Formulas from Geometry

Formulas for Area (A), Perimeter (P), Circumference (C), and Volume (V):

Square

$A = s^2$

$P = 4s$

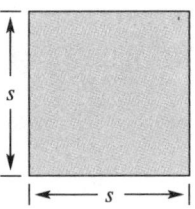

Rectangle

$A = lw$

$P = 2l + 2w$

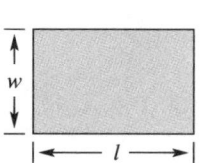

Circle

$A = \pi r^2$

$C = 2\pi r$

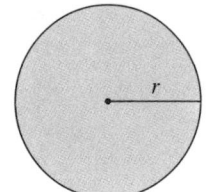

Triangle

$A = \dfrac{1}{2}bh$

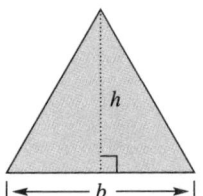

Trapezoid

$A = \dfrac{1}{2}h(b_1 + b_2)$

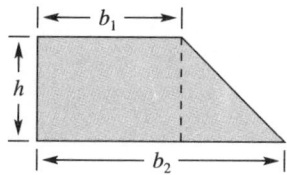

Parallelogram

$A = bh$

$P = 2a + 2b$

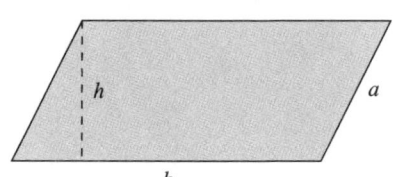

Pythagorean Theorem

$a^2 + b^2 = c^2$

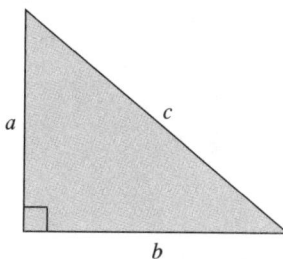

Cube

$V = s^3$

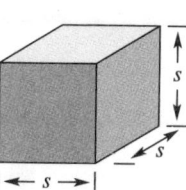

Rectangular Solid

$V = lwh$

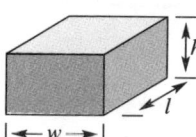

Circular Cylinder

$V = \pi r^2 h$

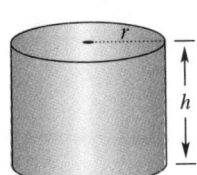

Sphere

$V = \dfrac{4}{3}\pi r^3$

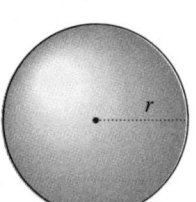

INSTRUCTOR'S ANNOTATED EDITION

INTERMEDIATE ALGEBRA

Graphs and Functions

INTERMEDIATE ALGEBRA
Graphs and Functions

ROLAND E. LARSON
The Pennsylvania State University, The Behrend College

ROBERT P. HOSTETLER
The Pennsylvania State University, The Behrend College

CAROLYN F. NEPTUNE
Johnson County Community College

with the assistance of
DAVID E. HEYD
The Pennsylvania State University, The Behrend College

D. C. HEATH AND COMPANY
Lexington, Massachusetts   Toronto

Address editorial correspondence to:

D. C. Heath
125 Spring Street
Lexington, MA 02173

ACQUISITIONS EDITOR: Ann Marie Jones

DEVELOPMENTAL EDITOR: Cathy Cantin

PRODUCTION EDITOR: Karen Carter

DESIGNER: Cornelia Boynton

ART EDITOR: Gary Crespo

PRODUCTION SUPERVISOR: Lisa Merrill

COVER: Paul Klee (1879–1940). *Pyramide, 1930.138.* Paul Klee Foundation, Museum of Fine Arts Berne. © 1991, Copyright by COSMOPRESS, Geneva, Switzerland.

COMPOSITION: Meridian Creative Group

TECHNICAL ART: Folium, Inc.; Illustrious, Inc.; Tech-Graphics, Inc.; Techsetters

PHOTO CREDITS: 1, Julie Houck/Stock, Boston; 39, Edward L. Miller/Stock, Boston; 88, Jan Halaska/Photo Researchers, Inc.; 185, Ron Sanford/AllStock; 193, The Bettmann Archive; 205, Brian Smith; 229, Jonathan Watts/Science Photo Library/Photo Researchers, Inc.; 238, John Running/Stock, Boston; 259, J. Carini/The Image Works; 282, John Head/Science Photo Library/Photo Researchers, Inc.; 300, D. and I. MacDonald/The Picture Cube; 312, Dawson Jones/Stock, Boston; 330, Topham/The Image Works; 351, Bob Daemmrich/Stock, Boston; 354, Courtesy of Lucasfilm, Ltd.; 359, Hiroji Kubota/Magnum Photos; 373, G. C. Kelley/Photo Researchers, Inc.; 384, Fridmar Damm/Leo de Wys Inc.; 430, Steve Goldberg/Monkmeyer Press Photo; 450, George Munday/Leo de Wys Inc.; 469, The Bettmann Archive; 517, Mimi Forsyth/Monkmeyer Press Photo; 595, Bob Daemmrich/Stock, Boston; 601, Bryce Flynn/Stock, Boston; 604, Martin Dohrn/Science Photo Library/Photo Researchers, Inc.; 623, Gamma-Liaison; 627, Dan Abernathy; 669, Scott Camazine/Photo Researchers, Inc.; 675, Fridmar Damm/Leo de Wys Inc.; 677, David Dul/Unicorn Stock Photo; 682, Peter Menzel/Stock, Boston; 720, David Young-Wolff/PhotoEdit; 725, Nita Winters/The Image Works; 733, Jim Schwabel/New England Stock Photo; 745, Dr. E. R. Degginger; 751, D. G. Arnold; 786, H. Armstrong Roberts, Inc.

Published simultaneously in Canada.

Printed in the United States of America.

International Standard Book Number: 0-669-33756-0

Library of Congress Catalog Number: 93-71003

10 9 8 7 6 5 4 3 2 1

Preface

The Instructor's Annotated Edition of *Intermediate Algebra: Graphs and Functions* contains all of the material that is in the student text, as well as answers to the even-numbered exercises. It also offers teaching strategies, anticipates common student errors, and gives additional examples and supplementary exercises. These notes appear in color in the margin adjacent to the relevant text. Thanks to Debra A. Landre of San Joaquin Delta College and Cheryl V. Roberts of Northern Virginia Community College–Manassas for the excellent set of annotations.

The student edition of the text contains answers to quizzes, tests, warm-ups, and odd-numbered exercises. It does not have answers to even-numbered exercises or instructor's notes.

Answers to Exercises

Chapter 1

Chapter Opener *(page 1)*

$1080°$, $135°$

SECTION 1.1 *(page 13)*

1. (a) 1, 4, 6   **(b)** -10, 0, 1, 4, 6
    (c) -10, $-\frac{2}{3}$, $-\frac{1}{4}$, 0, $\frac{5}{8}$, 1, 4, 6   **(d)** $-\sqrt{5}$, $\sqrt{3}$, 2π

2. (a) 8, 245   **(b)** 0, 8, 245
    (c) $-\frac{7}{2}$, $-\frac{3}{8}$, 0, $\frac{10}{3}$, 8, 245   **(d)** $-\sqrt{6}$, $-\frac{\pi}{2}$, $\sqrt{15}$

3. $2 < 5$

4. $-8 < 3$

5. $-7 < -2$

6. $-2 > -5$

7. $\frac{1}{3} > \frac{1}{4}$

8. $-\frac{3}{2} > -\frac{5}{2}$

9. $-\frac{5}{8} < \frac{1}{2}$

10. $-\pi < -3$

11. 6   **12.** 20   **13.** 19   **14.** 125   **15.** 8
16. 86   **17.** 39   **18.** 190   **19.** 225
20. 62   **21.** -16 **22.** $-\frac{3}{8}$ **23.** -85
24. -36.5 **25.** -34, 34   **26.** 52, 52
27. $\frac{3}{11}$, $\frac{3}{11}$ **28.** $-\frac{5}{32}$, $\frac{5}{32}$

29.

30.

31.

32.

33. $|3| < |-6|$ **34.** $|-8| > |5|$
35. $|-4| = |4|$ **36.** $|12| = |-12|$
37. $-|-2| > -|-3|$ **38.** $-5 = -|-5|$
39. $x < 0$ **40.** $t > 0$ **41.** $x \geq 0$
42. $u \geq 16$ **43.** $p < \$225$ **44.** $p \geq 30$
45. 45   **46.** 68   **47.** -19 **48.** -28
49. -20 **50.** 0.7   **51.** $\frac{5}{4}$
52. $\frac{4}{9}$ **53.** $\frac{1}{10}$ **54.** $\frac{1}{12}$
55. $\frac{105}{8}$ or $13\frac{1}{8}$ **56.** $-\frac{97}{6}$ or $-16\frac{1}{6}$
57. 60   **58.** -28 **59.** -28 **60.** -28
61. -30 **62.** -21 **63.** 32.13   **64.** 14.08
65. $\frac{1}{2}$ **66.** $-\frac{6}{13}$ **67.** $-\frac{1}{3}$ **68.** 2
69. 6   **70.** 2   **71.** $-\frac{5}{2}$ **72.** $-\frac{22}{5}$
73. $\frac{46}{17}$ **74.** $\frac{23}{16}$ **75.** 4.03   **76.** 10.37

77. $(4)(4)(4)$ **78.** $\left(\frac{2}{3}\right)\left(\frac{2}{3}\right)\left(\frac{2}{3}\right)\left(\frac{2}{3}\right)$

79. $\left(-\frac{4}{5}\right)\left(-\frac{4}{5}\right)\left(-\frac{4}{5}\right)\left(-\frac{4}{5}\right)\left(-\frac{4}{5}\right)\left(-\frac{4}{5}\right)$

80. $(-6)(-6)(-6)(-6)(-6)$ **81.** $\left(\frac{5}{8}\right)^4$

82. -7^3 **83.** $(-4)^6$ **84.** $(-7)^3$ **85.** 16
86. 9   **87.** -1 **88.** -243 **89.** -16
90. -9 **91.** -64 **92.** -243 **93.** $\frac{49}{64}$
94. $-\frac{64}{125}$ **95.** $-\frac{49}{64}$ **96.** $-\frac{64}{125}$ **97.** 9
98. 2   **99.** 7   **100.** 10   **101.** 5
102. 4   **103.** -2 **104.** -1 **105.** -1
106. 5   **107.** 3   **108.** 2   **109.** 36
110. 8   **111.** 57   **112.** 40   **113.** 135
114. 0   **115.** -7.8 **116.** 5   **117.** \$2533.56
118. \$10,800   **119.** $\frac{17}{180}$ **120.** $\frac{11}{70}$

121.

Day	Daily Gain or Loss
Tuesday	+$5
Wednesday	+$8
Thursday	−$5
Friday	+$16

122.

Year	Yearly Gain or Loss (in millions)
1989	+$0.5
1990	−$0.2
1991	+$1.3
1992	−$0.2
1993	+$0.9

123. 15 m^2 **124.** 20 in.^2 **125.** 36 ft^2

126. 128 cm^2 **127.** 6.125 ft^3

128. 40 bales/ton, 2940 ft^3

129. True. Every integer p can be written in the form of the rational number $p/1$.

130. False. Rationals include all integers *plus* all fractions of the form p/q where $q \neq 0$.

131. False. Given the integer 2, its reciprocal is the rational number $\frac{1}{2}$.

132. True. The reciprocal of a rational number p/q is q/p, a rational number because $p \neq 0$.

133. True. Any nonzero real number raised to an even power is positive.

134. False. A negative number raised to an odd power is negative.

USING A GRAPHING CALCULATOR   *(page 20)*

1. $(-4)^2$ **2.** $\dfrac{7}{-3-5}$ **3.** $(1+4)^3$

4. $3(5-2)$ **5.** 397.440   **6.** −704.969

7. 4764.250   **8.** 67.021   **9.** 15,379.913

10. 7.833   **11.** −0.218   **12.** 6.759

13. The calculator steps correspond to $0.05(1.24) + 2.36$ instead of $0.05(1.24 + 2.36)$.
Correct Calculator Steps:

.05 ⨯ (1.24 + 2.36) ENTER
Display: .18

14. The calculator steps correspond to $523 - 145 - 136$ instead of $523 - (145 - 136)$.
Correct Calculator Steps:

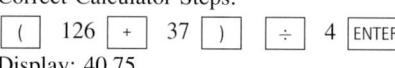
523 − (145 − 136) ENTER
Display: 514

15. The calculator steps correspond to $126 + \dfrac{37}{4}$ instead of $\dfrac{126 + 37}{4}$.
Correct Calculator Steps:

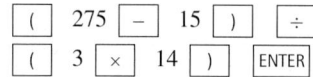
(126 + 37) ÷ 4 ENTER
Display: 40.75

16. The calculator steps correspond to $275 - \dfrac{15}{3}(14)$ instead of $\dfrac{275 - 15}{3(14)}$.
Correct Calculator Steps:

(275 − 15) ÷
(3 ⨯ 14) ENTER

Display: 6.19

SECTION 1.2   *(page 28)*

Warm-Up *(page 28)*

1. 225   **2.** −240   **3.** −150   **4.** 120

5. $\frac{8}{5}$ **6.** $-\frac{1}{4}$ **7.** $\frac{6}{5}$ **8.** $\frac{3}{10}$

9. −27   **10.** $\frac{25}{64}$

1. Commutative Property of Addition
2. Commutative Property of Multiplication
3. Commutative Property of Multiplication
4. Commutative Property of Addition
5. Associative Property of Multiplication
6. Additive Identity Property
7. Multiplicative Identity Property
8. Distributive Property
9. Associative Property of Addition
10. Associative Property of Multiplication
11. Distributive Property
12. Associative Property of Addition
13. Additive Identity Property
14. Multiplicative Identity Property
15. Additive Inverse Property

16. Multiplicative Inverse Property
17. Multiplicative Inverse Property
18. Distributive Property
19. Associative Property of Addition
20. Additive Inverse Property
21. Commutative Property of Multiplication
22. Distributive Property
23. Distributive Property
24. Multiplicative Inverse Property
25. $(3 \cdot 6)y$ 26. $-6 + 10$ 27. $-3(15)$
28. $(6 + 5) + y$ 29. $5 \cdot 6 + 5 \cdot z$ or $30 + 5z$
30. $8 \cdot 4 + y \cdot 4$ or $32 + 4y$ 31. $-x + 25$
32. 0 33. $x + 8$ 34. $8x$
35. (a) -10 (b) $\frac{1}{10}$ 36. (a) -18 (b) $\frac{1}{18}$
37. (a) 16 (b) $-\frac{1}{16}$ 38. (a) 52 (b) $-\frac{1}{52}$
39. (a) $-6z$ (b) $\frac{1}{6z}$ 40. (a) $-2y$ (b) $\frac{1}{2y}$
41. (a) $-x - 1$ or $-(x + 1)$ (b) $\frac{1}{x + 1}$
42. (a) $-(y - 4)$ or $-y + 4$ (b) $\frac{1}{y - 4}$
43. $x + (5 + 3)$ 44. $z + (6 + 10)$ 45. $(32 + 4) + y$
46. $(15 + 3) + x$ 47. $(3 \cdot 4)5$ 48. $10 \cdot (8 \cdot 5)$
49. $(6 \cdot 2)y$ 50. $(8 \cdot 3)x$
51. $20 \cdot a + 20 \cdot 5$ or $20a + 100$
52. $-3 \cdot y + (-3)8$ or $-3y - 24$
53. $5 \cdot 3x + 5 \cdot 4$ or $15x + 20$
54. $6 \cdot 2x + 6 \cdot 5$ or $12x + 30$
55. $x \cdot (-2) + 6 \cdot (-2)$ or $-2x - 12$
56. $z \cdot 12 + 10 \cdot 12$ or $12z + 120$
57. Given
   Addition Property of Equality
   Associative Property of Addition
   Additive Inverse Property
   Additive Identity Property
58. Given
   Addition Property of Equality
   Associative Property of Addition
   Additive Inverse Property
   Additive Identity Property
59. Given
   Addition Property of Equality
   Associative Property of Addition
   Additive Inverse Property
   Additive Identity Property
   Multiplication Property of Equality
   Associative Property of Multiplication
   Multiplicative Inverse Property
   Multiplicative Identity Property

60. Given
   Addition Property of Equality
   Associative Property of Addition
   Additive Inverse Property
   Additive Identity Property
   Multiplication Property of Equality
   Associative Property of Multiplication
   Multiplicative Inverse Property
   Multiplicative Identity Property
61. $3x + 15$ 62. $4x + 8$ 63. $-2x - 16$
64. $-9x - 36$ 65. 0 66. 1 67. 28
68. 25 69. 434 70. 245 71. 62.82
72. 239.4 73. $a(b + c) = ab + ac$
74. $a(b - c) = ab - ac$ 75. \$9.5 billion
76. \$13.4 billion

SECTION 1.3   *(page 40)*

Warm-Up *(page 40)*

1. 168 2. 20 3. $-\frac{21}{2}$ 4. 7
5. $\frac{72}{5}$ 6. $-\frac{19}{10}$
7. Multiplicative Inverse Property
8. Associative Property of Addition
9. Distributive Property
10. Additive Identity Property

1. $10x$, 5 2. $-16t^2$, 48 3. $-3y^2$, $2y$, -8
4. $25z^3$, $-4.8z^2$ 5. $4x^2$, $-3y^2$, $-5x$, $2y$
6. $14u^2$, $25uv$, $-3v^2$ 7. x^2, $-2.5x$, $-\frac{1}{x}$
8. $\frac{3}{t^2}$, $\frac{-4}{t}$, 6 9. 5 10. 4
11. $-\frac{3}{4}$ 12. -8.4
13. Commutative Property of Addition
14. Associative Property of Addition
15. Associative Property of Multiplication
16. Commutative Property of Multiplication
17. Multiplicative Inverse Property
18. Additive Inverse Property
19. Distributive Property
20. Additive Identity Property
21. Multiplicative Identity Property
22. Distributive Property
23. (a) $5x + 5 \cdot 6$ or $5x + 30$ (b) $(x + 6)5$
24. (a) $6(x + 1)$ (b) $6 + 6x$

25. (a) $(xy)6$ **(b)** $(6x)y$
26. (a) $3ab$ **(b)** $0 + 3ab$
27. (a) 0 **(b)** $(-4t^2) + 4t^2$
28. (a) $3 + (6-9)$ **(b)** 0 **29.** $(x \cdot x \cdot x) \cdot (x \cdot x \cdot x \cdot x)$
30. $(z \cdot z) \cdot (z \cdot z \cdot z \cdot z \cdot z)$ **31.** $(-2x)(-2x)(-2x)$
32. $(2y)(2y)(2y)$ **33.** $\left(\frac{y}{5}\right)\left(\frac{y}{5}\right)\left(\frac{y}{5}\right)\left(\frac{y}{5}\right)\left(\frac{y}{5}\right)$
34. $\left(\frac{3}{t}\right)\left(\frac{3}{t}\right)\left(\frac{3}{t}\right)\left(\frac{3}{t}\right)\left(\frac{3}{t}\right)$ **35.** $(5x)^4$

36. $y^3 y^4$ **37.** $x^3 y^3$ **38.** $(-9t)^6$ **39.** x^{12}
40. u^9 **41.** $27y^6$ **42.** $36x^8$ **43.** $16x^2$
44. $-64x^3$ **45.** $-125z^6$ **46.** $25z^6$ **47.** $6x^3 y^4$
48. $-10a^3 b^7$ **49.** $-54u^5 v^3$ **50.** $2000x^{10} y^4$
51. x^2 **52.** y^7 **53.** $81x^2$ **54.** $4y^2$
55. $\frac{4}{3}xy$ **56.** $-3x^2 y$ **57.** $\frac{16}{3}x^2 y^2$ **58.** $16a^4 b^4$
59. $-25y^6$ **60.** $54y^5$ **61.** $-3125z^8$ **62.** $18n^3$
63. $16a^4$ **64.** $16a^3$ **65.** $8xy$ **66.** $-z^4$
67. $\frac{2}{3}a^7 b$ **68.** $-\frac{9}{8}c^7 d^3$ **69.** $\frac{a^6}{27}$
70. $\frac{x^6}{y^4}$ **71.** $\frac{-4x^8}{25y^2}$ **72.** $-\frac{27a^9}{8b^{15}}$
73. x **74.** a^m **75.** x^{4n} **76.** a^{3k}
77. x^{n+4} **78.** y^m **79.** rs^{m+3} **80.** $x^{2n} y^{2m+2}$
81. $7x$ **82.** $2x^2$ **83.** $8y$ **84.** $14y$
85. $8x + 18y$ **86.** $-9a - \frac{2}{3}b$ **87.** $\frac{11}{2}z^2 + \frac{3}{2}z + 10$
88. $3y^3 - 6y^2 + 4y - 4$ **89.** $4u^2 v^2 + uv$
90. $5m^2 n^2 - 9mn$ **91.** $5a^2 b^2 - 2ab$
92. $2xy + 8$ **93.** $8x^2 + 4x - 12$
94. $8z^3 - 32z^2 + 16$ **95.** $-18y^2 + 3y + 6$
96. $5x^2 - 10y - 5$ **97.** $12x - 35$
98. $4x^2 - 3$ **99.** $-3y^2 - 7y - 7$
100. $-4a^2 + 13a + 34$ **101.** $-2b^2 + 4b - 36$
102. $-12x^2 - 100$ **103.** $9x^3 - 5$
104. $x^3 - 12$ **105.** $2y^3 + y^2 + y$
106. $ab^3 - 8ab$ **107.** $-x^2 y^2 - 2xy$
108. $z^6 + 3z^4 + 4z^3$ **109.** $-51a^7$
110. 0 **111.** $4x - 14$ **112.** $4x + 12$
113. (a) 3 **(b)** -10 **114. (a)** 7 **(b)** $-\frac{13}{7}$
115. (a) 7 **(b)** 7 **116. (a)** 0 **(b)** 0
117. (a) 0 **(b)** $\frac{3}{10}$ **118. (a)** Undefined   **(b)** $\frac{11}{2}$
119. (a) 13 **(b)** -36 **120. (a)** 7 **(b)** 7
121. (a) 0 **(b)** Undefined   **122. (a)** 3 **(b)** 0
123. (a) 210 **(b)** 140
124. (a) \$4250   **(b)** \$157.50
125. (a) $-7, -5, -3, -1, 1, 3$
   (b) A two-unit increase
   (c) A $\frac{3}{4}$-unit increase

126. (a) $-1, 2, 5, 8, 11, 14$
   (b) A three-unit increase
   (c) A $\frac{2}{3}$-unit increase
127. (a) \$2759.4009 million   **(b)** Appear the same
128. (a) \$4005.6241 million   **(b)** Equivalent
129. (a) No   **(b)** The bar graph
130.

$$b_1 h + \frac{1}{2}(b_2 - b_1)h = b_1 h + \left(\frac{1}{2}b_2 - \frac{1}{2}b_1\right)h$$
$$= b_1 h + \frac{1}{2}b_2 h - \frac{1}{2}b_1 h$$
$$= \left(b_1 h - \frac{1}{2}b_1 h\right) + \frac{1}{2}b_2 h$$
$$= \frac{1}{2}b_1 h + \frac{1}{2}b_2 h$$
$$= \frac{1}{2}h(b_1 + b_2)$$
$$= \frac{h}{2}(b_1 + b_2)$$

131. $\frac{57}{2}$ **132.** 1440 ft^2

CHAPTER 1 MID-CHAPTER QUIZ   *(page 45)*

1. $-\frac{3}{2} > -\frac{5}{2}$ **2.** 50 **3.** -68 **4.** $\frac{1}{24}$
5. 71.03 **6.** $\frac{2}{5}$ **7.** $\frac{27}{8}$ **8.** -2 **9.** $(5x)^4$
10. Distributive Property
11. Multiplicative Identity Property
12. Additive Inverse Property
13. $\frac{1}{x+1}$ **14.** $5x^4, -3x^2, 2, \frac{-1}{x^2}$ **15.** $4x^5$
16. $16x^4$ **17.** $\frac{y^6}{27}$ **18.** $\frac{3}{2}xy^2$
19. $-x^2 + 2xy$ **20.** $-6x^3 - 3x$
21. (a) 225 **(b)** 5 **22.** $6x + 16$
23. Loss of \$498,833.36

SECTION 1.4   *(page 55)*

Warm-Up   *(page 54)*

1. $23y - 69$ **2.** $12 - 30z$ **3.** $-10 + 4x$
4. $-25x + 20$ **5.** $32x, -10y, -7$
6. $-3s, 4t, 1$ **7.** $7x - 15$ **8.** $3y + 36$
9. $-4u$ **10.** $-5v - 18$

1. $10x - 4$, 1, 10 **2.** $3x^2 + 8$, 2, 3
3. $-3y^4 + 5$, 4, -3 **4.** $-3x^3 - 2x^2 - 3$, 3, -3
5. $4t^5 - t^2 + 6t + 3$, 5, 4 **6.** $-16z^2 + 8z$, 2, -16
7. -4, 0, -4 **8.** $-16t^2 + v_0 t$, 2, -16
9. Binomial **10.** Monomial
11. Trinomial **12.** Binomial
13. Exponent in y^{-3} is not nonnegative.
14. The term $4x^{1/3}$ has a fractional exponent.
15. ax^3 where $a \neq 0$ **16.** $-2x^4 + x^3 + 1$
17. $8x^2 + 4$ **18.** Any constant
19. (a) 16 (b) 0 (c) -16 (d) 16
20. (a) -4 (b) 0 (c) -4 (d) $\frac{9}{4}$
21. (a) -27 (b) -16 (c) 0 (d) $\frac{9}{16}$
22. (a) -1 (b) $-\frac{16}{27}$ (c) 0 (d) 7
23. $7x^2 + 3$ **24.** $3x^3 + x + 3$ **25.** $6x^2 - 7x + 8$
26. $4y^2 - y + 3$ **27.** $-2x^2 + 15$
28. $3s^2 + 26s - 32$ **29.** $2x^2 - 3x$
30. $4x^3 + 2x^2 + 9x - 6$ **31.** 4
32. $3v^2 + 2v - 2$ **33.** $x^2 - 3x + 2$
34. $2y^4 - 4$ **35.** $7x^3 + 2x$
36. $7y^2 - 9y + 2$ **37.** $4x^3 + x + 4$
38. $2y^4 + 2y^2$ **39.** $x^2 - 2x + 5$ **40.** 0
41. $-4x^3 - 2x + 13$ **42.** $1.2t^4 - 0.8t^2 + 1.4$
43. $-2y^4 - 5y + 4$ **44.** $4a^2 + 15$
45. $2x^3 - 2x + 3$ **46.** $11z^4 - 7z^2 + 12$
47. $q^2 + 15$ **48.** $-12s^2 - 6s - 5$
49. $7x^3 + 22x^2 + 4$ **50.** $26y^4 - 10y - 10$
51. $29s + 8$ **52.** $p^3 - p^2 - 3p + 9$
53. $3t^2 + 29$ **54.** $-5v - 1$
55. $3v^2 + 78v + 27$ **56.** $60x^2 - 80x + 19$
57. $16a^3$ **58.** $-18n^3$
59. $10y - 2y^2$ **60.** $10z^2 - 35z$
61. $8x^5 - 12x^4 + 20x^3$ **62.** $-9y^4 + 21y^3 - 9y^2$
63. $-10x^2 - 6x^4 + 14x^5$ **64.** $-33a^3 + 9a^2$
65. $x^2 + 3x - 28$ **66.** $y^2 + y - 6$
67. $2a^2 - 11a + 15$ **68.** $6x^2 + 5x + 1$
69. $6x^2 + 7xy + 2y^2$ **70.** $6x^2 - 7xy + 2y^2$
71. $48y^2 + 32y - 3$ **72.** $10t^2 - \frac{163}{2}t + 12$
73. $75x^3 + 30x^2$ **74.** $-12t^4 + 12t^2$
75. $-a^2 + 19a$ **76.** $2t^2 + 7t - 16$
77. $x^4 - 2x^3 - 3x^2 + 8x - 4$
78. $t^3 - 2t^2 - 14t + 3$
79. $2u^3 + 13u^2 + 11u - 20$ **80.** $x^4 - 2x^3 - 8x - 16$
81. $4x^4 - 4x^3 + 6x - 9$ **82.** $x^3 - 5x^2 + 10x - 6$
83. $a^3 + 15a^2 + 75a + 125$ **84.** $y^3 - 6y^2 + 12y - 8$
85. $x^3 + 27$ **86.** $x^3 - 64$
87. $t^4 - t^2 + 4t - 4$ **88.** $2y^4 + 3y^3 - 18y - 5$
89. $-6x^3 - 5x^2 + 4x - 3$ **90.** $6s^3 - 23s^2 + 38s - 24$
91. $x^2 - 4$ **92.** $16 - 9z^2$

93. $a^2 - 36c^2$ **94.** $64n^2 - m^2$
95. $4x^2 - \frac{1}{16}$ **96.** $\frac{4}{9}x^2 - 49$
97. $0.04t^2 - 0.25$ **98.** $16a^2 - 0.01b^2$
99. $x^2 + 10x + 25$ **100.** $x^2 + 18x + 81$
101. $25x^2 - 20x + 4$ **102.** $9x^2 - 48x + 64$
103. $4a^2 + 12ab + 9b^2$ **104.** $16x^2 - 40xy + 25y^2$
105. $x^2 - 2xy + y^2 + 4x - 4y + 4$
106. $x^2 + 2xy + y^2 - 8x - 8y + 16$
107. $u^2 - 2uv + 6u + v^2 - 6v + 9$
108. $4z^2 + 4yz + 4z + y^2 + 2y + 1$ **109.** $-8x - 7$
110. $k - 67$ **111.** $6y^2 - 32y + 36$
112. $-3b^2 - 18b + 1$ **113.** $12t$
114. $2a^2 + 72$ **115.** $8x^2 + 26x$ **116.** $1.2x^2$
117. (a) $A = (x + a)(x + b) = x^2 + ax + bx + ab$
(b) Because both expressions represent the area, the two must be equal. Therefore,
$$(x + a)(x + b) = x^2 + ax + bx + ab.$$
This statement illustrates the FOIL Method.
118. (a) $A = (x + a)^2 = x^2 + 2ax + a^2$
(b) Because both expressions represent the area, the two must be equal. Therefore,
$$(x + a)^2 = x^2 + 2ax + a^2.$$
This statement illustrates the special product for the square root of a binomial.
119. $1000 + 2000r + 1000r^2$
120. (a) $55,000 (b) $70,000 (c) $15,000
121. Dropped, 100 ft
122. Thrown upward, 0 ft
123. Thrown downward, 50 ft
124. Thrown upward, 300 ft
125. 224 ft, 216 ft, 176 ft
126. (a) $t = 2$ sec to $t = 3$ sec
(b) The object fell 8 ft in the interval from $t = 1$ sec to $t = 2$ sec, but the object fell 40 ft in the interval from $t = 2$ sec to $t = 3$ sec.
127. $4.29x^2$ **128.** $-14.14y^3$
129. $-7.148a^2 + 15.691a$ **130.** $12.823k^2 - 15.204k$
131. (a) $y = 31.43 - 0.55t + 0.013t^2$
1975—29.005 gal
1985—26.105 gal
(b) Decreasing

SECTION 1.5   *(page 67)*

Warm-Up   *(page 67)*

1. $24x^2 - 36x$ 2. $28y - 21y^2$
3. $-60x + 42x^2$ 4. $-2y^2 - 2y$ 5. $2t$
6. $2z^2 + 3z - 35$ 7. $121 - x^2$
8. $36r^2 - 25s^2$ 9. $x^3 - 8$ 10. $x^3 + 1$

1. 6   2. 2   3. $3x$ 4. $9x^3$
5. $6z^2$ 6. $15y$ 7. $14b^2$ 8. $4xy$
9. $21(x + 8)^2$ 10. $22(3 - y)$ 11. $x(1 - z)^2$
12. $2(x + 5)$ 13. $8(z - 1)$ 14. $5(x + 1)$
15. $2(2u + 5)$ 16. $-5(3t + 2)$ 17. $6(4x^2 - 3)$
18. $7(2z^3 + 3)$ 19. $x(2x + 1)$ 20. $-a(a^2 + 4)$
21. $7u(3u - 2)$ 22. $12y^2(3y^2 + 2)$
23. $1(11u^2 + 9)$ 24. $1(16x^2 - 3y^3)$
25. $3y(x^2y - 5)$ 26. $2uv(2 + 3uv)$
27. $4(7x^2 + 4x - 2)$ 28. $3(3 - 9y - 5y^2)$
29. $x^2(14x^2 + 21x + 9)$ 30. $y^2(17x^5y - x + 34)$
31. $-5(x - 2)$ 32. $-4(x^4 - 8)$
33. $-7(2x - 1)$ 34. $-5(x - 3)$
35. $-2(x^2 - 2x - 4)$ 36. $-6(x^2 - 2x + 3)$
37. $-1(4t^2 - 2t + 15)$ 38. $-1(5s^4 - 32s - 16)$
39. $(y - 3)(2y + 5)$ 40. $(s + 9)(7t - 6)$
41. $(t^2 + 1)(5t - 4)$ 42. $(a^2 - 3)(3a + 10)$
43. $a(a + 6)(1 - a)$ 44. $y(x - y)$
45. $(y - 6)(y + 2)$ 46. $(y + 3)(y + 4)$
47. $(x - 3)(x^2 + 2)$ 48. $(a - 5)(a^2 + 6)$
49. $(x + 25)(x + 1)$ 50. $(x - 7)(x + 1)$
51. $(x + 2)(x^2 + 1)$ 52. $(t - 11)(t^2 + 1)$
53. $(a - 4)(a^2 + 2)$ 54. $(s + 2)(3s^2 + 5)$
55. $(z + 3)(z^3 - 2)$ 56. $(2u - 1)(2u^3 - 3)$
57. $(c - 3)(d + 3)$ 58. $(u + v)(u - 4)$
59. $(x + 8)(x - 8)$ 60. $(y - 12)(y + 12)$
61. $(4y + 3z)(4y - 3z)$ 62. $(3z - 5w)(3z + 5w)$
63. $(x + 2y)(x - 2y)$ 64. $(3a - 5b)(3a + 5b)$
65. $(a^4 + 6)(a^4 - 6)$ 66. $(y^5 - 8)(y^5 + 8)$
67. $(10 - 3y)(10 + 3y)$ 68. $(25 - 7x)(25 + 7x)$
69. $(ab - 4)(ab + 4)$ 70. $(uv - 5)(uv + 5)$
71. $(a + 11)(a - 3)$ 72. $(x - 5)(x - 1)$
73. $(4 - z)(14 + z)$ 74. $(13 - y)(7 + y)$
75. $(x - 2)(x^2 + 2x + 4)$ 76. $(t - 3)(t^2 + 3t + 9)$
77. $(y + 4z)(y^2 - 4yz + 16z^2)$
78. $(z + 5w)(z^2 - 5wz + 25w^2)$
79. $(2t - 3)(4t^2 + 6t + 9)$
80. $(3s + 4)(9s^2 - 12s + 16)$ 81. $2(2 - 5x)(2 + 5x)$
82. $a(a - 4)(a + 4)$ 83. $(y + 3x)(y - 3x)(y^2 + 9x^2)$

84. $(u - 4v)(u + 4v)(u^2 + 16v^2)$
85. $2(x - 3)(x^2 + 3x + 9)$
86. $5(y - 5)(y^2 + 5y + 25)$ 87. 6399   88. 396
89. $(2x^n - 5)(2n^n + 5)$
90. $(3 - 2y^n)(3 + 2y^n)(9 + 4y^{2n})$
91. $x^2(3x + 4) - (3x + 4) = (x - 1)(x + 1)(3x + 4)$
    $3x(x^2 - 1) + 4(x^2 - 1) = (x - 1)(x + 1)(3x + 4)$
92. $2x^2(3x - 4) + 3(3x - 4) = (3x - 4)(2x^2 + 3)$
    $3x(2x^2 + 3) - 4(2x^2 + 3) = (2x^2 + 3)(3x - 4)$
93. $p = 800 - 0.25x$ 94. $w = 45 - l$ 95. $4x$
96. $4x + 12$ 97. $P(1 + rt)$ 98. $kx(Q - x)$
99. $\pi(R - r)(R + r)$

SECTION 1.6   *(page 80)*

Warm-Up   *(page 80)*

1. $x^2 + 2x + 1$ 2. $x^2 - 10x + 25$
3. $4 - 4y + y^2$ 4. $6 + y - 2y^2$
5. $10z^2 - 29z - 21$ 6. $2t^3 - 5t^2 - 12t$
7. $3x(x - 7)$ 8. $-(x - 3)(x^2 + 1)$
9. $(2t - 13)(2t + 13)$ 10. $(y - 4)(y^2 + 4y + 16)$

1. $(x + 2)^2$ 2. $(z + 3)^2$ 3. $(a - 6)^2$
4. $(y - 7)^2$ 5. $(5y - 1)^2$ 6. $(2z + 7)^2$
7. $(3b + 2)^2$ 8. $(2x - 1)^2$ 9. $(2x - y)^2$
10. $(m + 3n)^2$ 11. $(u + 4v)^2$ 12. $(2y + 5z)^2$
13. $x + 1$ 14. $a - 2$ 15. $y - 5$
16. $y + 2$ 17. $z - 2$ 18. $z + 6$
19. $(x + 3)(x + 1)$ 20. $(x + 5)(x + 2)$
21. $(x - 3)(x - 2)$ 22. $(x - 6)(x - 4)$
23. $(y + 10)(y - 3)$ 24. $(m - 5)(m + 2)$
25. $(t - 7)(t + 3)$ 26. $(x + 6)(x - 2)$
27. $(x - 12)(x - 8)$
28. $(u + 3v)(u + 2v)$ 29. $(x - 7y)(x + 5y)$
30. $(a - 10b)(a - 11b)$ 31. $5x + 3$ 32. $5x + 4$
33. $5a - 3$ 34. $5c - 4$ 35. $2y - 9$
36. $3y - 10$ 37. $(3x + 1)(x + 1)$
38. $(5x + 2)(x + 1)$ 39. $(2x + 3)(4x - 3)$
40. $(2t + 1)(4t - 5)$ 41. $(3b - 1)(2b + 7)$
42. $(2a - 5)(a - 4)$ 43. $(6x + y)(4x - 3y)$
44. $(5x + 4)(4x - 3)$ 45. Not factorable
46. $(5x - 3y)(2x + 3y)$ 47. $(2u - 5)(u + 7v)$
48. Not factorable   49. $(-1)(2x - 3)(x + 2)$
50. $-(3x + 2)(2x - 3)$ 51. $(-1)(4x + 1)(15x - 1)$

52. $(-1)(4x+1)(3x-2)$ **53.** $(3x+4)(x+2)$
54. $(2x+3)(x+3)$ **55.** $(2x-1)(3x+2)$
56. $(2x+3)(3x-5)$ **57.** $(3x-1)(5x-2)$
58. $(2x+3)(x+3)$ **59.** $3x^3(x-2)(x+2)$
60. $5(2y-3)(2y+3)$ **61.** $2t(5t-9)(t+2)$
62. $(4z-7)^2$ **63.** $(3x-2)(7x-2)$
64. $(1-x)(1+x)(6-x)$ **65.** $(3-z)(9+z)$
66. $3(t-2)(t^2+2t+4)$ **67.** $2(3x-1)(9x^2+3x+1)$
68. $v(v^2+3v+5)$ **69.** $3a(3a-b)(3a+b)$
70. $4mn(m+4n)(2m-3n)$ **71.** ±18 **72.** ±8
73. ±6 **74.** ±12 **75.** ±12 **76.** ±40
77. 16   **78.** 36   **79.** 9   **80.** 100   **81.** 25
82. 4   **83.** 14   **84.** (-3) **85.** $\pm9, \pm11, \pm19$
86. $\pm9, \pm15$ **87.** $\pm4, \pm20$ **88.** ±6
89. $\pm13, \pm14, \pm22, \pm41$ **90.** $\pm1, \pm7, \pm13, \pm29$
91. $8, -16$ **92.** $18, -36$ **93.** $2, -40$
94. $35, -28$ **95.** $-5, -32$ **96.** $-12, -6$
97. 2704   **98.** 1521   **99.** c
100. b   **101.** a   **102.** d
103. $(x+3)(x+1)$

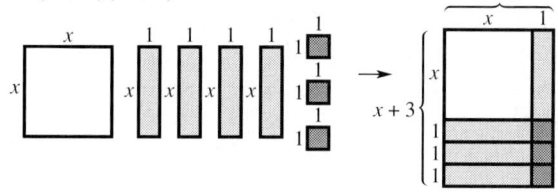

104. $(x+1)(x+4)$

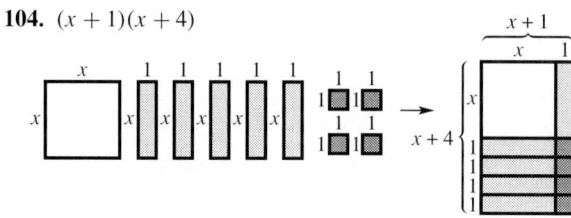

CHAPTER 1 REVIEW EXERCISES   *(page 84)*

1. $3 > -\frac{1}{8}$

2. $-2 > -8$

3. $-\frac{8}{5} < -\frac{2}{5}$

4. $8.4 > -\pi$

5. 7.2   **6.** 1.6   **7.** -7.2 **8.** 3.6
9. 11   **10.** -9 **11.** 230   **12.** -308
13. -38 **14.** 48,000   **15.** -4200 **16.** -960
17. 14   **18.** Undefined   **19.** 0   **20.** 16
21. $\frac{11}{21}$ **22.** $\frac{1}{2}$ **23.** $\frac{1}{6}$ **24.** $\frac{107}{96}$
25. $\frac{17}{8}$ **26.** $\frac{9}{4}$ **27.** $-\frac{1}{20}$ **28.** 1
29. 2   **30.** $-\frac{5}{2}$ **31.** $\frac{6}{5}$ **32.** $-\frac{8}{7}$
33. 5.25   **34.** 5   **35.** -216 **36.** -81
37. $\frac{1}{12}$ **38.** -36 **39.** 20   **40.** 40
41. Additive Inverse Property
42. Multiplicative Inverse Property
43. Distributive Property
44. Commutative Property of Multiplication
45. Associative Property of Addition
46. Associative Property of Multiplication
47. Commutative Property of Multiplication
48. Additive Identity Property
49. Multiplicative Inverse Property
50. Associative Property of Multiplication
51. Associative Property of Addition
52. Distributive Property
53. $4s-8t$ **54.** $-10x+20y$ **55.** $-3y^2+10y$
56. $3x^2+4xy$ **57.** $u-3v$ **58.** $-20+12j$
59. $5x-10$ **60.** $-7z+1$ **61.** $5x-y$
62. $20x-80$ **63.** $20u$ **64.** $33-6y$
65. $18b-15a$ **66.** $-2t^2+t$ **67.** x^6 **68.** $-2y^5$
69. $-64a^7$ **70.** $a^{11}b^2$ **71.** $8u^2v^2$ **72.** $\frac{y^6}{25}$
73. $162a^{13}b^6$ **74.** $-3x^3y^4$ **75.** (a) 0   (b) -3
76. (a) 0   (b) -133 **77.** (a) 15   (b) 3
78. (a) 0   (b) Not possible   **79.** $6x-x^2$
80. $\frac{9}{2}x+1$ **81.** $-9x^3+9x-4$ **82.** $-8z^2$
83. $6y^2-2y+15$ **84.** $6a^3+5$
85. $15x^2-11x-12$ **86.** $12y^4-7y^2-10$
87. $4x^3-5x+6$ **88.** $20s^3-9s^2-32s+15$
89. u^2-8u+7 **90.** 0
91. $16x^2-56x+49$ **92.** $64-48x+9x^2$
93. $25u^2-64$ **94.** $49a^2-16$
95. u^2-v^2-6u+9
96. $m^2-10m+25+2mn-10n+n^2$
97. $3x^2(2+5x)$ **98.** $4y(2-3y^3)$
99. $4a^2b^2(2a-3b)$ **100.** $27r^3s^2(4-5r)$
101. $7(x+5)(x+9)$ **102.** $(u-9v)u$
103. $(3a-10)(3a+10)$ **104.** $(b+30)(b-30)$
105. $(u-3)(u+15)$ **106.** $(y+1)(y-7)$
107. $(x-20)^2$ **108.** $(y+13)^2$ **109.** $(2s+9t)^2$
110. $(u-5v)^2$ **111.** $4a(1-4a)(1+4a)$
112. $3b(1+9b^2)$
113. $4(2x-3)(2x-1)$ **114.** $(x+3)(x+2)(x-2)$

115. $(2x + 1)(4x^2 - 2x + 1)$ **116.** $(t - 6)(t^2 + 6t + 36)$
117. $(2u - 7)^2$ **118.** $(x - 1)^2$
119. Not factorable
120. $(3x - 1)(x + 8)$ **121.** $t(t + 2)(4t - 1)$
122. $2h(3h + 1)(h - 4)$ **123.** ± 6
124. $\pm 5, \pm 7$ **125.** $\pm 1, \pm 7, \pm 13, \pm 29$
126. $\pm 5, \pm 19, \pm 31, \pm 65$
127. 55,578,932   **128.** 6,214,010
129. \$644   **130.** \$1203   **131.** 1
132. (a) 20,736   (b) (x^4) is the same as $(x^2)^2$
133. $10p^3 - 20p^4 + 10p^5$

CHAPTER 1 TEST   *(page 87)*

1. $-\frac{1}{2}$ **2.** $\frac{1}{6}$ **3.** $\frac{4}{27}$
4. $-\frac{27}{125}$ **5.** 11   **6.** 15
7. Associative Property of Multiplication
8. Multiplicative Inverse Property
9. $-5x$ **10.** $10x - 15$ **11.** $3x^4y^3$
12. $-2x^2 + 5x - 1$ **13.** $-2y^2 - 2y$
14. $8x^2 - 4x + 10$ **15.** $11t + 7$ **16.** $\dfrac{y^3}{8}$
17. $2x^2 + 7xy - 15y^2$ **18.** $6s^3 - 17s^2 + 26s - 21$
19. $16x^2 - 24x + 9$ **20.** $16 - a^2 - 2ab - b^2$
21. $6y(3y - 2)$ **22.** $(x - 2)(5x^2 - 6)$
23. $(3u - 1)^2$ **24.** $2(3x + 2)(x - 5)$
25. 640 ft^3 **26.** $x^2 + 26x$

Chapter 2

Chapter Opener   *(page 88)*

$s \approx 342, 119.44$ ft^2

SECTION 2.1   *(page 100)*

Warm-Up   *(page 99)*

1. $-2x$ **2.** $6x + 64$ **3.** $1 - 6x$
4. $-3x + 22$ **5.** $12y$ **6.** $0.02x + 100$
7. $2v(2v + 9)$ **8.** $(4x + 5)(4x - 5)$
9. $(x + 9)(x - 3)$ **10.** $-(4x - 3)(2x - 1)$

1. (a) No   (b) Yes   **2.** (a) Yes   (b) No
3. (a) No   (b) Yes   **4.** (a) No   (b) No
5. (a) Yes   (b) No   **6.** (a) No   (b) No
7. 15, 3, -5 **8.** 21, 21, 3   **9.** 5, 7, $-2x$, $-\frac{7}{2}$
10. 25, -15, -3, 5   **11.** 4   **12.** -2 **13.** $-\frac{2}{3}$
14. $\frac{5}{4}$ **15.** 2   **16.** 3   **17.** $\frac{1}{3}$
18. $-\frac{1}{2}$ **19.** 0   **20.** Not possible because $2 \neq 0$.
21. Not possible because $7 \neq 0$.   **22.** 0   **23.** 11
24. 3   **25.** -3 **26.** 10   **27.** -3 **28.** 100
29. $\frac{6}{5}$ **30.** 1   **31.** 50   **32.** -14 **33.** $\frac{19}{10}$
34. $-\frac{11}{30}$ **35.** 0   **36.** $-\frac{2}{11}$ **37.** $-\frac{10}{3}$ **38.** $\frac{24}{7}$
39. 72   **40.** $-\frac{48}{11}$ **41.** 23   **42.** 46   **43.** 0, 8
44. 0, -6 **45.** 3, -10 **46.** 3, -8
47. $-\frac{5}{2}, -\frac{1}{3}$ **48.** $\frac{3}{5}, 8$ **49.** 5, -2 **50.** 4, -3
51. 4   **52.** -2 **53.** -8 **54.** 6
55. $\frac{5}{4}, -5$ **56.** $-\frac{1}{7}, -\frac{1}{2}$ **57.** ± 5 **58.** ± 2
59. $-2, 6$ **60.** 3, -11 **61.** 0, $-\frac{5}{3}$ **62.** 0, $\frac{3}{2}$
63. $-10, 2$ **64.** ± 3 **65.** 5, -2 **66.** 9, -8
67. $-\frac{7}{2}, 5$ **68.** $-\frac{5}{3}, \frac{4}{3}$ **69.** $-7, 0$
70. 12, 3   **71.** -2 **72.** 3   **73.** -2
74. 8   **75.** $\frac{25}{3}$ **76.** 2   **77.** -20 **78.** $\frac{5}{2}$
79. 12   **80.** $-\frac{1}{2}$ **81.** ± 4 **82.** ± 11
83. $\frac{3}{2}$ **84.** $-\frac{3}{2}, -1$ **85.** $-\frac{1}{2}, 7$ **86.** $-\frac{1}{3}, 11$
87. $-6, 5$ **88.** $-12, 10$ **89.** 3.89   **90.** 14.58
91. 30.28   **92.** 0.16
93. (a)

t	1	1.5	2
Width	300	240	200
Length	300	360	400
Area	90,000	86,400	80,000

t	3	4	5
Width	150	120	100
Length	450	480	500
Area	67,500	57,600	50,000

(b) In a rectangle of fixed perimeter with length l equal to t times width w and $t \geq 1$, as t increases, w decreases, l increases, and the area A decreases. The maximum area occurs when the length and width are equal (when $t = 1$).

94. 2 sec   **95.** 6 hr   **96.** 1986
97. 15 ft × 22 ft   **98.** 20 sec
99. (a) $V = $ length · width · height
$V = x \cdot x \cdot 5 = 5x^2$

(b) 20, 80, 180, 320

(c) 22 in. × 22 in.

100. 20 in. × 20 in.

101. A linear equation in x is an equation that can be written in the form $ax + b = 0$, $a \neq 0$. A quadratic equation in x is an equation that can be written in the form
$ax^2 + bx + c = 0$, $a \neq 0$.

29. $r = \dfrac{-a_1 + S}{S}$ **30.** $m_2 = \dfrac{Fr^2}{km_1}$

31. 15 in.   **32.** 0.88 ft   **33.** 81 in.^{2}
34. Base 8 in.; Height: 12 in.
35. 2 ft   **36.** 3 m

37. (a) $p = \dfrac{A}{2\pi w}$ **(b)** $\dfrac{11}{2\pi}$ cm

38. (a) $r = \dfrac{P - 2x}{2\pi}$ **(b)** $r = \dfrac{35}{2\pi} \approx 5.57$

39. 7%   **40.** 5 years   **41.** \$6340.96
42. \$20,532.29   **43.** 2275 mi   **44.** 450 ft

45. $\dfrac{100}{11}$ hr   **46.** $\dfrac{125}{16}$ sec   **47.** $\dfrac{2000}{3}$ ft/sec

48. 55 mi/hr   **49.** $\frac{5}{17}$ hr   **50.** 500 sec
51. 12.3 mi/hr   **52.** 10.8 mi/hr   **53.** 30 days
54. 150 lb   **55.** $x = 0$, $x = -\dfrac{b}{a}$ **56.** 0, 1

SECTION 2.2   *(page 111)*

Warm-Up   *(page 110)*

1. $\frac{22}{5}$ **2.** −14   **3.** 5   **4.** $\frac{4}{3}$
5. 3   **6.** 6   **7.** −4, 2   **8.** $-\frac{3}{2}$
9. −12, 8   **10.** −4, 0

---

1. (a) $x = \dfrac{6 + 3y}{2}$ **(b)** $y = \dfrac{2x - 6}{3}$

2. (a) $x = \dfrac{3y - 15}{5}$ **(b)** $y = \dfrac{5x + 15}{3}$

3. (a) $x = \dfrac{10y - 11}{7}$ **(b)** $y = \dfrac{7x + 11}{10}$

4. (a) $x = \dfrac{42 - 7y}{12}$ **(b)** $y = \dfrac{42 - 12x}{7}$

5. (a) $x = -6y$ **(b)** $y = \frac{1}{6}x$

6. (a) $x = -\dfrac{1 + 27y}{3}$ **(b)** $y = -\dfrac{1 + 3x}{27}$

7. (a) $x = \dfrac{10 - 2y}{5}$ **(b)** $y = \dfrac{10 - 5x}{2}$

8. (a) $x = \dfrac{60 - 2y}{9}$ **(b)** $y = \dfrac{60 - 9x}{2}$

19. $R = \dfrac{E}{I}$ **20.** $C = \dfrac{S}{1 + r}$ **21.** $L = \dfrac{S}{1 - r}$

22. $h = \dfrac{2A}{b}$ **23.** $b = \dfrac{2A - ah}{h}$ **24.** $h = \dfrac{V}{\pi r^2}$

25. $r = \dfrac{A - P}{Pt}$ **26.** $P = \dfrac{A}{(1 + r/n)^{nt}}$

27. $n = \dfrac{S}{a_1 + a_n}$ **28.** $n = \dfrac{L + d - a}{d}$

SECTION 2.3   *(page 124)*

Warm-Up   *(page 124)*

1. $-6 < 2$

2. $\frac{11}{3} > 2$

3. $-3 > -7$

4. $-5 < 0$

5. $\frac{13}{3} > -\frac{3}{2}$

6. $\frac{3}{5} < \frac{13}{16}$

7. 3   **8.** 27   **9.** $\frac{33}{4}$ **10.** 3

1.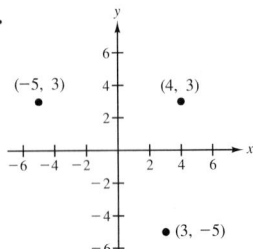

2.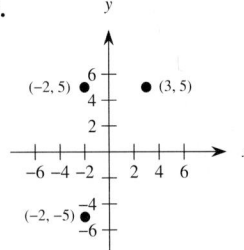

13.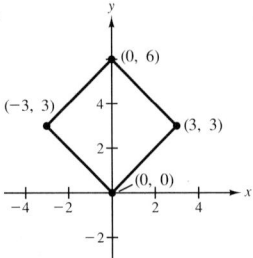

14.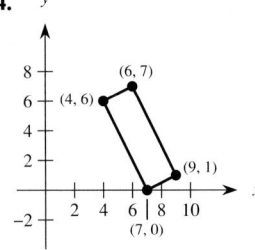

3.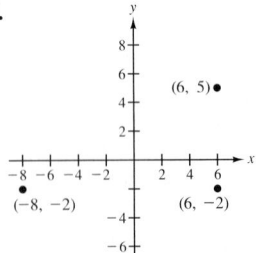

4.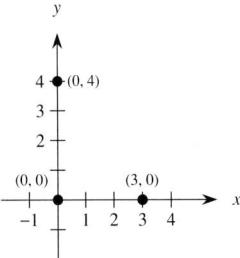

15.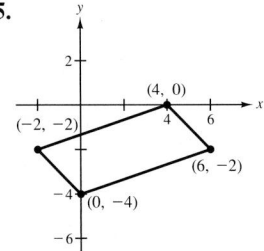

16.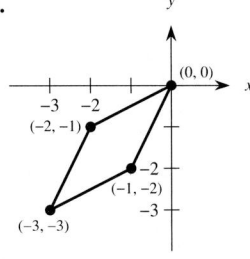

5.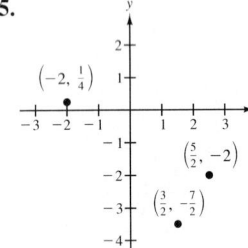

6.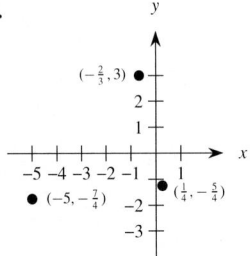

7. (a) Quadrant IV   **(b)** Quadrant II
8. (a) Quadrant I   **(b)** Quadrant III
9. A: $(4, -2)$ B: $(-3, -2.5)$ C: $(3, 0.5)$
10. A: $(-3, 2)$ B: $(4, -1)$ C: $\left(-\frac{1}{2}, -2\right)$

11.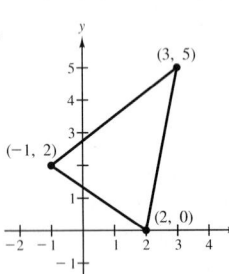

12.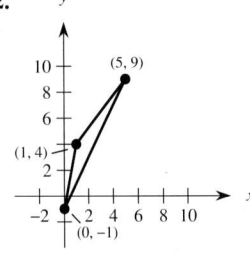

17. (a) Quadrant II, Quadrant III
  (b) If $y > 0$, the point would be located to the *left* of the vertical axis and *above* the horizontal axis in the second quadrant. If $y < 0$, the point would be located to the *left* of the vertical axis and *below* the horizontal axis in the third quadrant. Note: If $y = 0$, the point would be located on the x-axis *between* the second and third quadrants.

18. (a) Quadrant I, Quadrant II
  (b) If $x > 0$, the point would be located to the *right* of the vertical axis and *above* the horizontal axis in the first quadrant. If $x < 0$, the point would be located to the *left* of the vertical axis and *above* the horizontal axis in the second quadrant. Note: If $x = 0$, the point would be located on the y-axis *between* the first and second quadrants.

19. (a) Quadrant I, Quadrant III
  (b) The product xy is positive; therefore x and y have the same sign. If x and y are positive, the point would be located to the *right* of the vertical axis and *above* the horizontal axis in the first quadrant. If x and y are negative, the point would be located to the *left* of the vertical axis and *below* the horizontal axis in the third quadrant.

20. (a) Quadrant IV
  (b) All points in the fourth quadrant have positive x-coordinates and negative y-coordinates.

21. $(3, -4)$ **22.** $(-2, 5)$

23. $(-10, -10)$ **24.** $(7, -7)$

25.

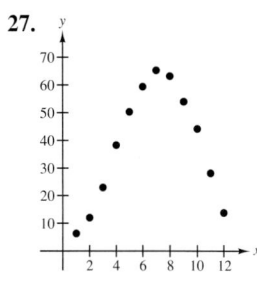

26.

27.

28.

29. **(a)** Yes   **(b)** No   **(c)** No   **(d)** Yes

30. **(a)** No   **(b)** Yes   **(c)** Yes   **(d)** No

31. **(a)** No   **(b)** Yes   **(c)** Yes   **(d)** No

32. **(a)** Yes   **(b)** No   **(c)** Yes   **(d)** No

33. **(a)** Yes   **(b)** No   **(c)** Yes   **(d)** Yes

34. **(a)** Yes   **(b)** Yes   **(c)** No   **(d)** No

35.

x	-2	0	2	4	6
$y = 5x - 1$	-11	-1	9	19	29

36.

x	-2	0	2	4	6
$y = \frac{3}{4}x + 2$	$\frac{1}{2}$	2	$\frac{7}{2}$	5	$\frac{13}{2}$

37.

x	-2	-1	0	1	2
$y = x^2 - x + 3$	9	5	3	3	5

38.

x	-6	-3	0	$\frac{3}{4}$	10
$y = \left\|\frac{4}{3}x - \frac{1}{3}\right\|$	$\frac{25}{3}$	$\frac{13}{3}$	$\frac{1}{3}$	$\frac{2}{3}$	13

39.

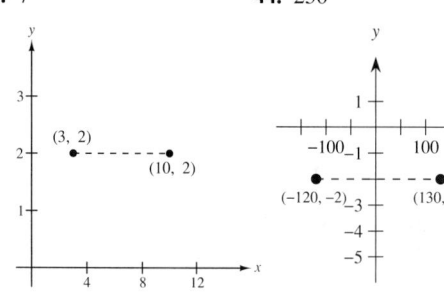

x	100	150	200
$y = 28x + 3000$	5800	7200	8600

x	250	300
$y = 28x + 3000$	10,000	11,400

40.

x	2	4	8	10	20
$y = 0.75x + 8$	9.5	11	14	15.5	23

41. 7   **42.** 7

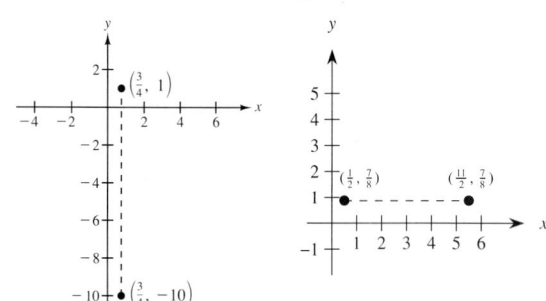

43. 7   **44.** 250

45. 11   **46.** 5

47. (a) $(x, y) = (10, 2)$ **(b)** 6, 8 **(c)** 10 **(d)** 10
48. (a) $(x, y) = (10, 1)$ **(b)** 5, 12 **(c)** 13 **(d)** 13
49. (a) $(x, y) = (4, -4)$ **(b)** 8, 7
   (c) $\sqrt{113}$ **(d)** $\sqrt{113}$
50. (a) $(x, y) = (6, 6)$ **(b)** 8, 6 **(c)** 10 **(d)** 10
51. 5 **52.** 13 **53.** 15 **54.** 17
55. 7.81 **56.** 16.55 **57.** 5.39 **58.** 1.41
59. Yes **60.** No **61.** No **62.** Yes
63. $2 + \sqrt{274} \approx 18.55$ ft
64. The shortest distance is 40 ft.
65. $(1, 4)$ **66.** $(2, 0)$

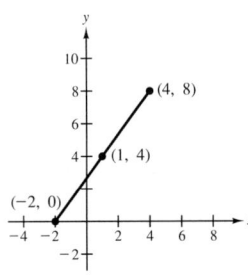

67. $\left(\frac{7}{2}, \frac{9}{2}\right)$ **68.** $\left(\frac{11}{2}, 3\right)$

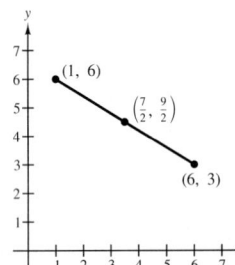
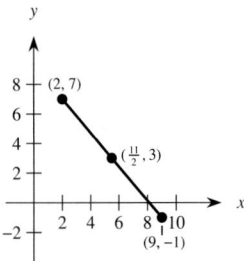

69. $(-1, 1), (3, 2), (0, 4)$
70. $(1, -1), (3, -4), (5, -1), (3, 2)$
71. Reflection about the y-axis

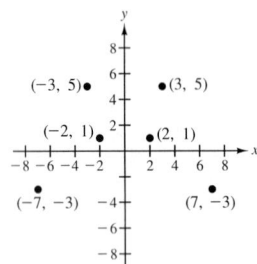

72. Reflection about the x-axis

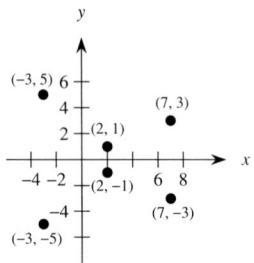

CHAPTER 2 MID-CHAPTER QUIZ *(page 129)*

1. 8 **2.** 2 **3.** 2 **4.** 12
5. -15, 16 **6.** -12, 6 **7.** $\frac{1}{2}$
8. $-\frac{2}{3}$, 2 **9.** $C = \dfrac{5F - 160}{9}$ **10.** $l = \dfrac{P - 2w}{2}$
11. $h = \dfrac{S - 2\pi r^2}{2\pi r}$
   $h = \dfrac{35 - 6\pi}{2\pi} \approx 2.57$ in.
12. 4 sec **13.** $\frac{15}{2}$ hr
14. (a) Quadrant II, Quadrant IV
   (b) The product xy is negative; therefore, x and y have opposite signs. If x is negative and y is positive, the point would be located to the *left* of the vertical axis and *above* the horizontal axis. Thus, the point would be in the second quadrant. If x is positive and y is negative, the point would be located to the *right* of the vertical axis and *below* the horizontal axis. Thus, the point would be in the fourth quadrant.

15.

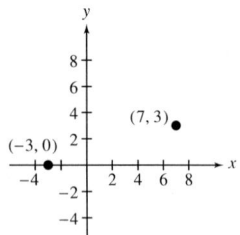

16. 10.44
17. (a) Yes **(b)** No **(c)** No **(d)** Yes

SECTION 2.4   *(page 143)*

Warm-Up   *(page 142)*

1.

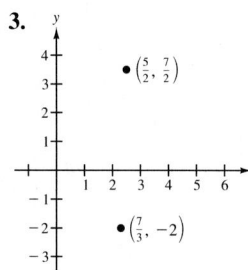

2.

3.

4.

5. $\frac{48}{5}$ **6.** 6   **7.** 13   **8.** $-\frac{11}{3}$

9. $\frac{130}{19} \approx 6.84$ **10.** 7

1.

x	-4	-1	0
y	14	5	2
(x, y)	$(-4, 14)$	$(-1, 5)$	$(0, 2)$

x	2	4
y	-4	-10
(x, y)	$(2, -4)$	$(4, -10)$

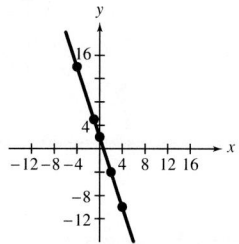

2.

x	-3	0	2	3	$\frac{21}{2}$
y	-4	-2	$-\frac{2}{3}$	0	5
(x, y)	$(-3, -4)$	$(0, -2)$	$\left(2, -\frac{2}{3}\right)$	$(3, 0)$	$\left(\frac{21}{2}, 5\right)$

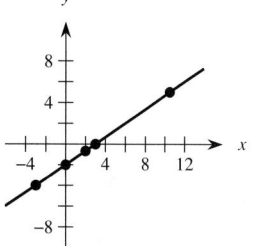

3.

x	± 2	-1	0
y	0	3	4
(x, y)	$(\pm 2, 0)$	$(-1, 3)$	$(0, 4)$

x	2	± 3
y	0	-5
(x, y)	$(2, 0)$	$(\pm 3, -5)$

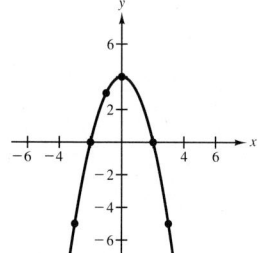

4.

x	-1	0	1	2	3
y	$-\frac{9}{2}$	-4	$-\frac{7}{2}$	0	$\frac{19}{2}$
(x, y)	$\left(-1, -\frac{9}{2}\right)$	$(0, -4)$	$\left(1, -\frac{7}{2}\right)$	$(2, 0)$	$\left(3, \frac{19}{2}\right)$

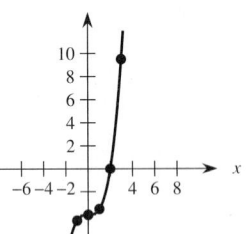

11. 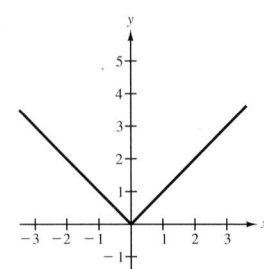   **12.**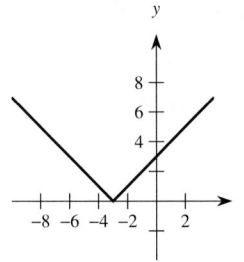

13. $(10, 0)$, $(0, 5)$ **14.** $(-4, 0)$, $(0, 6)$
15. $(-20, 0)$, $(0, 15)$ **16.** $(30, 0)$, $(0, 12)$
17. $(\pm 5, 0)$, $(0, -25)$ **18.** $(-1, 0)$, $(0, 1)$
19. $(0, 0)$ **20.** $(0, 0)$ **21.** No x-intercept, $(0, 3)$
22. $(3, 0)$, No y-intercept   **23.** $\left(\frac{2}{3}, 0\right)$, $(-1, 0)$, $(0, -2)$
24. $(3, 0)$, $(4, 0)$, $(0, 12)$

5. 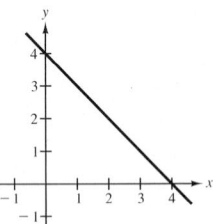   **6.**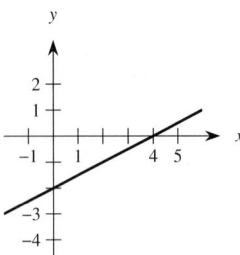

25. 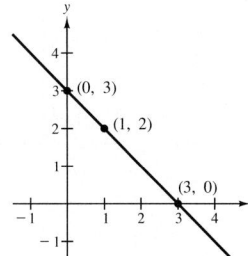   **26.**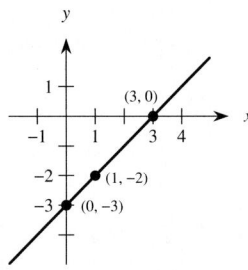

7. 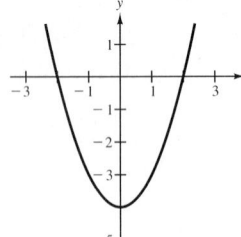   **8.**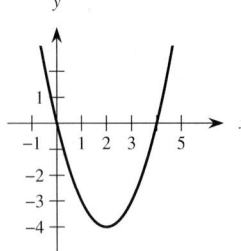

27. 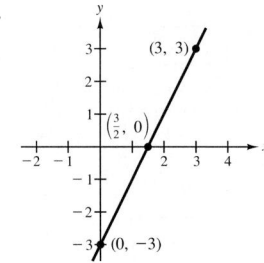   **28.**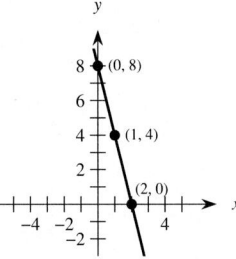

9. 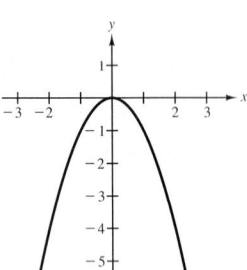   **10.**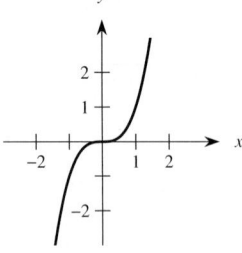

29. 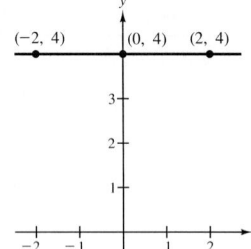   **30.**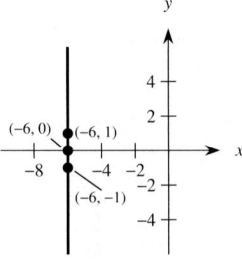

31.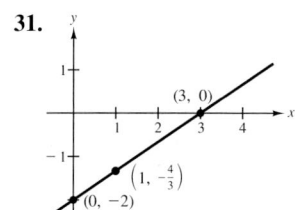

32.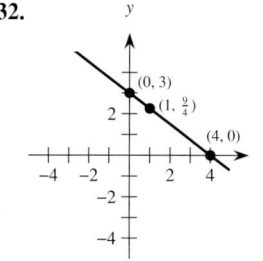

41.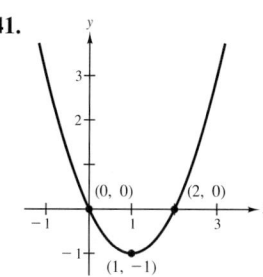

42.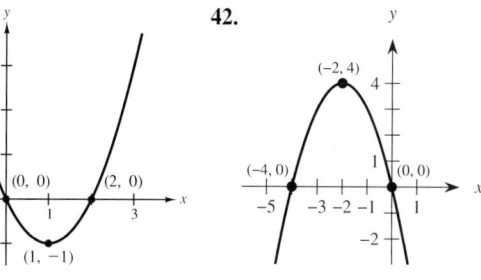

33.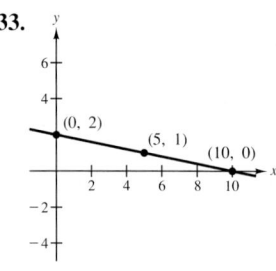

34.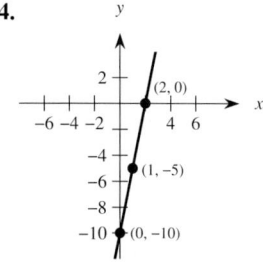

43.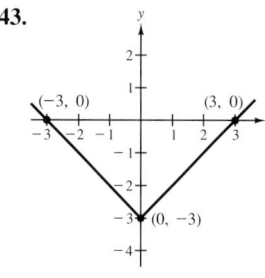

44.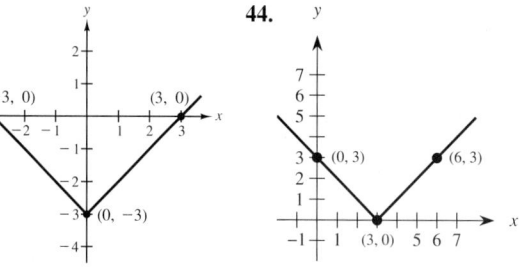

35.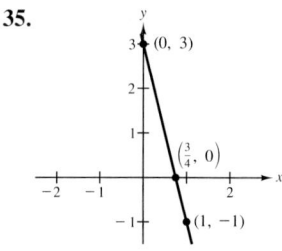

36.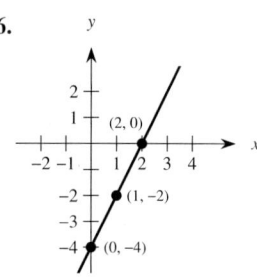

45. $x = 4$

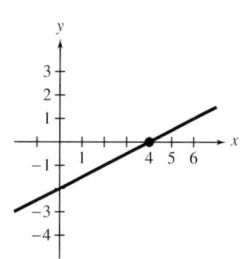

46. $x = 2$

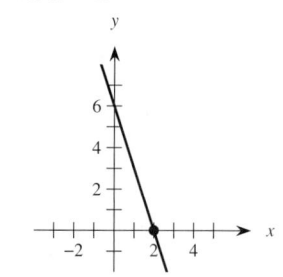

37.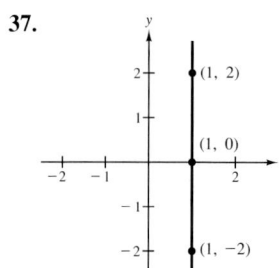

38.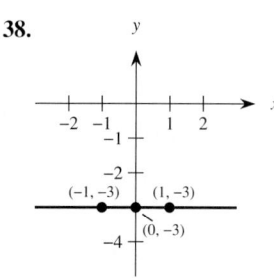

47. $x = \frac{9}{2}$

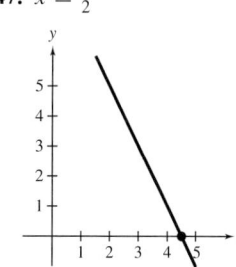

48. $x = -4$

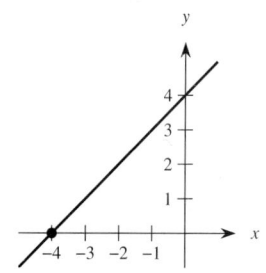

39.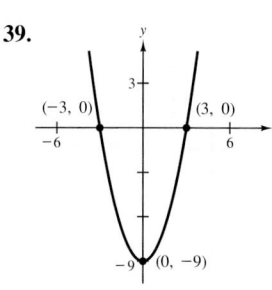

40.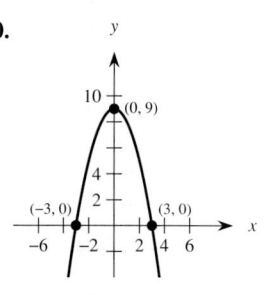

49. $x = \pm 2$

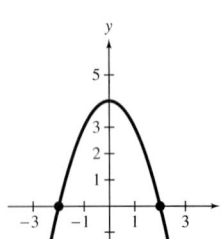

50. $x = 0$, $x = -2$

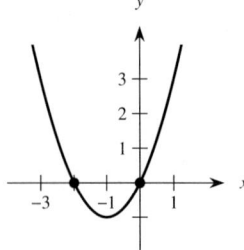

51. $x = 1$

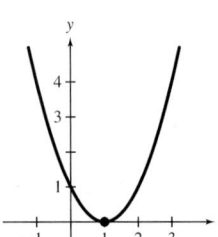

52. $x = 3$, $x = 1$

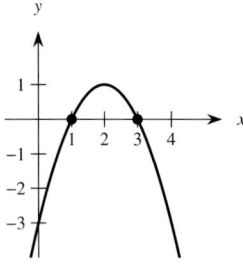

53.

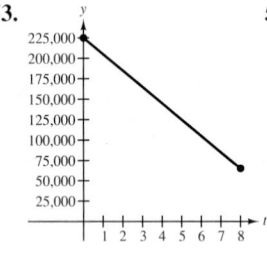

54.

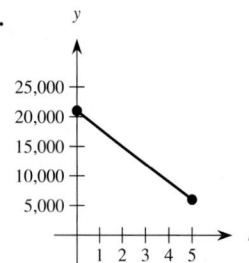

55. (b)

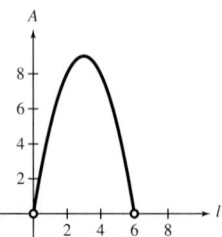

(c) 3 m × 3 m

56. (a) 2010   **(b)** 2010

57. (a)

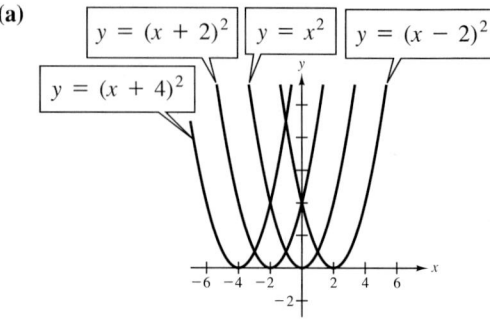

(b) The graph of $y = (x + c)^2$, $c > 0$, is obtained by shifting the graph of $y = x^2$ to the *left c* units. The graph of $y = (x - c)^2$, $c > 0$, is obtained by shifting the graph of $y = x^2$ to the *right c* units.

58. (a)

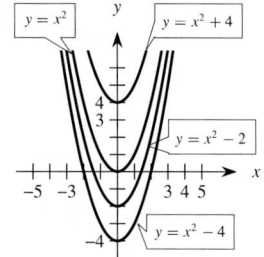

(b) The graph of $y = x^2 + c$, $c > 0$, is obtained by shifting the graph of $y = x^2$ *upward c* units. The graph of $y = x^2 - c$, $c > 0$, is obtained by shifting the graph of $y = x^2$ *downward c* units.

USING A GRAPHING CALCULATOR   (page 149)

1.

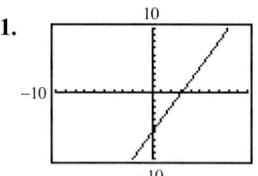

2.

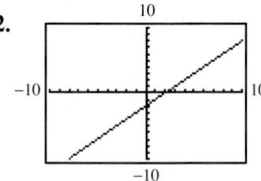

3.

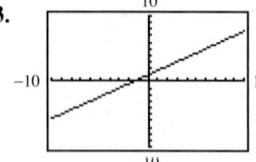

4.

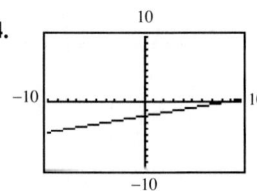

5.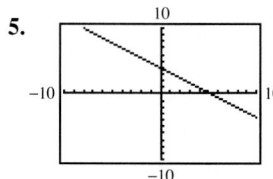

6.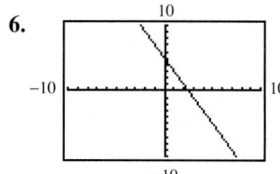

7.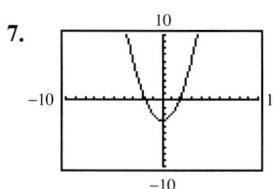

8.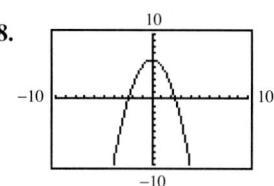

9.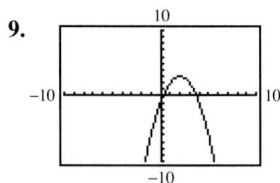

10.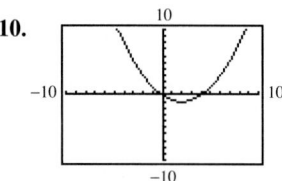

11.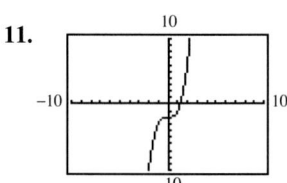

12.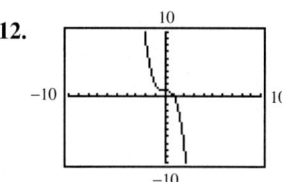

13.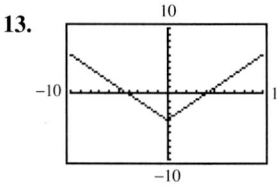

14.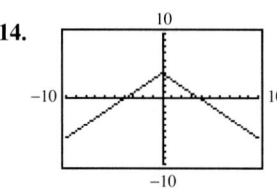

15.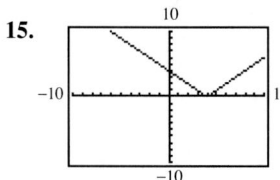

16.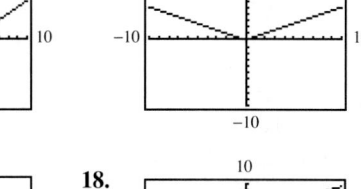

17.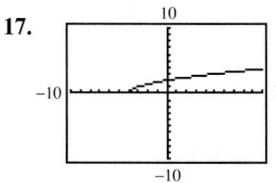

18.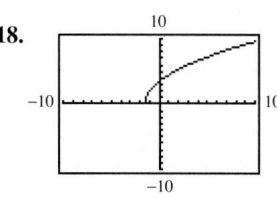

19.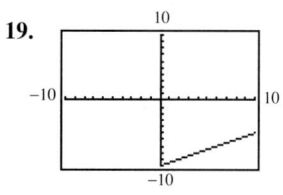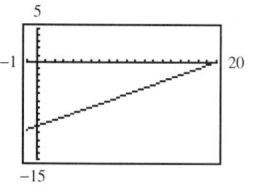

20.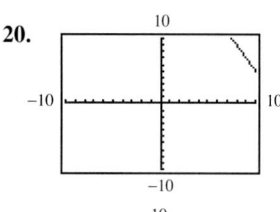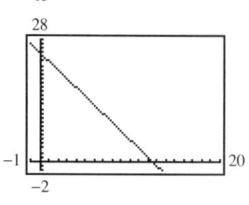

21.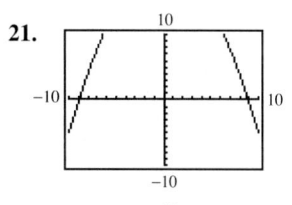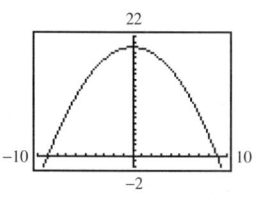

22.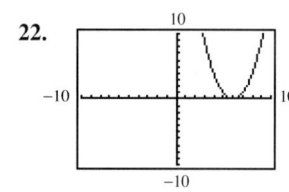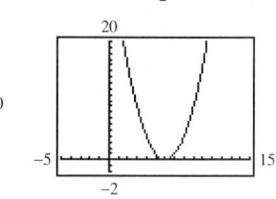

23.
```
RANGE
Xmin=-8
Xmax=8
Xscl=1
Ymin=-1
Ymax=17
Yscl=1
```

24.
```
RANGE
Xmin=-5
Xmax=15
Xscl=1
Ymin=-15
Ymax=5
Yscl=1
```

25.
```
RANGE
Xmin=-2
Xmax=10
Xscl=1
Ymin=-5
Ymax=5
Yscl=1
```

26.
```
RANGE
Xmin=-1
Xmax=8
Xscl=1
Ymin=-6
Ymax=10
Yscl=1
```

27.
```
RANGE
Xmin=-5
Xmax=5
Xscl=1
Ymin=-3
Ymax=6
Yscl=1
```

28.
```
RANGE
Xmin=-3
Xmax=3
Xscl=1
Ymin=-1
Ymax=18
Yscl=1
```

SECTION 2.5 *(page 161)*

Warm-Up *(page 161)*

1. $2h$ **2.** $3t$ **3.** $2x^2 - 16x + 27$
4. $-3x^2 - 12x - 2$ **5.** $-8x - 10$
6. $-2x^2 + 14x$
7. (a) -4 (b) 8 (c) 8 (d) 71
8. (a) 0 (b) 9 (c) 0 (d) 8.75
9. (a) 0 (b) Undefined   (c) 8 (d) -6
10. (a) 2 (b) -2 (c) $\frac{5}{2}$ (d) $\frac{101}{10}$

1. (a) Function from A to B
   (b) Not a function from A to B
   (c) Function from A to B
   (d) Not a function from A to B
2. (a) Not a function from A to B
   (b) Function from A to B
   (c) Not a function from A to B
   (d) Function from A to B
3. $y = 10x + 12$ **4.** $y = -8x + 3$
5. $y = \dfrac{-3x + 2}{7}$ **6.** $y = \dfrac{x + 3}{9}$
7. $y = -x^2 + 2x + 3$ **8.** $y = \dfrac{2x^2 + 4}{3}$
9. $y = \dfrac{-|x| + 4}{2}$ **10.** $y = \dfrac{1}{3}|x + 1|$
11. There are two values of y associated with one value of x.
12. There are two values of y associated with one value of x.
13. There are two values of y associated with one value of x.
14. There are two values of y associated with one value of x.
15. (a) 2 (b) -2 (c) k (d) $k + 1$
16. (a) 0 (b) -3 (c) m (d) $t + 2$
17. (a) $3, 3$ (b) $-4, -4$ (c) s, s (d) $s - 2, s - 2$
18. (a) 1 (b) -4 (c) h (d) $h - 8$
19. (a) 29 (b) 11 (c) $12a - 2$ (d) $12a + 5$
20. (a) 10 (b) $-\frac{1}{2}$ (c) $20 - 7t$ (d) $17 - 7t$
21. (a) 8 (b) $\frac{2}{9}$ (c) $\frac{1}{2}a^2 + 4a + 8$ (d) $8 + \frac{1}{2}a^2$
22. (a) 0 (b) 0 (c) $t^2 - 2t$ (d) $t^2 + 2t$
23. (a) 2 (b) 3 (c) $\sqrt{21}$ (d) $\sqrt{5z + 5}$
24. (a) 4 (b) 4 (c) -7 (d) $8 - |x - 6|$
25. (a) 0 (b) $-\dfrac{3}{2}$ (c) $-\dfrac{5}{2}$ (d) $\dfrac{3x + 12}{x - 1}$

26. (a) 1 (b) 1 (c) -1 (d) 0
27. (a) 2 (b) -2 (c) 10 (d) -8
28. (a) 0 (b) $\frac{3}{2}$ (c) 4 (d) 552
29. (a) 0 (b) $\frac{7}{4}$ (c) 3 (d) 0
30. (a) 0 (b) 1 (c) 0 (d) 11
31. (a) 2 (b) $\dfrac{2(x - 6)}{x}$
32. (a) $x^2 + 2x$ (b) $-2x + 5$
33. (a) $-x^2 - 4x - 3$ (b) $-2x - 2$
34. (a) $x^3 + 6x^2 + 12x + 8$ (b) $3x^2 + 6x + 4$
35. Domain: $\{0, 2, 4, 6\}$
   Range: $\{0, 1, 8, 27\}$
36. Domain: $\{-3, -1, 4, 10\}$
   Range: $\{-\frac{17}{2}, -\frac{5}{2}, 2, 11\}$
37. Domain: All real numbers r such that $r > 0$
   Range: All real numbers C such that $C > 0$
38. Domain: All real numbers s such that $s > 0$
   Range: All real numbers A such that $A > 0$
39. All real numbers x such that $x \neq 3$
40. All real numbers x such that $x \neq -4$
41. All real numbers x such that $x \neq 2, 1$
42. All real numbers x
43. All real numbers t such that $t \neq 0, -2$
44. All real numbers s such that $s \neq 6, 10$
45. All real numbers x
46. All real numbers x such that $x \neq 2, -2$
47. All real numbers x such that $x \geq 2$
48. All real numbers x such that $x \geq 5$
49. All real numbers x such that $x \geq 0$
50. All real numbers x **51.** All real numbers x
52. All real numbers x such that $x \geq 0$
53. All real numbers t **54.** All real numbers x
55. $\left\{(3, 150), (2, 100), (8, 400), (6, 300), \left(\frac{1}{2}, 25\right)\right\}$
56. $\{(1, 1), (2, 8), (3, 27), (4, 64), (5, 125), (6, 216)\}$
57. $\{(1986, \text{New York}), (1987, \text{Minnesota}),$
   $(1988, \text{Los Angeles}), (1989, \text{Oakland}),$
   $(1990, \text{Cincinnati}), (1991, \text{Minnesota}),$
   $(1992, \text{Toronto})\}$
58. $\{(1973, \text{Nixon}), (1977, \text{Carter}),$
   $(1981, \text{Reagan}), (1989, \text{Bush}),$
   $(1993, \text{Clinton})\}$
59. (a) 80.6 (b) 89.3 (c) 74.8 (d) 63.9
60. (a) 50 (b) 53 (c) 55 (d) 57
61. $P(x) = 4x$ **62.** $S(x) = 6x^2$
63. $V(x) = x^3$ **64.** $A(x) = (35 - 2x)^2$
65. $A(x) = (32 - x)^2$ **66.** $C(x) = 8000 + 1.95x$
67. (a) \$240   (b) \$320   (c) \$380   (d) \$440
68. Function   **69.** Function
70. Not a function   **71.** Not a function

SECTION 2.6   *(page 176)*

Warm-Up   *(page 176)*

1.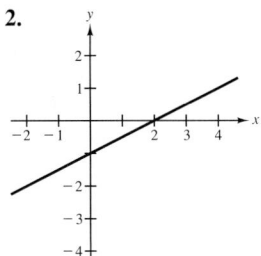

2.

3.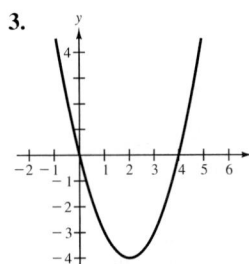

4.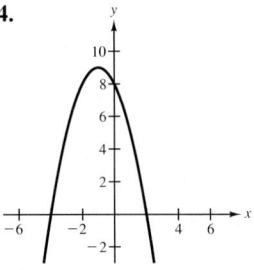

5.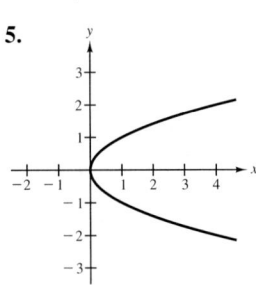

6.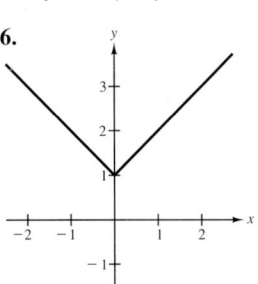

7. **(a).** 12   **(b)** $\frac{3}{16}$ **8. (a)** -7 **(b)** $-2x$

9. (a) $\frac{1}{3}$ **(b)** $\frac{c-6}{c+4}$

10. (a) $2\sqrt{3}$ **(b)** $\sqrt{t-1}$

---

1. c   **2.** b   **3.** e   **4.** a   **5.** f   **6.** d

7.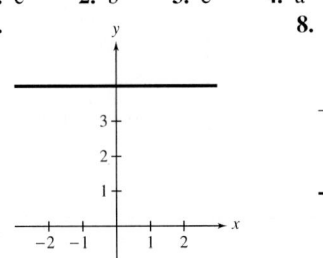

8.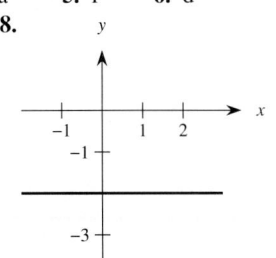

9.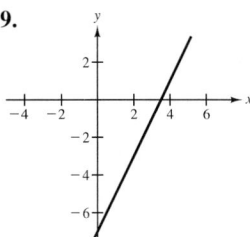

10.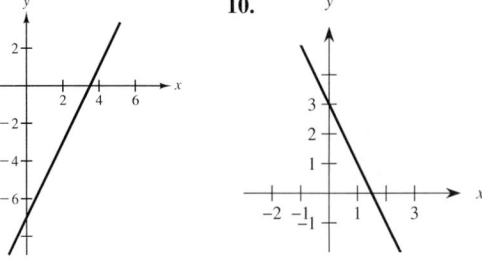

11.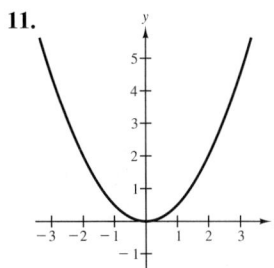

12.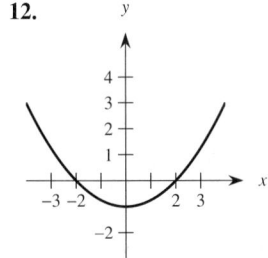

13.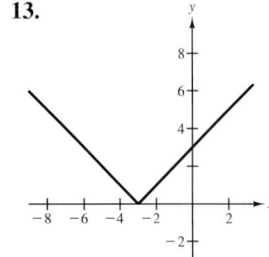

14.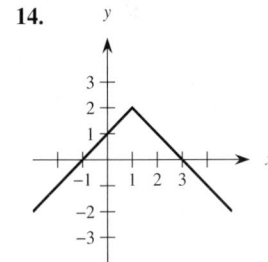

15.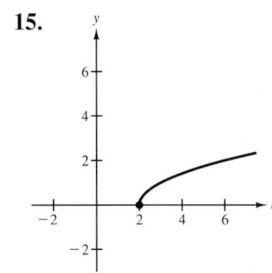

16.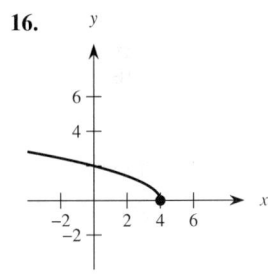

17.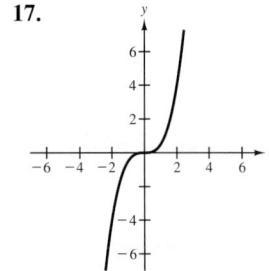

18.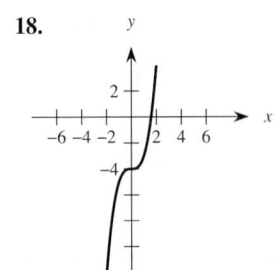

19.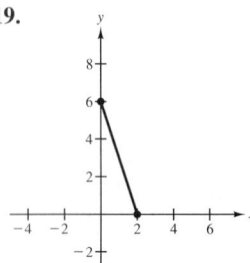

20.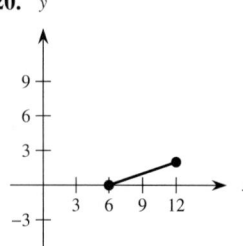

21.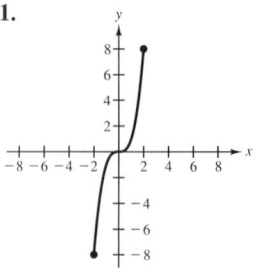

22.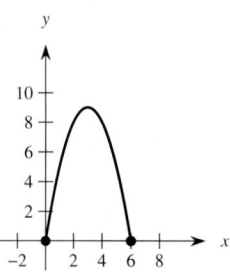

23. y is a function of x.
24. y is a function of x.
25. y is not a function of x.
26. y is a function of x.
27. y is a function of x.

28. y is not a function of x.

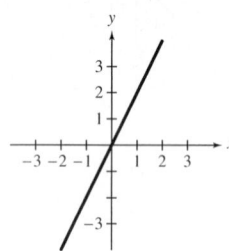

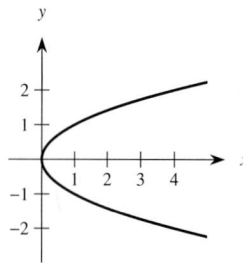

29. y is a function of x.

30. y is a function of x.

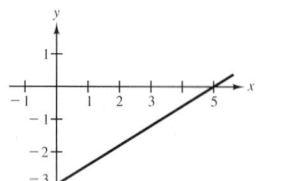

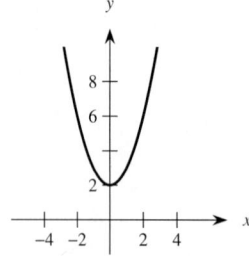

31. y is not a function of x.

32. y is not a function of x.

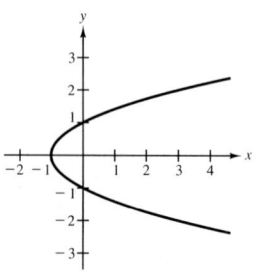

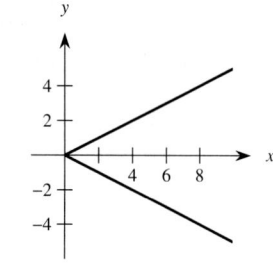

33. y is a function of x.

34. y is not a function of x.

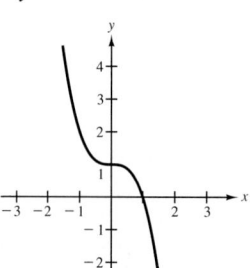

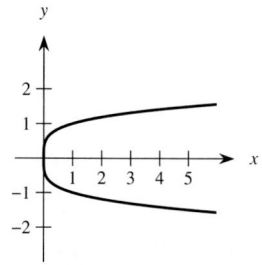

35. Vertical shift two units upward

36. Vertical shift four units downward

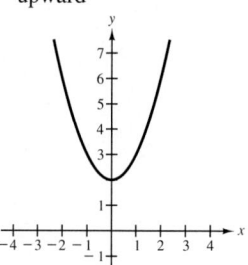

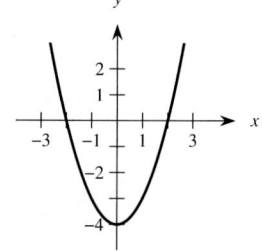

37. Horizontal shift two units to the left

38. Horizontal shift four units to the right

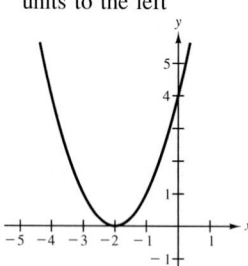

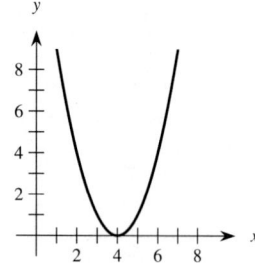

39. Reflection in the x-axis

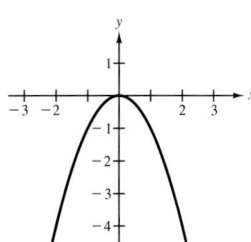

40. Reflection in the x-axis and a vertical shift four units upward

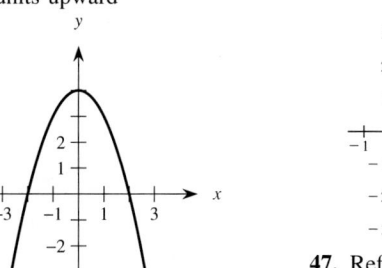

41. Horizontal shift three units to the right and a reflection in the x-axis

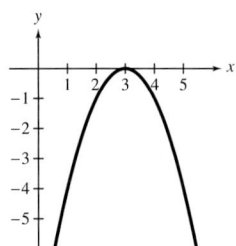

42. Horizontal shift one unit to the left, a reflection in the x-axis, and a vertical shift two units upward

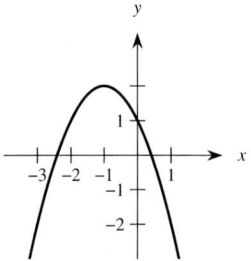

43. Vertical shift three units upward

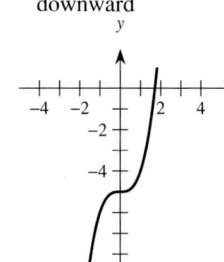

44. Vertical shift five units downward

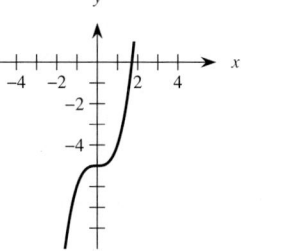

45. Horizontal shift three units to the right

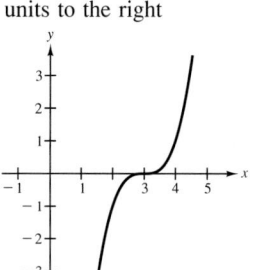

46. Horizontal shift two units to the left

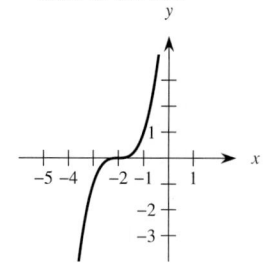

47. Reflection in the y-axis

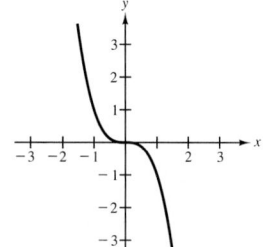

48. Reflection in the x-axis

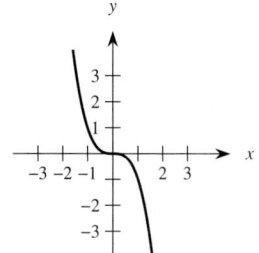

49. Reflection in the x-axis followed by a horizontal shift one unit to the right followed by a vertical shift two units upward

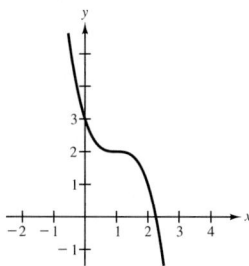

50. Vertical shift three units downward and a horizontal shift two units to the left

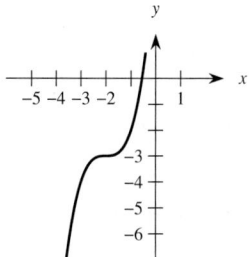

51. Horizontal shift five units to the right

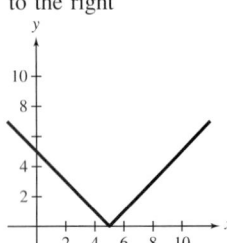

52. Vertical shift five units downward

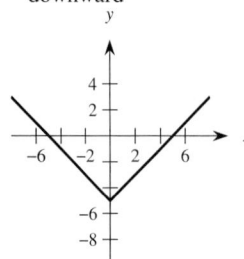

53. Reflection in the x-axis

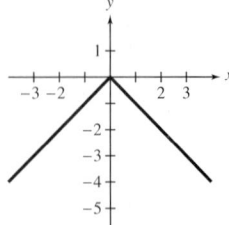

54. Reflection in the x-axis followed by a vertical shift five units upward

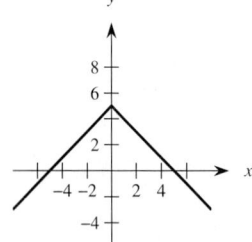

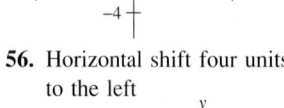

55. Vertical shift two units upward

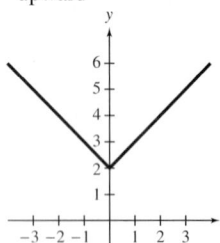

56. Horizontal shift four units to the left

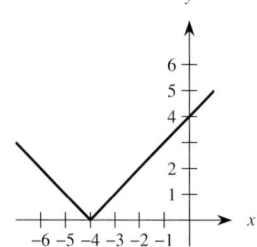

57. Horizontal shift three units to the left and a vertical shift one unit downward

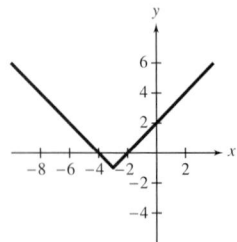

58. Horizontal shift four units to the right and two units upward

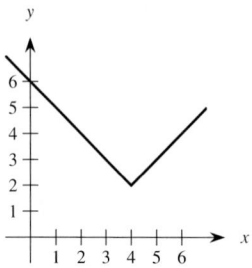

59. $y = -\sqrt{x}$ **60.** $y = \sqrt{x} + 1$
61. $y = \sqrt{x + 2}$ **62.** $y = \sqrt{x - 3}$
63. $y = \sqrt{-x} + 2$ **64.** $y = \sqrt{x + 3} + 2$

65.
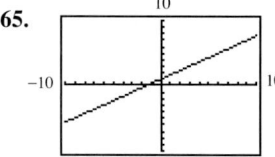

66.
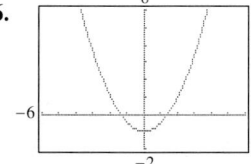

67.
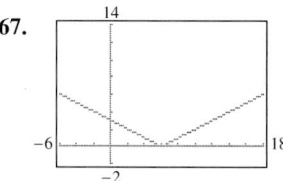

68.
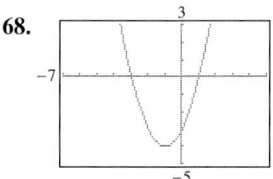

69.
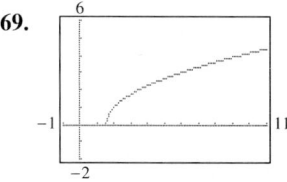

70.
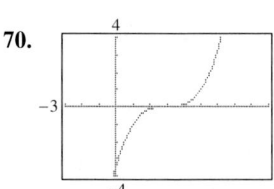

71. The graph of $f(x) = \sqrt{x - 3}$ indicates that only those x-values that are greater than or equal to 3 are included.
Domain: All real numbers $x \geq 3$.

72. The graph of $f(x) = \sqrt{x + 1}$ indicates that only those x-values that are greater than or equal to -1 are included.
Domain: All real numbers $x \geq -1$.

73. (a) $0 < x < 4$

(b)

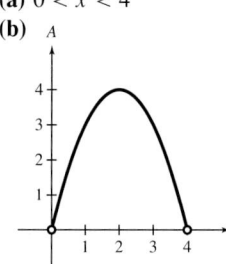

$x = 2$

74. (a) 4 ft

(b)

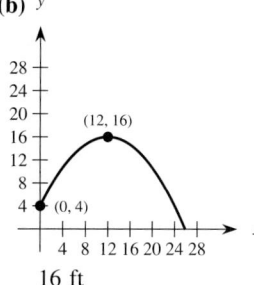

16 ft

(c) 26 ft

CHAPTER 2 REVIEW EXERCISES   *(page 180)*

1. (a) No   **(b)** Yes   **2. (a)** Yes   **(b)** No
3. (a) Yes   **(b)** No   **4. (a)** No   **(b)** Yes
5. 2   **6.** 8   **7.** 14   **8.** $-\frac{6}{5}$
9. -8.2 **10.** 2   **11.** $\frac{16}{3}$ **12.** $-\frac{7}{5}$
13. 2   **14.** $\frac{17}{2}$ **15.** $\frac{20}{17}$ **16.** $\frac{5}{7}$
17. 0, 3   **18.** 0, $-\frac{7}{4}$ **19.** 10, -10 **20.** 2, -8
21. 15, 10   **22.** $\frac{3}{2}$ **23.** $-\frac{4}{3}$, 2
24. 9, -4 **25.** 3.38   **26.** 5.14   **27.** 33.32
28. 2.41   **29.** $x = \frac{1}{2}(7y - 4)$ **30.** $v = \dfrac{2u - 9}{18}$

31. $h = \dfrac{V}{\pi r^2}$ **32.** $h = \dfrac{S - 2\pi r^2}{2\pi r}$

33.

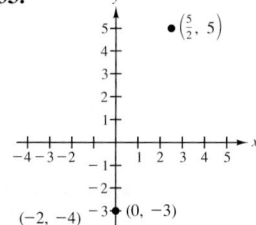

34.

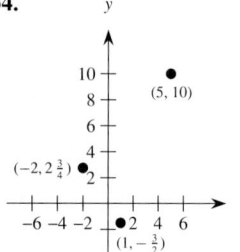

35. $d = 5$ **36.** $d = 4$

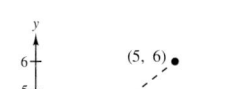

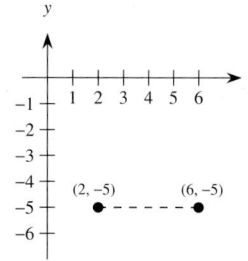

37. $d \approx 7.62$

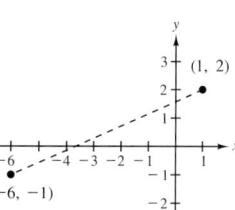

38. $d = 13$

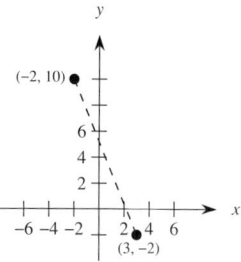

39.

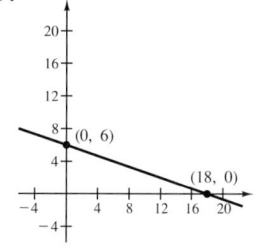

40.

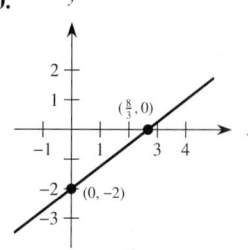

41.

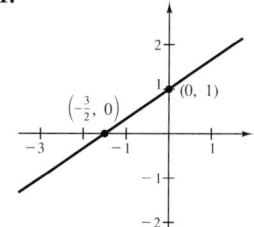

42.

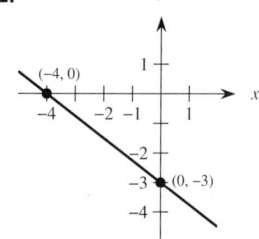

43.

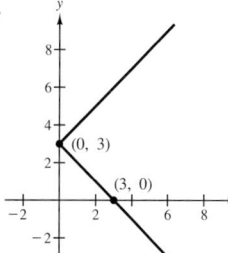

44.

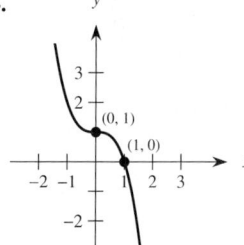

45. y is not a function of x.   **46.** y is a function of x.

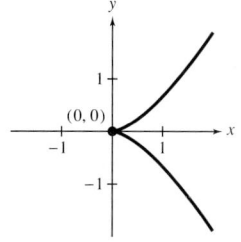

47. y is a function of x.   **48.** y is not a function of x.

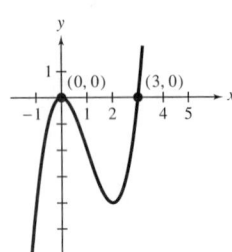

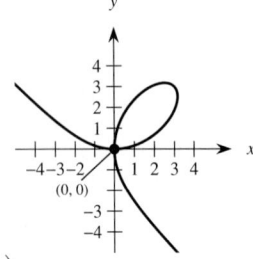

49. $(9, 0)$, $(6, 0)$ **50.** $\left(\frac{8}{5}, 0\right)$, $(0, -4)$

51. No x-intercept, $(0, -1)$, $(0, 1)$
52. $(-4, 0)$, $(0, -6)$
53. $(8, 0)$, $(-8, 0)$, $(0, 8)$ **54.** $(0, 0)$, $(2, 0)$
55. $(2, 0)$, $(0, 4)$ **56.** $(1, 0)$, $(5, 0)$, $(0, -5)$

57. (a) 29   (b) 3   (c) $\dfrac{36 - 5t}{2}$ (d) $4 - \dfrac{5}{2}(x + h)$

58. (a) 0   (b) 20   (c) 33   (d) $t^2 - 16$

59. (a) 3   (b) 0   (c) $\sqrt{2}$ (d) $\sqrt{5 - 5x}$

60. (a) 1   (b) 1   (c) $\dfrac{5}{4}$ (d) $\dfrac{|x + 2|}{4}$

61. (a) -3 (b) 2   (c) 0   (d) -7
62. (a) 2   (b) $-\frac{1}{8}$ (c) 0   (d) 5

63. (a) -2 (b) $\dfrac{2(6 - x)}{x}$

64. (a) 3   (b) $\dfrac{3x - 30}{x}$

65. (a) $(x + h)^2 = x^2 + 2xh + h^2$ (b) $2xh + h^2$
66. (a) 8   (b) 0   **67.** All real numbers x
68. All real number s such that $s \neq 0$
69. All real numbers x such that $x \neq 5$
70. All real numbers x such that $x \geq -4$
71. (a) Reflection of the graph of $f(x) = x^2$ in the x-axis with a left shift of one unit
(b)

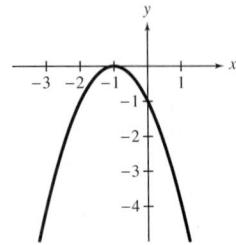

72. (a) A reflection of $y = x^2$ about the x-axis followed by a horizontal shift two units to the right and a vertical shift nine units upward
(b)

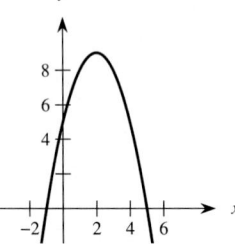

73. (a) The graph of $f(t) = |t|$ shifted two units to the right and one unit downward
(b)

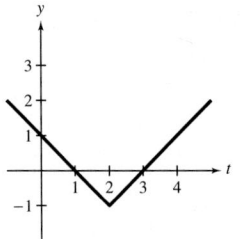

74. (a) A horizontal shift of $y = |x|$ one unit to the left and a vertical shift two units downward
(b)

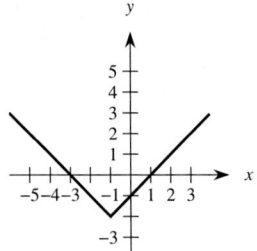

75. (a) The graph of $f(x) = x^3$ shifted downward two units
(b)

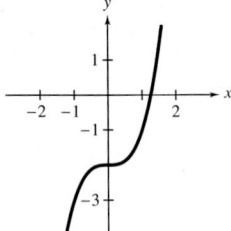

76. (a) A horizontal shift of $y = x^3$ two units to the right
(b)

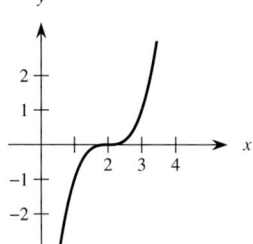

77. 7.96
78 (a) 16 ft/sec **(b)** $\frac{5}{2}$ sec **(c)** -16 ft/sec
79. 6 in. × 8 in. **80.** 20 ft × 35 ft

CHAPTER 2 TEST *(page 183)*

1. (a) No **(b)** Yes **2.** $r = \frac{1}{3}(2s - 1)$
3. 4 **4.** 4 **5.** 24
6. $\frac{19}{2}$ **7.** $-5, 1$ **8.** $-\frac{4}{3}, 3$
9. (a) Quadrant I, Quadrant III
(b) The product xy is positive, and therefore x and y have the same sign. If x and y are both positive, the point is located to the *right* of the vertical axis and *above* the horizontal axis in the first quadrant. If x and y are both negative, the point is located to the *left* of the vertical axis and *below* the horizontal axis in the third quadrant.
10. $\sqrt{73} \approx 8.54$
11. $(3, 0), (-1, 0), (0, -3)$
12.

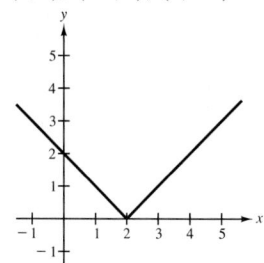

13. y is not a function of x.
14. (a) 16 **(b)** $3h$ **(c)** $3t + 1$ **(d)** $s - 2$

15.

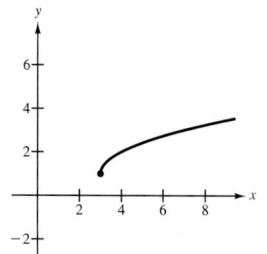

16. (a) $y = |x - 2|$ **(b)** $y = |x| - 2$ **(c)** $y = 2 - |x|$
17. 9 cm × 6 cm **18.** \$2000

Chapter 3

Chapter Opener *(page 185)*

1120 miles, 30 days

SECTION 3.1 *(page 198)*

> **Warm-Up** *(page 197)*
>
> **1.** $43 - 6x$ **2.** $u^2 - 16$ **3.** $2n - 3$
> **4.** $16n + 6$ **5.** -4 **6.** 27 **7.** 14
> **8.** -19 **9.** $\frac{940}{3}$ **10.** $\frac{1}{10}$

1. $8 + n$ **2.** $n - 6$ **3.** $15 - 3n$ **4.** $4n - 6$
5. $\frac{1}{3}n$ **6.** $\frac{7n}{5}$ **7.** $0.30L$ **8.** $0.40C$
9. $\frac{x}{6}$ **10.** $\frac{y}{3}$ **11.** $\frac{3 + 4x}{8}$
12. $10y - 35$ **13.** $|n - 5|$ **14.** $\left|\frac{y}{4}\right|$
15. The sum of three times a number and 2
16. The difference of four times a number and 5
17. Eight times the difference of a number and 5
18. Negative three times the sum of a number and 2
19. The ratio of a number to 8
20. Four-fifths of a number
21. The sum of a number and 10, all divided by 3
22. Twenty-five more than the quotient of a number and 6
23. $0.25n$ **24.** $10m + 25n$ **25.** $55t$
26. $5r$ **27.** $\frac{100}{r}$ **28.** $\frac{360}{t}$

29. $0.45y$ **30.** $0.65q$ **31.** $0.0125l$
32. $0.06L$ **33.** $0.80L$ **34.** $18 + 3n$
35. $8.25 + 0.60q$ **36.** $11.65 + 0.80q$ **37.** s^2
38. $6w$ **39.** $3.24l$ **40.** $0.375b^2$
41. $4n + 2$ **42.** $2n + 1$ **43.** $6w^2$ ft^2
44. $l^2 - 6l$ ft^2 **45.** 44, 46, 48   **46.** 15, 17
47. 10   **48.** 42

49.   Number $+$ 30 $=$ 82
$$x + 30 = 82$$
$$x = 52$$

50.   6 $\cdot$ Number -12 $=$ 300
$$6(x - 12) = 300$$
$$x = 62$$

51.   3 $\cdot$ Side $=$ Perimeter
$$3s = 129$$
$$s = 43$$

52.   Perimeter $= 2 \cdot$ Length
$+ 2 \cdot$ Width
$$64 = 2(3w) + 2w$$
$$w = 8, \ 3w = 24$$

53.   Cost for parts $+$ Labor cost per hour $\cdot$
Number of hours $=$ Total cost
$$275 + 35x = 380$$
$$x = 3$$

54.   Earnings $= 8 \cdot$ Hourly pay
$+ 0.75 \cdot$ Number of units
$$146 = 8(10) + 0.75x, \ x = 88$$

55.   Number of people laid off $=$
Percentage of work force $\cdot$
Number of employees
$$25 = p(160)$$
$$0.15625 = p \ \text{ or } \ 15.625\%$$

56. 7%

57.   Number of defective parts in sample $=$
Percentage of defective parts in sample $\cdot$
Size of sample
$$3 = 0.015x$$
$$200 = x$$

58. $22,391

59. $0.80 \cdot$ Possible points $- 372 = 8$
$$0.80x - 372 = 8$$
$$x = 475$$

60. 41.9%   **61.** 2004   **62.** 379   **63.** 7315
64. 87.9%   **65.** $\frac{3}{4}$ **66.** $\frac{10}{3}$ **67.** $\frac{1}{25}$
68. $\frac{5}{8}$ **69.** $\frac{85}{4}$ **70.** $\frac{3}{2}$ **71.** 4   **72.** $\frac{216}{7}$
73. $\frac{15}{12}$ **74.** $\frac{7}{4}$ **75.** 16   **76.** $\frac{20}{3}$
77. 20 pt   **78.** $10\frac{1}{2}$ cups   **79.** $1800
80. 46,400 votes   **81.** 350 mi   **82.** 112.5 min
83. 3   **84.** 5.09   **85.** 46.9 ft   **86.** 6.43 ft

SECTION 3.2   *(page 210)*

Warm-Up   *(page 210)*

1. $\frac{11}{4}$ **2.** 5   **3.** $\frac{5}{9}$ **4.** $\frac{10}{3}$ **5.** $\frac{2}{5}$
6. 20   **7.** -4 **8.** $-\frac{6}{5}$ **9.** 9   **10.** 50

---

1. $18.38, 40%   **2.** $29.47, 35%
3. $152.00, 65%   **4.** $419.25, 44%
5. $22,250.00, 21%   **6.** $12,550.00, 31%
7. $416.70, $191.70   **8.** $976.00, $244.00
9. $23.76, 48%   **10.** $39.27, 33%
11. $111.00, 63%   **12.** $210.45, 39%
13. $33.25, $61.75   **14.** $19.90, $3.98
15. $1145.00, 22%   **16.** $459.50, 44%
17. $54.15   **18.** $0.33/lb
19. Department store   **20.** Mail-order
21. 2.5 hr   **22.** 2 hr
23. 9 min, $2.06   **24.** 8 hr   **25.** 18.3%
26. 15.2%   **27.** 9%   **28.** $1034.40
29. Tax $=$ $267;   Total bill $=$ $4717;
Amount financed $=$ $3717

30. $580   **31.** 30 $0.15 stamps; 40 $0.30 stamps
32. 15 dimes; 35 quarters   **33.** 700   **34.** 100
35. 500—$3.00, 400—$2.74, 300—$2.48, 200—$2.22,
  100—$1.96, 0—$1.70
  (a) Decreases   **(b)** Decreases
  (c) The price of the mixture would be equal to the
    average price of the oats and the corn per bushel.
36. 5—$16, 4—$23.20, 3—$30.40, 2—$37.60,
  1—$44.80, 0—$52
  (a) Decreases   **(b)** Increases
  (c) The price of the alloy would be the average of the
    prices of metal A and metal B.
37. 4 roses   **38.** 75 lb at $3.88; 25 lb at $4.88
39. $30,000 at 8.5%; $10,000 at 10%
40. $1000 at 6.5%; $3000 at 9%
41. 50 gal at 20%;   50 gal at 60%
42. 6 liters at 50%; 4 liters at 75%
43. 8 qt at 15%;   16 qt at 60%
44. 13.75 gal at 60%; 41.25 gal at 80%
45. $\frac{5}{6}$ gal   **46.** 0.61 gal   **47.** 1440 mi
48. 375 m/min   **49.** $\frac{1}{3}$ hr   **50.** 3 hr, $2\frac{3}{4}$ hr
51. 43.6 mi/hr   **52.** $\frac{2}{3}$ hr; $3\frac{1}{3}$ mi
53. $83\frac{1}{3}$ mi/hr   **54.** 55.6 mi/hr   **55.** $\frac{8}{15}t$
56. $\frac{1}{3}$, $\frac{1}{4}$, $1\frac{5}{7}$ hr   **57.** 10 min   **58.** 6 hr
59. $\frac{1}{5}$, $\frac{1}{8}$, $\frac{40}{13}$, or $3\frac{1}{13}$ hr
60. 1985, $0.366/hr   **61.** 1985, $0.286
62. (a) 10.899 million barrels   **(b)** 8   **63.** 8.3%

SECTION 3.3   *(page 226)*

Warm-Up *(page 226)*

1. $-\frac{3}{4} > -5$ **2.** $-\frac{1}{5} > -\frac{1}{3}$ **3.** $\pi > -3$
4. $6 < \frac{13}{2}$ **5.** $\frac{7}{2}$ **6.** $\frac{48}{7}$ **7.** 40
8. 153   **9.** -12 **10.** 24

---

1. (a) Yes   **(b)** No   **(c)** Yes   **(d)** No
2. (a) No   **(b)** No   **(c)** Yes   **(d)** No
3. (a) No   **(b)** Yes   **(c)** Yes   **(d)** No
4. (a) Yes   **(b)** Yes   **(c)** No   **(d)** No
5. f   **6.** c   **7.** a   **8.** e   **9.** d   **10.** b

11. $(-5, 3]$

12. $[1, 4)$

13. $\left(0, \frac{3}{2}\right]$

14. $(-7, -3]$

15. $\left(-\infty, -\frac{5}{2}\right)$

16. $[-5, \infty)$

17. $x \leq 2$

18. $z > 4$

19. $x < \frac{11}{2}$

20. $x > \frac{5}{2}$

21. $x \leq -4$

22. $x \geq -4$

23. $x > 8$

24. $x < 10$

25. $x > 7$

26. $x \leq 6$

27. $x \geq 7$

28. $y \le 6$

29. $x > -\frac{2}{3}$

30. $x < \frac{7}{5}$

31. $x > \frac{9}{2}$

32. $x \le 2$

33. $2 < y$

34. $z < \frac{10}{3}$

35. $-10 \ge y$

36. $z \le 0$

37. $x \ge -12$

38. $x \ge \frac{12}{7}$

39. $\frac{5}{2} < x < 7$

40. $0 \ge x > -5$

41. $-\frac{3}{2} < x < \frac{9}{2}$

42. $5 \le x < 13$

43. $1 < x < 10$

44. $8 > x \ge 2$

45. $-\frac{3}{5} < x < -\frac{1}{5}$

46. $6 > x > -6$

47. $x \ge 0$ **48.** $y > -2$ **49.** $z \ge 2$
50. $450 \le x \le 500$ **51.** x is at least $\frac{5}{2}$.
52. t is less than 4.
53. z is greater than 0 and no more than π.
54. t is at least -4 but no more than 4.
55. \$2600   **56.** \$525
57. The average temperature in Miami is greater than the average temperature in New York.
58. Elevation of San Francisco is less than the elevation of Denver.
59. $m < 25{,}357.14286$ or $m \le 25{,}357$
60. 29,655 mi   **61.** $x \ge 31$ **62.** $x \ge 942$
63. The call must be less than 6.38 minutes. If a portion of a minute is billed as a full minute, then the call must be less than or equal to 6 minutes.
64. The call must be less than 15.26 minutes. If a portion of a minute is billed as a full minute, then the call must be less than or equal to 15 minutes.
65. $2 \le d \le 8$ **66.** $5 \le d \le 15$
67. $-8 < -t \le 5$ **68.** $-10 \le -x \le 3$
69. $3 \le n \le \frac{15}{2}$ **70.** $3 \le n \le 9$
71. $n > 21$ **72.** $n > 32$
73. 1970–1974   **74.** 1985, 1986
75. 1993   **76.** $a + b > c$
77. $10{,}000 \le b \le 120{,}000$
78. $1000 \le B \le 120{,}000$

CHAPTER 3 MID-CHAPTER QUIZ   *(page 230)*

1. $5m + 10n$ **2.** $L + 0.06L = 1.06L$ **3.** $\frac{4}{3}$
4. 262,263   **5.** 2 in. × 24 in.   **6.** \$563,952
7. Taxes: 7.12%
   Employee benefits: 10.98%
   Miscellaneous: 12.46%
   Insurance: 2.37%
   Supplies: 2.67%
   Utilities: 1.19%
   Rent: 3.86%
   Wages: 59.35%

8. $\frac{1}{50}$ **9.** 17.1 gal   **10.** $\frac{10}{3}$ **11.** $188

12. 10 hr   **13.** 6 ml   **14.** $\frac{18}{7}$ or $2\frac{4}{7}$ hr

15. $x \geq 4$

16. $t > \frac{1}{3}$

17. $-6 \leq x < 4$

18. $x \geq \frac{6}{5}$

19. $x > 33$

SECTION 3.4   *(page 240)*

Warm-Up   *(page 240)*

1. 15   **2.** 3.2   **3.** -72
4. -4 **5.** 18   **6.** -19.8
7. $-\frac{1}{2} < x < \frac{1}{2}$ **8.** $-\frac{1}{2} \leq x \leq \frac{3}{2}$
9. $-6 \leq x \leq 6$ **10.** $20 < x < 30$

---

1. 45, -45 **2.** 16, -16 **3.** 0
4. No solution   **5.** 21, 11   **6.** 200, 0
7. 11, -14 **8.** $\frac{2}{7}$, -2 **9.** $\frac{16}{3}$, 16
10. $\frac{16}{5}$, -2 **11.** No solution   **12.** -6, 14
13. $-\frac{11}{5}$, $\frac{17}{5}$ **14.** $-\frac{11}{4}$, $-\frac{29}{4}$ **15.** $-\frac{39}{2}$, $\frac{15}{2}$
16. $\frac{5}{8}$, $\frac{25}{8}$ **17.** 18.75, -6.25 **18.** 1.14, 4.93
19. -3, 7   **20.** $\frac{3}{4}$, $\frac{17}{2}$ **21.** 11, 13
22. $\frac{21}{2}$, $-\frac{29}{8}$ **23.** $-\frac{15}{4}$ **24.** -1
25. (a) Yes   **(b)** No   **(c)** No   **(d)** Yes
26. (a) No   **(b)** Yes   **(c)** Yes   **(d)** No
27. (a) No   **(b)** Yes   **(c)** Yes   **(d)** No
28. (a) Yes   **(b)** No   **(c)** No   **(d)** Yes
29. (a) Yes   **(b)** No   **(c)** No   **(d)** Yes
30. (a) No   **(b)** Yes   **(c)** Yes   **(d)** No
31. (a) No   **(b)** Yes   **(c)** Yes   **(d)** No
32. (a) Yes   **(b)** No   **(c)** No   **(d)** Yes

33.

34.

35.

36.

37. $-4 < y < 4$

38. $-6 < x < 6$

39. $-2 \leq y \leq 6$

40. $-3 \leq x \leq 9$

41. $-6 < y < 2$

42. $-9 < x < 3$

43. $y \leq -4$ or $y \geq 4$

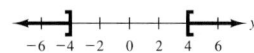

44. $x \leq -6$ or $x \geq 6$

45. $y < -2$ or $y > 6$

46. $x < -3$ or $x > 9$

47. $y \leq -6$ or $y \geq 2$

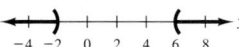

48. $x \le -9$ or $x \ge 3$

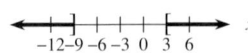

49. $-7 < x < 7$

50. $-\frac{9}{4} \le z \le \frac{9}{4}$

51. $-9 \le y \le 9$

52. $-8 < t < 8$

53. $x < -2$ or $x > \frac{2}{3}$

54. $x \le -1$ or $x \ge 2$

55. $x < 3$ or $x > 7$

56. $-5 < a < 3$

57. $-5 < x < 35$ **58.** $-5.\overline{3} \le t \le 16$

59. $-82 \le x \le 78$ **60.** $-104 < y < 136$

61. $t \le -\frac{15}{2}$ or $t \ge \frac{5}{2}$

62. $t < -2$ or $t > \frac{4}{3}$

63. $s > 23$ or $s < -17$

64. $a \le -38$ or $a \ge 26$

65. $z < -50$ or $z > 110$ **66.** No solution

67.

68.

69. $|x| \le 2$ **70.** $|x| < 4$

71. $|x - 10| < 3$ **72.** $|x - 75| \le 3$

73. $|x - 19| < 3$ **74.** $|x + 11| \le 2$

75. $|x| < 3$ **76.** $|x| > 2$ **77.** $|x - 5| > 6$

78. $|x - 16| < 5$ **79.** $|N - 222.5| \le 102.5$

80. $144 \le N \le 244$

$|N - 194| \le 50$

81. $-1 < x < 7$ **82.** $-4 < x < 2$

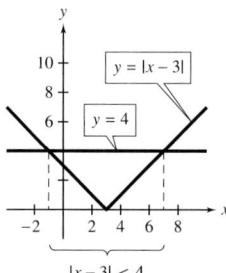

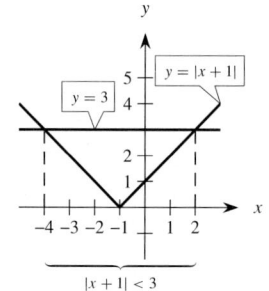

83. $x < -6$ or $x > 2$ **84.** $x < -4$ or $x > 6$

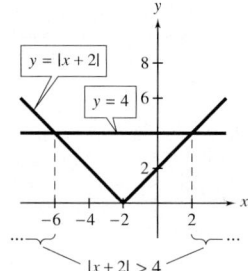

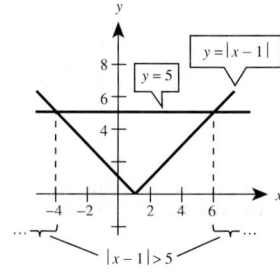

SECTION 3.5   *(page 255)*

Warm-Up   *(page 255)*

1. $\frac{1}{2}$ **2.** $\frac{1}{2}$ **3.** $-\frac{29}{2}$ **4.** $-\frac{1}{2}$

5. $\frac{16}{5}$ **6.** $-\frac{7}{3}$ **7.** $y = \frac{5}{8}(x - 1)$

8. $y = -\frac{9}{2}x$ **9.** $y = 2x - 1$

10. $y = \frac{1}{4}(3x + 20)$

1. $f(x) = 3x + 10$ **2.** $f(x) = -\frac{1}{2}x + 4$

3. $f(x) = -\frac{1}{4}x + \frac{45}{8}$ **4.** $f(x) = \frac{8}{9}x$

5. $f(x) = 4$ **6.** $f(x) = \frac{2}{3}x + \frac{13}{3}$

7. $(0, -6)$, $(3, 0)$ **8.** $(0, 6)$, $(2, 0)$

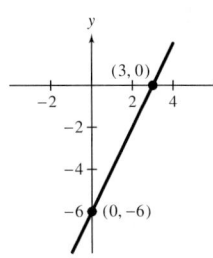   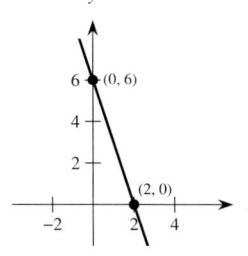

9. $(0, 1)$, $\left(\frac{4}{3}, 0\right)$ **10.** $(0, 2)$, $\left(-\frac{5}{2}, 0\right)$

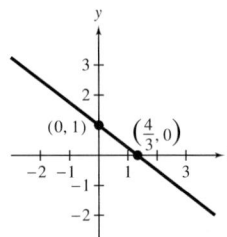   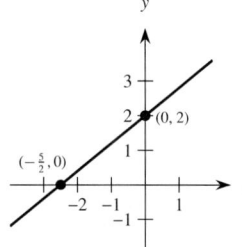

11. $\frac{2}{3}$ **12.** $-\frac{1}{2}$ **13.** -2

14. 4 **15.** Undefined   **16.** 0

17. (a) L_3 **(b)** L_2 **(c)** L_1

18. (a) L_2 **(b)** L_3 **(c)** L_1

19. $\frac{5}{7}$, rises   **20.** $\frac{4}{3}$, rises

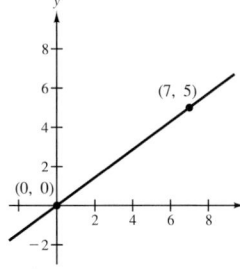   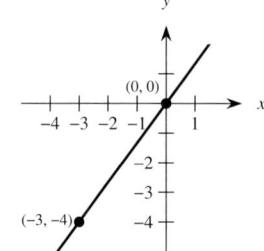

21. $\frac{1}{2}$, rises   **22.** $\frac{2}{3}$, rises

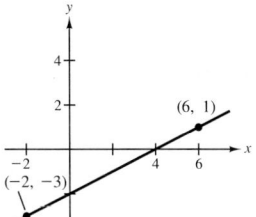   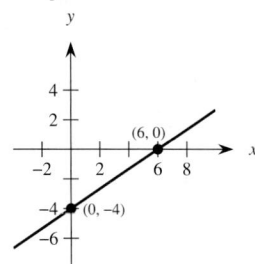

23. $-\frac{3}{2}$, falls   **24.** -4, falls

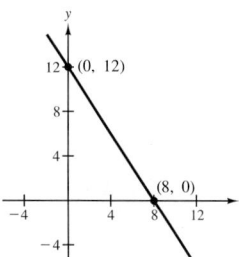   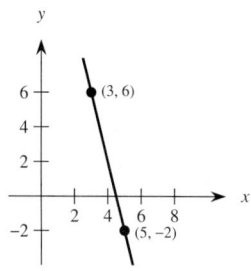

25. Undefined, vertical   **26.** $-\frac{8}{5}$, falls

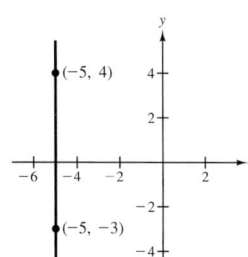   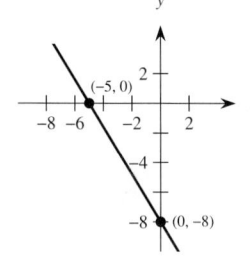

27. 0, horizontal   **28.** 2, rises

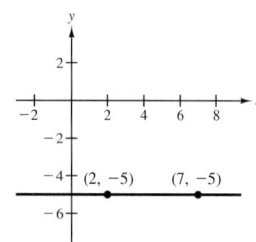   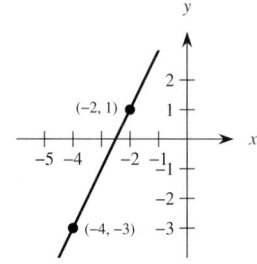

29. $-\frac{18}{17}$, falls   **30.** $\frac{8}{17}$, rises

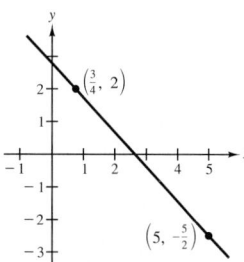   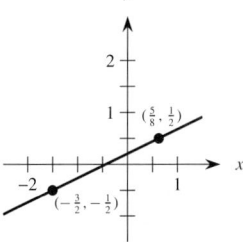

31. $-\frac{5}{6}$, falls

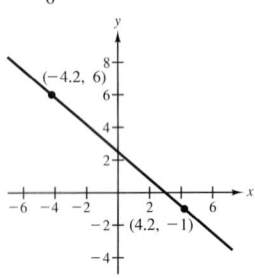

32. Undefined, vertical

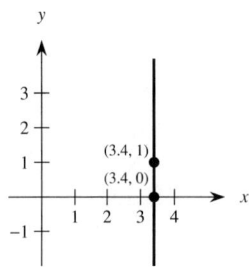

33. $\frac{29}{9}$, rises

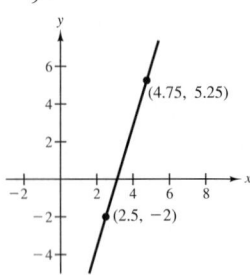

34. 0, horizontal

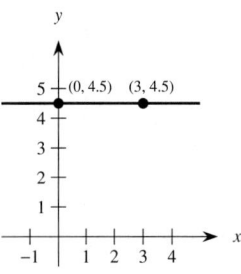

35. $\frac{5}{3}$, rises

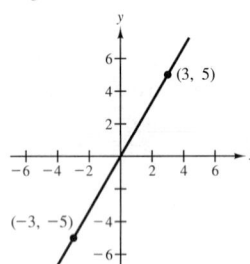

36. $\frac{6}{5}$, rises

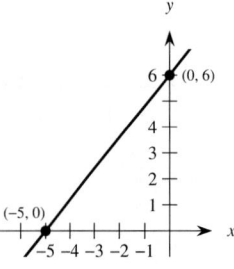

37. $x = 1$ **38.** $x = \frac{7}{3}$

39. $y = -15$ **40.** $y = -10$

41.

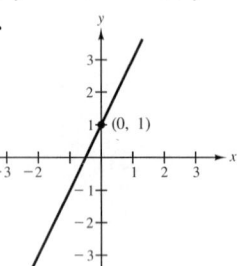

42.

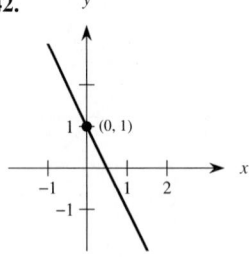

43.

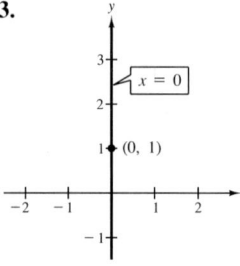

44.

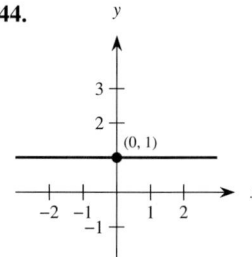

45.

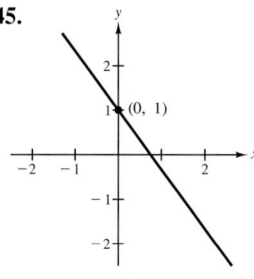

46.

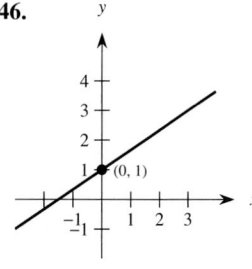

47.

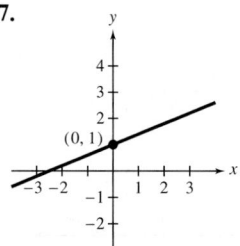

48.

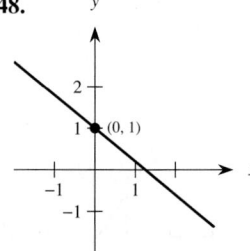

49.

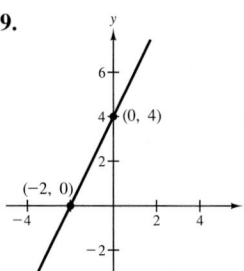

50.

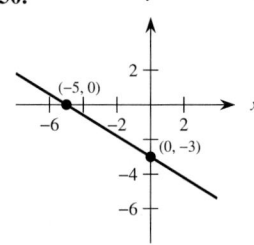

51.

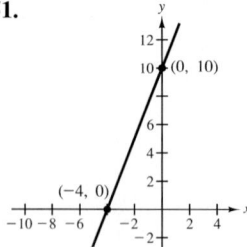

52.

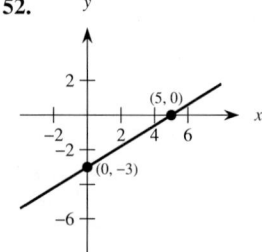

53. 3, $(0, -2)$

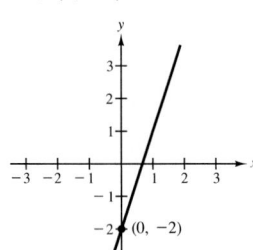

54. 1, $(0, -5)$

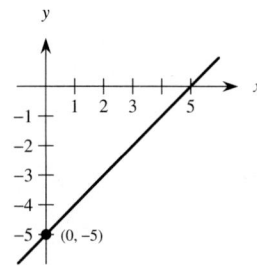

61. 0, $(0, 2)$

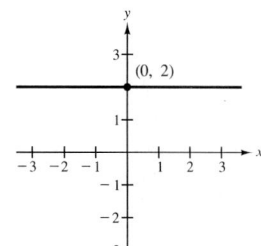

62. 0, $(0, -4)$

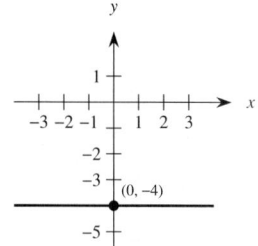

55. -1, $(0, 0)$

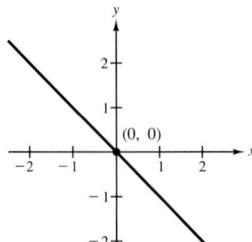

56. 1, $(0, 0)$

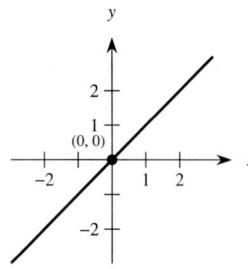

63. 5, $(0, -5)$

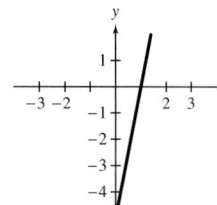

64. $-\frac{5}{6}$, $(0, 5)$

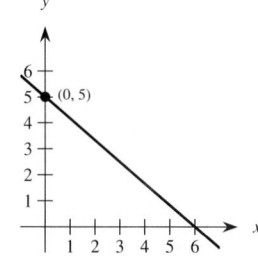

57. $-\frac{3}{2}$, $(0, 1)$

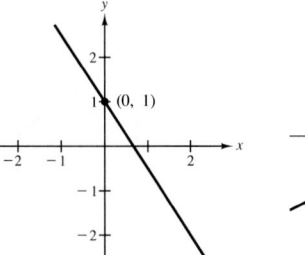

58. $\frac{1}{2}$, $(0, -1)$

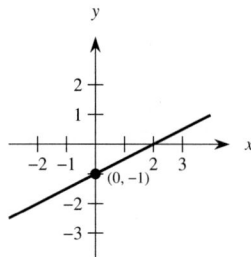

65. $(6, 2)$, $(10, 2)$ **66.** $(-4, 2)$, $(-4, -8)$
67. $(4, -1)$, $(5, 2)$ **68.** $(0, -3)$, $(1, -1)$
69. $(1, 2)$, $(2, 1)$ **70.** $(-1, 3)$, $(0, 0)$
71. $(-2, 4)$, $(1, 8)$ **72.** $(3, -2)$, $(7, -5)$
73. $(4, 0)$, $(4, -1)$ **74.** $(1, -2)$, $(0, -2)$

75.

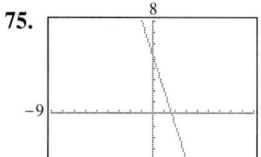

76.

59. $\frac{1}{4}$, $\left(0, \frac{1}{2}\right)$

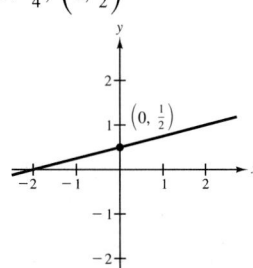

60. $-\frac{4}{3}$, $\left(0, \frac{1}{2}\right)$

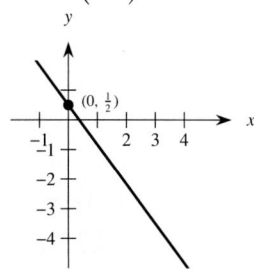

77.

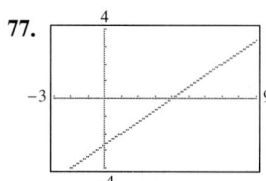

78.

79. Perpendicular   **80.** Neither
81. Parallel   **82.** Perpendicular

83. Parallel

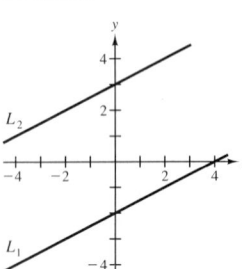

84. Parallel

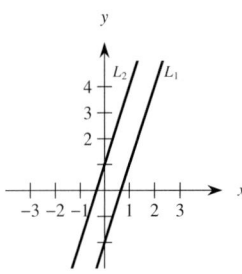

85. Perpendicular

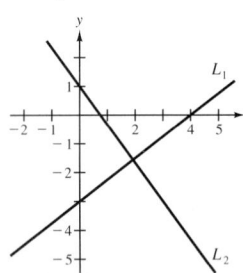

86. Perpendicular

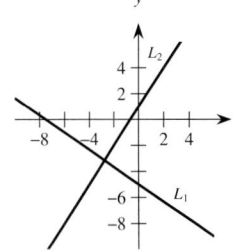

87. (a) 1988 and 1989   (b) 1985 and 1986

88. $\frac{22}{35}$% per year

89. (a)

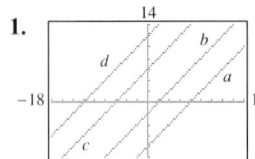

(b) $19.25 / day

90. 11.2   **91.** 16,667 ft   **92.** $\frac{45}{4}$

USING A GRAPHING CALCULATOR   (page 262)

1.

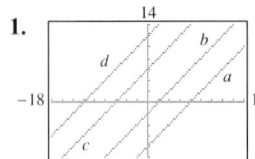

2.

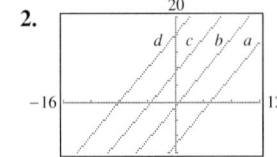

3.

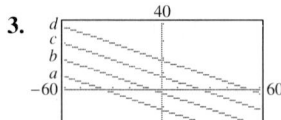

4.

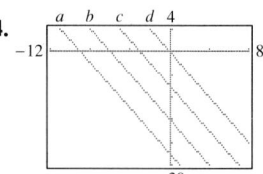

5.

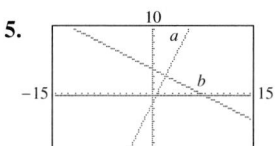

The lines are perpendicular because their slopes are negative reciprocals. Use a square setting for the viewing rectangle.

6.

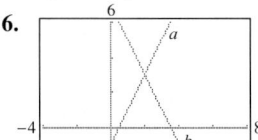

The lines are *not* perpendicular, so a square setting is not necessary.

7.

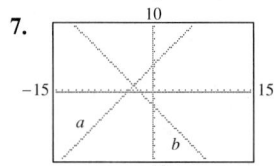

The lines are perpendicular because their slopes are negative reciprocals. Use a square setting for the viewing rectangle.

8.

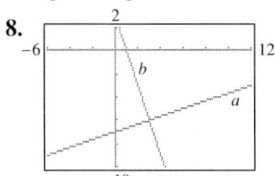

The lines are perpendicular because their slopes are negative reciprocals. Use a square setting for the viewing rectangle. (If you want to see all the intercepts on the screen, use a square setting such as -30 to 30 on the x-axis and -20 to 20 on the y-axis.)

9. No. Parallel lines appear parallel with square *or* nonsquare settings for the viewing rectangle.

10. Yes. Perpendicular lines only appear perpendicular with a *square* setting for the viewing rectangle.

11. No. Intersecting lines appear to intersect with square *or* nonsquare settings for the viewing rectangle.

12. Yes. A line with a slope of 2 only appears to have a slope of 2 with a *square* setting for the viewing rectangle.

SECTION 3.6 *(page 273)*

Warm-Up *(page 273)*

1. $-\frac{1}{4}$ 2. 1 3. Undefined 4. $-\frac{9}{10}$

5. 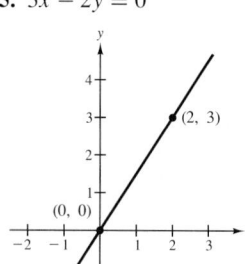 6.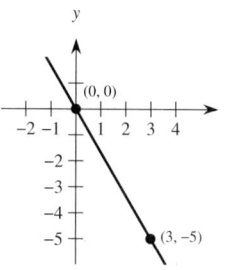

7. $y = -\frac{3}{5}x + \frac{21}{5}$ 8. $y = \frac{4}{3}x - 9$

9. $y = -4x + 18$ 10. $y = x + 9$

1. $m = 6$, $(-4, 3)$ 2. $m = -12$, $(1, -10)$

3. $m = \frac{3}{4}$, $\left(-2, -\frac{5}{8}\right)$ 4. $m = \frac{5}{6}$, $\left(\frac{9}{10}, \frac{3}{4}\right)$

5. $m = \frac{2}{3}$, $(0, -2)$ 6. $m = -5$, $(0, 12)$

7. $m = \frac{3}{2}$, $(0, 0)$ 8. $m = -\frac{5}{3}$, $\left(0, \frac{10}{3}\right)$

9. $m = 0$, $(0, 3)$ 10. Slope is undefined, $(-4, 0)$

11. $m = \frac{5}{2}$, $(0, 12)$ 12. $m = -\frac{3}{4}$, $(0, 4)$

13. $x - 2y - 12 = 0$ 14. $x + 3y - 27 = 0$

15. $4x + 2y - 3 = 0$ 16. $12x - 4y - 7 = 0$

17. $2x + y - 4 = 0$ 18. $3x - y = 0$

19. $2x + 3y - 10 = 0$ 20. $5x - 4y - 13 = 0$

21. $x - 2 = 0$ 22. $y - 5 = 0$

23. $8x - 6y + 9 = 0$

24. $6x + 2y + 8 = 0$ or $3x + y + 4 = 0$

25. $3x - 2y = 0$

26. $5x + 36 = 0$

27. $3x + 7y - 15 = 0$
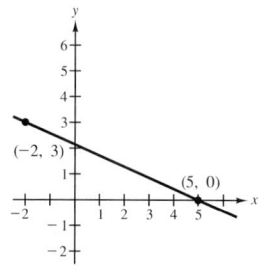

28. $3x - 5y + 3 = 0$
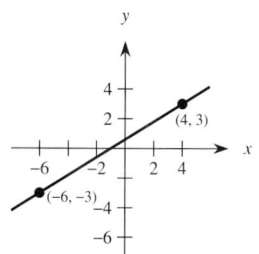

29. $x - 2y + 6 = 0$
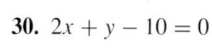

30. $2x + y - 10 = 0$

31. $y - 3 = 0$
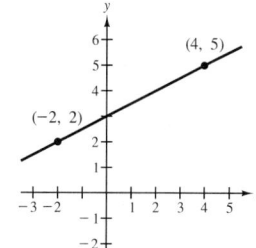

32. $x - 1 = 0$
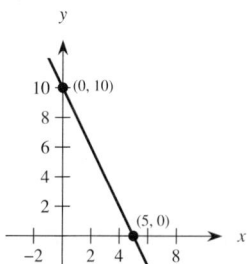

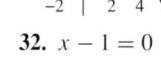
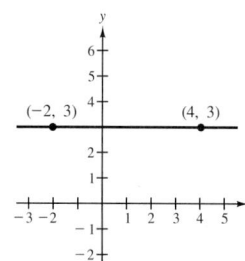
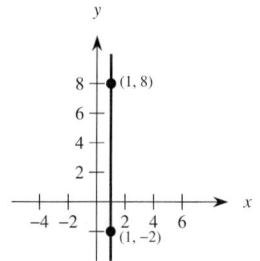

33. $14x + 6y - 39 = 0$ **34.** $x - 3y + 3 = 0$

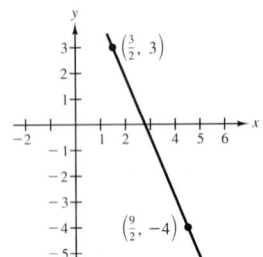   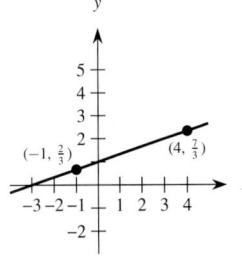

35. $10x + 90y - 27 = 0$ **36.** $2x - y + 1 = 0$

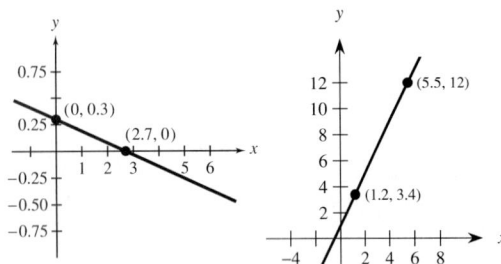

37. $f(x) = -\frac{2}{5}x$ **38.** $f(x) = -\frac{1}{8}x + \frac{37}{8}$

39. $f(x) = \frac{5}{2}x - 2$ **40.** $f(x) = 12$

41. $f(x) = \frac{7}{6}x + \frac{1}{4}$ **42.** $f(x) = -\frac{1}{2}x - \frac{19}{10}$

43. (a) $3x - y - 5 = 0$ **(b)** $x + 3y - 5 = 0$

44. (a) $x + 6y - 21 = 0$ **(b)** $6x - y + 22 = 0$

45. (a) $5x + 4y + 9 = 0$ **(b)** $4x - 5y + 40 = 0$

46. (a) $40x - 32y + 47 = 0$ **(b)** $16x + 26y - 55 = 0$

47. (a) $y - 2 = 0$ **(b)** $x + 1 = 0$

48. (a) $x - 3 = 0$ **(b)** $y + 4 = 0$

49.

x	0	50	100	500	1000
C	5000	6000	7000	15,000	25,000

50.

F	$-20°$	$-0.04°$	$20°$	$32°$	$100°$	$212°$
C	$-\frac{260°}{9}$	$-17.8°$	$-\frac{20°}{3}$	$0°$	$\frac{340°}{9}$	$100°$

51. $S = 1500 + 0.03M$ **52.** $C = 125 + 0.27x$

53. $S = 0.70L$ **54.** $V = -1475t + 7400$

55. $y = 72 + 5t$ **56.** $N = 1500 + 60t$, 2700

57. (a) $x = -\frac{1}{15}p + 80$ **(b)** 42   **(c)** 48

58. (a) $x = -10,000p + 14,000$ **(b)** 3000 cans
(c) 5000 cans

59. $C = 0.32t + 2.3$, \$7.1 billion

60. 0.05% or $\frac{1}{20}\%$ **61.** $s = 0.6t + 4.9$

62. $y = \frac{1}{8}x$; 8 ft, 16 ft, 24 ft, 32 ft

63. (a) and (b)

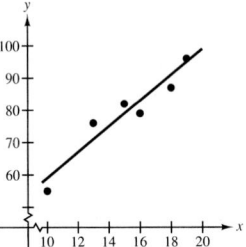

(c) $y = 4x + 19$
(d) 87

64. (a)

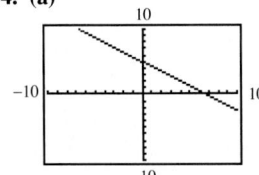

(b)

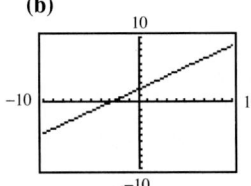

65. Parallel   **66.** Perpendicular

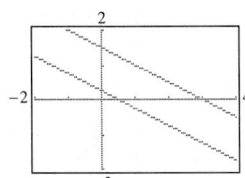   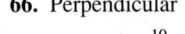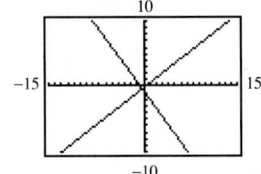

CHAPTER 3 REVIEW EXERCISES   *(page 277)*

1. $200 - 3n$ **2.** $100 + 15n$ **3.** $n^2 + 49$

4. $|n + 10|$ **5.** The sum of twice a number and 7

6. Three less than five times a number

7. The difference of a number and 5, all divided by 4

8. Negative three times the difference of a number and 10

9. $0.18I$ **10.** $8r$ **11.** $l(l - 5) = l^2 - 5l$

12. $3n + 6$ **13.** \$350.93   **14.** 84.2%

15. \$194.25   **16.** Department store   **17.** 15.5%

18. \$8333.33   **19.** 25 dimes, 35 quarters   **20.** 480

21. 6.35 hr   **22.** $\frac{45}{124}$ km/min   **23.** 2800 mi

24. $53\frac{1}{3}$ mi/hr   **25.** $\frac{15}{4}$ or $3\frac{3}{4}$ hr

26. $7\frac{1}{2}$ in. $\times 13\frac{1}{2}$ in.

27. \$20,000 at 8.5\%; \$30,000 at 10\%

28. \$71.23   **29.** $\frac{4}{3}$ **30.** $\frac{6}{5}$

31. $\frac{3}{20}$ **32.** $\frac{2}{1}$ **33.** $y = \frac{7}{2}$

34. $x = \frac{20}{3}$ **35.** $b = \frac{25}{2}$ **36.** $x = 5$

37. $1856.25   **38.** $\frac{15}{4}$ or $3\frac{3}{4}$ cups   **39.** 487.5 mi

40. 25 pints or 3.125 gal   **41.** 3   **42.** $\frac{14}{3}$

43. $z \le 10$ **44.** $x \ge 0$

45. $7 \le y < 14$ **46.** $V < 27$

47. $x > 3$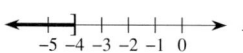

48. $x \le -4$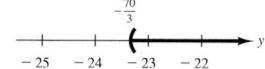

49. $y > -\frac{70}{3}$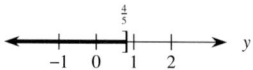

50. $y \le \frac{4}{5}$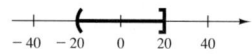

51. $-20 < x \le 20$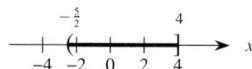

52. $-\frac{5}{2} < x \le 4$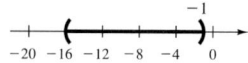

53. $-16 < x < -1$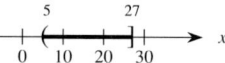

54. $5 < x \le 27$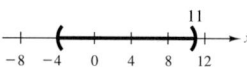

55. $-4 < x < 11$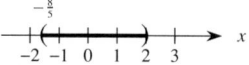

56. $-\frac{8}{5} < x < 2$

57. $x > 7$ or $x < 1$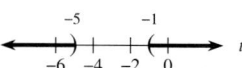

58. $t < -5$ or $t > -1$

59. $b > 5$ or $b < -9$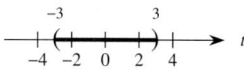

60. $-3 < t < 3$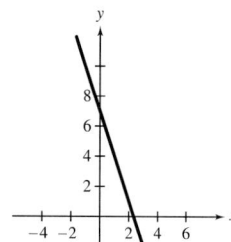

61. $-4, 8$ **62.** $-5, 2$ **63.** $\frac{1}{2}, 3$

64. $-\frac{7}{3}, -\frac{5}{7}$ **65.** $\frac{2}{7}$ **66.** $-\frac{13}{5}$

67. 0   **68.** Undefined

69. $-3, (0, 7)$ **70.** $2, (0, -3)$

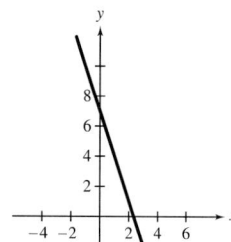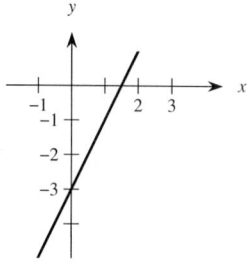

71. $\frac{5}{2}, (0, -5)$ **72.** $-\frac{3}{4}, (0, -2)$

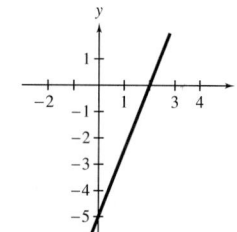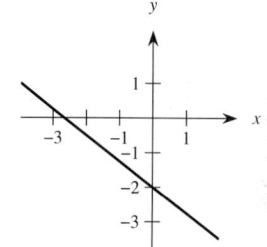

73. $2x - y - 6 = 0$ **74.** $3x - y + 10 = 0$

75. $4x + y = 0$ **76.** $2x + y - 8 = 0$

77. $2x + 3y - 17 = 0$ **78.** $9x - 6y + 10 = 0$

79. $x - 7 = 0$ **80.** $y - 5 = 0$

81. $x + 2y + 6 = 0$ **82.** $3x - 2y = 0$

83. $y - 10 = 0$ **84.** $x + 4 = 0$

85. $55x + 32y - 176 = 0$

86. $9x + 14y + 62 = 0$

87. $2x - 5 = 0$ **88.** $9x - 24y - 8 = 0$

89. (a) $x + 2y - 9 = 0$ (b) $2x - y + 7 = 0$

90. (a) $3x + y - 1 = 0$ (b) $x - 3y - 3 = 0$

91. (a) $8x - 6y + 15 = 0$ (b) $24x + 32y - 105 = 0$

92. (a) $x - 12 = 0$ (b) $y - 1 = 0$

93. (a) $C = 8.55x + 25{,}000$ (b) $P = 4.05x - 25{,}000$

CHAPTER 3 TEST   *(page 280)*

1. $5n - 8$ **2.** $0.6l^2$ **3.** $8000

4. $1466.67   **5.** $2\frac{1}{2}$ hr

6. $33\frac{1}{3}$ liters at 10%; $66\frac{2}{3}$ liters at 40%   **7.** $\frac{2}{3}$ hr

8. $x > 2$

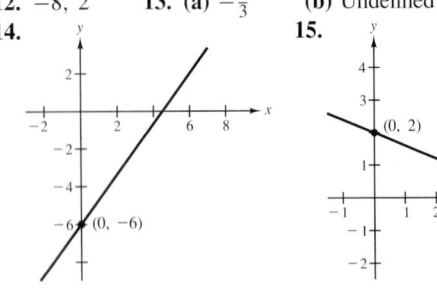

9. $-7 < x \le 1$

10. $1 \le x \le 5$

11. $x > -3$ or $x < -5$

12. $-8, 2$ **13. (a)** $-\frac{2}{3}$ **(b)** Undefined
14.   **15.**

$(0, 2)$

$(5, 0)$

$(0, -6)$

16. $y = -\frac{3}{2}x + 4$ **17.** $x - 2y - 55 = 0$
18. $\frac{3}{5}$ **19.** $V = -2250t + 16{,}000$, $2\frac{2}{3}$ years

CUMULATIVE TEST: CHAPTERS 1–3   *(page 281)*

1. 23 **2.** $-\frac{10}{27}$ **3.** $8a^8 b^7$
4. $2x^3 y - 11xy$ **5.** $t^2 - 9t$
6. $x^2 - 2xy + y^2 + 4x + 4$ **7.** $\frac{3}{2}$
8. $-\frac{3}{2}$ **9.** ± 8 **10.** $-\frac{1}{2}, 3$
11. y is a function of x.
12. All real numbers x such that $x \ge 2$
13. (a) 4 **(b)** $c^2 + 3c$ **14.** $3n - 8$
15. 10 and 11 or -9 and -8 **16.** $\$1408.75$
17. $\frac{14}{3}$ **18.** 6.5
19. $x \le -1$ or $x \ge 5$

20. $x \ge 103$ **21. (a)** 10 **(b)** $3x - 4y + 12 = 0$

22.

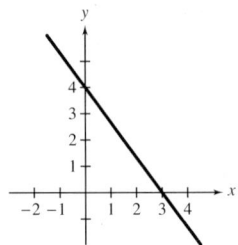

Chapter 4

Chapter Opener   *(page 282)*

Increasing; Compact disc sales
Decreasing; Long playing record sales
1986

SECTION 4.1   *(page 294)*

Warm-Up   *(page 294)*

1.
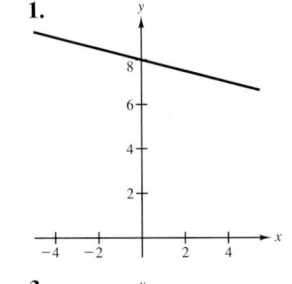

2.
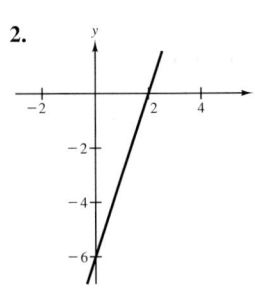

3.
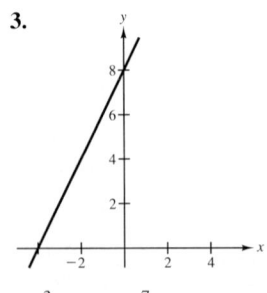

4.
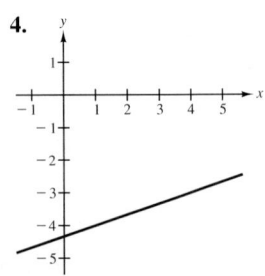

5. $\frac{3}{5}$ **6.** $\frac{7}{2}$ **7.** $2x - y = 0$
8. $x + y - 6 = 0$ **9.** $-\frac{2}{3}$ **10.** $\frac{9}{5}$

1. (a) Yes   **(b)** No   **2. (a)** No   **(b)** Yes
3. (a) No   **(b)** Yes   **4. (a)** Yes   **(b)** No
5. No solution   **6.** $(-2, 3)$
7. $\left(1, \frac{1}{3}\right)$ **8.** No solution
9. Infinite number of solutions   **10.** $(5, -3)$
11. $(-1, -1)$ **12.** $(0, 5)$

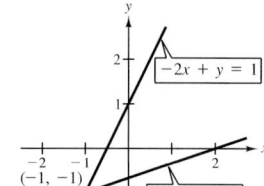
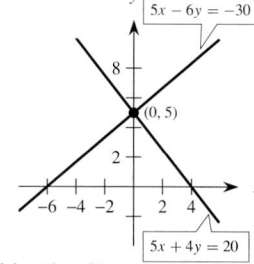

13. $(10, 0)$ **14.** $(6, -2)$

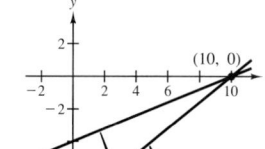
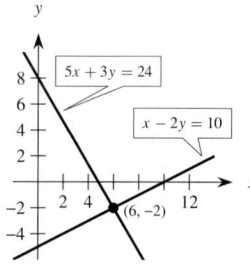

15. $(9, 12)$ **16.** $(4, 3)$

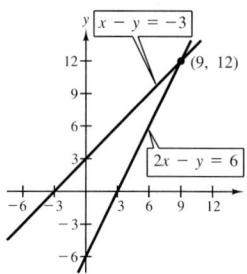
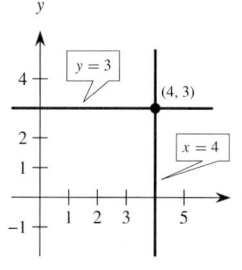

17. $\left(-\frac{3}{2}, \frac{5}{2}\right)$ **18.** $\left(1, -\frac{1}{4}\right)$

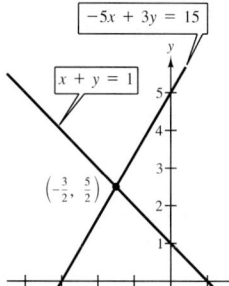
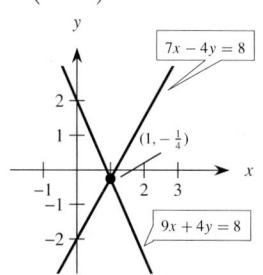

19. Infinite number of solutions   **20.** No solution
21. One solution   **22.** One solution
23. No solution   **24.** Infinite number of solutions
25. $\left(2992, \frac{798}{25}\right)$ **26.** $\left(276, \frac{3220}{11}\right)$
27. $(2, 1)$ **28.** $(2, 2)$ **29.** $(4, 3)$
30. $(6, 2)$ **31.** $\left(4, -\frac{1}{2}\right)$ **32.** $\left(\frac{5}{2}, \frac{3}{2}\right)$
33. $(1, 2)$ **34.** $\left(-\frac{20}{3}, -\frac{5}{3}\right)$
35. $(4, -2)$ **36.** $(-3, -1)$ **37.** $(7, 2)$
38. $(4, 2)$ **39.** $(-2, -1)$ **40.** No solution
41. No solution   **42.** $\left(4, \frac{4}{3}\right)$
43. Infinite number of solutions
44. Infinite number of solutions
45. $(5, 3)$ **46.** $(2, -3)$

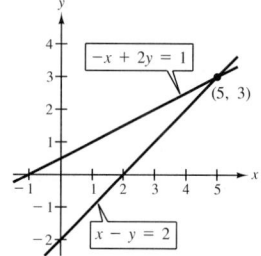
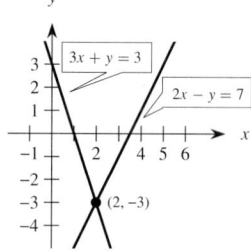

47. $(2, -2)$ **48.** $(4, 3)$

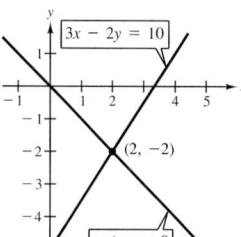

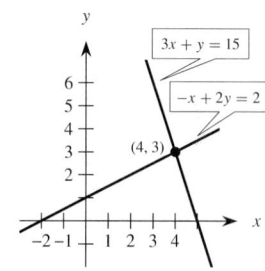

49. No solution   **50.** No solution

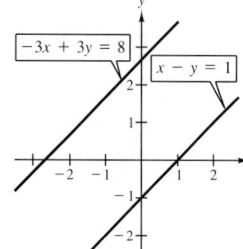
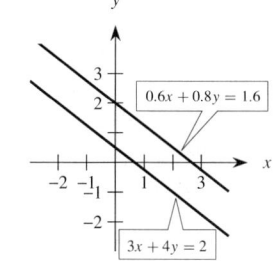

51. Infinite number of solutions   **52.** $\left(2, -\frac{3}{4}\right)$

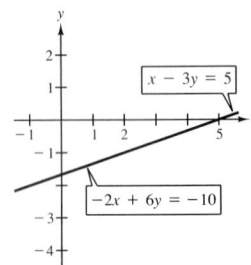

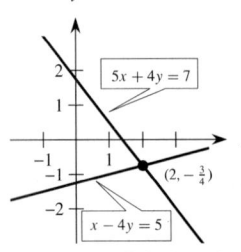

53. (3, 2)   **54.** (1, 5)   **55.** (−2, 5)   **56.** (1, −4)
57. (−1, −1)   **58.** (−3, 4)
59. (7, −2)   **60.** (10, 5)
61. No solution   **62.** Infinite number of solutions
63. $\left(\frac{3}{2}, 1\right)$ **64.** $\left(\frac{1}{4}, \frac{3}{4}\right)$
65. (−2, −1)   **66.** (20, 10)
67. Infinite number of solutions   **68.** No solution
69. $\left(\frac{25}{2}, -\frac{1}{5}\right)$ **70.** $\left(\frac{47}{6}, \frac{11}{3}\right)$ **71.** (3000, −2000)
72. (−2500, 1800)   **73.** (4, 3)   **74.** (−2, 6)
75. (2, 7)   **76.** (−3, 7)   **77.** (15, 10)
78. (−2, 7)   **79.** $\left(\frac{33}{4}, \frac{25}{2}\right)$ **80.** (9, 10)
81. 5 dimes, 10 quarters
82. 7 dimes, 25 quarters   **83.** 11 nickels, 14 dimes
84. 25 nickels, 25 quarters
85. Regular: \$1.11; Premium: \$1.22
86. 280 student tickets; 520 adult tickets
87. \$5.65 variety: 6 lb; \$8.95 variety: 4 lb
88. 70 tons of the \$75 per ton hay; 30 tons of the \$125 per ton hay
89. 40% solution: 12 l; 65% solution: 8 l
90. 80% solution: $\frac{100}{3}$ gal; 50% solution: $\frac{50}{3}$ gal
91. \$15,000 at 8%; \$5000 at 9.5%
92. \$4000 at 8.5%; \$8000 at 10%
93. 10,000 units   **94.** 2959 units   **95.** 63
96. Speed of plane: 540 mph; Speed of wind: 60 mph
97. (a) $y = \frac{3}{2}x - \frac{1}{6}$ **98. (a)** $y = -\frac{3}{2}x + \frac{19}{6}$
(b)   **(b)**

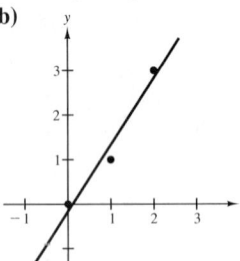

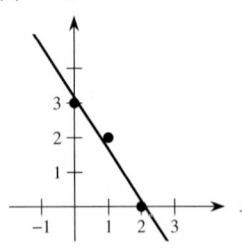

99. Depth: 10 ft; Length: 122 ft, 125 ft
100. $k = 27$
101. $\begin{cases} 3x + 2y = 22 \\ 2x - y = 3 \end{cases}$ **102.** $\begin{cases} x + y = 4 \\ 3x - y = -12 \end{cases}$

103. $\begin{cases} x + y = -3 \\ 7x - 3y = -1 \end{cases}$ **104.** $\begin{cases} 10x - y = 2 \\ -8x + 3y = 5 \end{cases}$

USING A GRAPHING CALCULATOR   *(page 303)*

1.

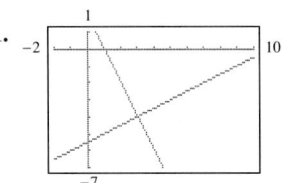

The solution appears to be approximately (3, −4).
Check:

$y = -2x + 2$ $y = \frac{1}{2}x - \frac{11}{2}$

$-4 \stackrel{?}{=} -2(3) + 2$ $-4 \geq \frac{1}{2}(3) - \frac{11}{2}$

$-4 = -4$ $-4 = -\frac{8}{2}$

The solution is (3, −4).

2.

The solution appears to be approximately (−4, −1).
Check:

$y = 3x + 11$ $y = \frac{1}{3}x + \frac{1}{3}$

$-1 \stackrel{?}{=} 3(-4) + 11$ $-1 \stackrel{?}{=} \frac{1}{3}(-4) + \frac{1}{3}$

$-1 = -1$ $-1 = -\frac{3}{3}$

The solution is (−4, −1).

3.

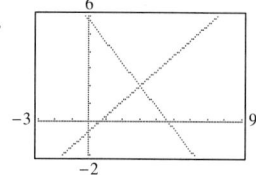

Rewrite: $\begin{cases} y = \frac{7}{8}x - \frac{5}{8} \\ y = -\frac{5}{4}x + \frac{23}{4} \end{cases}$

The solution appears to be approximately $(3, 2)$.
Check:

$$7x - 8y = 5 \qquad\qquad 5x + 4y = 23$$

$$7(3) - 8(2) \overset{?}{=} 5 \qquad 5(3) + 4(2) \overset{?}{=} 23$$

$$21 - 16 = 5 \qquad\qquad 15 + 8 = 23$$

The solution is $(3, 2)$.

4.

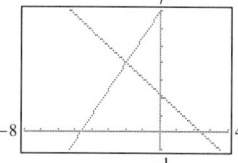

Rewrite: $\begin{cases} y = -\frac{10}{11}x + \frac{24}{11} \\ y = \frac{3}{2}x + 7 \end{cases}$

The solution appears to be approximately $(-2, 4)$.
Check:

$$10x + 11y = 24 \qquad\qquad 9x - 6y = -42$$

$$10(-2) + 11(4) \overset{?}{=} 24 \qquad 9(-2) - 6(4) \overset{?}{=} -42$$

$$-20 + 44 = 24 \qquad\qquad -18 - 24 = -42$$

The solution is $(-2, 4)$.

5. 1984

SECTION 4.2 *(page 314)*

Warm-Up *(page 313)*

1. $(5, 4)$ **2.** $(-1, 1)$ **3.** Inconsistent
4. Infinite number of solutions **5.** $(3, -2)$
6. $(-2, -2)$ **7.** $5x - z$ **8.** $34y - 5x$
9. $35z - 17y$ **10.** $7x + 26z$

1. $(22, -1, -5)$ **2.** $(-1, 2, 3)$ **3.** $(14, 3, -1)$
4. $(10, -14, 2)$ **5.** $(1, 2, 3)$ **6.** $(3, 2, 1)$
7. $(1, 2, 3)$ **8.** $(0, 4, -2)$ **9.** $(2, -3, -2)$
10. $(-2, 4, 1)$ **11.** No solution **12.** $(5, 2, -4)$
13. $(-4, 8, 5)$ **14.** $(5, -2, 0)$
15. $(3a + 1, 3a - 1, a)$, where a is any real number
16. $\left(\frac{3}{10}, \frac{2}{5}, 0\right)$ **17.** $(1, 0, -2)$
18. $(2, -1, 1)$ **19.** $(-4, 2, 3)$ **20.** $(0, -4, 5)$
21. $(1, -1, 2)$ **22.** $\left(\frac{1}{2}, \frac{13}{11}, \frac{35}{22}\right)$
23. $(-2a - 2, a + 1, a)$, where a is any real number
24. $(-a + 2, a - 1, a)$, where a is any real number
25. $\left(\frac{7}{5}a + 1, \frac{1}{5}a - a, a\right)$, where a is any real number
26. $\left(-\frac{1}{2}a + \frac{1}{2}, \frac{3}{5}a + \frac{2}{5}, a\right)$, where a is any real number
27. $\left(-\frac{1}{2}a + \frac{1}{4}, \frac{1}{2}a + \frac{5}{4}, a\right)$, where a is any real number
28. No solution **29.** $y = 2x^2 + 3x - 4$
30. $y = -x^2 + 2x + 5$
31. $y = x^2 - 4x + 3$ **32.** $y = -2x^2 + 5x - 1$
33. $x^2 + y^2 - 4x = 0$ **34.** $x^2 + y^2 - 6y = 0$
35. $x^2 + y^2 - 6x - 8y = 0$
36. $x^2 + y^2 - 3x - 2y = 0$
37. $s = -16t^2 + 144$ **38.** $s = -16t^2 + 64t$
39. $s = -16t^2 + 48t$ **40.** $s = -16t^2 + 60t + 10$
41. 20 gal of spray X; 18 gal of spray Y; 16 gal of spray Z
42. (a) No 10% solution, $8\frac{1}{3}$ liters of 20% solution, $1\frac{2}{3}$ liters of 50% solution.
(b) $6\frac{1}{4}$ liters of 10% solution, no 20% solution, $3\frac{3}{4}$ liters of 50% solution.
(c) 1 liter of 10% solution, 7 liters of 20% solution, 2 liters of 50% solution.
43. Strings: 50; Wind: 20; Percussion: 8
44. $y = \frac{1}{2}x^2 - \frac{1}{2}x$
45. $\begin{cases} x + y + z = 3 \\ x - y - 3z = 1 \\ 2x + 3y - z = -3 \end{cases}$
46. $\begin{aligned} 2x - 3y - 2z &= 9 \\ x + 2y &= 19 \\ 3y + 2z &= 1 \end{aligned}$

SECTION 4.3 *(page 327)*

Warm-Up *(page 327)*

1. $(2, 1)$ **2.** $(3, -2)$ **3.** $(5, 4)$

4. $\left(\frac{1}{2}, \frac{1}{3}\right)$ **5.** $(1, 2, 1)$ **6.** $(4, -1, 3)$

7. 9 **8.** $\frac{14}{5}$ **9.** $-\frac{19}{4}$ **10.** $-\frac{11}{6}$

---

1. 4×2 **2.** 3×3 **3.** 3×4 **4.** 3×5

5. $\begin{bmatrix} 4 & -5 & \vdots & -2 \\ -1 & 8 & \vdots & 10 \end{bmatrix}$ **6.** $\begin{bmatrix} 8 & 3 & \vdots & 25 \\ 3 & -9 & \vdots & 12 \end{bmatrix}$

7. $\begin{bmatrix} 1 & 10 & -3 & \vdots & 2 \\ 5 & -3 & 4 & \vdots & 0 \\ 2 & 4 & 0 & \vdots & 6 \end{bmatrix}$

8. $\begin{bmatrix} 9 & -3 & 20 & 1 & \vdots & 13 \\ 12 & 0 & -8 & 0 & \vdots & 5 \end{bmatrix}$

9. $\begin{cases} 4x + 3y = 8 \\ x - 2y = 3 \end{cases}$ **10.** $\begin{cases} 9x - 4y = 0 \\ 6x + y = -4 \end{cases}$

11. $\begin{cases} x \qquad + 2z = -10 \\ \quad 3y - z = 5 \\ 4x + 2y \qquad = 3 \end{cases}$

12. $\begin{cases} 5x + 8y + 2z \qquad = -1 \\ -2x + 15y + 5z + w = 9 \\ x + 6y - 7z \qquad = -3 \end{cases}$

13. $\begin{bmatrix} 1 & 4 & 3 \\ 0 & 2 & -1 \end{bmatrix}$ **14.** $\begin{bmatrix} 1 & 2 & \frac{8}{3} \\ 4 & -3 & 6 \end{bmatrix}$

15. $\begin{bmatrix} 1 & 1 & 4 & -1 \\ 0 & 5 & -2 & 6 \\ 0 & 3 & 20 & 4 \end{bmatrix}$ $\begin{bmatrix} 1 & 1 & 4 & -1 \\ 0 & 1 & -\frac{2}{5} & \frac{6}{5} \\ 0 & 3 & 20 & 4 \end{bmatrix}$

16. $\begin{bmatrix} 1 & 2 & 4 & \frac{3}{2} \\ 1 & -1 & -3 & 2 \\ 2 & 6 & 4 & 9 \end{bmatrix}$ $\begin{bmatrix} 1 & 2 & 4 & \frac{3}{2} \\ 0 & -3 & -7 & \frac{1}{2} \\ 0 & 2 & -4 & 6 \end{bmatrix}$

17. $\begin{bmatrix} 1 & 2 & 3 \\ 0 & 1 & 2 \end{bmatrix}$ **18.** $\begin{bmatrix} 1 & 3 & 6 \\ 0 & 1 & 9 \end{bmatrix}$

19. $\begin{bmatrix} 1 & -1 & -2 \\ 0 & 1 & \frac{9}{10} \end{bmatrix}$ **20.** $\begin{bmatrix} 1 & \frac{2}{3} & 2 \\ 0 & 1 & -\frac{21}{5} \end{bmatrix}$

21. $\begin{bmatrix} 1 & 1 & 0 & 5 \\ 0 & 1 & 2 & 0 \\ 0 & 0 & 1 & -1 \end{bmatrix}$ **22.** $\begin{bmatrix} 1 & 2 & -1 & 3 \\ 0 & 1 & -2 & 5 \\ 0 & 0 & 1 & -1 \end{bmatrix}$

23. $\begin{bmatrix} 1 & -1 & -1 & 1 \\ 0 & 1 & 6 & 3 \\ 0 & 0 & 1 & \frac{4}{5} \end{bmatrix}$ **24.** $\begin{bmatrix} 1 & -3 & 0 & -7 \\ 0 & 1 & 1 & 2 \\ 0 & 0 & 0 & 0 \end{bmatrix}$

25. $\begin{cases} x - 2y = 4 \\ y = -3 \end{cases}$ **26.** $\begin{cases} x + 5y = 0 \\ y = -1 \end{cases}$

$(-2, -3)$ $(5, -1)$

27. $\begin{cases} x - y + 2z = 4 \\ y - z = 2 \\ z = -2 \end{cases}$ **28.** $\begin{cases} x + 2y - 2z = -1 \\ y + z = 9 \\ z = -3 \end{cases}$

$(8, 0, -2)$ $(-31, 12, -3)$

29. $(3, 2)$ **30.** $(-1, 3)$ **31.** $(1, 1)$

32. $(0.2, 0.5)$ **33.** No solution

34. $(3a + 5, a)$ **35.** $(4, -3, 2)$

36. $(2, -3, 2)$ **37.** $(1, -1, 2)$

38. No solution   **39.** $(1, 2, -1)$

40. $(8, 10, 6)$ **41.** $(2a + 1, 3a + 2, a)$

42. $\left(-\frac{3}{2}a + \frac{3}{2}, \frac{1}{3}a + \frac{1}{3}, a\right)$

43. No solution   **44.** No solution

45. $\left(2, 5, \frac{5}{2}\right)$ **46.** $\left(1, \frac{2}{3}, -1\right)$

47. 8%: \$800,000; 9%: \$500,000; 12%: \$200,000

48. \$4000 invested at 5%, \$5000 invested at 6%, \$7000 invested at 7%

49. There are infinitely many solutions.
One possible allocation is: \$250,000 in certificates of deposit, \$125,000 in municipal bonds, and \$125,000 in blue-chip stocks.

50. $\left(-\frac{1}{2}a + 406,250, \frac{1}{2}a - 31,250, 125,000 - a, a\right)$,
where $62,500 \le a \le 125,000$

51. $y = x^2 + 2x + 4$ **52.** $y = -x^2 + 2x + 10$

53. $x^2 + y^2 - 5x - 3y + 6 = 0$ **54.** 5, 8, 20

CHAPTER 4 MID-CHAPTER QUIZ *(page 330)*

1. $(2, 1)$ **2.** No solution   **3.** $(1, 3)$

4. $(5, 3)$ **5.** $(9, 4)$

6. Infinite number of solutions

7. $(-18, 4, 2)$ **8.** $(-1, 5, 5)$

9. Their associated systems of equations have the same solution.

10. $(16, -5, 2)$

11. $\begin{cases} 2l + 2w = 90 \\ 2l - 3w = 0 \end{cases}$ **12.** $\begin{cases} x + y = 300 \\ x - 3y = 0 \end{cases}$

13. $\begin{cases} 0.94x + 0.92y + 0.80z = 20,144 \\ 0.04x + 0.02y + 0.04z = 766 \\ 0.02x + 0.06y + 0.16z = 1,990 \end{cases}$

14. $\begin{cases} 2x - y + z = 8 \\ x + 2y + 5z = 1 \\ x + z = 3 \end{cases}$

63. $3x - 5y = 0$ **64.** $2x + 6y - 10 = 0$
65. $7x - 6y - 28 = 0$ **66.** $2x + 3y - 8 = 0$
67. $y = 2x^2 - 6x + 1$ **68.** $y = x^2 - \frac{3}{2}x + 2$
69. $y = -3x^2 + 2x$ **70.** $y = -x^2 + 10$
71. 248   **72.** 105.625
73. $\left(\frac{51}{16}, -\frac{7}{16}, -\frac{13}{16}\right)$ **74.** $\left(-2, \frac{3}{2}, 2\right)$
75. (a) $\left(\frac{4k - 3}{2k - 1}, \frac{4k - 1}{2k - 1}\right)$ (b) $\frac{1}{2}$

SECTION 4.4   *(page 341)*

Warm-Up   *(page 341)*

1. 10   **2.** −56   **3.** −82   **4.** 153
5. 32   **6.** $-\frac{65}{3}$ **7.** −2
8. $\frac{9}{11}$ **9.** $\frac{6}{5}$ **10.** $-\frac{11}{3}$

1. 5   **2.** −11   **3.** 27   **4.** 14
5. 0   **6.** 0   **7.** 6   **8.** 0
9. −24   **10.** −13   **11.** −0.16   **12.** −0.72
13. −24   **14.** −48   **15.** −2   **16.** 6
17. −30   **18.** −20   **19.** 3   **20.** −2
21. 0   **22.** 0   **23.** −75   **24.** 151
25. −58   **26.** 0   **27.** 0   **28.** −28
29. −0.22   **30.** 0.002   **31.** $x - 5y + 2$
32. $-7x + 3y - 8$ **33.** (1, 2)   **34.** (−3, 4)
35. (2, −2)   **36.** $\left(\frac{1}{2}, \frac{1}{3}\right)$ **37.** $\left(\frac{3}{4}, -\frac{1}{2}\right)$
38. $D = 0$, cannot use Cramer's Rule
39. $D = 0$, cannot use Cramer's Rule
40. $\left(\frac{32}{7}, \frac{30}{7}\right)$ **41.** $\left(\frac{2}{3}, \frac{1}{2}\right)$ **42.** $\left(-\frac{1}{13}, \frac{8}{13}\right)$
43. (−1, 3, 2)   **44.** (5, 8, −2)   **45.** $\left(1, \frac{1}{2}, \frac{3}{2}\right)$
46. $\left(\frac{3}{4}, \frac{25}{28}, -\frac{73}{28}\right)$ **47.** (1, −2, 1)
48. $D = 0$, cannot use Cramer's Rule (no solution)
49. $D = 0$, cannot use Cramer's Rule
50. (0, −3, 2)   **51.** 7   **52.** 14   **53.** $\frac{31}{2}$
54. $\frac{33}{8}$ **55.** $\frac{53}{2}$ **56.** 15   **57.** 250 mi²
58. 3100 ft²   **59.** Collinear   **60.** Not collinear
61. Not collinear   **62.** Collinear

SECTION 4.5   *(page 349)*

Warm-Up   *(page 349)*

1.
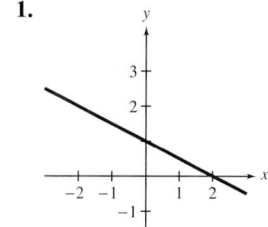

2.
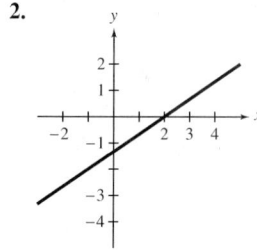

3.
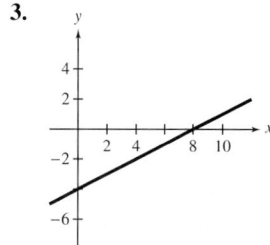

4.
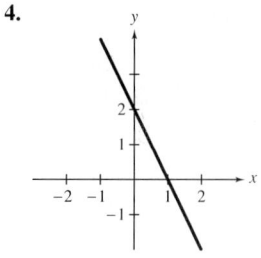

5.
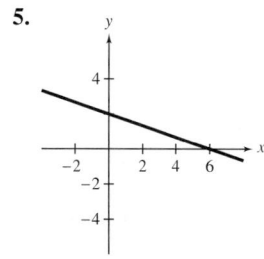

6.
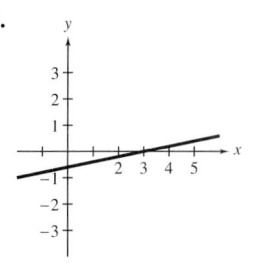

7.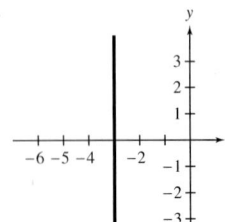

8.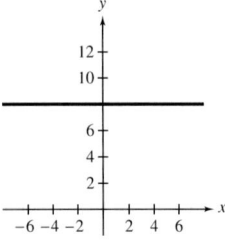

9.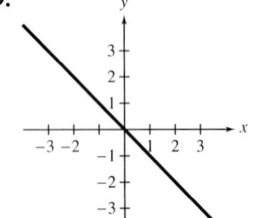

10.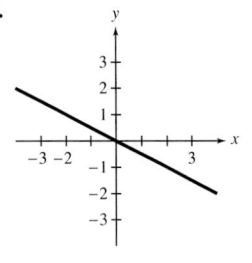

1. b   **2.** a   **3.** d   **4.** e   **5.** f   **6.** c

7. (a) Yes   **(b)** No   **(c)** Yes   **(d)** Yes
8. (a) No   **(b)** No   **(c)** Yes   **(d)** Yes
9. (a) No   **(b)** No   **(c)** Yes   **(d)** Yes
10. (a) Yes   **(b)** Yes   **(c)** No   **(d)** Yes
11. (a) Yes   **(b)** No   **(c)** No   **(d)** Yes
12. (a) No   **(b)** No   **(c)** Yes   **(d)** Yes
13. (a) No   **(b)** Yes   **(c)** No   **(d)** Yes
14. (a) No   **(b)** Yes   **(c)** Yes   **(d)** No

15.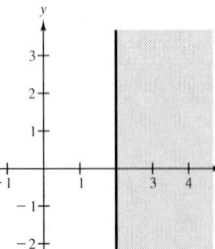

16.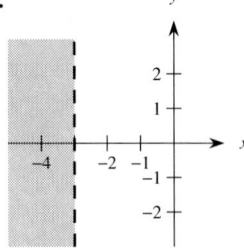

17.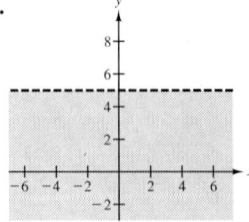

18.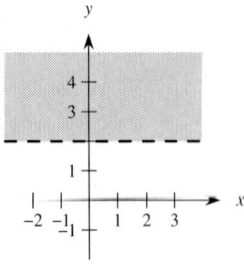

19.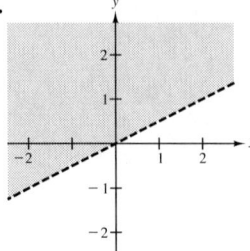

20.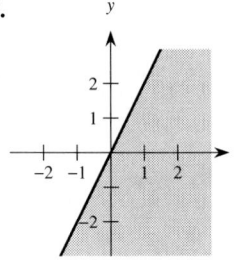

21.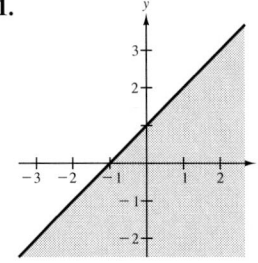

22.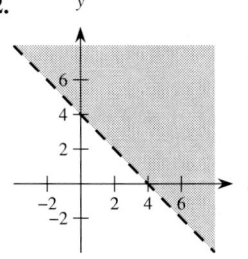

23.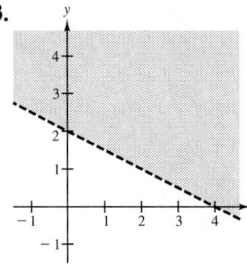

24.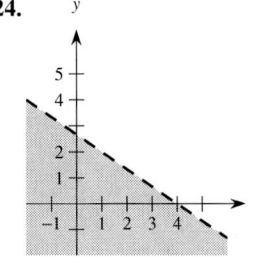

25.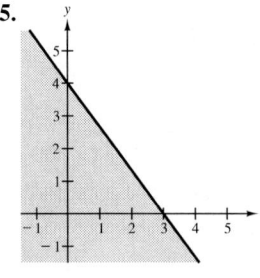

26.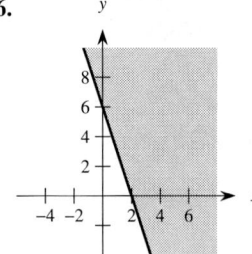

27.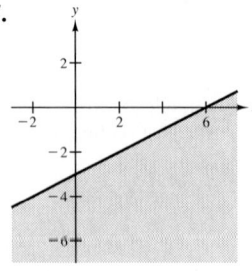

28.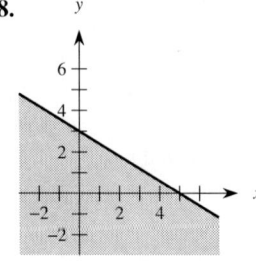

29.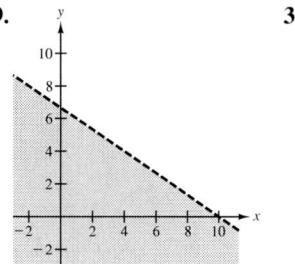

30.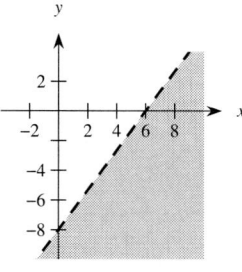

31. $3x + 4y > 17$ **32.** $-x + y < 2$ **33.** $y < 2$

34. $x < 2$ **35.** $x - 2y < 0$ **36.** $x + y < 0$

37. $2x + 2y \leq 500$ or $y \leq -x + 250$, $x \geq 0$, $y \geq 0$

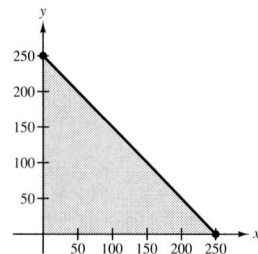

38. $10x + 15y \leq 1000$ or $y \leq -\frac{2}{3}x + \frac{200}{3}$, $x \geq 0$, $y \geq 0$

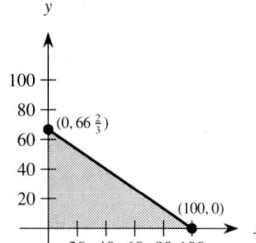

39. (a) and **(b)**

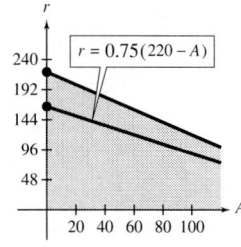

40. $0.40x + 0.80y \leq 10$
$$x \geq 0$$
$$y \geq 0$$

(3, 6), Yes

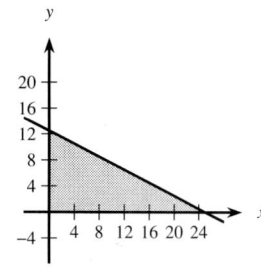

SECTION 4.6   *(page 361)*

Warm-Up   *(page 360)*

1. $-11x$ **2.** $-41v$ **3.** $16x - 3$
4. $4y - 2$ **5.** 4   **6.** 11
7. Perpendicular   **8.** Parallel
9. Parallel   **10.** Neither

1. c   **2.** b   **3.** f   **4.** e   **5.** a   **6.** d

7.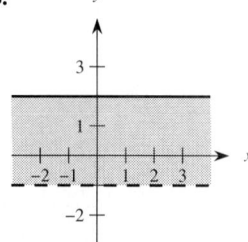

8.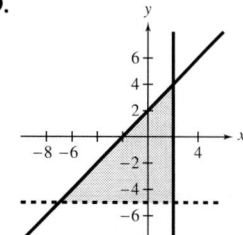

9.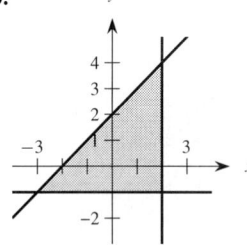

10.

11.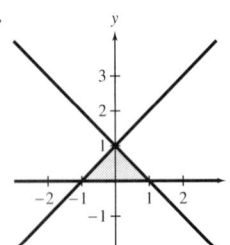

12.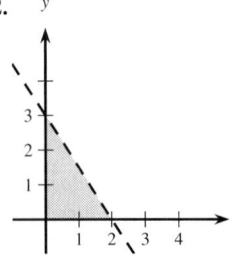

13.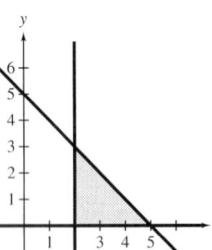

14.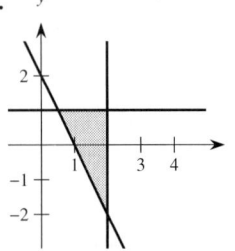

15.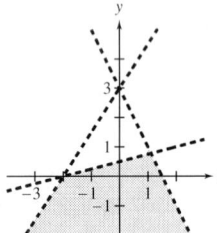

16.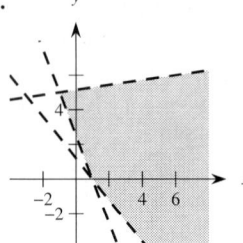

17.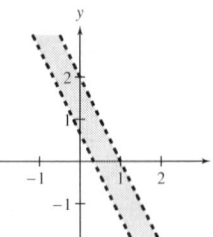

18.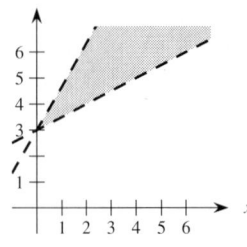

19.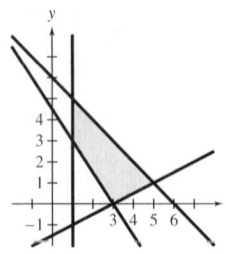

20.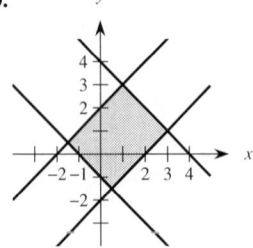

21. $x \geq 1$
$x \leq 8$
$y \geq -5$
$y \leq 3$

22. $y \leq 9$
$y \geq -1$
$y \geq x - 6$
$y \leq x + 3$

23. $y \leq \frac{9}{10}x + \frac{42}{5}$
$y \geq 3x$
$y \geq \frac{2}{3}x + 7$

24. $y \geq 2x - 4$
$y \leq 3x - 2$
$y \leq \frac{5}{4}x - \frac{1}{4}$

25. $x \leq 90$
$y \leq 0$
$y \geq -10$
$y \geq -\frac{1}{7}x$

26. $y \leq 22$
$y \geq 10$
$y \geq 2x - 24$
$y \geq -2x - 24$

27. $x + y \leq 20{,}000$
$x \geq 5000$
$y \geq 5000$
$y \geq 2x$

28. $x \geq 2y$
$800x + 1200y \leq 20{,}000$
$x \geq 4$
$y \geq 2$

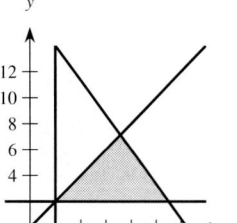

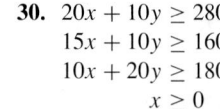

29. $x + y \geq 15{,}000$
$15x + 25y \geq 275{,}000$
$x \geq 8000$
$y \geq 4000$

30. $20x + 10y \geq 280$
$15x + 10y \geq 160$
$10x + 20y \geq 180$
$x \geq 0$
$y \geq 0$

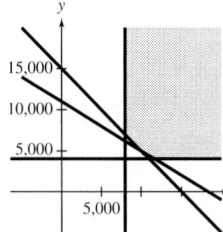

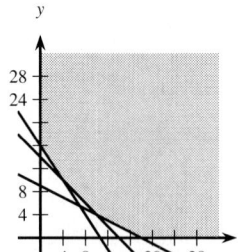

31. Minimum value at (0, 0): 0
Maximum value at (0, 6): 30
32. Minimum value at (0, 0): 0
Maximum value at (0, 4): 32

33. Minimum value at (0, 0): 0
Maximum value at (6, 0): 60
34. Minimum value at (0, 0): 0
Maximum value at (2, 0): 14
35. Minimum value at (0, 0): 0
Maximum value at (3, 4): 17
36. Minimum value at (0, 2): 6
Maximum value at (5, 3): 29
37. Minimum value at (0, 0): 0
Maximum value at (4, 0): 20
38. Minimum value at (3, 0): 3
Maximum value at (0, 4): 24
39. Minimum value at (0, 0): 0
Maximum value at (60, 20): 740
40. Minimum value at (0, 600): 21,000
Maximum value at (900, 0): 45,000
41. Minimum value at (0, 0): 0
Maximum value at (30, 45): 2325
42. Minimum value at (675, 0): 10,800
Maximum value at (0, 800): 16,000
43. Minimum value at (4, −1): −9
Maximum value at (−5, 3): 13

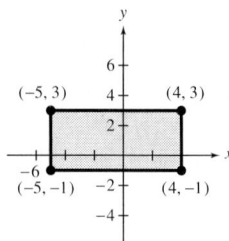

44. Minimum value at (−4, 1): −16
Maximum value at (2, 6): 34

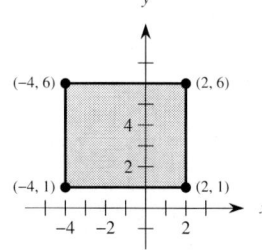

45. Minimum value at (12, 0): 12
Maximum value at (12, 10): 52

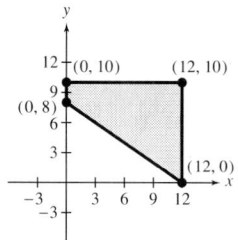

46. Minimum value at $\left(\frac{1}{4}, 0\right)$: $\frac{3}{2}$
Maximum value at (5, 19): 68

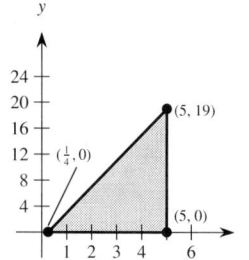

47. Minimum value at (0, 0): 0
Maximum value at (5, 0): 30

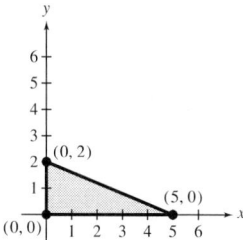

48. Minimum value at (0, 0): 0
Maximum value at (0, 8): 64

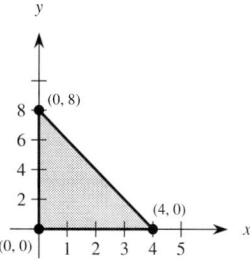

49. Minimum value at (0, 0): 0
Maximum value at (5, 0): 45

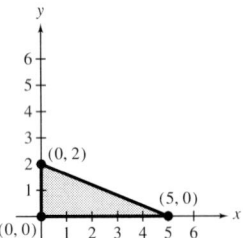

50. Minimum value at (0, 0): 0
Maximum value at (4, 0): 28

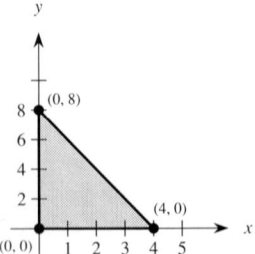

51. Minimum value at (5, 3): 35
No maximum value

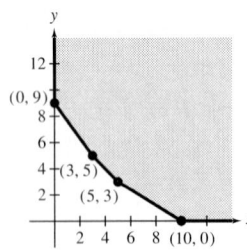

52. Minimum value at (0, 0): 0
Maximum value at (4, 1): 21

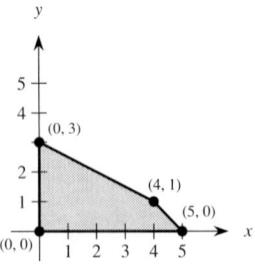

53. Minimum value at (10, 0): 20
No maximum value

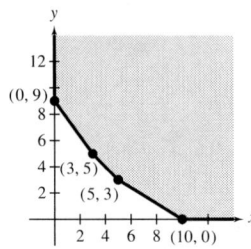

54. Minimum value at (0, 3): −3
Maximum value at (5, 0): 10

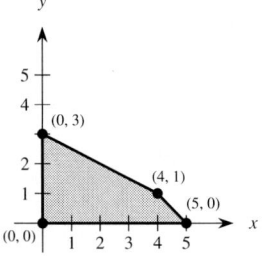

55. 750 units of Model A; 1000 units of Model B
56. 60 acres of crop A; 90 acres of crop B
57. 8 audits; 8 tax returns
58. $225,000 in Type A; $225,000 in Type B
59. 3 bags of Brand X; 6 bags of Brand Y
60. $3000

CHAPTER 4 REVIEW EXERCISES   *(page 366)*

1. (1, 1)

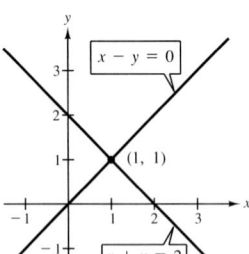

2. (3, 3)

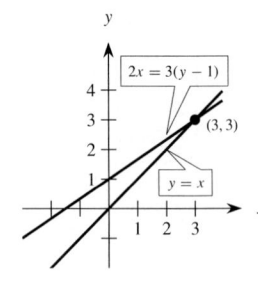

3. No solution

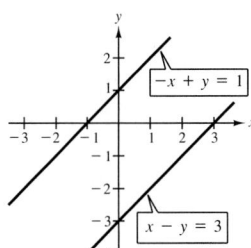

4. $(2, -3)$

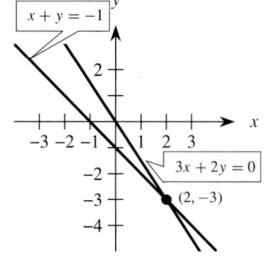

5. $(4, 8)$

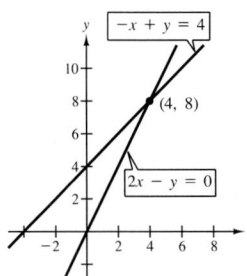

6. No solution

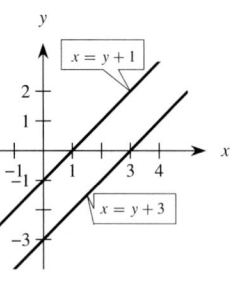

7. $(2, -1)$ **8.** $(8, 2)$ **9** No solution

10. $(-1, 4)$ **11.** $(-10, -5)$

12. Infinite number of solutions **13.** $(0, 0)$

14. $(1, -3)$ **15.** $\left(\frac{5}{2}, 3\right)$ **16.** $\left(-\frac{1}{2}, \frac{4}{5}\right)$

17. $(2, -3, 3)$ **18.** $(5, 2, -6)$

19. $(10, -12)$ **20.** $(-9, -4)$

21. $\left(\frac{3}{5}, \frac{1}{2}\right)$ **22.** $\left(-\frac{1}{5}, \frac{7}{10}\right)$

23. $\left(\frac{24}{5}, \frac{22}{5}, -\frac{8}{5}\right)$ **24.** $\left(\frac{1}{2}, -\frac{1}{3}, 1\right)$

25. $(-3, 7)$ **26.** $\left(\frac{1}{3}, -\frac{1}{2}\right)$

27. $D = 0$, cannot use Cramer's Rule

28. $\left(3, \frac{1}{3}\right)$ **29.** $(2, -3, 3)$

30. $D = 0$, cannot use Cramer's Rule **31.** 16

32. 24 **33.** 7 **34.** $\frac{25}{8}$ **35.** $x - 2y + 4 = 0$

36. $3x + 2y - 16 = 0$ **37.** $2x + 6y - 13 = 0$

38. $-3x + 1.5y - 2.7 = 0$

39. $\begin{cases} 3x + y = -2 \\ 6x + y = 0 \end{cases}$ **40.** $\begin{cases} x + y = 2 \\ 2x - y = -32 \end{cases}$

41. 16,667 units **42.** 96 m $\times$ 144 m

43. 75% solution: 40 gal; 50% solution: 60 gal

44. Shorter piece: 32 in.; Longer piece: 96 in.

45. \$9.95 tapes: 400; \$14.95 tapes: 250

46. $218\frac{3}{4}$ mph, $193\frac{3}{4}$ mph

47. \$8000 at 7%; \$5000 at 9%; \$7000 at 11%

48. 16, 20, 32 **49.** $y = 2x^2 + x - 6$

50. $y = 3x^2 + 11x - 20$

51. $x^2 + y^2 - 4x + 2y - 4 = 0$

52. $x^2 + y^2 - 2x + 4y - 20 = 0$

53.

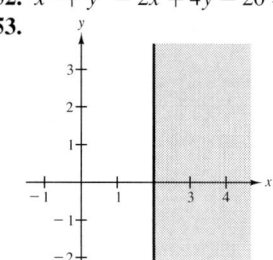

54.

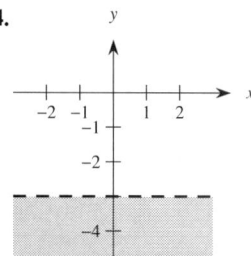

55.

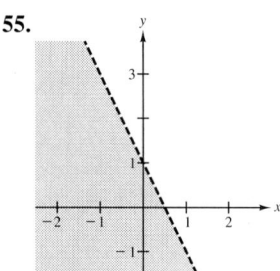

56.

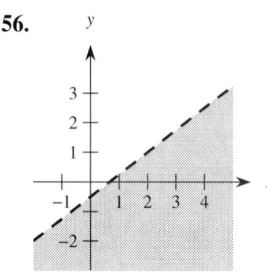

57.

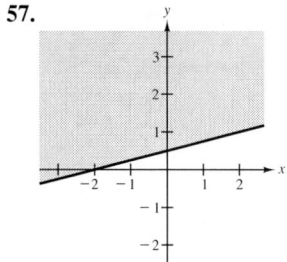

58.

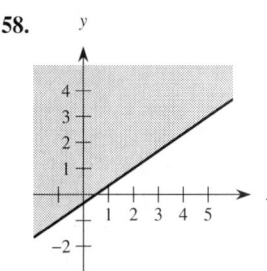

59.

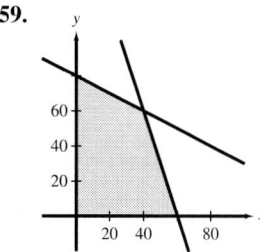

60.

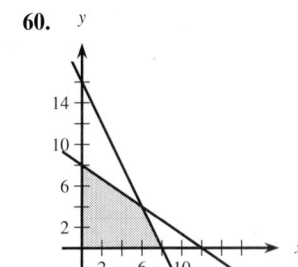

61. 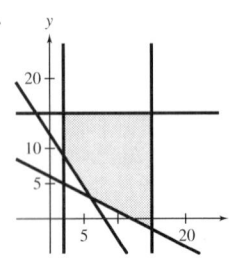   **62.**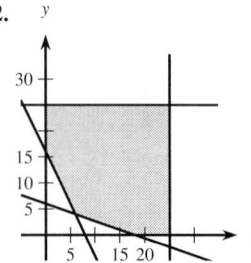

63. $\begin{cases} 2x + y \geq 7 \\ x - y \leq 2 \\ 2x + y \leq 22 \\ x - y \geq -4 \end{cases}$ **64.** $\begin{cases} x - y \geq -1 \\ 3x + y \leq 25 \\ x + 7y \geq 15 \end{cases}$

65. $\begin{cases} x + y \leq 1500 \\ x \geq 400 \\ y \geq 600 \end{cases}$ **66.** $\begin{cases} 20x + 30y \leq 24{,}000 \\ 12x + 8y \leq 12{,}400 \\ x \geq 0 \\ y \geq 0 \end{cases}$

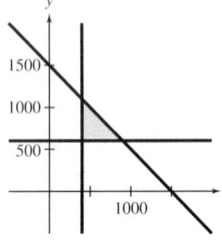   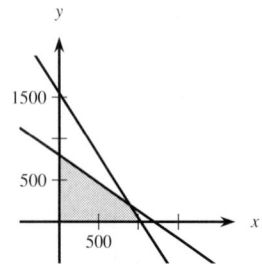

67. Maximum at (5, 8): 47
68. Minimum at (25, 50): 600
69. Minimum at (15, 0): 26.25
70. Maximum at (500, 500): 60,000
71. 5 units of Product A; 2 units of Product B
72. 3 bags of Brand X; 2 bags of Brand Y

CHAPTER 4 TEST   *(page 371)*

1. (b)   **2.** (3, 2)   **3.** (2, 4)
4. (−2, 2)   **5.** $\left(\frac{1}{4}, \frac{1}{3}\right)$ **6.** $(2, 2a - 1, a)$
7. (−1, 3, 3)   **8.** $\left(4, \frac{1}{7}\right)$ **9.** −62
10. $\begin{cases} x + 2y = -1 \\ x + y = 2 \end{cases}$ **11.** $\begin{cases} x + y = 200 \\ x = 4y \end{cases}$
            40 mi, 160 mi
12. $y = 2x^2 - 3x + 4$ **13.** 12

14. 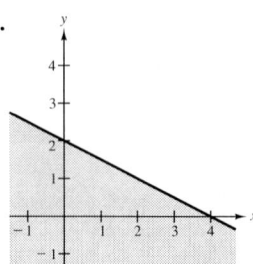   **15.**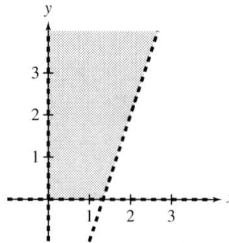

16.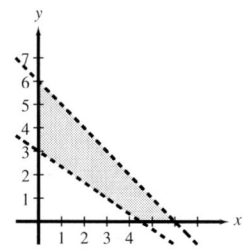

17. Minimum at (0, 0): 0
    Maximum at (3, 3): 48
18. Let x and y be the numbers of units of Model A and
    Model B, respectively. Then
    Objective function: $C = 100x + 150y$
    Constraints: $\begin{aligned} 3.5x + 8y &\leq 5600 \\ 2.5x + 2y &\leq 2000 \\ 1.3x + 0.9y &\leq 900 \\ x &\geq 0 \\ y &\geq 0 \end{aligned}$

Chapter 5

Chapter Opener   *(page 282)*

Hummingbird ≈ 62 times per second
California gull ≈ 4 times per second

SECTION 5.1   *(page 382)*

Warm-Up   *(page 382)*

1. −20   **2.** 70   **3.** −21   **4.** −27
5. $\frac{15}{26}$ **6.** $\frac{11}{23}$ **7.** $\frac{5}{8}$ **8.** $\frac{7}{12}$
9. $20a(x^2 + 3x + 4)$ **10.** $y^2(x - 8)(2x + 7)$

1. $(-\infty, 8) \cup (8, \infty)$ **2.** $(-\infty, 13) \cup (13, \infty)$

3. $(-\infty, -4) \cup (-4, \infty)$ **4.** $(-\infty, \infty)$

5. $(-\infty, \infty)$ **6.** $(-\infty, \infty)$

7. $(-\infty, -4) \cup (-4, 4) \cup (4, \infty)$

8. $(-\infty, 0) \cup (0, 4) \cup (4, \infty)$

9. $(-\infty, -1) \cup (-1, 5) \cup (5, \infty)$

10. $\left(-\infty, -\frac{3}{4}\right) \cup \left(-\frac{3}{4}, 2\right) \cup (2, \infty)$

11. $(-\infty, 2) \cup (2, 4) \cup (4, \infty)$

12. $(-\infty, -3) \cup (-3, 1) \cup (1, \infty)$

13. $\{1, 2, 3, 4, \ldots\}$ **14.** $\{1, 2, 3, 4, \ldots\}$

15. $\dfrac{x}{5}$ **16.** $\dfrac{4y}{3}$ **17.** $6y,\ y \neq 0$

18. $1,\ z \neq 0$ **19.** $\dfrac{6x}{5y^3},\ x \neq 0$ **20.** $\dfrac{4z}{15y^3},\ z \neq 0$

21. $x,\ x \neq 8,\ x \neq 0$ **22.** $\dfrac{a^2}{b^2(b-3)}$

23. $\dfrac{1}{2},\ x \neq \dfrac{3}{2}$ **24.** $\dfrac{y+9}{2},\ y \neq 9$

25. $-\dfrac{1}{3},\ x \neq 5$ **26.** $-(x+6)$ or $-x-6,\ x \neq 6$

27. $\dfrac{3y^2}{y^2+1},\ x \neq 0$ **28.** $x,\ 3xy \neq -1$

29. $\dfrac{y-8}{15},\ y \neq -8$ **30.** $x - 5z,\ x \neq -5z$

31. $\dfrac{1}{a+3}$ **32.** $u - 6,\ u \neq 6$

33. $\dfrac{x}{x-7}$ **34.** $\dfrac{z+11}{3},\ z \neq -11$

35. $\dfrac{y(y+2)}{y+6},\ y \neq 2$ **36.** $\dfrac{x}{x+3},\ x \neq 7$

37. $\dfrac{-1}{2x+3},\ x \neq 3$ **38.** $\dfrac{y+4}{y+6},\ y \neq -\dfrac{5}{2}$

39. $\dfrac{5x+4}{5x+2},\ x \neq \dfrac{1}{3}$ **40.** $\dfrac{8z-5}{7z-4},\ z \neq -\dfrac{4}{7}$

41. $\dfrac{5+3xy}{y^2},\ x \neq 0$ **42.** $\dfrac{2(u-3v)}{9},\ u \neq 0,\ v \neq 0$

43. $\dfrac{3(m-2n)}{m+2n}$ **44.** $\dfrac{x-y}{x+y},\ x \neq -2y$

45. $\dfrac{x^2+2xy+4y^2}{x+2y},\ x \neq 2y$

46. $\dfrac{x^2-3xz+9z^2}{x-2z},\ x \neq -3z$

47. $y + 5z,\ y^2 - 5yz + 25z^2$

48. $\dfrac{x^2-x+1}{y+1},\ x \neq -1$

49. $\dfrac{u-4}{v-2},\ v \neq 3$ **50.** $\dfrac{n+3}{m+n},\ m \neq n$

51. $\dfrac{3}{2} \neq 9$ **52.** $\dfrac{3}{5} \neq -4$

53. $\dfrac{16}{21} \neq \dfrac{3}{4}$ **54.** $\dfrac{9}{8} \neq \dfrac{1}{2}$

55. $x^n - 3,\ x \neq 0$ **56.** $\dfrac{1}{x+1},\ x \neq 0$

57. $x^n - 2,\ x^n \neq -2$ **58.** $\dfrac{x^n-3}{x},\ x^n \neq -4$

59.

-1	0	1	2	Undefined	4	5
-1	0	1	2	3	4	5

The two expressions are equivalent for all x-values except $x = 2$.

60.

3	4	Undefined	6	7	8	9
3	4	5	6	7	8	9

The two expressions are equal for all x-values except $x = 0$.

61. (a) 0 **(b)** $\dfrac{x-10}{4x}$ is undefined for $x = 0$

(c) $\dfrac{3}{2}$ **(d)** $\dfrac{1}{24}$

62. (a) 0 **(b)** 0

(c) $\dfrac{x^2-4x}{x^2-9}$ is undefined for $x = 3$

(d) $\dfrac{x^2-4x}{x^2-9}$ is undefined for $x = -3$

63. (a) $\dfrac{2500+9.25x}{x}$ **(b)** $\{1, 2, 3, 4, \ldots\}$

(c) \$34.25

64. $[0, 100)$ **65.** $\dfrac{1}{9},\ x \neq 0$

66. $\dfrac{x}{x+8}$ **67.** π

68. (a) Van $= 45(t+3)$; Car $= 60t$ **(b)** $\dfrac{4t}{3(t+3)}$

69. $\dfrac{107.1+12.64t+0.54t^2}{31.6+0.51t-0.14t^2} \times 1000$

70. 3389, 3762, 4197, 4703, 5296, and 5997

SECTION 5.2 *(page 392)*

Warm-Up *(page 392)*

1. $\frac{3}{10}$ 2. $\frac{1}{20}$ 3. $\frac{5}{56}$ 4. $-\frac{15}{112}$ 5. 5
6. $-\frac{4}{35}$ 7. $(x-1)(x-2)$ 8. $(x-1)(2x+7)$
9. $(x+1)(11x-5)$ 10. $(2x-7)^2$

1. $\frac{3x}{2}$, $x \neq 0$ 2. 30, $a \neq 0$ 3. $\frac{s^3}{6}$, $s \neq 0$

4. $\frac{8x^5}{21}$, $x \neq 0$ 5. $24u^2$, $u \neq 0$ 6. $\frac{5}{7}$, $x \neq 0$

7. 24 8. $\frac{23}{30}$

9. $\frac{2uv(u+v)}{3(3u+v)}$, $u \neq 0$ 10. 1, $x \neq -25$

11. -1, $r \neq 12$ 12. -1, $z \neq 8$, $z \neq -8$

13. $-\frac{x+8}{x^2}$, $x \neq \frac{3}{2}$ 14. $-\frac{x+14}{5x^2}$, $x \neq 10$

15. $\frac{2(r+2)}{11r}$, $r \neq 2$ 16. 2, $y \neq -3$, $y \neq 4$

17. $2t+5$, $t \neq 3$, $t \neq -2$ 18. $\frac{2(y+4)}{y^2(y-2)}$, $y \neq 4$

19. $\frac{xy(x+2y)}{(x-2y)}$ 20. $(u-2v)(u+2v)$, $u \neq 2v$

21. $\frac{(x-y)^2}{x+y}$, $x \neq -3y$ 22. $\frac{x^2+4y^2}{(x+2y)^2}$, $x \neq 2y$

23. $\frac{(u+v)^2}{(u-v)^2}$, $x \neq y$, $x \neq -y$

24. $\frac{(x+3)^2}{x}$, $x \neq 3$, $x \neq 4$

25. $\frac{(x-1)(2x+1)}{(3x-2)(x+2)}$, $x \neq 5$, $x \neq -5$, $x \neq -1$

26. $\frac{2t^2(l+2t)}{(2t-5)(t+2)^2}$, $t \neq -1$, $t \neq -3$, $t \neq 0$

27. $\frac{3y^2}{2ux^2}$, $v \neq 0$ 28. $\frac{4}{3x^3y^3}$ 29. $\frac{3}{2(a+b)}$

30. $\frac{(x^2+9)(x-2)}{x+3}$, $x \neq -2$, $x \neq 2$

31. $x^4y(x+2y)$, $x \neq 0$, $y \neq 0$, $x \neq -2y$

32. $\frac{y(x+y)}{(x-4)(x-y)}$, $x \neq 0$, $y \neq 0$

33. $\frac{3x}{10}$, $x \neq 0$ 34. $\frac{9u^3}{2v^6}$, $u \neq 0$

35. $-\frac{5x}{2}$, $x \neq 0$, $x \neq 5$

36. $\frac{x(x+1)}{2}$, $x \neq -7$, $x \neq -1$

37. $\frac{(x+3)(4x+1)}{(3x-1)(x-1)}$, $x \neq -3$, $x \neq -\frac{1}{4}$

38. $\frac{(x+5)(x+1)}{x^2}$, $x \neq 5$, $x \neq -1$

39. $\frac{(x+2)(x+3)}{x}$, $x \neq -2$, $x \neq 2$

40. $\frac{t+1}{t+3}$, $t \neq 3$, $t \neq 2$

41. $\frac{(x+1)(2x-5)}{x}$, $x \neq -1$, $x \neq -5$, $x \neq -\frac{2}{3}$

42. $\frac{5(t+5)}{4t^3}$, $t \neq -10$, $t \neq 10$

43. $\frac{x^4}{(x^n+1)^2}$, $x^n \neq -3$, $x^n \neq 3$, $x \neq 0$

44. $\frac{(x^n-8)(x^n-5)}{x^n(x^n+1)}$ 45. $\frac{2w^2+3w}{6}$

46. $\frac{2w^2-w}{6}$ 47. $\frac{x}{4(2x+1)}$ 48. $\frac{x}{2(2x+1)}$

49. $\frac{\pi x}{4(2x+1)}$ 50. $\frac{x}{4(2x+1)}$

SECTION 5.3 *(page 403)*

Warm-Up *(page 403)*

1. $\frac{7}{9}$ 2. $\frac{1}{2}$ 3. $\frac{3}{16}$ 4. $-\frac{3}{5}$ 5. $\frac{37}{36}$
6. $\frac{37}{48}$ 7. $-\frac{146}{225}$ 8. $\frac{181}{21}$ 9. 26 10. $\frac{5}{64}$

1. $-\frac{2}{9}$ 2. $\frac{2z^2-2}{3}$ 3. $\frac{2+y}{2}$ 4. $\frac{1}{3}$

5. $-\frac{3}{a}$ 6. $\frac{5+2z}{5z}$ 7. $\frac{x+6}{3}$ 8. $-\frac{x}{13}$

9. $-\frac{4}{3}$ 10. $-\frac{25}{9}$ 11. $\frac{1}{x-3}$, $x \neq 0$

12. 1, $x \neq -4$ 13. 1, $y \neq 6$
14. 5, $s \neq -\frac{5}{2}$ 15. $20x^3$ 16. $42t^5$
17. $15x^2(x+5)$ 18. $54y^3(y-3)^2$
19. $6x(x+2)(x-2)$
20. $2t^2(t-3)(t+3)(t^2+3t+9)$
21. x^2 22. $7(x-3)^2$ 23. $(u+1)$

24. $5t(3t-5)$ **25.** $-(x+2)$ **26.** $-x$

27. $\dfrac{2n^2(n+8)}{6n^2(n-4)}, \dfrac{10(n-4)}{6n^2(n-4)}$

28. $\dfrac{8s^2(s-1)}{s(s+2)^2(s-1)}, \dfrac{3(s+2)}{s(s+2)^2(s-1)}$

29. $\dfrac{2(x+3)}{x^2(x+3)(x-3)}, \dfrac{5x(x-3)}{x^2(x+3)(x-3)}$

30. $\dfrac{5t}{2t(t-3)^2}, \dfrac{8(t-3)}{2t(t-3)^2}$

31. $\dfrac{(x-8)(x-5)}{(x+5)(x-5)^2}, \dfrac{9x(x+5)}{(x+5)(x-5)^2}$

32. $\dfrac{3y^2}{y(y+3)(y-4)}, \dfrac{(y-4)^2}{y(y+3)(y-4)}$

33. $\dfrac{25-12x}{20x}$ **34.** $\dfrac{100b+1}{10b^2}$ **35.** $\dfrac{7(a+2)}{a^2}$

36. $\dfrac{3-4u}{18u^2}$ **37.** $0, x \neq 4$ **38.** $\dfrac{22}{2-5}$

39. $\dfrac{3(x+2)}{x-8}$ **40.** $\dfrac{1-y}{y-6}$

41. $1, x \neq \dfrac{2}{3}$ **42.** $\dfrac{y+3}{5y-3}$

43. $\dfrac{5(5x+22)}{x+4}$ **44.** $\dfrac{4(45-2x)}{x-10}$

45. $\dfrac{1}{2x(x-3)}$ **46.** $\dfrac{2}{x(x+2)}$

47. $\dfrac{x^2-7x-15}{(x+3)(x-2)}$ **48.** $\dfrac{7+4t}{t(t+1)}$

49. $\dfrac{x-2}{x(x+1)}$ **50.** $\dfrac{2(x+6)}{x(x-4)}$

51. $\dfrac{5(x+y)}{(x+5y)(x-5y)}$ **52.** $\dfrac{4(5x+3y)}{(2x+3y)(2x-3y)}$

53. $\dfrac{4}{x^2(x^2+1)}$ **54.** $\dfrac{4y+1}{2y^2}$

55. $\dfrac{x^2+3x+9}{x(x-3)(x+3)}$ **56.** $\dfrac{6}{(x+5)(x-6)}$

57. $\dfrac{4x}{(x-4)^2}$ **58.** $\dfrac{3x-7}{(x-2)^2}$

59. $\dfrac{2(4x^2+5x-3)}{x^2(x+3)}$ **60.** $\dfrac{4x^2-1}{2x(x+1)^2}$

61. $\dfrac{x}{x-1}, x \neq -6$ **62.** $\dfrac{6x+5}{2(x+10)(x+5)}$

63. $\dfrac{3x+2y+3}{(3x+2y)(2x-y)}$ **64.** $\dfrac{6x+11y}{(3x-y)(x+4y)}$

65. $\dfrac{10x-3}{(2x-1)(x-1)}$ **66.** $\dfrac{7x+3}{(1+x)(1-x)}$

67. $-\dfrac{5x+3}{3x(x-1)(3x+1)}$ **68.** $\dfrac{-x(8x-5)}{(x-2)(x-3)(5x+4)}$

69. $\dfrac{y-x}{xy}, x \neq -y$ **70.** $\dfrac{10x}{(x-y)(x+y)}$

71. $\dfrac{5u-2v}{(u-v)^2}$ **72.** $0, x \neq 0, y \neq 0$

73. $\dfrac{x}{2(3x+1)}$ **74.** $\dfrac{2x}{3(4x-1)}$

75. $\dfrac{4+3x}{4-3x}, x \neq 0$ **76.** $\dfrac{1-t}{1+t}, t \neq 0$

77. $-4x-1, x \neq 0, x \neq \dfrac{1}{4}$ **78.** $\dfrac{6-y}{y}, y \neq -6$

79. $\dfrac{3}{4}, x \neq 0, x \neq 3$ **80.** $\dfrac{(x-2)(x-1)}{x^2-3x+6}, x \neq 3$

81. $\dfrac{y+1}{y-3}, y \neq 0, y \neq 1$ **82.** $\dfrac{x^2-2}{2x}, x \neq -2$

83. $y-x, x \neq 0, y \neq 0, x \neq -y$

84. $\dfrac{x+y}{x-y}, x \neq 2y$ **85.** $\dfrac{5t}{12}$

86. $\dfrac{8t}{15}$ **87.** $\dfrac{5x}{24}$ **88.** $\dfrac{4x}{15}$

89. $\dfrac{x}{4}, \dfrac{x}{3}, \dfrac{5x}{12}$ **90.** $\dfrac{11x}{45}, \dfrac{13x}{45}$

91. (a) 12.7% **(b)** $r = \dfrac{288(NM-P)}{N(12P+NM)}, \; 12.7\%$

92. (a) 12% **(b)** $r = \dfrac{288(NM-P)}{N(12P+NM)}, \; 12\%$

93. $\dfrac{R_1 R_2}{R_1 + R_2}$ **94.** $\dfrac{4x+1}{x^2(x+1)}$

95. $(-4, 2, 2)$

$\dfrac{4}{x^3-x} = -\dfrac{4}{x} + \dfrac{2}{x+1} + \dfrac{2}{x-1}$

96. $\dfrac{1}{x-1} - \dfrac{1}{x+2}$

CHAPTER 5 MID-CHAPTER QUIZ   *(page 407)*

1. $\dfrac{5x}{2y^3}, x \neq 0$ **2.** $\dfrac{-1}{2x+1}, x \neq 4$

3. $36x^2(x+2)^2$ **4.** $5x(x+1)(x-1)$

5. $\dfrac{3(x+4)}{x^2}, x \neq 4$ **6.** $\dfrac{7(x+1)}{xy(x-2)}$

7. $\dfrac{x+5}{2x+3}, x \neq \dfrac{3}{2}, x \neq -5$ **8.** $\dfrac{x+6}{3x^2}$

9. $\dfrac{-5(4x+5)}{(x-6)(x+5)}$ **10.** $\dfrac{49x^2+24x-5}{6x(x+1)(x-1)}$

11. $\dfrac{4(x+2)}{1-16x}$, $x \neq 0$ **12.** $\dfrac{x+5}{x}$, $x \neq -3$, $x \neq 3$

13. $\dfrac{7}{6x(x+5)}$ **14.** $\dfrac{11t}{15}$

SECTION 5.4   *(page 417)*

Warm-Up   *(page 417)*

1. $\dfrac{1}{4}$ **2.** $\dfrac{29}{90}$ **3.** $-\dfrac{4x}{3}$ **4.** $\dfrac{t^2}{9}$ **5.** $\dfrac{2u}{9v^2}$

6. $-\dfrac{r^3}{7}$ **7.** $-15x^4 + 12x^3$ **8.** $70y^3 - 20y$

9. $2x^3 + 7x^2 - 3x - 4$ **10.** $8x^2 + 10x + 3$

---

1. $3z + 5$ **2.** $3x + 4$ **3.** $\frac{5}{2}z^2 + z - 3$

4. $\dfrac{u^2}{4} + \dfrac{u}{2} - \dfrac{3}{2}$ **5.** $7x^2 - 2x$, $x \neq 0$

6. $6a + 7$, $a \neq 0$ **7.** $-10z^2 - 6$, $z \neq 0$

8. $-3c^3 + 4c$, $c \neq 0$ **9.** $4z^2 + \frac{3}{2}z - 1$, $z \neq 0$

10. $2x^2 + \dfrac{8}{3}x - 6$, $x \neq 0$ **11.** $m^3 + 2m - \dfrac{7}{m}$

12. $-l + \dfrac{8}{l}$ **13.** $\dfrac{5}{2}x - 4 + \dfrac{7}{2}y$, $x \neq 0$, $y \neq 0$

14. $-7s^2 t + \dfrac{7}{2}t - \dfrac{9}{s^2}$, $t \neq 0$

15. $x + 1$, $x \neq -5$ **16.** $r - 1$, $r \neq 9$

17. $x - 5$, $x \neq 3$ **18.** $t - 12$, $t \neq 6$

19. $x + 10$, $x \neq -5$ **20.** $y - 8$, $y \neq -2$

21. $x + 7$, $x \neq 3$ **22.** $-x + 5$, $x \neq -1$

23. $y + 3$, $y \neq -\frac{1}{2}$ **24.** $5t + 4$, $t \neq \frac{3}{2}$

25. $6t - 5$, $t \neq \frac{5}{2}$ **26.** $-4u + 3$, $u \neq -\frac{5}{2}$

27. $4x - 1$, $x \neq -\frac{1}{4}$ **28.** $9y + 5$, $y \neq \frac{5}{9}$

29. $x^2 - 5x + 25$, $x \neq -5$ **30.** $x^2 + 3x + 9$, $x \neq 3$

31. $x^2 + 4$, $x \neq 2$ **32.** $x^2 - 4x - 12$, $x \neq -4$

33. $2 + \dfrac{5}{x+2}$ **34.** $6 - \dfrac{23}{2x+3}$

35. $x - 4 + \dfrac{32}{x+4}$ **36.** $y - 2 + \dfrac{12}{y+2}$

37. $5x - 8 + \dfrac{19}{x+2}$ **38.** $2x - 3 + \dfrac{14}{x+4}$

39. $4x + 3 - \dfrac{11}{3x+2}$ **40.** $2x + 1 + \dfrac{4}{4x-1}$

41. $\dfrac{6}{5}z + \dfrac{41}{25} + \dfrac{41}{25(5z-1)}$ **42.** $\dfrac{2}{3}y + \dfrac{5}{9} - \dfrac{25}{9(3y+5)}$

43. $2x^2 + x + 4 + \dfrac{6}{x-3}$ **44.** $5x^2 - 2x + 14 + \dfrac{6}{x+1}$

45. $x^5 + x^4 + x^3 + x^2 + x + 1$, $x \neq 1$

46. $x^2 + x + 1 + \dfrac{1}{x-1}$ **47.** $x + 2$

48. $x - 1 + \dfrac{-3x - 10}{2x^2 + 4x + 5}$

49. $x + 4$, $x^2 + 3x - 2 \neq 0$

50. $4x - 1$, $x^2 - 6x + 1 \neq 0$

51. $x^2 - 3x - 5$, $x^2 + x - 3 \neq 0$

52. $2x^2 - 4x + 5$, $x \neq -3$, $x \neq 1$

53. $x^{2n} + x^n + 4$, $x^n \neq -2$

54. $x^{2n} + 5$, $x^n \neq 1$ **55.** $(x - 7)(x - 8)$

56. $(x - 11)(x + 14)$ **57.** $(2a - 5)(a + 9)$

58. $(2t + 3)(t - 8)$ **59.** $(15x + 10)\left(x - \frac{4}{5}\right)$

60. $(18x - 24)\left(x + \frac{5}{6}\right)$ **61.** $(y^2 + 3y + 2)(y - 2)$

62. $(-3z^2 + 8z - 4)(z - 4)$ **63.** $(2t^2 + 5t - 6)(t + 5)$

64. $(5t^2 + 3t + 4)(t - 6)$ **65.** $(x^3 - 3x + 1)(x - 1)$

66. $(x^3 + 2x^2 + 4x + 8)(x - 2)$

67. $x^2 - x + 4 - \dfrac{17}{x+4}$

68. $x^3 - 2x^2 + 4x - 8 + \dfrac{16}{x+2}$

69. $x^3 - 2x^2 - 4x - 7 - \dfrac{4}{x-2}$

70. $2x^4 + 6x^3 + 15x^2 + 45x + 136 + \dfrac{408}{x-3}$

71. $5x^2 - 25x + 125 - \dfrac{613}{x+5}$

72. $8 + \dfrac{115}{x-10}$ **73.** $5x^2 + 14x + 56 + \dfrac{232}{x-4}$

74. $5x^2 - 10x + 26 - \dfrac{44}{x+2}$

75. $10x^3 + 10x^2 + 60x + 360 + \dfrac{1360}{x-6}$

76. $x^4 - 16x^3 + 48x^2 - 144x + 312 - \dfrac{856}{x+3}$

77. $0.1x + 0.82 + \dfrac{1.164}{x-0.2}$ **78.** $x^2 - x + 0.2 + \dfrac{2.2}{x+1}$

79. -8 **80.** -1188

81. $2x$, $x \neq 0$ **82.** $3xy$, $x \neq 0$, $y \neq 0$

83. $7uv$, $u \neq 0$, $v \neq 0$ **84.** $-2x + 7$, $x \neq 1$

85. Invalid. 5's are terms, not factors.
86. Invalid. 6's are terms, not factors.
87. Valid
88. Invalid. 8's are digits of the integers, not factors.
89. $x^3 - 5x^2 - 5x - 10$ **90.** $x^4 + 4x^3 + 3x^2 - 4x - 4$
91. $x^2 - 3$ **92.** $x^2 + 50x + 400$
93. $2x + 8$ **94.** $h^2 + 2h$
95. (a)

36	$\frac{51}{8}$	-12	16	148

(b)

36	$\frac{51}{8}$	-12	16	148

(c) The value of $f(c)$ is equal to the remainder when the polynomial is divided by $x - c$.
$f(4) = 66$

SECTION 5.5   *(page 428)*

Warm-Up   *(page 428)*

1. $(x - 5)(x + 2)$ **2.** $(x - 5)(x - 2)$
3. $x(x + 1)(x + 3)$ **4.** $(x^2 - 2)(x - 4)$
5.
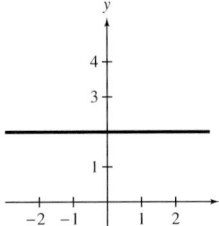
6.
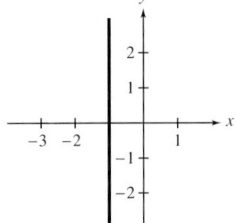
7.
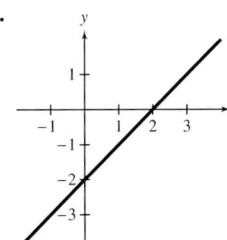
8.
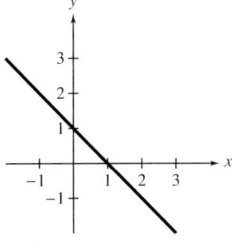
9. $5 + \dfrac{26}{x - 4}$ **10.** $-2 + \dfrac{7}{x + 2}$

1. e   **2.** f   **3.** a   **4.** b   **5.** c   **6.** d
7. Domain: all real numbers x such that $x \neq 5$
Vertical asymptote: $x = 5$
Horizontal asymptote: $y = 0$
8. Domain: all real numbers x such that $x \neq -5$
Vertical asymptote: $x = -5$
Horizontal asymptote: $y = 0$
9. Domain: all real numbers x such that $x \neq 0$
Vertical asymptote: $x = 0$
Horizontal asymptote: $y = 0$
10. Domain: all real numbers x such that $x \neq 2$
Vertical asymptote: $x = 2$
Horizontal asymptote: $y = 0$
11. Domain: all real numbers x such that $x \neq \pm 3$
Vertical asymptotes: $x = -3$, $x = 3$
Horizontal asymptote: $y = 0$
12. Domain: all real numbers x such that $x \neq 0$
Vertical asymptote: $x = 0$
Horizontal asymptote: $y = 0$
13. Domain: all real numbers x such that $x \neq 2$
Vertical asymptote: $x = 2$
Horizontal asymptote: $y = -1$
14. Domain: all real numbers x such that $x \neq -\frac{1}{2}$
Vertical asymptote: $x = -\frac{1}{2}$
Horizontal asymptote: $y = -\frac{5}{2}$
15. Domain: all real numbers x
Vertical asymptote: none
Horizontal asymptote: $y = 3$
16. Domain: all real numbers x
Vertical asymptote: none
Horizontal asymptote: $y = 3$
17. Domain: all real numbers x such that $x \neq \pm 1$
Vertical asymptotes: $x = -1$, $x = 1$
Horizontal asymptote: none
18. Domain: all real numbers x such that $x \neq -1$
Vertical asymptote: $x = -1$
Horizontal asymptote: None
19.
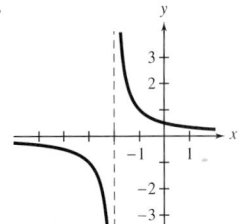
20.
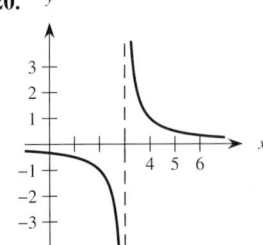

21.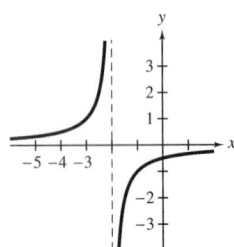

22.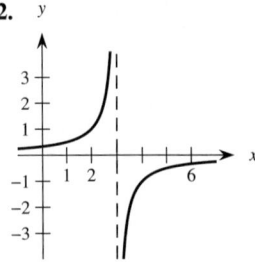

31.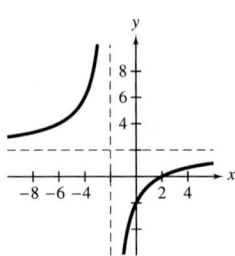

32.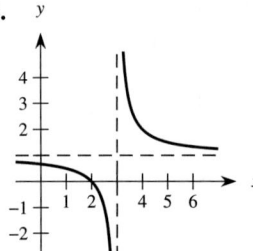

23.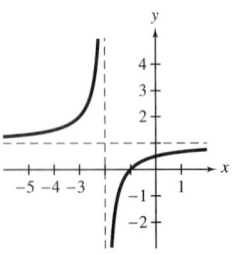

24.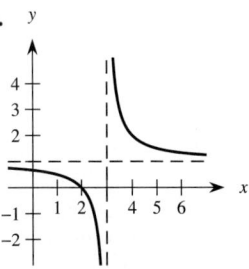

33.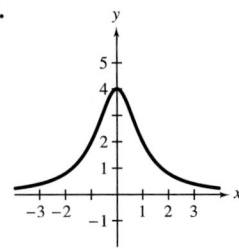

34.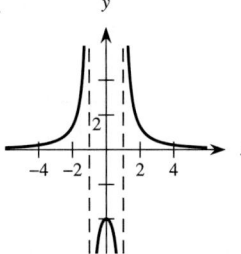

25.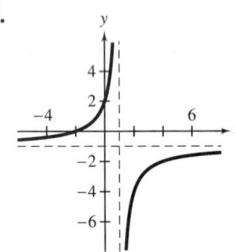

26.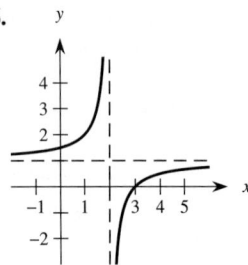

35.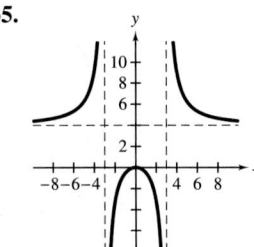

36.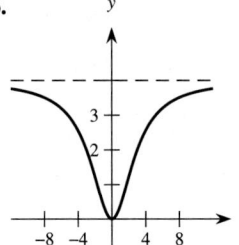

27.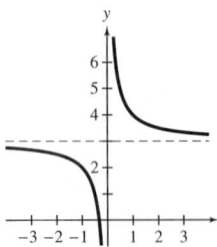

28.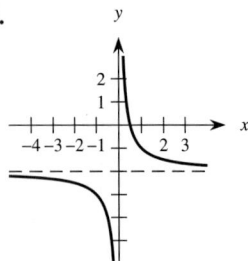

37.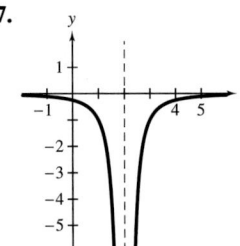

38.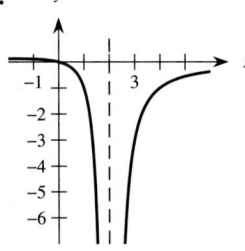

29.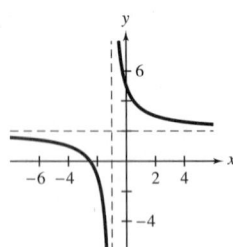

30.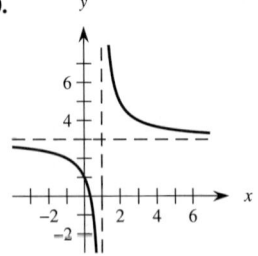

39.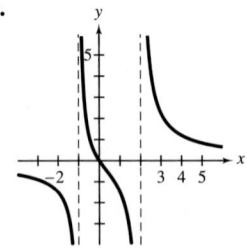

40.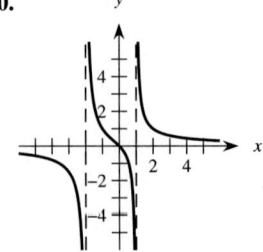

41. (a) 28.33 million dollars
(b) 170 million dollars
(c) 765 million dollars
(d) No

42. (a) $4411.76
(b) $25,000.00
(c) $225,000.00
(d) No

43. (a) 167,250,400
(b) 750; The limiting size of the population
(c)

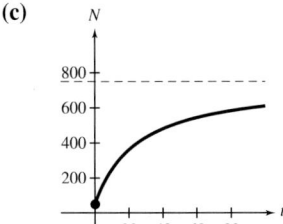

44. $\overline{C}(1000) = \$150.25$
$\overline{C}(10,000) = \$15.25$
$\overline{C}(100,000) = \$1.75$

45.

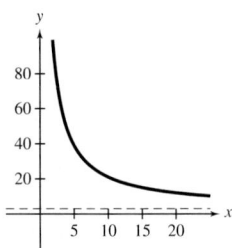

$$\overline{C} = \frac{3x + 180}{x}, \; x > 0, \; \overline{C} = 3$$

As the number of movies rented increases, the average cost per video approaches the rental price.

46. More than 60

3.

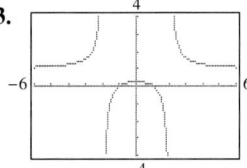

4.

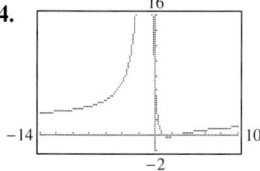

5.

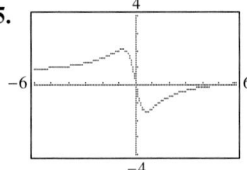

6.

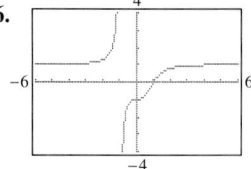

7. $g(x) = x + \dfrac{1}{x + 1}$

For large values of $|x|$, $g(x)$ is close to $f(x) = x$ because $\dfrac{1}{|x| + 1}$ is very small.

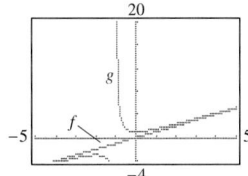

8.

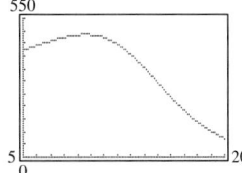

Sales peaked sometime during 1979.

USING A GRAPHING CALCULATOR   *(page 432)*

1.

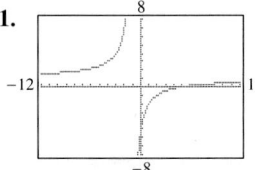

2.

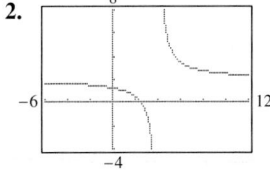

SECTION 5.6   *(page 441)*

Warm-Up   *(page 441)*

1. $\frac{16}{3}$ **2.** -19 **3.** 3   **4.** $\frac{5}{2}$ **5.** $\frac{2}{3}, -8$
6. $16, -\frac{3}{2}$ **7.** $0, \frac{21}{2}$ **8.** 5
9. $6, -7$ **10.** $0, 8$

1. (a) No   **(b)** No   **(c)** No   **(d)** Yes
2. (a) No   **(b)** Yes   **(c)** Yes   **(d)** No
3. (a) No   **(b)** Yes   **(c)** Yes   **(d)** No
4. (a) Yes   **(b)** No   **(c)** No   **(d)** No

5. $\frac{3}{2}$ **6.** 24   **7.** $\frac{1}{4}$ **8.** $\frac{15}{2}$ **9.** $-\frac{8}{3}$

10. $\frac{9}{2}$ **11.** 10   **12.** 20   **13.** $\frac{7}{4}$ **14.** 10

15. $\frac{1}{3}$ **16.** $-\frac{27}{4}$ **17.** 61   **18.** $-\frac{4}{5}$ **19.** $\frac{18}{5}$

20. 3   **21.** 3   **22.** 130   **23.** 3   **24.** $\frac{4}{5}$

25. $-\frac{11}{5}$ **26.** 5   **27.** $\frac{4}{3}$ **28.** -25

29. $-\frac{4}{15}$ **30.** 5   **31.** No solution

32. $-\frac{1}{6}$ **33.** No solution   **34.** $\frac{1}{3}$, 2

35. 1, $\frac{3}{2}$ **36.** No solution   **37.** 3

38. 20   **39.** 1   **40.** ± 10 **41.** ± 6

42. ± 30 **43.** ± 4 **44.** ± 15

45. 8, -9 **46.** -2, -10 **47.** 3, 13

48. 5, $-\frac{3}{2}$ **49.** 0, 2   **50.** -6, 7

51. 8, -3 **52.** $\frac{4}{3}$, 2   **53.** 2, 3

54. $-\frac{3}{2}$, 2   **55.** 8, $\frac{1}{8}$ **56.** 12, $\frac{1}{8}$

57. 40 mph   **58.** 50 mph   **59.** 15

60. 3000 units   **61.** 85%   **62.** 9 hr

63. 3 hr; $1\frac{7}{8}$ min; $1\frac{2}{3}$ hr

64. 2 days, $1\frac{16}{17}$ hr, $\dfrac{ab}{a+b}$ days

65. (a) $\dfrac{1}{20}$ min   **(b)** $\dfrac{x}{20}$ min   **(c)** $1\frac{3}{4}$ min

66. (a) $\dfrac{1}{15}$ min   **(b)** $\dfrac{x}{15}$ min   **(c)** $8\frac{2}{3}$ min

67. 15 hr; $22\frac{1}{2}$ hr   **68.** $16\frac{2}{3}$ hr; 25 hr

69. $11\frac{1}{4}$ hr   **70.** $9\frac{4}{5}$ hr; $12\frac{1}{4}$ hr   **71.** 1986

72. 1988   **73.** 1984   **74.** 1991

CHAPTER 5 REVIEW EXERCISES   *(page 445)*

1. $(-\infty, 8) \cup (8, \infty)$ **2.** $(-\infty, -12) \cup (-12, \infty)$
3. $(-\infty, 1) \cup (1, 6) \cup (6, \infty)$
4. $(-\infty, -4) \cup (-4, 0) \cup (0, 4) \cup (4, \infty)$

5. $\dfrac{2x^3}{5}$, $x \neq 0$, $y \neq 0$ **6.** $\dfrac{y^4}{14z^2}$, $y \neq 0$

7. $\dfrac{b-3}{6(b-4)}$ **8.** $\dfrac{2}{5a+13}$, $a \neq 0$

9. -9, $x \neq y$ **10.** $\dfrac{1}{x-4}$, $x \neq 3$

11. $\dfrac{y}{8x}$, $y \neq 0$ **12.** $60x^5y$, $x \neq 0$, $y \neq 0$

13. $12z(z-6)$, $z \neq -6$ **14.** $\dfrac{x+4}{2(x-4)}$

15. $-\frac{1}{4}$, $u \neq 0$, $v \neq 3$

16. $\dfrac{25x(x-1)}{x+5}$, $x \neq 0$, $x \neq 1$, $x \neq -1$

17. $3x^2$, $x \neq 0$ **18.** 0, $x \neq 0$, $y \neq 0$

19. $\dfrac{125y}{x}$, $y \neq 0$ **20.** $\dfrac{3}{2z^4}$

21. $\dfrac{x(x-1)}{x-7}$, $x \neq -1$, $x \neq 1$

22. $\dfrac{4}{3xy}$ **23.** $-\dfrac{7}{9}a$ **24.** $\dfrac{5y+11}{2y+1}$

25. $\dfrac{-48x+35}{48x}$ **26.** $\dfrac{17y}{24x}$

27. $\dfrac{4x+3}{(x+5)(x-12)}$ **28.** $\dfrac{x-22}{(x-10)(4-x)}$

29. $\dfrac{5x^3 - 5x^2 - 31x + 13}{(x+2)(x-3)}$

30. $\dfrac{31x - 78}{(x+6)(x-5)}$ **31.** $\dfrac{x+24}{x(x^2+4)}$

32. $\dfrac{4x}{(x+2)(x-5)}$ **33.** $\dfrac{6(x-9)}{(x+3)^2(x-3)}$

34. $\dfrac{9y^2 + 50y - 80}{y(y+5)(y-2)}$ **35.** $\dfrac{6(x+5)}{x(x+7)}$, $x \neq 5$, $x \neq -5$

36. $\frac{2}{5}$, $x \neq 2$, $x \neq \frac{4}{3}$

37. $\dfrac{3t^2}{5t-2}$, $t \neq 0$ **38.** $x - 1$, $x \neq 0$, $x \neq 2$

39. $\dfrac{-a^2 + a + 16}{(4a^2 + 16a + 1)(a-4)}$ **40.** $\dfrac{y-x}{xy}$, $x \neq -y$

41. $2x^2 - \dfrac{1}{2}$, $x \neq 0$ **42.** $2 + \dfrac{19}{5x-2}$

43. $2x^2 + x - 1 + \dfrac{1}{3x-1}$

44. $4x^3 + 7x^2 + 7x + 32 + \dfrac{64}{x-2}$

45. $x^2 - 2$, $x \neq 1$, $x \neq -1$

46. $3x^4 + 3x^2 + 3 + \dfrac{3}{x^2-1}$ **47.** $x^2 + 5x - 7$

48. $x^3 + 3x^2 - 2$ **49.** $x^3 + 3x^2 + 6x + 18 + \dfrac{29}{x-3}$

50. $2x^2 - x + \dfrac{11}{2} - \dfrac{\frac{19}{4}}{x+\frac{1}{2}}$

51. d   **52.** a   **53.** c   **54.** b

55.

56.

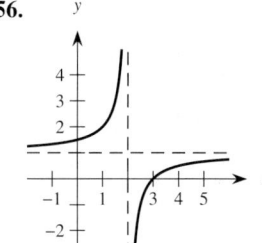

57.

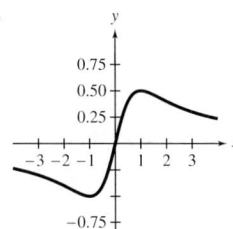

58.

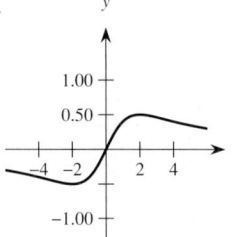

59.

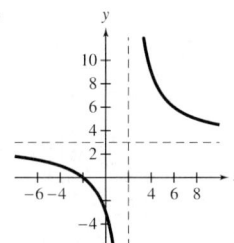

60.

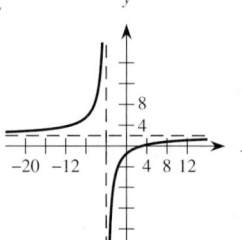

61.

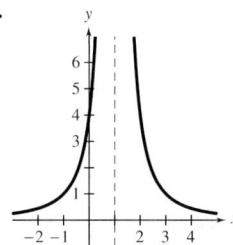

62.

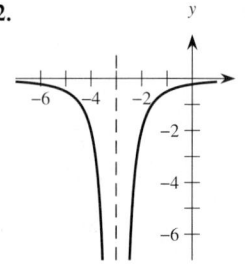

63.

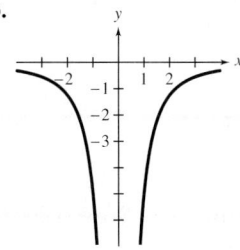

64.

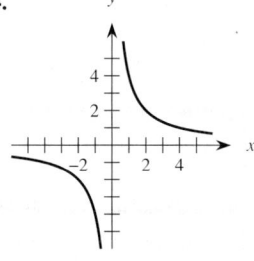

65.

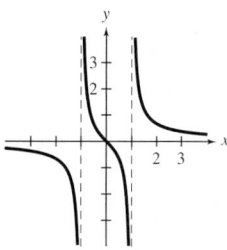

66.

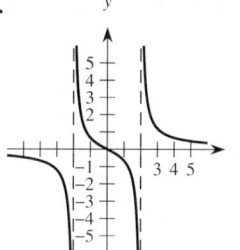

67.

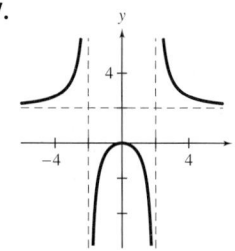

68.

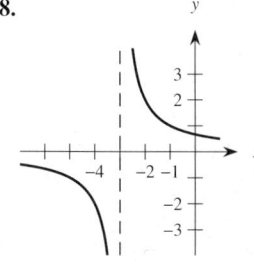

69. Horizontal asymptote: $y = \frac{1}{2}$

As x gets larger, the average cost per unit approaches $0.50.

70. $\overline{C}(1000) = \$100.90$
$\overline{C}(10,000) = \$10.90$
$\overline{C}(100,000) = \$1.90$

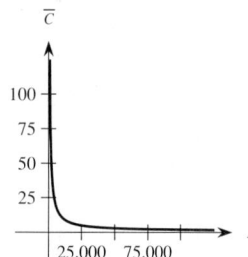

71. **(a)** 176 million dollars   **(b)** 528 million dollars
(c) 1584 million dollars   **(d)** No

72. **(a)** $N(5) = 304$ thousand fish
$N(10) \approx 453.3$ thousand fish
$N(25) \approx 702.2$ thousand fish
(b) The population is limited by the horizontal asymptote $N = 1200$ thousand fish.

73. -40 **74.** 3   **75.** $\frac{3}{2}$ **76.** 2

77. 5   **78.** $3, -\frac{16}{3}$ **79.** $6, -4$ **80.** $-2, -4$

81. $2, -6$ **82.** $3, -\frac{3}{2}$ **83.** $-2, 2$ **84.** $\frac{1}{2}, 4$

85. $-\frac{9}{5}, 3$ **86.** $0, 4$ **87.** 56 mph

88. 25   **89.** 4   **90.** 7   **91.** $6\frac{2}{3}$ min

CHAPTER 5 TEST *(page 449)*

1. $(\infty, -5) \cup (-5, 5) \cup (5, \infty)$
2. $x^3(x+3)^2(x-3)$ 3. $-\frac{1}{3}$, $x \neq 2$
4. $\frac{2a+3}{5}$, $a \neq 4$ 5. $\frac{5z}{3}$, $z \neq 0$
6. $\frac{4}{y+4}$, $y \neq 2$ 7. $\frac{(2x+3)^2}{x+1}$, $x \neq \frac{3}{2}$
8. $\frac{14y^2}{15}$, $x \neq 0$ 9. $\frac{x^3}{4}$, $x \neq -2$
10. 4, $x \neq -1$ 11. $-\frac{2x^2 - 2x - 1}{x+1}$
12. $\frac{5x^2 - 15x - 2}{(x+2)(x-3)}$ 13. $\frac{5x^3 + x^2 - 7x - 5}{x^2(x+1)^2}$
14. $-(3x+1)$ 15. $t^2 + 3 - \frac{6t-6}{t^2-2}$
16. $2x^3 + 6x^2 + 3x + 9 + \frac{20}{x-3}$

17.   18.

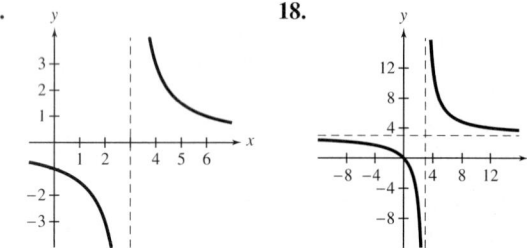

19. 22 20. $-1, -\frac{15}{2}$
21. No solution   22. $6\frac{2}{3}$ hr, 10 hr

Chapter 6

Chapter Opener *(page 450)*

≈ 29.47 yr

SECTION 6.1 *(page 459)*

Warm-Up *(page 459)*

1. a^{10} 2. $x^5 y^4$ 3. $9y^3$ 4. $4rs^2$
5. $-t^6$ 6. $16z^2$ 7. $36x^5 y^8$
8. $-128u^5 v^7$ 9. $\frac{16a^8}{81b^4}$ 10. $\frac{9x^2}{16y^6}$

1. $\frac{1}{25}$ 2. $\frac{1}{16}$ 3. $-\frac{1}{1000}$ 4. $-\frac{1}{400}$ 5. $-\frac{1}{243}$
6. 1 7. 64 8. -64 9. -32 10. $-\frac{1}{36}$
11. $\frac{3}{2}$ 12. $\frac{125}{64}$ 13. 1 14. $\frac{64}{25}$ 15. 1
16. $\frac{1}{4}$ 17. 729 18. $\frac{1}{125}$ 19. $100{,}000$
20. 10 21. $\frac{1}{64}$ 22. 16 23. $\frac{1}{16}$ 24. 125
25. $\frac{3}{16}$ 26. $\frac{35}{9}$ 27. $\frac{64}{121}$ 28. -6 29. $\frac{16}{15}$
30. 1 31. y^2 32. $\frac{1}{x^7}$ 33. x^6 34. $\frac{y}{x^3}$
35. x^6 36. $\frac{1}{s^6}$ 37. a 38. $\frac{2}{5u}$ 39. t^2
40. 1 41. $-\frac{12}{xy^3}$ 42. $\frac{3s^3}{10t^4}$ 43. $\frac{y^4}{9x^4}$
44. $-\frac{y^9}{64z^3}$ 45. $\frac{3x^5}{y^4}$ 46. $\frac{1}{y^2}$ 47. $\frac{81v^8}{u^6}$
48. $\frac{3x}{25y^3}$ 49. $\frac{x^{12} y^8}{16}$ 50. $\frac{64y^6}{x^{12}}$ 51. $\frac{1}{4x^4}$
52. $\frac{5184}{y^7}$ 53. $\frac{z^2}{16}$ 54. $\frac{x^x}{9z^4}$ 55. $\frac{125x^9}{y^{12}}$
56. $\frac{1}{625x^8 y^8}$ 57. 1 58. $\frac{a^6}{64b^9}$ 59. $\frac{10}{x}$
60. $\frac{x^2}{y^2}$ 61. $\frac{b^5}{a^5}$ 62. 1 63. $\frac{t^{32}}{s^{32}}$
64. $\frac{1}{u^{12} v^{12}}$ 65. $4096\, m^6$ 66. $\frac{16x^4}{81}$
67. $\frac{1}{x^8 y^{12}}$ 68. $x^8 y^{12}$ 69. $\frac{1}{a^4 b^4}$ 70. $\frac{x^5}{108}$
71. $\frac{v^2}{uv^2 + 1}$ 72. $\frac{x^2 + y^2}{x^2}$ or $1 + \frac{y^2}{x^2}$
73. $\frac{ab}{b-a}$ 74. $\frac{v-u}{v+u}$ 75. 5.75×10^7
76. -1.394×10^8 77. 9.461×10^{15} 78. 1×10^{-7}
79. 8.99×10^{-5} 80. 3.937×10^{-5} 81. $524{,}000{,}000$
82. $350{,}000{,}000$ 83. $13{,}000{,}000$

84. 0.000000001   **85.** 0.00000000048

86. 0.0009   **87.** 680,000

88. 125,000,000,000,000   **89.** 4000

90. 0.000005   **91.** 9,000,000,000,000,000

92. 1,600,000,000,000   **93.** 4.703×10^{11}

94. 3.682×10^{34} **95.** 3.463×10^{10}

96. 2.738×10^{20} **97.** $4x^8$ **98.** $36\pi x^8$

99. 80,000,000,000,000,000,000,000

100. 9.3×10^7 mi

101. 1.58×10^{-5} yr ≈ 8.3 min

102. 8.505×10^{17} m or $850,500,000,000,000,000

103. 0.333×10^6 or 333,000

104. Yes, $k \approx 1.0$ for each planet

105. $11,587.63

SECTION 6.2   *(page 471)*

Warm-Up   *(page 470)*

1. 169   **2.** -169 **3.** $\frac{8}{27}$ **4.** $\frac{3}{10}$

5. 48   **6.** $\frac{27}{8}$ **7.** $\frac{1}{x^3}$ **8.** $-\frac{15y^4}{x^2}$

9. $\frac{9y^2}{4x^2}$ **10.** $\frac{v^4}{u^5}$

---

1. 7   **2.** 24.5   **3.** 4.2   **4.** 6

5. Square root   **6.** Cube root   **7.** 9

8. Not a real number   **9.** $\frac{3}{4}$ **10.** 0.3

11. 8   **12.** -10 **13.** Not a real number

14. 12   **15.** $-\frac{2}{3}$ **16.** Not a real number

17. 0.4   **18.** -0.03 **19.** 5   **20.** -2

21. $-\frac{1}{4}$ **22.** -0.2 **23.** 3   **24.** 2

25. Not a real number   **26.** $-\frac{1}{5}$ **27.** -0.3

28. 2   **29.** 13   **30.** 5   **31.** 14   **32.** 10

33. 13   **34.** 20   **35.** -20 **36.** -35 **37.** 16

38. 2   **39.** $16^{1/2} = 4$ **40.** $81^{1/4} = 3$

41. $27^{2/3} = 9$ **42.** $\sqrt[3]{125} = 5$ **43.** $\sqrt[4]{256^3} = 64$

44. $\sqrt[5]{32^3} = 8$ **45.** 5   **46.** -11 **47.** 8

48. $\frac{1}{27}$ **49.** $\frac{1}{4}$ **50.** $\frac{1}{27}$ **51.** $\frac{4}{9}$ **52.** $\frac{4}{5}$

53. $\frac{3}{11}$ **54.** $\frac{10,000}{81}$ **55.** 1   **56.** 5

57. Irrational   **58.** Rational

59. Rational   **60.** Irrational

61. 8.5440   **62.** Not a real number

63. 1.0420   **64.** 5.1701   **65.** 0.0038

66. 97.4503   **67.** 4.3004   **68.** 5.4175

69. 66.7213   **70.** -8.4419 **71.** $|t|$ **72.** z

73. y^3 **74.** a^2 **75.** x^3 **76.** $z^{1/5}$

77. $\frac{9y^{3/2}}{x^{2/3}}$ **78.** $\frac{-8u^{9/5}}{v^{3/5}}$ **79.** $\frac{3y^2}{4z^{4/3}}$ **80.** $\frac{1}{a^{5/4}}$

81. $x^{1/4}$ **82.** $\frac{9m^{1/3}n^2}{16}$ **83.** $c^{1/2}$ **84.** $\frac{1}{k^{1/2}}$

85. $\sqrt[8]{y}$ **86.** $\sqrt[6]{2x}$ **87.** $2x^{3/2} - 3x^{1/2}$

88. $3x^{10/3} - 4x^{7/3} + 5x^{4/3}$

89. $1 + 5y$ **90.** $x - 9$

91. All real numbers $x \geq 0$

92. All real numbers $x \geq 0$

93. All real numbers x **94.** All real numbers x

95. All real numbers $x > 0$

96. All real numbers $x \neq 0$

97. 12.8%   **98.** 14.9%

99. 13 in. $\times$ 13 in. $\times$ 13 in.

100. 23 ft $\times$ 23 ft   **101.** 0.026 in.

102. **(a)** 1, 4, 5, 6, 9   **(b)** No

SECTION 6.3   *(page 480)*

Warm-Up   *(page 480)*

1. 100   **2.** -10 **3.** $-\frac{3}{5}$ **4.** $\frac{2}{5}$

5. $x^{11/6}$ **6.** $a^{1/4}$ **7.** $16x^4$

8. $4096x^2y^8$ **9.** $\frac{16}{x^{1/4}y^{1/4}}$ **10.** $\frac{8y^2}{z^{5/2}}$

---

1. $\sqrt{30}$ **2.** $\sqrt[5]{171}$ **3.** $\sqrt[3]{110}$ **4.** $\sqrt[4]{105}$

5. $\sqrt{\frac{15}{31}}$ **6.** $\sqrt[4]{\frac{85}{9}}$ **7.** $\sqrt[5]{\frac{152}{3}}$ **8.** $\sqrt[4]{\frac{633}{5}}$

9. $3\sqrt{35}$ **10.** $3\sqrt[3]{4}$ **11.** $3\sqrt[4]{11}$ **12.** $10\sqrt[5]{3}$

13. $\frac{\sqrt{35}}{3}$ **14.** $\frac{\sqrt[4]{165}}{2}$ **15.** $\frac{\sqrt[3]{11}}{10}$ **16.** $\frac{\sqrt[5]{2}}{3}$

17. $2\sqrt{5}$ **18.** $5\sqrt{2}$ **19.** $3\sqrt{3}$ **20.** $5\sqrt{5}$

21. 0.2   **22.** 0.5   **23.** $2\sqrt[3]{3}$ **24.** $3\sqrt[3]{2}$

25. $10\sqrt[4]{3}$ **26.** $2\sqrt[5]{3}$ **27.** $\frac{\sqrt{15}}{2}$ **28.** $\frac{\sqrt{5}}{6}$

29. $\frac{\sqrt[3]{35}}{4}$ **30.** $\frac{\sqrt[4]{5}}{2}$ **31.** $\frac{\sqrt[5]{15}}{3}$ **32.** $\frac{1}{10}$

33. 0.3   **34.** 0.4   **35.** $3x^2\sqrt{x}$ **36.** $8x\sqrt{x}$

37. $xy\sqrt[3]{x}$ **38.** $ab^2\sqrt[3]{a^2}$ **39.** $2xy\sqrt[5]{y}$

40. $2|uv|\sqrt[4]{8v^3}$ **41.** $\dfrac{\sqrt{13}}{5}$ **42.** $\dfrac{\sqrt{15}}{6}$

43. $\dfrac{2\sqrt[5]{x^2}}{y}$ **44.** $\dfrac{2z\sqrt[3]{2}}{y^2}$ **45.** $\dfrac{3a\sqrt[3]{2a}}{b^3}$

46. $\dfrac{\sqrt[4]{3u^2}}{2v^2}$ **47.** $4y^2\sqrt{3x}$ **48.** $\dfrac{3|x|\sqrt{2}}{z^3}$

49. $\dfrac{4a^2\sqrt{2}}{|b|}$ **50.** $2x\sqrt[3]{2x^2}$ **51.** $3x^2$

52. $\dfrac{5|x|\sqrt{3}}{y^2}$ **53.** $2x\sqrt[5]{3}$ **54.** 562

55. $\dfrac{2\sqrt[3]{2}}{3}$ **56.** $2xy^2\sqrt[3]{9x^2y}$ **57.** $\dfrac{\sqrt{3}}{3}$

58. $\dfrac{\sqrt{5}}{5}$ **59.** $4\sqrt{3}$ **60.** $\dfrac{\sqrt{10}}{2}$ **61.** $\dfrac{\sqrt[4]{20}}{2}$

62. $\dfrac{\sqrt[3]{45}}{5}$ **63.** $\dfrac{3\sqrt[3]{2}}{2}$ **64.** $5\sqrt[5]{2}$ **65.** $\dfrac{\sqrt{y}}{y}$

66. $\dfrac{\sqrt{2x}}{2x}$ **67.** $\dfrac{\sqrt{3x}}{x}$ **68.** $\dfrac{\sqrt{5c}}{|c|}$ **69.** $\dfrac{2\sqrt{x}}{x^2}$

70. $\dfrac{5\sqrt{2x}}{4x^3}$ **71.** $\dfrac{\sqrt[3]{18xy^2}}{3y}$ **72.** $\dfrac{\sqrt[3]{60x^2y}}{3y}$

73. $\dfrac{a^2\sqrt[3]{a^2b}}{b}$ **74.** $\dfrac{3u\sqrt[4]{2u}}{2}$ **75.** $\dfrac{2\sqrt{3b}}{b^2}$

76. $\dfrac{\sqrt{xy}}{xy}$ **77.** $\dfrac{b\sqrt[4]{ab}}{a}$ **78.** $\dfrac{\sqrt{5}}{y^2}$

79. $\dfrac{\sqrt[3]{2178x^2z^2}}{11z^2}$ **80.** $\dfrac{4t^3}{s^2}$ **81.** $2\sqrt{2}$

82. $-\dfrac{4}{5}\sqrt{5}$ **83.** $-5\sqrt[4]{7}-11\sqrt[4]{3}$

84. $5\sqrt[3]{17}+7\sqrt[3]{2}+\sqrt{2}$ **85.** $24\sqrt{2}-6$

86. $44\sqrt{2}$ **87.** $30\sqrt[3]{2}$ **88.** $5\sqrt[4]{3}$

89. These radical expressions cannot be combined.

90. $13\sqrt{x+1}$ **91.** $13\sqrt{y}$ **92.** $-t\sqrt[3]{2t}$

93. $(10-z)\sqrt[3]{z}$ **94.** $(10+6u)\sqrt[3]{3u^2}$ **95.** $\dfrac{2\sqrt{5}}{5}$

96. $\dfrac{3\sqrt{10}}{2}$ **97.** $\dfrac{9\sqrt{5}}{5}$ **98.** $\dfrac{10\sqrt{3x}}{3}$

99. 5.667 **100.** 1.586 **101.** 1.993 **102.** 0.0013

103. $\sqrt{7}+\sqrt{18}>\sqrt{7+18}$ **104.** $\sqrt{10}-\sqrt{6}<\sqrt{10-6}$

105. $5>\sqrt{3^2+2^2}$ **106.** $5=\sqrt{3^2+4^2}$

107. 89.44 cycles/sec **108.** 2.22 sec

109. $\approx$ 175 vibrations/sec $\approx$ 932 vibrations/sec

110. The frequencies of successive notes of the same letter double going from left to right.

111. 1

112. No. Rationalizing the denominator does not change the value of the expression.

CHAPTER 6 MID-CHAPTER QUIZ *(page 483)*

1. $\frac{1}{8}$ **2.** $\frac{1}{64}$ **3.** $\frac{16}{9}$ **4.** $\frac{1}{6}$ **5.** 3 **6.** 8

7. $-125x^3$ **8.** $12x^4$ **9.** $\dfrac{y^2}{2}$ **10.** $\dfrac{7x^{1/6}}{6}$

11. $6\sqrt{2}$ **12.** $\dfrac{2c\sqrt{10c}}{5d^3}$ **13.** $2x\sqrt[3]{2xy}$

14. $14\sqrt{2x}-3\sqrt{x}$ **15.** 1.3837×10^{-2}

16. 1.573×10^{10} **17.** $\dfrac{\sqrt{15y}}{5y}$ **18.** $\dfrac{\sqrt[3]{6t^2}}{2t}$

19. Factor of 4 **20.** 24 in. $\times$ 24 in. $\times$ 24 in.

SECTION 6.4 *(page 488)*

Warm-Up *(page 488)*

1. $34-5x$ **2.** $18x-28$ **3.** $t^2+\frac{2}{3}t+\frac{1}{9}$
4. $4x^2-20x+25$ **5.** $25x^2-16$
6. $2-\frac{23}{2}x-3x^2$ **7.** $2xy\sqrt[3]{2x}$
8. $6xy\sqrt{2y}$ **9.** $(x-4)\sqrt{3}$ **10.** $2\sqrt{7}x$

1. 4 **2.** $6\sqrt{3}$ **3.** $2\sqrt{5}-\sqrt{15}$ **4.** $\sqrt{55}-3\sqrt{11}$
5. $2\sqrt{10}+8\sqrt{2}$ **6.** $7\sqrt{2}+3\sqrt{7}$ **7.** -1 **8.** 6
9. 4 **10.** 46 **11.** $8\sqrt{5}+24$
12. $36-16\sqrt{5}$ **13.** $\sqrt{15}+3\sqrt{3}-5\sqrt{5}-15$
14. $2\sqrt{15}+6\sqrt{30}+6\sqrt{2}+36$ **15.** $y+4\sqrt{y}$
16. $5\sqrt{x}-x$ **17.** $9x+25$ **18.** $49-27t$
19. $2x+20\sqrt{2x}+100$ **20.** $25-10\sqrt{3v}+3v$
21. $45x-17\sqrt{x}-6$ **22.** $16u-19\sqrt{u}+3$
23. $x-y$ **24.** $9u-3v$ **25.** $2-7\sqrt[3]{4}$
26. $\sqrt[3]{45}-5\sqrt[3]{9}+5\sqrt[3]{5}-25$
27. $\sqrt[3]{4x^2}+10\sqrt[3]{2x}+25$
28. $y-5\sqrt[3]{y}+2\sqrt[3]{y^2}-10$
29. $2y-10\sqrt[3]{2y}+10\sqrt[3]{4y^2}-100$
30. $t+5\sqrt[3]{t^2}+\sqrt[3]{t}-3$
31. $(x+3)$ **32.** $(1-x)$ **33.** $(4-3x)$
34. $(5+4y)$ **35.** $(2u+\sqrt{2u}\,)$ **36.** $(3s-\sqrt{2}\,)$

37. $\dfrac{1 - 2\sqrt{x}}{3}$ **38.** $\dfrac{-1 + 9\sqrt{2y}}{6}$ **39.** $\dfrac{-1 + \sqrt{3y}}{4}$

40. $\dfrac{t + \sqrt{2t}}{-3}$ **41.** $2 - \sqrt{5}, -1$ **42.** $\sqrt{2} + 9, -79$

43. $\sqrt{11} + \sqrt{3}, 8$ **44.** $\sqrt{10} - \sqrt{7}, 3$ **45.** $\sqrt{x} + 3, x - 9$

46. $\sqrt{t} - 7, t - 49$ **47.** $\sqrt{2u} + \sqrt{3}, 2u - 3$

48. $\sqrt{5a} - \sqrt{2}, 5a - 2$ **49.** $\dfrac{\sqrt{22} + 2}{3}$

50. $\dfrac{2\sqrt{10} + 5}{5}$ **51.** $-4\sqrt{7} + 12$ **52.** $\dfrac{15 - 5\sqrt{5}}{2}$

53. $\dfrac{4\sqrt{7} + 11}{3}$ **54.** $\dfrac{12 - 7\sqrt{3}}{3}$

55. $\dfrac{x\sqrt{15} + x\sqrt{3}}{4}$ **56.** $\dfrac{6(y + 1)(y^2 - \sqrt{y})}{y(y^3 - 1)}$

57. $\dfrac{t\sqrt{5t} + t\sqrt{t}}{2}$ **58.** $\dfrac{5x\sqrt{x} + 5x\sqrt{2}}{x - 2}$

59. $\dfrac{2x - 9\sqrt{x} - 5}{4x - 1}$ **60.** $\dfrac{4t + 4\sqrt{t} + 1}{4t - 1}$

61. $192\sqrt{2}$ in.^{2} **62.** $\dfrac{500k\sqrt{k^2 + 1}}{1 + k^2}$

SECTION 6.5 *(page 496)*

Warm-Up *(page 496)*

1. 2 **2.** $\frac{9}{2}$ **3.** $\frac{27}{5}$ **4.** $\frac{1}{6}$
5. 2, -25 **6.** 12, -114 **7.** $x - 9$
8. $\sqrt{5u}$ **9.** $4t + 12\sqrt{t} + 9$ **10.** $25\sqrt{2}x$

1. Yes **2.** Yes **3.** Yes **4.** Yes **5.** Yes
6. No **7.** 400 **8.** 25 **9.** 49 **10.** 169
11. No solution **12.** No solution **13.** 525
14. 157 **15.** 90 **16.** $\frac{9}{2}$ **17.** $\frac{44}{3}$ **18.** $\frac{11}{5}$
19. $\frac{14}{25}$ **20.** $\frac{33}{4}$ **21.** $-\frac{2}{3}$ **22.** 2 **23.** 4
24. 7 **25.** No solution **26.** 5 **27.** 7
28. $\frac{78}{23}$ **29.** -15 **30.** No solution **31.** 8
32. 9 **33.** 1, 3 **34.** $-1, -2$ **35.** $\frac{1}{2}$
36. $\frac{13}{4}$ **37.** $\frac{4}{5}$ **38.** No solution **39.** ± 4
40. ± 1 **41.** 56.57 ft/sec **42.** 113.14 ft/sec
43. 56.25 ft **44.** 225 ft **45.** 64 ft
46. 144 ft **47.** 1.82 ft **48.** 0.46 ft

49. 500 units **50.** 26,250 passengers

51. $h^2 = \dfrac{s^2 - \pi^2 r^4}{\pi^2 r^2}$

USING A GRAPHING CALCULATOR *(page 501)*

1. 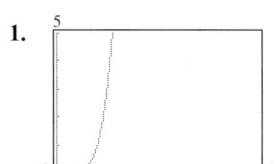 **2.**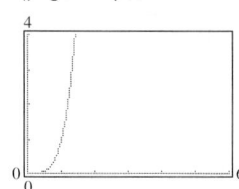

3. 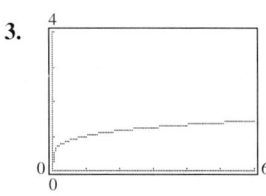 **4.**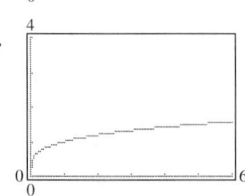

5. 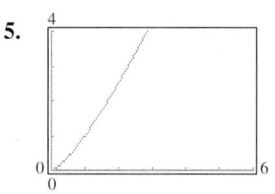 **6.**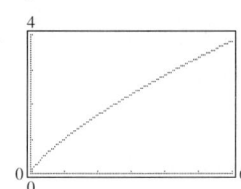

7. 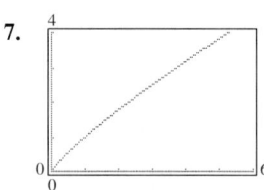 **8.**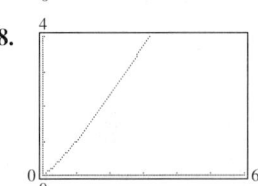

9.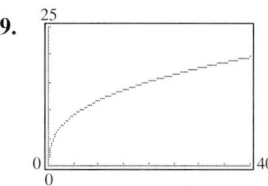

As the weight of the primate increases, the chest circumference increases. The *rate* of increase of the chest circumference is highest at lower weights, and the *rate* of growth of the chest circumference becomes lower as the weight continues to increase.

10. 23.2 lb

SECTION 6.6   *(page 509)*

Warm-Up   *(page 509)*

1. $4\sqrt{3}$ **2.** $6\sqrt{3}$ **3.** $23\sqrt{2}$ **4.** 150

5. $\dfrac{4\sqrt{10}}{5}$ **6.** $\dfrac{5(\sqrt{3}+1)}{4}$ **7.** $0, \dfrac{35}{2}$

8. $-4, \frac{5}{4}$ **9.** $-2, 9$ **10.** $-1, \frac{3}{2}$

---

1. $2i$ **2.** $3i$ **3.** $-12i$ **4.** $7i$ **5.** $\frac{2}{5}i$

6. $-\frac{6}{11}i$ **7.** $0.3i$ **8.** $0.02i$ **9.** $2\sqrt{2}\,i$

10. $5\sqrt{3}\,i$ **11.** $\sqrt{7}\,i$ **12.** $\sqrt{15}\,i$ **13.** $10i$

14. $-i$ **15.** $3\sqrt{2}\,i$ **16.** $13\sqrt{5}\,i$ **17.** -4

18. $-5\sqrt{6}$ **19.** $-3\sqrt{6}$ **20.** -0.44

21. $-2\sqrt{3}-3$ **22.** 6 **23.** $5\sqrt{2}-4\sqrt{5}$

24. $-\frac{7}{3}\sqrt{6}$ **25.** -6 **26.** -2 **27.** -16

28. -25 **29.** $a=3, b=-4$ **30.** $a=-8, b=6$

31. $a=-4, b=-2\sqrt{2}$ **32.** $a=-3, b=6$

33. $a=2, b=-2$ **34.** $a=2, b=\frac{9}{2}$

35. $10+4i$ **36.** $-6-5i$ **37.** $-14-40i$

38. $18+31i$ **39.** 13 **40.** $-70i$

41. $-14+20i$ **42.** $17+18i$ **43.** $\frac{13}{6}+\frac{3}{2}i$

44. $-6.15-9.3i$ **45.** $-3+49i$

46. $(-1-\sqrt{2})+(1-\sqrt{2})i$ **47.** -36

48. 20 **49.** $-36i$ **50.** $180i$

51. $27i$ **52.** -64 **53.** $-65-10i$

54. $80-60i$ **55.** $20-12i$ **56.** $-45-30i$

57. $-40-5i$ **58.** $-69+55i$ **59.** $-7-24i$

60. $48+14i$ **61.** $-21+20i$ **62.** $55-48i$

63. $-24-2\sqrt{3}\,i$ **64.** $-12+3i$ **65.** $2+11i$

66. $-9-46i$ **67.** $2-i, 5$ **68.** $3-2i, 13$

69. $5+\sqrt{6}\,i, 31$ **70.** $10+3i, 109$ **71.** $-2+8i, 68$

72. $-4-\sqrt{2}\,i, 18$ **73.** $-10i, 100$ **74.** $20, 400$

75. $1-\sqrt{3}\,i, 4$ **76.** $-3+\sqrt{5}\,i, 14$

77. $1.5-\sqrt{-0.25}, 2.5$ **78.** $3.2+\sqrt{-0.04}, 10.28$

79. $2+2i$ **80.** $6-2i$ **81.** $-\frac{24}{53}+\frac{84}{53}i$

82. $\frac{15}{4}+\frac{15}{4}i$ **83.** $-\frac{6}{5}+\frac{2}{5}i$ **84.** $\frac{3}{2}+\frac{5}{2}i$

85. $0-10i$ **86.** $\frac{1}{3}-\frac{1}{3}i$ **87.** $\frac{3}{8}-\frac{1}{2}i$

88. $-\frac{2}{3}+\frac{11}{12}i$ **89.** $\frac{8}{5}-\frac{1}{5}i$ **90.** $-\frac{9}{41}-\frac{40}{41}i$

91. $1-\frac{6}{5}i$ **92.** $\frac{47}{26}+\frac{27}{26}i$ **93.** $-\frac{53}{25}+\frac{29}{25}i$

94. $\frac{14}{29}-\frac{35}{29}i$ **95.** $2a+0i$ **96.** a^2+b^2

97. $0+2bi$ **98.** $2a^2-2b^2$

99. $(-1+2i)^2+2(-1+2i)+5$
$= 1-4i+4i^2-2+4i+5$
$= (-4+4)i+(1-4-2+5)$
$= 0$

100. $(2-3i)^2-4(2-3i)+13$
$= 4-12i+9i^2-8+12i+13$
$= (-12+12)i+(4-9-8+13)$
$= 0$

CHAPTER 6 REVIEW EXERCISES   *(page 512)*

1. $\frac{1}{72}$ **2.** $\frac{16}{625}$ **3.** $\frac{125}{8}$ **4.** 81

5. 3.6×10^7 **6.** 2.4×10^5 **7.** $8x$

8. $-108x^3$ **9.** $\dfrac{x^6}{y^8}$ **10.** $5y$ **11.** $\dfrac{1}{t^3}$

12. a^4 **13.** $\dfrac{27}{y^3}$ **14.** 1 **15.** 1.2 **16.** 0.4

17. $\frac{5}{6}$ **18.** $-\frac{8}{15}$ **19.** 12 **20.** 5

21. $11{,}414.13$ **22.** $18{,}380.16$ **23.** 10.63

24. 0.06 **25.** $16^{1/2}=4$ **26.** $\sqrt[4]{16}=2$

27. 81 **28.** $\frac{1}{9}$ **29.** 0.04 **30.** $32{,}554.94$

31. $x^{7/12}$ **32.** $4y$ **33.** $\dfrac{3}{x^{1/4}y^{2/5}}$ **34.** $\dfrac{24}{7}a^5b^3$

35. $6\sqrt{10}$ **36.** $\frac{5}{3}\sqrt{2}$ **37.** $0.5x^2\sqrt{y}$

38. $0.4s^3t\sqrt{t}$ **39.** $2ab\sqrt[3]{6b}$ **40.** $2uv\sqrt[4]{2v}$

41. $\dfrac{\sqrt{30}}{6}$ **42.** $\dfrac{\sqrt{15}}{10}$ **43.** $\dfrac{\sqrt{3x}}{2x}$

44. $\dfrac{2y\sqrt{10z}}{5z}$ **45.** $\dfrac{\sqrt[3]{4x^2}}{x}$ **46.** $\dfrac{2\sqrt[3]{2st}}{s}$

47. $-24\sqrt{10}$ **48.** $35\sqrt{2}+4\sqrt{3}$ **49.** $7\sqrt[4]{y+3}$

50. $12\sqrt{x}-2\sqrt[3]{x}$ **51.** $12\sqrt{5}+41$ **52.** $3-x$

53. $6\sqrt{2}+7\sqrt{6}-4\sqrt{3}-14$ **54.** $\dfrac{15\sqrt{x}-45}{x-9}$

55. $\dfrac{x+20\sqrt{x}+100}{x-100}$ **56.** $\dfrac{3s-2\sqrt{s}-8}{s-4}$

57. 225 **58.** No solution   **59.** 105 **60.** $\frac{6}{5}$

61. $-3, -5$ **62.** $\frac{9}{5}$ **63.** 5 **64.** 2 **65.** $\frac{3}{32}$

66. No solution   **67.** $4\sqrt{3}\,i$ **68.** $3+20\sqrt{5}\,i$

69. $\frac{3}{4}-\sqrt{3}\,i$ **70.** $-0.5+3.3i$ **71.** $8-3i$

72. $20-28i$ **73.** -90 **74.** $-70-40i$

75. 25 **76.** $11-60i$ **77.** $59+74i$

78. $-\frac{4}{5}i$ **79.** $\frac{9}{17}+\frac{2}{17}i$ **80.** $-\frac{7}{82}+\frac{19}{82}i$

81. 256 ft   **82.** 1.37 ft

CHAPTER 6 TEST   *(page 514)*

1. $\frac{3}{8}$ **2.** 3×10^{-5} **3.** $\frac{1}{9}$ **4.** 6

5. 3.2×10^{-5} **6.** 30,400,000

7. $\frac{3}{5t}$ **8.** $\frac{y^2}{xy^2 + 1}$ **9.** $x^{1/3}$ **10.** 25

11. $\frac{4}{3}\sqrt{2}$ **12.** $2\sqrt[3]{3}$ **13.** $-\frac{\sqrt{6}+9}{25}$

14. $\frac{\sqrt[3]{2}}{2}$ **15.** $-25\sqrt{3x}$ **16.** $5\sqrt{3x} + 3\sqrt{5}$

17. $16 - 8\sqrt{2x} + 2x$ **18.** $(4y + 3)$

19. No solution   **20.** 9   **21.** $2 - 2i$

22. $-5 - 12i$ **23.** $-8 + 4i$ **24.** $13 + 13i$

25. $-2 - 5i$ **26.** $-\frac{8}{41} + \frac{10}{41}i$ **27.** 100 ft

CUMULATIVE TEST: CHAPTERS 4-6   *(page 515)*

1. $(2, 1)$ **2.** $(3, -2)$ **3.** $(5, 4)$

4. $(1, -2, 3)$ **5.** 20   **6.** 8, 12, 24

7.

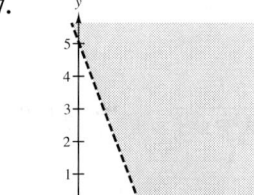

8. Minimum at $(3, 2)$: 19, Maximum at $(6, 2)$: 34

9. $\frac{x(x + 2)(x + 4)}{9(x - 4)}$, $x \neq 0$, $x \neq -4$

10. $\frac{3x + 5}{x(x + 3)}$ **11.** $x + y$, $x \neq y$, $x \neq 0$, $y \neq 0$

12. $x^2 - 3x + 9$

13.

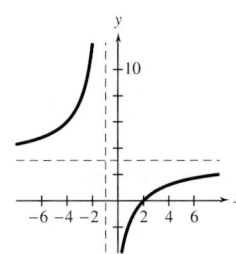

14. 2   **15.** $-\frac{2y^6}{3x^4}$ **16.** $t^{1/2}$ **17.** 1.6×10^7

18. $35\sqrt{5x}$ **19.** $2x - 6\sqrt{2x} + 9$ **20.** $\sqrt{10} + 2$

21. $-4 + 3\sqrt{2}\,i$ **22.** 6   **23.** 20 hr, 30 hr

Chapter 7

Chapter Opener   *(page 517)*

$t = 10$ sec

SECTION 7.1   *(page 526)*

Warm-Up   *(page 526)*

1. $(4x + 11)(4x - 11)$ **2.** $(3t - 4)^2$

3. $(x - 3)(5x + 2)$ **4.** $(x - 10)(x - 4)$

5. $y(2y - 1)(2y + 3)$ **6.** $4x(x^2 - 3x + 4)$

7. $-\frac{6}{5}$ **8.** $\frac{7}{2}$ **9.** $-9, \frac{15}{2}$ **10.** $\frac{8}{5}$, 12

1. 0, 3   **2.** 0, 3   **3.** 9, 12   **4.** 8, $\frac{3}{4}$ **5.** $\pm\frac{5}{2}$

6. $\pm\frac{11}{4}$ **7.** 6   **8.** $-\frac{4}{3}$ **9.** -30 **10.** $\frac{1}{2}, \frac{3}{4}$

11. 1, 6   **12.** $-1, -4$ **13.** $\frac{1}{2}, -\frac{5}{6}$ **14.** $\frac{3}{8}, -1$

15. $\frac{5}{3}, 6$ **16.** $\frac{3}{2}, 4$ **17.** ± 3 **18.** ± 5 **19.** ± 8

20. ± 12 **21.** ± 8 **22.** ± 13 **23.** $\pm\frac{4}{5}$

24. $\pm\frac{11}{3}$ **25.** $\pm\frac{15}{2}$ **26.** $\pm\frac{1}{4}$ **27.** 9, -17

28. 45, -5 **29.** 2.5, 3.5   **30.** $-1.1, -2.9$

31. $2 \pm \sqrt{7}$ **32.** $-8 \pm 2\sqrt{7}$ **33.** $\frac{-1 \pm 5\sqrt{2}}{2}$

34. $\frac{5 \pm 4\sqrt{3}}{3}$ **35.** $\frac{3 \pm 7\sqrt{2}}{4}$ **36.** $\frac{-11 \pm 10\sqrt{3}}{5}$

37. $\pm 6i$ **38.** $\pm 3i$ **39.** $\pm 2i$ **40.** $\pm 4i$

41. $3 \pm 5i$ **42.** $-5 \pm 9i$ **43.** $-\frac{4}{3} \pm 4i$

44. $\frac{3}{2} \pm \frac{5}{2}i$ **45.** $-6 \pm \frac{11}{3}i$ **46.** $4 \pm \frac{13}{2}i$

47. $1 \pm 3\sqrt{3}\,i$ **48.** $-\frac{3}{2} \pm \frac{3}{2}\sqrt{6}\,i$ **49.** $-1 \pm 0.2i$

50. $3 \pm 1.5i$ **51.** $\frac{2}{3} \pm \frac{1}{3}i$

52. $-\frac{5}{8} \pm \frac{7}{4}i$ **53.** $-\frac{7}{3} \pm \frac{\sqrt{38}}{3}i$ **54.** $\frac{5}{6} \pm \frac{2}{5}\sqrt{5}\,i$

55. $\pm 2, \pm 4$ **56.** $1, -2$ **57.** $\frac{1}{27}, -27$

58. $\frac{25}{4}, 4$ **59.** $\pm 2\sqrt{2}$ **60.** ± 2

61. ± 10 **62.** ± 20 **63.** $-5, 15$

64. $8, -32$ **65.** $5 \pm 10i$ **66.** $-12 \pm 20i$

67. $\pm\sqrt{15}, \pm\sqrt{5}\,i$ **68.** $\pm 2\sqrt{2}, \pm 4\sqrt{2}\,i$

69. $\pm 2, \pm 2i$ **70.** $\pm 3, \pm 3i$

71. $\pm\sqrt{3}, \pm 1$ **72.** $\pm 1, \pm 2\sqrt{2}\,i$

73. $x^2 - 3x - 10 = 0$ **74.** $3x^2 + 5x - 2 = 0$

75. $x^2 - 2x - 1 = 0$ **76.** $x^2 - 2x + 3 = 0$

77. 7.37 m   **78.** 4   **79.** 10

80. 24   **81.** 5.66   **82.** 3   **83.** 4.33

84. 21.63   **85.** 19.60 ft   **86.** 111.80 ft

87. 166 mi, 44 min   **88.** 74.25 mi, $\approx$ 14 hr, 51 min

89. 3.31 mi   **90.** 9.53 sec   **91.** 9 sec

92. 100 units, 140 units   **93.** 1980   **94.** 1986

SECTION 7.2   *(page 537)*

Warm-Up   *(page 537)*

1. $x^2 + 6x + 8$ **2.** $x^2 + 20x + 150$

3. $u^2 - u + \frac{1}{4}$ **4.** $v^2 - 24v + 44$

5. $\pm\frac{2}{7}$ **6.** $\pm\frac{3}{4}i$ **7.** $4 \pm 6i$

8. $-2 \pm \sqrt{5}$ **9.** $\frac{2}{3} \pm \frac{\sqrt{5}}{3}$ **10.** $\frac{3}{8}, 12$

---

1. 16   **2.** 36   **3.** 100   **4.** 1   **5.** $\frac{25}{4}$

6. $\frac{49}{4}$ **7.** $\frac{9}{25}$ **8.** $\frac{4}{9}$ **9.** $\frac{9}{100}$ **10.** $\frac{1}{36}$

11. 0.04   **12.** 5.0625   **13.** 0, 25   **14.** 0, -32

15. 1, 7   **16.** 2, -6 **17.** 4, -6 **18.** $-3, -9$

19. $-3, -4$ **20.** 2, -5 **21.** $-3, 6$

22. 9, -4 **23.** $\frac{3}{2}, 4$ **24.** $-\frac{1}{3}, 2$

25. $2 + \sqrt{7} \approx 4.65$ **26.** $3 + \sqrt{2} \approx 4.41$
$2 - \sqrt{7} \approx -0.65$ $3 - \sqrt{2} \approx 1.59$

27. $-2 + \sqrt{7} \approx 0.65$ **28.** $-3 + \sqrt{2} \approx -1.59$
$-2 - \sqrt{7} \approx -4.65$ $-3 - \sqrt{2} \approx -4.41$

29. $2 + \sqrt{3} \approx 3.73$ **30.** $5 + 2\sqrt{10} \approx 11.32$
$2 - \sqrt{3} \approx 0.27$ $5 - 2\sqrt{10} \approx -1.32$

31. $-1 + \sqrt{2}\,i \approx -1 + 1.41i$
$-1 - \sqrt{2}\,i \approx -1 - 1.41i$

32. $3 + \sqrt{3}\,i \approx 3 + 1.73i$
$3 - \sqrt{3}\,i \approx 3 - 1.73i$

33. $5 + 3\sqrt{3} \approx 10.20$
$5 - 3\sqrt{3} \approx -0.20$

34. $-4 + 2\sqrt{5} \approx 0.47$
$-4 - 2\sqrt{5} \approx -8.47$

35. $-10 + 3\sqrt{10} \approx -0.51$
$-10 - 3\sqrt{10} \approx -19.49$

36. $-3 + \sqrt{33} \approx 2.74$
$-3 - \sqrt{33} \approx -8.74$

37. $\dfrac{1 + 2\sqrt{7}}{3} \approx 2.10$
$\dfrac{1 - 2\sqrt{7}}{3} \approx -1.43$

38. $\dfrac{-2 + \sqrt{29}}{5} \approx 0.68$
$\dfrac{-2 - \sqrt{29}}{5} \approx -1.48$

39. $\dfrac{-5 + \sqrt{13}}{2} \approx -0.70$
$\dfrac{-5 - \sqrt{13}}{2} \approx -4.30$

40. $\dfrac{9 + \sqrt{85}}{2} \approx 9.11$
$\dfrac{9 - \sqrt{85}}{2} \approx -0.11$

41. $\dfrac{-3 + \sqrt{17}}{2} \approx 0.56$
$\dfrac{-3 - \sqrt{17}}{2} \approx -3.56$

42. $\dfrac{7 + \sqrt{13}}{2} \approx 5.30$
$\dfrac{7 - \sqrt{13}}{2} \approx 1.70$

43. $\dfrac{1}{2} + \dfrac{\sqrt{3}}{2}i \approx 0.5 + 0.87i$
$\dfrac{1}{2} - \dfrac{\sqrt{3}}{2}i \approx 0.5 - 0.87i$

44. $\dfrac{-1 + \sqrt{5}}{2} \approx 0.62$
$\dfrac{-1 - \sqrt{5}}{2} \approx -1.62$

45. $\dfrac{-4 + \sqrt{10}}{2} \approx -0.42$
$\dfrac{-4 - \sqrt{10}}{2} \approx -3.58$

46. $\dfrac{2 + \sqrt{7}}{3} \approx 1.55$
$\dfrac{2 - \sqrt{7}}{3} \approx -0.22$

47. $\dfrac{-9 + \sqrt{21}}{6} \approx -0.74$
$\dfrac{-9 - \sqrt{21}}{6} \approx -2.26$

48. $\dfrac{15 + \sqrt{85}}{10} \approx 2.42$
$\dfrac{15 - \sqrt{85}}{10} \approx 0.58$

49. $\dfrac{-1 + \sqrt{10}}{2} \approx 1.08$
$\dfrac{-1 - \sqrt{10}}{2} \approx -2.08$

50. $\dfrac{3}{8} + \dfrac{\sqrt{23}}{8}i \approx 0.38 + 0.60i$
$\dfrac{3}{8} - \dfrac{\sqrt{23}}{8}i \approx 0.38 - 0.60i$

51. $\dfrac{7 + \sqrt{57}}{2} \approx 7.27$
$\dfrac{7 - \sqrt{57}}{2} \approx -0.27$

52. $\dfrac{-4 + \sqrt{106}}{6} \approx 1.05$
$\dfrac{-4 - \sqrt{106}}{6} \approx -2.38$

53. $\dfrac{-5 + \sqrt{17}}{2} \approx -0.44$
$\dfrac{-5 - \sqrt{17}}{2} \approx -4.56$

54. $\dfrac{-5 + \sqrt{35}}{2} \approx 0.46$
$\dfrac{-5 - \sqrt{35}}{2} \approx -5.46$

55. 2.73, −0.73   **56.** 6.45, 1.55

57. 6.83   **58.** 6

59. (a) $x^2 + 8x$ (b) $x^2 + 8x + 16$ (c) $(x + 4)^2$

60.

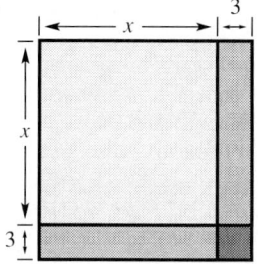

$x^2 + 6x$

$x^2 + 6x + 9 = (x + 3)^2$

61. 4 cm, 6 cm   **62.** 271 m, 129 m

63. 73.5 ft   **64.** 15 ft × $46\frac{2}{3}$ ft; 20 ft × 35 ft

65. 42 units, 58 units   **66.** 139 units, 861 units

SECTION 7.3   *(page 548)*

Warm-Up   *(page 548)*

1. 3, 7   **2.** 2, −16   **3.** 3, $-\frac{10}{3}$

4. $\frac{3}{5}$, 3   **5.** $\frac{-3 \pm \sqrt{17}}{2}$ **6.** $-1 \pm \frac{\sqrt{3}}{3}i$

7. 3   **8.** $\sqrt{11}i$ **9.** $4\sqrt{6}$ **10.** $3\sqrt{17}$

---

1. $2x^2 + 2x - 7 = 0$ **2.** $7x^2 + 15x - 5 = 0$

3. $-x^2 + 10x - 5 = 0$ **4.** $3x^2 + 8x - 15 = 0$

5. 4, 7   **6.** 9, 3   **7.** −2, −4   **8.** −2, −7

9. $-\frac{1}{2}$ **10** $-\frac{2}{3}$ **11.** $-\frac{3}{2}$ **12.** $\frac{5}{3}$ **13.** $-\frac{1}{2}, \frac{2}{3}$

14. $\frac{3}{5}, \frac{1}{2}$ **15.** −15, 20   **16.** −30, 10

17. 2 distinct imaginary solutions

18. 2 distinct irrational solutions

19. 2 distinct irrational solutions

20. 2 distinct imaginary solutions

21. 2 distinct imaginary solutions

22. 2 distinct rational solutions

23. 1 (repeated) rational solution

24. 2 distinct irrational solutions

25. $1 \pm \sqrt{5}$ **26.** $1 \pm \sqrt{7}$ **27.** $-2 \pm \sqrt{3}$

28. $-3 \pm \sqrt{5}$ **29.** $-3 \pm 2\sqrt{3}$ **30.** $-4 \pm 2\sqrt{5}$

31. $5 \pm \sqrt{2}$ **32.** $6 \pm \sqrt{7}$

33. $-\frac{3}{2} \pm \frac{\sqrt{3}}{2}i$ **34.** $\frac{1}{4} \pm \frac{\sqrt{7}}{4}i$

35. $\frac{1 \pm \sqrt{3}}{2}$ **36.** $\frac{-3 \pm \sqrt{5}}{4}$

37. $\frac{-2 \pm \sqrt{10}}{2}$ **38.** $-\frac{3}{4} \pm \frac{\sqrt{15}}{4}i$

39. $\frac{-1 \pm \sqrt{5}}{3}$ **40.** $\frac{2 \pm \sqrt{6}}{4}$

41. $\frac{1 \pm \sqrt{5}}{5}$ **42.** $\frac{-0.6 \pm \sqrt{2}}{2}$

43. $\frac{-1 \pm \sqrt{10}}{5}$ **44.** $\frac{-0.02 \pm \sqrt{0.003}}{0.03}$ **45.** ±13

46. $\pm 5\sqrt{6}$ **47.** 0, −15   **48.** 18, $-\frac{3}{2}$

49. $\frac{9}{5}, \frac{21}{5}$ **50.** $-\frac{11}{3}, -\frac{1}{3}$ **51.** $-4 \pm 4i$

52. $\pm \frac{7}{2}i$ **53.** $\frac{-5 \pm 5\sqrt{17}}{12}$ **54.** $\frac{15}{4} \pm \frac{15\sqrt{7}}{4}i$

55. 8, 16   **56.** $\frac{5}{2}, -\frac{11}{6}$

57. $-4 \pm 3i$ **58.** $-2 \pm \frac{\sqrt{7}}{2}$

59. 0.372, 3.228   **60.** 2.548, −2.748

61. 0.200, 99.800   **62.** 2.396, 0.361

63. 2.33, −1.08   **64.** 11.74, 0.26

65. 3.56   **66.** 4.41

67. (a) $c < 9$ (b) $c = 9$ (c) $c > 9$

68. (a) $c < 36$ (b) $c = 36$ (c) $c > 36$

69. (a) $c < 16$ (b) $c = 16$ (c) $c > 16$

70. (a) $c < 1$ (b) $c = 1$ (c) $c > 1$

71. (a) 2.5 sec   (b) $\frac{5 + 5\sqrt{3}}{4} \approx 3.4$ sec

72. 16, 18   **73.** 1968   **74.** 1981

CHAPTER 7 MID-CHAPTER QUIZ   *(page 551)*

1. 0, 9   **2.** $\pm\frac{5}{2}$ **3.** 4, −5   **4.** $-\frac{2}{3}, \frac{3}{4}$

5. $\pm 2\sqrt{2}$ **6.** 1, 19   **7.** $-3 \pm \sqrt{13}$

8. $\frac{3 \pm \sqrt{19}}{2}$ **9.** $\frac{1}{2}$, 1   **10.** $\frac{1 \pm \sqrt{11}\,i}{6}$

11. $-3 \pm 3i$ **12.** $\pm\sqrt{2}, \pm\sqrt{5}\,i$

13. −3, 5   **14.** −1, 5

15. (a) $c = \frac{49}{4}$ (b) $c < \frac{49}{4}$ (c) $c > \frac{49}{4}$

16. 200 units   **17.** 20 ft × 8 ft

USING A GRAPHING CALCULATOR   *(page 552)*

1. x-intercepts are -0.9 and 2.6
2. x-intercepts are -7.6 and 1.6
3. x-intercepts are -0.4 and -23.6
4. x-intercepts are -1.1 and 9.9
5. x-intercepts are -15.7 and 1.7
6. x-intercepts are -1.0 and -0.3
7. 10.9, 19.1 **8.** 10.1, 14.9

SECTION 7.4   *(page 562)*

Warm-Up   *(page 561)*

1. -7 **2.** $0, \frac{2}{15}$ **3.** $\frac{5}{2}, -1$ **4.** $10, -12$

5. $1, -17$ **6.** 1 **7.** $-\frac{3}{2} \pm \frac{\sqrt{11}}{2} i$

8. $\dfrac{-3 \pm \sqrt{29}}{2}$ **9.** $6, -1$ **10.** $3, -1$

1. $15, 16$ **2.** $33, 34$ **3.** $14, 16$ **4.** $15, 17$
5. $12, 13$ **6.** $14, 15$ **7.** 108 in.^{2} **8.** 96 in.^{2}
9. 70 ft   **10.** 60 cm   **11.** 64 in.   **12.** 210 in.
13. 180 km^2 **14.** 720 ft^2 **15.** 440 m   **16.** 90 ft
17. 50 ft $\times$ 250 ft or 125 ft $\times$ 100 ft
18. Height: 12 in.; Width: 24 in.
19. Base: 24 in.; Height: 16 in.
20. Base: 25 in.; Height: 50 in.
21. No.
  Area $= \frac{1}{2}(b_1 + b_2)h = \frac{1}{2}x\,[x + (550 - 2x)] = 43{,}560$
  This equation has no real solution.
22. Rectangular region: no; Circular region: yes
23. 8% **24.** 9.5% **25.** 18 doz, \$1.20/doz
26. 9 computers, \$3000   **27.** 48 **28.** 32
29. 15.86 mi   **30.** 15 in. $\times$ 36 in.
31. 400 mph   **32.** 45 mph
33. 9.1 hr, 11.1 hr   **34.** 10.7 hr, 13.7 hr
35. 3 sec   **36.** $6\frac{1}{4}$ sec   **37.** 9.5 sec   **38.** 7.8 sec
39. 4.7 sec   **40.** 1.6 sec   **41.** 1988
42. $t \approx 6$, which corresponds to the year 1986.

SECTION 7.5   *(page 573)*

Warm-Up   *(page 573)*

1. 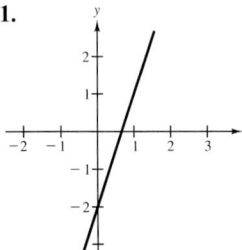   **2.**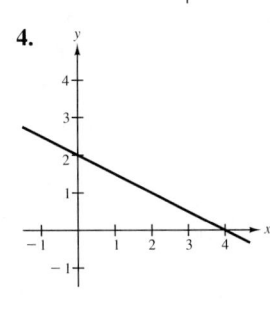

3.   **4.**

5. $\frac{9}{2}$ **6.** 20 **7.** $0, -8$ **8.** $\dfrac{-1 \pm \sqrt{41}}{4}$
9. $2x^2 + 20x + 20$ **10.** $-3x^2 + 12x - 19$

1. e   **2.** f   **3.** b   **4.** c   **5.** d   **6.** a
7. Up, $(0, 2)$ **8.** Down, $(-5, -3)$ **9.** Down, $(10, 4)$
10. Up, $(12, 3)$ **11.** Up, $(0, -6)$ **12.** Down, $(-1, 0)$
13. Up, $(3, 0)$ **14.** Down, $(-1, 3)$ **15.** $(-5, 0)$, $(5, 0)$
16. $(-7, 0)$, $(7, 0)$ **17.** $(0, 0)$, $(9, 0)$
18. $(0, 0)$, $(-4, 0)$ **19.** $\left(\frac{3}{2}, 0\right)$ **20.** $\left(-\frac{5}{2}, 0\right)$, $(2, 0)$
21. No x-intercepts   **22.** $(-2, 0)$, $(5, 0)$
23. $f(x) = (x - 2)^2 + 3$, $(2, 3)$
24. $g(x) = (x + 2)^2 - 4$, $(-2, -4)$
25. $f(x) = (x - 0)^2 + 2$, $(0, 2)$
26. $y = (x + 3)^2 - 14$, $(-3, -14)$
27. $y = -(x - 1)^2 - 6$, $(1, -6)$
28. $y = -(x + 5)^2 + 35$, $(-5, 35)$
29. $y = 2\left[x - \left(-\frac{3}{2}\right)\right]^2 - \frac{5}{2}$, $\left(-\frac{3}{2}, -\frac{5}{2}\right)$
30. $y = 3\left(x - \frac{1}{2}\right)^2 - \frac{39}{4}$, $\left(\frac{1}{2}, -\frac{39}{4}\right)$

31.
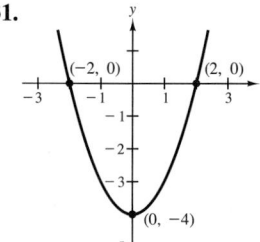

32.
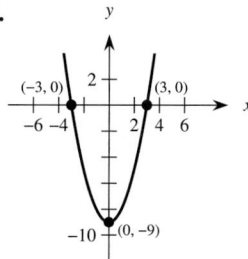

41.
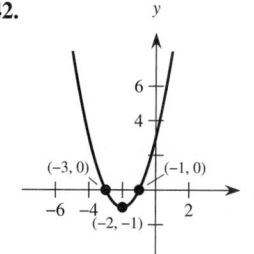

42.
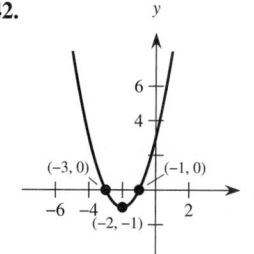

33.
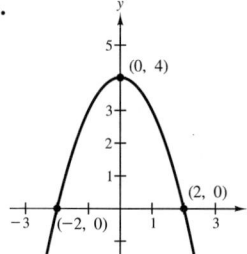

34.
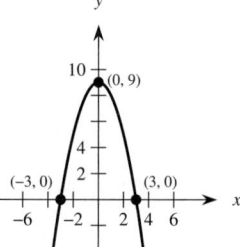

43.
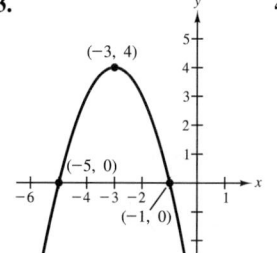

44.
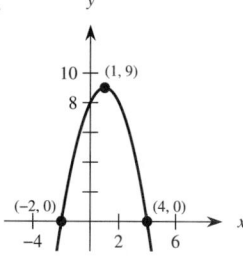

35.
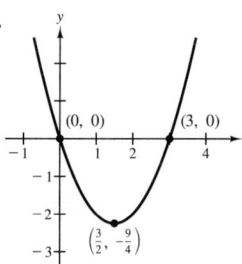

36.
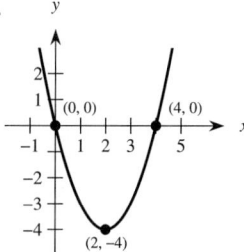

45.
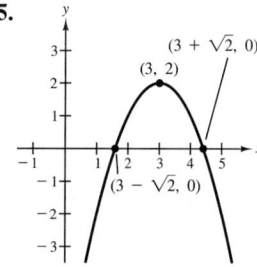

46.
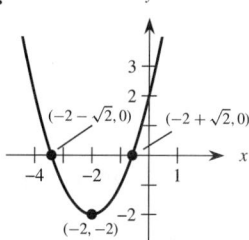

37.
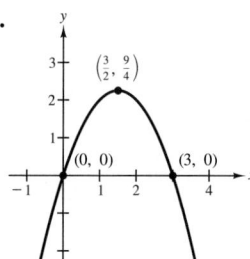

38.
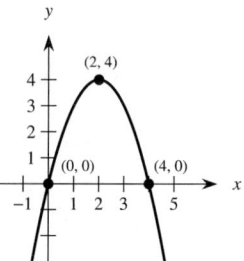

47.
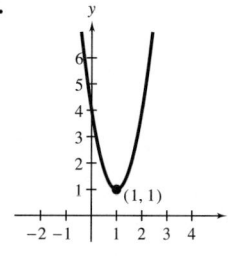

48.
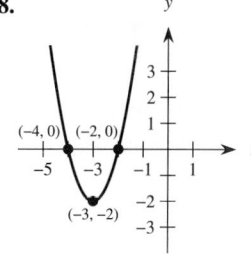

39.
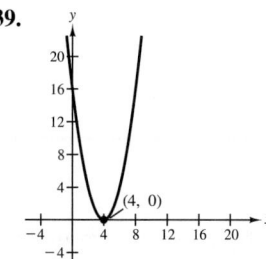

40.
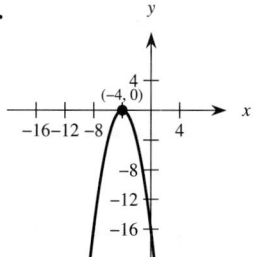

49.
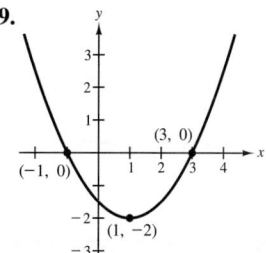

50.
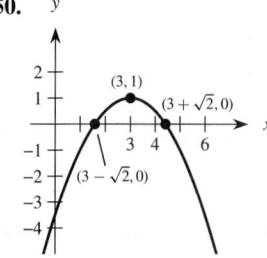

51.

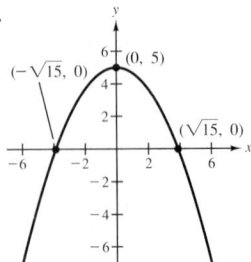

52.

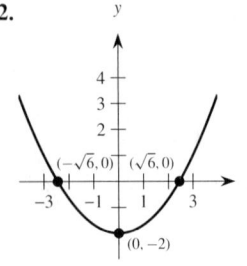

53.

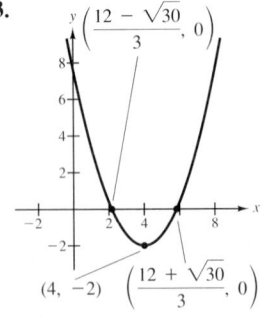

54.

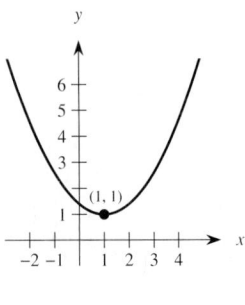

55. $x = -2,\ x = 2$

56. $x = 1,\ x = 5$

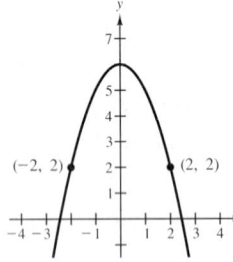

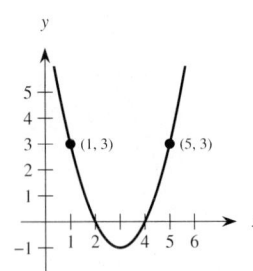

57. $x = 3 \pm \sqrt{2}$

58. $x = \dfrac{-2 \pm \sqrt{2}}{2}$

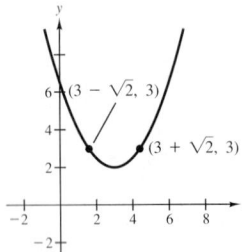

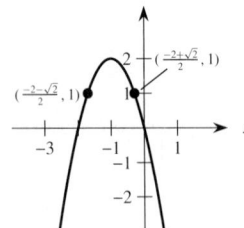

59. $y = -x^2 - 4x + 4$ **60.** $y = -x^2 + 4$

61. $y = x^2 - 4x$ **62.** $y = x^2 + 4x + 2$

63. $y = -2x^2 - 12x - 15$ **64.** $y = 2x^2 - 12x + 18$

65. $y = x^2 - 4x + 5$ **66.** $y = x^2 + 6x + 6$

67. $y = -x^2 - 6x - 5$ **68.** $y = -x^2 + 6x - 11$

69. $y = x^2 - 4x$ **70.** $y = x^2 + 4x$

71. $y = \frac{1}{2}x^2 - 3x + \frac{13}{2}$ **72.** $y = 5x^2 + 10x + 4$

73. $y = -4x^2 - 8x + 1$ **74.** $y = -4x^2 + 40x - 90$

75. $y = \frac{1}{25}x^2 - \frac{2}{5}x + 3$ **76.** $y = -\frac{1}{20}x^2 + 20$

77. (a)

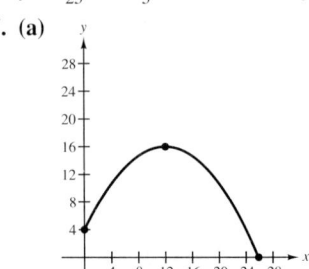

(b) 4 ft   **(c)** 16 ft   **(d)** $12 + 8\sqrt{3} \approx 25.9$ ft

78. 14 ft

79. (b) (50, 3375), 150 units

(c) Recommended only for orders between 100 and 150 units.

80.

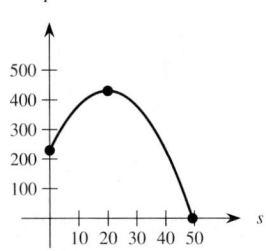

$2000

81. (a) 100 ft   **(b)** 400 ft

(c)

x	±100	±200	±300	±400	±500
y	16	64	144	256	400

82. $y = \frac{1}{2500}x^2$

83. Vertex: (2, 0.5)   **84.** Vertex: (2.5, 3)

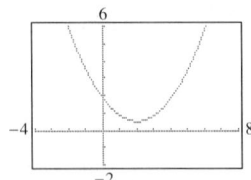

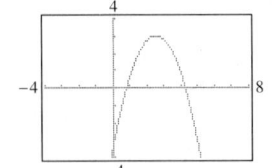

85. Vertex: $(-2, 4.9)$

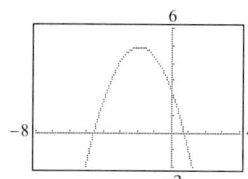

86. Vertex: $(5, 4.25)$

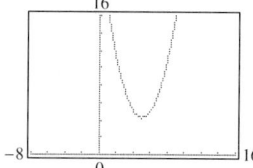

87. 50

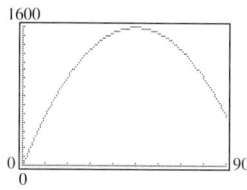

88. 20

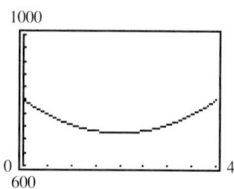

SECTION 7.6   *(page 588)*

Warm-Up   *(page 588)*

1. $x \le 5$

2. $x \ge -\frac{5}{3}$

3. $x < \frac{3}{2}$

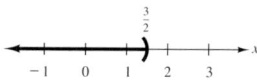

4. $x < 5$

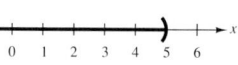

5. $1 < x < 5$

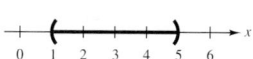

6. $x < 2, \ x > 8$

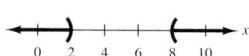

7. Negative   **8.** Zero
9. Positive   **10.** Negative

1. 3

2. $-\frac{3}{2}$

3. $0, \frac{5}{2}$

4. $0, 3$

5. $1, 3$

6. $-\frac{4}{3}, 2$

7. Negative: $(-\infty, 4)$
Positive: $(4, \infty)$

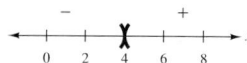

8. Positive: $(-\infty, 3)$
Negative: $(3, \infty)$

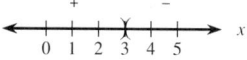

9. Negative: $(0, 4)$
Positive: $(-\infty, 0) \cup (4, \infty)$

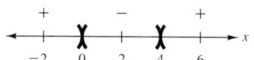

10. Negative: $(-\infty, 0) \cup (3, \infty)$
Positive: $(0, 3)$

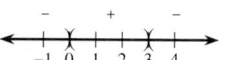

11. Negative: $(-1, 5)$
Positive: $(-\infty, -1) \cup (5, \infty)$

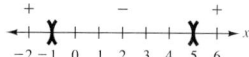

12. Positive: $\left(-\infty, \dfrac{2-\sqrt{10}}{2}\right) \cup \left(\dfrac{2+\sqrt{10}}{2}, \infty\right)$

Negative: $\left(\dfrac{2-\sqrt{10}}{2}, \dfrac{2+\sqrt{10}}{2}\right)$

13. $[-3, \infty)$

14. $(-\infty, 4)$

15. $(0, 2)$

16. $(-\infty, 0) \cup (6, \infty)$

17. $(-\infty, 0) \cup (2, \infty)$

18. $(0, 6)$

19. $(-\infty, -2] \cup [2, \infty)$

20. $[-3, 3]$

21. $[-5, 2]$

22. $(5, 10)$

23. $\left(-\infty, -\dfrac{1}{2}\right) \cup (4, \infty)$

24. $(-\infty, -2] \cup \left[\dfrac{2}{3}, \infty\right)$

25. No solution

26. $(-\infty, \infty)$

27. $(-\infty, \infty)$

28. $[-2, 8]$

29. $(-\infty, 2-\sqrt{2}) \cup (2+\sqrt{2}, \infty)$

30. $\left(-\infty, 4-\sqrt{5}\,\right] \cup [4+\sqrt{5}, \infty)$

31. No solution

32. $(-\infty, \infty)$

33. $(-\infty, 5-\sqrt{6}\,) \cup (5+\sqrt{6}, \infty)$

34. $\left(-\infty, -3-\sqrt{6}\,\right] \cup [-3+\sqrt{6}, \infty)$

35. $(-2, 0) \cup (2, \infty)$

36. $(-\infty, -2]$

37. 3

38. -2

39. $0, -5$

40. 2, 10

41. $(3, \infty)$

42. $(-\infty, 4)$

43. $(-\infty, 3)$

44. $(4, \infty)$

45. $[0, 4)$

46. $(-3, 1)$

47. $(-\infty, 3] \cup \left[\frac{11}{2}, \infty\right)$

48. $(-\infty, -5] \cup \left(-\frac{2}{3}, \infty\right)$

49. $(4, 7)$

50. $\left(-\infty, -\frac{7}{3}\right) \cup (-2, \infty)$

51. $(-\infty, 3) \cup [6, \infty)$

52. $(5, 6]$

53. $(3, 5)$ **54.** $\left(\dfrac{11 - \sqrt{71}}{4}, \dfrac{11 + \sqrt{71}}{4}\right)$

55. $(12, 20)$ **56.** $\left[25 - 5\sqrt{5}, 25 + 5\sqrt{5}\right]$

57. $r > 7.24\%$ **58.** $90{,}000 \le x \le 100{,}000$

CHAPTER 7 REVIEW EXERCISES   (page 591)

1. $0, -12$ **2.** $0, 18$ **3.** $-10, \frac{8}{3}$ **4.** $\frac{9}{2}, -\frac{4}{7}$

5. $\pm\frac{1}{2}$ **6.** ± 6 **7.** $-\frac{5}{2}$ **8.** $-\frac{4}{3}$

9. $-4, 5$ **10.** $-\frac{4}{3}, \frac{2}{3}$ **11.** $-9, 10$

12. $3, -1$ **13.** ± 100 **14.** $\pm 7\sqrt{2}$

15. ± 1.5 **16.** $\pm 2\sqrt{2}$ **17.** $-4, 36$

18. $-2.8, -3.2$ **19.** $3 \pm 2\sqrt{3}$ **20.** $-6 \pm \sqrt{30}$

21. $\frac{3}{2} \pm \frac{\sqrt{3}}{2}i$ **22.** $-\frac{1}{4} \pm \frac{\sqrt{17}}{4}$

23. $\dfrac{-5 \pm \sqrt{19}}{2}$ **24.** $\frac{1}{3} \pm \frac{\sqrt{5}}{3}i$

25. $5, -6$ **26.** $9, -8$ **27.** $3, -\frac{7}{2}$

28. $4, -\frac{5}{2}$ **29.** $\frac{10}{3} \pm \frac{5\sqrt{2}}{3}i$

30. $\dfrac{-2.5 \pm 0.5\sqrt{73}}{-2}$ **31.** $3 \pm 5\sqrt{10}$ **32.** $0, 36$

33. $-7, 12$ **34.** $-\frac{1}{3}$ **35.** $-2, 20$ **36.** $6 \pm 2\sqrt{15}$

37. $3 \pm i$ **38.** $7 \pm 2\sqrt{11}$ **39.** $-\frac{1}{2}, -1$

40. $\dfrac{-7 \pm \sqrt{249}}{10}$ **41.** $\dfrac{3 \pm \sqrt{17}}{2}$ **42.** $\dfrac{11 + \sqrt{13}}{2}$

43.   **44.**

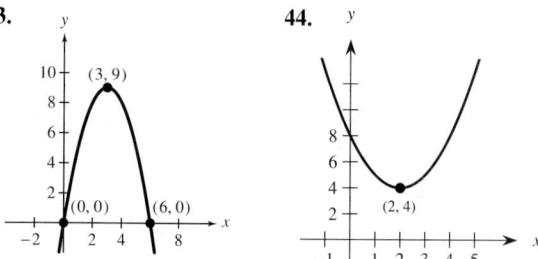

45.   **46.**

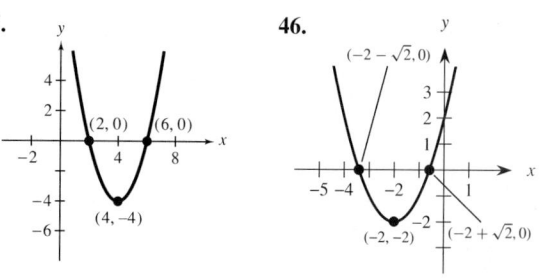

47. $y = -2x^2 + 12x - 13$ **48.** $y = 3x^2 + 12x + 15$

49. $y = \frac{1}{16}x^2 - \frac{5}{8}x + \frac{25}{16}$ **50.** $y = -x^2 - 4x + 1$

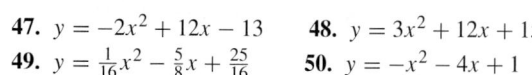

51. $(-\infty, 3)$

52. $(-2, \infty)$

53. $(0, 7)$

54. $(-\infty, 0] \cup [10, \infty)$

55. $(-\infty, -2] \cup [6, \infty)$

56. $(-\infty, -1) \cup (11, \infty)$

57. $\left(-4, \frac{5}{2}\right)$

58. $\left(-\infty, -\frac{4}{3}\right) \cup (2, \infty)$

59. $(-\infty, 0] \cup \left(\frac{7}{2}, \infty\right)$

60. $\left(1, \frac{9}{2}\right]$

61. 11, 12   **62.** 12, 13

63. $\dfrac{5\sqrt{2}}{2}$ sec   **64.** 6 sec

65. (a) $P = 2l + 2w = 48$ **(b)** $A = lw$
$\qquad\qquad l + 2 = 24 \qquad\qquad w = 24 - l$
$\qquad\qquad w = 24 - l \qquad\qquad A = l(24 - l)$
(c) 44, 80, 108, 128, 140, 144, 140, 128, 108
(d) The rectangle with the largest area has a length and
width of 12. This rectangle is a square.

66. 60 in. × 100 in.   **67.** 15 people   **68.** 60 mph

69. $\dfrac{9 + 3\sqrt{17}}{2} \approx 10.68$ hr

$\dfrac{9 + 3\sqrt{17}}{2} + 3 \approx 13.68$ hr

70. (a)

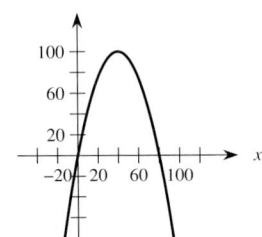

(b) 100 ft   **(c)** 80 ft
71. 1993
72. (a) $x^2 - 9 = 0$
(b) $2x^2 - 3x - 2 = 0$
(c) $x^2 + 6x + 5 = 0$
(d) $x^2 + 2x - 80 = 0$

CHAPTER 7 TEST *(page 594)*

1. $-5, 10$ **2.** $-\frac{3}{8}, 3$ **3.** 1.7, 2.3

4. $-3 \pm 9i$ **5.** $\dfrac{9}{4}$ **6.** $\dfrac{3 \pm \sqrt{3}}{2}$

7. -56; two distinct imaginary solutions

8. $\dfrac{4 \pm \sqrt{7}}{3}$ **9.** 7.41   **10.** $x^2 - x - 20 = 0$

11.

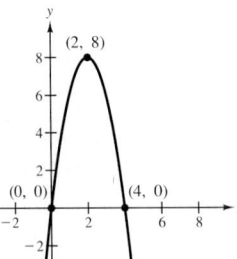

12. $y = \frac{2}{3}x^2 - 4x + 4$

13. $(0, 3)$

14. $(-\infty, -2] \cup [6, \infty)$

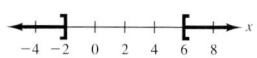

15. $\left(2, \frac{11}{4}\right)$

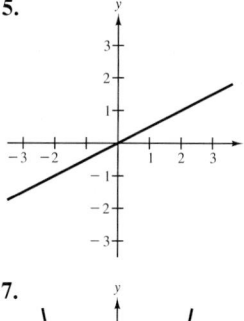

16. 14, 15   **17.** 12 ft × 20 ft
18. 50 mph   **19.** 1.58 sec   **20.** 120

Chapter 8

Chapter Opener   *(page 595)*

Width = 150 yd; Length = 180 yd

SECTION 8.1   *(page 602)*

Warm-Up   *(page 602)*

1. All real values x
2. All real values x such that $x \leq 10$
3. All real values x such that $x \neq 5$
4. All real values x such that $x = 5$ and $x \neq -5$

5.

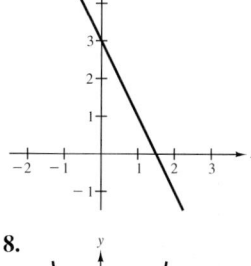

6.

7.

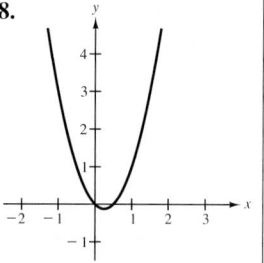

8.

9.

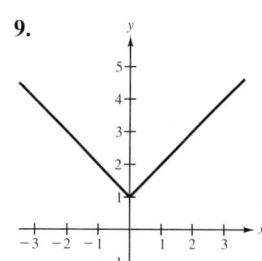

10.

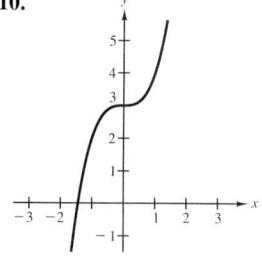

1. $(f + g)(x) = x^2 + 2x$, $(-\infty, \infty)$
$(f - g)(x) = 2x - x^2$, $(-\infty, \infty)$
$(fg)(x) = 2x^3$, $(-\infty, \infty)$
$\left(\dfrac{f}{g}\right)(x) = \dfrac{2}{x}$, $(-\infty, 0) \cup (0, \infty)$

2. $(f + g)(x) = x + 1 + \sqrt{x}$, $[0, \infty)$
$(f - g)(x) = x + 1 - \sqrt{x}$, $[0, \infty)$
$(fg)(x) = x^{2/3} + x^{1/2}$, $[0, \infty)$
$\left(\dfrac{f}{g}\right)(x) = \dfrac{x + 1}{\sqrt{x}}$, $(0, \infty)$

3. $(f + g)(x) = x^2 + 4x - 12$, $(-\infty, \infty)$
$(f - g)(x) = -x^2 + 4x + 6$, $(-\infty, \infty)$
$(fg)(x) = 4x^3 - 3x^2 - 36x + 27$, $(-\infty, \infty)$
$\left(\dfrac{f}{g}\right)(x) = \dfrac{4x - 3}{x^2 - 9}$,
$\qquad\qquad (-\infty, -3) \cup (-3, 3) \cup (3, \infty)$

4. $(f + g)(x) = x^3 + x^2 + 1$, $(-\infty, \infty)$
$(f - g)(x) = x^3 - x^2 - 1$, $(-\infty, \infty)$
$(fg)(x) = x^5 + x^3$, $(-\infty, \infty)$
$\left(\dfrac{f}{g}\right)(x) = \dfrac{x^3}{x^2 + 1}$, $(-\infty, \infty)$

5. $(f + g)(x) = \dfrac{1}{x} + 5x$, $(-\infty, 0) \cup (0, \infty)$

$(f - g)(x) = \dfrac{1}{x} - 5x$, $(-\infty, 0) \cup (0, \infty)$

$(fg)(x) = 5$, $(-\infty, 0) \cup (0, \infty)$

$\left(\dfrac{f}{g}\right)(x) = \dfrac{1}{5x^2}$, $(-\infty, 0) \cup (0, \infty)$

6. $(f + g)(x) = \dfrac{2x}{(x + 2)(x - 2)}$,
$(-\infty, -2) \cup (-2, 2) \cup (2, \infty)$

$(f - g)(x) = \dfrac{-4}{(x + 2)(x - 2)}$,
$(-\infty, -2) \cup (-2, 2) \cup (2, \infty)$

$(fg)(x) = \dfrac{1}{(x + 2)(x - 2)}$,
$(-\infty, -2) \cup (-2, 2) \cup (2, \infty)$

$\left(\dfrac{f}{g}\right)(x) = \dfrac{x - 2}{x + 2}$,
$(-\infty, -2) \cup (-2, 2) \cup (2, \infty)$

7. $(f + g)(x) = \sqrt{x + 4} + \sqrt{4 - x}$, $[-4, 4]$
$(f - g)(x) = \sqrt{x + 4} - \sqrt{4 - x}$, $[-4, 4]$
$(fg)(x) = \sqrt{16 - x^2}$, $[-4, 4]$
$\left(\dfrac{f}{g}\right)(x) = \dfrac{\sqrt{x + 4}}{\sqrt{4 - x}}$, $[-4, 4]$

8. $(f + g)(x) = \sqrt{x} + \sqrt{x^2 - 16}$, $[4, \infty)$
$(f - g)(x) = \sqrt{x} - \sqrt{x^2 - 16}$, $[4, \infty)$
$(fg)(x) = \sqrt{x^3 - 16x}$, $[4, \infty)$
$\left(\dfrac{f}{g}\right)(x) = \dfrac{\sqrt{x}}{\sqrt{x^2 - 16}}$, $[4, \infty)$

9. (a) 11 (b) 0 **10.** (a) −15 (b) 125

11. (a) 19 (b) −79

12. (a) 47
(b) $(f - g)(1)$ is not real [1 is not in the domain of $f(x)$]

13. (a) 10 (b) $\frac{2}{5}$ **14.** (a) 0 (b) 4

15. (a) $\frac{1}{3}$ (b) Undefined (division by zero)

16. (a) $\frac{8}{3}$ (b) Undefined (division by zero)

17. (a) $x^2 - 3$ (b) $(x - 3)^2$ (c) 13 (d) 16
18. (a) $x^3 + 5$ (b) $(x + 5)^3$ (c) 13 (d) 8
19. (a) $3|x - 1|$ (b) $3|x - 3|$ (c) 0 (d) 3
20. (a) $|2x + 5|$ (b) $2|x| + 5$ (c) 1 (d) 13
21. (a) $\sqrt{x + 5}$ (b) $\sqrt{x} + 5$ (c) 3 (d) 8
22. (a) $\sqrt{2x + 3}$ (b) $2\sqrt{x + 6} - 3$ (c) 3 (d) 1
23. (a) $\dfrac{1}{\sqrt{x} - 3}$ (b) $\dfrac{1}{\sqrt{x - 3}}$ (c) $\dfrac{1}{4}$ (d) $\dfrac{1}{3}$
24. (a) $\dfrac{4x^2}{1 - 4x^2}$ (b) $\dfrac{x^2 - 4}{4}$ (c) $-\dfrac{16}{15}$ (d) $-\dfrac{3}{4}$

25. $f \circ g$: $(-\infty, \infty)$, $g \circ f$: $(-\infty, \infty)$
26. $f \circ g$: $(-\infty, \infty)$, $g \circ f$: $(-\infty, \infty)$
27. $f \circ g$: $[2, \infty)$, $g \circ f$: $[0, \infty)$
28. $f \circ g$: $(-\infty, -\sqrt{5}] \cup [\sqrt{5}, \infty)$, $g \circ f$: $[5, \infty)$
29. $f \circ g$: $(-\infty, \infty)$, $g \circ f$: $(-\infty, -9) \cup (-9, \infty)$
30. $f \circ g$: $[0, 16) \cup (16, \infty)$,
$g \circ f$: $(-\infty, 0] \cup (4, \infty)$
31. $t^2 - 8t - 3$ **32.** $-\frac{2}{13}$ **33.** $t^2 - 2t + 3$
34. $5z^3 - 12z^2 - 9z$ **35.** 5 **36.** $2x + h - 3$
37. 10 **38.** 3 **39.** $5y^2 - 15y + 3$
40. $25z^2 + 15z$ **41.** (a) −1 (b) −2 (c) −2
42. (a) 2 (b) −3 (c) −3
43. −1 **44.** 1 **45.** −3 **46.** 1

47. 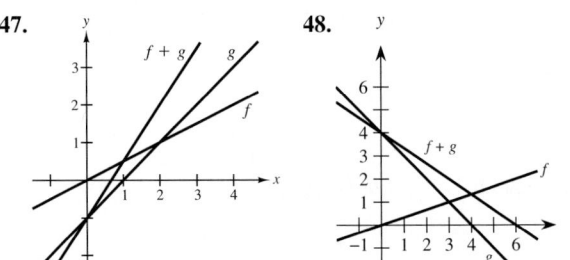 **48.**

49. 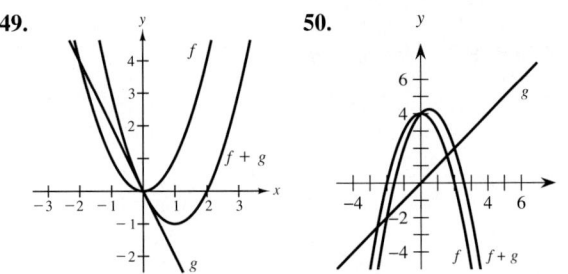 **50.**

51. $T(x) = \frac{3}{4}x + \frac{1}{15}x^2$

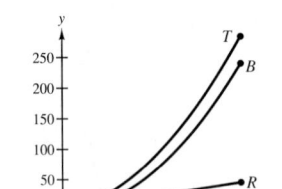

52. (a) $R_1 + R_2 = 750 + 0.78t - 0.8t^2$
$t = 5, 6, 7, 8, 9, 10$
(b) Total sales have been decreasing.

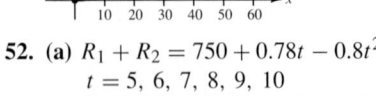

53. (a) $f(g(x)) = 0.02x - 200,000$
(b) $g(f(x)) = 0.02(x - 200,000)$
This part represents the bonus, because it gives 2% of sales over $200,000.

54. $A(r(t)) = 0.36\pi t^2$

55.

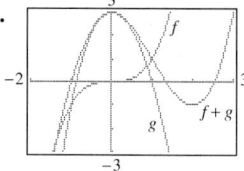

$f(x)$ is more significant in determining the magnitude of the sum.

56.

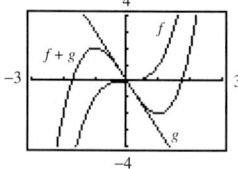

$f(x)$ is more significant in determining the magnitude of the sum.

SECTION 8.2 *(page 614)*

Warm-Up *(page 613)*

1. All real values x **2.** All real values x

3. Real values x such that $-4 \le x \le 4$

4. Real values x such that $x \ne -6$ and $x \ne 6$

5. $(f + g)(x) = x^2 + 3x - 9, \ (-\infty, \infty)$
$(f - g)(x) = x^2 - 3x - 9, \ (-\infty, \infty)$
$(fg)(x) = 3x^3 - 27x, \ (-\infty, \infty)$
$\left(\dfrac{f}{g}\right)(x) = \dfrac{x^2 - 9}{3x}, \ (-\infty, 0) \ \cup \ (0, \infty)$
$(f \circ g)(x) = 9x^2 - 9, \ (-\infty, \infty)$
$(g \circ f)(x) = 3(x^2 - 9), \ (-\infty, \infty)$

6. $(f + g)(x) = \frac{1}{2}x^2 + 2x - 3, \ (-\infty, \infty)$
$(f - g)(x) = -\frac{1}{2}x^2 + 2x - 3, \ (-\infty, \infty)$
$(fg)(x) = x^3 - \frac{3}{2}x^2, \ (-\infty, \infty)$
$\left(\dfrac{f}{g}\right)(x) = \dfrac{4x - 6}{x^2}, \ (-\infty, 0) \ \cup \ (0, \infty)$
$(f \circ g)(x) = x^2 - 3, \ (-\infty, \infty)$
$(g \circ f)(x) = \frac{1}{2}(2x - 3)^2, \ (-\infty, \infty)$

7. $(f + g)(x) = \sqrt[3]{x - 2} + x^2, \ (-\infty, \infty)$
$(f - g)(x) = \sqrt[3]{x - 2} - x^2, \ (-\infty, \infty)$
$(fg)(x) = x^2\sqrt[3]{x - 2}, \ (-\infty, \infty)$
$\left(\dfrac{f}{g}\right)(x) = \dfrac{\sqrt[3]{x - 2}}{x^2}, \ (-\infty, 0) \ \cup \ (0, \infty)$
$(f \circ g)(x) = \sqrt[3]{x^2 - 2}, \ (-\infty, \infty)$
$(g \circ f)(x) = \sqrt[3]{(x - 2)^2}, \ (-\infty, \infty)$

8. $(f + g)(x) = \sqrt{x} + x^2 - 4, \ [0, \infty)$
$(f - g)(x) = \sqrt{x} - x^2 + 4, \ [0, \infty)$
$(fg)(x) = \sqrt{x}(x^2 - 4), \ [0, \infty)$
$\left(\dfrac{f}{g}\right)(x) = \dfrac{\sqrt{x}}{x^2 - 4}, \ [0, 2) \ \cup \ (2, \infty)$
$(f \circ g)(x) = \sqrt{x^2 - 4}, \ (-\infty, -2] \cup [2, \infty)$
$(g \circ f)(x) = x - 4, \ [0, \infty)$

9. $(f + g)(x) = \dfrac{2(x - 1)}{x(x - 2)}, \ (-\infty, 0) \cup (0, 2) \ \cup \ (2, \infty)$
$(f - g)(x) = \dfrac{2}{x(2 - x)}, \ (-\infty, 0) \ \cup \ (0, 2) \cup (2, \infty)$
$(fg)(x) = \dfrac{1}{x(x - 2)}, \ (-\infty, 0) \ \cup \ (0, 2) \cup (2, \infty)$
$\left(\dfrac{f}{g}\right)(x) = \dfrac{x - 2}{x}, \ (-\infty, 0) \ \cup \ (0, 2) \ \cup \ (2, \infty)$
$(f \circ g)(x) = x - 2, \ (-\infty, 2) \ \cup \ (2, \infty)$
$(g \circ f)(x) = \dfrac{x}{1 - 2x},$
$\left(-\infty, \dfrac{1}{2}\right) \cup \left(\dfrac{1}{2}, 2\right) \ \cup \ (2, \infty)$

10. $(f + g)(x) = \dfrac{x^2 + 4x + 5}{x + 2}, \ (-\infty, -2) \cup (-2, \infty)$
$(f - g)(x) = \dfrac{x^2 + 4x + 3}{x + 2}, \ (-\infty, -2) \cup (-2, \infty)$
$(fg)(x) = 1, \ (-\infty, -2) \ \cup \ (-2, \infty)$
$\left(\dfrac{f}{g}\right)(x) = (x + 2)^2, \ (-\infty, -2) \cup (-2, \infty)$
$(f \circ g)(x) = \dfrac{2x + 5}{x + 2}, \ (-\infty, -2) \cup (-2, \infty)$
$(g \circ f)(x) = \dfrac{1}{x + 4}, \ (-\infty, -4) \cup (-4, \infty)$

1. $f^{-1}(x) = \dfrac{x}{5}$ 2. $f^{-1}(x) = 3x$

3. $f^{-1}(x) = x - 10$ 4. $f^{-1}(x) = x + 5$

5. $f^{-1}(x) = \sqrt[7]{x}$ 6. $f^{-1}(x) = \sqrt[5]{x}$

7. $f^{-1}(x) = x^3$ 8. $f^{-1}(x) = x^5$

9. $f(g(x)) = f\left(\dfrac{1}{10}x\right) = 10\left(\dfrac{1}{10}x\right) = x$

   $g(f(x)) = g(10x) = \dfrac{1}{10}(10x) = x$

10. $f(g(x)) = f\left(\dfrac{3x}{2}\right) = \dfrac{2}{3}\left(\dfrac{3x}{2}\right) = x$

    $g(f(x)) = g\left(\dfrac{2x}{3}\right) = \dfrac{3}{2}\left(\dfrac{2x}{3}\right) = x$

11. $f(g(x)) = f(x - 15) = (x - 15) + 15 = x$

    $g(f(x)) = g(x + 15) = (x + 15) - 15 = x$

12. $f(g(x)) = f(3 - x) = 3 - (3 - x) = x$

    $g(f(x)) = g(3 - x) = 3 - (3 - x) = x$

13. $f(g(x)) = f\left(\dfrac{1 - x}{2}\right)$

    $= 1 - 2\left(\dfrac{1 - x}{2}\right)$

    $= 1 - (1 - x) = x$

    $g(f(x)) = g(1 - 2x)$

    $= \dfrac{1 - (1 - 2x)}{2}$

    $= \dfrac{2x}{2} = x$

14. $f(g(x)) = f\left[\dfrac{1}{2}(x + 1)\right] = 2\left[\dfrac{1}{2}(x + 1)\right] - 1 = x$

    $g(f(x)) = g(2x - 1) = \dfrac{1}{2}[(2x - 1) + 1] = x$

15. $f(g(x)) = f\left[\dfrac{1}{3}(2 - x)\right] = 2 - 3\left[\dfrac{1}{3}(2 - x)\right]$

    $= 2 - (2 - x)$

    $= x$

    $g(f(x)) = g(2 - 3x) = \dfrac{1}{3}[2 - (2 - 3x)]$

    $= \dfrac{1}{3}(3x)$

    $= x$

16. $f(g(x)) = f[-4(x - 3)] = -\dfrac{1}{4}[-4(x - 3)] + 3 = x$

    $g(f(x)) = g\left(-\dfrac{1}{4}x + 3\right) = -4\left[\left(-\dfrac{1}{4}x + 3\right) - 3\right] = x$

17. $f(g(x)) = f(x^3 - 1) = \sqrt[3]{(x^3 - 1) + 1}$

    $= \sqrt[3]{x^3}$

    $= x$

    $g(f(x)) = g(\sqrt[3]{x + 1}) = (\sqrt[3]{x + 1})^3 - 1$

    $= x + 1 - 1$

    $= x$

18. $f(g(x)) = f(\sqrt[5]{x}) = (\sqrt[5]{x})^5 = x$

    $g(f(x)) = g(x^5) = \sqrt[5]{x^5} = x$

19. $f(g(x)) = f\left(\dfrac{1}{x}\right) = \dfrac{1}{(1/x)} = x$

    $g(f(x)) = g\left(\dfrac{1}{x}\right) = \dfrac{1}{(1/x)} = x$

20. $f(g(x)) = f\left(3 + \dfrac{1}{x}\right) = \dfrac{1}{(3 + 1/x) - 3} = x$

    $g(f(x)) = g\left(\dfrac{1}{x + 3}\right) = 3 + \dfrac{1}{1/(x - 3)} = x$

21. b   22. c   23. d   24. a

25.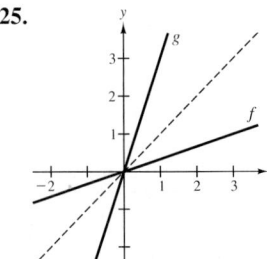

26.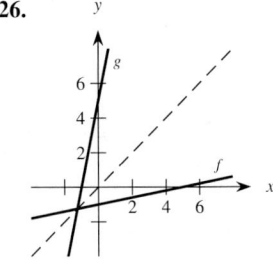

27.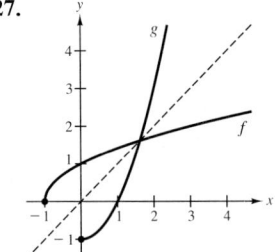

28.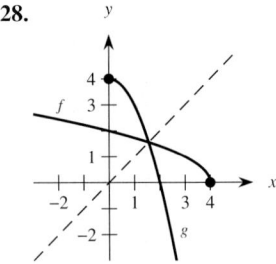

29.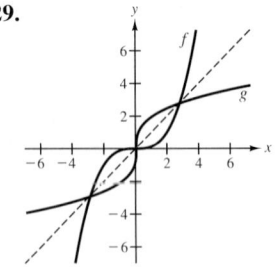

30.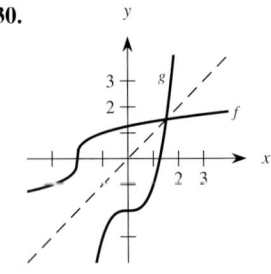

31. $f^{-1}(x) = \dfrac{x}{8}$ **32.** $f^{-1}(x) = 10x$

33. $g^{-1}(x) = x - 25$ **34.** $f^{-1}(x) = 7 - x$

35. $g^{-1}(x) = \dfrac{3 - x}{4}$ **36.** $g^{-1}(x) = \dfrac{x - 1}{6}$

37. $h^{-1}(x) = x^2,\ x \geq 0$

38. $h^{-1}(x) = x^2 - 5,\ x \geq 0$

39. $f^{-1}(t) = \sqrt[3]{t + 1}$ **40.** $h^{-1}(t) = \sqrt[5]{t}$

41. $g^{-1}(s) = \dfrac{5}{s}$ **42.** $f^{-1}(s) = 3 - \dfrac{2}{s}$

43. No   **44.** Yes   **45.** Yes   **46.** No

47. No   **48.** No   **49.** Yes   **50.** Yes

51. No   **52.** Yes   **53.** Yes   **54.** No

55. No   **56.** Yes

57. $x \geq 0,\ f^{-1}(x) = \sqrt[4]{x}$

58. $x \geq 0,\ f^{-1}(x) = \sqrt{9 - x}$

59. $x \geq 2,\ f^{-1}(x) = \sqrt{x} + 2$

60. $x \geq 2,\ f^{-1}(x) = |x| + 2$

61.

x	0	1	3	4
f^{-1}	6	4	2	0

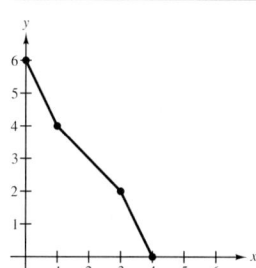

62.

x	-1	0	1	5
f^{-1}	-2	0	3	4

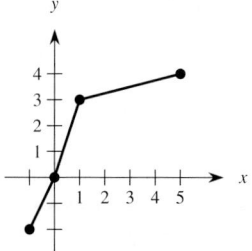

63.

x	-4	-2	2	3
f^{-1}	-2	-1	1	3

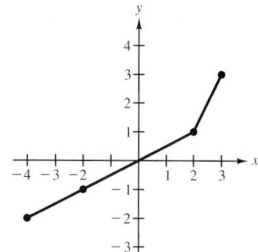

64.

x	-3	-2	0	6
f^{-1}	4	3	-1	-2

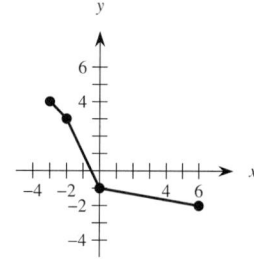

65. (a) $f^{-1}(x) = \dfrac{3 - x}{2}$ (b) $(f^{-1})^{-1}(x) = 3 - 2x$

66. (a) $(f \circ g)(x) = 4x + 24$

(b) $(f \circ g)^{-1}(x) = \dfrac{x - 24}{4}$

(c) $f^{-1}(x) = \dfrac{x}{4}$

$g^{-1}(x) = x - 6$

(d) $(g^{-1} \circ f^{-1})(x) = \dfrac{x - 24}{4}$

$(g^{-1} \circ f^{-1})(x) = (f \circ g)^{-1}(x)$

67. $f^{-1}(x) = \sqrt[3]{x - 1}$

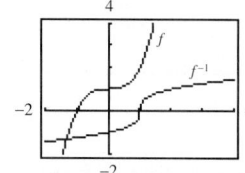

68. $f^{-1}(x) = \sqrt{x^2 + 4}$, $x \ge 0$

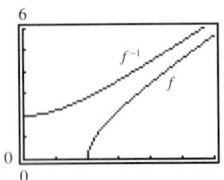

69. (a) $y = \frac{20}{13}(x - 9)$

(b) x: hourly wage; y: number of units produced

(c) 8

SECTION 8.3 *(page 625)*

Warm-Up *(page 625)*

1. $\frac{1}{3}$ **2.** $\frac{10}{3}$ **3.** 0.92

4. 513 **5.** 40 **6.** $\frac{400}{7}$

7.

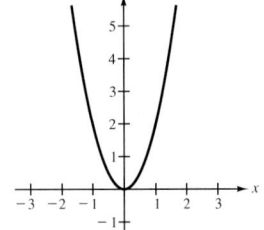

8.

9.

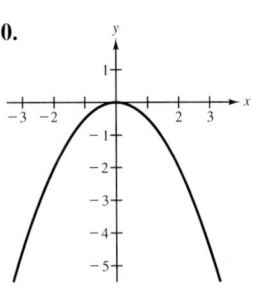

10.

1. $I = kV$ **2.** $C = kr$ **3.** $u = kv^2$

4. $V = kx^3$ **5.** $p = \frac{k}{d}$ **6.** $S = \frac{k}{v^2}$

7. $P = \frac{k}{\sqrt{1 + i}}$ **8.** $A = \frac{k}{t^4}$ **9.** $A = klw$

10. $V = khr^2$ **11.** $P = \frac{k}{V}$ **12.** $F = \frac{km_1m_2}{r^2}$

13. A varies jointly as b and h.

14. A varies directly as the square of r.

15. V varies jointly as the square of r and as h.

16. r varies directly as d and inversely as t.

17. $s = 5t$ **18.** $h = \frac{7}{3}r$ **19.** $F = \frac{5}{16}x^2$

20. $v = 6\sqrt{s}$ **21.** $n = \frac{48}{m}$ **22.** $q = \frac{75}{p}$

23. $F = \frac{25}{6}xy$ **24.** $V = \frac{1}{3}hb^2$

25. $d = \frac{120x^2}{r}$ **26.** $z = \frac{125x}{\sqrt{y}}$

27. \$4921.25 **28.** \$504

29. (a) 2 in. **(b)** 15 lb

30. (a) 1.2 in. **(b)** 25 lb

31. 18 lb **32.** 0.6 in. **33.** $208\frac{1}{3}$ ft

34. 576 ft **35.** 3072 W **36.** 0.61 mph

37. 324 lb **38.** 0.36 lb/in.^{2}, 116 lb

39. $p = \frac{114}{t}$, 17.5% **40.** $T = \frac{4000}{d}$, 0.91° C

41. 667 units **42.** 121.6 lb

43. $\frac{1}{4}$ **44.** 4 **45.** 3125 lb

46. No, k is different for each pizza. The 15-inch pizza is the best buy.

47.

x	2	4	6	8	10
$y = 1x^2$	4	16	36	64	100

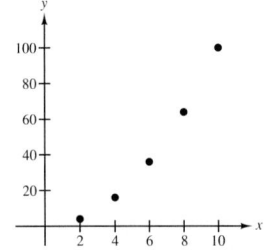

48.

x	2	4	6	8	10
$y = 2x^2$	8	32	72	128	200

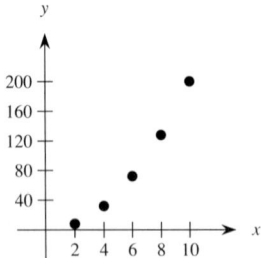

49.

x	2	4	6	8	10
$y = \frac{1}{2}x^2$	2	8	18	32	50

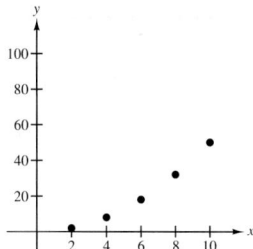

50.

x	2	4	6	8	10
$y = \frac{1}{4}x^2$	1	4	9	16	25

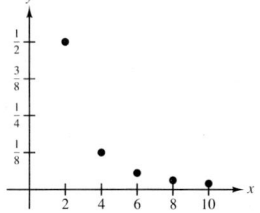

51.

x	2	4	6	8	10
$y = \frac{2}{x^2}$	$\frac{1}{2}$	$\frac{1}{8}$	$\frac{1}{18}$	$\frac{1}{32}$	$\frac{1}{50}$

52.

x	2	4	6	8	10
$y = \frac{5}{x^2}$	$\frac{5}{4}$	$\frac{5}{16}$	$\frac{5}{36}$	$\frac{5}{64}$	$\frac{1}{20}$

53.

x	2	4	6	8	10
$y = \frac{10}{x^2}$	$\frac{5}{2}$	$\frac{5}{8}$	$\frac{5}{18}$	$\frac{5}{32}$	$\frac{1}{10}$

54.

x	2	4	6	8	10
$y = \frac{20}{x^2}$	5	$\frac{5}{4}$	$\frac{5}{9}$	$\frac{5}{16}$	$\frac{1}{5}$

CHAPTER 8 MID-CHAPTER QUIZ (page 629)

1. $-x^2 + 2x - 8$, $(-\infty, \infty)$ 2. $x^2 + 2x - 8$, $(-\infty, \infty)$
3. $2x^3 - 8x^2$, $(-\infty, \infty)$ 4. $x^2 - 2x + 8$, $(-\infty, \infty)$
5. $\dfrac{2x - 8}{x^2}$, $(-\infty, 0) \cup (0, \infty)$

6. $\dfrac{x^2}{2x - 8}$, $(-\infty, 4) \cup (4, \infty)$

7. $2x^2 - 8$, $(-\infty, \infty)$ 8. $(2x - 8)^2$, $(-\infty, \infty)$
9. $C(x(t)) = 3000t + 750$, $0 \le t \le 40$
   Cost of units produced in t hours
10. $f(g(x)) = 100 - 15\left[\frac{1}{15}(100 - x)\right]$
    $= 100 - (100 - x) = x$
    $g(f(x)) = \frac{1}{15}[100 - (100 - 15x)]$
    $= \frac{1}{15}(15x) = x$
11. $f^{-1}(x) = \sqrt[3]{x + 8}$

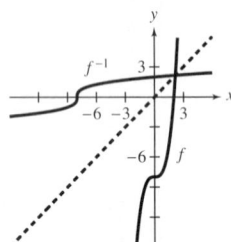

12. No. Let $x > 0$ 13. $z = 3t$ 14. $m = \dfrac{14}{n}$

15. $S = 5hr^2$ 16. $N = \dfrac{15t^2}{s}$ 17. 30 min

SECTION 8.4 (page 636)

Warm-Up (page 636)

1. $(3x - 2)(4x + 5)$ 2. $x(5x - 6)^2$
3. $z^2(12z + 5)(z + 1)$ 4. $(y + 5)(y^2 - 5y + 25)$
5. $(x + 3)(x + 2)(x - 2)$ 6. $(x + 2)(x^2 + 3)$
7. No real solution   8. $3 \pm \sqrt{5}$
9. $-\frac{1}{2} \pm \sqrt{3}$ 10. ± 3

1. Horizontal line   2. Horizontal line

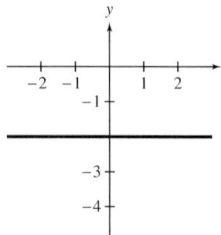   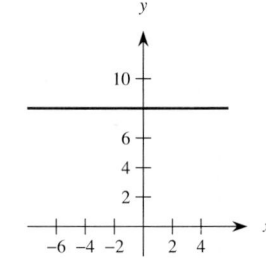

3. Line with slope -3 4. Line with slope $\frac{1}{2}$

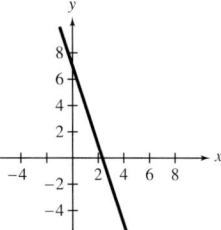   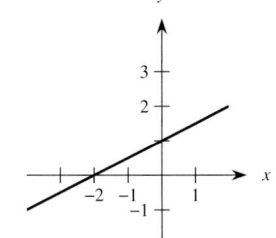

5. Parabola   6. Parabola

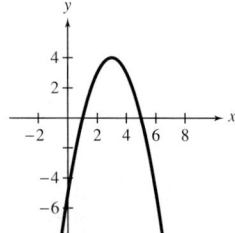   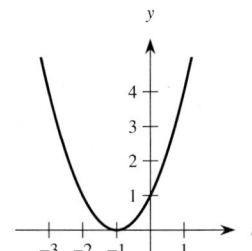

7. 4   8. 2   9. 1   10. 1   11. 3   12. 5
13. Horizontal translation 5 units to the right
14. Horizontal translation 3 units to the left
15. Vertical translation 3 units upward
16. Vertical translation 2 units downward
17. Reflection in the x-axis
18. Reflection in the x-axis

19. (a) 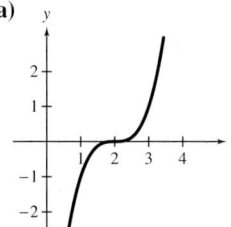 **(b)**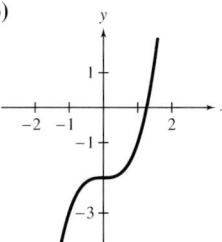

(c) 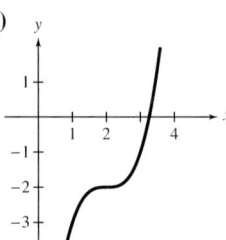 **(d)**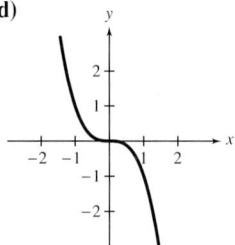

20. (a) 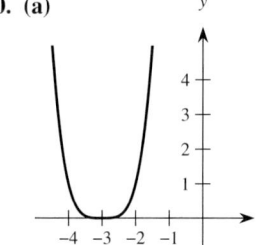 **(b)**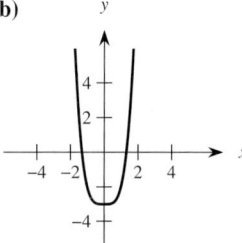

(c) 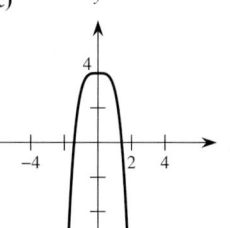 **(d)**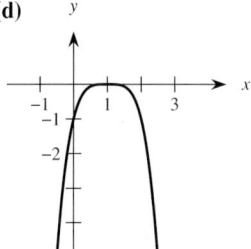

21. Rises to the left; Rises to the right
22. Falls to the left; Falls to the right
23. Falls to the left; Rises to the right
24. Rises to the left; Falls to the right
25. Rises to the left; Falls to the right
26. Falls to the left; Falls to the right

27. $(\pm 5, 0)$, $(0, -25)$ **28.** $(\pm 7, 0)$, $(0, 49)$
29. $(3, 0)$, $(0, 9)$ **30.** $(-5, 0)$, $(0, 25)$
31. $(-2, 0)$, $(1, 0)$, $(0, -2)$ **32.** $(2 \pm \sqrt{3}, 0)$, $(0, 3)$
33. $(2, 0)$, $(0, 0)$ **34.** $(-4, 0)$, $(0, 0)$, $(5, 0)$
35. $(\pm 1, 0)$, $\left(0, -\frac{1}{2}\right)$ **36.** $(\pm \sqrt{5}, 0)$, $(0, -40)$
37. $(0, 10)$ **38.** $(4, 0)$, $(\pm 5, 0)$, $(0, 100)$
39. e   **40.** c   **41.** b   **42.** f
43. a   **44.** g   **45.** d   **46.** h

47. 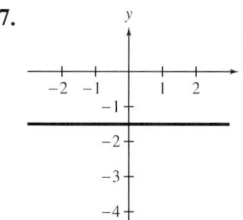 **48.**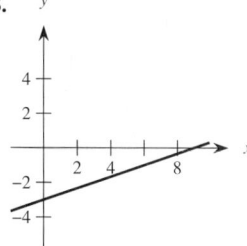

49. 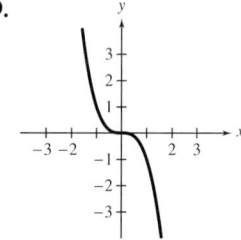 **50.**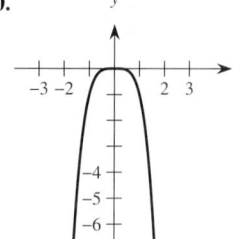

51. 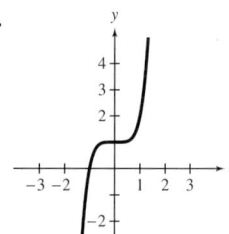 **52.**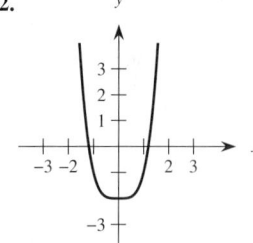

53. 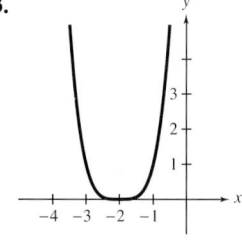 **54.**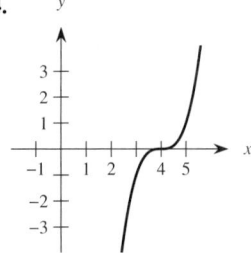

55.

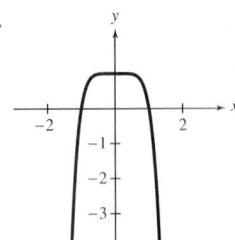

56.

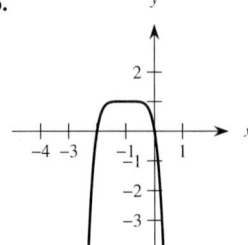

57.

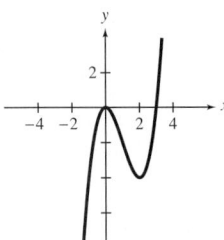

58.

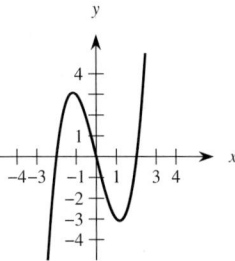

59.

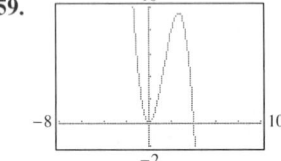

60.

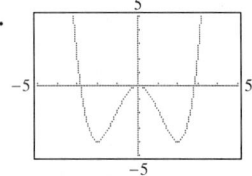

61.

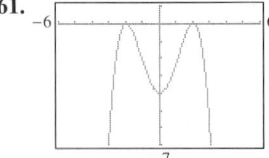

62.

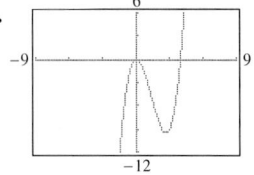

63.

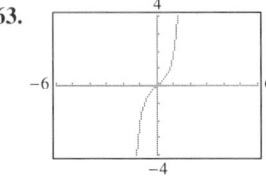

64.

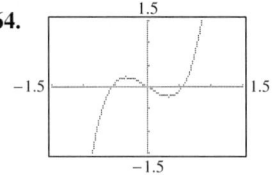

65. (b) $(0, 6)$
(c)

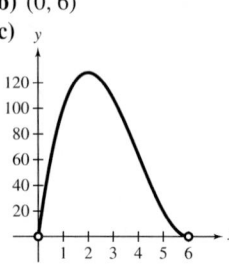

$x = 2$

66. (b) $(0, 3)$
(c)

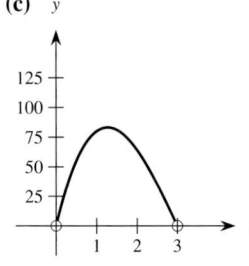

$x = 1.27$

SECTION 8.5   *(page 649)*

Warm-Up   *(page 649)*

1. $y = \frac{1}{3}(9 - 2x)$ **2.** $y = \pm\sqrt{36 - x^2}$

3. $y = \pm\sqrt{\dfrac{6 - 2x^2}{3}}$ **4.** $y = \pm\sqrt{2x^2 - 3}$

5. $15\sqrt{2}$ **6.** $15\sqrt{6}$

7.

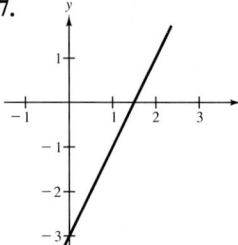

8.

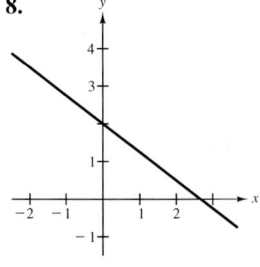

9.

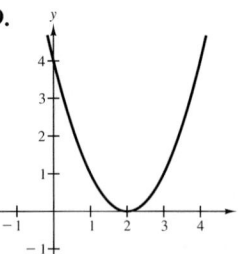

10.

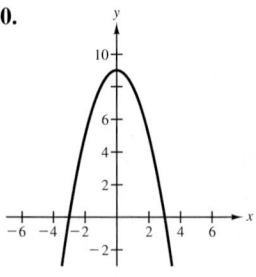

1. c   **2.** f   **3.** e   **4.** b   **5.** a   **6.** d

7. $x^2 + y^2 = 25$ **8.** $x^2 + y^2 = 49$

9. $x^2 + y^2 = \frac{4}{9}$ **10.** $x^2 + y^2 = \frac{25}{4}$

11. $x^2 + y^2 = 64$ **12.** $x^2 + y^2 = 4$

13. $x^2 + y^2 = 29$ **14.** $x^2 + y^2 = 17$

15. Center: $(0, 0)$ **16.** Center: $(0, 0)$
   $r = 4$ $r = 5$

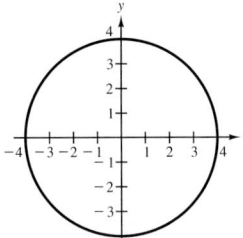

17. Center: $(0, 0)$ **18.** Center: $(0, 0)$
   $r = \frac{1}{2}$ $r = \frac{8}{3}$

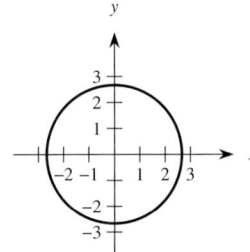

19. Center: $(0, 0)$ **20.** Center: $(0, 0)$
   $r = \frac{12}{5}$ $r = 2$

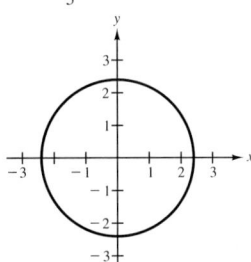
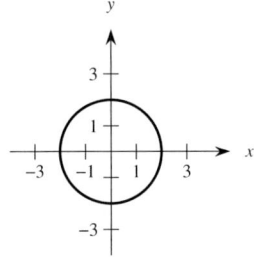

21. $\dfrac{x^2}{16} + \dfrac{y^2}{9} = 1$ **22.** $\dfrac{x^2}{16} + \dfrac{y^2}{1} = 1$

23. $\dfrac{x^2}{9} + \dfrac{y^2}{16} = 1$ **24.** $\dfrac{x^2}{1} + \dfrac{y^2}{25} = 1$

25. $\dfrac{x^2}{100} + \dfrac{y^2}{36} = 1$ **26.** $\dfrac{x^2}{225} + \dfrac{y^2}{625} = 1$

27. Vertices: $(\pm 4, 0)$ **28.** Vertices: $(\pm 5, 0)$
   Co-vertices: $(0, \pm 2)$ Co-vertices: $(0, \pm 3)$

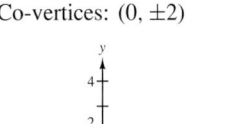
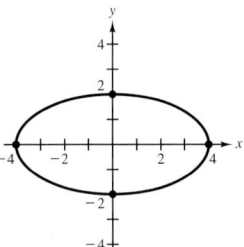

29. Vertices: $(0, \pm 4)$ **30.** Vertices: $(0, \pm 5)$
   Co-vertices: $(\pm 2, 0)$ Co-vertices: $(\pm 3, 0)$

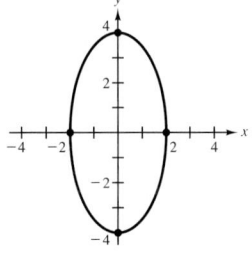
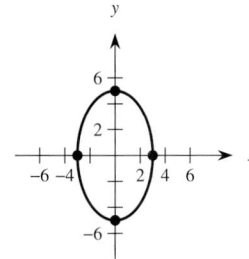

31. Vertices: $\left(\pm \frac{5}{3}, 0\right)$ **32.** Vertices: $(\pm 1, 0)$
   Co-vertices: $\left(0, \pm \frac{4}{3}\right)$ Co-vertices: $\left(0, \pm \frac{1}{2}\right)$

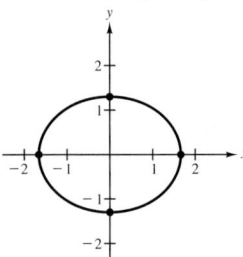

33. Vertices: $(0, \pm 2)$ **34.** Vertices: $(\pm 3, 0)$
   Co-vertices: $(\pm 1, 0)$ Co-vertices: $(0, \pm 2)$

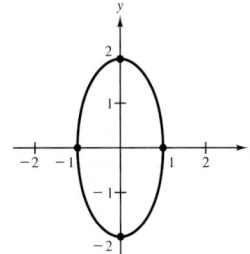
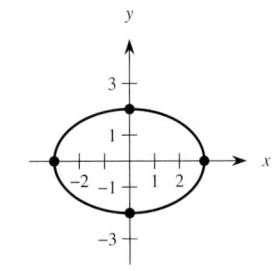

35. Vertices: $(\pm 3, 0)$
Asymptotes: $y = \pm x$

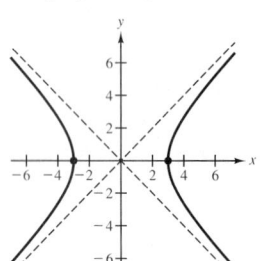

36. Vertices: $(\pm 1, 0)$
Asymptotes: $y = \pm x$

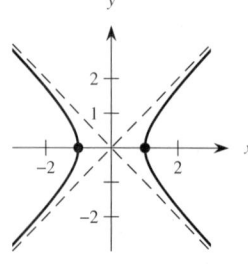

37. Vertices: $(0, \pm 3)$
Asymptotes: $y = \pm x$

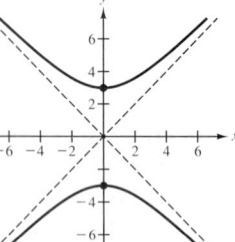

38. Vertices: $(0, \pm 1)$
Asymptotes: $y = \pm x$

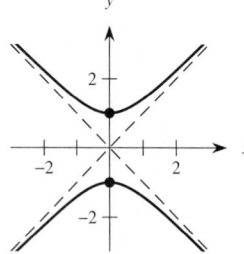

39. Vertices: $(\pm 3, 0)$
Asymptotes: $y = \pm \frac{5}{3} x$

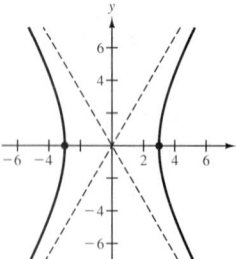

40. Vertices: $(\pm 2, 0)$
Asymptotes: $y = \pm \frac{3}{2} x$

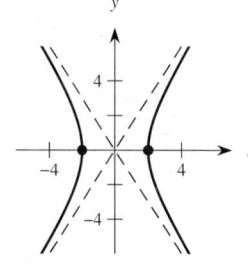

41. Vertices: $(0, \pm 3)$
Asymptotes: $y = \pm \frac{3}{5} x$

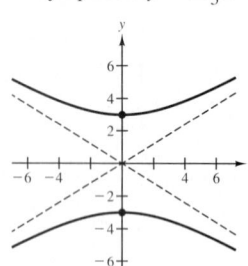

42. Vertices: $(0, \pm 2)$
Asymptotes: $y = \pm \frac{2}{3} x$

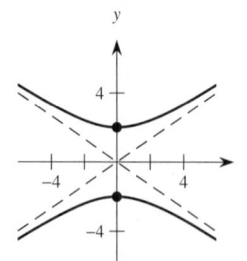

43. Vertices: $(\pm 1, 0)$
Asymptotes: $y = \pm \frac{3}{2} x$

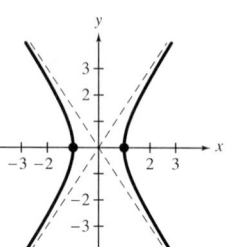

44. Vertices: $\left(0, \pm \frac{1}{2}\right)$
Asymptotes: $y = \pm \frac{1}{5} x$

45. Vertices: $(\pm 4, 0)$
Asymptotes: $y = \pm \frac{1}{2} x$

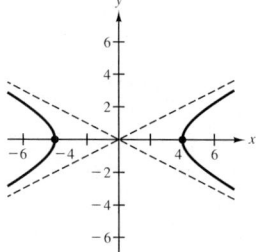

46. Vertices: $(0, \pm 3)$
Asymptotes: $y = \pm \frac{3}{2} x$

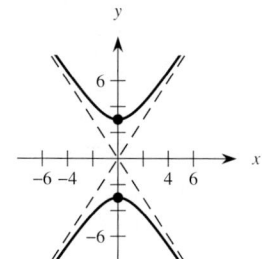

47. $\dfrac{x^2}{16} - \dfrac{y^2}{64} = 1$ **48.** $\dfrac{x^2}{4} - \dfrac{9y^2}{4} = 1$

49. $\dfrac{y^2}{16} - \dfrac{x^2}{64} = 1$ **50.** $\dfrac{y^2}{4} - \dfrac{9x^2}{4} = 1$

51. $\dfrac{x^2}{81} - \dfrac{y^2}{36} = 1$ **52.** $\dfrac{y^2}{25} - \dfrac{x^2}{25} = 1$

53. Parabola   **54.** Line   **55.** Ellipse
56. Circle   **57.** Hyperbola   **58.** Parabola
59. Circle   **60.** Circle   **61.** Line
62. Line   **63.** Parabola   **64.** Hyperbola
65. $x^2 + y^2 = 4500^2 = 20{,}250{,}000$
66. $x^2 + y^2 = 144$, 10.4 ft
67. Circle: $x^2 + y^2 = 25^2 = 625$
Height: $2y = 2\sqrt{625 - x^2}$
Area: $(2x)(2y) = 4x\sqrt{625 - x^2}$
68. $\sqrt{475} \approx 21.8$ ft   **69.** $\sqrt{304} \approx 17.4$ ft
70. $\dfrac{x^2}{16} + \dfrac{16y^2}{225} = 1$ or $\dfrac{16x^2}{225} + \dfrac{y^2}{16} = 1$

71.

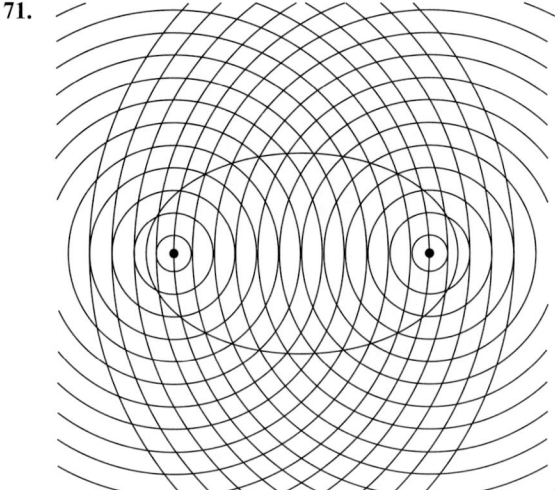

72.

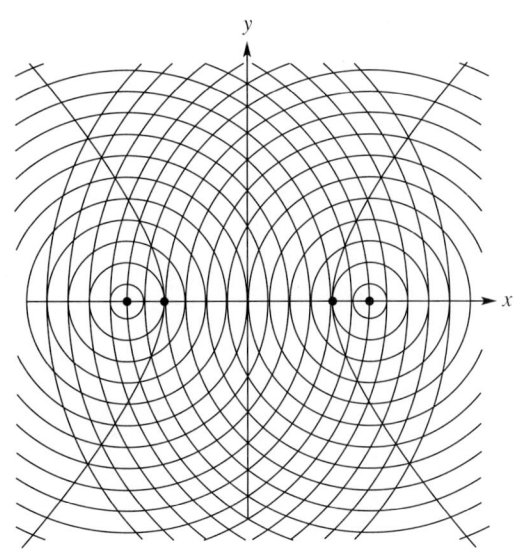

SECTION 8.6 *(page 660)*

Warm-Up *(page 660)*

1.

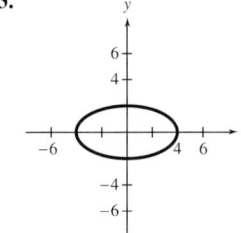

2.

3.

4.

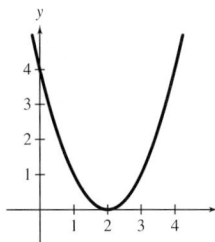

5.

6.

7. $\left(2, \frac{1}{2}\right)$ **8.** $(4, -2)$

9. $(1, -4, 3)$ **10.** $(4, 0, 6)$

1. $(-2, 4)$, $(1, 1)$ **2.** $(5, 0)$, $(3, 4)$
3. No real solution   **4.** $(3, 1)$
5. $(0, 3)$, $(-2, 0)$ **6.** No real solution

7. $(0, 0)$, $(1, 1)$, $(-1, -1)$ **8.** $(2, 4)$, $(-1, 1)$

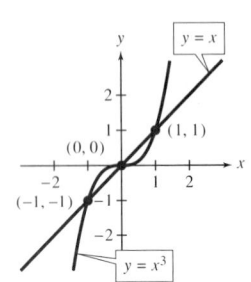

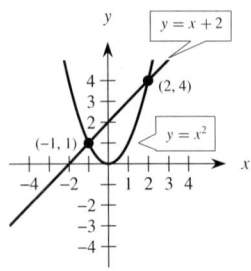

9. $(0, 0)$, $(2, 4)$ **10.** $(2, 4)$, $(-1, 7)$

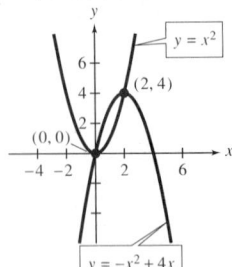

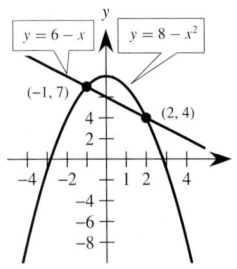
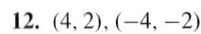

11. $(4, 2)$, $(9, 3)$ **12.** $(4, 2)$, $(-4, -2)$

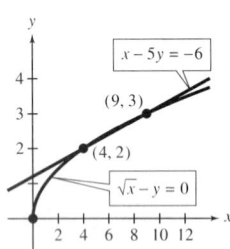

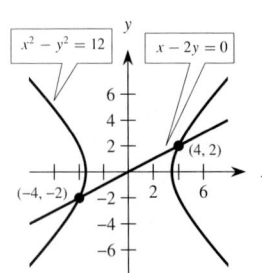

13. $(-3, 18)$, $(2, 8)$ **14.** $(-1, 5)$, $(-2, 20)$
15. $(2, 5)$, $(-3, 0)$ **16.** $(4, 2)$, $(1, -1)$
17. No real solution   **18.** No real solution
19. $(0, 5)$, $(-4, -3)$ **20.** $(-5, 12)$, $(12, -5)$
21. $(0, 2)$, $(3, 1)$ **22.** $(-1, -1)$, $(0, 0)$, $(1, 1)$
23. $(0, 4)$, $(3, 0)$ **24.** $(0, 0)$, $(2, 8)$, $(-2, 8)$
25. No real solution   **26.** $(5, 3)$, $(4, 0)$
27. $(10, 30)$, $(20, 15)$ **28.** $(10, 15)$, $(30, 5)$

29. $(\pm\sqrt{3}, -1)$ **30.** $\left(1, \pm 2\right)$, $\left(-\frac{1}{2}, \pm\frac{\sqrt{22}}{2}\right)$

31. $(2, \pm 2\sqrt{3})$, $(-1, \pm 3)$ **32.** $(\pm\sqrt{11}, -2)$, $(\pm 2\sqrt{2}, 1)$
33. $(\pm 2, \pm\sqrt{3})$ **34.** $(\pm 3, \pm 4)$

35. $\left(\pm\frac{2\sqrt{5}}{5}, \pm\frac{2\sqrt{5}}{5}\right)$ **36.** $\left(\pm\frac{2\sqrt{3}}{3}, \pm\frac{\sqrt{3}}{3}\right)$

37. $(0, 0)$, $(1, 1)$ **38.** $(-5.4, 5.9)$, $(7.3, 3.3)$

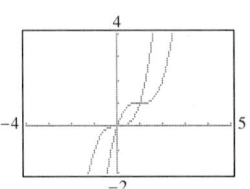

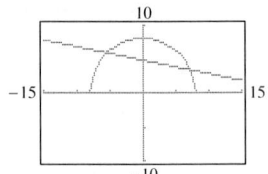

39. $\left(-\frac{3}{5}, -\frac{4}{5}\right)$ to $\left(\frac{4}{5}, -\frac{3}{5}\right)$
40. $(\pm 10, 0)$, $(6, 8)$ Yes, it is a right triangle.
41. Late 1987

USING A GRAPHING CALCULATOR   *(page 664)*

1.
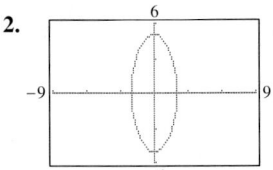

2.

3.
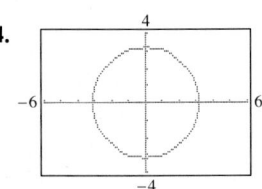

4.

5.
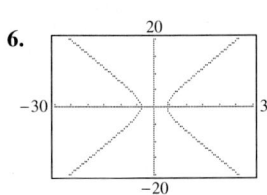

6.

7. $(-3, -1)$, $(1, -3)$ **8.** $(3, -1)$, $\left(-\frac{3}{15}, \frac{29}{13}\right)$
9. $(0, 0)$, $(2, 1)$
10. Asymptotes: $y = -\frac{4}{5}x$, $y = \frac{4}{5}x$

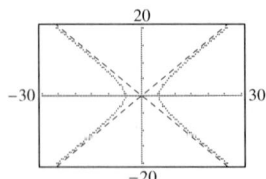

CHAPTER 8 REVIEW EXERCISES *(page 665)*

1. **(a)** 0   **(b)** 5   **(c)** 12   **(d)** −1
2. **(a)** $-\frac{3}{4}$ **(b)** 3   **(c)** $\frac{1}{27}$ **(d)** 2
3. **(a)** $-\dfrac{1}{3}$ **(b)** 83   **(c)** $-\dfrac{1}{48}$ **(d)** $-\dfrac{\sqrt{2}}{6}$
4. **(a)** 5   **(b)** 0   **(c)** 30   **(d)** 1
5. **(a)** $-\frac{8}{3}$ **(b)** −2   **(c)** 1   **(d)** $\frac{1}{10}$
6. **(a)** 0   **(b)** $-\frac{1}{3}$ **(c)** $-\frac{4}{7}$ **(d)** −3
7. **(a)** $x^2 + 2$ **(b)** $(x + 2)^2$ **(c)** 6   **(d)** 1
8. **(a)** $\sqrt[3]{x + 2}$ **(b)** $\sqrt[3]{x} + 2$ **(c)** 2   **(d)** 6
9. **(a)** $|x|$ **(b)** x **(c)** 5   **(d)** −1
10. **(a)** x **(b)** x **(c)** 1   **(d)** $\frac{1}{5}$
11. $f \circ g$: $[2, \infty)$, $g \circ f$: $[4, \infty)$
12. $f \circ g$: $(-\infty, -2) \cup (-2, 2) \cup (2, \infty)$
    $g \circ f$: $-\infty < x < 4, \ 4 < x < \infty$
13. No   **14.** Yes   **15.** Yes   **16.** No
17. $f^{-1}(x) = 4x$ **18.** $f^{-1}(x) = \dfrac{x + 3}{2}$
19. $h^{-1}(x) = x^2, \ x \geq 0$ **20.** $g^{-1}(x) = \sqrt{x - 2}$
21. Because it is possible to find a horizontal line that intersects the graph of f at more than one point, the function does not have an inverse.
22. $h^{-1}(t) = t$ **23.** $k = 6, \ y = 6\sqrt[3]{x}$
24. $k = 27, \ r = \dfrac{27}{s}$ **25.** $k = \dfrac{1}{18}, \ T = \dfrac{1}{18}rs^2$
26. $k = 750, \ D = \dfrac{750x^3}{y}$
27. **(a)** $k = \frac{1}{8}$ **(b)** 1953.125 kW
28. **(a)** 8 in.   **(b)** 62.5 lb
29. 150 lb   **30.** d will increase by a factor of 4
31. 945 units   **32.** 158 lb
33. Falls to the left; Falls to the right
34. Falls to the left; Rises to the right
35. Rises to the left; Rises to the right
36. Rises to the left; Falls to the right

37. (2, 0), (0, 8)

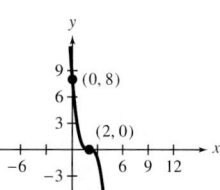

38. (−1, 0), (0, 1)

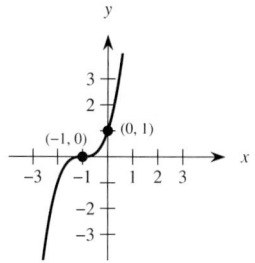

39. (0, 0), (−1, 0), (2, 0)

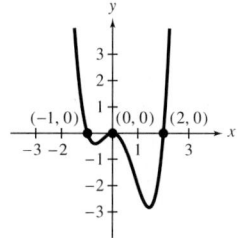

40. (0, 0), (±2, 0)

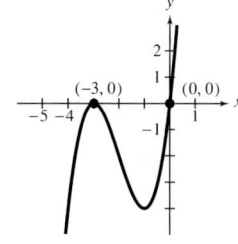

41. (0, 0), (−3, 0)

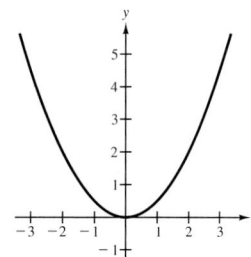

42. (0, 0), (±2, 0)

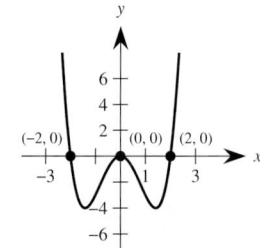

43. Parabola

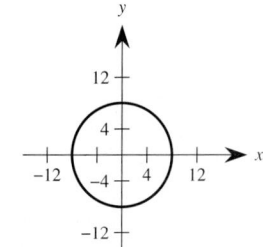

44. Circle

45. Hyperbola

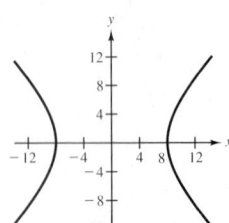

46. Ellipse

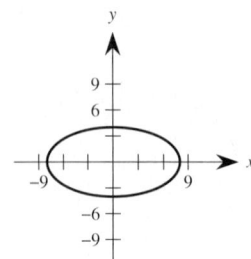

47. Ellipse

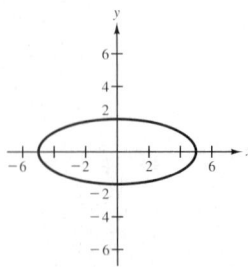

48. Hyperbola

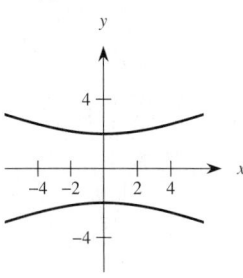

49. Circle

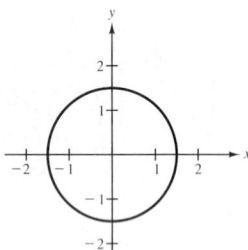

50. Ellipse

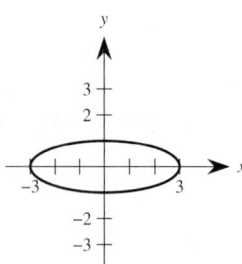

51. $\dfrac{x^2}{36} - \dfrac{y^2}{4} = 1$ **52.** $x^2 + y^2 = 100$

53. $\dfrac{x^2}{4} + \dfrac{y^2}{25} = 1$ **54.** $\dfrac{x^2}{100} + \dfrac{y^2}{36} = 1$

55. $x^2 + y^2 = 400$ **56.** $\dfrac{y^2}{16} - \dfrac{x^2}{4} = 1$

57. $x^2 + y^2 = 25,000,000$

58. $\dfrac{x^2}{20,250,000} + \dfrac{y^2}{25,000,000} = 1$

59. $(-1, 5), (-2, 20)$ **60.** No real solution

61. $(-1, 0), (0, -1)$

62. $(-5\sqrt{2}, 5\sqrt{2}), (5\sqrt{2}, -5\sqrt{2})$

63. $(0, 2), \left(-\dfrac{16}{5}, -\dfrac{6}{5}\right)$ **64.** $(0, -5), (-8, 3)$

65. $(\pm 5, 0)$ **66.** $(0, \pm 4)$

CHAPTER 8 TEST   *(page 668)*

1. -13 **2.** -30 **3.** $\frac{1}{3}$ **4.** 12
5. $[-5, 5]$ **6.** $f^{-1}(x) = 2x + 2$

7. $S = \dfrac{kx^2}{y}$ **8.** $v = \dfrac{1}{4}\sqrt{u}$ **9.** 240 m^2

10. Rises to the left; Falls to the right
11. $(3, 0), (0, 9)$

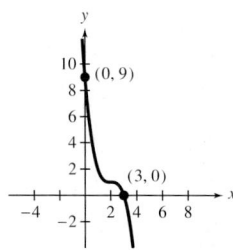

12. $x^2 + y^2 = 25$ **13.** $\dfrac{x^2}{9} + \dfrac{y^2}{100} = 1$

14. $\dfrac{x^2}{9} - \dfrac{y^2}{\frac{9}{4}} = 1$

15. (a)   **(b)**

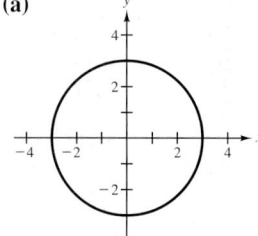   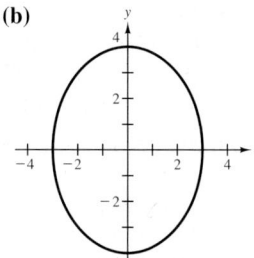

(c)   **(d)**

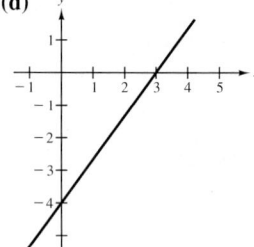

16. (a) $(2, -2), (6, -18)$ **(b)** $(6, 8), (8, 6)$

Chapter 9

Chapter Opener   *(page 669)*

$V_4 \approx 1.277$ $V_5 \approx 1.357$ Yes

SECTION 9.1   *(page 676)*

Warm-Up   *(page 676)*

1. $\frac{27}{64}$ **2.** 900   **3.** $\frac{1}{4}$ **4.** $\frac{1}{81}$ **5.** 5

6. $\frac{1}{9}$ **7.** 1   **8.** y^5 **9.** $13x$ **10.** $-\dfrac{8}{y^3}$

1. $\log_7 49 = 2$ **2.** $\log_5 625 = 4$
3. $\log_3 \frac{1}{9} = -2$ **4.** $\log_6 \frac{1}{216} = -3$
5. $\log_8 4 = \frac{2}{3}$ **6.** $\log_{81} 27 = \frac{3}{4}$
7. $5^2 = 25$ **8.** $10^4 = 10,000$ **9.** $4^{-2} = \frac{1}{16}$
10. $3^{-5} = \frac{1}{243}$ **11.** $36^{1/2} = 6$ **12.** $32^{2/5} = 4$
13. 3   **14.** 3   **15.** -2 **16.** -2 **17.** 0
18. There is no power to which 5 can be raised to obtain -6.
19. There is no power to which 2 can be raised to obtain -3.
20. 0   **21.** 1   **22.** 1   **23.** 3
24. -2 **25.** $\frac{1}{2}$ **26.** $\frac{1}{2}$ **27.** $\frac{3}{4}$
28. $\frac{3}{2}$ **29.** 4   **30.** 3   **31.** -2
32. There is no power to which 2 can be raised to obtain 0.
33. 2   **34.** -4 **35.** 0   **36.** 0   **37.** 1
38. -3 **39.** 1   **40.** 6   **41.** 1.4914
42. 3.7251   **43.** -0.0706 **44.** -0.4622
45. 0.7335   **46.** -0.0625 **47.** 3.2189
48. 1.8825   **49.** -0.2877 **50.** -0.3119
51. 0.0757   **52.** 0.0083
53.

r	0.07	0.08	0.09	0.10	0.11	0.12
t	9.9	8.7	7.7	6.9	6.3	5.8

54. 120 decibels   **55.** 53.4 in.
56. 282.8 mph

SECTION 9.2   *(page 684)*

Warm-Up   *(page 683)*

1. x^{20} **2.** $\dfrac{x^2}{16}$ **3.** $\dfrac{3}{2y^5}$ **4.** $\dfrac{4y^2}{9x^4}$
5. $\frac{1}{6}x^3 y$ **6.** 1   **7.** $4^3 = 64$ **8.** $3^{-4} = \frac{1}{81}$
9. $e^{2x} = e^{2x}$ **10.** $e^{1.6094\cdots} = 5$

1. 2   **2.** 4   **3.** 1
4. 1   **5.** 0   **6.** 0   **7.** -3
8. There is no power to which 3 can be raised to obtain -3.
9. 0.862   **10.** 1.293   **11.** 1.114   **12.** 1.976
13. 0.252   **14.** 0.935   **15.** 0.216   **16.** 0.228
17. 2.407   **18.** 1.419   **19.** 0   **20.** 2.586
21. $\log_3 11 + \log_3 x$ **22.** $\ln 5 + \ln x$ **23.** $3 \ln y$
24. $2 \log_7 x$ **25.** $\log_2 z - \log_2 17$
26. $\log_{10} 7 - \log_{10} y$ **27.** $-2 \log_5 x$ **28.** $\frac{1}{2} \log_2 s$
29. $\frac{1}{3} \log_3 (x+1)$ **30.** $-\frac{1}{2} \log_4 t$
31. $\ln 3 + 2 \ln x + \ln y$ **32.** $\ln y + 2 \ln(y-1)$
33. $2 \log_2 x - \log_2(x-3)$ **34.** $\frac{1}{2}(\log_5 x - \log_5 y)$
35. $\frac{1}{3}[\ln x + \ln(x+5)]$ **36.** $2[\ln 3 + \ln x + \ln(x-5)]$
37. $\log_2 3x$ **38.** $\log_5 6xy$ **39.** $\log_{10} \dfrac{4}{x}$
40. $\ln \dfrac{10x}{z}$ **41.** $\ln b^4,\ b > 0$ **42.** $\log_3 z^{10},\ z > 0$
43. $\log_5 (2x)^{-2},\ x > 0$ **44.** $\ln(x+3)^{-5}$
45. $\ln \sqrt[3]{2x+1}$ **46.** $\log_3 \dfrac{1}{\sqrt{5y}}$ **47.** $\log_3 2\sqrt{y}$
48. $\ln \dfrac{6}{z^3}$ **49.** $\ln \dfrac{x^2 y^3}{z},\ x > 0$ **50.** $\ln \dfrac{3^4}{x^2 y},\ x > 0$
51. $\ln(xy)^4,\ x > 0,\ y > 0$ **52.** $\ln \left(\dfrac{x}{x+1} \right)^2,\ x > 0$
53. True   **54.** True   **55.** False
56. False   **57.** False   **58.** True
59. $1 - \log_4 x$ **60.** $2 + \log_3 4$
61. $1 + \frac{1}{2} \log_5 2$ **62.** $\frac{1}{2} + \frac{1}{2} \log_2 11$
63. $2 + \ln 3$ **64.** $\ln 6 - 5$
65. $\ln 1 = 0$, $\ln 4 \approx 1.3862$, $\ln 6 \approx 1.7917$,
$\ln 8 \approx 2.0793$, $\ln 9 \approx 2.1972$, $\ln 10 \approx 2.3025$,
$\ln 12 \approx 2.4848$, $\ln 15 \approx 2.7080$, $\ln 16 \approx 2.7724$,
$\ln 18 \approx 2.8903$, $\ln 20 \approx 2.9956$
$\ln 7$, $\ln 11$, $\ln 13$, $\ln 14$, $\ln 17$, and $\ln 19$ cannot be
approximated using $\ln 2$, $\ln 3$, and $\ln 5$.

66. $\dfrac{\ln 3}{\ln 5} \neq \ln \dfrac{3}{5} = \ln 3 - \ln 5$

$0.6826 \neq -0.5108 = -0.5108$

67. $B = 10(\log_{10} I + 16)$, 60 decibels

68. $f(t) = 80 - 12\log_{10}(t + 1)$

74.27, 68.55

69. $E = 1.4\log_{10}\dfrac{C_2}{C_1}$ **70.** 0.4214

SECTION 9.3 *(page 694)*

Warm-Up *(page 693)*

1. x^8 **2.** 1 **3.** z^3 **4.** $\frac{3}{4}(x-3)^4$

5. $x^8 y^4$ **6.** $\dfrac{x^5}{32y^5}$ **7.** $\dfrac{x^4}{y^3}$ **8.** $\dfrac{1}{a^8}$

9. b^3 **10.** $2x$

---

1. (a) $\frac{1}{9}$ (b) 1 (c) 3

2. (a) $\frac{1}{5}$ (b) 5 (c) 125

3. (a) 9 (b) 1 (c) $\frac{1}{3}$

4. (a) 5 (b) $\frac{1}{5}$ (c) 0.062

5. (a) 500 (b) 250 (c) 56.657

6. (a) 1200 (b) 533.333 (c) 237.037

7. (a) 1000 (b) 1628.895 (c) 2653.298

8. (a) 7875.661 (b) 3029.948 (c) 918.058

9. (a) 0.368 (b) 1 (c) 1.649

10. (a) 543.656 (b) 1477.811 (c) 10,919.630

11. (a) 73.891 (b) 1.353 (c) 0.183

12. (a) 50 (b) 62.246 (c) 73.106

13. b **14.** a **15.** e **16.** d **17.** f **18.** c

19. 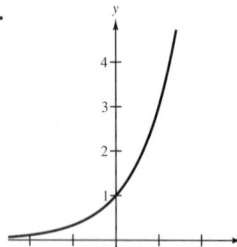 **20.**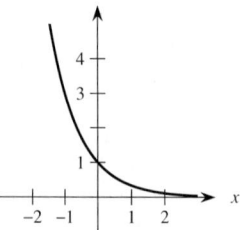

21. 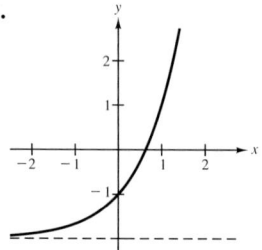 **22.**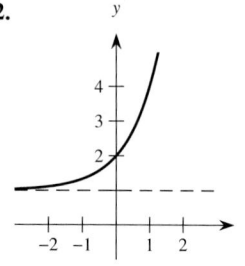

23. 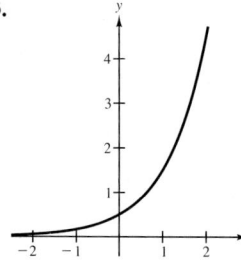 **24.**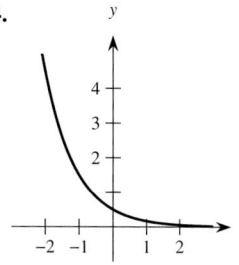

25. 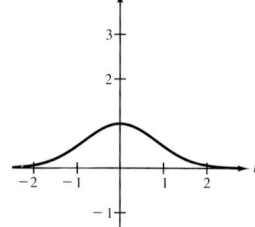 **26.**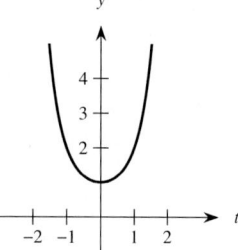

27. 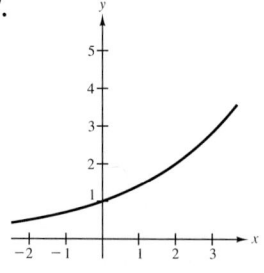 **28.**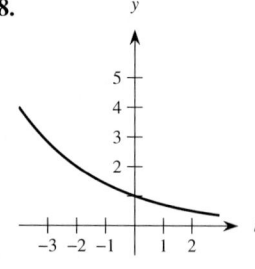

29. 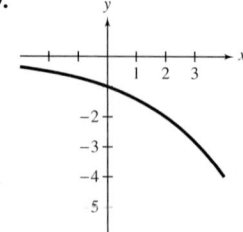 **30.**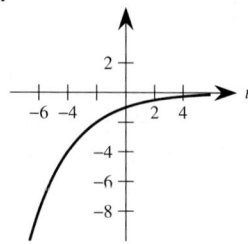

31.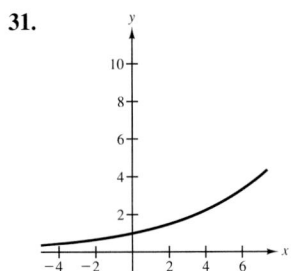

32.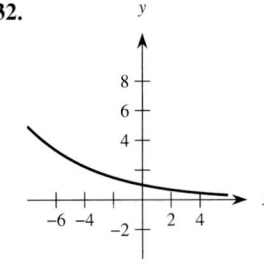

41.

n	1	4	12
A	\$466.10	\$487.54	\$492.68

n	365	Continuous compounding
A	\$495.22	\$495.30

42.

n	1	4	12
A	\$4734.72	\$4870.38	\$4902.71

n	365	Continuous compounding
A	\$4918.66	\$4818.21

33.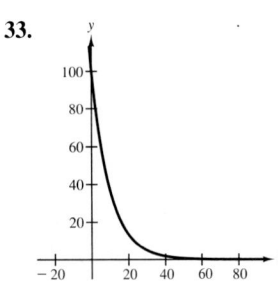

34.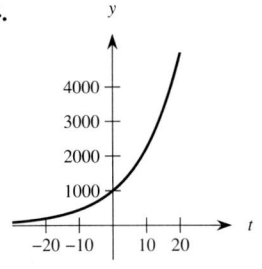

43.

n	1	4	12
A	\$226,296.28	\$259,889.34	\$268,503.32

n	365	Continuous compounding
A	\$272,841.24	\$272,990.75

35. (a) 272.184 million   **(b)** 288.627 million
36. (a) \$80,634.95   **(b)** \$161,629.89
37. \$9000

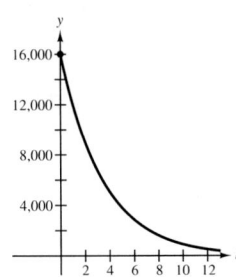

44.

n	1	4	12
A	\$1717.35	\$1723.32	\$1724.71

n	365	Continuous compounding
A	\$1725.39	\$1725.41

38.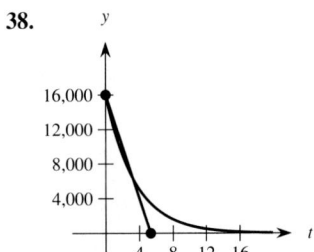

45.

n	1	4	12
A	\$152,203.13	\$167,212.90	\$170,948.62

n	365	Continuous compounding
A	\$172,813.72	\$172,877.82

After 2 years, the person selling the car would prefer the depreciation model $V(t) = 16,000 - 3000t$. After 4 years, the person selling the car would prefer the depreciation model $V(t) = 16,000 \left(\frac{3}{4}\right)^t$.

39. \$32.50   **40. (a)** \$22.04   **(b)** \$20.13

46.

n	1	4	12
A	$18,760.65	$20,993.96	$21,551.27

n	365	Continuous compounding
A	$21,829.69	$21,839.26

47.

n	1	4	12
P	$2541.75	$2498.00	$2487.98

n	365	Continuous compounding
P	$2483.09	$2482.93

48.

n	1	4	12
A	$17,843.09	$16,862.99	$16,641.28

n	365	Continuous compounding
A	$16,533.56	$16,529.89

49.

n	1	4	12
P	$18,429.30	$15,830.43	$15,272.04

n	365	Continuous compounding
P	$15,004.64	$14,995.58

50.

n	1	4	12
A	$2163.33	$2154.76	$2152.77

n	365	Continuous compounding
A	$2151.80	$2151.77

51.

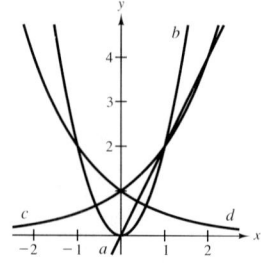

52. 1,073,741,824 pennies or $10,737,418.24

53.

t	0	25	50	75
h	2000 ft	1450 ft	950 ft	450 ft

Ground level: 97.5 sec

54.

t	0	50	100
h	3000 ft	1840 ft	740 ft

Ground level: 134 sec

55. 2.155 million

56.

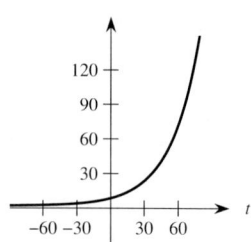

13.5 million

57.

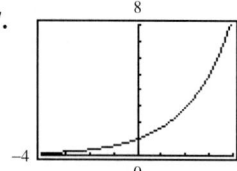

58.

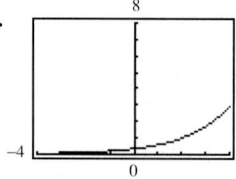

59.

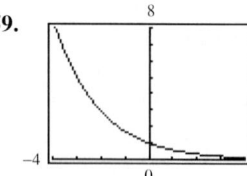

60.

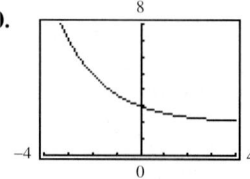

61. 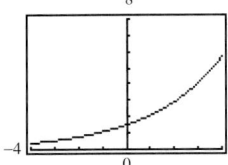 **62.**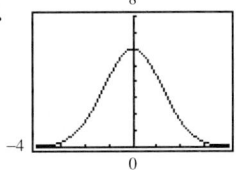

63.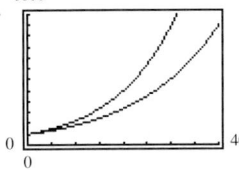

CHAPTER 9 MID-CHAPTER QUIZ   *(page 698)*

1. (a) $\log_5 125 = 3$ **(b)** $\ln 1 = 0$

2. (a) $16^{1/2} = 4$ **(b)** $5^{-2} = \frac{1}{25}$

3. 2   **4.** -2 **5. (a)** 2.7126   **(b)** 7.6578

6. $3[\ln x - \ln(x+3)]$ **7.** $\log_3\left[x^2(x+5)^4\right]$, $x > 0$

8. $3 - \log_5 x$ **9.** 864.665

10. (a) 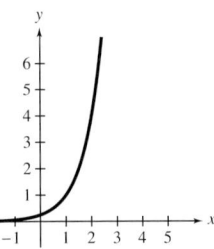   **(b)**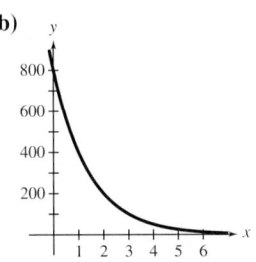

11. (a) \$2099.59   **(b)** \$2102.53   **12.** 4.6 lb/in.^{2}

SECTION 9.4   *(page 705)*

Warm-Up *(page 704)*

1. 4   **2.** 3   **3.** -3

4. 3   **5.** 0.368   **6.** 1.649

7. g is a vertical translation of f 4 units downward.

8. g is a horizontal translation of f 4 units to the right.

9. g is a reflection of f in the x-axis.

10. g is a reflection of f in the y-axis.

1. f is the inverse of g.   **2.** f is the inverse of g.

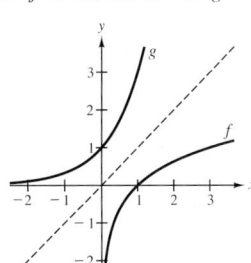   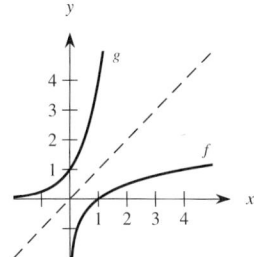

3. f is the inverse of g.   **4.** f is the inverse of g.

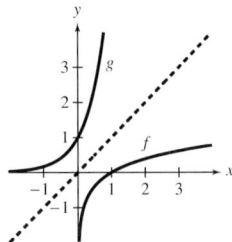   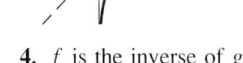

5. e   **6.** b   **7.** d   **8.** c   **9.** a   **10.** f

11. 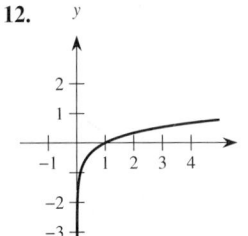   **12.**

13. 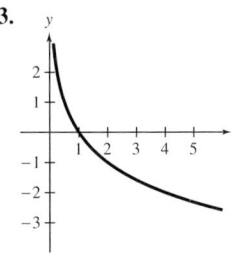   **14.**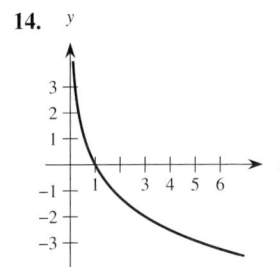

15.
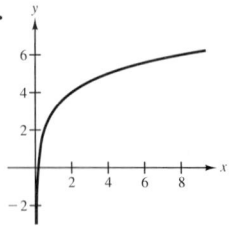

16.
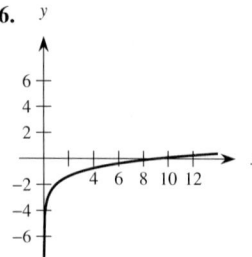

31.
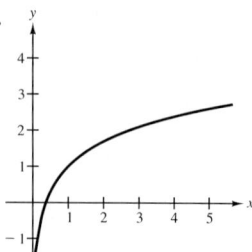

32.
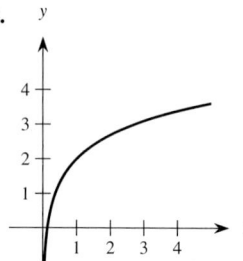

17.
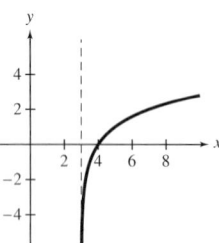

18.
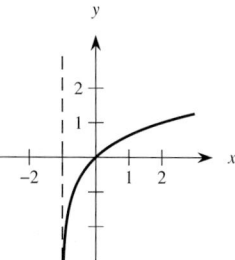

33.
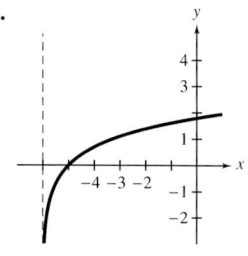

34.
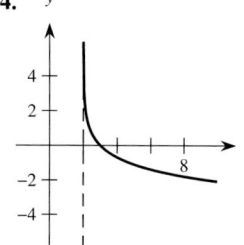

19.
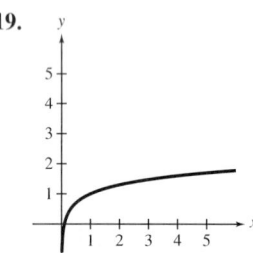

20.
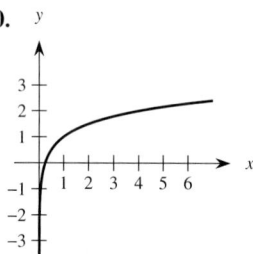

35.
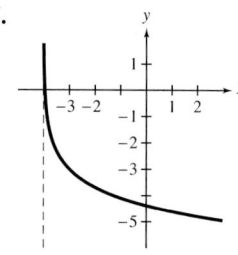

36.
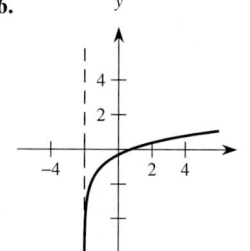

21. b   **22.** e   **23.** d   **24.** c   **25.** f   **26.** a

27.
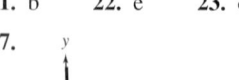

28.
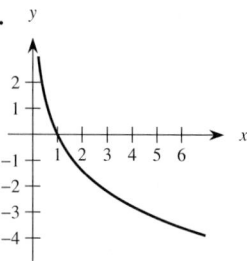

37.
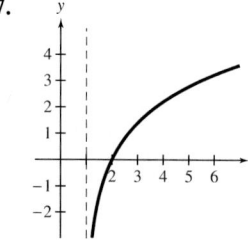

38.
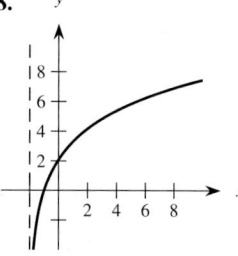

39. 2.3481   **40.** 3.8737   **41.** 1.7712
42. 0.7124   **43.** −0.4739   **44.** −0.2056
45. 2.6332   **46.** 1.6117   **47.** −2
48. 3.8227   **49.** 1.3481   **50.** 1.9360
51. $(0, \infty)$ **52.** $f^{-1}(x) = 10^x$
53. $3 \le f(x) \le 4$ or $[3, 4]$ **54.** $(0, 1)$
55. A factor of 10   **56.** $\dfrac{a}{b} = \dfrac{1000b}{b} - 1000$

29.
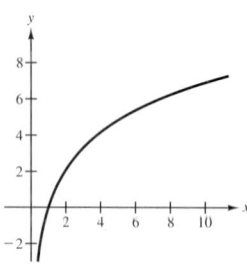

30.
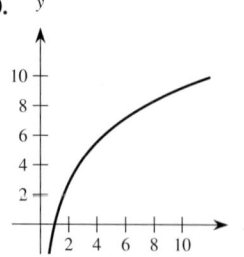

57.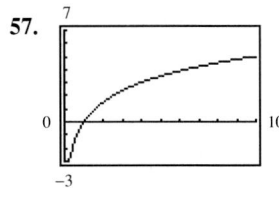

58.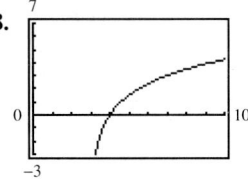

59.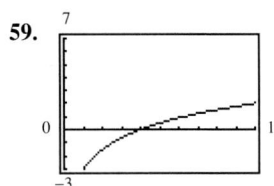

60.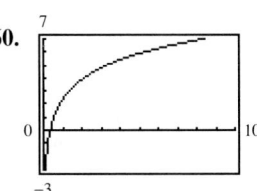

61.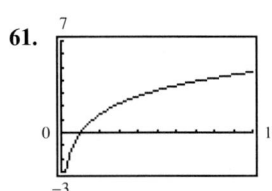

62.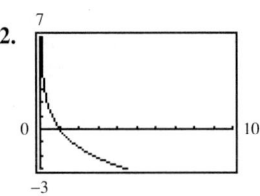

5.

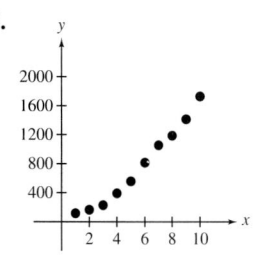

Model: $y = 100.487857(1.360675)^x$;
1991 sales: \$2974.5 million

SECTION 9.5 *(page 717)*

Warm-Up *(page 717)*

1. $\frac{5}{2}$ **2.** $-\frac{7}{2}$ **3.** 10, -2 **4.** 2, 3

5. $\frac{28}{\ln 5}$ **6.** 0 **7.** 3 **8.** 6

9. 0 **10.** 4

---

1. 5 **2.** 3 **3.** 8 **4.** -3 **5.** 3 **6.** 2
7. -3 **8.** -4 **9.** 4 **10.** 18 **11.** $\frac{22}{5}$
12. 10 **13.** 81 **14.** 125 **15.** 500,000
16. 11 **17.** 5.49 **18.** 3.33 **19.** 0.86
20. 3.94 **21.** 3.28 **22.** 2.87 **23.** 2.49
24. 12.28 **25.** 0.39 **26.** 4.01 **27.** 6.80
28. 12.22 **29.** 0.80 **30.** 9.16 **31.** 35.35
32. 7.39 **33.** 21.82 **34.** 61.40 **35.** 8.99
36. -0.35 **37.** 1 **38.** e **39.** 1000
40. 0.01 **41.** 7.91 **42.** 43.42 **43.** 8.17
44. 10.04 **45.** -0.78 **46.** 1.07
47. 31,622,771.6 **48.** 2000 **49.** ±20.09
50. 442,413.39 **51.** -2, 5 **52.** 0.62, -1.62
53. 7.70 yr **54.** 6.9% **55.** $10^{-8.5}$
56. 3.64 months
57. 1.0234 g/cm^3; 1.0241 g/cm^3 **58.** 262.5°

USING A GRAPHING CALCULATOR *(page 710)*

1.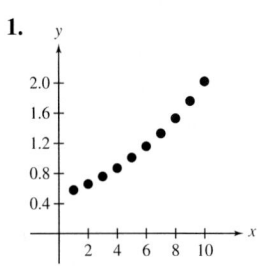

Model:
$y = 0.5013(1.14960^x)$

2.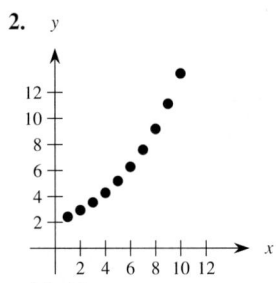

Model:
$y = 2.0005(1.2099)^x$

3.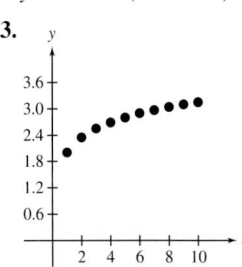

Model:
$y = 2.008 + 0.4993 \ln x$,
$y = 2.008 + 0.49931 \ln x$

4.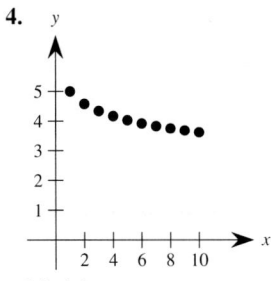

Model:
$y = 4.9985 - 0.6001 \ln x$

SECTION 9.6 *(page 726)*

Warm-Up *(page 726)*

1. $\frac{33}{4}$ **2.** $\frac{3}{5}, -1$ **3.** 41 **4.** $\frac{11}{2}$
5. 18 **6.** 69 **7.** 4.16 **8.** 11.31
9. 139.32 **10.** -1.39

---

1. 7% **2.** 10% **3.** 9%
4. 9.5% **5.** 8% **6.** 7.5%
7. 8.75 years **8.** 13.23 years **9.** 6.60 years
10. 7.64 years **11.** 9.24 years **12.** 11.55 years
13. Continuous **14.** Continuous **15.** Quarterly
16. Daily **17.** 8.33% **18.** 9.97%
19. 7.23% **20.** 8% **21.** 5.39%
22. 9.31% **23.** $1652.99 **24.** $3351.60
25. $626.46 **26.** $1492.79 **27.** $5496.57
28. $184,369.97 **29.** $320,250.81 **30.** $10,444.45
31. Total deposits: $7200.00;
  Total interest: $10,529.42
32. Total deposits: $14,400.00;
  Total interest: $91,143.80
33. $k = \frac{1}{2}\ln\frac{8}{3} \approx 0.4904$ **34.** $k = \frac{1}{5}\ln 3 \approx 0.2197$
35. $k = \frac{1}{3}\ln\frac{1}{2} \approx -0.2310$ **36.** $k = \frac{1}{7}\ln\frac{1}{2} \approx -0.0990$
37. $y = 9.6e^{0.0072t}$; 11.1 million
38. $y = 2.1e^{0.0142t}$; 2.8 million
39. $y = 3.3e^{0.0452t}$; 8.1 million
40. $y = 6.1e^{0.0478t}$; 15.9 million
41. 3.3 g **42.** 7.5 g **43.** 15,700 yr **44.** $9281
45. (a) $S = 10(1 - e^{-0.0575x})$ **(b)** 3300 units
46. (a) 1000 **(b)** 2642 **(c)** 5.9 years
47. 501 times as great **48.** 63 times as great
49. 7.04 **50.** 0.00002
51. 10,000,000 times **52.** 10 **53.** 8.995 billion
54. (a)

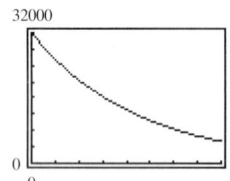

(b) $25,600
(c) 3.1 years

55. (a)

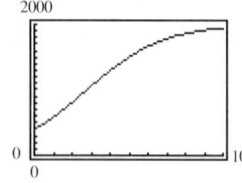

(b) 1298 units
(c) 3.2 years
(d) 2000 units

CHAPTER 9 REVIEW EXERCISES *(page 730)*

1. $\log_4 64 = 3$ **2.** $\log_{25} 125 = \frac{3}{2}$ **3.** 3
4. $\frac{1}{2}$ **5.** -2 **6.** -2 **7.** 7 **8.** -1
9. 0 **10.** -3 **11.** $\log_4 6 + 4\log_4 x$
12. $\log_{10} 2 - 3\log_{10} x$ **13.** $\frac{1}{5}\log_5(x + 2)$
14. $\frac{1}{3}(\ln x - \ln 5)$ **15.** $\ln(x + 2) - \ln(x - 2)$
16. $\ln x + 2\ln(x - 3)$
17. $\log_4 \dfrac{x}{10}$ **18.** $\log_2 y^5$
19. $\log_8 32x^3$ **20.** $4 + \ln x^8$, $x > 0$
21. $\ln \dfrac{9}{4x^2}$, $x > 0$ **22.** $\ln\left(\dfrac{1}{3y}\right)^{2/3}$
23. True **24.** False **25.** False
26. True **27.** True **28.** True
29. 1.7959 **30.** 0.55665 **31.** -0.4307
32. -0.2519 **33.** 1.0293 **34.** 3.1133
35. (a) $\frac{1}{8}$ **(b)** 2 **(c)** 4
36. (a) 4 **(b)** 1 **(c)** $\frac{1}{4}$
37. (a) 2.718 **(b)** 0.351 **(c)** 0.135
38. (a) 0 **(b)** -0.492 **(c)** -0.882
39. (a) 0 **(b)** 3 **(c)** -0.631
40. (a) -2 **(b)** -1 **(c)** 1.477
41. (a) 1 **(b)** -1.099 **(c)** 2.303
42. (a) 2 **(b)** 0.233 **(c)** 7.090
43. (a) -6 **(b)** 0 **(c)** 22.5
44. (a) 1 **(b)** 3 **(c)** 1.189
45. d **46.** f **47.** a **48.** b **49.** c **50.** e
51.

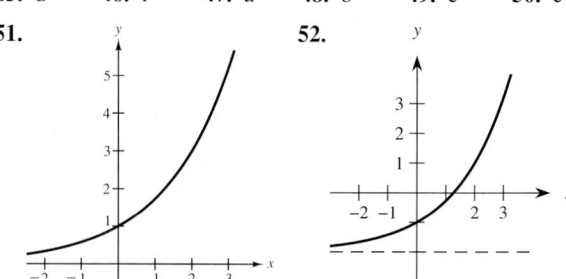

52.

53.

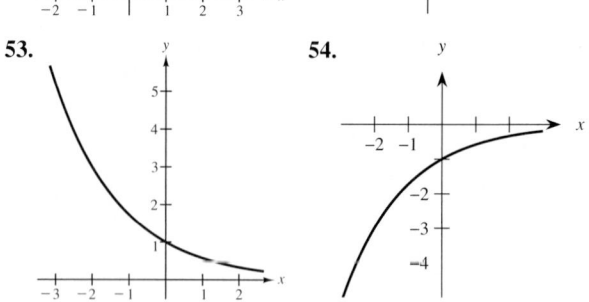

54.

55.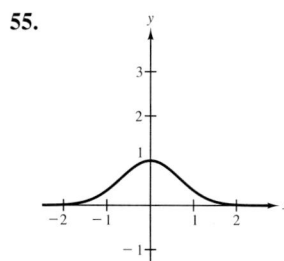

56.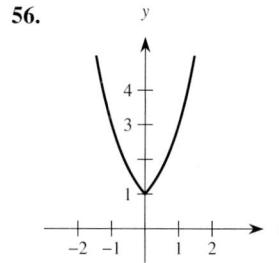

57.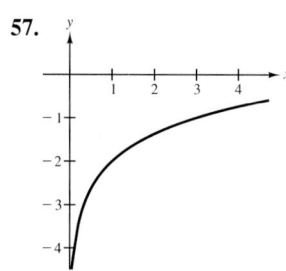

58.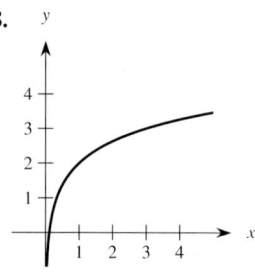

59.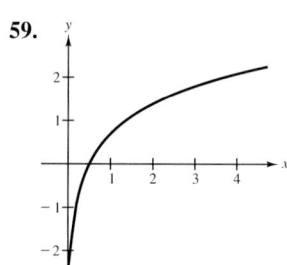

60.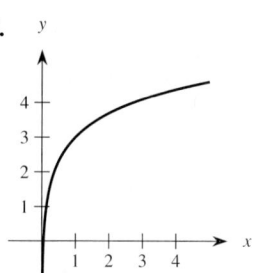

61.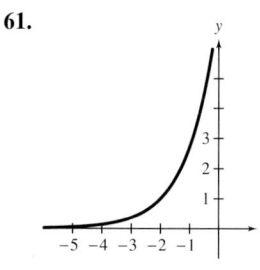

62.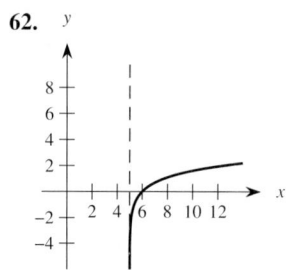

63. 1.585   **64.** −2.322   **65.** 2.132
66. −1.159   **67.** 6   **68.** 6   **69.** 1
70. 50   **71.** 243   **72.** 35   **73.** 5.66
74. 3.32   **75.** 6.23   **76.** 2.68   **77.** 32.99
78. −0.65   **79.** 15.81   **80.** 64   **81.** 1408.10
82. 0.61   **83.** 0.32   **84.** 452.99

85.

n	1	4	12
A	\$3806.13	\$4009.59	\$4058.25

n	365	Continuous compounding
A	\$4082.26	\$4083.08

86.

n	1	4	12
A	\$2154.40	\$2286.27	\$2317.63

n	365	Continuous compounding
A	\$2333.08	\$2333.61

87.

n	1	4	12
A	\$67,275.00	\$72,095.68	\$73,280.74

n	365	Continuous compounding
A	\$73,870.32	\$73,890.56

88.

n	1	4	12
A	\$2700	\$2706.08	\$2707.50

n	365	Continuous compounding
A	\$2708.19	\$2708.22

89.

n	1	4	12
P	\$2301.55	\$2103.50	\$2059.87

n	365	Continuous compounding
P	\$2038.82	\$2038.11

90.

n	1	4	12
A	\$943.40	\$942.18	\$941.91

n	365	Continuous compounding
A	\$941.77	\$941.76

91. 4.6 years   **92.** 8.7 years
93. 150 units   **94.** 0.000316 W/m^2
95.

t	0	1	5	10
$p(t)$	200	227	346	472

t	20	50	75
$p(t)$	579	600	600

The limiting size of the herd is 600.

96. (a) 301   **(b)** 3.85 years

CHAPTER 9 TEST   *(page 734)*

1. $5^3 = 125$ **2.** $\log_4 \frac{1}{16} = -2$ **3.** $\frac{1}{3}$
4. $\log_4 5 + 2\log_4 x - \frac{1}{2}\log_4 y$
5. $\ln \frac{x}{y^4}$, $y > 0$ **6.** $3 + \log_5 6$
7.

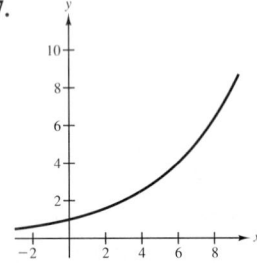

8. (a) \$8012.78   **(b)** \$8110.40

9. $g = f^{-1}$

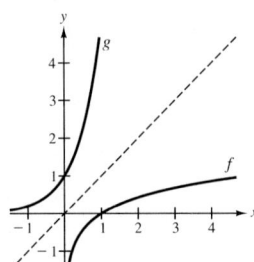

10. (a) 54   **(b)** 24   **(c)** 81   **(d)** 44.09
11. 64   **12.** 0.97   **13.** 13.73   **14.** 15.52
15. \$10,806.08   **16.** 7%   **17.** \$8469
18. (a) 600   **(b)** 1141   **(c)** 4.4 years

Chapter 10

Chapter Opener   *(page 735)*

\$320.71 after 20 years; \$1842.02 after 50 years

SECTION 10.1   *(page 743)*

Warm-Up *(page 743)*

1. $\frac{1}{7}$ **2.** 24   **3.** $\frac{3}{4}$ **4.** $\frac{13}{12}$

5. n **6.** $n - 1$ **7.** $\dfrac{2n - 3}{n - 5}$

8. $\dfrac{n + 1}{n}$ **9.** $\dfrac{n + 2}{n^2}$ **10.** $\dfrac{2n}{n^2 - 1}$

1. 2, 4, 6, 8, 10   **2.** 3, 6, 9, 12, 15
3. -2, 4, -6, 8, -10 **4.** 3, -6, 9, -12, 15
5. $\frac{1}{2}, \frac{1}{4}, \frac{1}{8}, \frac{1}{16}, \frac{1}{32}$ **6.** $\frac{1}{3}, \frac{1}{9}, \frac{1}{27}, \frac{1}{81}, \frac{1}{243}$
7. $\frac{1}{4}, -\frac{1}{8}, \frac{1}{16}, -\frac{1}{32}, \frac{1}{64}$ **8.** $-\frac{1}{3}, \frac{1}{9}, -\frac{1}{27}, \frac{1}{81}, -\frac{1}{243}$
9. $\frac{1}{2}, \frac{1}{3}, \frac{1}{4}, \frac{1}{5}, \frac{1}{6}$ **10.** $1, \frac{3}{5}, \frac{3}{7}, \frac{1}{3}, \frac{3}{11}$
11. $\frac{2}{5}, \frac{1}{2}, \frac{6}{11}, \frac{4}{7}, \frac{10}{17}$ **12.** $\frac{5}{7}, \frac{10}{11}, 1, \frac{20}{19}, \frac{25}{23}$
13. $-1, \frac{1}{4}, -\frac{1}{9}, \frac{1}{16}, -\frac{1}{25}$ **14.** $1, \dfrac{1}{\sqrt{2}}, \dfrac{1}{\sqrt{3}}, \dfrac{1}{2}, \dfrac{1}{\sqrt{5}}$
15. 2, 2, $\frac{4}{3}, \frac{2}{3}, \frac{4}{15}$ **16.** 1, 2, 3, 4, 5

17. $\frac{9}{2}, \frac{19}{4}, \frac{39}{8}, \frac{79}{16}, \frac{159}{32}$ **18.** $\frac{22}{3}, \frac{64}{9}, \frac{190}{27}, \frac{568}{81}, \frac{1702}{243}$

19. 0, 6, −6, 18, −30   **20.** 0, $\frac{1}{2}$, 0, $\frac{1}{8}$, 0

21. 5   **22.** 18   **23.** $\frac{1}{132}$ **24.** $\frac{1}{336}$

25. $\dfrac{1}{n+1}$ **26.** $(n+2)(n+1)$ **27.** $2n$

28. $(2n+2)(2n+1)$ **29.** 63   **30.** 50

31. 35   **32.** 102   **33.** $\frac{47}{60}$ **34.** $\frac{9}{5}$ **35.** −48

36. 100   **37.** $\frac{8}{9}$ **38.** $\frac{50}{21}$ **39.** $\frac{182}{243}$ **40.** $\frac{2059}{64}$

41. 112   **42.** 110   **43.** 852   **44.** $\frac{65}{4}$

45. 6.5793   **46.** 8.5015   **47.** $\displaystyle\sum_{k=1}^{5} k$ **48.** $\displaystyle\sum_{k=1}^{6} (k+7)$

49. $\displaystyle\sum_{k=1}^{5} 2k$ **50.** $\displaystyle\sum_{k=1}^{4} 6(k+3)$ **51.** $\displaystyle\sum_{k=1}^{10} \frac{1}{2k}$

52. $\displaystyle\sum_{k=1}^{50} \frac{3}{1+k}$ **53.** $\displaystyle\sum_{k=1}^{20} \frac{1}{k^2}$ **54.** $\displaystyle\sum_{k=0}^{12} \frac{1}{2^k}$

55. $\displaystyle\sum_{k=0}^{9} \frac{1}{(-3)^k}$ **56.** $\displaystyle\sum_{k=0}^{20} \left(-\frac{2}{3}\right)^k$ **57.** $\displaystyle\sum_{k=1}^{11} \frac{k}{k+1}$

58. $\displaystyle\sum_{k=0}^{9} \frac{2k+2}{3k+4}$ **59.** $\displaystyle\sum_{k=0}^{6} k!$ **60.** $\displaystyle\sum_{k=0}^{6} \frac{1}{k!}$

61. 3.6   **62.** 78.75   **63.** 0.8   **64.** 1.8

65. (a) \$502.92, \$505.85, \$508.80, \$511.77,
\$514.75, \$517.76, \$520.78, \$523.82
(b) \$2019.37

66. \$3796.88

67. $a_5 = 108°$, $a_6 = 120°$

At the point where any two hexagons and a pentagon
meet, the sum of the three angles is
$$a_5 + 2a_6 = 348° < 360°$$
Therefore, there is a gap of 12°.

68. 25.7°, 45°, 60°, 72°, 81.8°

SECTION 10.2   *(page 752)*

Warm-Up   *(page 752)*

1. $\frac{49}{20}$ **2.** $\frac{455}{72}$ **3.** 120   **4.** 40   **5.** 285

6. 1960   **7.** 77   **8.** $\frac{355}{2}$ **9.** 63   **10.** 144

---

1. 3   **2.** 8   **3.** −6   **4.** −400   **5.** $\frac{2}{3}$

6. $\frac{3}{4}$ **7.** Arithmetic, 2   **8.** Arithmetic, 4

9. Arithmetic, $\frac{3}{2}$ **10.** Arithmetic, 8

11. Arithmetic, −16   **12.** Not arithmetic

13. Not arithmetic   **14.** Arithmetic, $\frac{1}{6}$

15. Arithmetic, 0.8   **16.** Not arithmetic

17. Not arithmetic   **18.** Not arithmetic

19. 7, 10, 13, 16, 19   **20.** 1, 6, 11, 16, 21

21. 6, 4, 2, 0, −2   **22.** 90, 80, 70, 60, 50

23. $\frac{3}{2}$, 4, $\frac{13}{2}$, 9, $\frac{23}{2}$ **24.** $\frac{8}{3}$, $\frac{10}{3}$, 4, $\frac{14}{3}$, $\frac{16}{3}$

25. 4, $\frac{15}{4}$, $\frac{7}{2}$, $\frac{13}{4}$, 3   **26.** 36, 40, 44, 48, 52

27. d   **28.** b   **29.** c   **30.** a

31. $a_n = \frac{1}{2}n + \frac{5}{2}$ **32.** $a_n = 1.2n - 2.2$

33. $a_n = -25n + 1025$ **34.** $a_n = -8n + 72$

35. $a_n = 4n + 16$ **36.** $a_n = 6n + 10$

37. $a_n = -10n + 60$ **38.** $a_n = n + 9$

39. 5, 8, 11, 14, 17   **40.** 8, 15, 22, 29, 36

41. 9, 6, 3, 0, −3   **42.** 12, 6, 0, −6, −12

43. −10, −4, 2, 8, 14

44. −20, −24, −28, −32, −36

45. 210   **46.** 1860   **47.** 2700   **48.** 20,100

49. 19,500   **50.** 7900   **51.** 62,625   **52.** 120,200

53. 522   **54.** 1940   **55.** 900   **56.** 15,500

57. 12,200   **58.** 3000   **59.** 243   **60.** 66

61. 2850   **62.** 4455   **63.** 2550   **64.** 10,000

65. \$246,000   **66.** \$116.25   **67.** \$25.43

68. 126   **69.** 632 bales   **70.** 114   **71.** 1024 ft

SECTION 10.3   *(page 762)*

Warm-Up   *(page 762)*

1. $\frac{16}{81}$ **2.** $\frac{27}{125}$ **3.** $\frac{1}{72}$ **4.** 1   **5.** n

6. $n + 1$ **7.** $60n^6$ **8.** $16n^5$

9. $\dfrac{n}{2}$ **10.** $\dfrac{9}{5}n^2$

---

1. 3   **2.** −2   **3.** $-\frac{1}{2}$ **4.** $\frac{1}{3}$ **5.** $-\frac{3}{2}$

6. $\frac{2}{3}$ **7.** e **8.** 1.1   **9.** Not geometric

10. Geometric, 2   **11.** Geometric, $\frac{1}{2}$

12. Not geometric **13.** Geometric, $-\frac{2}{3}$

14. Geometric, -2 **15.** Geometric, 1.02

16. Geometric, 0.2 **17.** 4, 8, 16, 32, 64

18. 3, 12, 48, 192, 768 **19.** 6, 2, $\frac{2}{3}$, $\frac{2}{9}$, $\frac{2}{27}$

20. 4, 2, 1, $\frac{1}{2}$, $\frac{1}{4}$ **21.** 1, $-\frac{1}{2}$, $\frac{1}{4}$, $-\frac{1}{8}$, $\frac{1}{16}$

22. 32, -24, 18, $-\frac{27}{2}$, $\frac{81}{8}$

23. 1000, 1010, 1020.1, 1030.301, 1040.60401

24. 4000, 3960.396, 3921.184, 3882.361, 3843.921

25. $a_n 2(3)^{n-1}$ **26.** $a_n 5(4)^{n-1}$

27. $a_n 2^{n-1}$ **28.** $a_n(-5)^{n-1}$

29. $a_n = 4\left(-\frac{1}{2}\right)^{n-1}$ **30.** $a_n = 9\left(\frac{2}{3}\right)^{n-1}$

31. $a_n = 8\left(\frac{1}{4}\right)^{n-1}$ **32.** $a_n = 18\left(\frac{4}{9}\right)^{n-1}$

33. $\frac{3}{256}$ **34.** $\frac{2187}{2048}$ **35.** $48\sqrt{2}$ **36.** $500(1.06)^{39}$

37. $\frac{81}{64}$ **38.** 531,441 **39.** $\pm\frac{243}{32}$ **40.** ± 64

41. 1023 **42.** 364 **43.** 772.478 **44.** 35.989

45. 2.250 **46.** 6.400 **47.** 5727.500

48. 9981.756 **49.** $-14,762$ **50.** -4095

51. 16.000 **52.** 26.532 **53.** 152.095

54. 32.000 **55.** 1103.568 **56.** 25,905.652

57. (a) $250,000(0.75)^n$ (b) \$59,326.17

58. (a) $500,000(1.01)^n$ (b) 610,095

59. \$3,623,993 **60.** \$4,098,168 **61.** \$19,496.56

62. \$3600.53 **63.** \$105,428.44 **64.** \$455,865.06

65. (a) \$5,368,709.11 (b) \$10,737,418.23

66. (a) \$4,236,400,000 (b) \$12,709,000,000

67. 70.875 in.^{2} **68.** 72.969 in.^{2}

69. 1.476×10^{20}; No

CHAPTER 10 MID-CHAPTER QUIZ (page 765)

1. 25, 20, 15, 10, 5 **2.** 1, $-\frac{1}{4}$, $\frac{1}{16}$, $-\frac{1}{64}$, $\frac{1}{256}$

3. $\sum_{n=1}^{7}(6n-5)$ **4.** $\sum_{n=0}^{4}\left(\frac{4}{3}\right)^n$ **5.** 132

6. $n(n-1)$ **7.** Geometric, $r = 4$

8. Arithmetic, $d = -\frac{1}{2}$ **9.** $a_n = 9n - 6$

10. $a_n = 5(-0.5)^{n-1}$ **11.** 36 **12.** 88

13. 5.864×10^{12} **14.** $\frac{153}{32}$

15. (a) \$10,066.676, \$10,133.78, \$10,201.34,
\$10,269.35, \$10,337.81

(b) \$22,196.40

16. \$2737.50 **17.** \$141,021.55 million

SECTION 10.4 (page 772)

Warm-Up (page 771)

1. $-3x^2 + 6x$ **2.** $x^2 - 9$ **3.** $x^2 - 4x + 4$

4. $4x^2 + 12x + 9$ **5.** $\dfrac{5x^3}{y^2}$ **6.** $-8x^6 y^3$

7. 24 **8.** 8 **9.** 1320 **10.** 45

---

1. 15 **2.** 35 **3.** 252 **4.** 220 **5.** 1

6. 1 **7.** 1225 **8.** 200 **9.** 12,650 **10.** 8568

11. 455 **12.** 126 **13.** 53,130 **14.** 4060

15. 1, 9, 36, 84, 126, 126, 84, 36, 9, 1

16. 1, 10, 45, 120, 210, 252, 210, 120, 45, 10, 1

17. (a) 35 (b) 126 **18.** (a) 15 (b) 84

19. $x^4 + 4x^3 y + 6x^2 y^2 + 4xy^3 + y^4$

20. $u^6 + 6u^5 v + 15u^4 v^2 + 20u^3 v^3 + 15u^2 v^4 + 6uv^5 + v^6$

21. $x^5 + 5x^4 + 10x^3 + 10x^2 + 5x + 1$

22. $x^5 + 10x^4 + 40x^3 + 80x^2 + 80x + 32$

23. $x^6 + 18x^5 + 135x^4 + 540x^3 + 1215x^2 + 1458x + 729$

24. $x^4 + 16x^3 + 96x^2 + 256x + 256$

25. $x^5 - 5x^4 y + 10x^3 y^2 - 10x^2 y^3 + 5xy^4 - y^5$

26. $x^4 - 20x^3 + 150x^2 - 500x + 625$

27. $u^3 - 6u^2 v + 12uv^2 - 8v^3$

28. $32x^5 + 80x^4 y + 80x^3 y^2 + 40x^2 y^3 + 10xy^4 + y^5$

29. $81a^4 + 216a^3 b + 216a^2 b^2 + 96ab^3 + 16b^4$

30. $64u^3 - 144u^2 v + 108uv^2 - 27v^3$

31. $x^8 + 8x^7 y + 28x^6 y^2 + 56x^5 y^3 + 70x^4 y^4$
$\quad + 56x^3 y^5 + 28x^2 y^6 + 8xy^7 + y^8$

32. $r^7 - 7r^6 s + 21r^5 s^2 - 35r^4 s^3 + 35r^3 s^4$
$\quad\quad - 21r^2 s^5 + 7rs^6 - s^7$

33. $x^6 - 12x^5 + 60x^4 - 160x^3$
$\quad + 240x^2 - 192x + 64$

34. $32x^5 + 240x^4 + 720x^3 + 1080x^2 + 810x + 243$

35. 120 **36.** 5940 **37.** -1365

38. 120 **39.** 1760 **40.** $-64,481,508$

41. -4 **42.** $-38, 41i$

43. $2035 + 828i$ **44.** $-64,481,508$

45. $\frac{1}{32} + \frac{5}{32} + \frac{10}{32} + \frac{10}{32} + \frac{5}{32} + \frac{1}{32}$

46. $\frac{16}{81} + \frac{32}{81} + \frac{24}{81} + \frac{8}{81} + \frac{1}{81}$

47. $\frac{1}{256} + \frac{12}{256} + \frac{54}{256} + \frac{108}{256} + \frac{81}{256}$

48. $\frac{8}{125} + \frac{36}{125} + \frac{54}{125} + \frac{27}{125}$

49. 1.172   **50.** 1049.890
51. 510,568.785   **52.** 467.721
53. True; $_nC_m = \dfrac{n!}{m!(n-m)!} = {_nC_{n-m}}$
54. The nth determinant is equal to $_{n+1}C_{n-1}$.

SECTION 10.5   *(page 781)*

Warm-Up   *(page 781)*

1. 10   **2.** 120   **3.** 84   **4.** 1
5. 576   **6.** 720   **7.** 870   **8.** 1260
9. 4845   **10.** 24

1. 10   **2.** 10   **3.** 8   **4.** 8   **5.** 6   **6.** 3
7. 7   **8.** 14   **9.** 6   **10.** 14   **11.** 260
12. 2600   **13.** 6,760,000
14. (a) 900   (b) 720   (c) 400
15. 6   **16.** 12   **17.** 48   **18.** 18   **19.** 16
20. 18   **21.** XYZ, XZY, YXZ, YZX, ZXY, ZYX
22. ABCD, ABDC, ACBD, ACDB, ADBC, ADCB,
BACD, BADC, BCAD, BCDA, BDAC, BDCA,
CABD, CADB, CBAD, CBDA, CDAB, CDBA,
DABC, DACB, DBAC, DBCA, DCAB, DCBA
23. 120   **24.** 720   **25.** 40,320   **26.** 5040
27. {A, B}, {A, C}, {A, D}, {A, E}, {A, F}, {B, C},
{B, D}, {B, E}, {B, F}, {C, D}, {C, E}, {C, F},
{D, E}, {D, F}, {E, F}
28. {A, B, C}, {A, B, D}, {A, B, E}, {A, B, F},
{A, C, D}, {A, C, E}, {A, C, F}, {A, D, E},
{A, D, F}, {A, E, F}, {B, C, D}, {B, C, E},
{B, C, F}, {B, D, E}, {B, D, F}, {B, E, F},
{C, D, E}, {C, D, F}, {C, E, F}, {D, E, F}
29. 1140   **30.** 142,506   **31.** 126   **32.** 220
33. (a) 3   (b) 6   (c) 15   (d) 28
(e) 45   (f) 66
34. (a) 70   (b) 16   **35.** 30   **36.** 3003
37. 84   **38.** 21   **39.** 5   **40.** 9
41. 20   **42.** 35

SECTION 10.6   *(page 789)*

Warm-Up   *(page 789)*

1. $\frac{11}{12}$ **2.** $\frac{5}{8}$ **3.** $\frac{37}{30}$ **4.** $\frac{95}{144}$ **5.** $\frac{1}{24}$
6. $\frac{1}{120}$ **7.** $\frac{3}{14}$ **8.** $\frac{1}{6}$ **9.** $\frac{5}{8}$ **10.** $\frac{117}{125}$

1. {A, B, C, D, E, . . ., X, Y, Z}
2. {2, 3, 4, 5, 6, 7, 8, 9, 10, 11, 12}
3. {AB, AC, AD, AE, BC, BD, BE, CD, CE, DE}
4. {YYY, YYN, YNY, YNN, NYY, NNY, NYN, NNN}
5. $\frac{3}{8}$ **6.** $\frac{1}{2}$ **7.** $\frac{7}{8}$ **8.** $\frac{7}{8}$ **9.** $\frac{1}{2}$
10. $\frac{1}{13}$ **11.** $\frac{3}{13}$ **12.** $\frac{3}{26}$ **13.** $\frac{1}{6}$
14. 0   **15.** $\frac{5}{6}$ **16.** 1
17. (a) $\frac{1}{5}$ (b) $\frac{1}{3}$ (c) 1
18. (a) $\frac{1}{4}$ (b) $\frac{1}{2}$ (c) 1
19. (a) $\frac{1}{20}$ (b) $\frac{2}{5}$ (c) $\frac{1}{2}$ (d) $\frac{1}{4}$
20. 0.2   **21.** $\frac{14}{65}$ **22.** $\frac{1}{2}$ **23.** 0.9
24. 0.1; The sum of the two probabilities must be 1.
25. 0.4375   **26.** (a) $\dfrac{\pi}{4}$
27.

	Female	
	X	X
Male X	XX	XX
Y	XY	XY

Probability of a girl $= \frac{1}{2}$
Probability of a boy $= \frac{1}{2}$

28.

	A	o
B	AB	Bo
o	Ao	oo

Probability of each parent: A and B
Probability of Type A $= \frac{1}{4}$
Probability of Type B $= \frac{1}{4}$
Probability of Type AB $= \frac{1}{4}$
Probability of Type O $= \frac{1}{4}$

29. $\frac{1}{24}$ **30.** $\frac{1}{56}$ **31.** $\frac{1}{45}$ **32.** $\frac{1}{210}$
33. $\dfrac{1}{54,145}$ **34.** $\dfrac{33}{66,640}$ **35.** $1 - p$ **36.** 1

1. 10; 40% **2.** 100; 21% **3.** 1000; 5.5%

4. As M gets larger, the number of times each integer is chosen will be close to the number of times predicted by probability.

5. Experimental results will vary.

6. Experimental results will vary.

CHAPTER 10 REVIEW EXERCISES *(page 796)*

1. $\displaystyle\sum_{k=1}^{4}(5k-3)$ **2.** $\displaystyle\sum_{k=1}^{5}(90-10k)$ **3.** $\displaystyle\sum_{k=1}^{6}\frac{1}{3k}$

4. $\displaystyle\sum_{k=0}^{4}\left(\frac{1}{3}\right)^{k}$ **5.** 380 **6.** $\dfrac{1}{140{,}556}$

7. $n(n-1)(n-2)$ **8.** $\dfrac{1}{(n+1)n}$

9. 127, 122, 117, 112, 107 **10.** 5, 7, 9, 11, 13

11. $\frac{5}{4}$, 2, $\frac{11}{4}$, $\frac{7}{2}$, $\frac{17}{4}$ **12.** $\frac{2}{5}$, $-\frac{1}{5}$, $-\frac{4}{5}$, $-\frac{7}{5}$, -2

13. 10, 30, 90, 270, 810

14. 2, -10, 50, -250, 1250

15. 100, -50, 25, -12.5, 6.25

16. 12, 2, $\frac{1}{3}$, $\frac{1}{18}$, $\frac{1}{108}$

17. $4n+6$ **18.** $-2n+34$

19. $-50n+1050$ **20.** $8n+4$

21. $a_n=\left(-\frac{2}{3}\right)^{n-1}$ **22.** $a_n=100(1.07)^{n-1}$

23. $a_n=24(2)^{n-1}$ **24.** $a_n=16\left(-\frac{1}{4}\right)^{n-1}$

25. $a_n=12\left(-\frac{1}{2}\right)^{n-1}$ **26.** $a_n=3\left(\frac{1}{3}\right)^{n-1}$

27. 28 **28.** $-\frac{7}{12}$ **29.** $\frac{4}{5}$ **30.** $\frac{17}{15}$ **31.** 486

32. 450 **33.** $\frac{2525}{2}$ **34.** $\frac{3825}{2}$ **35.** 8190

36. 2730 **37.** 2.571 **38.** 453.320 **39.** 19.842

40. -2.205 **41.** 115,019.345 **42.** 442,592.56

43. 5100 **44.** 19,950 **45.** 462

46. (a) $a_n=120{,}000(0.70)^n$ (b) \$20,168.40

47. (a) $a_n=85{,}000(1.012)$ (b) 154,328

48. \$4,371,379.65

49. $x^{10}+10x^9+45x^8+120x^7+210x^6$
$+252x^5+210x^4+120x^3+45x^2+10x+1$

50. $u^9-9u^8v+36u^7v^2-84u^6v^3+126u^5v^4$
$-126u^4v^5+84u^3v^6-36u^2v^7+9uv^8-v^9$

51. $y^6-12y^5+60y^4-160y^3+240y^2$
$-192y+64$

52. $x^5+15x^4+90x^3+270x^2+405x+243$

53. $x^8-4x^7+7x^6-7x^5+\frac{35}{8}x^4$
$-\frac{7}{4}x^3+\frac{7}{16}x^2-\frac{1}{16}x+\frac{1}{256}$

54. $81x^4-216x^3y+216x^2y^2-96xy^3+16y^4$

55. 8 **56.** 21 **57.** 3003

58. 5040 **59.** $\frac{1}{3}$ **60.** $\frac{15}{16}$ **61.** $\frac{1}{24}$

62. No, rolling a 3 with one six-sided die has the greater probability of occurring.

63. 0.346 **64.** $\frac{1}{25}$, $\frac{7}{25}$

CHAPTER 10 TEST *(page 798)*

1. 1, $-\frac{2}{3}$, $\frac{4}{9}$, $-\frac{8}{27}$, $\frac{16}{81}$ **2.** 35

3. $\displaystyle\sum_{k=1}^{12}\frac{2}{3k+1}$ **4.** 12, 16, 20, 24, 28

5. $a_n=-100n+5100$ **6.** 3825

7. $-\frac{3}{2}$ **8.** $a_n=4\left(\frac{1}{2}\right)^{n-1}$

9. 765 **10.** \$47,868.33 **11.** 1140

12. $x^5-10x^4+40x^3-80x^2+80x-32$

13. 56 **14.** 26,000 **15.** 12,650

16. 0.25 **17.** $\frac{3}{26}$ **18.** $\frac{1}{6}$

CUMULATIVE TEST: CHAPTERS 7-10 *(page 799)*

1. $5 \pm 5\sqrt{2}\,i$ **2.** $\dfrac{-3 \pm \sqrt{3}}{3}$ **3.** 9

4. (a)

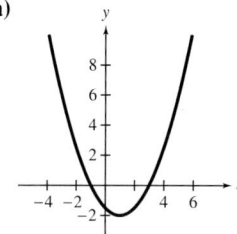

(b)

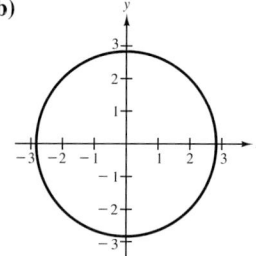

(c)

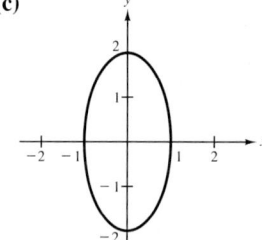

(d)

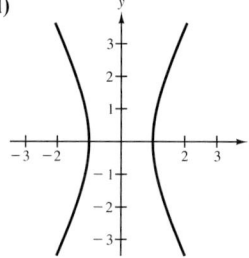

5. $y = \frac{2}{3}x^2 - 4x + 4$ **6.** 11 ft

7. (a) -5 **(b)** 51   **(c)** $\dfrac{\sqrt{3}}{5}$ **(d)** 3

8. $f^{-1}(x) = \frac{1}{3}(x + 2)$

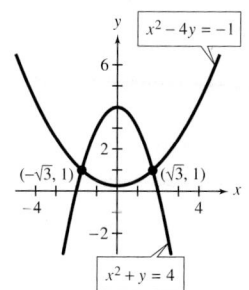

9. 128 ft
10. $(\pm\sqrt{3}, 1)$

11. -2
12.

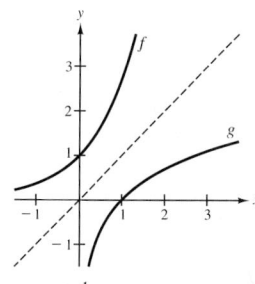

$g = f^{-1}$

13. $\log_2 \dfrac{(xy)^3}{z}$

14. (a) 3   **(b)** 12.182   **(c)** 18.013   **(d)** 0.867
15. 8.3%   **16.** 15.4 years   **17.** 800
18. $\frac{4}{81}$ **19.** -35 **20.** 48   **21.** 120
22. 0.535, 0.465   **23.** $\frac{5}{18}$

Appendix   Introduction to Logic

SECTION A.1   *(page A5)*

1. Statement   **2.** Nonstatement   **3.** Open statement
4. Nonstatement   **5.** Open statement   **6.** Statement
7. Open statement   **8.** Statement   **9.** Nonstatement
10. Statement   **11.** Open statement
12. Nonstatement   **13. (a)** True   **(b)** False
14. (a) False   **(b)** True   **15. (a)** True   **(b)** True
16. (a) True   **(b)** True   **17. (a)** False   **(b)** False
18. (a) True   **(b)** False   **19. (a)** True   **(b)** False
20. (a) False   **(b)** True
21. (a) The sun is not shining.
  (b) It is not hot.
  (c) The sun is shining and it is hot.
  (d) The sun is shining or it is hot.
22. (a) The car does not have a radio.
  (b) The car is not red.
  (c) The car has a radio and the car is red.
  (d) The car has a radio or the car is red.
23. (a) Lions are not mammals.
  (b) Lions are not carnivorous.
  (c) Lions are mammals and lions are carnivorous.
  (d) Lions are mammals or lions are carnivorous.
24. (a) Twelve is not less than fifteen.
  (b) Seven is not a prime number.
  (c) Twelve is less than fifteen and seven is a prime number.
  (d) Twelve is less than fifteen or seven is a prime number.

25. (a) The sun is not shining and it is hot.
 (b) The sun is not shining or it is hot.
 (c) The sun is shining and it is not hot.
 (d) The sun is shining or it is not hot.
26. (a) The car does not have a radio and the car is red.
 (b) The car does not have a radio or the car is red.
 (c) The car has a radio and the car is not red.
 (d) The car has a radio or the car is not red.
27. (a) Lions are not mammals and lions are carnivorous.
 (b) Lions are not mammals or lions are carnivorous.
 (c) Lions are mammals and lions are not carnivorous.
 (d) Lions are mammals or lions are not carnivorous.
28. (a) Twelve is not less than fifteen and seven is a prime number.
 (b) Twelve is not less than fifteen or seven is a prime number.
 (c) Twelve is less than fifteen and seven is not a prime number.
 (d) Twelve is less than fifteen or seven is not a prime number.
29. $p \wedge \sim q$ **30.** $\sim p \vee \sim q$ **31.** $\sim p \vee q$
32. $p \wedge q$ **33.** $\sim p \vee \sim q$ **34.** $p \wedge q$
35. $\sim p \wedge q$ **36.** $p \vee \sim q$ **37.** The bus is blue.
38. Frank is six feet tall.   **39.** x is not equal to 4.
40. x is equal to 4.   **41.** The Earth is flat.
42. The Earth is not flat.
43.

p	q	$\sim p$	$\sim p \wedge q$
T	T	F	F
T	F	F	F
F	T	T	T
F	F	T	F

44.

p	q	$\sim p$	$\sim p \vee q$
T	T	F	T
T	F	F	F
F	T	T	T
F	F	T	T

45.

p	q	$\sim p$	$\sim q$	$\sim p \vee \sim q$
T	T	F	F	F
T	F	F	T	T
F	T	T	F	T
F	F	T	T	T

46.

p	q	$\sim p$	$\sim q$	$\sim p \wedge \sim q$
T	T	F	F	F
T	F	F	T	F
F	T	T	F	F
F	F	T	T	T

47.

p	q	$\sim q$	$p \vee \sim q$
T	T	F	T
T	F	T	T
F	T	F	F
F	F	T	T

48.

p	q	$\sim q$	$p \wedge \sim q$
T	T	F	F
T	F	T	T
F	T	F	F
F	F	T	F

49. Not logically equivalent   **50.** Logically equivalent
51. Logically equivalent   **52.** Not logically equivalent
53. Logically equivalent   **54.** Not logically equivalent
55. Not logically equivalent   **56.** Logically equivalent
57. Logically equivalent   **58.** Not logically equivalent
59. Not a tautology   **60.** A tautology
61. A tautology   **62.** Not a tautology

63.

						Identical	
						↓	↓
p	q	$\sim p$	$\sim q$	$p \wedge q$	$\sim(p \wedge q)$	$\sim p \vee \sim q$	
T	T	F	F	T	F	F	
T	F	F	T	F	T	T	
F	T	T	F	F	T	T	
F	F	T	T	F	T	T	

SECTION A.2 *(page A13)*

1. **(a)** If the engine is running, then the engine is wasting gasoline.
   (b) If the engine is wasting gasoline, then the engine is running.
   (c) If the engine is not wasting gasoline, then the engine is not running.
   (d) If the engine is running, then the engine is not wasting gasoline.
2. **(a)** If the student is at school, then it is nine o'clock.
   (b) If it is nine o'clock, then the student is at school.
   (c) If it is not nine o'clock, then the student is not at school.
   (d) If the student is at school, then it is not nine o'clock.
3. **(a)** If the integer is even, then it is divisible by two.
   (b) If it is divisible by two, then the integer is even.
   (c) If it is not divisible by two, then the integer is not even.
   (d) If the integer is even, then it is not divisible by two.
4. **(a)** If the person is generous, then the person is rich.
   (b) If the person is rich, then the person is generous.
   (c) If the person is not rich, then the person is not generous.
   (d) If the person is generous, then the person is not rich.
5. $q \rightarrow p$ 6. $\sim q \rightarrow \sim p$ 7. $p \rightarrow q$ 8. $q \rightarrow p$
9. $p \rightarrow q$ 10. $p \rightarrow q$ **11.** True **12.** False
13. True **14.** True **15.** False **16.** True
17. True **18.** True **19.** True **20.** False
21. Converse: If you can see the eclipse, then the sky is clear.
   Inverse: If the sky is not clear, then you cannot see the eclipse.
   Contrapositive: If you cannot see the eclipse, then the sky is not clear.
22. Converse: If he is ineligible for the job, then the person is nearsighted.
   Inverse: If the person is not nearsighted, then he is eligible for the job.
   Contrapositive: If he is eligible for the job, then the person is not nearsighted.
23. Converse: If the deficit increases, then taxes were raised.
   Inverse: If taxes are not raised, then the deficit will not increase.
   Contrapositive: If the deficit will not increase, then the taxes are not raised.
24. Converse: If the company's profits will decrease, then wages are raised.
   Inverse: If wages are not raised, then the company's profits will not decrease.
   Contrapositive: If the company's profits will not decrease, then wages are not raised.
25. Converse: It is necessary to apply for the visa to have a birth certificate.
   Inverse: It is not necessary to have a birth certificate to not apply for the visa.
   Contrapositive: It is not necessary to apply for the visa to not have a birth certificate.
26. Converse: The sum of its digits is divisible by three only if the number is divisible by three.
   Inverse: The number is not divisible by three only if the sum of its digits is not divisible by three.
   Contrapositive: The sum of its digits is not divisible by three only if the number is not divisible by three.
27. Paul is not a junior and not a senior.
28. Jack is not a senior or he does not play varsity basketball.
29. The temperature increases and the metal rod will not expand.
30. The test fails and the project will not be halted.
31. We will go to the ocean and the weather forecast is not good.
32. We are going to win the game and not complete the pass on this play.
33. No student is in an extracurricular activity.
34. No odd integer is not a prime number.
35. Some contact sports are not dangerous.
36. Some members must not pay their dues prior to June 1.
37. Some children are allowed at the concert.
38. Some contestants are over the age of twelve.
39. No $20 bills are counterfeit.
40. No units are defective.

41.

p	q	$\sim q$	$p \rightarrow \sim q$	$\sim(p \rightarrow \sim q)$
T	T	F	F	T
T	F	T	T	F
F	T	F	T	F
F	F	T	T	F

42.

p	q	$\sim q$	$p \rightarrow q$	$\sim q \rightarrow (p \rightarrow q)$
T	T	F	T	T
T	F	T	F	F
F	T	F	T	T
F	F	T	T	T

43.

p	q	$q \rightarrow p$	$\sim(q \rightarrow p)$	$\sim(q \rightarrow p) \wedge q$
T	T	T	F	F
T	F	T	F	F
F	T	F	T	T
F	F	T	F	F

44.

p	q	$\sim p$	$\sim p \vee q$	$p \rightarrow (\sim p \vee q)$
T	T	F	T	T
T	F	F	F	F
F	T	T	T	T
F	F	T	T	T

45.

p	q	$\sim p$	$p \vee q$	$(p \vee q) \wedge (\sim p)$	$[(p \vee q) \wedge (\sim p)] \rightarrow q$
T	T	F	T	F	T
T	F	F	T	F	T
F	T	T	T	T	T
F	F	T	F	F	T

46.

p	q	$\sim q$	$p \rightarrow q$	$(p \rightarrow q) \wedge (\sim q)$	$[(p \rightarrow q) \wedge (\sim q)] \rightarrow p$
T	T	F	T	F	T
T	F	T	F	F	T
F	T	F	T	F	T
F	F	T	T	T	F

47.

p	q	$\sim p$	$\sim q$	$p \leftrightarrow (\sim q)$	$(p \leftrightarrow \sim q) \rightarrow \sim p$
T	T	F	F	F	T
T	F	F	T	T	F
F	T	T	F	T	T
F	F	T	T	F	T

48.

p	q	$\sim p$	$\sim q$	$p \vee \sim q$	$q \to \sim p$	$(p \vee \sim q) \leftrightarrow (q \to \sim p)$
T	T	F	F	T	F	F
T	F	F	T	T	T	T
F	T	T	F	F	T	F
F	F	T	T	T	T	T

49.

p	q	$\sim p$	$\sim q$	$q \to p$	$\sim p \to \sim q$
T	T	F	F	T	T
T	F	F	T	T	T
F	T	T	F	F	F
F	F	T	T	T	T

Identical

50.

p	q	$\sim p$	$\sim p \to q$	$p \vee q$
T	T	F	T	T
T	F	F	T	T
F	T	T	T	T
F	F	T	F	F

Identical

51.

p	q	$\sim q$	$p \to q$	$\sim(p \to q)$	$p \wedge \sim q$
T	T	F	T	F	F
T	F	T	F	T	T
F	T	F	T	F	F
F	F	T	T	F	F

Identical

52.

p	q	$p \vee q$	$(p \vee q) \to q$	$p \to q$
T	T	T	T	T
T	F	T	F	F
F	T	T	T	T
F	F	F	T	T

Identical

53.

p	q	$\sim p$	$\sim q$	$p \to q$	$(p \to q) \vee \sim q$	$p \vee \sim p$
T	T	F	F	T	T	T
T	F	F	T	F	T	T
F	T	T	F	T	T	T
F	F	T	T	T	T	T

Identical

54.

p	q	$\sim p$	$\sim q$	$\sim p \vee q$	$q \to (\sim p \vee q)$	$q \vee \sim q$
T	T	F	F	T	T	T
T	F	F	T	F	T	T
F	T	T	F	T	T	T
F	F	T	T	T	T	T

└── Identical ──┘

55.

p	q	$\sim p$	$\sim p \wedge q$	$p \to (\sim p \wedge q)$
T	T	F	F	F
T	F	F	F	F
F	T	T	T	T
F	F	T	F	T

└── Identical ──┘

56.

p	q	$\sim q$	$p \wedge q$	$\sim(p \wedge q)$	$\sim(p \wedge q) \to \sim q$	$p \vee \sim q$
T	T	F	T	F	T	T
T	F	T	F	T	T	T
F	T	F	F	T	F	F
F	F	T	F	T	T	T

└── Identical ──┘

57. (c)   **58.** (c)   **59.** (a)   **60.** (c)

61.   **62.**

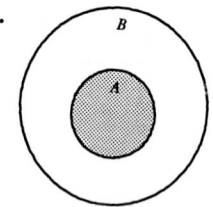

63.   **64.**

65.

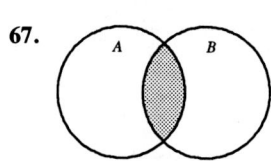

66.

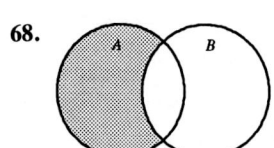

67.

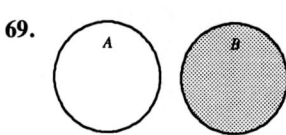

68.

69.

70.

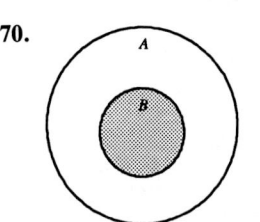

71. (a) Statement does not follow.
    (b) Statement follows.
72. (a) Statement follows.
    (b) Statement follows.
73. (a) Statement does not follow.
    (b) Statement does not follow.
74. (a) Statement does not follow.
    (b) Statement does not follow.

SECTION A.3 *(page A22)*

1.

p	q	$\sim p$	$\sim q$	$p \to \sim q$	$(p \to \sim q) \wedge q$	$[(p \to \sim q) \wedge q] \to \sim p$
T	T	F	F	F	F	T
T	F	F	T	T	F	T
F	T	T	F	T	T	T
F	F	T	T	T	F	T

2.

p	q	$p \leftrightarrow q$	$(p \leftrightarrow q) \wedge p$	$[(p \leftrightarrow q) \wedge p] \to q$
T	T	T	T	T
T	F	F	F	T
F	T	F	F	T
F	F	T	F	T

3.

p	q	$\sim p$	$p \vee q$	$(p \vee q) \wedge \sim p$	$[(p \vee q) \wedge \sim p] \to q$
T	T	F	T	F	T
T	F	F	T	F	T
F	T	T	T	T	T
F	F	T	F	F	T

4.

p	q	$\sim p$	$p \wedge q$	$(p \wedge q) \wedge \sim p$	$[(p \wedge q) \wedge \sim p] \to q$
T	T	F	T	F	T
T	F	F	F	F	T
F	T	T	F	F	T
F	F	T	F	F	T

5.

p	q	$\sim p$	$\sim q$	$(\sim p \to q)$	$(\sim p \to q) \wedge p$	$[(\sim p \to q) \wedge p] \to \sim q$
T	T	F	F	T	T	F
T	F	F	T	T	T	T
F	T	T	F	T	F	T
F	F	T	T	F	F	T

6.

p	q	$\sim p$	$\sim q$	$p \rightarrow q$	$(p \rightarrow q) \wedge \sim p$	$[(p \rightarrow q) \wedge \sim p] \rightarrow \sim q$
T	T	F	F	T	F	T
T	F	F	T	F	F	T
F	T	T	F	T	T	F
F	F	T	T	T	T	T

7.

p	q	$p \vee q$	$(p \vee q) \wedge q$	$[(p \vee q) \wedge q] \rightarrow p$
T	T	T	T	T
T	F	T	F	T
F	T	T	T	F
F	F	F	F	T

8.

p	q	$p \wedge q$	$\sim(p \wedge q)$	$\sim(p \wedge q) \wedge q$	$[\sim(p \wedge q) \wedge q] \rightarrow p$
T	T	T	F	F	T
T	F	F	T	F	T
F	T	F	T	T	F
F	F	F	T	F	T

9. Valid   **10.** Valid   **11.** Invalid   **12.** Invalid
13. Valid   **14.** Valid   **15.** Valid   **16.** Invalid
17. Invalid   **18.** Valid   **19.** Valid   **20.** Valid
21. Invalid   **22.** Invalid   **23.** (b)   **24.** (c)
25. (c)   **26.** (c)   **27.** (b)   **28.** (a)   **29.** (c)
30. (b)

31. Valid   **32.** Valid

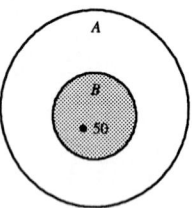

A: All numbers divisible by five
B: All numbers divisible by ten

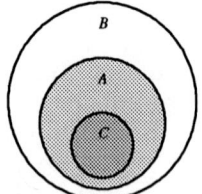

A: All human beings
B: Things that require adequate rest
C: All infants

33. Invalid

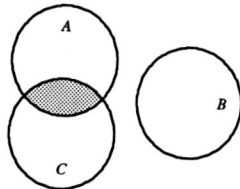

A: People eligible to vote
B: People under the age of 18
C: College students

34. Invalid

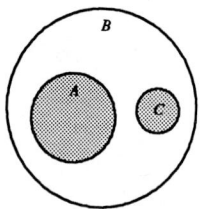

A: Every amateur radio operator
B: People who have a radio license
C: Jackie

35. Let p be the statement "Sue drives to work," let q represent "She will stop at the grocery store," and let r represent "She'll buy milk."

First write:

> Premise #1: $p \rightarrow q$
> Premise #2: $q \rightarrow r$
> Premise #3: p

Reorder the premises:

> Premise #3: p
> Premise #1: $p \rightarrow q$
> Premise #2: $q \rightarrow r$
> Conclusion: r

Then we can conclude r. That is, "Sue will get milk."

36. Let p represent "Bill is patient," let q represent "He will succeed," let r represent "Bill will get bonus pay."

First write:

> Premise #1: $p \rightarrow q$
> Premise #2: $q \rightarrow r$
> Premise #3: $\sim r$

Conclusion from Premise #1, Premise #2: $p \rightarrow r$
Conclusion from $p \rightarrow r$ and Premise #3: $\sim p$

That is, "Bill is not patient."

37. Let p represent "This is a good product," let q represent "We will buy it," and let r represent "The product was made by XYZ Corporation."

First write:

> Premise #1: $p \rightarrow q$
> Premise #2: $r \vee \sim q$
> Premise #3: $\sim r$

Note that $p \rightarrow q \equiv \sim q \rightarrow \sim p$, and reorder the premises:

> Premise #2: $r \vee \sim q$
> Premise #3: $\sim r$
> (Conclusion from Premise #2, Premise #3: $\sim q$)
> Premise #1: $\sim q \rightarrow \sim p$
> Conclusion: $\sim p$

Then we can conclude $\sim p$. That is, "It is not a good product."

38. Let p represent "The book is returned in two weeks," let q represent "There is a fine," and let r represent "You may not check out another book."

First write:

Premise #1: $p \rightarrow \sim q$
Premise #2: $q \vee r$
Premise #3: $\sim r$

Conclusion from Premise #2, Premise #3: q
Note $q \equiv \sim(\sim q)$
Conclusion from q, Premise #1: $\sim p$
That is, "The book was not returned within two weeks."

INSTRUCTOR'S ANNOTATED EDITION

INTERMEDIATE ALGEBRA

Graphs and Functions

INTERMEDIATE ALGEBRA
Graphs and Functions

ROLAND E. LARSON
The Pennsylvania State University, The Behrend College

ROBERT P. HOSTETLER
The Pennsylvania State University, The Behrend College

CAROLYN F. NEPTUNE
Johnson County Community College

with the assistance of
DAVID E. HEYD
The Pennsylvania State University, The Behrend College

D. C. HEATH AND COMPANY
Lexington, Massachusetts   Toronto

Address editorial correspondence to:

D. C. Heath
125 Spring Street
Lexington, MA 02173

ACQUISITIONS EDITOR: Ann Marie Jones

DEVELOPMENTAL EDITOR: Cathy Cantin

PRODUCTION EDITOR: Karen Carter

DESIGNER: Cornelia Boynton

ART EDITOR: Gary Crespo

PRODUCTION SUPERVISOR: Lisa Merrill

COVER: Paul Klee (1879–1940). *Pyramide, 1930.138.* Paul Klee Foundation, Museum of Fine Arts Berne. © 1991, Copyright by COSMOPRESS, Geneva, Switzerland.

COMPOSITION: Meridian Creative Group

TECHNICAL ART: Folium, Inc.; Illustrious, Inc.; Tech-Graphics, Inc.; Techsetters

PHOTO CREDITS: 1, Julie Houck/Stock, Boston; 39, Edward L. Miller/Stock, Boston; 88, Jan Halaska/Photo Researchers, Inc.; 185, Ron Sanford/AllStock; 193, The Bettmann Archive; 205, Brian Smith; 229, Jonathan Watts/Science Photo Library/Photo Researchers, Inc.; 238, John Running/Stock, Boston; 259, J. Carini/The Image Works; 282, John Head/Science Photo Library/Photo Researchers, Inc.; 300, D. and I. MacDonald/The Picture Cube; 312, Dawson Jones/Stock, Boston; 330, Topham/The Image Works; 351, Bob Daemmrich/Stock, Boston; 354, Courtesy of Lucasfilm, Ltd.; 359, Hiroji Kubota/Magnum Photos; 373, G. C. Kelley/Photo Researchers, Inc.; 384, Fridmar Damm/Leo de Wys Inc.; 430, Steve Goldberg/Monkmeyer Press Photo; 450, George Munday/Leo de Wys Inc.; 469, The Bettmann Archive; 517, Mimi Forsyth/Monkmeyer Press Photo; 595, Bob Daemmrich/Stock, Boston; 601, Bryce Flynn/Stock, Boston; 604, Martin Dohrn/Science Photo Library/Photo Researchers, Inc.; 623, Gamma-Liaison; 627, Dan Abernathy; 669, Scott Camazine/Photo Researchers, Inc.; 675, Fridmar Damm/Leo de Wys Inc.; 677, David Dul/Unicorn Stock Photo; 682, Peter Menzel/Stock, Boston; 720, David Young-Wolff/PhotoEdit; 725, Nita Winters/The Image Works; 733, Jim Schwabel/New England Stock Photo; 745, Dr. E. R. Degginger; 751, D. G. Arnold; 786, H. Armstrong Roberts, Inc.

Published simultaneously in Canada.

Printed in the United States of America.

International Standard Book Number: 0-669-33755-2

Library of Congress Catalog Number: 93-71003

10   9   8   7   6   5   4   3   2   1

Preface

Intermediate Algebra: Graphs and Functions has two basic goals: first, to help students develop proficiency in algebra; and, second, to show students how algebra can be used as a modeling language for real-life problems. To support these two functions, the text has several key pedagogical features.

Mathematics as Problem Solving Throughout the text, students are encouraged to consider multiple approaches to problem-solving — algebraic, graphical, and numerical. For real-life problems, students are encouraged to use the following approach: construct a verbal model, label variable and constant terms, construct an algebraic model, solve the algebraic model, answer the question, and check the answer in the original statement of the problem.

Mathematics as Communication The discussion problems at the end of each section offer students the opportunity to think, talk, and write about mathematics in different ways. Students are encouraged to draw new conclusions about the concepts presented and to develop a sense of how each topic studied fits into the whole concept of algebra.

Mathematics as Reasoning While the text stresses skill development in algebra, it does so in the broader context of developing an understanding of algebraic properties and principles. Many of the examples and exercises in the text ask students to explain the reasons for choosing a particular problem-solving approach.

Mathematical Connections Numerous applications are integrated throughout every section of the text — both as solved examples and as exercises. As a result, students will constantly use and review their problem-solving skills. The text applications cover a wide range of relevant topics, and many use real data. The examples and exercises in the text connect algebra not only to real life, but also to other branches of mathematics, such as geometry.

These and other features of the text are described in greater detail on the following pages.

Features of the Text

Linear Functions, Equations, and Inequalities

3.1   Applications of Linear Equations
3.2   More Applications: Consumer and Scientific Problems
3.3   Linear Inequalities in One Variable
3.4   Equations and Inequalities Involving Absolute Value
3.5   Slope: An Aid to Graphing Linear Functions
3.6   Equations of Lines

*E*ach year, monarch butterflies migrate from the northern United States to Mexico. The butterflies travel at an average rate of 80 miles per day. The distance d (in miles) traveled for a given time t (in days) is given by

(Distance) = (rate)(time)

$$d = 80t \, .$$

Use this model to appro the distance traveled by monarch butterfly in 2 entire trip covers a dist How long does it take t

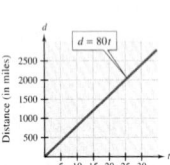

Features of the Text

Chapter Opener

Each chapter contains a list of the topics to be covered and a real-life application that helps motivate the chapter. This application poses one or two questions that are designed to pique students' curiosity.

Section Topics

Each section begins with a list of important topics that are covered in the section. These topics are also the subsection titles and can be used for easy reference and review by students.

Definitions and Rules

All of the important rules, formulas, and definitions are boxed for emphasis. Each is also titled for easy reference.

Notes

Notes appear after definitions and examples. Anticipating students' needs, the notes give additional insight, point out common errors, and describe generalizations.

SECTION 1.4

Operations with Polynomials

Basic Definitions   •   Adding and Subtracting Polynomials   •   Multiplying Polynomials   •   Special Products   •   Applications

Basic Definitions

An algebraic expression containing only terms of the form ax^k, where a is any real number and k is a nonnegative integer, is called a **polynomial in one variable** or simply a **polynomial.** Here are some examples of polynomials in one variable.

$$3x - 8, \quad x^4 + 3x^3 - x^2 - 8x + 1, \quad x^3 + 5, \quad \text{and} \quad 9x^5$$

In the term ax^k, a is called the **coefficient,** and k the **degree,** of the term. Note that the degree of the term ax is 1, and the degree of a constant term is zero. Because a polynomial is an algebraic *sum*, the coefficients take on the signs between the terms. For instance,

$$x^3 - 4x^2 + 3 = (1)x^3 + (-4)x^2 + (0)x + 3$$

has coefficients 1, -4, 0, and 3.

Polynomials are usually written in order of descending powers of the variable. This is referred to as **standard form.** For example, the standard form of $3x^2 - 5 - x^3 + 2x$ is

$$-x^3 + 3x^2 + 2x - 5. \qquad \text{Standard form}$$

The **degree of a polynomial** is defined as the degree of the term with the highest power, and the coefficient of this term is called the **leading coefficient** of the polynomial. For instance, the polynomial $-3x^4 + 4x^2 + x + 7$ is of fourth degree and its leading coefficient is -3.

Definition of a Polynomial in *x*	Let $a_n, \ldots, a_2, a_1, a_0$ be real numbers and let n be a *nonnegative integer.* A **polynomial in x** is an expression of the form $$a_n x^n + a_{n-1} x^{n-1} + \cdots + a_2 x^2 + a_1 x + a_0 \, ,$$ where $a_n \neq 0$. The polynomial is of **degree n**, and the number a_n is called the **leading coefficient.** The number a_0 is called the **constant term.**

NOTE   The following are *not* polynomials, for the reasons stated.

$$2x^{-1} + 5 \qquad \text{Exponent in } 2x^{-1} \text{ is not nonnegative.}$$
$$x^3 + 3x^{1/2} \qquad \text{Exponent in } 3x^{1/2} \text{ is } not \text{ an integer.}$$

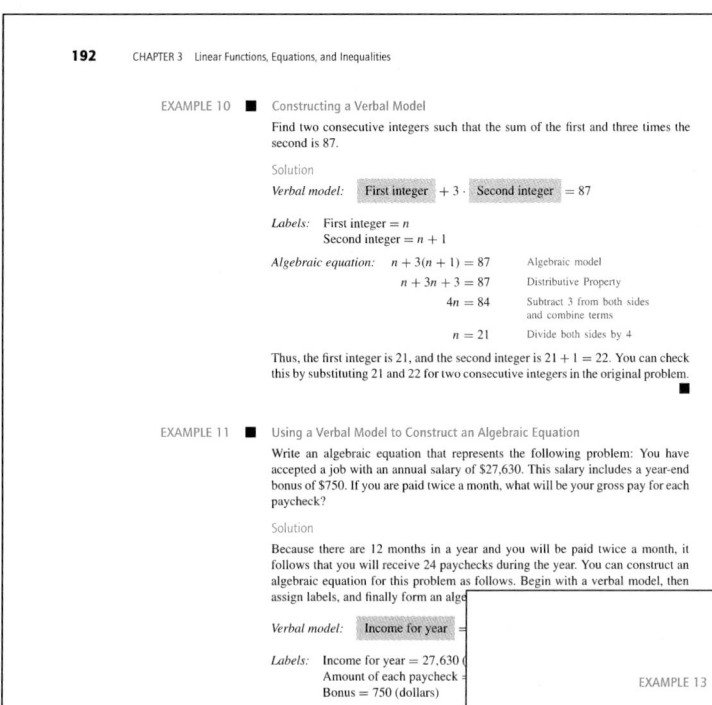

192 CHAPTER 3 Linear Functions, Equations, and Inequalities

EXAMPLE 10 ■ Constructing a Verbal Model

Find two consecutive integers such that the sum of the first and three times the second is 87.

Solution

Verbal model: First integer + 3 · Second integer = 87

Labels: First integer = n
Second integer = $n + 1$

Algebraic equation:
$n + 3(n + 1) = 87$ — Algebraic model
$n + 3n + 3 = 87$ — Distributive Property
$4n = 84$ — Subtract 3 from both sides and combine terms
$n = 21$ — Divide both sides by 4

Thus, the first integer is 21, and the second integer is $21 + 1 = 22$. You can check this by substituting 21 and 22 for two consecutive integers in the original problem. ■

EXAMPLE 11 ■ Using a Verbal Model to Construct an Algebraic Equation

Write an algebraic equation that represents the following problem: You have accepted a job with an annual salary of $27,630. This salary includes a year-end bonus of $750. If you are paid twice a month, what will be your gross pay for each paycheck?

Solution

Because there are 12 months in a year and you will be paid twice a month, it follows that you will receive 24 paychecks during the year. You can construct an algebraic equation for this problem as follows. Begin with a verbal model, then assign labels, and finally form an alge

Verbal model: Income for year =

Labels: Income for year = 27,630
Amount of each paycheck =
Bonus = 750 (dollars)

Algebraic equation: 27,630 = 24x

The algebraic equation for this prob
Using the techniques discussed in Se
If you do this, you will find that the s

You could solve the problems in E
instance, in Example 11 you could so
$750 from the annual salary of $27,63

Problem-Solving Process

Students are taught the following strategies — in keeping with the spirit of NCTM standards — for solving applied problems. (1) Construct a verbal model; (2) label variable and constant terms; (3) construct an algebraic model; (4) using the model, solve the problem; and (5) check the answer in the original statement of the problem.

Examples

Each of the nearly 900 examples was carefully chosen to illustrate a particular concept or problem-solving technique and to enhance students' understanding. The examples are titled for easy reference.

SECTION 1.3 Algebraic Expressions **39**

EXAMPLE 13 ■ Creating a Real-Life Model

For the years 1980 through 1990, the number B (in millions) of children's books that were sold in the United States can be modeled by $B = 18.7t + 119$, where $t = 0$ represents 1980. During the same years, the average price p of a children's book can be modeled by $p = 0.2t + 3.3$. These two models can be used to write a model for the total sales (in millions of dollars) of children's books for 1980 through 1990. Use the model to find the total sales in 1988. (*Source: Book Industry Times*)

Solution

Let S represent the yearly sales in millions of dollars.

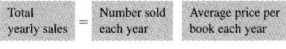

Total yearly sales = Number sold each year · Average price per book each year

$$S = (18.7t + 119)(0.2t + 3.3)$$

To find the total sales in 1988, substitute 8 for t in this model.

$S = (18.7t + 119)(0.2t + 3.3)$ — Total sales model
$= (18.7 \cdot 8 + 119)(0.2 \cdot 8 + 3.3)$ — Substitute 8 for t
$= (268.6)(4.9)$ — Perform operations within parentheses
$= 1316.14$ — Multiply

The total sales of children's books in 1988 was about $1316 million (or about $1.3 billion). The bar graph shown in Figure 1.9 shows the total sales of children's books from 1980 through 1990. ■

According to the NEA, over half of the adults in the United States read at least one book for enjoyment each year. Reading books to children increases the likelihood that the children will read for enjoyment as adults.

FIGURE 1.9

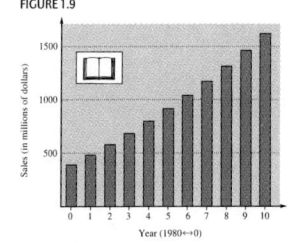

Applications

Real-world applications are integrated throughout the text both in examples and in exercises. These applications offer students a constant review of problem-solving skills and emphasize the relevance of the mathematics. Many of the applications use current, real data, and all are titled for reference.

Using a Graphing Calculator

Most graphing calculators have two graphing modes: a *connected mode* and a *dot mode* (see figures below). The connected mode works well for graphs of functions that are continuous (have no holes or breaks). The connected mode does not, however, work well for the graphs of many rational functions because they are often composed of two or more disconnected branches. To correct this problem, change the calculator to dot mode.

On the screen shown at left below, notice the vertical line at approximately $x = 1$. This line is *not* part of the graph—it is simply the calculator's attempt to connect the two branches of the graph.

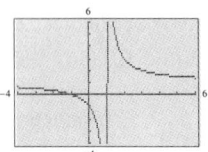 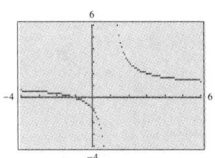

Connected Mode
Graph of $f(x) = \dfrac{x+1}{x-1}$

Dot Mode
Graph of $f(x) = \dfrac{x+1}{x-1}$

EXAMPLE 1 ■ Investigating Asymptotic Behavior

Some people think that the graph of a rational function cannot cross its horizontal asymptote. This, however, is not the case. For instance, consider the graph of

$$f(x) = \frac{2x^2 - 3x + 5}{x^2 + 1}$$

and its horizontal asymptote $y = 2$. From the screen at left, it appears that the graph crosses its horizontal asymptote when $x = 1$.

Solution

From the screen at left, it appears that the graph crosses its horizontal asymptote when $x = 1$. You can confirm this by evaluating

$$f(1) = \frac{2(1^2) - 3(1) + 5}{1^2 + 1} = \frac{4}{2}$$

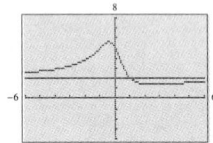

EXAMPLE 2 ■ Pollution Level of a Pond

Some organic waste has been dumped into a pond. One component of the decomposition process is oxidation, whereby oxygen dissolved in the pond water is combined with decomposing material. Let $L = 1$ represent the normal oxygen level in the pond, and let t represent the number of weeks that have elapsed since the waste was dumped. The oxygen level in the pond can be modeled by

$$L = \frac{t^2 - t + 1}{t^2 + 1}.$$

Sketch the graph of this model, and use the graph to explain how the oxygen level has changed during the 15 weeks since the waste was dumped.

Solution

The screen at left shows the graph of the model and the line $L = 1$ (the normal oxygen level). From the graph, you can see that the oxygen dropped to 50% of its normal level after 1 week. Then, during the next several weeks, the oxygen level gradually returned to normal. Now, at the end of the 15th week, the oxygen level has reached 93% of normal. ■

Oxygen Level of Pond

EXERCISES

In Exercises 1–6, find bounds for x and y such that the calculator screen displays the basic characteristics of the graph of the rational function.

1. $f(x) = \dfrac{x - 5}{x + 1}$ 　　**2.** $f(x) = \dfrac{2x - 5}{x - 4}$ 　　**3.** $f(x) = \dfrac{x^2 - 1}{x^2 - 4}$

4. $f(x) = \dfrac{2x^2 - 7x + 5}{x^2 + 2x + 1}$ 　　**5.** $f(x) = \dfrac{x^2 - 5x}{2x^2 + 1}$ 　　**6.** $f(x) = \dfrac{x^3 - 1}{x^3 + 1}$

7. Use a graphing calculator to sketch the graphs of f and g on the same screen. (Use a range of $-5 \le x \le 5$ and $-4 \le y \le 20$.)

$$f(x) = x \quad \text{and} \quad g(x) = \frac{x^2 + x + 1}{x + 1}$$

Use long division to rewrite $g(x)$. Then use the result to explain why the two graphs are close to each other for large values of $|x|$.

8. For the years 1975 to 1990, the number N (in millions) of long-playing albums sold in the United States can be approximated by the model

$$N = \frac{1000 - 74.41t + 1.45t^2}{2.94 - 0.336t + 0.0125t^2},$$

where $t = 5$ represents 1975. Sketch the graph of this function and use the graph to find the year in which the sales peaked. (*Source:* Recording Industry Association of America)

Graphing /Scientific Calculator

Each chapter contains a section devoted to the use of calculators and computers in problem solving. Exercises labeled ▦ in the text sections that follow give students the opportunity to use technology as a problem-solving tool.

Graphing

Graphing is introduced in Chapter 2. From that point on, students are encouraged to consider using graphs to reinforce algebraic solutions.

EXAMPLE 5 ■ Finding the Intercepts of a Graph

Find the intercepts and sketch the graph of $y = 2x - 3$.

Solution

To determine whether the graph has any x-intercepts, let y be zero and solve the resulting equation for x.

$y = 2x - 3$	Given equation
$0 = 2x - 3$	Let $y = 0$
$\dfrac{3}{2} = x$	Solve equation for x

Therefore, the graph has one x-intercept, which occurs at the point $\left(\frac{3}{2}, 0\right)$.

To determine whether the graph has any y-intercepts, let x be zero and solve the resulting equation for y.

$y = 2x - 3$	Given equation
$y = 2(0) - 3$	Let $x = 0$
$y = -3$	Solve equation for y

Therefore, the graph has one y-intercept, which occurs at the point $(0, -3)$.

To sketch the graph of the equation, make a table of values, as shown in Table 2.9. (Notice that the two intercepts are included in the table.) Finally, using the solution points in the table, sketch the graph of the equation, as shown in Figure 2.20.

TABLE 2.9

x	-1	0	1	$\frac{3}{2}$	2	3	4
$y = 2x - 3$	-5	-3	-1	0	1	3	5
Solution Points	$(-1, -5)$	$(0, -3)$	$(1, -1)$	$\left(\frac{3}{2}, 0\right)$	$(2, 1)$	$(3, 3)$	$(4, 5)$

FIGURE 2.20

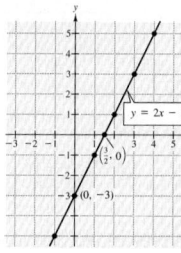

Mid-Chapter Quiz

Each chapter contains a mid-chapter quiz after the third section. This feature allows students to perform a self-assessment midway through the chapter. Answers to mid-chapter quizzes are given at the end of the text.

2

Mid-Chapter Quiz

Take this quiz as you would take a quiz in class. After you are done, check your work against the answers in the back of the book.

In Exercises 1–8, solve the equation.

1. $24 - 2x = x$
2. $3(x - 4) + 4 = -2(x - 1)$
3. $\dfrac{5x}{6} - \dfrac{2}{3} = 1$
4. $\dfrac{t}{4} - \dfrac{t}{6} = 1$
5. $(s - 16)(s + 15) = 0$
6. $y(y + 6) = 72$
7. $4x^2 - 4x + 1 = 0$
8. $3t^2 - 4t + 6 = 10$

9. Solve for C
   Temperature Conversion: $F = \frac{9}{5}C + 32$

10. Solve for l
    Perimeter of a Rectangle: $P = 2(l + w)$

11. The surface area of a cylinder (see figure) is given by $S = 2\pi rh + 2\pi r^2$.
    (a) Solve for h.
    (b) Find h is $S = 105$ square inches and $r = 3$ inches.

Figure for 11

12. An object is thrown upward from a height of 64 feet with an initial velocity of 48 feet per second. Find the time t for the object to reach the ground by solving the following equation.
    $$-16t^2 + 48t + 64 = 0$$

13. Two people can complete a task in t hours, where t must satisfy the equation
    $$\frac{t}{12} + \frac{t}{20} = 1.$$
    Find the required time t.

14. (a) Determine the quadrant(s) in which the point (x, y) is located if $xy < 0$.
    (b) Explain your reasoning.

In Exercises 15 and 16, use the points $(-3, 0)$ and $(7, 3)$.

15. Plot the points in a rectangular coordinate system.

16. Find the distance between the points. Round your result to two decimal places.

17. Determine whether the ordered paris are solution points to the equation $5x - 3y + 10 = 0$.

    (a) $(-2, 0)$ (b) $(0, 3)$ (c) $(1, -2)$ (d) $(-5, -5)$

210   CHAPTER 3   Linear Functions, Equations, and Inequalities

DISCUSSION PROBLEM ■ Red Herrings

Most applied problems in textbooks give precisely the right amount of information that is necessary to solve a given problem. In real life, however, you often must sort through the given information and discard information that is irrelevant to the problem. Such irrelevant information is called a **red herring.** Find the red herrings in the following problems.

(a) Suppose you are hired for a job that pays \$8 per hour. After a 90-day review, your salary will be increased to \$8.50 per hour. During your first two weeks on the job, you work 80 hours. How much will you be paid for your first two weeks of work?

(b) A person leaves home at noon and drives 50 miles before stopping to fill the car with gas. The person then continues driving until 3 P.M. At the beginning of the trip, the car's odometer reading is 45,768, and at the end of the trip, the reading is 45,930. Find the average speed of the car during the entire trip. ■

Warm-Up

The following warm-up exercises involve skills that were covered in earlier sections. You will use these skills in the exercise set for this section.

In Exercises 1–10, solve the equation.

1. $44 - 16x = 0$
2. $-4(x - 5) = 0$
3. $3[4 + 5(x - 1)] = 6x + 2$
4. $\frac{3x}{8} + \frac{3}{4} = 2$
5. $\frac{x}{3} + \frac{x}{2} = \frac{1}{3}$
6. $x - \frac{x}{4} = 15$
7. $\frac{5x}{4} + \frac{1}{2} = x - \frac{1}{2}$
9. $0.25x + 0.75(10 - x) = 3$

3.2 EXERCISES   means that a graphing utility can help solve the exercise or check your sol

In Exercises 1–8, find the missing quantities. (Assume that the markup rate is a based on the cost.)

Merchandise	Cost	Selling Price	Markup	Markup Rate
1. Wristwatch	\$45.95	\$64.33		
2. Bicycle	\$84.20	\$113.67		
3. Sleeping bag		\$250.80	\$98.80	

176   CHAPTER 2   Introduction to Equations, Graphs, and Functions

Warm-Up

The following warm-up exercises involve skills that were covered in earlier sections. You will use these skills in the exercise set for this section.

In Exercises 1–6, sketch the graph of the equation.

1. $y = 3 - 2x$
2. $y = \frac{1}{2}x - 1$
3. $y = (x - 2)^2 - 4$
4. $y = 9 - (x + 1)^2$
5. $x - y^2 = 0$
6. $y = |x| + 1$

In Exercises 7–10, evaluate the function at the indicated values.

Function	Function Value	
7. $f(x) = \frac{1}{3}x^2$	(a) $f(6)$	(b) $f\left(\frac{3}{4}\right)$
8. $f(x) = 3 - 2x$	(a) $f(5)$	(b) $f(x + 3) - f(3)$
9. $f(x) = \frac{x}{x + 10}$	(a) $f(5)$	(b) $f(c - 6)$
10. $f(x) = \sqrt{x - 4}$	(a) $f(16)$	(b) $f(t + 3)$

2.6 EXERCISES   means that a graphing utility can help you solve the exercise or check your solution.

In Exercises 1–6, match the function with its graph. [The graphs are labeled (a), (b), (c), (d), (e), and (f).]

1. $f(x) = 4 - 2x$
2. $f(x) = \frac{3}{2}x + 1$
3. $g(x) = \sqrt{1 - x}$
4. $g(x) = (x + 2)^2$
5. $h(x) = |x| - 2$
6. $h(x) = |x + 2|$

(a) (b) (c)

(d) (e) (f)

Discussion Problems

Discussion problems appear at the end of each section. They encourage students to think, talk, and write about mathematics, both individually and in groups.

Warm-Up

Each section (except for Section 1.1) contains a set of ten warm-up exercises that enables students to review and practice the previously learned skills necessary to master the new skills presented in the section. All warm-up exercises are answered at the end of the text.

Calculators and Computers

Two types of calculators are discussed: scientific and graphing. In examples and exercises, the graphing capability of graphing calculators and computer graphing software is investigated; however, coverage of this material is optional.

In Exercises 47–50, find (a) the missing coordinates of the one vertex of the right triangle, (b) the lengths of the vertical and horizontal sides of the triangle, (c) the length of the hypotenuse, and (d) the distance between the two given points.

47. $(10, 8), (2, 2)$

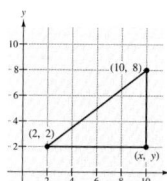

48. $(-2, 1), (10, 6)$

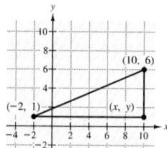

49. $(-3, -4), (4, 4)$

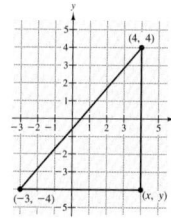

50. $(0, 6), (6, -2)$

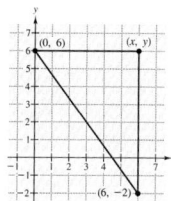

In Exercises 51–58, find the distance between the two points. If appropriate, round the result to two decimal places.

51. $(1, 3), (5, 6)$ **52.** $(3, 10), (15, 5)$

53. $(0, 0), (12, -9)$ **54.** $(-5, 0), (3, 15)$

55. $(-2, -3), (4, 2)$ **56.** $(-5, 4), (10, -3)$

57. $(1, 3), (3, -2)$ **58.** $\left(\frac{1}{2}, 1\right), \left(\frac{3}{2}, 2\right)$

In Exercises 59–62, determine whether the points are vertices of a right triangle.

59. $(2, 3), (2, 6), (6, 3)$

60. $(2, 4), (-1, 6), (-3, 1)$

61. $(8, 3), (5, 2), (1, 9)$

62. $(2, 4), (1, 1), (7, -1)$

63. *Housing Construction*   A house is 30 feet wide and the ridge of the roof is 7 feet above the top of the walls (see figure). Find the length of the rafters if they overhang the edge of the walls by 2 feet.

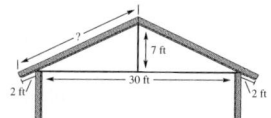

lgebra

rinomial.

$+ 169$ **109.** $4s^2 + 36st + 81t^2$ **110.** $u^2 - 10uv + 25v^2$

pletely, if possible.

$3b + 27b^3$ **113.** $8x(2x - 3) - 4(2x - 3)$

$8x^3 + 1$ **116.** $t^3 - 216$

$x^2 - 2x + 1$ **119.** $x^2 + x - 1$

$4t^3 + 7t^2 - 2t$ **122.** $6h^3 - 22h^2 - 8h$

at the trinomial is factorable.

123. $x^2 + bx + 5$ **124.** $x^2 + bx + 6$ **125.** $2x^2 + bx - 15$ **126.** $3x^2 + bx - 22$

Magazine Circulations   In Exercises 127 and 128, use the figure showing the circulations for the top five magazines in the United States for the first six months of 1990. (*Source:* Audit Bureau of Circulations)

Figure for 127 and 128

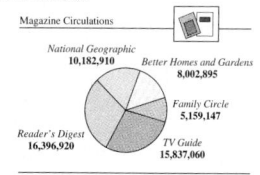

127. Determine the combined circulation for the five magazines.

128. What is the difference in circulation between *Reader's Digest* and *National Geographic*?

129. *Total Charge*   Suppose you purchased a product by making a down payment of $239 plus nine monthly payments of $45 each. What is the total amount you paid for the product?

130. *Total Charge*   Suppose you purchased a product by making a down payment of $387 plus 12 monthly payments of $68 each. What is the total amount you paid for the product?

131. *Exploratory Exercise*   Enter any number between 0 and 1 in a calculator. Take the square root of the number. Then take the square root of the result, and keep repeating the process. What number does the calculator display seem to be approaching?

132. *Calculator Experiment*   Use a calculator to calculate 12^4 in two ways.*

Scientific	Graphing
(a) 12 y^x 4 $=$	(a) 12 $\wedge$ 4 ENTER
(b) 12 x^2 x^2	(b) 12 x^2 x^2 ENTER

Why do these two methods give the same result?

133. *Probability*   The probability of three successes in five trials of an experiment is $10p^3(1 - p)^2$. Find this product.

*The graphing calculator keystrokes in this text correspond to the TI-81 and TI-82 graphing calculators from Texas Instruments. For other graphing calculators, the keystrokes may differ.

Exercise Sets

The nearly 7000 exercises contain numerous computational and applied problems dealing with a wide range of topics. Anticipating students' needs, these problems are carefully graded to increase in difficulty as students' problem-solving skills develop. Each pair of consecutive problems is similar, with the answer to the odd-numbered problem given at the end of the text. Exercise sets appear at the end of each text section. The opportunity to use calculators — to show patterns, to experiment, to calculate, or to create graphic models — is available with selected topics.

Graphics

The ability to visualize problems is a critical skill that students need in order to solve them. To encourage the development of this skill, the text has an abundance of figures, which are computer-generated for accuracy.

180 CHAPTER 2 Introduction to Equations, Graphs, and Functions

2 REVIEW EXERCISES

In Exercises 1–4, determine whether the values of the variable are solutions of the equation.

Equation	Values of the Variable
1. $45 - 7x = 3$	(a) $x = 3$ (b) $x = 6$
2. $3(5 - x) = 6 - x$	(a) $x = \frac{9}{2}$ (b) $x = -\frac{2}{3}$
3. $\frac{x}{7} + \frac{x}{5} = 1$	(a) $x = \frac{35}{12}$ (b) $x = 2$
4. $\frac{x+2}{6} = \frac{7}{2}$	(a) $x = -12$ (b) $x = 19$

In Exercises 5–24, solve the equation and check the result. (Some of the equations have no solutions.)

5. $17 - 7x = 3$

6. $3 + 6x = 51$

7. $4y - 6(y - 5) = 2$

8. $7x + 2(7 - x) = 8$

9. $1.4t + 2.1 = 0.9t - 2$

10. $8(x - 2) = 3(x - 2)$

11. $\frac{3x}{4} = 4$

12. $\frac{5x}{14} = \frac{1}{2}$

13. $\frac{4}{5}x - \frac{1}{10} = \frac{3}{2}$

14. $\frac{1}{4}s + \frac{3}{8} = \frac{5}{2}$

15. $\frac{v - 20}{-8} = 2v$

16. $x + \frac{2x}{5} = 1$

17. $10x(x - 3) = 0$

18. $3x(4x + 7) = 0$

19. $v^2 - 100 = 0$

20. $(x + 3)^2 - 25 = 0$

21. $x^2 - 25x = -150$

22. $4t^2 - 12t = -9$

23. $3s^2 - 2s - 8 = 0$

24. $z(5 - z) + 36 = 0$

In Exercises 25–28, solve the equation and round your answer to two decimal places. (A calculator may be helpful.)

25. $382x - 575 = 715$

26. $3.625x + 3.5 = 22.125$

27. $\frac{x}{2.33} = 14.302$

28. $\frac{7x}{3} + 2.5 = 8.125$

In Exercises 29–32, solve the equation for the specified variable.

29. $2x - 7y + 4 = 0$. Solve for x.

30. $\frac{2}{3}u - 4v = 2v + 3$. Solve for v.

31. $V = \pi r^2 h$. Solve for h.

32. $S = 2\pi r^2 + 2\pi rh$. Solve for h.

In Exercises 33 and 34, plot the points in a rectangular coordinate system.

33. $(0, -3)$, $(\frac{5}{2}, 5)$, $(-2, -4)$

34. $(1, -\frac{3}{2})$, $(-2, 2\frac{3}{4})$, $(5, 10)$

In Exercises 35–38, plot the points and find the distance between them. If appropriate, round the result to two decimal places.

35. $(1, 3)$, $(5, 6)$

36. $(2, -5)$, $(6, -5)$

37. $(-6, -1)$, $(1, 2)$

38. $(-2, 10)$, $(3, -2)$

Graphs, and Functions

$) - f(3)$

$)$

62. $h(x) = \begin{cases} x^3, & \text{if } x \le 1 \\ (x - 1)^2 + 1, & \text{if } x > 1 \end{cases}$

(a) $h(2)$ (b) $h\left(-\frac{1}{2}\right)$ (c) $h(0)$ (d) $h(4) - h(3)$

64. $f(x) = 3x$

(a) $\frac{f(x + 1) - f(1)}{x}$ (b) $\frac{f(x - 5) - f(5)}{x}$

66. $f(x) = 8$

(a) $f(x + h)$ (b) $\frac{f(x + h) - f(x)}{h}$

ion.

67. $h(x) = 4x^2 - 7$

68. $g(s) = \frac{5}{s}$

69. $f(x) = \frac{1}{5 - x}$

70. $G(x) = \sqrt{x + 4}$

In Exercises 71–76, (a) describe the transformation and (b) sketch the graph.

71. $g(x) = -(x + 1)^2$

72. $h(x) = 9 - (x - 2)^2$

73. $f(t) = |t - 2| - 1$

74. $f(x) = |x + 1| - 2$

75. $g(x) = x^3 - 2$

76. $h(x) = (x - 2)^3$

77. Rocker Arm Design Find the distance d between the centers of the two small bolt holes in the rocker arm shown in the figure. Round the result to two decimal places.

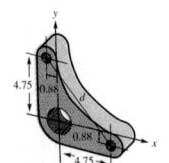

78. Velocity of a Ball The velocity of a ball thrown upward from ground level is given by $v = -32t + 80$, where t is the time in seconds and v is the velocity in feet per second.
(a) Find the velocity when $t = 2$.
(b) Find the time when the ball reaches its maximum height. (*Hint:* Find the time when $v = 0$.)
(c) Find the velocity when $t = 3$.

79. Dimensions of a Rectangle The area of the rectangle in the figure is 48 square inches. Find x.

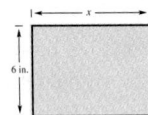

80. Dimensions of a Rectangle The perimeter of the rectangle in the figure is 110 feet. Find the dimensions of the rectangle.

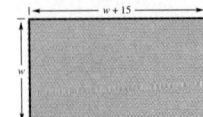

Review Exercises

A set of review exercises appears at the end of each chapter. Answers to all odd-numbered review exercises appear at the end of the text.

Geometry

Geometric formulas and concepts are reviewed throughout the text. For reference, common formulas are presented inside the front and back covers.

2

Chapter Test

Take this test as you would take a test in class. After you are done, check your work against the answers in the back of the book.

1. Determine whether the given value of x is a solution of the equation
$3(5 - 2x) - (3x - 1) = -2$.

(a) $x = -4$ (b) $x = 2$

2. Solve for r in the equation $2(r - s) = 5r - 4s + 1$.

In Exercises 3–8, solve the equation.

3. $6x - 5 = 19$ **4.** $15 - 7(1 - x) = 3(x + 8)$ **5.** $\frac{2x}{3} = \frac{x}{2} + 4$

6. $\frac{t - 5}{12} = \frac{3}{8}$ **7.** $(y + 2)^2 - 9 = 0$ **8.** $12 + 5y - 3y^2 = 0$

9. Determine the quadrant in which the point (x, y) lies if $xy > 0$. Explain your reasoning.

10. Find the distance between the points $(0, 9)$ and $(3, 1)$.

11. Find the x- and y-intercepts of the graph of the equation $y = x(x + 1) - 3(x + 1)$.

12. Sketch the graph of the equation $y = |x - 2|$.

13. Use the vertical line test on the graph in the figure to determine whether the equation
$y^2(4 - x) = x^3$ represents y as a function of x.

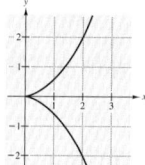

14. Evaluate the function $f(x) = 3x - 2$ at the indicated values.

(a) $f(6)$ (b) $f(x + h) - f(x)$

(c) $f(t + 1)$ (d) $f\left(\frac{x}{3}\right)$

15. Sketch the graph of the function $g(x) = \sqrt{x - 3} + 1$.

Chapter Test

Each chapter contains an end-of-chapter test. Answers to chapter tests are given at the end of the text.

1-3

Cumulative Test

Take this test as you would take a test in class. After you are done, check your work against the answers in the back of the book.

In Exercises 1 and 2, evaluate the expression.

1. $18 - (3 - 8)$ **2.** $-\frac{8}{45} \div \frac{12}{25}$

In Exercises 3–6, perform the indicated operations and simplify.

3. $(2a^2b)^3(-ab^2)^2$ **4.** $3xy(x^2 - 2) - xy(x^2 + 5)$ **5.** $t(3t - 1) - 2t(t + 4)$ **6.** $[2 + (x - y)]^2$

In Exercises 7–10, solve the equation.

7. $12 - 5(3 - x) = x + 3$ **8.** $1 - \frac{x + 2}{4} = \frac{7}{8}$ **9.** $y^2 - 64 = 0$ **10.** $2t^2 - 5t - 3 = 0$

11. Determine whether the equation $x - y^3 = 0$ represents y as a function of x.

12. Find the domain of the function $f(x) = \sqrt{x - 2}$.

13. Given $f(x) = x^2 - 3x$, find   (a) $f(4)$.
                                 (b) $f(c + 3)$.

14. Write an algebraic expression for the statement, "The number n is tripled and the product is decreased by 8."

15. Find two consecutive integers such that their sum is 89 less than their product.

16. The annual insurance premium for a policyholder is \$1225. Find the annual premium if the policyholder must pay a 15% surcharge because of a driving violation.

17. Solve the proportion $\frac{t - 1}{4} = \frac{11}{12}$.

18. Solve for the length x of the side of the second triangle by using the fact that corresponding sides of similar triangles are proportional.

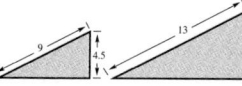

19. Solve and sketch the solution $|x - 2| \geq 3$.

20. The revenue from selling x units of a product is $R = 12.90x$. The cost of producing x units is $C = 8.50x + 450$. For a profit to be obtained, the revenue must be greater than the cost. For what values of x will this product produce a profit?

21. Consider the two points $(-4, 0)$ and $(4, 6)$.   (a) Find the distance between the points.
                                             (b) Find an equation of the line through the points.

22. Sketch a graph of the linear equation $4x + 3y - 12 = 0$.

Cumulative Test

Cumulative tests have been placed after Chapters 3, 6, and 10. These tests reinforce the message that is presented throughout the text — that mathematics is a continuing story and requires constant synthesis and review. Answers are given at the end of the text.

Supplements

Intermediate Algebra: Graphs and Functions by Larson, Hostetler, and Neptune is accompanied by a comprehensive supplements package for maximum teaching effectiveness and efficiency.

Instructor's Annotated Edition

Student Study and Solutions Guide by Carolyn F. Neptune, Johnson County Community College

Complete Solutions Guide by Carolyn F. Neptune, Johnson County Community College

Test Item File with Ready-made Tests by David C. Falvo, The Pennsylvania State University, The Behrend College

Intermediate Algebra: Graphs and Functions Videotapes by Dana Mosely, Valencia Community College

Intermediate Algebra: Graphs and Functions TUTOR by Timothy R. Larson and John R. Musser

Test-Generating Software (Macintosh, IBM)

Study Skills Videotapes by Paul Nolting

This complete supplements package offers ancillary materials for students and instructors and for classroom resources. Each item is keyed directly to the textbook for ease of use. For the convenience of software users, a technical support telephone number is available with all D.C. Heath software products: (617) 860-1218. The components of this comprehensive teaching and learning package are outlined on the following page.

INTERMEDIATE ALGEBRA: GRAPHS AND FUNCTIONS

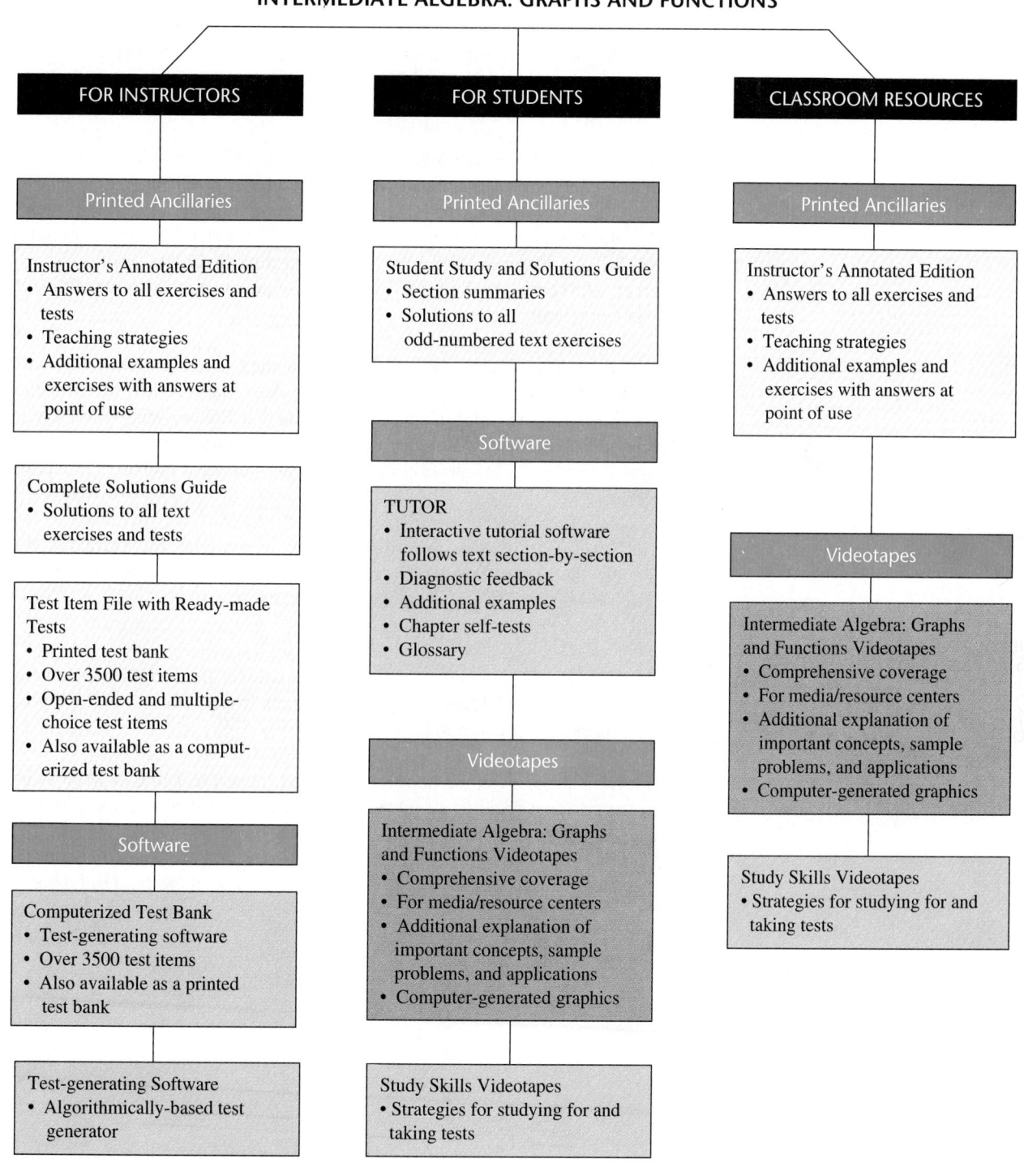

FOR INSTRUCTORS

Printed Ancillaries

Instructor's Annotated Edition
- Answers to all exercises and tests
- Teaching strategies
- Additional examples and exercises with answers at point of use

Complete Solutions Guide
- Solutions to all text exercises and tests

Test Item File with Ready-made Tests
- Printed test bank
- Over 3500 test items
- Open-ended and multiple-choice test items
- Also available as a computerized test bank

Software

Computerized Test Bank
- Test-generating software
- Over 3500 test items
- Also available as a printed test bank

Test-generating Software
- Algorithmically-based test generator

FOR STUDENTS

Printed Ancillaries

Student Study and Solutions Guide
- Section summaries
- Solutions to all odd-numbered text exercises

Software

TUTOR
- Interactive tutorial software follows text section-by-section
- Diagnostic feedback
- Additional examples
- Chapter self-tests
- Glossary

Videotapes

Intermediate Algebra: Graphs and Functions Videotapes
- Comprehensive coverage
- For media/resource centers
- Additional explanation of important concepts, sample problems, and applications
- Computer-generated graphics

Study Skills Videotapes
- Strategies for studying for and taking tests

CLASSROOM RESOURCES

Printed Ancillaries

Instructor's Annotated Edition
- Answers to all exercises and tests
- Teaching strategies
- Additional examples and exercises with answers at point of use

Videotapes

Intermediate Algebra: Graphs and Functions Videotapes
- Comprehensive coverage
- For media/resource centers
- Additional explanation of important concepts, sample problems, and applications
- Computer-generated graphics

Study Skills Videotapes
- Strategies for studying for and taking tests

Acknowledgments

We would like to thank the many people who have helped us prepare the text and supplements package. Their encouragement, criticisms, and suggestions have been invaluable to us.

Reviewers: Lionel Geller, Dawson College; William Grimes, Central Missouri State University; Rosalyn T. Jones, Albany State College; Debra A. Landre, San Joaquin Delta College; Myrna F. Manly, El Camino Community College; James I. McCullough, Arapahoe Community College; Katherine McLain, Cosumnes River College; Karen S. Norwood, North Carolina State University; Nora I. Schukei, University of South Carolina at Beaufort; Kay Stroope, Phillips County Community College.

A special thanks to all the people at D.C. Heath and Company who worked with us in the development of the text, especially Ann Marie Jones, Mathematics Acquisitions Editor; Cathy Cantin, Developmental Editor; Wing-Harn Chen, Developmental Assistant; Karen Carter, Production Editor; Cornelia Boynton, Designer; Carolyn Johnson, Editorial Associate; and Lisa Merrill, Production Supervisor.

David E. Heyd assisted us in writing the text and solving the exercises. We would also like to thank the staff at Larson Texts, Inc., who assisted with proofreading the manuscript; preparing and proofreading the art package; and checking and typesetting the supplements.

On a personal level, we are grateful to our spouses, Deanna Gilbert Larson, Eloise Hostetler, and Harold Neptune for their love, patience, and support. Also, a special thanks goes to R. Scott O'Neil.

If you have suggestions for improving the text, please feel free to write to us. Over the past two decades we have received many useful comments from both instructors and students, and we value these very much.

Roland E. Larson
Robert P. Hostetler
Carolyn F. Neptune

Contents

HOW TO STUDY ALGEBRA   xix

WHAT IS ALGEBRA?   xxii

CHAPTER 1   **Concepts of Elementary Algebra**   1

1.1   Operations with Real Numbers   2

USING A SCIENTIFIC CALCULATOR OR A GRAPHING CALCULATOR   18

1.2   Properties of Real Numbers   21

1.3   Algebraic Expressions   31

MID-CHAPTER QUIZ   45

1.4   Operations with Polynomials   46

1.5   Factoring Polynomials   59

1.6   Factoring Trinomials   70

REVIEW EXERCISES   84

CHAPTER TEST   87

CHAPTER 2   **Introduction to Equations, Graphs, and Functions**   88

2.1   Solving Linear and Quadratic Equations   89

2.2   Literal Equations and Formulas   103

2.3   The Rectangular Coordinate System   114

MID-CHAPTER QUIZ   129

2.4   Graphs of Equations   130

USING A GRAPHING CALCULATOR   145

2.5   Relations, Functions, and Function Notation   151

2.6   Graphs of Functions   166

REVIEW EXERCISES   180

CHAPTER TEST   183

CHAPTER 3   **Linear Functions, Equations, and Inequalities**   185

3.1   Applications of Linear Equations   186

3.2   More Applications: Consumer and Scientific Problems   203

3.3   Linear Inequalities in One Variable   216

MID-CHAPTER QUIZ   230

3.4   Equations and Inequalities Involving Absolute Value   232

3.5   Slope: An Aid to Graphing Linear Functions   243

USING A GRAPHING CALCULATOR   260

3.6   Equations of Lines   263

REVIEW EXERCISES   277

CHAPTER TEST   280

CUMULATIVE TEST: 1–3   281

CHAPTER **4**   Systems of Linear Equations and Inequalities   282

4.1   Systems of Linear Equations in Two Variables   283

USING A GRAPHING CALCULATOR   301

4.2   Systems of Linear Equations in Three Variables   304

4.3   Matrices and Systems of Linear Equations   317

MID-CHAPTER QUIZ   330

4.4   Linear Systems and Determinants   331

4.5   Graphs of Linear Inequalities in Two Variables   344

4.6   Systems of Linear Inequalities and Linear Programming   351

REVIEW EXERCISES   366

CHAPTER TEST   371

CHAPTER **5**   Rational Expressions and Rational Functions   373

5.1   Simplifying Rational Expressions   374

5.2   Multiplying and Dividing Rational Expressions   386

5.3   Adding and Subtracting Rational Expressions   394

MID-CHAPTER QUIZ   407

5.4   Dividing Polynomials   408

5.5   Graphing Rational Functions   420

USING A GRAPHING CALCULATOR   431

5.6   Solving Equations Involving Rational Expressions   433

REVIEW EXERCISES   445

CHAPTER TEST   449

CHAPTER **6**   Radicals and Complex Numbers   450

6.1   Integer Exponents and Scientific Notation   451

6.2   Rational Exponents and Radicals   462

6.3   Simplifying and Combining Radicals   474

MID-CHAPTER QUIZ   483

6.4   Multiplying and Dividing Radical Expressions   484

6.5   Solving Equations Involving Radicals   490

USING A GRAPHING CALCULATOR   499

6.6   Complex Numbers   502

REVIEW EXERCISES   512

CHAPTER TEST   514

CUMULATIVE TEST : 4-6   515

CHAPTER **7**   Quadratic Functions, Equations, and Inequalities   517

7.1   The Factoring and Square Root Methods   518

7.2   Completing the Square   530

7.3   The Quadratic Formula and the Discriminant   540

MID-CHAPTER QUIZ   551

USING A GRAPHING CALCULATOR   552

7.4   Applications of Quadratic Equations   555

7.5   Graphing Quadratic Functions   565

7.6   Nonlinear Inequalities in One Variable   578

REVIEW EXERCISES   591

CHAPTER TEST   594

CHAPTER **8**   Additional Functions and Relations   595

8.1   Combinations of Functions   596

8.2   Inverse Functions   605

8.3   Variation and Mathematical Models   618

MID-CHAPTER QUIZ   629

8.4   Polynomial Functions and Their Graphs   630

8.5   Circles, Ellipses, and Hyperbolas   639

8.6   Nonlinear Systems of Equations   653

USING A GRAPHING CALCULATOR   663

REVIEW EXERCISES   665

CHAPTER TEST   668

CHAPTER **9**   **Exponential and Logarithmic Functions**   **669**

9.1   Exponential and Logarithmic Expressions   670

9.2   Properties of Logarithms   678

9.3   Exponential Functions and Their Graphs   685

MID-CHAPTER QUIZ   698

9.4   Logarithmic Functions and Their Graphs   699

USING A GRAPHING CALCULATOR   708

9.5   Solving Exponential and Logarithmic Equations   712

9.6   Applications of Exponential and Logarithmic Functions   719

REVIEW EXERCISES   730

CHAPTER TEST   734

CHAPTER **10**   **Additional Topics in Algebra**   **735**

10.1   Sequences   736

10.2   Arithmetic Sequences   746

10.3   Geometric Sequences   755

MID-CHAPTER QUIZ   765

10.4   The Binomial Theorem   766

10.5   Counting Principles, Permutations, and Combinations   773

10.6   Probability   783

USING A RANDOM NUMBER GENERATOR   793

REVIEW EXERCISES   796

CHAPTER TEST   798

CUMULATIVE TEST: 7–10   799

Appendix   **Introduction to Logic**   **A1**

A.1   Introduction to Logic   A2

A.2   Implications, Quantifiers, and Venn Diagrams   A8

A.3   Logical Argument   A16

Answers to Odd-Numbered Exercises   **A27**

Index of Applications   **A98**

Index   **A101**

How to Study Algebra

Studying Mathematics

Studying mathematics requires a different approach from many other subjects because it is a linear process. In other words, the material learned on one day is built upon material learned previously. It is necessary to keep up with course-work day by day — there are no shortcuts.

Making a Plan

Make your own course plan right now! Determine the number of hours per week you need to spend on algebra. A good guideline is to study two to four hours for every hour in class. After your first major test you will know if your efforts were sufficient. If you did not make the grade you wanted, then you should increase your study time, improve your study efficiency, or both.

Preparing for Class

Before attending class, read the portion of the text that is to be covered, paying special attention to the definitions and rules that are boxed in blue. This practice takes a lot of self-discipline, but it pays off. If time does not permit, then you should at least review your previous day's notes. Going to class prepared will enable you to benefit much more from your instructor's presentation. Algebra, like most other technical subjects, is easier to understand the second or third time you hear it.

Attending Class

Attend every class. Arrive on time with your text, a pen or pencil, paper for notes, and your calculator. If you have to miss a class, get the notes from another student, get help from your tutor, or view the appropriate mathematics video-tape. You must learn the information that was taught in the missed class before attending the next class. Remember, learning mathematics is linear — do not miss class!

Participating in Class

As you read the text before class, write down any questions that you have about the material. Ask your instructor these questions during class to save yourself time and frustration with your homework.

Taking Notes

Take notes in class, especially on definitions, examples, concepts, and rules. Focus on the instructor's cues to indicate important material. Then, as soon after class as possible, read through your notes, adding any explanations that are necessary to make your notes understandable *to you*.

Doing the Homework

Learning algebra is like learning to play the piano or learning to play basketball. You cannot become skilled just by watching someone else do it. You must also do it yourself. The best time to do your homework is right after class, when the concepts are still fresh in your mind. Immediately doing the homework increases the chances of retaining the information in long-term memory.

Finding a Study Partner

When you get stuck on a problem, it may help to try to work with someone else. Even if you feel you are giving more help than you are getting, you will find that teaching others is also an excellent way to learn.

Building a Math Library

Start building a library of books that can help you with this course and future math courses. Consider using the *Student Study and Solutions Guide* that accompanies the text. Also, as you will probably be taking other math courses after you finish this course, we suggest that you keep the text. It will be a valuable reference book. Adding computer software and math videotapes is another way to build your mathematics library.

Keeping Up with the Work

Don't let yourself fall behind in the course. If you think that you are having trouble, seek help immediately. Ask your instructor, attend your school's tutoring service, talk with your study partner, use additional study aids such as videos or software tutorial — do something. If you are having trouble with the material in one chapter of your algebra text, there is a good chance that you will also have trouble with later chapters.

Getting Stuck

Everyone who has ever taken a math course has had this experience: you are working on a problem and cannot solve it, or you have solved it but your answer does not agree with the answer given at the end of the book. People have different approaches to this problem. You might ask for help, take a break to clear your thoughts, sleep on it, rework the problem, or reread the section in the text. Don't get frustrated or spend too much time on a single problem.

Keeping Your Skills Sharp

Before each exercise set in the text we have included a short set of *Warm-Up Exercises*. These exercises will help you review skills that you learned in previous exercises and retain them in long-term memory. These sets are designed to take only a few minutes to solve. We suggest working the entire set before you start each new exercise set. (All of the *Warm-Up Exercises* are answered at the end of the text.)

Checking Your Work

One of the nice things about algebra is that you can test your own solution. One way to do this is to plug the answers into the equation, then solve it to see if the numbers on each side of the equation are equal. Another way to check your work is to redo the problem on a separate sheet of paper to see if your two answers match. If they don't match, then compare each of the problem steps. Locating your systematic test errors can improve your test score. If, in addition to your "solving skills," you work on your "checking skills," you should find your test scores improving.

Preparing for Exams

Cramming for algebra exams seldom works. If you have kept up with the work and followed the suggestions given here, you should be almost ready for the exam. We have included three features that should help as a final preparation. Review the *Mid-Chapter Quiz*, work the *Review Exercises*, and set aside an hour to take the sample *Chapter Test*. Analyze the results from your *Chapter Test* to locate test-taking errors.

Taking Exams

Most instructors suggest that you do not study right up to the minute you are taking a test. This practice tends to make people anxious. The best cure for anxiousness during tests is to prepare well before taking the test. Once the test has begun, read the directions carefully and try to work at a reasonable pace. (You might want to read the entire test first, then work the problems in the order with which you feel most comfortable.) Hurrying tends to cause people to make careless errors. If you finish early, take a few moments to clear your thoughts and then take time to go over your work.

Learning from Mistakes

When you get an exam back, be sure to go over any errors that you might have made. Reviewing your test will save you from making the same systematic or conceptual errors over and over again. Don't be too quick to pass off an error as just a "dumb mistake." Take advantage of any mistakes by continually hunting for ways to improve your test-taking abilities.

What Is Algebra?

To some, algebra is manipulating symbols or performing mathematical operations with letters instead of numbers. To others, it is factoring, solving equations, or solving word problems. And to still others, algebra is a mathematical language that can be used to model real-world problems. In fact, algebra is all of these!

As you study this text, it is helpful to view algebra from the "big picture" — to see how the various rules, operations, and strategies fit together.

The rules of arithmetic are generalized through the use of symbols and letters to form the basic rules of algebra. The algebra rules are used to *rewrite* algebraic expressions and equations in new, more useful forms. The ability to rewrite algebraic expressions and equations is needed in all three major components of algebra — *simplifying* expressions, *solving* equations, and *graphing* functions. The following chart shows how this text fits into the "big picture" of algebra.

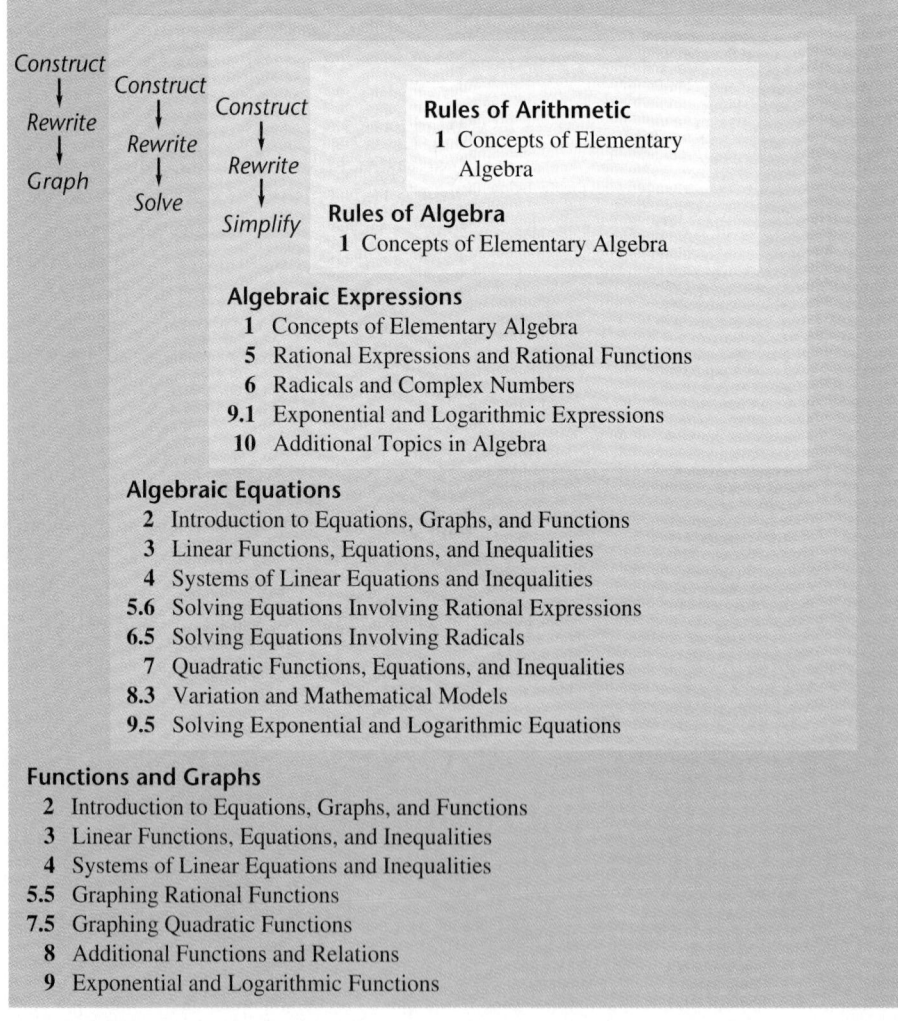

Construct
↓
Rewrite
↓
Graph

Construct
↓
Rewrite
↓
Solve

Construct
↓
Rewrite
↓
Simplify

Rules of Arithmetic
1 Concepts of Elementary Algebra

Rules of Algebra
1 Concepts of Elementary Algebra

Algebraic Expressions
1 Concepts of Elementary Algebra
5 Rational Expressions and Rational Functions
6 Radicals and Complex Numbers
9.1 Exponential and Logarithmic Expressions
10 Additional Topics in Algebra

Algebraic Equations
2 Introduction to Equations, Graphs, and Functions
3 Linear Functions, Equations, and Inequalities
4 Systems of Linear Equations and Inequalities
5.6 Solving Equations Involving Rational Expressions
6.5 Solving Equations Involving Radicals
7 Quadratic Functions, Equations, and Inequalities
8.3 Variation and Mathematical Models
9.5 Solving Exponential and Logarithmic Equations

Functions and Graphs
2 Introduction to Equations, Graphs, and Functions
3 Linear Functions, Equations, and Inequalities
4 Systems of Linear Equations and Inequalities
5.5 Graphing Rational Functions
7.5 Graphing Quadratic Functions
8 Additional Functions and Relations
9 Exponential and Logarithmic Functions

Concepts of Elementary Algebra

1.1   Operations with Real Numbers

1.2   Properties of Real Numbers

1.3   Algebraic Expressions

1.4   Operations with Polynomials

1.5   Factoring Polynomials

1.6   Factoring Trinomials

Each chapter begins with a list of topics to be covered and an introductory application featuring a mathematical model.

The sum s of the measures of the angles of an n-sided polygon is given by

$$s = 180° (n - 2).$$

Records of this formula date back to ancient Greece—even before the time of Euclid, who lived in the third century B.C. For a regular polygon, each angle has the same measure a, and is given by

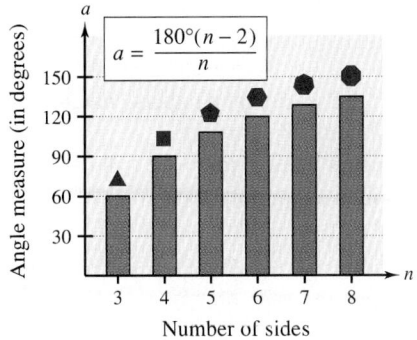

$$a = \frac{180°\,(n-2)}{n}.$$

Angle measure (in degrees)

$$a = \frac{180°(n-2)}{n}$$

Number of sides

The top face of the gemstone shown above is a regular octagon (n = 8). What is the sum of the measures of its angles? What is the measure of each of its angles?

1

The Real Numbers and Their Subsets

A **set** is a collection of objects.* For instance, the set {1, 2, 3} contains the three numbers 1, 2, and 3. In this text, a *pair* of braces { } always indicates that we are describing the members of a set. Parentheses ( ) and brackets [ ] are used to represent other ideas.

The set of numbers that is used in arithmetic is the set of **real numbers.** The term *real* distinguishes real numbers from *imaginary* numbers—a type of number that you will study later in this text.

One of the most commonly used subsets of real numbers is the set of **natural numbers** or **positive integers**

$$\{1,\ 2,\ 3,\ 4,\ \ldots\}.$$

The three dots indicate that the pattern continues (the set also contains the numbers 5, 6, 7, and so on).

Positive integers can be used to describe many quantities that you encounter in everyday life—you might be taking four classes this term, or you might be paying $420 per month for rent. But even in everyday life, positive integers cannot describe some concepts accurately. For instance, you could have a zero balance in your checking account, or the temperature could be $-10°$ (10 degrees below zero). To describe such quantities, you need to expand the set of positive integers to include **zero** and the **negative integers.** The positive integers and zero make up the set of **whole numbers.** The expanded set containing the whole numbers and the negative integers is called the set of **integers,** which is written as follows.

$$\underbrace{\{\ldots,\ -3,\ -2,\ -1,}_{\substack{\text{Negative} \\ \text{integers}}}\ \overbrace{0,}^{\text{Zero}}\ \underbrace{1,\ 2,\ 3,\ \ldots\}}_{\substack{\text{Positive} \\ \text{integers}}}$$

The set of integers is a **subset** of the set of real numbers, which is another way of saying that every integer is a real number.

Even when the entire set of integers is used, there are still many quantities in everyday life that cannot be described accurately. The costs of many items are not in whole-dollar amounts, but in parts of dollars, such as $1.19 or $39.98. You might work $8\frac{1}{2}$ hours, or you might miss the first *half* of a movie. To describe such

---

*Whenever a mathematical term is formally introduced in this text, the word will appear in boldface type. Be sure you understand the meaning of each new word—it is important that each new word become part of your mathematical vocabulary.

quantities, you can expand the set of integers to include **fractions.** The expanded set is called the set of **rational numbers.** Formally, a real number is called **rational** if it can be written as the ratio p/q of two integers, where $q \neq 0$ (the symbol $\neq$ means **does not equal**). For instance,

$$2 = \frac{2}{1}, \quad \frac{1}{3} = 0.333 \ldots, \quad \frac{1}{8} = 0.125, \quad \text{and} \quad \frac{125}{111} = 1.126126 \ldots$$

are rational numbers. Real numbers that cannot be written as ratios of two integers are called **irrational.** For instance, the numbers

$$\sqrt{2} = 1.4142135 \ldots \quad \text{and} \quad \pi = 3.1415926 \ldots$$

are irrational. The decimal representation of a rational number is either **terminating** or **repeating.** For instance, the decimal representation of $\frac{1}{4} = 0.25$ is terminating, and the decimal representation of $\frac{4}{11} = 0.363636\cdots = 0.\overline{36}$ is repeating. (The line over 36 indicates which digits repeat.)

The decimal representation of an irrational number neither terminates nor repeats. When performing operations with such numbers, you usually use a decimal approximation that has been **rounded.*** For instance, rounded to four decimal places, the decimal approximations of $\frac{2}{3}$ and π are

$$\frac{2}{3} \approx 0.6667 \quad \text{and} \quad \pi \approx 3.1416.$$

The symbol $\approx$ means **equals approximately.**

Figure 1.1 shows several commonly used subsets of real numbers and their relationships to each other.

FIGURE 1.1

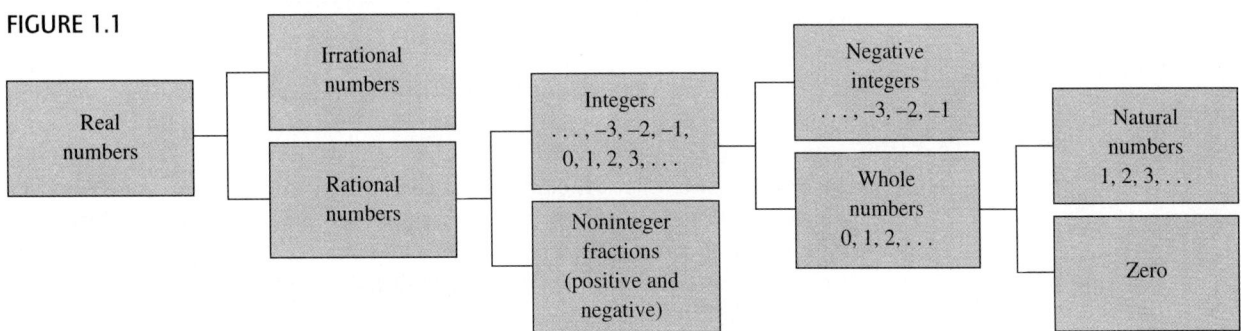

---

*The rounding rule we use in this text is to round *up* if the succeeding digit is 5 or more and round *down* if the succeeding digit is 4 or less. For example, if you wanted to round 7.35 to one decimal place, you would round up to 7.4. Similarly, if you wanted to round 2.364 to two decimal places, you would round down to 2.36.

EXAMPLE 1   ■   Identifying Real Numbers

Which of the numbers in the following set are (a) natural numbers, (b) integers, (c) rational numbers, and (d) irrational numbers?

$$\left\{-7, -\sqrt{3}, -1, -\frac{1}{5}, 0, \frac{3}{4}, \sqrt{2}, \pi, 5\right\}$$

Solution

(a) Natural numbers: $\{5\}$

(b) Integers: $\{-7, -1, 0, 5\}$

(c) Rational numbers: $\left\{-7, -1, -\frac{1}{5}, 0, \frac{3}{4}, 5\right\}$

(d) Irrational numbers: $\{-\sqrt{3}, \sqrt{2}, \pi\}$ ■

Order and Distance on the Real Number Line

The picture that is used to represent the real numbers is called the **real number line.** It consists of a horizontal line with a point (the **origin**) labeled as 0. Numbers to the left of 0 are **negative** and numbers to the right of 0 are **positive,** as shown in Figure 1.2.

FIGURE 1.2

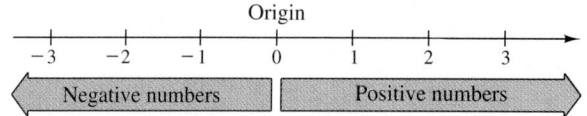

The real number zero is neither positive nor negative. Thus, when you want to talk about real numbers that might be positive or zero, you should use the term **nonnegative real number.**

Each point on the real number line corresponds to exactly one real number, and each real number corresponds to exactly one point on the real number line, as shown in Figure 1.3. When you draw the point (on the real number line) that corresponds to a real number, you are **plotting** the real number.

FIGURE 1.3

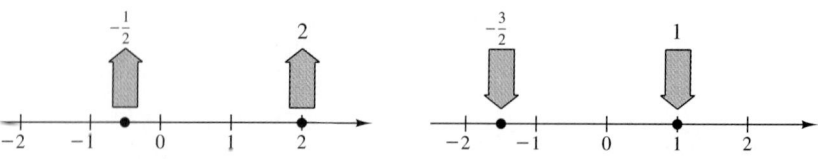

Each point on the real number line corresponds to a real number.

Each real number corresponds to a point on the real number line.

FIGURE 1.4

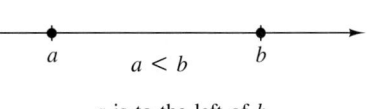

a is to the left of b.

The real number line provides you with a way of comparing any two real numbers. For instance, if you choose any two (different) numbers on the real number line, then one of the numbers must be to the left of the other number. The number to the left is **less than** the number to the right. Similarly, the number to the right is **greater than** the number to the left, as shown in Figure 1.4.

Definition of Order on the Real Number Line	If the real number a lies to the left of the real number b on the real number line, then we say that a is **less than** b and write $a < b.$ This relationship can also be described by saying that b is **greater than** a and by writing $b > a$. The symbol $a \leq b$ means that a is **less than or equal to** b, and the symbol $b \geq a$ means that b is **greater than or equal to** a. The symbols $<, >, \leq,$ and $\geq$ are called **inequality symbols.**

When you are asked to **order** two numbers, you are simply being asked to say which of the two numbers is greater.

EXAMPLE 2   ■   Ordering Real Numbers

The "point" of the arrow always points to the number on the left on the number line.

Determine the correct inequality symbol ($<$ or $>$) between the numbers in each of the following pairs.

(a) -3 ⬚ -5 (b) $\frac{1}{5}$ ⬚ $\frac{1}{3}$

Solution

(a) Because -3 lies to the right of -5 on the real number line, you can say that -3 is *greater than* -5, and write

$$-3 > -5.$$

See Figure 1.5(a).

(b) Because $\frac{1}{5}$ lies to the left of $\frac{1}{3}$ on the real number line, you can say that $\frac{1}{5}$ is *less than* $\frac{1}{3}$, and write

$$\frac{1}{5} < \frac{1}{3}.$$

See Figure 1.5(b).

FIGURE 1.5   (a)

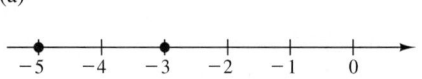

(b)

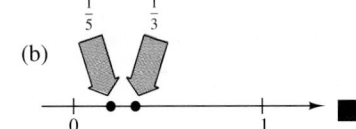

■

Once you know how to represent real numbers as points on the real number line, it is natural to talk about the **distance between two real numbers.** Specifically, if a and b are two real numbers such that $a \leq b$, then the distance between a and b is defined as $b - a$.

Definition of Distance Between Two Real Numbers

If a and b are two real numbers such that $a \leq b$, then the **distance between a and b is**

(Distance between a and b) $= b - a$.

Note from this definition that if $a = b$, then the distance between a and b is zero. If $a < b$, then the distance between a and b is positive.

EXAMPLE 3   ■   Finding the Distance Between Two Real Numbers

Find the distance between the real numbers in each of the following pairs.

(a) -2 and 3   (b) 0 and 4

Solution

(a) The distance between -2 and 3 is

$$3 - (-2) = 3 + 2 = 5,$$

as shown in Figure 1.6(a).

(b) The distance between 0 and 4 is

$$4 - 0 = 4,$$

as shown in Figure 1.6(b).

FIGURE 1.6   (a)

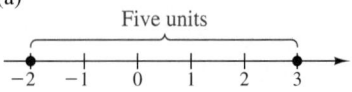
Five units

(b)

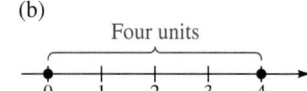
Four units

The Absolute Value of a Real Number

The distance between a real number a and 0 (the origin) is called the **absolute value** of a. Double vertical bars | | are used to denote absolute value. For example, $|-8|$ denotes the distance between -8 and 0. Thus, $|-8| = 8$.

| **Definition of Absolute Value of a Real Number** | The **absolute value** of a real number a is defined as the distance between a and 0 on the real number line. |

Rule   *Example*

1. If $a > 0$, then $|a| = a - 0 = a$. $|3| = 3$

2. If $a = 0$, then $|a| = 0 - 0 = 0$. $|0| = 0$

3. If $a < 0$, then $|a| = 0 - a = -a$. $|-2| = -(-2) = 2$

The absolute value of any real number is either positive or zero. Moreover, the only real number whose absolute value is zero is zero. That is, $|0| = 0$.

Commuting to and from work via the same route can be a good example for illustrating distance and absolute value. The distance between your workplace and your home is the same regardless of the direction (assuming the same route), and the real number value is never negative.

NOTE Be sure you see from this definition that the absolute value of a real number is never negative. For instance, if $a = -3$, then $|a| = |-3| = -(-3) = 3 = -a$.

Two real numbers are said to be **opposites** of each other if they lie the same distance from, but on opposite sides of, zero. For example, -2 is the opposite of 2. Because opposite numbers lie the same distance from zero on the real number line, they have the same absolute value. Thus, $|5| = 5$ and $|-5| = 5$.

EXAMPLE 4   ■   Evaluating Expressions That Involve Absolute Value

(a) $\left|\frac{3}{4}\right| = \frac{3}{4}$ (b) $|-3.2| = 3.2$ (c) $-|-6| = -(6) = -6$

Note that part (c) does not contradict the fact that the absolute value of a number cannot be negative. In this case, the minus sign in the answer comes from the minus sign in $-|-6|$ that is *outside* the absolute value bars.   ■

For any two real numbers a and b, exactly one of the following must be true: $a < b$, $a = b$, or $a > b$. This property of real numbers is called the **Law of Trichotomy.**

EXAMPLE 5   ■   Comparing Real Numbers

Determine the correct symbol ($<$, $>$, or $=$) between the real numbers in each pair.

(a) $|-9|$ ▨ $|9|$ (b) $|-2|$ ▨ 1 (c) -4 ▨ $-|-4|$

Solution

(a) $|-9| = |9|$, because both are equal to 9.

(b) $|-2| > 1$, because $|-2| = 2$ and 2 is greater than 1.

(c) $-4 = -|-4|$, because both numbers are equal to -4.   ■

When we defined the distance between two real numbers a and b as $b - a$, the definition included the restriction $a \leq b$. Using absolute value, you can generalize this definition as follows. If a and b are *any* two real numbers, then the distance between a and b is

(Distance between a and b) = $|b - a|$ or, equivalently, $|a - b|$.

For instance, the distance between -2 and 1 is

$$|-2 - 1| = |-3| = 3.$$

Operations with Real Numbers

There are four basic operations of arithmetic: addition, subtraction, multiplication, and division.

The result of **adding** two real numbers is called the **sum** of the two numbers, and the two real numbers are called **terms** of the sum. **Subtraction** of one real number from another can be described as *adding the opposite* of the second number to the first number. For instance,

$$7 - 5 = 7 + (-5) = 2 \quad \text{and} \quad 10 - (-13) = 10 + 13 = 23.$$

The result of subtracting one real number from another is called the **difference** of the two numbers.

EXAMPLE 6   ■   Adding and Subtracting Real Numbers

(a) $-84 + 14 = -70$

(b) $6 + (-13) + 10 = 3$

(c) $-13.8 - 7.02 = -13.8 + (-7.02) = -20.82$

(d) $\dfrac{1}{5} - \dfrac{2}{5} = \dfrac{1-2}{5} = -\dfrac{1}{5}$

(e) To add two fractions with *unlike denominators*, you must first rewrite one (or both) of the fractions so that they have the same denominator.

$$\frac{3}{8} + \frac{5}{12} = \frac{3(3)}{8(3)} + \frac{5(2)}{12(2)} \qquad \text{Common denominator is 24}$$

$$= \frac{9}{24} + \frac{10}{24} \qquad \text{Fractions have like denominators}$$

$$= \frac{9 + 10}{24} \qquad \text{Add numerators}$$

$$= \frac{19}{24}$$ ■

The result of **multiplying** two real numbers is called their **product,** and each of the two numbers is called a **factor** of the product. The product of zero and any other number is zero. For instance, if you multiply 0 and 4, you obtain $(0)(4) = 0$. Multiplication is denoted in a variety of ways. For instance,

$$7 \times 3, \quad 7 \cdot 3, \quad 7(3), \quad \text{and} \quad (7)(3)$$

all denote the product of "7 times 3," which you know is 21.

EXAMPLE 7 ■

Initiate a discussion about why a negative number times a negative number is a positive number. A pattern approach can be used to illustrate the product. Alternatively, assign the question as an essay.

Multiplying Real Numbers

(a) $(-5)(-7) = 35$

(b) $(-1.2)(0.4) = -0.48$

(c) To find the product of more than two numbers, find the product of their absolute values. If there is an *even* number of negative factors, then the product is positive. If there is an *odd* number of negative factors, then the product is negative. For instance, in the following product there are two negative factors, so the product must be positive, and you can write

$$5(-3)(-4)(7) = 420.$$

(d) To multiply two fractions, multiply their numerators and their denominators. For instance, the product of $\frac{2}{3}$ and $\frac{4}{5}$ is

$$\left(\frac{2}{3}\right)\left(\frac{4}{5}\right) = \frac{(2)(4)}{(3)(5)} = \frac{8}{15}. \qquad \blacksquare$$

The **reciprocal** of a nonzero real number a is defined to be the number that a must be multiplied by to obtain 1. For instance, the reciprocal of 3 is $\frac{1}{3}$ because

$$3\left(\frac{1}{3}\right) = 1.$$

Similarly, the reciprocal of $-\frac{4}{5}$ is $-\frac{5}{4}$ because

$$-\frac{4}{5}\left(-\frac{5}{4}\right) = 1.$$

In general, the reciprocal of a/b is b/a. Note that the reciprocal of a positive number is positive, and the reciprocal of a negative number is negative. Also, be sure you see that zero does not have a reciprocal because there is no number that can be multiplied by zero to obtain 1.

To divide one real number by a second (nonzero) real number, multiply the first number by the reciprocal of the second number. The result of dividing two real numbers is called the **quotient** of the numbers. The quotient of a and b can be written as

$$a \div b, \quad a/b, \quad \text{or} \quad \frac{a}{b}.$$

The number a is called the **numerator** (or **dividend**), and the number b is called the **denominator** (or **divisor**).

EXAMPLE 8   ■   Dividing Real Numbers

(a) $-30 \div 5 = -30\left(\dfrac{1}{5}\right) = -\dfrac{30}{5} = -6$

(b) $-\dfrac{9}{14} \div -\dfrac{1}{3} = -\dfrac{9}{14}\left(-\dfrac{3}{1}\right) = \dfrac{27}{14}$

(c) $\dfrac{5}{16} \div 2\dfrac{3}{4} = \dfrac{5}{16} \div \dfrac{11}{4} = \dfrac{5}{16}\left(\dfrac{4}{11}\right) = \dfrac{5(4)}{4(4)(11)} = \dfrac{5}{44}$

(d) $\dfrac{-\frac{2}{3}}{\frac{3}{5}} = -\dfrac{2}{3} \div \dfrac{3}{5} = -\dfrac{2}{3}\left(\dfrac{5}{3}\right) = -\dfrac{10}{9}$ ■

Note that division by zero is *not* defined. For instance, the expression $\frac{3}{0}$ is undefined.

Positive Integer Exponents

Let n be a positive integer and let a be a real number. Then the product of n factors of a is given by

$$a^n = \underbrace{a \cdot a \cdot a \cdots a}_{n \text{ factors}}.$$

In the **exponential form** a^n, a is called the **base** and n is the **exponent.** When you write the exponential form a^n, you are **raising a to the nth power.**

When a number is raised to the *first* power, you usually do not write the exponent 1. For instance, you would usually write 5 rather than 5^1.

EXAMPLE 9   ■   Evaluating Exponential Expressions

(a) $(-3)^4 = (-3)(-3)(-3)(-3) = 81$

(b) $-3^4 = -(3)(3)(3)(3) = -81$

(c) $-(-3)^4 = -(-3)(-3)(-3)(-3) = -81$

(d) $\left(\frac{2}{5}\right)^3 = \left(\frac{2}{5}\right)\left(\frac{2}{5}\right)\left(\frac{2}{5}\right) = \frac{8}{125}$

(e) $(-2)^5 = (-2)(-2)(-2)(-2)(-2) = -32$

(f) $-2^5 = -(2)(2)(2)(2)(2) = -32$ ■

Roots

When you multiply a number by itself, you are **squaring** the number. For instance, $5^2 = 25$. To undo this "squaring operation," you can take the **square root** of 25. Similarly, you can undo a cubing operation by taking a **cube root.**

Historical Note: In the 16th century, as many as four different symbols were used to indicate a square root. Other symbols were used for cube roots.

Number	Equal Factors	Root
$25 = 5^2$	$5 \cdot 5$	5 (square root)
$25 = (-5)^2$	$(-5)(-5)$	-5 (square root)
$9 = 3^2$	$3 \cdot 3$	3 (square root)
$-27 = (-3)^3$	$(-3)(-3)(-3)$	-3 (cube root)
$16 = 2^4$	$2 \cdot 2 \cdot 2 \cdot 2$	2 (fourth root)

In general, the **nth root** of a number is defined as follows.

Definition of *n*th Root of a Number

Let a and b be real numbers and let n be a positive integer such that $n \geq 2$. If

$$a = b^n,$$

then b is an ***n*th root of *a***. If $n = 2$, the root is a **square root,** and if $n = 3$, the root is a **cube root.**

By applying this definition, you can see that some numbers have more than one *n*th root. For example, both 5 and -5 are square roots of 25 because $25 = 5^2$ and $25 = (-5)^2$. Similarly, both 2 and -2 are fourth roots of 16 because $16 = 2^4$ and $16 = (-2)^4$. To avoid ambiguity, if a is a real number, then the **principal *n*th root of *a*** is defined as the *n*th root that has the same sign as a, and it is denoted by the **radical symbol**

$$\sqrt[n]{a}. \qquad \text{Principal } n\text{th root}$$

The positive integer n is called the **index** of the radical, and the number a is called the **radicand.** If $n = 2$, we omit the index and write $\sqrt{a}$ rather than $\sqrt[2]{a}$. "Having the same sign" means that the principal *n*th root of a is positive if a is positive and negative if a is negative. For example, $\sqrt{4} = 2$ and $\sqrt[3]{-8} = -2$.

EXAMPLE 10 ■ Finding the Principal *n*th Root of a Number

(a) $\sqrt{36} = 6$

(b) $\sqrt{0} = 0$

(c) The number -4 has no real square root because there is no real number that can be squared to produce -4. Thus, $\sqrt{-4}$ is not a real number. This same comment can be made about any negative number. That is, if a is negative, then $\sqrt{a}$ is not a real number.

(d) $\sqrt[3]{8} = 2$

(e) $\sqrt[3]{-27} = -3$

(f) $\sqrt[4]{256} = 4$ ■

Order of Operations

An established **order of operations** helps avoid confusion when one is evaluating numerical expressions.

Order of Operations

1. Perform operations that occur within grouping symbols such as parentheses or brackets.

2. Evaluate powers and roots.

3. Perform multiplications and divisions from left to right.

4. Perform additions and subtractions from left to right.

EXAMPLE 11   ■   Evaluating Expressions Using Order of Operations

$$(a)\ -4 + 2(-2 + 5)^2 = -4 + 2(3)^2 \qquad \text{Add within parentheses}$$
$$= -4 + 2(9) \qquad \text{Evaluate the power}$$
$$= -4 + 18 \qquad \text{Multiply}$$
$$= 14 \qquad \text{Add}$$

This is a good time to demonstrate informally the use of the calculator. Have students do each of the following by hand and then check their answers on the calculator.

$$3 + 8 \div 2 + 1 = 8$$
$$5 - 2^2 \cdot 3 + 7 = 0$$

$$(b)\ (-4 + 2)(-2 + 5)^2 = (-2)(3)^2 \qquad \text{Add within parentheses}$$
$$= (-2)(9) \qquad \text{Evaluate the power}$$
$$= -18 \qquad \text{Multiply}$$

$$(c)\ 5 - 12 \div 3 - 7 = 5 - 4 - 7 \qquad \text{Divide first}$$
$$= -6 \qquad \text{Subtract from left to right}$$

$$(d)\ (-3)(-2)^2 = (-3)(4) \qquad \text{Evaluate the power}$$
$$= -12 \qquad \text{Multiply}$$

$$(e)\ (-3)(-2^2) = (-3)(-4) \qquad \text{Evaluate the power}$$
$$= 12 \qquad \text{Multiply}$$ ■

DISCUSSION PROBLEM   ■   Adding Fractions

There are two basic ways to add fractions with different denominators. (a) Rewrite the fractions using a common denominator and then add using the rule for adding fractions with the same denominator. (b) Use the rule

$$\frac{a}{c} + \frac{b}{d} = \frac{ad + bc}{cd}.$$

Write a short paragraph describing the advantages and disadvantages of each technique. Which of the following solutions is more efficient?

(a) $\dfrac{3}{5} + \dfrac{7}{10} = \dfrac{3(2)}{5(2)} + \dfrac{7}{10}$

$\qquad\qquad = \dfrac{6}{10} + \dfrac{7}{10}$

$\qquad\qquad = \dfrac{13}{10}$

(b) $\dfrac{3}{5} + \dfrac{7}{10} = \dfrac{3(10) + 7(5)}{5(10)}$

$\qquad\qquad = \dfrac{65}{50}$

$\qquad\qquad = \dfrac{\cancel{5}(13)}{\cancel{5}(10)}$

$\qquad\qquad = \dfrac{13}{10}$ ∎

1.1 EXERCISES

1. Which of the real numbers in the following set are (a) natural numbers, (b) integers, (c) rational numbers, and (d) irrational numbers?

$$\left\{ -10,\ -\sqrt{5},\ -\dfrac{2}{3},\ -\dfrac{1}{4},\ 0,\ \dfrac{5}{8},\ 1,\ \sqrt{3},\ 4,\ 2\pi,\ 6 \right\}$$

2. Which of the real numbers in the following set are (a) natural numbers, (b) integers, (c) rational numbers, and (d) irrational numbers?

$$\left\{ -\dfrac{7}{2},\ -\sqrt{6},\ -\dfrac{\pi}{2},\ -\dfrac{3}{8},\ 0,\ \sqrt{15},\ \dfrac{10}{3},\ 8,\ 245 \right\}$$

In Exercises 3–10, plot the two real numbers on the real number line and determine the correct inequality symbol ($<$ or $>$) between the two numbers.

3. 2 ▨ 5

4. −8 ▨ 3

5. −7 ▨ −2

6. −2 ▨ −5

7. $\dfrac{1}{3}$ ▨ $\dfrac{1}{4}$

8. $-\dfrac{3}{2}$ ▨ $-\dfrac{5}{2}$

9. $-\dfrac{5}{8}$ ▨ $\dfrac{1}{2}$

10. $-\pi$ ▨ -3

In Exercises 11–18, find the distance between the pair of real numbers.

11. 4 and 10

12. 14 and −6

13. −12 and 7

14. 0 and 125

15. −8 and 0

16. −54 and 32

17. −6 and −45

18. −300 and −110

In Exercises 19–24, evaluate the quantity.

19. $|-225|$

20. $|62|$

21. $-|16|$

22. $-\left|\dfrac{3}{8}\right|$

23. $-|-85|$

24. $-|-36.5|$

In Exercises 25–28, find the opposite of the number and the absolute value of the number.

25. 34

26. −52

27. $-\dfrac{3}{11}$

28. $\dfrac{5}{32}$

In Exercises 29–32, plot the number and its opposite on the real number line. In each case, notice that the absolute value of the number is the distance between zero and each of the two numbers you have plotted.

29. 7

30. $\dfrac{7}{4}$

31. $-\dfrac{3}{5}$

32. −4.25

In Exercises 33–38, place the correct symbol ($<$, $>$, or $=$) between the pair of real numbers.

33. $|3|$ $|-6|$ **34.** $|-8|$ $|5|$ **35.** $|-4|$ $|4|$ **36.** $|12|$ $|-12|$

37. $-|-2|$ $-|-3|$ **38.** -5 $-|-5|$

In Exercises 39–44, write the statement using inequality notation.

39. x is negative.   **40.** t is positive.   **41.** x is nonnegative.   **42.** u is at least 16.

43. The price p of a coat will be less than \$225 during a sale.

44. The tire pressure p must be at least 30 pounds per square inch.

In Exercises 45–60, perform the indicated operation(s).

45. $13 + 32$ **46.** $-16 + 84$ **47.** $13 + (-32)$ **48.** $-22 - 6$

49. $4 - 16 + (-8)$ **50.** $5.8 - 6.2 + 1.1$ **51.** $\frac{3}{8} + \frac{7}{8}$ **52.** $\frac{5}{9} - \frac{1}{9}$

53. $\frac{3}{5} + \left(-\frac{1}{2}\right)$ **54.** $\frac{5}{6} - \frac{3}{4}$ **55.** $5\frac{3}{4} + 7\frac{3}{8}$ **56.** $8\frac{1}{2} - 24\frac{2}{3}$

57. $85 - |-25|$ **58.** $-36 + |-8|$ **59.** $-|-15.667| - 12.333$ **60.** $-\left|-15\frac{2}{3}\right| - 12\frac{1}{3}$

In Exercises 61–68, find the product without using a calculator.

61. $5(-6)$ **62.** $-7(3)$ **63.** $6.3(5.1)$ **64.** $(-4.4)(-3.2)$

65. $\left(-\frac{5}{8}\right)\left(-\frac{4}{5}\right)$ **66.** $\left(\frac{10}{13}\right)\left(-\frac{3}{5}\right)$ **67.** $-\frac{9}{8}\left(\frac{16}{27}\right)\left(\frac{1}{2}\right)$ **68.** $\frac{2}{3}\left(-\frac{18}{5}\right)\left(-\frac{5}{6}\right)$

In Exercises 69–74, find the quotient.

69. $\dfrac{-18}{-3}$ **70.** $-\dfrac{30}{-15}$ **71.** $-\dfrac{4}{5} \div \dfrac{8}{25}$ **72.** $\dfrac{\frac{-11}{12}}{\frac{5}{24}}$

73. $5\frac{3}{4} \div 2\frac{1}{8}$ **74.** $-3\frac{5}{6} \div -2\frac{2}{3}$

In Exercises 75 and 76, find the quotient and round your answer to two decimal places. (A calculator may be useful.)

75. $\dfrac{25.5}{6.325}$ **76.** $\dfrac{265.45}{25.6}$

In Exercises 77–80, write the quantity as a repeated multiplication problem.

77. 4^3 **78.** $\left(\frac{2}{3}\right)^4$ **79.** $\left(-\frac{4}{5}\right)^6$ **80.** $(-6)^5$

In Exercises 81–84, write the repeated multiplication problem using exponential notation.

81. $\left(\frac{5}{8}\right) \times \left(\frac{5}{8}\right) \times \left(\frac{5}{8}\right) \times \left(\frac{5}{8}\right)$

82. $-(7 \times 7 \times 7)$

83. $(-4)(-4)(-4)(-4)(-4)(-4)$

84. $(-7) \times (-7) \times (-7)$

In Exercises 85–96, evaluate the exponential expression.

85. $(-2)^4$ **86.** $(-3)^2$ **87.** $(-1)^3$ **88.** $(-3)^5$ **89.** -2^4 **90.** -3^2

91. -4^3 **92.** -3^5 **93.** $\left(-\frac{7}{8}\right)^2$ **94.** $\left(-\frac{4}{5}\right)^3$ **95.** $-\left(\frac{7}{8}\right)^2$ **96.** $-\left(\frac{4}{5}\right)^3$

In Exercises 97–108, find the root.

97. $\sqrt{81}$ **98.** $\sqrt{4}$ **99.** $\sqrt{49}$ **100.** $\sqrt{100}$ **101.** $\sqrt[3]{125}$ **102.** $\sqrt[3]{64}$

103. $\sqrt[3]{-8}$ **104.** $\sqrt[3]{-1}$ **105.** $-\sqrt[4]{1}$ **106.** $-\sqrt[3]{-125}$ **107.** $\sqrt[4]{81}$ **108.** $\sqrt[5]{32}$

In Exercises 109–116, evaluate the expression.

109. $16 - 5(6 - 10)$

110. $|-36| + 7(12 - 16)$

111. $45 + 3(16 \div 4)$

112. $72 - 8(6^2 \div 9)$

113. $5^3 + |-14 + 4|$

114. $|(-2)^5| - (25 + 7)$

115. $0.2(6 - 10)^3 + \sqrt{25}$

116. $\dfrac{5^3 - 50}{-15} + \sqrt[3]{1000}$

117. *Balance in an Account* During a given month, one deposit of $1236.45 was made in a checking account. There were four withdrawals of $25.62, $455, $125, and $715.95. The balance at the beginning of the month was $2618.68. Find the balance at the end of the month. (Disregard any interest that may have been earned.)

118. *Savings Plan* Suppose that you decide to save $50 per month for 18 years. How much money will you set aside during the 18 years?

In Exercises 119 and 120, determine the unknown fractional part of the **pie chart.**

119.

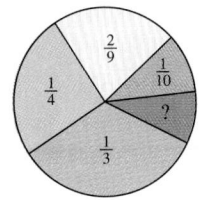

120.

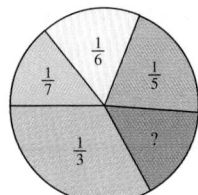

121. *Stock Values* On Monday you purchased $500 worth of stock. The value of the stock during the remainder of the week is shown in the **bar chart.** Use the chart to complete the table showing the daily gains and losses during the week.

Day	Daily Gain or Loss
Tuesday	
Wednesday	
Thursday	
Friday	

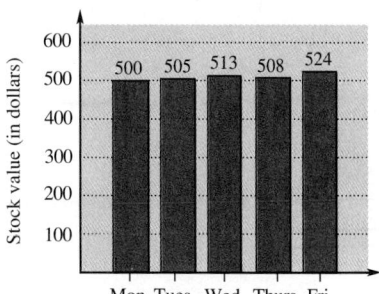

122. *Company Profits* The annual profit for a company (in millions of dollars) is shown in the bar chart. Complete the table, showing the annual profit gain or loss during each indicated year.

Year	Yearly Gain or Loss
1989	
1990	
1991	
1992	
1993	

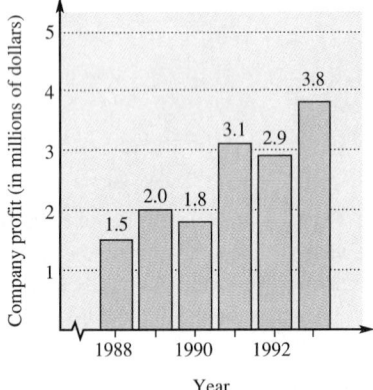

Area In Exercises 123–126, find the area of the figure. (The area of a rectangle is $A = lw$, and the area of a triangle is $A = \frac{1}{2}bh$.)

123.

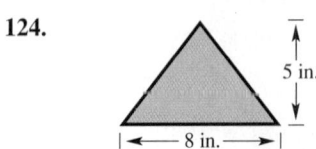

3 m

5 m

124.

5 in.

8 in.

125.

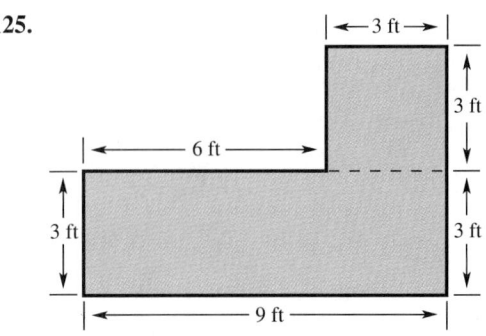

3 ft

3 ft

6 ft

3 ft

3 ft

9 ft

126.

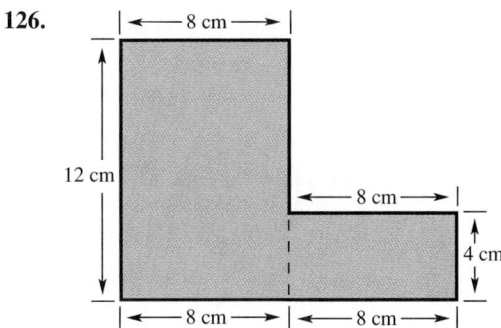

8 cm

12 cm

8 cm

4 cm

8 cm   8 cm

In Exercises 127 and 128, use the information that a bale of hay in the form of a rectangular solid has dimensions 14 inches by 18 inches by 42 inches and weighs approximately 50 pounds (see figure) to answer the questions.

Figure for 127 and 128

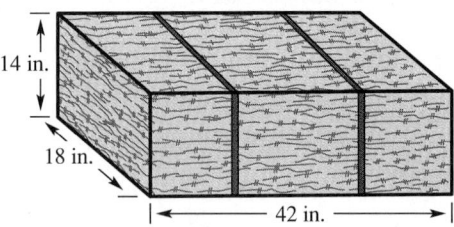

14 in.

18 in.

42 in.

127. *Volume* Find the volume of a bale of hay in cubic feet if the volume of a rectangular solid is the product of its length, width, and height. (Use the fact that 1728 in.^{3} = 1 ft^3.)

128. *Volume* Approximate the number of bales in a ton of hay. Then, approximate the volume of a stack of baled hay that weighs 12 tons.

True or False In Exercises 129–134, determine whether the statement is true or false. Explain your reasoning.

129. Every integer is a rational number.

130. Every rational number is an integer.

131. The reciprocal of every nonzero integer is an integer.

132. The reciprocal of every nonzero rational number is a rational number.

133. If a negative real number is raised to the 12th power, the result will be positive.

134. If a negative real number is raised to the 11th power, the result will be positive.

Using a Scientific Calculator or a Graphing Calculator

This book includes many examples and exercises that are best done with a calculator. Most of these can be done with a scientific calculator *or* a graphing calculator. (Some require a graphing calculator.) Occasionally, we will give sample keystrokes for calculations. When we do this, however, remember that the keystroke sequences listed in the text may not agree precisely with the steps required by *your* calculator. So be sure you are familiar with the use of the keys on your own calculator.*

EXAMPLE 1 ■ Evaluating Expressions on a Calculator

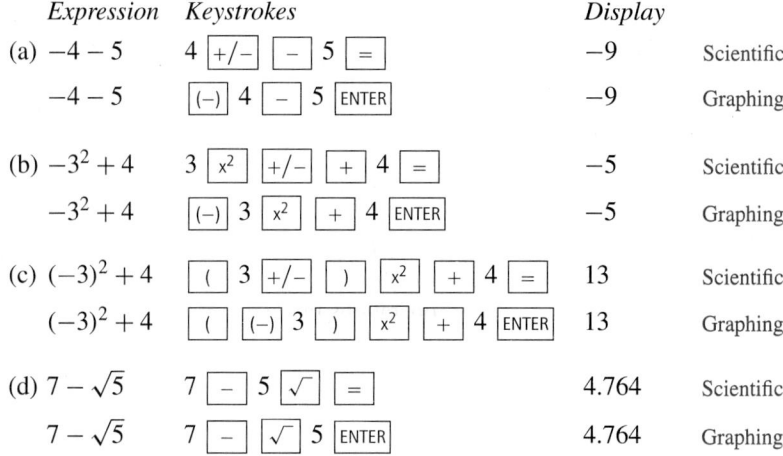

	Expression	*Keystrokes*	*Display*	
(a)	$-4 - 5$	4 +/− − 5 =	−9	Scientific
	$-4 - 5$	(−) 4 − 5 ENTER	−9	Graphing
(b)	$-3^2 + 4$	3 x² +/− + 4 =	−5	Scientific
	$-3^2 + 4$	(−) 3 x² + 4 ENTER	−5	Graphing
(c)	$(-3)^2 + 4$	(3 +/−) x² + 4 =	13	Scientific
	$(-3)^2 + 4$	((−) 3) x² + 4 ENTER	13	Graphing
(d)	$7 - \sqrt{5}$	7 − 5 √ =	4.764	Scientific
	$7 - \sqrt{5}$	7 − √ 5 ENTER	4.764	Graphing

In part (d) the display on most calculators will show more digits. For instance, your calculator might display 4.763932023. Rounded to three decimal places, the result is 4.764. ■

Incorrectly negating a number on the calculator is one of the most common mistakes students make. Be sure to stress proper use of calculators.

On a scientific calculator, notice the difference between the change sign key +/− and the subtraction key − . On a graphing calculator, the negation key (−) and the subtraction key − may not perform the same operations.

*The graphing calculator keystrokes given in this text correspond to the TI-81™ and TI-82™ graphing calculators from Texas Instruments. For other graphing calculators, the keystrokes may differ.

EXAMPLE 2 ■ Evaluating Expressions on a Calculator

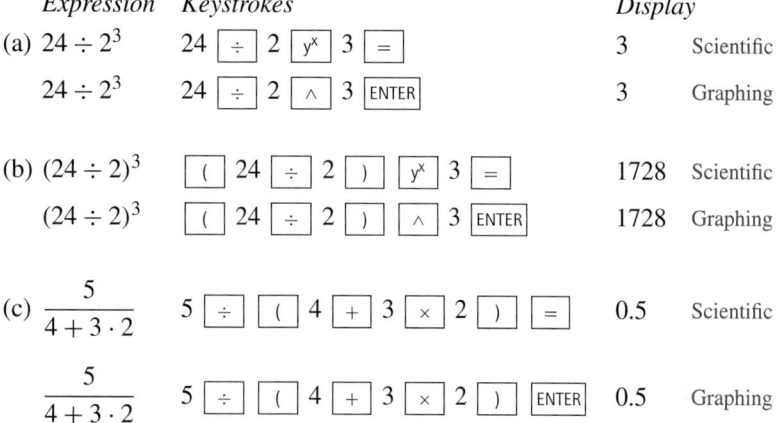

Many calculators follow the same order of operations that we use in algebra (as discussed in Section 1.1). Some calculators, however, use a different order of operations—so be sure that you understand the order of operations used by your own calculator. If you are unsure of the order of operations used by your calculator, you should insert enough parentheses to make sure the calculator performs the operations in the order you intend.

EXAMPLE 3 ■ Using Parentheses with a Calculator

Suppose you live in a state that charges 6% sales tax. You are buying two items that cost $24.95 and $36.95. The sales clerk uses a calculator to compute your sales tax as follows.

$$24.95 \boxed{+} 36.95 \boxed{\times} .06 \boxed{\text{ENTER}}$$

From the calculator display, the sales tax appears to be $27.17. Both you and the sales clerk realize that this is too much. What went wrong?

Solution

The order of operations used by the calculator was $24.95 + (36.95)(0.06) \approx 24.95 + 2.22 = 27.17$. The correct sales tax of $3.71 can be obtained by using the following calculator steps.

$$\boxed{(} \ 24.95 \boxed{+} 36.95 \boxed{)} \boxed{\times} .06 \boxed{\text{ENTER}}$$ ■

EXERCISES

In Exercises 1–4, write an expression that corresponds to the calculator keystrokes.

1. ((−) 4) ∧ 2 ENTER

2. 7 ÷ ((−) 3 − 5) ENTER

3. (1 + 4) ∧ 3 ENTER

4. 3 × (5 − 2) ENTER

In Exercises 5–12, use a calculator to evaluate the expression. Round the result to three decimal places.

5. $3(5.3 - 4.1^2)^2$

6. $(-3.7 - 5.2)^3$

7. $(0.21 + 5.23)^5$

8. $\frac{4}{3}\pi(4^2)$

9. $2075(1 + 0.65)^4$

10. $\dfrac{9.2 - 4.5}{0.6}$

11. $\dfrac{3.4}{-7.2 - 8.4}$

12. $\dfrac{1 + 3(4^2)}{7.25}$

In Exercises 13–16, a calculator was used incorrectly to evaluate the expression. Describe the error and correct it.

	Expression	*Calculator Steps*	*Display*
13.	$0.05(1.24 + 2.36)$	.05 × 1.24 + 2.36 ENTER	2.422
14.	$523 - (145 - 136)$	523 − 145 − 136 ENTER	242
15.	$\dfrac{126 + 37}{4}$	126 + 37 ÷ 4 ENTER	135.25
16.	$\dfrac{275 - 15}{3(14)}$	275 − 15 ÷ 3 × 14 ENTER	205

SECTION
1.2

Properties of Real Numbers
Mathematical Systems • *Basic Properties of Real Numbers* • *Additional Properties of Real Numbers*

Mathematical Systems

In this section, we will review the properties of real numbers. These properties make up the third component of what is called a **mathematical system.** These three components are a set of numbers, operations with the set of numbers, and properties of the numbers (and operations).

Figure 1.7 is a diagram that represents different mathematical systems. Note that the set of numbers for the system can vary. The set can consist of whole numbers, integers, rational numbers, or real numbers. (Later, you will see that the set can also consist of algebraic expressions.)

FIGURE 1.7

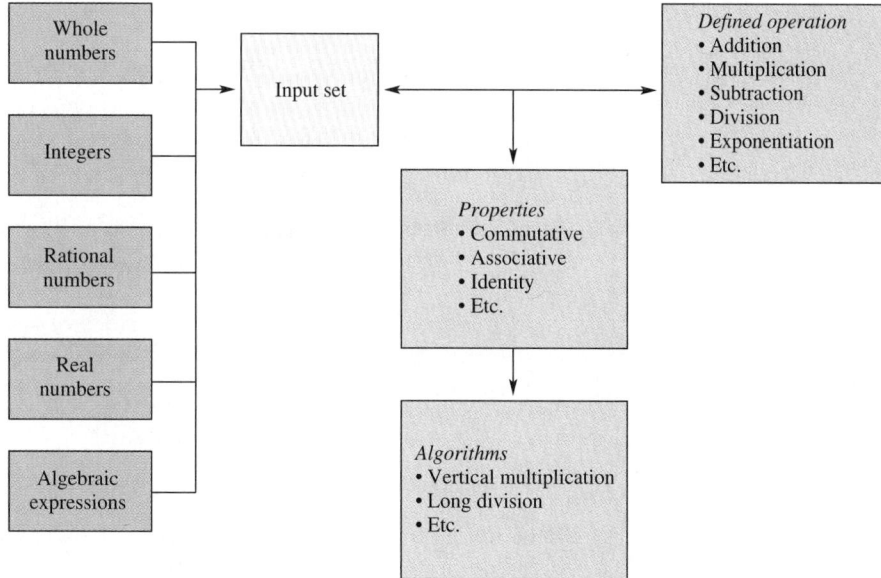

Basic Properties of Real Numbers

For the mathematical system that consists of the set of real numbers together with the operations of addition, subtraction, multiplication, and division, the resulting properties are called the **properties of real numbers.** For each property in the following list, we give a verbal description of the property, as well as one or two examples.

Properties of Real Numbers

Let a, b, and c be real numbers.

Property	*Verbal Description*
Closure Property of Addition and Multiplication	The sum and product of two real numbers are real numbers.
$\quad a + b$ is a real number $\quad ab$ is a real number	Examples: $1 + 5 = 6$, and 6 is a real number $\qquad\qquad 7 \cdot 3 = 21$, and 21 is a real number
Commutative Property of Addition	Two real numbers can be added in either order.
$\quad a + b = b + a$	Example: $2 + 6 = 6 + 2$
Commutative Property of Multiplication	Two real numbers can be multiplied in either order.
$\quad a \cdot b = b \cdot a$	Example: $3 \cdot (-5) = -5 \cdot 3$
Associative Property of Addition	When three real numbers are added, it makes no difference which two are added first.
$\quad (a+b)+c = a+(b+c)$	Example: $(1 + 7) + 4 = 1 + (7 + 4)$
Associative Property of Multiplication	When three real numbers are multiplied, it makes no difference which two are multiplied first.
$\quad (ab)c = a(bc)$	Example: $(4 \cdot 3) \cdot 9 = 4 \cdot (3 \cdot 9)$
Distributive Property	Multiplication distributes over addition.
$\quad a(b + c) = ab + ac$ $\quad (b + c)a = ba + ca$	Examples: $2(3 + 4) = 2 \cdot 3 + 2 \cdot 4$ $\qquad\qquad (3 + 4)2 = 3 \cdot 2 + 4 \cdot 2$
Additive Identity Property	The sum of zero and a real number equals the number itself.
$\quad a + 0 = 0 + a = a$	Example: $4 + 0 = 0 + 4 = 4$
Multiplicative Identity Property	The product of 1 and a real number equals the number itself.
$\quad a \cdot 1 = 1 \cdot a = a$	Example: $5 \cdot 1 = 1 \cdot 5 = 5$
Additive Inverse Property	The sum of a real number and its opposite is 0.
$\quad a + (-a) = 0$	Example: $5 + (-5) = 0$
Multiplicative Inverse Property	The product of a nonzero real number and its reciprocal is 1
$\quad a \cdot \dfrac{1}{a} = 1, \quad a \neq 0$	Example: $7 \cdot \dfrac{1}{7} = 1$

NOTE   Why are the operations of subtraction and division not listed in the preceding collection? It is because they fail to possess many of the properties described in the list. For instance, subtraction and division are not commutative. To see this, consider $4 - 3 \neq 3 - 4$ and $15 \div 5 \neq 5 \div 15$. Similarly, the examples $8 - (6 - 2) \neq (8 - 6) - 2$ and $20 \div (4 \div 2) \neq (20 \div 4) \div 2$ illustrate the fact that subtraction and division are not associative.

EXAMPLE 1   ■   **Identifying Properties of Real Numbers**

Name the property of real numbers that justifies each given statement.

(a) $4(a + 3) = 4 \cdot a + 4 \cdot 3$

(b) $6 \cdot \frac{1}{6} = 1$

(c) $-3 + (2 + b) = (-3 + 2) + b$

(d) $(b + 8) + 0 = b + 8$

Solution

(a) This statement is justified by the Distributive Property.

(b) This statement is justified by the Multiplicative Inverse Property.

(c) This statement is justified by the Associative Property of Addition.

(d) This statement is justified by the Additive Identity Property.   ■

EXAMPLE 2   ■   **Identifying Properties of Real Numbers**

The area of the rectangle in Figure 1.8 can be represented in two ways: as the area of a single rectangle, or as the sum of the areas of the two rectangles. Find this area in both ways. What property of real numbers does this demonstrate?

Solution

FIGURE 1.8

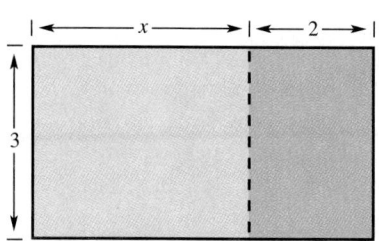

The area of the single rectangle with width = 3 and length = $x + 2$ is $A = 3(x + 2)$. The areas of the two rectangles are $A_1 = 3(x)$ and $A_2 = 3(2)$. Because the area of the single rectangle and the sum of the areas of the two rectangles are equal, you can say that

$$A = A_1 + A_2$$
$$3(x + 2) = 3(x) + 3(2)$$
$$= 3x + 6.$$

This is an example of the Distributive Property.   ■

EXAMPLE 3   ■   Using the Properties of Real Numbers

Complete the following statements using the specified properties of real numbers.

(a) Multiplicative Identity Property   (b) Associative Property of Addition

$$(4a)1 = \quad\quad$$ $$(b + 8) + 3 = \quad\quad$$

Review the difference between the
Commutative Properties and the
Associative Properties.

(c) Additive Inverse Property   (d) Distributive Property

$$0 = 5c + \quad\quad$$ $$4 \cdot b + 4 \cdot 5 = \quad\quad$$

Solution

(a) By the Multiplicative Identity Property, you can write

$$(4a)1 = 4a.$$

(b) By the Associative Property of Addition, you can write

$$(b + 8) + 3 = b + (8 + 3).$$

(c) By the Additive Inverse Property, you can write

$$0 = 5c + (-5c).$$

(d) By the Distributive Property, you can write

$$4 \cdot b + 4 \cdot 5 = 4(b + 5).$$ ■

Additional Properties of Real Numbers

Once you have determined the basic properties of a mathematical system (called
the **axioms** of the system), you can go on to develop other properties of the system.
These additional properties are often called **theorems,** and the formal arguments
that justify the theorems are called **proofs.** The following list summarizes several
additional properties of real numbers.

**Additional Properties of
Real Numbers**

Let a, b, and c be real numbers.

PROPERTIES OF EQUALITY

Property	*Verbal Description*
Addition Property of Equality If $a = b$, then $a + c = b + c$.	Adding a real number to both sides of a true equation produces another true equation.
Multiplication Property of Equality If $a = b$, then $ac = bc, c \neq 0$.	Multiplying both sides of a true equation by a nonzero real number produces another true equation.
Cancellation Property of Addition If $a + c = b + c$, then $a = b$.	Subtracting a real number from both sides of a true equation produces another true equation.

Cancellation Property of Multiplication

If $ac = bc$ and $c \neq 0$, then $a = b$.

Dividing both sides of a true equation by a nonzero real number produces another true equation.

PROPERTIES OF ZERO

Property	*Verbal Description*

Multiplication Property of Zero

$0 \cdot a = 0$

The product of zero and any real number is zero.

Division Property of Zero

$\dfrac{0}{a} = 0, \quad a \neq 0$

If zero is divided by any *nonzero* real number, the result is zero.

Division by Zero Is Undefined

$\dfrac{a}{0}$ is undefined.

We do not define division by zero.

PROPERTIES OF NEGATION

Property	*Verbal Description*

Multiplication by -1

$(-1)a = -a$

$(-1)(-a) = a$

The opposite of a real number a can be obtained by multiplying the real number by -1.

Placement of Minus Signs

$(-a)(b) = -(ab) = (a)(-b)$

The opposite of the product of two numbers is equal to the product of one of the numbers and the opposite of the other.

Product of Two Opposites

$(-a)(-b) = ab$

The product of the opposites of two real numbers is equal to the product of the two real numbers.

EXAMPLE 4   ■   Proof of a Property of Equality

Prove that if $a + c = b + c$, then $a = b$. (You may use the Addition Property of Equality.)

Solution

$$a + c = b + c \qquad \text{Given equation}$$
$$(a + c) + (-c) = (b + c) + (-c) \qquad \text{Addition Property of Equality}$$
$$a + [c + (-c)] = b + [c + (-c)] \qquad \text{Associative Property of Addition}$$
$$a + 0 = b + 0 \qquad \text{Additive Inverse Property}$$
$$a = b \qquad \text{Additive Identity Property}$$

■

EXAMPLE 5 ■ Proof of a Property of Negation

Prove that $(-1)a = -a$. (You may use any of the Properties of Equality and Properties of Zero.)

Solution

At first glance, it is a little difficult to see what you are being asked to prove. However, a good way to start is to consider carefully the definitions of each of the three numbers in the equation.

$a =$ given real number

$-1 =$ the additive inverse of 1

$-a =$ the additive inverse of a

By showing that $(-1)a$ has the same properties as the additive inverse of a, you will be showing that $(-1)a$ must be the additive inverse of a.

$$
\begin{aligned}
(-1)a + a &= (-1)a + (1)(a) && \text{Multiplicative Identity Property} \\
&= (-1 + 1)a && \text{Distributive Property} \\
&= (0)a && \text{Additive Inverse Property} \\
&= 0 && \text{Multiplication Property of Zero}
\end{aligned}
$$

Because you have shown that $(-1)a + a = 0$, you can now use the fact that $-a + a = 0$ to conclude that $(-1)a + a = -a + a$. From this, you can complete the proof as follows.

$$
\begin{aligned}
(-1)a + a &= -a + a && \text{Shown in first part of proof} \\
(-1)a &= -a && \text{Cancellation Property of Addition} \quad \blacksquare
\end{aligned}
$$

There are many other properties of real numbers that we could have included in the list. Many of these are properties that you use—even if you don't happen to know their formal names. Here are a few. The **Symmetric Property of Equality** states that if $a = b$, then $b = a$. The **Transitive Property of Equality** states that if $a = b$ and $b = c$, then $a = c$.

The list of additional properties of real numbers forms a very important part of algebra. Knowing the names of the properties is not especially important, but knowing how to use each property is extremely important. The next two examples show how several of the properties are used to solve common problems in algebra.

EXAMPLE 6 ■ Applying the Properties of Real Numbers

In the following list of equations, each equation can be justified on the basis of the previous equation using one of the properties of real numbers. Name the property of real numbers that justifies each step.

$$
\begin{aligned}
b + 2 &= 6 && \text{Given equation} \\
(b + 2) + (-2) &= 6 + (-2)
\end{aligned}
$$

$$b + [2 + (-2)] = 6 - 2$$
$$b + 0 = 4$$
$$b = 4$$

Solution

$b + 2 = 6$	Given equation
$(b + 2) + (-2) = 6 + (-2)$	Addition Property of Equality
$b + [2 + (-2)] = 4$	Associative Property of Addition
$b + 0 = 4$	Additive Inverse Property
$b = 4$	Additive Identity Property ■

You might begin by asking students to compare the first two lines and describe the change.

EXAMPLE 7 ■ **Applying the Properties of Real Numbers**

In the following list of equations, each equation can be justified on the basis of the previous equation using one of the properties of real numbers. Which property of real numbers justifies each step?

$3a = 9$	Given equation
$\left(\dfrac{1}{3}\right)(3a) = \left(\dfrac{1}{3}\right)(9)$	
$\left(\dfrac{1}{3} \cdot 3\right)(a) = 3$	
$(1)(a) = 3$	
$a = 3$	

Going from the second line to the third line requires the use of the Associative Property of Multiplication. Be aware that students may see the parentheses in the second line and attempt to use the Distributive Property. Point out that here we have $3 \cdot a$, *not* $3 + a$.

Solution

$3a = 9$	Given equation
$\left(\dfrac{1}{3}\right)(3a) = \left(\dfrac{1}{3}\right)(9)$	Multiplication Property of Equality
$\left(\dfrac{1}{3} \cdot 3\right)(a) = 3$	Associative Property of Multiplication
$(1)(a) = 3$	Multiplicative Inverse Property
$a = 3$	Multiplicative Identity Property ■

DISCUSSION PROBLEM ■ **You Be the Instructor**

One of the most common errors in algebra is illustrated in the following *incorrect* use of the Distributive Property.

$$-3(4 + x) = -12 + 3x$$

Suppose you are teaching an algebra class. What would you say to your class to help them avoid this type of error? ■

Warm-Up

The following warm-up exercises involve skills that were covered earlier. You will use these skills in the exercise set for this section.

In the student text, answers to all Warm-Up exercises can be found in the back of the text.

In Exercises 1–10, perform the indicated operation(s).

1. $625 + (-400)$ **2.** $-360 + 120$ **3.** $-2(225 - 150)$

4. $5(57 - 33)$ **5.** $7 \times \frac{8}{35}$ **6.** $-\frac{4}{15} \times \frac{15}{16}$

7. $\frac{3}{8} \div \frac{5}{16}$ **8.** $\frac{9}{10} \div 3$ **9.** $(12 - 15)^3$

10. $\left(\frac{5}{8}\right)^2$

1.2 EXERCISES

In Exercises 1–24, name the property of real numbers that justifies the statement.

1. $3 + (-5) = -5 + 3$ **2.** $-5(7) = 7(-5)$ **3.** $(5 + 10)(8) = 8(5 + 10)$

4. $25 + 35 = 35 + 25$ **5.** $5(2a) = (5 \cdot 2)a$ **6.** $5 + 0 = 5$

7. $7 \cdot 1 = 7$ **8.** $3(6 + b) = 3 \cdot 6 + 3 \cdot b$ **9.** $3 + (12 + 9) = (3 + 12) + 9$

10. $(-4 \cdot 10) \cdot 8 = -4(10 \cdot 8)$ **11.** $(8 + 5)(10) = 8 \cdot 10 + 5 \cdot 10$ **12.** $(16 + 8) + 5 = 16 + (8 + 5)$

13. $3x + 0 = 3x$ **14.** $8y \cdot 1 = 8y$ **15.** $-25 + 25 = 0$

16. $10x \cdot \dfrac{1}{10x} = 1, \ x \neq 0$ **17.** $\dfrac{1}{y} \cdot y = 1, \ y \neq 0$ **18.** $6(x + 3) = 6 \cdot x + 6 \cdot 3$

19. $(6 + x) + m = 6 + (x + m)$ **20.** $3x + (-3x) = 0$ **21.** $(x + 4)x = x(x + 4)$

22. $(x + 4)7 = 7x + 28$ **23.** $5(y + 10) = 5y + 50$ **24.** $(z - 3) \cdot \dfrac{1}{z - 3} = 1, \ z \neq 3$

In Exercises 25–34, use the property of real numbers to fill in the missing part of the statement.

25. *Associative Property of Multiplication:* $3(6y) = $

26. *Commutative Property of Addition:* $10 + (-6) = $

27. *Commutative Property of Multiplication:* $15(-3) = $

28. *Associative Property of Addition:* $6 + (5 + y) = $

29. *Distributive Property:* $5(6 + z) = $

30. *Distributive Property:* $(8 + y)(4) = $

31. *Commutative Property of Addition:* $25 + (-x) =$

32. *Additive Inverse Property:* $13x + (-13x) =$

33. *Multiplicative Identity Property:* $(x + 8) \cdot 1 =$

34. *Additive Identity Property:* $(8x) + 0 =$

In Exercises 35–42, give (a) the additive inverse and (b) the multiplicative inverse of the quantity.

35. 10

36. 18

37. -16

38. -52

39. $6z, z \neq 0$

40. $2y, y \neq 0$

41. $x + 1, \ x \neq -1$

42. $y - 4, \ y \neq 4$

In Exercises 43–50, rewrite the expression using the Associative Property of Addition or the Associative Property of Multiplication.

43. $(x + 5) + 3$

44. $(z + 6) + 10$

45. $32 + (4 + y)$

46. $15 + (3 + x)$

47. $3(4 \cdot 5)$

48. $(10 \cdot 8) \cdot 5$

49. $6(2y)$

50. $8(3x)$

In Exercises 51–56, rewrite the expression using the Distributive Property.

51. $20(a + 5)$

52. $-3(y + 8)$

53. $5(3x + 4)$

54. $6(2x + 5)$

55. $(x + 6)(-2)$

56. $(z + 10)(12)$

In Exercises 57–60, identify the property of real numbers that justifies each step.

57.
$$x + 5 = 3 \quad \text{Given}$$
$$(x + 5) + (-5) = 3 + (-5)$$
$$x + [5 + (-5)] = -2$$
$$x + 0 = -2$$
$$x = -2$$

58.
$$x - 8 = 20 \quad \text{Given}$$
$$(x - 8) + 8 = 20 + 8$$
$$x + (-8 + 8) = 28$$
$$x + 0 = 28$$
$$x = 28$$

59.
$$2x - 5 = 6 \quad \text{Given}$$
$$(2x - 5) + 5 = 6 + 5$$
$$2x + (-5 + 5) = 11$$
$$2x + 0 = 11$$
$$2x = 11$$
$$\tfrac{1}{2}(2x) = \tfrac{1}{2}(11)$$
$$\left(\tfrac{1}{2} \cdot 2\right)x = \tfrac{11}{2}$$
$$1 \cdot x = \tfrac{11}{2}$$
$$x = \tfrac{11}{2}$$

60.
$$3x + 4 = 10 \quad \text{Given}$$
$$(3x + 4) + (-4) = 10 + (-4)$$
$$3x + [4 + (-4)] = 6$$
$$3x + 0 = 6$$
$$3x = 6$$
$$\tfrac{1}{3}(3x) = \tfrac{1}{3}(6)$$
$$\left(\tfrac{1}{3} \cdot 3\right)x = 2$$
$$1 \cdot x = 2$$
$$x = 2$$

In Exercises 61–66, the right side of the equation is *not* equal to the left side. Change the right side so that it *does* equal the left side.

61. $3(x + 5) \neq 3x + 5$

62. $4(x + 2) \neq 4x + 2$

63. $-2(x + 8) \neq -2x + 16$

64. $-9(x + 4) \neq -9x + 36$

65. $3 \cdot \left(\frac{0}{3}\right) \neq 1$

66. $6 \cdot \left(\frac{1}{6}\right) \neq 0$

In Exercises 67–72, use the Distributive Property to perform the required arithmetic mentally. For example, suppose you work in an industry where the wage is \$14 per hour with "time and a half" for overtime. Thus, your hourly wage for overtime is given by $14(1.5) = 14\left(1 + \frac{1}{2}\right) = 14 + 7 = \$21$.

67. $16(1.75)$

68. $15\left(1\frac{2}{3}\right)$

69. $7(62) = 7(60 + 2)$

70. $5(49) = 5(50 - 1)$

71. $9(6.98) = 9(7 - 0.02)$

72. $12(19.95) = 12(20 - 0.05)$

73. *Area of a Rectangle* The figure shows two adjoining rectangles. Demonstrate the Distributive Property by filling in the blanks to express the total area of the two rectangles in two ways.

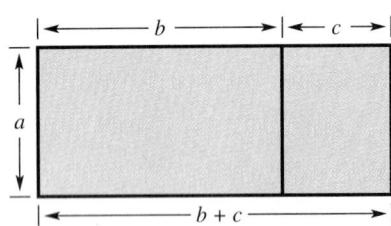

74. *Area of a Rectangle* The figure shows two adjoining rectangles. Demonstrate the "subtraction version" of the Distributive Property by filling in the blanks to express the area of the left rectangle in two ways.

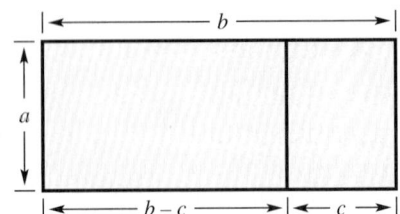

Sales Goals In Exercises 75 and 76, use the following equation, which approximates the sales goal for General Mills.

$$\text{Sales goal} = 7.0 + 0.1t^2$$

In this equation, the sales goal is measured in billions of dollars and t represents time with $t = 0$ corresponding to 1990 (see figure). (*Source*: General Mills 1992 Annual Report)

Figure for 75 and 76

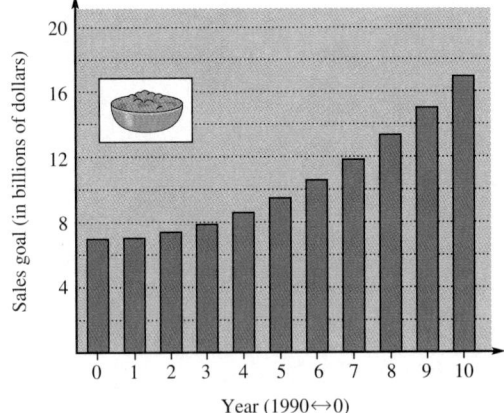

75. Use the graph to approximate the sales goal for 1995.

76. Use the graph to approximate the sales goal for 1998.

77. Prove that if $ac = bc$ and $c \neq 0$, then $a = b$.

78. Prove that $(-1)(-a) = a$.

Algebraic Expressions

Algebraic Expressions • Basic Rules of Algebra • Properties of Exponents • Simplifying Algebraic Expressions • Evaluating Algebraic Expressions

Algebraic Expressions

Historical Note: The French mathematician François Viéte (1540-1603) was the first to use letters to represent numbers. He used vowels to represent unknown quantities and consonants to represent known quantities.

This section marks our transition from arithmetic to algebra. One of the basic characteristics of algebra is the use of letters (or combinations of letters) to represent numbers. The letters used to represent the numbers are called **variables,** and combinations of letters and numbers are called **algebraic expressions.** Here are a few examples.

$$3x, \quad x + 2, \quad \frac{x}{x^2 + 1}, \quad 2x - 3y$$

Definition of an Algebraic Expression	A collection of letters (called **variables**) and real numbers (called **constants**) combined using the operations of addition, subtraction, multiplication, and division is called an **algebraic expression.**

The **terms** of an algebraic expression are those parts that are separated by *addition.* For example, the algebraic expression $x^2 - 3x + 6$ has three terms: x^2, $-3x$, and 6. Note that $-3x$ is a term, rather than $3x$, because

$$x^2 - 3x + 6 = x^2 + (-3x) + 6. \qquad \text{Think of subtraction as addition.}$$

The terms x^2 and $-3x$ are called the **variable terms** of the expression, and 6 is called the **constant term** of the expression. The numerical factor of a variable term is called the **coefficient** of the variable term. For instance, the coefficient of the variable term $-3x$ is -3, and the coefficient of the variable term x^2 is 1.

EXAMPLE 1 ■ Identifying Terms and Coefficients

Algebraic Expression	*Terms*	*Coefficients*
(a) $5x - \frac{1}{3}$	$5x, \quad -\frac{1}{3}$	$5, \quad -\frac{1}{3}$
(b) $4y + 6x - 9$	$4y, \quad 6x, \quad -9$	$4, \quad 6, \quad -9$

Basic Rules of Algebra

The properties of real numbers that were discussed in Section 1.2 can be used to rewrite algebraic expressions. The following list is similar to that given in Section 1.2, except that the examples involve algebraic expressions. In other words, the properties are true for variables and algebraic expressions as well as for real numbers.

Basic Rules of Algebra

Let a, b, and c be real numbers, variables, or algebraic expressions.

Property	*Example*
Commutative Property of Addition	
$a + b = b + a$	$5x + x^2 = x^2 + 5x$
Commutative Property of Multiplication	
$ab = ba$	$(3 + x)x^3 = x^3(3 + x)$
Associative Property of Addition	
$(a + b) + c = a + (b + c)$	$(-x + 6) + 3x^2 = -x + (6 + 3x^2)$
Associative Property of Multiplication	
$(ab)c = a(bc)$	$(5x \cdot 4y)(6) = (5x)(4y \cdot 6)$
Distributive Property	
$a(b + c) = ab + ac$	$2x(4 + 3x) = 2x \cdot 4 + 2x \cdot 3x$
$(a + b)c = ac + bc$	$(y + 6)y = y \cdot y + 6 \cdot y$
Additive Identity Property	
$a + 0 = 0 + a = a$	$4y^2 + 0 = 4y^2$
Multiplicative Identity Property	
$a \cdot 1 = 1 \cdot a = a$	$(-5x^3)(1) = -5x^3$
Additive Inverse Property	
$a + (-a) = 0$	$4x^2 + (-4x^2) = 0$
Multiplicative Inverse Property	
$a \cdot \dfrac{1}{a} = 1, \quad a \neq 0$	$(x^2 + 1)\left(\dfrac{1}{x^2 + 1}\right) = 1$

NOTE   Because subtraction is defined as "adding the opposite," the Distributive Property is also true for subtraction. For instance, the "subtraction form" of $a(b + c) = ab + ac$ is

$$a(b - c) = a[b + (-c)] = ab + a(-c) = ab - ac.$$

In addition to these basic rules, the properties of equality, zero, and negation given in the previous section are also valid for algebraic expressions. The next example illustrates the use of a variety of the basic rules and properties.

EXAMPLE 2   ■   Identifying the Basic Rules of Algebra

Identify the rule of algebra used in each of the following equations.

(a) $(5x^2)3 = 3(5x^2)$

(b) $(3x^2 + x) - (3x^2 + x) = 0$

(c) $(y - 6)3 + (y - 6)y = (y - 6)(3 + y)$

(d) $(5 + x^2) + 4x^2 = 5 + (x^2 + 4x^2)$

(e) $5x \cdot \dfrac{1}{5x} = 1, \quad x \neq 0$

Solution

(a) This equation illustrates the Commutative Property of Multiplication. In other words, you obtain the same result whether you multiply $5x^2$ by 3, or 3 by $5x^2$.

(b) This equation illustrates the Additive Inverse Property. In terms of subtraction, this property simply states that when any expression is subtracted from itself the result is zero.

The concept in Example 2(c) can sometimes be difficult for students.

(c) This equation illustrates the Distributive Property in reverse order.

$$ab + ac = a(b + c)$$ Distributive Property

$$(y - 6)3 + (y - 6)y = (y - 6)(3 + y)$$

Note in this case that $a = y - 6$, $b = 3$, and $c = y$.

(d) This equation illustrates the Associative Property of Addition. In other words, to form the sum $5 + x^2 + 4x^2$, it doesn't matter whether 5 and x^2 are added first or x^2 and $4x^2$ are added first.

(e) This equation illustrates the Multiplicative Inverse Property. Note that it is important that x be a nonzero number. If x were zero, the reciprocal of x would be undefined.   ■

Properties of Exponents

Historical Note: Originally, Arabian mathematicians used their words for colors to represent quantities (*cosa, censa, cubo*). These words were eventually abbreviated to *co, ce, cu*. René Descartes (1596-1650) simplified this even further by introducing the symbols x, x^2, and x^3.

When multiplying two exponential expressions that have the *same base,* you add exponents. To see why this is true, consider the product

$$a^3 \cdot a^2.$$

Because the first expression represents $a \cdot a \cdot a$ and the second represents $a \cdot a$, the product of the two expressions represents $a \cdot a \cdot a \cdot a \cdot a$, as follows.

$$a^3 \cdot a^2 = \underbrace{(a \cdot a \cdot a)}_{\substack{\text{Three} \\ \text{factors}}} \cdot \underbrace{(a \cdot a)}_{\substack{\text{Two} \\ \text{factors}}} = \underbrace{(a \cdot a \cdot a \cdot a \cdot a)}_{\substack{\text{Five} \\ \text{factors}}} = a^{3+2} = a^5$$

The properties of exponents are summarized as follows.

Properties of Exponents   Let m and n be positive integers, and let a and b represent real numbers, variables, or algebraic expressions.

Property	*Example*
1. $a^m \cdot a^n = a^{m+n}$	$x^2(x^4) = x^{2+4} = x^6$
2. $(ab)^m = a^m \cdot b^m$	$(2x)^3 = 2^3(x^3) = 8x^3$
3. $(a^m)^n = a^{mn}$	$(x^2)^3 = x^{2 \cdot 3} = x^6$
4. $\dfrac{a^m}{a^n} = a^{m-n}, \quad m > n, \ a \neq 0$	$\dfrac{x^6}{x^2} = x^{6-2} = x^4, \quad x \neq 0$
5. $\left(\dfrac{a}{b}\right)^m = \dfrac{a^m}{b^m}, \quad b \neq 0$	$\left(\dfrac{x}{2}\right)^3 = \dfrac{x^3}{2^3} = \dfrac{x^3}{8}$

The first and second properties can be extended to products involving three or more factors. For example,

$$a^m \cdot a^n \cdot a^k = a^{m+n+k} \quad \text{and} \quad (abc)^m = a^m b^m c^m.$$

The next two examples show how these rules can be used to rewrite products involving exponential forms.

EXAMPLE 3 ■   **Illustrating the Properties of Exponents**

(a) To multiply two exponential expressions that have the *same base,* add exponents.

$$x^2 \cdot x^3 = \underbrace{x \cdot x}_{2 \text{ factors}} \cdot \underbrace{x \cdot x \cdot x}_{3 \text{ factors}} = \underbrace{x \cdot x \cdot x \cdot x \cdot x}_{5 \text{ factors}} = x^{2+3} = x^5$$

(b) To raise the product of two factors to the *same power,* raise each factor to the power and multiply the results.

$$(3x)^3 = \underbrace{3x \cdot 3x \cdot 3x}_{3 \text{ factors}} = \underbrace{3 \cdot 3 \cdot 3}_{3 \text{ factors}} \cdot \underbrace{x \cdot x \cdot x}_{3 \text{ factors}} = 3^3 \cdot x^3 = 27x^3$$

(c) To raise an exponential expression to a power, multiply the powers.

$$(x^3)^2 = \underbrace{(x \cdot x \cdot x)}_{3 \text{ factors}} \cdot \underbrace{(x \cdot x \cdot x)}_{3 \text{ factors}} = \underbrace{(x \cdot x \cdot x \cdot x \cdot x \cdot x)}_{6 \text{ factors}} = x^{3 \cdot 2} = x^6 \quad ■$$

EXAMPLE 4 ■   **Illustrating the Properties of Exponents**

(a) To divide two exponential expressions that have the *same base,* subtract exponents.

$$\frac{x^4}{x^2} = \frac{\overbrace{x \cdot x \cdot x \cdot x}^{4 \text{ factors}}}{\underbrace{x \cdot x}_{2 \text{ factors}}} = x^{4-2} = x^2$$

(b) To raise the quotient of two expressions to the *same power,* raise each expression to the power and divide the results.

$$\left(\frac{x}{3}\right)^3 = \frac{x}{3} \cdot \frac{x}{3} \cdot \frac{x}{3} = \overbrace{\frac{x \cdot x \cdot x}{\underbrace{3 \cdot 3 \cdot 3}_{\text{3 factors}}}}^{\text{3 factors}} = \frac{x^3}{3^3} = \frac{x^3}{27}$$

■

EXAMPLE 5   ■   **Applying Properties of Exponents**

Use the properties of exponents to simplify the following expressions.

(a) $(x^2 y^4)(3x)$　　　　(b) $(-2y^2)^3$　　　　(c) $-2(y^2)^3$

It may be helpful to have students clearly identify the base of an exponential expression, which will help them distinguish between expressions such as -2^4 and $(-2)^4$.

Solution

(a) $(x^2 y^4)(3x) = 3(x^2 \cdot x)(y^4) = 3x^3 y^4$

(b) $(-2y^2)^3 = (-2)^3 (y^2)^3 = -8y^6$

(c) $-2(y^2)^3 = (-2)(y^{2 \cdot 3}) = -2y^6$

■

EXAMPLE 6   ■   **Applying Properties of Exponents**

Use the properties of exponents to simplify the following expressions.

(a) $\dfrac{14a^5 b^3}{7a^2 b^2}$　　　　(b) $\left(\dfrac{x^2}{2y}\right)^3$　　　　(c) $\dfrac{x^{n+2} y^{3n}}{x^2 y^n}$

Solution

(a) $\dfrac{14a^5 b^3}{7a^2 b^2} = 2(a^{5-2})(b^{3-2}) = 2a^3 b$

(b) $\left(\dfrac{x^2}{2y}\right)^3 = \dfrac{(x^2)^3}{(2y)^3} = \dfrac{x^6}{2^3 y^3} = \dfrac{x^6}{8y^3}$

(c) $\dfrac{x^{n+2} y^{3n}}{x^2 y^n} = x^{(n+2)-2} y^{3n-n} = x^n y^{2n}$

■

Simplifying Algebraic Expressions

One common use of the basic rules of algebra is to rewrite an algebraic expression in a simpler form. To **simplify** an algebraic expression generally means to *remove symbols of grouping* such as parentheses or brackets and *combine like terms.*

Two or more terms of an algebraic expression can be combined only if they are *like terms.* In an algebraic expression, two terms are said to be **like terms** if they are both constant terms or if they have the same variable factor(s). For example, the terms $4x$ and $-2x$ are like terms because they have the same variable factor. Similarly, $2x^2 y$, $-x^2 y$, and $\frac{1}{2}(x^2 y)$ are like terms. Note that $4x^2 y$ and $-x^2 y^2$ are not like terms because their variable factors are different.

To combine like terms in an algebraic expression, simply add their respective coefficients and attach the common variable factor. This is actually an application of the Distributive Property, as shown in Example 7.

EXAMPLE 7   ■   Combining Like Terms

Simplify the following expressions by combining like terms.

(a) $2x + 3x - 4$ (b) $-3 + 5 + 2y - 7y$ (c) $5x + 3y - 4x$

Solution

(a) $2x + 3x - 4 = (2 + 3)x - 4$ Distributive Property

$\qquad\qquad\quad\; = 5x - 4$ Simplest form

(b) $-3 + 5 + 2y - 7y = (-3 + 5) + (2 - 7)y$ Distributive Property

$\qquad\qquad\qquad\quad\; = 2 - 5y$ Simplest form

(c) $5x + 3y - 4x = 3y + 5x - 4x$ Commutative Property

$\qquad\qquad\quad\; = 3y + (5x - 4x)$ Associative Property

$\qquad\qquad\quad\; = 3y + (5 - 4)x$ Distributive Property

$\qquad\qquad\quad\; = 3y + x$ Simplest form   ■

As you gain experience with the rules of algebra, you may want to combine some of the steps in your work. For instance, you might feel comfortable listing only the following steps to solve Example 7(c).

$5x + 3y - 4x = 3y + (5x - 4x)$ Group like terms

$\qquad\qquad\quad = 3y + x$ Combine like terms

EXAMPLE 8   ■   Combining Like Terms

Simplify the following expressions by combining like terms.

(a) $7x + 7y - 4x - y$

(b) $2x^2 + 3x - 5x^2 - x$

(c) $3xy^2 - 4x^2y^2 + 2xy^2 + (xy)^2$

Solution

(a) $7x + 7y - 4x - y = (7x - 4x) + (7y - y)$ Group like terms

$\qquad\qquad\qquad\quad = 3x + 6y$ Combine like terms

(b) $2x^2 + 3x - 5x^2 - x = (2x^2 - 5x^2) + (3x - x)$ Group like terms

$\qquad\qquad\qquad\qquad = -3x^2 + 2x$ Combine like terms

(c) $3xy^2 - 4x^2y^2 + 2xy^2 + (xy)^2 = (3xy^2 + 2xy^2) + (-4x^2y^2 + x^2y^2)$

$\qquad\qquad\qquad\qquad\qquad\qquad = 5xy^2 - 3x^2y^2$ ■

EXAMPLE 9 ■ Removing Symbols of Grouping

Simplify the algebraic expression

$$3(x - 5) - (2x - 7).$$

Solution

$$
\begin{aligned}
3(x - 5) - (2x - 7) &= 3x - 15 - 2x + 7 && \text{Distributive Property}\\
&= (3x - 2x) + (-15 + 7) && \text{Group terms}\\
&= x - 8 && \text{Combine like terms} \quad\blacksquare
\end{aligned}
$$

NOTE A set of parentheses preceded by a *minus* sign can be removed by changing the sign of each term inside the parentheses. For instance,

$$8x - (5x - 4) = 8x - 5x + 4 = 3x + 4.$$

A set of parentheses preceded by a *plus* sign can be removed without changing the signs of the terms inside the parentheses. For instance,

$$8x + (5x - 4) = 8x + 5x - 4 = 13x - 4.$$

When removing symbols of grouping, remove the innermost symbols first and combine like terms. Then repeat the process for any remaining symbols of grouping.

EXAMPLE 10 ■ Removing Symbols of Grouping

Simplify the following algebraic expressions.

(a) $5x - 2x[3 + 2(x - 7)]$

(b) $-3x(5x^4) + (2x)^5$

Solution

$$
\begin{aligned}
\text{(a)} \quad 5x - 2x[3 + 2(x - 7)] &= 5x - 2x[3 + 2x - 14] && \text{Remove parentheses}\\
&= 5x - 2x[2x - 11] && \text{Combine like terms in}\\
& && \text{brackets}\\
&= 5x - 4x^2 + 22x && \text{Remove brackets}\\
&= -4x^2 + 27x && \text{Combine like terms}
\end{aligned}
$$

$$
\begin{aligned}
\text{(b)} \quad -3x(5x^4) + (2x)^5 &= -15x^5 + (2^5)(x^5)\\
&= -15x^5 + 32x^5\\
&= 17x^5 \quad\blacksquare
\end{aligned}
$$

Evaluating Algebraic Expressions

A common error is to confuse the operation of subtraction with the substitution of a negative number. Point out the use of parentheses when substituting a numerical value for a variable in an expression.

To **evaluate** an algebraic expression, substitute numerical values for each of the variables in the expression. Here are some examples.

Expression	Value of Variable	Substitute	Value of Expression
$3x + 2$	$x = 2$	$3(2) + 2$	$6 + 2 = 8$
$4x^2 + 2x - 1$	$x = -1$	$4(-1)^2 + 2(-1) - 1$	$4 - 2 - 1 = 1$
$2x(x + 4)$	$x = -2$	$2(-2)(-2 + 4)$	$2(-2)(2) = -8$

EXAMPLE 11   ■   **Evaluating Algebraic Expressions**

Evaluate the following algebraic expressions when $x = -2$ and $y = 5$.

(a) $2y - 3x$ (b) $5 + x^2$ (c) $5 - x^2$

Solution

Additional example. Evaluate $-x^4$ when $x = -1$.

Solution: $-(-1)^4 = -1$

(Note that $-(-1)^4 \neq +1^4$.)

(a) When $x = -2$ and $y = 5$, the expression $2y - 3x$ has a value of
$$2(5) - 3(-2) = 10 + 6 = 16.$$

(b) When $x = -2$, the expression $5 + x^2$ has a value of
$$5 + (-2)^2 = 5 + 4 = 9.$$

(c) When $x = -2$, the expression $5 - x^2$ has a value of
$$5 - (-2)^2 = 5 - 4 = 1.$$

■

EXAMPLE 12   ■   **Evaluating Algebraic Expressions**

Evaluate the following algebraic expressions when $x = 2$ and $y = -1$.

(a) $x^2 - 2xy + y^2$ (b) $|y - x|$ (c) $\dfrac{2xy}{5x + y}$

Solution

(a) When $x = 2$ and $y = -1$, the expression $x^2 - 2xy + y^2$ has a value of
$$2^2 - 2(2)(-1) + (-1)^2 = 4 + 4 + 1 = 9.$$

(b) When $x = 2$ and $y = -1$, the expression $|y - x|$ has a value of
$$|-1 - 2| = |-3| = 3.$$

(c) When $x = 2$ and $y = -1$, the expression $2xy/(5x + y)$ has a value of
$$\frac{2(2)(-1)}{5(2) - 1} = \frac{-4}{10 - 1} = -\frac{4}{9}.$$

■

EXAMPLE 13 ■ Creating a Real-Life Model

For the years 1980 through 1990, the number B (in millions) of children's books that were sold in the United States can be modeled by $B = 18.7t + 119$, where $t = 0$ represents 1980. During the same years, the average price p of a children's book can be modeled by $p = 0.2t + 3.3$. These two models can be used to write a model for the total sales (in millions of dollars) of children's books for 1980 through 1990. Use the model to find the total sales in 1988. (*Source: Book Industry Times*)

Solution

Let S represent the yearly sales in millions of dollars.

Total yearly sales	=	Number sold each year	·	Average price per book each year

$$S = (18.7t + 119)(0.2t + 3.3)$$

To find the total sales in 1988, substitute 8 for t in this model.

$S = (18.7t + 119)(0.2t + 3.3)$	Total sales model
$= (18.7 \cdot 8 + 119)(0.2 \cdot 8 + 3.3)$	Substitute 8 for t
$= (268.6)(4.9)$	Perform operations within parentheses
$= 1316.14$	Multiply

According to the NEA, over half of the adults in the United States read at least one book for enjoyment each year. Reading books to children increases the likelihood that the children will read for enjoyment as adults.

The total sales of children's books in 1988 was about $1316 million (or about $1.3 billion). The bar graph shown in Figure 1.9 shows the total sales of children's books from 1980 through 1990. ■

FIGURE 1.9

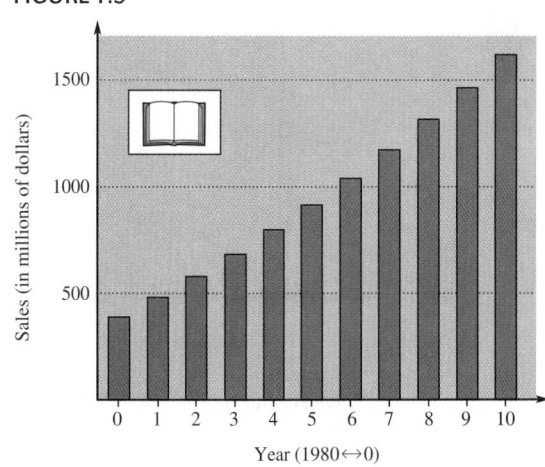

DISCUSSION PROBLEM   ■   Unit Analysis

In Example 13, the *unit* for the model for the yearly sales of children's books is *millions of dollars per year.* Explain how the following "unit analysis" diagram can be used to determine this unit. Then give other examples in which you can use a similar unit analysis.

$$\frac{\text{Millions of books}}{\text{Year}} \cdot \frac{\text{dollars}}{\text{book}} = \frac{\text{millions of dollars}}{\text{year}}$$

■

■

Warm-Up

The following warm-up exercises involve skills that were covered in earlier sections. You will use these skills in the exercise set for this section.

In Exercises 1–6, evaluate the quantity.

1. $6(3 + 5^2)$

2. $-5(6 - 10)$

3. $3(-4) + 4\left(\frac{3}{8}\right)$

4. $-|35 - 60| + 8(9 - 5)$

5. $\dfrac{360}{3^2 + 4^2}$

6. $\dfrac{13}{5} - \dfrac{9}{2}$

In Exercises 7–10, name the property of real numbers that justifies the statement.

7. $8x \cdot \dfrac{1}{8x} = 1$

8. $5 + (-3 + x) = (5 - 3) + x$

9. $-4(x + 10) = -4 \cdot x + (-4)(10)$

10. $3x + 0 = 3x$

■ **1.3**   **EXERCISES**

In Exercises 1–8, identify the terms of the algebraic expression.

1. $10x + 5$

2. $-16t^2 + 48$

3. $-3y^2 + 2y - 8$

4. $25z^3 - 4.8z^2$

5. $4x^2 - 3y^2 - 5x + 2y$

6. $14u^2 + 25uv - 3v^2$

7. $x^2 - 2.5x - \dfrac{1}{x}$

8. $\dfrac{3}{t^2} - \dfrac{4}{t} + 6$

In Exercises 9–12, determine the coefficient of the term.

9. $5y^3$

10. $4x^6$

11. $-\frac{3}{4}t^2$

12. $-8.4x$

In Exercises 13–22, identify the rule of algebra that is illustrated by the equation.

13. $4 - 3x = -3x + 4$

14. $(10 + x) - y = 10 + (x - y)$

15. $-5(2x) = (-5 \cdot 2)x$

16. $(x - 2)(3) = 3(x - 2)$

17. $(x + 5) \cdot \dfrac{1}{x + 5} = 1, \quad x \neq -5$

18. $(x^2 + 1) - (x^2 + 1) = 0$

19. $5(y^3 + 3) = 5y^3 + 5 \cdot 3$

20. $10x^3y + 0 = 10x^3y$

21. $(16t^4) \cdot 1 = 16t^4$

22. $-32(u^2 - 3u) = -32u^2 + 96u$

In Exercises 23–28, use the property to rewrite the expression.

23. (a) *Distributive Property:* $5(x + 6) = $ ▓▓

　　(b) *Commutative Property of Multiplication:* $5(x + 6) = $ ▓▓

24. (a) *Distributive Property:* $6x + 6 = $ ▓▓

　　(b) *Commutative Property of Addition:* $6x + 6 = $ ▓▓

25. (a) *Commutative Property of Multiplication:* $6(xy) = $ ▓▓

　　(b) *Associative Property of Multiplication:* $6(xy) = $ ▓▓

26. (a) *Additive Identity Property:* $3ab + 0 = $ ▓▓

　　(b) *Commutative Property of Addition:* $3ab + 0 = $ ▓▓

27. (a) *Additive Inverse Property:* $4t^2 + (-4t^2) = $ ▓▓

　　(b) *Commutative Property of Addition:* $4t^2 + (-4t^2) = $ ▓▓

28. (a) *Associative Property of Addition:* $(3 + 6) + (-9) = $ ▓▓

　　(b) *Additive Inverse Property:* $9 + (-9) = $ ▓▓

In Exercises 29–34, use the definition of exponents to write the expression as a repeated multiplication.

29. $x^3 \cdot x^4$

30. $z^2 \cdot z^5$

31. $(-2x)^3$

32. $(2y)^3$

33. $\left(\dfrac{y}{5}\right)^4$

34. $\left(\dfrac{3}{t}\right)^5$

In Exercises 35–38, write the expression using exponential notation.

35. $(5x)(5x)(5x)(5x)$

36. $(y \cdot y \cdot y)(y \cdot y \cdot y \cdot y)$

37. $(x \cdot x \cdot x)(y \cdot y \cdot y)$

38. $(-9t)(-9t)(-9t)(-9t)(-9t)(-9t)$

In Exercises 39–80, use the properties of exponents to simplify the expression.

39. $x^5 \cdot x^7$

40. $u^3 \cdot u^5 \cdot u$

41. $3^3y^4 \cdot y^2$

42. $6^2x^3 \cdot x^5$

43. $(-4x)^2$

44. $(-4x)^3$

45. $(-5z^2)^3$

46. $(-5z^3)^2$

47. $(2xy)(3x^2y^3)$

48. $(-5a^2b^3)(2ab^4)$

49. $(3uv)^2(-6u^3v)$

50. $(10x^2y)^3(2x^4y)$

51. $\dfrac{x^6}{x^4}$

52. $\dfrac{y^9}{y^2}$

53. $\dfrac{3^7x^5}{3^3x^3}$

54. $\dfrac{2^4y^5}{2^2y^3}$

55. $\dfrac{28x^2y^3}{21xy^2}$ **56.** $\dfrac{(-3xy)^3}{9xy^2}$ **57.** $\dfrac{(2xy)^5}{6(xy)^3}$ **58.** $\dfrac{4^3(ab)^6}{4(ab)^2}$

59. $(5y)^2(-y^4)$ **60.** $(3y)^3(2y^2)$ **61.** $-5z^4(-5z)^4$ **62.** $(-6n)(-3n^2)$

63. $(-2a)^2(-2a)^2$ **64.** $(-2a^2)(-8a)$ **65.** $\dfrac{(2x)^4y^2}{2x^3y}$ **66.** $(-z^2)(z^2)$

67. $\dfrac{6(a^3b)^3}{(3ab)^2}$ **68.** $\dfrac{(-3c^5d^3)^2}{(-2cd)^3}$ **69.** $\left(\dfrac{a^2}{3}\right)^3$ **70.** $\left(-\dfrac{x^3}{y^2}\right)^2$

71. $-\left(\dfrac{2x^4}{5y}\right)^2$ **72.** $-\left(\dfrac{3a^3}{2b^5}\right)^3$ **73.** $\dfrac{x^{n+1}}{x^n}$ **74.** $\dfrac{a^{m+3}}{a^3}$

75. $(x^n)^4$ **76.** $(a^3)^k$ **77.** $x^{n+1}\cdot x^3$ **78.** $y^{m-2}\cdot y^2$

79. $\dfrac{r^{n+2}s^{m+4}}{r^{n+1}s}$ **80.** $\left(\dfrac{x^{2n}y^{m+4}}{x^n y^3}\right)^2$

In Exercises 81–92, simplify the expression by combining like terms.

81. $3x + 4x$ **82.** $-2x^2 + 4x^2$ **83.** $9y - 5y + 4y$

84. $8y + 7y - y$ **85.** $3x - 2y + 5x + 20y$ **86.** $-2a + \frac{1}{3}b - 7a - b$

87. $8z^2 + \frac{3}{2}z - \frac{5}{2}z^2 + 10$ **88.** $-5y^3 + 3y - 6y^2 + 8y^3 + y - 4$

89. $2uv + 5u^2v^2 - uv - (uv)^2$ **90.** $3m^2n^2 - 4mn - n(5m) + 2(mn)^2$

91. $5(ab)^2 + 2ab - 4ab$ **92.** $3xy - xy + 8$

In Exercises 93–96, use the Distributive Property to rewrite the expression.

93. $4(2x^2 + x - 3)$ **94.** $8(z^3 - 4z^2 + 2)$ **95.** $-3(6y^2 - y - 2)$ **96.** $-5(-x^2 + 2y + 1)$

In Exercises 97–110, simplify the algebraic expression.

97. $10(x - 3) + 2x - 5$ **98.** $3(x^2 + 1) + x^2 - 6$

99. $-3(y^2 + 3y - 1) + 2(y - 5)$ **100.** $5(a + 6) - 4(a^2 - 2a - 1)$

101. $2[3(b - 5) - (b^2 + b + 3)]$ **102.** $4[5 - 3(x^2 + 10)]$

103. $2x(5x^2) - (x^3 + 5)$ **104.** $x(x^2 + 3) - 3(x + 4)$

105. $y^2(y + 1) + y(y^2 + 1)$ **106.** $2ab(b^2 - 3) - ab(b^2 + 2)$

107. $x(xy^2 + y) - 2xy(xy + 1)$ **108.** $z^2(z^4 - z^2) + 4z^3(z + 1)$

109. $-2a(3a^2)^3 + \dfrac{9a^8}{3a}$ **110.** $5y^3 + \dfrac{4y^5}{2y^2} - (7y)y^2$

111. *Geometry* Write an expression for the perimeter of the movie screen in the figure and simplify.

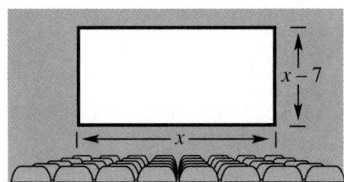

112. *Geometry* Write an expression for the perimeter of the triangle in the figure and simplify.

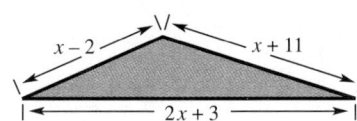

In Exercises 113–124, evaluate the algebraic expression for the specified value of the variable. If not possible, state the reason.

Expression   *Value*

113. $5 - 3x$ (a) $x = \frac{2}{3}$
                          (b) $x = 5$

114. $\frac{3}{2}x - 2$ (a) $x = 6$
                          (b) $x = -3$

115. $10 - |x|$ (a) $x = 3$
                          (b) $x = -3$

116. $2x^2 + 5x - 3$ (a) $x = \frac{1}{2}$
                          (b) $x = -3$

117. $\dfrac{x}{x^2 + 1}$ (a) $x = 0$
                          (b) $x = 3$

118. $5 - \dfrac{3}{x}$ (a) $x = 0$
                          (b) $x = -6$

119. $3x + 2y$ (a) $x = 1, \ y = 5$
                          (b) $x = -6, \ y = -9$

120. $x^2 - xy + y^2$ (a) $x = 2, \ y = -1$
                          (b) $x = -3, \ y = -2$

Expression   *Value*

121. $\dfrac{x}{x - y}$ (a) $x = 0, \ y = 10$
                          (b) $x = 4, \ y = 4$

122. $|y - x|$ (a) $x = 2, \ y = 5$
                          (b) $x = -2, \ y = -2$

123. Distance traveled: rt (a) $r = 40, \ t = 5\frac{1}{4}$
                          (b) $r = 35, \ t = 4$

124. Simple interest: Prt (a) $P = \$5000, \ r = 0.085,$ $t = 10$
                          (b) $P = \$750, \ r = 0.07,$ $t = 3$

125. *Exploratory Exercise*
   (a) Complete the table by evaluating the expression $2x - 5$.

x	-1	0	1	2	3	4
$2x - 5$						

   (b) From the table in part (a), determine the increase in the value of the expression for each one-unit increase in x.
   (c) Use the results of parts (a) and (b) to guess the increase in the algebraic expression $\frac{3}{4}x + 5$ for each one-unit increase in x.

126. *Exploratory Exercise*
   (a) Complete the table by evaluating the expression $3x + 2$.

x	-1	0	1	2	3	4
$3x + 2$						

   (b) From the table in part (a), determine the increase in the value of the expression for each one-unit increase in x.
   (c) Use the results of parts (a) and (b) to guess the increase in the algebraic expression $\frac{2}{3}x - 4$ for each one-unit increase in x.

Athletic Shoes In Exercises 127–129, use the equation, which approximates the annual sales of athletic and sports shoes in the United States from 1980 to 1989.

$$\text{Sales} = (41.78 - 0.40t + 0.76t^2 - 0.05t^3)^2$$

In this formula, the sales are given in millions of dollars and t represents the year, with $t = 0$ corresponding to 1980 (see figure). (*Source:* National Sporting Goods Association)

Figure for 127-129

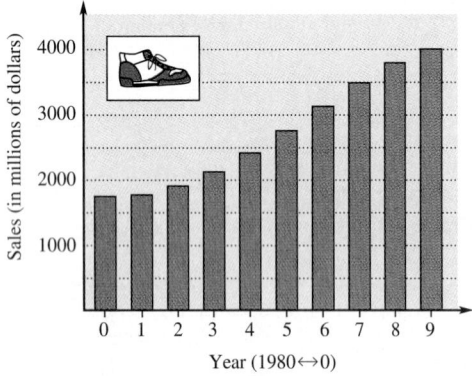

Year (1980↔0)

127. (a) Use a calculator to find the sales of athletic and sports shoes in 1985. (b) How does your answer compare with the bar graph?

128. (a) Use a calculator to find the sales of athletic and sports shoes in 1989. (b) How does your answer compare with the bar graph?

129. (a) From 1980 to 1989, did sales increase by the same amount each year? (b) Did you use the bar graph or the model to determine your answer?

130. *Area of a Trapezoid* The area of the trapezoid with parallel bases of length b_1 and b_2 and height h in the figure is

$$\frac{h}{2}(b_1 + b_2).$$

Use the Distributive Property to show that the area can also be expressed as

$$b_1h + \frac{1}{2}(b_2 - b_1)h.$$

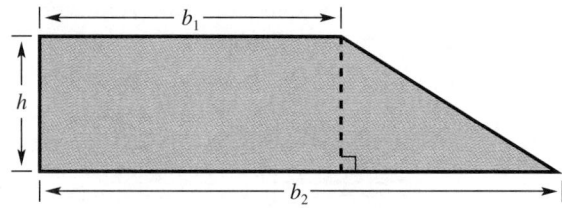

131. *Area of a Trapezoid* Use both formulas given in Exercise 130 to find the area of a trapezoid with $b_1 = 7, b_2 = 12$, and $h = 3$.

132. *Amount of Roofing Material* The roof shown in the figure is made up of two trapezoids and two triangles. Find the total area of the roof.

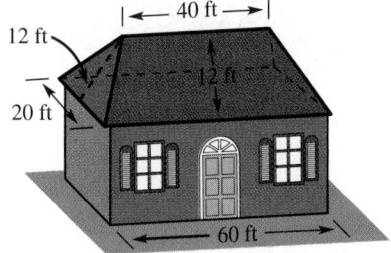

Mid-Chapter Quiz

Take this quiz as you would take a quiz in class. After you are done, check your work against the answers in the back of the book.

1. Place the correct symbol ($<$ or $>$) between the real numbers $-\frac{3}{2}$ and $-\frac{5}{2}$.

2. Find the distance between the real numbers 18 and -32.

3. Evaluate: $16 + (-84)$

4. Evaluate: $\frac{5}{8} + \frac{1}{4} - \frac{5}{6}$

5. Evaluate: $|-16.25| + 54.78$

6. Evaluate: $\frac{3}{4} \div \frac{15}{8}$

7. Evaluate: $-\left(-\frac{3}{2}\right)^3$

8. Evaluate: $\frac{2^4 - 6}{5} - \sqrt{16}$

9. Write the expression $(5x)(5x)(5x)(5x)$ using exponential notation.

10. Name the property of real numbers demonstrated by $5(x + 3) = 5 \cdot x + 5 \cdot 3$.

11. Name the property of real numbers demonstrated by $6y \cdot 1 = 6y$.

12. Name the property of real numbers demonstrated by $(z + 3) - (z + 3) = 0$.

13. Give the multiplicative inverse of the quantity $x + 1$, $x \neq -1$.

14. Identify the terms of the algebraic expression $5x^4 - 3x^2 + 2 - \dfrac{1}{x^2}$.

15. Simplify: $4x^2 \cdot x^3$

16. Simplify: $(-2x)^4$

17. Simplify: $\left(\dfrac{y^2}{3}\right)^3$

18. Simplify: $\dfrac{18x^2y^3}{12xy}$

19. Simplify: $4x^2 - 3xy + 5xy - 5x^2$

20. Simplify: $3[x - 2x(x^2 + 1)]$

21. Evaluate the expression $10x^2 - |5x|$ if (a) $x = 5$ and (b) $x = -1$.

22. Find an expression for the perimeter of the rectangle.

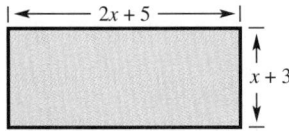

23. The midyear financial statement of a company showed a profit of $1,415,322.62. At the close of the year, the financial statement showed a profit for the year of $916,489.26. Find the profit or loss of the company for the second 6 months of the year.

45

SECTION**1.4**

Operations with Polynomials

Basic Definitions • Adding and Subtracting Polynomials • Multiplying Polynomials • Special Products • Applications

Basic Definitions

An algebraic expression containing only terms of the form ax^k, where a is any real number and k is a nonnegative integer, is called a **polynomial in one variable** or simply a **polynomial.** Here are some examples of polynomials in one variable.

$$3x - 8, \quad x^4 + 3x^3 - x^2 - 8x + 1, \quad x^3 + 5, \quad \text{and} \quad 9x^5$$

In the term ax^k, a is called the **coefficient,** and k the **degree,** of the term. Note that the degree of the term ax is 1, and the degree of a constant term is zero. Because a polynomial is an algebraic *sum,* the coefficients take on the signs between the terms. For instance,

$$x^3 - 4x^2 + 3 = (1)x^3 + (-4)x^2 + (0)x + 3$$

has coefficients 1, -4, 0, and 3.

Polynomials are usually written in order of descending powers of the variable. This is referred to as **standard form.** For example, the standard form of $3x^2 - 5 - x^3 + 2x$ is

$$-x^3 + 3x^2 + 2x - 5. \qquad \text{Standard form}$$

The **degree of a polynomial** is defined as the degree of the term with the highest power, and the coefficient of this term is called the **leading coefficient** of the polynomial. For instance, the polynomial $-3x^4 + 4x^2 + x + 7$ is of fourth degree and its leading coefficient is -3.

Definition of a Polynomial in x

Let $a_n, \ldots, a_2, a_1, a_0$ be real numbers and let n be a *nonnegative integer.* A **polynomial in x** is an expression of the form

$$a_n x^n + a_{n-1} x^{n-1} + \cdots + a_2 x^2 + a_1 x + a_0 ,$$

where $a_n \neq 0$. The polynomial is of **degree n**, and the number a_n is called the **leading coefficient.** The number a_0 is called the **constant term.**

NOTE The following are *not* polynomials, for the reasons stated.

$2x^{-1} + 5$ Exponent in $2x^{-1}$ is *not* nonnegative.

$x^3 + 3x^{1/2}$ Exponent in $3x^{1/2}$ is *not* an integer.

EXAMPLE 1 ■ Identifying Leading Coefficients and Degrees of Polynomials

Polynomial	Standard Form	Degree	Leading Coefficient
(a) $5x^2 - 2x^7 + 4 - 2x$	$-2x^7 + 5x^2 - 2x + 4$	7	-2
(b) $16 - 8x^3$	$-8x^3 + 16$	3	-8
(c) 10	10	0	10
(d) $5 + x^4 - 6x^3$	$x^4 - 6x^3 + 5$	4	1 ■

A polynomial with only one term is called a **monomial.** Polynomials with two *unlike* terms are called **binomials,** and those with three *unlike* terms are called **trinomials.** Here are some examples.

$5x^3$	Monomial
$-4x + 3$	Binomial
$2x^2 + 3x - 7$	Trinomial

NOTE The prefix *mono* means one, the prefix *bi* means two, and the prefix *tri* means three.

EXAMPLE 2 ■ Evaluating a Polynomial

Find the value of the following polynomial when $x = 4$.

$$x^3 - 5x^2 + 6x - 3$$

Solution

When $x = 4$, you have the following.

Value $= x^3 - 5x^2 + 6x - 3$	Given polynomial
$= 4^3 - 5(4^2) + 6(4) - 3$	Replace x by 4
$= 64 - 80 + 24 - 3$	Evaluate terms
$= 5$	Simplify ■

Adding and Subtracting Polynomials

To add two polynomials, simply combine like terms. This can be done in either a horizontal or a vertical format, as shown in Examples 3 and 4.

EXAMPLE 3 ■ Adding Polynomials Horizontally

$(2x^3 + x^2 - 5) + (x^2 + x + 6)$	Given polynomials
$= (2x^3) + (x^2 + x^2) + (x) + (-5 + 6)$	Group like terms
$= 2x^3 + 2x^2 + x + 1$	Standard form ■

EXAMPLE 4 ■ Using a Vertical Format to Add Polynomials

To use a vertical format to add polynomials, align the terms of the polynomials by their degrees, as follows.

$$
\begin{array}{l}
5x^3 + 2x^2 - x + 7 \\
\ 3x^2 - 4x + 7 \\
-x^3 + 4x^2 - 8 \\
\hline
4x^3 + 9x^2 - 5x + 6
\end{array}
$$ ■

To subtract one polynomial from another, change the sign of each of the terms of the polynomial that is being subtracted and then add the resulting like terms.

EXAMPLE 5 ■ Subtracting Polynomials Horizontally

$$(3x^3 - 5x^2 + 3) - (x^3 + 2x^2 - x - 4)$$ Given polynomials

$$= 3x^3 - 5x^2 + 3 - x^3 - 2x^2 + x + 4$$ Change signs and add

$$= (3x^3 - x^3) + (-5x^2 - 2x^2) + (x) + (3 + 4)$$ Group like terms

$$= 2x^3 - 7x^2 + x + 7$$ Standard form   ■

Be especially careful to get the correct signs when you are subtracting one polynomial from another. One of the most common mistakes in algebra is to forget to change signs correctly when subtracting one expression from another.

EXAMPLE 6 ■ Using a Vertical Format to Subtract Polynomials

Use a vertical format to perform the following operations.

$$(4x^4 - 2x^3 + 5x^2 - x + 8) - (3x^4 - 2x^3 + 3x - 4)$$

Solution

Encourage students to perform subtraction carefully.

$$
\begin{array}{ll}
(4x^4 - 2x^3 + 5x^2 - x + 8) & \Rightarrow \quad 4x^4 - 2x^3 + 5x^2 - x + 8 \\
-(3x^4 - 2x^3 + 3x - 4) & \Rightarrow \quad -3x^4 + 2x^3 - 3x + 4 \\
\hline
& x^4 + 5x^2 - 4x + 12
\end{array}
$$ ■

EXAMPLE 7 ■ Combining Polynomials

$$(2x^2 - 2x + 1) - [(x^2 + x - 3) + (-2x^2 - 4x + 5)]$$

$$= (2x^2 - 2x + 1) - [(x^2 - 2x^2) + (x - 4x) + (-3 + 5)]$$

$$= (2x^2 - 2x + 1) - [-x^2 - 3x + 2]$$

$$= 2x^2 - 2x + 1 + x^2 + 3x - 2$$

$$= (2x^2 + x^2) + (-2x + 3x) + (1 - 2)$$

$$= 3x^2 + x - 1$$ ■

Multiplying Polynomials

The simplest type of polynomial multiplication involves a monomial multiplier. The product is obtained by direct application of the Distributive Property.

EXAMPLE 8 ■   **Finding Products with Monomial Multipliers**

You might illustrate the use of the Commutative Property of Multiplication.

$$(2x - 7)(-3x) = (-3x)(2x - 7)$$
$$= -3x(2x) - 3x(-7)$$
$$= -6x^2 + 21x$$

(a) $(2x - 7)(-3x) = 2x(-3x) - 7(-3x)$ Distributive Property

$\qquad\qquad\qquad = -6x^2 + 21x$ Standard form

(b) $4x^2(3x - 2x^3 + 1) = 4x^2(3x) - 4x^2(2x^3) + 4x^2(1)$ Distributive Property

$\qquad\qquad\qquad\qquad = 12x^3 - 8x^5 + 4x^2$ Properties of exponents

$\qquad\qquad\qquad\qquad = -8x^5 + 12x^3 + 4x^2$ Standard form   ■

To multiply two binomials, you can use both (left and right) forms of the Distributive Property. For example, if you treat the binomial $(2x + 7)$ as a single quantity, you can multiply $(3x - 2)$ by $(2x + 7)$ as follows.

$$(3x - 2)(2x + 7) = 3x(2x + 7) - (2)(2x + 7)$$
$$= (3x)(2x) + (3x)(7) - (2)(2x) - (2)(7)$$
$$= 6x^2 + 21x - 4x - 14$$

| Product of **F**irst terms | Product of **O**uter terms | Product of **I**nner terms | Product of **L**ast terms |

$$= 6x^2 + 17x - 14$$

FOIL Diagram

$$(3x - 2)(2x + 7)$$

First
Outer
Inner
Last

With practice, you should be able to multiply two binomials without writing out all of the above steps. In fact, the four products in the boxes above suggest that you can put the product of two binomials in the FOIL form in just one step. This is called the **FOIL Method.** Note that the words *first, outer, inner,* and *last* refer to the positions of the terms in the original product.

EXAMPLE 9 ■   **Multiplying Binomials (Distributive Property)**

$$(x + 2)(x - 3) = x(x - 3) + 2(x - 3)$$ Right Distributive Property

$$= x^2 - 3x + 2x - 6$$ Left Distributive Property

$$= x^2 - x - 6$$ Combine like terms   ■

EXAMPLE 10 ■   **Multiplying Binomials (FOIL Method)**

$$\overset{\text{F}}{}\quad\overset{\text{O}}{}\quad\overset{\text{I}}{}\quad\overset{\text{L}}{}$$
$$(3x + 4)(2x + 1) = 6x^2 + 3x + 8x + 4$$

$$= 6x^2 + 11x + 4$$ Combine like terms   ■

To multiply two polynomials that have three or more terms, you can use the same basic principle that you use when multiplying monomials and binomials. That is, *each term of one polynomial must be multiplied by each term of the other polynomial.* This can be done using either a horizontal or a vertical format.

EXAMPLE 11 ■ Multiplying Polynomials (Horizontal Format)

$$(4x^2 - 3x - 1)(2x - 5)$$

$$= 4x^2(2x - 5) - 3x(2x - 5) - 1(2x - 5) \qquad \text{Distributive Property}$$

$$= 8x^3 - 20x^2 - 6x^2 + 15x - 2x + 5$$

$$= 8x^3 - 26x^2 + 13x + 5 \qquad \text{Combine like terms} \quad ■$$

When multiplying two polynomials, it is best to write each in standard form before using either the horizontal or vertical format. This is illustrated in the next example.

EXAMPLE 12 ■ Multiplying Polynomials (Vertical Format)

With a vertical format, line up like terms in the same vertical columns, much as you align digits in whole-number multiplication.

$$
\begin{array}{r}
4x^2 + x - 2 \qquad \text{Standard form}\\
\times \qquad -x^2 + 3x + 5 \qquad \text{Standard form}\\
\hline
20x^2 + 5x - 10 \qquad \Longleftarrow \quad 5(4x^2 + x - 2)\\
12x^3 + 3x^2 - 6x \qquad \Longleftarrow \quad 3x(4x^2 + x - 2)\\
-4x^4 - x^3 + 2x^2 \qquad \Longleftarrow \quad -x^2(4x^2 + x - 2)\\
\hline
-4x^4 + 11x^3 + 25x^2 - x - 10 \qquad\qquad ■
\end{array}
$$

EXAMPLE 13 ■ Multiplying Polynomials

Multiply $(x - 4)^3$.

Solution

$$(x - 4)^3 = (x - 4)(x - 4)(x - 4)$$

$$= [(x - 4)(x - 4)](x - 4) \qquad \text{Associative Property}$$

$$= [x^2 - 4x - 4x + 16](x - 4) \qquad \text{Find } (x - 4)^2$$

$$= [x^2 - 8x + 16](x - 4) \qquad \text{Combine like terms}$$

$$= x^2(x - 4) - 8x(x - 4) + 16(x - 4) \qquad \text{Distributive Property}$$

$$= x^3 - 4x^2 - 8x^2 + 32x + 16x - 64 \qquad \text{Distributive Property}$$

$$= x^3 - 12x^2 + 48x - 64 \qquad \text{Combine like terms} \quad ■$$

EXAMPLE 14 ■ An Area Model for Multiplying Polynomials

Use an area model (or algebra tiles) to show that

$$(x + 2)(2x + 1) = 2x^2 + 5x + 2.$$

Solution

Think of a rectangle whose sides have lengths $x + 2$ and $2x + 1$. The area of this rectangle is

$$(x + 2)(2x + 1). \qquad \text{Area = (width)(height)}$$

FIGURE 1.10

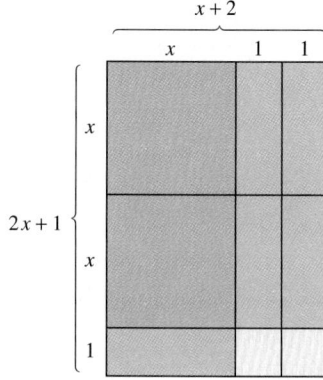

Another way to find the area is to add the areas of the rectangular parts, as shown in Figure 1.10. There are two squares whose sides are x, five rectangles whose sides are x and 1, and two squares whose sides are 1. The total area of these nine rectangles is

$$2x^2 + 5x + 2. \qquad \text{Area = sum of rectangular areas}$$

Because each method must produce the same area, you can conclude that

$$(x + 2)(2x + 1) = 2x^2 + 5x + 2. \qquad \blacksquare$$

Special Products

Some binomial products have special forms that occur frequently in algebra. For instance, the product $(x+3)(x-3)$ is called the **product of the sum and difference of two terms.** With such products, the two middle terms cancel, as follows.

$$
\begin{aligned}
(x + 3)(x - 3) &= x^2 - 3x + 3x - 9 && \text{Sum and difference of two terms}\\
&= x^2 - 9 && \text{Product has no middle term}
\end{aligned}
$$

Another common type of product is the **square of a binomial.** With this type of product, the middle term is always twice the product of the terms in the binomial.

$$
\begin{aligned}
(2x + 5)^2 &= (2x + 5)(2x + 5) && \text{Square of a binomial}\\
&= 4x^2 + 10x + 10x + 25 && \text{Outer and inner terms are equal}\\
&= 4x^2 + 20x + 25 && \text{Middle term is twice the product of the terms of the binomial}
\end{aligned}
$$

You should learn to recognize the patterns of these two special products. The general forms of these special products are listed in the following statements.

■ **Special Products**

Let u and v be real numbers, variables, or algebraic expressions. Then the following formulas are true.

Sum and Difference of Two Terms

$$(u + v)(u - v) = u^2 - v^2$$

Example

$$(3x - 4)(3x + 4) = 9x^2 - 16$$

Square of a Binomial

$$(u + v)^2 = u^2 + 2uv + v^2$$

Example

$$
\begin{aligned}
(4x + 9)^2 &= (4x)^2 + 2(4x)(9) + 9^2\\
&= 16x^2 + 72x + 81
\end{aligned}
$$

$$(u - v)^2 = u^2 - 2uv + v^2$$

Example

$$
\begin{aligned}
(x - 6)^2 &= x^2 - 2(x)(6) + 6^2\\
&= x^2 - 12x + 36
\end{aligned}
$$

Check that students do not forget the middle term. For example, $(a + b)^2 \neq a^2 + b^2$.

NOTE When a binomial is squared, the resulting middle term is always *twice* the product of the two terms.

$$(x + y)^2 = x^2 + \underbrace{2(xy)}_{\substack{\text{Twice the product} \\ \text{of the terms}}} + y^2$$

Terms

EXAMPLE 15 ■ **Finding Special Products**

(a) $(3x - 2)(3x + 2) = (3x)^2 - 2^2$ Special Product

               $= 9x^2 - 4$ Combine like terms

(b) $(2x - 7)^2 = (2x)^2 - 2(2x)(7) + 7^2$ Special Product

               $= 4x^2 - 28x + 49$ Combine like terms

Ask students to show geometrically that $(a + b)^2 = a^2 + 2ab + b^2$.

(c) $[(a - 2) + b]^2 = (a - 2)^2 + 2(a - 2)b + b^2$ Special Product

               $= a^2 - 4a + 4 + 2ab - 4b + b^2$ Combine like terms ■

Applications

There are many applications that require the evaluation of polynomials. One commonly used second-degree polynomial is called a **position polynomial.** This polynomial has the form

$$-16t^2 + v_0 t + s_0, \qquad \text{Position polynomial}$$

where t is time, measured in seconds. The value of this polynomial gives the height (in feet) of a free-falling object above the ground, assuming no air resistance. The coefficient of t, v_0, is called the **initial velocity** of the object, and the constant term, s_0, is called the **initial height** of the object. If the initial velocity is positive, then the object was projected upward (at $t = 0$), and if the initial velocity is negative, then the object was projected downward.

EXAMPLE 16 ■ **Finding the Height of a Free-Falling Object**

An object is thrown downward from the top of a 200-foot building. The initial velocity is -10 feet per second. Use the position polynomial

$$-16t^2 - 10t + 200$$

to find the height of the object when $t = 1$, $t = 2$, and $t = 3$ (see Figure 1.11).

Solution

When $t = 1$, the height of the object is

$$\text{Height} = -16(1^2) - 10(1) + 200$$
$$= -16 - 10 + 200$$
$$= 174 \text{ feet.}$$

FIGURE 1.11

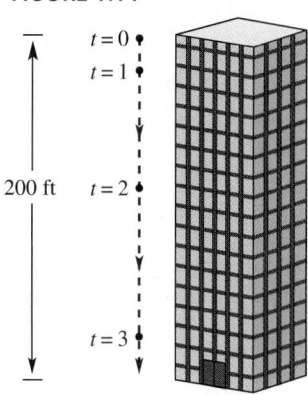

200 ft

$t = 0$
$t = 1$
$t = 2$
$t = 3$

When $t = 2$, the height of the object is

$$\text{Height} = -16(2^2) - 10(2) + 200$$
$$= -64 - 20 + 200$$
$$= 116 \ \text{feet.}$$

When $t = 3$, the height of the object is

$$\text{Height} = -16(3^2) - 10(3) + 200$$
$$= -144 - 30 + 200$$
$$= 26 \ \text{feet.}$$ ■

EXAMPLE 17   ■   Using Polynomial Models

The numbers of pounds of shrimp S and of lobster L consumed in the United States from 1980 to 1990 (see Figure 1.12) can be modeled by

$$S = 42.8t + 425 \qquad\qquad \text{Shrimp (millions of pounds)}$$
$$L = -1.5t^2 + 20.7t + 194.5, \qquad \text{Lobster (millions of pounds)}$$

where $t = 0$ represents 1980. Find a model that represents the total amount T of shrimp *and* lobster consumed from 1980 to 1990. Estimate the total amount T consumed in 1991.

FIGURE 1.12

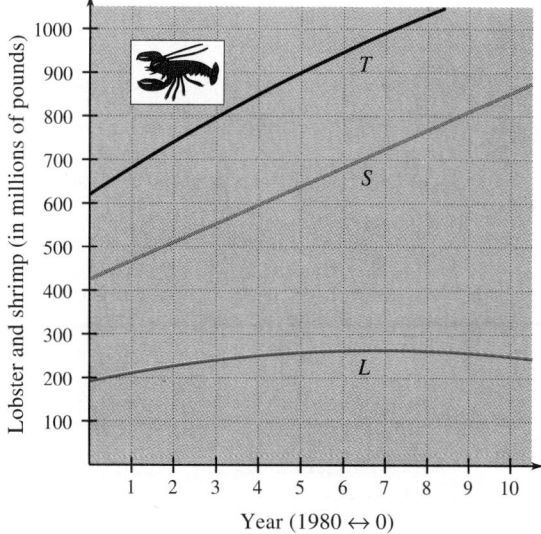

Lobster and shrimp (in millions of pounds)

Year (1980 ↔ 0)

Solution

The sum of the two polynomial models is

$$(42.8t + 425) + (-1.5t^2 + 20.7t + 194.5) = -1.5t^2 + 63.5t + 619.5.$$

The model for the total consumption of shrimp and lobster is

$$T = -1.5t^2 + 63.5t + 619.5. \qquad \text{Total (millions of pounds)}$$

Using this model, and substituting $t = 11$, you can estimate the 1991 consumption to be

$$T = -1.5(11^2) + 63.5(11) + 619.5 = 1136.5 \text{ million pounds.}$$

(*Source:* U. S. National Oceanic and Atmospheric Administration)   ■

DISCUSSION PROBLEM   ■   **The Position Polynomial**

Consider the following polynomials in the variable t to represent the height above ground of an object moving vertically up or down through the air.

$$-16t^2 + 80t + 30, \quad -16t^2 - 80t + 30, \quad -16t^2 + 30, \quad -16t^2 + 30t$$

Write a short paragraph to support your answer to each of the following questions.

(a) Which polynomial represents the height of an object thrown *downward*?

(b) Which polynomial represents the height of an object that is *dropped* from a height of 30 feet?

(c) Which polynomial represents the height of an object thrown *upward* from ground level?   ■

Warm-Up

The following warm-up exercises involve skills that were covered in earlier sections. You will use these skills in the exercise set for this section.

In Exercises 1–4, expand the expression.

1. $23(y - 3)$ **2.** $6(2 - 5z)$ **3.** $-\frac{2}{3}(15 - 6x)$ **4.** $-5(5x - 4)$

In Exercises 5 and 6, list the terms of the algebraic expression.

5. $32x - 10y - 7$ **6.** $-3s + 4t + 1$

In Exercises 7–10, simplify the algebraic expression.

7. $3(x - 5) + 4x$ **8.** $5(4 - y) + 8(y + 2)$

9. $-3(u - 6) - (u + 18)$ **10.** $(v - 3) - 3(2v + 5)$

| **1.4** | **EXERCISES** |

In Exercises 1–8, write the polynomial in standard form, and find its degree and leading coefficient.

1. $10x - 4$

2. $3x^2 + 8$

3. $5 - 3y^4$

4. $-3x^3 - 2x^2 - 3$

5. $6t + 4t^5 - t^2 + 3$

6. $8z - 16z^2$

7. -4

8. $v_0 t - 16t^2$
(v_0 is a constant)

In Exercises 9–12, determine whether the polynomial is a monomial, binomial, or trinomial.

9. $12 - 5y^2$

10. t^3

11. $x^3 + 2x^2 - 4$

12. $25 - 2u^2$

In Exercises 13 and 14, state why the algebraic expression is not a polynomial.

13. $y^{-3} - 2$

14. $x^3 + 4x^{1/3}$

In Exercises 15–18, give an example of a polynomial in one variable satisfying the conditions. (*Note:* Each problem has more than one correct answer.)

15. A monomial of degree 3

16. A trinomial of degree 4 and leading coefficient -2

17. A binomial of degree 2 and leading coefficient 8

18. A monomial of degree 0

In Exercises 19–22, find the value of the polynomial at the given values of the variable.

19. $x^3 - 12x$
 (a) $x = -2$ (b) $x = 0$ (c) $x = 2$ (d) $x = 4$

20. $\frac{1}{4}x^4 - 2x^2$
 (a) $x = -2$ (b) $x = 0$ (c) $x = 2$ (d) $x = 3$

21. $x^4 - 4x^3 + 16x - 16$
 (a) $x = -1$ (b) $x = 0$ (c) $x = 2$ (d) $x = \frac{5}{2}$

22. $3t^4 + 4t^3$
 (a) $t = -1$ (b) $t = -\frac{2}{3}$ (c) $t = 0$ (d) $t = 1$

In Exercises 23–28, perform the addition using a horizontal format.

23. $(2x^2 - 3) + (5x^2 + 6)$

24. $(3x^3 - 2x + 8) + (3x - 5)$

25. $(x^2 - 3x + 8) + (2x^2 - 4x) + 3x^2$

26. $(5y + 6) + (4y^2 - 6y - 3)$

27. Add $3x^2 + 8$ to $7 - 5x^2$.

28. Add $20s - 12s^2 - 32$ to $15s^2 + 6s$.

In Exercises 29–32, perform the addition using a vertical format.

29. $(5x^2 - 3x + 4) + (-3x^2 - 4)$

30. $(4x^3 - 2x^2 + 8x) + (4x^2 + x - 6)$

31. $(2b - 3) + (b^2 - 2b) + (7 - b^2)$

32. $(v^2 + v - 3) + (4v + 1) + (2v^2 - 3v)$

In Exercises 33–38, perform the subtraction using a horizontal format.

33. $(3x^2 - 2x + 1) - (2x^2 + x - 1)$

34. $(5y^4 - 2) - (3y^4 + 2)$

35. $(8x^3 - 4x^2 + 3x) - [(x^3 - 4x^2 + 5) + (x - 5)]$

36. $(5y^2 - 2y) - [(y^2 + y) - (3y^2 - 6y + 2)]$

37. Subtract $6x^3 - x + 11$ from $10x^3 + 15$.

38. Subtract $y^4 - y^2$ from $y^2 + 3y^4$.

In Exercises 39–42, perform the subtraction using a vertical format.

39. $(x^2 - x + 3) - (x - 2)$

40. $(z^2 - 5) - (z^2 - 5)$

41. $(-2x^3 - 15x + 25) - (2x^3 - 13x + 12)$

42. $(0.2t^4 - 0.5t^2) - (0.3t^2 - t^4 - 1.4)$

In Exercises 43–56, perform the indicated operations.

43. $(2 - 8y) + (-2y^4 + 3y + 2)$

44. $(3a^2 + 5a) + (7 - a^2 - 5a) + (2a^2 + 8)$

45. $-(2x^3 - 3) + (4x^3 - 2x)$

46. $(8z^4 - 7z^2) + 3(z^4 + 4)$

47. $(5q^2 - 3q + 5) - (4q^2 - 3q - 10)$

48. $(-10s^2 - 5) - (2s^2 + 6s)$

49. $(10x^2 - 11) - (-7x^3 - 12x^2 - 15)$

50. $(15y^4 - 18y - 18) - (-11y^4 - 8y - 8)$

51. $5s - [6s - (30s + 8)]$

52. $(p^3 + 4) - [(p^2 + 4) + (3p - 9)]$

53. $2(t^2 + 12) - 5(t^2 + 5) + 6(t^2 + 5)$

54. $-10(v + 2) + 8(v - 1) - 3(v - 9)$

55. $15v - 3(3v - v^2) + 9(8v + 3)$

56. $9(7x^2 - 3x + 3) - 4(15x + 2) - (3x^2 - 7x)$

In Exercises 57–76, perform the indicated multiplication and simplify.

57. $(-2a^2)(-8a)$

58. $(-6n)(3n^2)$

59. $2y(5 - y)$

60. $5z(2z - 7)$

61. $4x^3(2x^2 - 3x + 5)$

62. $3y^2(-3y^2 + 7y - 3)$

63. $-2x^2(5 + 3x^2 - 7x^3)$

64. $-3a^2(11a - 3)$

65. $(x + 7)(x - 4)$

66. $(y - 2)(y + 3)$

67. $(2a - 5)(a - 3)$

68. $(3x + 1)(2x + 1)$

69. $(2x + y)(3x + 2y)$

70. $(2x - y)(3x - 2y)$

71. $\left(4y - \frac{1}{3}\right)(12y + 9)$

72. $\left(5t - \frac{3}{4}\right)(2t - 16)$

73. $-3x(-5x)(5x + 2)$

74. $4t(-3t)(t^2 - 1)$

75. $5a(a + 2) - 3a(2a - 3)$

76. $(2t - 1)(t + 1) + 3(2t - 5)$

In Exercises 77–84, use a horizontal format to perform the indicated multiplication.

77. $(x^3 - 3x + 2)(x - 2)$

78. $(t + 3)(t^2 - 5t + 1)$

79. $(u + 5)(2u^2 + 3u - 4)$

80. $(x^2 + 4)(x^2 - 2x - 4)$

81. $(2x^2 - 3)(2x^2 - 2x + 3)$

82. $(x - 1)(x^2 - 4x + 6)$

83. $(a + 5)^3$

84. $(y - 2)^3$

In Exercises 85–90, use a vertical format to perform the indicated multiplication.

85. $(x + 3)(x^2 - 3x + 9)$

86. $(x - 4)(x^2 + 4x + 16)$

87. $(t^2 + t - 2)(t^2 - t + 2)$

88. $(y^2 + 3y + 5)(2y^2 - 3y - 1)$

89. $(2x + 3)(-3x^2 + 2x - 1)$

90. $(3s - 4)(2s^2 - 5s + 6)$

In Exercises 91–108, perform the indicated multiplication.

91. $(x + 2)(x - 2)$

92. $(4 + 3z)(4 - 3z)$

93. $(a - 6c)(a + 6c)$

94. $(8n - m)(8n + m)$

95. $\left(2x - \frac{1}{4}\right)\left(2x + \frac{1}{4}\right)$

96. $\left(\frac{2}{3}x + 7\right)\left(\frac{2}{3}x - 7\right)$

97. $(0.2t + 0.5)(0.2t - 0.5)$

98. $(4a - 0.1b)(4a + 0.1b)$

99. $(x + 5)^2$

100. $(x + 9)^2$

101. $(5x - 2)^2$

102. $(3x - 8)^2$

103. $(2a + 3b)^2$

104. $(4x - 5y)^2$

105. $[(x + 2) - y]^2$

106. $[(x - 4) + y]^2$

107. $[u - (v - 3)]^2$

108. $[2z + (y + 1)]^2$

In Exercises 109–114, perform the indicated operations and simplify.

109. $(x + 3)(x - 3) - (x^2 + 8x - 2)$

110. $(k - 8)(k + 8) - (k^2 - k + 3)$

111. $5y(y - 4) + (y - 6)^2$

112. $(b + 1)^2 - 4b(b + 5)$

113. $(t + 3)^2 - (t - 3)^2$

114. $(a + 6)^2 + (a - 6)^2$

Geometry In Exercises 115 and 116, find the area of the shaded portion of the figure.

115.

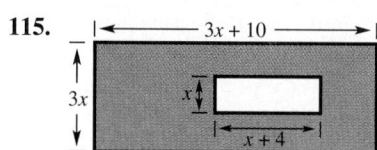

116.

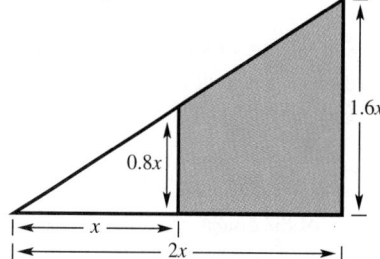

117. *Area of a Rectangle* (a) Find the area of the rectangle (see figure) in two ways. (b) Explain how the result is related to the FOIL Method.

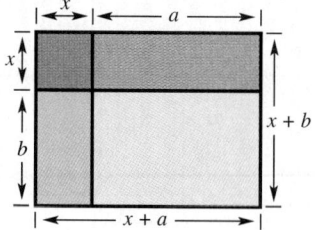

118. *Area of a Square* (a) Find the area of the square (see figure) in two ways. (b) Explain how the result is related to a special product.

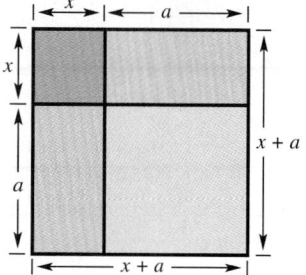

119. *Compounded Interest* After two years, an investment of \$1000 compounded annually at an interest rate r will yield an amount $1000(1 + r)^2$. Find this product.

120. *Profit* A manufacturer can produce and sell x radios per week. The total cost (in dollars) for producing the radios is

$$C = 8x + 15{,}000$$

and the total revenue is

$$R = 14x.$$

  (a) Find the cost of producing 5000 radios per week.
  (b) Find the total revenue from selling 5000 radios per week.
  (c) Find the profit obtained by selling 5000 radios per week.

Free-Falling Object In Exercises 121–124, use the position polynomial to determine whether the free-falling object was dropped, thrown upward, or thrown downward. Also determine the height of the object at time $t = 0$.

121. $-16t^2 + 100$ **122.** $-16t^2 + 50t$

123. $-16t^2 - 24t + 50$ **124.** $-16t^2 + 32t + 300$

125. *Free-Falling Object* An object is thrown upward from the top of a 200-foot building (see figure). The initial velocity is 40 feet per second. Use the position polynomial

$$\text{Height} = -16t^2 + 40t + 200$$

to find the height of the object when $t = 1$, $t = 2$, and $t = 3$.

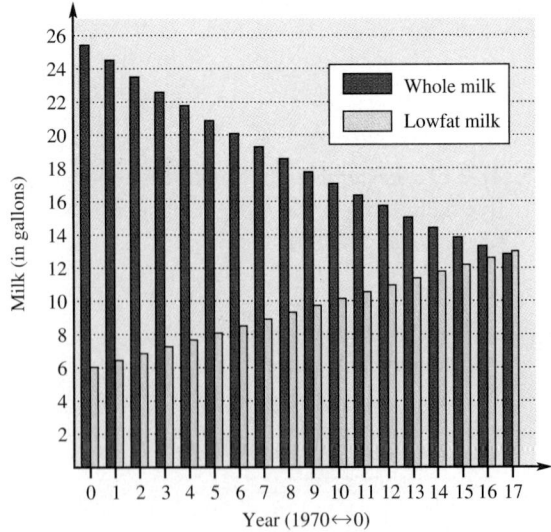

200 ft

126. (a) Use the result of Exercise 125 to determine in which of the following intervals the object fell farther: from $t = 1$ second to $t = 2$ seconds or from $t = 2$ seconds to $t = 3$ seconds.
  (b) Explain your reasoning.

In Exercises 127–130, use a calculator to perform the indicated operations.

127. $8.04x^2 - 9.37x^2 + 5.62x^2$

128. $-11.98y^3 + 4.63y^3 - 6.79y^3$

129. $(4.098a^2 + 6.349a) - (11.246a^2 - 9.342a)$

130. $(27.433k^2 - 19.018k) + (-14.61k^2 + 3.814k)$

131. *Consumption of Milk* The per capita consumption (average consumption per person) of whole milk and low-fat milk in the United States between 1970 and 1987 can be approximated by two polynomial models.

$$y = 25.4 - 0.96t + 0.013t^2 \qquad \text{Whole milk}$$
$$y = 6.03 + 0.41t \qquad\qquad\qquad \text{Lowfat milk}$$

In these models, y represents the average consumption per person in gallons and t represents the year, with $t = 0$ corresponding to 1970. (*Source:* USDA)

  (a) Find a polynomial model that represents the per capita consumption of milk (of both types) during this time period. Use this model to find the per capita consumption of milk in 1975 and in 1985.
  (b) Between 1970 and 1985, the per capita consumption of whole milk was decreasing and the per capita consumption of lowfat milk was increasing (see figure). Was the per capita consumption of milk (of both types) increasing or decreasing?

Milk (in gallons)

Whole milk

Lowfat milk

Year (1970↔0)

Factoring Polynomials

Factoring Polynomials with Common Factors • Factoring by Grouping • Factoring the Difference of Two Squares • Factoring the Sum or Difference of Two Cubes • Factoring Completely

Factoring Polynomials with Common Factors

Now we will switch from the process of multiplying polynomials to the *reverse* process—**factoring polynomials.** This section and the next section deal only with polynomials that have integer coefficients. Remember that in Section 1.4 you used the Distributive Property to *multiply* and *remove* parentheses, as follows.

Multiply

$$3x(4 - 5x) = 12x - 15x^2 \qquad \text{Distributive Property}$$

In this and the next section, you will use the Distributive Property in the reverse direction to *factor* and *create* parentheses.

Factor

$$12x - 15x^2 = 3x(4 - 5x) \qquad \text{Distributive Property}$$

Factoring an expression (by the Distributive Property) changes a *sum of terms* into a *product of factors.* Later you will see that this is an important strategy for solving equations and for simplifying algebraic expressions.

To be efficient in factoring an expression, you need to understand the concept of the **greatest common factor** of two (or more) integers or terms. Recall from arithmetic that every integer can be factored into a product of prime numbers. The **greatest common factor** of two or more integers is the greatest integer that is a factor of each number. For example, the greatest common factor of 18 and 42 is 6.

$$\left. \begin{array}{l} 18 = 2 \cdot 3 \cdot 3 \\ 42 = 2 \cdot 3 \cdot 7 \end{array} \right\} 2 \cdot 3 = 6 \text{ is common to both integers}$$

In a similar way, an algebraic term can be factored into a product of prime factors, as shown in Example 1.

EXAMPLE 1   ■   Finding the Greatest Common Factor

Find the greatest common factor of

$$6x^5, \quad 30x^4, \quad \text{and} \quad 12x^3.$$

Solution

From the factorizations

$$6x^5 = 2 \cdot 3 \cdot x \cdot x \cdot x \cdot x \cdot x = (6x^3)(x^2)$$

$$30x^4 = 2 \cdot 3 \cdot 5 \cdot x \cdot x \cdot x \cdot x = (6x^3)(5x)$$

$$12x^3 = 2 \cdot 2 \cdot 3 \cdot x \cdot x \cdot x = (6x^3)(2)$$

you can conclude that the greatest common factor is $6x^3$.   ■

Consider the three terms given in Example 1 as terms of the polynomial

$$6x^5 + 30x^4 + 12x^3.$$

The greatest common factor, $6x^3$, of these terms is called the **greatest common monomial factor** of the polynomial. When you use the Distributive Property to remove this factor from each term of the polynomial, you are **factoring out** the greatest common monomial factor.

$$6x^5 + 30x^4 + 12x^3 = 6x^3(x^2) + 6x^3(5x) + 6x^3(2) \qquad \text{Factor each term}$$

$$= 6x^3(x^2 + 5x + 2) \qquad \text{Factor out common monomial factor}$$

EXAMPLE 2   ■   Factoring Out the Greatest Common Monomial Factor

Factor out the greatest common monomial factor from the polynomial

$$24x^3 - 32x^2.$$

Solution

The greatest common factor of $24x^3$ and $32x^2$ is $8x^2$. Thus, you can factor the given polynomial as follows.

$$24x^3 - 32x^2 = (8x^2)(3x) - (8x^2)(4)$$

$$= 8x^2(3x - 4) \qquad ■$$

NOTE If a polynomial in x (with integer coefficients) has a greatest common monomial factor of the form ax^n, then the following statements must be true.

1. The coefficient a of the greatest common monomial factor must be the greatest integer that *divides* each of the coefficients in the polynomial.

2. The variable factor x^n of the greatest common monomial factor has the *least* power of x of all terms of the polynomial.

We usually consider the greatest common monomial factor of a polynomial to have a positive coefficient. However, sometimes it is convenient to factor a negative number out of a polynomial. You can see how this is done in the next example.

EXAMPLE 3   ■   A Negative Common Monomial Factor

Factor the polynomial $-3x^2 + 12x - 18$ in two ways.

(a) Factor out a 3.   (b) Factor out a -3.

Solution

(a) To factor out the common monomial factor of 3, write the following.

$$-3x^2 + 12x - 18 = 3(-x^2) + 3(4x) + 3(-6) \qquad \text{Factor out each term}$$

$$= 3(-x^2 + 4x - 6) \qquad \text{Factor out the polynomial}$$

(b) To factor -3 out of the polynomial, write the following.

$$-3x^2 + 12x - 18 = -3(x^2) + (-3)(-4x) + (-3)(6) \quad \text{Factor out each term}$$

$$= -3(x^2 - 4x + 6) \qquad \text{Factor out the polynomial} \quad ■$$

Factoring by Grouping

There are occasions when a common factor of a polynomial is not simply a monomial. For instance, the polynomial

$$x^2(2x - 3) + 4(2x - 3)$$

has the common *binomial* factor $(2x - 3)$. Factoring out this common factor produces

$$x^2(2x - 3) + 4(2x - 3) = (2x - 3)(x^2 + 4).$$

This type of factoring is part of a more general procedure called **factoring by grouping.**

EXAMPLE 4   ■   A Common Binomial Factor

Factor the polynomial

$$5x^2(6x - 5) - 2(6x - 5).$$

Solution

Each of the terms of this polynomial has a binomial factor of $(6x - 5)$. Factoring this binomial out of each term produces the following.

$$5x^2(6x - 5) - 2(6x - 5) = (6x - 5)(5x^2 - 2) \qquad\qquad ■$$

In Example 4, the given polynomial was already grouped, and so it was easy to determine the common binomial factor. In practice, you will have to do the grouping as well as the factoring. To see how this works, consider the expression

$$x^3 - 3x^2 - 5x + 15$$

and try to *factor* it. Note first that there is no common monomial factor to take out of all four terms. But suppose you *group* the first two terms together and the last two terms together. Then you have the following.

$$x^3 - 3x^2 - 5x + 15 = (x^3 - 3x^2) - (5x - 15) \qquad \text{Group terms}$$

$$= x^2(x - 3) - 5(x - 3) \qquad \text{Factor out common factor in each group}$$

$$= (x - 3)(x^2 - 5) \qquad \text{Factor out common binomial factor}$$

EXAMPLE 5 ■ Factoring by Grouping

Factor the polynomial $x^3 - 5x^2 + x - 5$.

Solution

Point out that when factoring a polynomial with four or more terms, students first should try factoring by grouping.

Note that the common factor must be exactly the same. For example, $x(x-3) - 5(x+3)$ cannot be factored.

By grouping the first two terms together, and the third and fourth terms together, you obtain the following.

$$x^3 - 5x^2 + x - 5 = (x^3 - 5x^2) + (x - 5)$$

$$= x^2(x - 5) + (x - 5)$$

$$= (x - 5)(x^2 + 1)$$

Be sure you see that the expression in the second line of the solution, $x^2(x - 5) + (x - 5)$, is only an *intermediate step* in the factoring process. This expression is itself *not* considered to be factored—of the four expressions listed in the solution, only the last is in factored form. ■

NOTE You can check to see that you have factored an expression correctly by multiplying out the factors and comparing the result with the original expression.

Factoring the Difference of Two Squares

Some polynomials have special forms that you should learn to recognize so that they can be factored easily. Here are some examples of forms that you should be able to recognize by the time you have completed this section.

$$x^2 - 9 = (x + 3)(x - 3) \qquad \text{Difference of two squares}$$

$$x^3 + 8 = (x + 2)(x^2 - 2x + 4) \qquad \text{Sum of two cubes}$$

$$x^3 - 1 = (x - 1)(x^2 + x + 1) \qquad \text{Difference of two cubes}$$

One of the easiest special polynomial forms to recognize and to factor is the form $u^2 - v^2$, called a **difference of two squares.** It factors according to the following pattern.

Difference of Two Squares

Let u and v be real numbers, variables, or algebraic expressions. Then the expression $u^2 - v^2$ can be factored as follows.

$$u^2 - v^2 = (u + v)(u - v)$$

$\uparrow$ Difference $\uparrow$ $\uparrow$ Opposite signs

To recognize perfect squares, look for coefficients that are squares of integers and for variables raised to *even* powers. Here are some examples.

Consider reinforcing the pattern using parentheses.

$81y^4 - 49 = (\quad)^2 - (\quad)^2$
$= (9y^2)^2 - (7)^2$
$= (9y^2 + 7)(9y^2 - 7)$

Given Polynomial		*Think: Difference of Squares*		*Factored Form*
$x^2 - 4$	$\Longrightarrow$	$x^2 - 2^2$	$\Longrightarrow$	$(x + 2)(x - 2)$
$4x^2 - 25$	$\Longrightarrow$	$(2x)^2 - 5^2$	$\Longrightarrow$	$(2x + 5)(2x - 5)$
$25 - 49x^2$	$\Longrightarrow$	$5^2 - (7x)^2$	$\Longrightarrow$	$(5 + 7x)(5 - 7x)$
$x^{10} - 16y^2$	$\Longrightarrow$	$(x^5)^2 - (4y)^2$	$\Longrightarrow$	$(x^5 + 4y)(x^5 - 4y)$

EXAMPLE 6 ■ Factoring the Difference of Two Squares

Factor the polynomial

$$x^2 - 64.$$

Solution

Because x^2 and 64 are both perfect squares, you can recognize this polynomial as the difference of two squares. Therefore, the polynomial factors as follows.

$x^2 - 64 = x^2 - 8^2$ Write as difference of two squares

$= (x + 8)(x - 8)$ Factored form   ■

EXAMPLE 7 ■ Factoring the Difference of Two Squares

Factor the polynomial

$$49x^2 - 81y^2.$$

Solution

Because $49x^2$ and $81y^2$ are both perfect squares, you can recognize this polynomial as the difference of two squares. Therefore, the polynomial factors as follows.

$49x^2 - 81y^2 = (7x)^2 - (9y)^2$ Write as difference of two squares

$= (7x + 9y)(7x - 9y)$ Factored form   ■

Remember that the rule $u^2 - v^2 = (u + v)(u - v)$ applies to polynomials or expressions in which u and v are themselves expressions. The next two examples illustrate this possibility.

EXAMPLE 8 ■ Factoring the Difference of Two Squares

Extension example:

$(3x - 1)^2 - (x + 6)^2$
$= [(3x - 1) + (x + 6)][(3x - 1) - (x + 6)]$
$= [3x - 1 + x + 6][3x - 1 - x - 6]$
$= (4x + 5)(2x - 7)$

$$(x + 2)^2 - 9 = (x + 2)^2 - 3^2 \qquad \text{Write as difference of two squares}$$
$$= [(x + 2) + 3][(x + 2) - 3] \qquad \text{Factored form}$$
$$= (x + 5)(x - 1) \qquad \text{Simplify} \qquad ■$$

Factoring the Sum or Difference of Two Cubes

The last type of special factoring discussed in this section is the sum or difference of two cubes. The patterns for these two special forms are summarized as follows. In these patterns, pay particular attention to the signs of the terms.

Sum or Difference of Two Cubes

Let u and v be real numbers, variables, or algebraic expressions. Then the expressions $u^3 + v^3$ and $u^3 - v^3$ can be factored as follows.

Like signs

1. $u^3 + v^3 = (u + v)(u^2 - uv + v^2)$

Unlike signs

Like signs

2. $u^3 - v^3 = (u - v)(u^2 + uv + v^2)$

Unlike signs

EXAMPLE 9 ■ Factoring Sums and Differences of Cubes

Factor the following polynomials.

(a) $x^3 - 125$
(b) $8y^3 + 1$

Solution

(a) This polynomial is the difference of two cubes, because x^3 is the cube of x and 125 is the cube of 5. Therefore, you can factor the polynomial as follows.

$$x^3 - 125 = x^3 - 5^3 \qquad \text{Write as difference of two cubes}$$
$$= (x - 5)(x^2 + 5x + 5^2) \qquad \text{Factored form}$$
$$= (x - 5)(x^2 + 5x + 25) \qquad \text{Simplify}$$

(b) This polynomial is the sum of two cubes, because $8y^3$ is the cube of $2y$ and 1 is its own cube. Therefore, you can factor the polynomial as follows.

$$8y^3 + 1 = (2y)^3 + 1^3 \qquad \text{Write as sum of two cubes}$$
$$= (2y + 1)[(2y)^2 - (2y)(1) + 1^2] \qquad \text{Factored form}$$
$$= (2y + 1)(4y^2 - 2y + 1) \qquad \text{Simplify} \qquad \blacksquare$$

It is easy to make arithmetic errors when applying the patterns for factoring the sum or difference of two cubes. When you use these patterns, you should check your work by multiplying the two factors. For instance, you can check the factors given in Example 9 as follows.

$$
\begin{array}{r}
x^2 + 5x + 25 \\
x - 5 \\
\hline
-5x^2 - 25x - 125 \\
x^3 + 5x^2 + 25x \\
\hline
x^3 \qquad\qquad - 125
\end{array}
\qquad
\begin{array}{r}
4y^2 - 2y + 1 \\
2y + 1 \\
\hline
4y^2 - 2y + 1 \\
8y^3 - 4y^2 + 2y \\
\hline
8y^3 \qquad\qquad + 1
\end{array}
$$

Factoring Completely

Remind students always to check first for a common monomial factor.

Sometimes the difference of two squares can be hidden by the presence of a common monomial factor. Remember that with *all* factoring techniques, you should first remove any common monomial factors. This is demonstrated in Example 10.

EXAMPLE 10 $\blacksquare$ Removing a Common Monomial Factor First

Factor the polynomial

$$125x^2 - 80.$$

Solution

The polynomial $125x^2 - 80$ has a common monomial factor of 5. After removing this factor, you can see that the remaining polynomial is the difference of two squares.

$$125x^2 - 80 = 5(25x^2 - 16) \qquad \text{Remove common monomial factor}$$
$$= 5[(5x)^2 - 4^2] \qquad \text{Write as difference of two squares}$$
$$= 5(5x + 4)(5x - 4) \qquad \text{Factored form} \qquad \blacksquare$$

The polynomial in Example 10 is said to be **completely factored** because none of its factors can be factored further using integer coefficients.

EXAMPLE 11 $\blacksquare$ Removing a Common Monomial Factor First

Factor the following polynomial completely.

$$3x^3 + 81$$

Solution

$$3x^3 + 81 = 3(x^3 + 27) \qquad\qquad \text{Remove common monomial factor}$$

$$= 3(x^3 + 3^3) \qquad\qquad \text{Write as sum of cubes}$$

$$= 3(x + 3)(x^2 - 3x + 9) \qquad \text{Factored form} \qquad \blacksquare$$

Which of the following is factored completely and correctly?

(a) $8x^3 - 64 = 8(x^3 - 8)$

$$= 8(x - 2)(x^2 + 2x + 4)$$

(b) $8x^3 - 64 = (2x - 4)(4x^2 + 8x + 16)$

Discuss the results.

To factor a polynomial completely, always check to see whether the factors obtained might themselves be factorable. That is, can any of the factors be factored? For instance, after factoring the polynomial $x^4 - 16$ once as the difference of two squares,

$$x^4 - 16 = (x^2)^2 - 4^2 = (x^2 + 4)(x^2 - 4),$$

you can see that the second factor is itself the difference of two squares. Thus, to factor the polynomial *completely*, you must continue the factoring process, as follows.

$$x^4 - 16 = (x^2 + 4)(x^2 - 4) = (x^2 + 4)(x + 2)(x - 2)$$

Other instances of "repeated factoring" are given in the next example.

EXAMPLE 12   ■   Factoring Completely

Factor the following polynomials completely.

(a) $x^4 - y^4$ (b) $81m^4 - 1$

Solution

(a) Recognizing $x^4 - y^4$ as a difference of two squares, you can write

$$x^4 - y^4 = (x^2)^2 - (y^2)^2 = (x^2 + y^2)(x^2 - y^2).$$

Note that the second factor $(x^2 - y^2)$ is itself a difference of two squares and you therefore obtain

$$x^4 - y^4 = (x^2 + y^2)(x^2 - y^2) = (x^2 + y^2)(x + y)(x - y).$$

(b) Recognizing $81m^4 - 1$ as a difference of two squares, you can write

$$81m^4 - 1 = (9m^2)^2 - (1)^2 = (9m^2 + 1)(9m^2 - 1).$$

Now, because the second factor $(9m^2 - 1)$ is itself a difference of two squares, you can continue the factoring process as follows.

$$81m^4 - 1 = (9m^2 + 1)(9m^2 - 1) = (9m^2 + 1)(3m + 1)(3m - 1) \qquad \blacksquare$$

Note in Example 12 that the *sum of two squares* does not factor further. A second-degree polynomial that is the sum of two squares, such as $x^2 + y^2$ or $9m^2 + 1$, is *not factorable.**

---

*In this text, when we say that a polynomial is not factorable, we mean that it is not factorable *using integer coefficients.*

DISCUSSION PROBLEM ■ Factoring Higher-Power Polynomials

Consider the following factorizations.

(a) $x^6 - 64 = (x^3)^2 - 8^2$

$= (x^3 - 8)(x^3 + 8)$

$= (x - 2)(x^2 + 2x + 4)(x + 2)(x^2 - 2x + 4)$

(b) $x^6 - 64 = (x^2)^3 - 4^3$

$= (x^2 - 4)(x^4 + 4x^2 + 16)$

$= (x - 2)(x + 2)(x^4 + 4x^2 + 16)$

Which factorization is more complete? Justify the apparent difference between these two results. ■

Warm-Up

The following warm-up exercises involve skills that were covered in earlier sections. You will use these skills in the exercise set for this section.

In Exercises 1–10, perform the indicated multiplication.

1. $12x(2x - 3)$ **2.** $7y(4 - 3y)$ **3.** $-6x(10 - 7x)$ **4.** $-2y(y + 1)$

5. $t(t^2 + 1) - t(t^2 - 1)$ **6.** $2z(z + 5) - 7(z + 5)$ **7.** $(11 - x)(11 + x)$

8. $(6r + 5s)(6r - 5s)$ **9.** $(x - 2)(x^2 + 2x + 4)$ **10.** $(x + 1)(x^2 - x + 1)$

1.5 EXERCISES

In Exercises 1–12, find the greatest common factor of the expressions.

1. 48, 90

2. 36, 150, 100

3. $3x^2$, $12x$

4. $27x^4$, $18x^3$

5. $30z^2$, $12z^3$

6. $45y$, $150y^3$

7. $28b^2$, $14b^3$, $42b^5$

8. $16x^2y$, $84xy^2$, $36x^2y^2$

9. $42(x + 8)^2$, $63(x + 8)^3$

10. $66(3 - y)$, $44(3 - y)^2$

11. $4x(1 - z)^2$, $x^2(1 - z)^3$

12. $2(x + 5)$, $8(x + 5)$

In Exercises 13–30, factor by removing the greatest common monomial factor. (Some of the polynomials have no common monomial factor other than 1 or −1.)

13. $8z - 8$

14. $5x + 5$

15. $4u + 10$

16. $-15t - 10$

17. $24x^2 - 18$

18. $14z^3 + 21$

19. $2x^2 + x$

20. $-a^3 - 4a$

21. $21u^2 - 14u$ **22.** $36y^4 + 24y^2$ **23.** $11u^2 + 9$ **24.** $16x^2 - 3y^3$

25. $3x^2y^2 - 15y$ **26.** $4uv + 6u^2v^2$ **27.** $28x^2 + 16x - 8$ **28.** $9 - 27y - 15y^2$

29. $14x^4 + 21x^3 + 9x^2$ **30.** $17x^5y^3 - xy^2 + 34y^2$

In Exercises 31–38, factor a negative real number from the polynomial and then write the polynomial factor in standard form.

31. $10 - 5x$ **32.** $32 - 4x^4$ **33.** $7 - 14x$ **34.** $15 - 5x$

35. $8 + 4x - 2x^2$ **36.** $12x - 6x^2 - 18$ **37.** $2t - 15 - 4t^3$ **38.** $16 + 32s^2 - 5s^4$

In Exercises 39–44, factor the given polynomial by factoring out the common binomial factor.

39. $2y(y - 3) + 5(y - 3)$ **40.** $7t(s + 9) - 6(s + 9)$ **41.** $5t(t^2 + 1) - 4(t^2 + 1)$

42. $3a(a^2 - 3) + 10(a^2 - 3)$ **43.** $a(a + 6) - a^2(a + 6)$ **44.** $(5x + y)(x - y) - 5x(x - y)$

In Exercises 45–58, factor the expression by grouping.

45. $y^2 - 6y + 2y - 12$ **46.** $y^2 + 3y + 4y + 12$ **47.** $x^3 - 3x^2 + 2x - 6$

48. $a^3 - 5a^2 + 6a - 30$ **49.** $x^2 + 25x + x + 25$ **50.** $x^2 - 7x + x - 7$

51. $x^3 + 2x^2 + x + 2$ **52.** $t^3 - 11t^2 + t - 11$ **53.** $a^3 - 4a^2 + 2a - 8$

54. $3s^3 + 6s^2 + 5s + 10$ **55.** $z^4 + 3z^3 - 2z - 6$ **56.** $4u^4 - 2u^3 - 6u + 3$

57. $cd + 3c - 3d - 9$ **58.** $u^2 + uv - 4u - 4v$

In Exercises 59–74, factor the difference of two squares.

59. $x^2 - 64$ **60.** $y^2 - 144$ **61.** $16y^2 - 9z^2$ **62.** $9z^2 - 25w^2$

63. $x^2 - 4y^2$ **64.** $81a^2 - b^6$ **65.** $a^8 - 36$ **66.** $y^{10} - 64$

67. $100 - 9y^2$ **68.** $625 - 49x^2$ **69.** $a^2b^2 - 16$ **70.** $u^2v^2 - 25$

71. $(a + 4)^2 - 49$ **72.** $(x - 3)^2 - 4$ **73.** $81 - (z + 5)^2$ **74.** $100 - (y - 3)^2$

In Exercises 75–80, factor the sum or difference of cubes.

75. $x^3 - 8$ **76.** $t^3 - 27$ **77.** $y^3 + 64z^3$

78. $z^3 + 125w^3$ **79.** $8t^3 - 27$ **80.** $27s^3 + 64$

In Exercises 81–86, factor the algebraic expression completely.

81. $8 - 50x^2$ **82.** $a^3 - 16a$ **83.** $y^4 - 81x^4$

84. $u^4 - 256v^4$ **85.** $2x^3 - 54$ **86.** $5y^3 - 625$

In Exercises 87 and 88, evaluate the quantity mentally using the following example as a model.

$$48 \cdot 52 = (50 - 2)(50 + 2) = 50^2 - 2^2 = 2496$$

87. $79 \cdot 81$ **88.** $18 \cdot 22$

In Exercises 89 and 90, factor the expression. (Assume $n > 0$.)

89. $4x^{2n} - 25$ **90.** $81 - 16y^{4n}$

In Exercises 91 and 92, show all the different groupings that can be used to factor completely the polynomial. Show that the various factorizations yield the same result.

91. $3x^3 + 4x^2 - 3x - 4$ **92.** $6x^3 - 8x^2 + 9x - 12$

93. *Revenue and Price* The revenue for selling x units of a product at a price of p dollars per unit is xp. For a particular commodity, the revenue is

$$R = 800x - 0.25x^2.$$

Factor the expression for the revenue and determine an expression for the price in terms of x.

94. *Width of a Rectangle* The area of a rectangle of length l is given by $45l - l^2$. Factor this expression to determine the width of the rectangle.

Geometry In Exercises 95 and 96, find the perimeter of the rectangle or square. The lengths of the sides are obtained by factoring the expression for the area.

95. Area $= x^2 - 49$

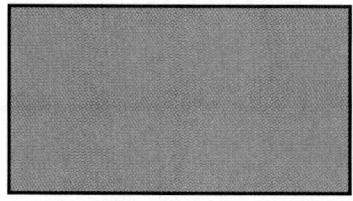

96. Area $= x(x + 3) + 3(x + 3)$

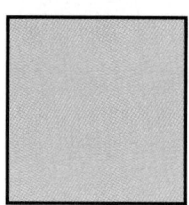

97. *Simple Interest* The total amount of money from a principal of P invested at $r\%$ simple interest for t years is $P + Prt$. Factor this expression.

98. *Chemical Reaction* The rate of change of a chemical reaction is

$$kQx - kx^2,$$

where Q is the amount of the original substance, x is the amount of substance formed, and k is a constant of proportionality. Factor the expression for this rate of change.

99. *Product Design* A washer on the drive train of a car has an inside radius of r centimeters and an outside radius of R centimeters (see figure). Find the area of one of the flat surfaces of the washer and express the area in factored form.

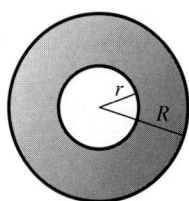

Factoring Trinomials

Factoring Perfect Square Trinomials • Factoring Trinomials of the Form $x^2 + bx + c$ • Factoring Trinomials of the Form $ax^2 + bx + c$ • Factoring Trinomials by Grouping (Optional) • Summary of Factoring

Factoring Perfect Square Trinomials

A **perfect square trinomial** is the square of a binomial. For instance,

$$x^2 + 6x + 9 = (x + 3)^2$$

is the square of the binomial $(x + 3)$, and

$$4x^2 - 20x + 25 = (2x - 5)^2$$

is the square of the binomial $(2x - 5)$. Perfect square trinomials come in two forms: one in which the middle term is positive, and the other in which the middle term is negative.

Perfect Square Trinomials

Let u and v represent real numbers, variables, or algebraic expressions. Then the following perfect square trinomials can be factored as indicated.

1. $u^2 + 2uv + v^2 = (u + v)^2$ 2. $u^2 - 2uv + v^2 = (u - v)^2$

           Same sign   Same sign

To recognize a perfect square trinomial, remember that the first and last terms must be perfect squares and positive, and that the middle term must be twice the product of u and v. (Note that the middle term can be positive or negative.)

EXAMPLE 1 ■ Factoring Perfect Square Trinomials

(a) $x^2 - 4x + 4 = x^2 - 2(2x) + 2^2 = (x - 2)^2$

(b) $16y^2 + 24y + 9 = (4y)^2 + 2(4y)(3) + 3^2 = (4y + 3)^2$

(c) $9x^2 - 30xy + 25y^2 = (3x)^2 - 2(3x)(5y) + (5y)^2 = (3x - 5y)^2$ ■

EXAMPLE 2 ■ Removing a Common Monomial Factor First

Remind students to check whether a common factor can be removed before trying other methods of factoring.

Factor the following trinomials.

(a) $3x^2 - 30x + 75$ (b) $16y^3 + 80y^2 + 100y$

Solution

(a) Begin by factoring out the common monomial factor of 3. The remaining polynomial is a perfect square trinomial.

$$3x^2 - 30x + 75 = 3(x^2 - 10x + 25) \qquad \text{Remove common monomial factor}$$

$$= 3(x - 5)^2 \qquad \text{Factor as perfect square trinomial}$$

(b) Begin by factoring out the common monomial factor of $4y$. The remaining polynomial is a perfect square trinomial.

$$16y^3 + 80y^2 + 100y = 4y(4y^2 + 20y + 25) \qquad \text{Remove common monomial factor}$$

$$= 4y(2y + 5)^2 \qquad \text{Factor as perfect square trinomial}$$

■

Factoring Trinomials of the Form $x^2 + bx + c$

Try covering the factored forms in the left-hand column below. Can you determine the factored forms from the trinomial forms?

Factored Form	**F**	**O**	**I**	**L**	*Trinomial Form*
$(x - 1)(x + 4) =$	x^2	$+ 4x -$	x	$- 4 =$	$x^2 + 3x - 4$
$(x - 3)(x - 2) =$	x^2	$- 2x -$	$3x$	$+ 6 =$	$x^2 - 5x + 6$
$(3x + 5)(x + 1) =$	$3x^2$	$+ 3x +$	$5x$	$+ 5 =$	$3x^2 + 8x + 5$

Your goal here is to factor trinomials of the form $x^2 + bx + c$. To begin, consider the following factorization.

$$x^2 + bx + c = (x + m)(x + n)$$

By multiplying the right-hand side, you obtain the following.

$$(x + m)(x + n) = x^2 + nx + mx + mn$$

$$= x^2 + \underbrace{(m + n)}_{\substack{\text{Sum of} \\ \text{terms}}}x + \underbrace{mn}_{\substack{\text{Product} \\ \text{of terms}}}$$

$$\downarrow \qquad \downarrow$$

$$= x^2 + \boxed{b}\, x + \boxed{c}$$

Thus, to *factor* a trinomial $x^2 + bx + c$ into a product of two binomials, you must find *factors* of c with a sum of b. Note how this strategy is used in the following examples.

EXAMPLE 3   ■   Factoring Trinomials

Factor the following trinomials.

(a) $x^2 - 2x - 8$ (b) $x^2 - 5x + 6$

Solution

(a) For this trinomial, you have $x^2 + bx + c = x^2 - 2x - 8$. Thus, $b = -2$ and $c = -8$, and you need to find two numbers with a product of -8 and a sum of -2. After considering the various possibilities, you can see that the numbers 2 and -4 work. Therefore, the factorization is as follows.

$$x^2 - 2x - 8 = (x +)(x +)$$
Think: You need factors of -8 with a sum of -2.

$$= (x + 2)(x - 4)$$
$(2)(-4) = -8,\ 2 - 4 = -2$

(b) For this trinomial, you have $x^2 + bx + c = x^2 - 5x + 6$. Thus, $b = -5$ and $c = 6$, and you need to find two numbers with a product of 6 and a sum of -5. After considering the various possibilities, you can see that the numbers -2 and -3 work. Therefore, the factorization is as follows.

$$x^2 - 5x + 6 = (x +)(x +)$$
Think: You need factors of 6 with a sum of -5.

$$= (x - 2)(x - 3)$$
$(-2)(-3) = 6,\ -2 - 3 = -5$

∎

NOTE When the constant term of the trinomial is positive, its factors must have *like* signs; otherwise, its factors have *unlike* signs.

When factoring a trinomial of the form $x^2 + bx + c$, if you have trouble finding two factors of c with a sum of b, it may be helpful to list all of the distinct pairs of factors and then choose the appropriate one from the list. For instance, consider the trinomial

$$x^2 - 5x - 24.$$

For this trinomial, you have $c = -24$ and $b = -5$. Thus, you need to find two factors of -24 with a sum of -5, as follows.

Factors of -24	*Sum of Factors*		
$(1)(-24)$	$1 - 24 = -23$		
$(-1)(24)$	$-1 + 24 = 23$		
$(2)(-12)$	$2 - 12 = -10$		
$(-2)(12)$	$-2 + 12 = 10$		
$(3)(-8)$	$3 - 8 = -5$	$\Longleftarrow$	Correct factors
$(-3)(8)$	$-3 + 8 = 5$		
$(4)(-6)$	$4 - 6 = -2$		
$(-4)(6)$	$-4 + 6 = 2$		

With experience, you will be able to narrow this list down *mentally* to only two or three possibilities whose sums can then be tested to determine the correct factorization, which is $x^2 - 5x - 24 = (x + 3)(x - 8)$.

EXAMPLE 4   ■   Factoring a Trinomial

Factor the trinomial $x^2 - 7x - 18$.

Solution

To factor this trinomial, you need to find two factors of -18 with a sum of -7. After testing some possibilities, you can see that 2 and -9 work. Thus, the factorization is as follows.

$$x^2 - 7x - 18 = (x +)(x +)$$

Think: You need factors of -18 with a sum of -7.

$$= (x + 2)(x - 9)$$

$(2)(-9) = -18, \ 2 - 9 = -7$ ■

In factoring the following, it may be useful to factor out a -1 first:
$$-x^2 + x + 20 = -(x^2 - x - 20)$$
$$= -(x - 5)(x + 4)$$
$$= (5 - x)(x + 4)$$
The last step is optional.

Applications of algebra sometimes involve trinomials that have a common monomial factor. To factor such trinomials completely, first factor out the common monomial factor. Then try to factor the resulting trinomial by the methods given in this section. For instance, the trinomial

$$5x^2 - 5x - 30 = 5(x^2 - x - 6) = 5(x - 3)(x + 2)$$

is factored completely.

EXAMPLE 5   ■   Factoring Completely

Factor the following trinomial completely.

$$4x^3 - 8x^2 - 60x$$

Solution

This trinomial has a common monomial factor of $4x$. Thus, you should start the factoring process by factoring $4x$ out of each term.

$$4x^3 - 8x^2 - 60x = 4x(x^2 - 2x - 15)$$

Factor out common factor

$$= 4x(x +)(x +)$$

Think: You need factors of -15 with a sum of -2

$$= 4x(x + 3)(x - 5)$$

$(3)(-5) = -15, \ 3 - 5 = -2$ ■

Factoring Trinomials of the Form $ax^2 + bx + c$

To factor a trinomial whose leading coefficient is not 1, use the following pattern.

Factors of a

$$ax^2 + bx + c = (x +)(x +)$$

Factors of c

The goal is to find a combination of factors of a and c such that the outer and inner products add up to the middle term bx. For instance, in the trinomial $6x^2 + 17x + 5$, you have $a = 6$, $c = 5$, and $b = 17$. After some experimentation, you can determine that the factorization is

$$\begin{array}{cccc} \textbf{F} & \textbf{O} & \textbf{I} & \textbf{L} \\ \downarrow & \downarrow & \downarrow & \downarrow \end{array}$$

$$(2x + 5)(3x + 1) = 6x^2 + 2x + 15x + 5 = 6x^2 + 17x + 5.$$

Note that the outer (O) and inner (I) products add up to $17x$.

EXAMPLE 6   ■   **Factoring a Trinomial of the Form $ax^2 + bx + c$**

Factor the trinomial

$$6x^2 + 5x - 4.$$

Solution

For this trinomial, you have $ax^2 + bx + c = 6x^2 + 5x - 4$, which implies that $a = 6$, $c = -4$, and $b = 5$. The possible factors of 6 are (1)(6) and (2)(3), and the possible factors of -4 are $(-1)(4)$, $(1)(-4)$, and $(2)(-2)$. By trying the *many* different combinations of these factors, you obtain the following list.

$$(x + 1)(6x - 4) = 6x^2 + 2x - 4$$

$$(x - 1)(6x + 4) = 6x^2 - 2x - 4$$

$$(x + 4)(6x - 1) = 6x^2 + 23x - 4$$

$$(x - 4)(6x + 1) = 6x^2 - 23x - 4$$

$$(x + 2)(6x - 2) = 6x^2 + 10x - 4$$

$$(x - 2)(6x + 2) = 6x^2 - 10x - 4$$

$$(2x + 1)(3x - 4) = 6x^2 - 5x - 4$$

$$(2x - 1)(3x + 4) = 6x^2 + 5x - 4 \quad \Longleftarrow \quad \text{Correct factorization}$$

$$(2x + 4)(3x - 1) = 6x^2 + 10x - 4$$

$$(2x - 4)(3x + 1) = 6x^2 - 10x - 4$$

$$(2x + 2)(3x - 2) = 6x^2 + 2x - 4$$

$$(2x - 2)(3x + 2) = 6x^2 - 2x - 4$$

Thus, you can conclude that the correct factorization is

$$6x^2 + 5x - 4 = (2x - 1)(3x + 4). \qquad\qquad ■$$

Progress check: Factor the following completely.

(a) $x^2 + 8x + 12$ (b) $6y^2 - 13y + 6$
(c) $2x^3 - 18x$

Answers: (a) $(x + 2)(x + 6)$
        (b) $(2y - 3)(3y - 2)$
        (c) $2x(x + 3)(x - 3)$

To help shorten the list of *possible* factorizations of a trinomial, use the following guidelines.

Guidelines for Limiting Possible Factorizations of a Trinomial	When factoring a trinomial of the form $ax^2 + bx + c$, consider the following guidelines.

1. If the trinomial has a *common monomial factor*, you should remove the monomial factor before trying to find binomial factors. For instance, the trinomial $12x^2 + 10x - 8$ has a common factor of 2. By removing this common factor, you obtain $12x^2 + 10x - 8 = 2(6x^2 + 5x - 4)$.

2. Do not switch the signs of the factors of c unless the middle term is correct except in sign. In Example 6, after determining that $(x + 4)(6x - 1)$ is not the correct factorization, for instance, it is unnecessary to test $(x - 4)(6x + 1)$.

3. Do not use binomial factors that have a common monomial factor. Such a factor cannot be correct, because the trinomial has no common monomial factor. (Any common monomial factor was already removed in step 1.) For instance, in Example 6, it is unnecessary to test $(x + 1)(6x - 4) = 6x^2 + 2x - 4$ because the factor $(6x - 4)$ has a common factor of 2.

With these suggestions, you could shorten the list given in Example 6 to the following three possible factorizations.

$$(x + 4)(6x - 1) = 6x^2 + 23x - 4$$

$$(2x + 1)(3x - 4) = 6x^2 - 5x - 4$$

$$(2x - 1)(3x + 4) = 6x^2 + 5x - 4 \quad \Longleftarrow \quad \text{Correct factorization}$$

Do you see why you can cut the list from 12 possible factorizations to only three?

EXAMPLE 7   ■   **Factoring a Trinomial of the Form $ax^2 + bx + c$**

Factor the trinomial $2x^2 - x - 21$.

Solution

For this trinomial, you have $a = 2$, which factors as $(1)(2)$, and $c = -21$, which factors as $(1)(-21)$, $(-1)(21)$, $(3)(-7)$, or $(-7)(3)$. You can test the possible factors as follows.

$$(2x + 1)(x - 21) = 2x^2 - 41x - 21$$

$$(2x + 21)(x - 1) = 2x^2 + 19x - 21$$

$$(2x + 3)(x - 7) = 2x^2 - 11x - 21$$

$$(2x + 7)(x - 3) = 2x^2 + \quad x - 21 \qquad \text{Middle term has incorrect sign}$$

$$(2x - 7)(x + 3) = 2x^2 - \quad x - 21 \quad \Longleftarrow \quad \text{Correct factorization}$$

Therefore, the correct factorization is

$$2x^2 - x - 21 = (2x - 7)(x + 3).$$

■

NOTE Remember that if the middle term is correct except in sign, you need only change the signs of the factors of c, as in Example 7.

EXAMPLE 8 ■ **Factoring a Trinomial of the Form** $ax^2 + bx + c$

Factor the trinomial $3x^2 + 11x + 10$.

Solution

For this trinomial, you have $a = 3$, which factors as $(1)(3)$, and $c = 10$, which factors as $(1)(10)$ or $(2)(5)$. You can test the possible factors as follows.

$$(x + 10)(3x + 1) = 3x^2 + 31x + 10$$
$$(x + 1)(3x + 10) = 3x^2 + 13x + 10$$
$$(x + 5)(3x + 2) = 3x^2 + 17x + 10$$
$$(x + 2)(3x + 5) = 3x^2 + 11x + 10 \quad \Longleftarrow \quad \text{Correct factorization}$$

Therefore, the correct factorization is

$$3x^2 + 11x + 10 = (x + 2)(3x + 5).$$ ■

Remember that if a trinomial has a common monomial factor, the common monomial factor should be removed first. The next two examples illustrate this technique.

EXAMPLE 9 ■ **Factoring Completely**

Factor the following trinomial completely.

$$8x^2y - 60xy + 28y$$

Solution

Begin by factoring out the common monomial factor $4y$.

$$8x^2y - 60xy + 28y = 4y(2x^2 - 15x + 7)$$

Now, for the new trinomial $2x^2 - 15x + 7$, you have $a = 2$ and $c = 7$. The possible factorizations of this trinomial are as follows.

$$(2x - 7)(x - 1) = 2x^2 - 9x + 7$$
$$(2x - 1)(x - 7) = 2x^2 - 15x + 7 \quad \Longleftarrow \quad \text{Correct factorization}$$

Therefore, the complete factorization of the original trinomial is

$$8x^2y - 60xy + 28y = 4y(2x^2 - 15x + 7) = 4y(2x - 1)(x - 7).$$ ■

When factoring a trinomial with a negative leading coefficient, first factor -1 out of the trinomial, as demonstrated in Example 10.

EXAMPLE 10   ■   **Factoring a Trinomial with a Negative Leading Coefficient**

Factor the trinomial $-3x^2 + 16x + 35$.

Solution

This trinomial has a negative leading coefficient, so you should begin by factoring (-1) out of the trinomial.

$$-3x^2 + 16x + 35 = (-1)(3x^2 - 16x - 35)$$

Now, for the new trinomial $3x^2 - 16x - 35$, you have $a = 3$ and $c = -35$. The possible factorizations of this trinomial are as follows.

$$(3x - 1)(x + 35) = 3x^2 + 104x - 35$$
$$(3x - 35)(x + 1) = 3x^2 - 32x - 35$$
$$(3x - 7)(x + 5) = 3x^2 + 8x - 35$$
$$(3x - 5)(x + 7) = 3x^2 + 16x - 35 \qquad \text{Middle term has incorrect sign}$$
$$(3x + 5)(x - 7) = 3x^2 - 16x - 35 \iff \text{Correct factorization}$$

Thus, the correct factorization is

$$-3x^2 + 16x + 35 = (-1)(3x + 5)(x - 7).$$

Alternative forms of this factorization include

$$(3x + 5)(-x + 7) \quad \text{and} \quad (-3x - 5)(x - 7). \qquad ■$$

Not all trinomials are factorable using only integers. For instance, to factor $x^2 + 3x + 5$, you need factors of 5 that add up to 3. This is not possible, because the only integer factors of 5 are 1 and 5, and their sum is not 3. Such a trinomial is *not factorable*. Try factoring $2x^2 - 3x + 2$ to see that it is also not factorable. Watch for other trinomials that are not factorable in the exercises for this section.

Factoring Trinomials by Grouping (Optional)

In this section, you have seen that factoring a trinomial can involve quite a bit of trial and error. An alternative technique that some people like is to use factoring by grouping. For instance, suppose you rewrite the trinomial $2x^2 + x - 15$ as

$$2x^2 + x - 15 = 2x^2 + 6x - 5x - 15.$$

Then, by grouping the first two terms and the third and fourth terms, you can factor the polynomial as follows.

$$2x^2 + x - 15 = 2x^2 + 6x - 5x - 15 \qquad \text{Rewrite middle term}$$
$$= (2x^2 + 6x) - (5x + 15) \qquad \text{Group terms}$$
$$= 2x(x + 3) - 5(x + 3) \qquad \text{Factor groups}$$
$$= (2x - 5)(x + 3) \qquad \text{Distributive Property}$$

The key to this method of factoring is knowing how to rewrite the middle term. In general, *to factor a trinomial $ax^2 + bx + c$ by grouping, choose factors of the product ac that add up to b and use these factors to rewrite the middle term.* This technique is illustrated in Example 11.

EXAMPLE 11 ■ Factoring a Trinomial by Grouping

Use factoring by grouping to factor the trinomial

$$2x^2 + 5x - 3.$$

Solution

In the trinomial $2x^2 + 5x - 3$, you have $a = 2$ and $c = -3$, which implies that the product ac is -6. Now, because -6 factors as $(6)(-1)$, and $6 - 1 = 5 = b$, you can rewrite the middle term as $5x = 6x - x$. This produces the following.

$$
\begin{aligned}
2x^2 + 5x - 3 &= 2x^2 + 6x - x - 3 && \text{Rewrite middle term}\\
&= (2x^2 + 6x) - (x + 3) && \text{Group terms}\\
&= 2x(x + 3) - (x + 3) && \text{Factor groups}\\
&= (x + 3)(2x - 1) && \text{Distributive Property}
\end{aligned}
$$

Therefore, the trinomial factors as

$$2x^2 + 5x - 3 = (x + 3)(2x - 1). \qquad \blacksquare$$

What do you think of this optional technique? Some people think that it is more efficient than the trial-and-error process, especially when the coefficients a and c have many factors.

Summary of Factoring

Ask students to write an essay about what it means to "factor completely."

Although the basic factoring techniques have been discussed one at a time, from this point on you must decide which technique to apply to any given problem situation. The following guidelines should assist you in this selection process.

Guidelines for Factoring Polynomials

1. Factor out any common factors.

2. Factor according to one of the special polynomial forms: difference of squares, sum or difference of cubes, or perfect square trinomials.

3. Factor trinomials, $ax^2 + bx + c$, using the methods for $a = 1$ or $a \neq 1$.

4. Factor by grouping—for polynomials with four terms.

5. Check to see if the factors themselves can be further factored.

6. Check the results by multiplying the factors.

EXAMPLE 12 ■ **Factoring Polynomials**

Factor the following polynomials.

(a) $3x^2 - 108$

(b) $4x^3 - 32x^2 + 64x$

(c) $6x^3 + 27x^2 - 15x$

(d) $x^3 - 3x^2 - 4x + 12$

Solution

(a) $3x^2 - 108 = 3(x^2 - 36)$ Factor out common factor

$\qquad\qquad\quad = 3(x + 6)(x - 6)$ Difference of two squares

(b) $4x^3 - 32x^2 + 64x = 4x(x^2 - 8x + 16)$ Factor out common factor

$\qquad\qquad\qquad\quad = 4x(x - 4)^2$ Perfect square trinomial

(c) $6x^3 + 27x^2 - 15x = 3x(2x^2 + 9x - 5)$ Factor out common factor

$\qquad\qquad\qquad\quad = 3x(2x - 1)(x + 5)$ Factor trinomial

(d) $x^3 - 3x^2 - 4x + 12 = (x^3 - 3x^2) + (-4x + 12)$ Group terms

$\qquad\qquad\qquad\quad = x^2(x - 3) - 4(x - 3)$ Factor out common factors

$\qquad\qquad\qquad\quad = (x - 3)(x^2 - 4)$ Distributive Property

$\qquad\qquad\qquad\quad = (x - 3)(x + 2)(x - 2)$ Difference of two squares ■

DISCUSSION PROBLEM ■ **You Be the Instructor**

Suppose you are making up a test on this chapter for your algebra class. The test has 20 problems and you decide that eight of the problems will come from Sections 1.5 and 1.6 (on factoring). Create eight factoring problems that you think would represent a fair test of your class's factoring skills. ■

Warm-Up

The following warm-up exercises involve skills that were covered in earlier sections. You will use these skills in the exercise set for this section.

In Exercises 1–6, perform the indicated multiplication.

1. $(x + 1)^2$ **2.** $(x - 5)^2$ **3.** $(2 - y)^2$

4. $(2 - y)(3 + 2y)$ **5.** $(5z + 3)(2z - 7)$ **6.** $t(t - 4)(2t + 3)$

In Exercises 7–10, factor the expression completely.

7. $3x^2 - 21x$ **8.** $-x^3 + 3x^2 - x + 3$

9. $4t^2 - 169$ **10.** $y^3 - 64$

1.6   EXERCISES

In Exercises 1–12, factor the perfect square trinomial.

1. $x^2 + 4x + 4$ **2.** $z^2 + 6z + 9$ **3.** $a^2 - 12a + 36$ **4.** $y^2 - 14y + 49$

5. $25y^2 - 10y + 1$ **6.** $4z^2 + 28z + 49$ **7.** $9b^2 + 12b + 4$ **8.** $4x^2 - 4x + 1$

9. $4x^2 - 4xy + y^2$ **10.** $m^2 + 6mn + 9n^2$ **11.** $u^2 + 8uv + 16v^2$ **12.** $4y^2 + 20yz + 25z^2$

In Exercises 13–18, fill in the missing factor.

13. $x^2 + 5x + 4 = (x + 4)(\qquad)$ **14.** $a^2 + 2a - 8 = (a + 4)(\qquad)$

15. $y^2 - y - 20 = (y + 4)(\qquad)$ **16.** $y^2 + 6y + 8 = (y + 4)(\qquad)$

17. $z^2 - 6z + 8 = (z - 4)(\qquad)$ **18.** $z^2 + 2z - 24 = (z - 4)(\qquad)$

In Exercises 19–30, factor the expression.

19. $x^2 + 4x + 3$ **20.** $x^2 + 7x + 10$ **21.** $x^2 - 5x + 6$ **22.** $x^2 - 10x + 24$

23. $y^2 + 7y - 30$ **24.** $m^2 - 3m - 10$ **25.** $t^2 - 4t - 21$ **26.** $x^2 + 4x - 12$

27. $x^2 - 20x + 96$ **28.** $u^2 + 5uv + 6v^2$ **29.** $x^2 - 2xy - 35y^2$ **30.** $a^2 - 21ab + 110b^2$

In Exercises 31–36, fill in the missing factor.

31. $5x^2 + 18x + 9 = (x + 3)(\qquad)$ **32.** $5x^2 + 19x + 12 = (x + 3)(\qquad)$

33. $5a^2 + 12a - 9 = (a + 3)(\qquad)$ **34.** $5c^2 + 11c - 12 = (c + 3)(\qquad)$

35. $2y^2 - 3y - 27 = (y + 3)(\qquad)$ **36.** $3y^2 - y - 30 = (y + 3)(\qquad)$

In Exercises 37–48, factor the algebraic expression, if possible. Some of the expressions are not factorable using integer coefficients.

37. $3x^2 + 4x + 1$

38. $5x^2 + 7x + 2$

39. $8x^2 + 6x - 9$

40. $8t^2 - 6t - 5$

41. $6b^2 + 19b - 7$

42. $2a^2 - 13a + 20$

43. $24x^2 - 14xy - 3y^2$

44. $20x^2 + x - 12$

45. $3y^2 + 4y + 12$

46. $10x^2 + 9xy - 9y^2$

47. $2u^2 + 9uv - 35v^2$

48. $r^2 - 9rs - 9s^2$

In Exercises 49–52, factor the trinomial.

49. $-2x^2 - x + 6$

50. $-6x^2 + 5x + 6$

51. $1 - 11x - 60x^2$

52. $2 + 5x - 12x^2$

In Exercises 53–58, factor the trinomial by grouping (see Example 11).

53. $3x^2 + 10x + 8$

54. $2x^2 + 9x + 9$

55. $6x^2 + x - 2$

56. $6x^2 - x - 15$

57. $15x^2 - 11x + 2$

58. $12x^2 - 28x + 15$

In Exercises 59–70, factor the algebraic expression completely.

59. $3x^5 - 12x^3$

60. $20y^2 - 45$

61. $10t^3 + 2t^2 - 36t$

62. $16z^2 - 56z + 49$

63. $4x(3x - 2) + (3x - 2)^2$

64. $6 - x - 6x^2 + x^3$

65. $36 - (z + 3)^2$

66. $3t^3 - 24$

67. $54x^3 - 2$

68. $v^3 + 3v^2 + 5v$

69. $27a^3 - 3ab^2$

70. $8m^3n + 20m^2n^2 - 48mn^3$

In Exercises 71–76, find two real numbers b such that the algebraic expression is a perfect square trinomial.

71. $x^2 + bx + 81$

72. $x^2 + bx + 16$

73. $9y^2 + by + 1$

74. $36z^2 + bz + 1$

75. $4x^2 + bx + 9$

76. $16x^2 + bxy + 25y^2$

In Exercises 77–82, find a real number c such that the algebraic expression is a perfect square trinomial.

77. $x^2 + 8x + c$

78. $x^2 + 12x + c$

79. $y^2 - 6y + c$

80. $z^2 - 20z + c$

81. $16a^2 + 40a + c$

82. $9t^2 - 12t + c$

In Exercises 83 and 84, fill in the missing number.

83. $x^2 + 12x + 50 = (x + 6)^2 + \ \rule{2em}{0.8em}\ $

84. $x^2 + 10x + 22 = (x + 5)^2 + \ \rule{2em}{0.8em}\ $

In Exercises 85–90, find all integers b such that the trinomial can be factored.

85. $x^2 + bx + 18$

86. $x^2 + bx + 14$

87. $x^2 + bx - 21$

88. $x^2 + bx - 7$

89. $5x^2 + bx + 8$

90. $3x^2 + bx - 10$

In Exercises 91–96, find *two* integers c such that the trinomial can be factored. (There are many correct answers.)

91. $x^2 + 6x + c$ **92.** $x^2 + 9x + c$ **93.** $x^2 - 3x + c$ **94.** $x^2 - 12x + c$

95. $t^2 - 4t + c$ **96.** $s^2 + s + c$

In Exercises 97 and 98, evaluate the quantity mentally using the following example as a model.

$$29^2 = (30 - 1)^2 = 30^2 - 2(30)(1) + 1^2 = 900 - 60 + 1 = 841$$

97. 52^2 **98.** 39^2

In Exercises 99–102, match the "geometric factoring model" with the correct factoring formula. [The models are labeled (a), (b), (c), and (d).]

99. $a^2 - b^2 = (a + b)(a - b)$ **100.** $a^2 + 2ab + b^2 = (a + b)^2$

101. $a^2 + 2a + 1 = (a + 1)^2$ **102.** $ab + a + b + 1 = (a + 1)(b + 1)$

(a)

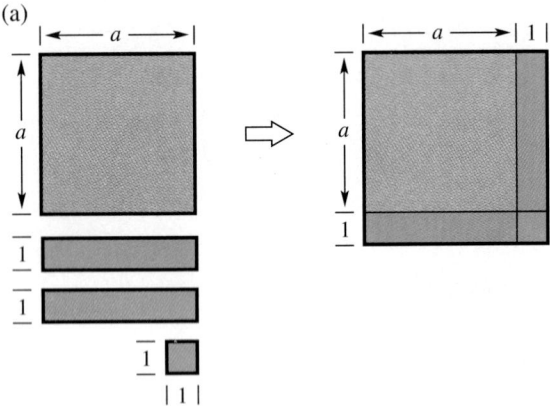

(b)

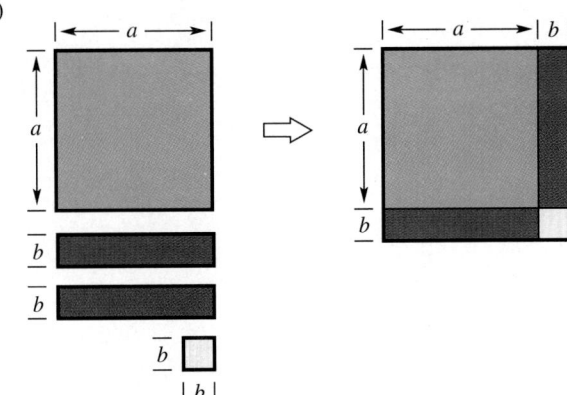

(c)

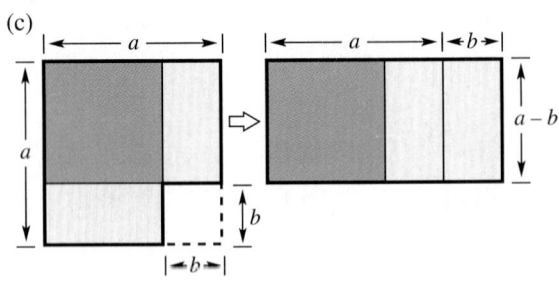

(d)

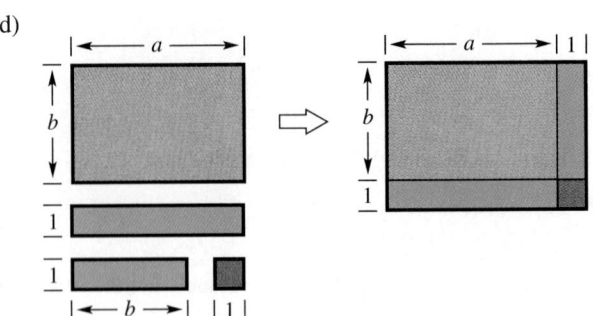

In Exercises 103 and 104, factor the trinomial and show the result using the concept of area, as illustrated in the figure.

$$x^2 + 3x + 2 = (x + 1)(x + 2)$$

Figure for 103 and 104

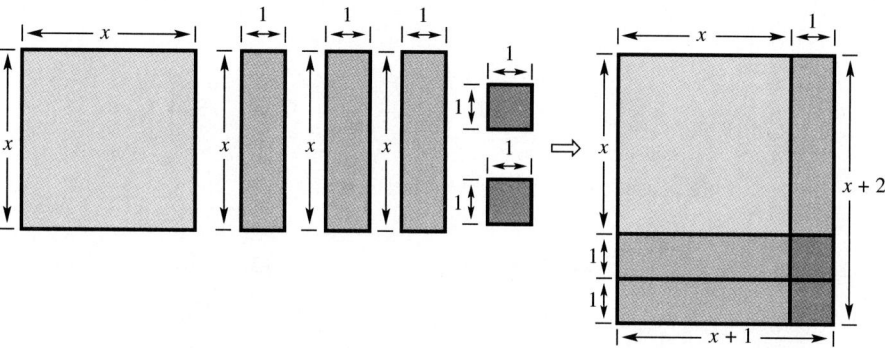

103. $x^2 + 4x + 3$

104. $x^2 + 5x + 4$

This is a good place for students to review "How to Study Algebra" on pages xix–xxi.

Urge students to use the Mid-Chapter Quiz, Review Exercises, and Chapter Test to review and practice the skills discussed in this chapter and to assess their mastery of these skills.

 REVIEW EXERCISES

In Exercises 1–4, show each real number as a point on the real number line and determine the correct inequality symbol ($<$ or $>$) between the real numbers.

1. $3 \rule{1cm}{0.4pt} -\frac{1}{8}$

2. $-2 \rule{1cm}{0.4pt} -8$

3. $-\frac{8}{5} \rule{1cm}{0.4pt} -\frac{2}{5}$

4. $8.4 \rule{1cm}{0.4pt} -\pi$

In Exercises 5–8, evaluate the expression.

5. $|-7.2|$

6. $|1.6|$

7. $-|-7.2|$

8. $|-3.6|$

In Exercises 9–40, evaluate the quantity. (If not possible, state the reason.) Write all fractions in reduced form.

9. $15 + (-4)$

10. $-12 + 3$

11. $340 - 115 + 5$

12. $-154 + 86 - 240$

13. $|-96| - |134|$

14. $16(3000)$

15. $120(-5)(7)$

16. $(-16)(-15)(-4)$

17. $\dfrac{-56}{-4}$

18. $\dfrac{85}{0}$

19. $\dfrac{45 - |-45|}{2}$

20. $\dfrac{288}{2 \cdot 3 \cdot 3}$

21. $\frac{4}{21} + \frac{7}{21}$

22. $\frac{21}{16} - \frac{13}{16}$

23. $-\frac{5}{6} + 1$

24. $\frac{21}{32} + \frac{11}{24}$

25. $8\frac{3}{4} - 6\frac{5}{8}$

26. $-2\frac{9}{10} + 5\frac{3}{20}$

27. $\dfrac{3}{8} \cdot \dfrac{-2}{15}$

28. $\frac{5}{21} \cdot \frac{21}{5}$

29. $-\frac{7}{15} \div -\frac{7}{30}$

30. $-\frac{2}{3} \div \frac{4}{15}$

31. $\dfrac{\frac{3}{4}}{\frac{5}{8}}$

32. $\dfrac{-\frac{4}{5}}{\frac{7}{10}}$

33. $-35 + 26.5 + 13.75$

34. $\frac{6.25}{1.25}$

35. $(-6)^3$

36. $-(-3)^4$

37. $\dfrac{3}{6^2}$

38. $-4(25 - 4^2)$

39. $120 - (5^2 \cdot 4)$

40. $45 - 45 \div 3^2$

In Exercises 41–46, name the property of real numbers that justifies the statement.

41. $13 - 13 = 0$

42. $7\left(\frac{1}{7}\right) = 1$

43. $7(9 + 3) = 7 \cdot 9 + 7 \cdot 3$

44. $15(4) = 4(15)$

45. $5 + (4 - y) = (5 + 4) - y$

46. $6(4z) = (6 \cdot 4)z$

In Exercises 47–52, identify the rule of algebra that is illustrated by the equation.

47. $(u - v)(2) = 2(u - v)$

48. $(x + y) + 0 = x + y$

49. $ab \cdot \dfrac{1}{ab} = 1, \quad ab \neq 0$

50. $x(yz) = (xy)z$

51. $(u + v) + w = u + (v + w)$

52. $4(x - y) = 4 \cdot x - 4 \cdot y$

In Exercises 53–58, expand the expression by using the Distributive Property.

53. $\frac{2}{3}(6s - 12t)$

54. $-5(2x - 4y)$

55. $-y(3y - 10)$

56. $x(3x + 4y)$

57. $-(-u + 3v)$

58. $(5 - 3j)(-4)$

In Exercises 59–66, simplify the algebraic expression.

59. $5(x - 4) + 10$

60. $15 - 7(z + 2)$

61. $3x - (y - 2x)$

62. $30x - (10x + 80)$

63. $-2(1 - 20u) + 2(1 - 10u)$

64. $9(11 - 2y) - 6(11 - 2y)$

65. $3[b + 5(b - a)]$

66. $-2t[8 - (6 - t)] + 5t$

In Exercises 67–74, simplify the expression by using the properties of exponents.

67. $x^2 \cdot x^3 \cdot x$

68. $y^3(-2y^2)$

69. $(-2a^2)^3(8a)$

70. $\dfrac{(a^5 b)^3}{a^4 b}$

71. $\dfrac{120u^5 v^3}{15u^3 v}$

72. $\left(\dfrac{-y^3}{5}\right)^2$

73. $(2ab^2)(3a^3 b)^4$

74. $(xy)(-3x^2 y^3)$

In Exercises 75–78, evaluate the algebraic expression for the specified value of the variable. If not possible, state the reason.

75. $x^2 - 2x - 3$
   (a) $x = 3$ (b) $x = 0$

76. $y^3 - 125$
   (a) $y = 5$ (b) $y = -2$

77. $3x^2 - y(x + 1)$
   (a) $x = 2, \quad y = -1$ (b) $x = -1, \quad y = 20$

78. $\dfrac{x}{y + 2}$
   (a) $x = 0, \quad y = 3$ (b) $x = 5, \quad y = -2$

In Exercises 79–90, perform the specified operations and simplify.

79. $(5x + 3x^2) + (x - 4x^2)$

80. $\left(\frac{1}{2}x + \frac{2}{3}\right) + \left(4x + \frac{1}{3}\right)$

81. $(-x^3 - 3x) - 4(2x^3 - 3x + 1)$

82. $(7z^2 + 6z) - 3(5z^2 + 2z)$

83. $3y^2 - [2y - 3(y^2 + 5)]$

84. $(16a^3 + 5a) - 5[a + (2a^3 - 1)]$

85. $(5x + 3)(3x - 4)$

86. $(3y^2 + 2)(4y^2 - 5)$

87. $(2x^2 - 3x + 2)(2x + 3)$

88. $(5s^2 + 4s - 3)(4s - 5)$

89. $2u(u - 7) - (u + 1)(u - 7)$

90. $(3v + 2)(-5v) + 5v(3v + 2)$

In Exercises 91–96, find the product by using the special product formulas.

91. $(4x - 7)^2$

92. $(8 - 3x)^2$

93. $(5u - 8)(5u + 8)$

94. $(7a + 4)(7a - 4)$

95. $[(u - 3) + v][(u - 3) - v]$

96. $[(m - 5) + n]^2$

In Exercises 97–102, factor the polynomial by removing the common factor.

97. $6x^2 + 15x^3$

98. $8y - 12y^4$

99. $8a^3 b^2 - 12a^2 b^3$

100. $108r^3 s^2 - 135r^4 s^2$

101. $28(x + 5) + 7(x + 5)^2$

102. $(u - 9v)(u - v) + v(u - 9v)$

In Exercises 103–106, factor the difference of two squares.

103. $9a^2 - 100$

104. $b^2 - 900$

105. $(u + 6)^2 - 81$

106. $(y - 3)^2 - 16$

In Exercises 107–110, factor the perfect square trinomial.

107. $x^2 - 40x + 400$ **108.** $y^2 + 26y + 169$ **109.** $4s^2 + 36st + 81t^2$ **110.** $u^2 - 10uv + 25v^2$

In Exercises 111–122, factor the expression completely, if possible.

111. $4a - 64a^3$ **112.** $3b + 27b^3$ **113.** $8x(2x - 3) - 4(2x - 3)$

114. $x^3 + 3x^2 - 4x - 12$ **115.** $8x^3 + 1$ **116.** $t^3 - 216$

117. $4u^2 - 28u + 49$ **118.** $x^2 - 2x + 1$ **119.** $x^2 + x - 1$

120. $3x^2 + 23x - 8$ **121.** $4t^3 + 7t^2 - 2t$ **122.** $6h^3 - 22h^2 - 8h$

In Exercises 123–126, find all integers b such that the trinomial is factorable.

123. $x^2 + bx + 5$ **124.** $x^2 + bx + 6$ **125.** $2x^2 + bx - 15$ **126.** $3x^2 + bx - 22$

Magazine Circulations In Exercises 127 and 128, use the figure showing the circulations for the top five magazines in the United States for the first six months of 1990. (*Source: Audit Bureau of Circulations*)

Figure for 127 and 128

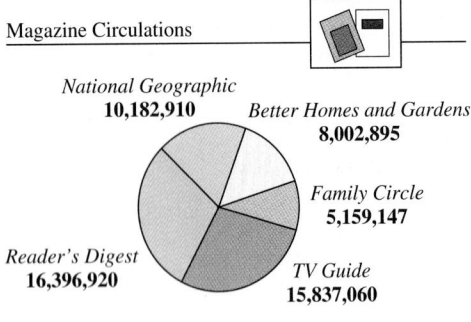

Magazine Circulations

National Geographic
10,182,910

Better Homes and Gardens
8,002,895

Family Circle
5,159,147

TV Guide
15,837,060

Reader's Digest
16,396,920

127. Determine the combined circulation for the five magazines.

128. What is the difference in circulation between *Reader's Digest* and *National Geographic*?

129. *Total Charge* Suppose you purchased a product by making a down payment of $239 plus nine monthly payments of $45 each. What is the total amount you paid for the product?

130. *Total Charge* Suppose you purchased a product by making a down payment of $387 plus 12 monthly payments of $68 each. What is the total amount you paid for the product?

131. *Exploratory Exercise* Enter any number between 0 and 1 in a calculator. Take the square root of the number. Then take the square root of the result, and keep repeating the process. What number does the calculator display seem to be approaching?

132. *Calculator Experiment* Use a calculator to calculate 12^4 in two ways.*

    Scientific   *Graphing*

(a) 12 [y^x] 4 [=]   (a) 12 [∧] 4 [ENTER]

(b) 12 [x^2] [x^2]   (b) 12 [x^2] [x^2] [ENTER]

Why do these two methods give the same result?

133. *Probability* The probability of three successes in five trials of an experiment is $10p^3(1 - p)^2$. Find this product.

*The graphing calculator keystrokes in this text correspond to the TI-81 and TI-82 graphing calculators from Texas Instruments. For other graphing calculators, the keystrokes may differ.

Chapter Test

1

Take this test as you would take a test in class. After you are done, check your work against the answers in the back of the book.

1. Evaluate: $\frac{2}{3} + \left(-\frac{7}{6}\right)$

2. Evaluate: $\left(-\frac{7}{16}\right)\left(-\frac{8}{21}\right)$

3. Evaluate: $\frac{5}{18} \div \frac{15}{8}$

4. Evaluate: $\left(-\frac{3}{5}\right)^3$

5. Evaluate: $\sqrt{25} + 3(36 \div 18)$

6. Evaluate: $\frac{4^2 - 6}{5} + 13$

7. Name the property of real numbers demonstrated by $(-3 \cdot 5) \cdot 6 = -3(5 \cdot 6)$.

8. Name the property of real numbers demonstrated by $3y \cdot \dfrac{1}{3y} = 1, \ y \neq 0$.

9. Give the additive inverse of the quantity $5x$.

10. Rewrite the expression $5(2x - 3)$ using the Distributive Property.

11. Simplify: $(3x^2y)(-xy)^2$

12. Simplify: $3x^2 - 2x - 5x^2 + 7x - 1$

13. Simplify: $(16 - y^2) - (16 + 2y + y^2)$

14. Simplify: $-2(2x^4 - 5) + 4x(x^3 + 2x - 1)$

15. Simplify: $4t - [3t - (10t + 7)]$

16. Simplify: $2y\left(\dfrac{y}{4}\right)^2$

17. Multiply: $(2x - 3y)(x + 5y)$

18. Multiply: $(2s - 3)(3s^2 - 4s + 7)$

19. Multiply: $(4x - 3)^2$

20. Multiply: $[4 - (a + b)][4 + (a + b)]$

21. Factor: $18y^2 - 12y$

22. Factor: $5x^3 - 10x^2 - 6x + 12$

23. Factor: $9u^2 - 6u + 1$

24. Factor: $6x^2 - 26x - 20$

25. A *cord* of wood is a pile 4 feet high, 4 feet wide, and 8 feet long. The volume of a rectangular solid is its length times its width times its height. Find the number of cubic feet in five cords of wood.

26. Find the area of the shaded region in the figure.

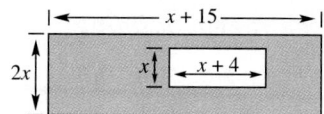

Depending on available resources, encourage students to take advantage of the supplementary materials available for each section and chapter of the text.

- Student Solutions Guide
- Videotapes
- Tutorial Software
- Improving Mathematics Study and Test Taking Skills Videotape

Instructors may also take advantage of:

- Complete Solutions Guide
- Test Item File and Resource Guide
- Computerized Testing Software

Introduction to Equations, Graphs, and Functions

2.1   Solving Linear and Quadratic Equations

2.2   Literal Equations and Formulas

2.3   The Rectangular Coordinate System

2.4   Graphs of Equations

2.5   Relations, Functions, and Function Notation

2.6   Graphs of Functions

The surface area S of a sphere is a function of the sphere's radius r:

$$S = (\text{circumference})(\text{diameter})$$
$$= (2\pi r)(2r) = 4\pi r^2.$$

The symbol π (the Greek letter pi) represents the ratio of the circumference of a circle to its diameter, which is approximately 3.1416. The geodesic dome shown is called **Spaceship Earth**. It is the theme building for Walt Disney World's Epcot Center. The building has a diameter of 165 feet. What is its approximate surface area?

$S = 4\pi r^2$

Surface area (in square feet)

250
200
150
100
50

1 2 3 4 5

Radius (in feet)

SECTION
2.1

Solving Linear and Quadratic Equations
Introduction • *Solving Linear Equations in Standard Form* • *Solving Linear Equations in Nonstandard Form* • *Quadratic Equations and the Zero-Factor Property* • *Solving Quadratic Equations by Factoring*

Introduction

In Chapter 1, you used the basic rules of algebra to rewrite and simplify algebraic *expressions.* In this section, you will use these same rules to rewrite and solve algebraic *equations.*

Remind students to *simplify* expressions and *solve* equations.

An **equation** is a statement that two mathematical expressions are equal. Some examples are

$$x = 4, \quad 4x + 3 = 15, \quad 2x - 8 = 2(x - 4), \quad \text{and} \quad x^2 - 16 = 0.$$

To **solve** an equation involving x means that you are to find all values of x for which the equation is true. Such values are called **solutions,** and the solutions are said to **satisfy** the equation. For instance, 4 is a solution of $x = 4$ because $4 = 4$ is a true statement. Similarly, 3 is a solution of $4x + 3 = 15$ because $4(3) + 3 = 15$ is a true statement.

The **solution set** of an equation is the set of all real numbers that are solutions of the equation. Sometimes, an equation will have the set of all real numbers as its solution set. Such an equation is called an **identity.** For instance, the equation

$$2x - 8 = 2(x - 4) \qquad \qquad \text{Identity}$$

is an identity because the equation is true for all real values of x. Try values such as 0, 1, -2, and 5 in this equation to see that each one is a solution.

An equation whose solution set is not the entire set of real numbers is called a **conditional equation.** For instance, the equation

$$x^2 - 16 = 0 \qquad \qquad \text{Conditional equation}$$

is a conditional equation because it has only two solutions, 4 and -4. An equation that has no solution is said to have an **empty solution set,** which is denoted by { } or Ø.

Examples 1 and 2 show how to **check** whether a given value of x is a solution of an equation.

EXAMPLE 1   ■   Checking a Solution of an Equation

Determine whether -3 is a solution of the equation

$$-3x - 5 = 4x + 16.$$

Solution

To check whether -3 is a solution of this equation, replace the variable x by the number -3. This must be done for all occurrences of x on *both sides* of the equation. After replacing the variable, simplify both sides of the equation. If both sides turn out to be the same number, then -3 is a solution, and the solution "checks." If the two sides turn out to be different numbers, then -3 is not a solution.

$-3x - 5 = 4x + 16$	Given equation
$-3(-3) - 5 \overset{?}{=} 4(-3) + 16$	Replace x by -3
$9 - 5 \overset{?}{=} -12 + 16$	Simplify
$4 = 4$	Solution checks

Because both sides of the equation turn out to be the same number, you can conclude that -3 is a solution of the given equation. ∎

NOTE When checking a solution, write a question mark over the equal sign to indicate that you are uncertain whether the "equation" is true.

EXAMPLE 2 ■ A Trial Solution That Does Not Check

Determine whether -2 is a solution of the equation

$$-3x - 5 = 4x + 16.$$

Solution

$-3x - 5 = 4x + 16$	Given equation
$-3(-2) - 5 \overset{?}{=} 4(-2) + 16$	Replace x by -2
$6 - 5 \overset{?}{=} -8 + 16$	Simplify
$1 \neq 8$	-2 is not a solution

Because the two sides of the equation turn out to be different after x is replaced by -2, you can conclude that -2 is not a solution of the given equation. ∎

It is helpful to think of an equation as having two sides that are "in balance." Consequently, when you try to solve an equation, you must be careful to maintain that balance by performing the same operation on both sides.

Two equations that have the same set of solutions are called **equivalent.** For instance, the equations $x = 3$ and $x - 3 = 0$ are equivalent because both have only one solution—the number 3. When any one of the four techniques in the following list is applied to an equation, the resulting equation is equivalent to the original equation.

Forming Equivalent Equations

A given equation can be transformed into an *equivalent equation* by one or more of the following steps.

	Given Equation	Equivalent Equation
1. Remove symbols of grouping, combine like terms, or reduce fractions on one or both sides of the equation.	$3x - x = 8$	$2x = 8$
2. Add (or subtract) the same quantity to (or from) *both* sides of the equation.	$x - 3 = 5$	$x = 8$
3. Multiply (or divide) *both* sides of the equation by the same *nonzero* quantity.	$3x = 9$	$x = 3$
4. Interchange the two sides of the equation.	$7 = x$	$x = 7$

Point out that students should not multiply or divide both sides of an equation by a variable, as the value of that variable could be zero.

In solving an equation, the four techniques for forming equivalent equations can be used to eliminate terms or factors in the equation. For example, to solve the equation $x + 4 = 2$, you need to eliminate the term $+4$ on the left side. This is accomplished by adding its opposite, -4, to both sides.

$$x + 4 = 2 \qquad \text{Given equation}$$
$$x + 4 + (-4) = 2 + (-4) \qquad \text{Add } -4 \text{ to both sides}$$
$$x + 0 = -2 \qquad \text{Combine like terms}$$
$$x = -2 \qquad \text{Solution}$$

Note in this solution that -4 was *added* to both sides of the equation to eliminate 4 on the left side. You could just as easily have *subtracted* 4 from both sides. Both techniques are legitimate—which one you choose is a matter of personal preference.

Solving Linear Equations in Standard Form

The most common type of equation in one variable is a **linear equation.**

Definition of Linear Equation

A **linear equation** in one variable x is an equation that can be written in the standard form

$$ax + b = c,$$

where a, b, and c are real numbers with $a \neq 0$. A linear equation in one variable is also called a **first-degree equation** because its variable has an (implied) exponent of one.

Remember that to *solve* an equation in x means to find the values of x that satisfy the equation. For a linear equation in the standard form $ax + b = c$, the goal is to **isolate** x by rewriting the standard equation in the form $x =$ (a number).

EXAMPLE 3   ■   Solving a Linear Equation in Standard Form

$$4x - 2 = 10 \qquad \text{Given equation}$$
$$4x - 2 + 2 = 10 + 2 \qquad \text{Add 2 to both sides}$$
$$4x = 12 \qquad \text{Combine like terms}$$
$$\frac{4x}{4} = \frac{12}{4} \qquad \text{Divide both sides by 4}$$
$$x = 3 \qquad \text{Solution}$$

Check

$$4x - 2 = 10 \qquad \text{Given equation}$$
$$4(3) - 2 \overset{?}{=} 10 \qquad \text{Replace } x \text{ by 3}$$
$$12 - 2 \overset{?}{=} 10$$
$$10 = 10 \qquad \text{Solution checks} \qquad ■$$

Encourage students to get into the habit of checking their answers. It will help them to remember to do so later when checking is critical, and it is a useful study skill when doing exercises or taking tests.

You know that 3 is a solution of the equation given in Example 3. But how can you be sure that the equation does not have other solutions? The answer is that a linear equation in standard form always has *exactly one* solution. Other types of equations can have no solutions or two or more solutions. For instance, the nonlinear equation $x^2 = 4$ has two solutions: 2 and -2.

As you gain experience with solving linear equations, you will probably find that you can perform some of the solution steps in your head. For instance, you might solve the equation given in Example 3 by writing only the following steps.

$$4x - 2 = 10 \qquad \text{Given equation}$$
$$4x = 12 \qquad \text{Add 2 to both sides}$$
$$x = 3 \qquad \text{Divide both sides by 4}$$

Remember, however, that you should not skip the final step—checking your solution.

Solving Linear Equations in Nonstandard Form

Linear equations often occur in nonstandard forms that contain symbols of grouping or like terms that are not combined. The next two examples show how to solve these linear equations.

EXAMPLE 4   ■   Solving a Linear Equation in Nonstandard Form

$$x + 2 = 2x - 6$$ Given equation

$$-2x + x + 2 = -2x + 2x - 6$$ Add $-2x$ to both sides

$$-x + 2 = -6$$ Combine like terms

$$-x + 2 - 2 = -6 - 2$$ Subtract 2 from both sides

$$-x = -8$$ Combine like terms

$$(-1)(-x) = (-1)(-8)$$ Multiply both sides by -1

$$x = 8$$ Solution

Thus, the solution is 8. Check this solution in the original equation.   ■

When solving linear equations that are in nonstandard form, you should be aware that it is possible for the equation to have *no solution*. For instance, the equation

$$2x - 4 = 2(x - 3)$$ Given equation

$$2x - 4 = 2x - 6$$ Distributive Property

$$-4 = -6$$ Subtract $2x$ from both sides

has no solution because there is no value of x for which -4 is equal to -6.

EXAMPLE 5   ■   Solving a Linear Equation in Nonstandard Form

$$6(y - 1) + 4y = 3(7y + 1)$$ Given equation

$$6y - 6 + 4y = 21y + 3$$ Remove symbols of grouping

$$10y - 6 = 21y + 3$$ Combine like terms

$$10y - 21y - 6 = 21y - 21y + 3$$ Subtract $21y$ from both sides

$$-11y - 6 = 3$$ Combine like terms

$$-11y - 6 + 6 = 3 + 6$$ Add 6 to both sides

$$-11y = 9$$ Combine like terms

$$\frac{-11y}{-11} = \frac{9}{-11}$$ Divide both sides by -11

$$y = -\frac{9}{11}$$ Solution

Sometimes students add or subtract terms from both sides of an equation before simplifying the original sides. Stress the usefulness of simplifying each side completely first to make the equation easier to manipulate.

Thus, the solution is $-\frac{9}{11}$. Check this solution in the original equation.   ■

NOTE In Example 5, you could round the solution to $y \approx -0.818$. In this rounded form, however, the solution will not check exactly in the original equation.

If a linear equation contains fractions, first clear the equation of fractions by multiplying both sides of the equation by the least common denominator of the fractions occurring in the equation. For instance, the least common denominator in the following example is 12.

EXAMPLE 6 ■   Solving a Linear Equation That Contains Fractions

$$\frac{x}{3} + \frac{3x}{4} = 2 \qquad \text{Given equation}$$

$$12\left(\frac{x}{3} + \frac{3x}{4}\right) = 12(2) \qquad \text{Multiply both sides by 12}$$

$$12 \cdot \frac{x}{3} + 12 \cdot \frac{3x}{4} = 12(2) \qquad \text{Distributive Property}$$

$$4x + 9x = 24 \qquad \text{Simplify}$$

$$13x = 24 \qquad \text{Combine like terms}$$

$$\frac{13x}{13} = \frac{24}{13} \qquad \text{Divide both sides by 13}$$

$$x = \frac{24}{13} \qquad \text{Solution}$$

Discuss the idea of trying alternative strategies for completing a mathematical task. Note the text suggestion here. In addition, another approach to Example 6 that students may suggest is to begin by finding a common denominator and then combining fractions. Point out how this process is generally less efficient than the text method.

Thus, the solution is $\frac{24}{13}$. Check this solution in the original equation. ■

The next example shows how to solve a linear equation involving decimals. The procedure is basically the same, but the arithmetic is somewhat messier.

EXAMPLE 7 ■   Solving a Linear Equation Involving Decimals

$$0.12x + 0.09(5000 - x) = 513 \qquad \text{Given equation}$$

$$0.12x + 450 - 0.09x = 513 \qquad \text{Distributive Property}$$

$$(0.12x - 0.09x) + 450 = 513 \qquad \text{Group like terms}$$

$$0.03x + 450 - 450 = 513 - 450 \qquad \text{Subtract 450 from both sides}$$

$$0.03x = 63 \qquad \text{Combine like terms}$$

$$\frac{0.03x}{0.03} = \frac{63}{0.03} \qquad \text{Divide both sides by 0.03}$$

$$x = 2100 \qquad \text{Solution}$$

Thus, the solution is 2100. Check this solution in the original equation. ■

A different approach to solving the linear equation in Example 7 would be to begin by multiplying both sides of the equation by 100. For this particular equation, this would clear the equation of decimals and produce the equivalent equation

$$12x + 9(5000 - x) = 51,300.$$

Try solving this equation to see that you obtain the same solution.

Quadratic Equations and the Zero-Factor Property

So far in this section, the equations have all been *first-degree* polynomial equations (*linear* equations). In the remainder of the section, you will study a technique for solving **second-degree** polynomial equations (*quadratic* equations). A **quadratic equation in x in standard form** is an equation that can be written in the form

$$ax^2 + bx + c = 0, \qquad a \neq 0.$$ Quadratic equation

In Sections 1.5 and 1.6, you reviewed techniques for factoring trinomials of the form $ax^2 + bx + c$. These skills can be combined with the **Zero-Factor Property** to *solve* quadratic equations.

■

Zero-Factor Property

Let u and v represent real numbers, variables, or algebraic expressions. If u and v are factors such that

$$uv = 0,$$

then $u = 0$ or $v = 0$. This property applies to three or more factors as well.

This is a good time to justify the time students spend on learning how to factor.

NOTE The Zero-Factor Property is just another way of saying that the only way the product of two (or more) real numbers can be zero is if one (or more) of the real numbers is zero.

The Zero-Factor Property is the primary property that is used to solve equations in algebra. For instance, to solve the equation

$$(x - 2)(x + 3) = 0,$$

you can use the Zero-Factor Property to conclude that either $(x - 2)$ or $(x + 3)$ must be zero. Setting the first factor equal to zero implies that $x = 2$ is a solution. That is,

$$x - 2 = 0 \quad \Longrightarrow \quad x = 2.$$

Similarly, setting the second factor equal to zero implies that $x = -3$ is a solution. That is,

$$x + 3 = 0 \quad \Longrightarrow \quad x = -3.$$

Thus, the equation $(x - 2)(x + 3) = 0$ has exactly two solutions: 2 and -3.

Solving Quadratic Equations by Factoring

In each of the examples that follow, note how factoring skills are combined with the Zero-Factor Property to solve equations.

NOTE Factoring and the Zero-Factor Property allow you to solve a quadratic equation by converting it into two *linear* equations (which you already know how to solve). This is a common strategy of algebra—to break down a given problem into simpler parts, each solvable by previously learned methods.

EXAMPLE 8 ■ **Using Factoring to Solve an Equation**

Solve the quadratic equation

$$x^2 - x - 12 = 0.$$

Solution

Begin by checking to see that the right side of the equation is zero. Next, factor the left side of the equation. Finally, apply the Zero-Factor Property to find the solutions.

$x^2 - x - 12 = 0$	Given equation in standard form
$(x + 3)(x - 4) = 0$	Factor left side of equation
$x + 3 = 0 \implies x = -3$	Set first factor equal to 0
$x - 4 = 0 \implies x = 4$	Set second factor equal to 0

Thus, the given equation has two solutions: -3 and 4.

Check First Solution

$x^2 - x - 12 = 0$	Given equation
$(-3)^2 - (-3) - 12 \stackrel{?}{=} 0$	Replace x by -3
$9 + 3 - 12 \stackrel{?}{=} 0$	
$0 = 0$	Solution checks

Check Second Solution

$x^2 - x - 12 = 0$	Given equation
$4^2 - 4 - 12 \stackrel{?}{=} 0$	Replace x by 4
$16 - 4 - 12 \stackrel{?}{=} 0$	
$0 = 0$	Solution checks ■

To use the Zero-Factor Property, a quadratic equation *must* be written in **standard form.** That is, the quadratic must be on the left side of the equation and zero must be the only term on the right side of the equation. For instance, to write the equation $x^2 - 3x = 18$ in standard form, you must subtract 18 from both sides of the equation, as follows.

$x^2 - 3x = 18$	Given equation (nonstandard form)
$x^2 - 3x - 18 = 18 - 18$	Subtract 18 from both sides
$x^2 - 3x - 18 = 0$	Standard form

EXAMPLE 9 ■ **Solving a Quadratic Equation**

Solve the quadratic equation

$$3x^2 + 5x = 12.$$

As students gain experience with solving equations, encourage them to perform some steps manually. For instance, students can be shown how to solve $3x - 4 = 0$ or $x + 3 = 0$ mentally.

Solution

$$3x^2 + 5x = 12 \qquad \text{Given equation}$$

$$3x^2 + 5x - 12 = 0 \qquad \text{Write in standard form}$$

$$(3x - 4)(x + 3) = 0 \qquad \text{Factor left side of equation}$$

$$3x - 4 = 0 \implies x = \frac{4}{3} \qquad \text{Set first factor equal to 0}$$

$$x + 3 = 0 \implies x = -3 \qquad \text{Set second factor equal to 0}$$

Therefore, the solutions are $\frac{4}{3}$ and -3. Check these solutions in the original equation. ∎

After converting a quadratic equation to standard form, you should check to see whether the left side of the equation has a common numerical factor. If it does, you can divide both sides of the equation by this factor without "losing" any of the solutions. For instance, each of the equations

$$2x^2 - 2x - 24 = 0, \quad 2(x^2 - x - 12) = 0, \quad \text{and} \quad x^2 - x - 12 = 0$$

has the same solutions.

In Examples 8 and 9, the given equations each involved a second-degree polynomial and each had *two different* solutions. Sometimes you will encounter a second-degree polynomial equation that has only one (repeated) solution. This occurs when the left side of the equation is a perfect square trinomial, as shown in Example 10.

EXAMPLE 10 ■ **A Quadratic Equation with a Repeated Solution**

Solve the quadratic equation

$$x^2 - 6x + 11 = 2.$$

Solution

$$x^2 - 6x + 11 = 2 \qquad \text{Given equation}$$

$$x^2 - 6x + 9 = 0 \qquad \text{Write in standard form}$$

$$(x - 3)^2 = 0 \qquad \text{Factor}$$

$$x - 3 = 0 \qquad \text{Set factor equal to 0}$$

$$x = 3 \qquad \text{Solution is 3}$$

Note that even though the left side of this equation has two factors, the two factors are the same. Thus, you can conclude that the only solution of the equation is 3. (This solution is called a **repeated solution.**) Check this solution in the original equation. ∎

Be sure you see that the Zero-Factor Property can be applied only to a product that is equal to *zero*. For instance, you cannot conclude from the equation

$$x(x - 1) = 6$$

that $x = 6$ and $x - 1 = 6$ yield solutions. Instead, you must first write the equation in standard form and then factor the left side, as follows.

$$x^2 - x - 6 = 0 \implies (x - 3)(x + 2) = 0$$

Now, from the factored form you can see that the solutions are 3 and -2.

EXAMPLE 11 ■ **Solving a Quadratic Equation**

Solve the quadratic equation

$$(x + 2)(x + 4) = 3.$$

Solution

$(x + 2)(x + 4) = 3$	Given equation
$x^2 + 6x + 8 = 3$	Multiply factors
$x^2 + 6x + 5 = 0$	Standard form
$(x + 1)(x + 5) = 0$	Factor left side of equation
$x + 1 = 0 \implies x = -1$	Set first factor equal to 0
$x + 5 = 0 \implies x = -5$	Set second factor equal to 0

Therefore, the equation has two solutions: -1 and -5. Check these solutions in the original equation. ■

You might again discuss Example 10, a quadratic equation with degree 2. We can expect at most two real solutions. Point out that we have two solutions, but they just happen to be the same so we don't need to repeat them.

Some quadratic equations do not have solutions that are real numbers. For instance, there is no real number that satisfies the equation $x^2 = -4$. The reason that this equation has no real solution is that there is no real number that can be multiplied by itself to produce -4. This type of equation will be discussed further in Chapter 7.

You might remember from an earlier course in algebra that a polynomial equation can have *at most* as many solutions as its degree. For instance, a second-degree equation can have zero, one, or two real solutions, but it cannot have three or more solutions.

EXAMPLE 12 ■ **Solving Linear and Quadratic Equations**

Describe the technique you would use to solve each of the following equations.

(a) $7x + 1 = 2x - 3$ (b) $7x + 1 = 2x^2 - 3$

Solution

(a) This is a linear equation, so you would solve it by isolating x. After doing this, you would find that the solution is $-\frac{4}{5}$.

(b) This is a quadratic equation. To solve it, you would write the equation in standard form, factor the quadratic, and apply the Zero-Factor Property to conclude that the solutions are 4 and $-\frac{1}{2}$. ■

DISCUSSION PROBLEM   ■   **You Be the Instructor**

Suppose you are teaching an algebra class and one of your students hands in the following problem.

$$2x - 3(x - 2) = 5x + 12$$
$$2x - 3x - 6 = 5x + 12$$
$$-x - 6 = 5x + 12$$
$$-6x = 6$$
$$x = -1$$

Check

$$2(-1) - 3[(-1) - 2] \overset{?}{=} 5(-1) + 12$$

$$-2 - 3(-3) \overset{?}{=} -5 + 12$$

$$-2 + 9 \overset{?}{=} 7$$

$$7 = 7$$

Is the answer correct? Are the solution steps correct? How much "partial credit" would you give the student for a correct answer that was obtained by incorrect means? ■

Warm-Up

The following warm-up exercises involve skills that were covered in earlier sections. You will use these skills in the exercise set for this section.

In Exercises 1–6, simplify the algebraic expression.

1. $-2(x - 3) - 6$

2. $6(x + 10) + 4$

3. $4 - 3(2x + 1)$

4. $5(x + 2) - 4(2x - 3)$

5. $24\left(\dfrac{y}{3} + \dfrac{y}{6}\right)$

6. $0.12x + 0.05(2000 - 2x)$

In Exercises 7–10, factor the algebraic expression.

7. $4v^2 + 18v$

8. $16x^2 - 25$

9. $x^2 + 6x - 27$

10. $-8x^2 + 10x - 3$

2.1   EXERCISES

In Exercises 1–6, determine whether the given values of the variable are solutions of the equation.

Equation	*Values*		*Equation*	*Values*	
1. $3x - 7 = 2$	(a) $x = 0$	(b) $x = 3$	**2.** $5x + 9 = 4$	(a) $x = -1$	(b) $x = 2$
3. $3x^2 - 2x = 21$	(a) $x = -3$	(b) $x = 3$	**4.** $10x - 3 = 7x$	(a) $x = 0$	(b) $x = -1$
5. $3x + 3 = 2(x - 4)$	(a) $x = -11$	(b) $x = 5$	**6.** $7x - 1 = 5(x + 5)^2$	(a) $x = 2$	(b) $x = -2$

In Exercises 7–10, complete the solution.

7.
$$3x + 15 = 0$$
$$3x + 15 - 15 = 0 - \boxed{}$$
$$3x = -15$$
$$\frac{3x}{3} = \frac{-15}{\boxed{}}$$
$$x = \boxed{}$$

8.
$$7x - 21 = 0$$
$$7x - 21 + 21 = 0 + \boxed{}$$
$$7x = 21$$
$$\frac{7x}{7} = \frac{\boxed{}}{7}$$
$$x = \boxed{}$$

9.
$$-2x + 5 = 12$$
$$-2x + 5 - \boxed{} = 12 - 5$$
$$-2x = \boxed{}$$
$$\frac{\boxed{}}{-2} = \frac{7}{-2}$$
$$x = \boxed{}$$

10.
$$25 - 3x = 10$$
$$25 - 3x - \boxed{} = 10 - 25$$
$$-3x = \boxed{}$$
$$\frac{-3x}{\boxed{}} = \frac{-15}{-3}$$
$$x = \boxed{}$$

In Exercises 11–42, solve the linear equation and check the result. (If not possible, state the reason.)

11. $3x = 12$

12. $-14x = 28$

13. $6x + 4 = 0$

14. $8z - 10 = 0$

15. $23x - 4 = 42$

16. $15x - 18 = 27$

17. $7 - 8x = 13x$

18. $2s - 16 = 34s$

19. $15t = 0$

20. $6a + 2 = 6a$

21. $-8t + 7 = -8t$

22. $4x = -12x$

23. $8(x - 8) = 24$

24. $6(x + 2) = 30$

25. $5 - (2y - 4) = 15$

26. $26 - (3x - 10) = 6$

27. $12(x + 3) = 7(x + 3)$

28. $-25(x - 100) = 16(x - 100)$

29. $8x - 3(x - 2) = 12$

30. $12 = 6(y + 1) - 8(1 - y)$

31. $\dfrac{u}{5} = 10$

32. $-\dfrac{z}{2} = 7$

33. $t - \dfrac{2}{5} = \dfrac{3}{2}$

34. $z + \dfrac{1}{15} = -\dfrac{3}{10}$

35. $\dfrac{t + 4}{14} = \dfrac{2}{7}$

36. $\dfrac{11x}{6} + \dfrac{1}{3} = 0$

37. $\dfrac{t}{5} - \dfrac{t}{2} = 1$

38. $\dfrac{t}{6} + \dfrac{t}{8} = 1$

39. $\dfrac{4u}{3} = \dfrac{5u}{4} + 6$ **40.** $\dfrac{-3x}{4} - 4 = \dfrac{x}{6}$ **41.** $0.3x + 1.5 = 8.4$ **42.** $16.3 - 0.2x = 7.1$

In Exercises 43–48, solve the quadratic equation.

43. $2x(x - 8) = 0$ **44.** $z(z + 6) = 0$ **45.** $(y - 3)(y + 10) = 0$ **46.** $17(t - 3)(t + 8) = 0$

47. $(2t + 5)(3t + 1) = 0$ **48.** $(5x - 3)(x - 8) = 0$

In Exercises 49–70, solve the quadratic equation by factoring.

49. $x^2 - 3x - 10 = 0$ **50.** $x^2 - x - 12 = 0$ **51.** $m^2 - 8m + 16 = 0$ **52.** $a^2 + 4a + 4 = 0$

53. $x^2 + 16x + 64 = 0$ **54.** $x^2 - 12x + 36 = 0$ **55.** $4x^2 + 15x - 25 = 0$ **56.** $14x^2 + 9x + 1 = 0$

57. $x^2 - 25 = 0$ **58.** $25z^2 - 100 = 0$ **59.** $(t - 2)^2 - 16 = 0$ **60.** $(s + 4)^2 - 49 = 0$

61. $9x^2 + 15x = 0$ **62.** $4x^2 - 6x = 0$ **63.** $x(x + 10) - 2(x + 10) = 0$ **64.** $u(u - 3) + 3(u - 3) = 0$

65. $x(x - 3) = 10$ **66.** $x(x - 1) = 72$ **67.** $t(2t - 3) = 35$ **68.** $3u(3u + 1) = 20$

69. $(a + 2)(a + 5) = 10$ **70.** $(x - 8)(x - 7) = 20$

In Exercises 71–88, solve the equation.

71. $8 - 5t = 20 + t$ **72.** $3y + 14 = y + 20$ **73.** $-4(t + 2) = 0$ **74.** $8(z - 8) = 0$

75. $2(x + 7) - 9 = 5(x - 4)$ **76.** $6[x - (5x - 7)] = 4 - 5x$ **77.** $\frac{1}{3}x + 1 = \frac{1}{12}x - 4$ **78.** $\frac{1}{9}x + \frac{1}{3} = \frac{11}{18}$

79. $1.2(x - 3) = 10.8$ **80.** $6.5(1 - 2x) = 13$ **81.** $3y^2 - 48 = 0$ **82.** $x^2 - 121 = 0$

83. $4z^2 - 12z + 9 = 0$ **84.** $16t^2 + 40t + 24 = 0$ **85.** $7 + 13x - 2x^2 = 0$ **86.** $11 + 32y - 3y^2 = 0$

87. $(x - 4)(x + 5) = 10$ **88.** $(u - 8)(u + 10) = 40$

In Exercises 89–92, solve the equation and round your answer to two decimal places. (A calculator may be helpful.)

89. $1.234x + 3 = 7.805$ **90.** $325x - 4125 = 612$ **91.** $\dfrac{x}{10.625} = 2.850$ **92.** $2x + \dfrac{4.7}{4} = \dfrac{3}{2}$

93. *Exploratory Exercise* The length of a certain rectangle is t times its width (see figure). Thus, the perimeter P is given by $P = 2w + 2(tw)$, where w is the width of the rectangle. The perimeter of the rectangle is 1200 meters.

(a) Complete the table giving the widths, lengths, and areas of the rectangle for the specified values of t.

(b) Use the table to write a short paragraph describing the relationship among the width, length, and area of a rectangle that has a *fixed* perimeter.

Figure for 93

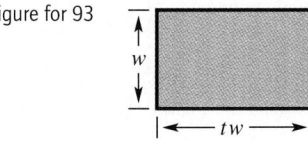

t	1	1.5	2	3	4	5
Width						
Length						
Area						

94. *Maximum Height of an Object* The velocity v of an object projected vertically with an initial velocity of 64 feet per second is given by $v = 64 - 32t$, where t is the time in seconds and air resistance is neglected. Find the time at which the maximum height of the object is obtained. (*Hint:* The maximum height is reached when $v = 0$.)

95. *Work Rate Problem* Two people can complete a task in t hours, where t must satisfy the equation

$$\frac{t}{10} + \frac{t}{15} = 1.$$

Find the required time t.

96. *Private School Expenditures* The total amount spent by primary and secondary private schools in the United States from 1975 to 1989 can be approximated by the linear equation

$$y = 7717.4 + 894.6t,$$

where y represents the amount spent in millions of dollars and t represents the year, with $t = 0$ corresponding to 1980 (see figure). According to this model, during which year did the total amount spent reach $13,085 million? (*Source:* U.S. National Center for Education Statistics)

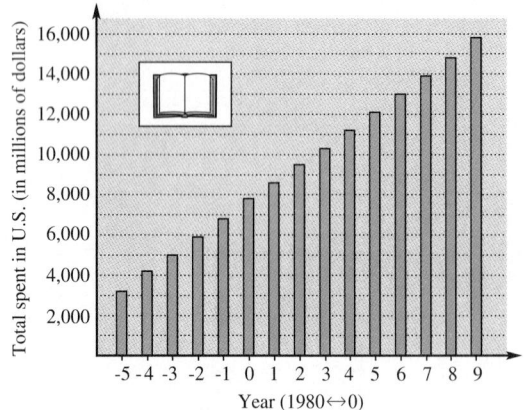

97. *Dimensions of a Rectangle* The rectangular floor of a storage shed has an area of 330 square feet. The length of the floor is 7 feet more than its width (see figure). Find the dimensions of the floor.

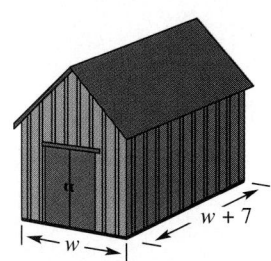

98. *Free-Falling Object* An object is dropped from a weather balloon 6400 feet above the ground. Find the time t for the object to reach the ground by solving the following equation.

$$-16t^2 + 6400 = 0$$

99. *Dimensions of an Open Box* An open box is to be made from a square piece of material by cutting 5-inch squares from each corner and turning up the sides (see figure). The volume V of a rectangular solid is the product of its length, width, and height, and x is the length of an edge of the square base.

(a) Show that the volume is given by $V = 5x^2$.
(b) Complete the table.

x	2	4	6	8
V				

(c) Find the size of the original piece of material if $V = 720$ cubic inches.

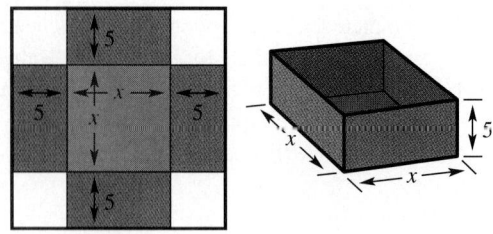

100. *Dimensions of an Open Box* An open box with a square base (see figure) is to be constructed from 880 square inches of material. What should be the dimensions of the base if the height of the box is to be 6 inches? (*Hint:* The surface area is given by $S = x^2 + 4xh$.)

101. In your own words, describe the difference between a linear equation in x and a quadratic equation in x.

Figure for 100

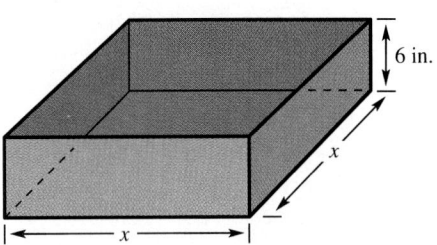

6 in.

x

x

| **SECTION** **2.2** | **Literal Equations and Formulas**
 Solving Literal Equations • *Using Formulas* |

Solving Literal Equations

A **literal equation** is an equation that has more than one variable. For instance, $5x + 2y = 7$ is a literal equation because it has two variables, x and y. The word *literal* comes from the Latin word for "letter." To **solve a literal equation** for one of its variables means to write an equivalent equation in which the "solved variable" is isolated on one side of the equation. For instance, you can solve the literal equation $5x + 2y = 7$ for x or for y, as follows.

Show that $x = \dfrac{7 - 2y}{5}$ is the same as $x = \dfrac{7}{5} - \dfrac{2}{5}y$. (Note also the discussion following Example 1.)

Solve for x

$$5x + 2y = 7$$
$$5x = 7 - 2y$$
$$x = \frac{7 - 2y}{5}$$

Solve for y

$$5x + 2y = 7$$
$$2y = 7 - 5x$$
$$y = \frac{7 - 5x}{2}$$

EXAMPLE 1 ■ Solving a Literal Equation

Solve the equation $5s - 2t = 8$ (a) for s and (b) for t.

Solution

(a) You can solve the equation for s as follows.

$$5s - 2t = 8 \qquad \text{Given equation}$$
$$5s = 8 + 2t \qquad \text{Add } 2t \text{ to both sides}$$
$$s = \frac{8 + 2t}{5} \qquad \text{Divide both sides by 5}$$

(b) You can solve the equation for t as follows.

$$5s - 2t = 8 \qquad \text{Given equation}$$

$$-2t = 8 - 5s \qquad \text{Subtract } 5s \text{ from both sides}$$

$$t = \frac{8 - 5s}{-2} \qquad \text{Divide both sides by } -2$$

$$t = \frac{5s - 8}{2} \qquad \text{Simplify} \qquad \blacksquare$$

There is no general agreement as to the "best" or "simplest" way to write solutions of literal equations. In Example 1(b), we rewrote the solution

$$t = \frac{8 - 5s}{-2} \quad \text{as} \quad t = \frac{5s - 8}{2}$$

because the latter form seems to be simpler (it involves fewer minus signs). This, however, is partly a matter of personal preference. For instance, you might prefer to keep the solution in the first form *or* you might prefer to write the solution as

$$t = \frac{5}{2}s - 4.$$

The actual form that you use is not particularly important. It is important, however, that you realize that many forms can be used *and* that you are able to convert from one form to another.

NOTE

$$t = \frac{8 - 5s}{-2} = \frac{(-1)(8 - 5s)}{(-1)(-2)} = \frac{-8 + 5s}{2} = \frac{5s - 8}{2}$$

and

$$t = \frac{8 - 5s}{-2} = \frac{8}{-2} + \frac{-5s}{-2} = -4 + \frac{5s}{2} = \frac{5}{2}s - 4.$$

EXAMPLE 2 ■ Solving a Literal Equation

Solve the literal equation $3y + 2x - 4 = 5y - 3x + 2$ for y.

Solution

$$3y + 2x - 4 = 5y - 3x + 2 \qquad \text{Given equation}$$

$$-2y + 2x - 4 = -3x + 2 \qquad \text{Subtract } 5y \text{ from both sides}$$

$$-2y - 4 = -5x + 2 \qquad \text{Subtract } 2x \text{ from both sides}$$

$$-2y = -5x + 6 \qquad \text{Add 4 to both sides}$$

$$y = \frac{-5x + 6}{-2} \qquad \text{Divide both sides by } -2$$

$$y = \frac{5x - 6}{2} \qquad \text{Simplify} \qquad \blacksquare$$

EXAMPLE 3 ■ **Solving a Literal Equation**

Solve the literal equation $A = P + Prt$ for P.

Solution

Notice how the Distributive Property is used to factor P out of the terms on the right side of the equation.

$A = P + Prt$	Given equation
$A = P(1 + rt)$	Distributive Property
$\dfrac{A}{1 + rt} = P$	Divide both sides by $1 + rt$ ■

Point out that this answer is not the same as $A/1 + A/rt$.

NOTE In Example 3, notice that it is not necessary to isolate the variable on the left side of the equation. Sometimes it is more convenient to isolate the variable on the right side.

Using Formulas

Encourage students to become familiar with common formulas, which are printed on the inside front and back covers of this text.

Many common types of geometric, scientific, and investment problems use ready-made equations called **formulas.** Knowing formulas such as those in the following lists will help you translate and solve a wide variety of real-life problems involving perimeter, area, volume, temperature, interest, and principal.

Common Formulas for Area and Perimeter

The formulas below give the area A and perimeter P (or circumference C) of the indicated figures.

Square	Rectangle	Circle	Triangle
$A = s^2$	$A = lw$	$A = \pi r^2$	$A = \frac{1}{2}bh$
$P = 4s$	$P = 2l + 2w$	$C = 2\pi r$	

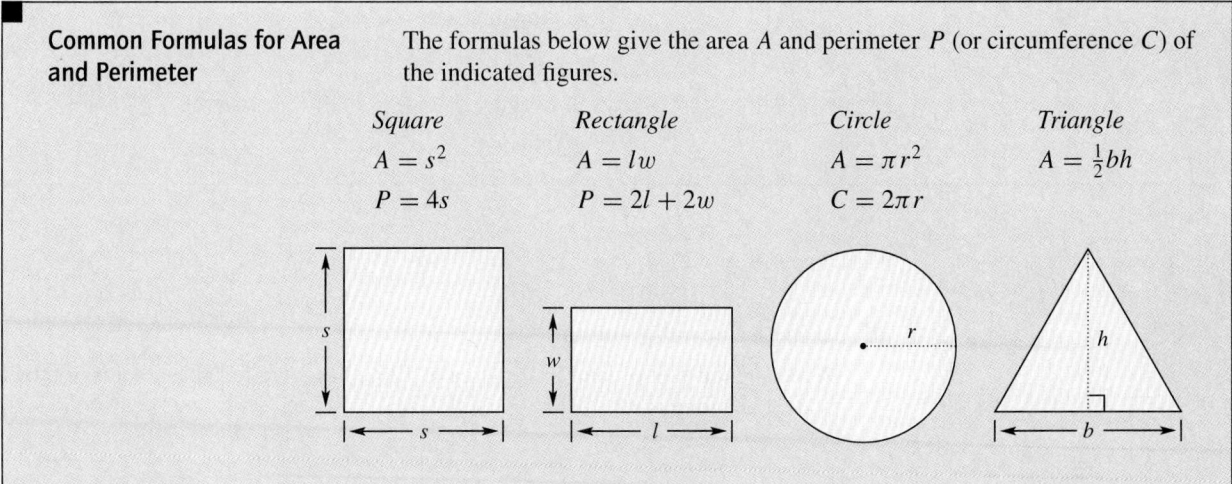

Common Formulas for Volume

The formulas below give the volume V of the indicated figures.

Cube	Rectangular Solid	Circular Cylinder	Sphere
$V = s^3$	$V = lwh$	$V = \pi r^2 h$	$V = \frac{4}{3}\pi r^3$

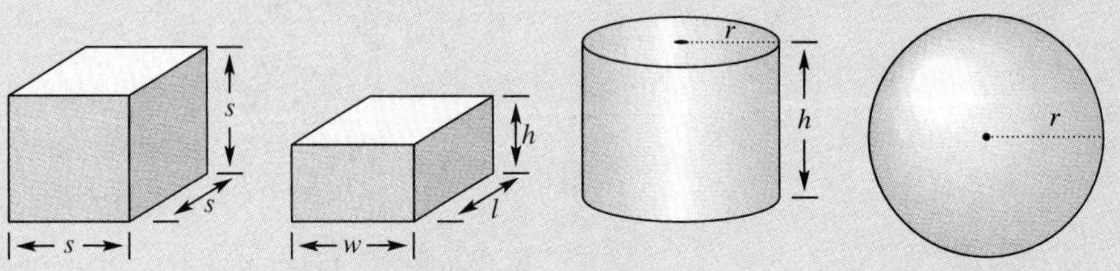

Miscellaneous Common Formulas

Temperature:

$F = \frac{9}{5}C + 32$

$F =$ degrees Fahrenheit $C =$ degrees Celsius

Simple Interest:

$I = Prt$

$I =$ interest $P =$ principal

$r =$ annual interest rate $t =$ time in years

Compound Interest:

$A = P\left(1 + \dfrac{r}{n}\right)^{nt}$

$A =$ balance $P =$ principal

$r =$ annual interest rate $n =$ compoundings per year

$t =$ time in years

Distance:

$d = rt$

$d =$ distance traveled $r =$ rate

$t =$ time

EXAMPLE 4 ◾ **Using a Geometric Formula**

A rectangular plot of land has an area of 120,000 square feet. The plot is 300 feet wide. How long is it?

Solution

In a problem such as this, it is helpful to begin by drawing a picture, as shown in Figure 2.1. To solve for the unknown length, begin with the formula for the area of a rectangle, as follows.

FIGURE 2.1

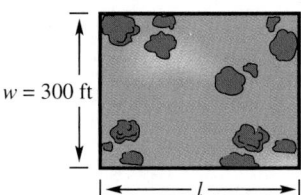

$w = 300$ ft

l

$$A = lw \qquad \text{Common formula}$$

$$120{,}000 = l(300) \qquad \text{Substitute 120,000 for } A \text{ and 300 for } w$$

$$\frac{120{,}000}{300} = l \qquad \text{Divide both sides by 300}$$

$$400 = l \qquad \text{Simplify}$$

Thus, the length of the rectangular plot is 400 feet. You can check this by multiplying 300 feet by 400 feet to obtain the given area of 120,000 square feet. ■

EXAMPLE 5 ■ Simple Interest

You deposit $8000 in an account paying simple interest. After 6 months, the account earns $300 in interest. What is the annual interest rate for this account?

Solution

$$I = Prt \qquad \text{Common formula}$$

$$300 = 8000(r)\left(\frac{1}{2}\right) \qquad \text{Substitute 300 for } I, \text{ 8000 for } P, \text{ and } \frac{1}{2} \text{ for } t$$

$$300 = 4000r \qquad \text{Simplify}$$

$$\frac{300}{4000} = r \qquad \text{Divide by 4000}$$

$$0.075 = r \qquad \text{Simplify}$$

Therefore, the annual interest rate is $r = 0.075$ (or 7.5%). Check this solution in the original problem. ■

EXAMPLE 6 ■ The Area of a Trapezoid

The area A of a trapezoid is given by the formula $A = \frac{1}{2}h(b_1 + b_2)$, where h is the height and b_1 and b_2 are the lengths of the upper and lower bases (see Figure 2.2).

FIGURE 2.2

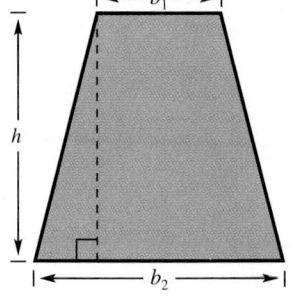

b_1

h

b_2

(a) Solve this formula for b_2.

(b) Use the result to find the length of the lower base of a trapezoid with a height of 6 inches, an upper base of 3 inches, and an area of 29 square inches.

Solution

(a)
$$A = \frac{1}{2}h(b_1 + b_2) \qquad \text{Formula for area}$$

$$2A = h(b_1 + b_2) \qquad \text{Multiply both sides by 2}$$

$$\frac{2A}{h} = (b_1 + b_2) \qquad \text{Divide both sides by } h$$

$$\frac{2A}{h} - b_1 = b_2 \qquad \text{Subtract } b_1 \text{ from both sides}$$

This gives the length of the lower base in terms of the height, the upper base, and the area.

(b) Substitute $h = 6$, $b_1 = 3$, and $A = 29$ into the formula for b_2.

$$b_2 = \frac{2A}{h} - b_1 \qquad\qquad \text{Formula for } b_2$$

$$= \frac{2(29)}{6} - 3 \qquad\qquad \text{Substitute 6 for } h, \text{ 3 for } b_1, \text{ and 29 for } A$$

$$\approx 6.67 \qquad\qquad \text{Use a calculator}$$

The length of the lower base is approximately 6.67 inches.   ∎

When evaluating a formula, be sure to follow the established *order of operations,* as discussed in Section 1.1. For instance, how would you evaluate the compound interest formula

$$A = P\left(1 + \frac{r}{n}\right)^{nt}$$

when $P = 1000$, $r = 0.06$, $n = 12$, and $t = 5$? By doing this properly, you should obtain a balance of $A = \$1348.85$.

EXAMPLE 7   ■   Finding the Principal

The formula $A = P + Prt$ gives the balance A in an account that earns a simple annual interest rate of r over t years with a deposit of P. How much money should you deposit in a savings account earning 6% interest annually to have a balance of $500 after $\frac{1}{2}$ year?

Solution

$$A = P + Prt \qquad\qquad \text{Given formula}$$

$$500 = P + P(0.06)\left(\frac{1}{2}\right) \qquad\qquad \text{Substitute values for } A, r, \text{ and } t$$

$$500 = P + P(0.03) \qquad\qquad \text{Simplify}$$

$$500 = P(1 + 0.03) \qquad\qquad \text{Factor out } P$$

$$\frac{500}{1.03} = P \qquad\qquad \text{Divide by 1.03}$$

$$485.44 \approx P \qquad\qquad \text{Principal}$$

Thus, you should deposit $485.44.   ∎

NOTE In Example 7, you could have first solved the given equation for P, as was done in Example 5, and then substituted values for A, r, and t. Try doing this to show that you get the same result for P.

The amount of horsepower produced by an engine whose pistons are turning a single crankshaft is

$$H = \left(\frac{1}{396,000}\right) PpArn,$$

where P is the average piston pressure, p is the piston stroke length, A is the piston area, r is the rotational speed of the crankshaft in revolutions per minute (rpm), and n is the number of pistons.

EXAMPLE 8   ■   Finding the Average Piston Pressure

(a) Solve the formula $H = \left(\dfrac{1}{396{,}000} \right) PpArn$ for P.

(b) A six-piston automobile engine is producing 150 horsepower at 2500 rpm. The piston stroke length is 3.3 inches, and the piston area is 10 square inches (see Figure 2.3). What is the average piston pressure (in pounds per square inch)?

Solution

(a)
$$H = \left(\frac{1}{396{,}000} \right) PpArn \qquad \text{Formula for horsepower}$$

$$396{,}000H = PpArn \qquad \text{Multiply both sides by 396,000}$$

$$\frac{396{,}000H}{pArn} = P \qquad \text{Divide both sides by factors } pArn$$

(b) Let $H = 150$, $A = 10$, $p = 3.3$, $r = 2500$, and $n = 6$.

$$P = \frac{396{,}000H}{pArn} = \frac{396{,}000(150)}{(3.3)(10)(2500)(6)} = \frac{59{,}400{,}000}{495{,}000} = 120$$

Thus, the average piston pressure is 120 pounds per square inch.

FIGURE 2.3

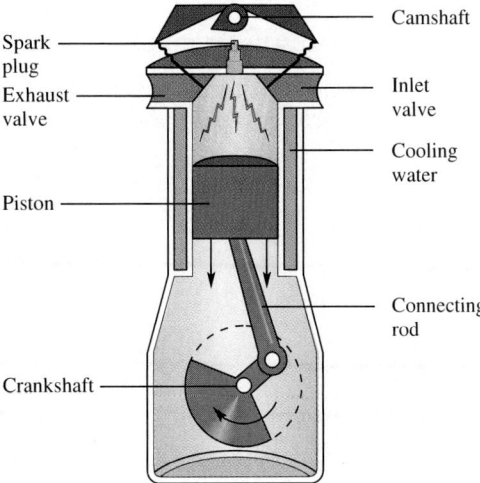

Camshaft

Spark plug

Exhaust valve

Inlet valve

Cooling water

Piston

Connecting rod

Crankshaft

■

When solving a formula for one of its variables, you may need to divide both sides of the equation by a variable quantity. (For instance, in Example 8a, both sides of the equation were divided by $pArn$.) When you do this, you are assuming that the variable quantity cannot be zero.

EXAMPLE 9 ■ Solving a Formula

The formula for distance is $d = rt$, where d is the distance traveled, r is the rate, and t is the time. Solve the formula for t. For which values of r is the solution valid?

Solution

$$d = rt \qquad \text{Formula for distance}$$

$$\frac{d}{r} = t \qquad \text{Divide both sides by } r$$

Because both sides of the equation were divided by r, it follows that r cannot be zero. Thus, the solution is valid for all nonzero values of r.   ■

DISCUSSION PROBLEM ■ Area of a Parallelogram

The formula for the area of a parallelogram is $A = bh$, where b is the length of the base and h is the height (see figure). Explain how this formula can be derived from the formula for the area of a triangle.

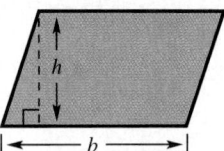

Warm-Up

The following warm-up exercises involve skills that were covered in earlier sections. You will use these skills in the exercise set for this section.

In Exercises 1–10, solve the equation.

1. $5x - 22 = 0$
2. $3x + 42 = 0$

3. $3t - 2(t + 5) = 2t - 15$
4. $2x + \dfrac{x}{2} = \dfrac{10}{3}$

5. $4y - 3 = 3y$
6. $3(z - 6) = 0$

7. $25(a + 4)(a - 2) = 0$
8. $4x^2 + 12x + 9 = 0$

9. $s(s + 4) = 96$
10. $4 - (b + 2)^2 = 0$

2.2 EXERCISES

In Exercises 1–8, solve the equation for (a) x and (b) y.

1. $2x - 3y = 6$

2. $-5x + 3y = 15$

3. $7x + 4 = 10y - 7$

4. $12(x - 2) + 7(y + 1) = 25$

5. $4[2x - 3(x + 2y)] = 0$

6. $-3[2(x + 4y) - (x - y)] = 1$

7. $\dfrac{x}{2} + \dfrac{y}{5} = 1$

8. $\dfrac{3x}{4} + \dfrac{y}{6} = 5$

In Exercises 9–18, verify that the solution can be written in the second form.

	Solution	*Rewritten Solution*		*Solution*	*Rewritten Solution*
9.	$\dfrac{y - 5}{-1}$	$5 - y$	**10.**	$\dfrac{x - 10}{3}$	$\dfrac{1}{3}(x - 10)$
11.	$\dfrac{3}{4}(8b + 4)$	$3(2b + 1)$	**12.**	$\dfrac{-7(3 - x)}{2}$	$\dfrac{7}{2}(x - 3)$
13.	$\dfrac{5x + 4}{2}$	$\dfrac{5}{2}x + 2$	**14.**	$\dfrac{3x - 7}{6}$	$\dfrac{1}{2}x - \dfrac{7}{6}$
15.	$\dfrac{2t - 5}{-2}$	$\dfrac{5}{2} - t$	**16.**	$\dfrac{15 - 12a}{-3}$	$4a - 5$
17.	$\dfrac{h + 4\pi}{2}$	$\dfrac{h}{2} + 2\pi$	**18.**	$u + 3uv$	$u(1 + 3v)$

In Exercises 19–30, solve for the indicated variable.

19. Solve for R

Ohm's Law: $E = IR$

20. Solve for C

Markup: $S = C + rC$

21. Solve for L

Discount: $S = L - rL$

22. Solve for h

Area of a Triangle: $A = \frac{1}{2}bh$

23. Solve for b

Area of a Trapezoid:
$A = \frac{1}{2}(a + b)h$

24. Solve for h

Volume of a Right Circular
Cylinder: $V = \pi r^2 h$

25. Solve for r

Investment at Simple Interest: $A = P + Prt$

26. Solve for P

Investment at Compound Interest: $A = P\left(1 + \dfrac{r}{n}\right)^{nt}$

27. Solve for n

Arithmetic Progression: $S = \dfrac{n}{2}(a_1 + a_n)$

28. Solve for n

Arithmetic Progression: $L = a + (n - 1)d$

29. Solve for r

Geometric Progression: $S = \dfrac{a_1}{1 - r}$

30. Solve for m_2

Newton's Law of Universal Gravitation: $F = \dfrac{km_1m_2}{r^2}$

In Exercises 31–34, use the geometric formulas given in this section to solve the problem.

31. Height of a Picture Frame A rectangular picture frame has a perimeter of 54 inches. The frame is 12 inches wide. How tall is it?

32. Height of a Picture Frame A rectangular picture frame has a perimeter of 3 feet. The width of the frame is 0.62 foot. Find the height of the frame.

33. Area of a Square Three squares are shown in the figure. The perimeter of square I is 12 square inches, and the area of square II is 36 square inches. Find the area of square III.

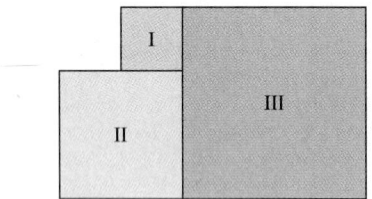

34. Dimensions of a Triangle The triangular cross section of a part being machined must have an area of 48 square inches (see figure). Find the base and height of the triangle if the height is $1\frac{1}{2}$ times the base.

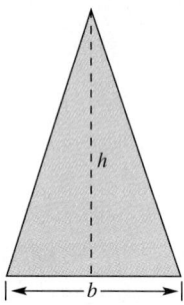

In Exercises 35 and 36, use the formulas for volume given in this section to solve the problem.

35. Test Chamber A sealed test chamber in the form of a cube has a volume of 8 cubic feet. Find the dimensions of the cube.

36. Weather Balloon A weather balloon in the form of a sphere has a volume of $\frac{4}{3}\pi(3^3)$ cubic meters. Find the radius of the balloon.

37. Area of a Circular Ring The area of a circular ring (see figure) is given by $A = 2\pi pw$.

    (a) Solve for p.

    (b) Find p given $A = 22$ square centimeters and $w = 2$ centimeters.

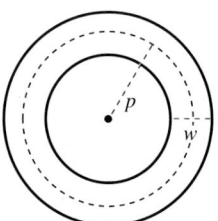

38. Perimeter of a Pond The perimeter of a pond (see figure) is given by $P = 2\pi r + 2x$.

    (a) Solve for r.

    (b) Find r if $P = 75$ feet and $x = 20$ feet.

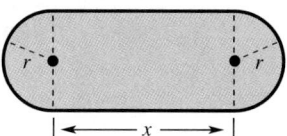

Simple Interest In Exercises 39 and 40, use the formula for simple interest.

39. Find the annual interest rate if the interest on an investment of $750 for 2 years is $105.

40. The interest on an investment of $2000 at 8% is $800. How long was the money invested?

Compound Interest In Exercises 41 and 42, use the formula for compound interest.

41. In 8 years, a child will need $12,000 for the first year of college. Find the principal that the parents must invest at 8% compounded monthly to yield this amount.

42. In 10 years, a couple plans to retire and buy a motor home. They estimate that this will require $50,000. Find the principal they must invest at 9% compounded quarterly to yield this amount.

Distance Problem In Exercises 43–48, determine the unknown distance, rate, or time.

	Distance, d	Rate, r	Time, t
43.		650 mi/hr	$3\frac{1}{2}$ hr
44.		45 ft/sec	10 sec
45.	1000 km	110 km/hr	
46.	250 ft	32 ft/sec	
47.	1000 ft		$\frac{3}{2}$ sec
48.	385 mi		7 hr

49. *Average Speed* Determine the time required for a space shuttle to travel a distance of 5000 miles when its average speed is 17,000 miles per hour (see figure).

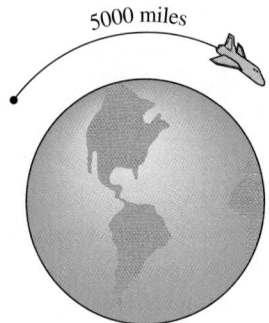

5000 miles

50. *Speed of Light* Determine the time for light to travel from the sun to the earth. The distance between the sun and the earth is 93,000,000 miles, and the speed of light is approximately 186,000 miles per second (see figure).

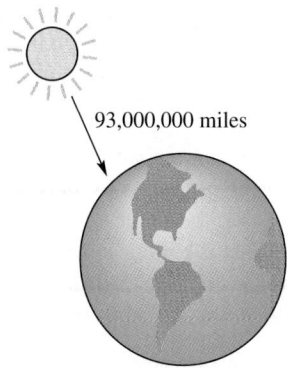

93,000,000 miles

New York City Marathon In Exercises 51 and 52, find the average speed of the record-holding runners in the New York City Marathon. The length of the course is 26 miles, 385 yards. (Note that 1 mile = 5280 feet = 1760 yards.)

Figure for 51 and 52

51. Men's record time: 2 hours, 8 minutes

52. Women's record time: 2 hours, $25\frac{1}{2}$ minutes

53. *Forager Honeybees* A forager honeybee spends about 3 weeks becoming accustomed to the immediate surroundings of its hive, and then spends the rest of its life collecting pollen and nectar. A forager's flight muscles can last only about 500 miles—after that the bee dies. The total number of miles T that a forager can fly in its lifetime L (in days) can be modeled by $T = m(L - 21)$, where m is the number of miles it flies each day. A hardworking forager honeybee can fly about 55 miles each day. About how long does it live?

54. *Pressure vs. Weight* The pressure P exerted on the floor by a person's shoe heel depends on the weight w of the person and the width of the heel H. The formula is

$$P \approx \frac{1.2w}{H^2},$$

where the weight is in pounds and the heel width is in inches. The heels on a man's shoes are 2 inches wide, and each exerts a pressure of 45 pounds per square inch. Estimate the man's weight.

55. Let a and b be real numbers, with $a \neq 0$. Find the solutions of $ax^2 + bx = 0$.

56. Let a be a nonzero real number. Find the solutions of $ax^2 - ax = 0$.

The Rectangular Coordinate System

The Rectangular Coordinate System   •   Ordered Pairs as Solutions of Equations   •   The Distance Between Two Points in the Plane

The Rectangular Coordinate System

Just as you can represent real numbers by points on the real number line, you can represent ordered pairs of real numbers by points in a plane. This plane is called a **rectangular coordinate system** or the **Cartesian plane,** after the French mathematician René Descartes (1596–1650).

Historical Note: Descartes introduced his analytical geometry in 1637. Most of the terminology we use comes from the Cartesian-coordinate system (linear, quadratic, etc.). Prior to this time there was not a distinct tie between algebra and geometry.

A rectangular coordinate system is formed by two real lines intersecting at a right angle, as shown in Figure 2.4(a). The horizontal number line is usually called the ***x*-axis,** and the vertical number line is usually called the **y-axis.** (The plural of axis is *axes.*) The point of intersection of the two axes is called the **origin,** and the axes separate the plane into four regions called **quadrants.**

FIGURE 2.4

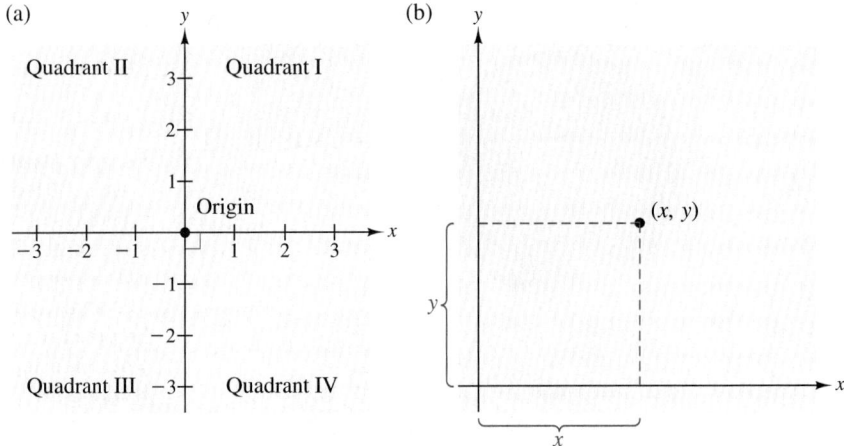

Each point in the plane corresponds to an **ordered pair** (x, y) of real numbers x and y, called the **coordinates** of the point. The first number (or ***x*-coordinate**) tells how far to the left or right the point is from the vertical axis, and the second number (or **y-coordinate**) tells how far up or down the point is from the horizontal axis, as shown in Figure 2.4(b).

NOTE   A positive x-coordinate implies that the point lies to the right of the vertical axis; a negative x-coordinate implies that the point lies to the left of the vertical axis; and an x-coordinate of zero implies that the point lies on the vertical axis. Similarly, a positive y-coordinate implies that the point lies above the horizontal axis, a negative y-coordinate implies that the point lies below the horizontal axis; and a y-coordinate of zero implies that the point lies on the horizontal axis.

Locating a given point in a plane is called **plotting** the point. Example 1 shows how this is done.

EXAMPLE 1   ■   **Plotting Points in a Rectangular Coordinate System**

Plot the points $(-2, 1)$, $(4, 0)$, $(3, -1)$, $(4, 3)$, $(0, 0)$, and $(-1, -3)$ in a rectangular coordinate system.

Solution

The point $(-2, 1)$ is two units to the *left* of the vertical axis and one unit *above* the horizontal axis. Similarly, the point $(4, 0)$ is four units to the *right* of the vertical axis and *on* the horizontal axis. (It is on the horizontal axis because its *y*-coordinate is zero.) The other four points can be plotted in a similar way, as shown in Figure 2.5.

FIGURE 2.5

Point out the sign of x and y in each of the four quadrants. Discuss questions similar to: If $(-a, b)$ is in quadrant IV, in which quadrant would we find $(a, -b)$?

Students tend to think that $(-a, b)$ must be in quadrant II. This discussion will help them realize that "$-a$" does not necessarily represent a negative value.

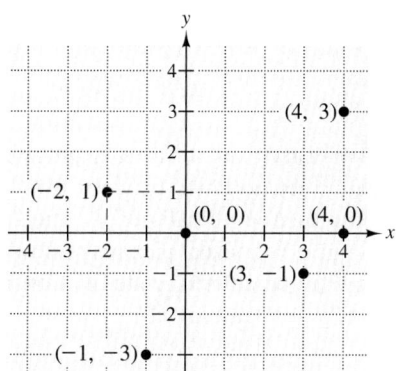

■

In Example 1, you were given the coordinates of several points and asked to plot the points in a rectangular coordinate system. Example 2 looks at the reverse problem. That is, you are given points in a rectangular coordinate system and are asked to determine their coordinates.

EXAMPLE 2   ■   **Finding Coordinates of Points**

Determine the coordinates for each of the points shown in Figure 2.6.

FIGURE 2.6

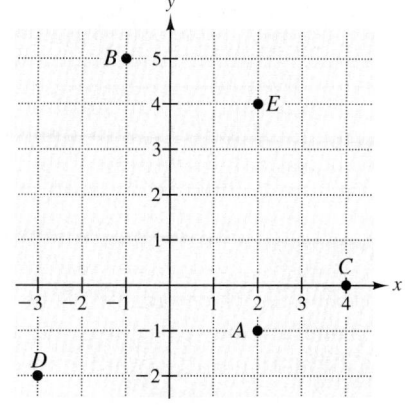

Solution

Point *A* lies two units to the *right* of the vertical axis and one unit *below* the horizontal axis. Therefore, point *A* must be given by the ordered pair $(2, -1)$. The coordinates of the other four points can be determined in a similar way; the results are summarized as follows.

Point	*Position*	*Coordinates*
A	2 units *right*, 1 unit *down*	$(2, -1)$
B	1 unit *left*, 5 units *up*	$(-1, 5)$
C	4 units *right*, 0 units *up*	$(4, 0)$
D	3 units *left*, 2 units *down*	$(-3, -2)$
E	2 units *right*, 4 units *up*	$(2, 4)$

■

Discuss how the rectangular coordinate system allows us to visualize relationships.

The primary value of a rectangular coordinate system is that it allows you to visualize relationships between two variables. Today, Descartes's ideas are commonly used in virtually every scientific and business-related field. Ordered pairs can give you visual pictures of how total sales are related to the price of an item, how temperature is related to time of day, how interest rates are related to months of a year, and so on.

EXAMPLE 3 ■ An Application of the Rectangular Coordinate System

The population of the state of California from 1975 through 1990 is listed in Table 2.1. Plot these points in a rectangular coordinate system. (Note that the population numbers given in the table are in thousands. Thus, the population in 1975 was 21,198 thousand, or 21,198,000.)

TABLE 2.1

Year	1975	1976	1977	1978	1979	1980	1981	1982
Population ($\times 1000$)	21,198	21,522	21,900	22,314	23,255	23,668	24,265	24,783

Year	1983	1984	1985	1986	1987	1988	1989	1990
Population ($\times 1000$)	25,308	26,358	26,981	26,997	27,653	28,314	28,607	29,839

Solution

Begin by choosing which variable will be plotted on the horizontal axis and which will be plotted on the vertical axis. For these data, it seems natural to plot the years on the horizontal axis (which means that the population must be plotted on the vertical axis). The points are shown in Figure 2.7. Note that the break in the *x*-axis indicates that the numbers between 0 and 1975 have been omitted. The break in the *y*-axis indicates that the numbers between 0 and 15,000 have been omitted.

FIGURE 2.7

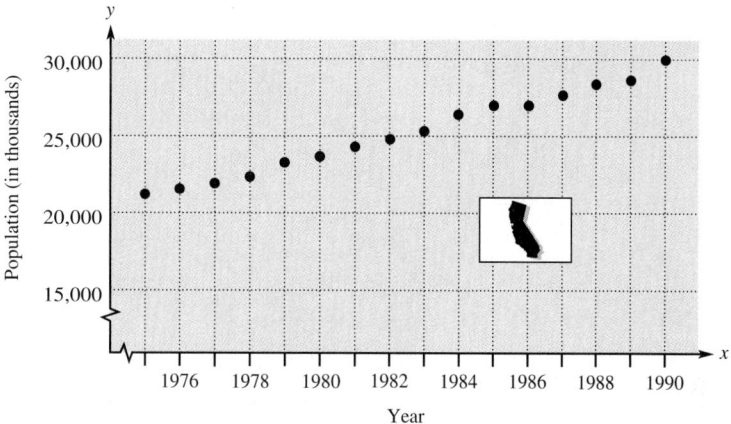

Ordered Pairs as Solutions of Equations

In Example 3, the relationship between the year and the population was given by a **table of values.** In mathematics, the relationship between the variables *x* and *y* is often given by an equation. From the equation, you can construct your own table of values. For instance, consider the equation

$$y = 3x + 2.$$

To construct a table of values for this equation, choose several *x*-values and then calculate the corresponding *y*-values. For example, if you choose $x = 1$, then the *y*-value is

$$y = 3(1) + 2 = 5.$$

The corresponding ordered pair $(x, y) = (1, 5)$ is called a **solution point** (or simply a **solution**) of the equation. Table 2.2 is a table of values (and their corresponding solution points) using *x*-values of $-3, -2, -1, 0, 1, 2,$ and 3.

TABLE 2.2

Choose x	Calculate y from $y = 3x + 2$	Solution Points
$x = -3$	$y = 3(-3) + 2 = -7$	$(-3, -7)$
$x = -2$	$y = 3(-2) + 2 = -4$	$(-2, -4)$
$x = -1$	$y = 3(-1) + 2 = -1$	$(-1, -1)$
$x = 0$	$y = 3(0) + 2 = 2$	$(0, 2)$
$x = 1$	$y = 3(1) + 2 = 5$	$(1, 5)$
$x = 2$	$y = 3(2) + 2 = 8$	$(2, 8)$
$x = 3$	$y = 3(3) + 2 = 11$	$(3, 11)$

Once you have constructed a table of values, you can get a visual idea of the relationship between the variables x and y by plotting the solution points in a rectangular coordinate system. For instance, the solution points shown in Table 2.2 are plotted in Figure 2.8.

FIGURE 2.8

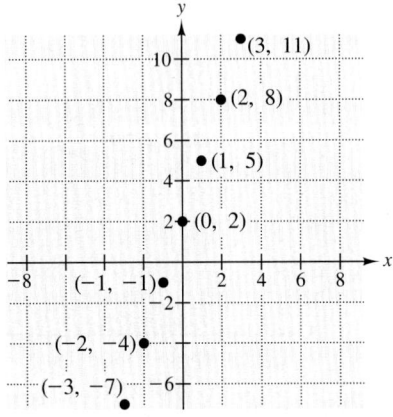

When making up a table of values for an equation, it is helpful to first solve the equation for y. For instance, the equation $5x + 3y = 4$ can be solved for y as follows.

$$5x + 3y = 4 \qquad\qquad \text{Given equation}$$

$$3y = -5x + 4 \qquad\qquad \text{Subtract } 5x \text{ from both sides}$$

$$y = -\frac{5}{3}x + \frac{4}{3} \qquad\qquad \text{Solve for } y$$

This procedure is demonstrated further in Example 4.

EXAMPLE 4   ■   Making a Table of Values

Make up a table of values showing five solution points for the equation

$$2x^2 - 3y = 5.$$

Then plot the solution points in the rectangular coordinate system. (Choose x-values of $-2, -1, 0, 1,$ and 2.)

Solution

Begin by solving the given equation for y.

$$2x^2 - 3y = 5$$

$$-3y = -2x^2 + 5$$

$$y = \frac{2}{3}x^2 - \frac{5}{3}$$

Now, using the equation $y = \frac{2}{3}x^2 - \frac{5}{3}$, construct the table of values, as shown in Table 2.3.

TABLE 2.3

Choose x	Calculate y from $y = \frac{2}{3}x^2 - \frac{5}{3}$	Solution Points
$x = -2$	$y = \frac{2}{3}(-2)^2 - \frac{5}{3} = \frac{8}{3} - \frac{5}{3} = 1$	$(-2, 1)$
$x = -1$	$y = \frac{2}{3}(-1)^2 - \frac{5}{3} = \frac{2}{3} - \frac{5}{3} = -1$	$(-1, -1)$
$x = 0$	$y = \frac{2}{3}(0)^2 - \frac{5}{3} = 0 - \frac{5}{3} = -\frac{5}{3}$	$\left(0, -\frac{5}{3}\right)$
$x = 1$	$y = \frac{2}{3}(1)^2 - \frac{5}{3} = \frac{2}{3} - \frac{5}{3} = -1$	$(1, -1)$
$x = 2$	$y = \frac{2}{3}(2)^2 - \frac{5}{3} = \frac{8}{3} - \frac{5}{3} = 1$	$(2, 1)$

Finally, from Table 2.3, plot the five solution points in the rectangular coordinate system, as shown in Figure 2.9.

FIGURE 2.9

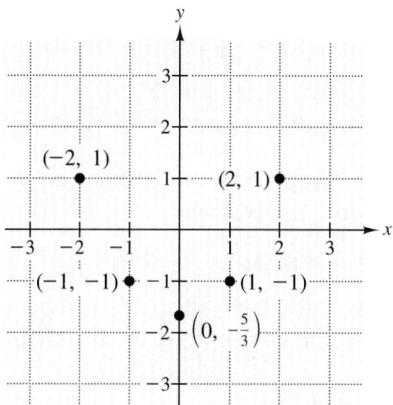

In the next example, you are given several ordered pairs and are asked to determine whether they are solutions of the given equation. To do this, you need to substitute the values of x and y into the equation. If the substitution produces a true equation, then the ordered pair (x, y) is a solution. Otherwise, the ordered pair is not a solution.

NOTE   When a substitution of (x, y) produces a true equation, the ordered pair (x, y) **satisfies** the equation.

EXAMPLE 5   ■   Verifying Solutions of an Equation

Determine which of the following ordered pairs are solutions of the equation $x^2 - 2y = 6$.

(a) $(2, 1)$ (b) $(0, -3)$ (c) $(-2, -5)$ (d) $\left(1, -\frac{5}{2}\right)$

Solution

(a) For the ordered pair $(x, y) = (2, 1)$, substitute $x = 2$ and $y = 1$ into the given equation.

$$x^2 - 2y = 6 \qquad\qquad \text{Given equation}$$
$$2^2 - 2(1) \overset{?}{=} 6 \qquad\qquad \text{Substitute } x = 2 \text{ and } y = 1$$
$$2 \neq 6 \qquad\qquad \text{Not a solution}$$

Because the substitution did not satisfy the given equation, you can conclude that the ordered pair $(2, 1)$ *is not* a solution of the given equation.

(b) The ordered pair $(0, -3)$ *is* a solution of the given equation because
$$0^2 - 2(-3) = 0 + 6 = 6.$$

(c) The ordered pair $(-2, -5)$ *is not* a solution of the given equation because
$$(-2)^2 - 2(-5) = 14 \neq 6.$$

(d) The ordered pair $\left(1, -\frac{5}{2}\right)$ *is* a solution of the given equation because
$$1^2 - 2\left(-\frac{5}{2}\right) = 1 - (-5) = 6. \qquad\qquad \blacksquare$$

The Distance Between Two Points in the Plane

You know from Section 1.1 that the distance d between two points a and b on the real number line is simply

$$d = |b - a|.$$

The same "absolute value rule" is used to find the distance between two points that lie on the same *vertical* or *horizontal* line, as shown in Example 6.

EXAMPLE 6   ■   Finding Horizontal and Vertical Distances

(a) Find the distance between the points $(2, -2)$ and $(2, 4)$.

(b) Find the distance between the points $(-3, -2)$ and $(2, -2)$.

Solution

(a) Because the x-coordinates are equal, you can visualize a vertical line through the points $(2, -2)$ and $(2, 4)$, as shown in Figure 2.10. The distance between these two points is given by the absolute value of the difference of their y-coordinates. That is,

$$\text{Vertical distance } = |4 - (-2)| = 6. \qquad\qquad \text{Subtract } y\text{-coordinates}$$

(b) Because the y-coordinates are equal, you can visualize a horizontal line through the points $(-3, -2)$ and $(2, -2)$, as shown in Figure 2.10. The distance between these two points is given by the absolute value of the difference of their x-coordinates. That is,

$$\text{Horizontal distance } = |2 - (-3)| = 5. \qquad\qquad \text{Subtract } x\text{-coordinates}$$

FIGURE 2.10

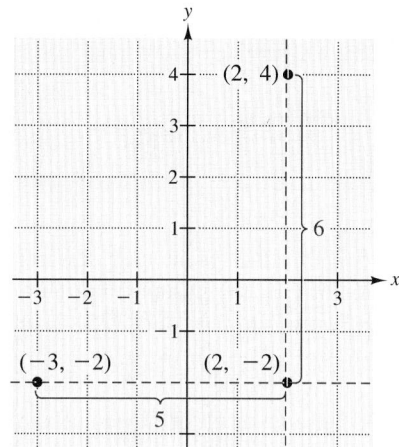

You might want to point out that the difference of y-coordinates can be taken in either order (see Solution (a)): $|4 - (-2)| = |-2 - 4|$.

FIGURE 2.11

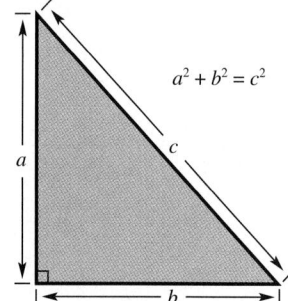

Encourage students to follow the development and to understand *why* the formula works. This understanding is critical to many areas in algebra and geometry.

FIGURE 2.12

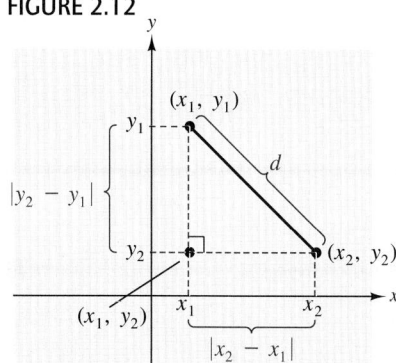

The technique applied in Example 6 can be used to develop a general formula for finding the distance between two points in the plane. This general formula will work for any two points, even if they do not lie on the same vertical or horizontal line. To develop the formula, you use the **Pythagorean Theorem,** which states that for a right triangle, the hypotenuse c and sides a and b are related by the formula $a^2 + b^2 = c^2$, as shown in Figure 2.11. (The converse is also true. That is, if $a^2 + b^2 = c^2$, then the triangle is a right triangle.)

To develop a general formula for the distance between two points (**Distance Formula**), let (x_1, y_1) and (x_2, y_2) represent two points in the plane (that do not lie on the same horizontal or vertical line). With these two points, a right triangle can be formed, as shown in Figure 2.12. Note that the third vertex of the triangle is (x_1, y_2). Because (x_1, y_1) and (x_1, y_2) lie on the same vertical line, the length of the vertical side of the triangle is $|y_2 - y_1|$. Similarly, the length of the horizontal side is $|x_2 - x_1|$. Thus, by the Pythagorean Theorem, the square of the distance between (x_1, y_1) and (x_2, y_2) is

$$d^2 = |x_2 - x_1|^2 + |y_2 - y_1|^2.$$

Because the distance d must be positive, you can choose the positive square root and write

$$d = \sqrt{|x_2 - x_1|^2 + |y_2 - y_1|^2}.$$

Finally, replacing $|x_2 - x_1|^2$ and $|y_2 - y_1|^2$ by the equivalent expressions $(x_2 - x_1)^2$ and $(y_2 - y_1)^2$ gives you the formula for the distance between two points in a rectangular coordinate plane.

The Distance Formula	The distance d between two points (x_1, y_1) and (x_2, y_2) in a coordinate plane is $$d = \sqrt{(x_2 - x_1)^2 + (y_2 - y_1)^2}.$$

Note that for the special case in which the two points lie on the same vertical or horizontal line, the Distance Formula still works. For instance, applying the Distance Formula to the points $(2, -2)$ and $(2, 4)$ produces

$$d = \sqrt{(2 - 2)^2 + [4 - (-2)]^2} = \sqrt{6^2} = 6,$$

which is the same result obtained in Example 6.

EXAMPLE 7   ■   **Finding the Distance Between Two Points**

Find the distance between the points $(-1, 2)$ and $(2, 4)$.

Solution

Let $(x_1, y_1) = (-1, 2)$ and $(x_2, y_2) = (2, 4)$, and apply the Distance Formula to obtain

$$d = \sqrt{[2 - (-1)]^2 + (4 - 2)^2}$$
$$= \sqrt{3^2 + 2^2}$$
$$= \sqrt{13}$$
$$\approx 3.61.$$

(See Figure 2.13.)

FIGURE 2.13

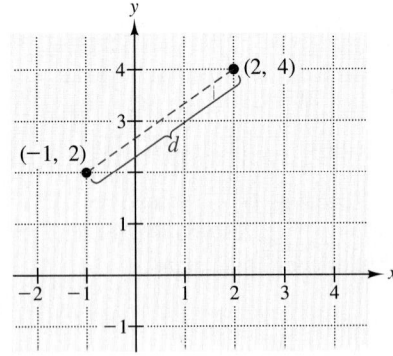

■

EXAMPLE 8   ■   An Application of the Distance Formula

Show that the points $(1, 2)$, $(3, 1)$, and $(4, 3)$ are vertices of a right triangle.

Solution

The three points are plotted in Figure 2.14. Using the Distance Formula, you can find the lengths of the three sides of the triangle.

$$d_1 = \sqrt{(3 - 1)^2 + (1 - 2)^2} = \sqrt{4 + 1} = \sqrt{5}$$

$$d_2 = \sqrt{(4 - 3)^2 + (3 - 1)^2} = \sqrt{1 + 4} = \sqrt{5}$$

$$d_3 = \sqrt{(4 - 1)^2 + (3 - 2)^2} = \sqrt{9 + 1} = \sqrt{10}$$

Because $d_1^2 + d_2^2 = 5 + 5 = 10 = d_3^2$, you can conclude from the Pythagorean Theorem that the triangle is a right triangle.

FIGURE 2.14

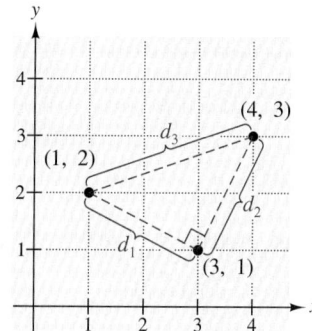

                                                                                ■

DISCUSSION PROBLEM   ■   Finding the Midpoint of a Line Segment

Plot the points $A = (-1, -3)$ and $B = (5, 2)$ in a rectangular coordinate system, and draw the line segment connecting the two points.

(a) Try to find the coordinates of the point C that is the midpoint of the line segment.

(b) Give a verbal explanation of how you arrived at your choice of coordinates for the point C.

(c) Write a paragraph describing how you could verify that the point C is the midpoint of the line segment. (Why is it not sufficient to show that the distance between A and C is the same as the distance between C and B?)

(d) Try to find a general formula for the coordinates of the midpoint of the line segment connecting the points (x_1, y_1) and (x_2, y_2).   ■

Warm-Up

The following warm-up exercises involve skills that were covered in earlier sections. You will use these skills in the exercise set for this section.

In Exercises 1–6, plot each real number on the real number line and place the correct inequality symbol ($<$ or $>$) between the real numbers.

1. -6 _____ 2 **2.** $\frac{11}{3}$ _____ 2 **3.** -3 _____ -7

4. -5 _____ 0 **5.** $\frac{13}{3}$ _____ $-\frac{3}{2}$ **6.** $\frac{3}{5}$ _____ $\frac{13}{16}$

In Exercises 7–10, evaluate the specified quantity.

7. $5 - |-2|$ **8.** $-15 - 3(4 - 18)$

9. $\dfrac{3 - 2(5 - 20)}{4}$ **10.** $\dfrac{|3 - 12|}{3}$

2.3   EXERCISES

In Exercises 1–6, plot the points in a rectangular coordinate system.

1. $(4, 3), (-5, 3), (3, -5)$ **2.** $(-2, 5), (-2, -5), (3, 5)$ **3.** $(-8, -2), (6, -2), (6, 5)$

4. $(0, 4), (0, 0), (3, 0)$ **5.** $\left(\frac{5}{2}, -2\right), \left(-2, \frac{1}{4}\right), \left(\frac{3}{2}, -\frac{7}{2}\right)$ **6.** $\left(-\frac{2}{3}, 3\right), \left(\frac{1}{4}, -\frac{5}{4}\right), \left(-5, -\frac{7}{4}\right)$

In Exercises 7 and 8, without plotting the points, determine the quadrant in which each point is located.

7. (a) $\left(3, -\frac{5}{8}\right)$ (b) $(-6.2, 8.05)$ **8.** (a) $(200, 1365.6)$ (b) $\left(-\frac{5}{11}, -\frac{3}{8}\right)$

In Exercises 9 and 10, approximate the coordinates of the plotted points.

9.
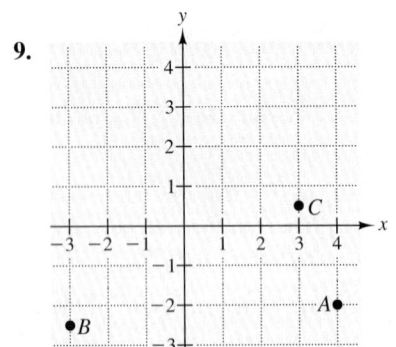

10.
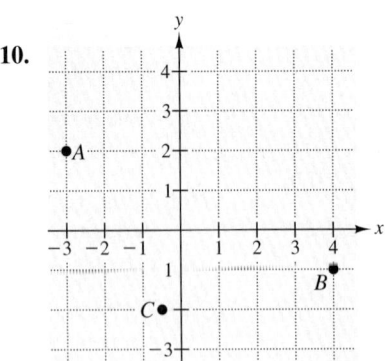

In Exercises 11–16, plot the points and connect them with line segments in such a way as to form the specified figure. (A **rhombus** is a parallelogram whose sides are all of the same length.)

11. *Triangle:* $(-1, 2), (2, 0), (3, 5)$

12. *Triangle:* $(1, 4), (0, -1), (5, 9)$

13. *Square:* $(0, 6), (3, 3), (0, 0), (-3, 3)$

14. *Rectangle:* $(7, 0), (9, 1), (4, 6), (6, 7)$

15. *Parallelogram:* $(4, 0), (6, -2), (0, -4), (-2, -2)$

16. *Rhombus:* $(-3, -3), (-2, -1), (-1, -2), (0, 0)$

In Exercises 17–20, (a) determine the quadrant or quadrants in which the point may be located in order to satisfy the given requirement. (b) Explain your reasoning.

17. $(-3, y)$, y is any real number

18. $(x, 5)$, x is any real number

19. (x, y), $xy > 0$

20. (x, y), $x > 0$, $y < 0$

In Exercises 21–24, find the coordinates of the indicated point.

21. The point is located three units to the right of the y-axis and four units below the x-axis.

22. The point is located two units to the left of the y-axis and five units above the x-axis.

23. The coordinates of the point are equal, and it is located in the third quadrant 10 units to the left of the y-axis.

24. The coordinates of the point are equal in magnitude and opposite in sign, and it is located seven units to the right of the y-axis.

In Exercises 25–28, plot the points whose coordinates are given in the table.

25. *Exam Scores* The table gives the time x in hours invested in concentrated study for five different algebra exams and the resulting score y.

x	5	2	3	6.5	4
y	81	71	88	92	86

26. *Fuel Efficiency* The table gives the speed x of a car in miles per hour and the approximate fuel efficiency y in miles per gallon.

x	50	55	60	65	70
y	28	26.4	24.8	23.4	22

27. *Average Temperature* The normal temperature y (in degrees Fahrenheit) for Duluth, Minnesota, for each month of the year is given in the table. (*Source: PC USA*) The months are numbered 1 through 12, with 1 corresponding to January.

x	y	x	y
1	6.3	7	65.3
2	12.0	8	63.2
3	22.9	9	54.0
4	38.3	10	44.2
5	50.3	11	28.2
6	59.4	12	13.8

28. *Net Earnings per Share of Stock* The net earnings per share of common stock in Procter & Gamble Company for the years 1988 through 1992 are given in the table. The time in years is given by x. (*Source: 1992 Annual Report of Procter & Gamble Company*)

x	y
1988	$1.49
1989	$1.78
1990	$2.25
1991	$2.46
1992	$2.62

In Exercises 29–34, determine whether the ordered pairs are solution points to the equation.

29. $y = 3x + 8$
  (a) $(3, 17)$ (b) $(-1, 10)$
  (c) $(0, 0)$ (d) $(-2, 2)$

30. $y = 10x - 7$
  (a) $(2, 10)$ (b) $(-2, -27)$
  (c) $(5, 43)$ (d) $(1, 5)$

31. $4y - 2x + 1 = 0$
  (a) $(0, 0)$ (b) $\left(\frac{1}{2}, 0\right)$
  (c) $\left(-3, -\frac{7}{4}\right)$ (d) $\left(1, -\frac{3}{4}\right)$

32. $5x - 2y + 50 = 0$
  (a) $(-10, 0)$ (b) $(-5, 5)$
  (c) $(0, 25)$ (d) $(20, -2)$

33. $x^2 - 3x + y = 0$
  (a) $(1, 2)$ (b) $(-2, 10)$
  (c) $(-3, -18)$ (d) $(0, 0)$

34. $y^2 + x^2 = 16$
  (a) $(4, 0)$ (b) $(0, -4)$
  (c) $(2, 2)$ (d) $(3, 7)$

In Exercises 35–38, complete the table by finding the indicated values of y.

35.

x	-2	0	2	4	6
$y = 5x - 1$					

36.

x	-2	0	2	4	6
$y = \frac{3}{4}x + 2$					

37.

x	-2	-1	0	1	2
$y = x^2 - x + 3$					

38.

x	-6	-3	0	$\frac{3}{4}$	10
$y = \left|\frac{4}{3}x - \frac{1}{3}\right|$					

39. *Cost* The cost y for producing x units is given by $y = 28x + 3000$. Complete the table to determine the costs for producing the specified numbers of units.

x	100	150	200	250	300
$y = 28x + 3000$					

40. *Hourly Wages* When an employee produces x units per hour, the wage y paid per hour is given by $y = 0.75x + 8$. Complete the table to determine the hourly wages for producing the specified numbers of units.

x	2	4	8	10	20
$y = 0.75x + 8$					

In Exercises 41–46, plot the points and find the distance between them.

41. $(3, -2), (3, 5)$

42. $(-2, 8), (-2, 1)$

43. $(3, 2), (10, 2)$

44. $(-120, -2), (130, -2)$

45. $\left(\frac{3}{4}, 1\right), \left(\frac{3}{4}, -10\right)$

46. $\left(\frac{1}{2}, \frac{7}{8}\right), \left(\frac{11}{2}, \frac{7}{8}\right)$

In Exercises 47–50, find (a) the missing coordinates of the one vertex of the right triangle, (b) the lengths of the vertical and horizontal sides of the triangle, (c) the length of the hypotenuse, and (d) the distance between the two given points.

47. $(10, 8), (2, 2)$

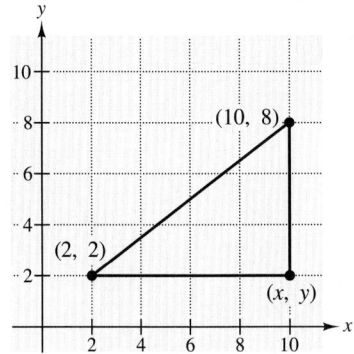

48. $(-2, 1), (10, 6)$

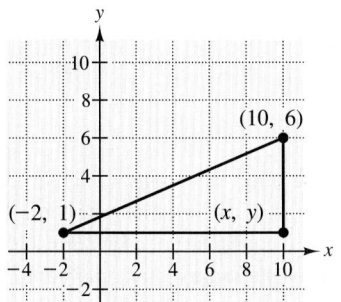

49. $(-3, -4), (4, 4)$

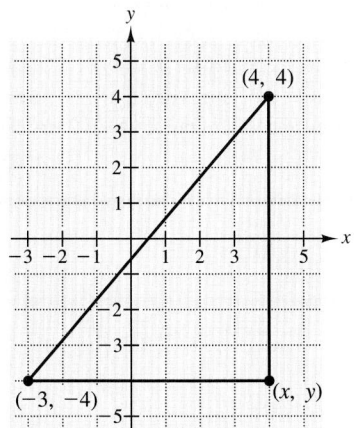

50. $(0, 6), (6, -2)$

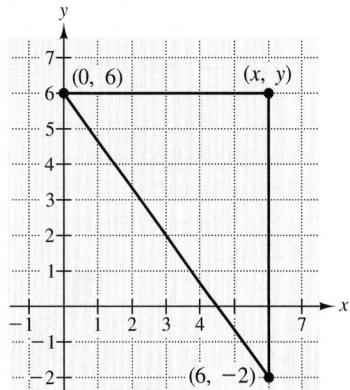

In Exercises 51–58, find the distance between the two points. If appropriate, round the result to two decimal places.

51. $(1, 3), (5, 6)$ **52.** $(3, 10), (15, 5)$

53. $(0, 0), (12, -9)$ **54.** $(-5, 0), (3, 15)$

55. $(-2, -3), (4, 2)$ **56.** $(-5, 4), (10, -3)$

57. $(1, 3), (3, -2)$ **58.** $\left(\frac{1}{2}, 1\right), \left(\frac{3}{2}, 2\right)$

In Exercises 59–62, determine whether the points are vertices of a right triangle.

59. $(2, 3), (2, 6), (6, 3)$

60. $(2, 4), (-1, 6), (-3, 1)$

61. $(8, 3), (5, 2), (1, 9)$

62. $(2, 4), (1, 1), (7, -1)$

63. *Housing Construction* A house is 30 feet wide and the ridge of the roof is 7 feet above the top of the walls (see figure). Find the length of the rafters if they overhang the edge of the walls by 2 feet.

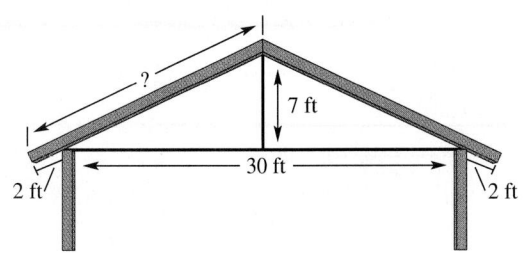

64. ***The Spider and the Fly*** A spider and a fly are on opposite walls of a rectangular room. The spider asks the fly if it can come over and "visit." The fly believes that the shortest distance, walking on one of the room's surfaces, is 42 feet. (That is the distance if the spider walks straight up the wall, over the ceiling, and straight down the opposite wall.) So the fly agrees to the visit, *provided* the spider can find a path that is shorter than 42 feet. Use the "unfolded" room and the Distance Formula to explain the fly's fatal miscalculation.

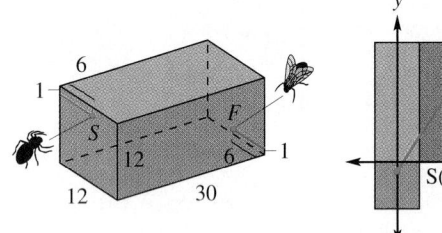

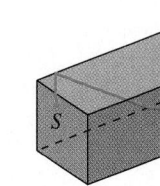

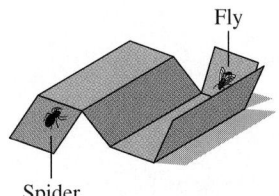

In Exercises 65–68, plot the given points and the midpoint of the line segment joining the given points. The coordinates of the midpoint of the line segment joining the points (x_1, y_1) and (x_2, y_2) are given by the following.

$$\text{Midpoint} = \left(\frac{x_1 + x_2}{2}, \frac{y_1 + y_2}{2} \right)$$

65. $(-2, 0), (4, 8)$

66. $(-3, -2), (7, 2)$

67. $(1, 6), (6, 3)$

68. $(2, 7), (9, -1)$

Shifting a Graph In Exercises 69 and 70, the figure is shifted to a new location in the plane. Find the coordinates of the vertices of the figure in its *new* location.

69.

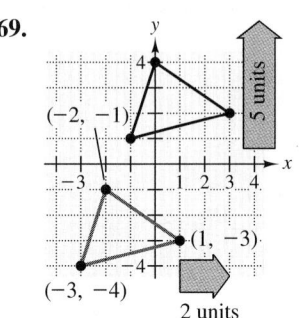

70. $(-5, 2)$

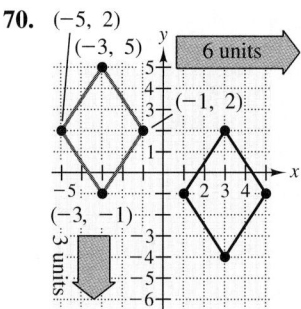

71. ***Exploratory Exercise*** Plot the points $(2, 1)$, $(-3, 5)$, and $(7, -3)$ in a rectangular coordinate system. Then change the sign of the x-coordinate of each point and plot the three new points in the same rectangular coordinate system. What inference can you make about the location of a point when the sign of the x-coordinate is changed?

72. ***Exploratory Exercise*** Plot the points $(2, 1)$, $(-3, 5)$, and $(7, -3)$ in a rectangular coordinate system. Then change the sign of the y-coordinate of each point and plot the three new points in the same rectangular coordinate system. What inference can you make about the location of a point when the sign of the y-coordinate is changed?

2

Mid-Chapter Quiz

Take this quiz as you would take a quiz in class. After you are done, check your work against the answers in the back of the book.

In Exercises 1–8, solve the equation.

1. $24 - 2x = x$

2. $3(x - 4) + 4 = -2(x - 1)$

3. $\dfrac{5x}{6} - \dfrac{2}{3} = 1$

4. $\dfrac{t}{4} - \dfrac{t}{6} = 1$

5. $(s - 16)(s + 15) = 0$

6. $y(y + 6) = 72$

7. $4x^2 - 4x + 1 = 0$

8. $3t^2 - 4t + 6 = 10$

9. Solve for C

Temperature Conversion: $F = \frac{9}{5}C + 32$

10. Solve for l

Perimeter of a Rectangle: $P = 2(l + w)$

11. The surface area of a cylinder (see figure) is given by $S = 2\pi rh + 2\pi r^2$.
(a) Solve for h.
(b) Find h is $S = 105$ square inches and $r = 3$ inches.

Figure for 11

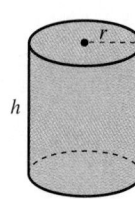

12. An object is thrown upward from a height of 64 feet with an initial velocity of 48 feet per second. Find the time t for the object to reach the ground by solving the following equation.

$$-16t^2 + 48t + 64 = 0$$

13. Two people can complete a task in t hours, where t must satisfy the equation

$$\frac{t}{12} + \frac{t}{20} = 1.$$

Find the required time t.

14. (a) Determine the quadrant(s) in which the point (x, y) is located if $xy < 0$.
(b) Explain your reasoning.

In Exercises 15 and 16, use the points $(-3, 0)$ and $(7, 3)$.

15. Plot the points in a rectangular coordinate system.

16. Find the distance between the points. Round your result to two decimal places.

17. Determine whether the ordered paris are solution points to the equation $5x - 3y + 10 = 0$.

(a) $(-2, 0)$ (b) $(0, 3)$ (c) $(1, -2)$ (d) $(-5, -5)$

Graphs of Equations

The Graph of an Equation • *Intercepts: An Aid to Sketching Graphs* • *The Connection Between Solutions and x-Intercepts* • *Applications*

You might begin this section by noting that when we *graph* equations involving two variables, we are looking for the set of all solutions—as well as for trends and a visual representation of the relationship.

The Graph of an Equation

The set of all solutions of an equation is called its **graph.** In Section 2.3, we took the first step in sketching the graph of an equation. That is, we constructed a table of values and plotted the corresponding solution points in a rectangular coordinate system. In this section, we will take the process one step further by connecting the solution points with a smooth curve or line.

To see how this works, make a sketch of the graph of the equation

$$y = -2x + 3.$$

To begin, construct a table of values, as shown in Table 2.4.

TABLE 2.4

x	-3	-2	-1	0	1	2	3
$y = -2x + 3$	9	7	5	3	1	-1	-3
Solution Points	$(-3, 9)$	$(-2, 7)$	$(-1, 5)$	$(0, 3)$	$(1, 1)$	$(2, -1)$	$(3, -3)$

Next, from the table of values, plot the solution points in a rectangular coordinate system, as shown in Figure 2.15(a). The seven points plotted in Figure 2.15(a) represent only a few of the points on the graph of the equation. To complete the sketch, try to determine whether the representative points form a pattern. In this case, all seven points appear to lie on a line, so you can complete the sketch by drawing a straight line through the points, as shown in Figure 2.15(b).

FIGURE 2.15

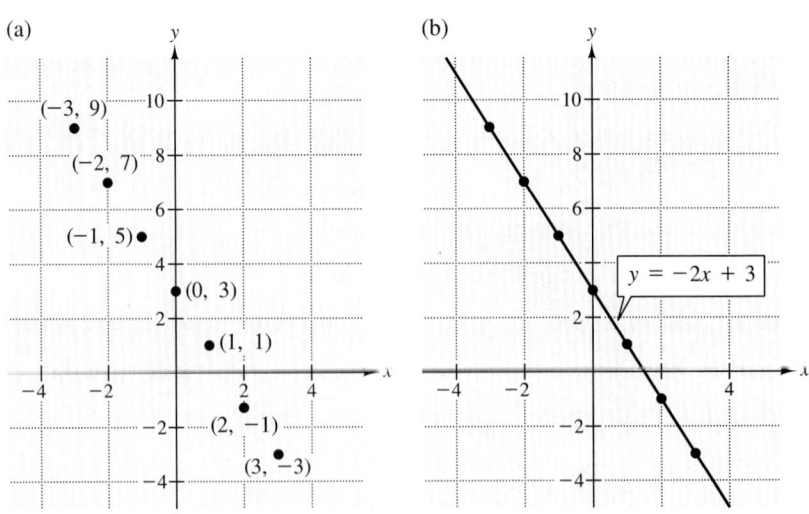

(a)   (b)

This method of sketching the graph of an equation is called the **point-plotting method.** The basic steps of the method are summarized as follows.

The Point-Plotting Method for Sketching a Graph	To sketch the graph of an equation by point plotting, use the following steps.
	1. If possible, rewrite the equation by isolating one of the variables.
	2. Make a table of values showing several solution points.
	3. Plot these points in a rectangular coordinate system.
	4. Connect the points with a smooth curve or line.

EXAMPLE 1 ■ **Sketching the Graph of an Equation**

Sketch the graph of the equation $3x - y = 2$.

Solution

Begin by rewriting the equation so that y is isolated on the left.

$$y = 3x - 2$$

Next, make a table of values, as shown in Table 2.5. (The choice of x-values to use in the table is somewhat arbitrary. However, the more x-values you choose, the easier it is to recognize a pattern.)

TABLE 2.5

x	-2	-1	0	1	2	3
$y = 3x - 2$	-8	-5	-2	1	4	7
Solution Points	$(-2, -8)$	$(-1, -5)$	$(0, -2)$	$(1, 1)$	$(2, 4)$	$(3, 7)$

Now, plot the six solution points. It appears that all six points lie on a line, so you can complete the sketch by drawing a line through these points, as shown in Figure 2.16.

FIGURE 2.16

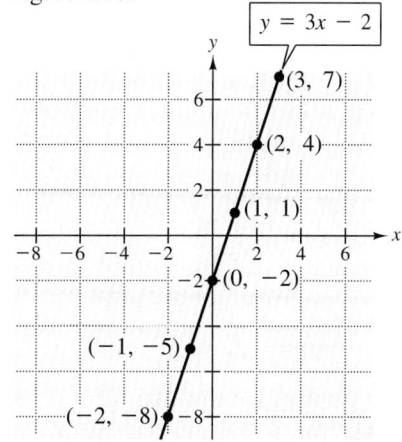

As mentioned in Example 1, the choice of x-values is arbitrary. Emphasize to students that if they chose different values for x, they would still generate the same line. Have them re-graph $y = 3x - 2$ by putting $x = 4, 5, 6, 7, 8$ in their tables. After they have finished, ask them to compare their graphs to the graph in Figure 2.16.

■

The equation in Example 1 is an example of a **linear equation** in two variables—it is of first degree in both variables and its graph is a straight line. (You will study this type of equation in Sections 3.5 and 3.6.) The graph of a nonlinear equation is not a straight line. The next two examples show how you can use the point-plotting method to sketch the graphs of nonlinear equations.

EXAMPLE 2 ■ Sketching the Graph of a Nonlinear Equation

Sketch the graph of the equation $-x^2 + 2x + y = 0$.

Solution

Begin by rewriting the equation so that y is isolated on the left.

$$y = x^2 - 2x$$

Next, make a table of values, as shown in Table 2.6. Watch out for the signs of the numbers when making such a table. For instance, when $x = -2$, the value of y is

$$y = (-2)^2 - 2(-2) = 4 + 4 = 8.$$

TABLE 2.6

x	-2	-1	0	1	2	3	4
$y = x^2 - 2x$	8	3	0	-1	0	3	8
Solution Points	$(-2, 8)$	$(-1, 3)$	$(0, 0)$	$(1, -1)$	$(2, 0)$	$(3, 3)$	$(4, 8)$

Plot the seven solution points. Finally, connect the points with a smooth curve, as shown in Figure 2.17.

FIGURE 2.17

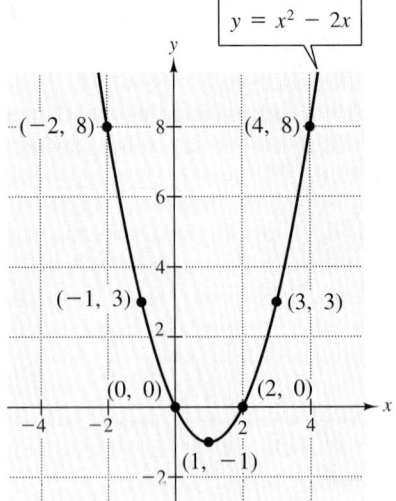

The graph of the equation given in Example 2 is called a **parabola.** You will study this type of graph in detail in Section 7.5.

EXAMPLE 3   ■   The Graph of an Equation Involving an Absolute Value

Sketch the graph of the equation $y = |x - 2|$.

Solution

This equation is already written in a form in which y is isolated on the left, so you can begin by making a table of values, as shown in Table 2.7. Be sure that you understand how the absolute value is evaluated. For instance, when $x = -2$, the value of y is

$$y = |-2 - 2| = |-4| = 4.$$

Similarly, when $x = 3$, the value of y is

$$y = |3 - 2| = |1| = 1.$$

TABLE 2.7

x	-2	-1	0	1	2	3	4	5
$y = |x - 2|$	4	3	2	1	0	1	2	3
Solution Points	$(-2, 4)$	$(-1, 3)$	$(0, 2)$	$(1, 1)$	$(2, 0)$	$(3, 1)$	$(4, 2)$	$(5, 3)$

Next, plot the eight solution points. It appears that the points lie in a "V-shaped" pattern, with the point $(2, 0)$ lying at the bottom of the "V." Following this pattern, connect the points to form the graph shown in Figure 2.18.

FIGURE 2.18

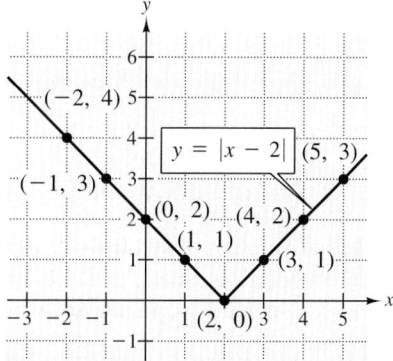

EXAMPLE 4   ■   The Graph of an Equation with Only One Variable

Sketch the graph of the equation $y = 3$.

Solution

When you make a table of values for this graph, you will notice that the value of y is 3, regardless of the value of x, as shown in Table 2.8 on the next page.

TABLE 2.8

Remind students that a graph is a visual representation of a relationship. In this case, at every point y is 3. See Table 2.8 and Figure 2.19.

x	-2	-1	0	1	2	3	4	5
$y = 3$	3	3	3	3	3	3	3	3
Solution Points	$(-2, 3)$	$(-1, 3)$	$(0, 3)$	$(1, 3)$	$(2, 3)$	$(3, 3)$	$(4, 3)$	$(5, 3)$

All eight solution points line on a horizontal line. You can sketch the graph of the equation by connecting the points with a horizontal line, as shown in Figure 2.19.

FIGURE 2.19

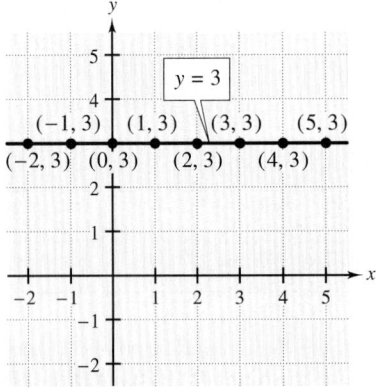

NOTE The graph of an equation of the form $y = a$ is a horizontal line, and the graph of an equation of the form $x = a$ is a vertical line. (More is said about this in Section 3.6.)

Intercepts: An Aid to Sketching Graphs

Definition of intercepts: The x-intercept can also be defined as only the x-value of the point at which the graph crosses the x-axis.

Two types of solution points that are especially useful are those having zero as either the x- or y-coordinate. Such points are called **intercepts** because they are the points at which the graph intersects the x- or y-axis.

Definition of Intercepts	The point $(a, 0)$ is called an **x-intercept** of the graph of an equation if it is a solution point of the equation. To find the x-intercepts, let y be zero and solve the equation for x.
	The point $(0, b)$ is called a **y-intercept** of the graph of an equation if it is a solution point of the equation. To find the y-intercepts, let x be zero and solve the equation for y.

EXAMPLE 5   ■   Finding the Intercepts of a Graph

Find the intercepts and sketch the graph of $y = 2x - 3$.

Illustrate the method used to find intercepts (0, ?) and (?, 0). Note the intercepts are in the table and labeled on the graph. (Be aware that students sometimes take the x-coordinate of the x-intercept and the y-coordinate of the y-intercept and incorrectly use them as coordinates of a single point.)

Solution

To determine whether the graph has any x-intercepts, let y be zero and solve the resulting equation for x.

$y = 2x - 3$	Given equation
$0 = 2x - 3$	Let $y = 0$
$\dfrac{3}{2} = x$	Solve equation for x

Therefore, the graph has one x-intercept, which occurs at the point $\left(\frac{3}{2}, 0\right)$.

To determine whether the graph has any y-intercepts, let x be zero and solve the resulting equation for y.

$y = 2x - 3$	Given equation
$y = 2(0) - 3$	Let $x = 0$
$y = -3$	Solve equation for y

Therefore, the graph has one y-intercept, which occurs at the point $(0, -3)$.

To sketch the graph of the equation, make a table of values, as shown in Table 2.9. (Notice that the two intercepts are included in the table.) Finally, using the solution points in the table, sketch the graph of the equation, as shown in Figure 2.20.

TABLE 2.9

x	-1	0	1	$\frac{3}{2}$	2	3	4
$y = 2x - 3$	-5	-3	-1	0	1	3	5
Solution Points	$(-1, -5)$	$(0, -3)$	$(1, -1)$	$\left(\frac{3}{2}, 0\right)$	$(2, 1)$	$(3, 3)$	$(4, 5)$

FIGURE 2.20

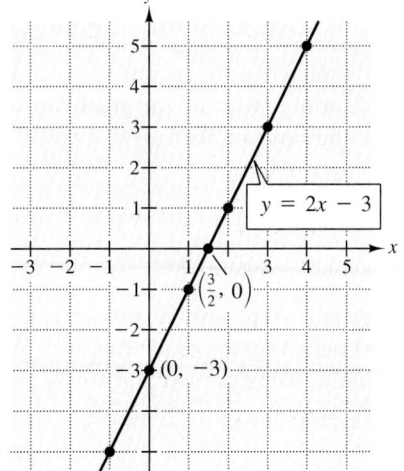

■

It is possible for a graph to have no intercepts or several intercepts. For instance, consider the three graphs in Figure 2.21.

FIGURE 2.21

No intercepts   One y-intercept   One y-intercept
                              No x-intercepts   Two x-intercepts

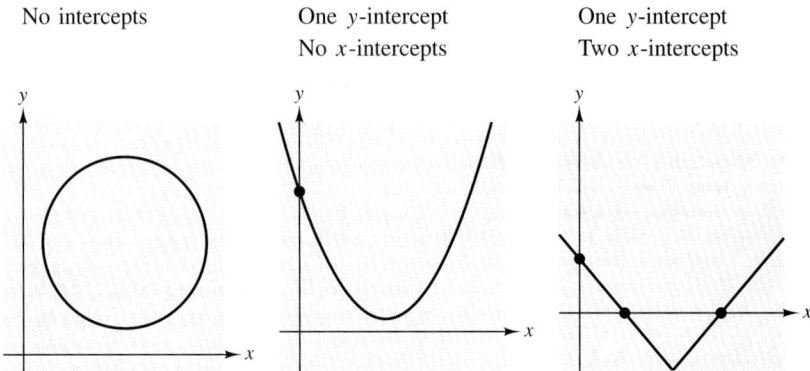

EXAMPLE 6 ■   **A Graph That Has Two x-Intercepts**

Find the intercepts and sketch the graph of $y = x^2 - 5x + 4$.

Solution

To determine whether the graph has any x-intercepts, let y be zero and solve the resulting equation for x.

$y = x^2 - 5x + 4$	Given equation
$0 = x^2 - 5x + 4$	Let $y = 0$
$0 = (x - 1)(x - 4)$	Factor

By setting each of these factors equal to zero, you can conclude that the equation $0 = x^2 - 5x + 4$ has two solutions: 1 and 4. Therefore, it follows that the graph has two x-intercepts, the points $(1, 0)$ and $(4, 0)$.

To determine whether the graph has any y-intercepts, let x be zero and solve the resulting equation for y.

$y = x^2 - 5x + 4$	Given equation
$y = 0^2 - 5(0) + 4$	Let $x = 0$
$y = 4$	Solve equation for y

Therefore, the graph has one y-intercept, which occurs at the point $(0, 4)$.

To sketch the graph of the equation, make a table of values, as shown in Table 2.10. Finally, using the solution points given in the table, sketch the graph of the equation, as shown in Figure 2.22.

TABLE 2.10

x	-1	0	1	2	3	4	5
$y = x^2 - 5x + 4$	10	4	0	-2	-2	0	4
Solution Points	$(-1, 10)$	$(0, 4)$	$(1, 0)$	$(2, -2)$	$(3, -2)$	$(4, 0)$	$(5, 4)$

FIGURE 2.22

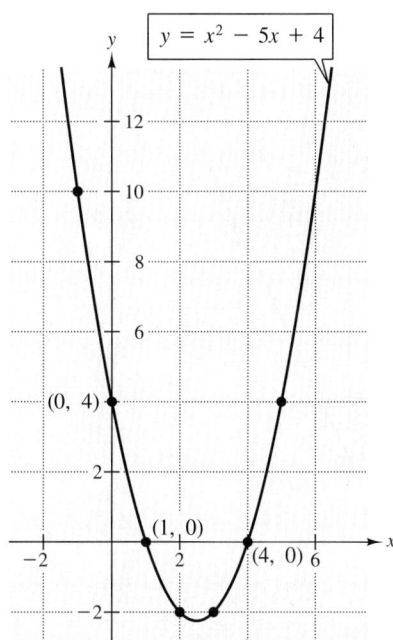

The Connection Between Solutions and *x*-Intercepts

You already know how to solve a linear equation in one variable. For instance, you could solve the equation $2x - 8 = 0$ as follows.

$2x - 8 = 0$	Given equation
$2x = 8$	Add 8 to both sides
$x = 4$	Divide both sides by 2

In Section 2.1, we emphasized the importance of checking this solution in the original equation. *You should continue to do this.* However, there is also a way to check a solution *graphically.*

To see how to do this, consider the line $y = 2x - 8$. The x-intercept of this line is the value of x when $y = 0$. In other words, the x-intercept of the line $y = 2x - 8$ is the solution of the equation $0 = 2x - 8$.

One-Variable Equation	*Two-Variable Equation*
The *solution* of $2x - 8 = 0$ is 4.	The *x-intercept* of $y = 2x - 8$ is 4.

Thus, to check graphically the solution of $2x - 8 = 0$, you can sketch the graph of $y = 2x - 8$ and observe the x-intercept, as shown in Figure 2.23.

FIGURE 2.23

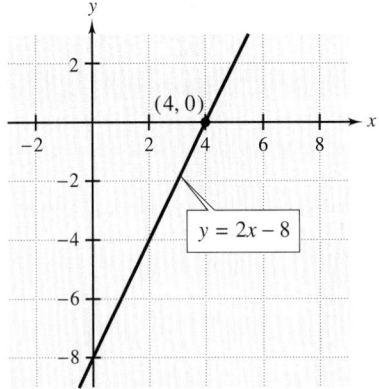

Using a Graphic Check of a Solution

The solution of an equation involving one variable x can be checked graphically with the following steps.

1. Write the equation so that all nonzero terms are on one side and zero is on the other side.

2. Sketch the graph of $y = $ (nonzero terms).

3. The solution of the one-variable equation is the x-intercept of the graph of the two-variable equation.

This connection between algebra and geometry represents one of the most wonderful discoveries ever made in mathematics. Before René Descartes introduced the coordinate plane in 1637, mathematicians had no easy way of "seeing" a solution of an algebraic equation.

EXAMPLE 7 ■ Using a Graphic Check of a Solution

Solve the equation $3x + 1 = -8$. Check your solution algebraically and graphically.

Solution

$$3x + 1 = -8 \qquad \text{Given equation}$$
$$3x = -9 \qquad \text{Subtract 1 from both sides}$$
$$x = -3 \qquad \text{Divide both sides by 3}$$

Graphic Check

Rewrite the equation so that all nonzero terms are on the left side.

$$3x + 1 = -8 \qquad \text{Given equation}$$
$$3x + 9 = 0 \qquad \text{Add 8 to both sides}$$

Now, sketch the graph of $y = 3x + 9$, as shown in Figure 2.24. Notice that the x-intercept is -3 (where $3x + 9 = 0$), which checks with the algebraic solution.

FIGURE 2.24

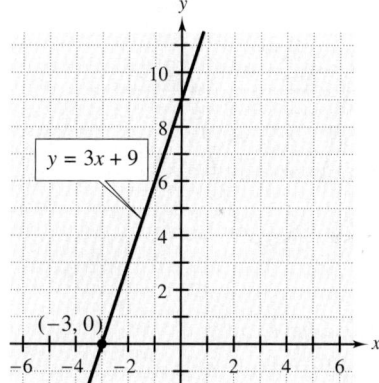

Algebraic Check

Substitute -3 for x in the original equation.

$$3(-3) + 1 \overset{?}{=} -8 \qquad \text{Substitute } -3 \text{ for } x$$
$$-9 + 1 \overset{?}{=} -8 \qquad \text{Simplify}$$
$$-8 = -8 \qquad \text{Solution checks}$$ ∎

EXAMPLE 8   ■   Using a Graphic Check of a Solution

Solve the equation $x^2 = 4$. Check your solution algebraically and graphically.

Solution

$$x^2 - 4 = 0 \qquad \text{Rewrite given equation}$$
$$(x + 2)(x - 2) = 0 \qquad \text{Factor as difference of two squares}$$
$$x + 2 = 0 \;\Rightarrow\; x = -2 \qquad \text{Set factors equal to 0}$$
$$x - 2 = 0 \;\Rightarrow\; x = 2$$

Graphic Check

Rewrite the equation so that all nonzero terms are on the left side.

$$x^2 = 4 \qquad\qquad \text{Given equation}$$

$$x^2 - 4 = 0 \qquad\qquad \text{Subtract 4 from both sides}$$

Now, sketch the graph of $y = x^2 - 4$, as shown in Figure 2.25. Notice that the x-intercepts are -2 and 2 (where $x^2 - 4 = 0$), which checks with the algebraic solution.

FIGURE 2.25

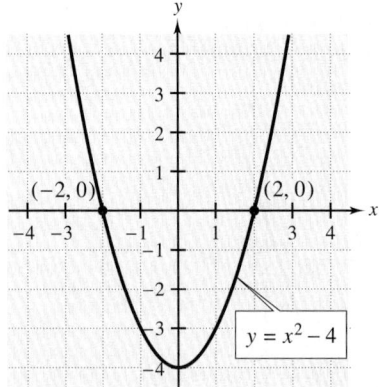

Algebraic Check

Check the first solution, $x = -2$, by substituting -2 for x in the original equation.

$$(-2)^2 \overset{?}{=} 4 \qquad\qquad \text{Substitute } -2 \text{ for } x$$

$$4 = 4 \qquad\qquad \text{Solution checks}$$

Check the second solution, $x = 2$, in a similar manner.

$$(2)^2 \overset{?}{=} 4 \qquad\qquad \text{Substitute 2 for } x$$

$$4 = 4 \qquad\qquad \text{Solution checks} \qquad\qquad \blacksquare$$

Applications

Newspapers and news magazines frequently use graphs to show month-by-month trends for rate of inflation, housing costs, wholesale prices, or the unemployment rate. Industrial firms and businesses use graphs to illustrate monthly or yearly production and sales statistics. Such graphs provide a simple geometric picture of the way one quantity changes with respect to another. The next example shows how a graph can help you visualize the concept of straight-line depreciation.

EXAMPLE 9   ■   Straight-Line Depreciation

A small business buys a new printing press for $65,000. For income tax purposes, the owner of the business decides to depreciate the printing press over a 10-year period. At the end of the 10 years, the salvage value of the printing press is expected to be $5000. Find an equation that relates the value of the printing press to the number of years it has been in service. Then sketch the graph of the equation. (Assume that the depreciation is the same each year. This type of depreciation is called **straight-line depreciation.**)

Solution

The total depreciation over the 10-year period is $65,000 − $5000 = $60,000. Because the same amount is depreciated each year, it follows that the annual depreciation is $60,000 ÷ 10 = $6000. Thus, after 1 year, the value of the printing press is

Value after 1 year = $65,000 − $6000 = $59,000.

By similar reasoning, we can see that the value after 2 years is

Value after 2 years = $65,000 − $6000(2) = $53,000.

Now, by letting y represent the value of the printing press after t years, we follow the pattern established for the first 2 years and conclude that the equation is

$$y = 65,000 - 6000t.$$

A sketch of the graph of this equation is shown in Figure 2.26.

FIGURE 2.26

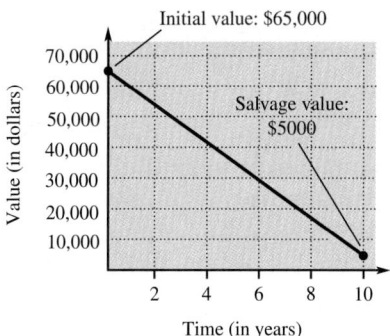

EXAMPLE 10   ■   Interpreting Graphs

The graph in Figure 2.27(a) on the next page shows the number of morning newspapers sold each weekday in the United States from 1978 through 1990, with $x = 0$ corresponding to 1980. The graph in Figure 2.27(b) shows the number of evening newspapers sold during the same years. What do these two graphs tell you about newspaper sales? (*Source:* Editor and Publisher Company)

FIGURE 2.27

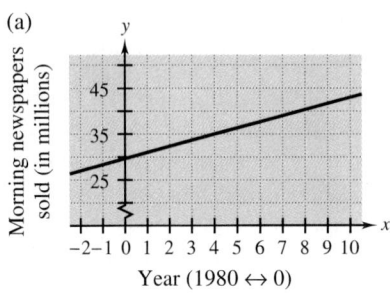 (a)

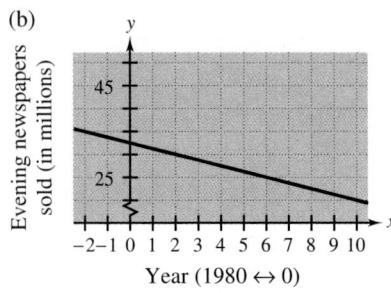 (b)

Give students additional practice reading graphs. Ask students to approximate the number of morning and evening newspapers that were sold in 1984, 1987, and 1989. (The answers are 1984—35 million, 27.5 million; 1987—39 million, 23 million; and 1989—42 million, 21.5 million.)

You might also take this opportunity to discuss increasing and decreasing lines.

Solution

The graph in Figure 2.27(a) shows that morning newspaper sales were *increasing* over the 12-year period, from about 27 million per day in 1978 to about 43 million per day in 1990. The graph in Figure 2.27(b) shows that evening newspaper sales were *decreasing* over the 12-year period, from about 35 million per day in 1978 to about 20 million per day in 1990. From these graphs, you can see that, during the 12-year period, Americans were switching from evening newspapers to morning newspapers. ■

DISCUSSION PROBLEM ■ **Vertical Shifts of Graphs**

Sketch the graphs of the following equations in the same rectangular coordinate system.

$$y = x^2, \quad y = x^2 + 1, \quad y = x^2 + 2, \quad y = x^2 + 3$$

Write a short paragraph describing the effect of the constant term a in the graph of the equation $y = x^2 + a$. ■

■ **Warm-Up**

The following warm-up exercises involve skills that were covered in earlier sections. You will use these skills in the exercise set for this section.

In Exercises 1–4, plot the points in a rectangular coordinate system.

1. $(-3, 2)$, $(5, -4)$ **2.** $(2, 8)$, $(7, -3)$

3. $\left(\frac{5}{2}, \frac{7}{2}\right)$, $\left(\frac{7}{3}, -2\right)$ **4.** $\left(-\frac{9}{4}, -\frac{1}{4}\right)$, $\left(-3, \frac{9}{2}\right)$

In Exercises 5 and 6, solve the equation.

5. $\frac{5}{8}x - 6 = 0$ **6.** $14 - \frac{7}{3}x = 0$

In Exercises 7–10, find the missing coordinate of the solution point.

7. $y = \frac{3}{5}x + 4$, $(15,)$ **8.** $y = 3 - \frac{5}{9}x$, $(12,)$

9. $y = 5.5 - 0.95x$, $(, -1)$ **10.** $y = 3 + 0.2x$, $(, 4.4)$

2.4   EXERCISES

In Exercises 1–4, complete the table and use the resulting solution points to sketch the graph of the equation.

1. $3x + y = 2$

x	-4			2	4
y		5	2		
(x, y)					

2. $2x - 3y = 6$

x	-3	0		3	
y			$-\frac{2}{3}$		5
(x, y)					

3. $y = 4 - x^2$

x		-1		2	
y	0		4		-5
(x, y)					

4. $y = \frac{1}{2}x^3 - 4$

x	-1		1		3
y		-4		0	
(x, y)					

In Exercises 5–12, sketch the graph of the equation.

5. $y = 4 - x$

6. $y = \frac{1}{2}x - 2$

7. $y = x^2 - 4$

8. $y = x^2 - 4x$

9. $y = -x^2$

10. $y = x^3$

11. $y = |x|$

12. $y = |x + 3|$

In Exercises 13–24, find the x- and y-intercepts (if any) of the graph of the equation.

13. $x + 2y = 10$

14. $3x - 2y + 12 = 0$

15. $y = \frac{3}{4}x + 15$

16. $y = 12 - \frac{2}{5}x$

17. $y = x^2 - 25$

18. $y = x^2 + 2x + 1$

19. $y = |x|$

20. $y = x^3$

21. $y = x^2 + 3$

22. $x = 3$

23. $y = 3x^2 + x - 2$

24. $y = x(x - 4) - 3(x - 4)$

In Exercises 25–44, sketch the graph of the equation and show the coordinates of three solution points (including intercepts).

25. $y = 3 - x$

26. $y = x - 3$

27. $y = 2x - 3$

28. $y = -4x + 8$

29. $y = 4$

30. $x = -6$

31. $2x - 3y = 6$

32. $3x + 4y = 12$

33. $x + 5y = 10$

34. $5x - y = 10$

35. $4x + y = 3$

36. $y - 2x = -4$

37. $x - 1 = 0$

38. $y + 3 = 0$

39. $y = x^2 - 9$

40. $y = 9 - x^2$

41. $y = x^2 - 2x$

42. $y = -x^2 - 4x$

43. $y = |x| - 3$

44. $y = |x - 3|$

In Exercises 45–52, solve the equation algebraically and check the result graphically.

45. $\frac{1}{2}x - 2 = 0$ **46.** $-3x + 6 = 0$ **47.** $-2(x - 1) = -7$ **48.** $2x - 1 = 3(x + 1)$

49. $4 - x^2 = 0$ **50.** $x^2 + 2x = 0$ **51.** $x^2 - 2x + 1 = 0$ **52.** $1 - (x - 2)^2 = 0$

53. *Straight-Line Depreciation* A manufacturing plant purchases a new molding machine that costs $225,000. The depreciated value y after t years is

$$y = 225{,}000 - 20{,}000t, \quad 0 \le t \le 8.$$

Sketch the graph of the equation over the given interval for t.

54. *Straight-Line Depreciation* A florist purchases a new van for delivering flowers. The cost of the van is $21,000. The depreciated value y after t years is

$$y = 21{,}000 - 3000t, \quad 0 \le t \le 5.$$

Sketch the graph of the equation over the given interval for t.

55. *Maximum Area* A rectangle of length l and width w has a perimeter of 12 meters (see figure).
  (a) Show that the width of the rectangle is $w = 6 - l$ and its area is $A = l(6 - l)$.
  (b) Sketch the graph of the equation for the area.
  (c) From the graph in part (b), estimate the dimensions of the rectangle that yield a maximum area.

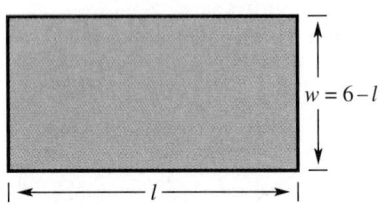

$w = 6 - l$

l

56. *Dentists in the United States* From 1980 to 1987, the number of dentists could be approximated by $D = 2857t + 141{,}250$, where t represents the year with $t = 0$ corresponding to 1980 (see figure).
  (a) Approximate from the graph the year when the United States will have 226,960 dentists.
  (b) Determine algebraically the year when the United States will have 226,960 dentists.

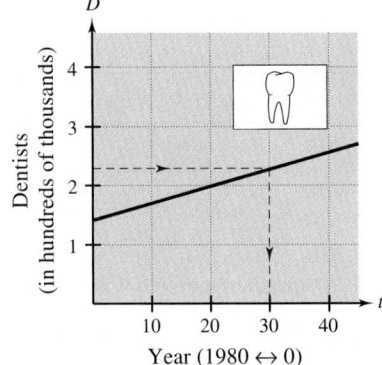

Year (1980 ↔ 0)

Exploratory Exercise In Exercises 57 and 58, (a) sketch the graph of each equation on the same set of coordinate axes. (b) What conclusions can you make by comparing the graphs?

57. $y = x^2$
  $y = (x - 2)^2$
  $y = (x + 2)^2$
  $y = (x + 4)^2$

58. $y = x^2$
  $y = x^2 - 2$
  $y = x^2 - 4$
  $y = x^2 + 4$

Using a Graphing Calculator

In Section 2.3, you studied the point-plotting method for sketching the graph of an equation. One of the disadvantages of the point-plotting method is that, to get a good idea about the shape of a graph, you need to plot *many* points. With only a few points, you could badly misrepresent the graph. For instance, consider the equation

$$y = \frac{1}{30}x(39 - 10x^2 + x^4).$$

Suppose you plotted only five points $(-3, -3)$, $(-1, -1)$, $(0, 0)$, $(1, 1)$, and $(3, 3)$, as shown in the figure on the left. From these five points, you might assume that the graph of the equation is a straight line. This, however, is not correct. By plotting several more points, you can see that the actual graph is not straight at all! (See the figure on the right.)

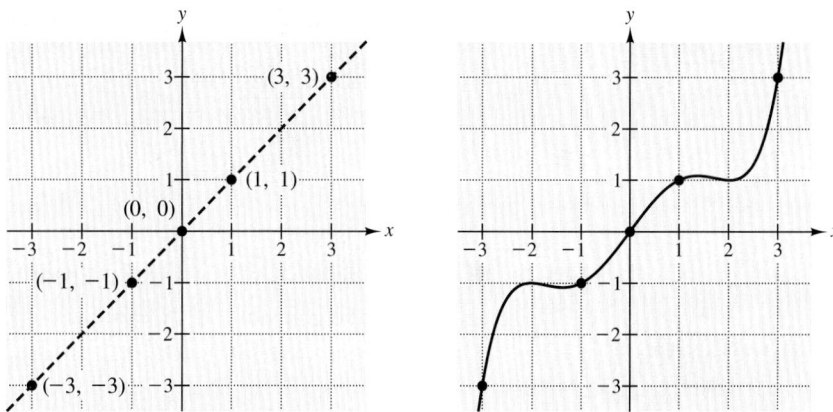

Thus, the point-plotting method leaves you with a dilemma. On the one hand, the method can be very inaccurate if only a few points are plotted. On the other hand, it is very time-consuming to plot a dozen (or more) points. Technology can help you solve this dilemma. Plotting several (even several hundred) points in a rectangular coordinate system is something that a computer or calculator can do easily.

The point-plotting method is the method used by *all* graphing packages for computers and *all* graphing calculators. Each computer or calculator screen is made up of a grid of hundreds or thousands of small areas called **pixels.** Screens that have many pixels per inch are said to have a higher **resolution** than screens that have fewer pixels.

145

There are many different types of graphing utilities—graphing calculators and software packages for computers. Here we describe the TI-81 and TI-82 graphing calculators from Texas Instruments. The steps used to sketch a simple graph are summarized in the following list.

Basic Graphing Steps for the TI-81 and TI-82 Graphing Calculators

To sketch the graph of an equation involving x and y on the TI-81 or TI-82, use the following steps. (Before performing these steps, you should set your calculator so that all of the standard defaults are active. On the TI-81, press ZOOM , set cursor to "Standard," and press ENTER .) On the TI-82, press ZOOM , set cursor to "Z standard," and press ENTER .

1. Rewrite the equation so that y is isolated on the left side of the equation.

2. Press the Y= key. Then enter the right side of the equation on the first line of the display. (The first line is labeled $Y_1 =$.)

3. Press the GRAPH key.

EXAMPLE 1 ■ Sketching the Graph of a Linear Equation

Sketch the graph of $x + 2y = 6$.

Solution

To begin, solve the given equation for y in terms of x.

$$x + 2y = 6 \qquad\qquad \text{Given equation}$$

$$2y = -x + 6 \qquad\qquad \text{Subtract } x \text{ from both sides}$$

$$y = -\frac{1}{2}x + 3 \qquad\qquad \text{Divide both sides by 2}$$

Next, after pressing the Y= key, enter the following keystrokes.

(−) X|T ÷ 2 + 3

The top row of the display should now be as follows.

$$Y_1 = \text{-X}/2 + 3$$

146

Now, press the GRAPH key. The screen should look like the one shown in the figure below. Notice that the calculator screen does not label the tick marks on the *x*-axis or the *y*-axis. To see what the tick marks represent, you can press RANGE on the TI-81, or WINDOW on the TI-82. If you set your calculator to the standard graphing defaults before solving Example 1, the screen should be as follows.

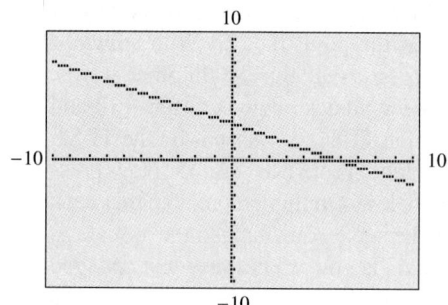

Xmin $= -10$	The minimum *x*-value is -10
Xmax $= 10$	The maximum *x*-value is 10
Xscl $= 1$	The *x*-scale is one unit per tick mark
Ymin $- 10$	The minimum *y*-value is -10
Ymax $= 10$	The maximum *y*-value is 10
Yscl $= 1$	The *y*-scale is one unit per tick mark ■

EXAMPLE 2 ■ **Sketching the Graph of an Equation Involving Absolute Value**

Sketch the graph of $y = |x - 3|$.

Solution

This equation is already written so that *y* is isolated on the left side of the equation. Thus, after pressing the Y= key, enter the following keystrokes.

ABS (X|T − 3)

The top row of the display should now be as follows.

$$Y_1 = \text{abs}(X-3)$$

Now, press the GRAPH key, and the screen should look like the one shown at the left. (To enter ABS on the TI-81 or TI-82, press 2nd and x⁻¹.) ■

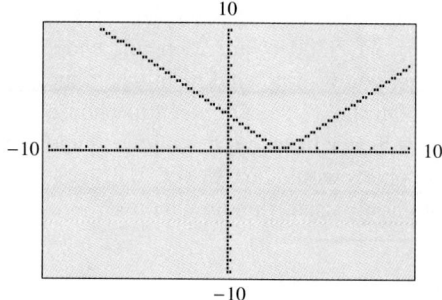

EXAMPLE 3 ■ Resetting the Scales on an Axis

Sketch the graph of $y = x^2 + 12$.

Solution

Press Y= and enter $y = x^2 + 12$ on the first line as X|T x^2 + 12. Press the GRAPH key, and you will notice that nothing appears on the screen (provided your calculator is set to standard defaults). The reason for this is that the lowest point on the graph of $y = x^2 + 12$ occurs at the point (0, 12). With *standard* range settings, you obtain a screen whose largest y-value is 10. In other words, none of the graph is visible on a screen whose y-values range between -10 and 10. To change these settings, press RANGE on the TI-81, or WINDOW on the TI-82, and change Ymax=10 to Ymax=30. Then change Yscl=1 to Yscl=5. Now press GRAPH, and you will obtain the graph shown below. On this graph, note that each tick mark on the y-axis represents five units because you changed the y-scale to 5. Also note that the highest point on the y-axis is now 30 because you changed the maximum value of y to 30. ■

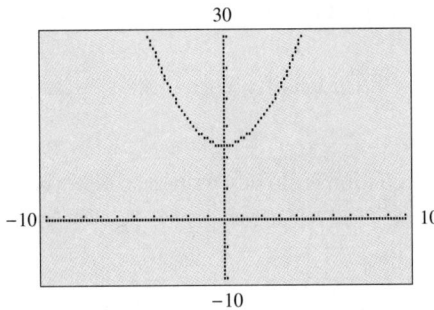

REMARK: A "friendly window" is one in which the x-coordinate of each pixel is an exact decimal value. To obtain a friendly window, you need to know the width of your calculator's display screen in pixels. (The TI-81 is 96 pixels wide; the TI-82, the Casio fx-7700G, and the Sharp EL-9300^{TM} are 95 pixels wide; and the TI-85 is 127 pixels wide.) Suppose that you want a window in which the x-coordinate of each pixel increases by 0.1. On the TI-81, you need to choose Xmin and Xmax so that (Xmax − Xmin) = 9.5. For instance, if Xmin = 0 and Xmax = 9.5, then the x-coordinates of the 96 pixels would be 0.0, 0.1, 0.2, . . ., 9.5. At the top of the next page are some commonly used friendly windows.

Pixel Increment	TI-81	TI-82, Casio fx-7700G, or Sharp EL-9300
1	Xmin = 0, Xmax = 95	Xmin = 0, Xmax = 94
1	Xmin = −45, Xmax = 50	Xmin = −44, Xmax = 50
0.5	Xmin = 0, Xmax = 47.5	Xmin = 0, Xmax = 47
0.5	Xmin = −23.5, Xmax = 24	Xmin = −23.5, Xmax = 23.5
0.2	Xmin = 0, Xmax = 19	Xmin = 0, Xmax = 18.8
0.2	Xmin = −9, Xmax = 10	Xmin = −9, Xmax = 9.8
0.1	Xmin = 0, Xmax = 9.5	Xmin = 0, Xmax = 9.4
0.1	Xmin = −4.7, Xmax = 4.8	Xmin = −4.7, Xmax = 4.7

If you want a friendly window also to be "square," then you should set the Ymin and Ymax so that (Ymax − Ymin) is two-thirds of (Xmax − Xmin).

EXERCISES

In Exercises 1–18, use a graphing calculator to graph the equation. Use the standard setting of the calculator.

```
Xmin=-10
Xmax=10
Xscl=1
Ymin=-10
Ymax=10
Yscl=1
```

1. $y = 2x - 6$

2. $y = x - 2$

3. $y = \frac{2}{3}x + 1$

4. $y = \frac{1}{4}x - 2$

5. $y = -\frac{3}{4}x + 4$

6. $y = -2x + 5$

7. $y = x^2 - 3$

8. $y = 6 - x^2$

9. $y = -x^2 + 4x - 1$

10. $y = \frac{1}{4}x^2 - x$

11. $y = x^3 - 2$

12. $y = 1 - x^3$

13. $y = |x| - 4$

14. $y = 4 - |x|$

15. $y = |x - 4|$

16. $y = \frac{1}{2}|x|$

17. $y = \sqrt{x + 4}$

18. $y = 2\sqrt{2x + 3}$

In Exercises 19–22, use a graphing calculator to graph the equation. Begin by using the standard setting of the calculator and then graph a second time using the specified setting.

19. $y = \frac{1}{2}x - 10$

20. $y = -2x + 25$

21. $y = 20 - \frac{1}{4}x^2$

22. $y = x^2 - 12x + 36$

```
Xmin=-1
Xmax=20
Xscl=1
Ymin=-15
Ymax=5
Yscl=1
```

```
Xmin=-1
Xmax=20
Xscl=1
Ymin=-2
Ymax=28
Yscl=1
```

```
Xmin=-10
Xmax=10
Xscl=1
Ymin=-2
Ymax=22
Yscl=1
```

```
Xmin=-5
Xmax=15
Xscl=1
Ymin=-2
Ymax=20
Yscl=1
```

In Exercises 23–28, find a setting on a graphing calculator such that the graph of the equation agrees with the given graph.

23. $y = -3x + 15$

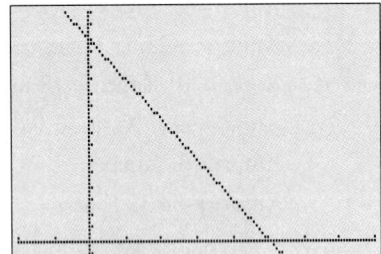

24. $y = \frac{2}{3}x - 8$

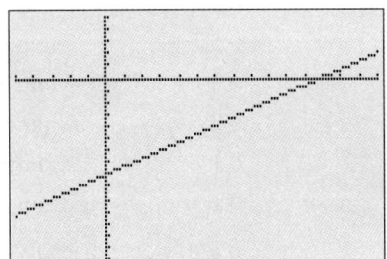

25. $y = -\frac{1}{8}x^2 + x$

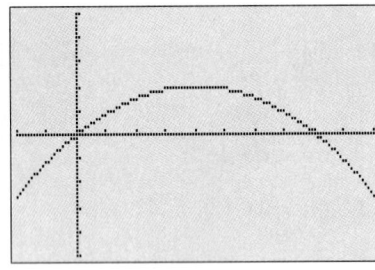

26. $y = 2x^2 - 16x + 26$

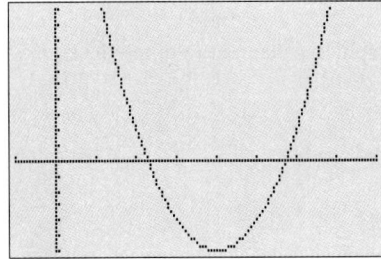

27. $y = x^3 - 3x + 2$

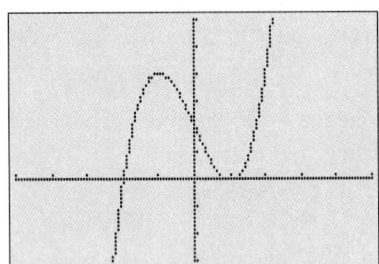

28. $y = (x^2 - 4)^2$

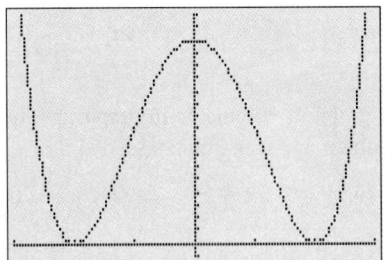

Relations, Functions, and Function Notation

*Relations   •   Functions   •   Function Notation   •   Finding the Domain and Range of a Function
•   Applications*

Relations

Many everyday occurrences involve two quantities that are related to each other by some rule of correspondence. Here are some examples.

1. The simple interest I earned on $1000 for 1 year is related to the annual percentage rate r by the formula $I = 1000r$.

2. The distance d traveled on a bicycle in 2 hours is related to the speed s by the formula $d = 2s$.

3. The area A of a circle is related to its radius r by the formula $A = \pi r^2$.

Not all correspondences between two quantities have simple mathematical formulas. For instance, you might match up quantities such as NFL starting quarterbacks with numbers of touchdown passes, or hours of the day with temperatures.

In mathematics, we define a relation between two variables x and y as a set of ordered pairs. The first number in each ordered pair is the x-value, and the second number is the corresponding y-value. The x-values are the inputs, or the domain, and the y-values are the outputs, or the range. Here is an example.

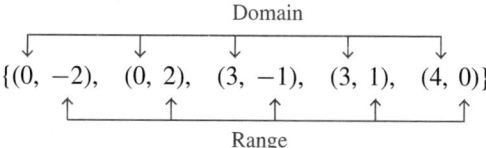

For this relation, the domain is the set {0, 3, 4} and the range is the set {−2, 2, −1, 1, 0}.

There are many ways to describe a relation: as a set of ordered pairs, as a graph, as a mapping diagram, as a table, as an equation, or as a verbal statement.

EXAMPLE 1   ■   Describing a Relation

The above relation was described as a set of ordered pairs. Here are some other ways to describe the same relation.

(a) *Graph*

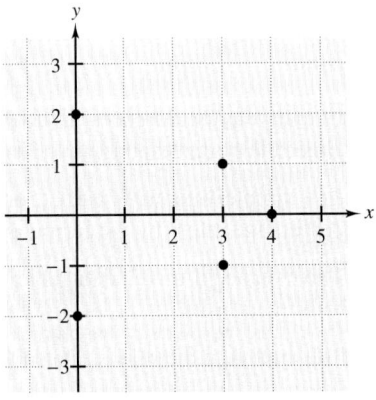

(b) *Mapping Diagram*

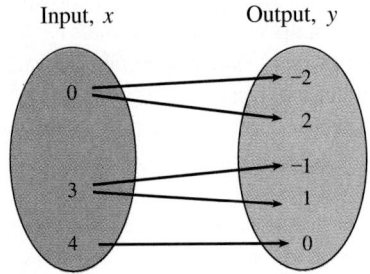

(c) *Table*

Input, x	0	0	3	3	4
Output, y	−2	2	−1	1	0

(d) *Equation*

$$y^2 = 4 - x, \quad \text{where} \quad x = 0,\ 3,\ 4$$

You can find the y-values that correspond to a given x-value by substituting the x-value in the equation. For instance, when $x = 0$, $y = \pm 2$.   ■

Functions

A relation between x and y is a **function** of x if each value of x corresponds to *exactly one* value of y. From this definition, you can see that the relation in Example 1 is *not* a function of x because some x-values correspond to two y-values. For instance, when $x = 0$, y has the values -2 and 2.

Definition of a Function	A **function** f from a set A to a set B is a rule of correspondence that assigns to each element x in the set A exactly one element y in the set B.
	The set A is called the **domain** (or set of inputs) of the function f, and the set B contains the **range** (or set of outputs) of the function.

The rule of correspondence for a function establishes a set of "input-output" ordered pairs of the form (x, y). In some cases, the rule may generate only a finite set of ordered pairs, whereas in other cases the rule may generate an infinite set of ordered pairs. Example 2 lists some examples of "input-output" ordered pairs that represent functions.

EXAMPLE 2   ■   Input-Output Ordered Pairs for Functions

Rule of Correspondence	*Set of Ordered Pairs*
(a) Winners of the Super Bowl from 1986 to 1993	Pairs: (Year, team) Set: {(1986, Bears), (1987, Giants), (1988, Redskins), (1989, 49ers), (1990, 49ers), (1991, Giants), (1992, Redskins), (1993, Cowboys)}
(b) $y = x - 2$	Pairs: (x, y) Set: {All points on the line given by $y = x - 2$}
(c) The squares of all positive integers that are less than 7	Pairs: (n, n^2) Set: {(1, 1), (2, 4), (3, 9), (4, 16), (5, 25), (6, 36)}
(d) The squares of all positive real numbers that are less than 7	Pairs: (x, x^2) Set: {All points (x, x^2) such that $0 < x < 7$}

Note that the sets in parts (a) and (c) have only finite numbers of ordered pairs, whereas the sets in parts (b) and (d) have infinite numbers of ordered pairs.   ■

EXAMPLE 3   ■   Representing a Function by a Set of Ordered Pairs

Find the set of ordered pairs that represents the function shown in Figure 2.28.

FIGURE 2.28   Function from Set A to Set B

Historical Note: Although the concept of function dates back to the Babylonians as early as 2000 B.C., the concept was fine-tuned in the mid-1600s. Descartes is often credited as the first to use the term *function*.

Set *A* is the domain
Input: 1, 2, 3, 4, 5, 6

Set *B* contains the range
Output: 4°, 9°, 12°, 13°, 15°

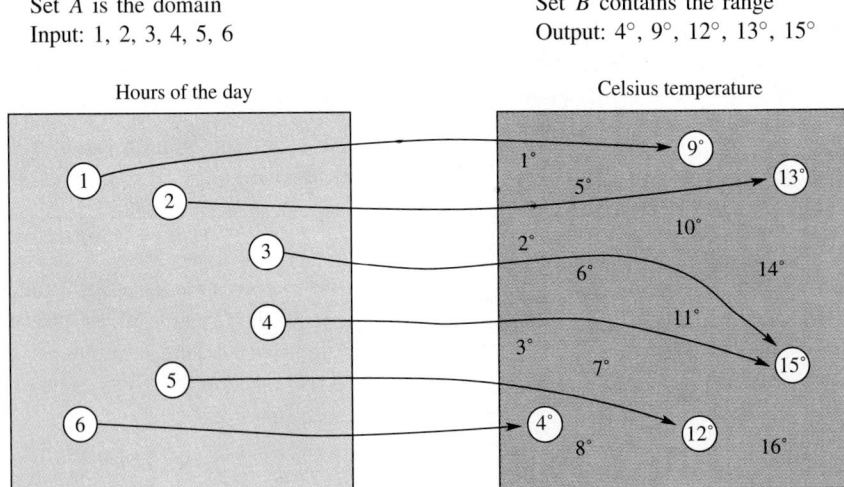

Hours of the day

Celsius temperature

Solution

From the figure, let the first coordinate of each ordered pair represent an input and the second coordinate represent the corresponding output. This produces the following set.

$\{(1, 9°), (2, 13°), (3, 15°), (4, 15°), (5, 12°), (6, 4°)\}$ ■

From Examples 2 and 3, you can construct the following list of characteristics of a function. Study this list carefully.

■
Characteristics of a Function

1. *Each* element in the domain *A* must be matched with an element in the range, which is contained in the set *B*.

2. Some elements in the set *B* may not be matched with any element in the domain *A*.

3. Two or more elements of the domain may be matched with the *same* element in the range.

4. *No* element of the domain may be matched with two different elements in the range.

These four characteristics are illustrated in Example 4.

EXAMPLE 4 ■   Test for Functions Represented by Ordered Pairs

Let $A = \{a, b, c\}$ and let $B = \{1, 2, 3, 4, 5\}$. Determine which of the following sets of ordered pairs or figures represent functions from set A to set B.

(a) $\{(a, 2), (b, 3), (c, 4)\}$ (b) $\{(a, 4), (b, 5)\}$

(c)

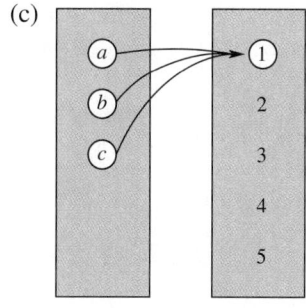

(d)

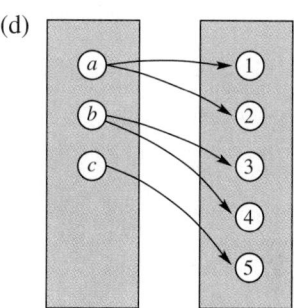

Solution

(a) This collection of ordered pairs *does* represent a function from A to B. Each element of A is matched with exactly one element of B.

(b) This collection of ordered pairs *does not* represent a function from A to B. Not all elements of A are matched with an element of B.

(c) This figure *does* represent a function from A to B. It does not matter that each element of A is matched with the same element in B.

(d) This figure *does not* represent a function from A to B. The element a in A is matched with *two* elements, 1 and 2, of B. This is also true of the element b. ■

Remind students that not all elements in set B must be used and that those elements from set B that are matched to elements from the domain, set A, form the set called the range. Have students identify the range in (a) and (c) in Example 4.

(The answers are (a) {2, 3, 4} and (c) {1}.)

Representing functions by sets of ordered pairs is a common practice in the study of *discrete mathematics*, which deals mainly with finite sets of data or with finite subsets of the set of real numbers. In algebra, however, it is more common to represent functions by equations or formulas involving two variables. For instance, the equation

$$y = x^2 \qquad \text{Squaring function}$$

represents the variable y as a function of the variable x. The variable x is called the **independent variable** and y is called the **dependent variable.** In this context, the domain of the function is the set of all *allowable* real values for the independent variable x, and the range of the function is the *resulting* set of all values taken on by the dependent variable y.

EXAMPLE 5 ■ Testing for Functions Represented by Equations

Determine which of the following equations represents y as a function of x.

(a) $y = x^2 + 1$ (b) $y^2 = x$

Solution

(a) From the equation

$$y = x^2 + 1,$$

you can see that for each value of x there corresponds just one value of y. For instance, when $x = 1$, y is determined to be $1^2 + 1 = 2$. Therefore, y *is* a function of x.

(b) From the equation

$$y^2 = x,$$

you can see that y is *not* a function of x because you can find two different y-values that correspond to the same x-value. For instance, when $x = 4$, y can be 2 or -2.   ■

Function Notation

When using an equation to represent a function, it is convenient to name the function so that it can be easily referenced. For example, you know from Example 5(a) that the equation $y = x^2 + 1$ describes y as a function of x. Suppose you give this function the name "f." Then you can use the following **function notation.**

Input	*Output*	*Equation*
x	$f(x)$	$f(x) = x^2 + 1$

Stress that $f(x)$ does *not* mean "f times x." We are simply giving the function the name f and indicating in which variable the function is written.

The symbol $f(x)$ is read as the **value of f at x** or simply "**f of x.**" Because $f(x)$ represents the y-value for a given x, you can write $y = f(x)$. Keep in mind that f is the *name* of the function, whereas $f(x)$ is the *value* of the function at x. For instance, the function given by

$$f(x) = 4 - 3x$$

has *function values* denoted by $f(-1)$, $f(0)$, $f(1)$, and so on. To find these values, substitute the specified input values into the given equation, as follows.

Stress the replacement concept.
$f(\quad) = 4 - 3(\quad).$

x-Value	*Function Value*
$x = -1$	$f(-1) = 4 - 3(-1) = 4 + 3 = 7$
$x = 0$	$f(0) = 4 - 3(0) = 4 - 0 = 4$
$x = 1$	$f(1) = 4 - 3(1) = 4 - 3 = 1$

The process of finding the value of $f(x)$ for a given value of x is called **evaluating a function.**

Although f is often used as a convenient function name and x as the independent variable, you can use other letters. For instance, $f(x) = 2x^2 - 3x + 5$, $f(t) = 2t^2 - 3t + 5$, and $g(s) = 2s^2 - 3s + 5$ all define the same function. In fact, the role of the independent variable in a function is simply that of a "placeholder." Consequently, the function given by $f(x) = 2x^2 - 3x + 5$ could be described by the form

$$f\,(\quad) = 2\,(\quad)^2 - 3\,(\quad) + 5,$$

where the parentheses are used in place of a letter. Therefore, to evaluate $f(-2)$, simply place -2 in each set of parentheses, as follows.

$$f(-2) = 2(-2)^2 - 3(-2) + 5$$
$$= 8 + 6 + 5$$
$$= 19$$

When evaluating a function, you aren't restricted to inserting only numerical values in the parentheses. For instance, the value of $f(3x)$ is

$$f(3x) = 2(3x)^2 - 3(3x) + 5$$
$$= 18x^2 - 9x + 5.$$

EXAMPLE 6   ■   Evaluating a Function

Let $f(x) = |2x - 5|$ and find the following.

(a) $f(1)$ (b) $f\left(\frac{5}{2}\right)$ (c) $f(4)$

Solution

(a) Replacing x by 1 produces
$$f(1) = |2(1) - 5| = |-3| = 3.$$

(b) Replacing x by $\frac{5}{2}$ produces
$$f\left(\frac{5}{2}\right) = \left|2\left(\frac{5}{2}\right) - 5\right| = |0| = 0.$$

(c) Replacing x by 4 produces
$$f(4) = |2(4) - 5| = |3| = 3.$$

Note that the values of $f(1)$ and $f(4)$ are the same. This does not contradict the definition of a function. Do you see why?   ■

EXAMPLE 7   ■   Evaluating a Function

Let $g(x) = 3x - x^2$ and find the following.

Mention the importance of this concept in calculus.

(a) $g(1)$ (b) $g(x + 1)$ (c) $g(x) + g(1)$

Solution

(a) Replacing x by 1 produces

$$g(1) = 3(1) - 1^2 = 3 - 1 = 2.$$

(b) Replacing x by $x + 1$ produces

$$\begin{aligned}
g(x + 1) &= 3(x + 1) - (x + 1)^2 \\
&= 3x + 3 - (x^2 + 2x + 1) \\
&= 3x + 3 - x^2 - 2x - 1 \\
&= -x^2 + x + 2.
\end{aligned}$$

Additional example. If $h(t) = -2t + 5$, find the following: (a) $h(a)$ (b) $h(-4a)$ (c) $-4h(a)$.

(d) True or false: $h(ka) = kh(a)$, where k is a constant.

(The answers are (a) $h(a) = -2a + 5$ (b) $h(-4a) = 8a + 5$ (c) $-4h(a) = 8a - 20$ (d) False, as $h(-4a) \neq -4h(a)$.)

(c) Using the result of part (a), you have

$$g(x) + g(1) = (3x - x^2) + 2 = -x^2 + 3x + 2.$$

By comparing the results of (b) and (c), you can see that $g(x + 1) \neq g(x) + g(1)$. In general, $g(a + b)$ is not equal to $g(a) + g(b)$. ∎

Sometimes a function is defined by more than one equation. An illustration of this is given in Example 8.

EXAMPLE 8   ■   A Function Defined by Two Equations

Evaluate the function

$$f(x) = \begin{cases} x^2 + 1, & \text{if } x < 0 \\ x - 2, & \text{if } x \geq 0 \end{cases}$$

at (a) $x = -1$, (b) $x = 0$, and (c) $x = 1$.

Solution

Point out that the absolute value function can be piecewise defined as follows.

$$f(x) = \begin{cases} x, & \text{if } x > 0 \\ 0, & \text{if } x = 0 \\ -x, & \text{if } x < 0 \end{cases}$$

(a) Because $x = -1 < 0$, use $f(x) = x^2 + 1$ to obtain

$$f(-1) = (-1)^2 + 1 = 2.$$

(b) Because $x = 0 \geq 0$, use $f(x) = x - 2$ to obtain

$$f(0) = 0 - 2 = -2.$$

(c) Because $x = 1 \geq 0$, use $f(x) = x - 2$ to obtain

$$f(1) = 1 - 2 = -1.$$ ∎

Finding the Domain and Range of a Function

The domain of a function may be explicitly described along with the function, or it may be *implied* by the expression used to define the function. The **implied domain** is the set of all real numbers for which you can find real number values for the function. For instance, the function given by

$$f(x) = \frac{1}{x^2 - 9}$$

has an implied domain that consists of all real values of x other than $x = \pm 3$. These two values are excluded from the domain because division by zero is undefined. Another common type of implied domain is that used to avoid even roots of negative numbers. For instance, the function given by $f(x) = \sqrt{x}$ is defined only for $x \geq 0$. Therefore, its implied domain is the set of all real numbers x such that $x \geq 0$.

EXAMPLE 9   ■   Finding the Domain and Range of a Function

Find the domain and range of each of the following functions.

(a) $\{(-3, 0), (-1, 2), (0, 4), (2, 4), (4, -1)\}$

(b) Area of a circle: $A = \pi r^2$

Solution

(a) The domain consists of all first coordinates in the set of ordered pairs, and the range consists of all second coordinates in the set of ordered pairs. Thus, the domain is $\{-3, -1, 0, 2, 4\}$ and the range is $\{0, 2, 4, -1\}$.

(b) For the area of a circle, you must choose positive values for the radius r. Thus, the domain is the set of all real numbers r such that $r > 0$. The range is therefore the set of all real numbers A such that $A > 0$.   ■

Note in Example 9(b) that the domain of a function can be implied by a physical context. For instance, from the equation $A = \pi r^2$, you would have no reason to restrict r to positive values. However, the fact that this function represents the area of a circle implies that the radius must be positive.

EXAMPLE 10   ■   Finding the Domain of a Function

Find the domain of each of the following functions.

(a) $f(x) = \dfrac{1}{x + 4}$

(b) $g(x) = \sqrt{x - 1}$

Solution

(a) For this function, you must exclude all values of x for which the denominator is 0. Thus, the domain of f consists of all real values of x such that $x \neq -4$.

(b) For this function, you must exclude all values of x for which the radicand $x - 1$ is negative. Thus, the domain consists of all x such that $x - 1 \geq 0$. In other words, the domain consists of all x such that $x \geq 1$.   ■

Applications

EXAMPLE 11 ■ **Finding an Equation to Represent a Function**

FIGURE 2.29

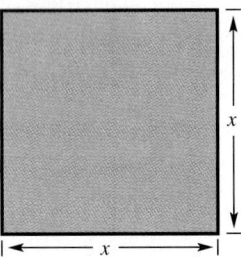

Is the area of a square a *function* of the length of one of its sides? If so, find an equation that represents this function.

Solution

Figure 2.29 shows a typical square. For this square, let the variable A represent the area of the square, and let the variable x represent the length of any one of its sides. (Note that by definition all sides of the square have the same length.) Because the area of a square is completely determined from the length of its side, you can see that A *is* a function of x. The equation that represents the function is

$$A = x^2. \qquad ■$$

EXAMPLE 12 ■ **The Average Life Span of Currency**

The average life span of United States currency is shown in Figure 2.30. Is the average life span a function of the value of the currency? (*Source:* Bureau of Engraving and Printing)

FIGURE 2.30

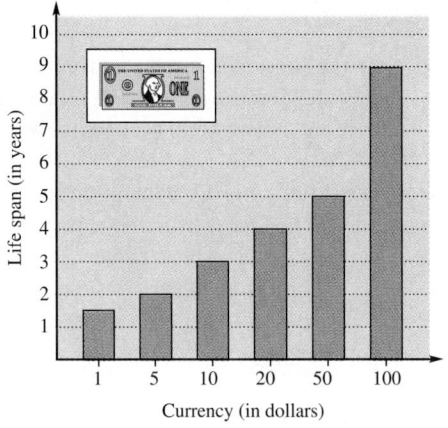

Currency (in dollars)

Solution

From the bar graph shown in the figure, you can see that the average life span *is* a function of the value of the currency. If you let the function be represented by f and let the value of the currency be represented by x, then you obtain the following function values.

$f(1) = 1.5$	A typical $1 bill lasts 1.5 years
$f(5) = 2$	A typical $5 bill lasts 2 years
$f(10) = 3$	A typical $10 bill lasts 3 years
$f(20) = 4$	A typical $20 bill lasts 4 years
$f(50) = 5$	A typical $50 bill lasts 5 years
$f(100) = 9$	A typical $100 bill lasts 9 years ■

DISCUSSION PROBLEM ■ **Identifying Functions from Mapping Diagrams**

Additional discussion problem: Ask students to identify functions from real life examples. Use relationships such as family members, age, social security numbers, and telephone numbers to illustrate functions and nonfunction relations.

Do the following mapping diagrams represent functions? Explain.

(a)
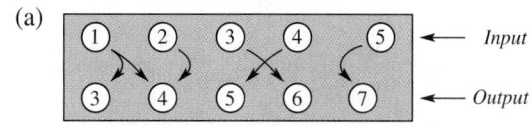

(b)
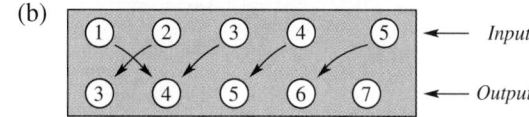

Warm-Up

The following warm-up exercises involve skills that were covered in earlier sections. You will use these skills in the exercise set for this section.

In Exercises 1–6, simplify the expression.

1. $2(x + h) - 3 - (2x - 3)$ **2.** $3(y + t) + 5 - (3y + 5)$ **3.** $2(x - 4)^2 - 5$

4. $-3(x + 2)^2 + 10$ **5.** $4 - 2[3 + 4(x + 1)]$ **6.** $5x + x[3 - 2(x - 3)]$

In Exercises 7–10, evaluate the expression at the given values of the variable. (If not possible, state the reason.)

7. $3x^2 - 4$ (a) $x = 0$ (b) $x = -2$ (c) $x = 2$ (d) $x = 5$

8. $9 - (x - 3)^2$ (a) $x = 0$ (b) $x = 3$ (c) $x = 6$ (d) $x = 2.5$

9. $\dfrac{x + 4}{x - 3}$ (a) $x = -4$ (b) $x = 3$ (c) $x = 4$ (d) $x = 2$

10. $x + \dfrac{1}{x}$ (a) $x = 1$ (b) $x = -1$ (c) $x = \dfrac{1}{2}$ (d) $x = 10$

2.5   EXERCISES

In Exercises 1 and 2, determine whether each relation (set of ordered pairs) represents a function from A to B.

1. $A = \{0, 1, 2, 3\}$ and $B = \{-2, -1, 0, 1, 2\}$
   (a) $\{(0, 1), (1, -2), (2, 0), (3, 2)\}$
   (b) $\{(0, -1), (2, 2), (1, -2), (3, 0), (1, 1)\}$
   (c) $\{(0, 0), (1, 0), (2, 0), (3, 0)\}$
   (d) $\{(0, 2), (3, 0), (1, 1)\}$

2. $A = \{1, 2, 3\}$ and $B = \{9, 10, 11, 12\}$
   (a) $\{(1, 10), (3, 11), (3, 12), (2, 12)\}$
   (b) $\{(1, 10), (2, 11), (3, 12)\}$
   (c) $\{(1, 10), (1, 9), (3, 11), (2, 12)\}$
   (d) $\{(3, 9), (2, 9), (1, 12)\}$

In Exercises 3–10, represent y as a function of x by solving the equation for y.

3. $10x - y + 12 = 0$ **4.** $8x + y - 3 = 0$ **5.** $3x + 7y - 2 = 0$ **6.** $x - 9y + 3 = 0$

7. $x^2 + y - 2x = 3$ **8.** $2x^2 - 3y + 4 = 0$ **9.** $|x| + 2y = 4$ **10.** $|x + 1| - 3y = 0$

In Exercises 11–14, show that both ordered pairs are solutions of the equation and explain why this implies that y is not a function of x.

11. $x^2 + y^2 = 25$, $(0, 5)$, $(0, -5)$ **12.** $x^2 + 4y^2 = 16$, $(0, 2)$, $(0, -2)$

13. $|y| = x + 2$, $(1, 3)$, $(1, -3)$ **14.** $|y - 2| = x$, $(2, 4)$, $(2, 0)$

In Exercises 15–18, fill in the blanks using the function and the specified value of the independent variable.

15. $f(x) = 3x + 5$

   (a) $f(2) = 3(\quad) + 5$

   (b) $f(-2) = 3(\quad) + 5$

   (c) $f(k) = 3(\quad) + 5$

   (d) $f(k + 1) = 3(\quad) + 5$

16. $f(x) = 3 - x^2$

   (a) $f(0) = 3 - (\quad)^2$

   (b) $f(-3) = 3 - (\quad)^2$

   (c) $f(m) = 3 - (\quad)^2$

   (d) $f(t + 2) = 3 - (\quad)^2$

17. $f(x) = \dfrac{x}{x + 2}$

   (a) $f(3) = \dfrac{(\quad)}{(\quad) + 2}$

   (b) $f(-4) = \dfrac{(\quad)}{(\quad) + 2}$

   (c) $f(s) = \dfrac{(\quad)}{(\quad) + 2}$

   (d) $f(s - 2) = \dfrac{(\quad)}{(\quad) + 2}$

18. $f(x) = \sqrt{x + 8}$

   (a) $f(1) = \sqrt{(\quad) + 8}$

   (b) $f(-4) = \sqrt{(\quad) + 8}$

   (c) $f(h) = \sqrt{(\quad) + 8}$

   (d) $f(h - 8) = \sqrt{(\quad) + 8}$

In Exercises 19–34, evaluate the function at the specified value of the independent variable and simplify when possible.

19. $f(x) = 12x - 7$ (a) $f(3)$ (b) $f\left(\frac{3}{2}\right)$ (c) $f(a) + f(1)$ (d) $f(a + 1)$

20. $f(x) = 3 - 7x$ (a) $f(-1)$ (b) $f\left(\frac{1}{2}\right)$ (c) $f(t) + f(-2)$ (d) $f(t - 2)$

21. $g(x) = \frac{1}{2}x^2$ (a) $g(4)$ (b) $g\left(\frac{2}{3}\right)$ (c) $g(4 + a)$ (d) $g(4) + g(a)$

22. $h(x) = x(x\ 2)$ (a) $h(2)$ (b) $h(0)$ (c) $h(t) + h(2)$ (d) $h(t + 2)$

23. $f(x) = \sqrt{x + 5}$ (a) $f(-1)$ (b) $f(4)$ (c) $f(16)$ (d) $f(5z)$

24. $g(x) = 8 - |x - 4|$ (a) $g(0)$ (b) $g(8)$ (c) $g(16) - g(-1)$ (d) $g(x - 2)$

25. $f(x) = \dfrac{3x}{x-5}$ 　　　　　 (a) $f(0)$ 　　 (b) $f\left(\dfrac{5}{3}\right)$ 　　 (c) $f(2) - f(-1)$ 　　 (d) $f(x+4)$

26. $g(x) = \dfrac{|x+1|}{x+1}$ 　　　　 (a) $g(2)$ 　　 (b) $g\left(-\dfrac{1}{3}\right)$ 　　 (c) $g(-4)$ 　　 (d) $g(3) + g(-5)$

27. $f(x) = \begin{cases} x+8, & \text{if } x < 0 \\ 10 - 2x, & \text{if } x \geq 0 \end{cases}$ 　 (a) $f(4)$ 　　 (b) $f(-10)$ 　　 (c) $f(0)$ 　　 (d) $f(6) - f(-2)$

28. $f(x) = \begin{cases} -x, & \text{if } x \leq 0 \\ x^2 - 3x, & \text{if } x > 0 \end{cases}$ 　 (a) $f(0)$ 　　 (b) $f\left(-\dfrac{3}{2}\right)$ 　　 (c) $f(4)$ 　　 (d) $f(-2) + f(25)$

29. $h(x) = \begin{cases} 4 - x^2, & \text{if } x \leq 2 \\ x - 2, & \text{if } x > 2 \end{cases}$ 　 (a) $h(2)$ 　　 (b) $h\left(-\dfrac{3}{2}\right)$ 　　 (c) $h(5)$ 　　 (d) $h(-3) + h(7)$

30. $f(x) = \begin{cases} x^2, & \text{if } x < 1 \\ x^2 - 3x + 2, & \text{if } x \geq 1 \end{cases}$ 　 (a) $f(1)$ 　　 (b) $f(-1)$ 　　 (c) $f(2)$ 　　 (d) $f(-3) + f(3)$

31. $f(x) = 2x + 5$ 　　 (a) $\dfrac{f(x+2) - f(2)}{x}$ 　　 (b) $\dfrac{f(x-3) - f(3)}{x}$

32. $f(x) = x^2 + 4$ 　　 (a) $f(x+1) - f(1)$ 　　 (b) $\dfrac{f(x-5) - f(x)}{5}$

33. $g(x) = 1 - x^2$ 　　 (a) $g(x+2)$ 　　 (b) $\dfrac{g(x+2) - g(x)}{2}$

34. $f(x) = x^3$ 　　 (a) $f(x+2)$ 　　 (b) $\dfrac{f(x+2) - f(x)}{2}$

In Exercises 35–38, find the domain and range of the function.

35. $\{(0, 0), (2, 1), (4, 8), (6, 27)\}$ 　　　　 **36.** $\left\{\left(-3, -\dfrac{17}{2}\right), \left(-1, -\dfrac{5}{2}\right), (4, 2), (10, 11)\right\}$

37. *Circumference of a Circle:* $C = 2\pi r$ 　　 **38.** *Area of a Square:* $A = s^2$

In Exercises 39–54, find the domain of the function.

39. $f(x) = \dfrac{2x}{x-3}$ 　　 **40.** $g(x) = \dfrac{x+5}{x+4}$ 　　 **41.** $g(x) = \dfrac{5x}{x^2 - 3x + 2}$ 　　 **42.** $h(x) = 4x - 3$

43. $f(t) = \dfrac{t+3}{t(t+2)}$ 　　 **44.** $g(s) = \dfrac{s-2}{(s-6)(s-10)}$ 　　 **45.** $h(x) = \dfrac{9}{x^2 + 1}$ 　　 **46.** $f(x) = \dfrac{4}{x^2 - 4}$

47. $f(x) = \sqrt{x-2}$ 　　 **48.** $f(x) = \sqrt{x-5}$ 　　 **49.** $H(x) = \sqrt{x}$ 　　 **50.** $G(x) = \sqrt[3]{x}$

51. $f(x) = \sqrt[3]{x+1}$ 　　 **52.** $h(x) = \sqrt[4]{x}$ 　　 **53.** $f(t) = t^2 + 4t - 1$ 　　 **54.** $f(x) = |x+3|$

In Exercises 55–58, determine the set of ordered pairs for the rule of correspondence.

55. The distance traveled in 3, 2, 8, 6, and $\frac{1}{2}$ hours if, in a given week, a salesperson travels a distance d in t hours at an average speed of 50 mph

56. The cubes of all positive integers less than or equal to 6

57. The winners of the World Series from 1986 to 1992

58. The men inaugurated as President of the United States in 1973, 1977, 1981, 1989, and 1993

59. *Beef Consumption* Use the graph to determine the second coordinate in the ordered pair, where the first coordinate (input) is the year and the second coordinate (output) is annual U.S. per capita beef consumption. (*Source:* U.S. Department of Agriculture)

(a) (1974,   )   (b) (1976,   )

(c) (1985,   )   (d) (1990,   )

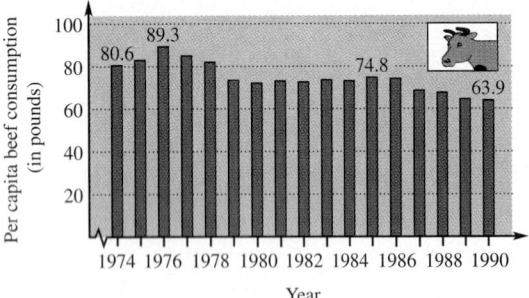

60. *State and Local Taxes* Use the graph to determine the second coordinate in the ordered pair, where the first coordinate (input) is the year and the second coordinate (output) is the average number of minutes per 8-hour workday an American works to earn enough to pay state and local taxes. (*Source:* Tax Foundation Facts and Figures on Government Finance)

(a) (1980,   )   (b) (1984,   )

(c) (1986,   )   (d) (1990,   )

Figure for 60

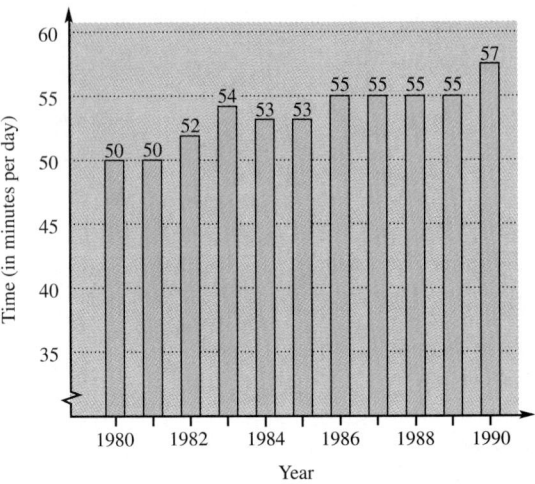

In Exercises 61–66, find an equation that represents the function. (Note that the variables do not necessarily have to be x and y.)

61. *Perimeter of a Square* Express the perimeter P of a square as a function of the length x of a side.

62. *Surface Area of a Cube* Express the surface area S of a cube as a function of the length x of one of the edges.

63. *Volume of a Cube* Express the volume V of a cube as a function of the length x of one of the edges.

64. *Area of a Square* Strips of width x are cut from the four sides of a square that is 35 inches on a side (see figure). Write the area A of the remaining square as a function of x.

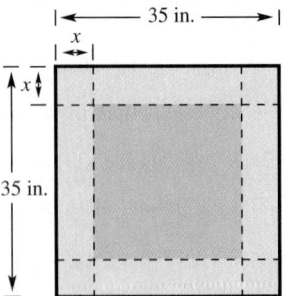

65. *Area of a Square* Strips of width x are cut from two adjacent sides of a square that is 32 inches on a side (see figure). Write the area A of the remaining square as a function of x.

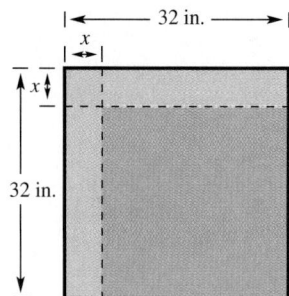

66. *Cost* The inventor of a new game believes that the variable cost for producing the game is $1.95 per unit and the fixed costs are $8000. If x is the number of games produced, express the total cost C as a function of x.

67. A wage earner is paid $8.00 per hour for regular time and time-and-a-half for overtime. The weekly wage function is

$$W(h) = \begin{cases} 8h, & 0 < h \le 40 \\ 12(h - 40) + 320, & h > 40 \end{cases}$$

where h represents the number of hours worked in the week. Evaluate the wage function for the following values of h.

(a) $h = 30$ hours   (b) $h = 40$ hours

(c) $h = 45$ hours   (d) $h = 50$ hours

In Exercises 68–71, determine whether or not the mapping diagram represents a function. Explain.

68.

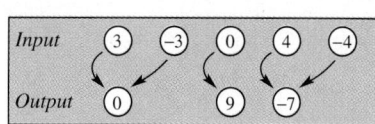

69.

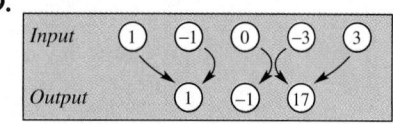

70.

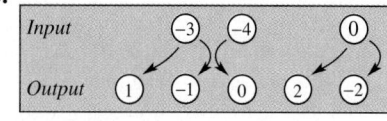

71.

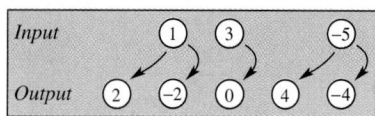

Graphs of Functions

The Graph of a Function   •   Vertical Line Test   •   Graphs of Basic Functions   •   Transformations of Graphs of Functions

The Graph of a Function

In Section 2.5, you studied several different types of functions. In this section, the discussion of functions is limited to those for which the independent variable and the dependent variable represent real numbers. The **graph** of such a function f is the set of ordered pairs $(x, f(x))$, where x is in the domain of f.

$x = x$-coordinate of the ordered pair

$f(x) = y$-coordinate of the ordered pair

Figure 2.31 shows a typical graph of a function. To sketch the graph of a function, you can use the same procedures discussed in Section 2.4 for sketching the graph of an equation.

FIGURE 2.31

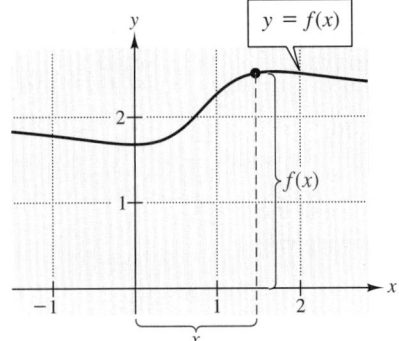

EXAMPLE 1   ■   Sketching the Graph of a Function

Sketch the graph of $f(x) = 2x - 1$.

Solution

Another way to write this equation is

$$y = 2x - 1.$$

Moreover, from your work in Section 2.4, you can construct a table of values and sketch the graph shown in Figure 2.32.

FIGURE 2.32

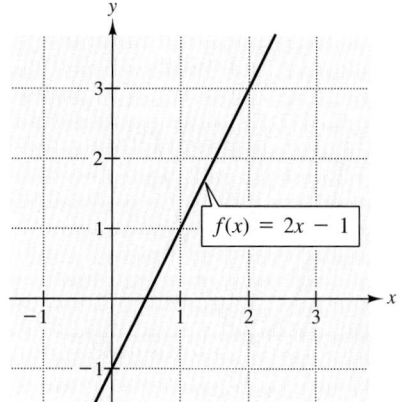

$f(x) = 2x - 1$

■

    In Example 1, the (implied) domain of the function is the set of all real numbers. When writing the equation of a function, we sometimes restrict its domain by writing a condition to the right of the equation. For instance, the domain of the function

$$f(x) = 4x + 5, \quad x \geq 0$$

is the set of all nonnegative real numbers. When sketching the graph of a function, be sure to look for any such restrictions.

EXAMPLE 2   ■   A Function with a Restricted Domain

Sketch the graph of the following function.

$$f(x) = x^2, \quad -2 < x \leq 2$$

Solution

You can begin by constructing a table of values, as shown in Table 2.11.

TABLE 2.11

x	-2	-1	0	1	2
$y = x^2$	4	1	0	1	4
Solution Points	$(-2, 4)$	$(-1, 1)$	$(0, 0)$	$(1, 1)$	$(2, 4)$

    Next, you can complete the graph by plotting the solution points in the table and connecting them with a smooth curve, as shown in Figure 2.33 on the next page. Note in this figure that the point $(-2, 4)$ is represented by an *open* dot, because this point is not part of the graph. The point $(2, 4)$ is represented by a *solid* dot, because this point is part of the graph.

FIGURE 2.33

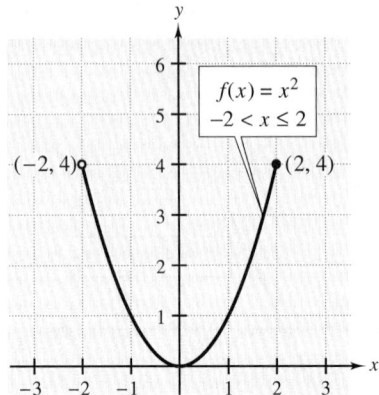

$$f(x) = x^2$$
$$-2 < x \le 2$$

$(-2, 4)$ $(2, 4)$

Determining the range of a function is often more difficult than determining its domain. Once you have sketched the graph of the function, however, you can use the graph to help determine the range. In Figure 2.33, for instance, you can see that the range of the function

$$f(x) = x^2, \quad -2 < x \le 2$$

is given by $0 \le y \le 4$. Remember that the domain of a function is the set of all allowable x-values, and the range is the resulting set of y-values.

Vertical Line Test

By the definition of a function, at most one y-value corresponds to a given x-value. This definition implies that a vertical line can intersect the graph of a function at most once. This important observation provides a convenient visual test for a function. This test is called the **vertical line test for functions.**

Vertical Line Test for Functions	A set of points in a rectangular coordinate system is the graph of y as a function of x if and only if no vertical line intersects the graph at more than one point.

The vertical line test gives you an easy way to determine whether an equation represents y as a function of x. If the graph of an equation has the property that no vertical line intersects the graph at two (or more) points, then the equation represents y as a function of x. On the other hand, if you can find a vertical line that intersects the graph at two (or more) points, then the equation does not represent y as a function of x (see Figure 2.34).

FIGURE 2.34   (a) Graph of a function of x (b) Not a graph of a function of x
No vertical line intersects   Vertical line intersects twice.
more than once.

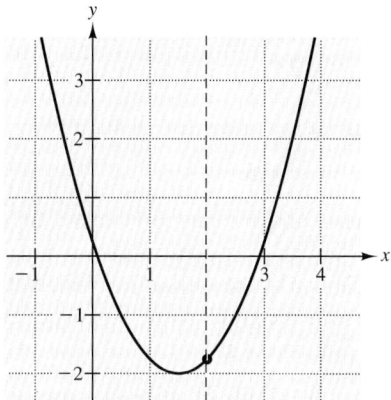

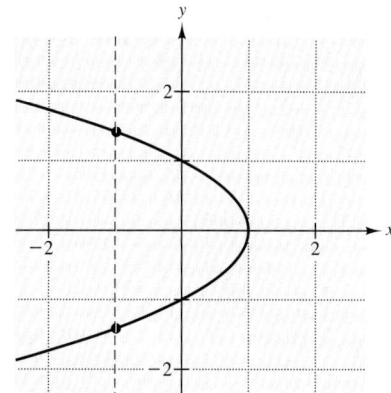

EXAMPLE 3   ■   **Using the Vertical Line Test for Functions**

Use the vertical line test to determine whether the equations whose graphs are shown in Figure 2.35 represent y as a function of x.

FIGURE 2.35   (a)   (b)

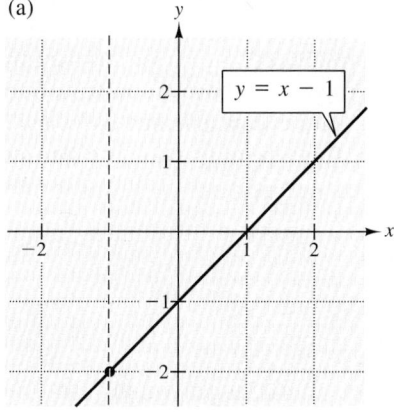

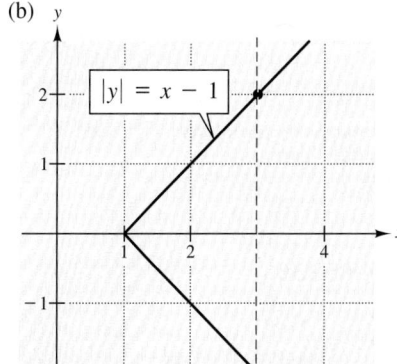

Solution

(a) For this equation, there is no vertical line that intersects the graph twice. Therefore, the equation $y = x - 1$ *does* represent y as a function of x.

(b) For this equation, there *are* some vertical lines that intersect the graph twice. Therefore, the equation $|y| = x - 1$ *does not* represent y as a function of x.

■

EXAMPLE 4   ■   **Testing for y as a Function of x**

Which of the following equations represent y as a function of x? Explain your reasoning.

(a) $x^2 = y - 5$ (b) $y^2 = x - 5$ (c) $y = 2$

Solution

(a) Begin by rewriting the equation in the form

$$y = x^2 + 5.$$

Because you are able to isolate y on the left side of the equation, it appears that the equation does represent y as a function of x. You can confirm this conclusion by the vertical line test. Notice in Figure 2.36(a) that no vertical line intersects the graph more than once. Therefore, this equation *does* represent y as a function of x.

(b) The graph of this equation is shown in Figure 2.36(b). Note that it is possible for a vertical line to intersect the graph at more than one point. Therefore, this equation *does not* represent y as a function of x. Remember that if an equation represents y as a function of x, then no two different solution points of the equation can have the same x-value. For this equation, you see that the points $(6, 1)$ and $(6, -1)$ are both solution points of the equation. The fact that these two points have the same x-value is another way that you can tell that the equation $y^2 = x - 5$ does not represent y as a function of x.

(c) At first glance, you might be tempted to say that this equation does not represent y as a function of x, because x does not appear in the equation. However, in Figure 2.36(c), you can see that the graph of this equation is simply a horizontal line (two units above the x-axis). Because this graph passes the vertical line test, you can conclude that the equation $y = 2$ *does* represent y as a function of x. This type of function is called a **constant function.**

Ask students to determine the domain and range of the function $y = 2$.

(The answers are domain, all real numbers, and range {2}.)

FIGURE 2.36

(a)

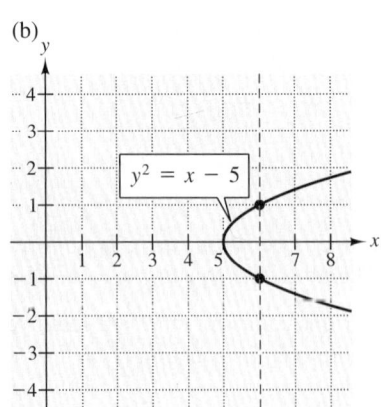

(b)

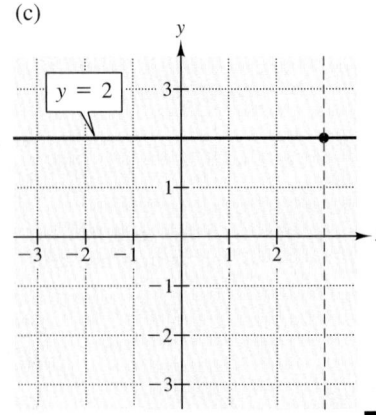

(c)

Graphs of Basic Functions

To become good at sketching the graphs of functions, it helps to be familiar with the graphs of some basic functions. The functions shown in Figure 2.37, and variations of them, occur frequently in applications.

FIGURE 2.37

(a) Constant function

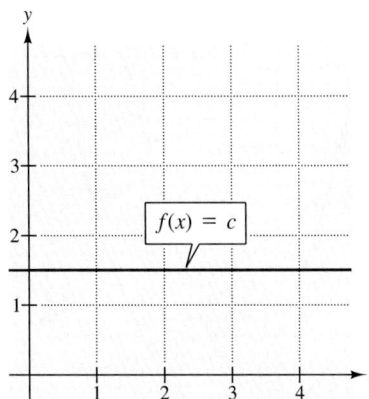

(b) Identity function

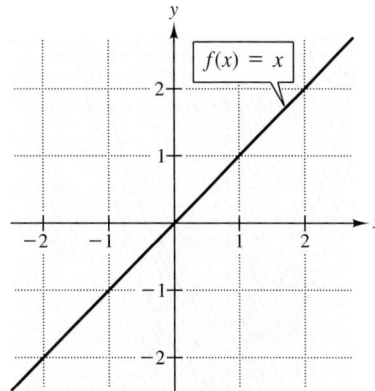

(c) Absolute value function

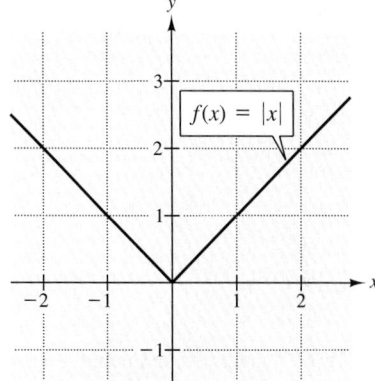

(d) Square root function

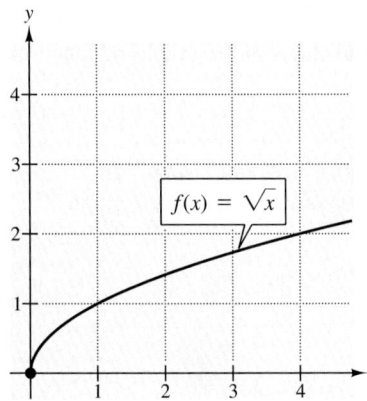

(e) Squaring function

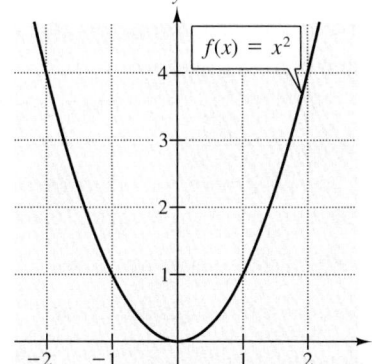

(f) Cubing function

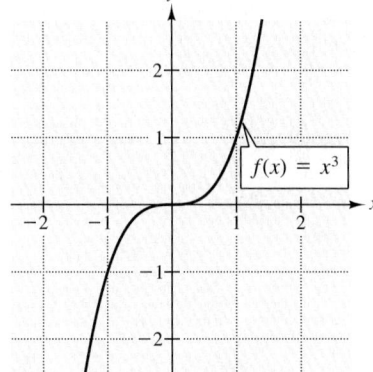

Transformations of Graphs of Functions

Let students discover these relationships by graphing several functions on the same plane.

Many functions have graphs that are simple transformations of the basic graphs shown in Figure 2.37. For example, you can obtain the graph of $h(x) = x^2 + 1$ by shifting the graph of $f(x) = x^2$ *upward* one unit, as shown in Figure 2.38 on the next page. In function notation, h and f are related as follows.

$$h(x) = x^2 + 1 = f(x) + 1 \qquad\qquad \text{Upward shift of 1}$$

Similarly, you can obtain the graph of $g(x) = (x - 1)^2$ by shifting the graph of

$f(x) = x^2$ to the *right* one unit, as shown in Figure 2.39. In this case, the functions g and f have the following relationship.

$$g(x) = (x - 1)^2 = f(x - 1)$$

Right shift of 1

FIGURE 2.38

Vertical shift: one unit up

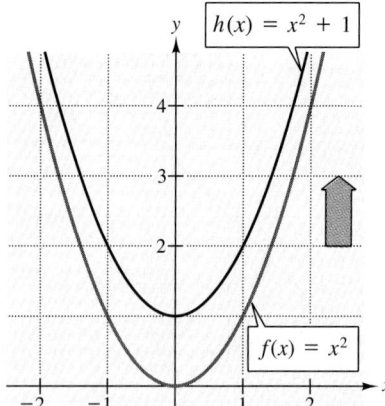

FIGURE 2.39

Horizontal shift: one unit right

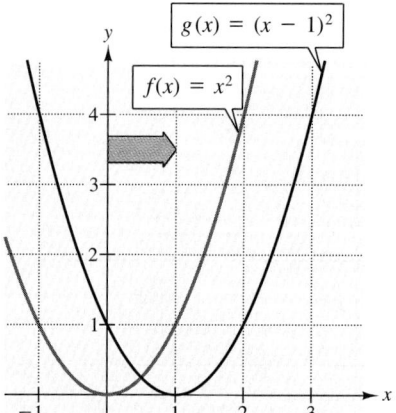

Have students do additional graphs $u(x) = x^2 - 1$ and $v(x) = (x + 1)^2$ and compare them to the graphs in Figures 2.38 and 2.39. Have students distinguish between $u(x) = x^2 - 1$ and $g(x) = (x - 1)^2$ by comparing the graphs of each one.

The various types of **horizontal** and **vertical shifts** of the graphs of functions are summarized as follows.

Vertical and Horizontal Shifts

Let c be a positive real number. **Vertical** and **horizontal shifts** of the graph of the function $y = f(x)$ are represented as follows.

1. Vertical shift c units **upward**: $h(x) = f(x) + c$

2. Vertical shift c units **downward**: $h(x) = f(x) - c$

3. Horizontal shift c units to the **right**: $h(x) = f(x - c)$

4. Horizontal shift c units to the **left**: $h(x) = f(x + c)$

EXAMPLE 5 ■ Shifts of the Graphs of Functions

Use the graph of $f(x) = x^2$ to sketch the graph of each of the following functions.

(a) $g(x) = x^2 - 2$ (b) $h(x) = (x + 3)^2$

Solution

(a) Relative to the graph of $f(x) = x^2$, the graph of $g(x) = x^2 - 2$ represents a *downward shift* of two units, as shown in Figure 2.40(a).

(b) Relative to the graph of $f(x) = x^2$, the graph of $h(x) = (x + 3)^2$ represents a *left shift* of three units, as shown in Figure 2.40(b).

FIGURE 2.40   (a) Vertical shift: two units down   (b) Horizontal shift: three units left

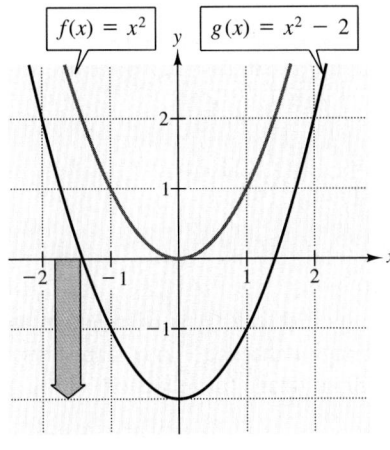
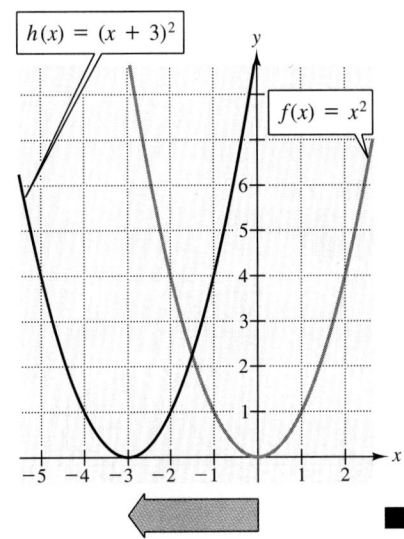

Some graphs can be obtained from a *combination* of vertical and horizontal shifts, as shown in part (b) of the next example.

EXAMPLE 6 ■ Shifts of the Graphs of Functions

Use the graph of $f(x) = x^3$ to sketch the graph of each of the following functions.

(a) $g(x) = x^3 + 2$ (b) $h(x) = (x - 1)^3 + 2$

Solution

(a) Relative to the graph of $f(x) = x^3$, the graph of $g(x) = x^3 + 2$ represents an *upward shift* of two units, as shown in Figure 2.41(a) on the next page.

(b) Relative to the graph of $f(x) = x^3$, the graph of $h(x) = (x - 1)^3 + 2$ represents a *right shift* of one unit, followed by an *upward shift* of two units, as shown in Figure 2.41(b).

FIGURE 2.41   (a) Vertical shift: two units up   (b) Horizontal shift: one unit right
Vertical shift: two units up

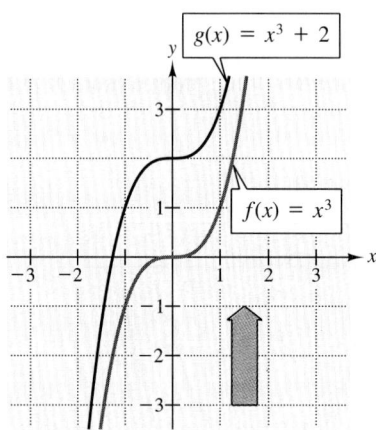

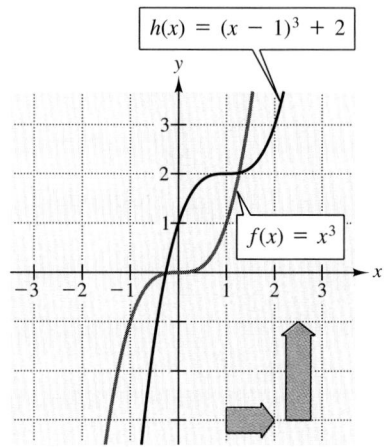

FIGURE 2.42

Reflection

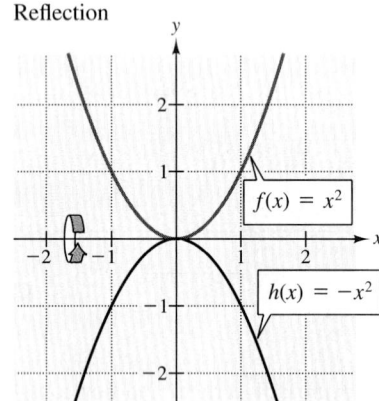

The second basic type of transformation is called a **reflection.** For instance, if you imagine that the x-axis represents a mirror, then the graph of $h(x) = -x^2$ is the mirror image (or reflection) of the graph of $f(x) = x^2$, as shown in Figure 2.42.

| **Reflections in the Coordinate Axes** | Reflections, in the coordinate axes, of the graph of $y = f(x)$ are represented as follows. |

1. **Reflection in the x-axis:** $h(x) = -f(x)$

2. **Reflection in the y-axis:** $h(x) = f(-x)$

EXAMPLE 7   ■   Reflections of the Graphs of Functions

Use the graph of $f(x) = \sqrt{x}$ to sketch the graph of each of the following functions.

(a) $g(x) = -\sqrt{x}$ (b) $h(x) = \sqrt{-x}$

Solution

(a) Relative to the graph of $f(x) = \sqrt{x}$, the graph of $g(x) = -\sqrt{x}$ represents a *reflection in the x-axis*, as shown in Figure 2.43(a).

(b) Relative to the graph of $f(x) = \sqrt{x}$, the graph of $h(x) = \sqrt{-x}$ represents a *reflection in the y-axis*, as shown in Figure 2.43(b).

FIGURE 2.43   (a) Reflection in x-axis   (b) Reflection in y-axis

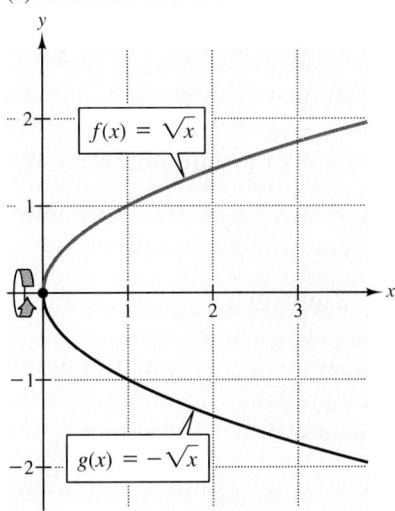

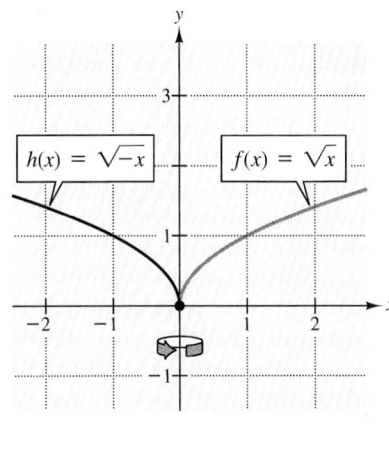

DISCUSSION PROBLEM ■ **Representing an Equation by More than One Function**

The graph of the equation $x^2 + y^2 = 1$ is a circle whose center is the origin and whose radius is 1, as shown in Figure 2.44(a). From the graph, you can see that the equation does not represent y as a function of x, because the graph does not pass the vertical line test for functions. Can you think of a way to write *two* different functions to represent this graph? Try finding one function to represent the upper semicircle and another function to represent the lower semicircle (see Figure 2.44b).

FIGURE 2.44   (a)   (b)

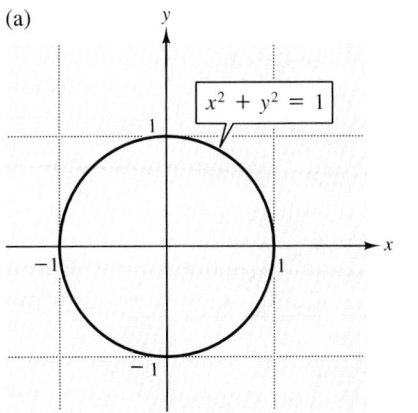

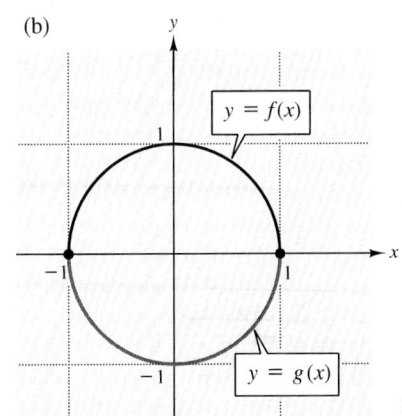

Warm-Up

The following warm-up exercises involve skills that were covered in earlier sections. You will use these skills in the exercise set for this section.

In Exercises 1–6, sketch the graph of the equation.

1. $y = 3 - 2x$ **2.** $y = \frac{1}{2}x - 1$ **3.** $y = (x - 2)^2 - 4$

4. $y = 9 - (x + 1)^2$ **5.** $x - y^2 = 0$ **6.** $y = |x| + 1$

In Exercises 7–10, evaluate the function at the indicated values.

Function	Function Value	
7. $f(x) = \frac{1}{3}x^2$	(a) $f(6)$	(b) $f\left(\frac{3}{4}\right)$
8. $f(x) = 3 - 2x$	(a) $f(5)$	(b) $f(x + 3) - f(3)$
9. $f(x) = \dfrac{x}{x + 10}$	(a) $f(5)$	(b) $f(c - 6)$
10. $f(x) = \sqrt{x - 4}$	(a) $f(16)$	(b) $f(t + 3)$

2.6   EXERCISES

 means that a graphing utility can help you solve the exercise or check your solution.

In Exercises 1–6, match the function with its graph. [The graphs are labeled (a), (b), (c), (d), (e), and (f).]

1. $f(x) = 4 - 2x$ **2.** $f(x) = \frac{3}{2}x + 1$ **3.** $g(x) = \sqrt{1 - x}$

4. $g(x) = (x + 2)^2$ **5.** $h(x) = |x| - 2$ **6.** $h(x) = |x + 2|$

(a)
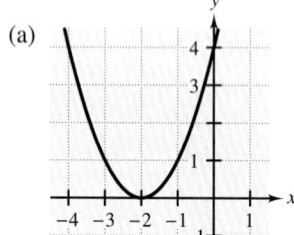

(b)
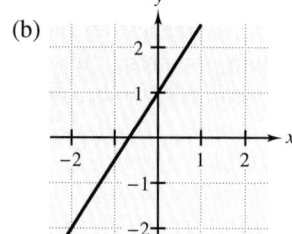

(c)
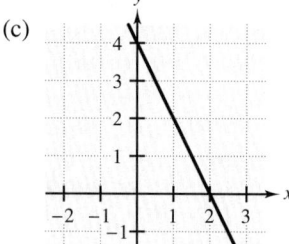

(d)
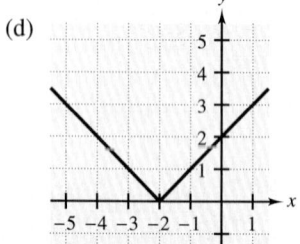

(e)
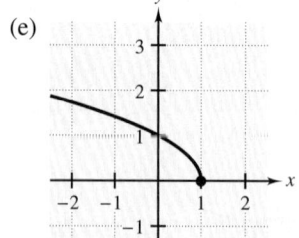

(f)
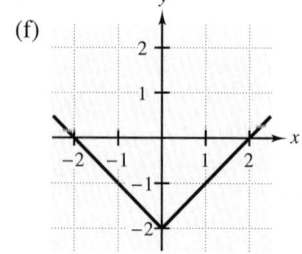

In Exercises 7–22, sketch the graph of the function.

7. $g(x) = 4$

8. $f(x) = -2$

9. $f(x) = 2x - 7$

10. $f(x) = 3 - 2x$

11. $g(x) = \frac{1}{2}x^2$

12. $h(x) = \frac{1}{4}x^2 - 1$

13. $f(x) = |x + 3|$

14. $g(x) = 2 - |x - 1|$

15. $f(t) = \sqrt{t - 2}$

16. $h(x) = \sqrt{4 - x}$

17. $g(s) = \frac{1}{2}s^3$

18. $f(x) = x^3 - 4$

19. $f(x) = 6 - 3x,$

$0 \le x \le 2$

20. $f(x) = \frac{1}{3}x - 2,$

$6 \le x \le 12$

21. $h(x) = x^3,$

$-2 \le x \le 2$

22. $h(x) = x(6 - x),$

$0 \le x \le 6$

In Exercises 23–26, determine whether y is a function of x.

23. $y = (x + 2)^2$

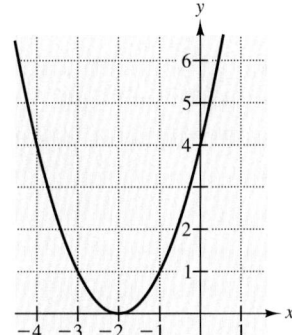

24. $x - 2y^2 = 0$

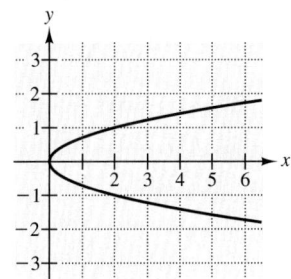

25. $x^2 + y^2 = 16$

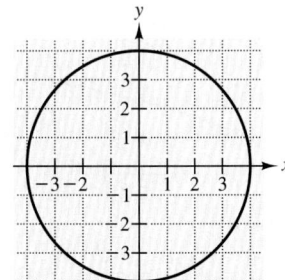

26. $y = x^3 - 3$

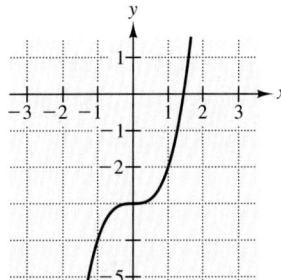

In Exercises 27–34, sketch the graph of the equation. Use the vertical line test to determine whether y is a function of x.

27. $y = 2x$

28. $y^2 = x$

29. $3x - 5y = 15$

30. $y = x^2 + 2$

31. $y^2 = x + 1$

32. $2|y| = x$

33. $y = 1 - x^3$

34. $x = y^4$

In Exercises 35–42, describe the transformation of the graph of $f(x) = x^2$ and sketch a graph of h.

35. $h(x) = x^2 + 2$ **36.** $h(x) = x^2 - 4$ **37.** $h(x) = (x + 2)^2$ **38.** $h(x) = (x - 4)^2$

39. $h(x) = -x^2$ **40.** $h(x) = -x^2 + 4$ **41.** $h(x) = -(x - 3)^2$ **42.** $h(x) = 2 - (x + 1)^2$

 In Exercises 43–50, describe the transformation of the graph of $f(x) = x^3$ and sketch a graph of h.

43. $h(x) = x^3 + 3$ **44.** $h(x) = x^3 - 5$ **45.** $h(x) = (x - 3)^3$ **46.** $h(x) = (x + 2)^3$

47. $h(x) = (-x)^3$ **48.** $h(x) = -x^3$ **49.** $h(x) = 2 - (x - 1)^3$ **50.** $h(x) = (x + 2)^3 - 3$

 In Exercises 51–58, describe the transformation of the graph of $f(x) = |x|$ and sketch the graph of h.

51. $h(x) = |x - 5|$ **52.** $h(x) = |x| - 5$ **53.** $h(x) = -|x|$ **54.** $h(x) = 5 - |x|$

55. $h(x) = |x| + 2$ **56.** $h(x) = |x + 4|$ **57.** $h(x) = |x + 3| - 1$ **58.** $h(x) = |x - 4| + 2$

In Exercises 59–64, use the graph of $f(x) = \sqrt{x}$ to write an equation that represents the transformation of the given graph.

59.
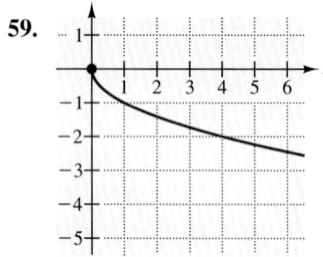

60.
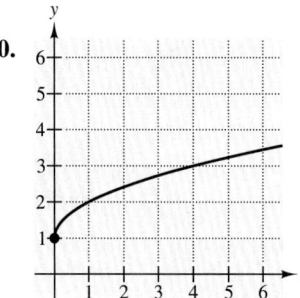

61.
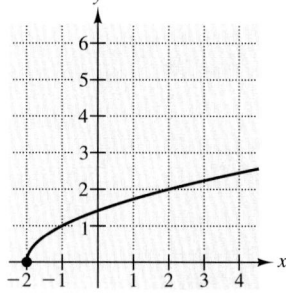

62.
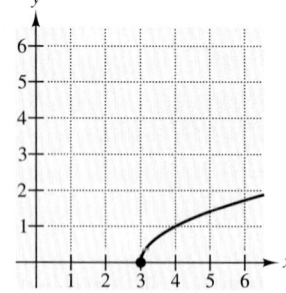

63.
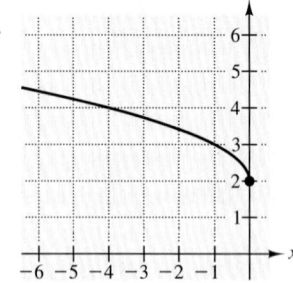

64.
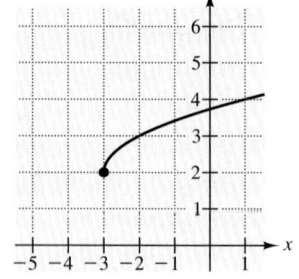

In Exercises 65–70, sketch the graph of the function using a graphing calculator.

65. $f(x) = \frac{2}{3}x + 1$ **66.** $f(x) = \frac{1}{2}x^2 - 1$ **67.** $g(x) = \frac{1}{2}|x - 6|$

68. $h(x) = x^2 + 2x - 3$ **69.** $y = \sqrt{2x - 3}$ **70.** $y = \frac{1}{8}(x - 3)^3$

In Exercises 71 and 72, explain how the graph of f can be used to determine its domain.

71. $f(x) = \sqrt{x - 3}$ **72.** $f(x) = \sqrt{x + 1}$

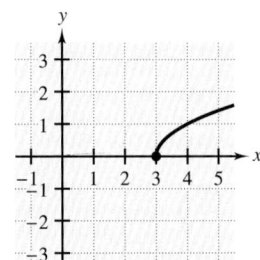   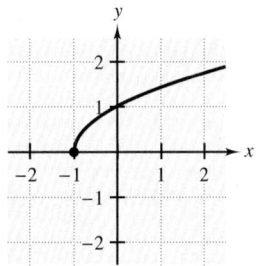

 73. *Area* A rectangle with perimeter 8 inches has length x. The area of the rectangle is given by the function $A(x) = x(4 - x)$.

(a) Determine the domain of the function.

(b) Sketch a graph of the function and estimate the value of x such that the area of the rectangle is maximum.

 74. *Trajectory of a Ball* The height y (in feet) of a ball thrown by a child is

$$y = -\frac{1}{12}x^2 + 2x + 4,$$

where x is the horizontal distance (in feet) from where the ball is thrown (see figure).

(a) How high was the ball when it left the child's hand? (*Hint:* Find y when $x = 0$.)

(b) Use the graph of the function to estimate the maximum height of the ball.

(c) Use the graph of the function to estimate how far from the child the ball struck the ground.

Figure for 74

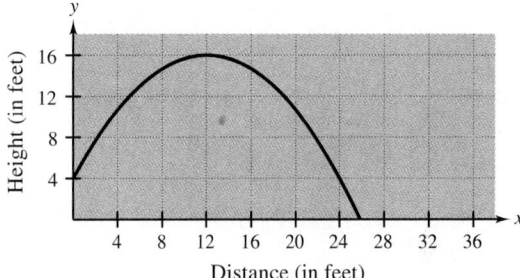

2 REVIEW EXERCISES

In Exercises 1–4, determine whether the values of the variable are solutions of the equation.

Equation	*Values of the Variable*	
1. $45 - 7x = 3$	(a) $x = 3$	(b) $x = 6$
2. $3(5 - x) = 6 - x$	(a) $x = \frac{9}{2}$	(b) $x = -\frac{2}{3}$
3. $\frac{x}{7} + \frac{x}{5} = 1$	(a) $x = \frac{35}{12}$	(b) $x = 2$
4. $\frac{x+2}{6} = \frac{7}{2}$	(a) $x = -12$	(b) $x = 19$

In Exercises 5–24, solve the equation and check the result. (Some of the equations have no solutions.)

5. $17 - 7x = 3$ **6.** $3 + 6x = 51$ **7.** $4y - 6(y - 5) = 2$ **8.** $7x + 2(7 - x) = 8$

9. $1.4t + 2.1 = 0.9t - 2$ **10.** $8(x - 2) = 3(x - 2)$ **11.** $\frac{3x}{4} = 4$ **12.** $-\frac{5x}{14} = \frac{1}{2}$

13. $\frac{4}{5}x - \frac{1}{10} = \frac{3}{2}$ **14.** $\frac{1}{4}s + \frac{3}{8} = \frac{5}{2}$ **15.** $\frac{v - 20}{-8} = 2v$ **16.** $x + \frac{2x}{5} = 1$

17. $10x(x - 3) = 0$ **18.** $3x(4x + 7) = 0$ **19.** $v^2 - 100 = 0$ **20.** $(x + 3)^2 - 25 = 0$

21. $x^2 - 25x = -150$ **22.** $4t^2 - 12t = -9$ **23.** $3s^2 - 2s - 8 = 0$ **24.** $z(5 - z) + 36 = 0$

In Exercises 25–28, solve the equation and round your answer to two decimal places. (A calculator may be helpful.)

25. $382x - 575 = 715$ **26.** $3.625x + 3.5 = 22.125$ **27.** $\frac{x}{2.33} = 14.302$ **28.** $\frac{7x}{3} + 2.5 = 8.125$

In Exercises 29–32, solve the equation for the specified variable.

29. $2x - 7y + 4 = 0$, Solve for x. **30.** $\frac{2}{3}u - 4v = 2v + 3$, Solve for v.

31. $V = \pi r^2 h$, Solve for h. **32.** $S = 2\pi r^2 + 2\pi rh$, Solve for h.

In Exercises 33 and 34, plot the points in a rectangular coordinate system.

33. $(0, -3)$, $\left(\frac{5}{2}, 5\right)$, $(-2, -4)$ **34.** $\left(1, -\frac{3}{2}\right)$, $\left(-2, 2\frac{3}{4}\right)$, $(5, 10)$

In Exercises 35–38, plot the points and find the distance between them. If appropriate, round the result to two decimal places.

35. $(1, 3)$, $(5, 6)$ **36.** $(2, -5)$, $(6, -5)$ **37.** $(-6, -1)$, $(1, 2)$ **38.** $(-2, 10)$, $(3, -2)$

In Exercises 39–44, sketch the graph of the equation and label the x- and y-intercepts.

39. $y = 6 - \frac{1}{3}x$

40. $y = \frac{3}{4}x - 2$

41. $3y - 2x - 3 = 0$

42. $3x + 4y + 12 = 0$

43. $x = |y - 3|$

44. $y = 1 - x^3$

In Exercises 45–48, label the intercepts on the given graph and use the vertical line test to determine whether y is a function of x.

45. $9y^2 = 4x^3$

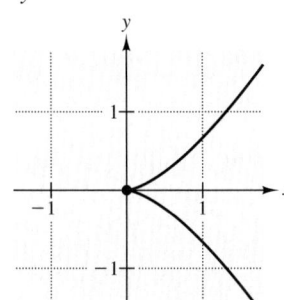

46. $y = 4x^3 - x^4$

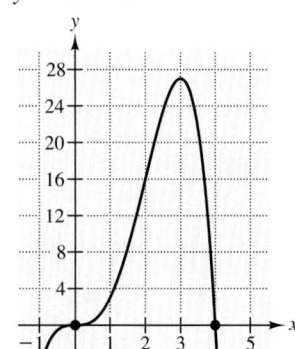

47. $y = x^2(x - 3)$

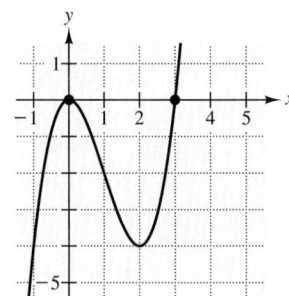

48. $x^3 + y^3 - 6xy = 0$

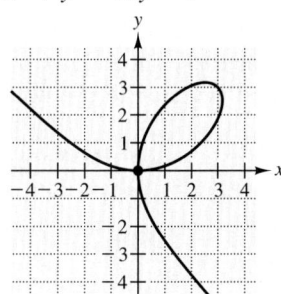

In Exercises 49–56, find the x- and y-intercepts of the graph of the equation.

49. $y = 6 - \frac{2}{3}x$

50. $y = \frac{5}{3}x - 4$

51. $y^2 - x^2 = 1$

52. $3x + 2y + 12 = 0$

53. $y = 8 - |x|$

54. $y = \frac{1}{2}x(2 - x)$

55. $y = (x - 2)^2$

56. $y = 4 - (x - 3)^2$

In Exercises 57–66, evaluate the function at the specified value of the independent variable and simplify when possible.

57. $f(x) = 4 - \frac{5}{2}x$

    (a) $f(-10)$ (b) $f\left(\frac{2}{5}\right)$ (c) $f(t) + f(-4)$ (d) $f(x+h)$

58. $h(x) = x(x - 8)$

    (a) $h(8)$ (b) $h(10)$ (c) $h(-3)$ (d) $h(t + 4)$

59. $f(t) = \sqrt{5 - t}$

    (a) $f(-4)$ (b) $f(5)$ (c) $f(3)$ (d) $f(5z)$

60. $g(x) = \dfrac{|x + 4|}{4}$

    (a) $g(0)$ (b) $g(-8)$ (c) $g(2) - g(-5)$ (d) $g(x - 2)$

61. $f(x) = \begin{cases} -3x, & \text{if } x \le 0 \\ 1 - x^2, & \text{if } x > 0 \end{cases}$

(a) $f(2)$ (b) $f\left(-\frac{2}{3}\right)$ (c) $f(1)$ (d) $f(4) - f(3)$

62. $h(x) = \begin{cases} x^3, & \text{if } x \le 1 \\ (x-1)^2 + 1, & \text{if } x > 1 \end{cases}$

(a) $h(2)$ (b) $h\left(-\frac{1}{2}\right)$ (c) $h(0)$ (d) $h(4) - h(3)$

63. $f(x) = 3 - 2x$

(a) $\dfrac{f(x+2) - f(2)}{x}$ (b) $\dfrac{f(x-3) - f(3)}{x}$

64. $f(x) = 3x$

(a) $\dfrac{f(x+1) - f(1)}{x}$ (b) $\dfrac{f(x-5) - f(5)}{x}$

65. $g(x) = x^2$

(a) $g(x+h)$ (b) $g(x+h) - g(x)$

66. $f(x) = 8$

(a) $f(x+h)$ (b) $\dfrac{f(x+h) - f(x)}{h}$

In Exercises 67–70, find the domain of the function.

67. $h(x) = 4x^2 - 7$

68. $g(s) = \dfrac{5}{s}$

69. $f(x) = \dfrac{1}{5-x}$

70. $G(x) = \sqrt{x+4}$

In Exercises 71–76, (a) describe the transformation and (b) sketch the graph.

71. $g(x) = -(x+1)^2$

72. $h(x) = 9 - (x-2)^2$

73. $f(t) = |t-2| - 1$

74. $f(x) = |x+1| - 2$

75. $g(x) = x^3 - 2$

76. $h(x) = (x-2)^3$

77. *Rocker Arm Design* Find the distance d between the centers of the two small bolt holes in the rocker arm shown in the figure. Round the result to two decimal places.

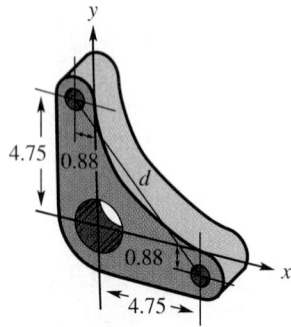

78. *Velocity of a Ball* The velocity of a ball thrown upward from ground level is given by $v = -32t + 80$, where t is the time in seconds and v is the velocity in feet per second.

(a) Find the velocity when $t = 2$.

(b) Find the time when the ball reaches its maximum height. (*Hint:* Find the time when $v = 0$.)

(c) Find the velocity when $t = 3$.

79. *Dimensions of a Rectangle* The area of the rectangle in the figure is 48 square inches. Find x.

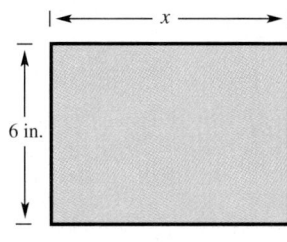

80. *Dimensions of a Rectangle* The perimeter of the rectangle in the figure is 110 feet. Find the dimensions of the rectangle.

2

Chapter Test

Take this test as you would take a test in class. After you are done, check your work against the answers in the back of the book.

1. Determine whether the given value of x is a solution of the equation
   $3(5 - 2x) - (3x - 1) = -2$.

   (a) $x = -4$ (b) $x = 2$

2. Solve for r in the equation $2(r - s) = 5r - 4s + 1$.

In Exercises 3–8, solve the equation.

3. $6x - 5 = 19$

4. $15 - 7(1 - x) = 3(x + 8)$

5. $\dfrac{2x}{3} = \dfrac{x}{2} + 4$

6. $\dfrac{t - 5}{12} = \dfrac{3}{8}$

7. $(y + 2)^2 - 9 = 0$

8. $12 + 5y - 3y^2 = 0$

9. Determine the quadrant in which the point (x, y) lies if $xy > 0$. Explain your reasoning.

10. Find the distance between the points $(0, 9)$ and $(3, 1)$.

11. Find the x- and y-intercepts of the graph of the equation $y = x(x + 1) - 3(x + 1)$.

12. Sketch the graph of the equation $y = |x - 2|$.

13. Use the vertical line test on the graph in the figure to determine whether the equation
    $y^2(4 - x) = x^3$ represents y as a function of x.

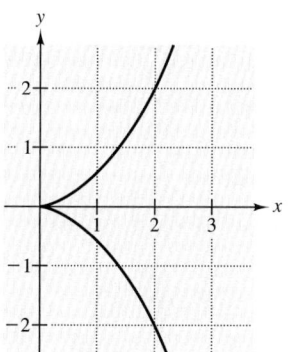

14. Evaluate the function $f(x) = 3x - 2$ at the indicated values.

    (a) $f(6)$ (b) $f(x + h) - f(x)$

    (c) $f(t + 1)$ (d) $f\left(\dfrac{s}{3}\right)$

15. Sketch the graph of the function $g(x) = \sqrt{x - 3} + 1$.

16. Use the graph of $f(x) = |x|$ to write an equation that represents the transformation of the given graph.

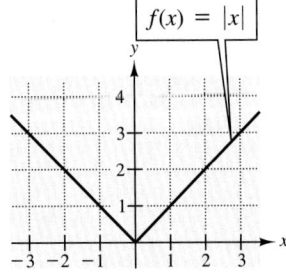

$f(x) = |x|$

(a)

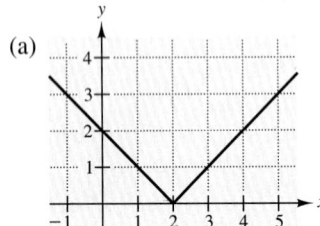

(b)

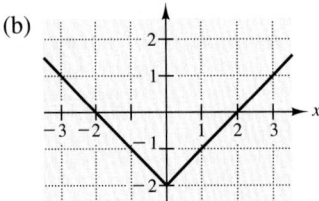

(c)

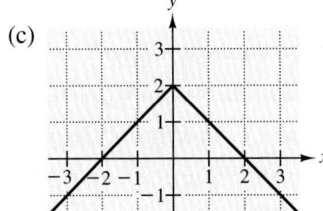

17. The length of a rectangle is $1\frac{1}{2}$ times its width. Find the dimensions of the rectangle if its area is 54 square centimeters.

18. Find the principal required to earn $300 in simple interest in 2 years if the annual interest rate is 7.5%.

Linear Functions, Equations, and Inequalities

3.1   Applications of Linear Equations

3.2   More Applications: Consumer and Scientific Problems

3.3   Linear Inequalities in One Variable

3.4   Equations and Inequalities Involving Absolute Value

3.5   Slope: An Aid to Graphing Linear Functions

3.6   Equations of Lines

ach year, monarch butterflies migrate from the northern United States to Mexico. The butterflies travel at an average rate of 80 miles per day. The distance d (in miles) traveled for a given time t (in days) is given by

$$(\text{Distance}) = (\text{rate})(\text{time})$$

$$d = 80t .$$

Use this model to approximate the distance traveled by a monarch butterfly in 2 weeks. The entire trip covers a distance of about 2400 miles. How long does it take the butterflies to make the trip?

$d = 80t$

Distance (in miles)

2500
2000
1500
1000
500

5 10 15 20 25 30

Time (in days)

Translating Phrases

So far in the text, you have been asked primarily to simplify *given* expressions and solve *given* equations. In this section, you will study ways to *construct* algebraic expressions and equations. When you translate a verbal phrase into an algebraic expression *or* translate a verbal sentence into an algebraic equation, it helps to watch for key words and phrases that indicate the four different operations of arithmetic.

Translating Key Words and Phrases	Key Words and Phrases	Verbal Description	Algebraic Expression
	Addition:		
	Sum, plus, greater, increased by, more than, exceeds, total of	The sum of 5 and x. Seven more than y.	$5 + x$ $y + 7$
	Subtraction:		
	Difference, minus, less than, decreased by, subtracted from, reduced by, the remainder	b is subtracted from 4. Three less than z.	$4 - b$ $z - 3$
	Multiplication:		
	Product, multiplied by, twice, times, percent of	Two times x.	$2x$
	Division:		
	Quotient, divided by, ratio, per	The ratio of x to 8.	$\dfrac{x}{8}$

EXAMPLE 1 ■ Translating Verbal Phrases Containing Specified Variables

Translate each of the following phrases into an algebraic expression.

(a) Seven more than three times x

(b) Four less than the product of 6 and n

(c) y decreased by the sum of 5 and x^2

(d) Four times the sum of y and 9

Solution

(a) *Verbal description:* Seven more than three times x
 Algebraic expression: $3x + 7$ Think: 7 added to what?

(b) *Verbal description:* Four less than the product of 6 and n
 Algebraic expression: $6n - 4$ Think: 4 subtracted from what?

(c) *Verbal description:* y decreased by the sum of 5 and x^2
 Algebraic expression: $y - (5 + x^2)$ Think: What is subtracted from y?

(d) *Verbal description:* Four times the sum of y and 9
 Algebraic expression: $4(y + 9)$ Think: 4 multiplied by what? ■

In most applications of algebra, the variables are not specified and it is your task to assign variables to the *appropriate* quantities. Although similar to the problems in Example 1, the problems in the next example may seem more difficult because variables have not been assigned to the unknown quantities.

EXAMPLE 2 ■ Translating Verbal Phrases Having No Specified Variables

Translate each of the following phrases into a variable expression.

(a) Four decreased by the product of 2 and a number

(b) A number increased by the sum of 8 and another number

(c) Four subtracted from a number and the result divided by 9.

This approach makes translation easier for students when they are solving application problems.

Solution

(a) *Verbal description:* Four decreased by the product of 2 and a number
 Label: x = the number
 Algebraic expression: $4 - 2x$ Think: What is subtracted from 4?

(b) *Verbal description:* A number increased by the sum of 8 and another number
 Labels: x = one number; y = the other number
 Algebraic expression: $x + (8 + y)$ Think: A number added to what?

(c) *Verbal description:* Four subtracted from a number and the result divided by 9
 Label: x = the number
 Algebraic expression: $\dfrac{x - 4}{9}$ Think: What is divided by 9? ■

A good way to learn algebra is to do it both forward and backward. In the next two examples, algebraic expressions are translated into verbal form. Keep in mind that other key words could be used to describe the operations in each expression. These two examples use key words and phrases that keep the verbal expressions clear and concise.

EXAMPLE 3   ■   Translating Algebraic Expressions into Verbal Form

Without using a variable, write a verbal description for each of the following algebraic expressions.

(a) $5x - 10$ (b) $5(x - 10)$

Solution

(a) *Algebraic expression:* $5x - 10$
    Primary operation: Subtraction
    Terms: $5x$ and 10
    Verbal description: Ten less than the product of 5 and a number

(b) *Algebraic expression:* $5(x - 10)$
    Primary operation: Multiplication
    Factors: 5 and $(x - 10)$
    Verbal description: A number decreased by 10 and the difference
                          multiplied by 5   ■

EXAMPLE 4   ■   Translating Algebraic Expressions into Verbal Form

Without using a variable, write a verbal description for each of the following algebraic expressions.

(a) $3 + \dfrac{x}{4}$ (b) $\dfrac{3 + x}{4}$

Solution

(a) *Algebraic expression:* $3 + \dfrac{x}{4}$

    Primary operation: Addition

    Terms: 3 and $\dfrac{x}{4}$

    Verbal description: Three added to the quotient of a number and 4

(b) *Algebraic expression:* $\dfrac{3 + x}{4}$

    Primary operation: Division
    Numerator, denominator: Numerator is $3 + x$, denominator is 4
    Verbal description: The sum of 3 and a number, all divided by 4   ■

Hidden Operations

Most real-life problems do not contain verbal expressions that clearly identify the arithmetic operations involved in the problems. You need to rely on common sense, past experience, and the physical nature of the problem to identify the operations hidden in the problem statement. For instance, products are often overlooked in verbal expressions. Watch for hidden products in the next two examples.

EXAMPLE 5 ■ Discovering Hidden Products

A cash register contains n nickels and d dimes. Write an expression for this amount of money in cents.

Solution

The concept of multiplying value by quantity is an important one as it can be applied later to a variety of applications.

The amount of money is a sum of products.

Verbal model: | Value of nickel | · | Number of nickels | $+$ | Value of dime | · | Number of dimes |

Labels: Value of nickel $= 5$ (cents/nickel)
Number of nickels $= n$ (nickels)
Value of dime $= 10$ (cents/dime)
Number of dimes $= d$ (dimes)

Algebraic expression: $5n + 10d$ (cents)   ■

EXAMPLE 6 ■ Discovering Hidden Products

Write an expression showing how far a person can ride a bicycle in t hours if the person travels at a constant rate of 12 miles per hour.

Solution

For this problem, you must rely on the *formula* for distance. This formula tells you that (distance) $=$ (rate)(time).

Verbal model: | Rate | · | Time |

Labels: Rate $= 12$ (miles per hour)
Time $= t$ (hours)

Algebraic expression: $12t$ (miles)   ■

NOTE In Example 6, the final answer is listed in terms of miles. This makes sense in the following way.

$$12 \, \frac{\text{miles}}{\cancel{\text{hour}}} \cdot t \, \cancel{\text{hours}}$$

Note that the hours "divide out," leaving miles as the unit of measure in the answer. This technique is called *unit analysis,* and it can be very helpful in determining the final unit of measure.

EXAMPLE 7 ■ Discovering Hidden Operations

Suppose you paid x dollars for an automobile. Your total cost included a 7% sales tax. Write an expression, in dollars, for the total cost of the automobile.

Solution

The total cost is a sum.

Verbal model: | Cost of automobile | $+$ | Tax |

Labels: Cost of automobile $= x$ (dollars)
Sales tax rate $= 0.07$ (decimal form)
Sales tax $= 0.07x$ (dollars)

Algebraic expression: $x + 0.07x = (1 + 0.07)x = 1.07x$ (dollars)   ■

Hidden operations are often involved when variable names (labels) are assigned to *two* unknown quantities. For example, suppose two numbers add up to 18 and one of the numbers is assigned the variable label x. What expression in x can be used to represent the second number? Consider a specific case first, and then apply it to a general case.

Specific case: If the first number is 7, then the second number is $18 - 7 = 11$.

General case: If the first number is x, then the second number is $18 - x$.

Note the use of *subtraction* to label the second number.

EXAMPLE 8   ■   **Using Hidden Operations to Label Unknown Quantities**

A person's weekly salary is d dollars. Write a label for the annual salary for this person.

Solution

There are 52 weeks in a year.

Encourage students to choose arbitrary values for a specific case before generalizing with variables. This is a valuable problem-solving strategy, which will ensure that their approach is feasible.

Specific case: If the weekly salary is $300, then the annual salary is 52(300) dollars.

General case: If the weekly salary is d dollars, then the annual salary is $52 \cdot d$ or $52d$ dollars.   ■

In mathematics it is useful to know how to represent special types of integers algebraically. Two integers are called **consecutive integers** if they differ by 1. Hence, for any integer n, the next two larger consecutive integers are $n + 1$ and $(n + 1) + 1$ or $n + 2$. Thus, you can denote three consecutive integers by n, $n + 1$, and $n + 2$.

Labels for Integers	Let *n* represent an integer.
	1. $\{n, n+1, n+2, \ldots\}$ denotes a set of *consecutive* integers.
	2. If *n* is an even integer, then $\{n, n+2, n+4, \ldots\}$ denotes a set of *consecutive even integers.*
Be sure to give examples of both consecutive even *and* consecutive odd integers.	3. If *n* is an odd integer, then $\{n, n+2, n+4, \ldots\}$ denotes a set of *consecutive odd integers.*

Introduction to Problem Solving

So far in this section, you have translated *verbal phrases* into *algebraic expressions.* When algebra is used to solve real-life problems, you usually have to carry the process one step further—by translating *verbal sentences* into *algebraic equations.* The process of translating phrases or sentences into algebraic expressions or equations is called **mathematical modeling.**

A good approach to mathematical modeling is to use two stages. The first stage is to form a *verbal model* by using the given verbal description of the problem. Then, after assigning labels to the unknown quantities in the verbal model, you can form a *mathematical model* or *algebraic equation.*

Verbal models help students organize and picture relationships, which can then be translated into equations.

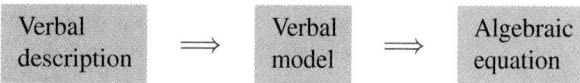

When you are trying to construct a verbal model, it is helpful to look for a hidden equality—a statement that two algebraic expressions are equal. These two expressions might be stated explicitly as being equal, or they might be known to be equal (based on prior knowledge or experience).

EXAMPLE 9 ■   Constructing a Verbal Model

Six times the sum of a number and 5 is 72. What is the number?

Solution

Verbal model: $6\left(\boxed{\text{A number}} + 5\right) = 72$

Labels: Number = *x*

Algebraic equation:	$6(x + 5) = 72$	Algebraic model
	$6x + 30 = 72$	Distributive Property
	$6x = 42$	Subtract 30 from both sides
	$x = 7$	Divide both sides by 6

Thus, the number is 7. You can check this by substituting 7 for "a number" in the original problem.   ■

EXAMPLE 10 ■ Constructing a Verbal Model

Find two consecutive integers such that the sum of the first and three times the second is 87.

Solution

Verbal model: | First integer | + 3 · | Second integer | = 87

Labels: First integer = n
Second integer = $n + 1$

Algebraic equation:

$n + 3(n + 1) = 87$	Algebraic model
$n + 3n + 3 = 87$	Distributive Property
$4n = 84$	Subtract 3 from both sides and combine terms
$n = 21$	Divide both sides by 4

Thus, the first integer is 21, and the second integer is $21 + 1 = 22$. You can check this by substituting 21 and 22 for two consecutive integers in the original problem. ■

EXAMPLE 11 ■ Using a Verbal Model to Construct an Algebraic Equation

Write an algebraic equation that represents the following problem: You have accepted a job with an annual salary of $27,630. This salary includes a year-end bonus of $750. If you are paid twice a month, what will be your gross pay for each paycheck?

Solution

Because there are 12 months in a year and you will be paid twice a month, it follows that you will receive 24 paychecks during the year. You can construct an algebraic equation for this problem as follows. Begin with a verbal model, then assign labels, and finally form an algebraic equation.

Verbal model: | Income for year | = | 24 paychecks | + | Bonus

Labels: Income for year = 27,630 (dollars)
Amount of each paycheck = x (dollars)
Bonus = 750 (dollars)

Algebraic equation: $27,630 = 24x + 750$

The algebraic equation for this problem is a *linear equation* in the variable x. Using the techniques discussed in Section 2.1, you can solve this equation for x. If you do this, you will find that the solution is $x = \$1120$. ■

You could solve the problems in Examples 9, 10, and 11 *without* algebra. For instance, in Example 11 you could solve the problem by subtracting the bonus of $750 from the annual salary of $27,630 and dividing the result by 24 pay periods.

These examples are presented so that you can practice writing algebraic versions of the problem-solving skills that you already possess. Your goal in this section is to be able to solve problems by common sense *and* to use this reasoning to write algebraic versions of the problems. Later, you will encounter more complicated problems for which algebra will be a necessary part of the solutions.

EXAMPLE 12 ▪ A Percent Application

A real estate agency receives a commission of $8092.50 for the sale of a $124,500 house. What percent commission is this?

Solution

Verbal model: | Commission | = | Percent (decimal form) | · | Sale price |

Labels: Commission = 8092.50 (dollars)
Percent = p (decimal form)
Sale price = 124,500 (dollars)

Algebraic equation: $8092.50 = p(124,500)$

$$\frac{8092.50}{124,500} = p$$

$$0.065 = p$$

Therefore, the real estate agency receives a commission of 6.5%. Check this solution in the original problem.

EXAMPLE 13 ▪ The First Transcontinental Railroad

The Central Pacific Company averaged 8.75 miles of track per month. The Union Pacific Company averaged 20 miles of track per month. The drawing below shows the two companies meeting in Promontory, Utah, as the track was completed. When was the track completed? How many miles of track did each company build?

In 1862, the United States Congress granted construction rights to two railroad companies to build a railroad connecting Omaha, Nebraska with Sacramento, California. The entire railroad was about 1590 miles long. The Central Pacific Company began building eastward from Sacramento in 1863. Twenty-four months later, the Union Pacific Company began building westward from Omaha.

Solution

Verbal model:

Total miles of track	=	Miles per month	·	Number of months	+	Miles per month	·	Number of months

Central Pacific   Union Pacific

Labels: Total length of track $= 1590$ (miles)
Central Pacific rate $= 8.75$ (miles per month)
Central Pacific time $= t$ (months)
Union Pacific rate $= 20$ (miles per month)
Union Pacific time $= t - 24$ (months)

Algebraic equation:

$$1590 = 8.75t + 20(t - 24)$$ Linear model

$$1590 = 8.75t + 20t - 480$$ Distributive Property

$$2070 = 28.75t$$ Add 480 to both sides.

$$72 = t$$ Divide both sides by 28.75.

The construction took 72 months (6 years) from the time the Central Pacific Company began. Thus, the track was completed in 1869. The numbers of miles of track built by the two companies were as follows.

Central Pacific: (8.75 miles per month)(72 months) $= 630$ miles

Union Pacific: (20 miles per month)(48 months) $= 960$ miles ■

Rates, Ratios, and Proportions

In real-life applications, the quotient a/b is called a **rate** if a and b have different units, and is called a **ratio** if a and b have the same unit. Note the *order* implied by a ratio. The ratio of a to b means a/b, whereas the ratio of b to a means b/a.

EXAMPLE 14 ■ Expressing Rates and Ratios

(a) You have driven 110 miles in 2 hours. Your average rate for the trip can be expressed as

Stress the differences between rate and ratio in this problem.

$$\text{Rate} = \frac{110 \text{ miles}}{2 \text{ hours}} = 55 \text{ miles per hour.}$$

(b) You have driven 110 miles in 2 hours and your friend has driven 100 miles in the same length of time. The ratio of the distance you have traveled to the distance your friend has traveled is

$$\text{Ratio} = \frac{110 \text{ miles}}{100 \text{ miles}} = \frac{11}{10}.$$ ■

When comparing two *measurements* by means of a ratio, be sure to use the same unit of measurement in both the numerator and the denominator.

EXAMPLE 15 ■   Comparing Measurements

Find ratios for comparing the relative sizes of the following. Use the same unit of measurement in the numerator and denominator.

(a) 200 cents to 5 dollars

(b) 28 months to $1\frac{1}{2}$ years

Solution

Another comparison: Find a ratio to compare the relative size of 3 pounds to 10 ounces.

$$\frac{3 \text{ lb}}{10 \text{ oz}} = \frac{48 \text{ oz}}{10 \text{ oz}} = \frac{24}{5}$$

(a) Because 200 cents is the same as 2 dollars, the ratio is

$$\frac{200 \text{ cents}}{5 \text{ dollars}} = \frac{2 \text{ dollars}}{5 \text{ dollars}} = \frac{2}{5}.$$

Note that if you had converted 5 dollars to 500 cents, you would have obtained the same ratio, because

$$\frac{200 \text{ cents}}{5 \text{ dollars}} = \frac{200 \text{ cents}}{500 \text{ cents}} = \frac{200}{500} = \frac{2}{5}.$$

(b) Because $1\frac{1}{2}$ years = 18 months, the ratio is

$$\frac{28 \text{ months}}{1\frac{1}{2} \text{ years}} = \frac{28 \text{ months}}{18 \text{ months}} = \frac{28}{18} = \frac{14}{9}. \qquad\blacksquare$$

A **proportion** is a statement that equates two ratios (or two rates). For example, if the ratio of a to b is the same as the ratio of c to d, you can write the proportion as

$$\frac{a}{b} = \frac{c}{d}.$$

In a typical application, you know values for three of the letters (quantities) and are required to find the value of the fourth. The quantities a and d are called the **extremes** of the proportion, and the quantities b and c are called the **means** of the proportion. In a proportion, the product of the means is equal to the product of the extremes.

$$ad = bc$$

Rewriting a proportion in this form is called **cross multiplying.**

EXAMPLE 16 ■   Solving a Proportion

The ratio of 8 to a certain number is the same as the ratio of 5 to 2. What is the number?

Solution

Let x represent the unknown number and write the following proportion.

$$\frac{8}{x} = \frac{5}{2}$$

You can solve this equation for x by cross multiplying to obtain

$$16 = 5x$$

$$\frac{16}{5} = x.$$

Thus, the number is $\frac{16}{5}$. Check this solution in the original problem. ■

EXAMPLE 17 ■ An Application of Proportion

Suppose you are driving from Arizona to New York, a trip of 2750 miles. You begin the trip with a full tank of gas, and after traveling 424 miles you refill the tank for $22. How much should you plan to spend on gasoline for the entire trip? (Assume that the driving conditions are similar throughout the trip.)

Solution

Verbal model: $\dfrac{\text{Dollars for trip}}{\text{Miles for trip}} = \dfrac{\text{Dollars for tank}}{\text{Miles for tank}}$

Labels: Entire trip: Cost of gasoline $= x$ (dollars),
Distance $= 2750$ (miles)

First part of trip: Cost of gasoline $= 22$ (dollars),
Distance $= 424$ (miles)

Proportion: $\dfrac{x}{2750} = \dfrac{22}{424}$

$$x = (2750)\left(\frac{22}{424}\right)$$

$$x \approx 142.69$$

Thus, you should plan to spend approximately $142.69 for gasoline on the trip. Check this solution in the original problem. ■

As a summary of what we have been doing and for future reference in work with applied problems, use the following approach to solve word problems.

Strategy for Solving Word Problems

To solve a word problem, use the following steps.

1. *Search* for the hidden equality—two expressions said to be equal or known to be equal. A sketch may be helpful.

2. *Write* a verbal model that equates these two expressions.

3. *Assign* numbers to the known quantities and letters (or algebraic expressions) to the variable quantities.

4. *Rewrite* the verbal model as an algebraic equation using the assigned labels.

5. *Solve* the resulting algebraic equation.

6. *Check* to see that the answer satisfies the word problem as stated.

DISCUSSION PROBLEM ■ Checking the Sensibility of an Answer

Activity: Ask a student to choose a positive whole number and keep the number a secret. Then give the following directions.

	Translation
Directions	(n = the number)
1. Add 6.	$n + 6$
2. Double this number.	$2(n + 6) =$ $2n + 12$
3. Subtract 10.	$2n + 2$
4. Divide this number by 2.	$n + 1$

Ask the student for the final number. You now know the original number, because it is one less than the final value. Discuss in algebraic terms why this game works.

When solving problems related to real-life situations, you should always ask yourself whether the answer makes sense. Write a short paragraph explaining why the following answers are suspicious.

(a) A problem asks you to find the sales tax on a car, and your answer is $4547.61.

(b) A problem asks you to find the number of square feet in a kitchen, and your answer is 736 square feet.

(c) A problem asks you to find the interest on a $2000 deposit that is left in a savings account for one month, and your answer is $105.75.

(d) A problem asks you to find the time it takes for a passenger plane to travel from New York to London, and your answer is $1\frac{1}{2}$ hours.   ■

Warm-Up

The following warm-up exercises involve skills that were covered in earlier sections. You will use these skills in the exercise set for this section.

In Exercises 1–4, use the rules of algebra to simplify the expression.

1. $25 - 6(x - 3)$ **2.** $-2(u + 8) + u(u + 2)$

3. $3(2n + 1) - 2(2n + 3)$ **4.** $5(2n) + 3(2n + 2)$

In Exercises 5–10, solve the equation. (If not possible, state the reason.)

5. $4 - \dfrac{x}{2} = 6$ **6.** $\dfrac{x}{3} + 1 = 10$

7. $8(x - 14) = 0$ **8.** $12(3 - x) = 5 - 7(2x + 1)$

9. $0.75x = 235$ **10.** $(1 + r)500 = 550$

3.1   EXERCISES

In Exercises 1–14, translate the statement into an algebraic expression.

1. The sum of 8 and a number n

2. Six less than a number n

3. Fifteen decreased by three times a number n

4. Six less than four times a number n

5. One-third of a number n

6. Seven-fifths of a number n

7. Thirty percent of the list price L

8. Forty percent of the cost C

9. The quotient of a number x and 6

10. The ratio of y to 3

11. The sum of 3 and four times a number x, all divided by 8

12. The product of a number y and 10 is decreased by 35.

13. The absolute value of the difference between a number n and 5

14. The absolute value of the quotient of y and 4

In Exercises 15–22, write a verbal description of the algebraic expression without using the variable.

15. $3x + 2$ 16. $4x - 5$ 17. $8(x - 5)$ 18. $-3(x + 2)$

19. $\dfrac{y}{8}$ 20. $\dfrac{4x}{5}$ 21. $\dfrac{x + 10}{3}$ 22. $25 + \dfrac{x}{6}$

In Exercises 23–42, write an algebraic expression that represents the specified quantity in the verbal statement and simplify if possible.

23. The amount of money (in dollars) represented by n quarters

24. The amount of money (in cents) represented by m dimes and n quarters

25. The distance traveled in t hours at an average speed of 55 miles per hour

26. The distance traveled in 5 hours at an average speed of r miles per hour

27. The time required to travel 100 miles at an average speed of r miles per hour

28. The average rate of speed for a journey of 360 miles in t hours

29. The amount of antifreeze in a cooling system containing y gallons of coolant that is 45% antifreeze

30. The amount of water in q quarts of a food product that is 65% water

31. The amount of wage tax due for a taxable income of I dollars that is taxed at the rate of 1.25%

32. The amount of sales tax on a purchase valued at L dollars if the tax rate is 6%

33. The sale price of a coat that has a list price of L dollars if it is a "20% off" sale

34. The total cost for a family to stay one night at a campground if the charge is $18 for the parents plus $3 for each of the n children

35. The total hourly wage for an employee when the base pay is $8.25 per hour and an additional $0.60 is paid for each of q units produced per hour

36. The total hourly wage for an employee when the base pay is $11.65 per hour and an additional $0.80 is paid for each of q units produced per hour

37. The area of a square with sides of length s (see figure)

38. The perimeter of a rectangle of width w and length $2w$ (see figure)

Figure for 37

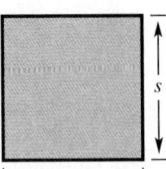

Figure for 38

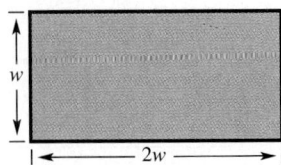

39. The perimeter of a rectangle of length l and width $0.62l$ (see figure)

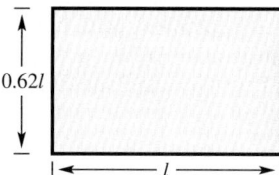

40. The area of a triangle with base b and height $0.75b$ (see figure)

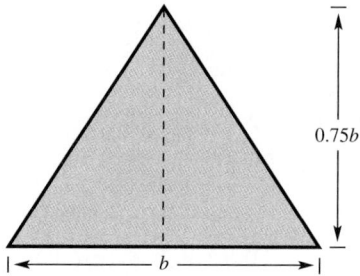

41. The sum of a number n and three times the number, all increased by 2

42. The sum of two consecutive integers, the first of which is n

43. *Advertising Space* An advertising banner has a width of w and length of $6w$, where w is measured in feet (see figure). Find an algebraic expression that represents the area of the banner. What is the unit of measure for the area?

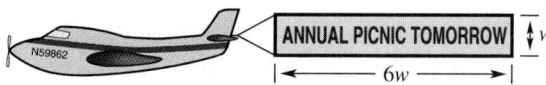

44. *Billiard Table* The top of a billiard table has a length of l and a width of $l - 6$, where l is measured in feet (see figure). Find an algebraic expression that represents the area of the top of the billiard table. What is the unit of measure for the area?

Figure for 44

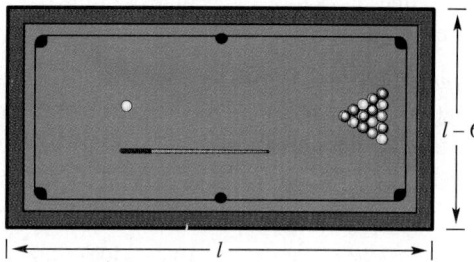

Number Problems In Exercises 45–48, solve the number problem.

45. The sum of three consecutive even integers is 138. Find the integers.

46. The difference of five times an odd integer and three times the next consecutive odd integer is 24. Find the integers.

47. Eight times the sum of a number and 6 is 128. What is the number?

48. When the sum of a number and 18 is divided by 5, the quotient is 12. Find the number.

In Exercises 49–60, construct a verbal model and write an algebraic equation that represents the problem. Then solve the equation.

49. *Number Problem* Find a number such that the sum of that number and 30 is 82.

50. *Number Problem* Find a number such that six times the difference of the number and 12 is 300.

51. *Dimensions of a Triangle* The Slow Moving Vehicle (SMV) sign (see figure) is an equilateral triangle. Find the length of the sides of the triangle if its perimeter is 129 centimeters.

52. *Dimensions of a Rectangle* The length of a rectangle is three times its width. The perimeter of the rectangle is 64 inches. Find the dimensions of the rectangle.

53. *Repair Time* The bill for the repair of an automobile is $380. Included in this bill is a charge of $275 for parts. If the remainder of the bill is for labor at a rate of $35 per hour, how many hours were spent in repairing the car?

54. *Units Produced* Suppose you have a job on an assembly line for which you are paid $10 per hour plus $0.75 per unit assembled. Find the number of units produced in an 8-hour day if your earnings for the day are $146.

55. *Company Layoff* Because of slumping sales, a small company laid off 25 of its 160 employees. What percentage of the work force was laid off?

56. *Real Estate Commission* A real estate agency receives a commission of $9100 for the sale of a $130,000 house. This commission is what percentage of the selling price?

57. *Defective Parts* A quality control engineer reported that 1.5% of a sample of parts were defective. Find the size of the sample if the engineer detected three defective parts.

58. *Price Inflation* The price of a new van is approximately 115% of what it was 3 years ago. What was the approximate price 3 years ago if the current price is $25,750?

59. *Course Grade* Suppose you missed a B in your mathematics course by 8 points. If your point total for the course was 372, how many points were possible in the course? (Assume that you needed 80% of the course total for a B.)

60. *Room Dimensions* The floor of a room measuring 10 feet by 12 feet is partially covered by a circular rug with a radius of 4 feet, as shown in the figure. What percentage of the floor is covered by the rug?

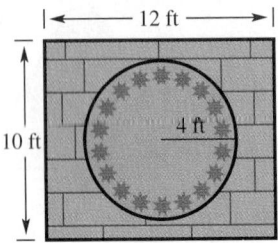

In Exercises 61–64, use the double bar graph to answer the questions. The graph shows the percentages of respondents of faculty and staff at The Pennsylvania State University who are subject to common health-related risk factors. There were 3789 female and 3083 male respondents in the study. (*Source:* The Pennsylvania State University)

Figure for 61-64

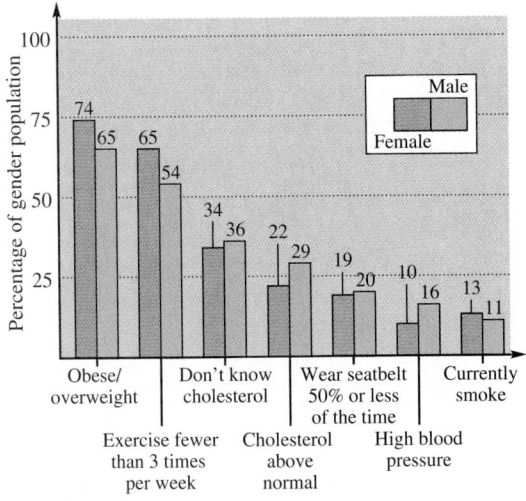

61. Estimate the number of males in the sample who are overweight.

62. Estimate the number of females in the sample who have high blood pressure.

63. Estimate the number of female employees at the university if 51.8% returned their survey forms.

64. Approximate the percentage of respondents who are nonsmokers.

In Exercises 65–68, express the ratio as a fraction. Use the same unit in both the numerator and denominator, and write your answer in reduced form.

65. 36 inches to 48 inches

66. 5 pounds to 24 ounces

67. 40 milliliters to 1 liter

68. 125 centimeters to 2 meters

69. *Compression Ratio* The **compression ratio** of a cylinder is the ratio of its expanded volume to its compressed volume (see figure). The expanded volume of one cylinder of a small diesel engine is 425 cubic centimeters, and its compressed volume is 20 cubic centimeters. Find the compression ratio of this engine.

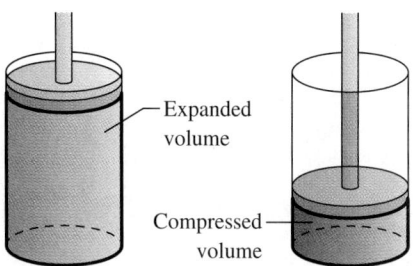

Expanded volume

Compressed volume

70. *Gear Ratio* The **gear ratio** of two gears is the ratio of the number of teeth in one gear to the number of teeth in the second gear (see figure). If two gears in a gearbox have 60 teeth and 40 teeth, find the ratio of the larger gear to the smaller.

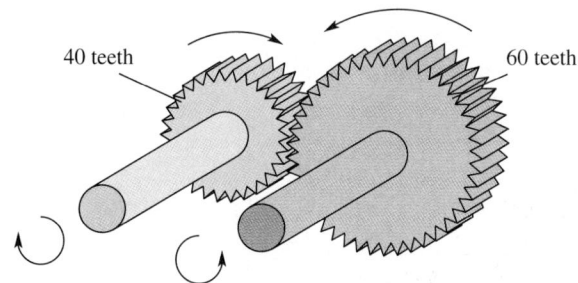

40 teeth   60 teeth

In Exercises 71–76, solve the proportion.

71. $\dfrac{x}{6} = \dfrac{2}{3}$

72. $\dfrac{y}{36} = \dfrac{6}{7}$

73. $\dfrac{4}{5} = \dfrac{6}{t}$

74. $\dfrac{8}{7} = \dfrac{2}{x}$

75. $\dfrac{y+5}{6} = \dfrac{y-2}{4}$

76. $\dfrac{z-3}{3} = \dfrac{z+8}{12}$

77. *Fuel Mixture* The gasoline-to-oil ratio of a two-cycle engine is 40 to 1. Determine the amount of gasoline required to produce a mixture that contains one-half pint of oil.

78. *Recipe Proportions* Three cups of flour are required to make 1 batch of cookies. How many cups are required to make $3\frac{1}{2}$ batches?

79. *Property Tax* The taxes on property with an assessed value of $75,000 are $1125. Find the taxes on property with an assessed value of $120,000.

80. *Public Opinion Poll* In a public opinion poll, 870 people from a sample of 1500 indicated that they would vote for a specific candidate. Assuming this poll to be a correct indicator of the electorate, how many votes can the candidate expect to receive from a total of 80,000 votes cast?

81. *Map Scale* Use the map scale in the figure to *approximate* the straight-line distance from Los Angeles to San Francisco.

0   100
miles

San Francisco

CALIFORNIA

Los Angeles

82. *Pumping Time* A pump can fill a 500-gallon tank in 45 minutes. Determine the time required to fill a 1250-gallon tank with this pump.

Similar Triangles In Exercises 83 and 84, solve for the length x of the side of the triangle on the right by using the fact that corresponding sides of similar triangles are proportional. (*Note:* In each pair, assume that the triangles are similar.)

83.

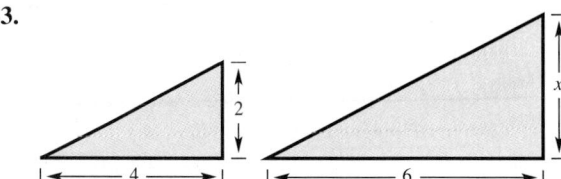

2   x

4   6

84.

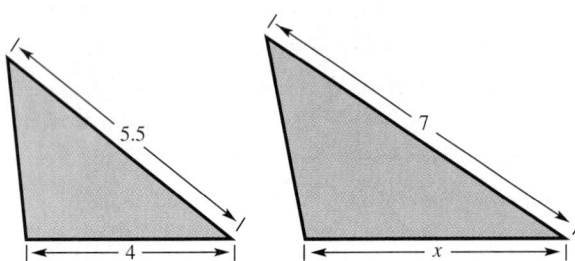

85. ***Tree Height*** A man who is 6 feet tall walks directly toward the tip of the shadow of a tree. When the man is 75 feet from the tree, he starts forming his own shadow beyond the shadow of the tree. Find the height of the tree, if the length of the shadow of the tree beyond this point is 11 feet.

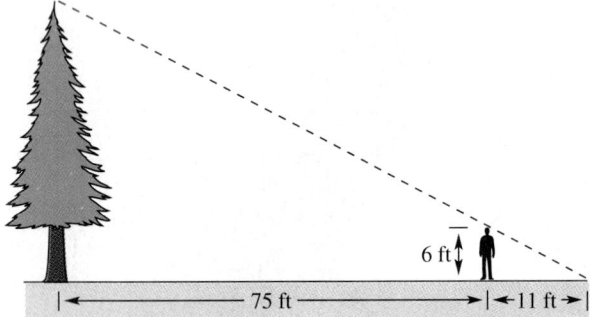

86. ***Shadow Length*** Find the length of the shadow of a man who is 6 feet tall and is standing 15 feet from a streetlight that is 20 feet high (see figure).

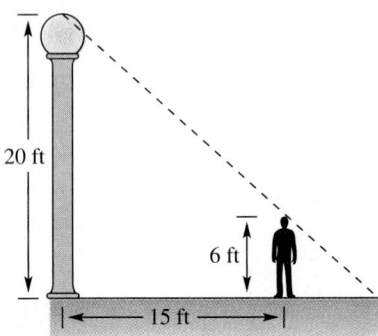

More Applications: Consumer and Scientific Problems
Consumer Problems • *Mixture Problems* • *Rate Problems*

Consumer Problems

Many consumer problems have a mathematical model that involves the sum of a fixed term and a variable term. The variable term is often a hidden product in which one of the factors is a percentage or some other type of rate. Watch for these occurrences in the discussion and examples that follow.

The **markup** on a consumer item is the difference between the amount a retailer pays for an item (**cost**) and the **price** at which the retailer sells the item. A verbal model for this relationship is as follows.

$$\boxed{\text{Selling price}} = \boxed{\text{Cost}} + \boxed{\text{Markup}}$$

In any such problem, the markup may be known or it may be expressed as a *percentage of the cost*. This percentage is called the **markup rate.**

$$\boxed{\text{Markup}} = \boxed{\text{Markup rate}} \cdot \boxed{\text{Cost}}$$

EXAMPLE 1 ■ Finding the Markup Rate

A clothing store sells a pair of jeans for $42. If the cost of the jeans is $16.80, what is the markup rate?

Solution

Mention that the phrases *markup rate*, *discount rate*, and *percentage rate* all imply an answer written in the form of a percentage.

Verbal model: $\boxed{\text{Selling price}} = \boxed{\text{Cost}} + \boxed{\text{Markup}}$

Labels: Selling price $= 42$ (dollars)
Cost $= 16.80$ (dollars)
Markup rate $= p$ (decimal form)
Markup $= 16.80p$ (dollars)

Point out the alternative of first multiplying both sides of the equation by 10 to eliminate the use of decimals.

Algebraic equation:

$$42 = 16.8 + p(16.8)$$
$$42 - 16.8 = p(16.8)$$
$$25.2 = p(16.8)$$
$$\frac{25.2}{16.8} = p$$
$$1.5 = p$$

Thus, because $p = 1.5$, it follows that the markup rate is 150%. (Remember that when you convert the decimal form $p = 1.5$ to percentage form, you obtain 150%.) Check this solution in the original problem. ■

As consumers, we all like a bargain! Many consumers often delay buying an item until it goes on sale at a discounted price. The model for this situation is

$$\boxed{\text{Sale price}} \;=\; \boxed{\text{List price}} \;-\; \boxed{\text{Discount}} \;,$$

where the **discount** is given in dollars. The **discount rate** is the percentage of the list price that yields the discount. That is,

$$\boxed{\text{Discount}} \;=\; \boxed{\text{Discount rate}} \;\cdot\; \boxed{\text{List price}} \;.$$

EXAMPLE 2 ■ Finding the Discount and the Discount Rate

A compact disc player is marked down from its list price of $820 to a sale price of $574. What is the discount rate?

Solution

Verbal model: $\boxed{\text{Discount}} = \boxed{\dfrac{\text{Discount}}{\text{rate}}} \cdot \boxed{\dfrac{\text{List}}{\text{price}}}$

Labels: Discount $= 820 - 574 = 246$ (dollars)
List price $= 820$ (dollars)
Discount rate $= p$ (decimal form)

Algebraic equation: $246 = p(820)$

$$\frac{246}{820} = p$$

$$0.30 = p$$

Thus, the discount rate is 30%. Check this solution in the original problem. ■

EXAMPLE 3 ■ Finding the Hours of Labor

An auto repair bill of $338 lists $170 for parts and the rest for labor. If the labor rate is $28 per hour, how many hours did it take to repair the auto?

Solution

Verbal model: $\boxed{\dfrac{\text{Total}}{\text{bill}}} = \boxed{\dfrac{\text{Price}}{\text{of parts}}} + \boxed{\dfrac{\text{Price}}{\text{of labor}}}$

Labels: Total bill $= 338$ (dollars)
Price of parts $= 170$ (dollars)
Number of hours of labor $= x$ (hours)
Hourly rate for labor $= 28$ (dollars per hour)
Price of labor $= 28x$ (dollars)

Algebraic equation:

$$338 = 170 + 28x$$

$$168 = 28x$$

$$\frac{168}{28} = x$$

$$6 = x$$

Thus, it took 6 hours (of labor) to repair the auto. Check this solution in the original problem. ■

BakeryCorp, a Miami-based supplier of baked goods to restaurants, was started in 1986 by three brothers: (clockwise from left) Luis, José, and Juan Lacal. In 7 years, sales have grown to more than $1,250,000. Information provided by *Your Company Magazine* and *BakeryCorp* [Miami, FL; call tollfree (800) 521-4345].

EXAMPLE 4 ■ Finding the Number of Hours Worked

In 1991, *BakeryCorp* employees earned between $4.50 and $7.00 per hour for the first 40 hours per week and time and a half for overtime. Write an expression that models the weekly earnings of a *BakeryCorp* employee who earns $7 per hour and works overtime each week. How many hours would the employee work in a week to earn $332.50?

Solution

Verbal model:	Hourly wage	·	40 hours	+ 1.5 ·	Hourly wage	·	Overtime hours	=	Weekly earnings

Labels: Hourly wage = 7 (dollars per hour)
Total hours worked in week = t (hours)
Overtime hours = $t - 40$ (hours)
Total earnings = 332.50 (dollars)

Algebraic equation:

$$7(40) + 1.5(7)(t - 40) = 332.50$$

$$280 + 10.5t - 420 = 332.50$$

$$10.5t = 472.50$$

$$t = 45$$

The employee would work 45 hours during the week in question. ■

Mixture Problems

Many real-life problems involve combinations of two or more quantities that make up a new or different quantity. We call such problems **mixture problems.** They are usually composed of the sum of two or more "hidden products" that fit the following verbal model.

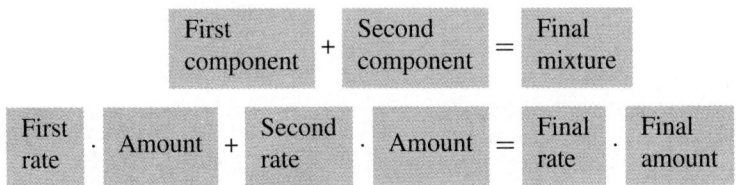

EXAMPLE 5 ■ A Mixture Problem

A nursery wants to mix two types of lawn seed. One type sells for $10 per pound, and the other type sells for $15 per pound. To obtain 20 pounds of a mixture that sells for $12 per pound, how many pounds of each type of seed are needed?

Solution

Verbal model: $\boxed{\text{Total cost of \$10 seed}} + \boxed{\text{Total cost of \$15 seed}} = \boxed{\text{Total cost of \$12 seed}}$

Labels: $10 seed: Cost = 10 (dollars per pound), Amount = x (pounds)
$15 seed: Cost = 15 (dollars per pound), Amount = $20 - x$ (pounds)
$12 seed: Cost = 12 (dollars per pound), Amount = 20 (pounds)

Algebraic equation:

$$10x + 15(20 - x) = 12(20)$$
$$10x + 300 - 15x = 240$$
$$-5x = -60$$
$$x = 12$$

Therefore, the mixture should contain 12 pounds of the $10 seed and $20 - 12 = 8$ pounds of the $15 seed. Check this solution in the original problem. ■

Note that each of these total costs is determined by multiplying the value by the quantity as in Section 3.1. Mention also that each of the three parts of the verbal model is measuring cost.

There are two additional guidelines that you should remember when working mixture problems. First, when you set up your verbal model, be sure to check that you are working with the *same type of units* in each part of the model. For instance, in Example 5, note that each of the three parts of the verbal model are measuring cost. (If two parts were measuring cost and the other part pounds, you would know that the model was incorrect.)

Second, when you have found the solution, go back to the original statement of the problem and check to see that the solution makes sense—both algebraically and logically. For instance, in Example 5, you might perform a check as follows.

$$\overbrace{\left(\begin{smallmatrix}\$10\text{ per}\\\text{pound}\end{smallmatrix}\right)\left(\begin{smallmatrix}12\\\text{pounds}\end{smallmatrix}\right)}^{\$10\text{ seed}} + \overbrace{\left(\begin{smallmatrix}\$15\text{ per}\\\text{pound}\end{smallmatrix}\right)\left(\begin{smallmatrix}8\\\text{pounds}\end{smallmatrix}\right)}^{\$15\text{ seed}} = \overbrace{\left(\begin{smallmatrix}\$12\text{ per}\\\text{pound}\end{smallmatrix}\right)\left(\begin{smallmatrix}20\\\text{pounds}\end{smallmatrix}\right)}^{\$12\text{ seed}}$$

$$\$120 + \$120 = \$240$$

EXAMPLE 6 ■   A Solution Mixture Problem

A pharmacist needs to strengthen a 20% alcohol solution so that it contains 36% alcohol. How much pure alcohol should be added to 240 milliliters of the 20% solution (see Figure 3.1)?

FIGURE 3.1

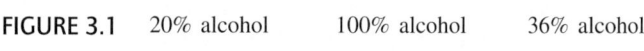

20% alcohol   100% alcohol   36% alcohol

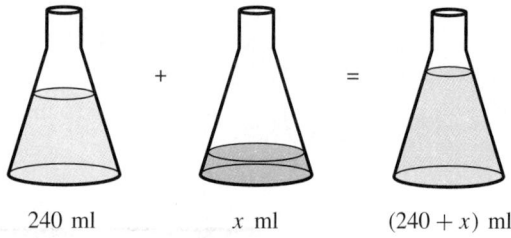

+   =

240 ml   *x* ml   (240 + *x*) ml

Solution

You might mention that if we were *diluting* this solution rather than strengthening it, we would add a solution containing 0% alcohol.

Verbal model: | Amount of alcohol in original solution | + | Amount of alcohol in pure solution | = | Amount of alcohol in final solution |

Labels: Original solution: Percent alcohol = 0.20 (decimal form)
                   Amount of alcohol = 240 (milliliters)

   Pure alcohol: Percent alcohol = 1.00 (decimal form)
                   Amount of alcohol = *x* (milliliters)

   Final solution: Percent alcohol = 0.36 (decimal form)
                   Amount of alcohol = 240 + *x* (milliliters)

Algebraic equation: $0.20(240) + 1.00(x) = 0.36(240 + x)$

$$48 + x = 86.4 + 0.36x$$

$$0.64x = 38.4$$

$$x = \frac{38.4}{0.64} = 60 \text{ milliliters}$$

Thus, the pharmacist should add 60 milliliters of pure alcohol to the original solution. Check this solution in the original problem.   ■

Rate Problems

Time-dependent problems such as those involving distance can be classified as **rate problems.** They fit the verbal model

$$\boxed{\text{Distance}} = \boxed{\text{Rate}} \cdot \boxed{\text{Time}}.$$

For instance, if you are traveling at a constant (or average) rate of 55 miles per hour for 45 minutes, then the total distance traveled is given by

$$\left(55 \, \frac{\text{miles}}{\text{hour}}\right)\left(\frac{45}{60} \, \text{hour}\right) = 41.25 \text{ miles}.$$

As with all problems involving application, be sure to check that the units in the verbal model make sense. For instance, in this problem, the rate is given in *miles per hour*. Therefore, in order for the solution to be given in *miles*, you must convert the time (from minutes) to *hours*. In the model, you think of the two "hours" as cancelling as follows.

$$\left(55 \, \frac{\text{miles}}{\cancel{\text{hour}}}\right)\left(\frac{45}{60} \, \cancel{\text{hour}}\right) = 41.25 \text{ miles}$$

EXAMPLE 7 ■ Distance-Rate Problems

Students are traveling in two cars to a football game 150 miles away. The first car leaves on time and travels at an average speed of 48 miles per hour. The second car starts $\frac{1}{2}$ hour later and travels at an average speed of 58 miles per hour. At these speeds, how long will it take the second car to catch up to the first car?

Caution students *not* to express a half hour as 30 *minutes* when using rates expressed in miles per *hour*.

Solution

Verbal model: $\boxed{\text{Distance}} = \boxed{\text{Rate}} \cdot \boxed{\text{Time}}$

Labels: $d_1 =$ Distance of first car $= 48t$ (miles)
$d_2 =$ Distance of second car $= 58\left(t - \frac{1}{2}\right)$ (miles)

Algebraic equation: $d_1 = d_2$

$$48t = 58\left(t - \frac{1}{2}\right)$$
$$48t = 58t - 29$$
$$29 = 10t$$
$$2.9 = t$$

After the first car travels for 2.9 hours, the second car catches up to it. Thus, it takes the second $2.9 - 0.5 = 2.4$ hours to catch up to the first car. ■

In work problems, the **rate of work** is the *reciprocal* of the time needed to do the entire job. For instance, if it takes 5 hours to complete a job, then the per-hour work rate is

$$\frac{1}{5} \text{ job per hour.}$$

Similarly, if it takes $3\frac{1}{2}$ minutes to complete a job, then the per-minute rate is

$$\frac{1}{3\frac{1}{2}} = \frac{1}{\frac{7}{2}} = \frac{2}{7} \text{ job per minute.}$$

EXAMPLE 8 ■ Work Rate Problem

Consider two machines in a paper manufacturing plant. Machine 1 can produce 2000 pounds of paper in 4 hours. Machine 2 is newer and can produce 2000 pounds of paper in $2\frac{1}{2}$ hours. How long will it take the two machines working together to produce 2000 pounds of paper?

Solution

The verbal model can also be written as $1/t = 1/4 + 2/5$. In this case, $1/t$ = the portion of the work done together in one hour; $1/4$ = the portion of the work done by Machine 1 in one hour; and $2/5$ = the portion of the work done by Machine 2 in one hour.

Verbal model:

Work done	$=$	Portion done by machine 1	$+$	Portion done by machine 2

Labels: Both machines: Work done = 1 (job)
Machine 1: Rate = $\frac{1}{4}$ (job per hour), Time = t (hours)
Machine 2: Rate = $\frac{2}{5}$ (job per hour), Time = t (hours)

Algebraic equation:

$$1 = \left(\frac{1}{4}\right)(t) + \left(\frac{2}{5}\right)(t)$$

$$1 = \left(\frac{1}{4} + \frac{2}{5}\right)(t)$$

$$1 = \left(\frac{13}{20}\right)(t)$$

$$\frac{1}{\frac{13}{20}} = t$$

$$\frac{20}{13} = t$$

Thus, it would take $\frac{20}{13}$ hours (or about 1.54 hours) for both machines to complete the job. Check this solution in the original problem. ■

Note in Example 8 that the "2000 pounds" of paper was unnecessary information. We simply represented the 2000 pounds as "one complete job." This type of unnecessary information in an applied problem is sometimes called a *red herring*.

DISCUSSION PROBLEM   ■   Red Herrings

Challenge problem: If $3x - 7 = 17$, what is $2x + 1$? Answer: 17

Most applied problems in textbooks give precisely the right amount of information that is necessary to solve a given problem. In real life, however, you often must sort through the given information and discard information that is irrelevant to the problem. Such irrelevant information is called a **red herring.** Find the red herrings in the following problems.

(a) Suppose you are hired for a job that pays $8 per hour. After a 90-day review, your salary will be increased to $8.50 per hour. During your first two weeks on the job, you work 80 hours. How much will you be paid for your first two weeks of work?

(b) A person leaves home at noon and drives 50 miles before stopping to fill the car with gas. The person then continues driving until 3 P.M. At the beginning of the trip, the car's odometer reading is 45,768, and at the end of the trip, the reading is 45,930. Find the average speed of the car during the entire trip. ■

Warm-Up

The following warm-up exercises involve skills that were covered in earlier sections. You will use these skills in the exercise set for this section.

In Exercises 1–10, solve the equation.

1. $44 - 16x = 0$

2. $-4(x - 5) = 0$

3. $3[4 + 5(x - 1)] = 6x + 2$

4. $\dfrac{3x}{8} + \dfrac{3}{4} = 2$

5. $\dfrac{x}{3} + \dfrac{x}{2} = \dfrac{1}{3}$

6. $x - \dfrac{x}{4} = 15$

7. $\dfrac{5x}{4} + \dfrac{1}{2} = x - \dfrac{1}{2}$

8. $\dfrac{3}{2}(z + 5) - \dfrac{1}{4}(z + 24) = 0$

9. $0.25x + 0.75(10 - x) = 3$

10. $0.60x + 0.40(100 - x) = 50$

3.2   **EXERCISES**

▦ means that a graphing utility can help you solve the exercise or check your solution.

In Exercises 1–8, find the missing quantities. (Assume that the markup rate is a percentage based on the cost.)

Merchandise	Cost	Selling Price	Markup	Markup Rate
1. Wristwatch	$45.95	$64.33		
2. Bicycle	$84.20	$113.67		
3. Sleeping bag		$250.80	$98.80	

Merchandise	Cost	Selling Price	Markup	Markup Rate
4. VCR		$603.72	$184.47	
5. Van		$26,922.50	$4672.50	
6. Tractor		$16,440.50	$3890.50	
7. Camera	$225.00			85.2%
8. Refrigerator	$732.00			$33\frac{1}{3}\%$

In Exercises 9–16, find the missing quantities. (Assume that the discount rate is a percentage based on the list price.)

Merchandise	List Price	Sale Price	Discount	Discount Rate
9. Wheel alignment	$49.50	$25.74		
10. Earrings	$119.00	$79.73		
11. Coffee service	$300.00		$189.00	
12. Luggage	$345.00		$134.55	
13. Athletic shoes	$95.00			65%
14. Gallon of paint		$15.92		20%
15. Camcorder		$893.10	$251.90	
16. Typewriter		$257.32	$202.18	

17. **Tire Cost** An auto store gives the list price of a tire as $79.42. During a promotional sale, the store is selling four tires for the price of three. The store needs a markup on cost of 10% during the sale. What is the cost to the store of each tire?

18. **Price per Pound** The produce manager of a supermarket pays $22.60 for a 100-pound box of bananas. From past experience, the manager estimates that 10% of the bananas will spoil before they are sold. At what price per pound should the bananas be sold to give the supermarket an average markup rate on cost of 30%?

19. **Comparison Shopping** A department store is offering a discount of 20% on a sewing machine with a list price of $279.95. A mail-order catalog offers the same machine for $228.95 plus $4.32 for shipping. Which is the better buy?

20. **Comparison Shopping** A hardware store is offering a discount of 15% on a 4000-watt generator with a list

price of $699.99. A mail-order company has the same generator for $549.95 plus $14.32 for shipping. Which is the better buy?

21. **Labor Charges** An auto repair bill of $216.37 lists $136.37 for parts and the rest for labor. If the labor rate is $32 per hour, how many hours did it take to repair the auto?

22. **Labor Charges** An appliance repair store charges $35 for the first half hour of a service call. For each additional half hour of labor, there is a charge of $18. Find the length of a service call for which the charge is $89.

23. **Long Distance Rates** The weekday rate for a telephone call is $0.75 for the first minute plus $0.55 for each additional minute. Determine the length of a call that cost $5.15. What would have been the cost of the call if it had been made during the weekend, when there is a 60% discount?

24. **Overtime Hours** Last week you earned $740. If you are paid $14.50 per hour for the first 40 hours and $20 for each hour over 40, how many hours of overtime did you work?

25. **Tip Rate** A customer left a total of $10 for a meal that cost $8.45. Determine the tip rate.

26. **Tip Rate** A customer left a total of $40 for a meal that cost $34.73. Determine the tip rate.

27. **Commission Rate** Determine the commission rate for an employee who earned $450 in commissions for sales of $5000.

28. **Insurance Premiums** The annual insurance premium for a policyholder is $862. Find the annual premium if the policyholder must pay a 20% surcharge because of an accident.

29. **Amount Financed** A customer bought a lawn tractor that cost $4450 plus 6% sales tax. Find the amount of the sales tax and the total bill. Find the amount that was financed if a down payment of $1000 was made.

30. **Weekly Pay** The weekly salary of an employee is $250 plus a 6% commission on the employee's total sales. Find the weekly pay for a week in which the sales are $5500.

31. **Number of Stamps** You have a set of 70 stamps with a total value of $16.50. If this set includes $0.15 stamps and $0.30 stamps, find the number of each type.

32. **Coin Problem** A person has 50 coins in dimes and quarters with a combined value of $10.25. Determine the number of coins of each type.

33. **Opinion Poll** Fourteen hundred people were surveyed in an opinion poll. Political candidates A and B received approximately the same preference, but Candidate C was preferred by twice the number of people as either A or B. Determine the number of people in the sample that preferred Candidate C.

34. **Ticket Sales** Ticket sales for a play total $2200. There are three times as many adult tickets sold, and the prices of the tickets for adults and children are $6 and $4, respectively. Find the number of children's tickets sold.

35. **Feed Mixture** A rancher must purchase 500 bushels of a feed mixture for cattle. She is considering oats and corn, which cost $1.70 per bushel and $3.00 per bushel, respectively. Complete the table, where x is the number of bushels of oats in the mixture.

Oats x	Corn $500 - x$	Price/Bushel of the Mixture
0		
100		
200		
300		
400		
500		

(a) How does the increase in the number of bushels of oats affect the number of bushels of corn in the mixture?

(b) How does the increase in the number of bushels of oats affect the price per bushel of the mixture?

(c) If there were an equal number of bushels of oats and corn in the mixture, how would the price of the mixture be related to the price of each component?

36. **Metal Mixture** A metallurgist is making 5 ounces of an alloy from metal A, which costs $52 per ounce, and metal B, which costs $16 per ounce. Complete the table, where x is the number of ounces of metal A in the alloy.

Metal A x	Metal B $5 - x$	Price/Ounce of the Alloy
0		
1		
2		
3		
4		
5		

(a) How does the increase in the number of ounces of metal A in the alloy affect the number of ounces of metal B in the alloy?

(b) How does the increase in the number of ounces of metal A in the alloy affect the price of the alloy?

(c) If there were equal amounts of metal A and metal B in the alloy, how would the price of the alloy be related to the price of each component?

37. *Floral Arrangement* A floral shop creates a mixed arrangement of roses, at a cost of $1.25 each, and carnations, at a cost of $0.75 each. An arrangement of 1 dozen flowers costs $11.00. Determine the number of roses per dozen flowers in the arrangement.

38. *Nut Mixture* A grocer mixes two kinds of nuts costing $3.88 per pound and $4.88 per pound, to make 100 pounds of a mixture costing $4.13 per pound. How many pounds of each kind of nut were put into the mixture?

39. *Simple Interest* An inheritance of $40,000 is divided into two investments earning 8.5% and 10% simple interest. How much is in each investment if the total interest for 1 year is $3550?

40. *Simple Interest* Four thousand dollars is divided into two investments earning $6\frac{1}{2}\%$ and 9% simple interest. How much is in each investment if the total interest for 1 year is $335?

Mixture Problem In Exercises 41–44, determine the number of units of solution 1 and solution 2 needed to obtain the desired amount and concentration of the final solution. (The percentage indicates how much alcohol is in an alcohol/water mixture.)

	Solution 1	Solution 2	Final Solution	Amount of Final Solution
41.	20%	60%	40%	100 gal
42.	50%	75%	60%	10 L
43.	15%	60%	45%	24 qt
44.	60%	80%	75%	55 gal

45. *Antifreeze Coolant* The cooling system on a truck contains 5 gallons of coolant that is 40% antifreeze. How much must be withdrawn and replaced with 100% antifreeze to bring the coolant in the system to 50% antifreeze?

46. *Fuel Mixture* Suppose you mix gasoline and oil to obtain $2\frac{1}{2}$ gallons of mixture for an engine. The mixture is 40 parts gasoline and 1 part two-cycle oil. How much gasoline must be added to bring the mixture to 50 parts gasoline and 1 part oil?

47. *Flying Distance* Two planes leave an airport at approximately the same time and fly in opposite directions. How far apart are the planes after $1\frac{1}{3}$ hours if their speeds are 480 miles per hour and 600 miles per hour?

48. *Average Speed* An Olympic runner completes a 5000-meter race in 13 minutes and 20 seconds. What is the average speed of the runner?

49. *Travel Time* Two cars start at a given point and travel in the same direction at average speeds of 30 miles per hour and 45 miles per hour (see figure). How much time must elapse before the two cars are 5 miles apart?

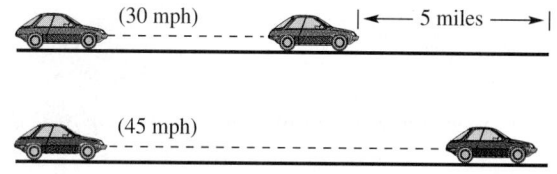

50. *Travel Time* On the first part of a 317-mile trip, a sales representative averaged 58 miles per hour. The sales representative averaged only 52 miles per hour on the last part of the trip because of the increased volume of traffic (see figure). Find the amount of driving time at each speed if the trip took 5 hours and 45 minutes.

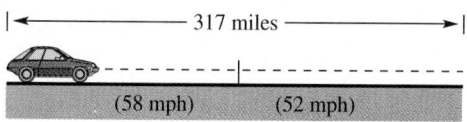

51. *Average Speed* A truck driver traveled at an average speed of 48 miles per hour on a 100-mile trip to pick up a load of freight. On the return trip with the truck fully loaded, the average speed was 40 miles per hour. Find the average speed for the round trip.

52. ***Running Time*** A jogger leaves a given point on a fitness trail, running at a rate of 4 miles per hour. After 10 minutes, a second jogger leaves from the same location, running at 5 miles per hour. How long will it take the second jogger to overtake the first, and how far will they have run?

Environmental Hazard In Exercises 53 and 54, use the following information. You are attending a wedding in Cedar City, Utah. As the bride and groom prepare to go to the wedding reception, helium balloons are released into the air. The wind is blowing as shown on the map in the figure. The radius of the blue circle is 1000 miles, and the radius of the green circle is 2000 miles. After 24 hours, the balloons fall to the earth due to the loss of helium.

Figure for 53 and 54

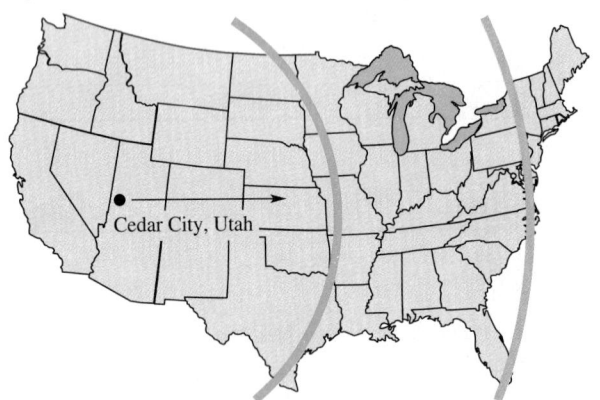

53. Approximately what wind speed will allow the balloons to reach the Atlantic Ocean?

54. Suppose the balloons remain aloft for 36 hours. Approximately what wind speed will allow the balloons to reach the Atlantic Ocean?

55. ***Work Rate Problem*** After two people work together for t hours on a common task, the fractional part of the job done by each of the two workers is $t/3$ and $t/5$. What fractional part of the task has been completed?

56. ***Work Rate Problem*** Suppose you can mow a lawn in 3 hours and your friend can mow it in 4 hours. What fractional part of the lawn can each of you mow in 1 hour? How long will it take both of you to mow the lawn together?

57. ***Work Rate Problem*** It takes 30 minutes for a pump to remove an amount of water from a basement. A larger pump can remove the same amount of water in half the time. If both pumps were operating together, how long would it take to remove the water?

58. ***Work Rate Problem*** It takes 90 minutes to remove an amount of water from a basement when two pumps of different sizes are operating together. It takes 2 hours to remove the same amount of water when only the larger pump is running. How long would it take to remove the same amount of water if only the smaller pump were running?

59. ***Work Rate Problem*** You can complete a typing project in 5 hours. A friend of yours would take 8 hours to complete the same project. What fractional part of the task can be accomplished by each typist in 1 hour? If you both work on the project together, in how many hours can it be completed?

60. ***Average Salary*** The average salary for bus drivers at public schools in the United States between 1975 and 1989 can be approximated by the linear model

$$y = 5.45 + 0.366t,$$

where y represents the salary (in dollars per hour) and t represents the year, with $t = 0$ corresponding to 1980 (see figure). During which year was the average salary $7.28? What was the average annual raise for bus drivers during this 15-year period?

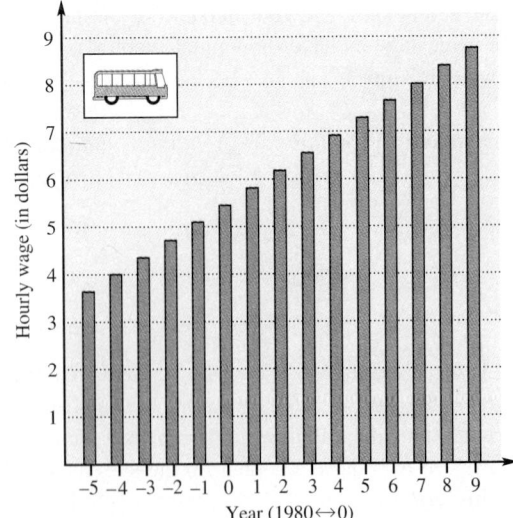

61. *Average Salary* The average salary for cafeteria workers at public schools in the United States between 1975 and 1989 can be approximated by the linear model

$$y = 3.96 + 0.286t,$$

where y represents the salary (in dollars per hour) and t represents the year, with $t = 0$ corresponding to 1980 (see figure). During which year was the average salary $5.39? What was the average annual raise for cafeteria workers during this 15-year period?

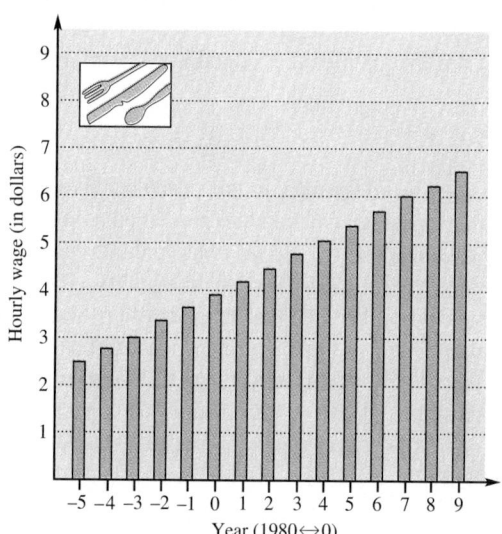

62. *Energy Use* The figure shows how 17.3 million barrels of oil are consumed each day in the United States. (a) How many barrels are used in the transportation sector? (b) We use approximately how many times as much oil in transportation as in heating and cooking? (*Source: American Petroleum Institute, Department of Energy*)

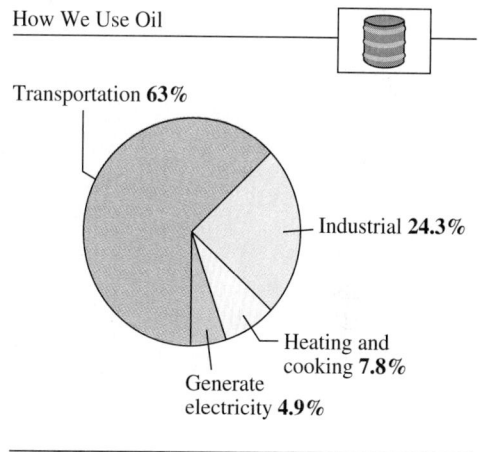

63. *Generation of Electricity* Use the figure to approximate the percentage of electricity generated in the United States that is produced by hydroelectric plants. (*Source: Energy Information Administration; Annual Energy Review*)

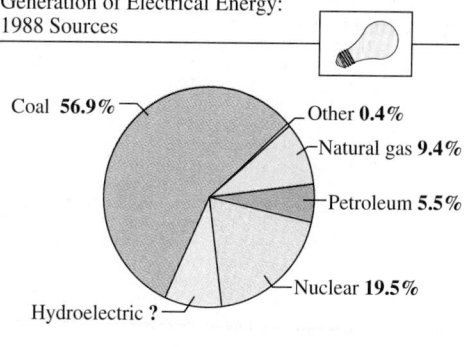

<table>
<tr><td>**SECTION**
3.3</td><td>

Linear Inequalities in One Variable

Inequalities and Intervals on the Real Number Line • *Properties of Inequalities* • *Solving Linear Inequalities* • *Solving Compound Linear Inequalities* • *Applications of Linear Inequalities*
</td></tr>
</table>

Inequalities and Intervals on the Real Number Line

Historical Note: The first recorded use of the inequality symbols < and > is in Artis Analyticae Praxis, published in 1631 and written by the English mathematician Thomas Harriot (1560-1621).

Simple inequalities were introduced in Section 1.1 to *order* the real numbers. There you compared two real numbers using the inequality symbols $<$, $\leq$, $>$, and $\geq$. For instance, because -5 lies to the left of 3 on the real number line, you can say that -5 is less than 3 and write $-5 < 3$. In this section you will study **algebraic inequalities,** which are inequalities that contain one or more variable terms. Some examples are

$$x \leq 4, \quad z \geq -3, \quad y + 2 > 7, \quad \text{and} \quad 4x - 6 < 3x + 8.$$

As with an equation, you **solve an inequality** in the variable x by finding all values of x for which the inequality is true. Such values are called **solutions,** and the solutions are said to **satisfy** the inequality. The set of all real numbers that are solutions of an inequality is called the **solution set** of the inequality.

The set of all points on the real number line that represent the solution set is called the **graph** of the inequality. Graphs of many types of inequalities consist of intervals on the real number line. The four different types of **bounded** intervals on the real number line are summarized as follows. (Note that a parenthesis is used to denote the inequality symbols $<$ and $>$, and a bracket is used to denote the inequality symbols $\leq$ and $\geq$.)

Bounded Intervals on the Real Number Line

Let a and b be real numbers such that $a < b$. The following intervals on the real number line are called **bounded intervals.** The numbers a and b are called the **endpoints** of each interval.

Assure students that mastering the use of interval notation will help them in the future.

Notation	Interval Type	Inequality	Graph
$[a, b]$	Closed	$a \leq x \leq b$	
(a, b)	Open	$a < x < b$	
$[a, b)$	Half-open	$a \leq x < b$	
$(a, b]$	Half-open	$a < x \leq b$	

Note that a closed interval contains both of its endpoints, a half-open interval contains only one of its endpoints, and an open interval does not contain either of its endpoints.

EXAMPLE 1   ■   Sketching the Graph of a Bounded Inequality

Sketch the graphs of the following inequalities, and write the interval notation for each.

(a) $-3 < x \leq 1$ (b) $0 < x < 2$ (c) $1 \leq x \leq 5$

FIGURE 3.2

(a)

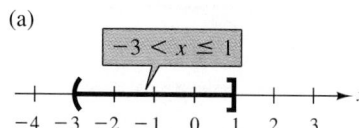

Solution

(a) The graph of this inequality is the set of all numbers on the real line that are greater than -3 and less than or equal to 1, as shown in Figure 3.2(a). The interval notation for this graph is

$(-3, 1]$.   Half-open interval

This interval does not contain the left endpoint, -3, but it does contain the right endpoint, 1.

(b)

(b) The graph of this inequality is the set of all numbers on the real line that are greater than 0 and less than 2, as shown in Figure 3.2(b). The interval notation for this graph is

$(0, 2)$.   Open interval

This interval does not contain the left endpoint, 0, or the right endpoint, 2.

(c)

(c) The graph of this inequality is the set of all numbers on the real line that are greater than or equal to 1 and less than or equal to 5, as shown in Figure 3.2(c). The interval notation for this graph is

$[1, 5]$.   Closed interval

This interval contains its left endpoint, 1, and its right endpoint, 5.   ■

The **length** of the interval $[a, b]$ is simply the distance between its endpoints: $b - a$. The lengths of (a, b), $(a, b]$, and $[a, b)$ are the same. The reason that these four types of intervals are called **bounded** is that each has a finite length. An interval that *does not* have a finite length is called **unbounded** (or **infinite**). The five basic types of unbounded intervals are summarized in the following list.

Unbounded Intervals on the Real Number Line

Let a and b be real numbers. The following intervals on the real number line are called **unbounded intervals.**

Notation	Interval Type	Inequality	Graph
$[a, \infty)$	Half-open	$a \leq x$	
(a, ∞)	Open	$a < x$	
$(-\infty, b]$	Half-open	$x \leq b$	
$(-\infty, b)$	Open	$x < b$	
$(-\infty, \infty)$	Entire real line		

The symbols ∞ (**positive infinity**) and $-\infty$ (**negative infinity**) do not represent real numbers. They are simply convenient symbols used to describe the unboundedness of an interval such as $(1, \infty)$.

EXAMPLE 2 ■ Sketching the Graph of an Unbounded Interval

Sketch the graphs of the following inequalities, and write the interval notation for each.

(a) $-2 < x$ (b) $x \leq 2$

FIGURE 3.3

(a)

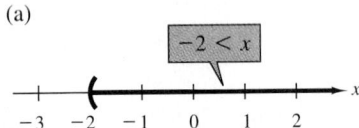

Solution

(a) The graph of this inequality is the set of all numbers on the real line that are greater than -2, as shown in Figure 3.3(a). The interval notation for this graph is

     $(-2, \infty)$.   Open, unbounded interval

This interval does not contain its left endpoint, -2. Moreover, it has no right endpoint (because it is unbounded to the right).

(b)

(b) The graph of this inequality is the set of all numbers on the real line that are less than or equal to 2, as shown in Figure 3.3(b). The interval notation for this graph is

     $(-\infty, 2]$.   Half-open, unbounded interval

This interval has no left endpoint, but it does have a right endpoint, 2, which is included in the interval. ■

Note that the inequality $-2 < x$ in Example 2(a) could be written as $x > -2$. Similarly, the inequality $x \leq 2$ could be written as $2 \geq x$.

Properties of Inequalities

The procedures for solving linear inequalities in one variable are much like those for solving linear equations. To isolate the variable, you can use **Properties of Inequalities.** These properties are similar to the Properties of Equality, but there are two important exceptions. When both sides of an inequality are multiplied or divided by a negative number, the direction of the inequality symbol must be reversed. Here is an example.

$$-2 < 5 \qquad \text{Given inequality}$$
$$(-3)(-2) > (-3)(5) \qquad \text{Multiply both sides by } -3 \text{ and reverse the inequality}$$
$$6 > -15$$

Two inequalities that have the same solution set are called **equivalent.** The following list describes various operations that can be used to create equivalent inequalities.

Properties of Inequalities

Addition and Subtraction Properties

Adding (or subtracting) the same quantity to (or from) both sides of an inequality produces an equivalent inequality.

If $a < b$, then $a + c < b + c$.   If $a < b$, then $a - c < b - c$.

Multiplication and Division Properties (Positive quantities)

Multiplying (or dividing) both sides of an inequality by a *positive* quantity produces an equivalent inequality.

If $a < b$ and $c > 0$, then $ac < bc$.   If $a < b$ and $c > 0$, then $\dfrac{a}{c} < \dfrac{b}{c}$.

Multiplication and Division Properties (Negative quantities)

Multiplying (or dividing) both sides of an inequality by a *negative* quantity produces an equivalent inequality in which the inequality symbol is reversed.

If $a < b$ and $c < 0$, then $ac > bc$.   If $a < b$ and $c < 0$, then $\dfrac{a}{c} > \dfrac{b}{c}$.

Transitive Property

Consider three quantities for which the first quantity is less than the second, and the second is less than the third. It follows that the first quantity must be less than the third.

If $a < b$ and $b < c$, then $a < c$.

NOTE Each of the above properties is true if the symbol $<$ is replaced by $\leq$. Moreover, the letters a, b, and c can be real numbers, variables, or variable expressions. Note that you cannot multiply or divide both sides of an inequality by zero.

Solving Linear Inequalities

The Properties of Inequalities can be applied to solve inequalities. The only type of inequality you will work with in this section is a **linear inequality** in a single variable. For instance, the inequalities

$$2x + 3 > 4, \quad 3x - 4 \leq 2x + 5, \quad \text{and} \quad 3 \geq -5x + 4$$

are all linear inequalities in the variable x, because the (implied) exponent of x in each inequality is 1. (The inequality $x^2 < 4$ is *not* linear in the variable x, because x has an exponent of 2.)

As you read through each of the following examples, pay special attention to the steps in which the inequality symbol is reversed. Remember that when you multiply or divide an inequality by a negative number, you must reverse the inequality symbol.

EXAMPLE 3   ■   Solving a Linear Inequality

Solve the following linear inequality and sketch the graph of its solution set.

$$x + 6 < 9$$

Solution

$x + 6 < 9$	Given inequality
$x + 6 - 6 < 9 - 6$	Subtract 6 from both sides
$x < 3$	Solution set

Thus, the solution set consists of all real numbers that are less than 3. The interval notation for this solution set is

$$(-\infty, 3). \qquad \text{Solution set}$$

The graph of this solution set is shown in Figure 3.4.

■

FIGURE 3.4

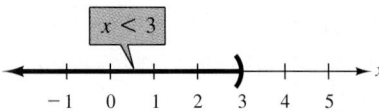

Checking the solution set of an inequality is not as simple as checking the solutions of an equation. (There are usually too many x-values to substitute back into the original inequality.) You can, however, get an indication of the validity of a solution set by substituting a few convenient values of x. For instance, in Example 3, the solution of $x + 6 < 9$ was found to be $x < 3$. Try checking that $x = 0$ satisfies the original inequality, whereas $x = 4$ does not.

EXAMPLE 4 ■ Solving a Linear Inequality

Solve the following linear inequality and sketch the graph of its solution set.

$$2y - 3 \leq -9$$

Solution

$2y - 3 \leq -9$	Given inequality
$2y - 3 + 3 \leq -9 + 3$	Add 3 to both sides
$2y \leq -6$	Combine like terms
$\dfrac{2y}{2} \leq \dfrac{-6}{2}$	Divide both sides by (positive) 2
$y \leq -3$	Solution set

Thus, the solution set consists of all real numbers that are less than or equal to -3. The interval notation for this solution set is

$$(-\infty, -3].$$ Solution set

The graph of this solution set is shown in Figure 3.5.

FIGURE 3.5

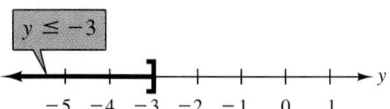

■

EXAMPLE 5 ■ Solving a Linear Inequality

Solve the following linear inequality and sketch the graph of its solution set.

$$8 - 3x \leq 20$$

Solution

$8 - 3x \leq 20$	Given inequality
$8 - 8 - 3x \leq 20 - 8$	Subtract 8 from both sides
$-3x \leq 12$	Combine like terms
$\dfrac{-3x}{-3} \geq \dfrac{12}{-3}$	Divide both sides by -3 and reverse inequality symbol
$x \geq -4$	Solution set

Thus, the solution set consists of all real numbers that are greater than or equal to -4. The interval notation for this solution set is

$$[-4, \infty).$$ Solution set

The graph of this solution set is shown in Figure 3.6.

FIGURE 3.6

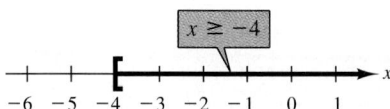

■

The next example involves a linear inequality in which the variable occurs more than once.

EXAMPLE 6   ■   Solving a Linear Inequality

Solve the following linear inequality and sketch the graph of its solution set.

$$9x - 4 > 5x + 2$$

Solution

$9x - 4 > 5x + 2$	Given inequality
$9x - 4 + 4 > 5x + 2 + 4$	Add 4 to both sides
$9x > 5x + 6$	Simplify
$9x - 5x > 6$	Subtract $5x$ from both sides
$4x > 6$	Combine like terms
$\dfrac{4x}{4} > \dfrac{6}{4}$	Divide both sides by (positive) 4
$x > \dfrac{3}{2}$	

Thus, the solution set consists of all real numbers that are greater than $\frac{3}{2}$. The interval notation for this solution set is

$$\left(\frac{3}{2}, \infty\right). \qquad \text{Solution set}$$

The graph of this solution set is shown in Figure 3.7.

■

FIGURE 3.7

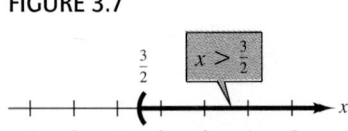

When one (or more) of the terms in a linear inequality involve constant denominators, you can multiply both sides of the equation by the least common denominator. This clears the inequality of fractions. For instance, as a first step in solving the linear inequality

$$10 - \frac{3x}{2} \geq x - 5,$$

you should multiply both sides of the inequality by 2. (Try solving this inequality— you should find that the solution is $x \leq 6$.) This procedure is further demonstrated in Example 7.

EXAMPLE 7   ■   Solving a Linear Inequality

Solve the following linear inequality and sketch the graph of its solution set.

$$\frac{2x}{3} + 12 < \frac{x}{6} + 18$$

Example: $\dfrac{x}{-4} - 2 > \dfrac{3x}{2} + 1$

Solution: $x + 8 > -6x - 4$
$$x + 6x > -4 - 8$$
$$7x < -12$$
$$x < -\dfrac{12}{7}$$

FIGURE 3.8

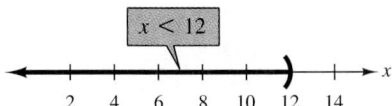

Historical Note: John Wallis (1616-1703), an English mathematician, was the first to use ∞ as a symbol for infinity. He was a man of many talents: he was the king's chaplain; he devised a system for teaching deaf-mutes; and he was an expert in cryptology.

Solution

$\dfrac{2x}{3} + 12 < \dfrac{x}{6} + 18$	Given inequality
$4x + 72 < x + 108$	Multiply both sides by 6
$4x - x < 108 - 72$	Subtract x and 72 from both sides
$3x < 36$	Combine like terms
$x < 12$	Divide both sides by 3

Thus, the solution set consists of all real numbers that are less than 12. The interval notation for this solution set is

$$(-\infty, 12).$$ Solution set

The graph of this solution set is shown in Figure 3.8.

∎

Solving Compound Linear Inequalities

Sometimes it is convenient to write two inequalities as a **compound inequality.** For instance, you can write the two inequalities $-7 \leq 5x - 2$ and $5x - 2 < 8$ more simply as

$$-7 \leq 5x - 2 < 8.$$

This form allows you to solve the two given inequalities together, as demonstrated in Example 8.

EXAMPLE 8   ∎   **Solving a Compound Inequality**

Solve the following compound inequality and sketch the graph of its solution set.

$$-7 \leq 5x - 2 < 8$$

Remind students to operate on all three parts.

Solution

$-7 \leq 5x - 2 < 8$	Given inequality
$-7 + 2 \leq 5x - 2 + 2 < 8 + 2$	Add 2 to all three parts
$-5 \leq 5x < 10$	Combine like terms
$\dfrac{-5}{5} \leq \dfrac{5x}{5} < \dfrac{10}{5}$	Divide each part by 5
$-1 \leq x < 2$	Solution set

FIGURE 3.9

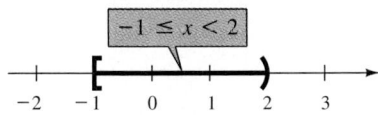

Thus, the solution set consists of all real numbers that are greater than or equal to -1 and less than 2. The interval notation for this solution set is

$$[-1, 2).$$ Solution set

The graph of this solution set is shown in Figure 3.9.

∎

Show students how they can obtain this solution by graphing $-1 \leq x$ and $x < 2$. The intersection of these two inequalities is $-1 \leq x < 2$. See Figure 3.9.

The compound inequality in Example 8 could have been solved in two parts as follows.

$$-7 \leq 5x - 2 \quad \text{and} \quad 5x - 2 < 8$$
$$-5 \leq 5x \qquad\qquad\quad 5x < 10$$
$$-1 \leq x \qquad\qquad\quad x < 2$$

The solution set consists of all real numbers that satisfy *both* inequalities. In other words, the solution set is the set of all values of x for which $-1 \leq x < 2$.

Applications of Linear Inequalities

Linear inequalities in real-life problems arise from statements that involve phrases such as "at least," "no more than," "minimum value," and so on.

Before looking at some applications, consider the following translations of verbal statements into inequalities.

EXAMPLE 9 ■   Translating Verbal Statements

	Verbal Statement	*Inequality*
(a)	x is at most 3.	$x \leq 3$
(b)	x is no more than 3.	$x \leq 3$
(c)	x is at least 3.	$x \geq 3$
(d)	x is more than 3.	$x > 3$
(e)	x is less than 3.	$x < 3$

■

EXAMPLE 10 ■   An Application of an Inequality

An overnight delivery service will not accept any package whose girth exceeds 132 inches. (The girth of a rectangular package is the sum of its length and the perimeter of its cross section.) Suppose you are sending a rectangular package that has square cross sections. If the length of the package is 68 inches, what is the maximum width of one of the sides?

Solution

Begin by making a sketch. In Figure 3.10, notice that the length of the package is 68 inches, and each side is x inches long.

FIGURE 3.10

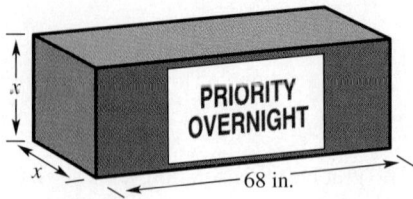

Verbal model: Girth $\leq$ 132 inches

Labels: Width of a side $= x$ (inches)
Length $= 68$ (inches)
Girth $= 4x + 68$ (inches)

Inequality: $4x + 68 \leq 132$
$$4x \leq 64$$
$$x \leq 16 \text{ inches}$$

Thus, the width of each side of the package must be less than or equal to 16 inches.   ∎

EXAMPLE 11   ∎   Finding the Freezing and Boiling Points

You have added enough antifreeze to your car's cooling system to lower the freezing point to $-35°$ C and raise the boiling point to $125°$ C. The coolant will remain a liquid as long as the temperature C (in degrees Celsius) satisfies the inequality $-35 < C < 125$. Write an inequality that describes the liquid state of the coolant in degrees Fahrenheit.

Solution

Let F represent the temperature in degrees Fahrenheit and use the formula $C = \frac{5}{9}(F - 32)$.

$-35 <$ C < 125		Rewrite original inequality
$-35 < \dfrac{5}{9}(F - 32) < 125$		Replace C by $\frac{5}{9}(F - 32)$
$-63 <$ $F - 32$ < 225		Multiply each expression by $\frac{9}{5}$
$-31 <$ F < 257		Add 32 to each expression

In most parts of the United States, people add antifreeze to automobile cooling systems. The antifreeze lowers the freezing point of the coolant and raises the boiling point. (The changes in freezing and boiling points depend on the amount and type of antifreeze that is added.)

The coolant will remain a liquid as long as the temperature stays between $-31°$ F and $257°$ F.   ∎

DISCUSSION PROBLEM   ∎   You Be the Instructor

Writing assignment: Why can't you multiply (or divide) both sides of an equation or an inequality by zero?

Suppose you are teaching an algebra class and one of your students asks the following question: "Why is it that when you add a *negative* number to both sides of an equation you don't have to reverse the inequality, but when you multiply or divide both sides of an inequality by a *negative* number you do have to reverse the inequality?" How would you answer this question?   ∎

Warm-Up

The following warm-up exercises involve skills that were covered in earlier sections. You will use these skills in the exercise set for this section.

In Exercises 1–4, place the correct inequality symbol ($<$ or $>$) between the two real numbers.

1. $-\frac{3}{4}$ ___ -5 **2.** $-\frac{1}{5}$ ___ $-\frac{1}{3}$

3. π ___ -3 **4.** 6 ___ $\frac{13}{2}$

In Exercises 5–10, solve the equation.

5. $-2n + 12 = 5$ **6.** $16 + 7z = 64$ **7.** $\dfrac{12 + x}{4} = 13$

8. $20 - \dfrac{x}{9} = 3$ **9.** $55 - 4(3 - y) = -5$ **10.** $8(t - 24) = 0$

3.3   EXERCISES

▦ means that a graphing utility can help you solve the exercise or check your solution.

In Exercises 1–4, determine whether each given value of x satisfies the inequality.

Inequality   *Value*

1. $7x - 10 > 0$ (a) $x = 3$ (b) $x = -2$ (c) $x = \frac{5}{2}$ (d) $x = \frac{1}{2}$

2. $3x + 2 < \dfrac{7x}{5}$ (a) $x = 0$ (b) $x = 4$ (c) $x = -4$ (d) $x = -1$

3. $0 < \dfrac{x + 5}{6} < 2$ (a) $x = 10$ (b) $x = 4$ (c) $x = 0$ (d) $x = -6$

4. $-2 < \dfrac{3 - x}{2} \le 2$ (a) $x = 0$ (b) $x = 3$ (c) $x = 9$ (d) $x = -12$

In Exercises 5–10, match the inequality with its graph. [The graphs are labeled (a), (b), (c), (d), (e), and (f).]

5. $-1 < x \le 2$ **6.** $-1 < x \le 1$ **7.** $-1 \le x \le 1$

8. $-1 < x < 2$ **9.** $-2 \le x < 1$ **10.** $-2 < x < 1$

(a)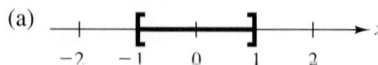

(b)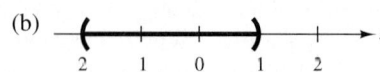

(c)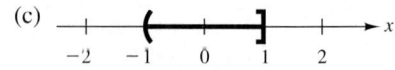

(d)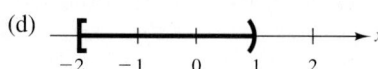

(e)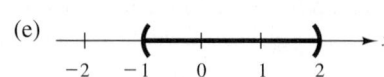

(f)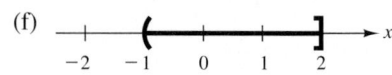

In Exercises 11–16, sketch the graph of the inequality and write the interval notation for the inequality.

11. $-5 < x \leq 3$

12. $4 > x \geq 1$

13. $\frac{3}{2} \geq x > 0$

14. $-7 < x \leq -3$

15. $x < -\frac{5}{2}$

16. $x \geq -5$

In Exercises 17–46, solve the inequality and sketch the solution on the real number line.

17. $x + 7 \leq 9$

18. $z - 4 > 0$

19. $4x < 22$

20. $2x > 5$

21. $-9x \geq 36$

22. $-6x \leq 24$

23. $-\frac{3}{4}x < -6$

24. $-\frac{1}{5}x > -2$

25. $2x - 5 > 9$

26. $3x + 4 \leq 22$

27. $5 - x \leq -2$

28. $1 - y \geq -5$

29. $5 - 3x < 7$

30. $12 - 5x > 5$

31. $3x - 11 > -x + 7$

32. $21x - 11 \leq 6x + 19$

33. $16 < 4(y + 2) - 5(2 - y)$

34. $4[z - 2(z + 1)] < 2 - 7z$

35. $-3(y + 10) \geq 4(y + 10)$

36. $2(4 - z) \geq 8(1 + z)$

37. $\dfrac{x}{6} - \dfrac{x}{4} \leq 1$

38. $\dfrac{x + 3}{6} + \dfrac{x}{8} \geq 1$

39. $0 < 2x - 5 < 9$

40. $-4 \leq 2 - 3(x + 2) < 11$

41. $-3 < \dfrac{2x - 3}{2} < 3$

42. $0 \leq \dfrac{x - 5}{2} < 4$

43. $1 > \dfrac{x - 4}{-3} > -2$

44. $-\dfrac{2}{3} < \dfrac{x - 4}{-6} \leq \dfrac{1}{3}$

45. $\dfrac{2}{5} < x + 1 < \dfrac{4}{5}$

46. $-1 < -\dfrac{x}{6} < 1$

In Exercises 47–50, use inequality notation to denote each of the statements.

47. x is nonnegative.

48. y is more than -2.

49. z is at least 2.

50. x is at least 450 but no more than 500.

In Exercises 51–54, write a statement describing the set of real numbers that satisfies the inequality.

51. $x \geq \frac{5}{2}$

52. $t < 4$

53. $0 < z \leq \pi$

54. $-4 \leq t \leq 4$

55. ***Travel Budget*** A student group has $4500 budgeted for a field trip. The cost of transportation for the trip is $1900. To stay within the budget, all other costs C must be no more than what amount?

56. ***Monthly Budget*** Suppose that you budget $1200 a month for your total expenses. Rent is $400 per month and food costs are $275. To stay within your budget, all other costs C must be no more than what amount?

57. ***Comparing Average Temperatures*** The average temperature in Miami is greater than the average temperature in Washington, D.C., and the average temperature in Washington, D.C. is greater than the average temperature in New York City. How does the average temperature in Miami compare with the average temperature in New York City?

58. *Comparing Elevations* The elevation (above sea level) of San Francisco is less than the elevation of Dallas, and the elevation of Dallas is less than the elevation of Denver. How does the elevation of San Francisco compare with the elevation of Denver?

59. *Operating Costs* A utility company has a fleet of vans. The annual operating cost per van is

$$C = 0.28m + 2900,$$

where m is the number of miles traveled by a van in a year. What number of miles will yield an annual operating cost that is less than $10,000?

60. *Operating Costs* A fuel company has a fleet of trucks. The annual operating cost per truck is

$$C = 0.58m + 7800,$$

where m is the number of miles traveled by a truck in a year. What number of miles will yield an annual operating cost that is less than $25,000?

61. *Profit* The revenue for selling x units of a product is

$$R = 89.95x.$$

The cost of producing x units is

$$C = 61x + 875.$$

To obtain a profit, the revenue must be greater than the cost. For what values of x will this product produce a profit?

62. *Profit* The revenue for selling x units of a product is

$$R = 105.45x.$$

The cost of producing x units is

$$C = 78x + 25,850.$$

To obtain a profit, the revenue must be greater than the cost. For what values of x will this product produce a profit?

63. *Long-Distance Charges* The cost for a long-distance telephone call is $0.96 for the first minute and $0.75 for each additional minute. If the total cost of the call cannot exceed $5, find the interval of time that is available for the call.

64. *Long-Distance Charges* The cost for a long-distance telephone call is $1.45 for the first minute and $0.95 for each additional minute. If the total cost of the call cannot exceed $15, find the interval of time that is available for the call.

65. *Distance* If you live 5 miles from school and your friend lives 3 miles from you (see figure), then the distance d that your friend lives from school is in what interval?

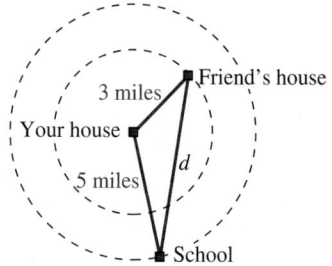

66. *Distance* If you live 10 miles from your work and your work is 5 miles from school, then the distance d that you live from school is in what interval?

67. If $-5 \le t < 8$, then $-t$ must be in what interval?

68. If $-3 \le x \le 10$, then $-x$ must be in what interval?

69. Four times a number n must be at least 12 and no more than 30. What interval contains this number?

70. Five times a number n must be at least 15 and no more than 45. What interval contains this number?

71. Determine all real numbers n such that $\frac{1}{3}n$ must be greater than 7.

72. Determine all real numbers n such that $\frac{1}{4}n$ must be greater than 8.

Air Pollutant Emissions In Exercises 73 and 74, use the following equation, which gives the approximate amount of air pollutant emissions of lead in the continental United States from 1970 to 1986 (see figure at top of next page).

$$y = 199.85 - 11.88t$$

In this model, y represents the amount of pollutant in thousands of metric tons (a metric ton is 2204.6 pounds) and t represents the year, with $t = 0$ corresponding to 1970. (*Source:* U.S. Environmental Protection Agency)

Figure for 73 and 74

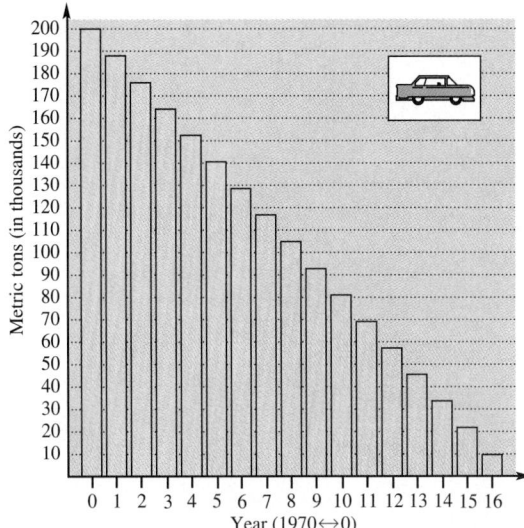

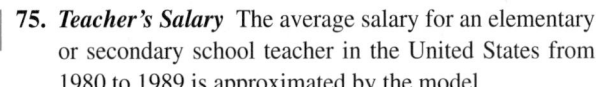

73. During which years from 1970 to 1986 was the air pollutant emission of lead greater than 140.45 thousand metric tons? (This corresponds to an emission of 309,636,070 pounds per year.)

74. During which years from 1970 to 1986 was the air pollutant emission of lead less than 33.53 thousand metric tons? (This corresponds to an emission of 73,920,238 pounds per year.)

75. *Teacher's Salary* The average salary for an elementary or secondary school teacher in the United States from 1980 to 1989 is approximated by the model

$$\text{Salary} = 16,116 + 1496t, \qquad 0 \le t,$$

where the salary is in dollars and the time t represents the calendar year, with $t = 0$ corresponding to 1980 (see figure). Assuming that this model is correct, when did the average salary exceed \$35,000? (*Source:* National Education Association)

If a bat flies into your room and you scream, will it hear you? If you scare the bat and it screams, will you hear it?

Figure for 75

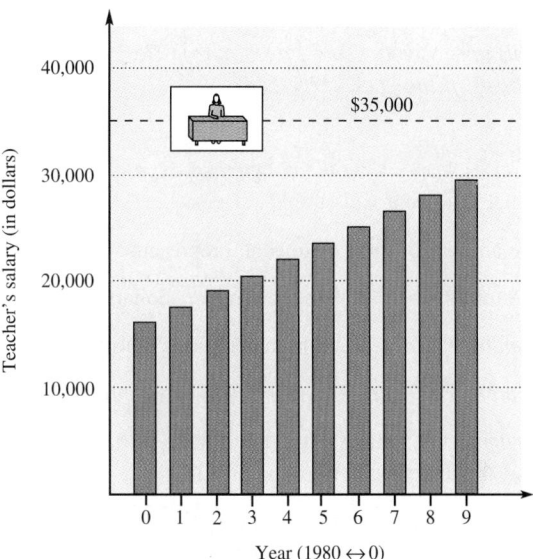

76. *Geometry* If a, b, and c are the lengths of the sides of the triangle in the figure, then $a + b$ must be greater than what value?

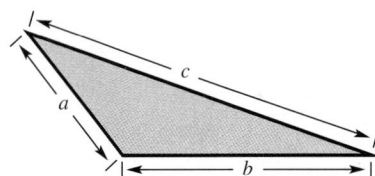

Bats In Exercises 77 and 78, use the following information. The range of a human's voice frequency h (in cycles per second) is about $85 \le h \le 1100$. The range of a human's hearing frequency H is about $20 \le H \le 20,000$.

77. The relationship between a human's voice frequency h and a bat's voice frequency b is

$$h = 85 + \frac{203}{22,000}(b - 10,000).$$

Find the range of a bat's voice frequency.

78. The relationship between a human's hearing frequency H and a bat's hearing frequency B is

$$H = 20 + \frac{999}{5950}(B - 1000).$$

Find the range of a bat's hearing frequency.

Mid-Chapter Quiz

Take this quiz as you would take a quiz in class. After you are done, check your work against the answers in the back of the book.

In Exercises 1 and 2, write an algebraic expression that represents the specified quantity in the verbal statement and simplify if possible.

1. The amount of money (in cents) represented by m nickels and n dimes

2. The total bill on an item that costs L dollars when the sales tax rate is 6%

3. *Number Problem* Find a number such that the product of that number and 12 is 16.

4. *Number Problem* The sum of two consecutive integers is 525. Find the two integers.

5. *Perimeter* A rectangle has a perimeter of 52 inches, and its length is 12 times its width. Find the dimensions of the rectangle.

6. *Projected Expenses* From January through May, a company's expenses have totaled $234,980. If the monthly expenses continue at this rate, what will be the total expenses for the year?

7. *Monthly Expenses* The expenses for a small company for January are shown in the pie chart. What is the percentage of the total monthly expenses for each budget item?

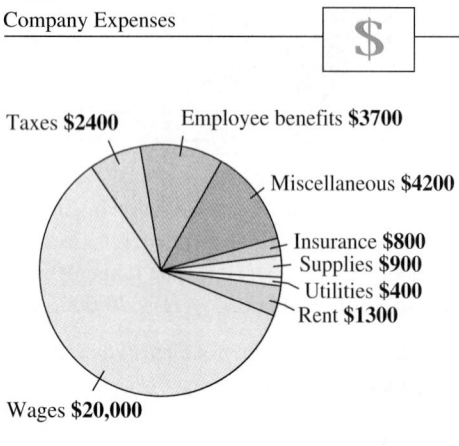

Company Expenses $

Taxes **$2400**
Employee benefits **$3700**
Miscellaneous **$4200**
Insurance **$800**
Supplies **$900**
Utilities **$400**
Rent **$1300**
Wages **$20,000**

8. *State Income Tax* The amount of state income tax withheld from your paycheck each week is $12.50 when your gross pay is $625. Find the ratio of tax to gross pay.

9. *Fuel Usage* A tractor uses 5 gallons of diesel fuel to plow for 105 minutes. Assuming that conditions remain the same, determine the number of gallons of fuel used in 6 hours.

10. **Similar Triangles** Solve for the length x of the side of the triangle on the right in the figure by using the fact that corresponding sides of similar triangles are proportional. (*Note:* Assume that the two triangles are similar.)

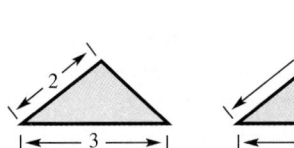 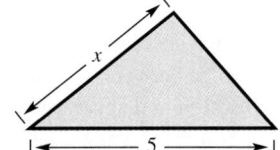

11. **Sale Price** The list price of a camera is $235. If the camera is on sale for 20% off the list price, find the sale price.

12. **Overtime Hours** An employee is paid a regular hourly rate of $12.00 and an overtime rate of $18.00 (for each hour over 40 hours in a week). During a given week, the employee earned $660. How many hours of overtime did the employee work?

13. **Alcohol Solution** A pharmacist needs to strengthen a 10% alcohol solution so that it contains 15% alcohol. How much pure alcohol should be added to 100 milliliters of the 10% solution? (Round your answer to the nearest milliliter.)

14. **Work Rate Problem** Find the time required for two people working together to complete a task if individually it takes them 4.5 hours and 6 hours.

In Exercises 15–18, solve the inequality and sketch the solution on the real number line.

15. $3x - 4 \geq 8$

16. $8 - 3t < 7$

17. $-5 < 3 - 2x \leq 15$

18. $\dfrac{x}{3} + \dfrac{x}{2} \geq 1$

19. **Profit** The revenue obtained from the sale of x units of a product is $R = 12x$. The cost for producing x units is $C = 8x + 132$. For a profit to be made, the revenue must be greater than the cost. For what values of x will this product produce a profit?

Equations and Inequalities Involving Absolute Value

Solving Equations Involving Absolute Value • *Solving Inequalities Involving Absolute Value* • *Connections with Graphs of Absolute Value Functions*

Solving Equations Involving Absolute Value

You know from Section 1.1 that the absolute value of a real number can be described geometrically as the *distance* between the number and zero. By placing a variable inside the absolute value sign and specifying a distance, you can form an **absolute value equation.** For instance, using x as the variable and 3 as the distance, you obtain the absolute value equation

$$|x| = 3.$$

The only solutions of this equation are -3 and 3, because these are the only two real numbers whose distance from zero is 3 (see Figure 3.11). In other words, the absolute value equation $|x| = 3$ has exactly two solutions: $x = -3$ and $x = 3$.

FIGURE 3.11

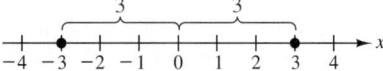

Solving an Absolute Value Equation

Let x be a variable or a variable expression and let a be a real number such that $a \geq 0$. The solution of the equation $|x| = a$ is given by $x = -a$ or $x = a$. That is,

$$|x| = a \quad \Longrightarrow \quad x = -a \quad \text{or} \quad x = a.$$

NOTE The strategy for solving absolute value equations is to *rewrite* the equation in *equivalent forms* that can be solved by previously learned methods. This is a common strategy in mathematics. That is, when you encounter a new type of problem, you try to rewrite the problem so that it can be solved by previously learned techniques.

EXAMPLE 1 ■ Solving Absolute Value Equations

Sometimes you can explain the concept of solving equations of the form $|x| = a$ with a variation of the question, "Of what number can you take the absolute value to get a?"

Solve the following absolute value equations.

(a) $|x| = 10$ (b) $|x| = 0$ (c) $|y| = -1$

Solution

(a) This absolute value equation is equivalent to the following two linear equations.

$$x = -10 \quad \text{or} \quad x = 10$$

Thus, the given equation has two solutions: -10 and 10.

(b) This absolute value equation is equivalent to the equations

$$x = -0 \quad \text{or} \quad x = 0.$$

Because both of these equations are the same, you can conclude that the given equation has only one solution: 0.

(c) This absolute value equation has *no solution* because it is not possible for the absolute value of a real number to be negative. ■

EXAMPLE 2 ■ **Solving Absolute Value Equations**

Solve the absolute value equation

$$|x - 2| = 4.$$

Solution

$|x - 2| = 4$		Given equation
$x - 2 = -4 \quad$ or $\quad x - 2 = 4$		Equivalent equations
$x - 2 + 2 = -4 + 2 \qquad x - 2 + 2 = 4 + 2$		Add 2 to both sides
$x = -2 \qquad\qquad x = 6$		Solutions

Thus, the given equation has two solutions: -2 and 6. You can check the solutions as follows.

Check

$|x - 2| = 4 \qquad |x - 2| = 4$		Given equation
$|-2 - 2| \overset{?}{=} 4 \qquad |6 - 2| \overset{?}{=} 4$		Replace x by -2 and 6
$|-4| \overset{?}{=} 4 \qquad |4| \overset{?}{=} 4$		Simplify
$4 = 4 \qquad\qquad 4 = 4$		Both solutions check ■

EXAMPLE 3 ■ **Solving Absolute Value Equations**

Solve the absolute value equation

$$|3x + 4| = 10.$$

Solution

$|3x + 4| = 10$		Given equation
$3x + 4 = -10 \quad$ or $\quad 3x + 4 = 10$		Equivalent equations
$3x + 4 - 4 = -10 - 4 \qquad 3x + 4 - 4 = 10 - 4$		Subtract 4 from both sides
$3x = -14 \qquad\qquad 3x = 6$		Combine like terms
$x = -\dfrac{14}{3} \qquad\qquad x = 2$		Solutions

Thus, the given equation has two solutions: $-\frac{14}{3}$ and 2. Check these solutions in the original equation. ■

EXAMPLE 4   ■   Solving Absolute Value Equations

Solve the absolute value equation

$$|x + 1| = -2.$$

Solution

This equation is tricky. If you were to use the technique shown in the first three examples, you would solve the two equations $x + 1 = -(-2)$ and $x + 1 = -2$. The resulting solutions would be $x = 1$ and $x = -3$. After checking, however, you would discover that neither of these numbers is a solution! Can you see what went wrong? Notice in the original equation that the right-hand side is negative. This type of equation,

$$|ax + b| = \text{(negative number)},$$

never has a solution, because the absolute value of a real number cannot be negative. Thus, the given equation has no solution.   ■

The equation in the next example is not given in the **standard form**

$$|ax + b| = c, \qquad c \geq 0.$$

Notice that the first step in solving such an equation is to write it in standard form.

EXAMPLE 5   ■   An Absolute Value Equation in Nonstandard Form

Solve the absolute value equation

$$|2x - 1| + 3 = 8.$$

Solution

$|2x - 1| + 3 = 8$	Given equation
$|2x - 1| = 5$	Standard form
$2x - 1 = -5$ or $2x - 1 = 5$	Equivalent equations
$2x = -4$ $\qquad$ $2x = 6$	Add 1 to both sides
$x = -2$ $\qquad$ $x = 3$	Solutions

Thus, the given equation has two solutions: -2 and 3. Check these solutions in the original equation.   ■

If two algebraic expressions are equal in absolute value, then they must either be *equal* to each other, or they must be the *opposites* of each other. Thus, to solve equations of the form

$$|ax + b| = |cx + d|,$$

form the two linear equations

$$\overbrace{ax + b = -(cx + d)}^{\text{Expressions opposite}} \quad \text{and} \quad \overbrace{ax + b = cx + d}^{\text{Expressions equal}}.$$

This technique is used in the next two examples.

EXAMPLE 6   ■   **Solving an Equation Involving Absolute Value**

Solve the equation

$$|3x - 4| = |7x - 16|.$$

Solution

$$|3x - 4| = |7x - 16| \qquad \text{Given equation}$$

$$3x - 4 = -(7x - 16) \quad \text{or} \quad 3x - 4 = 7x - 16 \qquad \text{Equivalent equations}$$

$$3x - 4 = -7x + 16 \qquad\qquad 3x = 7x - 12$$

$$10x = 20 \qquad\qquad\qquad -4x = -12$$

$$x = 2 \qquad\qquad\qquad\quad x = 3 \qquad \text{Solutions}$$

Thus, the given equation has two solutions: 2 and 3. Check these solutions in the original equation.   ■

When solving equations of the form $|ax + b| = |cx + d|$, it is possible that one of the following equations will not have a solution.

$$ax + b = -(cx + d), \qquad ax + b = cx + d$$

This is the case in the next example.

EXAMPLE 7   ■   **Solving an Equation Involving Absolute Value**

Solve the following equation.

$$|x + 5| = |x + 11|$$

Solution

By equating the expression $x + 5$ to the opposite of $x + 11$, you obtain

$$x + 5 = -(x + 11)$$

$$x + 5 = -x - 11$$

$$x = -x - 16$$

$$2x = -16$$

$$x = -8.$$

Thus, $x = -8$ is one solution. However, by setting the two expressions equal to each other, you obtain

$$x + 5 = x + 11$$

$$x = x + 6$$

$$0 = 6,$$

which makes no sense. Therefore, the original equation has only one solution: -8. Check this solution in the original equation.   ■

Solving Inequalities Involving Absolute Value

To see how to solve inequalities involving absolute value, consider the following comparisons.

Equation or Inequality	*Geometric Solution*	*Graph*
$|x| = 2$	Values of x that lie two units from 0: $x = -2$ or $x = 2$	
$|x| < 2$	Values of x that lie *less than* two units from 0: $-2 < x < 2$	
$|x| > 2$	Values of x that lie *more than* two units from 0: $x < -2$ or $x > 2$	

These comparisons suggest the following rule for solving inequalities involving absolute value.

Solving an Absolute Value Inequality

Let x be a variable or an algebraic expression and let a be a real number such that $a > 0$.

1. The solutions of $|x| < a$ are all values of x that lie between $-a$ and a. That is,

$$|x| < a \quad \text{if and only if} \quad -a < x < a.$$

2. The solutions of $|x| > a$ are all values of x that are less than $-a$ or greater than a. That is,

$$|x| > a \quad \text{if and only if} \quad x < -a \text{ or } x > a.$$

These two rules are also valid if $<$ is replaced by $\leq$ and $>$ is replaced by $\geq$.

EXAMPLE 8 ■ Solving an Absolute Value Inequality

Additional examples:

Inequality	Solution
1. $|2x + 1| \geq 3$	$(-\infty, -2] \cup [1, \infty)$
2. $|7 - x| < 2$	$(5, 9)$
3. $|3x - 2| > -3$	$(-\infty, \infty)$
4. $|5 + 4x| < -1$	no solutions

Solve the following inequality and sketch the graph of its solution set.

$$|x - 5| < 2$$

Solution

$|x - 5| < 2$	Given inequality
$-2 < x - 5 < 2$	Equivalent inequalities
$-2 + 5 < x - 5 + 5 < 2 + 5$	Add 5 to all three parts
$3 < x < 7$	Solution set

Thus, the solution set consists of all real numbers that are greater than 3 and less than 7. The interval notation for this solution set is

$(3, 7)$.   Solution set

The graph of this solution set is shown in Figure 3.12.

FIGURE 3.12

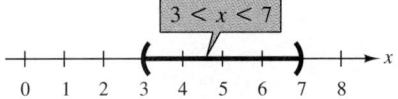

In Example 8, note that absolute value inequalities of the form $|x| < a$ can be solved using a "compound inequality" (see Example 8 in Section 3.3). Inequalities of the form $|x| > a$ cannot be solved using a compound inequality. Instead, you must solve two separate inequalities, as demonstrated in the next example.

EXAMPLE 9 ■ **Solving an Absolute Value Inequality**

Solve the following inequality and sketch the graph of its solution set.

$|3x - 4| \geq 5$

Solution

Reinforce the use of *or* in this problem.

$|3x - 4| \geq 5$			Given inequality
$3x - 4 \leq -5$	or	$3x - 4 \geq 5$	Equivalent inequalities
$3x - 4 + 4 \leq -5 + 4$		$3x - 4 + 4 \geq 5 + 4$	Add 4 to both sides
$3x \leq -1$		$3x \geq 9$	Combine like terms
$\dfrac{3x}{3} \leq \dfrac{-1}{3}$		$\dfrac{3x}{3} \geq \dfrac{9}{3}$	Divide both sides by 3
$x \leq -\dfrac{1}{3}$		$x \geq 3$	Solution set

Thus, the solution set consists of all real numbers that are less than or equal to $-\frac{1}{3}$ or greater than or equal to 3. The interval notation for this solution set is

$$\left(-\infty, -\frac{1}{3}\right] \cup [3, \infty).$$ Solution set

The symbol $\cup$ is called a **union** symbol, and it is used to denote the combining of two sets. The graph of this solution set is shown in Figure 3.13.

FIGURE 3.13

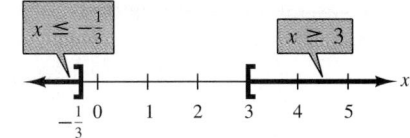

The pit organs of a rattlesnake can detect temperature changes as small as 0.005° F. By moving its head back and forth, a rattlesnake can detect warm prey, even in the dark.

EXAMPLE 10   ■   Creating a Model

FIGURE 3.14

To test the accuracy of a rattlesnake's "pit-organ sensory system," a biologist blindfolded a rattlesnake and presented the snake with a warm "target." Of 36 strikes, the snake was on target 17 times. In fact, the snake was within 5 degrees of the target for 30 of the strikes. Let A represent the number of degrees by which the snake is off target. Then $A = 0$ represents a strike that is aimed directly at the target. Positive values of A represent strikes to the right of the target and negative values of A represent strikes to the left of the target. Use the diagram shown in Figure 3.14 to write an absolute-value inequality that describes the interval in which the 36 strikes occurred.

Solution

From the diagram, you can see that the snake was never off by more than 15 degrees in either direction. As a compound inequality, this can be represented by

$$-15 \leq A \leq 15.$$

As an absolute value inequality, the interval in which the strikes occurred can be represented by

$$|A| \leq 15.$$ ■

Connections with Graphs of Absolute Value Functions

Each of the absolute value equations and inequalities in this section has contained only *one* variable. This section concludes with a look at a technique that uses the graphs of equations with *two* variables to perform a graphic check of solutions. The basic procedure is illustrated in Example 11.

EXAMPLE 11   ■   Performing a Graphic Check

Solve the inequality $|x - 5| < 2$. Then show how the graph of $y = |x - 5|$ can be used to check your solution graphically.

Solution

The algebraic solution of this inequality is shown in Example 8. From that example, you know that the solution is

$$3 < x < 7.$$

To check this solution graphically, sketch the graphs of

$$y = |x - 5| \quad \text{and} \quad y = 2$$

Note that each of the previous examples involving absolute value inequalities could be solved analytically and graphically.

on the same coordinate plane, as shown in Figure 3.15. In the figure, notice that for all values of x between 3 and 7, the graph of $y = |x - 5|$ lies *below* the line given by $y = 2$. Thus, you can conclude that

$$|x - 5| < 2$$

for all x such that $3 < x < 7$.

FIGURE 3.15

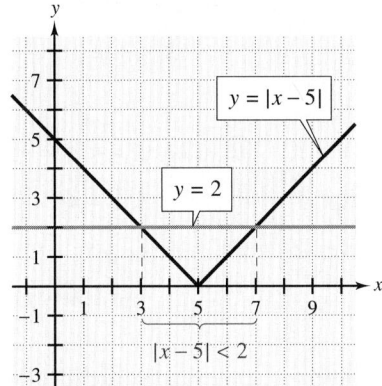

NOTE   Try using Figure 3.15 to solve graphically the inequality $|x - 5| > 2$. From the graph, can you see why the solution of this inequality involves *two* separate linear inequalities?

DISCUSSION PROBLEM   ■   A Challenging Problem

The following inequality is not covered by the standard solution techniques in this section. Try to find the solution set of this inequality and then write a short paragraph describing the procedure that you used.

$$|x - 4| \le \frac{x}{2}$$

■   **Warm-Up**

The following warm-up exercises involve skills that were covered in earlier sections. You will use these skills in the exercise set for this section.

In Exercises 1–6, evaluate the expression.

1. $|-15|$ **2.** $|-3.2|$ **3.** $-|72|$

4. $-|-4|$ **5.** $|14 - 32|$ **6.** $-|46.8 - 27|$

In Exercises 7–10, solve the inequality.

7. $-4 < 10x + 1 < 6$ **8.** $-2 \le 1 - 2x \le 2$

9. $-3 \le \dfrac{x}{2} \le 3$ **10.** $-5 < x - 25 < 5$

3.4   EXERCISES

▦   means that a graphing utility can help you solve the exercise or check your solution.

In Exercises 1–24, solve the equation. (Some of the equations have no solutions.)

1. $|t| = 45$ **2.** $|s| = 16$ **3.** $|h| = 0$ **4.** $|x| = -82$

5. $|x - 16| = 5$ **6.** $|z - 100| = 100$ **7.** $|2s + 3| = 25$ **8.** $|7a + 6| = 8$

9. $|32 - 3y| = 16$ **10.** $|3 - 5x| = 13$ **11.** $|3x + 4| = -16$ **12.** $|20 - 5t| = 50$

13. $|5x - 3| + 8 = 22$ **14.** $4|x + 5| = 9$ **15.** $\left|\frac{2}{3}x + 4\right| = 9$ **16.** $\left|\frac{3}{2} - \frac{4}{5}x\right| = 1$

17. $|0.32x - 2| = 4$ **18.** $|3.2 - 1.054x| = 2$ **19.** $|x + 8| = |2x + 1|$ **20.** $|10 - 3x| = |x + 7|$

21. $|45 - 4x| = |32 - 3x|$ **22.** $|5x + 4| = |3x + 25|$ **23.** $|2x + 3| = |2x + 12|$ **24.** $|x - 3| = |x + 5|$

In Exercises 25–32, determine whether the given values of the variable are solutions of the inequality.

Inequality	*Values*			*Inequality*	*Values*		
25. $|x| < 3$	(a) $x = 2$	(b) $x = -4$		**26.** $|x| \le 5$	(a) $x = -7$	(b) $x = -4$	
	(c) $x = 4$	(d) $x = -1$			(c) $x = 4$	(d) $x = 9$	
27. $|x| \ge 3$	(a) $x = 2$	(b) $x = -4$		**28.** $|x| > 5$	(a) $x = -7$	(b) $x = -4$	
	(c) $x = 4$	(d) $x = -1$			(c) $x = 4$	(d) $x = 9$	
29. $|x - 7| < 3$	(a) $x = 9$	(b) $x = -4$		**30.** $|x - 3| \le 5$	(a) $x = 16$	(b) $x = 3$	
	(c) $x = 11$	(d) $x = 6$			(c) $x = -2$	(d) $x = -3$	
31. $|x - 7| \ge 3$	(a) $x = 9$	(b) $x = -4$		**32.** $|x - 3| > 5$	(a) $x = 16$	(b) $x = 3$	
	(c) $x = 11$	(d) $x = 6$			(c) $x = -2$	(d) $x = -3$	

In Exercises 33–36, sketch a graph that shows the real numbers that satisfy the statement.

33. All real numbers greater than -2 *and* less than 5

34. All real numbers greater than or equal to 3 *and* less than 10

35. All real numbers less than or equal to 4 *or* greater than 7

36. All real numbers less than -6 *or* greater than or equal to 6

In Exercises 37–56, solve the inequality and sketch the graph of the solution on a number line.

37. $|y| < 4$

38. $|x| < 6$

39. $|y - 2| \le 4$

40. $|x - 3| \le 6$

41. $|y + 2| < 4$

42. $|x + 3| < 6$

43. $|y| \ge 4$

44. $|x| \ge 6$

45. $|y - 2| > 4$

46. $|x - 3| > 6$

47. $|y + 2| \ge 4$

48. $|x + 3| \ge 6$

49. $|2x| < 14$

50. $|4z| \le 9$

51. $\left|\dfrac{y}{3}\right| \le 3$

52. $\left|\dfrac{t}{2}\right| < 4$

53. $|3x + 2| > 4$

54. $|2x - 1| \ge 3$

55. $|x - 5| + 3 > 5$

56. $|a + 1| - 4 < 0$

In Exercises 57–66, solve the inequality.

57. $|0.2x - 3| < 4$

58. $|1.5t - 8| \le 16$

59. $\dfrac{|x + 2|}{10} \le 8$

60. $\dfrac{|y - 16|}{4} < 30$

61. $|6t + 15| \ge 30$

62. $|3t + 1| > 5$

63. $\dfrac{|s - 3|}{5} > 4$

64. $\dfrac{|a + 6|}{2} \ge 16$

65. $\left|\dfrac{z}{10} - 3\right| > 8$

66. $\left|\dfrac{x}{8} + 1\right| < 0$

67. *Temperature* The operating temperature for an electronic device must satisfy the inequality

$$|t - 72| < 10,$$

where t is the temperature in degrees Fahrenheit. Sketch the graph of the solution of the inequality on a number line.

68. *Time Study* A time study was conducted to determine the length of time required to perform a particular task in a manufacturing process. The time required by approximately two-thirds of the workers in the study satisfied the inequality

$$\left|\dfrac{t - 15.6}{1.9}\right| < 1,$$

where t is the time in minutes. Sketch the graph of the solution of the inequality on a number line.

In Exercises 69–74, write an inequality (involving absolute values) to represent the given interval.

69.
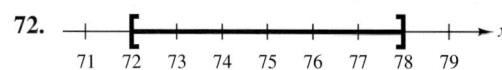

70.

71.

72.

73.

15 16 17 18 19 20 21 22 23

74.

−15 −14 −13 −12 −11 −10 −9 −8 −7

In Exercises 75–78, write an inequality (involving absolute values) to represent the verbal statement.

75. The set of all real numbers x whose distance from 0 is less than 3

76. The set of all real numbers x whose distance from 0 is more than 2

77. The set of all real numbers x whose distance from 5 is more than 6

78. The set of all real numbers x whose distance from 16 is less than 5

American Doctors In Exercises 79 and 80, use the following information. The table and bar graph show the number N of practicing medical doctors per 100,000 people in each of the 50 states in 1988. (*Source:* U.S. Bureau of Census)

AK 138	HI 225	ME 173	NJ 234	SD 138
AL 151	IA 145	MI 180	NM 173	TN 187
AR 144	ID 120	MN 212	NV 158	TX 169
AZ 191	IL 210	MO 190	NY 307	UT 177
CA 242	IN 151	MS 125	OH 191	VA 204
CO 202	KS 169	MT 150	OK 145	VT 244
CT 293	KY 161	NC 179	OR 196	WA 206
DE 191	LA 184	ND 167	PA 227	WI 181
FL 203	MA 322	NE 168	RI 244	WV 164
GA 167	MD 325	NH 186	SC 156	WY 134

Figure for 79 and 80

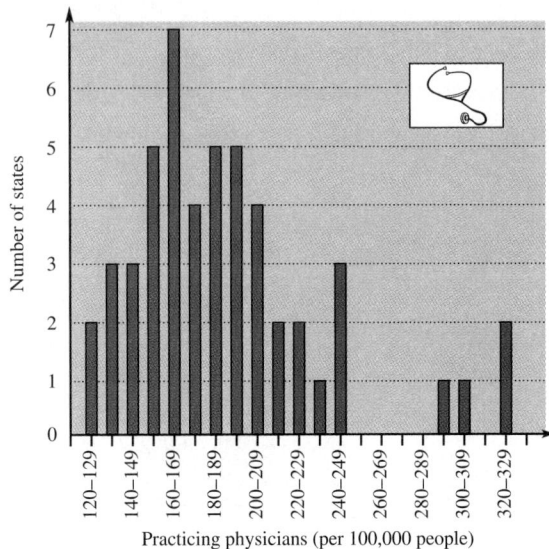

79. The range of N within the 50 states is given by the compound inequality

$$120 \le N \le 325.$$

Write an absolute value inequality that is equivalent to this compound inequality.

80. Drop the five lowest and five highest values of N. Complete the following statement: *Eighty percent of the values of N can be described by the compound inequality*

$$\boxed{} \le N \le \boxed{}$$

Write an absolute value inequality that is equivalent to the compound inequality.

In Exercises 81–84, solve the inequality and check the solution graphically (see Example 11).

81. $|x - 3| < 4$ **82.** $|x + 1| < 3$

83. $|x + 2| > 4$ **84.** $|x - 1| > 5$

Slope: An Aid to Graphing Linear Functions

Linear Functions • The Slope of a Line • Slope: An Aid to Sketching Lines • Parallel and Perpendicular Lines

Linear Functions

A **linear function in** x is a function of the form

$$f(x) = ax + b,$$

where $a \neq 0$. The name *linear* is used to describe this type of function because the graph of a linear function is a line. Linear functions can be written using function notation *or* as linear equations involving x and y. For instance,

$$f(x) = 3x + 2 \quad \text{and} \quad y = 3x + 2$$

represent the same linear function.

EXAMPLE 1   ■   Using Function Notation

The equation $8x - 6y - 9 = 0$ is a linear equation in x and y. Solve this equation for y. Then use function notation to write y as a function of x.

Solution

Begin by solving for y. Then substitute $f(x)$ for y.

$8x - 6y - 9 = 0$	Original equation
$8x - 6y = 9$	Add 9 to both sides
$-6y = -8x + 9$	Subtract 8x from both sides
$y = \dfrac{-8x}{-6} + \dfrac{9}{-6}$	Divide both sides by -6
$y = \dfrac{4}{3}x - \dfrac{3}{2}$	Simplify
$f(x) = \dfrac{4}{3}x - \dfrac{3}{2}$	Substitute $f(x)$ for y

■

EXAMPLE 2   ■   Using Intercepts to Sketch the Graph of a Linear Function

Find the intercepts of the linear function

$$f(x) = -2x + 4.$$

Then, use the intercepts to sketch the graph of the function.

Solution

To find the y-intercept, let $x = 0$.

$$f(0) = -2(0) + 4 = 4$$

Thus, the graph crosses the y-axis at the point $(0, 4)$. To find the x-intercept, let $f(x) = 0$ and solve for x.

$0 = -2x + 4$	Let $f(x) = 0$
$2x = 4$	Add $2x$ to both sides
$x = 2$	Divide both sides by 2

Thus, the graph crosses the x-axis at the point $(2, 0)$. You now know two points on the graph of the function. Moreover, because the graph of a linear function is a line, you can sketch the line by plotting the two points and drawing the line that passes through them, as shown in Figure 3.16.

FIGURE 3.16

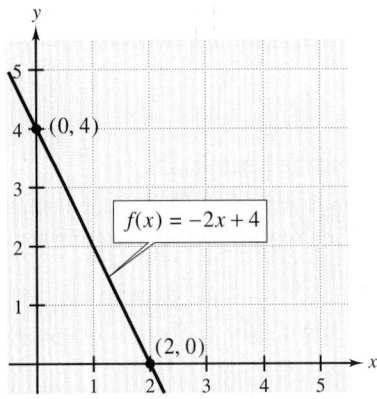

The Slope of a Line

The **slope** of a nonvertical line represents the number of units a line rises or falls vertically for each unit of horizontal change from left to right. For example, the line in Figure 3.17 rises two units for each unit of horizontal change from left to right.

Historical Note: The French verb meaning to mount, to climb, or to rise is *monter*. Because Descartes was largely responsible for the development of analytical geometry, his use of *m*—short for *monter*—to indicate the slope became the accepted term among European mathematicians.

FIGURE 3.17

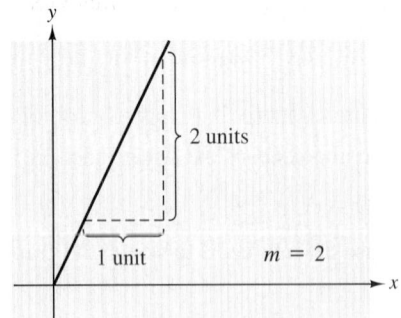

Definition of the Slope of a Line	The **slope** m of the nonvertical line passing through the points (x_1, y_1) and (x_2, y_2) is

$$m = \frac{y_2 - y_1}{x_2 - x_1} = \frac{\text{change in } y}{\text{change in } x},$$

where $x_1 \neq x_2$ (see Figure 3.18).

FIGURE 3.18

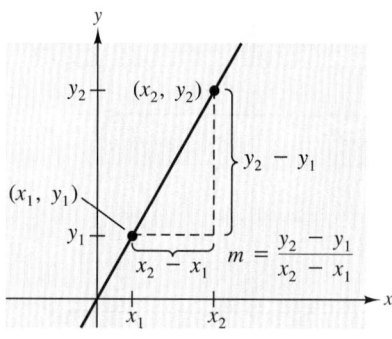

It may be helpful to show an example of each of the correct forms. The slope of the line through $(-1, 3)$ and $(4, -2)$ is

$$m = \frac{-2 - 3}{4 - (-1)} = \frac{-5}{5} = -1 \quad \text{or}$$

$$m = \frac{3 - (-2)}{-1 - 4} = \frac{5}{-5} = -1.$$

When using the formula for the slope of a line, remember that the *order of subtraction* is important. Given two points on a line, you are free to label either one of them as (x_1, y_1), and the other as (x_2, y_2). However, once this has been done, you must form the numerator and denominator using the same order of subtraction.

$$\underbrace{m = \frac{y_2 - y_1}{x_2 - x_1}}_{\text{Correct}} \qquad \underbrace{m = \frac{y_1 - y_2}{x_1 - x_2}}_{\text{Correct}} \qquad \underbrace{m = \frac{y_2 - y_1}{x_1 - x_2}}_{\text{Incorrect}}$$

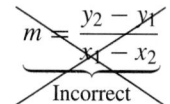

EXAMPLE 3   ■   Finding the Slope of a Line Passing Through Two Points

(a) The slope of the line through $(x_1, y_1) = (3, 4)$ and $(x_2, y_2) = (1, -2)$ is

$$m = \frac{y_2 - y_1}{x_2 - x_1} \quad \Longleftarrow \quad \text{Difference in } y\text{-values}$$
$$\phantom{m = \frac{y_2 - y_1}{x_2 - x_1}} \quad \Longleftarrow \quad \text{Difference in } x\text{-values}$$

$$= \frac{-2 - 4}{1 - 3}$$

$$= \frac{-6}{-2}$$

$$= 3.$$

(b) The slope of the line through $(-2, 4)$ and $(3, 4)$ is

$$m = \frac{4 - 4}{3 - (-2)} = \frac{0}{5} = 0.$$

(c) The slope of the line through $(2, -3)$ and $(0, 1)$ is

$$m = \frac{1 - (-3)}{0 - 2} = \frac{4}{-2} = -2.$$

The graphs of these three lines are shown in Figure 3.19.

FIGURE 3.19

(a) Positive slope

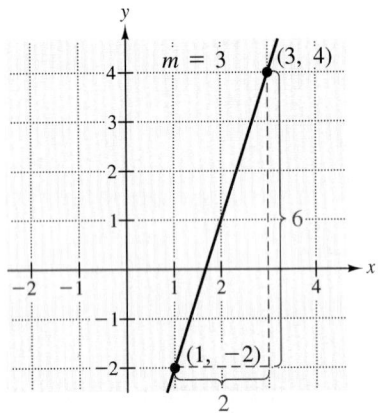

(b) Zero slope

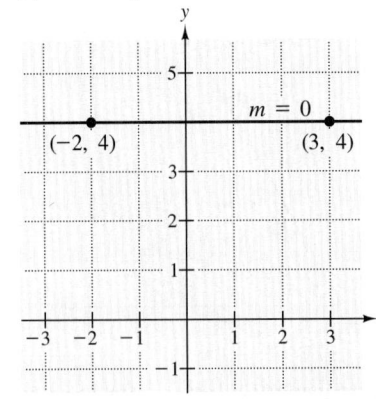

(c) Negative slope

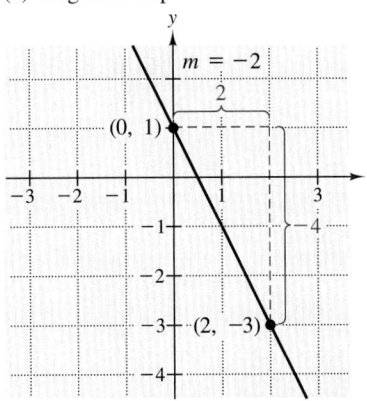

∎

The definition of slope does not apply to vertical lines. For instance, consider the points $(2, 3)$ and $(2, -1)$ on the vertical line shown in Figure 3.20. Applying the formula for slope, you have

$$\frac{-1 - 3}{2 - 2}. \qquad \text{Undefined division by zero}$$

FIGURE 3.20

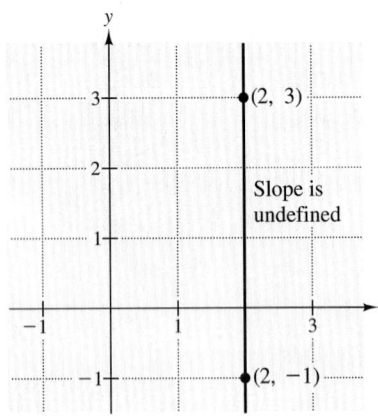

Because division by zero is not defined, the slope of a vertical line is not defined.

From the slopes of the lines shown in Figures 3.19 and 3.20, we make the following generalizations about the slope of a line.

1. A line with positive slope ($m > 0$) *rises* from left to right.

2. A line with negative slope ($m < 0$) *falls* from left to right.

3. A line with zero slope ($m = 0$) is *horizontal.*

4. A line with undefined slope is *vertical.*

EXAMPLE 4   ■   Using Slope to Describe Lines

Problem: Given the points $(2w, 3)$ and $(w, -2)$, what is the value of w if the slope of the line connecting these two points is equal to 2?

Solution: $\dfrac{-2 - 3}{w - 2w} = \dfrac{-5}{-w} = \dfrac{5}{w} = 2,$

so $w = \dfrac{5}{2}.$

Use slope to determine whether the line through each of the pairs of points rises, falls, is horizontal, or is vertical.

(a) $(3, 3), \quad (4, 1)$

(b) $(-2, 3), \quad (-2, -1)$

(c) $(-1, -3), \quad (2, 2)$

(d) $(-1, 2), \quad (2, 2)$

Solution

(a) Because

$$m = \frac{1 - 3}{4 - 3} = \frac{-2}{1} = -2 < 0, \qquad \text{Negative slope}$$

the line falls (from left to right).

(b) Because

$$m = \frac{-1 - 3}{-2 - (-2)} = \frac{-4}{0}, \qquad \text{Undefined slope}$$

is undefined, the line is vertical.

(c) Because

$$m = \frac{2 - (-3)}{2 - (-1)} = \frac{5}{3} > 0, \qquad \text{Positive slope}$$

the line rises (from left to right).

(d) Because

$$m = \frac{2 - 2}{2 - (-1)} = \frac{0}{3} = 0, \qquad \text{Zero slope}$$

the line is horizontal.

The graphs of these four lines are shown in Figure 3.21.

FIGURE 3.21

(a) Line falls
    Negative slope

(b) Vertical line
    Undefined slope

(c) Line rises
    Positive slope

(d) Horizontal line
    Zero slope

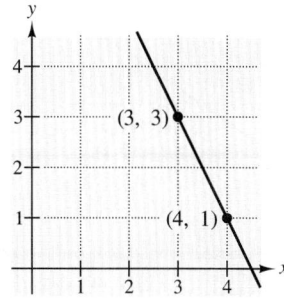
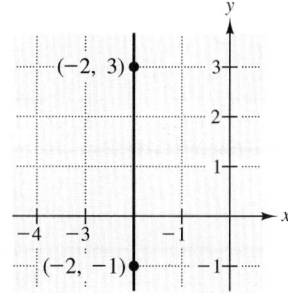
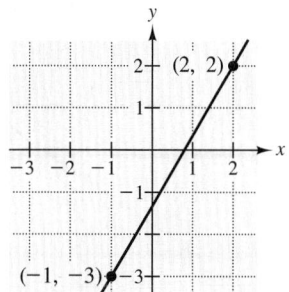
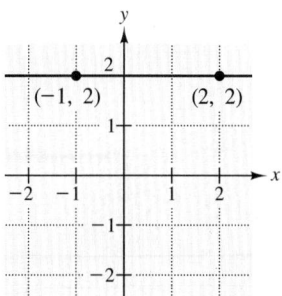

■

Any two points on a line can be used to calculate its slope. This is demonstrated in the next example.

EXAMPLE 5 ■ **Finding the Slope of a Line**

Sketch the line given by $2x - 3y = 6$. Then find the slope of the line. (Choose two different pairs of points on the line and show that the same slope is obtained from either pair.)

Solution

Begin by solving the given equation for y.

$$y = \frac{2}{3}x - 2$$

Then construct a table of values as shown in Table 3.1.

TABLE 3.1

x	-3	0	3	6
$y = \frac{2}{3}x - 2$	-4	-2	0	2
Solution Points	$(-3, -4)$	$(0, -2)$	$(3, 0)$	$(6, 2)$

The line is shown in Figure 3.22(a) and (b). Any two points on the line can be used to calculate the slope of a line. Here are two examples.

Points: $(-3, -4)$ and $(0, -2)$ *Points*: $(6, 2)$ and $(3, 0)$

$$m = \frac{-2 - (-4)}{0 - (-3)} = \frac{2}{3} \qquad m = \frac{0 - 2}{3 - 6} = \frac{-2}{-3} = \frac{2}{3}$$

Note that the same slope is obtained using either set of points. (Try other pairs of points on the line to confirm that you also obtain a slope of $m = \frac{2}{3}$.)

FIGURE 3.22   (a)

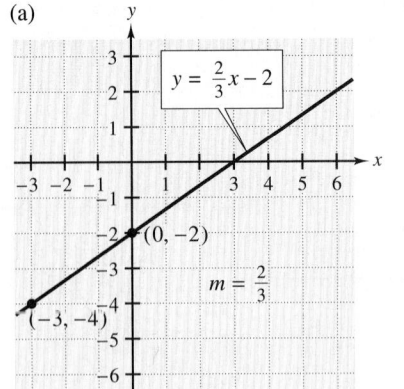

(b)

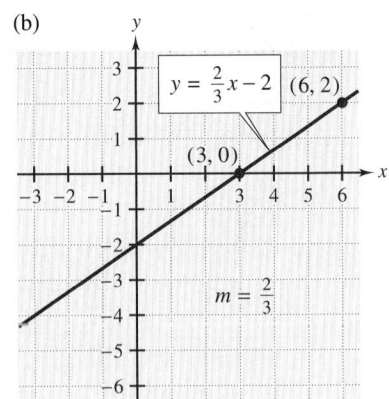

Slope: An Aid to Sketching Lines

As you learned earlier, before making a table of values for an equation, it is helpful to solve the equation for y. When you do this for a linear equation, you obtain some very useful information. Consider the results of Example 5.

$$y = \boxed{\frac{2}{3}}\, x + \boxed{-2}$$

$\qquad\qquad\uparrow\qquad\qquad\uparrow$

$\qquad\quad$ Slope $\qquad$ y-intercept $(0,\ -2)$

This is called the **slope-intercept** form of the equation of the line because the coefficient of x is the slope of the line and the constant term, -2, is the y-intercept of the line.

Slope-Intercept Form of the Equation of a Line	The graph of the equation $$y = mx + b$$ is a line with a slope of m and a y-intercept of $(0, b)$.

When sketching a line, remember that you need only two points to determine the line.

EXAMPLE 6 ■   Using the Slope and y-Intercept to Sketch a Line

Use the slope and y-intercept to sketch the graph of $y = \frac{1}{2}x + 1$.

Solution

The equation is already in slope-intercept form.

$$y = mx + b$$
$$y = \frac{1}{2}x + 1$$

Thus, the slope of the line is $m = \frac{1}{2}$ and the y-intercept is $(0, b) = (0, 1)$. Knowing this information, you can sketch the graph of the line as follows. First, plot the y-intercept. Then, using a slope of $\frac{1}{2}$,

$$m = \frac{1}{2} = \frac{\text{change in } y}{\text{change in } x},$$

locate a second point on the line by moving one unit up and two units to the right, as shown in Figure 3.23. Finally, obtain the graph by drawing the line that passes through the two points.

FIGURE 3.23

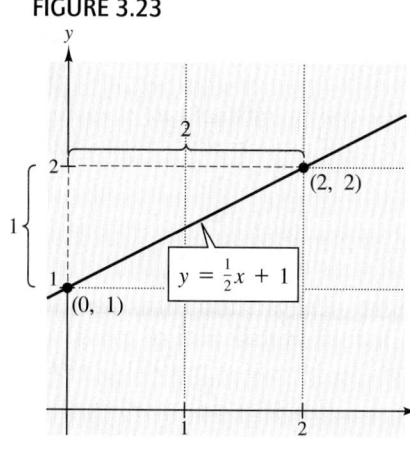

■

EXAMPLE 7   ■   Using the Slope and *y*-Intercept to Sketch a Line

Use the slope and *y*-intercept to sketch the graph of $4x + 3y - 2 = 0$.

Solution

Begin by writing the equation in slope-intercept form.

$$4x + 3y - 2 = 0 \qquad \text{Given equation}$$
$$3y = -4x + 2$$
$$y = -\frac{4}{3}x + \frac{2}{3} \qquad \text{Slope-intercept form}$$

From this form, you can see that the slope is $m = -\frac{4}{3}$ and the *y*-intercept is $(0, b) = \left(0, \frac{2}{3}\right)$. To sketch the line, first plot the *y*-intercept. Then, using a slope of $-\frac{4}{3}$,

$$m = \frac{-4}{3} = \frac{\text{change in } y}{\text{change in } x},$$

locate a second point on the line by moving four units down and three units to the right, as shown in Figure 3.24. (Note that we moved down because the slope is negative.)

FIGURE 3.24

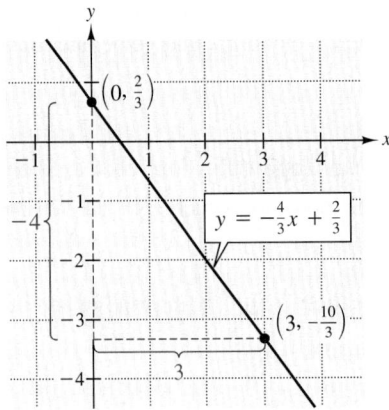

In real-life problems, slope is often used to describe a **constant rate of change** or an **average rate of change.** In such cases, units of measure are assigned, such as miles per hour or dollars per year.

EXAMPLE 8   ■   Slope as a Rate of Change

Discuss the pros and cons of predicting what percentage of engineering doctorates will be awarded to women in the years 1995, 2000, and 2040.

In 1980, 3.5% of the doctorates in engineering were awarded to women. By 1990, 8.5% were being earned by women. Find the average rate of change in the percentage of engineering doctorates earned by women from 1980 to 1990. (*Source*: National Research Council)

Solution

Let p represent the percentage of engineering doctorates earned by women and let t represent the year. The two given data points are represented by (t_1, p_1) and (t_2, p_2).

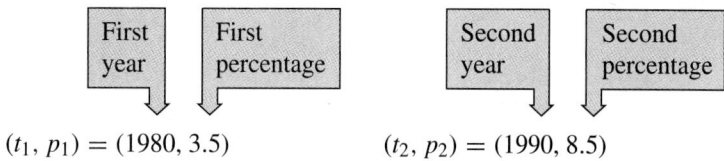

First year	First percentage		Second year	Second percentage

$(t_1, p_1) = (1980, 3.5)$ $(t_2, p_2) = (1990, 8.5)$

Now, use the formula for slope to find the average rate of change.

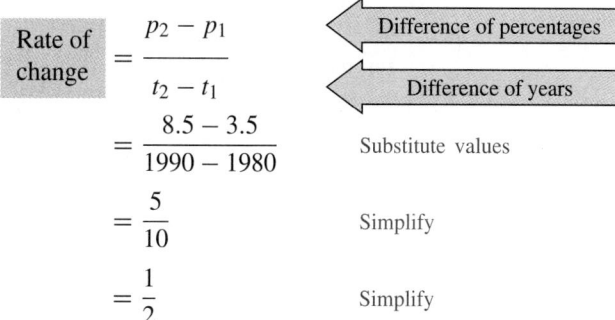

$$\text{Rate of change} = \frac{p_2 - p_1}{t_2 - t_1} \qquad \text{Difference of percentages} \qquad \text{Difference of years}$$

$$= \frac{8.5 - 3.5}{1990 - 1980} \qquad \text{Substitute values}$$

$$= \frac{5}{10} \qquad \text{Simplify}$$

$$= \frac{1}{2} \qquad \text{Simplify}$$

From 1980 through 1990, the *average rate of change* in the percentage of engineering doctorates awarded to women was $\frac{1}{2}\%$ per year. (The exact changes in percentage varied from one year to the next, as shown in the scatter plot in Figure 3.25.)

FIGURE 3.25

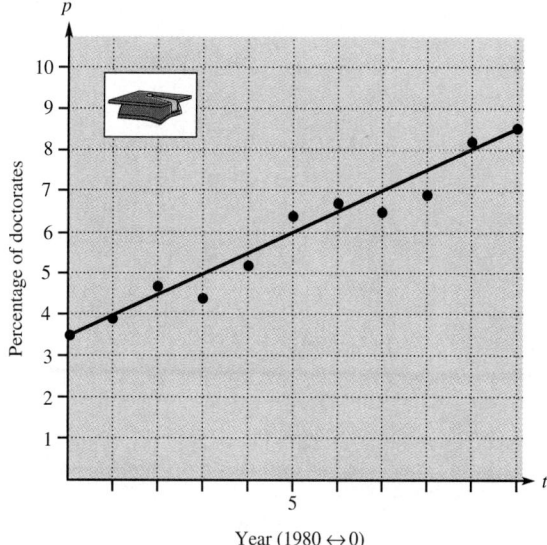

Year (1980 ↔ 0)

Parallel and Perpendicular Lines

You know from geometry that two lines in a plane are *parallel* if they do not intersect. What this means in terms of their slopes is suggested by the following example.

EXAMPLE 9   ■   **Lines That Have the Same Slope**

In the same rectangular coordinate system, sketch the two lines given by the equations

$$y = -2x \quad \text{and} \quad y = -2x + 3.$$

Solution

For the line given by

$$y = -2x,$$

the slope is $m = -2$ and the y-intercept is $(0, 0)$. For the line

$$y = -2x + 3,$$

the slope is also $m = -2$ and the y-intercept is $(0, 3)$. The graphs of these two lines are shown in Figure 3.26.

FIGURE 3.26

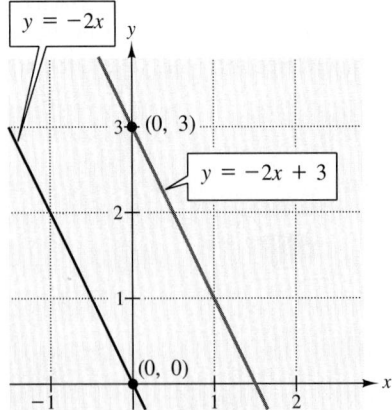

In Example 9, notice that the two lines have the same slope *and* appear to be parallel. The following rule tells us that this is always the case.

Parallel Lines	Two distinct nonvertical lines are parallel if and only if they have the same slope.

NOTE The phrase "if and only if" in this rule is used in mathematics as a way to write two statements in one. The first statement says that *if two distinct nonvertical lines have the same slope, then they must be parallel.* The second statement says that *if two distinct nonvertical lines are parallel, then they must have the same slope.*

From geometry, you know that two lines in a plane are *perpendicular* if they intersect at right (90°) angles. Two *nonvertical* lines are perpendicular if their slopes are negative reciprocals of each other. For instance, the line given by $y = 2x$ has a slope of $m_1 = 2$, and the line given by $y = -\frac{1}{2}x + 1$ has a slope of $m_2 = -\frac{1}{2}$. Note that the slope of the second line is the negative reciprocal of the slope of the first line. That is,

$$m_2 = -\frac{1}{2} = -\frac{1}{m_1}.$$

From Figure 3.27, it appears that the two lines are perpendicular. This result is generalized in the following rule.

FIGURE 3.27

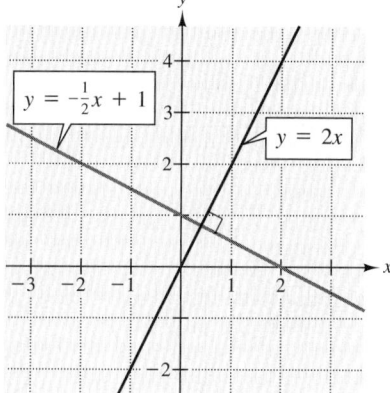

Perpendicular Lines

Consider two nonvertical lines whose slopes are m_1 and m_2. The two lines are perpendicular if and only if their slopes are *negative reciprocals* of each other. That is,

$$m_1 = -\frac{1}{m_2}.$$

EXAMPLE 10 ■ **Determining Whether Lines are Parallel or Perpendicular**

Determine whether the following pairs of lines are parallel or perpendicular, or neither.

(a) Line 1: $y = -2x - 5$ Line 2: $y = \frac{1}{2}x + 1$

(b) Line 1: $y = \frac{1}{2}x + 3$ Line 2: $y = \frac{1}{2}x - 1$

Solution

Also mention that $m_1 m_2 = -1$.

(a) For this pair of lines, the first line has a slope of $m_1 = -2$, and the second line has a slope of $m_2 = \frac{1}{2}$. Because these slopes are negative reciprocals of each other, the two lines must be perpendicular, as shown in Figure 3.28.

(b) Each of these two lines has a slope of $m = \frac{1}{2}$. Therefore, the two lines must be parallel, as shown in Figure 3.29.

FIGURE 3.28   **FIGURE 3.29**

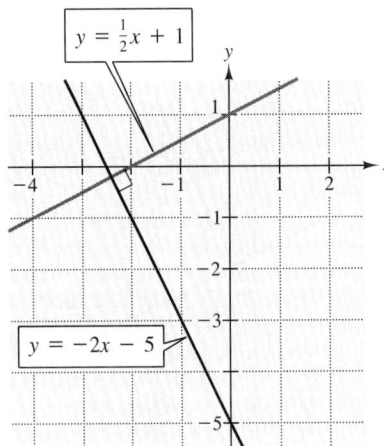

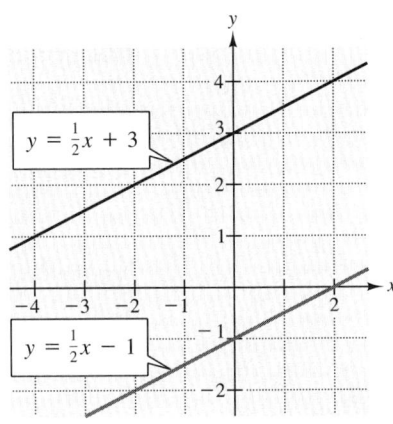

■

DISCUSSION PROBLEM ■ **An Application of Slope**

In 1980, a college had an enrollment of 4000 students. By 1990, the enrollment had increased to 6000 students.

(a) What was the average annual change in enrollment from 1980 to 1990?

(b) Use the average annual change in enrollment to estimate the enrollment in 1984, 1988, and 1992.

Year	1980	1984	1988	1990	1992
Enrollment	4000			6000	

(c) Sketch the line represented by the data given in the table in part (b). What is the slope of this line?

(d) Write a short paragraph that compares the concepts of *slope* and *average rate of change*.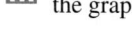

Warm-Up

The following warm-up exercises involve skills that were covered in earlier sections. You will use these skills in the exercise set for this section.

In Exercises 1–4, evaluate the quantity.

1. $\dfrac{8-3}{12-2}$

2. $\dfrac{-3-(-6)}{11-5}$

3. $\dfrac{14-(-15)}{-13-(-11)}$

4. $\dfrac{-3-7}{0-(-20)}$

In Exercises 5 and 6, find $-1/m$ for the given value of m.

5. $m = -\frac{5}{16}$

6. $m = \frac{3}{7}$

In Exercises 7–10, solve for y in terms of x.

7. $5x - 8y = 5$

8. $9x + 2y = 0$

9. $y - (-7) = 2[x - (-3)]$

10. $y - 14 = \frac{3}{4}(x - 12)$

3.5   **EXERCISES**

means that a graphing utility can help you solve the exercise or check your solution.

In Exercises 1–6, use function notation to write y as a function of x.

1. $3x - y + 10 = 0$

2. $x + 2y - 8 = 0$

3. $0.2x + 0.8y - 4.5 = 0$

4. $\frac{2}{3}x - \frac{3}{4}y = 0$

5. $y - 4 = 0$

6. $y - 3 = \frac{2}{3}(x + 2)$

In Exercises 7–10, find the intercepts of the linear function and use the intercepts to sketch the graph of the function.

7. $f(x) = 2x - 6$

8. $g(x) = -3x + 6$

9. $g(x) = -\frac{3}{4}x + 1$

10. $f(x) = 0.8x + 2$

In Exercises 11–16, estimate the slope of the line from its graph.

11.

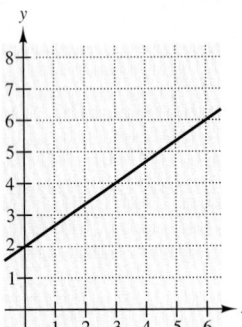

12.

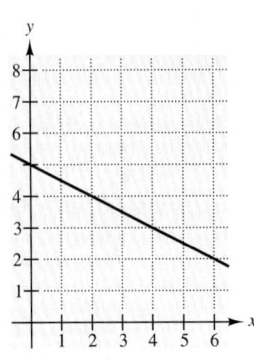

13.

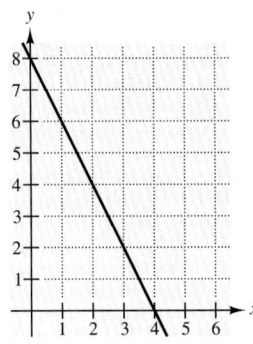

14.

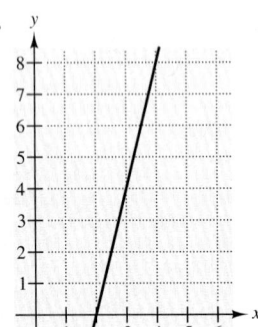

15.

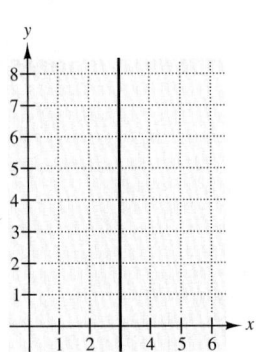

16.

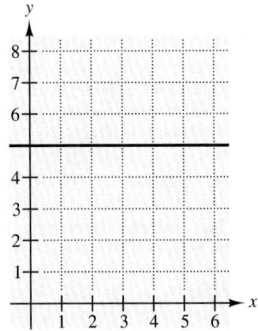

In Exercises 17 and 18, identify the line in the graph that has the specified slope m.

17.

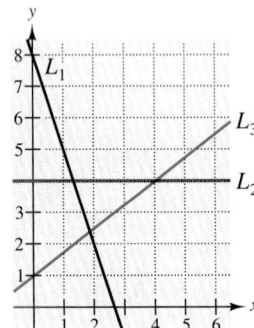

(a) $m = \frac{3}{4}$

(b) $m = 0$

(c) $m = -3$

18.

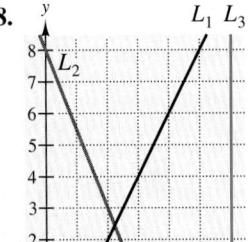

(a) $m = -\frac{5}{2}$

(b) m is undefined.

(c) $m = 2$

In Exercises 19–36, plot the points and find the slope (if possible) of the line passing through each pair of points. State whether the line rises, falls, is horizontal, or is vertical.

19. $(0, 0), (7, 5)$

20. $(0, 0), (-3, -4)$

21. $(-2, -3), (6, 1)$

22. $(0, -4), (6, 0)$

23. $(0, 12), (8, 0)$

24. $(3, 6), (5, -2)$

25. $(-5, -3), (-5, 4)$

26. $(0, -8), (-5, 0)$

27. $(2, -5), (7, -5)$

28. $(-2, 1), (-4, -3)$

29. $\left(\frac{3}{4}, 2\right), \left(5, -\frac{5}{2}\right)$

30. $\left(-\frac{3}{2}, -\frac{1}{2}\right), \left(\frac{5}{8}, \frac{1}{2}\right)$

31. $(4.2, -1), (-4.2, 6)$

32. $(3.4, 0), (3.4, 1)$

33. $(2.5, -2), (4.75, 5.25)$

34. $(0, 4.5), (3, 4.5)$

35. $(-3, -5), (3, 5)$

36. $(-5, 0), (0, 6)$

In Exercises 37–40, find the unknown coordinate of the indicated point such that the line through the two points will have the specified slope.

37. $(4, 5), (x, 7), m = -\frac{2}{3}$

38. $(x, -2), (5, 0), m = \frac{3}{4}$

39. $(-3, y), (9, 3), m = \frac{3}{2}$

40. $(-3, 20), (2, y), m = -6$

In Exercises 41–48, sketch the graph of a line through the point $(0, 1)$ having the given slope.

41. $m = 2$

42. $m = -2$

43. m is undefined.

44. $m = 0$

45. $m = -\frac{4}{3}$

46. $m = \frac{2}{3}$

47. $m = 0.4$

48. $m = -0.8$

In Exercises 49–52, plot the x- and y-intercepts and sketch the graph of the line.

49. $2x - y + 4 = 0$

50. $3x + 5y + 15 = 0$

51. $-5x + 2y - 20 = 0$

52. $3x - 5y - 15 = 0$

In Exercises 53–64, determine the slope and y-intercept of the line. Then sketch the line.

53. $3x - y - 2 = 0$

54. $x - y - 5 = 0$

55. $x + y = 0$

56. $x - y = 0$

57. $3x + 2y - 2 = 0$

58. $x - 2y - 2 = 0$

59. $x - 4y + 2 = 0$

60. $8x + 6y - 3 = 0$

61. $y - 2 = 0$

62. $y + 4 = 0$

63. $x - 0.2y - 1 = 0$

64. $0.5x + 0.6y - 3 = 0$

In Exercises 65–74, a point on a line and the slope of the line are given. Find two additional points on the line. (There are many correct answers to each problem.)

65. $(5, 2), m = 0$

66. $(-4, 3), m$ is undefined.

67. $(3, -4), m = 3$

68. $(-1, -5), m = 2$

69. $(0, 3), m = -1$

70. $(-2, 6), m = -3$

71. $(-5, 0), m = \frac{4}{3}$

72. $(-1, 1), m = -\frac{3}{4}$

73. $(4, 2), m$ is undefined.

74. $(2, -2), m = 0$

In Exercises 75–78, use a graphing calculator to graph the linear function.

75. $f(x) = -3x + 5$

76. $f(x) = 0.7x - 2$

77. $g(t) = \frac{2}{3}t - \frac{8}{3}$

78. $h(s) = -\frac{5}{2}s + \frac{3}{4}$

In Exercises 79–82, determine whether the lines L_1 and L_2 passing through the given pairs of points are parallel, or perpendicular, or neither.

79. L_1: $(1, 3), (2, 1)$
L_2: $(0, 0), (4, 2)$

80. L_1: $(-3, -3), (1, 7)$
L_2: $(0, 4), (5, -2)$

81. L_1: $(-2, 0), (4, 4)$
L_2: $(1, -2), (4, 0)$

82. L_1: $(-5, 3), (3, 0)$
L_2: $(1, 2), \left(3, \frac{22}{3}\right)$

In Exercises 83–86, sketch the graphs of the pair of equations on the same rectangular coordinate system, and state the relationship that exists between the resulting lines.

83. L_1: $y = \frac{1}{2}x - 2$
L_2: $y = \frac{1}{2}x + 3$

84. L_1: $y = 3x - 2$
L_2: $y = 3x + 1$

85. L_1: $y = \frac{3}{4}x - 3$
L_2: $y = -\frac{4}{3}x + 1$

86. L_1: $y = -\frac{2}{3}x - 5$
L_2: $y = \frac{3}{2}x + 1$

87. *Earnings per Share* The graph shows the earnings per share of common stock for Xerox Corporation for the years 1983 through 1989 (*Source:* Xerox Corporation). Between which two years did earnings

(a) Decrease most rapidly?

(b) Increase most rapidly?

88. *Law Enforcement* The percentage of female law enforcement officers in the United States rose from 9.4 in 1983 to 13.8 in 1990 (see figure). What was the average rate of change in the percentage of women in law enforcement over this period? (*Source:* U.S. Department of Labor)

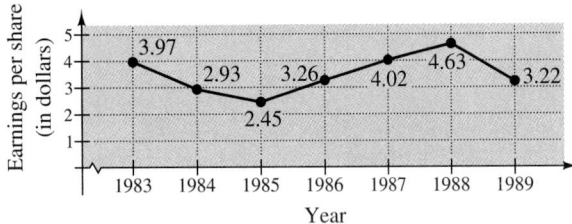

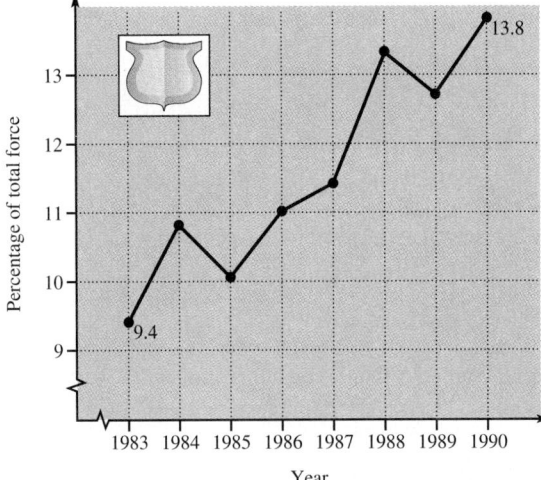

89. *Spending Your Paycheck* Consider the ordered pair (x, y), where y represents your pocket money and x is time, with $x = 0$ corresponding to payday. On payday, you have $200. Eight days later, you have $46.

   (a) Plot the two points modeling the given information and sketch a line segment connecting the points.
   (b) At what average rate did you spend your paycheck over the 8-day period?

90. *Leaning Tower of Pisa* When it was built, the Leaning Tower of Pisa in Italy was 180 feet tall. Since then, one side of the base has sunk 1 foot into the ground, causing the top of the tower to lean 16 feet off center (see figure). Approximate the slope of the side of the tower.

The tower leans because it was built on a layer of unstable soil—a mixture of clay, sand, and water.

91. *Road Grade* When driving down a mountain road, you notice warning signs indicating that it is a 12% grade. This means that the slope of the road is $-\frac{12}{100}$. Determine the amount of horizontal change in your position that has occurred if you note from elevation markers that you have descended 2000 feet vertically (see figure).

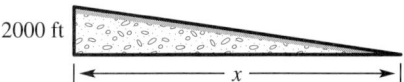

92. *Height of an Attic* The "rise-to-run" ratio used in determining the steepness of the roof on a house is 3 to 4. Determine the maximum height h of the attic of the house if the house is 30 feet wide (see figure).

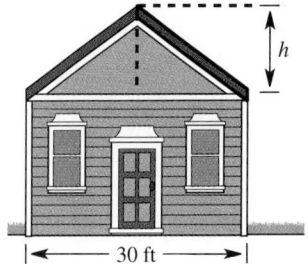

Using a Graphing Calculator

Most graphing calculators allow you to sketch more than one graph on the same display screen.

EXAMPLE 1 ■   **Sketching More than One Graph on the Same Screen**

Historical Note: Technologically, we have made significant progress in calculator development. Blaise Pascal (1623-1662), the French mathematician, invented the first mechanical calculator in 1642 at the young age of 18. He wanted to help his father, an accountant of sorts, to compute more easily. The Pascaline calculator is currently on display at the Museum of Technology in Paris.

Sketch the graphs of the following equations on the same display screen. Choose a viewing rectangle that shows the x-intercepts and y-intercepts of each graph.

(a) $y = -\frac{1}{2}x - 6$ \qquad\qquad (b) $y = -\frac{1}{2}x - 2$

(c) $y = -\frac{1}{2}x + 2$ \qquad\qquad (d) $y = -\frac{1}{2}x + 6$

Solution

There are two ways to determine a viewing rectangle that will show all the intercepts. One way is to make a list of the intercepts and then choose a viewing rectangle that includes them all. The other way is simply to experiment with various viewing rectangles. For instance, you might begin with the standard viewing rectangle, as shown below at the left. In this display, you can see that two x-intercepts are not shown. By enlarging the viewing rectangle slightly, you can see all eight intercepts, as shown below at the right.

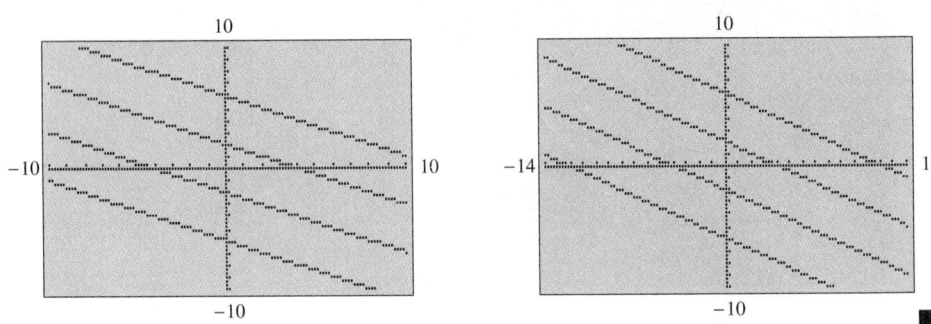

In Example 1, notice that all four lines are parallel (each has a slope of $-\frac{1}{2}$). Moreover, the lines will appear parallel on *any* viewing rectangle that shows the lines. This is *not* the case with perpendicular lines. In order for perpendicular lines to appear to meet at right angles, you must use a **square setting** in which the distance between consecutive tick marks is the same on the x- and y-axes. On most graphing calculators, the height of the display screen is two-thirds of its width. This means that the setting will be "square" if

$$\frac{\text{Ymax} - \text{Ymin}}{\text{Xmax} - \text{Xmin}} = \frac{2}{3}. \qquad \text{Square setting}$$

EXAMPLE 2 ■ Using a Square Setting

Sketch the lines represented by

$$y = x + 1 \quad \text{and} \quad y = -x + 3$$

on the same display screen. The two lines are perpendicular (they have slopes of $m_1 = 1$ and $m_2 = -1$). Choose a viewing rectangle in which the lines appear to meet at right angles.

Solution

If you use a standard viewing rectangle, you will obtain the display shown below at the left. In this display, notice that the two lines do not appear to be perpendicular. (The angle of intersection that opens upward appears to be greater than the angle that opens to the right.) By changing the graphing calculator range to a square setting, as shown below at the right, you obtain a display in which the two lines appear to be perpendicular.

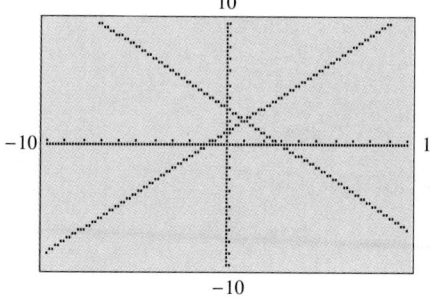

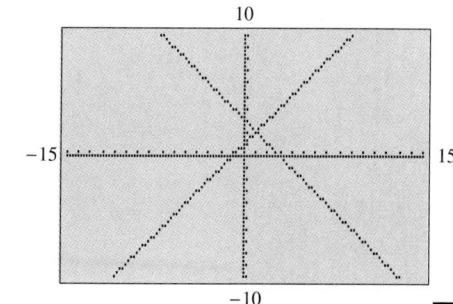

■

EXERCISES

In Exercises 1–4, sketch all four lines on the same display screen. Use a viewing rectangle that shows all intercepts of the graphs.

1. (a) $y = x - 8$
(b) $y = x - 2$
(c) $y = x + 6$
(d) $y = x + 12$

2. (a) $y = 2x - 10$
(b) $y = 2x - 1$
(c) $y = 2x + 7$
(d) $y = 2x + 16$

3. (a) $y = -\frac{1}{3}x - 12$
(b) $y = -\frac{1}{3}x - 2$
(c) $y = -\frac{1}{3}x + 8$
(d) $y = -\frac{1}{3}x + 18$

4. (a) $y = -2x - 18$
(b) $y = -2x - 12$
(c) $y = -2x - 6$
(d) $y = -2x$

In Exercises 5–8, sketch both lines on the same display screen. Use a viewing rectangle that shows the point at which the lines intersect. If the lines are perpendicular, use a viewing rectangle in which they appear to meet at right angles.

5. (a) $y = 2x - 1$
(b) $y = -\frac{1}{2}x + 4$

6. (a) $y = 2x - 1$
(b) $y = -2x + 7$

7. (a) $y = x + 4$
(b) $y = -x - 2$

8. (a) $y = \frac{1}{3}x - 7$
(b) $y = -3x + 3$

In Exercises 9–12, state whether having a viewing rectangle with a square or "nonsquare" setting affects the appearance of the given lines.

9. Parallel lines appear to be parallel.

10. Perpendicular lines appear to be perpendicular.

11. Intersecting lines appear to intersect.

12. A line with a slope of 2 appears to have a slope of 2.

Equations of Lines
Point-Slope Equation of a Line • *Horizontal and Vertical Lines* • *Summary of Equations of Lines* •
Applications

Point-Slope Equation of a Line

There are two basic types of problems in coordinate geometry.

1. Given an algebraic equation, sketch its graph.

2. Given some geometric conditions about a graph, construct its algebraic equation.

The first type of problem can be thought of as moving from algebra to geometry, whereas the second type can be thought of as moving the other way—from geometry to algebra. In Section 3.5, you were working with the first type of problem. In this section, you will study the second type. That is, you will be given certain facts about a straight line and will be asked to write an equation for the line.

If you know the slope of a line *and* the coordinates of one point on the line, then you can find an equation for the line.

EXAMPLE 1   ■   **Finding an Equation of a Line Given Its Slope and a Point on the Line**

Discuss the number of lines having a slope of $\frac{4}{3}$. Then discuss the number of lines that pass through $(-2, 1)$. Point out that only one line satisfies both conditions.

A line has a slope of $\frac{4}{3}$ and passes through the point $(-2, 1)$. Find an equation for this line.

Solution

Using the methods of the previous section, sketch the graph of the line, as shown in Figure 3.30. You know that the slope of a line is the same through any two points on the line. Thus, to find an equation of the line, let (x, y) represent *any* point on the line. Now, using the representative point (x, y) and the given point $(-2, 1)$, it follows that the slope of the line is

FIGURE 3.30

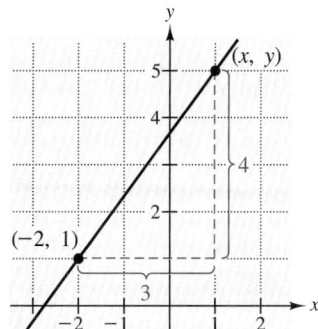

$$m = \frac{y - 1}{x - (-2)}. \quad \begin{array}{l} \Longleftarrow \text{ Difference in } y\text{-values} \\ \Longleftarrow \text{ Difference in } x\text{-values} \end{array}$$

Because the slope of the line is $m = \frac{4}{3}$, you can rewrite this equation as follows.

$$\frac{4}{3} = \frac{y - 1}{x + 2} \qquad \text{Slope formula}$$

$$4(x + 2) = 3(y - 1) \qquad \text{Cross multiply}$$

$$4x + 8 = 3y - 3$$

$$4x - 3y = -11 \qquad \text{Equation of line}$$

Therefore, you have found an equation for the line. It is $4x - 3y = -11$.   ■

The procedure in Example 1 can be used to derive a *formula* for the equation of a line, given its slope and a point on the line. In Figure 3.31, let (x_1, y_1) be a given point on the line whose slope is m. If (x, y) is *any other* point on the line, then it follows that

$$\frac{y - y_1}{x - x_1} = m.$$

This equation in variables x and y can be rewritten in the form

$$y - y_1 = m(x - x_1),$$

which is called the **point-slope form** of the equation of a line.

FIGURE 3.31

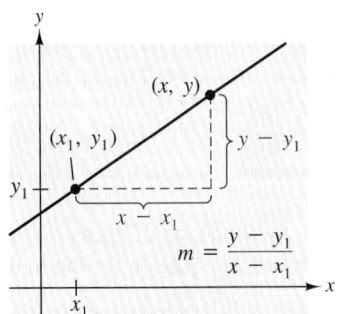

Point-Slope Form of the Equation of a Line

The **point-slope form** of the equation of the line that passes through the point (x_1, y_1) and has a slope of m is

$$y - y_1 = m(x - x_1).$$

EXAMPLE 2   ■   The Point-Slope Form of the Equation of a Line

FIGURE 3.32

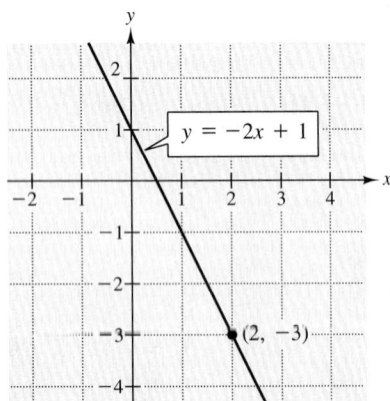

Find an equation of the line that passes through the point $(2, -3)$ and has a slope of -2.

Solution

Using the point-slope form with $(x_1, y_1) = (2, -3)$ and $m = -2$, you can write the following.

$y - y_1 = m(x - x_1)$	Point-slope form
$y - (-3) = -2(x - 2)$	Substitute $y_1 = -3$, $x_1 = 2$, and $m = -2$
$y + 3 = -2x + 4$	
$y = -2x + 1$	Equation of line

The graph of this line is shown in Figure 3.32.   ■

The point-slope form can be used to find the equation of a line passing through two points (x_1, y_1) and (x_2, y_2). First, use the formula for the slope of the line passing through two points.

$$m = \frac{y_2 - y_1}{x_2 - x_1}$$

Then, once you know the slope, use the point-slope form to obtain the equation

$$y - y_1 = \frac{y_2 - y_1}{x_2 - x_1}(x - x_1).$$

This is sometimes called the **two-point form** of the equation of a line.

EXAMPLE 3   ■   Finding an Equation of a Line Passing Through Two Points

Find an equation of the line that passes through the points $(4, 2)$ and $(-2, 3)$.

Solution

If you let $(x_1, y_1) = (4, 2)$ and $(x_2, y_2) = (-2, 3)$, then you can apply the formula for the slope of a line passing through two points as follows.

$$m = \frac{y_2 - y_1}{x_2 - x_1} = \frac{3 - 2}{-2 - 4} = -\frac{1}{6}$$

FIGURE 3.33

Now, using the point-slope form, you can find the equation of the line.

$$y - y_1 = m(x - x_1)$$

$$y - 2 = -\frac{1}{6}(x - 4)$$

$$y - 2 = -\frac{1}{6}x + \frac{2}{3}$$

$$y = -\frac{1}{6}x + \frac{8}{3}$$

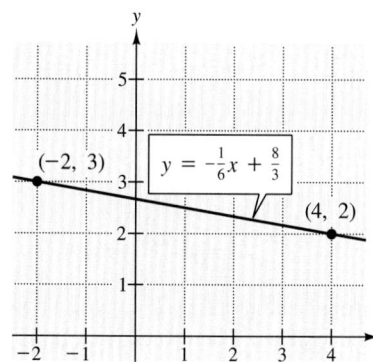

The graph of this line is shown in Figure 3.33.

■

In Example 3, it did not matter which of the two points you labeled as (x_1, y_1) and which as (x_2, y_2). Try switching these labels to $(x_1, y_1) = (-2, 3)$ and $(x_2, y_2) = (4, 2)$ and reworking the problem to confirm that you obtain the same equation.

Horizontal and Vertical Lines

From the slope-intercept form of the equation of a line, you can see that a horizontal line ($m = 0$) has an equation of the form

$$y = (0)x + b \quad \text{or} \quad y = b. \qquad \text{Horizontal line}$$

This is consistent with the fact that each point on a horizontal line through $(0, b)$ has a y-coordinate of b, as shown in Figure 3.34.

Similarly, each point on a vertical line through $(a, 0)$ has an x-coordinate of a, as shown in Figure 3.35. Hence, a vertical line has an equation of the form

$$x = a. \qquad \text{Vertical line}$$

FIGURE 3.34

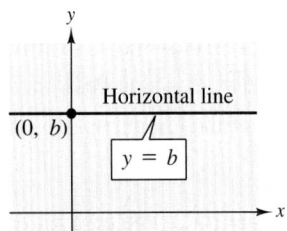

FIGURE 3.35

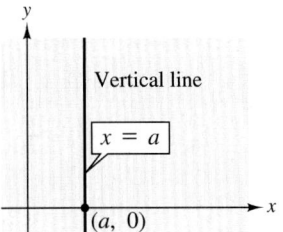

EXAMPLE 4 ■ Writing Equations of Horizontal and Vertical Lines

Write an equation for each of the following lines.

(a) Vertical line through $(-2, 4)$

(b) Horizontal line through $(-2, 4)$

(c) Line passing through $(-2, 3)$ and $(3, 3)$

(d) Line passing through $(-1, 2)$ and $(-1, 3)$

Solution

(a) Because this line is vertical and passes through the point $(-2, 4)$, you know that every point on the line has an x-coordinate of -2. Therefore, the equation of the line is
$$x = -2.$$

(b) Because this line is horizontal and passes through the point $(-2, 4)$, you know that every point on the line has a y-coordinate of 4. Therefore, the equation of the line is
$$y = 4.$$

(c) The line through $(-2, 3)$ and $(3, 3)$ is horizontal. Thus, its equation is
$$y = 3.$$

(d) The line through $(-1, 2)$ and $(-1, 3)$ is vertical. Thus, its equation is
$$x = -1.$$ ■

Summary of Equations of Lines

The equation of a vertical line cannot be written in the slope-intercept form because the slope of a vertical line is undefined. However, *every* line has an equation that can be written in the **general form**

$$ax + by + c = 0, \qquad \text{General form}$$

where a and b are not *both* zero. For instance, $y = -5x + 2$ can be written in general form as $5x + y - 2 = 0$. Similarly, $y - 4 = \frac{2}{3}(x + 5)$ can be written in general form as $2x - 3y + 22 = 0$.

If $a = 0$ (and $b \neq 0$), the equation can be reduced to the form $y = -c/b$, which represents a horizontal line. If $b = 0$ (and $a \neq 0$), the general equation can be reduced to the form $x = -c/a$, which represents a vertical line.

For convenience, the equations of lines discussed in Sections 3.5 and 3.6 are summarized in the following list.

Summary of Equations of Lines

Writing assignment: Have students describe what happens to the graph of $y = mx + b$ when the value of the coefficient on x is increased.

1. Slope of line through (x_1, y_1) and (x_2, y_2): $m = \dfrac{y_2 - y_1}{x_2 - x_1}$

2. General form of equation of line: $ax + by + c = 0$

3. Equation of vertical line: $x = a$

4. Equation of horizontal line: $y = b$

5. Slope-intercept form of equation of line: $y = mx + b$

6. Point-slope form of equation of line: $y - y_1 = m(x - x_1)$

7. Parallel lines have *equal* slopes: $m_1 = m_2$

8. Perpendicular lines have *negative reciprocal* slopes: $m_2 = -\dfrac{1}{m_1}$

EXAMPLE 5 ■   Equations of Parallel Lines

An application: To avoid midair collisions, flight controllers instruct pilots to take off and land in directions parallel to other planes.

Find an equation of the line that passes through the point $(3, -2)$ and is parallel to the line $x - 4y = 6$, as shown in Figure 3.36, at the top of the next page.

FIGURE 3.36

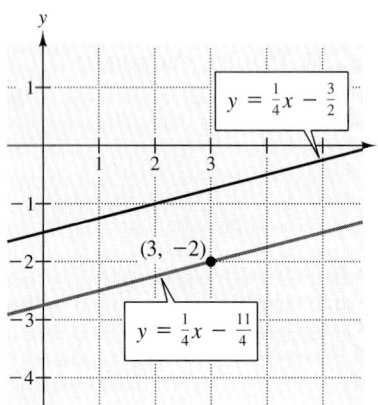

Solution

To begin, write the given equation in slope-intercept form.

$$x - 4y = 6 \qquad\qquad \text{Given equation}$$
$$-4y = -x + 6$$
$$4y = x - 6$$
$$y = \frac{1}{4}x - \frac{3}{2} \qquad\qquad \text{Slope-intercept form}$$

Therefore, the given line has a slope of $m = \frac{1}{4}$. Because any line parallel to the given line must also have a slope of $\frac{1}{4}$, the required line through $(3, -2)$ has the following equation.

$$y - y_1 = m(x - x_1)$$
$$y - (-2) = \frac{1}{4}(x - 3)$$
$$y + 2 = \frac{1}{4}x - \frac{3}{4}$$
$$y = \frac{1}{4}x - \frac{11}{4}$$

■

EXAMPLE 6   ■   Equations of Perpendicular Lines

Find an equation of the line that passes through the point $(-2, 3)$ and is perpendicular to the line $5x - 2y = 7$, as shown in Figure 3.37.

FIGURE 3.37

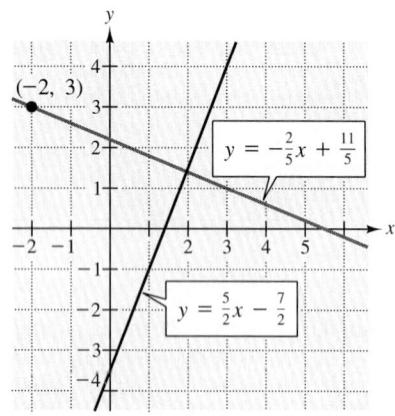

Solution

To begin, write the given equation in slope-intercept form.

$$5x - 2y = 7 \qquad\qquad \text{Given equation}$$
$$-2y = -5x + 7$$
$$2y = 5x - 7$$
$$y = \frac{5}{2}x - \frac{7}{2} \qquad\qquad \text{Slope-intercept form}$$

Thus, the given line has a slope of $m = \frac{5}{2}$, and any line perpendicular to this line must have a slope of $-\frac{2}{5}$ (the negative reciprocal of $\frac{5}{2}$). Therefore, the equation of the required line through $(-2, 3)$ has the following form.

$$y - y_1 = m(x - x_1)$$

$$y - 3 = -\frac{2}{5}[x - (-2)]$$

$$y - 3 = -\frac{2}{5}(x + 2)$$

$$y - 3 = -\frac{2}{5}x - \frac{4}{5}$$

$$y = -\frac{2}{5}x + \frac{11}{5}$$

■

Applications

Linear equations are used frequently as mathematical models in business. The next example gives you one idea of how useful such a model can be.

EXAMPLE 7 ■ Computer Software Sales

The total sales of a new computer software company were $500,000 for the second year and $1,000,000 for the fourth year. Using only this information, what would you estimate the total sales to be during the fifth year?

FIGURE 3.38

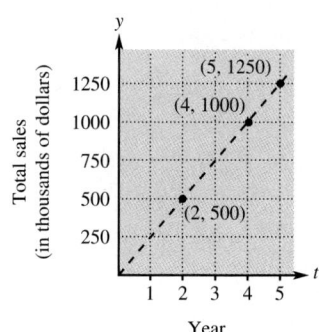

Year

Solution

To solve this problem, you can use use a *linear model,* with y representing the total sales (in thousands) and t representing the year. That is, in Figure 3.38, let $(2, 500)$ and $(4, 1000)$ be two points on the line representing the total sales for the company. The slope of the line passing through these points is

$$m = \frac{1000 - 500}{4 - 2} = 250.$$

Now, using the point-slope form, find the equation of the line as follows.

$$y - y_1 = m(t - t_1)$$

$$y - 500 = 250(t - 2)$$

$$y - 500 = 250t - 500$$

$$y = 250t$$

Finally, you can estimate the total sales during the fifth year ($t = 5$) to be

$$y = 250(5) = \$1250 \text{ thousand} = \$1,250,000.$$

■

The estimation method illustrated in Example 7 is called **linear extrapolation.** Note in Figure 3.39 that for linear extrapolation, the estimated point lies to the right of the given points. When the estimated point lies *between* two given points, the procedure is called **linear interpolation.**

FIGURE 3.39   Linear extrapolation   Linear interpolation

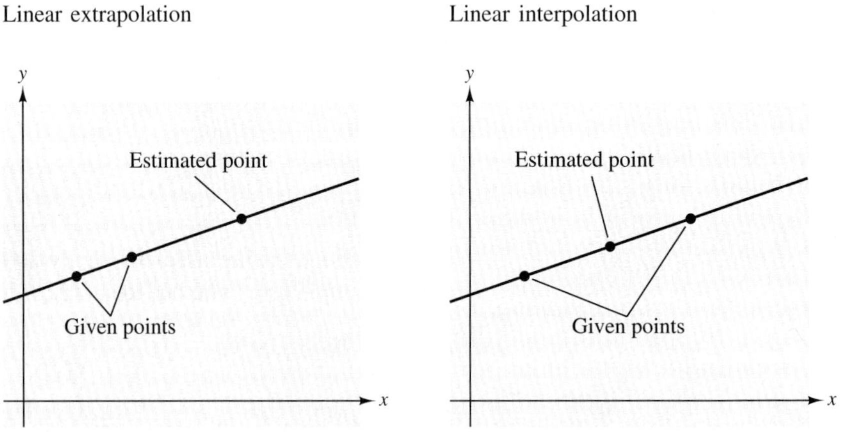

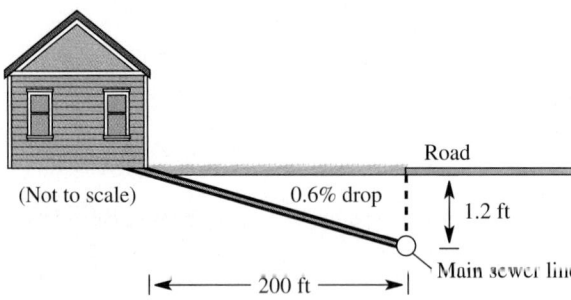

EXAMPLE 8 ■ Slope of a Sewer Line

A city ordinance requires a minimum drop of 0.6% in a home sewer line running to the main sewer line (along the street). This means that for each foot from the house to the street, the sewer line must drop 0.006 foot, as shown in Figure 3.40.

(a) Write an equation that expresses the minimum vertical drop y in terms of the horizontal length x of the pipe.

(b) How much drop is needed for a house that is 200 feet from the main line?

FIGURE 3.40

Solution

(a) In this case the slope is

$$m = -0.006,$$

because the line *drops* 0.006 foot vertically for each foot of horizontal length. Because a pipe of zero length would have a zero drop, it follows that $(x_1, y_1) = (0, 0)$ is a point that satisfies the equation relating x and y. Therefore, using the point-slope form, you can write the following equation.

$$y - y_1 = m(x - x_1)$$
$$y - 0 = -0.006(x - 0)$$
$$y = -0.006x$$

(b) The drop in a line for a house that is 200 feet from the street would have to be at least

$$y = -0.006(200) = -1.2 \text{ feet.}$$ ■

EXAMPLE 9   ■   **Writing a Linear Model**

In 1970, 130 African-American women held elected offices in the United States. By 1990, the number had increased to 1950. Write a linear model for the number n of African-American women who held elected offices between 1970 and 1990. (Let $t = 0$ represent 1970.) (*Source:* Joint Center for Political and Economic Studies)

Solution

One way to create such a model is to interpret the given information as two points on a line. Using $n = 130$ in 1970 ($t = 0$), you can determine that one point on the line is

(0, 130).   One point on the line

Using $n = 1950$ in 1990 ($t = 20$), you can determine that a second point on the line is

(20, 1950).   Second point on the line

(See Figure 3.41 at the top of the next page.) Thus, the slope of the line is

$$m = \frac{1950 - 130}{20 - 0} = \frac{1820}{20} = 91.$$

Because the n-intercept of the line is $b = 130$, you can conclude that an equation of the line is

$$n = mt + b$$ Slope-intercept form—use t and n.
$$n = 91t + 130.$$

FIGURE 3.41

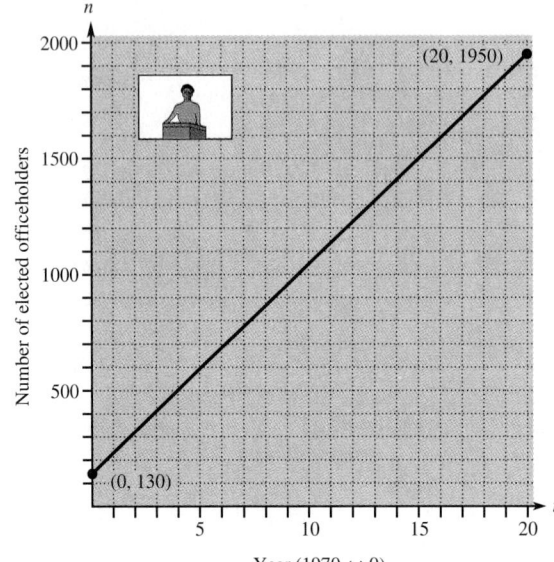

In 1973, Barbara Jordan became the first African-American woman from a southern state to serve in the United States Congress. From 1970 to 1990, the number of African-American women holding elective office increased at a rate that was approximately linear.

DISCUSSION PROBLEM ■   Misleading Graphs

Graphs can help you visualize relationships between two variables, but they can also be misused to imply results that are not correct. The two graphs in Figure 3.42 represent the *same* points. Which graph is misleading and why?

FIGURE 3.42

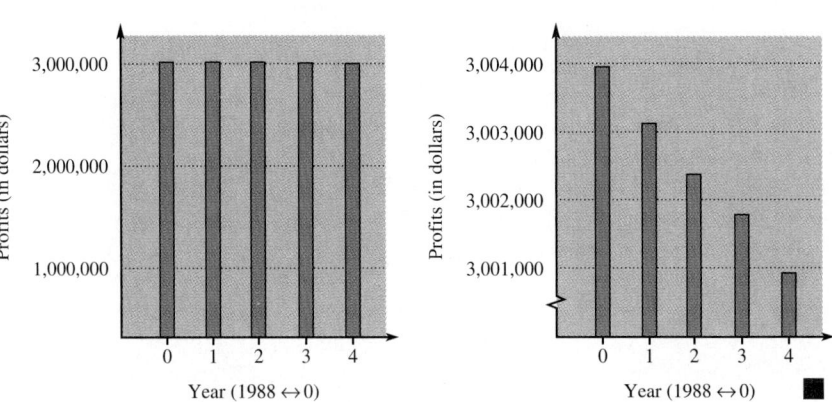

Warm-Up

The following warm-up exercises involve skills that were covered in earlier sections. You will use these skills in the exercise set for this section.

In Exercises 1–4, determine the slope of the line passing through the points.

1. $(-5, 2), (3, 0)$

2. $(-8, -4), (6, 10)$

3. $(-3, -2), (-3, 5)$

4. $\left(\frac{5}{6}, \frac{3}{2}\right), \left(\frac{5}{3}, \frac{3}{4}\right)$

In Exercises 5 and 6, sketch the lines through the given point with the indicated slopes. Make the sketches on the same set of coordinate axes.

	Point	*Slopes*			
5.	$(3, 2)$	(a) 0	(b) 14	(c) $\frac{3}{2}$	(d) $-\frac{1}{4}$
6.	$(-2, 4)$	(a) 2	(b) -2	(c) $\frac{1}{2}$	(d) undefined

In Exercises 7–10, use function notation to write y as a linear function of x.

7. $y - 3 = -\frac{3}{5}(x - 2)$

8. $y + 5 = \frac{4}{3}(x - 3)$

9. $y - (-2) = \dfrac{5 - (-3)}{4 - 6}(x - 5)$

10. $y - 7 = \dfrac{1 - 3}{0 - 2}[x - (-2)]$

3.6   EXERCISES

means that a graphing utility can help you solve the exercise or check your solution.

In Exercises 1–12, from the equation of the line, determine the slope (if possible) of the line and the coordinates of one point through which it passes.

1. $y - 3 = 6(x + 4)$

2. $y + 10 = -12(x - 1)$

3. $y + \frac{5}{8} = \frac{3}{4}(x + 2)$

4. $y - \frac{3}{4} = \frac{5}{6}\left(x - \frac{9}{10}\right)$

5. $y = \frac{2}{3}x - 2$

6. $y = -5x + 12$

7. $3x - 2y = 0$

8. $0.5x + 0.3y - 1 = 0$

9. $y - 3 = 0$

10. $x + 4 = 0$

11. $5x - 2y + 24 = 0$

12. $3x + 4y - 16 = 0$

In Exercises 13–24, find the general form of the equation of the line passing through the given point with the specified slope.

13. $(0, -6), \ m = \frac{1}{2}$

14. $(0, 9), \ m = -\frac{1}{3}$

15. $\left(0, \frac{3}{2}\right), \ m = -2$

16. $\left(0, -\frac{7}{4}\right), \ m = 3$

17. $(-2, 8), \ m = -2$

18. $(0, 0), \ m = 3$

19. $(5, 0), \ m = -\frac{2}{3}$

20. $(-3, -7), \ m = \frac{5}{4}$

21. $(2, -1), \ m$ is undefined.

22. $(-8, 5), \ m = 0$

23. $\left(\frac{3}{4}, \frac{5}{2}\right), \ m = \frac{4}{3}$

24. $\left(-\frac{3}{2}, \frac{1}{2}\right), \ m = -3$

In Exercises 25–36, find the general form of the equation of the line through the given points and sketch a graph of the line.

25. $(0, 0), (2, 3)$

26. $(0, 0), (3, -5)$

27. $(-2, 3), (5, 0)$

28. $(-6, -3), (4, 3)$

29. $(-2, 2), (4, 5)$

30. $(0, 10), (5, 0)$

31. $(-2, 3), (4, 3)$

32. $(1, -2), (1, 8)$

33. $\left(\frac{3}{2}, 3\right), \left(\frac{9}{2}, -4\right)$

34. $\left(4, \frac{7}{3}\right), \left(-1, \frac{2}{3}\right)$

35. $(0, 0.3), (2.7, 0)$

36. $(1.2, 3.4), (5.5, 12)$

In Exercises 37–42, find an equation of the line passing through the two given points. Use function notation to write y as a linear function of x.

37. $(-5, 2), (5, -2)$

38. $(5, 4), (-3, 5)$

39. $\left(1, \frac{1}{2}\right), \left(\frac{3}{2}, \frac{7}{4}\right)$

40. $(-2, 12), (6, 12)$

41. $(1.5, 2), (7.5, 9)$

42. $(-5, 0.6), (3, -3.4)$

 In Exercises 43–48, write equations of the lines through the given point (a) parallel to the given line and (b) perpendicular to the given line.

43. $(2, 1), \ 6x - 2y = 3$

44. $(-3, 4), \ x + 6y = 12$

45. $(-5, 4), \ 5x + 4y = 24$

46. $\left(\frac{5}{8}, \frac{9}{4}\right), \ -5x + 4y = 0$

47. $(-1, 2), \ y + 5 = 0$

48. $(3, -4), \ x - 10 = 0$

49. *Cost* The cost in dollars of producing x units of a certain product is $C = 20x + 5000$. Use this equation to complete the table.

x	0		100		1000
C		6000		15,000	

50. *Temperature Conversion* The linear relationship between the Fahrenheit and Celsius temperature scales is $C = \frac{5}{9}F - \frac{160}{9}$. Use this equation to complete the table.

F	$-20°$		$20°$		$100°$	
C		$-17.8°$		$0°$		$100°$

51. *Sales Commission* The salary of a sales representative is $1500 per month plus a 3% commission on total monthly sales. Write a linear equation giving the salary S in terms of the monthly sales M.

52. *Reimbursed Expenses* A sales representative is reimbursed $125 per day for lodging and meals plus $0.27 per mile driven. Write the daily cost C to the company in terms of x, the number of miles driven.

53. *Discount Price* A store is offering a 30% discount on all items in its inventory. Write a linear equation giving the sale price S for an item in terms of its list price L.

54. *Straight-Line Depreciation* A small business purchases a photocopier for $7400. It is estimated that after 4 years its depreciated value will be $1500. Assuming straight-line depreciation, write a linear equation giving the value V of the copier in terms of time t.

55. *Study Time* Suppose you can get a score of 72 on an exam with no additional study. For each additional hour of study, you can increase the score by five points. Write a linear equation giving your estimated score y in terms of the number of hours t of additional study.

56. *College Enrollment* A small college had an enrollment of 1500 students in 1980. During the next 10 years, the enrollment increased by approximately 60 students per year. Write a linear equation giving the enrollment N in terms of the year t. (Let $t = 0$ correspond to the year 1980.) If this constant rate of growth continues, predict the enrollment in the year 2000.

57. ***Rental Occupancy*** A real estate office handles an apartment complex with 50 units. When the rent per unit is $450 per month, all 50 units are occupied. However, when the rent is $525 per month, the average number of occupied units drops to 45. Assume that the relationship between the monthly rent p and the demand x is linear.
 (a) Write an equation of the line giving the demand x in terms of the rent p.
 (b) *(Linear Extrapolation)* Use this equation to predict the number of units occupied if the rent is raised to $570.
 (c) *(Linear Interpolation)* Use this equation to predict the number of units occupied if the rent is lowered to $480.

58. ***Soft Drink Sales*** When soft drinks cost $0.80 per can at a football game, approximately 6000 cans were sold. When the price was raised to $1.00 per can, the demand dropped to 4000. Assume that the relationship between the price p and demand x is linear.
 (a) Write an equation of the line giving the demand x in terms of the price p.
 (b) *(Linear Extrapolation)* Use the equation in part (a) to predict the number of cans of soft drinks sold if the price is raised to $1.10.
 (c) *(Linear Interpolation)* Use the equation in part (a) to predict the number of cans of soft drinks sold if the price is $0.90.

Running the IRS In Exercises 59 and 60, use the graph, which shows the cost C (in billions of dollars) of running the Internal Revenue Service from 1980 to 1990. (*Source:* Internal Revenue Service)

59. Using the costs for 1980 and 1990, write a linear model for the average cost, letting $t = 0$ represent 1980. Estimate the cost of running the IRS in 1995.

60. In 1989, the IRS collected about $9.5 trillion. Use the model written in Exercise 59 to approximate the percentage of this amount spent to run the IRS.

Figure for 59 and 60

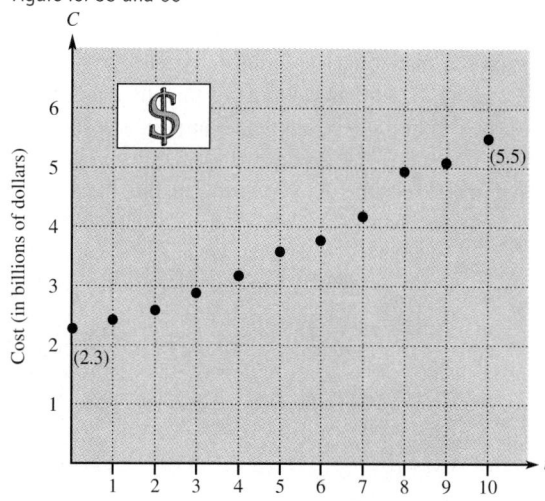

61. ***Bookstore Sales*** For the years 1986 through 1991, retail sales of American bookstores increased at a rate that was approximately linear (see figure). In 1986, the retail sales were $4.9 billion. In 1991, the retail sales were $7.9 billion. Write a linear model for the retail sales s (in billions of dollars) of American bookstores from 1986 to 1991. Let $t = 0$ represent 1986. (*Source:* American Booksellers Association)

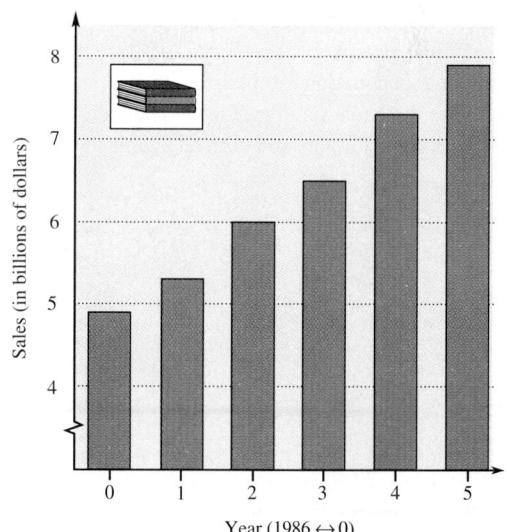

62. *Depth Markers* A swimming pool is 40 feet long, 20 feet wide, 4 feet deep at the shallow end, and 9 feet deep at the deep end. Imagine the side of the pool positioned on the rectangular coordinate system as shown in the figure, and find an equation of the line representing the edge of the inclined bottom of the pool. Use this equation to determine the distance from the deep end at which markers must be placed to indicate that the depth of the pool has changed by 1 foot.

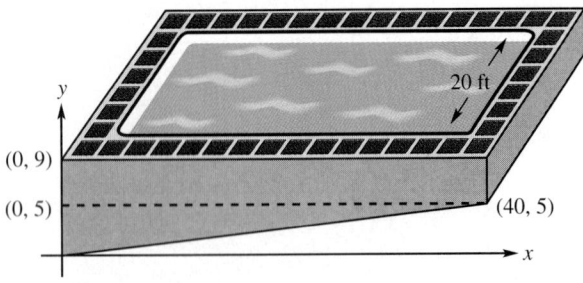
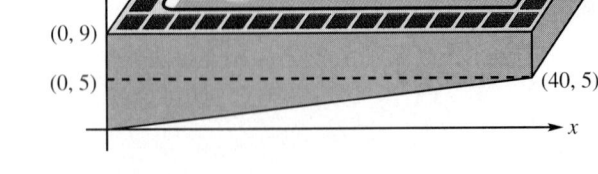

63. *Best-Fitting Line* An instructor gives 20-point quizzes and 100-point exams in a mathematics course. The average quiz and test scores for six students are given as ordered pairs (x, y), where x is the average quiz score and y is the average test score. The ordered pairs are $(18, 87)$, $(10, 55)$, $(19, 96)$, $(16, 79)$, $(13, 76)$, and $(15, 82)$.

(a) Plot the points.

(b) Use a ruler to sketch the best-fitting line through the points.

(c) Find an equation for the line sketched in part (b).

(d) Use the equation in part (c) to estimate the average test score for a person with an average quiz score of 17.

64. Use a graphing calculator to graph each equation.

(a) $y = -\frac{3}{4}x + 5$

(b) $6x - 9y + 18 = 0$

In Exercises 65 and 66, use a graphing calculator to graph both equations on the same screen and determine whether the lines are parallel or perpendicular. (Press ZOOM and move the cursor to "Square" so that the tick marks are equally spaced on each axis.)*

65. $x + 2y - 3 = 0$

$-2x - 4y + 1 = 0$

66. $3x - 4y - 1 = 0$

$4x + 3y + 2 = 0$

*The graphing calculator keystrokes in this text correspond to the TI-81 and TI-82 graphing calculators from Texas Instruments. For other calculators, the keystrokes may differ.

REVIEW EXERCISES

In Exercises 1–4, translate the phrase into a mathematical expression. (Let n represent the arbitrary real number.)

1. Two hundred decreased by 3 times a number

2. One hundred increased by the product of 15 and a real number

3. The sum of the square of a real number and 49

4. The absolute value of the sum of a real number and 10

In Exercises 5–8, write a verbal description of the algebraic expression without using the variable.

5. $2y + 7$

6. $5u - 3$

7. $\dfrac{x - 5}{4}$

8. $-3(a - 10)$

In Exercises 9–12, write an algebraic expression that represents the quantity given by the verbal statement.

9. The amount of income tax on a taxable income of I dollars when the tax rate is 18%

10. The distance traveled on a trip of 8 hours at an average speed of r miles per hour

11. The area of a rectangle whose length is l inches and whose width is 5 units less than the length

12. The sum of three consecutive odd integers, the first of which is n

13. *Retail Price* A camera that costs a retailer $259.95 is marked up by 35%. Find the price to the consumer.

14. *Markup Rate* A calculator with a cost to the consumer of $175 costs the retailer $95. Find the markup rate.

15. *Sale Price* The list price of a coat is $259. Find the sale price of the coat if it is reduced by 25%.

16. *Comparison Shopping* A mail-order catalog offers an attaché case with a list price of $99.97 plus $4.50 for shipping and handling. A local department store offers the same attaché case for $125.95. If the department store has a special 20 %-off sale on the case, which is the better buy?

17. *Tip Rate* A customer leaves a total of $25 for a meal that costs $21.65. Determine the tip rate.

18. *Sales Goal* The weekly salary of an employee is $150 plus a 6% commission on total sales. The employee needs a minimum salary of $650 per week. How much must be sold to produce this salary?

19. *Coin Problem* You have 60 coins in dimes and quarters with a combined value of $11.25. Determine the number of coins of each type.

20. *Opinion Poll* Twelve hundred people were surveyed in an opinion poll. Political candidates A and B received the same preference, but candidate C was preferred by $1\frac{1}{3}$ times the number of people who preferred candidate A (or B). Determine the number of people who preferred candidate C.

21. *Travel Time* Determine the time required for a bus to travel 330 miles if its average speed is 52 miles per hour.

22. *Average Speed* An Olympic cross-country skier completed the 15-kilometer event in 41 minutes and 20 seconds. What was the average speed for this skier?

23. *Distance* Determine the distance an air force jet can travel in $2\frac{1}{3}$ hours if its average speed is 1200 miles per hour.

24. *Average Speed* For 2 hours of a 400-mile trip, your average speed is only 40 miles per hour. Determine the average speed that must be maintained for the remainder of the trip if you want your average speed for the entire trip to be 50 miles per hour.

25. *Work Rate Problem* Find the time required for two people working together to complete a task if individually it takes them 10 hours and 6 hours.

26. *Dimensions of a Rectangle* The width of a rectangle is 6 inches less than its length. Find the dimensions of the rectangle if its perimeter is 42 inches.

27. *Simple Interest* An inheritance of $50,000 is divided between two investments earning 8.5% and 10% simple interest. How much is in each investment if the total interest for 1 year is $4700?

28. *Simple Interest* You invest $1000 in a certificate of deposit that has an annual simple interest rate of 7%. After 6 months, the interest is computed and added to the principal. During the second 6 months, the interest is computed using the original investment plus the interest earned during the first 6 months. What is the total interest earned during the first year of the investment?

In Exercises 29–32, express the ratio as a fraction in reduced form. (Use the same units for the numerator and denominator.)

29. 16 feet to 4 yards

30. 3 quarts to 5 pints

31. 45 seconds to 5 minutes

32. 3 meters to 150 centimeters

In Exercises 33–36, solve the proportion.

33. $\dfrac{7}{8} = \dfrac{y}{4}$

34. $\dfrac{x}{16} = \dfrac{5}{12}$

35. $\dfrac{b}{15} = \dfrac{5}{6}$

36. $\dfrac{x+1}{3} = \dfrac{x-1}{2}$

37. *Property Tax* The tax on property with an assessed value of $80,000 is $1350. Find the tax on property with an assessed value of $110,000. (Assume the same tax rate.)

38. *Recipe Enlargement* One and a half cups of milk are required to make one batch of pudding. How much is required to make $2\frac{1}{2}$ batches?

39. *Map Scale* One-third inch represents 50 miles on a map. Approximate the distance between two cities that are $3\frac{1}{4}$ inches apart on the map.

40. *Fuel Mixture* The gasoline-to-oil ratio for a lawn mower engine is 50 to 1. Determine the amount of gasoline required if one-half pint of the oil is used in producing the mixture.

Similar Triangles In Exercises 41 and 42, solve for the length x of the side of the triangle on the right by using the fact that corresponding sides of similar triangles are proportional. (*Note:* In each pair, assume that the triangles are similar.)

41.

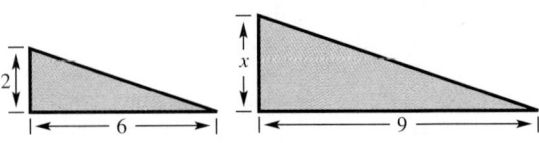

42.

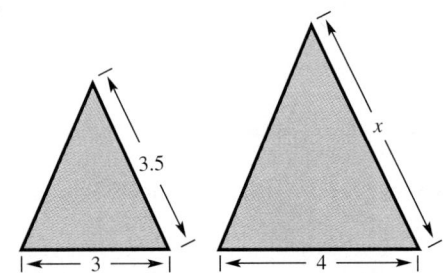

In Exercises 43–46, write an inequality for the statement.

43. z is no more than 10.

44. x is nonnegative.

45. y is at least 7 but less than 14.

46. The volume V is less than 27 cubic feet.

In Exercises 47–60, solve the inequality and sketch the solution on the real number line.

47. $5x + 3 > 18$

48. $-11x \geq 44$

49. $\dfrac{1}{3} - \dfrac{1}{2}y < 12$

50. $3(2 - y) \geq 2(1 + y)$

51. $-4 < \dfrac{x}{5} \leq 4$

52. $-13 \leq 3 - 4x < 13$

53. $5 > \dfrac{x+1}{-3} > 0$

54. $12 \geq \dfrac{x-3}{2} > 1$

55. $|2x - 7| < 15$

56. $|5x - 1| < 9$

57. $|x - 4| > 3$

58. $|t + 3| > 2$

59. $|b + 2| - 6 > 1$

60. $\left|\dfrac{t}{3}\right| < 1$

In Exercises 61–64, solve the equation.

61. $|x - 2| - 2 = 4$

62. $|2x + 3| = 7$

63. $|3x - 4| = |x + 2|$

64. $|5x + 6| = |2x - 1|$

In Exercises 65–68, find the slope of the line through the given points.

65. $(-1, 1),\ (6, 3)$

66. $(-2, 5),\ (3, -8)$

67. $(-1, 3),\ (4, 3)$

68. $(7, 2),\ (7, 8)$

In Exercises 69–72, determine the slope and y-intercept of the line; then sketch the line.

69. $3x + y = 7$

70. $2x - y = 3$

71. $5x - 2y - 10 = 0$

72. $3x + 4y + 8 = 0$

In Exercises 73–80, find the general form of the equation of the line passing through the given point with the specified slope.

73. $(1, -4),\ m = 2$

74. $(-5, -5),\ m = 3$

75. $(-1, 4),\ m = -4$

76. $(5, -2),\ m = -2$

77. $\left(\frac{5}{2}, 4\right),\ m = -\frac{2}{3}$

78. $\left(-2, -\frac{4}{3}\right),\ m = \frac{3}{2}$

79. $(7, 8),\ m$ is undefined

80. $(-6, 5),\ m = 0$

In Exercises 81–88, find the general form of the equation of the line passing through the two given points.

81. $(-6, 0),\ (0, -3)$

82. $(-2, -3),\ (4, 6)$

83. $(0, 10),\ (6, 10)$

84. $(-4, 3),\ (-4, 6)$

85. $(0, 5.5), (3.2, 0)$

86. $(-10, 2),\ (4, -7)$

87. $\left(\frac{5}{2}, 0\right),\ \left(\frac{5}{2}, 5\right)$

88. $\left(\frac{4}{3}, \frac{1}{6}\right),\ \left(4, \frac{7}{6}\right)$

In Exercises 89–92, find equations of the lines through the indicated point (a) parallel to the given line, and (b) perpendicular to the given line.

89. $(-1, 5),\ 2x + 4y = 1$

90. $\left(\frac{3}{5}, -\frac{4}{5}\right),\ 3x + y = 2$

91. $\left(\frac{3}{8}, 3\right),\ 4x - 3y = 12$

92. $(12, 1),\ 5x = 3$

93. *Cost and Profit* A company produces a product for which the variable cost is \$8.55 per unit and the fixed costs are \$25,000. The product is sold for \$12.60, and the company can sell all it produces.

(a) Write the cost C as a linear equation in terms of x, the number of units produced.

(b) Write the profit P as a linear equation in terms of x, the number of units sold.

Take this test as you would take a test in class. After you are done, check your work against the answers in the back of the book.

1. Translate into an algebraic expression the statement, "The product of a number n and 5 is decreased by 8."

2. Write an algebraic expression for the area of a rectangle of length l and width $0.6l$.

3. A store is offering a 20% discount on all items in its inventory. Find the list price of a tractor that has a sale price of $6400.

4. The tax on property with an assessed value of $90,000 is $1200. Find the tax on property with an assessed value of $110,000. (Assume the same tax rate.)

5. The bill (including parts and labor) for the repair of a home appliance was $165. The cost for parts was $85. How many hours were spent to repair the appliance if the cost of labor was $16 per half hour?

6. Two antifreeze solutions—one of 10% concentration and of 40% concentration—are mixed to create 100 liters of a 30% solution. Determine the numbers of liters of the 10% solution and the 40% solutions that are required.

7. Two cars start from the same point at a given time and travel in the same direction at average speeds of 40 miles per hour and 55 miles per hour. How much time must elapse before the two cars are 10 miles apart?

In Exercises 8–11, solve the inequality and sketch its solution.

8. $1 + 2x > 7 - x$ 　　 9. $0 \le \dfrac{1 - x}{4} < 2$ 　　 10. $|x - 3| \le 2$ 　　 11. $|x + 4| > 1$

12. Solve the equation $|x + 3| + 1 = 6$.

13. Find the slope (if possible) of the line passing through each pair of (a) $(-4, \ 7), \ (2, \ 3)$
    (b) $(3, \ -2), \ (3, \ 6)$

14. Sketch the graph of a line passing through the point $(0, \ -6)$ with slope $m = \frac{4}{3}$.

15. Plot the x- and y-intercepts and sketch the graph of the line given by the equation $2x + 5y - 10 = 0$.

16. Find the slope-intercept form of the equation of the line passing through the point $(0, 4)$ with slope $m = -\frac{3}{2}$.

17. Find the general form of the equation of the line passing through the points $(25, \ -15)$ and $(75, \ 10)$.

18. Find the slope of a line perpendicular to the line given by the equation $5x + 3y - 9 = 0$.

19. A small business purchases a car for $16,000. Its depreciated value after 4 years is estimated to be $7000. Assuming straight-line depreciation, write a linear equation giving the value V of the car in terms of t. Determine the time t when the car's value will be $10,000.

1-3

Cumulative Test

Take this test as you would take a test in class. After you are done, check your work against the answers in the back of the book.

In Exercises 1 and 2, evaluate the expression.

1. $18 - (3 - 8)$

2. $-\frac{8}{45} \div \frac{12}{25}$

In Exercises 3–6, perform the indicated operations and simplify.

3. $(2a^2b)^3(-ab^2)^2$

4. $3xy(x^2 - 2) - xy(x^2 + 5)$

5. $t(3t - 1) - 2t(t + 4)$

6. $[2 + (x - y)]^2$

In Exercises 7–10, solve the equation.

7. $12 - 5(3 - x) = x + 3$

8. $1 - \dfrac{x + 2}{4} = \dfrac{7}{8}$

9. $y^2 - 64 = 0$

10. $2t^2 - 5t - 3 = 0$

11. Determine whether the equation $x - y^3 = 0$ represents y as a function of x.

12. Find the domain of the function $f(x) = \sqrt{x - 2}$.

13. Given $f(x) = x^2 - 3x$, find   (a) $f(4)$.
                                               (b) $f(c + 3)$.

14. Write an algebraic expression for the statement, "The number n is tripled and the product is decreased by 8."

15. Find two consecutive integers such that their sum is 89 less than their product.

16. The annual insurance premium for a policyholder is $1225. Find the annual premium if the policyholder must pay a 15% surcharge because of a driving violation.

17. Solve the proportion $\dfrac{t - 1}{4} = \dfrac{11}{12}$.

18. Solve for the length x of the side of the second triangle by using the fact that corresponding sides of similar triangles are proportional.

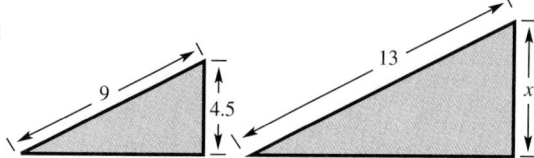

19. Solve and sketch the solution $|x - 2| \geq 3$.

20. The revenue from selling x units of a product is $R = 12.90x$. The cost of producing x units is $C = 8.50x + 450$. For a profit to be obtained, the revenue must be greater than the cost. For what values of x will this product produce a profit?

21. Consider the two points $(-4, 0)$ and $(4, 6)$.   (a) Find the distance between the points.
                                                             (b) Find an equation of the line through the points.

22. Sketch a graph of the linear equation $4x + 3y - 12 = 0$.

Systems of Linear Equations and Inequalities

4.1   Systems of Linear Equations in Two Variables

4.2   Systems of Linear Equations in Three Variables

4.3   Matrices and Systems of Linear Equations

4.4   Linear Systems and Determinants

4.5   Graphs of Linear Inequalities in Two Variables

4.6   Systems of Linear Inequalities and Linear Programming

rom 1985 through 1990, the number of compact discs (in millions) sold in the United States and the number of long-playing albums (in millions) sold in the United States can be modeled by

$$y = -205.5 + 45.3t, \quad 5 \le t \le 10, \quad \text{Compact disc sales}$$
$$y = 331.4 - 32.7t, \quad 5 \le t \le 10, \quad \text{Long-playing album sales}$$

where t represents the year, with t = 5 corresponding to 1985.
Which type of sales was increasing? Which was decreasing? During which year did the sales of compact discs overtake the sales of long-playing albums?

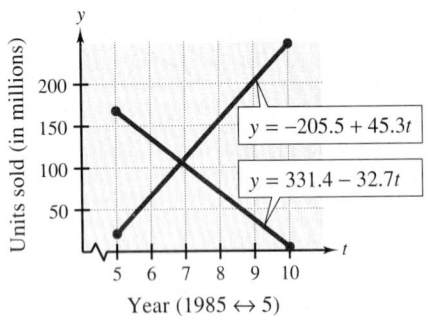

Source of data: Recording Industry Association of America, Inc.

Systems of Linear Equations in Two Variables

Systems of Equations • *Solving Systems of Equations by Graphing* • *Solving Systems of Equations by Substitution* • *The Method of Elimination* • *Applications*

Systems of Equations

Up to this point in the text, most problems have involved a single equation with one or two variables. Many problems in business and science involve a **system of equations** that consists of two or more equations, each involving two or more variables. A **solution** of a system of equations in two variables x and y is an ordered pair (x, y) of real numbers that satisfies *each* equation in the system. When you find the set of all solutions of the system of equations, you are **solving the system of equations.**

EXAMPLE 1   ■   Checking Solutions of a System of Equations

Check whether (a) the ordered pair $(3, 3)$ and (b) the ordered pair $(4, 2)$ are solutions of the following system of equations.

Ask students how they intuitively view a common solution for two equations. Stress that a solution to a system of equations must satisfy each equation in the system.

$$\begin{cases} x + y = 6 & \text{Equation 1} \\ 2x - 5y = -2 & \text{Equation 2} \end{cases}$$

Solution

(a) To determine whether the ordered pair $(3, 3)$ is a solution of the given system of equations, you must substitute $x = 3$ and $y = 3$ into *each* of the given equations. Substituting into Equation 1 produces

$$3 + 3 = 6. \qquad \text{Replace } x \text{ by 3 and } y \text{ by 3}$$

Similarly, substituting into Equation 2 produces

$$2(3) - 5(3) \neq -2. \qquad \text{Replace } x \text{ by 3 and } y \text{ by 3}$$

Because the ordered pair $(3, 3)$ fails to check in *both* equations, you can conclude that it is *not* a solution of the given system of equations.

(b) By substituting the coordinates of the ordered pair $(4, 2)$ into the two given equations, you can determine that this pair is a solution of the first equation,

$$4 + 2 = 6, \qquad \text{Replace } x \text{ by 4 and } y \text{ by 2}$$

and is also a solution of the second equation,

$$2(4) - 5(2) = -2. \qquad \text{Replace } x \text{ by 4 and } y \text{ by 2}$$

Therefore, $(4, 2)$ *is* a solution of the given system of equations.   ■

Solving Systems of Equations by Graphing

You can gain insight about the location and number of solutions of a system of equations by sketching the graph of each equation in the same coordinate plane. The solutions of the system correspond to the **points of intersection** of the graphs.

EXAMPLE 2   ■   The Graphical Method of Solving a System

Use the graphical method to solve the following system of equations.

$$\begin{cases} 2x + 3y = 7 & \text{Equation 1} \\ 2x - 5y = -1 & \text{Equation 2} \end{cases}$$

Solution

Because both equations in the given system are linear, you know that they have graphs that are straight lines. To sketch these lines, write each equation in slope-intercept form, as follows.

$$y = -\frac{2}{3}x + \frac{7}{3} \qquad\qquad \text{Equation 1}$$

$$y = \frac{2}{5}x + \frac{1}{5} \qquad\qquad \text{Equation 2}$$

The lines corresponding to these two equations are shown in Figure 4.1. From this figure, it appears that the two lines intersect in a single point, and that the coordinates of the point are approximately (2, 1). A check will show that this point is a solution of each of the original equations.

FIGURE 4.1

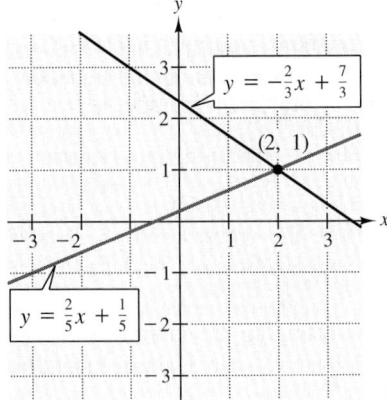

Use the system $3x - 2y = -4$, $9x + 4y = 13$ to point out the possible inaccuracy or need for approximation when using the graphical method.

Solution: $\left(\frac{1}{3}, \frac{5}{2}\right)$

To solve the system or to check the solution to the system with graphing technology, see page 301.

It is possible for a system of linear equations to have exactly one solution, infinitely many solutions, or no solution. To see why this is true, consider the following graphical interpretations of a system of two linear equations in two variables.

Graphical Interpretation of Solutions

For a system of two linear equations in two variables, the number of solutions is given by one of the following.

Number of Solutions	*Graphical Interpretation*
1. Exactly one solution	The two lines intersect at one point.
2. Infinitely many solutions	The two lines coincide (are identical).
3. No solution	The two lines are parallel.

These three possibilities are shown in Figure 4.2.

FIGURE 4.2

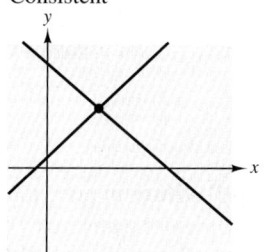

Consistent

Two lines that intersect: single point of intersection

Consistent

Two lines that coincide: infinitely many points of intersection

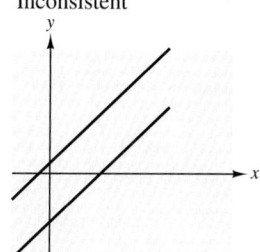

Inconsistent

Two parallel lines: no point of intersection

A system of linear equations is called **consistent** if it has at least one solution, and it is called **inconsistent** if it has no solution.

Solving Systems of Equations by Substitution

One way to solve a system of two equations in two variables algebraically is to convert the system to *one* equation in *one* variable by an appropriate substitution. This procedure is illustrated in Example 3.

EXAMPLE 3   ■

The Method of Substitution: One-Solution Case

Solve the following system of equations.

$$\begin{cases} -x + y = 3 & \text{Equation 1} \\ 3x + y = -1 & \text{Equation 2} \end{cases}$$

Caution students to avoid the common error of forgetting to find the value for *both* variables.

Solution

Begin by solving for y in the first equation.

$$y = x + 3 \qquad \text{Solve for } y \text{ in Equation 1}$$

FIGURE 4.3

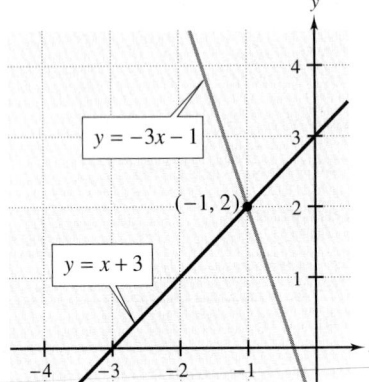

Next, substitute this expression for y into Equation 2.

$$3x + y = -1 \qquad \text{Equation 2}$$
$$3x + (x + 3) = -1 \qquad \text{Replace } y \text{ by } x + 3$$
$$4x = -4$$
$$x = -1 \qquad \text{Solve for } x$$

At this point, you know that the x-coordinate of the solution is -1. To find the y-coordinate, back-substitute the x-value into the revised Equation 1.

$$y = x + 3 \qquad \text{Revised Equation 1}$$
$$y = -1 + 3 \qquad \text{Replace } x \text{ by } -1$$
$$y = 2 \qquad \text{Solve for } y$$

Thus, the solution is $(-1, 2)$. Check to see that it satisfies both original equations. You can also check the solution graphically, as shown in Figure 4.3. ■

The term **back-substitute** implies that you work backwards. After finding a value for one of the variables, substitute that value back into one of the equations in the original (or revised) system to find the value of the other variable.

When using the method of substitution, it does not matter which variable you choose to solve for first. Whether you solve for y first or x first, you will obtain the same solution. When making your choice, you should choose the variable that is easier to work with. For instance, in the system

$$\begin{cases} 3x - 2y = 1 & \text{Equation 1} \\ x + 4y = 3 & \text{Equation 2} \end{cases}$$

it is easier to solve for x first (in the second equation).

The steps for using the method of substitution to solve a system of two equations involving two variables are summarized as follows.

■ Method of Substitution

To solve a system of two equations in two variables, use the following steps.

1. Solve one of the equations for one variable in terms of the other variable.

2. Substitute the expression found in step 1 into the other equation to obtain an equation of one variable.

3. Solve the equation obtained in step 2.

4. Back-substitute the solution from step 3 into the expression obtained in step 1 to find the value of the other variable.

5. Check the solution to ensure that it satisfies *both* of the original equations.

EXAMPLE 4   ■   Method of Substitution: No-Solution Case

Solve the following system of equations.

$$\begin{cases} 2x - 2y = 0 & \text{Equation 1} \\ x - y = 1 & \text{Equation 2} \end{cases}$$

Solution

Begin by solving for y in the second equation.

$$y = x - 1 \qquad \text{Solve for } y \text{ in Equation 2}$$

Next, substitute this expression for y into Equation 1.

$$2x - 2y = 0 \qquad \text{Equation 1}$$
$$2x - 2(x - 1) = 0 \qquad \text{Replace } y \text{ by } x - 1$$
$$2x - 2x + 2 = 0$$
$$2 = 0 \qquad \text{False statement}$$

Because the substitution process produced a false statement ($2 = 0$), you can conclude that the original system of equations has no solution. This is confirmed graphically in Figure 4.4, which shows that the two lines are parallel.   ■

FIGURE 4.4

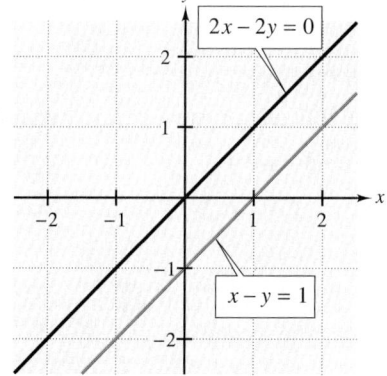

The Method of Elimination

A third method of solving a system of linear equations is called the **method of elimination.** The key step in the method of elimination is to obtain, for one of the variables, coefficients that differ only in sign, so that by *adding* the two equations this variable will be eliminated. The following system contains such coefficients for x.

$$\begin{array}{rl} 2x + 7y = 16 & \text{Equation 1} \\ \underline{-2x - 3y = -8} & \text{Equation 2} \\ 4y = 8 & \text{Add equations} \end{array}$$

Note that by adding the two equations, you eliminate the variable x and obtain a single equation in y. Solving this equation for y produces $y = 2$, which can be back-substituted into one of the original equations to solve for x. The method of elimination is summarized as follows.

The Method of Elimination

To use the **method of elimination** to solve a system of two linear equations in x and y, use the following steps.

1. Obtain coefficients for x (or y) that differ only in sign by multiplying all terms of one or both equations by suitable constants.

2. Add the equations to eliminate one variable and solve the resulting equation.

3. Back-substitute the value obtained in step 2 into either of the original equations and solve for the other variable.

4. Check your solution in both of the original equations.

To obtain coefficients (for one of the variables) that differ only in sign, you often need to multiply one or both of the equations by a suitable constant. This is demonstrated in the next two examples.

EXAMPLE 5 ■ The Method of Elimination

Point out that when multiplying an equation by a non-zero constant, all terms on each side of the equation are multiplied by this constant. The resulting equation is equivalent to the original, and therefore the solution set of the system is not affected.

Solve the following system of linear equations.

$$\begin{cases} 4x - 5y = 13 & \text{Equation 1} \\ 3x - y = 7 & \text{Equation 2} \end{cases}$$

Solution

For this system, you can obtain coefficients of y that differ only in sign by multiplying the second equation by -5.

$$\begin{array}{llll} 4x - 5y = 13 & \Longrightarrow & 4x - 5y = 13 & \text{Equation 1} \\ 3x - y = 7 & \Longrightarrow & -15x + 5y = -35 & \text{Multiply Equation 2 by } -5 \\ & & \overline{-11x = -22} & \text{Add equations} \end{array}$$

Thus, you see that $x = 2$. By back-substituting this value of x into the second equation, you can solve for y.

$$\begin{array}{ll} 3x - y = 7 & \text{Equation 2} \\ 3(2) - y = 7 & \text{Replace } x \text{ by } 2 \\ -y = 1 & \text{Subtract 6 from both sides} \\ y = -1 & \text{Solve for } y \end{array}$$

Therefore, the solution is $(2, -1)$. Check to see that this solution satisfies both original equations. The solution is verified graphically in Figure 4.5. ■

FIGURE 4.5

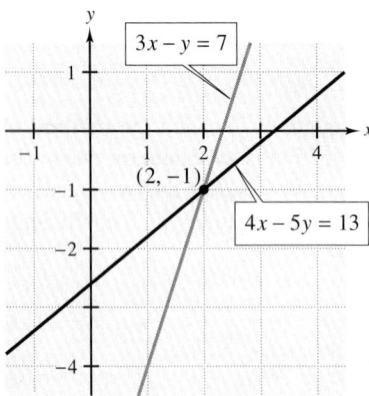

EXAMPLE 6   ■   The Method of Elimination

Solve the following system of linear equations.

$$\begin{cases} 5x + 3y = 9 & \text{Equation 1} \\ 2x - 4y = 14 & \text{Equation 2} \end{cases}$$

Solution

You can obtain coefficients of y that differ only in sign by multiplying the first equation by 4 and the second equation by 3.

FIGURE 4.6

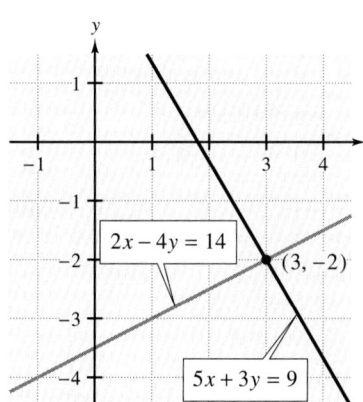

$$\begin{array}{lll} 5x + 3y = 9 & \implies & 20x + 12y = 36 \quad \text{Multiply Equation 1 by 4} \\ 2x - 4y = 14 & \implies & \underline{6x - 12y = 42} \quad \text{Multiply Equation 2 by 3} \\ & & 26x = 78 \quad \text{Add equations} \end{array}$$

From this equation, you can see that $x = 3$. By back-substituting this value of x into the second equation, you can solve for y, as follows.

$$\begin{array}{ll} 2x - 4y = 14 & \text{Equation 2} \\ 2(3) - 4y = 14 & \text{Replace } x \text{ by 3} \\ -4y = 8 & \text{Subtract 6 from both sides} \\ y = -2 & \text{Solve for } y \end{array}$$

Therefore, the solution is $(3, -2)$. Check this solution in both of the original equations. The solution is verified graphically in Figure 4.6.   ■

EXAMPLE 7   ■   The Method of Elimination: No-Solution Case

Solve the following system of linear equations.

FIGURE 4.7

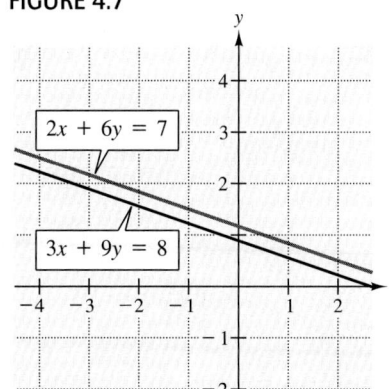

$$\begin{cases} 3x + 9y = 8 & \text{Equation 1} \\ 2x + 6y = 7 & \text{Equation 2} \end{cases}$$

Solution

To obtain coefficients of x that differ only in sign, multiply the first equation by 2 and the second equation by -3.

$$\begin{array}{lll} 3x + 9y = 8 & \implies & 6x + 18y = 16 \quad \text{Multiply Equation 1 by 2} \\ 2x + 6y = 7 & \implies & \underline{-6x - 18y = -21} \quad \text{Multiply Equation 2 by } -3 \\ & & 0 = -5 \quad \text{False statement} \end{array}$$

Because there are no values of x and y for which $0 = -5$, you can conclude that the system is inconsistent and has no solution. The lines corresponding to the two equations of this system are shown in Figure 4.7. Note that the two lines are parallel, and therefore have no point of intersection.   ■

Example 8 shows how the method of elimination works with a system that has infinitely many solutions.

EXAMPLE 8   ■   The Method of Elimination: Many-Solutions Case

Solve the following system of linear equations.

$$\begin{cases} -2x + 6y = 3 & \text{Equation 1} \\ 4x - 12y = -6 & \text{Equation 2} \end{cases}$$

Solution

To obtain coefficients of x that differ only in sign, multiply the first equation by 2.

$$\begin{array}{rclcrcl} -2x + 6y &=& 3 & \Longrightarrow & -4x + 12y &=& 6 \quad \text{Multiply Equation 1 by 2} \\ 4x - 12y &=& -6 & \Longrightarrow & 4x - 12y &=& -6 \quad \text{Equation 2} \\ \hline & & & & 0 &=& 0 \quad \text{Add equations} \end{array}$$

Because the two equations turned out to be equivalent, you can conclude that the system has infinitely many solutions. The solution set consists of all points (x, y) lying on the line $-2x + 6y = 3$, as shown in Figure 4.8.

FIGURE 4.8

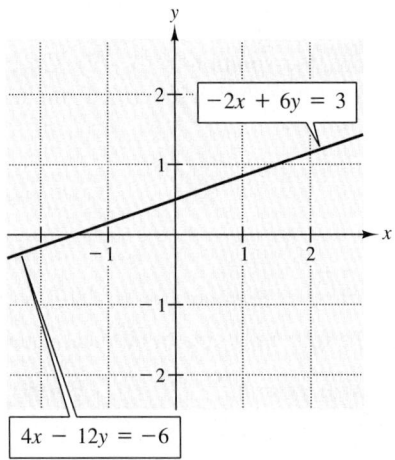

The next example shows how the method of elimination works with a system of linear equations having decimal coefficients.

EXAMPLE 9   ■   Solving a Linear System Having Decimal Coefficients

Solve the following system of linear equations.

$$\begin{cases} 0.02x - 0.05y = -0.38 & \text{Equation 1} \\ 0.03x + 0.04y = 1.04 & \text{Equation 2} \end{cases}$$

Solution

Because the coefficients in this system have two decimal places, it is convenient to begin by multiplying each equation by 100. (This produces a system in which the coefficients are all integers.)

$$\begin{cases} 2x - 5y = -38 & \text{Revised Equation 1} \\ 3x + 4y = 104 & \text{Revised Equation 2} \end{cases}$$

Now, to obtain coefficients of x that differ only in sign, multiply the first equation by 3 and the second equation by -2.

$$\begin{array}{rcl} 2x - 5y = -38 & \Rightarrow & 6x - 15y = -114 \quad \text{Multiply Equation 1 by 3} \\ 3x + 4y = 104 & \Rightarrow & \underline{-6x - 8y = -208} \quad \text{Multiply Equation 2 by } -2 \\ & & \qquad -23y = -322 \quad \text{Add equations} \end{array}$$

Thus, the value of y is

$$y = \frac{-322}{-23} = 14.$$

Back-substituting this value into revised Equation 2 produces the following.

$$\begin{array}{ll} 3x + 4y = 104 & \text{Equation 2} \\ 3x + 4(14) = 104 & \text{Replace } y \text{ by } 14 \\ 3x = 48 & \\ x = 16 & \text{Solve for } x \end{array}$$

Therefore, the solution is (16, 14). Check this solution in both of the original equations in the system. ∎

Applications

EXAMPLE 10 ∎ A Mixture Problem

A company with two stores buys six large delivery vans and five small delivery vans. The first store receives four of the large vans and two of the small vans for a total cost of $160,000. The second store receives two of the large vans and three of the small vans for a total cost of $128,000. What is the cost of each type of van?

Solution

The two unknowns in this problem are the costs of the two types of vans.

Verbal model: $4 \left(\boxed{\begin{array}{c} \text{Cost of} \\ \text{large van} \end{array}} \right) + 2 \left(\boxed{\begin{array}{c} \text{Cost of} \\ \text{small van} \end{array}} \right) = \boxed{\$160{,}000}$

$\qquad\qquad\quad 2 \left(\boxed{\begin{array}{c} \text{Cost of} \\ \text{large van} \end{array}} \right) + 3 \left(\boxed{\begin{array}{c} \text{Cost of} \\ \text{small van} \end{array}} \right) = \boxed{\$128{,}000}$

Labels: $x =$ Cost of large van (dollars)

$y =$ Cost of small van (dollars)

System of equations: $\begin{cases} 4x + 2y = 160{,}000 & \text{Equation 1} \\ 2x + 3y = 128{,}000 & \text{Equation 2} \end{cases}$

To solve this system of linear equations, you can use the method of elimination. To obtain coefficients of x that differ only in sign, multiply the second equation by -2.

$$\begin{array}{rcl} 4x + 2y = 160{,}000 & \Longrightarrow & 4x + 2y = 160{,}000 \\ 2x + 3y = 128{,}000 & \Longrightarrow & \underline{-4x - 6y = -256{,}000} \\ & & -4y = -96{,}000 \end{array}$$

Thus, the cost of each small van is $y = \$24{,}000$. By back-substituting this value into Equation 1, you can find the cost of each large van.

$$\begin{array}{ll} 4x + 2y = 160{,}000 & \text{Equation 1} \\ 4x + 2(24{,}000) = 160{,}000 & \text{Replace } y \text{ by } 24{,}000 \\ 4x = 112{,}000 & \\ x = 28{,}000 & \text{Solve for } x \end{array}$$

Thus, the cost of each large van is $x = \$28{,}000$, and the cost of each small van is $y = \$24{,}000$. Check this solution in the original problem. ■

EXAMPLE 11 ■ An Application

A total of \$12,000 was invested in two funds paying 6% and 8% simple interest. If the interest for 1 year is \$880, how much of the \$12,000 was invested in each fund?

Solution

Verbal model: | Amount of money invested in 6% account | $+$ | Amount of money invested in 8% account | $=$ | Total amount invested |

| Interest from 6% account | $+$ | Interest from 8% account | $=$ | Total interest |

Labels: $x =$ Amount invested at 6% (dollars)

$y =$ Amount invested at 8% (dollars)

System of equations: $\begin{cases} x + y = 12{,}000 & \text{Equation 1} \\ 0.06x + 0.08y = 880 & \text{Equation 2} \end{cases}$

To solve this system, solve the first equation for y to obtain $y = 12,000 - x$. Next, substitute this expression into the second equation to obtain

$$0.06x + 0.08(12,000 - x) = 880$$
$$0.06x + 960 - 0.08x = 880$$
$$-0.02x = -80$$
$$x = 4000.$$

Back-substituting this value for x into the revised first equation produces

$$y = 12,000 - 4000 = 8000.$$

Therefore, you can conclude that $4000 was invested in the 6% fund and $8000 was invested in the 8% fund.   ■

The total cost C of producing x units of a product usually has two components — the initial cost and the cost per unit. When enough units have been sold so that the total revenue R equals the total cost, the sales have reached the **break-even point.** You can find this break-even point by setting C equal to R and solving for x. In other words, the break-even point corresponds to the point of intersection of the cost and revenue graphs.

EXAMPLE 12   ■   An Application: Break-Even Analysis

A small business invests $14,000 in equipment to produce a product. Each unit of the product costs $0.80 to produce and is sold for $1.50. How many items must be sold before the business breaks even?

Solution

The total cost C of producing x units is

$$C = \overbrace{0.80x}^{\substack{\text{Cost} \\ \text{per unit}}} + \overbrace{14,000}^{\substack{\text{Initial} \\ \text{cost}}},$$

FIGURE 4.9

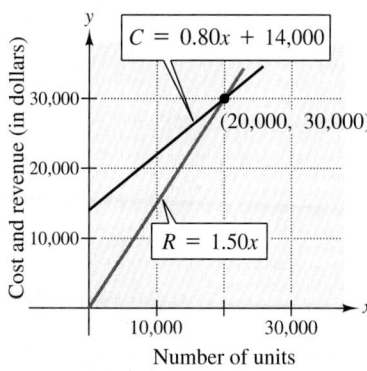

and the revenue R obtained by selling x units is

$$R = \overbrace{1.50x}^{\substack{\text{Price} \\ \text{per unit}}}.$$

Because the break-even point occurs when $R = C$, you have

$$1.50x = 0.80x + 14,000$$
$$0.7x = 14,000$$
$$x = 20,000.$$

Therefore, the business must sell 20,000 units before it breaks even. Note in Figure 4.9 that sales less than the break-even point correspond to a loss for the business, whereas sales greater than the break-even point correspond to a profit.   ■

DISCUSSION PROBLEM   ■   Creating a System of Equations with a Known Solution

Additional problem: For what value
of k does the following system have
no solution?

$2x - 3y = 4$

$4x - ky = 1$

Answer: $k = 6$

Suppose you are tutoring another student and want to create several systems of
equations that the student can use for practice. To begin, you want to find some
equations that have relatively simple solutions. For instance, try to create a system
of equations that has $(3, -2)$ as a solution. Write a short paragraph describing the
process you used to create the system.   ■

Warm-Up

*The following warm-up exercises involve skills that were covered in earlier sections.
You will use these skills in the exercise set for this section.*

In Exercises 1–4, sketch the graph of the equation.

1. $y = -\frac{1}{4}x + 8$ **2.** $y = 3(x - 2)$

3. $y - 2 = 2(x + 3)$ **4.** $y + 3 = \frac{1}{3}(x - 4)$

In Exercises 5 and 6, solve the equation.

5. $2y - 3(4y - 2) = 5$ **6.** $8x + 6(3 - 2x) = 4$

In Exercises 7 and 8, find an equation of the line passing through the two points.

7. $(-1, -2), \ (3, 6)$ **8.** $(1, 5), \ (6, 0)$

In Exercises 9 and 10, determine the slope of the line.

9. $4x + 6y = 4$ **10.** $-9x + 5y = 10$

4.1   EXERCISES

⊞ means that a graphing utility can help you
solve the exercise or check your solution.

 In Exercises 1–4, determine whether each ordered pair is a solution of the system of equations.

1. $\begin{cases} x + 2y = 9 \\ -2x + 3y = 10 \end{cases}$ (a) $(1, 4)$ (b) $(3, -1)$

2. $\begin{cases} 5x - 4y = 34 \\ x - 2y = 8 \end{cases}$ (a) $(0, 3)$ (b) $(6, -1)$

3. $\begin{cases} -2x + 7y = 46 \\ 3x + y = 0 \end{cases}$ (a) $(5, 8)$ (b) $(-2, 6)$

4. $\begin{cases} -5x - 2y = 23 \\ x + 4y = -19 \end{cases}$ (a) $(-3, -4)$ (b) $(-5, 1)$

In Exercises 5–10, use the graphs of the equations to determine whether the system has any solutions. Find any solutions that exist.

5. $\begin{cases} x + y = 4 \\ x + y = -1 \end{cases}$

6. $\begin{cases} -x + y = 5 \\ x + 2y = 4 \end{cases}$

7. $\begin{cases} 5x - 3y = 4 \\ 2x + 3y = 3 \end{cases}$

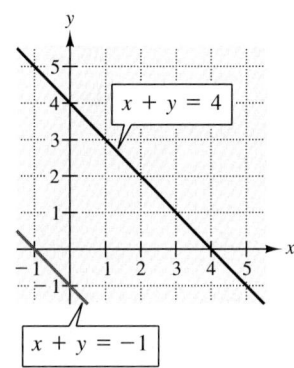

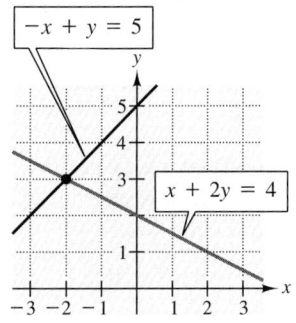

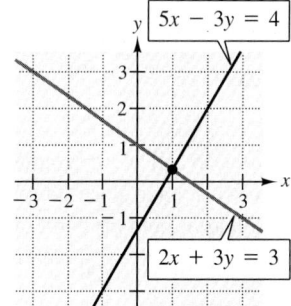

8. $\begin{cases} 2x - y = 4 \\ -4x + 2y = -12 \end{cases}$

9. $\begin{cases} -x + 2y = 5 \\ 2x - 4y = -10 \end{cases}$

10. $\begin{cases} x = 5 \\ y = -3 \end{cases}$

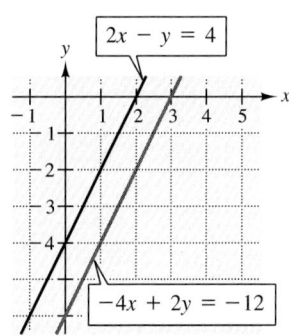

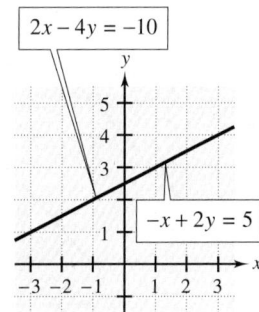

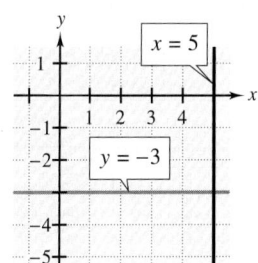

In Exercises 11–18, sketch the graphs of the given equations and find any solutions of the system of equations.

11. $\begin{cases} -2x + y = 1 \\ x - 3y = 2 \end{cases}$

12. $\begin{cases} 5x - 6y = -30 \\ 5x + 4y = 20 \end{cases}$

13. $\begin{cases} 2x - 5y = 20 \\ 4x - 5y = 40 \end{cases}$

14. $\begin{cases} 5x + 3y = 24 \\ x - 2y = 10 \end{cases}$

15. $\begin{cases} x - y = -3 \\ 2x - y = 6 \end{cases}$

16. $\begin{cases} x = 4 \\ y = 3 \end{cases}$

17. $\begin{cases} -5x + 3y = 15 \\ x + y = 1 \end{cases}$

18. $\begin{cases} 9x + 4y = 8 \\ 7x - 4y = 8 \end{cases}$

In Exercises 19–24, write each equation of the given system of linear equations in slope-intercept form and use this form to determine if the system has one solution, an infinite number of solutions, or no solution. (It is not necessary to sketch the graphs of the equations.)

19. $\begin{cases} 4x - 5y = 3 \\ -8x + 10y = -6 \end{cases}$

20. $\begin{cases} 4x - 5y = 3 \\ -8x + 10y = 14 \end{cases}$

21. $\begin{cases} -2x + 5y = 3 \\ 5x + 2y = 8 \end{cases}$

22. $\begin{cases} x + 10y = 12 \\ -2x + 5y = 2 \end{cases}$

23. $\begin{cases} -10x + 15y = 25 \\ 2x - 3y = -24 \end{cases}$

24. $\begin{cases} 4x - 5y = 28 \\ -2x + 2.5y = -14 \end{cases}$

In Exercises 25 and 26, the graphs of the two equations appear to be parallel. Yet, when each equation is written in slope-intercept form, we find that the slopes are not equal and thus the graphs intersect. Find the point of intersection of the two lines.

25. $\begin{cases} x - 100y = -200 \\ 3x - 275y = 198 \end{cases}$

26. $\begin{cases} 35x - 33y = 0 \\ 12x - 11y = 92 \end{cases}$

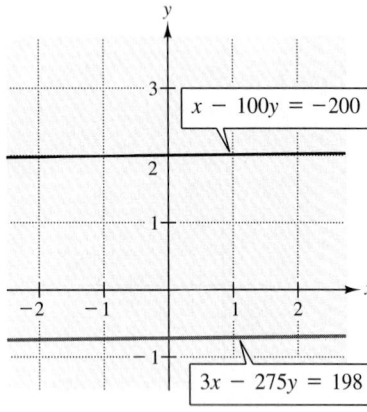

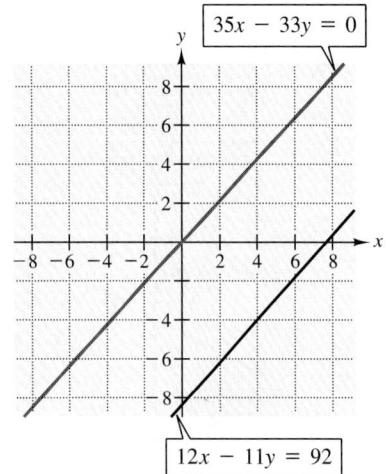

In Exercises 27–32, solve the system of equations by the method of substitution.

27. $\begin{cases} x - 2y = 0 \\ 3x + 2y = 8 \end{cases}$

28. $\begin{cases} x - y = 0 \\ 5x - 2y = 6 \end{cases}$

29. $\begin{cases} x = 4 \\ x - 2y = -2 \end{cases}$

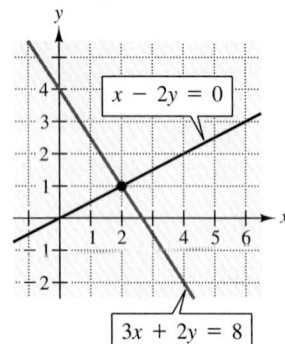

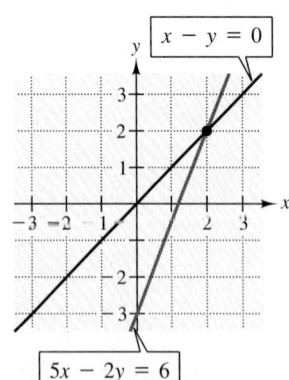

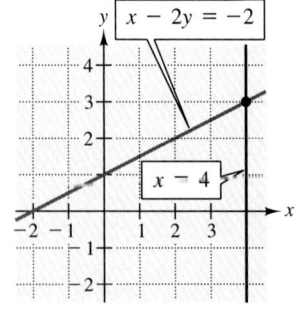

30. $\begin{cases} y = 2 \\ x - 6y = -6 \end{cases}$

31. $\begin{cases} 7x + 8y = 24 \\ x - 8y = 8 \end{cases}$

32. $\begin{cases} x - 3y = -2 \\ 5x + 3y = 17 \end{cases}$

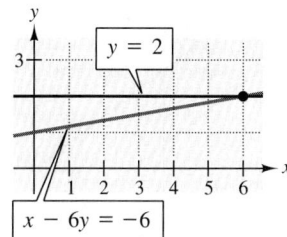

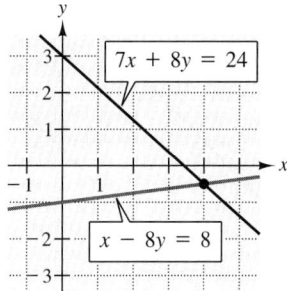

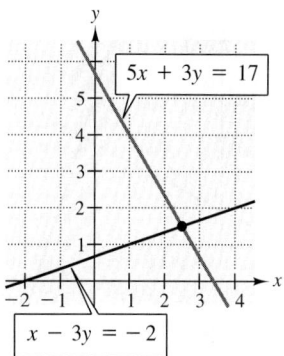

In Exercises 33–44, solve the system by the method of substitution.

33. $\begin{cases} x + y = 3 \\ 2x - y = 0 \end{cases}$

34. $\begin{cases} -x + y = 5 \\ x - 4y = 0 \end{cases}$

35. $\begin{cases} x + y = 2 \\ x - 4y = 12 \end{cases}$

36. $\begin{cases} x - 2y = -1 \\ x - 5y = 2 \end{cases}$

37. $\begin{cases} x + 6y = 19 \\ x - 7y = -7 \end{cases}$

38. $\begin{cases} x - 5y = -6 \\ 4x - 3y = 10 \end{cases}$

39. $\begin{cases} -13x + 16y = 10 \\ 5x + 16y = -26 \end{cases}$

40. $\begin{cases} 10x + 4y = 29 \\ 5x + 2y = 13 \end{cases}$

41. $\begin{cases} 12x - 8y = -15 \\ 18x - 12y = 9 \end{cases}$

42. $\begin{cases} 5x - 24y = -12 \\ 17x - 24y = 36 \end{cases}$

43. $\begin{cases} 2x - y = 2 \\ 6x - 3y = 6 \end{cases}$

44. $\begin{cases} -x + 4y = -8 \\ 0.5x - 2y = 4 \end{cases}$

In Exercises 45–52, solve the system by elimination. Identify and label each line with the proper equation and label the point of intersection (if any).

45. $\begin{cases} -x + 2y = 1 \\ x - y = 2 \end{cases}$

46. $\begin{cases} 3x + y = 3 \\ 2x - y = 7 \end{cases}$

47. $\begin{cases} x + y = 0 \\ 3x - 2y = 10 \end{cases}$

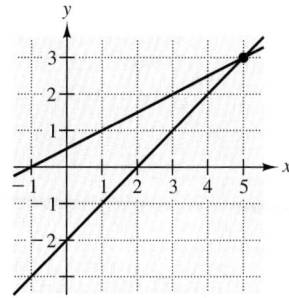

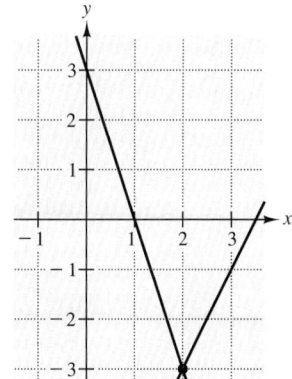

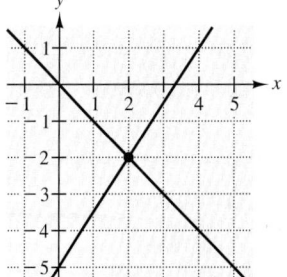

48. $\begin{cases} -x + 2y = 2 \\ 3x + y = 15 \end{cases}$

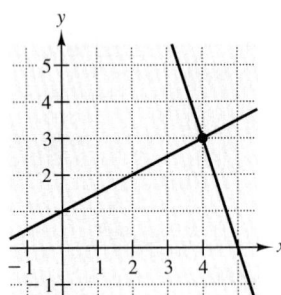

49. $\begin{cases} x - y = 1 \\ -3x + 3y = 8 \end{cases}$

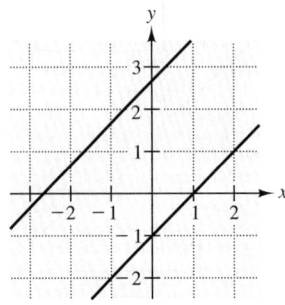

50. $\begin{cases} 3x + 4y = 2 \\ 0.6x + 0.8y = 1.6 \end{cases}$

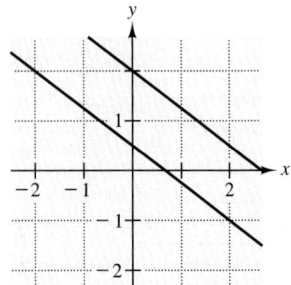

51. $\begin{cases} x - 3y = 5 \\ -2x + 6y = -10 \end{cases}$

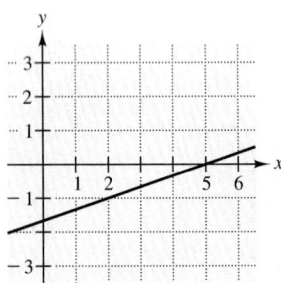

52. $\begin{cases} x - 4y = 5 \\ 5x + 4y = 7 \end{cases}$

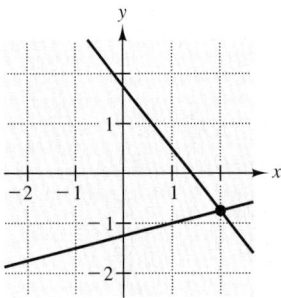

In Exercises 53–72, solve the system by the method of elimination.

53. $\begin{cases} 3x - 2y = 5 \\ x + 2y = 7 \end{cases}$

54. $\begin{cases} -x + 2y = 9 \\ x + 3y = 16 \end{cases}$

55. $\begin{cases} 4x + y = -3 \\ -4x + 3y = 23 \end{cases}$

56. $\begin{cases} -3x + 5y = -23 \\ 2x - 5y = 22 \end{cases}$

57. $\begin{cases} x - 3y = 2 \\ 3x - 7y = 4 \end{cases}$

58. $\begin{cases} 7r - s = -25 \\ 2r + 5s = 14 \end{cases}$

59. $\begin{cases} 2u + 3v = 8 \\ 3u + 4v = 13 \end{cases}$

60. $\begin{cases} 4x - 3y = 25 \\ -3x + 8y = 10 \end{cases}$

61. $\begin{cases} 12x - 5y = 2 \\ -24x + 10y = 6 \end{cases}$

62. $\begin{cases} -2x + 3y = 9 \\ 6x - 9y = -27 \end{cases}$

63. $\begin{cases} \frac{2}{3}r - s = 0 \\ 10r + 4s = 19 \end{cases}$

64. $\begin{cases} x - y = -\frac{1}{2} \\ 4x - 48y = -35 \end{cases}$

65. $\begin{cases} 0.7u - v = -0.4 \\ 0.3u - 0.8v = 0.2 \end{cases}$

66. $\begin{cases} 0.15x - 0.35y = -0.5 \\ -0.12x + 0.25y = 0.1 \end{cases}$

67. $\begin{cases} 5x + 7y = 25 \\ x + 1.4y = 5 \end{cases}$

68. $\begin{cases} 12b - 13m = 2 \\ -6b + 6.5m = -2 \end{cases}$

69. $\begin{cases} 2x = 25 \\ 4x - 10y = 52 \end{cases}$

70. $\begin{cases} 6x - 6y = 25 \\ 3y = 11 \end{cases}$

71. $\begin{cases} 2x + 3y = 0 \\ 3x + 5y = -1000 \end{cases}$

72. $\begin{cases} 0.4u + v = 800 \\ 0.7u + 2v = 1850 \end{cases}$

In Exercises 73–80, use the method of substitution or the method of elimination, whichever is more convenient, to solve the system.

73. $\begin{cases} \frac{3}{2}x + 2y = 12 \\ \frac{1}{4}x + y = 4 \end{cases}$

74. $\begin{cases} 4x + y = -2 \\ -6x + y = 18 \end{cases}$

75. $\begin{cases} y = 5x - 3 \\ y = -2x + 11 \end{cases}$

76. $\begin{cases} 3x + 2y = 5 \\ y = 2x + 13 \end{cases}$

77. $\begin{cases} 2x - y = 20 \\ y = x - 5 \end{cases}$

78. $\begin{cases} 3x - 2y = -20 \\ 5x + 6y = 32 \end{cases}$

79. $\begin{cases} 2x - 3y = -21 \\ 4y = 50 \end{cases}$

80. $\begin{cases} 5x + 2y = 65 \\ -4x + 3y = -6 \end{cases}$

Coin Problems In Exercises 81–84, determine how many coins of each type will yield the given value.

Coins	Value
81. 15 dimes and quarters	$3.00
82. 32 dimes and quarters	$6.95
83. 25 nickels and dimes	$1.95
84. 50 nickels and quarters	$7.50

85. Gasoline Mixture Twelve gallons of regular unleaded gasoline plus 8 gallons of premium unleaded gasoline cost $23.08. The price of premium unleaded is 11 cents more per gallon than the price of regular unleaded. Find the price per gallon for each grade of gasoline.

86. Ticket Sales Eight hundred tickets were sold for a theater production, and the receipts for the performance were $8600. The tickets for adults and students sold for $12.50 and $7.50, respectively. How many of each kind of ticket were sold?

87. Nut Mixture Ten pounds of mixed nuts sell for $6.97 per pound. The mixture contains two kinds of nuts, with one variety priced at $5.65 per pound and the other at $8.95 per pound. How many pounds of each variety of nuts were used in the mixture?

88. Hay Mixture How many tons of hay at $125 per ton and of hay at $75 per ton must be purchased to have 100 tons of hay with a value of $90 per ton?

89. Alcohol Mixture How many liters of a 40% alcohol solution must be mixed with a 65% solution to obtain 20 liters of a 50% solution?

90. Acid Mixture Fifty gallons of 70% acid solution are obtained by mixing an 80% solution with a 50% solution. How many gallons of each solution must be used to obtain the desired (70%) mixture?

91. Simple Interest A combined total of $20,000 is invested in two bonds that pay 8% and 9.5% simple interest. The annual interest is $1675. How much is invested in each bond?

92. Simple Interest A combined total of $12,000 is invested in two bonds that pay 8.5% and 10% simple interest. The annual interest is $1140. How much is invested in each bond?

93. Break-Even Analysis A small business invests $8000 in equipment for producing a product. Each unit of the product costs $1.20 to produce and is sold for $2.00. How many items must be sold before the business breaks even?

94. Break-Even Analysis A business invests $50,000 in equipment for producing a product. Each unit of the product costs $19.05 to produce and is sold for $35.95. How many items must be sold before the business breaks even?

95. Number Problem The sum of the digits of a given two-digit number is 9. If the digits are reversed, the number is decreased by 27. Find the number.

96. Air Speed An airplane flying into a headwind travels the 3000-mile flying distance between two cities in 6 hours and 15 minutes. On the return flight, the distance is traveled in 5 hours. Find the speed of the plane in still air and the speed of the wind, assuming that both remain constant throughout the round trip.

97. *Best-Fitting Line* The slope and y-intercept of the line $y = mx + b$ that best fits the three noncollinear points $(0, 0)$, $(1, 1)$ and $(2, 3)$ can be determined using the following system of linear equations.

$$\begin{cases} 3b + 5m = 7 \\ 3b + 3m = 4 \end{cases}$$

(a) Solve the system and find the equation of the best fitting line.

(b) Plot the three points and sketch the graph of the best fitting line.

98. *Best-Fitting Line* The slope and y-intercept of the line $y = mx + b$ that best fits the three noncollinear points $(0, 3)$, $(1, 2)$ and $(2, 0)$ can be determined using the following system of linear equations.

$$\begin{cases} 3b + 5m = 2 \\ 3b + 3m = 5 \end{cases}$$

(a) Solve the system and find the equation of the best fitting line.

(b) Plot the three points and sketch the graph of the best fitting line.

99. *Vietnam Veterans Memorial* "The Wall" in Washington, D.C., designed by Maya Ling Lin when she was a student at Yale University, has two vertical, triangular sections of black granite with a common side. The top of each section is level with the ground. The bottoms of the two sections can be modeled by the equations $y = \frac{2}{25}x - 10$ and $y = -\frac{5}{61}x - 10$ when the x-axis is superimposed on the top of the wall. Each unit in the coordinate system represents 1 foot. How deep is the memorial at the point where the two sections meet? How long is each section?

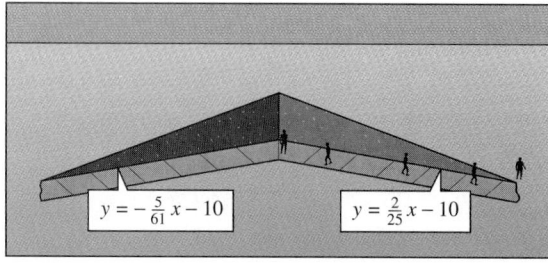

100. Determine the value of k such that the following system is inconsistent.

$$\begin{cases} 12x - 18y = 5 \\ -18x + ky = 10 \end{cases}$$

In Exercises 101–104, use the result of the discussion problem in this section (page 294) to create a system of equations having the given solution. (*Note:* There are many correct answers to each problem.)

101. $(4, 5)$ **102.** $(-2, 6)$

103. $(-1, -2)$ **104.** $\left(\frac{1}{2}, 3\right)$

The Vietnam Veterans Memorial honors all the men and women who fought and died in Vietnam.

Using a Graphing Calculator

A graphing calculator or computer can be used to sketch the graph of a system of linear equations. Here, we show how to use Texas Instruments TI-81 and TI-82 and a Casio fx-7700G™ to graph the following system.

$$\begin{cases} 3x - 4y = 2 & \text{Equation 1} \\ 5x + 6y = 16 & \text{Equation 2} \end{cases}$$

The graph of the system is shown below. Note that the two lines appear to intersect at the point $(2, 1)$. To check this, substitute 2 for x and 1 for y in each equation.

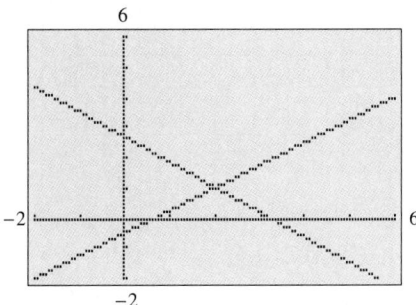

The first step in using a graphing calculator to sketch the graph of a linear system is to solve each equation for y.

$$\begin{cases} y = \dfrac{3}{4}x - \dfrac{1}{2} & \text{Equation 1} \\ y = -\dfrac{5}{6}x + \dfrac{8}{3} & \text{Equation 2} \end{cases}$$

Next, find a range setting that shows the point of intersection. You may have to try a few before you obtain a *viewing window* that shows the portion of the graph you want to see. For this particular system, you can use a range setting in which both x and y vary between -2 and 6. The keystrokes to use for the TI-81 or TI-82 and the Casio fx-7700G are shown on the next page.

Some uses of linear systems as real-life models call for *exact* solutions. For others, however, *approximate* solutions are sufficient. For instance, suppose you have found linear models that represent two different trends (such as population and sales) over a period of time. It is not reasonable to expect such a model to produce an answer such as "August 16, 1992." Rather, you expect the model to be accurate enough to produce an answer such as "sometime in 1992" or "about midyear in 1992."

301

TI-81 or TI-82

RANGE or WINDOW (Set range.)

    Xmin=-2

    Xmax=6

    Xscl=1

    Ymin=-2

    Ymax=6

    Yscl=1

Y= (3 ÷ 4) X|T

    − (1 ÷ 2) ENTER

    (−) (5 ÷ 6) X|T

    + (8 ÷ 3)

:$Y_1=(3/4)X-(1/2)$

:$Y_2=-(5/6)X+(8/3)$

GRAPH

Casio fx-7700G

Range (Set range.)

    Xmin=-2

    max=6

    scl=1

    Ymin=-2

    max=6

    scl=1

EXE Range

SHIFT F5 EXE

Graph (3 ÷ 4) X,θ,T

    − (1 ÷ 2)

Graph $Y=(3÷4)X-(1÷2)$ EXE

Graph (−) (5 ÷ 6) X,θ,T

    + (8 ÷ 3)

Graph $Y=-(5÷6)X+(8÷3)$ EXE

EXAMPLE 1 ■ Approximating the Solution of a Linear System

The total exports and imports of Puerto Rico (in millions of dollars) for 1970 through 1988 are represented by the linear models

$$\begin{cases} T = 688t + 370 & \text{Exports} \\ T = 538t + 2530, & \text{Imports} \end{cases}$$

where t is the year, with $t = 0$ corresponding to 1970. Use a graphing calculator to approximate the year in which Puerto Rico's exports began to exceed its imports.

Solution

Graphing calculators use x and y as variables, so you first need to rewrite the equations using x and y.

$$\begin{cases} y = 688x + 370 & \text{Exports} \\ y = 538x + 2530 & \text{Imports} \end{cases}$$

302

After entering these equations in the calculator and setting an appropriate range, you can obtain the graph shown below. From the graph, you can see that the exports began to exceed the imports around 1984. (The *trace* feature can help you approximate the point of intersection.)

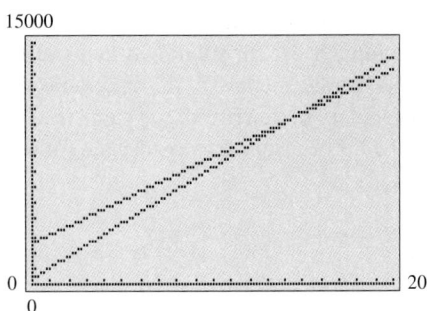

EXERCISES

In Exercises 1–4, use a graphing calculator to sketch the graph of the system. Approximate the solution and check your answer algebraically.

1. $\begin{cases} y = -2x + 2 \\ y = \frac{1}{2}x - \frac{11}{2} \end{cases}$

2. $\begin{cases} y = 3x + 11 \\ y = \frac{1}{3}x + \frac{1}{3} \end{cases}$

3. $\begin{cases} 7x - 8y = 5 \\ 5x + 4y = 23 \end{cases}$

4. $\begin{cases} 10x + 11y = 24 \\ 9x - 6y = -42 \end{cases}$

5. **Per Capita Income** For 1980 through 1990, the per capita incomes y in Nevada and New Hampshire can be modeled by

$$\begin{cases} y = 886t + 10{,}848 \qquad \text{Nevada} \\ y = 1233t + 9{,}150, \qquad \text{New Hampshire} \end{cases}$$

where t is the year, with $t = 0$ corresponding to 1980. Use a graphing calculator to approximate the year in which the per capita income of New Hampshire surpassed that of Nevada. (*Source:* Bureau of Economic Analysis)

Systems of Linear Equations in Three Variables
Row-Echelon Form • *The Method of Gaussian Elimination* • *Applications*

Row-Echelon Form

The method of elimination introduced in Section 4.1 can be applied to a system of linear equations in more than two variables. In fact, this method is easily adapted to computer use for solving systems of linear equations with dozens of variables.

When the method of elimination is used to solve a system of linear equations, the goal is to rewrite the system in a form to which back-substitution can be applied. For instance, consider the following two systems of linear equations.

$$\begin{cases} x - 2y + 2z = 9 \\ -x + 3y = -4 \\ 2x - 5y + z = 10 \end{cases} \qquad \begin{cases} x - 2y + 2z = 9 \\ y + 2z = 5 \\ z = 3 \end{cases}$$

The system on the right is said to be in **row-echelon form,** which means that it has a "stair-step" pattern with leading coefficients of 1.

Which of these two systems do you think is easier to solve? After comparing the two systems, it should be clear that it is easier to solve the system on the right. Example 1 shows how to use back-substitution to solve this system.

EXAMPLE 1   ■   Using Back-Substitution

Solve the following system of linear equations.

$$\begin{cases} x - 2y + 2z = 9 & \text{Equation 1} \\ y + 2z = 5 & \text{Equation 2} \\ z = 3 & \text{Equation 3} \end{cases}$$

Solution

From Equation 3, you know the value of z. To solve for y, substitute 3 for z in Equation 2 to obtain

$$y + 2(3) = 5 \qquad \text{Substitute 3 for } z$$
$$y = -1. \qquad \text{Solve for } y$$

Finally, substitute -1 for y and 3 for z in Equation 1 to obtain

$$x - 2(-1) + 2(3) = 9 \qquad \text{Substitute } -1 \text{ for } y, 3 \text{ for } z$$
$$x = 1. \qquad \text{Solve for } x$$

Thus, the solution is $x = 1$, $y = -1$, and $z = 3$, which can also be written as the **ordered triple** $(1, -1, 3)$.   ■

The Method of Gaussian Elimination

Two systems of equations are called **equivalent** if they have the same solution set. To solve a system that is not in row-echelon form, first convert it to an *equivalent* system that is in row-echelon form. Example 2 shows how this is done, by applying the method of elimination to a system of two linear equations.

EXAMPLE 2 ■ The Method of Gaussian Elimination

Use the method of elimination to solve the following system of linear systems.

$$\begin{cases} 3x - 2y = -1 & \text{Equation 1} \\ x - y = 0 & \text{Equation 2} \end{cases}$$

The initial goal is a coefficient of 1 for x in the first equation. Point out that the easiest way to achieve the goal is to switch the first and second equations.

Solution

$$\begin{cases} x - y = 0 \\ 3x - 2y = -1 \end{cases}$$ Interchange Equations 1 and 2.

$$\begin{cases} x - y = 0 \\ y = -1 \end{cases}$$ Adding -3 times Equation 1 to Equation 2 produces a new Equation 2.

Now, using back-substitution on the row-echelon system, you can find that the solution is $y = -1$ and $x = -1$. ■

The three basic types of operations that can be performed on a system of linear equations to produce an equivalent system are summarized as follows.

Operations That Lead to Equivalent Systems of Equations	Each of the following **row operations** on a system of linear equations produces an *equivalent* system of linear equations.

1. Interchange two equations.

2. Multiply one of the equations by a nonzero constant.

3. Add a multiple of one of the equations to another equation to replace the latter equation.

As shown in Example 2, rewriting a system of linear equations in row-echelon form usually involves a *chain* of equivalent systems, each of which is obtained by using one of the three basic row operations. This process is called **Gaussian elimination,** after the German mathematician Carl Friedrich Gauss (1777–1855). Example 3 shows how to use Gaussian elimination to solve the system that was given at the beginning of this section.

EXAMPLE 3   ■   **Using Gaussian Elimination to Solve a System**

Solve the following system of linear equations.

$$\begin{cases} x - 2y + 2z = 9 & \text{Equation 1} \\ -x + 3y = -4 & \text{Equation 2} \\ 2x - 5y + z = 10 & \text{Equation 3} \end{cases}$$

Solution

Stress the importance of using a systematic procedure, especially for more complicated problems.

Although there are several ways to begin, your goal is to develop a systematic procedure that can be applied to large systems. Work from the upper left corner, saving the x in the upper left position and eliminating the other x's from the first column.

$$\begin{cases} x - 2y + 2z = 9 \\ y + 2z = 5 \\ 2x - 5y + z = 10 \end{cases}$$

Adding Equation 1 to Equation 2 produces a new Equation 2.

$$\begin{cases} x - 2y + 2z = 9 \\ y + 2z = 5 \\ {-y} - 3z = -8 \end{cases}$$

Adding -2 times Equation 1 to Equation 3 produces a new Equation 3.

Now that all but the first x have been eliminated from the first column, you can go to work on the second column. (You need to eliminate y from Equation 3.)

$$\begin{cases} x - 2y + 2z = 9 \\ y + 2z = 5 \\ {-z} = -3 \end{cases}$$

Adding Equation 2 to Equation 3 produces a new Equation 3.

Finally, you need a coefficient of 1 for z in Equation 3.

$$\begin{cases} x - 2y + 2z = 9 \\ y + 2z = 5 \\ z = 3 \end{cases}$$

Multiplying Equation 3 by -1 produces a new Equation 3.

This is the same system that was solved in Example 1. As in that example, you can conclude that the solution is

$$x = 1, \quad y = -1, \quad \text{and} \quad z = 3. \qquad\qquad ■$$

NOTE   The two systems of linear equations given in Examples 1 and 3 are equivalent to each other.

In Example 3, you can check the solution $x = 1$, $y = -1$, and $z = 3$ as follows.

Equation 1: $(1) - 2(-1) + 2(3) = 9$ Check solution in each

Equation 2: $-(1) + 3(-1) = -4$ equation of original system

Equation 3: $2(1) - 5(-1) + (3) = 10$

EXAMPLE 4   ■   Using Elimination to Solve a System

Solve the following system of linear equations.

$$\begin{cases} 4x + y - 3z = 11 & \text{Equation 1} \\ 2x - 3y + 2z = 9 & \text{Equation 2} \\ x + y + z = -3 & \text{Equation 3} \end{cases}$$

Additional problem:

$$5x - 2y + 8z = -4$$

$$x + 3y - 6z = 9$$

$$2x + y + 4z = 6$$

Answer: $x = 0$, $y = 4$, and $z = \frac{1}{2}$

Solution

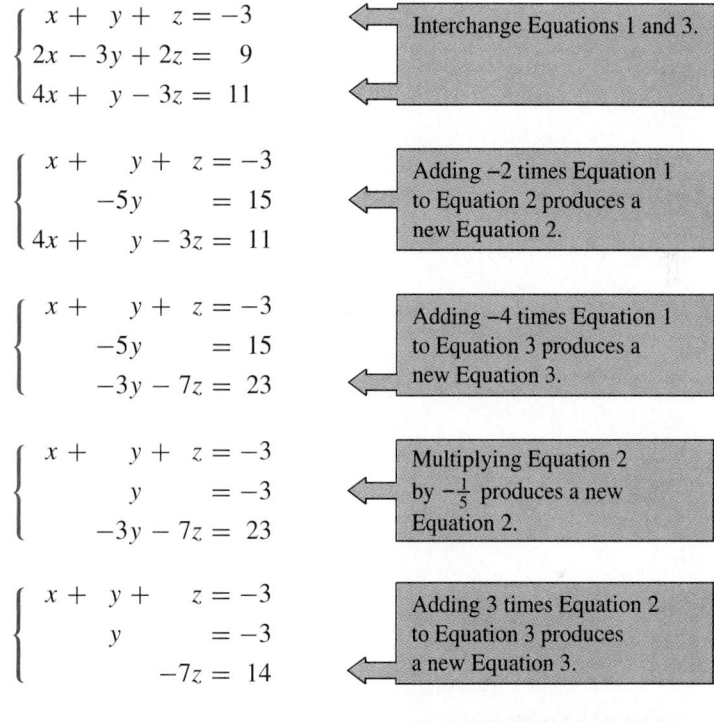

$$\begin{cases} x + y + z = -3 \\ 2x - 3y + 2z = 9 \\ 4x + y - 3z = 11 \end{cases}$$
Interchange Equations 1 and 3.

$$\begin{cases} x + y + z = -3 \\ -5y = 15 \\ 4x + y - 3z = 11 \end{cases}$$
Adding -2 times Equation 1 to Equation 2 produces a new Equation 2.

$$\begin{cases} x + y + z = -3 \\ -5y = 15 \\ -3y - 7z = 23 \end{cases}$$
Adding -4 times Equation 1 to Equation 3 produces a new Equation 3.

$$\begin{cases} x + y + z = -3 \\ y = -3 \\ -3y - 7z = 23 \end{cases}$$
Multiplying Equation 2 by $-\frac{1}{5}$ produces a new Equation 2.

$$\begin{cases} x + y + z = -3 \\ y = -3 \\ -7z = 14 \end{cases}$$
Adding 3 times Equation 2 to Equation 3 produces a new Equation 3.

$$\begin{cases} x + y + z = -3 \\ y = -3 \\ z = -2 \end{cases}$$
Multiplying Equation 3 by $-\frac{1}{7}$ produces a new Equation 3.

Now that the system of equations is in row-echelon form, you can see that $z = -2$ and $y = -3$. Moreover, by back-substituting these values into Equation 1, you can determine that $x = 2$. Therefore, the solution is

$$x = 2, \quad y = -3, \quad \text{and} \quad z = -2. \qquad ■$$

The next example concerns an *inconsistent* system—one that has no solution. The key to recognizing an inconsistent system is that, at some stage in the elimination process, you obtain an absurdity such as $0 = 7$.

EXAMPLE 5   ■   An Inconsistent System

Solve the following system of linear equations.

$$\begin{cases} x - 3y + z = 1 & \text{Equation 1} \\ 2x - y - 2z = 2 & \text{Equation 2} \\ x + 2y - 3z = -1 & \text{Equation 3} \end{cases}$$

Solution

$$\begin{cases} x - 3y + z = 1 \\ 5y - 4z = 0 \\ x + 2y - 3z = -1 \end{cases}$$

> Adding –2 times Equation 1 to Equation 2 produces a new Equation 2.

$$\begin{cases} x - 3y + z = 1 \\ 5y - 4z = 0 \\ 5y - 4z = -2 \end{cases}$$

> Adding –1 times Equation 1 to Equation 3 produces a new Equation 3.

$$\begin{cases} x - 3y + z = 1 \\ 5y - 4z = 0 \\ 0 = -2 \end{cases}$$

> Adding –1 times Equation 2 to Equation 3 produces a new Equation 3.

Because "Equation" 3 $(0 = -2)$ is a false statement, you can conclude that this system is inconsistent and therefore has no solution. Moreover, because this system is equivalent to the original system, you can conclude that the original system also has no solution.   ■

As with a system of linear equations in two variables, the solution(s) to a system of linear equations in more than two variables must fall into one of three categories.

The Number of Solutions of a System of Linear Equations

For a system of linear equations, exactly one of the following is true.

1. There is exactly one solution.

2. There are infinitely many solutions.

3. There is no solution.

Solutions of equations with three variables can be pictured with a *three-dimensional coordinate system*. The graph of a linear equation in three variables is a *plane*. Thus, the graph of a system of three linear equations in three variables consists of three planes. When these planes intersect in a single point, the system has exactly one solution. When the three planes have no point in common, the system has no solution. When the three planes contain a common line, the system has infinitely many solutions (see Figure 4.10).

FIGURE 4.10

Exactly one solution   No solution   Infinitely many solutions

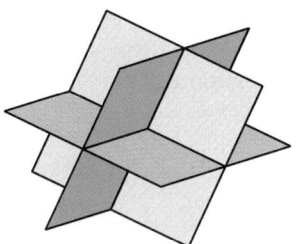 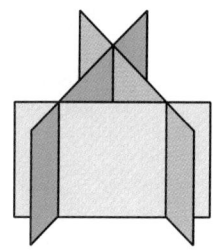 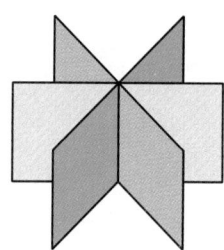

When a system of equations has no solution, you can simply state that it is *inconsistent*. If a system has exactly one solution, you can list the value of each variable. However, for systems that have infinitely many solutions, listing the solutions is a little awkward. For example, you might give the solutions to a system in three variables as

$$(a, a + 1, 2a), \qquad \text{where } a \text{ is any real number.}$$

This means that for each real number a, you have a valid solution to the system. A few of the infinitely many possible solutions are found by letting $a = -1, 0, 1$, and 2 to obtain $(-1, 0, -2)$, $(0, 1, 0)$, $(1, 2, 2)$, and $(2, 3, 4)$, respectively. Now consider the solutions represented by

$$(b - 1, b, 2b - 2), \qquad \text{where } b \text{ is any real number.}$$

A few possible solutions are $(-1, 0, -2)$, $(0, 1, 0)$, $(1, 2, 2)$, and $(2, 3, 4)$, which are found by letting $b = 0, 1, 2$, and 3, respectively. Note that both descriptions result in the same collection of solutions. Thus, when comparing descriptions of an infinite solution set, keep in mind that there is more than one way to describe the set.

EXAMPLE 6   ■   A System with Infinitely Many Solutions

Solve the system of linear equations.

$$\begin{cases} x + y - 3z = -1 & \text{Equation 1} \\ y - z = 0 & \text{Equation 2} \\ -x + 2y = 1 & \text{Equation 3} \end{cases}$$

Solution

Begin by rewriting the system in row-echelon form, as follows.

$$\begin{cases} x + y - 3z = -1 \\ y - z = 0 \\ 3y - 3z = 0 \end{cases}$$

Adding Equation 1 to Equation 3 produces a new Equation 3.

$$\begin{cases} x + y - 3z = -1 \\ \quad\quad y - z = 0 \\ \quad\quad\quad\quad 0 = 0 \end{cases}$$

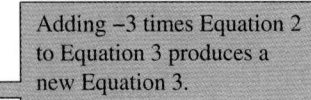

Adding –3 times Equation 2 to Equation 3 produces a new Equation 3.

This means that Equation 3 depends on Equations 1 and 2 in the sense that it gives you no additional information about the variables. Thus, the original system is equivalent to the system

$$\begin{cases} x + y - 3z = -1 \\ \quad\quad y - z = 0. \end{cases}$$

In this last equation, solve for y in terms of z to obtain $y = z$. By back-substituting for y in the previous equation, you can find x in terms of z, as follows.

$$x + z - 3z = -1$$
$$x - 2z = -1$$
$$x = 2z - 1$$

Finally, if we let $z = a$, the solutions to the given system are all of the form

$$x = 2a - 1, \quad y = a, \quad z = a,$$

where a is a real number. Thus, every ordered triple of the form

$$(2a - 1, a, a), \qquad a \text{ is a real number}$$

is a solution of the system. ∎

Applications

This section concludes with three applications involving systems of linear equations in three variables. The first example shows how to fit a parabola through three given points in the plane.

EXAMPLE 7 ■ An Application: Curve Fitting

Find a quadratic equation

$$y = ax^2 + bx + c$$

whose graph passes through the points $(-1, 3)$, $(1, 1)$, and $(2, 6)$.

Additional problems: Find a quadratic equation whose graph passes through

(a) $(0, 1)$, $(-2, 13)$, and $(3, 28)$

(b) $(1, 2)$, $(-1, 6)$, and $(4, 11)$

Answers: (a) $y = 3x^2 + 1$
(b) $y = x^2 - 2x + 3$

Solution

Because the graph of $y = ax^2 + bx + c$ passes through the points $(-1, 3)$, $(1, 1)$, and $(2, 6)$, you have the following.

When $x = -1, y = 3$: $a(-1)^2 + b(-1) + c = 3$

When $x = 1, y = 1$: $a(\ 1)^2 + b(\ 1) + c = 1$

When $x = 2, y = 6$: $a(\ 2)^2 + b(\ 2) + c = 6$

This produces the following system of linear equations in variables a, b, and c.

$$\begin{cases} a - b + c = 3 \\ a + b + c = 1 \\ 4a + 2b + c = 6 \end{cases}$$

FIGURE 4.11

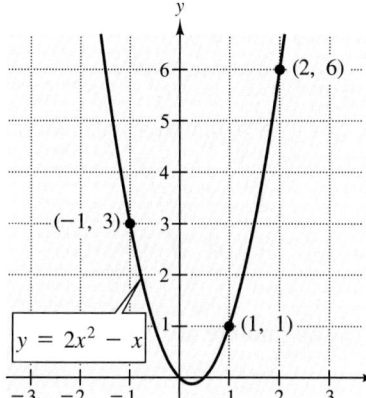

The solution to this system turns out to be

$$a = 2, \quad b = -1, \quad \text{and} \quad c = 0.$$

Thus, the equation of the parabola passing through the three given points is

$$y = 2x^2 - x,$$

as shown in Figure 4.11.

■

EXAMPLE 8   ■   An Application: Moving Object

You might ask students to account for the fact that the object reached a height of 164 feet <u>twice</u>.

The height at time t of an object that is moving in a (vertical) line with constant acceleration a is given by the **position equation**

$$s = \frac{1}{2}at^2 + v_0t + s_0.$$

The height s is measured in feet, the time t is measured in seconds, v_0 is the initial velocity (at time $t = 0$), and s_0 is the initial height. Find the values of a, v_0, and s_0, if $s = 164$ feet at 1 second, $s = 180$ feet at 2 seconds, and $s = 164$ feet at 3 seconds.

Solution

By substituting the three values of t and s into the position equation, you obtain three linear equations in a, v_0, and s_0.

When $t = 1$: $\frac{1}{2}a(1^2) + v_0(1) + s_0 = 164$

When $t = 2$: $\frac{1}{2}a(2^2) + v_0(2) + s_0 = 180$

When $t = 3$: $\frac{1}{2}a(3^2) + v_0(3) + s_0 = 164$

By multiplying the first and third equations by 2, this system can be rewritten

$$\begin{cases} a + 2v_0 + 2s_0 = 328 \\ 2a + 2v_0 + \quad s_0 = 180 \\ 9a + 6v_0 + 2s_0 = 328, \end{cases}$$

and you can apply elimination to obtain

$$\begin{cases} a + \quad 2v_0 + 2s_0 = \quad 328 \\ \quad -2v_0 - 3s_0 = -476 \\ \quad 2s_0 = \quad 232. \end{cases}$$

From the third equation, you can determine that $s_0 = 116$, so that back-substitution into the second equation yields

$$-2v_0 - 3(116) = -476$$
$$-2v_0 = -128$$
$$v_0 = 64.$$

Finally, by back-substituting $s_0 = 116$ and $v_0 = 64$ into the first equation, you have

$$a + 2(64) + 2(116) = 328$$
$$a = -32.$$

Thus, the position equation for this object is

$$s = -16t^2 + 64t + 116.$$ ∎

Explain to students how the need to understand the concepts of velocity and acceleration led to new developments in calculus by Newton and Leibniz.

EXAMPLE 9 ■ Solving a Mixture Problem

Three iron *alloys* contain the following percentages of carbon, chromium, and iron.

Alloy	Carbon	Chromium	Iron
Alloy X	1%	0%	99%
Alloy Y	1%	15%	84%
Alloy Z	4%	3%	93%

You have 15 tons of carbon, 39 tons of chromium, and 546 tons of iron. How much of the three types of alloy can you make?

The properties of pure iron can be altered by adding other elements to form iron alloys. Elements that are often added are carbon, chromium, silicon, and nickel.

Solution

Let x, y, and z represent the amounts of the three iron alloys. You can model this situation with the following linear system.

$$\begin{cases} 0.01x + 0.01y + 0.04z = 15 \\ 0.15y + 0.03z = 39 \\ 0.99x + 0.84y + 0.93z = 546 \end{cases}$$

Carbon

Chromium

Iron

After applying several elementary row operations, you can obtain the following system, which is in row-echelon form.

$$\begin{cases} x + y + 4z = 1500 \\ y + \dfrac{1}{5}z = 260 \\ z = 300 \end{cases}$$

Using substitution, you can find the solution to be $x = 100$, $y = 200$, and $z = 300$. Thus, you can make 100 tons of alloy X, 200 tons of alloy Y, and 300 tons of alloy Z. (Alloy X is a type of wrought iron, alloy Y is a type of stainless steel, and alloy Z is a type of cast iron.) ∎

DISCUSSION PROBLEM ■   **Too Much Information**

Set up a system of linear equations that represents the following number problem.
   The sum of three numbers is 9. The second number is twice the first number. The third number is half the sum of the first two numbers. The difference between the first and third numbers is 4. What are the numbers?
   Can this problem be solved? Explain why or why not. ∎

Warm-Up

The following warm-up exercises involve skills that were covered in earlier sections. You will use these skills in the exercise set for this section.

In Exercises 1–6, solve the system of linear equations.

1. $\begin{cases} x + 3y = 17 \\ 2x + y = 14 \end{cases}$ **2.** $\begin{cases} 4x + 2y = -2 \\ 3x - 2y = -5 \end{cases}$ **3.** $\begin{cases} -\frac{1}{2}x + \frac{3}{2}y = 2 \\ 2x - 6y = 1 \end{cases}$

4. $\begin{cases} \frac{2}{3}x + \frac{5}{6}y = 3 \\ 4x + 5y = 18 \end{cases}$ **5.** $\begin{cases} 2x + 5y = -4 \\ 3x + 6y = -3 \end{cases}$ **6.** $\begin{cases} 4x - 3y = -2 \\ 5x - 4y = -2 \end{cases}$

In Exercises 7–10, simplify the expression.

7. $(2x - 3y + 5z) + 3(x + y - 2z)$ **8.** $(x + 4y + 3z) - 3(2x - 10y + z)$

9. $-4(6x+5y-20z)+3(8x+y-15z)$ **10.** $5(2x - 3y + 4z) - 3(x - 5y - 2z)$

4.2   EXERCISES

In Exercises 1–4, use back-substitution to solve the system of linear equations.

1. $\begin{cases} x - 2y + 4z = 4 \\ 3y - z = 2 \\ z = -5 \end{cases}$
2. $\begin{cases} 5x + 4y - z = 0 \\ 10y - 3z = 11 \\ z = 3 \end{cases}$
3. $\begin{cases} x - 2y + 4z = 4 \\ y = 3 \\ y + z = 2 \end{cases}$
4. $\begin{cases} x = 10 \\ 3x + 2y = 2 \\ x + y + 2z = 0 \end{cases}$

In Exercises 5–28, use Gaussian elimination to solve the system of linear equations.

5. $\begin{cases} x + z = 4 \\ y = 2 \\ 4x + z = 7 \end{cases}$
6. $\begin{cases} x = 3 \\ -x + 3y = 3 \\ y + 2z = 4 \end{cases}$
7. $\begin{cases} x + y + z = 6 \\ 2x - y + z = 3 \\ 3x - z = 0 \end{cases}$
8. $\begin{cases} x + y + z = 2 \\ -x + 3y + 2z = 8 \\ 4x + y = 4 \end{cases}$

9. $\begin{cases} x + y + z = -3 \\ 4x + y - 3z = 11 \\ 2x - 3y + 2z = 9 \end{cases}$
10. $\begin{cases} x - y + 2z = -4 \\ 3x + y - 4z = -6 \\ 2x + 3y - 4z = 4 \end{cases}$
11. $\begin{cases} x + 2y + 6z = 5 \\ -x + y - 2z = 3 \\ x - 4y - 2z = 1 \end{cases}$
12. $\begin{cases} x + 6y + 2z = 9 \\ 3x - 2y + 3z = -1 \\ 5x - 5y + 2z = 7 \end{cases}$

13. $\begin{cases} 2x + 2z = 2 \\ 5x + 3y = 4 \\ 3y - 4z = 4 \end{cases}$
14. $\begin{cases} 6y + 4z = -12 \\ 3x + 3y = 9 \\ 2x - 3z = 10 \end{cases}$
15. $\begin{cases} 2x - 3y + 3z = 5 \\ x - 3z = 1 \\ 4x - 6y + 6z = 10 \end{cases}$
16. $\begin{cases} 2x + y + 3z = 1 \\ 2x + 6y + 8z = 3 \\ 6x + 8y + 18z = 5 \end{cases}$

17. $\begin{cases} 2x - 4y + z = 0 \\ 3x + 2z = -1 \\ -6x + 3y + 2z = -10 \end{cases}$
18. $\begin{cases} 3x - y - 2z = 5 \\ 2x + y + 3z = 6 \\ 6x - y - 4z = 9 \end{cases}$
19. $\begin{cases} y + z = 5 \\ 2x + 4z = 4 \\ 2x - 3y = -14 \end{cases}$

20. $\begin{cases} 5x + 2y = -8 \\ z = 5 \\ 3x - y + z = 9 \end{cases}$
21. $\begin{cases} 0.2x + 1.3y + 0.6z = 0.1 \\ 0.1x + 0.3z = 0.7 \\ 2x + 10y + 8z = 8 \end{cases}$
22. $\begin{cases} 0.3x - 0.1y + 0.2z = 0.35 \\ 2x + y - 2z = -1 \\ 2x + 4y + 3z = 10.5 \end{cases}$

23. $\begin{cases} x + 4y - 2z = 2 \\ x + y + z = -1 \\ 5x + 7y + 3z = -3 \end{cases}$
24. $\begin{cases} 2x + 3y - z = 1 \\ x + z = 2 \\ x - 3y + 4z = 5 \end{cases}$
25. $\begin{cases} x - 2y - z = 3 \\ 2x + y - 3z = 1 \\ x + 8y - 3z = -7 \end{cases}$

26. $\begin{cases} 2x + z = 1 \\ 5y - 3z = 2 \\ 6x + 20y - 9z = 11 \end{cases}$
27. $\begin{cases} 3x + y + z = 2 \\ 4x + 2z = 1 \\ 5x - y + 3z = 0 \end{cases}$
28. $\begin{cases} 2x + 3z = 4 \\ 5x + y + z = 2 \\ 11x + 3y - 3z = 0 \end{cases}$

Curve Fitting In Exercises 29–32, find the equation of the parabola $y = ax^2 + bx + c$ that passes through the given points.

29. $(0, -4)$, $(1, 1)$, $(2, 10)$

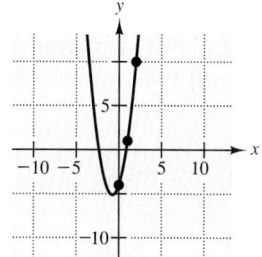

30. $(0, 5)$, $(1, 6)$, $(2, 5)$

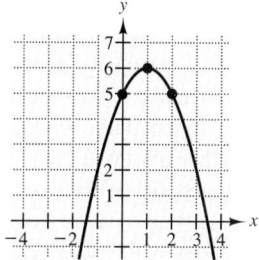

31. $(1, 0)$, $(2, -1)$, $(3, 0)$

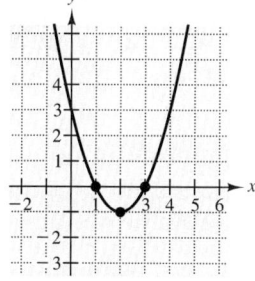

32. $(1, 2)$, $(2, 1)$, $(3, -4)$

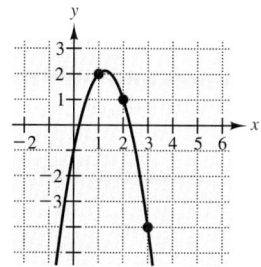

Curve Fitting In Exercises 33–36, find the equation of the circle $x^2 + y^2 + Dx + Ey + F = 0$ that passes through the given points.

33. $(0, 0)$, $(2, -2)$, $(4, 0)$

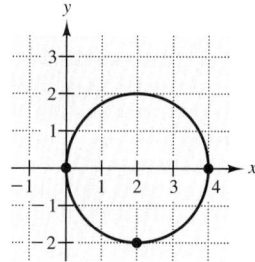

34. $(0, 0)$, $(0, 6)$, $(-3, 3)$

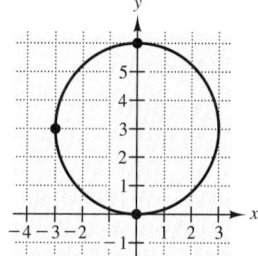

35. $(3, -1)$, $(-2, 4)$, $(6, 8)$

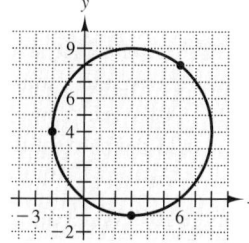

36. $(0, 0)$, $(0, 2)$, $(3, 0)$

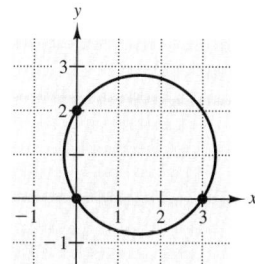

In Exercises 37–40, find the position equation $s = \frac{1}{2}at^2 + v_0t + s_0$ for an object that has the indicated heights at the specified times.

37. $s = 128$ feet at $t = 1$ second
$s = 80$ feet at $t = 2$ seconds
$s = 0$ feet at $t = 3$ seconds

38. $s = 48$ feet at $t = 1$ second
$s = 64$ feet at $t = 2$ seconds
$s = 48$ feet at $t = 3$ seconds

39. $s = 32$ feet at $t = 1$ second
$s = 32$ feet at $t = 2$ seconds
$s = 0$ feet at $t = 3$ seconds

40. $s = 10$ feet at $t = 0$ second
$s = 54$ feet at $t = 1$ second
$s = 46$ feet at $t = 3$ seconds

41. *Crop Spraying* A mixture of 12 gallons of chemical A, 16 gallons of chemical B, and 26 gallons of chemical C is required to kill a certain destructive crop insect. Commercial spray X contains 1, 2, and 2 parts of chemicals A, B, and C, respectively. Commercial spray Y contains only chemical C. Commercial spray Z contains only chemicals A and B in equal amounts. How much of each type of commercial spray is needed to obtain the desired mixture?

42. *Chemistry* A chemist needs 10 liters of a 25% acid solution. The solution is to be mixed from three solutions whose concentrations are 10%, 20%, and 50%. How many liters of each solution should the chemist use to satisfy the following?

(a) Use as little as possible of the 50% solution.
(b) Use as much as possible of the 50% solution.
(c) Use 2 liters of the 50% solution.

43. *School Orchestra* The table shows the percentages of each section of the North High School orchestra that were chosen to participate in the city orchestra, the county orchestra, and the state orchestra. Thirty members of the city orchestra, 17 members of the county orchestra, and 10 members of the state orchestra are from North High School. How many members are in each section of North High's orchestra?

Orchestra	String	Wind	Percussion
City Orchestra	40%	30%	50%
County Orchestra	20%	25%	25%
State Orchestra	10%	15%	25%

44. *Exploratory Exercise* The total numbers of diagonals of regular polygons with three, four, and five sides are three, six, and ten, as shown in the figure. Find a quadratic function $y = ax^2 + bx + c$ that fits these data. Then check to see if it gives the correct answer for a polygon with six sides.

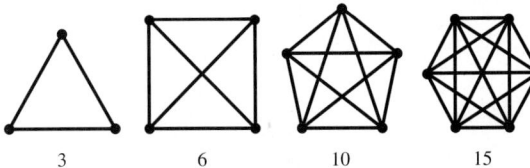

3   6   10   15

In Exercises 45 and 46, find a system of linear equations in three variables with integer coefficients that has the given point as a solution. (*Note:* There are many correct answers.)

45. $(4, -3, 2)$ **46.** $(5, 7, -10)$

Matrices and Systems of Linear Equations
Matrices • *Elementary Row Operations* • *Using Matrices to Solve a System of Linear Equations*

Matrices

Students may initially confuse rows with columns. Ask them to think of the (vertical) columns in front of the Parthenon.

In this section, you will study a streamlined technique for solving systems of linear equations. This technique involves the use of a rectangular array of real numbers called a **matrix.** (The plural of matrix is *matrices.*) Here is an example of a matrix.

$$\begin{bmatrix} 3 & -2 & 4 & 1 \\ 0 & 1 & -1 & 2 \\ 2 & 0 & -3 & 0 \end{bmatrix}$$

Note that this matrix has three rows and four columns. This means that its **order** is 3×4, which is read as "3 by 4." Each number in the matrix is called an **entry** of the matrix. Example 1 gives some other examples of matrices.

EXAMPLE 1   ■   Examples of Matrices

The following matrices have the indicated orders.

(a) Order: 2×3 (b) Order: 2×2 (c) Order: 3×2

$$\begin{bmatrix} 1 & -2 & 4 \\ 0 & 1 & -2 \end{bmatrix} \qquad \begin{bmatrix} 0 & 0 \\ 0 & 0 \end{bmatrix} \qquad \begin{bmatrix} 1 & -3 \\ -2 & 0 \\ 4 & -2 \end{bmatrix}$$

A matrix with the same number of rows as columns is called a **square matrix**. For instance, the 2×2 matrix in part (b) is square.   ■

A matrix derived from a system of linear equations (each written in standard form, with the constant term on the right) is called the **augmented matrix** of the system. Moreover, the matrix derived from the coefficients of the system (but which does not include the constant terms) is called the **coefficient matrix** of the system. Here is an example of a system of equations and its coefficient and augmented matrices.

System	*Coefficient Matrix*	*Augmented Matrix*

$$\begin{cases} x - 4y + 3z = 5 \\ -x + 3y - z = -3 \\ 2x \quad\;\; - 4z = 6 \end{cases} \qquad \begin{bmatrix} 1 & -4 & 3 \\ -1 & 3 & -1 \\ 2 & 0 & -4 \end{bmatrix} \qquad \begin{bmatrix} 1 & -4 & 3 & \vdots & 5 \\ -1 & 3 & -1 & \vdots & -3 \\ 2 & 0 & -4 & \vdots & 6 \end{bmatrix}$$

In both matrices, note the use of 0 for the missing y-variable in the third equation, and also note the fourth column of constant terms in the augmented matrix.

When forming either the coefficient matrix or the augmented matrix of a system, you should begin by vertically aligning the variables in the equations.

Given System	*Align Variables*	*Form Augmented Matrix*
$\begin{cases} x + 3y = 9 \\ -y + 4z = -2 \\ x - 5z = 0 \end{cases}$	$\begin{aligned} x + 3y &= 9 \\ -y + 4z &= -2 \\ x \qquad - 5z &= 0 \end{aligned}$	$\begin{bmatrix} 1 & 3 & 0 & \vdots & 9 \\ 0 & -1 & 4 & \vdots & -2 \\ 1 & 0 & -5 & \vdots & 0 \end{bmatrix}$

EXAMPLE 2 ■ **Forming Coefficient Matrices and Augmented Matrices**

Form the coefficient matrix and augmented matrix for each of the following systems of linear equations.

(a) $\begin{cases} -x + 5y = 2 \\ 7x - 2y = -6 \end{cases}$ (b) $\begin{cases} 3x + 2y - z = 1 \\ x + 2z = -3 \\ -2x - y = 4 \end{cases}$ (c) $\begin{cases} x = 3y - 1 \\ 2y - 5 = 9x \end{cases}$

Solution

(a)
System	*Coefficient Matrix*	*Augmented Matrix*
$\begin{cases} -x + 5y = 2 \\ 7x - 2y = -6 \end{cases}$	$\begin{bmatrix} -1 & 5 \\ 7 & -2 \end{bmatrix}$	$\begin{bmatrix} -1 & 5 & \vdots & 2 \\ 7 & -2 & \vdots & -6 \end{bmatrix}$

(b) To begin, rewrite the given system so that the variables are aligned vertically.

System	*Coefficient Matrix*	*Augmented Matrix*
$\begin{cases} 3x + 2y - z = 1 \\ x \quad + 2z = -3 \\ -2x - y = 4 \end{cases}$	$\begin{bmatrix} 3 & 2 & -1 \\ 1 & 0 & 2 \\ -2 & -1 & 0 \end{bmatrix}$	$\begin{bmatrix} 3 & 2 & -1 & \vdots & 1 \\ 1 & 0 & 2 & \vdots & -3 \\ -2 & -1 & 0 & \vdots & 4 \end{bmatrix}$

Be sure students understand that all equations must be written in standard form before they are translated to the augmented matrix.

(c) To begin, rewrite the given system so that the variables are aligned vertically and the constant terms are on the right.

System	*Coefficient Matrix*	*Augmented Matrix*
$\begin{cases} x - 3y = -1 \\ -9x + 2y = 5 \end{cases}$	$\begin{bmatrix} 1 & -3 \\ -9 & 2 \end{bmatrix}$	$\begin{bmatrix} 1 & -3 & \vdots & -1 \\ -9 & 2 & \vdots & 5 \end{bmatrix}$ ■

EXAMPLE 3 ■ **Forming a Linear System from Its Augmented Matrix**

Write a system of linear equations that is represented by each of the following augmented matrices.

(a) $\begin{bmatrix} 3 & -5 & \vdots & 4 \\ -1 & 2 & \vdots & 0 \end{bmatrix}$ (b) $\begin{bmatrix} 1 & \frac{3}{2} & \vdots & 2 \\ 0 & 1 & \vdots & -3 \end{bmatrix}$ (c) $\begin{bmatrix} 2 & 0 & -8 & \vdots & 1 \\ -1 & 1 & 1 & \vdots & 2 \\ 5 & -1 & 7 & \vdots & 3 \end{bmatrix}$

Solution

(a) $\begin{cases} 3x - 5y = 4 \\ -x + 2y = 0 \end{cases}$ (b) $\begin{cases} x + \frac{3}{2}y = 2 \\ \phantom{x + \frac{3}{2}}y = -3 \end{cases}$ (c) $\begin{cases} 2x - 8z = 1 \\ -x + y + z = 2 \\ 5x - y + 7z = 3 \end{cases}$ ■

Elementary Row Operations

In Section 4.2, you studied three operations that can be used on a system of linear equations to produce an equivalent system.

1. Interchange two equations.

2. Multiply an equation by a nonzero constant.

3. Add a multiple of an equation to another equation.

In matrix terminology, these three operations correspond to **elementary row operations.** An elementary row operation on an augmented matrix of a given system of linear equations produces a new augmented matrix corresponding to a new (but equivalent) system of linear equations. Two matrices are said to be **row-equivalent** if one can be obtained from the other by a sequence of elementary row operations.

Elementary Row Operations

Any of the following **elementary row operations** performed on an augmented matrix will produce a matrix that is row-equivalent to the original matrix.

1. Interchange two rows.

2. Multiply a row by a nonzero constant.

3. Add a multiple of a row to another row.

Although elementary row operations are simple to perform, they involve a lot of arithmetic. Because it is easy to make a mistake, you should get in the habit of noting the elementary row operations performed in each step so that you can go back and check your work. People use different schemes to denote which elementary row operation has been performed. The scheme used in this book is to write an abbreviated version of the row operation to the left of the row that has been changed, as shown in Example 4.

EXAMPLE 4 ■ **Elementary Row Operations**

(a) Interchange the first and second rows.

Original Matrix

$$\begin{bmatrix} 0 & 1 & 3 & 4 \\ -1 & 2 & 0 & 3 \\ 2 & -3 & 4 & 1 \end{bmatrix}$$

New Row-Equivalent Matrix

$$\begin{matrix} \to R_2 \\ \hookrightarrow R_1 \end{matrix} \begin{bmatrix} -1 & 2 & 0 & 3 \\ 0 & 1 & 3 & 4 \\ 2 & -3 & 4 & 1 \end{bmatrix}$$

(b) Multiply the first row by $\frac{1}{2}$.

Original Matrix

$$\begin{bmatrix} 2 & -4 & 6 & -2 \\ 1 & 3 & -3 & 0 \\ 5 & -2 & 1 & 2 \end{bmatrix}$$

New Row-Equivalent Matrix

$$\frac{1}{2}R_1 \to \begin{bmatrix} 1 & -2 & 3 & -1 \\ 1 & 3 & -3 & 0 \\ 5 & -2 & 1 & 2 \end{bmatrix}$$

(c) Add -2 times the first row to the third row.

Original Matrix

$$\begin{bmatrix} 1 & 2 & -4 & 3 \\ 0 & 3 & -2 & -1 \\ 2 & 1 & 5 & -2 \end{bmatrix}$$

New Row-Equivalent Matrix

$$-2R_1 + R_3 \to \begin{bmatrix} 1 & 2 & -4 & 3 \\ 0 & 3 & -2 & -1 \\ 0 & -3 & 13 & -8 \end{bmatrix}$$

Note that the elementary row operation is written beside the row that you are
changing.   ■

In Section 4.2, you used Gaussian elimination with back-substitution to solve
a system of linear equations. The next example demonstrates the matrix version
of Gaussian elimination. The two methods are essentially the same. The basic
difference is that with matrices you do not need to keep writing the variables.

EXAMPLE 5 ■ **Using Elementary Row Operations to Solve a System of Linear Equations**

System

$$\begin{cases} x - 2y + 2z = 9 \\ -x + 3y \quad\quad = -4 \\ 2x - 5y + z = 10 \end{cases}$$

Associated Augmented Matrix

$$\begin{bmatrix} 1 & -2 & 2 & \vdots & 9 \\ -1 & 3 & 0 & \vdots & -4 \\ 2 & -5 & 1 & \vdots & 10 \end{bmatrix}$$

Add the first equation to the
second equation.

$$\begin{cases} x - 2y + 2z = 9 \\ y + 2z = 5 \\ 2x - 5y + z = 10 \end{cases}$$

Add the first row to the
second row $(R_1 + R_2)$.

$$R_1 + R_2 \to \begin{bmatrix} 1 & -2 & 2 & \vdots & 9 \\ 0 & 1 & 2 & \vdots & 5 \\ 2 & -5 & 1 & \vdots & 10 \end{bmatrix}$$

Add -2 times the first equation to the third equation.

$$\begin{cases} x - 2y + 2z = 9 \\ y + 2z = 5 \\ -y - 3z = -8 \end{cases}$$

Add -2 times the first row to the third row $(-2R_1 + R_3)$.

$$-2R_1 + R_3 \;\rightarrow\; \begin{bmatrix} 1 & -2 & 2 & \vdots & 9 \\ 0 & 1 & 2 & \vdots & 5 \\ 0 & -1 & -3 & \vdots & -8 \end{bmatrix}$$

Add the second equation to the third equation.

$$\begin{cases} x - 2y + 2z = 9 \\ y + 2z = 5 \\ -z = -3 \end{cases}$$

Add the second row to the third row $(R_2 + R_3)$.

$$R_2 + R_3 \;\rightarrow\; \begin{bmatrix} 1 & -2 & 2 & \vdots & 9 \\ 0 & 1 & 2 & \vdots & 5 \\ 0 & 0 & -1 & \vdots & -3 \end{bmatrix}$$

Multiply the third equation by -1.

$$\begin{cases} x - 2y + 2z = 9 \\ y + 2z = 5 \\ z = 3 \end{cases}$$

Multiply the third row by -1.

$$-R_3 \;\rightarrow\; \begin{bmatrix} 1 & -2 & 2 & \vdots & 9 \\ 0 & 1 & 2 & \vdots & 5 \\ 0 & 0 & 1 & \vdots & 3 \end{bmatrix}$$

At this point, you can use back-substitution to find that the solution is $x = 1$, $y = -1$, and $z = 3$, as in Example 1 in Section 4.2. ∎

The last matrix in Example 5 is said to be in **row-echelon form.** The term *echelon* refers to the stair-step pattern formed by the nonzero elements of the matrix. To be in this form, a matrix must have the following properties.

Definition of Row-Echelon Form of a Matrix

A matrix in **row-echelon form** has the following properties.

1. All rows consisting entirely of zeros occur at the bottom of the matrix.

2. For each row that does not consist entirely of zeros, the first nonzero entry is 1 (called a **leading 1**).

3. For two successive (nonzero) rows, the leading 1 in the higher row is farther to the left than the leading 1 in the lower row.

EXAMPLE 6   ∎   Row-Echelon Form

The following matrices are in row-echelon form.

(a) $\begin{bmatrix} 1 & 2 & -1 & 4 \\ 0 & 1 & 0 & 3 \\ 0 & 0 & 1 & -2 \end{bmatrix}$ (b) $\begin{bmatrix} 0 & 1 & 0 & 3 \\ 0 & 0 & 1 & 3 \\ 0 & 0 & 0 & 0 \end{bmatrix}$

(c) $\begin{bmatrix} 1 & 2 & -3 & 4 \\ 0 & 1 & 1 & -1 \\ 0 & 0 & 1 & -3 \end{bmatrix}$ (d) $\begin{bmatrix} 1 & 2 & -1 & 2 \\ 0 & 1 & 2 & -4 \\ 0 & 0 & 0 & 0 \end{bmatrix}$ ∎

Using Matrices to Solve a System of Linear Equations

The procedure used in Example 5 is outlined in the following guidelines for using matrices to solve a system of linear equations.

Gaussian Elimination with Back-Substitution

To use matrices and Gaussian elimination to solve a system of linear equations, use the following steps.

1. Write the augmented matrix of the system of linear equations.

2. Use elementary row operations to rewrite the augmented matrix in row-echelon form.

3. Write the system of linear equations corresponding to the matrix in row-echelon form, and use back-substitution to find the solution.

Gaussian elimination with back-substitution works well for solving systems of linear equations with a computer. For this algorithm, the order in which the elementary row operations are performed is important. You should operate from *left to right by columns,* using elementary row operations to obtain zeros in all entries directly below the leading 1's.

EXAMPLE 7 ■ Gaussian Elimination with Back-Substitution

Solve the following system of linear equations.

$$\begin{cases} 2x - 3y = -2 \\ x + 2y = 13 \end{cases}$$

Solution

The augmented matrix for this system is as follows.

$$\begin{bmatrix} 2 & -3 & \vdots & -2 \\ 1 & 2 & \vdots & 13 \end{bmatrix}$$

Begin by obtaining a leading 1 in the upper left corner by interchanging the first and second rows, and then proceed to obtain zeros elsewhere in the first column.

$$\begin{matrix} \to R_2 \\ \to R_1 \end{matrix} \begin{bmatrix} 1 & 2 & \vdots & 13 \\ 2 & -3 & \vdots & -2 \end{bmatrix}$$ First column has leading 1 in upper left corner

$$-2R_1 + R_2 \to \begin{bmatrix} 1 & 2 & \vdots & 13 \\ 0 & -7 & \vdots & -28 \end{bmatrix}$$ First column has a zero under its leading 1

$$-\tfrac{1}{7}R_2 \to \begin{bmatrix} 1 & 2 & \vdots & 13 \\ 0 & 1 & \vdots & 4 \end{bmatrix}$$ Second column has leading 1 in second row

The system of linear equations that corresponds to the (row-echelon) matrix is as follows.

$$\begin{cases} x + 2y = 13 \\ y = 4 \end{cases}$$

Using back-substitution, you can find that the solution of the system is $x = 5$ and $y = 4$. Check this solution in the original system of linear equations. ■

EXAMPLE 8 ■ Gaussian Elimination with Back-Substitution

Solve the following system of linear equations.

$$\begin{cases} 3x + 3y = 9 \\ 2x - 3z = 10 \\ 6y + 4z = -12 \end{cases}$$

Solution

$$\begin{bmatrix} 3 & 3 & 0 & \vdots & 9 \\ 2 & 0 & -3 & \vdots & 10 \\ 0 & 6 & 4 & \vdots & -12 \end{bmatrix} \quad \text{Augmented matrix for given system of linear equations}$$

$$\tfrac{1}{3}R_1 \rightarrow \begin{bmatrix} 1 & 1 & 0 & \vdots & 3 \\ 2 & 0 & -3 & \vdots & 10 \\ 0 & 6 & 4 & \vdots & -12 \end{bmatrix} \quad \text{First column has leading 1 in upper left corner}$$

$$-2R_1 + R_2 \rightarrow \begin{bmatrix} 1 & 1 & 0 & \vdots & 3 \\ 0 & -2 & -3 & \vdots & 4 \\ 0 & 6 & 4 & \vdots & -12 \end{bmatrix} \quad \text{First column has zeros under its leading 1}$$

$$-\tfrac{1}{2}R_2 \rightarrow \begin{bmatrix} 1 & 1 & 0 & \vdots & 3 \\ 0 & 1 & \tfrac{3}{2} & \vdots & -2 \\ 0 & 6 & 4 & \vdots & -12 \end{bmatrix} \quad \text{Second column has leading 1 in second row}$$

$$-6R_2 + R_3 \rightarrow \begin{bmatrix} 1 & 1 & 0 & \vdots & 3 \\ 0 & 1 & \tfrac{3}{2} & \vdots & -2 \\ 0 & 0 & -5 & \vdots & 0 \end{bmatrix} \quad \text{Second column has zeros under its leading 1}$$

$$-\tfrac{1}{5}R_3 \rightarrow \begin{bmatrix} 1 & 1 & 0 & \vdots & 3 \\ 0 & 1 & \tfrac{3}{2} & \vdots & -2 \\ 0 & 0 & 1 & \vdots & 0 \end{bmatrix} \quad \text{Third column has leading 1 in third row}$$

The system of linear equations that corresponds to this (row-echelon) matrix is as follows.

$$\begin{cases} x + y \phantom{+ \tfrac{3}{2}z} = 3 \\ y + \dfrac{3}{2}z = -2 \\ z = 0 \end{cases}$$

Using back-substitution, you can find that the solution is $x = 5$, $y = -2$, and $z = 0$. Check this solution in the original system of linear equations. ■

So far in this section, each of the systems of linear equations has had only one solution. Remember from your earlier work in this chapter that it is possible for a system of linear equations to have no solution *or* an infinite number of solutions.

EXAMPLE 9   ■   A System with No Solution

Solve the following system of linear equations.

$$\begin{cases} 6x - 10y = -4 \\ 9x - 15y = 5 \end{cases}$$

Solution

$$\begin{bmatrix} 6 & -10 & \vdots & -4 \\ 9 & -15 & \vdots & 5 \end{bmatrix}$$ Augmented matrix for given system of equations

$\frac{1}{6}R_1 \rightarrow$ $\begin{bmatrix} 1 & -\frac{5}{3} & \vdots & -\frac{2}{3} \\ 9 & -15 & \vdots & 5 \end{bmatrix}$ First column has leading 1 in upper left corner

$-9R_1 + R_2 \rightarrow$ $\begin{bmatrix} 1 & -\frac{5}{3} & \vdots & -\frac{2}{3} \\ 0 & 0 & \vdots & 11 \end{bmatrix}$ First column has a zero under its leading 1

The system of linear equations that corresponds to the (row-echelon) matrix is

$$\begin{cases} x - \dfrac{5}{3}y = -\dfrac{2}{3} \\ \phantom{x - \dfrac{5}{3}y} 0 = 11. \end{cases}$$

Because the second "equation" in this system is not true for any values of x and y, you can conclude that the original system of equations has no solution.   ■

EXAMPLE 10   ■   A System with Infinitely Many Solutions

Solve the following system of linear equations.

$$\begin{cases} 12x - 6y = -3 \\ -8x + 4y = 2 \end{cases}$$

Solution

$$\begin{bmatrix} 12 & -6 & \vdots & -3 \\ -8 & 4 & \vdots & 2 \end{bmatrix}$$ Augmented matrix for given system of equations

$\frac{1}{12}R_1 \rightarrow$ $\begin{bmatrix} 1 & -\frac{1}{2} & \vdots & -\frac{1}{4} \\ -8 & 4 & \vdots & 2 \end{bmatrix}$ First column has leading 1 in upper left corner

$8R_1 + R_2 \rightarrow$ $\begin{bmatrix} 1 & -\frac{1}{2} & \vdots & -\frac{1}{4} \\ 0 & 0 & \vdots & 0 \end{bmatrix}$ First column has a zero under its leading 1

The system of linear equations that corresponds to the (row-echelon) matrix is

$$\begin{cases} x - \dfrac{1}{2}y = -\dfrac{1}{4} \\ \qquad\quad 0 = \quad 0. \end{cases}$$

Because the second equation in this system adds no information, you can conclude that the system of equations has an infinite number of solutions, represented by all points (x, y) on the line $x - \frac{1}{2}y = -\frac{1}{4}$. Because this line can be written as $x = -\frac{1}{4} + \frac{1}{2}y$, you can write the solution set as

$$\left(-\frac{1}{4} + \frac{1}{2}a, a\right), \qquad \text{where } a \text{ is any real number.} \qquad \blacksquare$$

Point out the difference between the solution to Example 10 and the solution to Example 9. Emphasize that $0 = 11$ is *never* true (no solution), while $0 = 0$ is *always* true (infinitely many solutions).

Investors in the stock market often try to keep *diversified portfolios* of stocks. This means they will own some stocks that they believe will pay high returns, and others that are likely to pay average or even below-average returns. Typically, the stocks that have the potential for higher returns also involve greater risk.

EXAMPLE 11 ■ Dividing an Investment Among Three Stocks

You have $30,000 to invest in the three stocks listed below. You want an average annual return of 9%. Because stock X is a high risk stock, you want the combined investment in stock Y and stock Z to be four times the amount invested in stock X. How much should you invest in each type of stock?

Stock	Expected Return
Stock X	12%
Stock Y	9%
Stock Z	8%

Solution

Let x, y, and z be the three investment amounts.

$$\begin{cases} x + \quad y + \quad z = 30{,}000 & \text{Total investment is \$30,000} \\ 0.12x + 0.09y + 0.08z = \quad 2{,}700 & \text{Average return is } 0.09(30{,}000) \\ 4x - \quad y - \quad z = \qquad 0 & (y+z) \text{ is 4 times } x \end{cases}$$

The augmented matrix for this system is as follows.

$$\begin{bmatrix} 1 & 1 & 1 & \vdots & 30{,}000 \\ 0.12 & 0.09 & 0.08 & \vdots & 2{,}700 \\ 4 & -1 & -1 & \vdots & 0 \end{bmatrix}$$

Using Gaussian elimination, you can obtain the following matrix (which is in row-echelon form).

$$\begin{bmatrix} 1 & 1 & 1 & \vdots & 30{,}000 \\ 0 & 1 & \frac{4}{3} & \vdots & 30{,}000 \\ 0 & 0 & 1 & \vdots & 18{,}000 \end{bmatrix}$$

Using back-substitution, you can determine that the solution is $x = \$6000$, $y = \$6000$, and $z = \$18{,}000$. ∎

DISCUSSION PROBLEM ■ Computer Software

This system is inconsistent.

There are many computer software programs that will solve systems of linear equations. One is called MATRIXPAD and is available from D. C. Heath and Company. Use this software (or some other program) to solve the examples given in this section. For example, a screen from MATRIXPAD showing the solution to Example 8, Section 4.2, is given below.

```
T-Register   3 x 3 ( 0 constants)   Det= 0
Z-Register   3 x 3 ( 0 constants)   Det= 0

Y-Register   3 x 4 ( 1 constants)

       1          2          2        328

       2          2          1        180

       9          6          2        328


X-Register   3 x 4 ( 1 constants)

       1          0          0         -32

       0          1          0          64

       0          0          1         116


Help   Quit   Arithmetic (Rational)    eChelon form (Reduced)    learN (Off)
```

Warm-Up

The following warm-up exercises involve skills that were covered in earlier sections. You will use these skills in the exercise set for this section.

In Exercises 1–6, solve the system of equations.

1. $\begin{cases} x - 3y = -1 \\ -3x + 2y = -4 \end{cases}$
2. $\begin{cases} 2x + y = 4 \\ -x - y = -1 \end{cases}$
3. $\begin{cases} x - 3y = -7 \\ -2x + y = -6 \end{cases}$

4. $\begin{cases} 2x + 3y = 2 \\ 4x + 9y = 5 \end{cases}$
5. $\begin{cases} x - y = -1 \\ x + 2y - 2z = 3 \\ 3x - y + 2z = 3 \end{cases}$
6. $\begin{cases} 2x + y - 2z = 1 \\ x - z = 1 \\ 3x + 3y + z = 12 \end{cases}$

In Exercises 7–10, evaluate the expression.

7. $3 - \frac{3}{2}(-4)$

8. $-2 + \frac{4}{5}(6)$

9. $5(-2) + 7\left(\frac{3}{4}\right)$

10. $-\frac{2}{3}(5) - \frac{1}{2}(-3)$

4.3 EERCISES

▦ means that a graphing utility can help you solve the exercise or check your solution.

In Exercises 1–4, determine the order of the matrix.

1. $\begin{bmatrix} 3 & -2 \\ -4 & 0 \\ 2 & -7 \\ -1 & -3 \end{bmatrix}$

2. $\begin{bmatrix} 4 & 0 & -5 \\ -1 & 8 & 9 \\ 0 & -3 & 4 \end{bmatrix}$

3. $\begin{bmatrix} 5 & -8 & 32 & -4 \\ 7 & 15 & 28 & -16 \\ -12 & 65 & -10 & 2 \end{bmatrix}$

4. $\begin{bmatrix} -2.3 & 4.5 & -25.1 & 7.5 & 1 \\ 11.8 & 82.3 & -19.6 & 4.0 & -2 \\ 0 & -9.3 & 24.0 & 8.7 & -5 \end{bmatrix}$

In Exercises 5–8, form the augmented matrix for the system of linear equations.

5. $\begin{cases} 4x - 5y = -2 \\ -x + 8y = 10 \end{cases}$
6. $\begin{cases} 8x + 3y = 25 \\ 3x - 9y = 12 \end{cases}$
7. $\begin{cases} x + 10y - 3z = 2 \\ 5x - 3y + 4z = 0 \\ 2x + 4y = 6 \end{cases}$
8. $\begin{cases} 9w - 3x + 20y + z = 13 \\ 12w - 8y = 5 \end{cases}$

In Exercises 9–12, write the system of linear equations represented by the augmented matrix. (Use variables x, y, z, and w.)

9. $\left[\begin{array}{cc:c} 4 & 3 & 8 \\ 1 & -2 & 3 \end{array}\right]$

10. $\left[\begin{array}{cc:c} 9 & -4 & 0 \\ 6 & 1 & -4 \end{array}\right]$

11. $\left[\begin{array}{ccc:c} 1 & 0 & 2 & -10 \\ 0 & 3 & -1 & 5 \\ 4 & 2 & 0 & 3 \end{array}\right]$

12. $\left[\begin{array}{cccc:c} 5 & 8 & 2 & 0 & -1 \\ -2 & 15 & 5 & 1 & 9 \\ 1 & 6 & -7 & 0 & -3 \end{array}\right]$

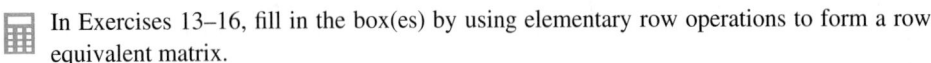 In Exercises 13–16, fill in the box(es) by using elementary row operations to form a row-equivalent matrix.

13. $\begin{bmatrix} 1 & 4 & 3 \\ 2 & 10 & 5 \end{bmatrix}$

$\begin{bmatrix} 1 & 4 & 3 \\ 0 & \boxed{} & -1 \end{bmatrix}$

14. $\begin{bmatrix} 3 & 6 & 8 \\ 4 & -3 & 6 \end{bmatrix}$

$\begin{bmatrix} 1 & \boxed{} & \frac{8}{3} \\ 4 & -3 & 6 \end{bmatrix}$

15. $\begin{bmatrix} 1 & 1 & 4 & -1 \\ 3 & 8 & 10 & 3 \\ -2 & 1 & 12 & 6 \end{bmatrix}$

$\begin{bmatrix} 1 & 1 & 4 & -1 \\ 0 & 5 & \boxed{} & \boxed{} \\ 0 & 3 & \boxed{} & \boxed{} \end{bmatrix}$

$\begin{bmatrix} 1 & 1 & 4 & -1 \\ 0 & 1 & \boxed{} & \boxed{} \\ 0 & 3 & 20 & 4 \end{bmatrix}$

16. $\begin{bmatrix} 2 & 4 & 8 & 3 \\ 1 & -1 & -3 & 2 \\ 2 & 6 & 4 & 9 \end{bmatrix}$

$\begin{bmatrix} 1 & \boxed{} & \boxed{} & \boxed{} \\ 1 & -1 & -3 & 2 \\ 2 & 6 & 4 & 9 \end{bmatrix}$

$\begin{bmatrix} 1 & 2 & 4 & \frac{3}{2} \\ 0 & \boxed{} & -7 & \frac{1}{2} \\ 0 & 2 & \boxed{} & \boxed{} \end{bmatrix}$

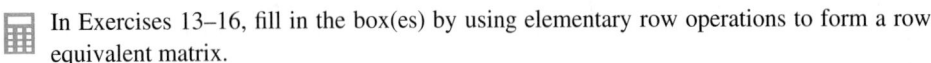 In Exercises 17–24, convert the matrix to row-echelon form. (*Note:* There is more than one correct answer.)

17. $\begin{bmatrix} 1 & 2 & 3 \\ 2 & -1 & -4 \end{bmatrix}$

18. $\begin{bmatrix} 1 & 3 & 6 \\ -4 & -9 & 3 \end{bmatrix}$

19. $\begin{bmatrix} 4 & 6 & 1 \\ -2 & 2 & 4 \end{bmatrix}$

20. $\begin{bmatrix} 3 & 2 & 6 \\ 2 & 3 & -3 \end{bmatrix}$

21. $\begin{bmatrix} 1 & 1 & 0 & 5 \\ -2 & -1 & 2 & -10 \\ 3 & 6 & 7 & 14 \end{bmatrix}$

22. $\begin{bmatrix} 1 & 2 & -1 & 3 \\ 3 & 7 & -5 & 14 \\ -2 & -1 & -3 & 8 \end{bmatrix}$

23. $\begin{bmatrix} 1 & -1 & -1 & 1 \\ 4 & -4 & 1 & 8 \\ -6 & 8 & 18 & 0 \end{bmatrix}$

24. $\begin{bmatrix} 1 & -3 & 0 & -7 \\ -3 & 10 & 1 & 23 \\ 4 & -10 & 2 & -24 \end{bmatrix}$

 In Exercises 25–28, write the system of linear equations represented by the augmented matrix. Then use back-substitution to find the solution. (Use variables x, y, and z.)

25. $\left[\begin{array}{cc:c} 1 & -2 & 4 \\ 0 & 1 & -3 \end{array}\right]$

26. $\left[\begin{array}{cc:c} 1 & 5 & 0 \\ 0 & 1 & -1 \end{array}\right]$

27. $\left[\begin{array}{ccc:c} 1 & -1 & 2 & 4 \\ 0 & 1 & -1 & 2 \\ 0 & 0 & 1 & -2 \end{array}\right]$

28. $\left[\begin{array}{ccc:c} 1 & 2 & -2 & -1 \\ 0 & 1 & 1 & 9 \\ 0 & 0 & 1 & -3 \end{array}\right]$

In Exercises 29–46, use matrices to solve the system of linear equations.

29. $\begin{cases} x + 2y = 7 \\ 3x + y = 11 \end{cases}$

30. $\begin{cases} 2x + 6y = 16 \\ 2x + 3y = 7 \end{cases}$

31. $\begin{cases} 6x - 4y = 2 \\ 5x + 2y = 7 \end{cases}$

32. $\begin{cases} 2x - y = -0.1 \\ 3x + 2y = 1.6 \end{cases}$

33. $\begin{cases} -x + 2y = 1.5 \\ 2x - 4y = 3 \end{cases}$

34. $\begin{cases} x - 3y = 5 \\ -2x + 6y = -10 \end{cases}$

35. $\begin{cases} x - 3z = -2 \\ 3x + y - 2z = 5 \\ 2x + 2y + z = 4 \end{cases}$

36. $\begin{cases} x - 2y = z - 6 \\ y + 4z = 5 \\ 4x + 2y + 3z = 8 \end{cases}$

37. $\begin{cases} x - 3y + 2z = 8 \\ \quad\;\; 2y - z = -4 \\ x \qquad\;\;\; + z = 3 \end{cases}$
38. $\begin{cases} \quad\; 2y + z = 3 \\ -4y - 2z = 0 \\ x + \;\; y + z = 2 \end{cases}$
39. $\begin{cases} 2x + 4y \qquad\;\; = 10 \\ 2x + 2y + 3z = \;\; 3 \\ -3x + \;\; y + 2z = -3 \end{cases}$
40. $\begin{cases} 2x - \;\; y + 3z = 24 \\ \quad\;\; 2y - \;\; z = 14 \\ 7x - 5y \qquad\;\; = 6 \end{cases}$

41. $\begin{cases} x + y - 5z = 3 \\ x \qquad - 2z = 1 \\ 2x - y - \;\; z = 0 \end{cases}$
42. $\begin{cases} 2x \qquad\;\; + 3z = 3 \\ 4x - 3y + 7z = 5 \\ 8x - 9y + 15z = 9 \end{cases}$
43. $\begin{cases} 2x \qquad + 4z = 1 \\ x + \;\; y + 3z = 0 \\ x + 3y + 5z = 0 \end{cases}$
44. $\begin{cases} 3x + \;\; y - 2z = 2 \\ 6x + 2y - 4z = 1 \\ -3x - \;\; y + 2z = 1 \end{cases}$

45. $\begin{cases} 2x + \;\; y - 2z = 4 \\ 3x - 2y + 4z = 6 \\ -4x + \;\; y + 6z = 12 \end{cases}$
46. $\begin{cases} 3x + 3y + \;\; z = 4 \\ 2x + 6y + \;\; z = 5 \\ -x - 3y + 2z = -5 \end{cases}$

 47. *Simple Interest* A corporation borrowed $1,500,000 to expand its product line. Some of the money was borrowed at 8%, some at 9%, and the remainder at 12%. The annual interest payment to the lenders was $133,000. If the amount borrowed at 8% was four times the amount borrowed at 12%, how much was borrowed at each rate?

 48. *Investments* An inheritance of $16,000 was divided among three investments that earn simple interest. The interest rates for the three investments were 5%, 6%, and 7%. Find the amount placed in each investment if the 5% and 6% investments were $3000 and $2000 less than the 7% investment, respectively.

Investment Portfolio In Exercises 49 and 50, consider an investor with a portfolio totaling $500,000 that is to be allocated among the following types of investments: (1) certificates of deposit, (2) municipal bonds, (3) blue-chip stocks, and (4) growth or speculative stocks. Describe several ways to allocate the $500,000 among the various investments.

49. The certificates of deposit yield 10% annually, and the municipal bonds pay 8% annually. Over a 5-year period, the investor expects the blue-chip stocks to return 12% annually and the growth stocks to return 13% annually. The investor wants a combined annual return of 10% with only one-fourth of the portfolio invested in stocks.

50. The certificates of deposit yield 9% annually, and the municipal bonds pay 5% annually. Over a 5-year period, the investor expects the blue-chip stocks to return 12% annually, and the growth stocks to return 14% annually. The investor wants a combined annual return of 10% with only one-fourth of the portfolio invested in stocks.

 Curve Fitting In Exercises 51 and 52, find the equation of the parabola $y = ax^2 + bx + c$ that passes through the given points.

51.
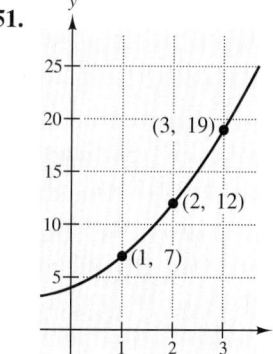

52.
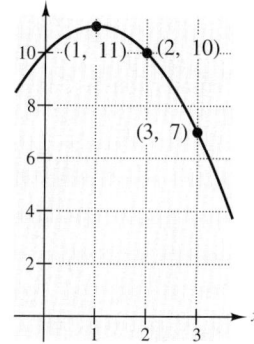

53. *Curve Fitting* Find the equation of the circle $x^2 + y^2 + Dx + Ey + F = 0$ that passes through the points $(1, 1)$, $(3, 3)$, and $(4, 2)$.

 54. *Number Problem* The sum of 3 positive numbers is 33. The second number is three greater than the first, and the third is four times the first. Find the three numbers.

Take this quiz as you would take a quiz in class. After you are done, check your work against the answers in the back of the book.

In Exercises 1 and 2, solve the system of equations graphically.

1. $\begin{cases} x - 2y = 0 \\ 2x + y = 5 \end{cases}$

2. $\begin{cases} x - y = -1 \\ 2x - 2y = 3 \end{cases}$

In Exercises 3 and 4, solve the system by the method of substitution.

3. $\begin{cases} x + y = 4 \\ -x + 4y = 11 \end{cases}$

4. $\begin{cases} 3x - 2y = 9 \\ 2x + y = 13 \end{cases}$

In Exercises 5–8, solve the system by the method of elimination or Gaussian elimination.

5. $\begin{cases} -2x + 7y = 10 \\ x - 3y = -3 \end{cases}$

6. $\begin{cases} x - 2y = 1 \\ -3x + 6y = -3 \end{cases}$

7. $\begin{cases} x + 9y + z = 20 \\ x + 10y - 2z = 18 \\ 3x + 27y + 2z = 58 \end{cases}$

8. $\begin{cases} 2x + 6y - 4z = 8 \\ 3x + 10y - 7z = 12 \\ -2x - 6y + 5z = -3 \end{cases}$

9. What is meant by saying that two augmented matrices are row-equivalent?

10. Use an augmented matrix to solve the system.

$$\begin{cases} x + 3y + z = 3 \\ x + 5y + 5z = 1 \\ 2x + 6y + 3z = 8 \end{cases}$$

In Exercises 11–13, set up the system of equations that models the problem. (It is not necessary to solve the system.)

11. **Geometry** The perimeter of a rectangle is 90 meters. Find the dimensions of the rectangle if its length is $1\frac{1}{2}$ times its width.

12. **Driving Distance** Two people share the driving on a 300-mile trip. One person drives three times as far as the other. Find the distance each person drives.

13. **Gold Alloys** Gold jewelry is seldom made of pure gold because pure gold is soft and expensive. Instead, gold is mixed with other metals to produce harder, less expensive gold alloys. The amount of gold (by weight) in an alloy is measured in karats. Anything made of 24-karat gold is 100% pure gold. An 18-karat gold mixture is 75% gold, and so on. Three gold alloys contain the percentages of gold, copper, and silver shown in the table. You have 20,144 grams of gold, 766 grams of copper, and 1990 grams of silver. How much of each alloy can you make?

	Gold	Copper	Silver
Alloy X	94%	92%	80%
Alloy Y	4%	2%	4%
Alloy Z	2%	6%	16%

14. Find a linear system of equations in three variables that has $(2, -3, 1)$ as a solution. (Note: There are many correct answers.)

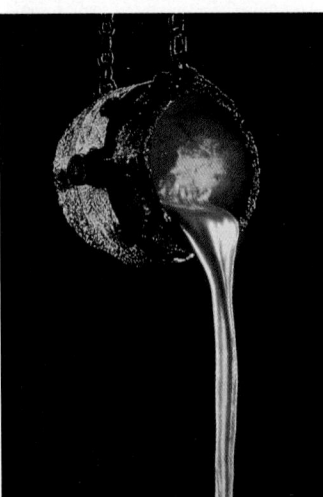

High-purity gold is mixed with other metals such as silver and copper to make jewelry.

Linear Systems and Determinants

The Determinant of a Matrix • *Cramer's Rule for Solving a System of Linear Equations* • *Applications*

The Determinant of a Matrix

In Section 4.3, we defined a *square* matrix as one that has the same number of rows and columns. Associated with each square matrix is a real number called its **determinant.**

Historically, the use of determinants arose from special number patterns that occur when systems of linear equations are solved. For instance, the system

$$\begin{cases} a_1 x + b_1 y = c_1 \\ a_2 x + b_2 y = c_2 \end{cases}$$

has a solution given by

$$x = \frac{c_1 b_2 - c_2 b_1}{a_1 b_2 - a_2 b_1} \quad \text{and} \quad y = \frac{a_1 c_2 - a_2 c_1}{a_1 b_2 - a_2 b_1},$$

provided $a_1 b_2 - a_2 b_1 \neq 0$. Note that the denominator of each fraction is the same; we call this denominator the **determinant** of the coefficient matrix of the system.

Historical Note: An ancient Chinese method of solving systems of equations involved the use of bamboo rods to represent coefficients. In 1683, Seki Kowa (1642-1708), the Japanese mathematician, advanced this method by rearranging the rods in a way that resembles our use of determinant notation today.

Coefficient Matrix	*Determinant*
$A = \begin{bmatrix} a_1 & b_1 \\ a_2 & b_2 \end{bmatrix}$	$\det(A) = a_1 b_2 - a_2 b_1$

The determinant of a matrix can also be denoted by vertical bars (rather than square brackets) on both sides of the matrix, as indicated in the following definition.

Definition of the Determinant of a 2 × 2 Matrix

The **determinant** of the matrix

$$A = \begin{bmatrix} a_1 & b_1 \\ a_2 & b_2 \end{bmatrix}$$

is given by

$$\det(A) = \begin{vmatrix} a_1 & b_1 \\ a_2 & b_2 \end{vmatrix} = a_1 b_2 - a_2 b_1.$$

A convenient method for remembering the formula for the determinant of a 2×2 matrix is shown in the following diagram.

$$\det(A) = \begin{vmatrix} a_1 & b_1 \\ a_2 & b_2 \end{vmatrix} = a_1b_2 - a_2b_1.$$

Note that the determinant is given by the difference of the products of the two diagonals of the matrix.

EXAMPLE 1 ■ **The Determinant of a 2 × 2 Matrix**

Find the determinant of each of the following matrices.

Historical Note: In a letter written in 1693 to Guillaume de L'Hospital, Gottfried Wilhelm von Leibniz (1646-1716) gave a written notation for determinants in what is considered to be the first formal use of them.

(a) $A = \begin{bmatrix} 2 & -3 \\ 1 & 4 \end{bmatrix}$ (b) $B = \begin{bmatrix} -1 & 2 \\ 2 & -4 \end{bmatrix}$ (c) $C = \begin{bmatrix} 1 & 3 \\ 2 & 5 \end{bmatrix}$

Solution

(a) $\det(A) = \begin{vmatrix} 2 & -3 \\ 1 & 4 \end{vmatrix} = 2(4) - 1(-3) = 8 + 3 = 11$

(b) $\det(B) = \begin{vmatrix} -1 & 2 \\ 2 & -4 \end{vmatrix} = (-1)(-4) - 2(2) = 4 - 4 = 0$

(c) $\det(C) = \begin{vmatrix} 1 & 3 \\ 2 & 5 \end{vmatrix} = 1(5) - 2(3) = -1$ ■

NOTE The determinant of a matrix can be positive, zero, or negative, as in Example 1.

The determinant of a 3×3 matrix is more difficult to evaluate than the determinant of a 2×2 matrix. To help with the evaluation, we use a technique called **expansion by minors.** This technique allows you to write the determinant of a 3×3 matrix in terms of three 2×2 determinants. The **minor** of an entry in a 3×3 matrix is the determinant of the 2×2 matrix that remains after deleting the row and column in which the entry occurs. Here are some examples.

Given Determinant	*Entry*	*Minor of Entry*	*Value of Minor*
$\begin{vmatrix} 1 & -1 & 3 \\ 0 & 2 & 5 \\ -2 & 4 & -7 \end{vmatrix}$	1	$\begin{vmatrix} 2 & 5 \\ 4 & -7 \end{vmatrix}$	$2(-7) - 4(5) = -34$
$\begin{vmatrix} 1 & -1 & 3 \\ 0 & 2 & 5 \\ -2 & 4 & -7 \end{vmatrix}$	-1	$\begin{vmatrix} 0 & 5 \\ -2 & -7 \end{vmatrix}$	$0(-7) - (-2)(5) = 10$
$\begin{vmatrix} 1 & -1 & 3 \\ 0 & 2 & 5 \\ -2 & 4 & -7 \end{vmatrix}$	5	$\begin{vmatrix} 1 & -1 \\ -2 & 4 \end{vmatrix}$	$1(4) - (-2)(-1) = 2$

To evaluate the determinant of a 3×3 matrix, you can expand by minors along an *entire* row (or column) according to the following scheme.

Expanding by Minors

The determinant of the 3×3 matrix

$$A = \begin{bmatrix} a_1 & b_1 & c_1 \\ a_2 & b_2 & c_2 \\ a_3 & b_3 & c_3 \end{bmatrix}$$

is given by

$$\begin{vmatrix} a_1 & b_1 & c_1 \\ a_2 & b_2 & c_2 \\ a_3 & b_3 & c_3 \end{vmatrix} = (a_1)(\text{minor of } a_1) - (b_1)(\text{minor of } b_1) + (c_1)(\text{minor of } c_1)$$

$$= a_1 \begin{vmatrix} b_2 & c_2 \\ b_3 & c_3 \end{vmatrix} - b_1 \begin{vmatrix} a_2 & c_2 \\ a_3 & c_3 \end{vmatrix} + c_1 \begin{vmatrix} a_2 & b_2 \\ a_3 & b_3 \end{vmatrix}.$$

This particular pattern is called **expanding by minors** along the first row. A similar pattern can be used to expand by minors along any row or column.

FIGURE 4.12

Sign pattern for 3×3 matrix

$$\begin{bmatrix} + & - & + \\ - & + & - \\ + & - & + \end{bmatrix}$$

The *signs* of the terms used in the expansion by minors follow the alternating pattern shown in Figure 4.12. For instance, the signs used to expand by minors along the second row are $-$, $+$, $-$, as follows.

$$\begin{vmatrix} a_1 & b_1 & c_1 \\ a_2 & b_2 & c_2 \\ a_3 & b_3 & c_3 \end{vmatrix} = -(a_2)(\text{minor of } a_2) + (b_2)(\text{minor of } b_2) - (c_2)(\text{minor of } c_2)$$

EXAMPLE 2 ■ Finding the Determinant of a 3×3 Matrix

Find the determinant of the matrix

$$A = \begin{bmatrix} -1 & 1 & 2 \\ 0 & 2 & 3 \\ 3 & 4 & 2 \end{bmatrix}.$$

Solution

Explain how the sign pattern shown in Figure 4.12 is used here.

By expanding by minors along the *first column*, you obtain the following.

$$\det(A) = \begin{vmatrix} -1 & 1 & 2 \\ 0 & 2 & 3 \\ 3 & 4 & 2 \end{vmatrix}$$

$$= (-1)\begin{vmatrix} 2 & 3 \\ 4 & 2 \end{vmatrix} - (0)\begin{vmatrix} 1 & 2 \\ 4 & 2 \end{vmatrix} + (3)\begin{vmatrix} 1 & 2 \\ 2 & 3 \end{vmatrix}$$

$$= (-1)(4 - 12) - (0)(2 - 8) + (3)(3 - 4)$$

$$= 8 - 0 - 3$$

$$= 5$$

■

Note in the expansion in Example 2 that a zero entry will always yield a zero term in the expansion by minors. Thus, when you are evaluating the determinant of a matrix, you should choose to expand along the row or column that has the most zero entries.

EXAMPLE 3 ■ Finding the Determinant of a 3 × 3 Matrix

Find the determinant of the matrix

$$A = \begin{bmatrix} 1 & 2 & 1 \\ 3 & 0 & 2 \\ 4 & 0 & -1 \end{bmatrix}.$$

Solution

By expanding by minors along the *second column*, you obtain the following.

$$\det(A) = \begin{vmatrix} 1 & 2 & 1 \\ 3 & 0 & 2 \\ 4 & 0 & -1 \end{vmatrix}$$

$$= -(2)\begin{vmatrix} 3 & 2 \\ 4 & -1 \end{vmatrix} + (0)\begin{vmatrix} 1 & 1 \\ 4 & -1 \end{vmatrix} - (0)\begin{vmatrix} 1 & 1 \\ 3 & 2 \end{vmatrix}$$

$$= -(2)(-3 - 8) + 0 - 0$$

$$= 22$$ ■

NOTE If you have a calculator that is capable of working with matrices, try using it to find the determinant of the matrix in Example 3. (The actual keystrokes used will depend on the type of calculator.)

Cramer's Rule for Solving a System of Linear Equations

Historical Note: Gabriel Cramer invented determinants independently and in 1750 published his method of using them to solve linear systems of equations. He did not, however, use the notation that we use today.

So far in this chapter, four methods for solving a system of linear equations have been discussed: graphing, substitution, elimination with equations, and elimination with matrices. We now look at one more method, called **Cramer's Rule,** named after Gabriel Cramer (1704–1752). This rule uses determinants to write the solution of a system of linear equations. To see how Cramer's Rule works, look again at the solution described at the beginning of this section. There, you saw that the system

$$\begin{cases} a_1x + b_1y = c_1 \\ a_2x + b_2y = c_2 \end{cases}$$

has a solution given by

$$x = \frac{c_1b_2 - c_2b_1}{a_1b_2 - a_2b_1} \quad \text{and} \quad y = \frac{a_1c_2 - a_2c_1}{a_1b_2 - a_2b_1},$$

provided $a_1b_2 - a_2b_1 \neq 0$.

Each numerator and denominator in this solution can be expressed as a determinant, as follows.

$$x = \frac{c_1 b_2 - c_2 b_1}{a_1 b_2 - a_2 b_1} = \frac{\begin{vmatrix} c_1 & b_1 \\ c_2 & b_2 \end{vmatrix}}{\begin{vmatrix} a_1 & b_1 \\ a_2 & b_2 \end{vmatrix}}, \qquad y = \frac{a_1 c_2 - a_2 c_1}{a_1 b_2 - a_2 b_1} = \frac{\begin{vmatrix} a_1 & c_1 \\ a_2 & c_2 \end{vmatrix}}{\begin{vmatrix} a_1 & b_1 \\ a_2 & b_2 \end{vmatrix}}$$

Relative to the original system, the denominator for x and y is simply the determinant of the *coefficient* matrix of the system. We denote this determinant by D. The numerators for x and y are denoted by D_x and D_y, respectively. They are formed by using the column of constants as replacements for the coefficients of x and y, as follows.

Coefficient Matrix	D	D_x	D_y
$\begin{bmatrix} a_1 & b_1 \\ a_2 & b_2 \end{bmatrix}$	$\begin{vmatrix} a_1 & b_1 \\ a_2 & b_2 \end{vmatrix}$	$\begin{vmatrix} c_1 & b_1 \\ c_2 & b_2 \end{vmatrix}$	$\begin{vmatrix} a_1 & c_1 \\ a_2 & c_2 \end{vmatrix}$

Cramer's Rule can be extended to a system of three linear equations in three variables, as indicated in the following explanation.

Cramer's Rule

1. For the system of linear equations

$$\begin{cases} a_1 x + b_1 y = c_1 \\ a_2 x + b_2 y = c_2, \end{cases}$$

the solution is given by

$$x = \frac{D_x}{D} = \frac{\begin{vmatrix} c_1 & b_1 \\ c_2 & b_2 \end{vmatrix}}{\begin{vmatrix} a_1 & b_1 \\ a_2 & b_2 \end{vmatrix}}, \qquad y = \frac{D_y}{D} = \frac{\begin{vmatrix} a_1 & c_1 \\ a_2 & c_2 \end{vmatrix}}{\begin{vmatrix} a_1 & b_1 \\ a_2 & b_2 \end{vmatrix}}, \qquad \text{provided } D \neq 0.$$

2. For the system of linear equations

$$\begin{cases} a_1 x + b_1 y + c_1 z = d_1 \\ a_2 x + b_2 y + c_2 z = d_2 \\ a_3 x + b_3 y + c_3 z = d_3, \end{cases}$$

the solution is given by

$$x = \frac{D_x}{D} = \frac{\begin{vmatrix} d_1 & b_1 & c_1 \\ d_2 & b_2 & c_2 \\ d_3 & b_3 & c_3 \end{vmatrix}}{\begin{vmatrix} a_1 & b_1 & c_1 \\ a_2 & b_2 & c_2 \\ a_3 & b_3 & c_3 \end{vmatrix}}, \qquad y = \frac{D_y}{D} = \frac{\begin{vmatrix} a_1 & d_1 & c_1 \\ a_2 & d_2 & c_2 \\ a_3 & d_3 & c_3 \end{vmatrix}}{\begin{vmatrix} a_1 & b_1 & c_1 \\ a_2 & b_2 & c_2 \\ a_3 & b_3 & c_3 \end{vmatrix}}, \qquad z = \frac{D_z}{D} = \frac{\begin{vmatrix} a_1 & b_1 & d_1 \\ a_2 & b_2 & d_2 \\ a_3 & b_3 & d_3 \end{vmatrix}}{\begin{vmatrix} a_1 & b_1 & c_1 \\ a_2 & b_2 & c_2 \\ a_3 & b_3 & c_3 \end{vmatrix}},$$

provided $D \neq 0$.

EXAMPLE 4   ■   **Using Cramer's Rule for a 2 × 2 System**

Use Cramer's Rule to solve the following system of linear equations.

$$\begin{cases} 4x - 2y = 10 \\ 3x - 5y = 11 \end{cases}$$

Solution

To begin, find the determinant of the coefficient matrix.

$$D = \begin{vmatrix} 4 & -2 \\ 3 & -5 \end{vmatrix} = -20 - (-6) = -14$$

Because this determinant is not zero, you can apply Cramer's Rule to find the solution, as follows.

$$x = \frac{D_x}{D} = \frac{\begin{vmatrix} 10 & -2 \\ 11 & -5 \end{vmatrix}}{-14} = \frac{(-50) - (-22)}{-14} = \frac{-28}{-14} = 2$$

$$y = \frac{D_y}{D} = \frac{\begin{vmatrix} 4 & 10 \\ 3 & 11 \end{vmatrix}}{-14} = \frac{44 - 30}{-14} = \frac{14}{-14} = -1$$

Therefore, the solution is $(2, -1)$. Try checking this solution in the original system of equations.   ■

EXAMPLE 5   ■   **Using Cramer's Rule for a 3 × 3 System**

Use Cramer's Rule to solve the following system of linear equations.

$$\begin{cases} -x + 2y - 3z = 1 \\ 2x \quad\quad + z = 0 \\ 3x - 4y + 4z = 2 \end{cases}$$

Solution

To begin, find the determinant of the coefficient matrix.

$$D = \begin{vmatrix} -1 & 2 & -3 \\ 2 & 0 & 1 \\ 3 & -4 & 4 \end{vmatrix} = -(2)\begin{vmatrix} 2 & -3 \\ -4 & 4 \end{vmatrix} + (0)\begin{vmatrix} -1 & -3 \\ 3 & 4 \end{vmatrix} - (1)\begin{vmatrix} -1 & 2 \\ 3 & -4 \end{vmatrix}$$

$$= -2(-4) + 0 - (-2)$$

$$= 10$$

Because this determinant is not zero, you can apply Cramer's Rule to find the solution, as follows.

$$x = \frac{D_x}{D} = \frac{\begin{vmatrix} 1 & 2 & -3 \\ 0 & 0 & 1 \\ 2 & -4 & 4 \end{vmatrix}}{10} = \frac{8}{10} = \frac{4}{5}$$

$$y = \frac{D_y}{D} = \frac{\begin{vmatrix} -1 & 1 & -3 \\ 2 & 0 & 1 \\ 3 & 2 & 4 \end{vmatrix}}{10} = \frac{-15}{10} = -\frac{3}{2}$$

$$z = \frac{D_z}{D} = \frac{\begin{vmatrix} -1 & 2 & 1 \\ 2 & 0 & 0 \\ 3 & -4 & 2 \end{vmatrix}}{10} = \frac{-16}{10} = -\frac{8}{5}$$

Therefore, the solution is $\left(\frac{4}{5}, -\frac{3}{2}, -\frac{8}{5}\right)$. Try checking this solution in the original system of equations.   ∎

When using Cramer's Rule, remember that the method does not apply if the determinant of the coefficient matrix is zero. For instance, the system

$$\begin{cases} 6x - 10y = -4 \\ 9x - 15y = 5 \end{cases}$$

has no solution (see Example 9 in Section 4.3), and the determinant of the coefficient matrix of this system is

$$D = \begin{vmatrix} 6 & -10 \\ 9 & -15 \end{vmatrix} = -90 - (-90) = 0.$$

Similarly, the system

$$\begin{cases} 12x - 6y = -3 \\ -8x + 4y = 2 \end{cases}$$

has infinitely many solutions (see Example 10 in Section 4.3), and the determinant of the coefficient matrix of this system is

$$D = \begin{vmatrix} 12 & -6 \\ -8 & 4 \end{vmatrix} = 48 - 48 = 0.$$

Writing assignment: Have students describe the different methods of solving linear systems of equations. Ask them to comment on the advantages and disadvantages of each method.

NOTE   Because of the many calculations involving determinants, Cramer's Rule is usually considered to be less efficient than using elimination to solve a system of linear equations. Moreover, Cramer's Rule is not as general as the elimination method because Cramer's Rule requires that the coefficient matrix of the system be square *and* that the system have exactly one solution.

Applications

Using determinants with Cramer's Rule to solve a system of linear equations is not as efficient as using the elimination method. *However,* determinants have many other practical applications. Three of these are presented in this section. The first gives a formula for finding the area of a triangle with vertices given by three points in a rectangular coordinate system.

■ Area of a Triangle

The area of a triangle with vertices (x_1, y_1), (x_2, y_2), and (x_3, y_3) is given by

$$\text{Area} = \pm \frac{1}{2} \begin{vmatrix} x_1 & y_1 & 1 \\ x_2 & y_2 & 1 \\ x_3 & y_3 & 1 \end{vmatrix},$$

where the symbol $\pm$ indicates that the appropriate sign should be chosen to yield a positive area.

EXAMPLE 6 ■ Finding the Area of a Triangle

Find the area of the triangle with vertices $(2, 0)$, $(1, 3)$, and $(3, 2)$, as shown in Figure 4.13.

FIGURE 4.13

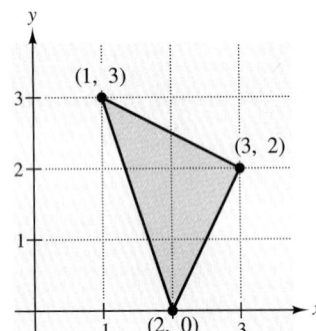

Solution

Choose $(x_1, y_1) = (2, 0)$, $(x_2, y_2) = (1, 3)$, and $(x_3, y_3) = (3, 2)$. Then, to find the area of the triangle, evaluate the determinant

$$\begin{vmatrix} x_1 & y_1 & 1 \\ x_2 & y_2 & 1 \\ x_3 & y_3 & 1 \end{vmatrix} = \begin{vmatrix} 2 & 0 & 1 \\ 1 & 3 & 1 \\ 3 & 2 & 1 \end{vmatrix}$$

$$= 2 \begin{vmatrix} 3 & 1 \\ 2 & 1 \end{vmatrix} - 0 \begin{vmatrix} 1 & 1 \\ 3 & 1 \end{vmatrix} + 1 \begin{vmatrix} 1 & 3 \\ 3 & 2 \end{vmatrix}$$

$$= 2(1) - 0 + 1(-7)$$

$$= -5.$$

Using this value, you can conclude that the area of the triangle is

$$\text{Area} = -\frac{1}{2} \begin{vmatrix} 2 & 0 & 1 \\ 1 & 3 & 1 \\ 3 & 2 & 1 \end{vmatrix} = -\frac{1}{2}(-5) = \frac{5}{2}.$$

■

NOTE To see the benefit of the "determinant formula for area," you should try finding the area of the triangle in Example 6 using the standard formula: area $= \frac{1}{2}$(base)(height).

FIGURE 4.14

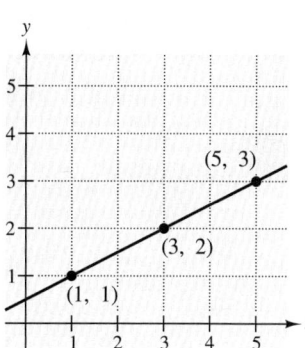

Suppose the three points in Example 6 had been on the same line. What would have happened had we applied the area formula to three such points? The answer is that the determinant would have been zero. Consider, for instance, the three collinear points $(1, 1)$, $(3, 2)$, and $(5, 3)$, as shown in Figure 4.14. The area of the "triangle" that has these three points as vertices is

$$\frac{1}{2} \begin{vmatrix} 1 & 1 & 1 \\ 3 & 2 & 1 \\ 5 & 3 & 1 \end{vmatrix} = \frac{1}{2} \left(1 \begin{vmatrix} 2 & 1 \\ 3 & 1 \end{vmatrix} - 1 \begin{vmatrix} 3 & 1 \\ 5 & 1 \end{vmatrix} + 1 \begin{vmatrix} 3 & 2 \\ 5 & 3 \end{vmatrix} \right)$$

$$= \frac{1}{2}[-1 - (-2) + (-1)] = 0.$$

This result is generalized as follows.

Test for Collinear Points

Three points (x_1, y_1), (x_2, y_2), and (x_3, y_3) are collinear (lie on the same line) if and only if

$$\begin{vmatrix} x_1 & y_1 & 1 \\ x_2 & y_2 & 1 \\ x_3 & y_3 & 1 \end{vmatrix} = 0.$$

EXAMPLE 7   ■   **Testing for Collinear Points**

Determine whether the points $(-2, -2)$, $(1, 1)$, and $(7, 5)$ lie on the same line (see Figure 4.15).

FIGURE 4.15

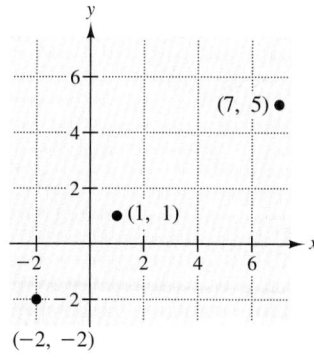

Solution

Letting $(x_1, y_1) = (-2, -2)$, $(x_2, y_2) = (1, 1)$, and $(x_3, y_3) = (7, 5)$, you have

$$\begin{vmatrix} x_1 & y_1 & 1 \\ x_2 & y_2 & 1 \\ x_3 & y_3 & 1 \end{vmatrix} = \begin{vmatrix} -2 & -2 & 1 \\ 1 & 1 & 1 \\ 7 & 5 & 1 \end{vmatrix}$$

$$= -2 \begin{vmatrix} 1 & 1 \\ 5 & 1 \end{vmatrix} - (-2) \begin{vmatrix} 1 & 1 \\ 7 & 1 \end{vmatrix} + 1 \begin{vmatrix} 1 & 1 \\ 7 & 5 \end{vmatrix}$$

$$= -2(-4) - (-2)(-6) + 1(-2)$$

$$= -6.$$

Because the value of this determinant is not zero, you can conclude that the three points do not lie on the same line. ■

The test for collinear points can be adapted to another use. That is, if you are given two points in a rectangular coordinate system, you can find the equation of the line passing through the two points as follows.

Two-Point Form of the Equation of a Line

An equation of the line passing through the distinct points (x_1, y_1) and (x_2, y_2) is given by

$$\begin{vmatrix} x & y & 1 \\ x_1 & y_1 & 1 \\ x_2 & y_2 & 1 \end{vmatrix} = 0.$$

EXAMPLE 8   ■   Finding an Equation of the Line Passing Through Two Points

Find an equation of the line passing through the two points $(-2, 1)$ and $(3, -2)$.

Solution

Applying the determinant formula for the equation of the line passing through these two points produces

$$\begin{vmatrix} x & y & 1 \\ -2 & 1 & 1 \\ 3 & -2 & 1 \end{vmatrix} = 0.$$

To evaluate this determinant, expand by minors along the first row to obtain the following.

$$x \begin{vmatrix} 1 & 1 \\ -2 & 1 \end{vmatrix} - y \begin{vmatrix} -2 & 1 \\ 3 & 1 \end{vmatrix} + 1 \begin{vmatrix} -2 & 1 \\ 3 & -2 \end{vmatrix} = 3x + 5y + 1 = 0$$

Therefore, an equation of the line is

$$3x + 5y + 1 = 0.$$ ■

DISCUSSION PROBLEM   ■   Determinant of a 3 × 3 Matrix

There is an alternative method that is commonly used for evaluating the determinant of a 3×3 matrix A. (This method *only* works for 3×3 matrices.) To apply this method, copy the first and second columns of A to form fourth and fifth columns. The determinant of A is then obtained by adding the products of three diagonals and subtracting the products of three diagonals, as shown in the diagram at the top of the next page.

$$A = \begin{bmatrix} 0 & 2 & 1 \\ 3 & -1 & 2 \\ 4 & -4 & 1 \end{bmatrix}$$

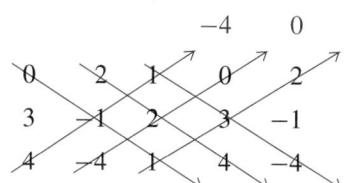

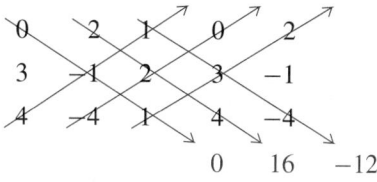

Thus, by adding the lower three products and subtracting the upper three products, you can find the determinant of A to be

$$\det(A) = 0 + 16 - 12 - (-4) - 0 - 6 = 2.$$

Try using this technique to find the determinant of the matrices in Examples 2 and 3. Do you think this method is easier than expanding by minors? ■

Warm-Up

The following warm-up exercises involve skills that were covered in earlier sections. You will use these skills in the exercise set for this section.

In Exercises 1–10, evaluate the quantity.

1. $2(25) - 5(8)$

2. $6(-8) - 4(2)$

3. $3(-2) - 7(4) + 8(-6)$

4. $10\left(\frac{3}{2}\right) - 6(-8) + (-3)(-30)$

5. $-4\left[-1(8) - 2\left(\frac{3}{4}\right) + (-4)\left(-\frac{3}{8}\right)\right]$

6. $\frac{5}{6}[10(-4) - (-1)(-6) + 4(5)]$

7. $\dfrac{2(-2) - 8(3)}{8\left(\frac{3}{2}\right) - 6\left(\frac{-1}{3}\right)}$

8. $\dfrac{5(3) - 4(-3)}{-6(2) - 5(-9)}$

9. $\dfrac{4(4) - 5(6) + 2(10)}{-5(1) - 10(3) + 8(5)}$

10. $\dfrac{-12\left(\frac{1}{4}\right) - 2(8) + 3(-1)}{3(2) - 4(15) + 10(6)}$

 4.4   EXERCISES

⊞ means that a graphing utility can help you solve the exercise or check your solution.

⊞ In Exercises 1–12, find the determinant of the matrix.

1. $\begin{bmatrix} 2 & 1 \\ 3 & 4 \end{bmatrix}$

2. $\begin{bmatrix} -3 & 1 \\ 5 & 2 \end{bmatrix}$

3. $\begin{bmatrix} 5 & 2 \\ -6 & 3 \end{bmatrix}$

4. $\begin{bmatrix} 2 & -2 \\ 4 & 3 \end{bmatrix}$

5. $\begin{bmatrix} 5 & -4 \\ -10 & 8 \end{bmatrix}$

6. $\begin{bmatrix} 4 & -3 \\ 0 & 0 \end{bmatrix}$

7. $\begin{bmatrix} 2 & 6 \\ 0 & 3 \end{bmatrix}$

8. $\begin{bmatrix} -2 & 3 \\ 6 & -9 \end{bmatrix}$

9. $\begin{bmatrix} -7 & 6 \\ \frac{1}{2} & 3 \end{bmatrix}$

10. $\begin{bmatrix} \frac{2}{3} & \frac{5}{6} \\ 14 & -2 \end{bmatrix}$

11. $\begin{bmatrix} 0.3 & 0.5 \\ 0.5 & 0.3 \end{bmatrix}$

12. $\begin{bmatrix} -1.2 & 4.5 \\ 0.4 & -0.9 \end{bmatrix}$

In Exercises 13–16, evaluate the determinant of the matrix in six different ways, by expanding by minors along each row and column.

13. $\begin{bmatrix} 2 & 3 & -1 \\ 6 & 0 & 0 \\ 4 & 1 & 1 \end{bmatrix}$
14. $\begin{bmatrix} 10 & 2 & -4 \\ 8 & 0 & -2 \\ 4 & 0 & 2 \end{bmatrix}$
15. $\begin{bmatrix} 1 & 1 & 2 \\ 3 & 1 & 0 \\ -2 & 0 & 3 \end{bmatrix}$
16. $\begin{bmatrix} 2 & 1 & 3 \\ 1 & 4 & 4 \\ 1 & 0 & 2 \end{bmatrix}$

In Exercises 17–32, evaluate the determinant of the matrix. Expand by minors on the row or column that appears to make the computation easiest.

17. $\begin{bmatrix} 2 & 4 & 6 \\ 0 & 3 & 1 \\ 0 & 0 & -5 \end{bmatrix}$
18. $\begin{bmatrix} 2 & 3 & 1 \\ 0 & 5 & -2 \\ 0 & 0 & -2 \end{bmatrix}$
19. $\begin{bmatrix} -2 & 2 & 3 \\ 1 & -1 & 0 \\ 0 & 1 & 4 \end{bmatrix}$
20. $\begin{bmatrix} 3 & 2 & 2 \\ 2 & 2 & 2 \\ -4 & 4 & 3 \end{bmatrix}$

21. $\begin{bmatrix} 1 & 4 & -2 \\ 3 & 6 & -6 \\ -2 & 1 & 4 \end{bmatrix}$
22. $\begin{bmatrix} 2 & -1 & 0 \\ 4 & 2 & 1 \\ 4 & 2 & 1 \end{bmatrix}$
23. $\begin{bmatrix} -3 & 2 & 1 \\ 4 & 5 & 6 \\ 2 & -3 & 1 \end{bmatrix}$
24. $\begin{bmatrix} -3 & 4 & 2 \\ 6 & 3 & 1 \\ 4 & -7 & -8 \end{bmatrix}$

25. $\begin{bmatrix} 1 & 4 & -2 \\ 3 & 2 & 0 \\ -1 & 4 & 3 \end{bmatrix}$
26. $\begin{bmatrix} 6 & 8 & -7 \\ 0 & 0 & 0 \\ 4 & -6 & 22 \end{bmatrix}$
27. $\begin{bmatrix} 4 & 3 & 3 \\ -2 & 6 & 6 \\ 3 & 1 & 1 \end{bmatrix}$
28. $\begin{bmatrix} \frac{1}{2} & \frac{3}{2} & \frac{1}{2} \\ 4 & 8 & 10 \\ -2 & -6 & 12 \end{bmatrix}$

29. $\begin{bmatrix} 0.1 & 0.2 & 0.3 \\ -0.3 & 0.2 & 0.2 \\ 5 & 4 & 4 \end{bmatrix}$
30. $\begin{bmatrix} -0.4 & 0.4 & 0.3 \\ 0.2 & 0.2 & 0.2 \\ 0.3 & 0.2 & 0.2 \end{bmatrix}$
31. $\begin{bmatrix} x & y & 1 \\ 3 & 1 & 1 \\ -2 & 0 & 1 \end{bmatrix}$
32. $\begin{bmatrix} x & y & 1 \\ -2 & -2 & 1 \\ 1 & 5 & 1 \end{bmatrix}$

In Exercises 33–50, use Cramer's Rule to solve the system of equations. (If not possible, state the reason.)

33. $\begin{cases} x + 2y = 5 \\ -x + y = 1 \end{cases}$

34. $\begin{cases} 2x - y = -10 \\ 3x + 2y = -1 \end{cases}$

35. $\begin{cases} 3x + 4y = -2 \\ 5x + 3y = 4 \end{cases}$

36. $\begin{cases} 18x + 12y = 13 \\ 30x + 24y = 23 \end{cases}$

37. $\begin{cases} 20x + 8y = 11 \\ 12x - 24y = 21 \end{cases}$

38. $\begin{cases} 13x - 6y = 17 \\ 26x - 12y = 8 \end{cases}$

39. $\begin{cases} -0.4x + 0.8y = 1.6 \\ 2x - 4y = 5 \end{cases}$

40. $\begin{cases} -0.4x + 0.8y = 1.6 \\ 0.2x + 0.3y = 2.2 \end{cases}$

41. $\begin{cases} 3u + 6v = 5 \\ 6u + 14v = 11 \end{cases}$

42. $\begin{cases} 3x_1 + 2x_2 = 1 \\ 2x_1 + 10x_2 = 6 \end{cases}$

43. $\begin{cases} 4x - y + z = -5 \\ 2x + 2y + 3z = 10 \\ 5x - 2y + 6z = 1 \end{cases}$

44. $\begin{cases} 4x - 2y + 3z = -2 \\ 2x + 2y + 5z = 16 \\ 8x - 5y - 2z = 4 \end{cases}$

45. $\begin{cases} 3x + 4y + 4z = 11 \\ 4x - 4y + 6z = 11 \\ 6x - 6y = 3 \end{cases}$

46. $\begin{cases} 14x_1 - 21x_2 - 7x_3 = 10 \\ -4x_1 + 2x_2 - 2x_3 = 4 \\ 56x_1 - 21x_2 + 7x_3 = 5 \end{cases}$

47. $\begin{cases} 3a + 3b + 4c = 1 \\ 3a + 5b + 9c = 2 \\ 5a + 9b + 17c = 4 \end{cases}$

48. $\begin{cases} 2x + 3y + 5z = 4 \\ 3x + 5y + 9z = 7 \\ 5x + 9y + 17z = 13 \end{cases}$

49. $\begin{cases} 5x - 3y + 2z = 2 \\ 2x + 2y - 3z = 3 \\ x - 7y + 8z = -4 \end{cases}$

50. $\begin{cases} 3x + 2y + 5z = 4 \\ 4x - 3y - 4z = 1 \\ -8x + 2y + 3z = 0 \end{cases}$

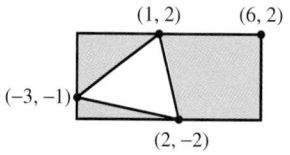

Area of a Triangle In Exercises 51–54, use a determinant to find the area of the triangle with the given vertices.

51. $(0, 0)$, $(3, 1)$, $(1, 5)$

52. $(-2, -3)$, $(2, -3)$, $(0, 4)$

53. $(-2, 1)$, $(3, -1)$, $(1, 6)$

54. $\left(0, \frac{1}{2}\right)$, $\left(\frac{5}{2}, 0\right)$, $(4, 3)$

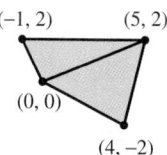

Area of a Region In Exercises 55 and 56, find the area of the shaded region of the figure.

55.

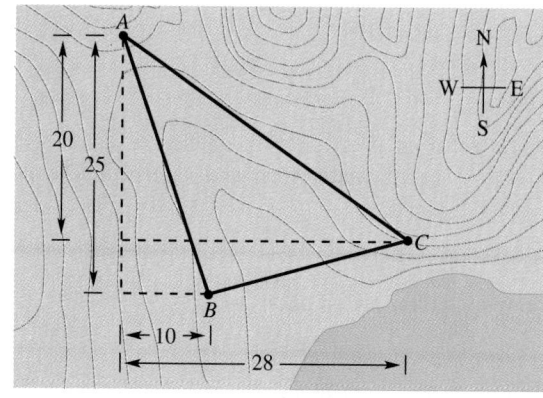

56.

57. ***Area of a Region*** A large region of forest has been infested by gypsy moths. The region is roughly triangular, as shown in the figure. From the northernmost vertex A of the region, the distance to the other vertices is 25 miles south and 10 miles east (for vertex B), and 20 miles south and 28 miles east (for vertex C). Approximate the number of square miles in this region.

58. ***Area of a Region*** Suppose you purchased a triangular tract of land, as shown in the figure. To estimate the number of square feet in the tract, you start at one vertex and walk 65 feet east and 50 feet north to the second vertex. Then, from the second vertex, you walk 85 feet west and 30 feet north to the third vertex. How many square feet are there in the tract of land?

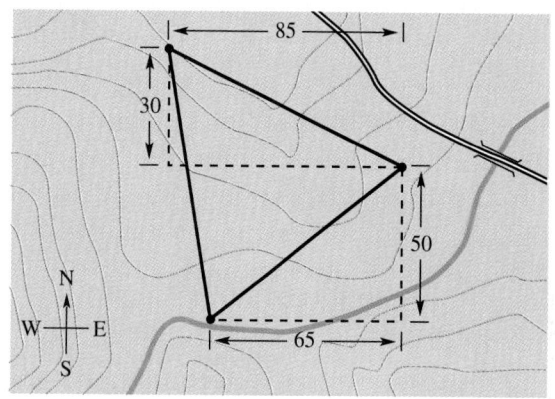

Collinear Points In Exercises 59–62, use a determinant to determine whether the given points are collinear.

59. $(-1, 11)$, $(0, 8)$, $(2, 2)$

60. $(-1, -1)$, $(1, 9)$, $(2, 13)$

61. $\left(-2, \frac{1}{3}\right)$, $(2, 1)$, $\left(3, \frac{1}{5}\right)$

62. $\left(0, \frac{1}{2}\right)$, $\left(1, \frac{7}{6}\right)$, $\left(9, \frac{13}{2}\right)$

Equation of a Line In Exercises 63–66, use a determinant to find an equation of the line through the given points.

63. $(0, 0)$, $(5, 3)$

64. $(-4, 3)$, $(2, 1)$

65. $(10, 7)$, $(-2, -7)$

66. $\left(-\frac{1}{2}, 3\right)$, $\left(\frac{5}{2}, 1\right)$

Curve Fitting In Exercises 67–70, use Cramer's Rule to find the equation of the parabola $y = ax^2 + bx + c$ that passes through the given points.

67. $(0, 1)$, $(1, -3)$, $(-2, 21)$

68. $(2, 3)$, $\left(-1, \frac{9}{2}\right)$, $(-2, 9)$

69. $(1, -1)$, $(-1, -5)$, $\left(\frac{1}{2}, \frac{1}{4}\right)$

70. $(-2, 6)$, $(1, 9)$, $(3, 1)$

In Exercises 71 and 72, use the matrix capabilities of a graphing calculator to evaluate the determinant of the matrix.

71. $\begin{bmatrix} 5 & -3 & 2 \\ 7 & 5 & -7 \\ 0 & 6 & -1 \end{bmatrix}$ **72.** $\begin{bmatrix} -\frac{1}{2} & -1 & 6 \\ 8 & -\frac{1}{4} & -4 \\ 1 & 2 & 1 \end{bmatrix}$

 In Exercises 73 and 74, use the matrix capabilities of a graphing calculator and Cramer's Rule to solve the system of equations.

73. $\begin{cases} 3x - 2y + 3z = 8 \\ x + 3y + 6z = -3 \\ x + 2y + 9z = -5 \end{cases}$

74. $\begin{cases} 6x + 4y - 8z = -22 \\ -2x + 2y + 3z = 13 \\ -2x + 2y - z = 5 \end{cases}$

75. (a) Use Cramer's Rule to solve the system of linear equations.

$$\begin{cases} kx + (1-k)y = 1 \\ (1-k)x + ky = 3 \end{cases}$$

(b) For what value of k can Cramer's Rule not be used?

Graphs of Linear Inequalities in Two Variables

Linear Inequalities in Two Variables • *The Graph of a Linear Inequality in Two Variables*

Linear Inequalities in Two Variables

A **linear inequality** in variables x and y is an inequality that can be written in one of the following forms (where a and b are not both zero).

$$ax + by < c \qquad ax + by \le c$$
$$ax + by > c \qquad ax + by \ge c$$

Here are some examples.

$$x - y > -3, \quad 4x - 3y < 7, \quad x \le 2, \quad \text{and} \quad y \ge -4$$

An ordered pair (x_1, y_1) is a **solution** of a linear inequality in x and y if the inequality is true when x_1 and y_1 are substituted for x and y, respectively.

EXAMPLE 1 ■ Verifying Solutions of Linear Inequalities

Determine whether the following points are solutions of $2x - 3y \ge -2$.

(a) $(0, 0)$
(b) $(2, 2)$
(c) $(0, 1)$

Solution

(a) To determine whether the point $(0, 0)$ is a solution of the inequality, substitute the coordinates of the point into the inequality as follows.

$$2x - 3y \geq -2 \qquad\qquad \text{Given inequality}$$

$$2(0) - 3(0) \overset{?}{\geq} -2 \qquad\qquad \text{Replace } x \text{ by } 0 \text{ and } y \text{ by } 0$$

$$0 \geq -2 \qquad\qquad \text{Inequality is satisfied}$$

Because the inequality is satisfied, you can conclude that the point $(0, 0)$ is a solution.

(b) By substituting the coordinates of the point $(2, 2)$ into the given inequality, you obtain $2(2) - 3(2) = -2$. Because -2 is greater than or equal to -2, you can conclude that the point $(2, 2)$ is also a solution of the inequality.

(c) By substituting the coordinates of the point $(0, 1)$ into the given inequality, you obtain $2(0) - 3(1) = -3$. Because -3 is less than -2, you can conclude that the point $(0, 1)$ is *not* a solution of the inequality.   ■

The Graph of a Linear Inequality in Two Variables

The **graph** of an inequality is the collection of all solution points of the inequality. To sketch the graph of a linear inequality such as $4x - 3y < 12$, begin by sketching the graph of the *corresponding linear equation* $4x - 3y = 12$. This graph is made with a *dashed* line for the inequalities $<$ and $>$ and with a *solid* line for the inequalities $\leq$ and $\geq$. The graph of the equation (corresponding to a given linear inequality) separates the plane into two regions, called **half-planes.** In each half-plane, one of the following *must* be true.

1. All points in the half-plane are solutions of the inequality.

2. No point in the half-plane is a solution of the inequality.

Thus, you can determine whether the points in an entire half-plane satisfy the inequality by simply testing *one* point in the region. This graphing procedure is summarized as follows.

Sketching the Graph of a Linear Inequality in Two Variables

1. Replace the inequality sign by an equal sign, and sketch the graph of the resulting equation. (Use a dashed line for $<$ or $>$ and a solid line for $\leq$ or $\geq$.)

2. Test one point in each of the half-planes formed by the graph in step 1. If the point satisfies the inequality, then shade the entire half-plane to denote that every point in the region satisfies the inequality.

EXAMPLE 2 ■ Sketching the Graph of a Linear Inequality

Sketch the graphs of the following linear inequalities.

(a) $x \geq -3$ (b) $y < 4$

Solution

(a) The graph of the corresponding equation $x = -3$ is a vertical line. The points that satisfy the inequality $x \geq -3$ are those lying to the right of (or on) this line, as shown in Figure 4.16.

(b) The graph of the corresponding equation $y = 4$ is a horizontal line. The points that satisfy the inequality $y < 4$ are those lying below this line, as shown in Figure 4.17.

FIGURE 4.16   **FIGURE 4.17**

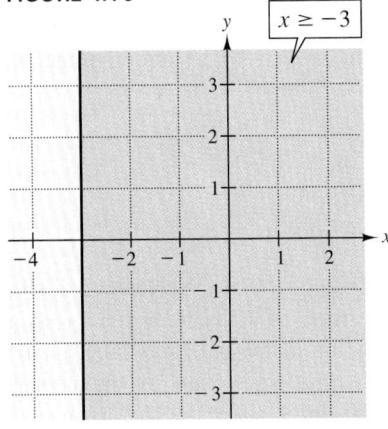

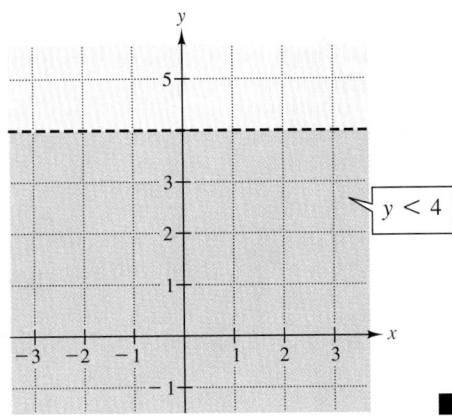

EXAMPLE 3 ■ Sketching the Graph of a Linear Inequality

Sketch the graph of the linear inequality

$$x + y > 3.$$

Solution

The graph of the corresponding equation $x + y = 3$ is a line, as shown in Figure 4.18. Because the origin $(0, 0)$ *does not* satisfy the inequality, the graph consists of the half-plane lying *above* the line. (Try checking a point above the line. Regardless of which point you choose, you will see that it satisfies the inequality.)

FIGURE 4.18

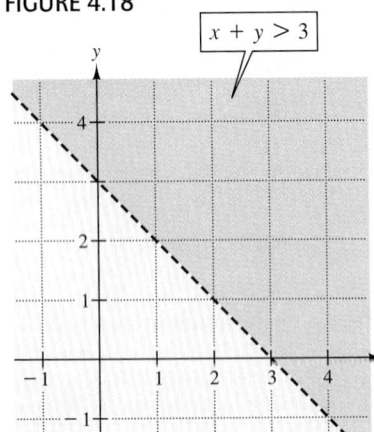

▶ Point out that initially trying the point $(0, 0)$ is usually the easiest, but should be done only if the line does not pass through the origin.

For a linear inequality in two variables, you can sometimes simplify the graphing procedure by writing the inequality in *slope-intercept* form. For instance, by writing $x + y > 3$ in the form

$$y > -x + 3,$$

you can see that the solution points lie *above* the line $y = -x + 3$, as shown in Figure 4.18. Similarly, by writing the inequality $4x - 3y > 12$ in the form

$$y < \frac{4}{3}x - 4,$$

you can see that the solutions lie *below* the line $y = \frac{4}{3}x - 4$, as shown in Figure 4.19.

FIGURE 4.19

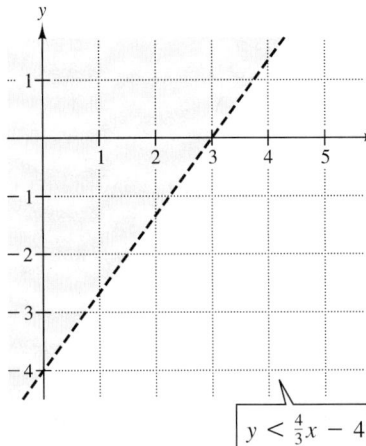

$y < \frac{4}{3}x - 4$

EXAMPLE 4   ■   Sketching the Graph of a Linear Inequality

Use the slope-intercept form of a linear equation as an aid in sketching the graph of the inequality

$$2x - 3y \leq 15.$$

Solution

To begin, rewrite the inequality in slope-intercept form.

$2x - 3y \leq 15$	Given inequality
$-3y \leq -2x + 15$	Subtract $2x$ from both sides
$y \geq \frac{2}{3}x - 5$	Divide by -3 and reverse inequality

From this form, you can conclude that the solution is the half-plane lying *on* or *above* the line

$$y = \frac{2}{3}x - 5.$$

The graph is shown in Figure 4.20.

FIGURE 4.20

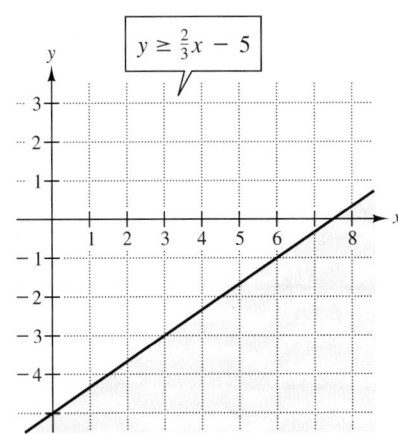

$y \geq \frac{2}{3}x - 5$

■

EXAMPLE 5 ■ Finding Ocean Treasure

You are on a treasure-diving ship that is hunting for gold and silver coins. Objects collected by the divers are placed in a wire basket. One of the divers signals you to reel in the basket. It feels like it contains about 50 pounds of material! If gold coins weigh about $\frac{1}{2}$ ounce each and silver coins weigh about $\frac{1}{4}$ ounce each, what are the different amounts of coins that you could be reeling in?

Solution

At 16 ounces per pound, the basket contains 800 ounces.

FIGURE 4.21

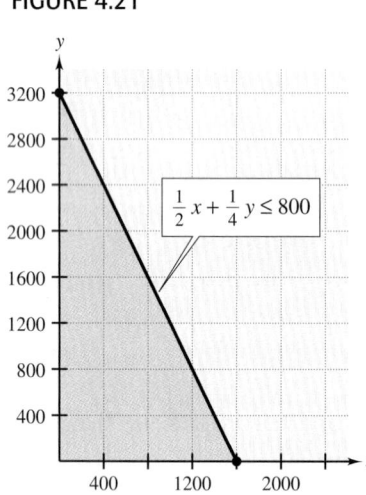

Verbal model:	Weight per coin	·	Number of gold coins	+	Weight per coin	·	Number of silver coins	≤	Weight in basket

Labels: Weight per gold coin $= \frac{1}{2}$ (ounce per coin)

Number of gold coins $= x$ (coins)

Weight per silver coin $= \frac{1}{4}$ (ounce per coin)

Number of silver coins $= y$ (coins)

Weight in basket $= 800$ (ounces)

Inequality: $\frac{1}{2}x + \frac{1}{4}y \leq 800$ Linear model

One solution is $(1600, 0)$: all gold coins. Another solution is $(0, 3200)$: all silver coins. Unfortunately, there are many other solutions including $(0, 0)$: 50 pounds of something, but no gold or silver coins. All possible solutions are shown in Figure 4.21. ■

DISCUSSION PROBLEM ■ A Nonlinear Inequality

Write a short paragraph describing how you would sketch the graph of the nonlinear inequality

$$y < x^2 + 1.$$

■

Warm-Up

The following warm-up exercises involve skills that were covered in earlier sections. You will use these skills in the exercise set for this section.

In Exercises 1–10, sketch the graph of the equation.

1. $x + 2y = 2$

2. $2x - 3y = 4$

3. $y = \frac{1}{2}(x - 2) - 3$

4. $y = -2(x + 1) + 4$

5. $x + 3y = 6$

6. $0.1x - 0.5y = 0.3$

7. $x = -3$

8. $y = 8$

9. $x + y = 0$

10. $9x + 18y = 0$

4.5 EXERCISES

means that a graphing utility can help you solve the exercise or check your solution.

In Exercises 1–6, match the inequality with its graph. [The graphs are labeled (a), (b), (c), (d), (e), and (f).]

1. $y \geq -2$

2. $x < -2$

3. $3x - 2y < 0$

4. $3x - 2y > 0$

5. $x + y < 4$

6. $x + y \leq 4$

(a)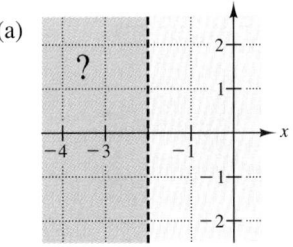

(b)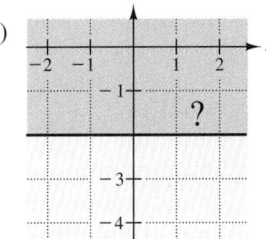

(c)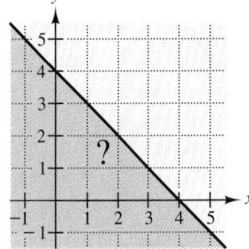

(d)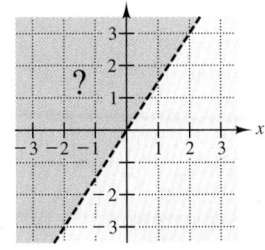

(e)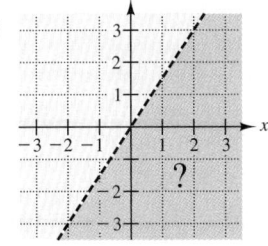

(f)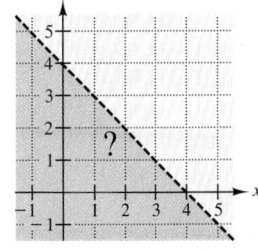

In Exercises 7–14, determine whether the points are solutions of the inequality.

7. $x - 2y < 4$ (a) (0, 0)   (b) (2, −1)   **8.** $x + y < 3$ (a) (0, 6)   (b) (4, 0)

(c) (3, 4)   (d) (5, 1)   (c) (0, −2)   (d) (1, 1)

9. $3x + y \geq 10$ (a) (1, 3)   (b) (−3, 1)   **10.** $-3x + 5y \geq 6$ (a) (2, 8)   (b) (−10, −3)

(c) (3, 1)   (d) (2, 15)   (c) (0, 0)   (d) (3, 3)

11. $y > 0.2x - 1$ (a) (0, 2)   (b) (6, 0)   **12.** $y < -3.5x + 7$ (a) (1, 5)   (b) (5, −1)

(c) (4, −1)   (d) (−2, 7)   (c) (−1, 4)   (d) $\left(0, \frac{4}{3}\right)$

13. $y \leq 3 - |x|$ (a) (−1, 4)   (b) (2, −2)   **14.** $y \geq |x - 3|$ (a) (0, 0)   (b) (1, 2)

(c) (6, 0)   (d) (5, −2)   (c) (4, 10)   (d) (5, −1)

In Exercises 15–30, sketch the graph of the linear inequality.

15. $x \geq 2$ **16.** $x < -3$ **17.** $y < 5$ **18.** $y > 2$

19. $y > \frac{1}{2}x$ **20.** $y \leq 2x$ **21.** $y \leq x + 1$ **22.** $y > 4 - x$

23. $y - 1 > -\frac{1}{2}(x - 2)$ **24.** $y - 2 < -\frac{2}{3}(x - 1)$ **25.** $\frac{x}{3} + \frac{y}{4} \leq 1$ **26.** $\frac{x}{2} + \frac{y}{6} \geq 1$

27. $x - 2y \geq 6$ **28.** $3x + 5y \leq 15$ **29.** $0.2x + 0.3y < 2$ **30.** $x - 0.75y > 6$

In Exercises 31–36, write an inequality for the shaded region shown in the figure.

31. (−1, 5)

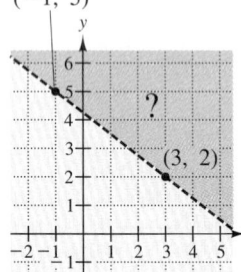

32.

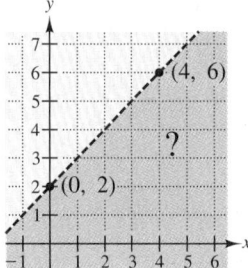

33.

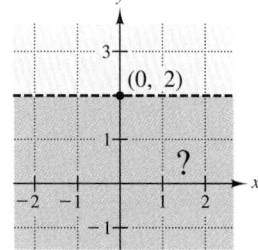

34.

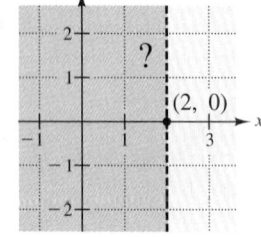

35.

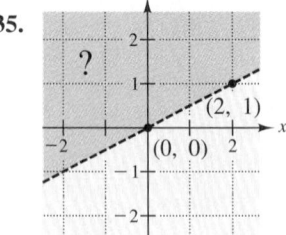

36.

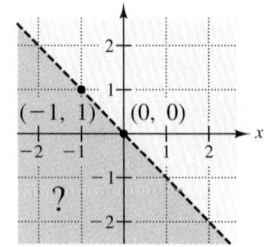

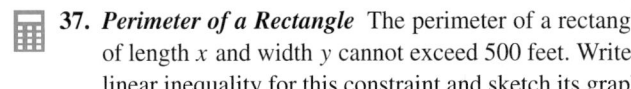

37. *Perimeter of a Rectangle* The perimeter of a rectangle of length x and width y cannot exceed 500 feet. Write a linear inequality for this constraint and sketch its graph.

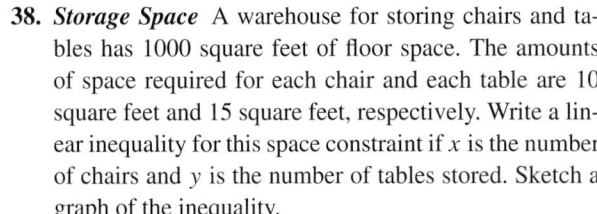

38. *Storage Space* A warehouse for storing chairs and tables has 1000 square feet of floor space. The amounts of space required for each chair and each table are 10 square feet and 15 square feet, respectively. Write a linear inequality for this space constraint if x is the number of chairs and y is the number of tables stored. Sketch a graph of the inequality.

39. *Getting a Workout* The maximum heart rate r (in beats per minute) of a person in normal health is related to the person's age A (in years). The relationship between r and A is given by the inequality $r \leq 220 - A$.

    (a) Sketch a graph of the inequality with A measured along the horizontal axis and r measured along the vertical axis.

    (b) Physiologists recommend that during a workout a person strive to increase his or her heart rate to 75% of the maximum rate for the person's age. Sketch the graph of $r = 0.75(220 - A)$ on the same set of coordinate axes used in part (a).

Jogging is an aerobic exercise because it helps the body efficiently process large amounts of oxygen by increasing heart and breathing rates.

40. *Pizza and Soda Pop* You and some friends go out for pizza. Together you have $26. You want to order two large pizzas with cheese at $8 each. Each additional topping costs $0.40, and each small soft drink costs $0.80. Write an inequality that represents the number of toppings x and the number of drinks y that your group can afford. Sketch a graph of this inequality. What are the coordinates for an order of six soft drinks and two large pizzas with cheese, each with three additional toppings? Is this a solution of the inequality? (Assume that there is no sales tax.)

SECTION 4.6   Systems of Linear Inequalities and Linear Programming

Systems of Linear Inequalities in Two Variables • *Linear Programming*

Systems of Linear Inequalities in Two Variables

Many practical problems in business, science, and engineering involve **systems of linear inequalities.** This type of system arises in problems that have *constraint* statements that contain phrases such as "more than," "less than," "at least," "no more than," "a minimum of," and "a maximum of."

A **solution** of a system of linear inequalities in x and y is a point (x, y) that satisfies each inequality in the system. For instance, the point $(2, 4)$ is a solution of the following system because $x = 2$ and $y = 4$ satisfy each of the inequalities in the system.

$$\begin{cases} x + y \leq 10 \\ 3x - y \leq 2 \end{cases}$$

To sketch the graph of a system of inequalities in two variables, first sketch (in the same coordinate system) the graph of each individual inequality. The **solution set** is the region that is *common* to every graph in the system.

EXAMPLE 1 ■ Sketching the Graph of a System of Linear Inequalities

Sketch the graph of the following system of linear inequalities.

$$\begin{cases} 2x - y \le 5 \\ x + 2y \ge 2 \end{cases}$$

Solution

FIGURE 4.22

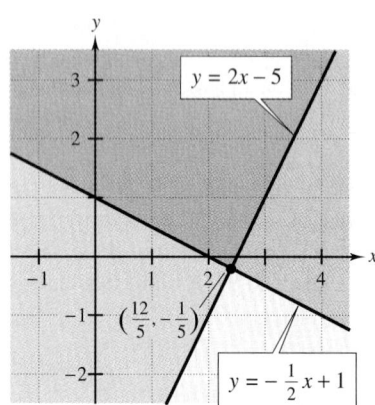

Begin by sketching the graph of each inequality. The graph of $2x - y \le 5$ consists of all points on or above the line $y = 2x - 5$. The graph of $x + 2y \ge 2$ consists of all points on or above the line $y = -\frac{1}{2}x + 1$. The graph of the *system* of linear inequalities consists of the wedge-shaped region that is common to the two half-planes (representing the individual inequalities), as shown in Figure 4.22. ■

In Figure 4.22, note that the two borderlines of the region

$$y = 2x - 5 \quad \text{and} \quad y = -\frac{1}{2}x + 1$$

intersect at the point $\left(\frac{12}{5}, -\frac{1}{5}\right)$. Such a point is called a **vertex** of the region. The region shown in the figure has only one vertex. Some regions, however, have several vertices. When you are sketching the graph of a system of linear inequalities, it is helpful to find and label any vertices of the region.

Here are some guidelines that may help you sketch the graph of a system of linear inequalities.

Graphing a System of Linear Inequalities

1. Sketch the line that corresponds to each inequality. (Use dashed lines for inequalities with < or > and solid lines for inequalities with ≤ or ≥.)

2. Lightly shade the half-plane that is the graph of each linear inequality. (Colored pencils may help you distinguish the different half-planes.)

3. The graph of the system is the intersection of the half-planes. (If you have used colored pencils, it is the region that has been shaded with *every* color.)

EXAMPLE 2 ■ Graphing a System of Linear Inequalities

Sketch the graph of the following system of linear inequalities.

$$\begin{cases} y < 4 \\ y > 1 \end{cases}$$

FIGURE 4.23

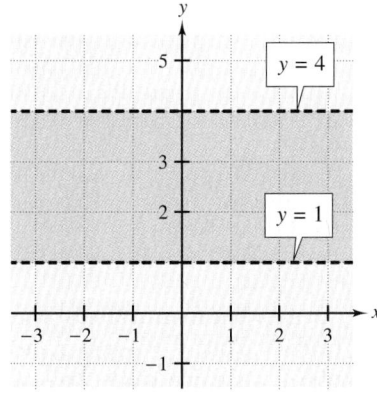

Solution

The graph of the first inequality is the half-plane *below* the horizontal line

$$y = 4. \qquad \text{Upper boundary}$$

The graph of the second inequality is the half-plane *above* the horizontal line

$$y = 1. \qquad \text{Lower boundary}$$

The graph of the system is the horizontal band that lies *between* the two horizontal lines (where $y < 4$ *and* $y > 1$), as shown in Figure 4.23.

◼

EXAMPLE 3   ◼   **Sketching the Graph of a System of Linear Inequalities**

Sketch the graph of the following system of linear inequalities and label the vertices.

$$\begin{cases} x + y < 4 \\ x > -3 \\ y \geq 2 \end{cases}$$

FIGURE 4.24

Solution

The graph of $x + y < 4$ consists of all points below the line $y = -x + 4$, the graph of $x > -3$ consists of all points to the right of the vertical line $x = -3$, and the graph of $y \geq 2$ consists of all points on or above the line $y = 2$. Notice in Figure 4.24 that the solution set of the system is the triangular region that is common to all three half-planes. To find the vertices of this triangular region, solve *pairs* of equations that represent the three boundaries of the region.

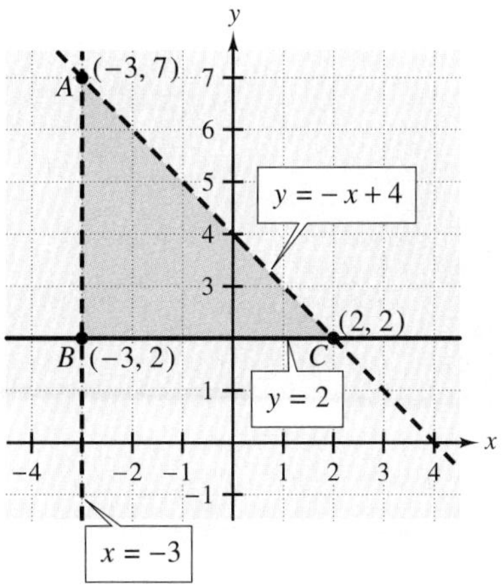

Vertex A: $(-3, 7)$

Solution of the system

$$\begin{cases} x + y = 4 \\ x = -3 \end{cases}$$

Vertex B: $(-3, 2)$

Solution of the system

$$\begin{cases} x = -3 \\ y = 2 \end{cases}$$

Vertex C: $(2, 2)$

Solution of the system

$$\begin{cases} x + y = 4 \\ y = 2 \end{cases}$$

◼

EXAMPLE 4   ■   Sketching the Graph of a System of Linear Inequalities

Sketch the graph of the following system of linear inequalities, and label the vertices.

$$\begin{cases} x + y \le 5 \\ 3x + 2y \le 12 \\ x \ge 0 \\ y \ge 0 \end{cases}$$

Solution

FIGURE 4.25

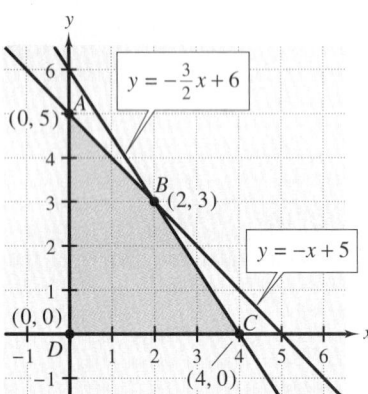

Begin by sketching the half-planes represented by the four linear inequalities. The graph of $x + y \le 5$ is the half-plane lying on or below the line $y = -x + 5$, the graph of $3x + 2y \le 12$ is the half-plane lying on or below the line $y = -\frac{3}{2}x + 6$, the graph of $x \ge 0$ is the half-plane lying to the right of the y-axis, and the graph of $y \ge 0$ is the half-plane lying above the x-axis. As shown in Figure 4.25, the region that is common to all four of these half-planes is a four-sided polygon. The vertices of the region are found as follows.

Vertex A: $(0, 5)$	*Vertex B:* $(2, 3)$	*Vertex C:* $(4, 0)$	*Vertex D:* $(0, 0)$
Solution of the system	**Solution of the system**	**Solution of the system**	**Solution of the system**
$\begin{cases} x + y = 5 \\ x = 0 \end{cases}$	$\begin{cases} x + y = 5 \\ 3x + 2y = 12 \end{cases}$	$\begin{cases} 3x + 2y = 12 \\ y = 0 \end{cases}$	$\begin{cases} x = 0 \\ y = 0 \end{cases}$ ■

Film studios create backgrounds, or *mattes,* by painting on glass (left) and blacking out unwanted areas. Actors are filmed on a real set, with opposite areas blacked out (middle). The two films are then combined (right). ("INDIANA JONES AND THE TEMPLE OF DOOM"™ⓒ Lucasfilm Ltd. [LFL] 1984. All Rights Reserved.)

EXAMPLE 5   ■   Finding the Boundaries of a Region

Find a system of inequalities that defines the region shown in Figure 4.26.

FIGURE 4.26

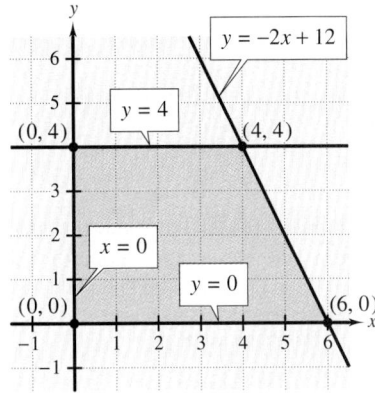

Solution

Three of the boundaries of the region are horizontal or vertical—they are easy to find. To find the diagonal boundary line, use the techniques from Section 3.6 to find the equation of the line passing through the points $(4, 4)$ and $(6, 0)$. The equation is $y = -2x + 12$. The system of linear inequalities that describes the region is as follows.

$$\begin{cases} y \leq 4 & \text{Region lies on and below line } y = 4 \\ y \geq 0 & \text{Region lies on and above } x\text{-axis} \\ x \geq 0 & \text{Region lies on and to the right of } y\text{-axis} \\ y \leq -2x + 12 & \text{Region lies on and below line } y = -2x + 12 \end{cases}$$

■

Linear Programming

Systems of linear inequalities are used extensively in business and economics to solve **optimization problems.** The word "optimize" means to find the greatest or least. Many optimization problems can be solved using a technique called **linear programming.** A two-variable linear programming problem consists of the following.

1. An **objective function** that expresses the quantity to be maximized (or minimized).

2. A system of **constraint linear inequalities** whose solution set represents the set of **feasible solutions.**

FIGURE 4.27

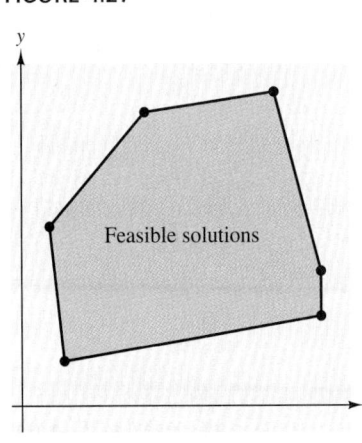

Feasible solutions

The solution of a linear programming problem is found by determining which point in the set of feasible solutions yields the optimal value of the objective function. For example, consider a linear programming problem in which you are asked to maximize the value of

$$C = ax + by \qquad \text{Objective function}$$

subject to a set of constraints that determine the region indicated in Figure 4.27. It can be shown that if there is an optimal solution, it must occur at one of the vertices of the region. In other words, *you can find the maximum value by testing C at each of the vertices,* as illustrated in Example 6.

EXAMPLE 6 ■ Solving a Linear Programming Problem

Find the minimum value and maximum value of

$$C = 4x + 5y \qquad \text{Objective quantity}$$

subject to the following constraints.

$$\begin{cases} x & \geq 0 \\ & y \geq 0 \\ x + y \leq 6 \end{cases} \qquad \text{Constraints}$$

FIGURE 4.28

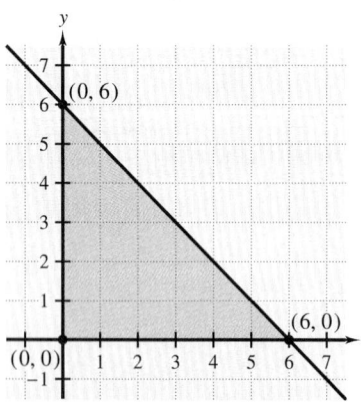

Solution

The graph of the constraint inequalities is shown in Figure 4.28. The three vertices are $(0, 0)$, $(6, 0)$ and $(0, 6)$. To find the minimum and maximum values of C, evaluate $C = 4x + 5y$ at each of the three vertices.

At $(0, 0)$: $C = 4(0) + 5(0) = 0$ Minimum value of C

At $(6, 0)$: $C = 4(6) + 5(0) = 24$

At $(0, 6)$: $C = 4(0) + 5(6) = 30$ Maximum value of C

The minimum value of C is 0. It occurs when $x = 0$ and $y = 0$. The maximum value of C is 30. It occurs when $x = 0$ and $y = 6$. ■

In Example 6, try evaluating C at other *feasible points* in the graph of the constraints. No matter which point you choose, the value of C will be greater than or equal to 0 and less than or equal to 30. For instance, at the point $(3, 2)$ the value of C is 22.

The steps used in Example 6 are outlined as follows.

Guidelines for Solving a Linear Programming Problem

To solve a linear programming problem involving two variables, use the following steps.

1. Sketch the region corresponding to the system of constraints. (The points inside or on the boundary of the region are called feasible solutions.)

2. Find the vertices of the region.

3. Test the objective function at each of the vertices and select the values of the variables that optimize the objective function. For a bounded region, both a minimum and a maximum value will exist. (For an unbounded region, *if* an optimal solution exists, it will occur at a vertex.)

These guidelines will work whether the objective function is to be maximized *or* minimized. For instance, in Example 6 the same test was used to find the maximum value of C and the minimum value of C.

EXAMPLE 7   ■   **Solving a Linear Programming Problem**

Find the minimum value and maximum value of

$$C = 3x + 4y \qquad \text{Objective quantity}$$

subject to the following constraints.

$$\begin{cases} x & \geq 2 \\ x & \leq 5 \\ & y \geq 1 \\ & y \leq 6 \end{cases} \qquad \text{Constraints}$$

FIGURE 4.29

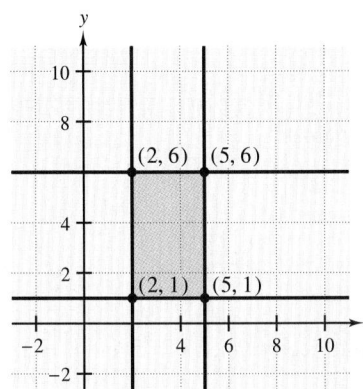

Solution

The graph of the constraint inequalities is shown in Figure 4.29. The four vertices are (2, 1), (2, 6), (5, 1) and (5, 6). To find the minimum and maximum values of C, evaluate $C = 3x + 4y$ at each of the four vertices.

At (2, 1): $C = 3(2) + 4(1) = 10$ Minimum value of C

At (2, 6): $C = 3(2) + 4(6) = 30$

At (5, 1): $C = 3(5) + 4(1) = 19$

At (5, 6): $C = 3(5) + 4(6) = 39$ Maximum value of C

The minimum value of C is 10. It occurs when $x = 2$ and $y = 1$. The maximum value of C is 39. It occurs when $x = 5$ and $y = 6$.   ■

EXAMPLE 8   ■   **Solving a Linear Programming Problem**

Find the minimum value and maximum value of

$$C = 6x + 7y \qquad \text{Objective quantity}$$

subject to the following constraints.

$$\begin{cases} x & \geq 0 \\ & y \geq 0 \\ 4x + 3y \geq 24 \\ x + 3y \geq 15 \end{cases} \qquad \text{Constraints}$$

FIGURE 4.30

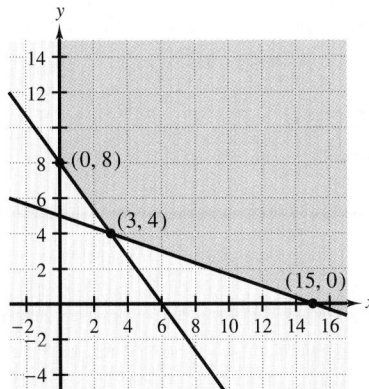

Solution

The graph of the system of linear inequalities is shown in Figure 4.30. The three vertices are (0, 8), (3, 4), and (15, 0). To find the minimum and maximum values of C, evaluate $C = 6x + 7y$ at each of the three vertices.

At (0, 8): $C = 6(0) + 7(8) = 56$

At (3, 4): $C = 6(3) + 7(4) = 46$ Minimum value of C

At (15, 0): $C = 6(15) + 7(0) = 90$

The minimum value of C is 46. It occurs when $x = 3$ and $y = 4$. There is no maximum value (the graph of the constraints is unbounded).   ■

EXAMPLE 9 ■ Solving a Linear Programming Problem

Find the maximum value of

$$C = 3x + 2y \qquad \text{Objective function}$$

subject to the following constraints.

$$\begin{cases} x + 2y \le 4 \\ x - \ y \le 1 \\ x \qquad \ge 0 \\ \qquad y \ge 0 \end{cases} \qquad \text{Constraints}$$

FIGURE 4.31

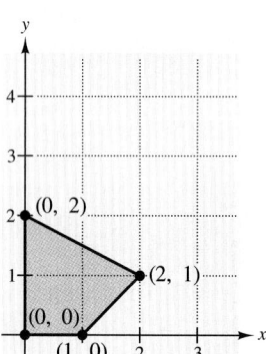

Solution

The constraints form the region shown in Figure 4.31. At the four vertices of this region, the objective function has the following values.

At $(0, 0)$: $C = 3(0) + 2(0) = 0$

At $(1, 0)$: $C = 3(1) + 2(0) = 3$

At $(2, 1)$: $C = 3(2) + 2(1) = 8$ Maximum value of C

At $(0, 2)$: $C = 3(0) + 2(2) = 4$

Thus, the maximum value of C is 8, and this value occurs when $x = 2$ and $y = 1$.

■

EXAMPLE 10 ■ Finding the Maximum Profit

You own a bicycle manufacturing plant and can assemble bicycles using two processes. The hours of unskilled labor, machine time, and skilled labor *per bicycle* are given below. You can use up to 4200 hours of unskilled labor and up to 2400 hours each of machine time and skilled labor. Process A earns a profit of $45 per bike, and process B earns a profit of $50 per bike. How many bicycles should you assemble by each process to obtain a maximum profit?

	Unskilled Labor	Machine Time	Skilled Labor
Hours for Process A	3	1	2
Hours for Process B	3	2	1

In most countries, the number of bicycles is greater than the number of automobiles.

Solution

Let a and b represent the number assembled by each process. Because you want a maximum profit P, the objective quantity is

$$P = 45a + 50b.$$ Profit: $45 per bike for process A

$50 per bike for process B

The constraints are as follows.

$$\begin{cases} 3a + 3b \leq 4200 & \text{Unskilled labor: Up to 4200 hours} \\ a + 2b \leq 2400 & \text{Machine time: Up to 2400 hours} \\ 2a + b \leq 2400 & \text{Skilled labor: Up to 2400 hours} \\ a \qquad \geq 0 & \text{Cannot produce a negative amount} \\ b \geq 0 & \text{Cannot produce a negative amount} \end{cases}$$

The region that represents the feasible solutions is shown in Figure 4.32. The profits at the vertices of the region are as follows.

At (0, 1200): $P = \$60,000$
At (400, 1000): $P = \$68,000$ Maximum profit
At (1000, 400): $P = \$65,000$
At (1200, 0): $P = \$54,000$
At (0, 0): $P = \quad \$0$

The maximum profit is obtained by making 400 bicycles by process A and 1000 bicycles by process B.   ■

FIGURE 4.32

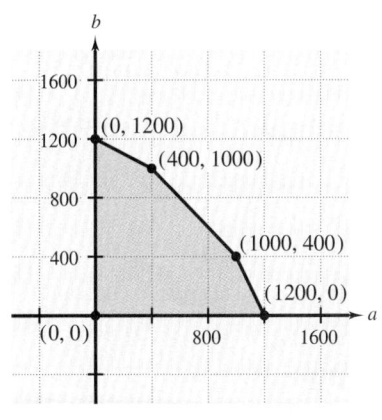

DISCUSSION PROBLEM ■ Creating a Linear Programming Problem

Write a linear programming problem whose constraints determine the region shown in Figure 4.33. Then find an objective function that is a maximum at the vertex (4, 4).

FIGURE 4.33

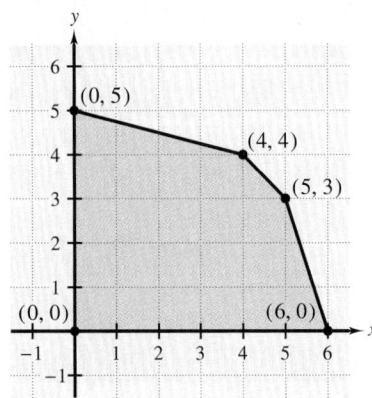

Warm-Up

The following warm-up exercises involve skills that were covered in earlier sections. You will use these skills in the exercise set for this section.

In Exercises 1–4, simplify the expression.

1. $(4x + 3y) - 3(5x + y)$ **2.** $(-15u + 4v) + 5(3u - 9v)$

3. $2x + (2x - 3) + 12x$ **4.** $y - (y + 2) + 4y$

In Exercises 5 and 6, solve the given equation.

5. $5x - (2x - 5) = 17$ **6.** $4t - 3(t + 1) = 8$

In Exercises 7–10, determine whether the lines represented by the pair of equations are parallel or perpendicular, or neither.

7. $\begin{cases} 3x - 4y = -10 \\ 4x + 3y = 11 \end{cases}$ **8.** $\begin{cases} 6x - 18y = 8 \\ -2x + 6y = 1 \end{cases}$

9. $\begin{cases} 0.3x - 0.5y = 1 \\ -1.2x + 2y = 2 \end{cases}$ **10.** $\begin{cases} 2x + y = 2 \\ 3x + 4y - 1 \end{cases}$

4.6 EXERCISES

⊞ means that a graphing utility can help you solve the exercise or check your solution.

In Exercises 1–6, match the system of inequalities with its graph. [The graphs are labeled (a), (b), (c), (d), (e), and (f).]

1. $\begin{cases} y > x \\ x > -3 \\ y \le 0 \end{cases}$
2. $\begin{cases} y \le 4 \\ y > -2 \end{cases}$
3. $\begin{cases} y < x \\ y > -3 \\ x \le 0 \end{cases}$
4. $\begin{cases} x \le 3 \\ y < 1 \\ y > -x + 1 \end{cases}$
5. $\begin{cases} y > -1 \\ x \ge -3 \\ y \le -x + 1 \end{cases}$
6. $\begin{cases} y > -4 \\ y \le 2 \end{cases}$

(a)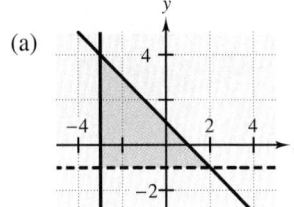

(b)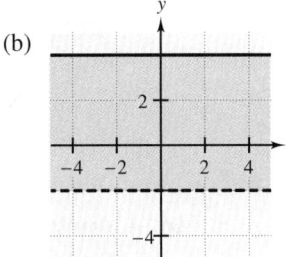

(c)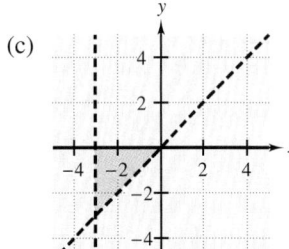

(d)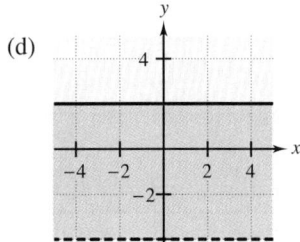

(e)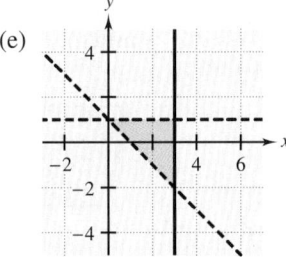

(f)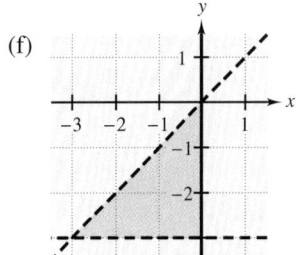

In Exercises 7–20, sketch a graph of the solution of the system of inequalities.

7. $\begin{cases} x < 3 \\ x > -2 \end{cases}$
8. $\begin{cases} y > -1 \\ y \le 2 \end{cases}$
9. $\begin{cases} y > -5 \\ x \le 2 \\ y \le x + 2 \end{cases}$
10. $\begin{cases} y \le -1 \\ x \le 2 \\ y \le x + 2 \end{cases}$

11. $\begin{cases} x + y \le 1 \\ -x + y \le 1 \\ y \ge 0 \end{cases}$
12. $\begin{cases} 3x + 2y < 6 \\ x \ge 0 \\ y \ge 0 \end{cases}$
13. $\begin{cases} x + y \le 5 \\ x \ge 2 \\ y \ge 0 \end{cases}$
14. $\begin{cases} 2x + y \ge 2 \\ x \le 2 \\ y \le 1 \end{cases}$

15. $\begin{cases} -3x + 2y < 6 \\ x - 4y > -2 \\ 2x + y < 3 \end{cases}$
16. $\begin{cases} x - 7y > -36 \\ 5x + 2y > 5 \\ 6x + 5y > 6 \end{cases}$
17. $\begin{cases} 2x + y < 2 \\ 6x + 3y > 2 \end{cases}$
18. $\begin{cases} x - 2y < -6 \\ 5x - 3y > -9 \end{cases}$

19. $\begin{cases} x \ge 1 \\ x - 2y \le 3 \\ 3x + 2y \ge 9 \\ x + y \le 6 \end{cases}$
20. $\begin{cases} x + y \le 4 \\ x + y \ge -1 \\ x - y \ge -2 \\ x - y \le 2 \end{cases}$

In Exercises 21–24, write a system of linear inequalities that describes the shaded region.

21.

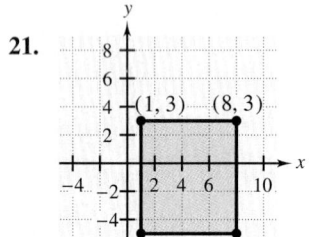

22.

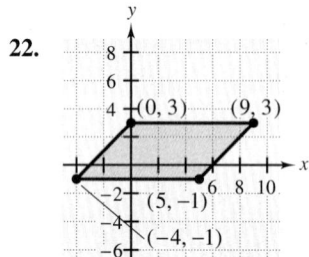

23.

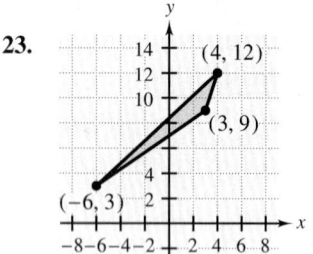

24.

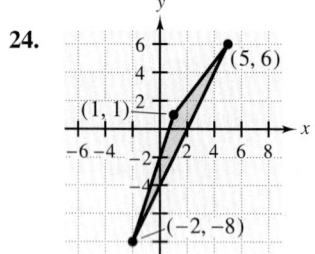

25. *Swimming Safety* The figure shows a cross section of a roped-off swimming area at a beach. Write a system of inequalities that describes the cross section. (Each unit in the coordinate system represents 1 foot.)

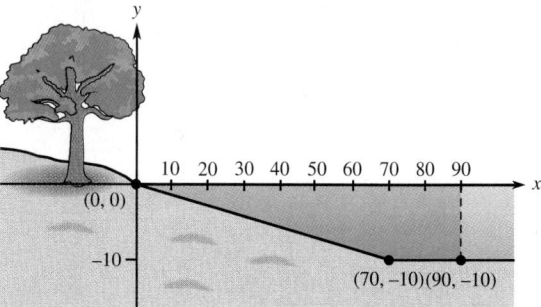

26. *A View of the Chorus* The figure shows the chorus platform on a stage. Write a system of inequalities that describes the part of the audience that can see the full chorus. (Each unit in the coordinate system represents 1 meter.)

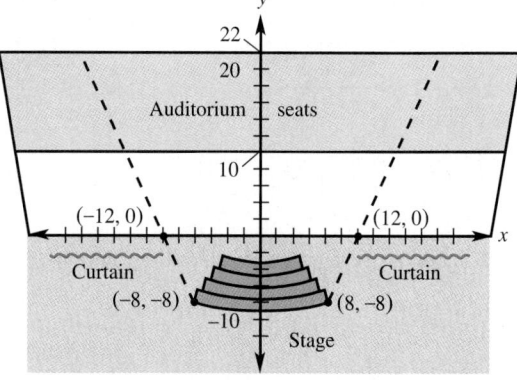

27. *Investment* A person plans to invest up to $20,000 in two different interest-bearing accounts. Each account is to contain at least $5000. Moreover, one account should have at least twice the amount that is in the other account. Write a system of inequalities to describe the various amounts that can be deposited in each account, and sketch the graph of the system.

28. *Computer Inventory* A store sells two models of a certain brand of computer. Because of the demand, it is necessary to stock at least twice as many units of model A as units of model B. The costs to the store for model A and model B are $800 and $1200, respectively. The management does not want more than $20,000 in computer inventory at any one time, and it wants at least four model A computers and two model B computers in inventory at all times. Write a system of inequalities describing all possible inventory levels, and sketch the graph of the system.

 29. *Concert Ticket Sales* Two types of tickets are to be sold for a concert. One type costs $15 per ticket and the other type costs $25 per ticket. The promoter of the concert must sell at least 15,000 tickets, including at least 8000 of the $15 tickets and at least 4000 of the $25 tickets. Moreover, the gross receipts must total at least $275,000 in order for the concert to be held. Write a system of inequalities describing the different numbers of tickets that can be sold, and sketch the graph of the system.

 30. *Diet Supplement* A dietitian is asked to design a special diet supplement using two different foods. Each ounce of food X contains 20 units of calcium, 15 units of iron, and 10 units of vitamin B. Each ounce of food Y contains 10 units of calcium, 10 units of iron, and 20 units of vitamin B. The minimum daily requirements in the diet are 280 units of calcium, 160 units of iron, and 180 units of vitamin B. Write a system of inequalities describing the different amounts of food X and food Y that can be used in the diet, and sketch the graph of the system.

In Exercises 31–42, find the minimum and maximum values of the objective function, subject to the constraints. (For each exercise, the graph of the region determined by the constraints is provided.)

31. Objective function: $C = 4x + 5y$

Constraints:
$$\begin{cases} x & \geq 0 \\ & y \geq 0 \\ x + y & \leq 6 \end{cases}$$

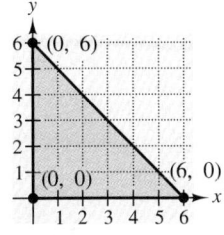

32. Objective function: $C = 2x + 8y$

Constraints:
$$\begin{cases} x & \geq 0 \\ & y \geq 0 \\ 2x + y & \leq 4 \end{cases}$$

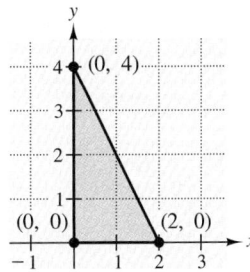

33. Objective function: $C = 10x + 6y$

Constraints: (See Exercise 31.)

34. Objective function: $C = 7x + 3y$

Constraints: (See Exercise 32.)

35. Objective function: $C = 3x + 2y$

Constraints:
$$\begin{cases} x & \geq 0 \\ & y \geq 0 \\ x + 3y & \leq 15 \\ 4x + y & \leq 16 \end{cases}$$

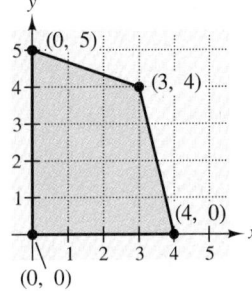

36. Objective function: $C = 4x + 3y$

Constraints:
$$\begin{cases} x & \geq & 0 \\ 2x + 3y & \geq & 6 \\ 3x - 2y & \leq & 9 \\ x + 5y & \leq & 20 \end{cases}$$

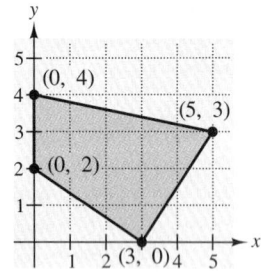

37. Objective function: $C = 5x + 0.5y$

Constraints: (See Exercise 35.)

38. Objective function: $C = x + 6y$

Constraints: (See Exercise 36.)

39. Objective function: $C = 10x + 7y$

Constraints:
$$\begin{cases} 0 \leq x \leq 60 \\ 0 \leq y \leq 45 \\ 5x + 6y \leq 420 \end{cases}$$

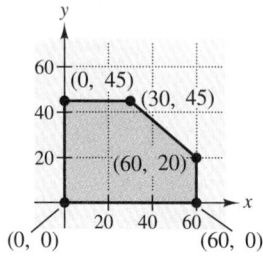

40. Objective function: $C = 50x + 35y$

Constraints:
$$\begin{cases} x & \geq & 0 \\ y & \geq & 0 \\ 8x + 9y & \leq & 7200 \\ 8x + 9y & \geq & 5400 \end{cases}$$

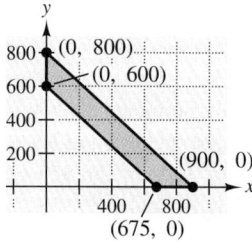

41. Objective function: $C = 25x + 35y$

Constraints: (See Exercise 39.)

42. Objective function: $C = 16x + 20y$

Constraints: (See Exercise 40.)

In Exercises 43–54, sketch the region determined by the constraints. Then find the minimum and maximum values of the objective function, subject to the constraints.

43. Objective function: $C = -2x + y$

Constraints:
$$\begin{cases} x & \geq -5 \\ x & \leq 4 \\ y \geq -1 \\ y \leq 3 \end{cases}$$

44. Objective function: $C = 5x + 4y$

Constraints:
$$\begin{cases} x & \leq 2 \\ x & \geq -4 \\ y \geq 1 \\ y \leq 6 \end{cases}$$

45. Objective function: $C = x + 4y$

Constraints:
$$\begin{cases} x & \geq 0 \\ x & \leq 12 \\ y \leq 10 \\ 2x + 3y \geq 24 \end{cases}$$

46. Objective function: $C = 6x + 2y$

Constraints: $\begin{cases} x & \geq 0 \\ x & \leq 5 \\ & y \geq 0 \\ 4x - y \geq 1 \end{cases}$

47. Objective function: $C = 6x + 10y$

Constraints: $\begin{cases} x & \geq 0 \\ & y \geq 0 \\ 2x + 5y \leq 10 \end{cases}$

48. Objective function: $C = 7x + 8y$

Constraints: $\begin{cases} x & \geq 0 \\ & y \geq 0 \\ x + \frac{1}{2}y \leq 4 \end{cases}$

49. Objective function: $C = 9x + 4y$

Constraints: (See Exercise 47.)

50. Objective function: $C = 7x + 2y$

Constraints: (See Exercise 48.)

51. Objective function: $C = 4x + 5y$

Constraints: $\begin{cases} x & \geq 0 \\ & y \geq 0 \\ 4x + 3y \geq 27 \\ x + y \geq 8 \\ 3x + 5y \geq 30 \end{cases}$

52. Objective function: $C = 4x + 5y$

Constraints: $\begin{cases} x & \geq 0 \\ & y \geq 0 \\ 2x + 2y \leq 10 \\ x + 2y \leq 6 \end{cases}$

53. Objective function: $C = 2x + 7y$

Constraints: (See Exercise 51.)

54. Objective function: $C = 2x - y$

Constraints: (See Exercise 52.)

55. *Maximum Profit* A manufacturer produces two models of bicycles. The amounts of time (in hours) required for assembling, painting, and packaging each model are as follows.

	Assembling	Painting	Packaging
Model A	2	4	1
Model B	2.5	1	0.75

The total amounts of time available for assembling, painting, and packaging are 4000 hours, 4800 hours, and 1500 hours, respectively. The profits per unit for the two models are $45 (model A) and $50 (model B). How many of each type should be produced to obtain maximum profit?

56. *Maximum Profit* A fruit grower has 150 acres of land available to raise two crops, A and B. It takes 1 day to trim an acre of crop A and 2 days to trim an acre of crop B, and there are 240 days per year available for trimming. It takes 0.3 day to pick an acre of crop A and 0.1 day to pick an acre of crop B, and there are 30 days per year available for picking. Find the number of acres of each fruit that should be planted to maximize profit, assuming that the profit is $140 per acre for crop A and $235 per acre for crop B.

57. *Maximum Revenue* An accounting firm has 900 hours of staff time and 100 hours of review time available each week. The firm charges $2000 for an audit and $300 for a tax return. Each audit requires 100 hours of staff time and 10 hours of review time. Each tax return requires 12.5 hours of staff time and 2.5 hours of review time. What numbers of audits and tax returns will generate maximum revenue?

58. *Investments* An investor has up to $450,000 to invest in two types of investments. Type A investments pay 6% annually, and type B investments pay 10% annually. To have a well-balanced portfolio, the investor imposes the following conditions. At least one-half of the total portfolio is to be allocated to type A investments and at least one-fourth of the portfolio to type B investments. How much should be allocated to each type of investment to obtain a maximum return?

59. *Minimum Cost* A farming cooperative mixes two brands of cattle feed. Brand X costs $25 per bag and contains two units of nutritional element A, two units of element B, and two units of element C. Brand Y costs $20 per bag and contains one unit of nutritional element A, nine units of element B, and three units of element C. Find the number of bags of each brand that should be mixed to produce a mixture having a minimum cost. The minimum requirements of nutrients A, B, and C are 12 units, 36 units, and 24 units, respectively.

60. *Magician* Your school has contracted with a magician to perform at the school. The school has guaranteed an attendance of at least 1000 and total ticket receipts of at least $4800. The tickets are $4 for students and $6 for nonstudents, of which the magician receives $2.50 and $4.50, respectively. What is the minimum amount of money the magician could receive?

4 REVIEW EXERCISES

In Exercises 1–6, sketch the graphs of the equations and find any solutions of the system of equations.

1. $\begin{cases} x + y = 2 \\ x - y = 0 \end{cases}$

2. $\begin{cases} 2x = 3(y - 1) \\ y = x \end{cases}$

3. $\begin{cases} x - y = 3 \\ -x + y = 1 \end{cases}$

4. $\begin{cases} x + y = -1 \\ 3x + 2y = 0 \end{cases}$

5. $\begin{cases} 2x - y = 0 \\ -x + y = 4 \end{cases}$

6. $\begin{cases} x = y + 3 \\ x = y + 1 \end{cases}$

In Exercises 7–12, solve the system of linear equations by the method of substitution.

7. $\begin{cases} 2x + 3y = 1 \\ x + 4y = -2 \end{cases}$

8. $\begin{cases} 3x - 7y = 10 \\ -2x + y = -14 \end{cases}$

9. $\begin{cases} -5x + 2y = 4 \\ 10x - 4y = 7 \end{cases}$

10. $\begin{cases} 5x + 2y = 3 \\ 2x + 3y = 10 \end{cases}$

11. $\begin{cases} 3x - 7y = 5 \\ 5x - 9y = -5 \end{cases}$

12. $\begin{cases} 24x - 4y = 20 \\ 6x - y = 5 \end{cases}$

In Exercises 13–18, solve the system of linear equations by the method of elimination or Gaussian elimination.

13. $\begin{cases} x + y = 0 \\ 2x + y = 0 \end{cases}$

14. $\begin{cases} 4x + y = 1 \\ x - y = 4 \end{cases}$

15. $\begin{cases} 2x - y = 2 \\ 6x + 8y = 39 \end{cases}$

16. $\begin{cases} 0.2x + 0.3y = 0.14 \\ 0.4x + 0.5y = 0.20 \end{cases}$

17. $\begin{cases} -x + y + 2z = 1 \\ 2x + 3y + z = -2 \\ 5x + 4y + 2z = 4 \end{cases}$

18. $\begin{cases} 2x + 3y + z = 10 \\ 2x - 3y - 3z = 22 \\ 4x - 2y + 3z = -2 \end{cases}$

In Exercises 19–24, use matrices and elementary row operations to solve the system of linear equations.

19. $\begin{cases} 5x + 4y = 2 \\ -x + y = -22 \end{cases}$

20. $\begin{cases} 2x - 5y = 2 \\ 3x - 7y = 1 \end{cases}$

21. $\begin{cases} 0.2x - 0.1y = 0.07 \\ 0.4x - 0.5y = -0.01 \end{cases}$

22. $\begin{cases} 2x + y = 0.3 \\ 3x - y = -1.3 \end{cases}$

23. $\begin{cases} x + 2y + 6z = 4 \\ -3x + 2y - z = -4 \\ 4x + 2z = 16 \end{cases}$

24. $\begin{cases} 2x_1 + 3x_2 + 3x_3 = 3 \\ 6x_1 + 6x_2 + 12x_3 = 13 \\ 12x_1 + 9x_2 - x_3 = 2 \end{cases}$

 In Exercises 25–30, solve the system of linear equations by using Cramer's Rule. (If not possible, state the reason.)

25. $\begin{cases} 7x + 12y = 63 \\ 2x + 3y = 15 \end{cases}$

26. $\begin{cases} 12x + 42y = -17 \\ 30x - 18y = 19 \end{cases}$

27. $\begin{cases} 3x - 2y = 16 \\ 12x - 8y = -5 \end{cases}$

28. $\begin{cases} 4x + 24y = 20 \\ -3x + 12y = -5 \end{cases}$

29. $\begin{cases} -x + y + 2z = 1 \\ 2x + 3y + z = -2 \\ 5x + 4y + 2z = 4 \end{cases}$

30. $\begin{cases} 2x_1 + x_2 + 2x_3 = 4 \\ 2x_1 + 2x_2 = 5 \\ 2x_1 - x_2 + 6x_3 = 2 \end{cases}$

 Area of a Triangle In Exercises 31–34, use a determinant to find the area of the triangle with the given vertices.

31. $(1, 0), (5, 0), (5, 8)$

32. $(-4, 0), (4, 0), (0, 6)$

33. $(1, 2), (4, -5), (3, 2)$

34. $\left(\frac{3}{2}, 1\right), \left(4, -\frac{1}{2}\right), (4, 2)$

Equation of a Line In Exercises 35–38, use a determinant to find the equation of the line through the given points.

35. $(-4, 0), (4, 4)$

36. $(2, 5), (6, -1)$

37. $\left(-\frac{5}{2}, 3\right), \left(\frac{7}{2}, 1\right)$

38. $(-0.8, 0.2), (0.7, 3.2)$

In Exercises 39 and 40, create a system of equations having the given solution. (*Note:* Each problem has many correct answers.)

39. $\left(\frac{2}{3}, -4\right)$

40. $(-10, 12)$

41. ***Break-Even Analysis*** A small business invests $25,000 in equipment to produce a product. Each unit of the product costs $3.75 to produce and is sold for $5.25. How many items must be sold before the business breaks even?

42. ***Dimensions of a Rectangle*** The perimeter of a rectangle is 480 meters, and its length is 150% of its width. Find the dimensions of the rectangle.

43. ***Acid Mixture*** One hundred gallons of a 60% acid solution is obtained by mixing a 75% solution with a 50% solution. How many gallons of each solution must be used to obtain the desired mixture?

44. ***Rope Length*** Suppose you must cut a rope that is 128 inches long into two pieces such that one piece is three times longer than the other. Find the length of each piece.

45. ***Cassette Tape Sales*** Suppose you are the manager of a music store and are going over receipts for the previous week's sales. Six hundred and fifty cassette tapes of two different types were sold. One type of cassette sold for $9.95, and the other sold for $14.95. The total cassette receipts were $7717.50. The cash register that was supposed to record the number of each type of cassette sold malfunctioned. Can you recover the information? If so, how many of each type of cassette were sold?

46. *Flying Speeds* Two planes leave Pittsburgh and Philadelphia at the same time, each going to the other city. Because of the wind, one plane flies 25 miles per hour faster than the other. Find the ground speed of each plane if the cities are 275 miles apart and the planes pass one another (at different altitudes) after 40 minutes of flying time.

 47. *Investments* An inheritance of $20,000 is divided among three investments yielding $1818 in interest per year. The interest rates for the three investments are 7%, 9%, and 11%. Find the amount placed in each investment if the second and third are $3000 and $1000 less than the first, respectively.

 48. *Number Problem* The sum of three positive numbers is 68. The second number is four greater than the first, and the third is twice the first. Find the three numbers.

 Curve Fitting In Exercises 49 and 50, find an equation of the parabola $y = ax^2 + bx + c$ that passes through the given points.

49. $(0, -6), (1, -3), (2, 4)$

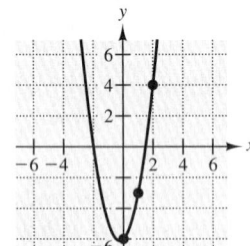

50. $(-5, 0), (1, -6), (2, 14)$

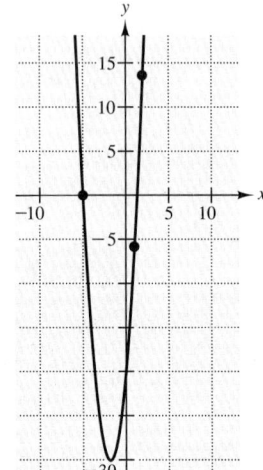

Curve Fitting In Exercises 51 and 52, find an equation of the circle $x^2 + y^2 + Dx + Ey + F = 0$ that passes through the given points.

51. $(2, 2), (5, -1), (-1, -1)$

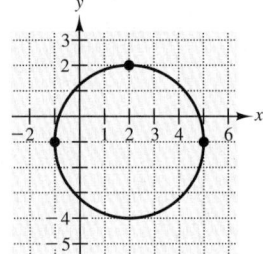

52. $(4, 2), (1, 3), (-2, -6)$

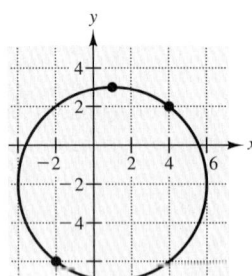

In Exercise 53–58, sketch the half-plane consisting of the solutions of the inequality.

53. $x - 2 \geq 0$

54. $y + 3 < 0$

55. $2x + y < 1$

56. $3x - 4y > 2$

57. $x \leq 4y - 2$

58. $(y - 3) \geq \frac{2}{3}(x - 5)$

In Exercises 59–62, sketch a graph of the solution set of the system of inequalities.

59. $\begin{cases} x + 2y \leq 160 \\ 3x + y \leq 180 \\ x \geq 0 \\ y \geq 0 \end{cases}$

60. $\begin{cases} 2x + 3y \leq 24 \\ 2x + y \leq 16 \\ x \geq 0 \\ y \geq 0 \end{cases}$

61. $\begin{cases} 3x + 2y \geq 24 \\ x + 2y \geq 12 \\ 2 \leq x \leq 15 \\ y \leq 15 \end{cases}$

62. $\begin{cases} 2x + y \geq 16 \\ x + 3y \geq 18 \\ 0 \leq x \leq 25 \\ 0 \leq y \leq 25 \end{cases}$

In Exercises 63 and 64, derive a set of inequalities to describe the region.

63. Parallelogram with vertices at $(1, 5)$, $(3, 1)$, $(6, 10)$, $(8, 6)$

64. Triangle with vertices at $(1, 2)$, $(6, 7)$, $(8, 1)$

 In Exercises 65 and 66, determine a system of inequalities that models the description, and sketch a graph of the solution of the system.

65. *Fruit Distribution* A Pennsylvania fruit grower has up to 1500 bushels of apples that are to be divided between markets in Harrisburg and Philadelphia. These two markets need at least 400 bushels and 600 bushels, respectively.

66. *Inventory Costs* A warehouse operator has up to 24,000 square feet of floor space in which to store two products. Each unit of product I requires 20 square feet of floor space and costs $12 per day to store. Each unit of product II requires 30 square feet of floor space and costs $8 per day to store. The total storage cost per day cannot exceed $12,400.

 In Exercises 67–70, find the required optimum value of the objective function subject to the indicated constraints.

67. Maximize the objective function:

$$C = 3x + 4y$$

Constraints: $\begin{cases} x \geq 0 \\ y \geq 0 \\ 2x + 5y \leq 50 \\ 4x + y \leq 28 \end{cases}$

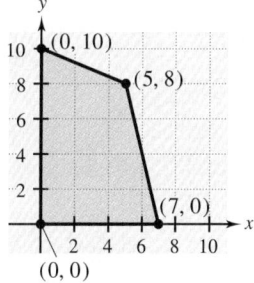

68. Minimize the objective function:

$$C = 10x + 7y$$

Constraints:
$$\begin{cases} x & \geq & 0 \\ & y \geq & 0 \\ 2x + y \geq & 100 \\ x + y \geq & 75 \end{cases}$$

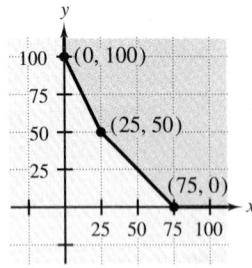

69. Minimize the objective function:

$$C = 1.75x + 2.25y$$

Constraints:
$$\begin{cases} x & \geq & 0 \\ & y \geq & 0 \\ 2x + & y \geq 25 \\ 3x + 2y \geq 45 \end{cases}$$

70. Maximize the objective function:

$$C = 50x + 70y$$

Constraints:
$$\begin{cases} x & \geq & 0 \\ & y \geq & 0 \\ x + 2y \leq & 1500 \\ 5x + 2y \leq & 3500 \end{cases}$$

71. *Maximum Profit* A manufacturer produces products A and B, yielding profits of $18 and $24 per unit, respectively. Each product must go through two processes, for which the required times per unit are shown in the table.

Process	Hours for Product A	Hours for Product B	Hours Available per Day
I	4	2	24
II	1	2	9

Find the daily production level for each unit so as to maximize profit.

72. *Minimum Cost* A pet supply company mixes two brands of dry dog food. Brand X costs $15 per bag and contains eight units of nutritional element A, one unit of nutritional element B, and two units of nutritional element C. Brand Y costs $30 per bag and contains two units of nutritional element A, one unit of nutritional element B, and seven units of nutritional element C. Each bag of dog food must contain at least 16 units, 5 units, and 20 units of nutritional elements A, B, and C, respectively. Find the number of bags of brands X and Y that should be mixed to produce a mixture meeting the minimum nutritional requirements and having a minimum cost.

Take this test as you would take a test in class. After you are done, check your work against the answers in the back of the book.

1. Determine which ordered pair is a solution of the system of equations.

$$\begin{cases} 2x - 2y = 1 \\ -x + 2y = 0 \end{cases}$$ (a) $(3, -4)$
(b) $\left(1, \frac{1}{2}\right)$

2. Solve the system of equations graphically.

$$\begin{cases} x - 2y = -1 \\ 2x + 3y = 12 \end{cases}$$

3. Solve the system of equations by the method of substitution.

$$\begin{cases} 5x - y = 6 \\ 4x - 3y = -4 \end{cases}$$

4. Solve the system of equations by the method of elimination.

$$\begin{cases} 3x - 4y = -14 \\ -3x + y = 8 \end{cases}$$

5. Solve the system of equations by the method of elimination.

$$\begin{cases} 8x + 3y = 3 \\ 4x - 6y = -1 \end{cases}$$

6. Use Gaussian elimination to solve the system of equations.

$$\begin{cases} x + 2y - 4z = 0 \\ 3x + y - 2z = 5 \\ 3x - y + 2z = 7 \end{cases}$$

7. Use matrices to solve the system of equations.

$$\begin{cases} x \qquad - 3z = -10 \\ \quad - 2y + 2z = 0 \\ x - 2y \qquad = -7 \end{cases}$$

8. Use Cramer's Rule to solve the system of equations.

$$\begin{cases} 2x - 7y = 7 \\ 3x + 7y = 13 \end{cases}$$

9. Evaluate the determinant of the matrix.

$$\begin{bmatrix} 3 & -2 & 0 \\ -1 & 5 & 3 \\ 2 & 7 & 1 \end{bmatrix}$$

10. Find a system of linear equations with integer coefficients that has the solution $(5, -3)$. (*Note:* This problem has many correct answers.)

11. Two people share the driving for a 200-mile trip. One person drives four times as far as the other. Write a system of linear equations that models this problem. Find the distance each person drives.

12. Find the equation of the parabola $y = ax^2 + bx + c$ that passes through the points $(0, 4)$, $(1, 3)$, and $(2, 6)$.

13. Use a determinant to find the area of the triangle with vertices $(0, 0)$, $(5, 4)$, and $(6, 0)$.

14. Sketch the graph of the inequality $x + 2y \leq 4$.

In Exercises 15 and 16, sketch the graph of the solution set of the system of inequalities.

15. $$\begin{cases} 3x - y < 4 \\ x \qquad > 0 \\ \quad y > 0 \end{cases}$$

16. $$\begin{cases} x + y < 6 \\ 2x + 3y > 9 \\ x \qquad \geq 0 \\ \quad y \geq 0 \end{cases}$$

17. Find the minimum and maximum values of the objective function, subject to the indicated constraints.

Objective function: $C = 5x + 11y$

Constraints:
$$\begin{cases} x & \geq & 0 \\ & y \geq & 0 \\ x + 3y & \leq & 12 \\ 3x + 2y & \leq & 15 \end{cases}$$

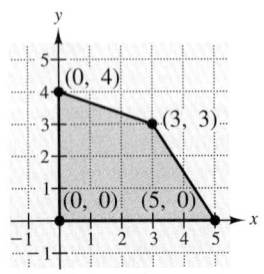

18. A manufacturer produces two models of a product. The amounts of time (in hours) required for assembling, finishing, and packaging each model are as follows.

	Assembling	Finishing	Packaging
Model A	3.5	2.5	1.3
Model B	8	2	0.9

The total amounts of time available for assembling, finishing, and packaging are 5600 hours, 2000 hours, and 900 hours, respectively. The profits per unit for the two models are $100 (model A) and $150 (model B). Find the objective function and the constraints that model the problem. (Do not solve.)

Rational Expressions and Rational Functions

5.1   Simplifying Rational Expressions

5.2   Multiplying and Dividing Rational Expressions

5.3   Adding and Subtracting Rational Expressions

5.4   Dividing Polynomials

5.5   Graphing Rational Functions

5.6   Solving Equations Involving Rational Expressions

The number of times per second n that a bird flaps its wings during normal flight is inversely proportional to its wing length x (in centimeters). This relationship can be modeled by the rational function

$$n = \frac{216}{x}.$$

Flaps per second

$n = \frac{216}{x}$

Wing length (in centimeters)

The bee hummingbird shown at right has a wing length of about 3.5 centimeters. Approximately how many times per second does it flap its wings? A California gull has a wing length of about 55 centimeters. Approximately how many times per second does it flap its wings?

Simplifying Rational Expressions

The Domain of a Rational Expression • *Simplifying Rational Expressions* • *Applications*

The Domain of a Rational Expression

A fraction whose numerator and denominator are polynomials is called a **rational expression.** Some examples are

$$\frac{3}{x+4}, \quad \frac{2x}{x^2-4x+4}, \quad \text{and} \quad \frac{x^2-5x}{x^2+2x-3}.$$

Because division by zero is undefined, the denominator of a rational expression cannot be zero. Therefore, in your work with rational expressions, you must assume that all real number values of the variable that make the denominator zero are excluded. For the three fractions above, $x = -4$ is excluded from the first fraction, $x = 2$ from the second, and both $x = 1$ and $x = -3$ from the third. The set of *usable* values of the variable is called the **domain** of the rational expression.

Emphasize the importance of determining the domain of a rational expression.

| **Definition of a Rational Expression** | Let u and v be polynomials. The algebraic expression $$\frac{u}{v}$$ is called a **rational expression.** The **domain** of this rational expression is the set of all real numbers for which $v \neq 0$. |

EXAMPLE 1 ■ Finding the Domain of a Rational Expression

Find the domain of each of the following rational expressions.

A rational expression is undefined if the *denominator* is equal to 0. Students ofent infer that this means the *variable in the denominator is 0*. Distinguish these two rules by the following examples:

(a) $\frac{4-n}{n}$

Domain: $(-\infty, 0) \cup (0, \infty)$

(b) $\frac{4-n}{n+6}$

Domain: $(-\infty, -6) \cup (-6, \infty)$

(a) $\dfrac{4}{x-2}$ (b) $\dfrac{2x+5}{8}$ (c) $3x^2 + 2x - 5$

Solution

(a) The denominator is zero when $x - 2 = 0$ or $x = 2$. Therefore, the domain is all real values of x such that $x \neq 2$. In interval notation, you can write the domain as

Domain $= (-\infty, 2) \cup (2, \infty)$.

Recall that the **union symbol** $\cup$ denotes the combining of two sets.

(b) The denominator, 8, is never zero; hence, the domain is the set of *all* real numbers. In interval notation, you can write the domain as

Domain $= (-\infty, \infty)$.

(c) Note that any polynomial is also a rational expression, because you can consider its denominator to be 1. For instance, in this case you can write

$$3x^2 + 2x - 5 = \frac{3x^2 + 2x - 5}{1}.$$

Because the denominator of this expression is not zero for any value of x, it follows that the domain is the set of all real numbers.   ■

NOTE   The domain of *any* polynomial is the set of all real numbers.

EXAMPLE 2   ■   Finding the Domain of a Rational Expression

Find the domain of each of the following rational expressions.

(a) $\dfrac{3x}{x^2 - 25}$ (b) $\dfrac{x^2 + 3x}{x^2 + 5x - 6}$

Solution

(a) The denominator of this rational expression is

$$x^2 - 25 = (x + 5)(x - 5).$$

Because this denominator is zero when $x = -5$ or when $x = 5$, the domain of the fraction is all real values of x such that $x \neq -5$ and $x \neq 5$.

(b) The denominator of this rational expression is

$$x^2 + 5x - 6 = (x + 6)(x - 1).$$

Because this denominator is zero when $x = -6$ or when $x = 1$, the domain of the fraction is all real values of x such that $x \neq -6$ and $x \neq 1$.   ■

In applications involving rational expressions, it is often necessary to restrict the domain further. To indicate such a restriction, the domain is written to the right of the fraction. For instance, the domain of the rational expression

$$\frac{x^2 + 20}{x + 4}, \qquad x > 0$$

is the set of all positive real numbers, indicated by the inequality $x > 0$.

EXAMPLE 3   ■   An Application Involving a Restricted Domain

Suppose you have started a small manufacturing business. The initial investment for the business was $120,000. The cost of each unit that you manufacture is $15. Thus, your total cost of producing x units is

$$\text{Total cost} = 15x + 120{,}000.$$

Your average cost per unit depends on the number of units produced. If you produce 100 units, your average cost per unit is

Point out that in solving applications we often have additional restrictions on the domain that are based on real-life conditions of the problem.

$$\frac{15(100) + 120{,}000}{100} = \$1215.$$

The average cost per unit decreases as the number of units increases. If you produce 1000 units, then your average cost per unit is

$$\frac{15(1000) + 120{,}000}{1000} = \$135.$$

In general, if you produce x units, your average cost per unit is

$$\text{Average cost per unit} = \frac{15x + 120{,}000}{x}.$$

What is the domain of this rational expression?

Solution

If you were simply considering the expression $(15x + 120{,}000)/x$ as a mathematical quantity, you would say that the domain is all real values of x such that $x \neq 0$. However, because this fraction is a mathematical model representing a real-life situation, you must consider which values of x make sense in real life. For this model, the variable x represents the number of units that you produce. Assuming that you cannot produce a fractional number of units, the domain is the set of positive integers. That is,

Domain $= \{1,\ 2,\ 3,\ 4,\ \ldots\}.$ ■

Simplifying Rational Expressions

The rules for operating with rational expressions are like those for operating with numerical fractions. As with numerical fractions, a rational expression is **simplified** or **in reduced form** if its numerator and denominator have no factors in common (other than ± 1). To reduce fractions, you can apply the following law.

Reduction Law for Fractions

Let u, v, and w represent numbers, variables, or algebraic expressions such that $v \neq 0$ and $w \neq 0$.

$$\frac{u\cancel{w}}{v\cancel{w}} = \frac{u}{v}$$

Be sure you see that this reduction law allows you to divide out only *factors,* not *terms*. For instance, consider the following.

$$\frac{\cancel{2} \cdot x}{\cancel{2}(x+5)} \qquad \text{You CAN divide out common } factor \text{ 2}$$

$$\frac{3+x}{3+2x} \qquad \text{You CANNOT divide out common } term \text{ 3}$$

$$\frac{\cancel{x}(x^2+6)}{\cancel{x} \cdot x} \qquad \text{You CAN divide out common } factor \text{ } x$$

$$\frac{(\cancel{x+2})(x^2+6)}{x(\cancel{x+2})} \qquad \text{You CAN divide out common } factor \text{ } (x+2)$$

$$\frac{x+4}{x} \qquad \text{You CANNOT divide out common } term \text{ } x$$

Using the reduction law to simplify a rational expression requires two steps. First, you must completely factor the numerator and denominator. Then, apply the reduction law to divide out any *factors* that are common to both the numerator and denominator. Thus, your success in simplifying rational expressions actually lies in your ability to factor completely the polynomials in both the numerator and denominator. You may want to review the factoring techniques discussed in Sections 1.5 and 1.6. For convenience, the guidelines for factoring polynomials are listed below.

Guidelines for Factoring Polynomials

1. Factor out any common factors.
2. Factor according to one of the special polynomial forms: difference of squares, sum or difference of cubes, or perfect square trinomials.
3. Factor trinomials, $ax^2 + bx + c$, using the methods for $a = 1$ or $a \neq 1$.
4. Factor by grouping—for polynomials with four terms.
5. Check to see if the factors themselves can be factored further.
6. Check the results by multiplying the factors.

EXAMPLE 4 ■ Simplifying a Rational Expression

Simplify the rational expression

$$\frac{2x^3 - 6x}{6x^2}.$$

Solution

Begin by completely factoring both the numerator and denominator. If the numerator and denominator have any common factors, divide them out to obtain the simplified form. If the numerator and denominator do not have any common factors, then the rational expression is already in simplified form.

$$\frac{2x^3 - 6x}{6x^2} = \frac{2x(x^2 - 3)}{6x^2}$$ Factor numerator and denominator

$$= \frac{2x(x^2 - 3)}{3(2x)(x)}$$ Divide out common factor $2x$

$$= \frac{x^2 - 3}{3x}$$ Simplified form ∎

EXAMPLE 5 ■ Adjusting the Domain After Simplifying

$$\frac{x^2 + 2x - 15}{4x - 12} = \frac{(x + 5)(x - 3)}{4(x - 3)}$$ Factor numerator and denominator

$$= \frac{(x + 5)(x - 3)}{4(x - 3)}$$ Divide out common factor $(x - 3)$

$$= \frac{x + 5}{4}, \quad x \neq 3$$ Simplified form

The simplification process used in this example has a tricky technicality. This technicality arises from the fact that dividing the numerator and denominator of a rational expression by their common factor can change the domain of the expression. For instance, in this example, the domain of the original expression is all real values of x such that $x \neq 3$. Thus, for you to equate the original expression to the simplified expression, you must restrict the domain of the simplified expression to exclude 3. ∎

In this text, when simplifying a rational expression, we follow the convention of listing *by the simplified expression* all values of x that must be specifically excluded from the domain in order to make the domains of the simplified and original expressions agree. For instance, in the next example the restriction $x \neq 4$ must be listed with the simplified expression in order to make the two domains agree. (Note that the value of -2 is excluded from both domains, so it is not necessary to list this value.)

EXAMPLE 6 ■ Simplifying a Rational Expression

$$\frac{x^3 - 16x}{x^2 - 2x - 8} = \frac{x(x^2 - 16)}{(x + 2)(x - 4)}$$ Partially factor

$$= \frac{x(x + 4)(x - 4)}{(x + 2)(x - 4)}$$ Factor completely

$$= \frac{x(x + 4)(x - 4)}{(x + 2)(x - 4)}$$ Divide out common factor $(x - 4)$

$$= \frac{x(x + 4)}{x + 2}, \quad x \neq 4$$ Simplified form ∎

Be sure to factor *completely* the numerator and denominator of an algebraic fraction before concluding that there is no common factor. This may involve a change in sign to see if further reduction is possible. Here are two examples.

$$x - 3 = -(3 - x) \quad \text{and} \quad 5 - 2x = -(2x - 5)$$

Watch for this in the next example.

EXAMPLE 7   ■   Simplification Involving a Change of Sign

$$\frac{2x^2 - 9x + 4}{12 + x - x^2} = \frac{(2x - 1)(x - 4)}{(4 - x)(3 + x)} \qquad \text{Factor completely}$$

$$= \frac{(2x - 1)(x - 4)}{-(x - 4)(3 + x)} \qquad (4 - x) = -(x - 4)$$

$$= \frac{(2x - 1)(x - 4)}{-(x - 4)(3 + x)} \qquad \text{Divide out common factor } (x - 4)$$

$$= -\frac{2x - 1}{3 + x}, \qquad x \neq 4 \qquad \text{Simplified form}$$

Be sure you see that when the common factor $(x - 4)$ is divided out, the minus sign stays. In the simplified form of the fraction, we usually like to move the minus sign out in front of the fraction. However, this is merely a personal preference. Any of the following forms is legitimate.

Although all of these forms are equivalent, you may want to discuss your preference with students.

$$-\frac{2x - 1}{3 + x} = \frac{-(2x - 1)}{3 + x} = \frac{-2x + 1}{3 + x} = \frac{2x - 1}{-(3 + x)} = \frac{2x - 1}{-3 - x} = \frac{1 - 2x}{3 + x} \quad ■$$

In the next two examples, the reduction law is applied to simplify rational expressions that involve more than one variable.

EXAMPLE 8   ■   Simplifying a Rational Expression Involving Two Variables

$$\frac{3xy + y^2}{2y} = \frac{y(3x + y)}{2y} \qquad \text{Factor numerator and denominator}$$

$$= \frac{y(3x + y)}{2y} \qquad \text{Divide out common factor } y$$

$$= \frac{3x + y}{2}, \qquad y \neq 0 \qquad \text{Simplified form} \qquad ■$$

EXAMPLE 9   ■   Simplifying a Rational Expression Involving Two Variables

Simplify the rational expression

$$\frac{2x^2 + 2xy - 4y^2}{5x^3 - 5xy^2}.$$

Solution

$$\frac{2x^2 + 2xy - 4y^2}{5x^3 - 5xy^2} = \frac{2(x^2 + xy - 2y^2)}{5x(x^2 - y^2)} \qquad \text{Partially factor}$$

Remind students to factor completely both denominator and numerator.

$$= \frac{2(x - y)(x + 2y)}{5x(x - y)(x + y)} \qquad \text{Completely factor}$$

$$= \frac{2(\cancel{x - y})(x + 2y)}{5x(\cancel{x - y})(x + y)} \qquad \text{Divide out common factor } (x - y)$$

$$= \frac{2(x + 2y)}{5x(x + y)}, \qquad x \neq y \quad \text{Simplified form} \qquad \blacksquare$$

NOTE As you study the examples and work the exercises in this and the following three sections, keep in mind that you are *rewriting expressions in simpler forms*. You are not solving equations. Equal signs in the steps of the simplification process only indicate that the new form of the expression is *equivalent* to the previous one.

Applications

The geometric model shown in Figure 5.1 can be used to find an algebraic model for the ratio of surface area to volume for a person or animal. The model is created from six rectangular boxes. The dimensions, surface area S and volume V of each box are listed with the figure.

$$\text{Each Arm and Leg:} \quad S = x^2 + 4(6x^2) = 25x^2$$
$$V = 6x^3$$
$$\text{Head:} \quad S = 5(4x^2) = 20x^2$$
$$V = 8x^3$$
$$\text{Trunk:} \quad S = 2(16x^2) + 4(24x^2) - 4x^2 - 4x^2 = 120x^2$$
$$V = 96x^3$$

FIGURE 5.1

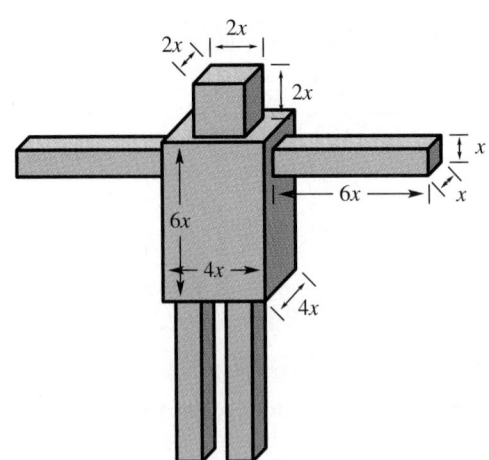

EXAMPLE 10   ■   Comparing Surface Area to Volume

Find a rational expression that represents the ratio of surface area to volume for the geometric model shown on page 380. Simplify the expression. Then evaluate the simplified expression for several values of x, where x is measured in feet.

Solution

$$\frac{\text{Surface area}}{\text{Volume}} = \frac{4(25x^2) + 20x^2 + 120x^2}{4(6x^3) + 8x^3 + 96x^3}$$

$$= \frac{240x^2}{128x^3}$$

$$= \frac{15}{8x}$$

In the following table notice that small animals have large ratios and large animals have small ratios.

Animal	Mouse	Squirrel	Cat	Human	Elephant
x	$\frac{1}{24}$	$\frac{1}{12}$	$\frac{1}{6}$	$\frac{2}{5}$	2
Ratio	45	22.5	11.25	4.7	0.9

■

The air resistance of an animal that is falling through the air depends on the ratio of the animal's surface area to its volume. The greater the ratio, the greater the air resistance. For instance, if a mouse and a human each falls 30 feet, the mouse will encounter a greater air resistance, which means that it will hit the ground at a lower speed.

DISCUSSION PROBLEM   ■   Equivalent Expressions

(a) Evaluate the two expressions for the x-values listed in the following table.

x-values	-2	-1	0	1	2	3	4
$\dfrac{x^2 - x - 6}{x - 3}$							
$x + 2$							

(b) Write a paragraph describing the equivalence (or nonequivalence) of these two expressions. Support your argument with appropriate algebra from this section *and* a discussion of the domains of the two expressions.   ■

Warm-Up

The following warm-up exercises involve skills that were covered in earlier sections. You will use these skills in the exercise set for this section.

In Exercises 1–4, find the missing real number that makes the two fractions equivalent.

1. $-\dfrac{5}{6} = \dfrac{}{24}$

2. $\dfrac{3}{7} = \dfrac{30}{}$

3. $\dfrac{4}{3} = \dfrac{-28}{}$

4. $-\dfrac{9}{16} = \dfrac{}{48}$

In Exercises 5–8, write the fraction in reduced form.

5. $\dfrac{45}{78}$ **6.** $\dfrac{55}{115}$ **7.** $\dfrac{160}{256}$ **8.** $\dfrac{189}{324}$

In Exercises 9 and 10, factor the algebraic expression.

9. $20ax^2 + 60ax + 80a$ **10.** $2x^2y^2 - 9xy^2 - 56y^2$

5.1   EXERCISES

In Exercises 1–12, find the domain of the rational expression.

1. $\dfrac{5}{x-8}$

2. $\dfrac{9}{x-13}$

3. $\dfrac{7x}{x+4}$

4. $x^4 + 2x^2 - 5$

5. $\dfrac{4}{x^2+9}$

6. $\dfrac{y^2-3}{7}$

7. $\dfrac{5t}{t^2-16}$

8. $\dfrac{z+2}{z(z-4)}$

9. $\dfrac{u^2}{u^2-4u-5}$

10. $\dfrac{y+5}{4y^2-5y-6}$

11. $\dfrac{6x+5}{x(x-2)-4(x-2)}$

12. $\dfrac{r+1}{4-(r+1)^2}$

In Exercises 13 and 14, determine the restricted domain of the mathematical model.

13. *Inventory Cost* The inventory cost when x units of a product are ordered from a supplier is

$$\frac{0.25x + 2000}{x}.$$

14. *Average Cost* The average cost for a manufacturer to produce x units of a product is

$$\frac{1.35x + 4570}{x}.$$

In Exercises 15–50, write the expression in reduced form.

15. $\dfrac{5x}{25}$

16. $\dfrac{32y}{24}$

17. $\dfrac{12y^2}{2y}$

18. $\dfrac{15z^3}{15z^3}$

19. $\dfrac{18x^2y}{15xy^4}$

20. $\dfrac{16y^2z^2}{60y^5z}$

21. $\dfrac{x^2(x-8)}{x(x-8)}$ **22.** $\dfrac{a^2b(b-3)}{b^3(b-3)^2}$ **23.** $\dfrac{2x-3}{4x-6}$ **24.** $\dfrac{y^2-81}{2y-18}$ **25.** $\dfrac{5-x}{3x-15}$ **26.** $\dfrac{x^2-36}{6-x}$

27. $\dfrac{3xy^2}{xy^2+x}$ **28.** $\dfrac{x+3x^2y}{3xy+1}$ **29.** $\dfrac{y^2-64}{5(3y+24)}$ **30.** $\dfrac{x^2-25z^2}{x+5z}$

31. $\dfrac{a+3}{a^2+6a+9}$ **32.** $\dfrac{u^2-12u+36}{u-6}$ **33.** $\dfrac{x^2-7x}{x^2-14x+49}$ **34.** $\dfrac{z^2+22z+121}{3z+33}$

35. $\dfrac{y^3-4y}{y^2+4y-12}$ **36.** $\dfrac{x^2-7x}{x^2-4x-21}$ **37.** $\dfrac{3-x}{2x^2-3x-9}$ **38.** $\dfrac{2y^2+13y+20}{2y^2+17y+30}$

39. $\dfrac{15x^2+7x-4}{15x^2+x-2}$ **40.** $\dfrac{56z^2-3z-20}{49z^2-16}$ **41.** $\dfrac{5xy+3x^2y^2}{xy^3}$ **42.** $\dfrac{4u^2v-12uv^2}{18uv}$

43. $\dfrac{3m^2-12n^2}{m^2+4mn+4n^2}$ **44.** $\dfrac{x^2+xy-2y^2}{x^2+3xy+2y^2}$ **45.** $\dfrac{x^3-8y^3}{x^2-4y^2}$ **46.** $\dfrac{x^3+27z^3}{x^2+xz-6z^2}$

47. $\dfrac{y^3+125z^3}{y^2-5yz+25z^2}$ **48.** $\dfrac{x^3+1}{x+xy+y+1}$ **49.** $\dfrac{uv-3u-4v+12}{v^2-5v+6}$ **50.** $\dfrac{mn+3m-n^2-3n}{m^2-n^2}$

In Exercises 51–54, demonstrate that the two quantities are *not* equivalent by evaluating each side when $x = 10$.

51. $\dfrac{x-4}{4} \neq x-1$ **52.** $\dfrac{x-4}{x} \neq -4$ **53.** $\dfrac{3x+2}{4x+2} \neq \dfrac{3}{4}$ **54.** $\dfrac{1-x}{2-x} \neq \dfrac{1}{2}$

In Exercises 55–58, write the fraction in reduced form. (Assume that n is a positive integer.)

55. $\dfrac{x^{n+1}-3x}{x}$ **56.** $\dfrac{x^{2n}}{x^{2n+1}+x^{2n}}$ **57.** $\dfrac{x^{2n}-4}{x^n+2}$ **58.** $\dfrac{x^{2n}+x^n-12}{x^{n+1}+4x}$

Exploratory Exercises   In Exercises 59 and 60, evaluate the two expressions for the x-values listed in the table. Comment on the domains and equivalence of the two expressions.

59.

x	-2	-1	0	1	2	3	4
$\dfrac{x^2-x-2}{x-2}$							
$x+1$							

60.

x	-2	-1	0	1	2	3	4
$\dfrac{x^2+5x}{x}$							
$x+5$							

In Exercises 61 and 62, evaluate the expression at each value of *x*. (If not possible, state the reason.)

	Expression	*Value*	
61.	$\dfrac{x-10}{4x}$	(a) $x = 10$	(b) $x = 0$
		(c) $x = -2$	(d) $x = 12$
62.	$\dfrac{x^2-4x}{x^2-9}$	(a) $x = 0$	(b) $x = 4$
		(c) $x = 3$	(d) $x = -3$

63. *Average Cost* A machine shop has a setup cost of $2500 for the production of a new product. The cost for labor and material used to produce each unit is $9.25.
   (a) Write an algebraic fraction that gives the average cost per unit when *x* units are produced.
   (b) Determine the domain of the fraction in part (a).
   (c) Find the average cost per unit when $x = 100$ units are produced.

64. *Pollution Removal* The cost in dollars of removing *p* percent of the air pollutants from the stack emissions of a utility company is given by the rational expression

$$\frac{80{,}000p}{100 - p}.$$

Determine the domain of the rational expression.

In Exercises 65 and 66, find the ratio of the shaded portion of the figure to the total area of the figure.

65.

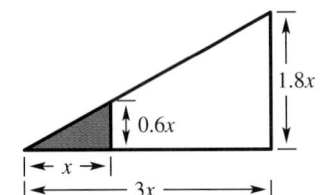

66.

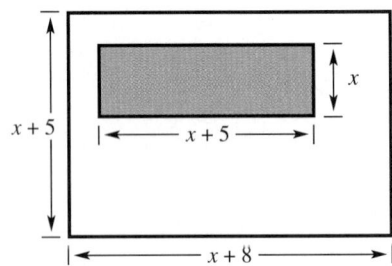

67. *Swimming Pools* One swimming pool is circular and the other is rectangular. The rectangular pool's width is three times its depth. Its length is 6 feet more than its width. The circular pool has a diameter that is twice the width of the rectangular pool, and it is 2 feet deeper. Find the ratio of the circular pool's volume to the rectangular pool's volume.

In 1989, more people went swimming than participated in any other recreational activity. (*Source:* National Sporting Goods Association)

68. *Distance Traveled* A van started on a trip and traveled at an average speed of 45 miles per hour. Three hours later, a car started on the same trip and traveled at an average speed of 60 miles per hour (see figure).
   (a) Find the distance each vehicle had traveled when the car had been on the road for t hours.
   (b) Use the result of part (a) to determine the ratio of the distance the car had traveled to the distance the van had traveled.

Figure for 68

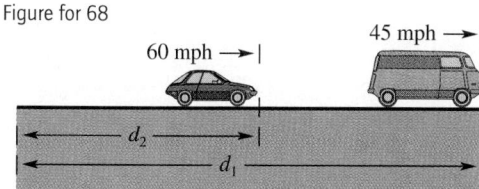

Cost of Medicare In Exercises 69 and 70, use the following polynomial models, which give the projected cost of Medicare between 1990 and 1995 and the U.S. population age 65 and older (see figure).

$$C = 107.1 + 12.64t + 0.54t^2 \quad \text{Cost of Medicare}$$

$$P = 31.6 + 0.51t - 0.14t^2 \quad \text{Population}$$

Figure for 69 and 70

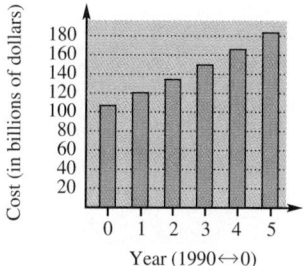

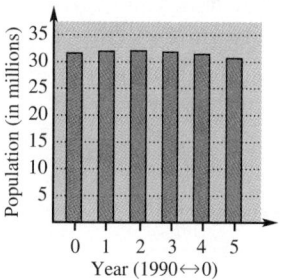

In these models, C represents the total annual cost of Medicare (in billions of dollars), P represents the U.S. population (in millions) age 65 and older, and t represents the year, with $t = 0$ corresponding to 1990. (*Sources:* Congressional Budget Office and U.S. Bureau of Census)

69. Find a rational model that represents the average cost of Medicare per person age 65 and older during the years 1990 to 1995.

70. Use the model found in Exercise 69 to complete the table showing the average cost of Medicare per person age 65 and older.

Year	Average Cost
1990	
1991	
1992	
1993	
1994	
1995	

SECTION

5.2

Multiplying and Dividing Rational Expressions

Multiplying Rational Expressions • *Dividing Rational Expressions* • *Compound Fractions*

Multiplying Rational Expressions

Remind students at the beginning that the restrictions on the original expression are $x \neq 2$ and $x \neq 4$.

The rule for multiplying rational expressions is the same as the rule for multiplying numerical fractions. That is, you *multiply numerators, multiply denominators, and write the new fraction in reduced form.* Here is an example.

$$\frac{x+3}{x-2} \cdot \frac{x-2}{x-4} = \frac{(x+3)(x-2)}{(x-2)(x-4)}$$ Multiply numerators and denominators

$$= \frac{(x+3)(x-2)}{(x-2)(x-4)}$$ Divide out common factor

$$= \frac{x+3}{x-4}, \qquad x \neq 2$$ Simplified form

Multiplying Rational Expressions

Let u, v, w, and z be real numbers, variables, or algebraic expressions such that $v \neq 0$ and $z \neq 0$.

$$\frac{u}{v} \cdot \frac{w}{z} = \frac{uw}{vz}$$

To recognize common factors in products of fractions, you should write the numerators and denominators in completely factored form, as demonstrated in the following examples.

EXAMPLE 1 ■ Multiplying Rational Expressions

Perform the following multiplication of rational expressions.

$$\frac{4x^3 y}{3x y^4} \cdot \frac{-6x^2 y^2}{10x^4}$$

Solution

$$\frac{4x^3 y}{3x y^4} \cdot \frac{-6x^2 y^2}{10x^4} = \frac{-24x^5 y^3}{30x^5 y^4}$$ Multiply numerators and denominators

$$= \frac{-4(6)(x^5)(y^3)}{5(6)(x^5)(y^3)(y)}$$ Divide out common factors

$$= -\frac{4}{5y}, \qquad x \neq 0$$ Simplified form   ■

EXAMPLE 2   ■   Multiplying Rational Expressions

Perform the following multiplication of rational expressions.

$$\frac{x}{5x^2 - 20x} \cdot \frac{x - 4}{2x^2 + x - 3}$$

Solution

$$\frac{x}{5x^2 - 20x} \cdot \frac{x - 4}{2x^2 + x - 3}$$

$$= \frac{x(x - 4)}{5x(x - 4)(x - 1)(2x + 3)} \qquad \text{Factor and multiply}$$

$$= \frac{\cancel{x}(\cancel{x - 4})(1)}{5\cancel{x}(\cancel{x - 4})(x - 1)(2x + 3)} \qquad \text{Divide out common factors}$$

$$= \frac{1}{5(x - 1)(2x + 3)}, \qquad x \neq 0, \; x \neq 4 \qquad \text{Simplified form} \qquad ■$$

EXAMPLE 3   ■   Multiplying Rational Expressions

$$\frac{4x^2 - 4x}{x^2 + 2x - 3} \cdot \frac{x^2 + x - 6}{4x}$$

$$= \frac{4x(x - 1)(x + 3)(x - 2)}{(x - 1)(x + 3)(4x)} \qquad \text{Factor and multiply}$$

$$= \frac{\cancel{4x}(\cancel{x - 1})(\cancel{x + 3})(x - 2)}{(\cancel{x - 1})(\cancel{x + 3})(\cancel{4x})} \qquad \text{Divide out common factors}$$

$$= x - 2, \qquad x \neq 0, \; x \neq 1, \; x \neq -3 \qquad \text{Simplified form} \qquad ■$$

The rule for multiplying fractions can be extended to cover products involving expressions that are not in fractional form. To do this, rewrite the nonfractional expression as a fraction with a denominator of 1. Here is a simple example.

$$\frac{x + 3}{x - 2} \cdot (5x) = \frac{x + 3}{x - 2} \cdot \frac{5x}{1}$$

$$= \frac{(x + 3)(5x)}{x - 2}$$

$$= \frac{5x(x + 3)}{x - 2}$$

EXAMPLE 4   ■   Multiplying Rational Expressions

Point out that any polynomial expression can be rewritten as a rational expression with denominator equal to 1.

$$\frac{3x}{2x^2 - 9x + 10} \cdot (2x - 5)$$

$$= \frac{3x}{2x^2 - 9x + 10} \cdot \frac{2x - 5}{1} \qquad \text{Write in fractional form}$$

$$= \frac{3x(2x - 5)}{(2x - 5)(x - 2)} \qquad \text{Factor and multiply}$$

$$= \frac{3x(2x\!\!\!\!/\,5)}{(2x\!\!\!\!/\,5)(x - 2)} \qquad \text{Divide out common factor}$$

$$= \frac{3x}{x - 2}, \qquad x \neq \frac{5}{2} \qquad \text{Simplified form} \qquad ■$$

The next two examples show how to divide out factors that differ only in sign.

EXAMPLE 5   ■   Multiplying Rational Expressions

In Examples 5 and 6, remind students that $a - b = -(b - a)$. This step is critical in the simplification process.

$$\frac{x^3 - 125}{x^2 + 5x + 25} \cdot \frac{x + 4}{10 - 2x}$$

$$= \frac{(x - 5)(x^2 + 5x + 25)}{x^2 + 5x + 25} \cdot \frac{x + 4}{2(5 - x)} \qquad \text{Factor completely}$$

$$= \frac{(x - 5)(x^2 + 5x + 25)}{x^2 + 5x + 25} \cdot \frac{x + 4}{-2(x - 5)} \qquad 2(5 - x) = -2(x - 5)$$

$$= \frac{(x - 5)(x^2 + 5x + 25)(x + 4)}{(x^2 + 5x + 25)(-2)(x - 5)} \qquad \begin{array}{l}\text{Multiply numerators and}\\ \text{denominators}\end{array}$$

$$= \frac{(x\!\!\!\!/\,5)(x^2\!\!\!\!/\,+5x\!\!\!\!/\,+25)(x + 4)}{(x^2\!\!\!\!/\,+5x\!\!\!\!/\,+25)(-2)(x\!\!\!\!/\,5)} \qquad \text{Divide out common factors}$$

$$= -\frac{x + 4}{2}, \qquad x \neq 5 \qquad \text{Simplified form} \qquad ■$$

EXAMPLE 6   ■   Multiplying Rational Expressions

$$\frac{y - x}{x^2 - y^2} \cdot \frac{x^2 - xy - 2y^2}{3x - 6y}$$

$$= \frac{y - x}{(x + y)(x - y)} \cdot \frac{(x - 2y)(x + y)}{3(x - 2y)} \qquad \text{Factor completely}$$

$$= \frac{(-1)(x - y)}{(x + y)(x - y)} \cdot \frac{(x - 2y)(x + y)}{3(x - 2y)} \qquad (y - x) = (-1)(x - y)$$

$$= \frac{(-1)(x - y)(x - 2y)(x + y)}{(x + y)(x - y)(3)(x - 2y)} \qquad \begin{array}{l}\text{Multiply numerators and}\\ \text{denominators}\end{array}$$

$$= \frac{(-1)(x-y)(x-2y)(x+y)}{(x+y)(x-y)(3)(x-2y)}$$ Divide out common factors

$$= -\frac{1}{3}, \qquad x \neq y, \ x \neq -y, \ x \neq 2y$$ Simplified form   ■

The rule for multiplying rational expressions can be extended to cover products of three or more fractions. This procedure is illustrated in Example 7.

EXAMPLE 7 ■ Multiplying Three Rational Expressions

$$\frac{x^2 - 3x + 2}{x + 2} \cdot \frac{3x}{x - 2} \cdot \frac{2x + 4}{x^2 - 5x}$$

$$= \frac{(x - 1)(x - 2)(3)(x)(2)(x + 2)}{(x + 2)(x - 2)(x)(x - 5)}$$ Factor and multiply

$$= \frac{(x - 1)(x-2)(3)(x)(2)(x+2)}{(x+2)(x-2)(x)(x - 5)}$$ Divide out common factors

$$= \frac{6(x - 1)}{x - 5}, \qquad x \neq 0, \ x \neq 2, \ x \neq -2$$ Simplified form   ■

Dividing Rational Expressions

To divide two rational expressions, you simply *invert the divisor and multiply.* That is, you multiply the first fraction by the *reciprocal* of the second. For instance, to perform the division

$$\frac{x}{x + 3} \div \frac{4}{x - 1},$$

invert the fraction $4/(x - 1)$ and multiply, as follows.

$$\frac{x}{x + 3} \div \frac{4}{x - 1} = \frac{x}{x + 3} \cdot \frac{x - 1}{4} = \frac{x(x - 1)}{(x + 3)(4)} = \frac{x(x - 1)}{4(x + 3)}, \qquad x \neq 1$$

Dividing Rational Expressions

Let u, v, w, and z be real numbers, variables, or algebraic expressions such that $v \neq 0$, $w \neq 0$, and $z \neq 0$.

$$\frac{u}{v} \div \frac{w}{z} = \frac{u}{v} \cdot \frac{z}{w} = \frac{uz}{vw}$$

EXAMPLE 8 ■ Dividing Rational Expressions

Perform the following division.

$$\frac{3x}{x + 2} \div \frac{x - 6}{x + 2}$$

Solution

$$\frac{3x}{x+2} \div \frac{x-6}{x+2} = \frac{3x}{x+2} \cdot \frac{x+2}{x-6}$$ Invert and multiply

$$= \frac{(3x)(x+2)}{(x+2)(x-6)}$$ Multiply numerators and denominators

$$= \frac{(3x)(x+2)}{(x+2)(x-6)}$$ Divide out common factor

$$= \frac{3x}{x-6}, \qquad x \neq -2$$ Simplified form ∎

EXAMPLE 9 ∎   **Dividing Rational Expressions**

$$\frac{2x}{3x-12} \div \frac{x^2-2x}{x^2-6x+8}$$

Solution

$$\frac{2x}{3x-12} \div \frac{x^2-2x}{x^2-6x+8} = \frac{2x}{3x-12} \cdot \frac{x^2-6x+8}{x^2-2x}$$ Invert and multiply

$$= \frac{(2x)(x-2)(x-4)}{(3)(x-4)(x)(x-2)}$$ Factor and multiply

$$= \frac{(2x)(x-2)(x-4)}{(3)(x-4)(x)(x-2)}$$ Divide out common factors

$$= \frac{2}{3}, \quad x \neq 0, \ x \neq 2, \ x \neq 4$$ Simplified form ∎

Compound Fractions

Problems involving the division of two rational expressions are sometimes written as **compound fractions.** The rules for dividing fractions still apply in such cases. For instance, consider the following compound fraction.

Compound fractions are given more extensive treatment in the next section where these ideas are extended to the sum or difference of rational expressions.

$$\frac{\left(\dfrac{x-1}{5}\right)}{\left(\dfrac{x+3}{x}\right)}$$

} Numerator fraction

→ Main fraction line

} Denominator fraction

(Note that for compound fractions we make the main fraction line slightly longer than the fraction lines in the numerator and denominator.) To perform the division implied by this compound fraction, invert the denominator and multiply:

$$\frac{\left(\dfrac{x-1}{5}\right)}{\left(\dfrac{x+3}{x}\right)} = \frac{x-1}{5} \cdot \frac{x}{x+3} = \frac{x(x-1)}{5(x+3)}, \qquad x \neq 0.$$

EXAMPLE 10 ■ Simplifying a Compound Fraction

Simplify the following compound fraction.

$$\frac{\left(\dfrac{x^2 + 2x - 3}{x - 3}\right)}{4x + 12}$$

Solution

Begin by converting the denominator to fraction form.

$$\frac{\left(\dfrac{x^2 + 2x - 3}{x - 3}\right)}{4x + 12} = \frac{\left(\dfrac{x^2 + 2x - 3}{x - 3}\right)}{\left(\dfrac{4x + 12}{1}\right)} \qquad \text{Rewrite denominator}$$

$$= \frac{x^2 + 2x - 3}{x - 3} \cdot \frac{1}{4x + 12} \qquad \text{Invert and multiply}$$

$$= \frac{(x - 1)(x + 3)}{(x - 3)(4)(x + 3)} \qquad \text{Factor and multiply}$$

$$= \frac{(x - 1)(x\!\!\!/\!+3)}{(x - 3)(4)(x\!\!\!/\!+3)} \qquad \text{Divide out common factor}$$

$$= \frac{x - 1}{4(x - 3)}, \qquad x \neq -3 \qquad \text{Simplified form} \qquad ■$$

DISCUSSION PROBLEM ■ You Be the Instructor

Error analysis is practiced here. Finding the error in others' work may encourage students to check their own work.

Suppose you are the instructor in an algebra course and one of your students turns in the following solution. Is the solution correct? If not, find the error(s) and write a correct solution.

$$\frac{9 - 4x^2}{6x^2 + 7x - 3} \cdot \frac{6x - 2}{2x - 3} \div \frac{3x + 1}{3x - 1}$$

$$= \frac{(3 + 2x)(3 - 2x)}{(3x + 1)(2x + 3)} \cdot \frac{2(3x - 1)}{2x - 3} \cdot \frac{3x + 1}{3x - 1}$$

$$= \frac{(2x + 3)(-1)(2x - 3)}{(3x + 1)(2x + 3)} \cdot \frac{2(3x - 1)}{(2x - 3)} \cdot \frac{3x + 1}{3x - 1}$$

$$= \frac{(2x + 3)(-1)(2x - 3)(2)(3x - 1)(3x + 1)}{(3x + 1)(2x + 3)(2x - 3)(3x - 1)}$$

$$= -2$$

Warm-Up

The following warm-up exercises involve skills that were covered in earlier sections. You will use these skills in the exercise set for this section.

In Exercises 1–6, evaluate the quantity and write your answer in reduced form.

1. $\dfrac{3}{16} \cdot \dfrac{8}{5}$

2. $\dfrac{5}{12} \cdot \dfrac{9}{75}$

3. $-\dfrac{13}{35} \cdot \dfrac{-25}{104}$

4. $\dfrac{-225}{-448} \div \dfrac{-105}{28}$

5. $\dfrac{14}{3} \div \dfrac{42}{45}$

6. $\dfrac{-8}{105} \div \dfrac{2}{3}$

In Exercises 7–10, factor the algebraic expression.

7. $x^2 - 3x + 2$

8. $2x^2 + 5x - 7$

9. $11x^2 + 6x - 5$

10. $4x^2 - 28x + 49$

5.2 EXERCISES

In Exercises 1–44, perform the indicated operation.

1. $\dfrac{7x^2}{3} \cdot \dfrac{9}{14x}$

2. $\dfrac{6}{5a} \cdot (25a)$

3. $\dfrac{8s^3}{9s} \cdot \dfrac{6s^2}{32s}$

4. $\dfrac{3x^4}{7x} \cdot \dfrac{8x^2}{9}$

5. $16u^4 \cdot \dfrac{12}{8u^2}$

6. $\dfrac{25}{8x} \cdot \dfrac{8x}{35}$

7. $\dfrac{8}{3 + 4x} \cdot (9 + 12x)$

8. $\dfrac{1 - 3x}{4} \cdot \dfrac{46}{15 - 45x}$

9. $\dfrac{8u^2v}{3u + v} \cdot \dfrac{u + v}{12u}$

10. $\dfrac{x + 25}{8} \cdot \dfrac{8}{x + 25}$

11. $\dfrac{12 - r}{3} \cdot \dfrac{3}{r - 12}$

12. $\dfrac{8 - z}{8 + z} \cdot \dfrac{z + 8}{z - 8}$

13. $\dfrac{(2x - 3)(x + 8)}{x^3} \cdot \dfrac{x}{3 - 2x}$

14. $\dfrac{x + 14}{x^3(10 - x)} \cdot \dfrac{x(x - 10)}{5}$

15. $\dfrac{6r}{r - 2} \cdot \dfrac{r^2 - 4}{33r^2}$

16. $\dfrac{5y - 20}{5y + 15} \cdot \dfrac{2y + 6}{y - 4}$

17. $\dfrac{2t^2 - t - 15}{t + 2} \cdot \dfrac{t^2 - t - 6}{t^2 - 6t + 9}$

18. $\dfrac{y^2 - 16}{2y^3} \cdot \dfrac{4y}{y^2 - 6y + 8}$

19. $(x^2 - 4y^2) \cdot \dfrac{xy}{(x - 2y)^2}$

20. $(u - 2v)^2 \cdot \dfrac{u + 2v}{u - 2v}$

21. $\dfrac{x^2 + 2xy - 3y^2}{(x + y)^2} \cdot \dfrac{x^2 - y^2}{x + 3y}$

22. $\dfrac{x - 2y}{x + 2y} \cdot \dfrac{x^2 + 4y^2}{x^2 - 4y^2}$

23. $\dfrac{xu - yu + xv - yv}{xu + yu - xv - yv} \cdot \dfrac{xu + yu + xv + yv}{xu - yu - xv + yv}$

24. $\dfrac{x^3 + 3x^2 - 4x - 12}{x^3 - 3x^2 - 4x + 12} \cdot \dfrac{x^2 - 9}{x}$

25. $\dfrac{x+5}{x-5} \cdot \dfrac{2x^2-9x-5}{3x^2+x-2} \cdot \dfrac{x^2-1}{x^2+7x+10}$

26. $\dfrac{t^2+4t+3}{2t^2-t-10} \cdot \dfrac{t}{t^2+3t+2} \cdot \dfrac{2t^2+4t^3}{t^2+3t}$

27. $\dfrac{7xy^2}{10u^2v} \div \dfrac{21x^3}{45uv}$

28. $\dfrac{25x^2y}{60x^3y^2} \div \dfrac{5x^4y^3}{16x^2y}$

29. $\dfrac{3(a+b)}{4} \div \dfrac{(a+b)^2}{2}$

30. $\dfrac{x^2+9}{5(x+2)} \div \dfrac{x+3}{5(x^2-4)}$

31. $\dfrac{(x^3y)^2}{(x+2y)^2} \div \dfrac{x^2y}{(x+2y)^3}$

32. $\dfrac{x^2-y^2}{2x^2-8x} \div \dfrac{(x-y)^2}{2xy}$

33. $\dfrac{\left(\dfrac{x^2}{12}\right)}{\left(\dfrac{5x}{18}\right)}$

34. $\dfrac{\left[\dfrac{(3u^2v)^2}{6v^3}\right]}{\left[\dfrac{(uv^3)^2}{3uv}\right]}$

35. $\dfrac{\left(\dfrac{25x^2}{x-5}\right)}{\left(\dfrac{10x}{5-x}\right)}$

36. $\dfrac{\left(\dfrac{5x}{x+7}\right)}{\left(\dfrac{10}{x^2+8x+7}\right)}$

37. $\dfrac{16x^2+8x+1}{3x^2+8x-3} \div \dfrac{4x^2-3x-1}{x^2+6x+9}$

38. $\dfrac{x^2-25}{x} \div \dfrac{x^3-5x^2}{x^2+x}$

39. $\dfrac{x(x+3)-2(x+3)}{x^2-4} \div \dfrac{x}{x^2+4x+4}$

40. $\dfrac{t^3+t^2-9t-9}{t^2-5t+6} \div \dfrac{t^2+6t+9}{t-2}$

41. $\dfrac{2x^2+5x-25}{3x^2+5x+2} \cdot \dfrac{3x^2+2x}{x+5} \div \left(\dfrac{x}{x+1}\right)^2$

42. $\dfrac{t^2-100}{4t^2} \cdot \dfrac{t^3-5t^2-50t}{t^4+10t^3} \div \dfrac{(t-10)^2}{5t}$

43. $x^3 \cdot \dfrac{x^{2n}-9}{x^{2n}+4x^n+3} \div \dfrac{x^{2n}-2x^n-3}{x}$, n is a positive integer.

44. $\dfrac{x^{n+1}-8x}{x^{2n}+2x^n+1} \cdot \dfrac{x^{2n}-4x^n-5}{x} \div x^n$, n is a positive integer.

Area In Exercises 45 and 46, determine the area of the shaded region of the figure.

45.

| $\frac{2w+3}{3}$ | $\frac{2w+3}{3}$ | $\frac{2w+3}{3}$ |

46.

| $\frac{2w-1}{2}$ | $\frac{2w-1}{2}$ |

Probability In Exercises 47–50, consider an experiment in which a marble is tossed into a rectangular box with dimensions x inches by $2x+1$ inches. The probability that the marble will come to rest in the *unshaded* portion of the box is equal to the ratio of the unshaded area to the total area of the figure. Find this probability.

47.

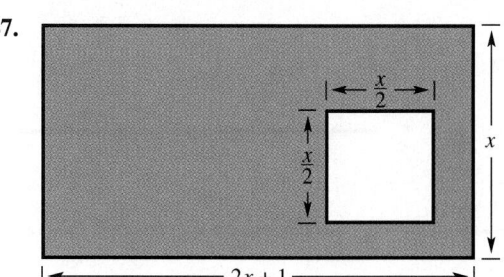

48.

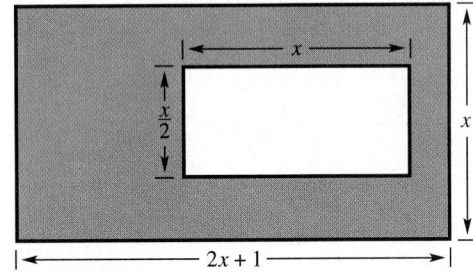

50.

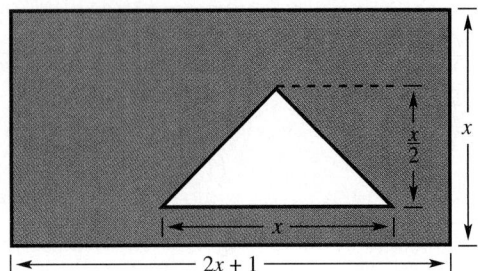

49.

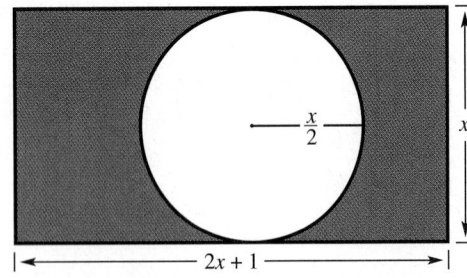

SECTION	**Adding and Subtracting Rational Expressions**
5.3	*Combining Rational Expressions with Like Denominators* • *The Least Common Multiple of Polynomials* • *Combining Rational Expressions with Unlike Denominators* • *Compound Fractions* • *Applications*

Combining Rational Expressions with Like Denominators

As with numerical fractions, the procedure used to add (or subtract) two rational expressions depends upon whether the expressions have *like* or *unlike* denominators. To add (or subtract) two rational expressions with *like* denominators, you simply combine their numerators and place the result over the common denominator. Here are some examples.

$$\frac{2}{7} + \frac{4}{7} = \frac{2+4}{7} = \frac{6}{7}$$

$$\frac{4}{3x} - \frac{5}{3x} = \frac{4-5}{3x} = -\frac{1}{3x}$$

$$\frac{3x+2}{3} - \frac{x}{3} = \frac{(3x+2)-(x)}{3} = \frac{2x+2}{3}$$

$$\frac{2x}{x+3} - \frac{4}{x+3} = \frac{2x-4}{x+3}$$

In each of these cases, one of the following rules was used.

Adding or Subtracting Rational Expressions with Like Denominators

Let u, v, and w be real numbers, variables, or variable expressions, with $w \neq 0$.

1. $\dfrac{u}{w} + \dfrac{v}{w} = \dfrac{u+v}{w}$ Add fractions with like denominators

2. $\dfrac{u}{w} - \dfrac{v}{w} = \dfrac{u-v}{w}$ Subtract fractions with like denominators

---

EXAMPLE 1 ■ **Adding Rational Expressions with Like Denominators**

Perform the following addition of rational expressions.

$$\frac{x}{4} + \frac{5-x}{4}$$

Solution

For these two rational expressions, the denominators are the same. Therefore, you can add the rational expressions by simply adding their numerators, as follows.

$$\frac{x}{4} + \frac{5-x}{4} = \frac{x + (5-x)}{4} = \frac{5}{4}$$ ■

EXAMPLE 2 ■ **Subtracting Rational Expressions with Like Denominators**

Perform the following subtraction of rational expressions.

$$\frac{7}{2x-3} - \frac{3x}{2x-3}$$

Solution

Because the denominators are the same, you can use the rule for subtracting rational expressions with like denominators, as follows.

$$\frac{7}{2x-3} - \frac{3x}{2x-3} = \frac{7-3x}{2x-3}$$ ■

After adding or subtracting two (or more) rational expressions, you should check the resulting fraction to see if it can be simplified. This procedure is illustrated in the next example.

EXAMPLE 3 ■ Subtracting Rational Expressions and Simplifying

$$\frac{x}{x^2 - 2xy - 3y^2} - \frac{3y}{x^2 - 2xy - 3y^2}$$

$$= \frac{x - 3y}{x^2 - 2xy - 3y^2} \qquad \text{Subtract numerators}$$

$$= \frac{x - 3y}{(x - 3y)(x + y)} \qquad \text{Factor completely}$$

$$= \frac{\cancel{x - 3y}(1)}{\cancel{(x - 3y)}(x + y)} \qquad \text{Divide out common factor}$$

$$= \frac{1}{x + y}, \qquad x \neq 3y \qquad \text{Simplified form} \qquad ■$$

The rules for adding and subtracting rational expressions with like denominators can be extended to cover sums and differences involving three or more rational expressions. Example 4 demonstrates this procedure.

EXAMPLE 4 ■ Combining Three Rational Expressions with Like Denominators

Encourage students to use parentheses in the first step of combining. This will help ensure that parentheses will be used for subtraction when necessary.

$$\frac{x^2 - 26}{x - 5} - \frac{2x + 4}{x - 5} + \frac{10 + x}{x - 5} = \frac{(x^2 - 26) - (2x + 4) + (10 + x)}{x - 5}$$

$$= \frac{x^2 - 26 - 2x - 4 + 10 + x}{x - 5}$$

$$= \frac{x^2 - x - 20}{x - 5}$$

$$= \frac{\cancel{(x - 5)}(x + 4)}{\cancel{x - 5}}$$

$$= x + 4, \qquad x \neq 5 \qquad ■$$

The Least Common Multiple of Polynomials

To add or subtract rational expressions with *unlike* denominators, you must first rewrite each expression using the **least common multiple** of the denominators of the individual expressions. The least common multiple of two (or more) polynomials is the simplest polynomial that is a multiple of each of the original polynomials. Here are some examples.

     Polynomials   *Least Common Multiple*

     27 and 18 $\overbrace{3 \cdot 3 \cdot 3}^{27} \cdot 2 = 3^3 \cdot 2 = 54$

$$\underbrace{}_{18}$$

Polynomials	*Least Common Multiple*
x and $x - 4$	$x \cdot (x - 4) = x(x - 4)$

x^2 and $x(x - 3)$ $\overbrace{x \cdot x}^{x^2} \cdot \underbrace{(x - 3)}_{x(x-3)} = x^2(x - 3)$

$x^2 - 4$ and $x + 2$ $\overbrace{(x + 2) \cdot (x - 2)}^{x^2 - 4} = x^2 - 4$
$\phantom{x^2 - 4 \text{ and } x + 2 \quad} \underbrace{}_{x+2}$

The examples above illustrate the following guidelines for finding the least common multiple of two or more polynomials.

Guidelines for Finding the Least Common Multiple of Polynomials

1. Factor each polynomial completely.

2. List all the *different* factors of the polynomials.

3. For each factor, determine the highest power (of that factor) that occurs in the polynomials.

4. The least common multiple is the product of the different factors, each raised to the appropriate power.

EXAMPLE 5 ■ Finding Least Common Multiples

Find the least common multiple of each of the following sets of polynomials.

(a) $6x, \ 2x^2$ (b) $x^2 - x, \ 2x - 2$ (c) $3x^2 + 6x, \ x^2 + 4x + 4$

Solution

Additional problems: Find the least common multiple of each of the following.

(a) $30y^3$ and $45y^2$

(b) $x^2 - 5x + 6$ and $x^2 - 4$

(c) $12x - 24$ and $8x^2 - 8x - 16$

Answers:

(a) $90y^3$

(b) $(x - 2)(x - 3)(x + 2)$

(c) $24(x - 2)(x + 1)$

(a) These two polynomials factor as follows.
$$6x = 2 \cdot 3 \cdot x \quad \text{and} \quad 2x^2 = 2 \cdot x^2$$
The different factors are 2, 3, and x. Using the highest powers of these factors, the least common multiple is $6x^2$.

(b) These two polynomials factor as follows.
$$x^2 - x = x(x - 1) \quad \text{and} \quad 2x - 2 = 2(x - 1)$$
The different factors are 2, x, and $x - 1$. Using the highest powers of these factors, the least common multiple is $2x(x - 1)$.

(c) These two polynomials factor as follows.
$$3x^2 + 6x = 3x(x + 2) \quad \text{and} \quad x^2 + 4x + 4 = (x + 2)^2$$
The different factors are 3, x, and $x + 2$. Using the highest powers of these factors, the least common multiple is $3x(x + 2)^2$. ■

Combining Rational Expressions with Unlike Denominators

To add or subtract rational expressions with *unlike* denominators, you must first rewrite the rational expressions so that they have *like* denominators. The (like) denominator that you use is the least common multiple of the original denominators, and it is called the **least common denominator** of the original rational expressions. Once the rational expressions have been written with like denominators, you simply add or subtract the rational expressions using the rules given at the beginning of this section. This procedure is demonstrated in the next four examples.

EXAMPLE 6   ■   Adding Rational Expressions with Unlike Denominators

Perform the following addition of rational expressions.

$$\frac{7}{6x} + \frac{5}{8x}$$

Solution

By factoring the denominators, $6x = 2 \cdot 3 \cdot x$ and $8x = 2^3 \cdot x$, you can conclude that the least common denominator is $24x$.

$$\frac{7}{6x} + \frac{5}{8x} = \frac{7(4)}{6x(4)} + \frac{5(3)}{8x(3)} \qquad \text{Rewrite fractions using least common denominator}$$

$$= \frac{28}{24x} + \frac{15}{24x} \qquad \text{Like denominators}$$

$$= \frac{28 + 15}{24x} \qquad \text{Add fractions}$$

$$= \frac{43}{24x} \qquad \text{Simplified form}$$ ■

EXAMPLE 7   ■   Subtracting Rational Expressions with Unlike Denominators

Perform the following subtraction of rational expressions.

$$\frac{3}{x - 3y} - \frac{5}{x + 2y}$$

Solution

The only factors of the denominators are $(x - 3y)$ and $(x + 2y)$. Therefore, the least common denominator is $(x - 3y)(x + 2y)$.

$$\frac{3}{x - 3y} - \frac{5}{x + 2y}$$

$$= \frac{3(x + 2y)}{(x - 3y)(x + 2y)} - \frac{5(x - 3y)}{(x - 3y)(x + 2y)} \qquad \text{Rewrite fractions using least common denominator}$$

$$= \frac{3x + 6y}{(x - 3y)(x + 2y)} - \frac{5x - 15y}{(x - 3y)(x + 2y)} \qquad \text{Like denominators}$$

$$= \frac{(3x + 6y) - (5x - 15y)}{(x - 3y)(x + 2y)}$$ Subtract fractions

$$= \frac{3x + 6y - 5x + 15y}{(x - 3y)(x + 2y)}$$ Remove parentheses

$$= \frac{-2x + 21y}{(x - 3y)(x + 2y)}$$ Simplified form   ∎

EXAMPLE 8   ∎   Adding Rational Expressions with Unlike Denominators

Perform the following addition of rational expressions.

$$\frac{6x}{x^2 - 4} + \frac{3}{2 - x}$$

Solution

The factors in the denominators are $x^2 - 4 = (x + 2)(x - 2)$ and $2 - x$. Using the fact that $2 - x = (-1)(x - 2)$, you can rewrite the second fraction by multiplying its numerator and denominator by -1.

$$\frac{6x}{x^2 - 4} + \frac{3}{2 - x} = \frac{6x}{(x + 2)(x - 2)} + \frac{(-1)3}{(-1)(2 - x)}$$

$$= \frac{6x}{(x + 2)(x - 2)} + \frac{-3}{x - 2}$$

Now, using the least common denominator of $(x + 2)(x - 2)$, you can add as follows.

$$\frac{6x}{(x + 2)(x - 2)} + \frac{-3}{x - 2}$$

$$= \frac{6x}{(x + 2)(x - 2)} + \frac{-3(x + 2)}{(x + 2)(x - 2)}$$ Rewrite fractions using least common denominator

$$= \frac{6x}{(x + 2)(x - 2)} + \frac{-3x - 6}{(x + 2)(x - 2)}$$ Like denominators

$$= \frac{6x + (-3x - 6)}{(x + 2)(x - 2)}$$ Add fractions

$$= \frac{6x - 3x - 6}{(x + 2)(x - 2)}$$ Remove parentheses

$$= \frac{3x - 6}{(x + 2)(x - 2)}$$

$$= \frac{3(x - 2)}{(x + 2)(x - 2)}$$ Divide out common factors

$$= \frac{3}{x + 2}, \quad x \neq 2$$ Simplified form   ∎

The next example shows how a least common denominator can be used to combine three rational expressions.

EXAMPLE 9 ■ **Combining Rational Expressions with Unlike Denominators**

Combine the following rational expressions by performing the indicated operations.

$$\frac{2x - 5}{6x + 9} - \frac{4}{2x^2 + 3x} + \frac{1}{x}$$

Point out that you could use any common denominator, but that the use of the *least* common denominator simplifies the process.

Solution

The denominators factor as $6x + 9 = 3(2x + 3)$, $2x^2 + 3x = x(2x + 3)$, and x. Therefore, you can conclude that the least common denominator is $3x(2x + 3)$.

$$\frac{2x - 5}{6x + 9} - \frac{4}{2x^2 + 3x} + \frac{1}{x}$$

$$= \frac{(2x - 5)(x)}{3(2x + 3)(x)} - \frac{(4)(3)}{x(2x + 3)(3)} + \frac{1(3)(2x + 3)}{(x)(3)(2x + 3)} \quad \text{Rewrite fractions using least common denominator}$$

$$= \frac{2x^2 - 5x}{3x(2x + 3)} - \frac{12}{3x(2x + 3)} + \frac{6x + 9}{3x(2x + 3)} \quad \text{Like denominators}$$

$$= \frac{2x^2 - 5x - 12 + 6x + 9}{3x(2x + 3)} \quad \text{Add fractions}$$

$$= \frac{2x^2 + x - 3}{3x(2x + 3)} \quad \text{Combine like terms}$$

$$= \frac{(x - 1)(2x + 3)}{3x(2x + 3)} \quad \text{Factor numerator}$$

$$= \frac{(x - 1)(2x + 3)}{3x(2x + 3)} \quad \text{Divide out common factor}$$

$$= \frac{x - 1}{3x}, \quad x \neq -\frac{3}{2} \quad \text{Simplified form} \qquad ■$$

Compound Fractions

Compound fractions can have numerators or denominators that are sums or differences of fractions. Here are some examples.

$$\frac{\left(\dfrac{4}{x} + 5\right)}{\left(2 - \dfrac{3}{x}\right)} \quad \text{and} \quad \frac{\left(\dfrac{1}{x + 4} - \dfrac{2}{x}\right)}{\left(\dfrac{3}{x} + \dfrac{3}{2x^2}\right)}$$

To simplify a compound fraction, begin by combining its numerator and its denominator into a single fraction. Then, divide as demonstrated in the previous section (by inverting the denominator and multiplying).

EXAMPLE 10 ■ Simplifying a Compound Fraction

$$\frac{\left(\dfrac{x}{4}+\dfrac{3}{2}\right)}{\left(2-\dfrac{3}{x}\right)} = \frac{\left(\dfrac{x}{4}+\dfrac{6}{4}\right)}{\left(\dfrac{2x}{x}-\dfrac{3}{x}\right)}$$ Find least common denominators

$$= \frac{\left(\dfrac{x+6}{4}\right)}{\left(\dfrac{2x-3}{x}\right)}$$ Add fractions in numerator and in denominator

$$= \frac{x+6}{4}\cdot\frac{x}{2x-3}$$ Invert and multiply

$$= \frac{x(x+6)}{4(2x-3)}, \qquad x \neq 0$$ Simplified form   ■

Note two methods for simplifying are shown. Students who had trouble adding and subtracting rational expressions may find this alternative method useful.

Another way to simplify the compound fraction given in Example 10 is to multiply the numerator and denominator by the least common denominator of *every* fraction in the numerator and denominator. For this fraction, notice what happens when we multiply the numerator and denominator by $4x$.

$$\frac{\left(\dfrac{x}{4}+\dfrac{3}{2}\right)}{\left(2-\dfrac{3}{x}\right)} = \frac{\left(\dfrac{x}{4}+\dfrac{3}{2}\right)}{\left(2-\dfrac{3}{x}\right)}\cdot\frac{4x}{4x} = \frac{\dfrac{x}{4}(4x)+\dfrac{3}{2}(4x)}{2(4x)-\dfrac{3}{x}(4x)} = \frac{x^2+6x}{8x-12}, \qquad x \neq 0$$

EXAMPLE 11 ■ Compound Fractions
Simplify the following compound fraction.

$$\frac{\left(\dfrac{2}{x+2}\right)}{\left(\dfrac{1}{x+2}+\dfrac{2}{x}\right)}$$

Solution

To simplify this compound fraction, multiply the numerator and denominator by the least common denominator of all fractions occurring in the numerator and denominator, which in this case is $x(x+2)$.

If we choose to combine the denominator into a single fraction, we have

$$\frac{\dfrac{2}{x+2}}{\left(\dfrac{1}{x+2}+\dfrac{2}{x}\right)} = \frac{\dfrac{2}{x+2}}{\left(\dfrac{x+2(x+2)}{x(x+2)}\right)} =$$

$$\frac{\dfrac{2}{x+2}}{\dfrac{3x+4}{x(x+2)}} = \frac{2}{(x+2)} \cdot \frac{x(x+2)}{3x+4} =$$

$$\frac{2x}{3x+4}, \quad x \neq 0,\ -2$$

Discuss which method is easier and why.

$$\frac{\left(\dfrac{2}{x+2}\right)}{\left(\dfrac{1}{x+2}+\dfrac{2}{x}\right)} = \frac{\left(\dfrac{2}{x+2}\right)(x)(x+2)}{\dfrac{1}{x+2}(x)(x+2)+\dfrac{2}{x}(x)(x+2)} \qquad \begin{array}{l}\text{Multiply numerator}\\\text{and denominator by}\\ x(x+2)\end{array}$$

$$= \frac{2x}{x+2(x+2)}$$

$$= \frac{2x}{3x+4}, \qquad x \neq -2,\ x \neq 0 \qquad \text{Simplified form} \quad ■$$

Applications

Marriage (and divorce) rates are often given in terms of the number of marriages (or divorces) per 1000 people. Since 1900, the marriage rate has remained relatively constant: the year with the lowest rate was 1958 (a rate of 8.4), and the year with the highest rate was 1945 (a rate of 12.2). The divorce rate, however, has increased greatly since 1900—from a low of 0.7 in 1900 to a high of 5.3 around 1980. (*Source:* U.S. Bureau of Census)

EXAMPLE 12   ■   Subtracting Rational Expressions

For the years 1975 through 1990, the marriage rate M (per 1000 people) and the divorce rate D (per 1000 people) can be modeled by

$$M = 9.6 + \frac{-2t+48}{t^2-25.2t+178} \qquad \text{Marriage rate}$$

and

$$D = 4.4 + \frac{2t+13}{t^2-17t+106}, \qquad \text{Divorce rate}$$

where $t = 5$ represents 1975. Find a model that gives the difference R between marriages and divorces (per 1000 people per year). (You do not need to simplify the model.)

FIGURE 5.2

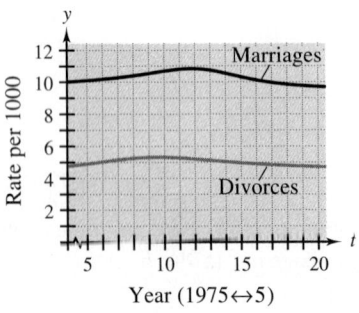

Year (1975↔5)

Solution

You can find the model for R by subtracting the model for D from the model for M.

$$R = M - D$$

$$= \left(9.6 + \frac{-2t+48}{t^2-25.2t+178}\right) - \left(4.4 + \frac{2t+13}{t^2-17t+106}\right)$$

$$= 5.2 + \frac{-2t+48}{t^2-25.2t+178} - \frac{2t+13}{t^2-17t+106}$$

The graphs of the two models are shown in Figure 5.2.   ■

DISCUSSION PROBLEM ■ You Be the Instructor

Suppose you are the instructor in an algebra course and one of your students turns in the following solution. Is the solution correct? If not, find the error(s) and write a correct solution.

$$\frac{4}{x-1} - \frac{4}{x} + \frac{x+3}{x^2-1} = \frac{4(x+1)}{x(x^2-1)} - \frac{4(x^2-1)}{x(x^2-1)} + \frac{x(x+3)}{x(x^2-1)}$$

$$= \frac{4x+4}{x(x^2-1)} - \frac{4x^2-4}{x(x^2-1)} + \frac{x^2+3x}{x(x^2-1)}$$

$$= \frac{4x+4-4x^2-4+x^2+3x}{x(x^2-1)}$$

$$= \frac{-3x^2+7x}{x(x^2-1)}$$

$$= \frac{x(-3x+7)}{x(x^2-1)}$$

$$\neq \frac{-3x+7}{x^2-1}, \qquad x \neq 0$$

■

Warm-Up

The following warm-up exercises involve skills that were covered in earlier sections. You will use these skills in the exercise set for this section.

In Exercises 1–10, perform the indicated operations.

1. $\frac{4}{9} + \frac{3}{9}$ **2.** $\frac{3}{20} + \frac{7}{20}$ **3.** $\frac{11}{32} - \frac{5}{32}$

4. $\frac{7}{10} - \frac{13}{10}$ **5.** $\frac{3}{4} + \frac{5}{18}$ **6.** $\frac{11}{24} + \frac{5}{16}$

7. $-\frac{8}{9} + \frac{18}{75}$ **8.** $\frac{27}{3} - \frac{8}{21}$ **9.** $\frac{28}{5} \cdot \frac{65}{14}$

10. $\frac{15}{16} \div 12$

5.3   EXERCISES

In Exercises 1–14, perform the indicated operation and write your answer in reduced form.

1. $\frac{x}{9} - \frac{x+2}{9}$

2. $\frac{z^2}{3} + \frac{z^2-2}{3}$

3. $\frac{4-y}{4} + \frac{3y}{4}$

4. $\frac{10x^2+1}{3} - \frac{10x^2}{3}$

5. $\frac{2}{3a} - \frac{11}{3a}$

6. $\frac{16+z}{5z} - \frac{11-z}{5z}$

7. $\dfrac{2x+5}{3} + \dfrac{1-x}{3}$

8. $\dfrac{6x}{13} - \dfrac{7x}{13}$

9. $\dfrac{3y}{3} - \dfrac{3y-3}{3} - \dfrac{7}{3}$

10. $\dfrac{-16u}{9} - \dfrac{27-16u}{9} + \dfrac{2}{9}$

11. $\dfrac{2x-1}{x(x-3)} + \dfrac{1-x}{x(x-3)}$

12. $\dfrac{5x-1}{x+4} + \dfrac{5-4x}{x+4}$

13. $\dfrac{3y-22}{y-6} - \dfrac{2y-16}{y-6}$

14. $\dfrac{7s-5}{2s+5} + \dfrac{3(s+10)}{2s+5}$

In Exercises 15–20, find the least common multiple of the pair of polynomials.

15. $5x^2$, $20x^3$

16. $14t^2$, $42t^5$

17. $15x^2$, $3(x+5)$

18. $18y^3$, $27y(y-3)^2$

19. $6(x^2-4)$, $2x(x+2)$

20. $t^3 + 3t^2 + 9t$, $2t^2(t^2-9)$

In Exercises 21–26, find the missing algebraic expression that makes the two fractions equivalent.

21. $\dfrac{7}{3y} = \dfrac{7x^2}{3y\,(\quad)}$, $x \neq 0$

22. $\dfrac{2x}{x-3} = \dfrac{14x(x-3)^2}{(x-3)\,(\quad)}$

23. $\dfrac{3u}{7v} = \dfrac{3u\,(\quad)}{7v(u+1)}$, $u \neq -1$

24. $\dfrac{3t+5}{t} = \dfrac{(3t+5)\,(\quad)}{5t^2(3t-5)}$, $t \neq \dfrac{5}{3}$

25. $\dfrac{13x}{x-2} = \dfrac{13x\,(\quad)}{4-x^2}$, $x \neq -2$

26. $\dfrac{x^2}{10-x} = \dfrac{x^2\,(\quad)}{x^2-10x}$, $x \neq 0$

In Exercises 27–32, find the least common denominator of the two fractions, and rewrite each fraction using the least common denominator.

27. $\dfrac{n+8}{3n-12}$, $\dfrac{10}{6n^2}$

28. $\dfrac{8s}{(s+2)^2}$, $\dfrac{3}{s^3+s^2-2s}$

29. $\dfrac{2}{x^2(x-3)}$, $\dfrac{5}{x(x+3)}$

30. $\dfrac{5t}{2t(t-3)^2}$, $\dfrac{4}{t(t-3)}$

31. $\dfrac{x-8}{x^2-25}$, $\dfrac{9x}{x^2-10x+25}$

32. $\dfrac{3y}{y^2-y-12}$, $\dfrac{y-4}{y^2+3y}$

In Exercises 33–72, perform the specified operation and simplify.

33. $\dfrac{5}{4x} - \dfrac{3}{5}$

34. $\dfrac{10}{b} + \dfrac{1}{10b^2}$

35. $\dfrac{7}{a} + \dfrac{14}{a^2}$

36. $\dfrac{1}{6u^2} - \dfrac{2}{9u}$

37. $\dfrac{20}{x-4} + \dfrac{20}{4-x}$

38. $\dfrac{15}{2-t} - \dfrac{7}{t-2}$

39. $\dfrac{3x}{x-8} - \dfrac{6}{8-x}$

40. $\dfrac{1}{y-6} + \dfrac{y}{6-y}$

41. $\dfrac{3x}{3x-2} + \dfrac{2}{2-3x}$

42. $\dfrac{y}{5y-3} - \dfrac{3}{3-5y}$

43. $25 + \dfrac{10}{x+4}$

44. $\dfrac{100}{x-10} - 8$

45. $-\dfrac{1}{6x} + \dfrac{1}{6(x-3)}$

46. $\dfrac{1}{x} - \dfrac{1}{x+2}$

47. $\dfrac{x}{x+3} - \dfrac{5}{x-2}$

48. $\dfrac{3}{t(t+1)} + \dfrac{4}{t}$

49. $\dfrac{3}{x+1} - \dfrac{2}{x}$

50. $\dfrac{5}{x-4} - \dfrac{3}{x}$

51. $\dfrac{3}{x-5y} + \dfrac{2}{x+5y}$

52. $\dfrac{7}{2x-3y} + \dfrac{3}{2x+3y}$

53. $\dfrac{4}{x^2} - \dfrac{4}{x^2+1}$

54. $\dfrac{2}{y} + \dfrac{1}{2y^2}$

55. $\dfrac{x}{x^2-9} + \dfrac{3}{x(x-3)}$

56. $\dfrac{x}{x^2-x-30} - \dfrac{1}{x+5}$

57. $\dfrac{4}{x-4} + \dfrac{16}{(x-4)^2}$

58. $\dfrac{3}{x-2} - \dfrac{1}{(x-2)^2}$

59. $\dfrac{4}{x} - \dfrac{2}{x^2} + \dfrac{4}{x+3}$

60. $\dfrac{5}{2(x+1)} - \dfrac{1}{2x} - \dfrac{3}{2(x+1)^2}$

61. $\dfrac{x+2}{x-1} - \dfrac{2}{x+6} - \dfrac{14}{x^2+5x-6}$

62. $\dfrac{x}{x^2+15x+50} + \dfrac{7}{2(x+10)} - \dfrac{3}{2(x+5)}$

63. $\dfrac{3}{6x^2+xy-2y^2} + \dfrac{1}{2x-y}$

64. $\dfrac{3}{3x^2+11xy-4y^2} + \dfrac{1}{(x+4y)^2}$

65. $\dfrac{4x}{1-3x+2x^2} + \dfrac{3}{x-1}$

66. $\dfrac{4x}{1-x^2} - \dfrac{3}{x-1}$

67. $\dfrac{1}{x} + \dfrac{1}{3(x-1)} - \dfrac{4x}{3x^2-2x-1}$

68. $\dfrac{2}{x-2} - \dfrac{3}{x-3} - \dfrac{3x}{5x^2-6x-8}$

69. $\dfrac{y}{x^2+xy} - \dfrac{x}{xy+y^2}$

70. $\dfrac{5}{x+y} + \dfrac{5}{x-y}$

71. $\dfrac{3u}{u^2-2uv+v^2} + \dfrac{2}{u-v}$

72. $\dfrac{1}{x} - \dfrac{3}{y} + \dfrac{3x-y}{xy}$

In Exercises 73–84, simplify the compound fraction.

73. $\dfrac{\dfrac{1}{2}}{\left(3+\dfrac{1}{x}\right)}$

74. $\dfrac{\dfrac{2}{3}}{\left(4-\dfrac{1}{x}\right)}$

75. $\dfrac{\left(\dfrac{4}{x}+3\right)}{\left(\dfrac{4}{x}-3\right)}$

76. $\dfrac{\left(\dfrac{1}{t}-1\right)}{\left(\dfrac{1}{t}+1\right)}$

77. $\dfrac{\left(16x-\dfrac{1}{x}\right)}{\left(\dfrac{1}{x}-4\right)}$

78. $\dfrac{\left(\dfrac{36}{y}-y\right)}{6+y}$

79. $\dfrac{\left(3+\dfrac{9}{x-3}\right)}{\left(4+\dfrac{12}{x-3}\right)}$

80. $\dfrac{\left(x+\dfrac{2}{x-3}\right)}{\left(x+\dfrac{6}{x-3}\right)}$

81. $\dfrac{\left(1-\dfrac{1}{y^2}\right)}{\left(1-\dfrac{4}{y}+\dfrac{3}{y^2}\right)}$

82. $\dfrac{\left(\dfrac{x+1}{x+2}-\dfrac{1}{x}\right)}{\left(\dfrac{2}{x+2}\right)}$

83. $\dfrac{\left(\dfrac{y}{x}-\dfrac{x}{y}\right)}{\left(\dfrac{x+y}{xy}\right)}$

84. $\dfrac{\left(x-\dfrac{2y^2}{x-y}\right)}{x-2y}$

85. Work Rate After two people have worked together on a common task for t hours, the fractional parts of the job done by the two workers are $t/4$ and $t/6$. What fractional part of the task has been completed?

86. Work Rate After two people have worked together on a common task for t hours, the fractional parts of the job done by the two workers are $t/3$ and $t/5$. What fractional part of the task has been completed?

87. Average of Two Numbers Determine the average of the two real numbers $x/4$ and $x/6$.

88. Average of Two Numbers Determine the average of the two real numbers $x/3$ and $x/5$.

89. Equal Parts Find three real numbers x_1, x_2, and x_3 that divide the real number line between $x/6$ and $x/2$ into four equal parts.

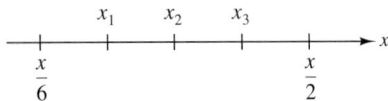

90. Equal Parts Find two real numbers x_1 and x_2 that divide the real number line between $x/5$ and $x/3$ into three equal parts.

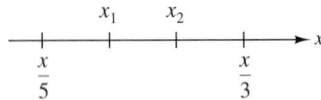

Monthly Payment In Exercises 91 and 92, use the following formula, which gives the approximate annual percentage rate r of a monthly installment loan.

$$r = \frac{\left[\dfrac{24(NM - P)}{N}\right]}{\left(P + \dfrac{NM}{12}\right)}$$

In this formula, N is the total number of payments, M is the monthly payment, and P is the amount financed.

91. (a) Approximate the annual percentage rate for a 4-year car loan of $15,000 that has monthly payments of $400.
   (b) Simplify the expression for the annual percentage rate r, and then redo part (a).

92. (a) Approximate the annual percentage rate for a 5-year car loan of $18,000 that has monthly payments of $400.
   (b) Simplify the expression for the annual percentage rate r, and then redo part (a).

93. Parallel Resistance When two resistors are connected in parallel (see figure), the total resistance is

$$\frac{1}{\left(\dfrac{1}{R_1} + \dfrac{1}{R_2}\right)}.$$

Simplify this compound fraction.

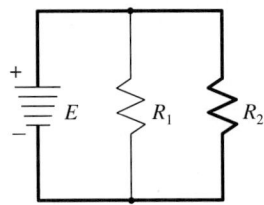

94. Determine whether the following solution is correct. If it is not, find and correct any errors.

$$\frac{2}{x} - \frac{3}{x+1} + \frac{x+1}{x^2} = \frac{2x(x+1) - 3x^2 + (x+1)^2}{x^2(x+1)}$$
$$= \frac{2x^2 + x - 3x^2 + x^2 + 1}{x^2(x+1)}$$
$$= \frac{x+1}{x^2(x+1)}$$
$$= \frac{1}{x^2}$$

95. Rewriting a Fraction The fraction $4/(x^3 - x)$ can be written as a sum of three fractions, as follows.

$$\frac{4}{x^3 - x} = \frac{A}{x} + \frac{B}{x+1} + \frac{C}{x-1}$$

The numbers A, B, and C are the solutions of the system

$$\begin{cases} A + B + C = 0 \\ \quad\ -B + C = 0 \\ -A \qquad\qquad = 4. \end{cases}$$

Solve the system and verify that the sum of the three resulting fractions is the original fraction.

Mid-Chapter Quiz

Take this quiz as you would take a quiz in class. After you are done, check your work against the answers in the back of the book.

In Exercises 1 and 2, write the expression in reduced form.

1. $\dfrac{45x^2 y}{18xy^4}$

2. $\dfrac{x-4}{4+7x-2x^2}$

In Exercises 3 and 4, find the least common multiple of the polynomials.

3. $12x^2,\ 18x(x+2),\ 9(x+2)^2$

4. $x^2-1,\ x+1,\ 5x$

In Exercises 5–10, perform the indicated operation and simplify the result.

5. $\dfrac{12}{x-4} \cdot \dfrac{x^2-16}{4x^2}$

6. $\dfrac{x^2-x-2}{2x^2 y} \div \dfrac{x^2-4x+4}{14x}$

7. $(2x-3) \div \dfrac{4x^2-9}{x+5}$

8. $\dfrac{4}{2x^2} + \dfrac{1}{3x}$

9. $\dfrac{3x^2}{x^2-x-30} + \dfrac{3x+5}{6-x}$

10. $\dfrac{10x}{3x^2-3} + \dfrac{4}{x-1} + \dfrac{5}{6x}$

In Exercises 11 and 12, simplify the compound fraction.

11. $\dfrac{\left(\dfrac{4}{x}+2\right)}{\left(\dfrac{1}{2x}-8\right)}$

12. $\dfrac{\left(\dfrac{4}{x^2-9}+\dfrac{2}{x-3}\right)}{\left(\dfrac{1}{x+3}+\dfrac{1}{x-3}\right)}$

13. Find the ratio of the area of the shaded portion of the figure to the total area of the figure.

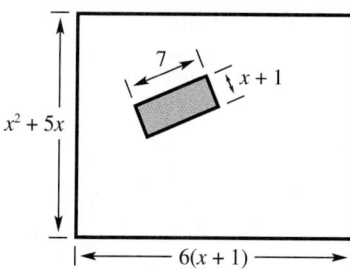

14. *Work Rate* After two people have worked together on a common task for t hours, the fractional parts of the job done by the two workers are $2t/5$ and $t/3$. What fractional part of the task has been completed?

Dividing Polynomials

Dividing a Polynomial by a Monomial   •   *Long Division*   •   *Synthetic Division*   •   *Factoring and Division*

Dividing a Polynomial by a Monomial

To divide a polynomial by a monomial, you can reverse the procedure used to add (or subtract) two rational expressions. For example, you can add the expressions 2 and $1/x$ as follows.

$$2 + \frac{1}{x} = \frac{2x}{x} + \frac{1}{x} = \frac{2x + 1}{x}$$

By reversing this process, you can divide the polynomial $2x + 1$ by the monomial x. That is,

$$\frac{2x + 1}{x} = \frac{2x}{x} + \frac{1}{x} = 2 + \frac{1}{x}.$$

Dividing a Polynomial by a Monomial

Let u, v, and w be real numbers, variables, or algebraic expressions such that $w \neq 0$.

1. $\dfrac{u + v}{w} = \dfrac{u}{w} + \dfrac{v}{w}$ 2. $\dfrac{u - v}{w} = \dfrac{u}{w} - \dfrac{v}{w}$

You might compare these rules to those at the beginning of Section 5.3.

Note that when a polynomial is divided by a monomial, the resulting expressions are usually reduced to simplest form.

EXAMPLE 1   ■   Dividing a Polynomial by a Monomial

Perform the following division and simplify.

$$\frac{12x^2 - 20x + 8}{4x}$$

Solution

$$\frac{12x^2 - 20x + 8}{4x} = \frac{12x^2}{4x} - \frac{20x}{4x} + \frac{8}{4x}$$

$$= \frac{3(4x)(x)}{4x} - \frac{5(4x)}{4x} + \frac{2(4)}{4x}$$

$$= 3x - 5 + \frac{2}{x}$$

Note that the result is not a polynomial.

■

EXAMPLE 2   ■   Dividing a Polynomial by a Monomial

Perform the following division and simplify.

$$\frac{5y^3 + 8y^2 - 10y}{5y^2}$$

Solution

$$\frac{5y^3 + 8y^2 - 10y}{5y^2} = \frac{5y^3}{5y^2} + \frac{8y^2}{5y^2} - \frac{10y}{5y^2}$$

$$= \frac{5y^2(y)}{5y^2} + \frac{8y^2}{5y^2} - \frac{2(5y)}{5y(y)}$$

$$= y + \frac{8}{5} - \frac{2}{y}$$

■

EXAMPLE 3   ■   Dividing a Polynomial by a Monomial

Perform the following division and simplify.

$$\frac{-12x^3y^2 + 6x^2y + 9y^3}{3xy^2}$$

Solution

$$\frac{-12x^3y^2 + 6x^2y + 9y^3}{3xy^2} = -\frac{12x^3y^2}{3xy^2} + \frac{6x^2y}{3xy^2} + \frac{9y^3}{3xy^2}$$

$$= -\frac{4(3x)(x^2)(y^2)}{(3x)(y^2)} + \frac{2(3x)(x)(y)}{3x(y)(y)} + \frac{3y(3y^2)}{x(3y^2)}$$

$$= -4x^2 + \frac{2x}{y} + \frac{3y}{x}$$

■

Long Division

In Section 5.1, you saw how to divide one polynomial by another by factoring and dividing out common factors. For instance, you can divide $(x^2 - 2x - 3)$ by $(x - 3)$ as follows.

$$(x^2 - 2x - 3) \div (x - 3) = \frac{x^2 - 2x - 3}{x - 3} = \frac{(x + 1)(x - 3)}{x - 3} = x + 1, \qquad x \neq 3$$

This procedure works well for polynomials that are easily factored. To divide other polynomials, you can use a more *general* procedure, which follows a "long division algorithm." This algorithm is similar to that used for dividing positive integers. For instance, the following diagram shows how a division algorithm is used to divide 6584 by 28.

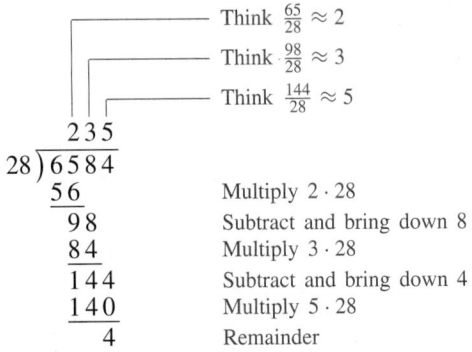

$$
\begin{array}{c}
\quad\ \ \text{Think } \tfrac{65}{28} \approx 2 \\
\quad\ \ \text{Think } \tfrac{98}{28} \approx 3 \\
\quad\ \ \text{Think } \tfrac{144}{28} \approx 5
\end{array}
$$

$$
\begin{array}{r}
235 \\
28\,\overline{)6584} \\
56 \\
\overline{98} \\
84 \\
\overline{144} \\
140 \\
\overline{4}
\end{array}
$$

Multiply $2 \cdot 28$

Subtract and bring down 8

Multiply $3 \cdot 28$

Subtract and bring down 4

Multiply $5 \cdot 28$

Remainder

The result of this division can be written as

$$
6584 \div 28 = 235 + \frac{4}{28} = 235 + \frac{1}{7}.
$$

In this division problem, the number 6854 is called the **dividend,** 28 is called the **divisor,** 235 is called the **quotient,** and 4 is called the **remainder.**

The next several examples show how the long division algorithm can be extended to cover division of one polynomial by another.

EXAMPLE 4   ■   Long Division Algorithm for Polynomials

Use the long division algorithm to perform the following division.

$$
(x^2 + 2x + 4) \div (x - 1)
$$

Solution

$$
\begin{array}{c}
\text{Think } \tfrac{x^2}{x} = x \\
\text{Think } \tfrac{3x}{x} = 3
\end{array}
$$

$$
\begin{array}{r}
x + 3 \\
x - 1\,\overline{)x^2 + 2x + 4} \\
x^2 - x \\
\overline{3x + 4} \\
3x - 3 \\
\overline{7}
\end{array}
$$

Multiply $x(x - 1)$

Subtract and bring down 4

Multiply $3(x - 1)$

Subtract

Considering the remainder as a fractional part of the divisor, you can write the result as

$$
\underbrace{\frac{\overbrace{x^2 + 2x + 4}^{\text{Dividend}}}{\underbrace{x - 1}_{\text{Divisor}}}}_{} = \overbrace{x + 3}^{\text{Quotient}} + \frac{\overbrace{7}^{\text{Remainder}}}{\underbrace{x - 1}_{\text{Divisor}}}.
$$

■

You can check a long division problem by multiplying. For instance, you can check the results of Example 4 as follows.

$$\frac{x^2 + 2x + 4}{x - 1} \stackrel{?}{=} x + 3 + \frac{7}{x - 1} \qquad \text{Result to be checked}$$

$$x^2 + 2x + 4 \stackrel{?}{=} (x + 3)(x - 1) + 7 \qquad \text{Multiply both sides by } (x - 1)$$

$$x^2 + 2x + 4 \stackrel{?}{=} (x^2 + 2x - 3) + 7$$

$$x^2 + 2x + 4 = x^2 + 2x + 4 \qquad \text{Result checks}$$

EXAMPLE 5   ■   Long Division Algorithm for Polynomials

Use the long division algorithm to perform the following division.

$$(3x^3 + 10x^2 + 6x - 4) \div (x + 2)$$

Solution

$$
\begin{array}{r}
\text{Think } \frac{3x^3}{x} = 3x^2 \\
\text{Think } \frac{4x^2}{x} = 4x \\
\text{Think } -\frac{2x}{x} = -2 \\
\end{array}
$$

$$
\begin{array}{r}
3x^2 + 4x - 2 \\
x + 2 \overline{\smash{\big)}\, 3x^3 + 10x^2 + 6x - 4} \\
\underline{3x^3 + 6x^2} \\
4x^2 + 6x \\
\underline{4x^2 + 8x} \\
-2x - 4 \\
\underline{-2x - 4} \\
0 \\
\end{array}
$$

Multiply $3x^2(x + 2)$
Subtract and bring down $6x$
Multiply $4x(x + 2)$
Subtract and bring down -4
Multiply $-2(x + 2)$
Remainder

Thus, you have

$$\frac{3x^3 + 10x^2 + 6x - 4}{x + 2} = 3x^2 + 4x - 2, \qquad x \neq -2.$$

Note that the remainder in this problem is zero. In such cases the denominator (or divisor) **divides evenly** into the numerator (or dividend).

Check

You can check this result by multiplying, as follows.

$$
\begin{array}{r}
3x^2 + 4x - 2 \\
\times \qquad\qquad x + 2 \\
\hline
6x^2 + 8x - 4 \\
3x^3 + 4x^2 - 2x \\
\hline
3x^3 + 10x^2 + 6x - 4 \\
\end{array}
$$

■

Errors are frequently made in the subtraction steps of the long division algorithm, so be especially careful each time you subtract. When using the long division algorithm for polynomials, be sure that both the divisor and the dividend are written in standard form before you begin the division process. This procedure is illustrated in Example 6.

EXAMPLE 6 ■ Writing in Standard Form Before Dividing

Use the long division algorithm to perform the following division.

$$(-13x^3 + 10x^4 + 8x - 7x^2 + 4) \div (3 - 2x)$$

Solution

Begin by writing the divisor and dividend in *standard polynomial form* (with decreasing powers of x).

$$
\begin{array}{r}
-5x^3 - x^2 + 2x - 1 \\
-2x+3 \overline{)\ 10x^4 - 13x^3 - 7x^2 + 8x + 4} \\
\underline{10x^4 - 15x^3} \\
2x^3 - 7x^2 \\
\underline{2x^3 - 3x^2} \\
-4x^2 + 8x \\
\underline{-4x^2 + 6x} \\
2x + 4 \\
\underline{2x - 3} \\
7
\end{array}
$$

Thus, you have

$$\frac{10x^4 - 13x^3 - 7x^2 + 8x + 4}{-2x+3} = -5x^3 - x^2 + 2x - 1 + \frac{7}{-2x+3}.$$

Check this result by multiplying. ■

When the dividend is missing some powers of x, the long division algorithm requires that you account for the missing powers, as shown in Example 7.

EXAMPLE 7 ■ Accounting for Missing Powers of x

Divide $(x^3 - 2)$ by $(x - 1)$.

Solution

Because there are no x^2-terms or x-terms in the dividend, you line up the subtractions by using *zero* coefficients (or by leaving spaces) for these missing terms.

$$
\require{enclose}
\begin{array}{r}
x^2 + \ x + 1 \\
x - 1 \enclose{longdiv}{x^3 + 0x^2 + 0x - 2} \\
\underline{x^3 - \ x^2} \\
x^2 + 0x \\
\underline{x^2 - \ x} \\
x - 2 \\
\underline{x - 1} \\
-1
\end{array}
$$

Thus, you can write

$$\frac{x^3 - 2}{x - 1} = x^2 + x + 1 - \frac{1}{x - 1}.$$ ∎

In each of the long division examples presented so far, the divisor has been a first-degree polynomial. The long division algorithm works just as well with polynomial divisors of degree 2 or more, as shown in Example 8.

EXAMPLE 8   ■   A Second-Degree Divisor

Use the long division algorithm to perform the following division.

$$\frac{x^4 + 6x^3 + 6x^2 - 10x - 3}{x^2 + 2x - 3}$$

Solution

$$
\require{enclose}
\begin{array}{r}
x^2 + \ 4x + 1 \\
x^2 + 2x - 3 \enclose{longdiv}{x^4 + 6x^3 + 6x^2 - 10x - 3} \\
\underline{x^4 + 2x^3 - 3x^2} \\
4x^3 + 9x^2 - 10x \\
\underline{4x^3 + 8x^2 - 12x} \\
x^2 + \ 2x - 3 \\
\underline{x^2 + \ 2x - 3} \\
0
\end{array}
$$

Thus, $(x^2 + 2x - 3)$ divides evenly into $(x^4 + 6x^3 + 6x^2 - 10x - 3)$, and you have

$$\frac{x^4 + 6x^3 + 6x^2 - 10x - 3}{x^2 + 2x - 3} = x^2 + 4x + 1, \qquad x \neq 1, \ x \neq -3.$$ ∎

Synthetic Division

There is a nice shortcut for division by polynomials of the form $(x - k)$. This shortcut is called **synthetic division**, and it is outlined for a third-degree polynomial as follows.

Synthetic Division for a Third-Degree Polynomial

To use synthetic division to divide $(ax^3 + bx^2 + cx + d)$ by $(x - k)$, use the following pattern.

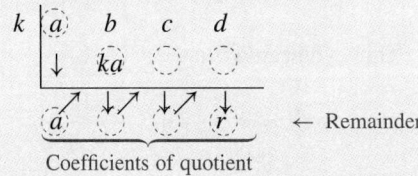

Vertical pattern: Add terms.
Diagonal pattern: Multiply by k.

Be sure you see that synthetic division works *only* for divisors of the form $(x - k)$. Moreover, the degree of the quotient is always one less than the degree of the dividend. This process is demonstrated in the next two examples.

EXAMPLE 9 ■ Using Synthetic Division

Use synthetic division to divide $(x^3 + 3x^2 - 4x - 10)$ by $(x - 2)$.

Solution

The coefficients of the dividend form the top row of the synthetic division table, and, because you are dividing by $(x - 2)$, you should write 2 at the upper left in the table. To begin the algorithm, bring down the first coefficient. Then, multiply this coefficient by 2, write the result in the second row of the table, and add the two numbers in the second column, as follows.

$$
\begin{array}{r|rrrr}
2 & 1 & 3 & -4 & -10 \\
  &   & 2 &   &   \\
\hline
  & 1 & 5 &   &
\end{array}
$$

By continuing this pattern, you obtain the following completed synthetic division table.

$$
\begin{array}{r|rrrr}
2 & 1 & 3 & -4 & -10 \\
  &   & 2 & 10 & 12 \\
\hline
  & 1 & 5 & 6 & 2
\end{array}
$$
← Remainder

Finally, from the bottom row of the table, you can determine that the quotient is $(1)x^2 + (5)x + (6)$ and the remainder is 2. Thus, the result of the division problem is

$$\frac{x^3 + 3x^2 - 4x - 10}{x - 2} = x^2 + 5x + 6 + \frac{2}{x - 2}.$$

■

If the remainder in a synthetic division problem turns out to be zero, you can conclude that the divisor divides *evenly* into the dividend, as shown in Example 10.

EXAMPLE 10   ■   **Using Synthetic Division**

Use synthetic division to divide $(x^4 - 10x^2 - 2x + 3)$ by $(x + 3)$.

Solution

The coefficients of the dividend form the top row of the synthetic division table, and, because you are dividing by $(x + 3)$, you should write -3 at the upper left in the table. The completed synthetic division table is as follows. (Note that a zero is included in place of the missing term in the dividend.)

$$
\begin{array}{r|rrrrr}
-3 & 1 & 0 & -10 & -2 & 3 \\
   &   & -3 & 9 & 3 & -3 \\
\hline
   & 1 & -3 & -1 & 1 & \textcircled{0}
\end{array}
$$
$\leftarrow$ Remainder

Because the remainder is zero, you can conclude that

$$\frac{x^4 - 10x^2 - 2x + 3}{x + 3} = x^3 - 3x^2 - x + 1, \qquad x \neq -3.$$

■

Factoring and Division

One of the uses of synthetic division (or long division) is in factoring of polynomials. This is demonstrated in Example 11.

EXAMPLE 11   ■   **Factoring a Polynomial**

Factor completely the polynomial

$$x^3 - 7x + 6.$$

Use the fact that $(x - 1)$ is one of the factors.

Solution

Remind students about the placeholder to account for missing powers of x.

Because you are given one of the factors, you should divide this factor into the given polynomial. (In this case we use synthetic division, but long division could have been used also.)

$$
\begin{array}{r|rrrr}
1 & 1 & 0 & -7 & 6 \\
 & & 1 & 1 & -6 \\
\hline
 & 1 & 1 & -6 & \boxed{0} \leftarrow \text{Remainder}
\end{array}
$$

Remind students to factor completely.

Thus, you can conclude that

$$\frac{x^3 - 7x + 6}{x - 1} = x^2 + x - 6, \qquad x \neq 1$$

which implies that

$$x^3 - 7x + 6 = (x - 1)(x^2 + x - 6).$$

Thus, you have used division to factor the polynomial $(x^3 - 7x + 6)$ into the product of a first-degree polynomial and a second-degree polynomial. To complete the factorization, factor the second-degree polynomial as follows.

$$
\begin{aligned}
x^3 - 7x + 6 &= (x - 1)(x^2 + x - 6) \\
&= (x - 1)(x + 3)(x - 2)
\end{aligned}
$$
∎

DISCUSSION PROBLEM ■   **Factoring and Problem Solving**

Consider the following function.

$$f(x) = 2x^4 + 7x^3 - 4x^2 - 27x - 18 = (x - 2)(x + 3)(2x + 3)(x + 1)$$

(a) Use the given complete factorization to find all the solutions of the following polynomial equation.

$$2x^4 + 7x^3 - 4x^2 - 27x - 18 = 0$$

(b) Evaluate the given function at the x-values listed in the following table.

x	-3	-2	$-\frac{3}{2}$	-1	0	2	3
$f(x)$							

(c) Use synthetic division to divide $f(x)$ by the divisors in the following table. In each case, list the remainder in the table.

x	-3	-2	$-\frac{3}{2}$	-1	0	2	3
Divisors	$x + 3$	$x + 2$	$x + \frac{3}{2}$	$x + 1$	x	$x - 2$	$x - 3$
Remainders							

(d) Compare the function values in the first table with the remainders in the second table. What conclusion can you draw? ■

■

Warm-Up

The following warm-up exercises involve skills that were covered in earlier sections. You will use these skills in the exercise set for this section.

In Exercises 1–6, rewrite the fraction in reduced form.

1. $\dfrac{36}{144}$ **2.** $\dfrac{58}{180}$ **3.** $\dfrac{-16x^2}{12x}$

4. $\dfrac{5t^4}{45t^2}$ **5.** $\dfrac{6u^2v}{27uv^3}$ **6.** $\dfrac{14r^4s^2}{-98rs^2}$

In Exercises 7 and 8, find the indicated product.

7. $-3x^3(5x - 4)$ **8.** $10y(7y^2 - 2)$

In Exercises 9 and 10, perform the indicated subtraction.

9. $(3x^3 + 5x^2 - 2x - 1) - (x^3 - 2x^2 + x + 3)$

10. $(7x^2 + 6) - (-x^2 - 10x + 3)$

5.4   EXERCISES

In Exercises 1–54, perform the specified division.

1. $\dfrac{6z + 10}{2}$ **2.** $\dfrac{9x + 12}{3}$ **3.** $\dfrac{10z^2 + 4z - 12}{4}$ **4.** $\dfrac{4u^2 + 8u - 24}{16}$

5. $(7x^3 - 2x^2) \div x$ **6.** $(6a^2 + 7a) \div a$ **7.** $\dfrac{50z^3 + 30z}{-5z}$ **8.** $\dfrac{18c^4 - 24c^2}{-6c}$

9. $\dfrac{8z^3 + 3z^2 - 2z}{2z}$ **10.** $\dfrac{6x^4 + 8x^3 - 18x^2}{3x^2}$ **11.** $\dfrac{m^4 + 2m^2 - 7}{m}$ **12.** $\dfrac{l^2 - 8}{-l}$

13. $(5x^2y - 8xy + 7xy^2) \div 2xy$ **14.** $(-14s^4t^2 + 7s^2t^2 - 18t) \div 2s^2t$ **15.** $\dfrac{x^2 + 6x + 5}{x + 5}$

16. $\dfrac{r^2 - 10r + 9}{r - 9}$ **17.** $\dfrac{x^2 - 8x + 15}{x - 3}$ **18.** $\dfrac{t^2 - 18t + 72}{t - 6}$

19. $(x^2 + 15x + 50) \div (x + 5)$ **20.** $(y^2 - 6y - 16) \div (y + 2)$ **21.** Divide $21 - 4x - x^2$ by $3 - x$.

22. Divide $5 + 4x - x^2$ by $1 + x$.   **23.** Divide $2y^2 + 7y + 3$ by $2y + 1$.   **24.** Divide $10t^2 - 7t - 12$ by $2t - 3$.

25. $(12t^2 - 40t + 25) \div (2t - 5)$ **26.** $(15 - 14u - 8u^2) \div (5 + 2u)$ **27.** $\dfrac{16x^2 - 1}{4x + 1}$

28. $\dfrac{81y^2 - 25}{9y - 5}$ **29.** $\dfrac{x^3 + 125}{x + 5}$ **30.** $\dfrac{x^3 - 27}{x - 3}$

31. $\dfrac{x^3 - 2x^2 + 4x - 8}{x - 2}$

32. $\dfrac{x^3 - 28x - 48}{x + 4}$

33. $(2x + 9) \div (x + 2)$

34. $(12x - 5) \div (2x + 3)$

35. $\dfrac{x^2 + 16}{x + 4}$

36. $\dfrac{y^2 + 8}{y + 2}$

37. $\dfrac{5x^2 + 2x + 3}{x + 2}$

38. $\dfrac{2x^2 + 5x + 2}{x + 4}$

39. $\dfrac{12x^2 + 17x - 5}{3x + 2}$

40. $\dfrac{8x^2 + 2x + 3}{4x - 1}$

41. $\dfrac{6z^2 + 7z}{5z - 1}$

42. $\dfrac{2y^2 + 5y}{3y + 5}$

43. $\dfrac{2x^3 - 5x^2 + x - 6}{x - 3}$

44. $\dfrac{5x^3 + 3x^2 + 12x + 20}{x + 1}$

45. Divide $x^6 - 1$ by $x - 1$.

46. Divide x^3 by $x - 1$.　　**47.** $(x^3 + 4x^2 + 7x + 6) \div (x^2 + 2x + 3)$　　**48.** $(2x^3 + 2x^2 - 2x - 15) \div (2x^2 + 4x + 5)$

49. $\dfrac{x^3 + 7x^2 + 10x - 8}{x^2 + 3x - 2}$

50. $\dfrac{4x^3 - 25x^2 + 10x - 1}{x^2 - 6x + 1}$

51. $\dfrac{x^4 - 2x^3 - 11x^2 + 4x + 15}{x^2 + x - 3}$

52. $\dfrac{2x^4 - 9x^2 + 22x - 15}{x^2 + 2x - 3}$

53. $\dfrac{x^{3n} + 3x^{2n} + 6x^n + 8}{x^n + 2}$, n is a positive integer

54. $\dfrac{x^{3n} - x^{2n} + 5x^n - 5}{x^n - 1}$, n is a positive integer

In Exercises 55–66, use synthetic division to divide the polynomials. Then use the result to write the dividend in factored form.

55. $\dfrac{x^2 - 15x + 56}{x - 8}$

56. $\dfrac{x^2 + 3x - 154}{x + 14}$

57. $\dfrac{2a^2 + 13a - 45}{a + 9}$

58. $\dfrac{24 + 13t - 2t^2}{8 - t}$

59. $\dfrac{15x^2 - 2x - 8}{x - \dfrac{4}{5}}$

60. $\dfrac{18x^2 - 9x - 20}{x + \dfrac{5}{6}}$

61. $\dfrac{y^3 + y^2 - 4y - 4}{y - 2}$

62. $\dfrac{-3z^3 + 20z^2 - 36z + 16}{z - 4}$

63. $\dfrac{2t^3 + 15t^2 + 19t - 30}{t + 5}$

64. $\dfrac{5t^3 - 27t^2 - 14t - 24}{t - 6}$

65. $\dfrac{x^4 - x^3 - 3x^2 + 4x - 1}{x - 1}$

66. $\dfrac{x^4 - 16}{x - 2}$

In Exercises 67–78, use synthetic division to divide the polynomials.

67. $\dfrac{x^3 + 3x^2 - 1}{x + 4}$

68. $\dfrac{x^4}{x + 2}$

69. $\dfrac{x^4 - 4x^3 + x + 10}{x - 2}$

70. $\dfrac{2x^5 - 3x^3 + x}{x - 3}$

71. $\dfrac{5x^3 + 12}{x + 5}$

72. $\dfrac{8x + 35}{x \quad 10}$

73. $\dfrac{5x^3 - 6x^2 + 8}{x - 4}$

74. $\dfrac{5x^3 + 6x + 8}{x + 2}$

75. $\dfrac{10x^4 - 50x^3 - 800}{x - 6}$

76. $\dfrac{x^5 - 13x^4 - 120x + 80}{x + 3}$

77. $\dfrac{0.1x^2 + 0.8x + 1}{x - 0.2}$

78. $\dfrac{x^3 - 0.8x + 2.4}{x + 1}$

In Exercises 79 and 80, find the constant c such that the denominator will divide evenly into the numerator.

79. $\dfrac{x^3 + 2x^2 - 4x + c}{x - 2}$

80. $\dfrac{x^4 - 3x^2 + c}{x + 6}$

In Exercises 81–84, simplify the algebraic expression.

81. $\dfrac{4x^4}{x^3} - 2x$

82. $\dfrac{15x^3 y}{10x^2} + \dfrac{3xy^2}{2y}$

83. $\dfrac{8u^2 v}{2u} + \dfrac{3(uv)^2}{uv}$

84. $\dfrac{x^2 + 2x - 3}{x - 1} - (3x - 4)$

In Exercises 85–88, determine whether the reduction shown is valid or invalid. If it is invalid, state what is wrong.

85. $\dfrac{5 + 12}{5} = \dfrac{\cancel{5} + 12}{\cancel{5}} = 12$

86. $\dfrac{6 - 3}{6 + 11} = \dfrac{\cancel{6} - 3}{\cancel{6} + 11} = \dfrac{-3}{11}$

87. $\dfrac{9 \cdot 12}{19 \cdot 9} = \dfrac{\cancel{9} \cdot 12}{19 \cdot \cancel{9}} = \dfrac{12}{19}$

88. $\dfrac{28}{83} = \dfrac{2\cancel{8}}{\cancel{8}3} = \dfrac{2}{3}$

89. When a polynomial is divided by $x - 6$, the quotient is $x^2 + x + 1$ and the remainder is -4. Find the polynomial.

90. When a polynomial is divided by $x + 3$, the quotient is $x^3 + x^2 - 4$ and the remainder is 8. Find the polynomial.

91. *Dimensions of a Rectangle* The area of a rectangle is $2x^3 + 3x^2 - 6x - 9$ (see figure). Find its length if its width is $2x + 3$.

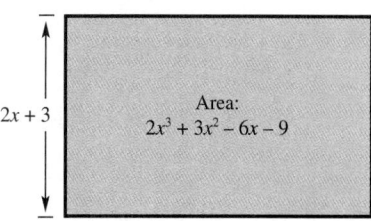

$2x + 3$

Area:
$2x^3 + 3x^2 - 6x - 9$

92. *Dimensions of a House* A rectangular house has a volume of

$$x^3 + 55x^2 + 650x + 2000$$

cubic feet (the space in the attic is not counted). The height of the house is $x + 5$ (see figure). Find the number of square feet of floor space *on the first floor* of the house.

$x + 5$

In Exercises 93 and 94, you are given the expression for the volume of the solid shown. Find the expression for the missing dimension.

93. $V = x^3 + 18x^2 + 80x + 96$

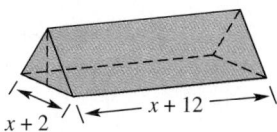

$x + 2$ $x + 12$

94. $V = h^4 + 3h^3 + 2h^2$

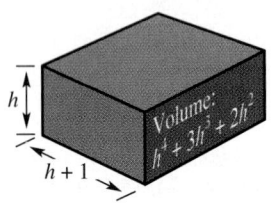

95. *Exploratory Exercise* Consider the function

$$f(x) = x^3 + 4x^2 - 15x - 2.$$

(a) Evaluate the function at the x-values listed in the following table.

x	-2	$-\frac{1}{2}$	1	3	5
$f(x)$					

(b) Use synthetic division to divide $f(x)$ by the divisors in the following table. In each case, list the remainder in the table.

x	-2	$-\frac{1}{2}$	1	3	5
Divisors	$x + 2$	$x + \frac{1}{2}$	$x - 1$	$x - 3$	$x - 5$
Remainders					

(c) What conclusion can you draw as you compare the function values in the first table with the remainders in the second? Use your conclusion and synthetic division to find $f(4)$.

<table>
<tr><td>**SECTION**
5.5</td><td>**Graphing Rational Functions**
The Domain of a Rational Function • *Horizontal and Vertical Asymptotes* • *Using Asymptotes to Graph a Rational Function*</td></tr>
</table>

The Domain of a Rational Function

A **rational function** is a function of the form

$$f(x) = \frac{p(x)}{q(x)}, \qquad q(x) \neq 0,$$

where $p(x)$ and $q(x)$ are polynomials in x. The *domain* of a rational function consists of the values of x for which the denominator $q(x)$ is not zero. For instance, the domain of

$$f(x) = \frac{x + 2}{x - 1}$$

is all real numbers except $x = 1$. When sketching the graph of a rational function, you should give special attention to the shape of the graph near x-values that are not in the domain.

EXAMPLE 1   ■   Sketching the Graph of a Rational Function

Sketch the graph of $f(x) = \dfrac{x+2}{x-1}$.

Solution

Begin by noticing that the domain is all real numbers except $x = 1$. Next, construct a table of values, including x-values that are close to 1 on the left *and* the right.

x-Values to the Left of 1

x	-3	-2	-1	0	0.5	0.9
$f(x)$	0.25	0	-0.5	-2	-5	-29

FIGURE 5.3

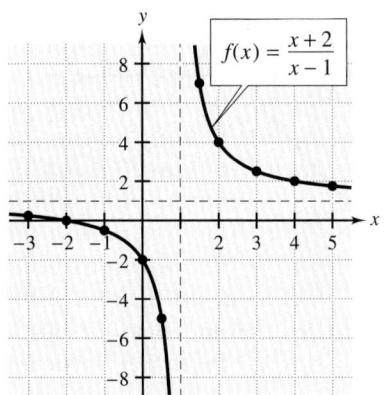

x-Values to the Right of 1

x	1.1	1.5	2	3	4	5
$f(x)$	31	7	4	2.5	2	1.75

Plot the points to the left of 1 and connect them with a smooth curve, as shown in Figure 5.3. Do the same for the points to the right of 1. *Do not* connect the two portions of the graph, which are called **branches.**   ■

For the graph of the function in Example 1, as x approaches 1 from the left, the values of $f(x)$ approach negative infinity. As x approaches 1 from the right, the values of $f(x)$ approach positive infinity.

Horizontal and Vertical Asymptotes

An **asymptote** of a graph is a line to which the graph becomes arbitrarily close as $|x|$ or $|y|$ increases without bound. In other words, if a graph has an asymptote, then it is possible to move far enough from the origin so that there is almost no difference between the graph and the asymptote.

The graph in Example 1 has two asymptotes: the line $x = 1$ is a **vertical asymptote,** and the line $y = 1$ is a **horizontal asymptote.** Other examples of asymptotes are shown in Figure 5.4 at the top of the next page.

FIGURE 5.4

Graph of $f(x) = \dfrac{1}{x^2 + 1}$

Horizontal asymptote: $y = 0$
Vertical asymptote: None

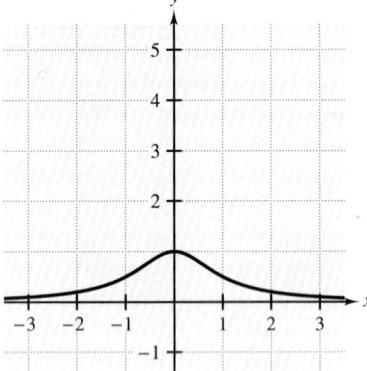

Graph of $f(x) = \dfrac{2x^2}{x^2 - 4}$

Horizontal asymptote: $y = 2$
Vertical asymptotes: $x = 2,\ x = -2$

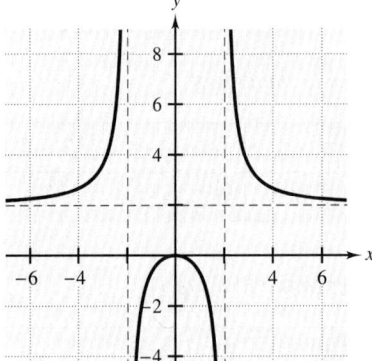

Graph of $f(x) = \dfrac{x^3 + 1}{x^2}$

Horizontal asymptote: None
Vertical asymptote: $x = 0$

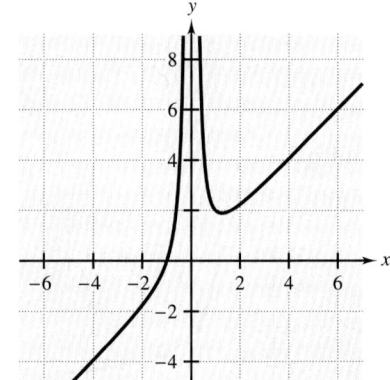

As you can see from these examples, the graph of a rational function may have no horizontal or vertical asymptotes, or it may have several. Here are some guidelines for finding the horizontal and vertical asymptotes of a rational function.

Horizontal and Vertical Asymptotes

Let $f(x) = p(x)/q(x)$ where $p(x)$ and $q(x)$ have *no common factors*.

1. The graph of f has a vertical asymptote at *each* real solution of the equation $q(x) = 0$.

2. The graph of f has at most one horizontal asymptote.

   - If the degree of $p(x)$ is less than the degree of $q(x)$, then the line $y = 0$ is a horizontal asymptote.

   - If the degree of $p(x)$ is equal to the degree of $q(x)$, then the line $y = a/b$ is a horizontal asymptote, where a is the leading coefficient of $p(x)$ and b is the leading coefficient of $q(x)$.

   - If the degree of $p(x)$ is greater than the degree of $q(x)$, then the graph has no horizontal asymptote.

EXAMPLE 2   ■   Finding Horizontal and Vertical Asymptotes

Find all horizontal and vertical asymptotes of the graph of

$$f(x) = \frac{2x}{3x^2 + 1}.$$

Solution

For this rational function, the degree of the numerator is less than the degree of the denominator. This implies that the graph of the function has the line

$$y = 0 \text{ (the } x\text{-axis)} \qquad \text{Horizontal asymptote}$$

as a horizontal asymptote, as shown in Figure 5.5. To find any vertical asymptotes, set the denominator equal to zero and solve the resulting equation for x.

$$3x^2 + 1 = 0 \qquad \text{Set denominator equal to zero}$$

This equation has no (real) solution. Thus, the graph has no vertical asymptotes.

FIGURE 5.5

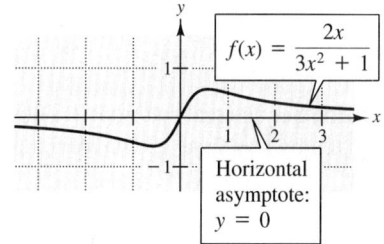

EXAMPLE 3   ■   Finding Horizontal and Vertical Asymptotes

Find all horizontal and vertical asymptotes of the graph of

$$f(x) = \frac{2x^2}{x^2 - 1}.$$

FIGURE 5.6

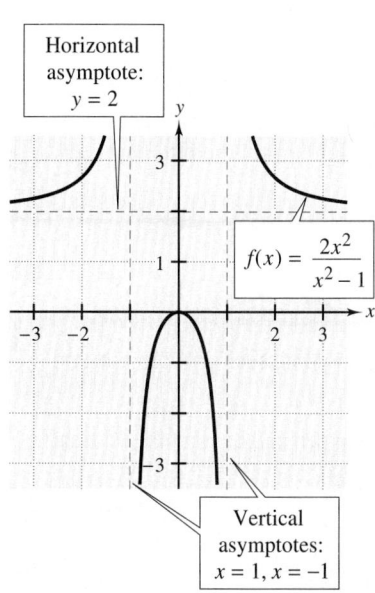

Horizontal asymptote: $y = 2$

$f(x) = \dfrac{2x^2}{x^2 - 1}$

Vertical asymptotes: $x = 1, x = -1$

Solution

For this rational function, the degree of the numerator is equal to the degree of the denominator. The leading coefficient of the numerator is 2, and the leading coefficient of the denominator is 1. Thus, the graph of f has the line

$$y = \frac{2}{1} = 2 \qquad \text{Horizontal asymptote}$$

as a horizontal asymptote, as shown in Figure 5.6. To find any vertical asymptotes, set the denominator equal to zero and solve the resulting equation for x.

$$x^2 - 1 = 0 \qquad \text{Set denominator equal to zero}$$

This equation has two real solutions: -1 and 1. Thus, the graph has two vertical asymptotes: the lines $x = -1$ and $x = 1$.

EXAMPLE 4 ■ Finding Horizontal and Vertical Asymptotes

Find all horizontal and vertical asymptotes of the graph of

$$f(x) = \frac{x^3}{10x - 20}.$$

FIGURE 5.7

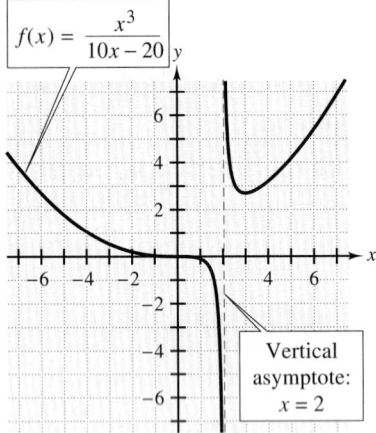

$f(x) = \dfrac{x^3}{10x - 20}$

Vertical asymptote: $x = 2$

Solution

For this rational function, the degree of the numerator is greater than the degree of the denominator. This implies that the graph has no horizontal asymptote. To find any vertical asymptotes, set the denominator equal to zero and solve the resulting equation for x.

$$10x - 20 = 0 \qquad \text{Set denominator equal to zero}$$

This equation has one real solution: 2. Thus, the graph has one vertical asymptote: the line given by $x = 2$, as shown in Figure 5.7.

■

Using Asymptotes to Graph a Rational Function

To sketch the graph of a rational function, we suggest the following guidelines.

Guidelines for Graphing Rational Functions

Let $f(x) = p(x)/q(x)$, where $p(x)$ and $q(x)$ are polynomials with no common factors.

1. Find and plot the y-intercept (if any) by evaluating $f(0)$.

2. Set the numerator equal to zero and solve the equation for x. The (real) solutions represent the x-intercepts of the graph. Plot these intercepts.

3. Set the denominator equal to zero and solve the equation for x. The (real) solutions represent the vertical asymptotes of the graph. Sketch these asymptotes.

4. Find and sketch the horizontal asymptotes of the graph.

5. Plot at least one point both between and beyond each x-intercept and vertical asymptote.

6. Use smooth curves to complete the graph between and beyond the vertical asymptotes.

EXAMPLE 5 ■ Sketching the Graph of a Rational Function

Sketch the graph of $f(x) = \dfrac{2}{x-3}$.

Solution

This organized process will be beneficial to students as the process becomes more complicated.

Begin by noting that the numerator and denominator have no common factors. Following the guidelines for sketching the graph of a rational function, you obtain the following.

FIGURE 5.8

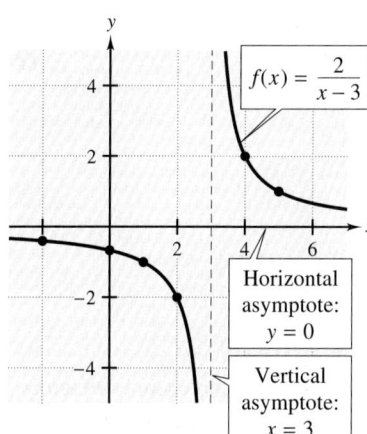

$f(x) = \dfrac{2}{x-3}$

Horizontal asymptote: $y = 0$

Vertical asymptote: $x = 3$

y-intercept:	Evaluate the function when $x = 0$ to obtain $f(0) = -\frac{2}{3}$. From this, you can conclude that the y-intercept is $\left(0, -\frac{2}{3}\right)$.
x-intercept:	Setting the numerator equal to zero produces $2 = 0$. Because this equation has no solution, you can conclude that the graph has no x-intercept.
Vertical asymptote:	Setting the denominator equal to zero produces $x - 3 = 0$. This equation has $x = 3$ as its only solution. Thus, you can conclude that the graph has one vertical asymptote: the line $x = 3$.
Horizontal asymptote:	The line $y = 0$ is a horizontal asymptote of the graph because the degree of the numerator is less than the degree of the denominator.
Additional points:	

x	-2	1	2	4	5
$f(x)$	$-\frac{2}{5}$	-1	-2	2	1

By plotting the intercepts, the asymptotes, and the additional points from the table, you can obtain the graph shown in Figure 5.8. ■

EXAMPLE 6 ■ Sketching the Graph of a Rational Function

Sketch the graph of $f(x) = \dfrac{2x-1}{x}$.

Solution

Begin by noting that the numerator and denominator have no common factors. Following the guidelines for sketching the graph of a rational function, you obtain the following.

y-intercept:	The graph has no y-intercept because $x = 0$ is not in the domain of the function.
x-intercept:	Setting the numerator equal to zero produces $2x - 1 = 0$. Because the solution of this equation is $x = \frac{1}{2}$, you can conclude that the graph has one x-intercept: $\left(\frac{1}{2}, 0\right)$.

FIGURE 5.9

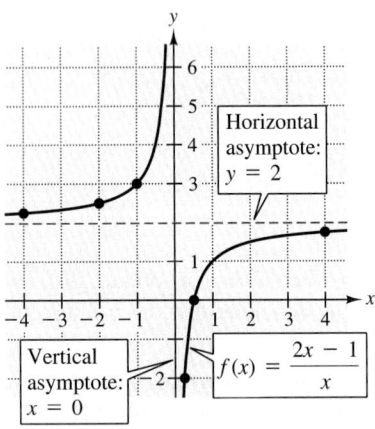

Vertical asymptote: Setting the denominator equal to zero produces $x = 0$. This equation has $x = 0$ as its only solution. Thus, you can conclude that the graph has one vertical asymptote: the line $x = 0$.

Horizontal asymptote: The degree of the numerator is equal to the degree of the denominator. Thus, the graph has a horizontal asymptote. Because the ratio of the leading coefficients is 2, the equation of the horizontal asymptote is $y = 2$.

Additional points:

x	-4	-2	-1	$\frac{1}{4}$	4
$f(x)$	$\frac{9}{4}$	$\frac{5}{2}$	3	-2	$\frac{7}{4}$

By plotting the intercepts, the asymptotes, and the additional points from the table, you can obtain the graph shown in Figure 5.9. ∎

EXAMPLE 7 ■   Sketching the Graph of a Rational Function

Other examples to graph:

(a) $f(x) = \dfrac{2}{x^2 - 4}$

(b) $f(x) = \dfrac{x}{x^2 + 2x - 15}$

Sketch the graph of $f(x) = \dfrac{2}{x^2 + 1}$.

Solution

Begin by noting that the numerator and denominator have no common factors. Following the guidelines for sketching the graph of a rational function, you obtain the following.

FIGURE 5.10

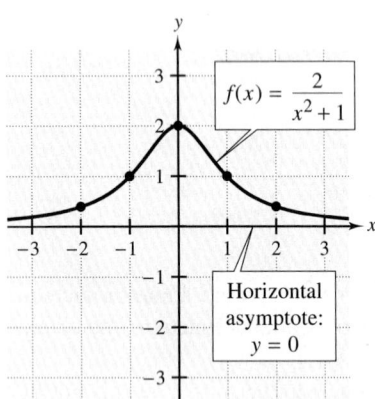

y-intercept: Evaluate the function when $x = 0$ to obtain $f(0) = 2$. From this, you can conclude that the y-intercept is $(0, 2)$.

x-intercept: Setting the numerator equal to zero produces $2 = 0$. Because this equation has no solution, you can conclude that the graph has no x-intercept.

Vertical asymptote: Setting the denominator equal to zero produces $x^2 + 1 = 0$. Because this equation has no solution, you can conclude that the graph has no vertical asymptote.

Horizontal asymptote: The line $y = 0$ is a horizontal asymptote of the graph because the degree of the numerator is less than the degree of the denominator.

Additional points:

x	-2	-1	1	2
$f(x)$	$\frac{2}{5}$	1	1	$\frac{2}{5}$

By plotting the intercepts, the asymptotes, and the additional points from the table, you can obtain the graph shown in Figure 5.10. ∎

EXAMPLE 8 ■ Finding the Average Cost

As a fund-raising project, a club is publishing a calendar. The cost of photography, typesetting, and printing is $850. In addition to these "one-time" charges, the *unit cost* of printing each calendar is $3.25. Let x represent the number of calendars that the club has printed. Write a model that represents the average cost per calendar.

Solution

The total cost C of printing x calendars is

$$C = 3.25x + 850.$$

The average cost per calendar A for printing x calendars is

$$A = \frac{3.25x + 850}{x}.$$

From the graph shown in Figure 5.11, notice that the average cost decreases as the number of calendars increases. This is further confirmed by constructing a table that compares the average cost A (in dollars per calendar) with the number of calendars printed x, as shown in the table.

Number Printed x	50	100	200	1000	2500	5000
Average Cost A	$20.25	$11.75	$7.50	$4.10	$3.59	$3.42

FIGURE 5.11

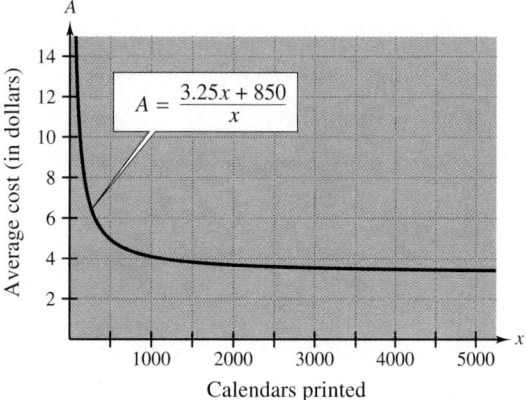

DISCUSSION PROBLEM ■ More About the Average Cost

In Example 8, what is the horizontal asymptote of the graph of the average cost function? What is the significance of this asymptote in the problem? Is it possible to sell enough calendars to obtain an average cost of $3.00 per calendar? Explain your reasoning. ■

Warm-Up

The following warm-up exercises involve skills that were covered in earlier sections. You will use these skills in the exercise set for this section.

In Exercises 1–4, factor the polynomial.

1. $x^2 - 3x - 10$

2. $x^2 - 7x + 10$

3. $x^3 + 4x^2 + 3x$

4. $x^3 - 4x^2 - 2x + 8$

In Exercises 5–8, sketch the graph of the equation.

5. $y = 2$

6. $x = -1$

7. $y = x - 2$

8. $y = -x + 1$

In Exercises 9 and 10, use long division to write the rational expression as the sum of a constant and a rational expression.

9. $\dfrac{5x + 6}{x - 4}$

10. $\dfrac{-2x + 3}{x + 2}$

5.5   EXERCISES

means that a graphing utility can help you solve the exercise or check your solution.

 In Exercises 1–6, match the rational function with its graph. [The graphs are labeled (a), (b), (c), (d), (e), and (f).]

1. $f(x) = \dfrac{2}{x + 1}$

2. $f(x) = \dfrac{1}{x - 4}$

3. $f(x) = \dfrac{x + 1}{x}$

4. $f(x) = \dfrac{1 - 2x}{x}$

5. $f(x) = \dfrac{x - 2}{x - 1}$

6. $f(x) = -\dfrac{x + 2}{x + 1}$

(a)
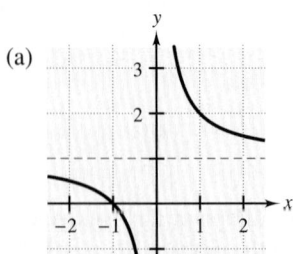

(b)
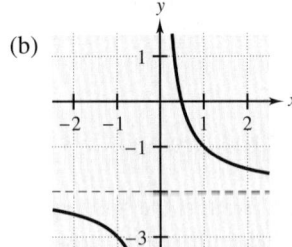

(c)
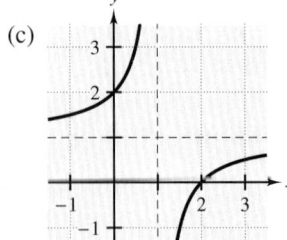

(d)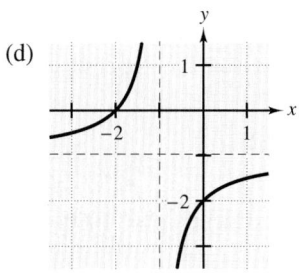

(e)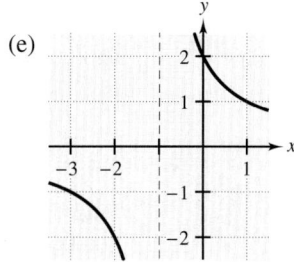

(f)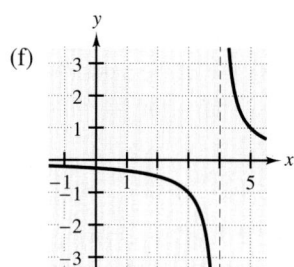

In Exercises 7–18, find the domain of the function, and identify any horizontal or vertical asymptotes.

7. $g(x) = \dfrac{1}{x - 5}$

8. $g(x) = \dfrac{3}{x + 5}$

9. $f(x) = \dfrac{1}{x^2}$

10. $f(x) = \dfrac{4}{(x - 2)^3}$

11. $h(x) = \dfrac{x}{x^2 - 9}$

12. $h(x) = \dfrac{3x^2 - 1}{x^3}$

13. $f(x) = \dfrac{2 + x}{2 - x}$

14. $f(x) = \dfrac{1 - 5x}{1 + 2x}$

15. $f(x) = \dfrac{3x^2 + 1}{x^2 + 9}$

16. $f(x) = \dfrac{3x^2 + x - 5}{x^2 + 1}$

17. $f(x) = \dfrac{x^3}{x^2 - 1}$

18. $f(x) = \dfrac{2x^2}{x + 1}$

In Exercises 19–40, sketch the graph of the function.

19. $f(x) = \dfrac{1}{x + 2}$

20. $f(x) = \dfrac{1}{x - 3}$

21. $h(x) = \dfrac{-1}{x + 2}$

22. $g(x) = \dfrac{1}{3 - x}$

23. $f(x) = \dfrac{x + 1}{x + 2}$

24. $f(x) = \dfrac{x - 2}{x - 3}$

25. $f(x) = \dfrac{2 + x}{1 - x}$

26. $f(x) = \dfrac{3 - x}{2 - x}$

27. $f(x) = \dfrac{3x + 1}{x}$

28. $f(x) = \dfrac{1 - 2x}{x}$

29. $C(x) = \dfrac{5 + 2x}{1 + x}$

30. $P(x) = \dfrac{1 - 3x}{1 - x}$

31. $g(x) = \dfrac{2x - 4}{x + 2}$

32. $h(x) = \dfrac{x - 2}{x - 3}$

33. $f(x) = \dfrac{4}{x^2 + 1}$

34. $g(x) = \dfrac{4}{x^2 - 1}$

35. $h(x) = \dfrac{4x^2}{x^2 - 9}$

36. $f(x) = \dfrac{4x^2}{x^2 + 9}$

37. $f(x) = -\dfrac{1}{(x - 2)^2}$

38. $g(x) = -\dfrac{x}{(x - 2)^2}$

39. $f(x) = \dfrac{3x}{x^2 - x - 2}$

40. $f(x) = \dfrac{2x}{x^2 + x - 2}$

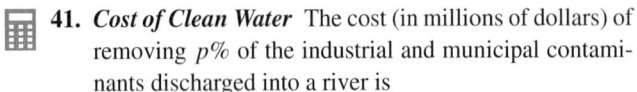

41. *Cost of Clean Water* The cost (in millions of dollars) of removing $p\%$ of the industrial and municipal contaminants discharged into a river is

$$C = \frac{255p}{100 - p}, \quad 0 \le p < 100.$$

(a) Find the cost of removing 10%.
(b) Find the cost of removing 40%.
(c) Find the cost of removing 75%.
(d) According to this model, would it be possible to remove 100% of the pollutants?

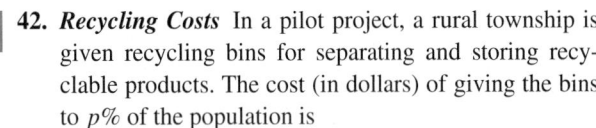

42. *Recycling Costs* In a pilot project, a rural township is given recycling bins for separating and storing recyclable products. The cost (in dollars) of giving the bins to $p\%$ of the population is

$$C = \frac{25,000p}{100 - p}, \quad 0 \le p < 100.$$

(a) Find the cost of giving the bins to 15% of the population.
(b) Find the cost of giving the bins to 50%.
(c) Find the cost of giving the bins to 90%.
(d) According to this model, would it be possible to give bins to 100% of the residents?

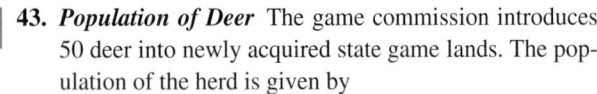

43. *Population of Deer* The game commission introduces 50 deer into newly acquired state game lands. The population of the herd is given by

$$N = \frac{10(5 + 3t)}{1 + 0.04t}, \quad 0 \le t,$$

where t is time in years.

(a) Find the population when t is 5, 10, and 25.
(b) Determine the horizontal asymptote of the rational function. What does it represent in the population model?
(c) Sketch a graph of the function.

44. *Average Cost* The cost of producing x units is $C = 150,000 + 0.25x$, and therefore the average cost per unit is

$$\overline{C} = \frac{C}{x} = \frac{150,000 + 0.25x}{x}, \quad 0 < x.$$

Find the average cost of producing $x = 1000$, $x = 10,000$, and $x = 100,000$ units.

Renting Videos In Exercises 45 and 46, use the following information. You have purchased a VCR for $180. You have also joined a video rental store where you can rent movies for $3 each.

45. Write a model that represents your average cost per movie (including the price of the VCR). Sketch the graph of this model. What is the horizontal asymptote and what does it represent?

46. Suppose the cost of admission to a movie theater is $6. How many videos must you rent to make your average cost per video less than $6? (*Hint:* Use the graph in Exercise 45 to solve the problem graphically.)

According to Tom Heymann in "On an Average Day," in 1990, Americans rented about 6 million movie videos each day, about twice the number of tickets sold at movie theaters.

Using a Graphing Calculator

Most graphing calculators have two graphing modes: a *connected mode* and a *dot mode* (see figures below). The connected mode works well for graphs of functions that are continuous (have no holes or breaks). The connected mode does not, however, work well for the graphs of many rational functions because they are often composed of two or more disconnected branches. To correct this problem, change the calculator to dot mode.

On the screen shown at left below, notice the vertical line at approximately $x = 1$. This line is *not* part of the graph—it is simply the calculator's attempt to connect the two branches of the graph.

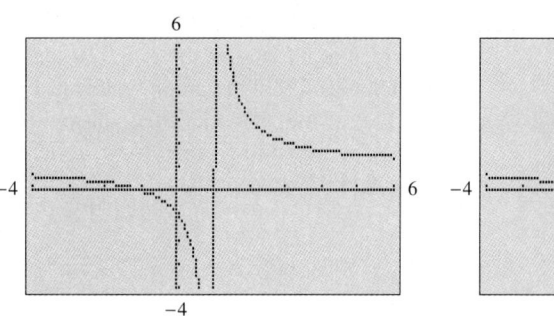

Connected Mode

Graph of $f(x) = \dfrac{x+1}{x-1}$

Dot Mode

Graph of $f(x) = \dfrac{x+1}{x-1}$

EXAMPLE 1 ■ Investigating Asymptotic Behavior

Some people think that the graph of a rational function cannot cross its horizontal asymptote. This, however, is not true. Use a graphing calculator to sketch the graph of

$$f(x) = \frac{2x^2 - 3x + 5}{x^2 + 1}$$

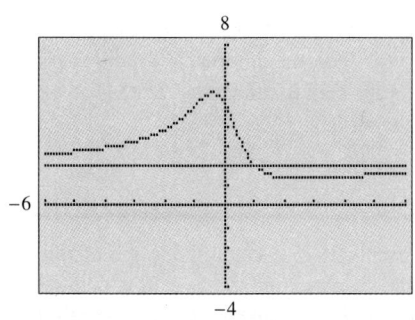

and its horizontal asymptote $y = 2$ on the screen. Find the x-value at which the graph crosses its horizontal asymptote.

Solution

From the screen at left, it appears that the graph crosses the horizontal line $y = 2$ when $x = 1$. You can confirm this by substituting 1 for x in the function.

$$f(1) = \frac{2(1^2) - 3(1) + 5}{1^2 + 1} = \frac{4}{2} = 2$$

■

431

EXAMPLE 2 ■ Pollution Level of a Pond

Some organic waste has been dumped into a pond. One component of the decomposition process is oxidation, whereby oxygen dissolved in the pond water is combined with decomposing material. Let $L = 1$ represent the normal oxygen level in the pond, and let t represent the number of weeks that have elapsed since the waste was dumped. The oxygen level in the pond can be modeled by

$$L = \frac{t^2 - t + 1}{t^2 + 1}.$$

Sketch the graph of this model, and use the graph to explain how the oxygen level has changed during the 15 weeks since the waste was dumped.

Solution

The screen at left shows the graph of the model and the line $L = 1$ (the normal oxygen level). From the graph, you can see that the oxygen dropped to 50% of its normal level after 1 week. Then, during the next several weeks, the oxygen level gradually returned to normal. Now, at the end of the 15th week, the oxygen level has reached 93% of normal. ■

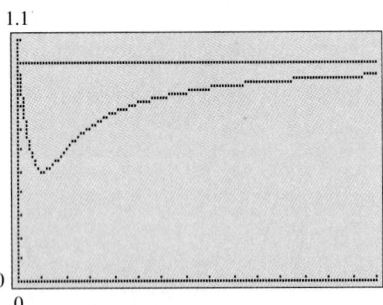

Oxygen Level of Pond

EXERCISES

In Exercises 1–6, find bounds for x and y such that the calculator screen displays the basic characteristics of the graph of the rational function.

1. $f(x) = \dfrac{x - 5}{x + 1}$

2. $f(x) = \dfrac{2x - 5}{x - 4}$

3. $f(x) = \dfrac{x^2 - 1}{x^2 - 4}$

4. $f(x) = \dfrac{2x^2 - 7x + 5}{x^2 + 2x + 1}$

5. $f(x) = \dfrac{x^2 - 5x}{2x^2 + 1}$

6. $f(x) = \dfrac{x^3 - 1}{x^3 + 1}$

7. Use a graphing calculator to sketch the graphs of f and g on the same screen. (Use a range of $-5 \le x \le 5$ and $-4 \le y \le 20$.)

$$f(x) = x \qquad \text{and} \qquad g(x) = \frac{x^2 + x + 1}{x + 1}$$

Use long division to rewrite $g(x)$. Then use the result to explain why the two graphs are close to each other for large values of $|x|$.

8. For the years 1975 to 1990, the number N (in millions) of long-playing albums sold in the United States can be approximated by the model

$$N = \frac{1000 - 74.41t + 1.45t^2}{2.94 - 0.336t + 0.0125t^2},$$

where $t = 5$ represents 1975. Sketch the graph of this function and use the graph to find the year in which the sales peaked. (*Source:* Recording Industry Association of America)

432

SECTION
5.6

Solving Equations Involving Rational Expressions

Equations Containing Constant Denominators   •   *Equations Containing Variable Denominators*
•   *Applications*

Equations Containing Constant Denominators

In Section 2.1, you learned how to solve equations containing fractions with *constant* denominators. That procedure is reviewed here because it is the basis for solving more general equations involving fractions. Recall from Section 2.1 that you can "clear an equation of fractions" by multiplying both sides of the equation by the least common denominator of the fractions in the equation. Note how this is done in Example 1.

EXAMPLE 1   ■   **An Equation Containing Constant Denominators**

Historical Note: The Rhind Papyrus is a major source of knowledge about ancient Egyptian mathematics. In one of its many documents, we find the *rule of false position*. Problem 24 states "A quantity and its $\frac{1}{7}$ added together become 19. What is the quantity?" It is solved as follows:

$$x + \frac{x}{7} = 19$$

Guess that x is, for example, 14.

$$14 + \tfrac{14}{7} = 14 + 2 = 16$$

As $19 = 16\left(\tfrac{19}{16}\right)$, the solution is $14\left(\tfrac{16}{19}\right)$ or $\tfrac{133}{8}$. Try the following from the Rhind Papyrus (using either the rule of false position or an equation involving rational expressions): Problem 25: A quantity and its $\frac{1}{2}$ added together become 16. What is the quantity? (Solution: $\frac{32}{3}$)

Solve the equation $\dfrac{3}{5} = \dfrac{x}{2}$.

Solution

Begin by multiplying both sides of the equation by the least common denominator, 10.

$\dfrac{3}{5} = \dfrac{x}{2}$	Given equation
$10\left(\dfrac{3}{5}\right) = 10\left(\dfrac{x}{2}\right)$	Multiply both sides by 10
$6 = 5x$	Simplify fractions
$\dfrac{6}{5} = x$	Solution

Check

$\dfrac{3}{5} = \dfrac{x}{2}$	Given equation
$\dfrac{3}{5} \overset{?}{=} \dfrac{\frac{6}{5}}{2}$	Replace x by $\frac{6}{5}$
$\dfrac{3}{5} \overset{?}{=} \left(\dfrac{6}{5}\right)\left(\dfrac{1}{2}\right)$	Invert and multiply
$\dfrac{3}{5} = \dfrac{3}{5}$	Solution checks

■

Note in Example 1 that the equation involves only two fractions, one on each side of the equation. To solve this type of equation, you can use cross multiplication. For instance, in Example 1, you could have solved the equation as follows.

Remind students that cross multiplication can be used only if we have one single fraction equal to another with no other terms involved.

$$\frac{3}{5} = \frac{x}{2} \qquad \text{Given equation}$$

$$3(2) = 5x \qquad \text{Cross multiply}$$

$$6 = 5x$$

$$\frac{6}{5} = x \qquad \text{Solution}$$

Although this cross-multiplication technique can be a little quicker than multiplying by the least common denominator, remember that *it can only be used with equations that have a single fraction on each side of the equation.*

EXAMPLE 2 ■ An Equation Containing Constant Denominators

Solve the equation $\dfrac{x}{6} = 7 - \dfrac{x}{12}$.

Solution

Begin by multiplying both sides of the equation by the least common denominator, 12.

$$\frac{x}{6} = 7 - \frac{x}{12} \qquad \text{Given equation}$$

$$12\left(\frac{x}{6}\right) = 12\left(7 - \frac{x}{12}\right) \qquad \text{Multiply both sides by 12}$$

$$2x = 84 - x \qquad \text{Distribute and simplify}$$

$$3x = 84 \qquad \text{Add } x \text{ to both sides}$$

$$x = 28 \qquad \text{Solution}$$

Therefore, the solution of the equation is 28. Check this solution in the original equation. ■

EXAMPLE 3 ■ An Equation Containing Constant Denominators

Solve the equation $\dfrac{x+2}{6} - \dfrac{x-4}{8} = \dfrac{2}{3}$.

Solution

Begin by multiplying both sides of the equation by the least common denominator, 24.

$$\frac{x+2}{6} - \frac{x-4}{8} = \frac{2}{3}$$ Given equation

$$24\left(\frac{x+2}{6} - \frac{x-4}{8}\right) = 24\left(\frac{2}{3}\right)$$ Multiply both sides by 24

$$4(x+2) - 3(x-4) = 8(2)$$ Distribute and simplify

$$4x + 8 - 3x + 12 = 16$$ Distributive Property

$$x = -4$$ Solution

Therefore, the solution of the equation is -4. Check this solution in the original equation.   ■

EXAMPLE 4   ■   **An Equation Containing Constant Denominators**

Solve the equation $\dfrac{x^2}{3} + \dfrac{x}{2} = \dfrac{5}{6}$.

Solution

Begin by multiplying both sides of the equation by the least common denominator, 6.

$$\frac{x^2}{3} + \frac{x}{2} = \frac{5}{6}$$ Given equation

$$6\left(\frac{x^2}{3} + \frac{x}{2}\right) = 6\left(\frac{5}{6}\right)$$ Multiply both sides by 6

$$2x^2 + 3x = 5$$ Distribute and simplify

$$2x^2 + 3x - 5 = 0$$ Write in standard form

$$(2x + 5)(x - 1) = 0$$ Factor

$$2x + 5 = 0 \quad \Longrightarrow \quad x = -\frac{5}{2}$$ Set first factor equal to 0

$$x - 1 = 0 \quad \Longrightarrow \quad x = 1$$ Set second factor equal to 0

Therefore, the equation has two solutions: $-\frac{5}{2}$ and 1. Check these solutions in the original equation.   ■

Equations Containing Variable Denominators

As stated in Section 5.1, you should always *exclude* those values of a variable that make the denominator of an algebraic fraction zero. This is especially critical when you are solving equations that contain variable denominators. You will see why in the examples that follow.

EXAMPLE 5 ■ An Equation Containing Variable Denominators

Solve the following equation.

$$\frac{2}{x} + \frac{5}{3} = \frac{7}{x}$$

Solution

Begin by multiplying both sides of the equation by the least common denominator, $3x$.

$$\frac{2}{x} + \frac{5}{3} = \frac{7}{x}$$ Given equation

$$3x\left(\frac{2}{x} + \frac{5}{3}\right) = 3x\left(\frac{7}{x}\right)$$ Multiply both sides by $3x$

$$6 + 5x = 21, \qquad x \neq 0$$ Distribute and simplify

$$5x = 15$$ Combine like terms

$$x = 3$$ Solution

Therefore, the solution appears to be 3. You can check this solution in the original equation, as follows.

FIGURE 5.12

Check

$$\frac{2}{x} + \frac{5}{3} = \frac{7}{x}$$ Given equation

$$\frac{2}{3} + \frac{5}{3} \overset{?}{=} \frac{7}{3}$$ Replace x by 3

$$\frac{7}{3} = \frac{7}{3}$$ Solution checks

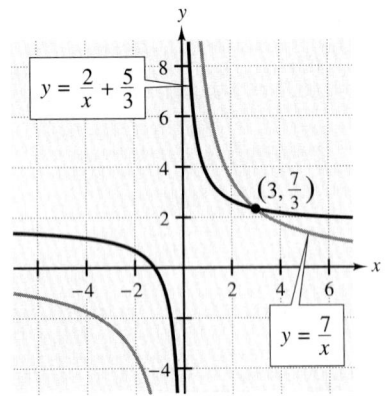

After checking, you can conclude that the solution of the equation is 3. Another way to check the solution is to sketch the graph of "each side" of the equation, as shown in Figure 5.12. In the figure, note that the two graphs intersect when $x = 3$.

■

Note that two ways for checking solutions are presented.

Throughout the text, we have emphasized the importance of checking solutions. Up to this point, the main reason for checking has been to make sure that you don't make errors in the solution process. In the next example, you will see that there is another reason for checking solutions in the *original* equation. That is, even with no mistakes in the solution process, it can happen that a "trial solution" does not satisfy the original equation. This type of "solution" is called **extraneous.** (An extraneous solution of an equation does not, by definition, satisfy its original equation, and must therefore *not* be listed as an actual solution.)

EXAMPLE 6 ■ **An Equation with No Solution**

Solve the equation $\dfrac{5x}{x-2} = 7 + \dfrac{10}{x-2}$.

Solution

As an aid to checking solutions, you might have students determine the domain *before* beginning the problem.

Begin by multiplying both sides of the equation by the least common denominator, $x - 2$.

$$\frac{5x}{x-2} = 7 + \frac{10}{x-2} \qquad \text{Given equation}$$

$$(x-2)\left(\frac{5x}{x-2}\right) = (x-2)\left(7 + \frac{10}{x-2}\right) \qquad \text{Multiply both sides by } x-2$$

$$5x = 7(x-2) + 10, \qquad x \ne 2 \qquad \text{Distribute and simplify}$$

$$5x = 7x - 4 \qquad \text{Group like terms}$$

$$-2x = -4 \qquad \text{Subtract } 7x \text{ from both sides}$$

$$x = 2 \qquad \text{Trial solution}$$

At this point, the solution appears to be 2. However, by performing the following check, you can see that this "trial solution" is extraneous.

Check

$$\frac{5x}{x-2} = 7 + \frac{10}{x-2} \qquad \text{Given equation}$$

$$\frac{5(2)}{2-2} \overset{?}{=} 7 + \frac{10}{2-2} \qquad \text{Replace } x \text{ by 2}$$

$$\frac{10}{0} \overset{?}{=} 7 + \frac{10}{0} \qquad \text{Division by zero is undefined}$$

Because the check resulted in *division by zero*, you can conclude that 2 is extraneous. Therefore, the given equation has no solution. Figure 5.13 shows that the graphs of "each side" of the equation,

$$y = \frac{5x}{x-2} \quad \text{and} \quad y = 7 + \frac{10}{x-2},$$

have no point of intersection. ■

FIGURE 5.13

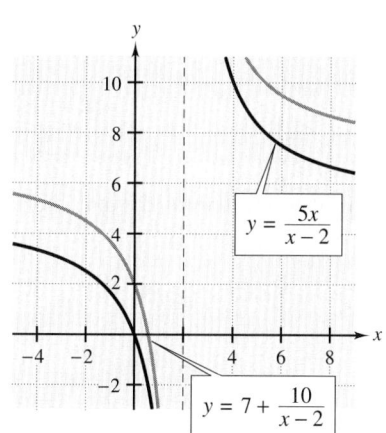

$$y = \frac{5x}{x-2}$$

$$y = 7 + \frac{10}{x-2}$$

NOTE In Example 6, can you see why $x = 2$ is extraneous? By looking back at the original equation, you can see that 2 is excluded from the domains of both of the fractions that occur in the equation.

EXAMPLE 7 ■ **An Equation Containing Variable Denominators**

Solve the equation $\dfrac{4}{x-2} + \dfrac{3x}{x+1} = 3$.

Solution

Begin by multiplying both sides of the equation by the least common denominator, $(x-2)(x+1)$.

$$\frac{4}{x-2} + \frac{3x}{x+1} = 3 \qquad \text{Given equation}$$

$$(x-2)(x+1)\left(\frac{4}{x-2} + \frac{3x}{x+1}\right) = 3(x-2)(x+1) \qquad \begin{array}{l}\text{Multiply both sides by} \\ (x-2)(x+1)\end{array}$$

$$4(x+1) + 3x(x-2) = 3(x^2 - x - 2), \qquad \text{Distribute and simplify}$$
$$x \neq 2, \; x \neq -1$$

$$4x + 4 + 3x^2 - 6x = 3x^2 - 3x - 6 \qquad \text{Distributive Property}$$

$$x = -10 \qquad \text{Solution}$$

After checking, you can conclude that the solution is -10. Check this solution in the original equation. ■

EXAMPLE 8 ■ **An Equation That Has Two Solutions**

Solve the equation $\dfrac{3x}{x+1} = \dfrac{12}{x^2 - 1} + 2$.

Solution

Begin by multiplying both sides of the equation by the least common denominator, $(x+1)(x-1)$ or $x^2 - 1$.

$$\frac{3x}{x+1} = \frac{12}{x^2 - 1} + 2 \qquad \text{Given equation}$$

$$(x^2 - 1)\left(\frac{3x}{x+1}\right) = (x^2 - 1)\left(\frac{12}{x^2 - 1} + 2\right) \qquad \text{Multiply both sides by } x^2 - 1$$

$$(x-1)(3x) = 12 + 2(x^2 - 1), \quad x \neq \pm 1 \qquad \text{Distribute and simplify}$$

$$3x^2 - 3x = 12 + 2x^2 - 2 \qquad \text{Distributive Property}$$

$$x^2 - 3x - 10 = 0 \qquad \text{Standard form}$$

$$(x+2)(x-5) = 0 \qquad \text{Factor}$$

$$x + 2 = 0 \implies x = -2 \qquad \text{Set first factor equal to 0}$$

$$x - 5 = 0 \implies x = 5 \qquad \text{Set second factor equal to 0}$$

After checking *each* solution, you can conclude that the equation has two solutions: -2 and 5. Check these solutions in the original equation. ■

Applications

EXAMPLE 9   ■   Average Cost

A manufacturing plant can produce x units of a certain item for $26 per unit *plus* an initial investment of $80,000. How many units must be produced in order for the average cost per unit to be $30?

FIGURE 5.14

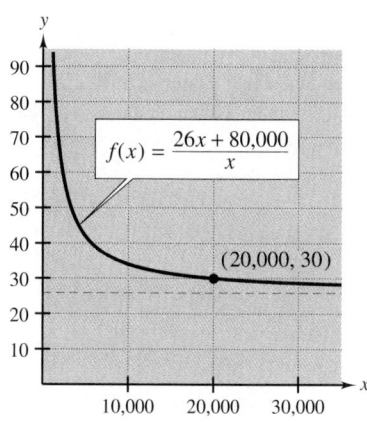

$$f(x) = \frac{26x + 80,000}{x}$$

(20,000, 30)

Solution

Verbal model: | Average cost per unit | = | Total cost | ÷ | Number of units |

Labels: Number of units $= x$ (units)
Total cost $= 26x + 80,000$ (dollars)
Average cost per unit $= 30$ (dollars per unit)

Equation: $30 = \dfrac{26x + 80,000}{x}$

$30x = 26x + 80,000, \qquad x \neq 0$

$4x = 80,000$

$x = 20,000$

Therefore, in order for the average cost per unit to be $30, the plant must manufacture 20,000 units. Check this solution in the original problem. Notice in Figure 5.14 that the average cost per unit drops as the number of units increases.   ■

EXAMPLE 10   ■   Comparing Two Rates

With only the cold water valve open, it takes 8 minutes to fill the tub of an automatic washer. With both the hot and cold water valves fully open, it takes only 5 minutes to fill the tub. How long will it take to fill the tub with only the hot water valve open?

Solution

Verbal model: | Rate for cold water | + | Rate for hot water | = | Rate for warm water |

Labels: Warm water: Time $= 5$ (minutes)
Rate $= \dfrac{1}{5}$ (tub per minute)

Cold water: Time $= 8$ (minutes)
Rate $= \dfrac{1}{8}$ (tub per minute)

Hot water: Time $= t$ (minutes)
Rate $= \dfrac{1}{t}$ (tub per minute)

In answering the question(s) to an application problem in words, use mixed fractions where appropriate. $\frac{40}{3}$ minutes is much more difficult to conceptualize than $13\frac{1}{3}$ minutes.

Equation:
$$\frac{1}{8} + \frac{1}{t} = \frac{1}{5}$$

$$40t\left(\frac{1}{8} + \frac{1}{t}\right) = 40t\left(\frac{1}{5}\right)$$

$$5t + 40 = 8t$$

$$40 = 3t$$

$$\frac{40}{3} = t$$

Thus, it will take about $13\frac{1}{3}$ minutes to fill the tub with hot water alone. Check this solution in the original problem. ■

EXAMPLE 11   ■   Finding a Batting Average

In this year's playing season, a baseball player has been up to bat 140 times and has hit safely 35 times. Thus, the player's batting average is $\frac{35}{140} = 0.250$. How many consecutive times must the player hit safely to obtain a batting average of 0.300?

Solution

Verbal model: Batting average $=$ Total hits $\div$ Total times at bat

Labels: Current times at bat $= 140$ (times at bat)
Current hits $= 35$ (hits)
Additional consecutive hits $= x$ (hits)
Batting average $= 0.300$ (hits per times at bat)

Equation:
$$0.300 = \frac{x + 35}{x + 140}$$

$$0.300(x + 140) = x + 35$$

$$0.3x + 42 = x + 35$$

$$7 = 0.7x$$

$$10 = x$$

Thus, the player must hit safely for the next 10 times at bat. After that, the player's batting average will be $\frac{45}{150} = 0.300$. Check this solution in the original problem. ■

DISCUSSION PROBLEM   ■   **Extraneous Solutions**

Consider the steps in the following solution.

$$\frac{x^2 + 2x + 2}{x - 1} = \frac{2x + 3}{x - 1}$$ Given equation

$$x^2 + 2x + 2 = (2x + 3), \qquad x \neq 1$$ Multiply both sides by $(x - 1)$

$$x^2 - 1 = 0$$

$$(x - 1)(x + 1) = 0$$ Factor

$$x + 1 = 0 \implies x = -1$$ Set first factor equal to zero

$$x - 1 = 0 \implies x = 1$$ Set second factor equal to zero

Which of these two possible solutions is extraneous? At which stage in the solution can you determine that this x-value cannot be a solution?   ■

Warm-Up

The following warm-up exercises involve skills that were covered in earlier sections. You will use these skills in the exercise set for this section.

In Exercises 1–10, solve the equation.

1. $3x = 16$ **2.** $-5x = 95$ **3.** $15x + 3 = 48$

4. $125 - 50x = 0$ **5.** $(3x - 2)(x + 8) = 0$ **6.** $(2x + 3)(x - 16) = 0$

7. $x(2x - 21) = 0$ **8.** $x(10 - x) = 25$

9. $x^2 + x - 42 = 0$ **10.** $t^2 - 8t = 0$

5.6   **EXERCISES**

▦   means that a graphing utility can help you solve the exercise or check your solution.

In Exercises 1–4, determine whether the given value of x is a solution to the equation.

Equation	Value		Equation	Value	
1. $\dfrac{x}{3} - \dfrac{x}{5} = \dfrac{4}{3}$	(a) $x = 0$	(b) $x = -1$	**2.** $x = 4 + \dfrac{21}{x}$	(a) $x = 0$	(b) $x = -3$
	(c) $x = \dfrac{1}{8}$	(d) $x = 10$		(c) $x = 7$	(d) $x = -1$
3. $\dfrac{x}{4} + \dfrac{3}{4x} = 1$	(a) $x = -1$	(b) $x = 1$	**4.** $5 - \dfrac{1}{x - 3} = 2$	(a) $x = \dfrac{10}{3}$	(b) $x = -\dfrac{1}{3}$
	(c) $x = 3$	(d) $x = 2$		(c) $x = 0$	(d) $x = 1$

In Exercises 5–14, solve the equation.

5. $\dfrac{x}{4} = \dfrac{3}{8}$

6. $\dfrac{x}{10} = \dfrac{12}{5}$

7. $\dfrac{t}{2} = \dfrac{1}{8}$

8. $\dfrac{y}{5} = \dfrac{3}{2}$

9. $\dfrac{z+2}{3} = \dfrac{z}{12}$

10. $5 + \dfrac{y}{3} = y + 2$

11. $\dfrac{4t}{3} = 15 - \dfrac{t}{6}$

12. $\dfrac{x}{3} + \dfrac{x}{6} = 10$

13. $\dfrac{h+2}{5} - \dfrac{h-1}{9} = \dfrac{2}{3}$

14. $\dfrac{u-2}{6} + \dfrac{2u+5}{15} = 3$

In Exercises 15–54, solve the equation. (Be sure to check for extraneous solutions.)

15. $\dfrac{7}{x} = 21$

16. $\dfrac{9}{t} = -\dfrac{4}{3}$

17. $\dfrac{9}{25-y} = -\dfrac{1}{4}$

18. $\dfrac{2}{u+4} = \dfrac{5}{8}$

19. $5 - \dfrac{12}{a} = \dfrac{5}{3}$

20. $\dfrac{6}{b} + 22 = 24$

21. $\dfrac{12}{y+5} + \dfrac{1}{2} = 2$

22. $\dfrac{7}{8} - \dfrac{16}{t-2} = \dfrac{3}{4}$

23. $\dfrac{5}{x} = \dfrac{25}{3(x+2)}$

24. $\dfrac{10}{x+4} = \dfrac{15}{4(x+1)}$

25. $\dfrac{8}{3x+5} = \dfrac{1}{x+2}$

26. $\dfrac{500}{3x+5} = \dfrac{50}{x-3}$

27. $\dfrac{3}{x+2} - \dfrac{1}{x} = \dfrac{1}{5x}$

28. $\dfrac{12}{x+5} + \dfrac{5}{x} = \dfrac{20}{x}$

29. $\dfrac{4}{2x+3} + \dfrac{17}{5(2x+3)} = 3$

30. $\dfrac{2}{6q+5} - \dfrac{3}{4(6q+5)} = \dfrac{1}{28}$

31. $\dfrac{10}{x(x-2)} + \dfrac{4}{x} = \dfrac{5}{x-2}$

32. $3\left(\dfrac{1}{x} + 4\right) = 2 + \dfrac{4}{3x}$

33. $\dfrac{x}{x+4} + \dfrac{4}{x+4} + 2 = 0$

34. $\dfrac{8}{2x+4} - \dfrac{3x+1}{3x^2+2} = \dfrac{2}{x+2}$

35. $\dfrac{2x^2-5}{x^2-4} + \dfrac{6}{x+2} = \dfrac{4x-7}{x-2}$

36. $\dfrac{2}{x^2-6x+8} = \dfrac{1}{x-4} + \dfrac{2}{x-2}$

37. $\dfrac{1}{x-5} + \dfrac{1}{x+5} = \dfrac{x+3}{x^2-25}$

38. $\dfrac{2}{x-10} - \dfrac{3}{x-2} = \dfrac{6}{x^2-12x+20}$

39. $\dfrac{20x^2-x-4}{2x^2-9x-18} = \dfrac{4x}{x-6} - \dfrac{2x-3}{2x+3}$

40. $\dfrac{3}{4} = \dfrac{75}{t^2}$

41. $\dfrac{1}{2} = \dfrac{18}{x^2}$

42. $\dfrac{1}{6} = \dfrac{150}{z^2}$

43. $\dfrac{32}{t} = 2t$

44. $\dfrac{45}{u} = \dfrac{u}{5}$

45. $x + 1 = \dfrac{72}{x}$

46. $t + 4\left(\dfrac{2t+5}{t+4}\right) = 0$

47. $1 = \dfrac{16}{y} - \dfrac{39}{y^2}$

48. $\dfrac{15}{x^2} + \dfrac{7}{x} = 2$

49. $\dfrac{1}{x-1} + \dfrac{3}{x+1} = 2$

50. $\dfrac{x+42}{x} = x$

51. $x - \dfrac{24}{x} = 5$

52. $\dfrac{3x}{2} + \dfrac{4}{x} = 5$

53. $\dfrac{x}{2} = \dfrac{\left(2 - \dfrac{3}{x}\right)}{\left(1 - \dfrac{1}{x}\right)}$

54. $\dfrac{2x}{3} = \dfrac{\left(1 + \dfrac{2}{x}\right)}{\left(1 + \dfrac{1}{x}\right)}$

55. *Number Problem* Find a number such that the sum of the number and its reciprocal is $\frac{65}{8}$.

56. *Number Problem* Find a number such that the sum of two times the number and three times its reciprocal is $\frac{97}{4}$.

57. *Wind Speed* A plane has a speed of 300 miles per hour in still air. Find the speed of the wind if the plane traveled a distance of 680 miles with a tail wind in the same time it took to travel 520 miles into a head wind.

58. *Average Speed* During the first 200 miles of a 320-mile trip, you travel at an average speed of r miles per hour. Because of the difference of speed limits among states, you increased your average speed by 10 miles per hour during the last part of the trip. Find the lower (earlier) average speed if the total time for the trip was 6 hours.

59. *Partnership Costs* A group of outdoor enthusiasts plan to buy a piece of property for $78,000 by sharing the cost equally. To ease the financial burden, they look for three additional partners so that the cost per person can be reduced by $1300. Determine the number of partners they would like to have in the group.

60. *Average Cost* The average cost for producing x units of a product is

$$\text{Average cost} = 1.50 + \frac{4200}{x}.$$

Determine the number of units that must be produced so that the average cost is $2.90.

61. *Pollution Removal* The cost C in dollars of removing $p\%$ of the air pollution in the stack emissions of a utility company that burns coal to generate electricity is

$$C = \frac{120,000p}{100 - p}.$$

Determine the percentage of the stack emissions that can be removed for $680,000.

62. *Population Growth* A biologist introduces 100 insects into a culture. The population P of the culture is approximated by the model

$$P = \frac{500(1 + 3t)}{5 + t},$$

where t is the time in hours. Find the time required for the population to increase to 1000 insects.

Work Rate In Exercises 63 and 64, complete the table by finding the times required for two individuals working together to complete three different tasks. The first two columns in the table give the times required for the two individuals working alone to complete the tasks. (Assume that when they work together their individual rates are not changed.)

63.

Person #1	Person #2	Together
6 hours	6 hours	
3 minutes	5 minutes	
5 hours	$2\frac{1}{2}$ hours	

64.

Person #1	Person #2	Together
4 days	4 days	
$5\frac{1}{2}$ hours	3 hours	
a days	b days	

65. *Photocopy Rate* A photocopier produces copies at a rate of 20 pages per minute.
(a) Determine the time required to copy one page.
(b) Determine the time required to copy x pages.
(c) Determine the time required to copy 35 pages.

66. *Pumping Rate* The rated volume of a pump is 15 gallons per minute.
(a) Determine the time required to pump 1 gallon.
(b) Determine the time required to pump x gallons.
(c) Determine the time required to pump 130 gallons.

67. *Work Rate* One landscaper works $1\frac{1}{2}$ times as fast as a second landscaper. Find their individual times for a certain job if it takes them 9 hours working together.

68. *Work Rate* The two landscapers in Exercise 67 begin a job working together. After 4 hours, the slower of the two is sent off to do another job, and it takes the faster worker an additional 10 hours to complete the original job. How long would it have taken each worker, working alone, to complete the original job?

69. *Swimming Pool* The flow rate of one pipe is $1\frac{1}{4}$ times as high as that of a second pipe. If a swimming pool can be filled in 5 hours using both pipes, find the time required to fill the pool using only the pipe with the lower flow rate.

70. *Swimming Pool* A swimming pool is being filled using two pipes that have the same relative flow rates as those in Exercise 69. After 1 hour, the pipe with the higher flow rate is shut off, and it takes an additional 10 hours to fill the pool. How long would it have taken to fill the pool using only the pipe with the lower flow rate? How long would it have taken using only the pipe with the higher flow rate?

Car Rental Revenues In Exercises 71 and 72, use the following model, which approximates the total revenue for the car rental industry in the United States for the years 1984 to 1990 (see figure).

$$y = \frac{1}{2.39 - 0.326t + 0.015t^2}$$

In this model, y represents the total revenue in billions of dollars and t represents the year, with $t = 0$ corresponding to 1980. (*Source:* Alex Brown and Sons)

Figure for 71 and 72

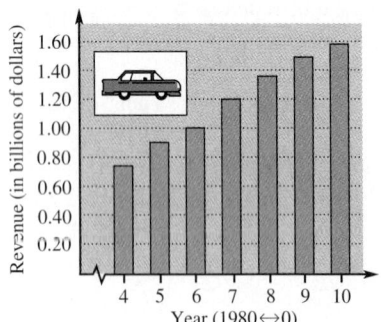

Year (1980 ↔ 0)

71. Use the bar graph to determine the year in which the total revenue approximated \$1.03 billion. Then use the model to verify your answer.

72. Use the bar graph to determine the year in which the total revenue approximated \$1.35 billion. Then use the model to verify your answer.

In Exercises 73 and 74, use the following information. For 1980 and 1990, the number P of conventional pianos sold in the United States can be modeled by

$$P = 223{,}720 - 9480t,$$

where $t = 0$ represents 1980 (see figure). The number K of electronic pianos and keyboards sold can be modeled by

$$K = 27{,}820 + 8710t.$$

Figure for 73 and 74

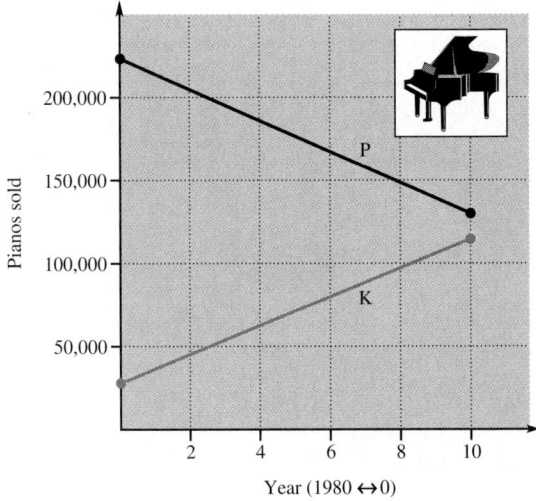

Year (1980 ↔ 0)

73. In what year was the ratio of conventional pianos sold to electronic pianos and keyboards sold equal to 3?

74. In what year does it appear that the number of electronic pianos and keyboards sold surpassed the number of conventional pianos sold?

5 REVIEW EXERCISES

In Exercises 1–4, find the domain of the rational expression.

1. $\dfrac{3y}{y-8}$

2. $\dfrac{t+4}{t+12}$

3. $\dfrac{u}{u^2-7u+6}$

4. $\dfrac{x-12}{x(x^2-16)}$

In Exercises 5–10, rewrite the fraction in reduced form.

5. $\dfrac{6x^4y^2}{15xy^2}$

6. $\dfrac{2(y^3z)^2}{28(yz^2)^2}$

7. $\dfrac{5b-15}{30b-120}$

8. $\dfrac{4a}{10a^2+26a}$

9. $\dfrac{9x-9y}{y-x}$

10. $\dfrac{x+3}{x^2-x-12}$

In Exercises 11–40, perform the indicated operation and simplify your answer.

11. $\dfrac{7}{8}\cdot\dfrac{2x}{y}\cdot\dfrac{y^2}{14x^2}$

12. $\dfrac{15(x^2y)^3}{3y^3}\cdot\dfrac{12y}{x}$

13. $\dfrac{60z}{z+6}\cdot\dfrac{z^2-36}{5}$

14. $\dfrac{1}{6}(x^2-16)\cdot\dfrac{3}{x^2-8x+16}$

15. $\dfrac{u}{u-3}\cdot\dfrac{3u-u^2}{4u^2}$

16. $x^2\cdot\dfrac{x+1}{x^2-x}\cdot\dfrac{(5x-5)^2}{x^2+6x+5}$

17. $\dfrac{\left(\dfrac{6}{x}\right)}{\left(\dfrac{2}{x^3}\right)}$

18. $\dfrac{0}{\left(\dfrac{5x^2}{2y}\right)}$

19. $25y^2\div\dfrac{xy}{5}$

20. $\dfrac{6}{z^2}\div 4z^2$

21. $\dfrac{x^2-7x}{x+1}\div\dfrac{x^2-14x+49}{x^2-1}$

22. $\left(\dfrac{6x}{y^2}\right)^2\div\left(\dfrac{3x}{y}\right)^3$

23. $\dfrac{4a}{9}-\dfrac{11a}{9}$

24. $\dfrac{2(3y+4)}{2y+1}+\dfrac{3-y}{2y+1}$

25. $\dfrac{15}{16x}-\dfrac{5}{24x}-1$

26. $-\dfrac{3y}{8x}+\dfrac{7y}{6x}-\dfrac{y}{12x}$

27. $\dfrac{1}{x+5}+\dfrac{3}{x-12}$

28. $\dfrac{2}{x-10}+\dfrac{3}{4-x}$

29. $5x+\dfrac{2}{x-3}-\dfrac{3}{x+2}$

30. $4-\dfrac{4x}{x+6}+\dfrac{7}{x-5}$

31. $\dfrac{6}{x}-\dfrac{6x-1}{x^2+4}$

32. $\dfrac{5}{x+2}+\dfrac{25-x}{x^2-3x-10}$

33. $\dfrac{5}{x+3}-\dfrac{4x}{(x+3)^2}-\dfrac{1}{x-3}$

34. $\dfrac{8}{y}-\dfrac{3}{y+5}+\dfrac{4}{y-2}$

35. $\dfrac{\left(\dfrac{6x^2}{x^2+2x-35}\right)}{\left(\dfrac{x^3}{x^2-25}\right)}$

36. $\dfrac{\left[\dfrac{24-18x}{(2-x)^2}\right]}{\left(\dfrac{60-45x}{x^2-4x+4}\right)}$

37. $\dfrac{3t}{\left(5-\dfrac{2}{t}\right)}$

38. $\dfrac{\left(x - 3 + \dfrac{2}{x}\right)}{\left(1 - \dfrac{2}{x}\right)}$

39. $\dfrac{\left(\dfrac{1}{a^2 - 16} - \dfrac{1}{a}\right)}{\left(\dfrac{1}{a^2 + 4a} + 4\right)}$

40. $\dfrac{\left(\dfrac{1}{x^2} - \dfrac{1}{y^2}\right)}{\left(\dfrac{1}{x} + \dfrac{1}{y}\right)}$

In Exercises 41–46, perform the indicated division.

41. $(4x^3 - x) \div 2x$

42. $(10x + 15) \div (5x - 2)$

43. $\dfrac{6x^3 + x^2 - 4x + 2}{3x - 1}$

44. $\dfrac{4x^4 - x^3 - 7x^2 + 18x}{x - 2}$

45. $\dfrac{x^4 - 3x^2 + 2}{x^2 - 1}$

46. $\dfrac{3x^6}{x^2 - 1}$

In Exercises 47–50, use synthetic division to perform the indicated division.

47. $\dfrac{x^3 + 7x^2 + 3x - 14}{x + 2}$

48. $\dfrac{x^4 - 2x^3 - 15x^2 - 2x + 10}{x - 5}$

49. $(x^4 - 3x^2 - 25) \div (x - 3)$

50. $(2x^3 + 5x - 2) \div \left(x + \dfrac{1}{2}\right)$

In Exercises 51–54, match the function with its graph. [The graphs are labeled (a), (b), (c), and (d).]

51. $f(x) = \dfrac{5}{x - 6}$

(a)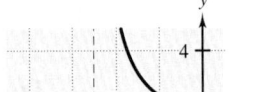

(b)

52. $f(x) = \dfrac{6}{x + 5}$

53. $f(x) = \dfrac{6x}{x - 5}$

54. $f(x) = \dfrac{2x}{x + 6}$

(c)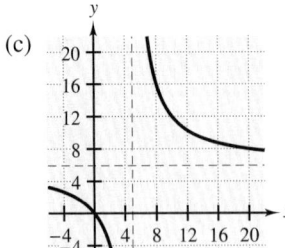

(d)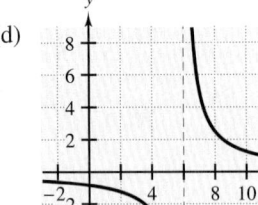

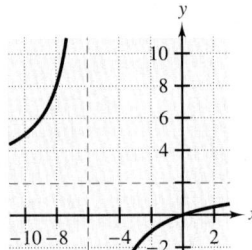

In Exercises 55–68, sketch the graph of the rational function.

55. $g(x) = \dfrac{2+x}{1-x}$

56. $h(x) = \dfrac{x-3}{x-2}$

57. $f(x) = \dfrac{x}{x^2+1}$

58. $f(x) = \dfrac{2x}{x^2+4}$

59. $P(x) = \dfrac{3x+6}{x-2}$

60. $s(x) = \dfrac{2x-6}{x+4}$

61. $h(x) = \dfrac{4}{(x-1)^2}$

62. $g(x) = \dfrac{-2}{(x+3)^2}$

63. $f(x) = \dfrac{-5}{x^2}$

64. $f(x) = \dfrac{4}{x}$

65. $y = \dfrac{x}{x^2-1}$

66. $y = \dfrac{2x}{x^2-4}$

67. $y = \dfrac{2x^2}{x^2-4}$

68. $y = \dfrac{2}{x+3}$

69. *Average Cost* A business has a cost of $C = 0.5x + 500$ for producing x units. The average cost per unit is

$$\overline{C} = \frac{C}{x} = \frac{0.5x + 500}{x}, \quad 0 < x.$$

Find the horizontal asymptote and state what it represents in the model.

 70. *Average Cost* The cost of producing x units is C and therefore the average cost per unit is

$$\overline{C} = \frac{C}{x} = \frac{100{,}000 + 0.9x}{x}, \quad 0 < x.$$

Sketch a graph of the average cost function and find the average cost of producing $x = 1000$, $x = 10{,}000$, and $x = 100{,}000$ units.

71. *Seizure of Illegal Drugs* The cost in millions of dollars for the federal government to seize $p\%$ of a certain illegal drug as it enters the country is

$$C = \frac{528p}{100 - p}, \quad 0 \le p < 100.$$

(a) Find the cost of seizing 25%.
(b) Find the cost of seizing 50%.
(c) Find the cost of seizing 75%.
(d) According to this model, would it be possible to seize 100% of the drug?

72. *Population of Fish* The Parks and Wildlife Commission introduces 80,000 fish into a large man-made lake. The population of the fish in thousands is

$$N = \frac{20(4 + 3t)}{1 + 0.05t}, \quad 0 \le t,$$

where t is time in years.

(a) Find the population when t is 5, 10, and 25.
(b) Determine the horizontal asymptote and state what it represents in this model.

In Exercises 73–86, solve the equation.

73. $\dfrac{3x}{8} = -15$

74. $\dfrac{t+1}{8} = \dfrac{1}{2}$

75. $3\left(8 - \dfrac{12}{t}\right) = 0$

76. $\dfrac{1}{3y-4} = \dfrac{6}{4(y+1)}$

77. $\dfrac{2}{y} - \dfrac{1}{3y} = \dfrac{1}{3}$

78. $8\left(\dfrac{6}{x} - \dfrac{1}{x+5}\right) = 15$

79. $r = 2 + \dfrac{24}{r}$

80. $\dfrac{3}{y+1} - \dfrac{8}{y} = 1$

81. $\dfrac{2}{x} - \dfrac{x}{6} = \dfrac{2}{3}$

82. $\dfrac{2x}{x+3} - \dfrac{3}{x} = 0$

83. $\dfrac{12}{x^2+x-12} - \dfrac{1}{x-3} = -1$

84. $\dfrac{3}{x-1} + \dfrac{6}{x^2-3x+2} = 2$

85. $\dfrac{5}{x^2-4} - \dfrac{6}{x-2} = -5$

86. $\dfrac{3}{x^2-9} + \dfrac{4}{x+3} = 1$

87. *Average Speed* Suppose you drove 56 miles (one way) on a service call for your company. On the return trip, which took 10 minutes less than the original trip, your average speed was 8 miles per hour faster. What was your average speed on the return trip?

88. *Batting Average* In this year's playing season, a baseball player has been up to bat 150 times and has hit safely 45 times. Thus, the player's batting average is $\frac{45}{150} = 0.300$. How many consecutive times must the player hit safely to obtain a batting average of 0.400?

89. *Forming a Partnership* A group of people agree to share equally in the cost of a $60,000 piece of machinery. If they could find two more people to join the group, each person's share of the cost would decrease by $5000. How many people are presently in the group?

90. *Forming a Partnership* An individual is planning to start a small business that will require $28,000 before any income can be generated. Because it is difficult to borrow for new ventures, the individual wants a group of friends to divide the cost equally and, in return, receive a share of the future profits. The person has found some investors, but three more are needed so that the price per person will be $1200 less. How many investors are currently in the group?

91. *Work Rate* Suppose that in 12 minutes your supervisor can do a task that you require 15 minutes to complete. Determine the time required to complete the task if you work together.

Take this test as you would take a test in class. After you are done, check your work against the answers in the back of the book.

1. Find the domain of the rational expression $\dfrac{3y}{y^2 - 25}$.

2. Find the least common denominator of the fractions $\dfrac{3}{x^2}$, $\dfrac{x}{x-3}$, $\dfrac{2x}{x^3(x+3)}$, $\dfrac{10}{x^2 + 6x + 9}$.

3. Simplify the rational expression $\dfrac{2-x}{3x-6}$.

4. Simplify the rational expression $\dfrac{2a^2 - 5a - 12}{5a - 20}$.

In Exercises 5–12, perform the specified operation and simplify.

5. $\dfrac{4z^3}{5} \cdot \dfrac{25}{12z^2}$

6. $\dfrac{y^2 + 8y + 16}{2(y-2)} \cdot \dfrac{8y - 16}{(y+4)^3}$

7. $(4x^2 - 9) \cdot \dfrac{2x + 3}{2x^2 - x - 3}$

8. $\dfrac{(2xy^2)^3}{15} \div \dfrac{12x^3}{21}$

9. $\dfrac{\left(\dfrac{3x}{x+2}\right)}{\left(\dfrac{12}{x^3 + 2x^2}\right)}$

10. $\dfrac{4}{x+1} + \dfrac{4x}{x+1}$

11. $2x + \dfrac{1 - 4x^2}{x+1}$

12. $\dfrac{5x}{x+2} - \dfrac{2}{x^2 - x - 6}$

13. Perform the specified operations and simplify: $\dfrac{3}{x} - \dfrac{5}{x^2} + \dfrac{2x}{x^2 + 2x + 1}$

14. Simplify: $\dfrac{\left(9x - \dfrac{1}{x}\right)}{\left(\dfrac{1}{x} - 3\right)}$

In Exercises 15 and 16, perform the specified division.

15. $\dfrac{t^4 + t^2 - 6t}{t^2 - 2}$ (Long division)

16. $\dfrac{2x^4 - 15x^2 - 7}{x - 3}$ (Synthetic division)

In Exercises 17 and 18, sketch the graph of the function.

17. $f(x) = \dfrac{3}{x-3}$

18. $g(x) = \dfrac{3x}{x-3}$

In Exercises 19–21, solve the equation (if possible).

19. $\dfrac{3}{h+2} = \dfrac{1}{8}$

20. $\dfrac{2}{x+5} - \dfrac{3}{x+3} = \dfrac{1}{x}$

21. $\dfrac{1}{x+1} + \dfrac{1}{x-1} = \dfrac{2}{x^2 - 1}$

22. One painter works $1\frac{1}{2}$ times as fast as a second painter. Find their individual times for painting a room if, working together, they can complete the job in 4 hours.

6 Radicals and Complex Numbers

6.1   Integer Exponents and Scientific Notation

6.2   Rational Exponents and Radicals

6.3   Simplifying and Combining Radicals

6.4   Multiplying and Dividing Radical Expressions

6.5   Solving Equations Involving Radicals

6.6   Complex Numbers

*J*ohannes Kepler (1571–1630) discovered a
mathematical model for the elliptical orbits of the
planets about the sun. His findings are summarized in
Kepler's Laws of Interplanetary Motion. One of Kepler's laws
relates the average distance x (in astronomical units) of a
planet from the sun to the time t (in years) that the planet takes
to make one revolution about the sun. This
relationship is given by

$$t = x^{3/2}.$$

The photo of Saturn shown above was taken by
Voyager I. The average distance between Saturn
and the sun is 9.541 astronomical units. How
long does it take Saturn to revolve about the sun?

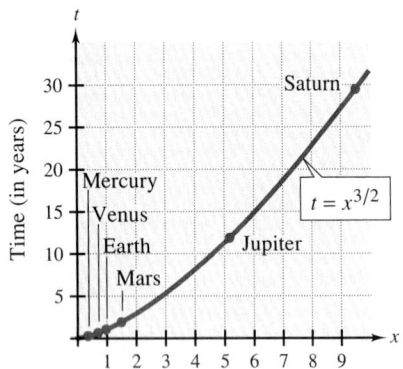

Integer Exponents and Scientific Notation

Integer Exponents   •   *Scientific Notation*

Integer Exponents

So far in the text, all exponents have been positive integers. In this section, we extend the concept of an exponent to include zero and negative integers.

If a is a real number such that $a \neq 0$, then a^0 is defined as 1. Moreover, if m is an integer, then a^{-m} is defined as the reciprocal of a^m.

Definition of Zero Exponents and Negative Exponents

Let a be a real number such that $a \neq 0$, and let m be an integer.

1. $a^0 = 1, \quad a \neq 0$ 2. $a^{-m} = \dfrac{1}{a^m}, \quad a \neq 0$

These definitions of zero and negative exponents are consistent with the properties of exponents given in Section 1.3. For instance, consider the following.

$$1 = \frac{x^6}{x^6} = x^{6-6} = x^0, \quad x \neq 0$$

$$\frac{1}{x^4} = \frac{x^2}{x^6} = x^{2-6} = x^{-4}, \quad x \neq 0$$

EXAMPLE 1 ■ **Zero Exponents and Negative Exponents**

Rewrite the following numbers without using zero exponents or negative exponents.

(a) 3^0 (b) 0^0 (c) 2^{-1} (d) 3^{-2}

A patterned approach may benefit some students. Have them complete the following pattern and relate the results to the appropriate definitions.

$2^3 = 8\} \div 2$
$2^2 = 4\} \div 2$
$2^1 = 2$
$2^0 =$
$2^{-1} =$

Solution

(a) $3^0 = 1$

(b) 0^0 is undefined. Remember that when raising a number to the zero power, you must be sure that the number is not zero.

(c) $2^{-1} = \dfrac{1}{2^1} = \dfrac{1}{2}$

(d) $3^{-2} = \dfrac{1}{3^2} = \dfrac{1}{9}$ ■

The following properties of exponents are valid for all integer exponents, including integer exponents that are zero or negative.

Properties of Exponents

Let m and n be integers, and let a and b represent real numbers, variables, or algebraic expressions.

Property

1. $a^m \cdot a^n = a^{m+n}$

2. $(ab)^m = a^m \cdot b^m$

3. $(a^m)^n = a^{mn}$

4. $\dfrac{a^m}{a^n} = a^{m-n}, \quad a \neq 0$

5. $\left(\dfrac{a}{b}\right)^m = \dfrac{a^m}{b^m}, \quad b \neq 0$

6. $\left(\dfrac{a}{b}\right)^{-m} = \left(\dfrac{b}{a}\right)^m, \quad \begin{array}{l} a \neq 0, \\ b \neq 0 \end{array}$

7. $a^0 = 1, \quad a \neq 0$

8. $a^{-m} = \dfrac{1}{a^m}, \quad a \neq 0$

Example

$x^4(x^3) = x^{4+3} = x^7$

$(3x)^2 = 3^2(x^2) = 9x^2$

$(x^2)^3 = x^{2 \cdot 3} = x^6$

$\dfrac{x^3}{x} = x^{3-1} = x^2, \quad x \neq 0$

$\left(\dfrac{x}{3}\right)^2 = \dfrac{x^2}{3^2} = \dfrac{x^2}{9}$

$\left(\dfrac{x}{3}\right)^{-2} = \left(\dfrac{3}{x}\right)^2 = \dfrac{3^2}{x^2} = \dfrac{9}{x^2}, \quad x \neq 0$

$(x^2 + 1)^0 = 1$

$x^{-2} = \dfrac{1}{x^2}, \quad x \neq 0$

EXAMPLE 2 ■ **Using Properties of Exponents**

Rewrite the following expressions using only positive exponents. (For each expression, assume that $x \neq 0$.)

(a) $2x^{-1}$ (b) $(2x)^{-1}$ (c) $\dfrac{3}{x^{-2}}$ (d) $\dfrac{1}{(3x)^{-2}}$

Ask students to compare these expressions involving zero exponents.

$2x^0 = 2$
$(2x)^0 = 1$
$2^0 x y^0 = x$

Solution

(a) $2x^{-1} = 2(x^{-1}) = 2\left(\dfrac{1}{x}\right) = \dfrac{2}{x}$

(b) $(2x)^{-1} = \dfrac{1}{(2x)^1} = \dfrac{1}{2x}$

(c) $\dfrac{3}{x^{-2}} = \dfrac{3}{\left(\dfrac{1}{x^2}\right)} = 3\left(\dfrac{x^2}{1}\right) = 3x^2$

Have students simplify $1/(3x^{-2})$ and compare their answers to Example 2d.

(d) $\dfrac{1}{(3x)^{-2}} = \dfrac{1}{\left[\dfrac{1}{(3x)^2}\right]} = \dfrac{1}{\left(\dfrac{1}{3^2 x^2}\right)} = \dfrac{1}{\left(\dfrac{1}{9x^2}\right)} = (1)\left(\dfrac{9x^2}{1}\right) = 9x^2$

■

As you become accustomed to working with negative exponents, you will probably not write as many steps as are shown in Example 2. To rewrite a fraction involving exponents, we like to use the following informal rule. *To move a factor from the numerator to the denominator or vice versa, change the sign of its exponent.* For instance, in the expression

$$\frac{3}{x^{-2}},$$

you can eliminate the negative exponent by "moving" the factor x^{-2} to the numerator and changing the exponent to 2. That is,

$$\frac{3}{x^{-2}} = 3x^2.$$

EXAMPLE 3   ■   **Using Properties of Exponents**

Rewrite the following expressions using only positive exponents. (For each expression, assume that $x \neq 0$ and $y \neq 0$.)

(a) $(-5x^{-3})^2$

(b) $-\left(\dfrac{7x}{y^2}\right)^{-2}$

Solution

(a) $(-5x^{-3})^2 = (-5)^2(x^{-3})^2 = 25x^{-6} = \dfrac{25}{x^6}$

(b) $-\left(\dfrac{7x}{y^2}\right)^{-2} = -\left(\dfrac{y^2}{7x}\right)^2 = -\dfrac{(y^2)^2}{(7x)^2} = -\dfrac{y^4}{49x^2}$ ■

EXAMPLE 4   ■   **Using Properties of Exponents**

Rewrite the following expressions using only positive exponents. (For each expression, assume that $x \neq 0$ and $y \neq 0$.)

(a) $\left(\dfrac{8x^{-1}y^4}{4x^3y^2}\right)^{-3}$

(b) $\dfrac{3xy^0}{x^2(5y)^0}$

Solution

Point out that students can simplify within parentheses first.

(a) $\left(\dfrac{8x^{-1}y^4}{4x^3y^2}\right)^{-3} = \left(\dfrac{2y^2}{x^4}\right)^{-3} = \left(\dfrac{x^4}{2y^2}\right)^3 = \dfrac{x^{12}}{2^3y^6} = \dfrac{x^{12}}{8y^6}$

(b) $\dfrac{3xy^0}{x^2(5y)^0} = \dfrac{3x(1)}{x^2(1)} = \dfrac{3}{x}$ ■

EXAMPLE 5   ■   **Using Properties of Exponents**

Rewrite the following expressions using only positive exponents. (For each expression, assume that $x \neq 0$ and $y \neq 0$.)

(a) $(5x^{-2}y^{-5})^{-2}(3xy^{-4})$

(b) $\dfrac{4x^{-5}y^3}{(7x^{-2}y)^{-1}}$

Solution

(a) $(5x^{-2}y^{-5})^{-2}(3xy^{-4}) = (5^{-2}x^4y^{10})(3xy^{-4}) = \dfrac{x^4y^{10}(3x)}{5^2y^4} = \dfrac{3x^5y^6}{25}$

(b) $\dfrac{4x^{-5}y^3}{(7x^{-2}y)^{-1}} = \dfrac{4x^{-5}y^3}{7^{-1}x^2y^{-1}} = \dfrac{4y^3 \cdot 7y}{x^2x^5} = \dfrac{28y^4}{x^7}$ ■

When you use the properties of exponents to rewrite an expression using only positive exponents, there is more than one way in which the steps of rewriting may be performed. However, the final results will be the same.

For instance, in Example 5(b), you could rewrite the expression as follows.

$$\dfrac{4x^{-5}y^3}{(7x^{-2}y)^{-1}} = (4x^{-5}y^3)(7x^{-2}y) = 28x^{-7}y^4 = \dfrac{28y^4}{x^7}$$

Notice that the final results are the same, although the steps are different.

Note the difference between the expressions

$$\dfrac{a^{-2}b}{c} \quad \text{and} \quad \dfrac{a^{-2}+b}{c}.$$

That is,

$$\dfrac{a^{-2}b}{c} = \dfrac{b}{a^2c}$$

but

$$\dfrac{a^{-2}+b}{c} = \dfrac{\dfrac{1}{a^2}+b}{c} = \dfrac{\dfrac{1+a^2b}{a^2}}{c} = \dfrac{1+a^2b}{a^2} \cdot \dfrac{1}{c} = \dfrac{1+a^2b}{a^2c}.$$

Have students simplify $(a^{-2}+b^{-2})/c^2$ and $(a^{-2}b^{-2})/c^2$. Compare the results. Answers: $(b^2 + a^2)/(a^2b^2c^2)$ and $1/(a^2b^2c^2)$

Scientific Notation

Exponents provide an efficient way of writing and computing with very large (or very small) numbers. For instance, a drop of water contains more than 33 billion billion molecules—that is, 33 followed by 18 zeros.

33,000,000,000,000,000,000

It is convenient to write such numbers in **scientific notation.** This notation has the form $c \times 10^n$, where $1 \leq c < 10$ and n is an integer. Thus, the number of molecules in a drop of water can be written in scientific notation as

$$3.3 \times 10,000,000,000,000,000,000 = 3.3 \times 10^{19}.$$

The *positive* exponent 19 indicates that the number is large (10 or more) and that the decimal point has been moved 19 places. A *negative* exponent in scientific notation indicates that the number is *small* (less than 1). For instance, the mass (in grams) of one electron is approximately

$$9.0^{-28} = 0.0000000000000000000000000009 \, .$$
                            28 decimal places

EXAMPLE 6 ■ **Converting from Decimal Notation to Scientific Notation**

Write each of the following real numbers in scientific notation.

(a) 0.0000684   (b) 937,200,000

Solution

(a) This decimal number is less than 1. Thus, to write the number in scientific notation, you should move the decimal point as follows.

$$0.0000684 = 6.84 \times 10^{-5}$$
          Five places

(b) This number is larger than 10. Thus, to write the number in scientific notation, you should move the decimal point as follows.

$$938,200,000.0 = 9.372 \times 10^{8}$$
          Eight places
                                                                        ■

EXAMPLE 7 ■ **Converting from Scientific Notation to Decimal Notation**

Convert each of the following numbers from scientific notation to decimal notation.

(a) 2.486×10^2 (b) 1.81×10^{-6}

Solution

(a) For this number, the exponent of 10 is a positive 2. Thus, to write the number in decimal notation, you should move the decimal point two places, as follows.

$$2.486 \times 10^2 = 248.6$$
                Two places

(b) For this number, the exponent of 10 is a negative 6. Thus, to write the number in decimal notation, you should move the decimal point six places, as follows.

$$1.81 \times 10^{-6} = 0.00000181$$
                Six places
                                                                        ■

To convert decimal numbers to scientific notation or vice versa, you can make the following associations.

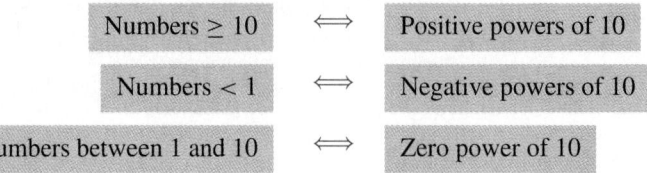

Numbers ≥ 10	$\Longleftrightarrow$	Positive powers of 10
Numbers < 1	$\Longleftrightarrow$	Negative powers of 10
Numbers between 1 and 10	$\Longleftrightarrow$	Zero power of 10

To demonstrate these three possibilities, have students write 2600, 0.00026, and 2.6 in scientific notation. Answers: 2.6×10^3, 2.6×10^{-4}, and 2.6×10^0

Most scientific calculators automatically switch to scientific notation when showing large (or small) numbers that exceed the display range. Try multiplying $86,500,000 \times 6000$. If your calculator follows standard conventions, its display should be

$$\boxed{5.19 \quad 11} \quad \text{or} \quad \boxed{5.19 \quad E \quad 11}.$$

Discuss how to operate a calculator using scientific notation. Remind students that the number on the right of the display is the exponent of base 10.

This means that $c = 5.19$ and the exponent of 10 is $n = 11$. Therefore, the number is 5.19×10^{11}.

To *enter* numbers in scientific notation, your calculator should have an exponential entry key labeled $\boxed{EE}$ or $\boxed{EXP}$. If you were to perform the preceding multiplication using scientific notation, you could begin by writing

$$86,500,000 \times 6000 = (8.65 \times 10^7)(6.0 \times 10^3)$$

and then entering the following.*

$8.65 \boxed{EXP} 7 \boxed{\times} 6 \boxed{EXP} 3 \boxed{=}$ Scientific

$8.65 \boxed{EE} 7 \boxed{\times} 6 \boxed{EE} 3 \boxed{ENTER}$ Graphing

EXAMPLE 8 ■ Using Scientific Notation with a Calculator

Use a calculator to perform the following operations.

(a) $65,000 \times 3,400,000,000$ (b) $0.000000348 \div 870$

Solution

(a) Because $65,000 = 6.5 \times 10^4$ and $3,400,000,000 = 3.4 \times 10^9$, you can multiply the two numbers using the following calculator steps.

$6.5 \boxed{EXP} 4 \boxed{\times} 3.4 \boxed{EXP} 9 \boxed{=}$ Scientific

$6.5 \boxed{EE} 4 \boxed{\times} 3.4 \boxed{EE} 9 \boxed{ENTER}$ Graphing

*The graphing calculator keystrokes in this text correspond to the TI-81 and TI-82 graphing calculators from Texas Instruments. For other calculators, the keystrokes may differ.

After these keystrokes have been entered, the calculator display should read
2.21 14 or 2.21 E 14 . Therefore, the product of the two numbers is

$(6.5 \times 10^4)(3.4 \times 10^9) = 2.21 \times 10^{14} = 221,000,000,000,000.$

(b) Because $0.000000348 = 3.48 \times 10^{-7}$ and $870 = 8.7 \times 10^2$, you can divide the two numbers as follows.

3.48 [EXP] [+/−] 7 [÷] 8.7 [EXP] 2 [=]   Scientific

3.48 [EE] [(−)] 7 [÷] 8.7 [EE] 2 [ENTER]   Graphing

After these keystrokes have been entered, the calculator display should read
4.0 −10 or 4 E −10 . Therefore, the quotient of the two numbers is

$$\frac{3.48 \times 10^{-7}}{8.7 \times 10^2} = 4.0 \times 10^{-10} = 0.0000000004. \qquad \blacksquare$$

EXAMPLE 9   ■   **Using Scientific Notation and Properties of Exponents**

Use scientific notation and the properties of exponents to simplify the following expression. Then evaluate the expression.

$$\frac{(2,400,000,000)(0.00000345)}{(0.00007)(3800)}$$

Solution

Begin by rewriting each number in scientific notation and simplifying.

$$\frac{(2,400,000,000)(0.00000345)}{(0.00007)(3800)} = \frac{(2.4 \times 10^9)(3.45 \times 10^{-6})}{(7.0 \times 10^{-5})(3.8 \times 10^3)}$$

$$= \frac{(2.4)(3.45)(10^3)}{(7)(3.8)(10^{-2})}$$

$$= \frac{(8.28)(10^5)}{26.6}$$

$$\approx 0.3112782(10^5)$$

$$= 31,127.82 \qquad \blacksquare$$

EXAMPLE 10 ■ Average Amount Spent on Swimwear

The U.S. population in 1990 was 250 million. Use the information shown in Figure 6.1 to find the amount that an average American spent on swimwear in 1990. (*Source: NSGA Sports Business Today,* 1990)

Solution

From the bar graph shown in Figure 6.1, you know that the total amount spent on swimwear in 1990 was $1.8 billion. To find the average amount spent on swimwear in 1990, divide the total amount spent by the number of people.

$$\frac{\text{Average amount}}{\text{per person}} = \frac{\text{Total amount}}{\text{Number of people}}$$

$$= \frac{1.8 \text{ billion}}{250 \text{ million}}$$

$$= \frac{1.8 \times 10^9}{2.5 \times 10^8}$$

$$= \frac{1.8}{2.5} \times 10^1$$

$$= 7.2$$

Thus, the average amount spent per person was $7.20.

FIGURE 6.1

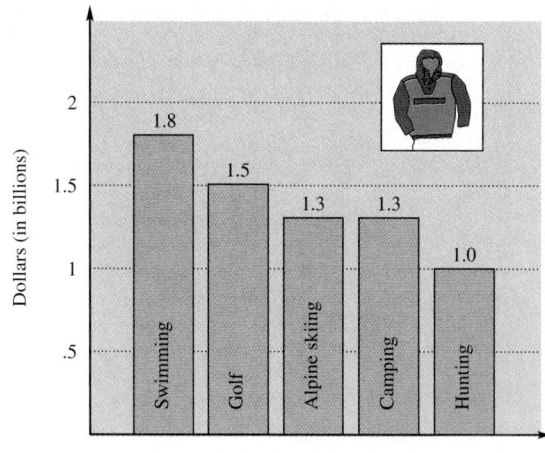

Dollars (in billions)

Swimming 1.8, Golf 1.5, Alpine skiing 1.3, Camping 1.3, Hunting 1.0

Sport

DISCUSSION PROBLEM ■ Limiting Value of an Expression

(a) Use a calculator to complete the following table, and write a short paragraph describing the value of the expression x^{-1} as the value of x grows larger and larger.

x	10^0	10^1	10^2	10^3	10^4	10^5	10^6
x^{-1}							

(b) Use a calculator to complete the following table, and write a short paragraph describing the value of the expression x^{-1} as the value of x gets closer and closer to zero (through positive values).

x	10^0	10^{-1}	10^{-2}	10^{-3}	10^{-4}	10^{-5}	10^{-6}
x^{-1}							

Warm-Up

The following warm-up exercises involve skills that were covered in earlier sections. You will use these skills in the exercise set for this section.

In Exercises 1–10, simplify the expression.

1. $a^4 \cdot a^6$

2. $x^2 \cdot y^4 \cdot x^3$

3. $\dfrac{81y^5}{3^2 y^2}, \quad y \neq 0$

4. $\dfrac{64r^2 s^4}{16r s^2}, \quad r \neq 0, \ s \neq 0$

5. $(-t^2)^3$

6. $(4z)^2$

7. $(-3x^2 y^3)^2 \cdot (4xy^2)$

8. $(-4uv^2)^3 \cdot (2u^2 v)$

9. $\left(\dfrac{2a^2}{3b}\right)^4, \quad b \neq 0$

10. $\left(\dfrac{3x}{4y^3}\right)^2, \quad y \neq 0$

6.1   EXERCISES

In Exercises 1–30, evaluate the expression.

1. 5^{-2}

2. 2^{-4}

3. -10^{-3}

4. -20^{-2}

5. $(-3)^{-5}$

6. 25^0

7. $\dfrac{1}{4^{-3}}$

8. $\dfrac{1}{-8^{-2}}$

9. $\dfrac{1}{(-2)^{-5}}$

10. $-\dfrac{1}{6^2}$

11. $\left(\frac{2}{3}\right)^{-1}$ **12.** $\left(\frac{4}{5}\right)^{-3}$ **13.** $\left(\frac{3}{16}\right)^{0}$ **14.** $\left(-\frac{5}{8}\right)^{-2}$ **15.** $27 \cdot 3^{-3}$

16. $4^2 \cdot 4^{-3}$ **17.** $\dfrac{3^4}{3^{-2}}$ **18.** $\dfrac{5^{-1}}{5^2}$ **19.** $\dfrac{10^3}{10^{-2}}$ **20.** $\dfrac{10^{-5}}{10^{-6}}$

21. $(2^{-3})^2$ **22.** $(-4^{-1})^{-2}$ **23.** $(4^2 \cdot 4^{-1})^{-2}$ **24.** $(5^3 \cdot 5^{-4})^{-3}$ **25.** $2^{-3} + 2^{-4}$

26. $4 - 3^{-2}$ **27.** $\left(\frac{3}{4} + \frac{5}{8}\right)^{-2}$ **28.** $\left(\frac{1}{2} - \frac{2}{3}\right)^{-1}$ **29.** $(5^0 - 4^{-2})^{-1}$ **30.** $(32 + 4^{-3})^0$

In Exercises 31–74, rewrite the expression using only positive exponents and simplify. (Assume that any variables in the expression are nonzero.)

31. $y^4 \cdot y^{-2}$ **32.** $x^{-2} \cdot x^{-5}$ **33.** $\dfrac{1}{x^{-6}}$ **34.** $\dfrac{x^{-3}}{y^{-1}}$ **35.** $\dfrac{x^4}{x^{-2}}$

36. $\dfrac{s^{-3}}{s^3}$ **37.** $\dfrac{a^{-6}}{a^{-7}}$ **38.** $\dfrac{6u^{-2}}{15u^{-1}}$ **39.** $\dfrac{(4t)^0}{t^{-2}}$ **40.** $\dfrac{(5u)^4}{(5u)^4}$

41. $(-3x^{-3}y^2)(4x^2y^{-5})$ **42.** $(5s^5t^{-5})\left(\dfrac{3s^{-2}}{50t^{-1}}\right)$ **43.** $(3x^2y^{-2})^{-2}$

44. $(-4y^{-3}z)^{-3}$ **45.** $\dfrac{6^2x^3y^{-3}}{12x^{-2}y}$ **46.** $\dfrac{2^{-4}y^{-1}z^{-3}}{4^{-2}yz^{-3}}$

47. $\left(\dfrac{3u^2v^{-1}}{3^3u^{-1}v^3}\right)^{-2}$ **48.** $\left(\dfrac{5^2x^3y^{-3}}{125xy}\right)\left(\dfrac{5x}{3y}\right)^{-1}$ **49.** $[(2x^{-3}y^{-2})^2]^{-2}$

50. $\left[\left(\dfrac{2x^2}{4y}\right)^{-3}\right]^2$ **51.** $(2x^2)^{-2}$ **52.** $\left(\dfrac{4}{y}\right)^3\left(\dfrac{3}{y}\right)^4$

53. $\left(\dfrac{4}{z}\right)^{-2}$ **54.** $\left(\dfrac{3z^2}{x}\right)^{-2}$ **55.** $\left(\dfrac{x^{-3}y^4}{5}\right)^{-3}$

56. $\left(\dfrac{5x^2}{y^{-2}}\right)^{-4}$ **57.** $(5x^2y^4z^6)^3(5x^2y^4z^6)^{-3}$ **58.** $(4a^{-2}b^3)^{-3}$

59. $\left(\dfrac{x}{10}\right)^{-1}$ **60.** $[(x^2y^{-2})^{-1}]^{-1}$ **61.** $\left(\dfrac{a^{-2}}{b^{-2}}\right)\left(\dfrac{b}{a}\right)^3$

62. $\left(\dfrac{a^{-3}}{b^{-3}}\right)\left(\dfrac{a}{b}\right)^3$ **63.** $(s^4t^{-4})^{-4}(s^{-4}t^4)^4$ **64.** $[(uv)^{-3}]^4$

65. $(4m)^3\left(\dfrac{1}{4m}\right)^{-3}$ **66.** $\left(\dfrac{2x}{3}\right)^2\left(\dfrac{4x^2}{9}\right)$ **67.** $[(x^4y^6)^{-1}]^2$

68. $[(x^{-4}y^{-6})^{-1}]^2$ **69.** $(ab)^{-2}(a^2b^2)^{-1}$ **70.** $\left(\dfrac{2}{x}\right)^{-2}\left(\dfrac{3}{x}\right)^{-3}$

71. $(u + v^{-2})^{-1}$ **72.** $x^{-2}(x^2 + y^2)$ **73.** $\dfrac{a + b}{ba^{-1} - ab^{-1}}$

74. $\dfrac{u^{-1} - v^{-1}}{u^{-1} + v^{-1}}$

In Exercises 75–80, write the number in scientific notation.

75. *Land Area of Earth:* 57,500,000 square miles

76. *Ocean Area of Earth:* 139,400,000 square miles

77. *Light-Year:* 9,461,000,000,000,000 kilometers

78. *Thickness of a Soap Bubble:* 0.0000001 meter

79. *Relative Density of Hydrogen:* 0.0000899

80. *One Micron (Millionth of a Meter):* 0.00003937 inch

In Exercises 81–86, write the number in decimal notation.

81. *U.S. Daily Coca-Cola Consumption:* 5.24×10^8 servings

82. *Number of Air Sacs in Lungs:* 3.5×10^8

83. *Interior Temperature of the Sun:* 1.3×10^7 degrees Celsius

84. *Width of an Air Molecule:* 9.0×10^{-9}

85. *Charge of an Electron:* 4.8×10^{-10} electrostatic units

86. *Width of a Human Hair:* 9.0×10^{-4} meters

In Exercises 87–92, evaluate the quantity without using a calculator.

87. $(2 \times 10^9)(3.4 \times 10^{-4})$

88. $(5 \times 10^4)^3$

89. $\dfrac{3.6 \times 10^9}{9 \times 10^5}$

90. $\dfrac{2.5 \times 10^{-3}}{5 \times 10^2}$

91. $(4,500,000)(2,000,000,000)$

92. $\dfrac{64,000,000}{0.00004}$

In Exercises 93–96, evaluate the quantity using a calculator. (Write your answer in scientific notation.)

93. $\dfrac{1.357 \times 10^{12}}{(4.2 \times 10^2)(6.87 \times 10^{-3})}$

94. $(8.67 \times 10^4)^7$

95. $\dfrac{(0.0000565)(2,850,000,000,000)}{0.00465}$

96. $\dfrac{(5,000,000)^3(0.000037)^2}{(0.005)^4}$

In Exercises 97 and 98, find an expression for the area of the figure.

97.

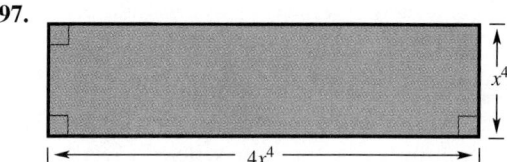

98.

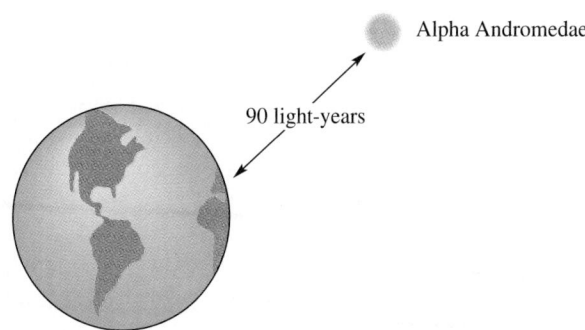

99. *Copper Electrons* A cube of copper that measures 1 centimeter on an edge has approximately 8×10^{22} free electrons. Write this real number in decimal form.

100. *Distance to the Sun* The distance from the earth to the sun is approximately 93 million miles. Write this distance in scientific notation.

101. *Light-Year* One light-year (the distance light can travel in 1 year) is approximately 9.45×10^{15} meters. Approximate the time required for light to travel from the sun to the earth if that distance is approximately 1.49×10^{11} meters.

102. *Distance to a Star* Determine the distance (in meters) to the star Alpha Andromedae if it is 90 light-years away from the earth (see figure). (See Exercise 101 for the definition of a light-year.)

103. *Masses of Earth and Sun* The masses of the earth and the sun are approximately 5.975×10^{24} kg and 1.99×10^{30} kg, respectively. The mass of the sun is approximately how many times that of the earth?

104. *Exploratory Exercise* In 1619, Johannes Kepler, a German astronomer, discovered that the period T (in years) of each planet in our solar system is related to the planet's mean distance R (in astronomical units) from the sun by the equation

$$\frac{T^2}{R^3} = k.$$

Test Kepler's equation for the nine planets in our solar system, using the table. Do you get approximately the same value of k for each planet? (Astronomical units relate the other planets' periods and mean distances to the earth's period and mean distance.)

Planet	T	R
Mercury	0.241	0.387
Venus	0.615	0.723
Earth	1.000	1.000
Mars	1.881	1.523
Jupiter	11.861	5.203
Saturn	29.457	9.541
Uranus	84.008	19.190
Neptune	164.784	30.086
Pluto	248.350	39.508

105. *Federal Debt* In 1989, the population of the United States was 247,350,000 people, and the federal debt was 2866.2 billion dollars. Use these two numbers to determine the amount each person would have had to pay to remove the debt.

SECTION	**Rational Exponents and Radicals**
6.2	*Roots and Radicals* ● *Rational Exponents* ● *Radicals and Calculators*

Roots and Radicals

In Section 1.1, we reviewed the use of radical notation to represent nth roots of real numbers. Recall from that section that b is called an nth root of a if $a = b^n$. Also recall that the principal nth root of a real number is defined as follows.

Principal nth Root of a Number	Let a be a real number that has at least one (real number) nth root. The **principal nth root of a** is the nth root that has the same sign as a, and it is denoted by the **radical symbol** $\sqrt[n]{a}.$ Principal nth root The positive integer n is called the **index** of the radical, and the number a is called the **radicand.** If $n = 2$, the index is omitted, and the root is written $\sqrt{a}$ rather than $\sqrt[2]{a}$.

Therefore, $\sqrt{49} = 7$, $\sqrt[3]{1000} = 10$, and $\sqrt[5]{-32} = -2$.

You need to be aware of the following properties of nth roots. (Remember that for nth roots n is an integer that is greater than or equal to 2.)

1. If a is a positive real number and n is *even,* then a has exactly two (real) nth roots. These roots are denoted by

$$\sqrt[n]{a} \quad \text{and} \quad -\sqrt[n]{a}.$$

2. If a is any real number and n is *odd,* then a has exactly one (real) nth root, which is denoted by $\sqrt[n]{a}$.

3. If a is a negative real number and n is *even,* then a has no (real) nth root.

Integers such as 1, 4, 9, 16, 49, and 81 are called **perfect squares** because they have integer square roots. Similarly, integers such as 1, 8, 27, 64, and 125 are called **perfect cubes** because they have integer cube roots.

EXAMPLE 1   ■   Evaluating Expressions Involving Radicals

Evaluate the following radical expressions.

(a) $\sqrt{144}$ (b) $-\sqrt{81}$ (c) $-\sqrt[3]{-64}$ (d) $\sqrt[4]{1}$

Solution

Have students compare $-\sqrt{-16}$ with the expression $-\sqrt[3]{-64}$ in (c). Answer: $-\sqrt{-16} = -(4i) = -4i$

(a) $\sqrt{144} = 12$

(b) $-\sqrt{81} = -9$

(c) $-\sqrt[3]{-64} = -(-4) = 4$

(d) $\sqrt[4]{1} = 1$ ■

Raising a number to the nth power and taking the principal nth root of a number can be thought of as *inverse* operations. For instance,

$$\left(\sqrt{4}\right)^2 = 2^2 = 4 \quad \text{and} \quad \sqrt{2^2} = \sqrt{4} = 2.$$

Similarly,

$$\left(\sqrt[3]{27}\right)^3 = 3^3 = 27 \quad \text{and} \quad \sqrt[3]{3^3} = \sqrt[3]{27} = 3.$$

For *even* roots involving *negative* numbers, this pattern breaks down a little. However, using absolute value signs, you can make the following generalizations.

Inverse Properties of *n*th Powers and *n*th Roots

Let a be a real number, and let n be an integer such that $n \geq 2$.

1. If a has a principal nth root, then
$$\left(\sqrt[n]{a}\right)^n = a.$$

2. If n is *odd*, then
$$\sqrt[n]{a^n} = a.$$

If n is *even*, then
$$\sqrt[n]{a^n} = |a|.$$

EXAMPLE 2 ■ Evaluating Radical Expressions

Evaluate the following radical expressions.

(a) $\sqrt[3]{5^3}$　　(b) $\sqrt[3]{(-2)^3}$　　(c) $\left(\sqrt{7}\right)^2$　　(d) $\sqrt{(-3)^2}$

Solution

(a) Because the index of the radical is odd, you can write
$$\sqrt[3]{5^3} = 5.$$

(b) The index of this radical is also odd, and you can write
$$\sqrt[3]{(-2)^3} = -2.$$

(c) Using the inverse property of powers and roots, you can write
$$\left(\sqrt{7}\right)^2 = 7.$$

(d) Because the index of this radical is even, you must include absolute value signs, and write
$$\sqrt{(-3)^2} = |-3| = 3.$$

Note that $\sqrt{(-3)^2} = \sqrt{9} = 3.$ ■

Rational Exponents

Rational exponents such as $4^{1/2}$ and $8^{1/3}$ are defined as follows.

<table>
<tr><td>

Definition of Rational Exponents

</td><td>

Let a be a real number, and let n be an integer such that $n \geq 2$. If the principal nth root of a exists, then $a^{1/n}$ is defined as

$$a^{1/n} = \sqrt[n]{a}.$$

Moreover, if m is a positive integer that has no common factor with n, then

$$a^{m/n} = \left(a^{1/n}\right)^m = \left(\sqrt[n]{a}\right)^m \quad \text{and} \quad a^{m/n} = (a^m)^{1/n} = \sqrt[n]{a^m}.$$

</td></tr>
</table>

Ask students to simplify $27^{4/3}$ using both methods and to decide which is easier. Students should be familiar with both evaluation techniques.

The numerator of a rational exponent denotes the *power* to which the base is raised, and the denominator denotes the *root* to be taken. Moreover, it doesn't matter in which order the two operations are performed, provided the nth root exists. Here is an example.

$$8^{2/3} = \left(\sqrt[3]{8}\right)^2 = 2^2 = 4 \qquad \text{Cube root, then second power}$$
$$8^{2/3} = \sqrt[3]{8^2} = \sqrt[3]{64} = 4 \qquad \text{Second power, then cube root}$$

The properties of exponents listed in Section 6.1 also apply to rational exponents (provided the roots indicated by the denominators exist). Those properties are listed here again, with different examples.

<table>
<tr><td>

Properties of Exponents

</td><td>

Let r and s be rational numbers, and let a and b be real numbers, variables, or algebraic expressions. If the roots indicated by the rational exponents exist, then the following properties are true.

Property	*Example*
1. $a^r \cdot a^s = a^{r+s}$	$4^{1/2}(4^{1/3}) = 4^{5/6}$
2. $(ab)^r = a^r \cdot b^r$	$(2x)^{1/2} = 2^{1/2}(x^{1/2})$
3. $(a^r)^s = a^{rs}$	$(x^3)^{1/3} = x$
4. $\dfrac{a^r}{a^s} = a^{r-s}, \quad a \neq 0$	$\dfrac{x^2}{x^{1/2}} = x^{2-(1/2)} = x^{3/2}$
5. $\left(\dfrac{a}{b}\right)^r = \dfrac{a^r}{b^r}, \quad b \neq 0$	$\left(\dfrac{x}{3}\right)^{1/3} = \dfrac{x^{1/3}}{3^{1/3}}$
6. $\left(\dfrac{a}{b}\right)^{-r} = \left(\dfrac{b}{a}\right)^r, \quad a \neq 0, \ b \neq 0$	$\left(\dfrac{x}{4}\right)^{-1/2} = \left(\dfrac{4}{x}\right)^{1/2} = \dfrac{2}{x^{1/2}}$
7. $a^{-r} = \dfrac{1}{a^r}, \quad a \neq 0$	$4^{-1/2} = \dfrac{1}{4^{1/2}} = \dfrac{1}{2}$
8. $a^0 = 1, \quad a \neq 0$	$(10x^3)^0 = 1$

</td></tr>
</table>

EXAMPLE 3   ■   **Evaluating Expressions with Rational Exponents**

Evaluate the following expressions.

(a) $8^{4/3}$ (b) $(4^2)^{3/2}$ (c) $25^{-3/2}$

Solution

(a) $8^{4/3} = \left(\sqrt[3]{8}\right)^4 = 2^4 = 16$

(b) $(4^2)^{3/2} = 4^{2(3/2)} = 4^3 = 64$

(c) $25^{-3/2} = \dfrac{1}{25^{3/2}} = \dfrac{1}{\left(\sqrt{25}\right)^3} = \dfrac{1}{5^3} = \dfrac{1}{125}$ ■

EXAMPLE 4   ■   **Evaluating Expressions with Rational Exponents**

Evaluate the following expressions.

(a) $\left(\frac{64}{125}\right)^{2/3}$ (b) $-9^{1/2}$ (c) $(-9)^{1/2}$

Solution

(a) $\left(\dfrac{64}{125}\right)^{2/3} = \left(\sqrt[3]{\dfrac{64}{125}}\right)^2 = \left(\dfrac{\sqrt[3]{64}}{\sqrt[3]{125}}\right)^2 = \left(\dfrac{4}{5}\right)^2 = \dfrac{16}{25}$

(b) $-9^{1/2} = -\sqrt{9} = -3$

(c) $(-9)^{1/2} = \sqrt{-9}$ is not a real number.   ■

EXAMPLE 5   ■   **Using Properties of Exponents**

Rewrite the following expressions using rational exponents.

(a) $x\sqrt[4]{x^3}$ (b) $\dfrac{\sqrt[3]{x^2}}{\sqrt{x^3}}$

Solution

(a) $x\sqrt[4]{x^3} = x(x^{3/4}) = x^{1+(3/4)} = x^{7/4}$

(b) $\dfrac{\sqrt[3]{x^2}}{\sqrt{x^3}} = \dfrac{x^{2/3}}{x^{3/2}} = x^{(2/3)-(3/2)} = x^{-5/6} = \dfrac{1}{x^{5/6}}$ ■

EXAMPLE 6   ■   **Using Properties of Exponents**

Use the properties of exponents to simplify the following expressions.

(a) $x^{3/5} \cdot x^{2/3}$ (b) $(x^{3/5})^{2/3}$ (c) $(6x^{3/5}y^{-3/2})^2, \quad y \neq 0$

Stress to students that the basic rules of exponents apply even if the exponents are negative or fractions. Give a comparison example such as:

$$x^2 \cdot x^4 = x^6$$
$$x^{-2} \cdot x^4 = x^2$$
$$x^{1/2} \cdot x^{1/4} = x^{3/4}$$

Solution

(a) $x^{3/5} \cdot x^{2/3} = x^{(3/5)+(2/3)} = x^{19/15}$

(b) $(x^{3/5})^{2/3} = x^{(3/5)(2/3)} = x^{2/5}$

(c) $(6x^{3/5}y^{-3/2})^2 = 6^2 x^{(3/5)2} y^{(-3/2)2} = \dfrac{36x^{6/5}}{y^3}$ ■

EXAMPLE 7   ■   **Using Properties of Exponents**

Use the properties of exponents to simplify the following expressions.

(a) $\sqrt{\sqrt[3]{x}}$

(b) $\dfrac{(2x-1)^{4/3}}{\sqrt[3]{2x-1}}, \quad x \neq \dfrac{1}{2}$

Solution

(a) $\sqrt{\sqrt[3]{x}} = \sqrt{x^{1/3}} = (x^{1/3})^{1/2} = x^{1/6} = \sqrt[6]{x}$

(b) $\dfrac{(2x-1)^{4/3}}{\sqrt[3]{2x-1}} = \dfrac{(2x-1)^{4/3}}{(2x-1)^{1/3}}$

$= (2x-1)^{(4/3)-(1/3)}$

$= 2x - 1, \quad x \neq \frac{1}{2}$ ■

Radicals and Calculators

Remind students that perfect n^{th} powers may not be obvious with some fractions until the fractions are reduced.

$$\sqrt[3]{\dfrac{54}{128}} = \sqrt[3]{\dfrac{27}{64}} = \sqrt[3]{\dfrac{3^3}{4^3}} = \sqrt[3]{\left(\dfrac{3}{4}\right)^4} = \dfrac{3}{4}$$

Some radicals, such as $\sqrt{16}$ and $\sqrt[3]{\dfrac{8}{27}}$, can be evaluated mentally because they are perfect nth powers ($\sqrt{16} = 4$ and $\sqrt[3]{\dfrac{8}{27}} = \dfrac{2}{3}$). These types of radicals are rational numbers. Most radicals, however, are not "perfect." They are irrational numbers that are best evaluated with a calculator. For instance, $\sqrt{7} = 2.6457513\ldots$.

There are two methods of evaluating radicals on most calculators. For square roots, you can use the *square root key* $\boxed{\checkmark}$. For other roots, you should first convert the radical to exponential form and then use the *exponential key* $\boxed{y^x}$ or $\boxed{\wedge}$.

NOTE For cube roots, check to see whether your calculator has a cube root key. For instance, the TI-81 and TI-82 have such a key as part of their "MATH" menu.

EXAMPLE 8 ■ Evaluating a Square Root with a Calculator

Use a calculator to evaluate $\sqrt{5}$. Round your answer to three decimal places.

Solution

Using the following calculator steps*

| 5 | √ | Scientific |

| √ | 5 | ENTER | Graphing |

results in a calculator display of 2.236068. Rounded to three decimal places, you have

$$\sqrt{5} \approx 2.236.$$

Try entering $5^{1/2}$ on your calculator to see that it yields the same result. ■

EXAMPLE 9 ■ Evaluating a Fifth Root with a Calculator

Use a calculator to evaluate $\sqrt[5]{25}$. Round your answer to three decimal places.

Solution

Begin by writing $\sqrt[5]{25}$ in exponential form.

$$\sqrt[5]{25} = 25^{1/5}$$

Alternatively, in this case you might show that $\sqrt[5]{25} = 25^{1/5} = 25^{0.2} \approx 1.904$.

Then use the following calculator steps.

The calculator display should read 1.9036539. Thus, you have

$$\sqrt[5]{25} \approx 1.904.$$

■

EXAMPLE 10 ■ Evaluating Radicals with a Calculator

Use a calculator to evaluate the following expressions. Round your answers to three decimal places.

(a) $\sqrt[3]{-4}$ (b) $(1.4)^{-2/5}$

Solution

(a) Most scientific calculators do not have a cube root key. Thus, to find the cube root of a negative number with a scientific calculator, use the fact that

$$\sqrt[3]{-4} = \sqrt[3]{(-1)(4)} = \sqrt[3]{-1}\sqrt[3]{4} = -\sqrt[3]{4} = -(4^{1/3})$$

*The graphing calculator keystrokes in this text correspond to the TI-81 and TI-82 graphing calculators from Texas Instruments. For other calculators, the keystrokes may differ.

and attach the negative sign of the radicand at the end of the keystroke sequence.

4 $\boxed{y^x}$ $\boxed{(}$ 1 $\boxed{\div}$ 3 $\boxed{)}$ $\boxed{=}$ $\boxed{+/-}$ Scientific

$\boxed{\sqrt[3]{}}$ $\boxed{(-)}$ 4 $\boxed{\text{ENTER}}$ Graphing

The calculator display is -1.5874011, which implies that

$$\sqrt[3]{-4} \approx -1.587.$$

Note: To enter $\sqrt[3]{}$ on a TI-81 or TI-82, press $\boxed{\text{MATH}}$, cursor to $\sqrt[3]{}$, and press $\boxed{\text{ENTER}}$.

Have students evaluate $(-4)^{3/2}$ on their calculators. Ask students why their displays give an E or Error message.

(b) Using the following keystroke sequence

1.4 $\boxed{y^x}$ $\boxed{(}$ 2 $\boxed{\div}$ 5 $\boxed{+/-}$ $\boxed{)}$ $\boxed{=}$ Scientific

1.4 $\boxed{\wedge}$ $\boxed{(}$ $\boxed{(-)}$ 2 $\boxed{\div}$ 5 $\boxed{)}$ $\boxed{\text{ENTER}}$ Graphing

results in a calculator display of 0.8740751. Thus, you have

$$(1.4)^{-2/5} \approx 0.874. \qquad\blacksquare$$

The *Olympias* was built in 1987. It is a reconstruction of a trireme (a Greek galley ship). The ship's triple set of oars was operated by volunteers.

EXAMPLE 11 ■ Finding the Speed of a Ship

The speed s (in knots) of the *Olympias* was found to be related to the power P (in kilowatts) generated by the rowers according to the model

$$s = \sqrt[3]{\frac{100P}{3}}.$$

The volunteer crew of the *Olympias* was able to generate maximum power of 10.5 kilowatts. What was the ship's greatest speed? (*Source: Scientific American*)

FIGURE 6.2

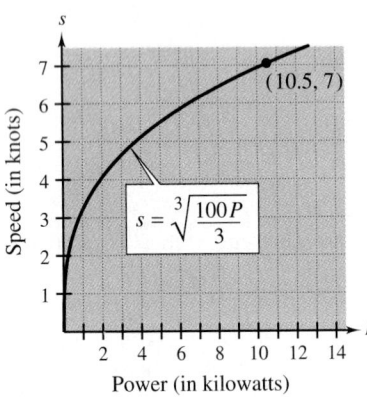

Power (in kilowatts)

Solution

To find the greatest speed, use the model and substitute 10.5 for P.

$$s = \sqrt[3]{\frac{100P}{3}}$$ Cube root model

$$= \sqrt[3]{\frac{100(10.5)}{3}}$$ Substitute 10.5 for P

$$\approx 7$$ Use a calculator

Thus, the greatest speed attained by the *Olympias* was about 7 knots (about 8 miles per hour). The graph in Figure 6.2 compares the speed s of the ship with the power P generated by the rowers.

■

DISCUSSION PROBLEM ■ **The Domain of a Radical Expression**

Describe the domain of each of the following functions. Write a short paragraph justifying each of your answers.

(a) $f(x) = \sqrt{x}$ (b) $f(x) = \dfrac{1}{\sqrt{x}}$

(c) $f(x) = \sqrt[3]{x}$ (d) $f(x) = \dfrac{1}{\sqrt[3]{x}}$ ▪

Warm-Up

The following warm-up exercises involve skills that were covered in earlier sections. You will use these skills in the exercise set for this section.

In Exercises 1–6, evaluate the expression.

1. $(-13)^2$ **2.** -13^2 **3.** $\left(\frac{2}{3}\right)^3$ **4.** $\frac{5}{6}\left(\frac{3}{5}\right)^2$ **5.** $\dfrac{3}{4^{-2}}$ **6.** $\left(\frac{2}{3}\right)^{-3}$

In Exercises 7–10, simplify the expression.

7. $(x^3 \cdot x^{-2})^{-3}$ **8.** $(5x^{-4}y^5)(-3x^2y^{-1})$

9. $\left(\dfrac{2x}{3y}\right)^{-2}$ **10.** $\left(\dfrac{7u^{-4}}{3v^{-2}}\right)\left(\dfrac{14u}{6v^2}\right)^{-1}$

6.2   EXERCISES

In Exercises 1–6, fill in each blank with the appropriate real number or word.

1. Because $7^2 = 49$, [blank] is a square root of 49.

2. Because $24.5^2 = 600.25$, [blank] is a square root of 600.25.

3. Because $4.2^3 = 74.088$, [blank] is a cube root of 74.088.

4. Because $6^4 = 1296$, [blank] is a fourth root of 1296.

5. Because $45^2 = 2025$, 45 is a [blank] root of 2025.

6. Because $12^3 = 1728$, 12 is a [blank] root of 1728.

In Exercises 7–10, find the indicated square root of the real number without using a calculator. (If not possible, state the reason.)

7. $\sqrt{81}$ **8.** $\sqrt{-144}$ **9.** $\sqrt{\frac{9}{16}}$ **10.** $\sqrt{0.09}$

In Exercises 11–30, evaluate the indicated nth root of the real number without using a calculator. (If not possible, state the reason.)

11. $\sqrt{64}$ **12.** $-\sqrt{100}$ **13.** $\sqrt{-100}$ **14.** $\sqrt{144}$ **15.** $-\sqrt{\frac{4}{9}}$

16. $\sqrt{-\frac{9}{25}}$ **17.** $\sqrt{0.16}$ **18.** $-\sqrt{0.0009}$ **19.** $\sqrt[3]{125}$ **20.** $\sqrt[3]{-8}$

21. $\sqrt[3]{-\frac{1}{64}}$ **22.** $-\sqrt[3]{0.008}$ **23.** $\sqrt[4]{81}$ **24.** $\sqrt[5]{32}$ **25.** $-\sqrt[4]{-625}$

26. $-\sqrt[4]{\frac{1}{625}}$ **27.** $\sqrt[5]{-0.00243}$ **28.** $\sqrt[6]{64}$ **29.** $\sqrt{49 - 4(2)(-15)}$ **30.** $\sqrt{\frac{75}{3}}$

In Exercises 31–38, use the inverse properties of nth powers and nth roots to evaluate the radical expression.

31. $(\sqrt{14})^2$ **32.** $\sqrt{10^2}$ **33.** $\sqrt{(-13)^2}$ **34.** $\sqrt[4]{(-20)^4}$ **35.** $(\sqrt[3]{-20})^3$

36. $(\sqrt[3]{-35})^3$ **37.** $\sqrt[5]{(16)^5}$ **38.** $\sqrt[6]{(2)^6}$

In Exercises 39–44, fill in the missing description.

	Radical Form	Rational Exponent Form		Radical Form	Rational Exponent Form
39.	$\sqrt{16} = 4$	[blank]	**40.**	$\sqrt[4]{81} = 3$	[blank]
41.	$\sqrt[3]{27^2} = 9$	[blank]	**42.**	[blank]	$125^{1/3} = 5$
43.	[blank]	$256^{3/4} = 64$	**44.**	[blank]	$32^{3/5} = 8$

In Exercises 45–56, evaluate the quantity without using a calculator.

45. $25^{1/2}$ **46.** $-121^{1/2}$ **47.** $16^{3/4}$ **48.** $81^{-3/4}$ **49.** $32^{-2/5}$

50. $243^{-3/5}$ **51.** $\left(\frac{8}{27}\right)^{2/3}$ **52.** $\left(\frac{256}{625}\right)^{1/4}$ **53.** $\left(\frac{121}{9}\right)^{-1/2}$ **54.** $\left(\frac{27}{1000}\right)^{-4/3}$

55. 3.4^0 **56.** $0.008^{-1/3}$

In Exercises 57–60, determine whether the square root is a rational or an irrational number.

57. $\sqrt{6}$ **58.** $\sqrt{\frac{9}{16}}$ **59.** $\sqrt{900}$ **60.** $\sqrt{72}$

In Exercises 61–70, use a calculator to approximate the quantity accurate to four decimal places. (If not possible, state the reason.)

61. $\sqrt{73}$ **62.** $\sqrt{-532}$ **63.** $\dfrac{8-\sqrt{35}}{2}$ **64.** $\dfrac{-5+\sqrt{3215}}{10}$ **65.** $1698^{-3/4}$

66. $962^{2/3}$ **67.** $\sqrt[4]{342}$ **68.** $\sqrt[3]{159}$ **69.** $\sqrt[3]{545^2}$ **70.** $\sqrt[5]{-35^3}$

In Exercises 71–86, use the properties of exponents to simplify the expression.

71. $\sqrt{t^2}$ **72.** $\sqrt[3]{z^3}$ **73.** $\sqrt[3]{y^9}$ **74.** $\sqrt[4]{a^8}$

75. $x^{2/3} \cdot x^{7/3}$ **76.** $z^{3/5} \cdot z^{-2/5}$ **77.** $(3x^{-1/3}y^{3/4})^2$ **78.** $(-2u^{3/5}v^{-1/5})^3$

79. $\dfrac{18y^{4/3}z^{-1/3}}{24y^{-2/3}z}$ **80.** $\dfrac{a^{3/4} \cdot a^{1/2}}{a^{5/2}}$ **81.** $\left(\dfrac{x^{1/4}}{x^{1/6}}\right)^3$ **82.** $\left(\dfrac{3m^{1/6}n^{1/3}}{4n^{-2/3}}\right)^2$

83. $(c^{3/2})^{1/3}$ **84.** $(k^{-1/3})^{3/2}$ **85.** $\sqrt{\sqrt[4]{y}}$ **86.** $\sqrt[3]{\sqrt{2x}}$

In Exercises 87–90, perform the indicated multiplication.

87. $x^{1/2}(2x-3)$ **88.** $x^{4/3}(3x^2-4x+5)$ **89.** $y^{-1/3}(y^{1/3}+5y^{4/3})$ **90.** $(x^{1/2}-3)(x^{1/2}+3)$

In Exercises 91–96, determine the domain of the function.

91. $f(x)=3\sqrt{x}$ **92.** $g(x)=\sqrt[4]{x}$ **93.** $g(x)=\sqrt[5]{x}$ **94.** $h(x)=\sqrt[3]{x^3}$

95. $f(x)=\dfrac{2}{\sqrt[4]{x}}$ **96.** $f(x)=\dfrac{10}{\sqrt[5]{x}}$

Declining Balances Depreciation In Exercises 97 and 98, find the annual depreciation rate r for the given depreciation problem. To find the annual depreciation rate by the **declining balances method,** use the formula

$$r = 1 - \left(\frac{S}{C}\right)^{1/n},$$

where n is the useful life of the item (in years), S is the salvage value (in dollars), and C is the original cost (in dollars).

97. A truck with an original cost of \$75,000 is depreciated over an 8-year period, as shown at the top of the next page.

Figure for 97

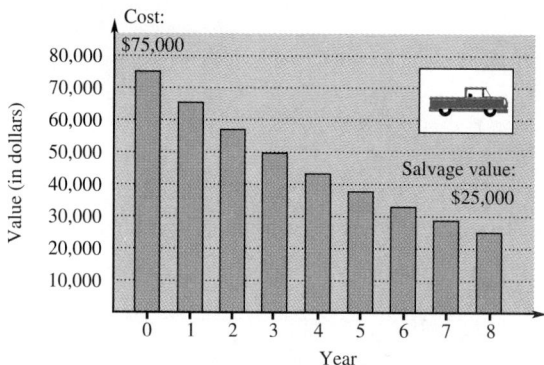

98. A printing press with an original cost of $125,000 is depreciated over a 10-year period, as shown in the figure.

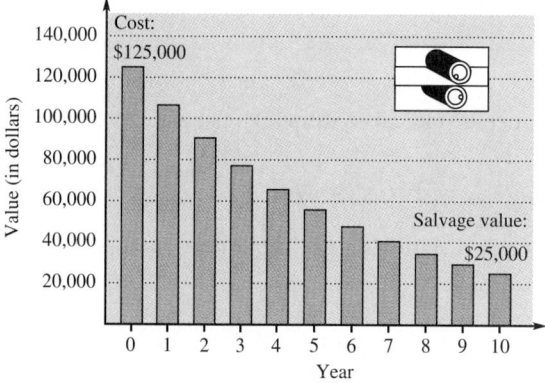

99. *Microwave Oven* The usable space in a particular microwave oven is in the form of a cube (see figure). The sales brochure indicates that the interior volume of the oven is 2197 cubic inches. Find the inside dimensions of the oven.

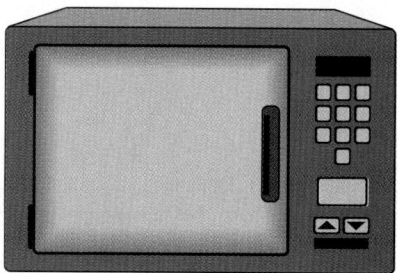

100. *Dimensions of a Room* Find the dimensions of a piece of carpet for a classroom with 529 square feet of floor space, assuming the floor is square.

101. *Velocity of a Stream* A stream of water moving at the rate of v feet per second can carry particles $0.03\sqrt{v}$ inches in diameter. Find the size of particles that can be carried by a stream flowing at the rate of $\frac{3}{4}$ foot per second.

102. *Exploratory Exercise* Find all possible "last digits" of perfect squares. (For instance, the last digit of 81 is 1, and the last digit of 64 is 4.) Is it possible that 4,322,788,987 is a perfect square?

Simplifying and Combining Radicals

Simplifying Radicals • *Combining Radicals*

Simplifying Radicals

The following two properties of radicals can be used to simplify radical expressions.

Properties of Radicals

Let u and v be real numbers, variables, or algebraic expressions. If the nth roots of u and v are real, then the following properties are true.

1. $\sqrt[n]{uv} = \sqrt[n]{u}\sqrt[n]{v}$ Multiplication Property

2. $\sqrt[n]{\dfrac{u}{v}} = \dfrac{\sqrt[n]{u}}{\sqrt[n]{v}}, \qquad v \neq 0$ Division Property

You can use these properties of radicals to simplify radical expressions, as follows.

$$\sqrt{48} = \sqrt{(16)(3)} = \sqrt{16}\sqrt{3} = \sqrt{4^2}\sqrt{3} = 4\sqrt{3}$$

This simplification process is called **removing perfect square factors from the radical.**

EXAMPLE 1 ■ Simplifying Radicals with Constant Radicands

For each of the following radical expressions, factor the radicand and simplify the radical by removing as many factors as possible.

(a) $\sqrt{75}$ (b) $\sqrt{72}$ (c) $\sqrt{162}$

Solution

(a) $\sqrt{75} = \sqrt{25(3)} = \sqrt{5^2(3)} = \sqrt{5^2}\sqrt{3} = 5\sqrt{3}$

(b) $\sqrt{72} = \sqrt{36(2)} = \sqrt{6^2(2)} = \sqrt{6^2}\sqrt{2} = 6\sqrt{2}$

(c) $\sqrt{162} = \sqrt{81(2)} = \sqrt{9^2(2)} = \sqrt{9^2}\sqrt{2} = 9\sqrt{2}$ ■

Removing *variable* factors from a square root radical is a little tricky. For instance, it is not valid to say that $\sqrt{x^2}$ is equal to x unless you happen to know that x is nonnegative. Without knowing anything about x, the only way you can simplify $\sqrt{x^2}$ is to include absolute value signs when you remove x from the radical. Thus, you have

$$\sqrt{x^2} = |x|.$$

When simplifying the expression $\sqrt{x^3}$, you do not need to include absolute value signs, because the domain of this expression does not include negative numbers. Thus, you can write

$$\sqrt{x^3} = \sqrt{x^2(x)} = x\sqrt{x}.$$

EXAMPLE 2 ■ Simplifying Radicals with Variable Radicands

For each of the following radical expressions, factor the radicand and simplify the radical by removing as many factors as possible.

(a) $\sqrt{25x^2}$ (b) $\sqrt{12x^3}, \quad x \geq 0$ (c) $\sqrt{144x^4}$

Solution

(a) $\sqrt{25x^2} = \sqrt{5^2 x^2} = \sqrt{5^2}\sqrt{x^2} = 5|x|$ $\sqrt{x^2} = |x|$

(b) $\sqrt{12x^3} = \sqrt{2^2 x^2 (3x)} = 2x\sqrt{3x}$ $\sqrt{2^2}\sqrt{x^2} = 2x, \quad x \geq 0$

(c) $\sqrt{144x^4} = \sqrt{12^2 (x^2)^2} = 12x^2$ $\sqrt{12^2}\sqrt{(x^2)^2} = 12|x^2| = 12x^2$ ■

In the same way that perfect squares can be removed from square root radicals, perfect nth powers can be removed from nth root radicals. This is demonstrated in Example 3.

EXAMPLE 3 ■ Simplifying Radicals

For each of the following radical expressions, factor the radicand and simplify the radical by removing as many perfect nth powers as possible.

(a) $\sqrt[3]{40}$ (b) $\sqrt[4]{x^5}, \quad x \geq 0$ (c) $\sqrt[3]{54x^3 y^5}$

Solution

(a) $\sqrt[3]{40} = \sqrt[3]{8(5)} = \sqrt[3]{2^3(5)} = 2\sqrt[3]{5}$ $\sqrt[3]{2^3} = 2$

(b) $\sqrt[4]{x^5} = \sqrt[4]{x^4(x)} = x\sqrt[4]{x}$ $\sqrt[4]{x^4} = x \quad x \geq 0$

(c) $\sqrt[3]{54x^3 y^5} = \sqrt[3]{27x^3 y^3 (2y^2)}$ $\sqrt[3]{3^3}\sqrt[3]{x^3}\sqrt[3]{y^3} = 3xy$

$\quad\quad\quad\quad = \sqrt[3]{3^3 x^3 y^3 (2y^2)}$

$\quad\quad\quad\quad = 3xy\sqrt[3]{2y^2}$ ■

Removing factors from a radical is only one of three techniques that are used to simplify radicals. All three techniques are summarized at the top of the next page.

Simplifying Radical Expressions	A radical expression is said to be in simplest form if all three of the following are true.

1. All possible factors have been removed from the radical.

2. No radical contains a fraction.

3. No denominator of a fraction contains a radical.

You can use a technique called **rationalizing the denominator.** This technique involves multiplying both the numerator and the denominator by a factor that creates a perfect nth power in the denominator. For square roots, you want a perfect square in the denominator; for cube roots, you want a perfect cube; and so on.

EXAMPLE 4   ■   Rationalizing the Denominator

Simplify the following expressions so that their denominators are free of radicals.

(a) $\sqrt{\dfrac{3}{5}}$
(b) $\dfrac{4}{\sqrt[3]{9}}$
(c) $\dfrac{8}{3\sqrt{18}}$

Solution

Students are usually comfortable with the rationalization of Example 4a. They tend to have difficulty deciding what to multiply by when they get to problems like Examples 4b and 4c. Have students multiply $4/\sqrt[3]{9}$ by $\sqrt[3]{9}/\sqrt[3]{9}$ and compare the results to Example 4b. Have students multiply $8/\left(3\sqrt{18}\right)$ by $\sqrt{18}/\sqrt{18}$ and compare the steps to Example 4c.

(a) $\sqrt{\dfrac{3}{5}} = \dfrac{\sqrt{3}}{\sqrt{5}} = \dfrac{\sqrt{3}}{\sqrt{5}} \cdot \dfrac{\sqrt{5}}{\sqrt{5}} = \dfrac{\sqrt{15}}{\sqrt{5^2}} = \dfrac{\sqrt{15}}{5}$ Multiply by $\sqrt{5}$ to create a perfect square in the denominator

(b) $\dfrac{4}{\sqrt[3]{9}} = \dfrac{4}{\sqrt[3]{9}} \cdot \dfrac{\sqrt[3]{3}}{\sqrt[3]{3}} = \dfrac{4\sqrt[3]{3}}{\sqrt[3]{3^3}} = \dfrac{4\sqrt[3]{3}}{3}$ Multiply by $\sqrt[3]{3}$ to create a perfect cube in the denominator

(c) $\dfrac{8}{3\sqrt{18}} = \dfrac{8}{3\sqrt{18}} \cdot \dfrac{\sqrt{2}}{\sqrt{2}} = \dfrac{8\sqrt{2}}{3\sqrt{36}} = \dfrac{8\sqrt{2}}{3(6)} = \dfrac{4\sqrt{2}}{9}$ Multiply by $\sqrt{2}$ to create a perfect square in the denominator   ■

EXAMPLE 5   ■   Rationalizing the Denominator

Simplify the following expressions so that their denominators are free of radicals.

(a) $\sqrt{\dfrac{8x}{12y^5}}$
(b) $\sqrt[3]{\dfrac{54x^6y^3}{5z^2}}$

Solution

(a) $\sqrt{\dfrac{8x}{12y^5}} = \sqrt{\dfrac{2x}{3y^5}} = \dfrac{\sqrt{2x}}{\sqrt{3y^5}} \cdot \dfrac{\sqrt{3y}}{\sqrt{3y}} = \dfrac{\sqrt{6xy}}{\sqrt{9y^6}} = \dfrac{\sqrt{6xy}}{3y^3}$

(b) $\sqrt[3]{\dfrac{54x^6y^3}{5z^2}} = \dfrac{\sqrt[3]{(3^3)(2)(x^6)(y^3)}}{\sqrt[3]{5z^2}} \cdot \dfrac{\sqrt[3]{25z}}{\sqrt[3]{25z}} = \dfrac{3x^2y\sqrt[3]{50z}}{\sqrt[3]{5^3z^3}} = \dfrac{3x^2y\sqrt[3]{50z}}{5z}$ ■

Combining Radicals

Two or more radical expressions are *alike* if they have the same radicands and the same indices. For instance, $\sqrt{2}$ and $3\sqrt{2}$ are alike, but $\sqrt{3}$ and $\sqrt[3]{3}$ are not alike. Two radical expressions that are alike can be added or subtracted by simply adding or subtracting their coefficients. For instance, the sum of $\sqrt{2}$ and $3\sqrt{2}$ is

$$\sqrt{2} + 3\sqrt{2} = (1+3)\sqrt{2} = 4\sqrt{2}.$$

EXAMPLE 6 ■ **Combining Radicals**

Simplify the following radical expressions by combining radicals that are alike.

(a) $\sqrt{7} + 5\sqrt{7} - 2\sqrt{7}$

(b) $6\sqrt{x} - \sqrt[3]{4} - 5\sqrt{x} + 2\sqrt[3]{4}$

(c) $3\sqrt[3]{x} + \sqrt[3]{8x}$

Solution

(a) $\sqrt{7} + 5\sqrt{7} - 2\sqrt{7} = (1+5-2)\sqrt{7} = 4\sqrt{7}$

Consider an analogy. For example, $6\sqrt{x} - \sqrt[3]{4} - 5\sqrt{x} + 2\sqrt[3]{4} = \sqrt{x} + \sqrt[3]{4}$ is similar to simplifying $6a - b - 5a + 2b$ to $a + b$, where $a = \sqrt{x}$ and $b = \sqrt[3]{4}$. Stress that only radicals that are *alike* can be combined.

(b) $6\sqrt{x} - \sqrt[3]{4} - 5\sqrt{x} + 2\sqrt[3]{4} = 6\sqrt{x} - 5\sqrt{x} - \sqrt[3]{4} + 2\sqrt[3]{4}$

$$= (6-5)\sqrt{x} + (-1+2)\sqrt[3]{4}$$

$$= \sqrt{x} + \sqrt[3]{4}$$

(c) As written, the radical expressions $3\sqrt[3]{x}$ and $\sqrt[3]{8x}$ are not alike. However, by simplifying the second expression, they can be made to be alike.

$$3\sqrt[3]{x} + \sqrt[3]{8x} = 3\sqrt[3]{x} + 2\sqrt[3]{x} = (3+2)\sqrt[3]{x} = 5\sqrt[3]{x}$$ ■

Notice in Example 6(c) that *before* concluding that two radicals cannot be combined, you should check to see that they are written in simplest form.

EXAMPLE 7 ■ **Simplifying Radical Expressions**

Simplify the following radical expressions.

(a) $\sqrt{45x} + 3\sqrt{20x}$ (b) $5\sqrt{x^3} - x\sqrt{4x}$ (c) $\sqrt[3]{54y^5} + 4\sqrt[3]{2y^2}$

Solution

(a) $\sqrt{45x} + 3\sqrt{20x} = \sqrt{9(5x)} + 3\sqrt{4(5x)} = 3\sqrt{5x} + 6\sqrt{5x} = 9\sqrt{5x}$

(b) $5\sqrt{x^3} - x\sqrt{4x} = 5x\sqrt{x} - 2x\sqrt{x} = 3x\sqrt{x}$

(c) $\sqrt[3]{54y^5} + 4\sqrt[3]{2y^2} = \sqrt[3]{27y^3(2y^2)} + 4\sqrt[3]{2y^2}$

$$= 3y\sqrt[3]{2y^2} + 4\sqrt[3]{2y^2} = (3y+4)\sqrt[3]{2y^2}$$ ■

In some instances, it may be necessary to rationalize denominators before combining radicals. This is demonstrated in Examples 8 and 9.

EXAMPLE 8   ■   **Rationalizing Denominators Before Simplifying**

Simplify the following radical expressions.

(a) $\sqrt{7} - \dfrac{5}{\sqrt{7}}$

(b) $\dfrac{\sqrt{12}}{6} + \sqrt{\dfrac{1}{3}} - \dfrac{2\sqrt{3}}{3}$

Solution

(a) $\sqrt{7} - \dfrac{5}{\sqrt{7}} = \sqrt{7} - \left(\dfrac{5}{\sqrt{7}} \cdot \dfrac{\sqrt{7}}{\sqrt{7}} \right)$

$\qquad\qquad = \sqrt{7} - \dfrac{5\sqrt{7}}{7}$

$\qquad\qquad = \left(1 - \dfrac{5}{7} \right) \sqrt{7}$

$\qquad\qquad = \dfrac{2}{7}\sqrt{7}$

(b) $\dfrac{\sqrt{12}}{6} + \sqrt{\dfrac{1}{3}} - \dfrac{2\sqrt{3}}{3} = \dfrac{2\sqrt{3}}{6} + \left(\dfrac{1}{\sqrt{3}} \cdot \dfrac{\sqrt{3}}{\sqrt{3}} \right) - \dfrac{2\sqrt{3}}{3}$

$\qquad\qquad = \dfrac{\sqrt{3}}{3} + \dfrac{\sqrt{3}}{3} - \dfrac{2\sqrt{3}}{3}$

$\qquad\qquad = \left(\dfrac{1}{3} + \dfrac{1}{3} - \dfrac{2}{3} \right) \sqrt{3}$

$\qquad\qquad = 0$

    ■

EXAMPLE 9   ■   **Rationalizing Denominators Before Simplifying**

Simplify the radical expression

$$\dfrac{1}{\sqrt{2x}} + \dfrac{\sqrt{8x}}{x}.$$

Solution

$\dfrac{1}{\sqrt{2x}} + \dfrac{\sqrt{8x}}{x} = \left(\dfrac{1}{\sqrt{2x}} \cdot \dfrac{\sqrt{2x}}{\sqrt{2x}} \right) + \dfrac{2\sqrt{2x}}{x}$

$\qquad\qquad = \dfrac{\sqrt{2x}}{2x} + \dfrac{2\sqrt{2x}}{x}$

$\qquad\qquad = \left(\dfrac{1}{2x} + \dfrac{2}{x} \right) \sqrt{2x}$

$\qquad\qquad = \left(\dfrac{1}{2x} + \dfrac{4}{2x} \right) \sqrt{2x}$

$\qquad\qquad = \dfrac{5}{2x}\sqrt{2x}$

    ■

EXAMPLE 10   ■   Finding the Frequencies of Musical Notes

The musical note A-440 (above middle C) has a frequency of 440 vibrations per second. The frequency F of any note can be found by the model

$$F = 440 \sqrt[12]{2^n},$$

where n represents the number of notes above or below A-440, as indicated in Figure 6.3. Two notes are said to *harmonize* (sound pleasing to the ear) if the ratio of their frequencies is an integer or a simple rational number. Show that any two notes that are an octave apart harmonize.

FIGURE 6.3

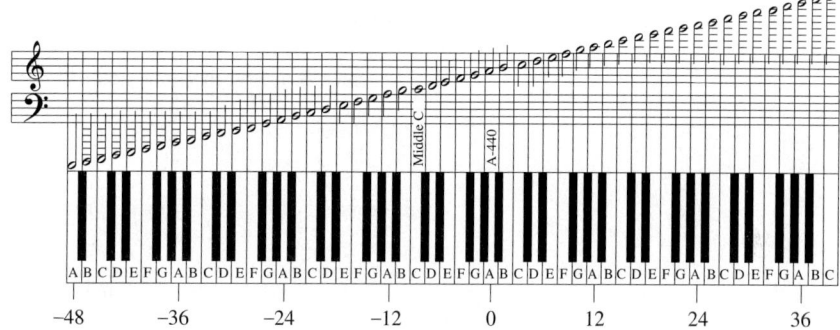

Solution

Notes that are an octave apart differ by 12 notes. (The "oct" in octave refers to the number of white keys on a piano.) Thus, you can represent the two notes by n and $n + 12$. The ratio of the frequencies of the two notes is as follows.

$$\frac{\text{Frequency of higher note}}{\text{Frequency of lower note}} = \frac{440 \sqrt[12]{2^{n+12}}}{440 \sqrt[12]{2^n}}$$

$$= \frac{\sqrt[12]{2^{n+12}}}{\sqrt[12]{2^n}}$$

$$= \sqrt[12]{\frac{2^{n+12}}{2^n}}$$

$$= \sqrt[12]{2^{12}}$$

$$= 2$$

Thus, the ratio of the two frequencies is 2—which means that the two notes harmonize.   ■

DISCUSSION PROBLEM ■ **Properties of Radicals**

Show how this example also demonstrates that $\sqrt{a^2 + b^2} \neq a + b$.

In general, $\sqrt{a + b}$ is *not* equal to $\sqrt{a} + \sqrt{b}$. One convenient way to demonstrate this is to let $a = 9$ and $b = 16$. Then

$$\sqrt{9 + 16} = \sqrt{25} = 5, \quad \text{whereas} \quad \sqrt{9} + \sqrt{16} = 3 + 4 = 7.$$

Can you find another example in which a, b, and $a + b$ are all perfect squares?

■

Warm-Up

The following warm-up exercises involve skills that were covered in earlier sections. You will use these skills in the exercise set for this section.

In Exercises 1–4, find the indicated nth root without using a calculator.

1. $\sqrt{10,000}$ **2.** $-\sqrt[3]{1000}$ **3.** $\sqrt[3]{-\frac{27}{125}}$ **4.** $\sqrt[4]{\frac{16}{625}}$

In Exercises 5–10, simplify the expression.

5. $x^{5/3} \cdot x^{1/6}$ **6.** $\dfrac{a^{7/4}}{a^{3/2}}$ **7.** $(2x)^{3/2}(2x)^{5/2}$

8. $(-4x^{1/3}y^{4/3})^6$ **9.** $\dfrac{32x^{1/4}y^{3/4}}{2x^{1/2}y}$ **10.** $\dfrac{8y^{3/2}z^{-2}}{y^{-1/2}z^{1/2}}$

6.3 EXERCISES

In Exercises 1–8, write the expression as a single radical.

1. $\sqrt{3} \cdot \sqrt{10}$ **2.** $\sqrt[5]{9} \cdot \sqrt[5]{19}$ **3.** $\sqrt[3]{11} \cdot \sqrt[3]{10}$ **4.** $\sqrt[4]{35} \cdot \sqrt[4]{3}$

5. $\dfrac{\sqrt{15}}{\sqrt{31}}$ **6.** $\dfrac{\sqrt[3]{85}}{\sqrt[3]{9}}$ **7.** $\dfrac{\sqrt[5]{152}}{\sqrt[5]{3}}$ **8.** $\dfrac{\sqrt[4]{633}}{\sqrt[4]{5}}$

In Exercises 9–16, simplify the expression.

9. $\sqrt{9 \cdot 35}$ **10.** $\sqrt[3]{27 \cdot 4}$ **11.** $\sqrt[4]{81 \cdot 11}$ **12.** $\sqrt[5]{100,000 \cdot 3}$

13. $\sqrt{\frac{35}{9}}$ **14.** $\sqrt[4]{\frac{165}{16}}$ **15.** $\sqrt[3]{\frac{11}{1000}}$ **16.** $\sqrt[5]{\frac{2}{243}}$

In Exercises 17–34, simplify the radical.

17. $\sqrt{20}$ **18.** $\sqrt{50}$ **19.** $\sqrt{27}$ **20.** $\sqrt{125}$ **21.** $\sqrt{0.04}$ **22.** $\sqrt{0.25}$

23. $\sqrt[3]{24}$ **24.** $\sqrt[3]{54}$ **25.** $\sqrt[4]{30,000}$ **26.** $\sqrt[5]{96}$ **27.** $\sqrt{\frac{15}{4}}$ **28.** $\sqrt{\frac{5}{36}}$

29. $\sqrt[3]{\frac{35}{64}}$ **30.** $\sqrt[4]{\frac{5}{16}}$ **31.** $\sqrt[5]{\frac{15}{243}}$ **32.** $\sqrt[3]{\frac{1}{1000}}$ **33.** $\sqrt[4]{0.0081}$ **34.** $\sqrt[3]{0.064}$

In Exercises 35–56, simplify the expression.

35. $\sqrt{9x^5}$ **36.** $\sqrt{64x^3}$ **37.** $\sqrt[3]{x^4y^3}$ **38.** $\sqrt[3]{a^5b^6}$ **39.** $\sqrt[5]{32x^5y^6}$ **40.** $\sqrt[4]{128u^4v^7}$

41. $\sqrt{\frac{13}{25}}$ **42.** $\sqrt{\frac{15}{36}}$ **43.** $\sqrt[5]{\frac{32x^2}{y^5}}$ **44.** $\sqrt[3]{\frac{16z^3}{y^6}}$ **45.** $\sqrt[3]{\frac{54a^4}{b^9}}$ **46.** $\sqrt[4]{\frac{3u^2}{16v^8}}$

47. $\sqrt{48xy^4}$ **48.** $\sqrt{\frac{18x^2}{z^6}}$ **49.** $\sqrt{\frac{32a^4}{b^2}}$ **50.** $\sqrt[4]{16x^5}$ **51.** $\sqrt[4]{(3x^2)^4}$ **52.** $\sqrt{75x^2y^{-4}}$

53. $\sqrt[5]{96x^5}$ **54.** $\sqrt[4]{562^4}$ **55.** $\sqrt[3]{\frac{16}{27}}$ **56.** $\sqrt[3]{72x^5y^7}$

In Exercises 57–80, rationalize the denominator and simplify further, if possible.

57. $\sqrt{\frac{1}{3}}$ **58.** $\sqrt{\frac{1}{5}}$ **59.** $\frac{12}{\sqrt{3}}$ **60.** $\frac{5}{\sqrt{10}}$ **61.** $\sqrt[4]{\frac{5}{4}}$ **62.** $\sqrt[3]{\frac{9}{25}}$

63. $\frac{6}{\sqrt[3]{32}}$ **64.** $\frac{10}{\sqrt[5]{16}}$ **65.** $\frac{1}{\sqrt{y}}$ **66.** $\frac{1}{\sqrt{2x}}$ **67.** $\sqrt{\frac{3}{x}}$ **68.** $\sqrt{\frac{5}{c}}$

69. $\sqrt{\frac{4}{x^3}}$ **70.** $\frac{5}{\sqrt{8x^5}}$ **71.** $\sqrt[3]{\frac{2x}{3y}}$ **72.** $\sqrt[3]{\frac{20x^2}{9y^2}}$ **73.** $\frac{a^3}{\sqrt[3]{ab^2}}$ **74.** $\frac{3u^2}{\sqrt[4]{8u^3}}$

75. $\frac{6}{\sqrt{3b^3}}$ **76.** $\frac{1}{\sqrt{xy}}$ **77.** $\frac{b^2}{\sqrt[4]{a^3b^3}}$ **78.** $\sqrt{\frac{5}{y^4}}$ **79.** $\sqrt[3]{\frac{18x^2}{11z^4}}$ **80.** $\frac{4t^5}{\sqrt{(st)^4}}$

In Exercises 81–94, combine the radical expressions, if possible.

81. $3\sqrt{2} - \sqrt{2}$ **82.** $\frac{2}{5}\sqrt{5} - \frac{6}{5}\sqrt{5}$ **83.** $\sqrt[4]{3} - 5\sqrt[4]{7} - 12\sqrt[4]{3}$

84. $9\sqrt[3]{17} + 7\sqrt[3]{2} - 4\sqrt[3]{17} + \sqrt{2}$ **85.** $12\sqrt{8} - 3\sqrt[3]{8}$ **86.** $4\sqrt{32} + 7\sqrt{32}$

87. $2\sqrt[3]{54} + 12\sqrt[3]{16}$ **88.** $4\sqrt[4]{48} - \sqrt[4]{243}$ **89.** $5\sqrt{x} - 3\sqrt[4]{x}$

90. $3\sqrt{x+1} + 10\sqrt{x+1}$ **91.** $\sqrt{25y} + \sqrt{64y}$ **92.** $\sqrt[3]{16t^4} - \sqrt[3]{54t^4}$

93. $10\sqrt[3]{z} - \sqrt[3]{z^4}$ **94.** $5\sqrt[3]{24u^2} + 2\sqrt[3]{81u^5}$

In Exercises 95–98, perform the indicated addition or subtraction and simplify your answer.

95. $\sqrt{5} - \frac{3}{\sqrt{5}}$ **96.** $\sqrt{10} + \frac{5}{\sqrt{10}}$ **97.** $\sqrt{20} - \sqrt{\frac{1}{5}}$ **98.** $\frac{x}{\sqrt{3x}} + \sqrt{27x}$

Technology In Exercises 99–102, use a calculator to evaluate the expression. Then simplify the expression and use a calculator to evaluate the simplified expression. Compare the results.

99. $\sqrt[3]{14} \cdot \sqrt[3]{13}$

100. $\sqrt[8]{8} \cdot \sqrt[8]{5}$

101. $(5^{6/7})^{1/2}$

102. $(6^{-3})^{5/3}$

In Exercises 103–106, place the correct inequality or equality symbol ($<$, $>$, or $=$) between the numbers.

103. $\sqrt{7} + \sqrt{18}$ ▢ $\sqrt{7 + 18}$

104. $\sqrt{10} - \sqrt{6}$ ▢ $\sqrt{10 - 6}$

105. 5 ▢ $\sqrt{3^2 + 2^2}$

106. 5 ▢ $\sqrt{3^2 + 4^2}$

107. ***Vibrating String*** The frequency f in cycles per second of a vibrating string is

$$f = \frac{1}{100}\sqrt{\frac{400 \times 10^6}{5}}.$$

Use a calculator to approximate this number. (Round your answer to two decimal places.)

108. ***Period of a Pendulum*** The period T in seconds of a pendulum (see figure) is

$$T = 2\pi\sqrt{\frac{L}{32}},$$

where L is the length of the pendulum in feet. Find the period of a pendulum whose length is 4 feet. (Round your answer to two decimal places.)

Figure for 108

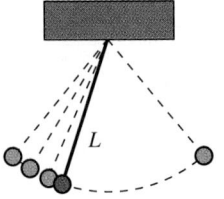

Musical Notes In Exercises 109 and 110, use the following information. The musical note A-440 (above middle C) has a frequency of 440 vibrations per second. The frequency F of any note can be found by the equation

$$F = 440 \cdot 2^{n/12},$$

where n represents the number of black and white keys below or above the given note and A-440 (see figure below).

109. Approximate the lowest and highest frequencies of a trumpet.

110. Describe the pattern of frequencies of successive notes of the same letter.

111. ***Exploratory Exercise*** Enter any positive real number in your calculator and find its square root. Then repeatedly take the square root of the result.

$$\sqrt{x\sqrt{\sqrt{x}}}, \quad \sqrt{\sqrt{\sqrt{\sqrt{x}}}}, \quad \dots$$

What real number does the display appear to be approaching?

112. ***Exploratory Exercise*** Square the real number $5/\sqrt{3}$ and note that the radical is eliminated from the denominator. Is this equivalent to rationalizing the denominator? Why or why not?

Figure for 109 and 110

⊢——— Cornet/Trumpet ———⊣

⊢——— Alto mellophone ———⊣

⊢——— French horn ———⊣

⊢— Trombone/Euphonium —⊣

⊢——— Bass tuba ———⊣

$-48 \quad\quad 36 \quad\quad -24 \quad\quad -12 \quad\quad 0 \quad\quad 12 \quad\quad 24 \quad\quad 36$

Mid-Chapter Quiz

Take this quiz as you would take a quiz in class. After you are done, check your work against the answers in the back of the book.

In Exercises 1–6, evaluate the expression.

1. 2^{-3}

2. $\dfrac{8^{-5}}{8^{-3}}$

3. $\left(\frac{3}{4}\right)^{-2}$

4. $36^{-1/2}$

5. $\sqrt[3]{\frac{81}{3}}$

6. $16^{3/4}$

In Exercises 7–14, simplify the expression.

7. $(-5x)^3$

8. $3x^5(4x^{-1})$

9. $\dfrac{(3y)^0}{2y^{-2}}$

10. $\dfrac{21x^{2/3}}{18x^{1/2}}$

11. $\sqrt{72}$

12. $\sqrt{\dfrac{40c^3}{25d^6}}$

13. $\sqrt[3]{16x^4y}$

14. $10\sqrt{2x} - 3\sqrt{x} + \sqrt{32x}$

15. One point (printer's measure) is equal to 0.013837 inch. Write this number in scientific notation.

16. Use a calculator to evaluate $(3.2 \times 10^3)^4(1.5 \times 10^{-4})$ and write the answer in scientific notation.

17. Rationalize the denominator: $\dfrac{3}{\sqrt{15y}}$

18. Rationalize the denominator: $\sqrt[3]{\dfrac{3}{4t}}$

19. The area of a circle is $A = \pi r^2$. When the radius of a circle is doubled, by what factor is the area increased? Justify your answer algebraically.

20. Find the dimensions of a cube that has a volume of 13,824 cubic inches (see figure).

Multiplying and Dividing Radical Expressions

Multiplying Radical Expressions • *Dividing Radical Expressions*

Multiplying Radical Expressions

You can multiply two radical expressions by using the Distributive Property and the Multiplication Property of Radicals. Recall from Section 6.3 that the product of two radicals is

$$\sqrt[n]{a}\,\sqrt[n]{b} = \sqrt[n]{ab},$$

where a and b are real numbers having nth roots that are also real numbers.

EXAMPLE 1 ■ Multiplying Radicals

$$\sqrt{3}(2 + \sqrt{5}) = 2\sqrt{3} + \sqrt{3}\sqrt{5} \qquad \text{Distributive Property}$$
$$= 2\sqrt{3} + \sqrt{15} \qquad \text{Multiplication Property of Radicals} \quad ■$$

EXAMPLE 2 ■ Multiplying Radicals

(a) $\sqrt{2}(4 - \sqrt{8}) = 4\sqrt{2} - \sqrt{2}\sqrt{8}$ Distributive Property

$\qquad\qquad\qquad = 4\sqrt{2} - \sqrt{16}$ Multiplication Property of Radicals

$\qquad\qquad\qquad = 4\sqrt{2} - 4$ Simplify

(b) $\sqrt{6}(\sqrt{12} - \sqrt{3}) = \sqrt{6}\sqrt{12} - \sqrt{6}\sqrt{3}$ Distributive Property

$\qquad\qquad\qquad\quad = \sqrt{72} - \sqrt{18}$ Multiplication Property of Radicals

$\qquad\qquad\qquad\quad = 6\sqrt{2} - 3\sqrt{2}$ Find perfect square factors

$\qquad\qquad\qquad\quad = 3\sqrt{2}$ Simplify   ■

EXAMPLE 3 ■ Multiplying Radicals

Students sometimes have more diffi-culty using the FOIL method when radicals are involved. You may want to use an additional example such as:

$$\left(4 + 2\sqrt{5}\right)\left(3 - 4\sqrt{5}\right)$$
$$= 12 - 16\sqrt{5} + 6\sqrt{5} - 8(5)$$
$$= -28 - 10\sqrt{5}$$

Have students verify their results using their calculators.

(a) $(2\sqrt{7} - 4)(\sqrt{7} + 1) = \overset{\text{F}}{\overbrace{2(\sqrt{7})^2}} + \overset{\text{O}}{\overbrace{2\sqrt{7}}} - \overset{\text{I}}{\overbrace{4\sqrt{7}}} - \overset{\text{L}}{\overbrace{4}}$ FOIL Method

$\qquad\qquad\qquad\qquad = 2(7) + (2 - 4)\sqrt{7} - 4$ Combine like radicals

$\qquad\qquad\qquad\qquad = 10 - 2\sqrt{7}$ Combine like terms

(b) $(3 - \sqrt{x})(1 + \sqrt{x}) = 3 + 3\sqrt{x} - \sqrt{x} - (\sqrt{x})^2$ FOIL Method

$\qquad\qquad\qquad\qquad = 3 + 2\sqrt{x} - x$ ■

EXAMPLE 4   ■   Multiplying Radicals

Looking ahead to solving equations involving radicals, have your students simplify $\left(\sqrt{3}+\sqrt{x}\right)^2$.

(a) $\left(2-\sqrt{5}\right)\left(2+\sqrt{5}\right) = 2^2 - \left(\sqrt{5}\right)^2$ Special product

$$= 4 - 5$$

$$= -1$$

(b) $\left(\sqrt{3}+\sqrt{x}\right)\left(\sqrt{3}-\sqrt{x}\right) = \left(\sqrt{3}\right)^2 - \left(\sqrt{x}\right)^2$ Special product

$$= 3 - x, \qquad x \geq 0 \qquad ■$$

In Example 4, the expressions $\left(2-\sqrt{5}\right)$ and $\left(2+\sqrt{5}\right)$ are called **conjugates** of each other. The product of two conjugates is the difference of two squares that is given by the special product $(a+b)(a-b) = a^2 - b^2$. Here are some other examples.

Ask students what pattern they can observe (*i.e.*, when you multiply a binomial with radical terms by its conjugate, the result is free of radicals.)

Expression	*Conjugate*	*Product*
$\left(1-\sqrt{3}\right)$	$\left(1+\sqrt{3}\right)$	$(1)^2 - \left(\sqrt{3}\right)^2 = 1 - 3 = -2$
$\left(\sqrt{5}+\sqrt{2}\right)$	$\left(\sqrt{5}-\sqrt{2}\right)$	$\left(\sqrt{5}\right)^2 - \left(\sqrt{2}\right)^2 = 5 - 2 = 3$
$\left(\sqrt{10}-3\right)$	$\left(\sqrt{10}+3\right)$	$\left(\sqrt{10}\right)^2 - (3)^2 = 10 - 9 = 1$
$\sqrt{x}+2$	$\sqrt{x}-2$	$\left(\sqrt{x}\right)^2 - (2)^2 = x - 4, \quad x \geq 0$

Dividing Radical Expressions

To simplify a *quotient* involving radicals, you rationalize the denominator. For single-term denominators, use the rationalizing process described in Section 6.3. For instance, to divide $\sqrt{3}$ by $\sqrt{5}$, you can multiply both the numerator and denominator by $\sqrt{5}$, as follows.

$$\frac{\sqrt{3}}{\sqrt{5}} = \frac{\sqrt{3}}{\sqrt{5}} \cdot \frac{\sqrt{5}}{\sqrt{5}} = \frac{\sqrt{15}}{5}$$

You can rationalize a denominator involving two terms by multiplying both the numerator and denominator by the *conjugate of the denominator.* Here is an example.

$$\frac{\sqrt{3}}{1-\sqrt{5}} = \frac{\sqrt{3}}{1-\sqrt{5}} \cdot \frac{1+\sqrt{5}}{1+\sqrt{5}}$$

$$= \frac{\sqrt{3}\left(1+\sqrt{5}\right)}{\left(1-\sqrt{5}\right)\left(1+\sqrt{5}\right)}$$

$$= \frac{\sqrt{3}+\sqrt{15}}{1-5}$$

$$= \frac{\sqrt{3}+\sqrt{15}}{-4}$$

EXAMPLE 5   ■   Simplifying Quotients Involving Radicals

Simplify the expression $\dfrac{4}{2 - \sqrt{3}}$ by rationalizing its denominator.

Solution

At this point students may still be tempted to rationalize $4/\left(2 - \sqrt{3}\right)$ by multiplying $\sqrt{3}/\sqrt{3}$. Ask them why this doesn't rationalize the denominator.

$$\frac{4}{2 - \sqrt{3}} = \frac{4}{2 - \sqrt{3}} \cdot \frac{2 + \sqrt{3}}{2 + \sqrt{3}} \qquad \text{Multiply by conjugate}$$

$$= \frac{4(2 + \sqrt{3})}{2^2 - (\sqrt{3})^2} \qquad \text{Special product}$$

$$= \frac{8 + 4\sqrt{3}}{4 - 3}$$

$$= 8 + 4\sqrt{3} \qquad \text{Simplest form} \qquad ■$$

EXAMPLE 6   ■   Simplifying Quotients Involving Radicals

(a) $\dfrac{5\sqrt{2}}{\sqrt{7} + \sqrt{2}} = \dfrac{5\sqrt{2}}{\sqrt{7} + \sqrt{2}} \cdot \dfrac{\sqrt{7} - \sqrt{2}}{\sqrt{7} - \sqrt{2}}$ $\qquad$ Multiply by conjugate

$$= \frac{5(\sqrt{14} - \sqrt{4})}{(\sqrt{7})^2 - (\sqrt{2})^2}$$

$$= \frac{5(\sqrt{14} - 2)}{7 - 2}$$

$$= \frac{\cancel{5}(\sqrt{14} - 2)}{\cancel{5}} \qquad \text{Divide out common factor}$$

$$= \sqrt{14} - 2$$

(b) $\dfrac{6}{\sqrt{x} - 2} = \dfrac{6}{\sqrt{x} - 2} \cdot \dfrac{\sqrt{x} + 2}{\sqrt{x} + 2}$ $\qquad$ Multiply by conjugate

$$= \frac{6(\sqrt{x} + 2)}{(\sqrt{x})^2 - (2)^2}$$

$$= \frac{6\sqrt{x} + 12}{x - 4} \qquad\qquad ■$$

EXAMPLE 7   ■   Dividing Radical Expressions

Perform the indicated division and simplify your answer.

$$(2 - \sqrt{3}) \div (\sqrt{6} + \sqrt{2})$$

Solution

$$\frac{2 - \sqrt{3}}{\sqrt{6} + \sqrt{2}} = \frac{2 - \sqrt{3}}{\sqrt{6} + \sqrt{2}} \cdot \frac{\sqrt{6} - \sqrt{2}}{\sqrt{6} - \sqrt{2}} \qquad \text{Multiply by conjugate}$$

$$= \frac{2\sqrt{6} - 2\sqrt{2} - \sqrt{18} + \sqrt{6}}{\left(\sqrt{6}\right)^2 - \left(\sqrt{2}\right)^2}$$

$$= \frac{3\sqrt{6} - 2\sqrt{2} - 3\sqrt{2}}{6 - 2} \qquad \sqrt{18} = 3\sqrt{2}$$

$$= \frac{3\sqrt{6} - 5\sqrt{2}}{4} \qquad \text{Simplest form} \qquad ■$$

DISCUSSION PROBLEM   ■   The Golden Section

The ratio of the width of the Temple of Hephaestus to its height (see Figure 6.4) is approximately

$$\frac{w}{h} \approx \frac{2}{\sqrt{5} - 1}.$$

Have students form the ratio of length to width of 3 × 5 and 5 × 8 index cards. Then ask them to convert these ratios to decimal form and compare the results to the golden section.

This number is called the **golden section.** Early Greeks believed that the most aesthetically pleasing rectangles were those with sides of this ratio. Rationalize the denominator of this number. Approximate your answer, rounded to two decimal places.

FIGURE 6.4

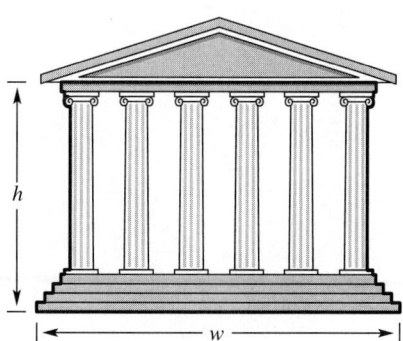

Warm-Up

The following warm-up exercises involve skills that were covered in earlier sections. You will use these skills in the exercise set for this section.

In Exercises 1–6, perform the indicated operations and simplify.

1. $19 - 5(x - 3)$ **2.** $32 + 6(3x - 10)$ **3.** $\left(t + \frac{1}{3}\right)^2$

4. $(2x - 5)^2$ **5.** $(5x - 4)(5x + 4)$ **6.** $\left(\frac{1}{2} - 3x\right)(4 + x)$

In Exercises 7–10, simplify the expression. (Assume that all variables are positive.)

7. $\sqrt[3]{16x^4 y^3}$ **8.** $\sqrt{72x^2 y^3}$

9. $\sqrt{3x^2} - 2\sqrt{12}$ **10.** $\dfrac{14x}{\sqrt{7}}$

6.4   EXERCISES

In Exercises 1–30, perform the indicated multiplication and simplify the result. (Assume that all variables are positive.)

1. $\sqrt{2} \cdot \sqrt{8}$

2. $\sqrt{6} \cdot \sqrt{18}$

3. $\sqrt{5}(2 - \sqrt{3})$

4. $\sqrt{11}(\sqrt{5} - 3)$

5. $\sqrt{2}(\sqrt{20} + 8)$

6. $\sqrt{7}(\sqrt{14} + 3)$

7. $(\sqrt{3} + 2)(\sqrt{3} - 2)$

8. $(\sqrt{15} + 3)(\sqrt{15} - 3)$

9. $(2\sqrt{2} + \sqrt{4})(2\sqrt{2} - \sqrt{4})$

10. $(4\sqrt{3} + \sqrt{2})(4\sqrt{3} - \sqrt{2})$

11. $(\sqrt{20} + 2)^2$

12. $(4 - \sqrt{20})^2$

13. $(\sqrt{5} + 3)(\sqrt{3} - 5)$

14. $(\sqrt{30} + 6)(\sqrt{2} + 6)$

15. $\sqrt{y}(\sqrt{y} + 4)$

16. $\sqrt{x}(5 - \sqrt{x})$

17. $(3\sqrt{x} - 5)(3\sqrt{x} + 5)$

18. $(7 - 3\sqrt{3t})(7 + 3\sqrt{3t})$

19. $(10 + \sqrt{2x})^2$

20. $(5 - \sqrt{3v})^2$

21. $(9\sqrt{x} + 2)(5\sqrt{x} - 3)$

22. $(16\sqrt{u} - 3)(\sqrt{u} - 1)$

23. $(\sqrt{x} + \sqrt{y})(\sqrt{x} - \sqrt{y})$

24. $(3\sqrt{u} + \sqrt{3v})(3\sqrt{u} - \sqrt{3v})$

25. $\sqrt[3]{4}(\sqrt[3]{2} - 7)$

26. $(\sqrt[3]{9} + 5)(\sqrt[3]{5} - 5)$

27. $(\sqrt[3]{2x} + 5)^2$

28. $(\sqrt[3]{y} + 2)(\sqrt[3]{y^2} - 5)$

29. $(\sqrt[3]{2y} + 10)(\sqrt[3]{4y^2} - 10)$

30. $(\sqrt[3]{t} + 1)(\sqrt[3]{t^2} + 4\sqrt[3]{t} - 3)$

In Exercises 31–36, insert the missing factor.

31. $5x\sqrt{3} + 15\sqrt{3} = 5\sqrt{3}()$

32. $x\sqrt{7} - x^2\sqrt{7} = x\sqrt{7}()$

33. $4\sqrt{12} - 2x\sqrt{27} = 2\sqrt{3}()$

34. $5\sqrt{50} + 10y\sqrt{8} = 5\sqrt{2}()$

35. $6u^2 + \sqrt{18u^3} = 3u()$

36. $12s^3 - \sqrt{32s^4} = 4s^2()$

In Exercises 37–40, write the fraction in reduced form.

37. $\dfrac{4 - 8\sqrt{x}}{12}$

38. $\dfrac{-3 + 27\sqrt{2y}}{18}$

39. $\dfrac{-2y + \sqrt{12y^3}}{8y}$

40. $\dfrac{-t^2 - \sqrt{2t^3}}{3t}$

In Exercises 41–48, determine the conjugate of the expression and then find the product of the expression and its conjugate. (Assume that all variables are positive.)

41. $2 + \sqrt{5}$

42. $\sqrt{2} - 9$

43. $\sqrt{11} - \sqrt{3}$

44. $\sqrt{10} + \sqrt{7}$

45. $\sqrt{x} - 3$

46. $\sqrt{t} + 7$

47. $\sqrt{2u} - \sqrt{3}$

48. $\sqrt{5a} + \sqrt{2}$

In Exercises 49–60, rationalize the denominator of the expression. (Assume that all variables are positive.)

49. $\dfrac{6}{\sqrt{22} - 2}$

50. $\dfrac{3}{2\sqrt{10} - 5}$

51. $\dfrac{8}{\sqrt{7} + 3}$

52. $\dfrac{10}{\sqrt{9} + \sqrt{5}}$

53. $\left(\sqrt{7} + 2\right) \div \left(\sqrt{7} - 2\right)$

54. $\left(5 - 3\sqrt{3}\right) \div \left(3 + \sqrt{3}\right)$

55. $\dfrac{3x}{\sqrt{15} - \sqrt{3}}$

56. $\dfrac{6(y + 1)}{y^2 + \sqrt{y}}$

57. $\dfrac{2t^2}{\sqrt{5t} - \sqrt{t}}$

58. $\dfrac{5x}{\sqrt{x} - \sqrt{2}}$

59. $\left(\sqrt{x} - 5\right) \div \left(2\sqrt{x} - 1\right)$

60. $\left(2\sqrt{t} + 1\right) \div \left(2\sqrt{t} - 1\right)$

61. *Strength of a Wooden Beam* The rectangular cross section of a wooden beam cut from a log 24 inches in diameter will have maximum strength if its width w and height h are

$$w = 8\sqrt{3} \quad \text{and} \quad h = \sqrt{24^2 - \left(8\sqrt{3}\right)^2}.$$

Find the area of the rectangular cross section and express the area in simplest form.

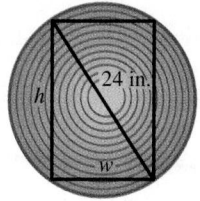

62. *Force to Move a Block* The force required to slide a steel block weighing 500 pounds across a milling machine is

$$\frac{500k}{\left(\dfrac{1}{\sqrt{k^2 + 1}} + \dfrac{k^2}{\sqrt{k^2 + 1}}\right)},$$

where k is a constant determined by the amount of friction (see figure). Simplify this expression of force.

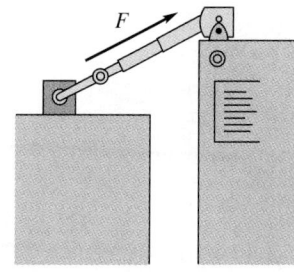

Solving Equations Involving Radicals

Solving Equations Involving Radicals   •   *Applications of Radicals*

Solving Equations Involving Radicals

Solving equations involving radicals is somewhat like solving equations that contain fractions—you try to get rid of the radicals and obtain a polynomial equation. Then you solve the polynomial equation using the standard procedures. The following property plays a key role.

Raising Both Sides of an Equation to the *n*th Power	Let u and v be real numbers, variables, or algebraic expressions, and let n be a positive integer. If $u = v$, then it follows that $u^n = v^n$. This operation is called **raising both sides of an equation to the *n*th power.**

You will see in this section that raising both sides of an equation to the *n*th power often introduces *extraneous* solutions. So when you use this procedure, it is critical that you check each solution in the *original* equation.

EXAMPLE 1   ■   **Solving an Equation Having One Radical**

$$\sqrt{x} - 8 = 0 \qquad \text{Given equation}$$
$$\sqrt{x} = 8 \qquad \text{Isolate radical}$$
$$\left(\sqrt{x}\right)^2 = 8^2 \qquad \text{Square both sides}$$
$$x = 64 \qquad \text{Solution}$$

Check

$$\sqrt{x} - 8 = 0 \qquad \text{Given equation}$$
$$\sqrt{64} - 8 \overset{?}{=} 0 \qquad \text{Replace } x \text{ by } 64$$
$$8 - 8 = 0 \qquad \text{Solution checks}$$

Therefore, the equation has one solution: $x = 64$. Another way to verify the correctness of the solution is to use a graph. For instance, in Figure 6.5, note that the graph of $y = \sqrt{x} - 8$ crosses the x-axis when $x = 64$.

FIGURE 6.5

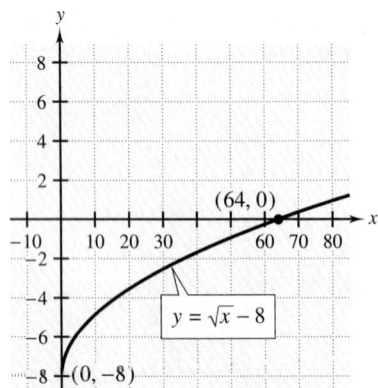

EXAMPLE 2 ■ Solving an Equation Having One Radical

$$\sqrt[3]{2x+1} - 2 = 3 \qquad \text{Given equation}$$
$$\sqrt[3]{2x+1} = 5 \qquad \text{Isolate radical}$$
$$\left(\sqrt[3]{2x+1}\right)^3 = 5^3 \qquad \text{Cube both sides}$$
$$2x + 1 = 125 \qquad \text{Simplify}$$
$$2x = 124$$
$$x = 62 \qquad \text{Solution}$$

Check

$$\sqrt[3]{2x+1} - 2 = 3 \qquad \text{Given equation}$$
$$\sqrt[3]{2(62)+1} - 2 \overset{?}{=} 3 \qquad \text{Replace } x \text{ by } 62$$
$$\sqrt[3]{125} - 2 \overset{?}{=} 3$$
$$5 - 2 = 3 \qquad \text{Solution checks}$$

Therefore, the equation has one solution: $x = 62$. ■

EXAMPLE 3 ■ Solving an Equation Having One Radical

$$\sqrt{3x} + 6 = 0 \qquad \text{Given equation}$$
$$\sqrt{3x} = -6 \qquad \text{Isolate radical}$$
$$\left(\sqrt{3x}\right)^2 = (-6)^2 \qquad \text{Square both sides}$$
$$3x = 36$$
$$x = 12 \qquad \text{Trial solution}$$

Check

$$\sqrt{3x} + 6 = 0 \qquad \text{Given equation}$$
$$\sqrt{3(12)} + 6 \overset{?}{=} 0 \qquad \text{Replace } x \text{ by } 12$$
$$6 + 6 \neq 0 \qquad \text{Solution does not check}$$

FIGURE 6.6

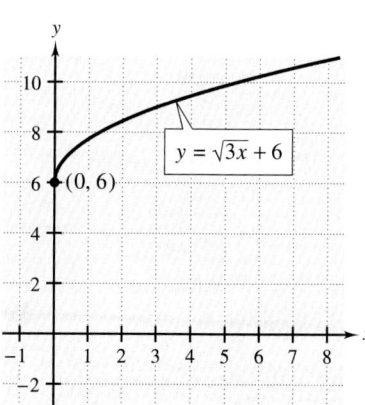

Therefore, the equation has no solution. In Figure 6.6, note that the graph of $y = \sqrt{3x} + 6$ has no x-intercepts. This visually confirms that the equation $\sqrt{3x} + 6 = 0$ has no solution.

■

EXAMPLE 4 ■ **Solving an Equation Having Two Radicals**

$$\sqrt{5x+3} = \sqrt{x+11}$$ Given equation

$$\left(\sqrt{5x+3}\right)^2 = \left(\sqrt{x+11}\right)^2$$ Square both sides

$$5x+3 = x+11$$

$$4x = 8$$

$$x = 2$$ Solution

Check

$$\sqrt{5x+3} = \sqrt{x+11}$$ Given equation

$$\sqrt{5(2)+3} \stackrel{?}{=} \sqrt{2+11}$$ Replace x by 2

$$\sqrt{13} = \sqrt{13}$$ Solution checks

Therefore, the equation has one solution: $x = 2$. ■

EXAMPLE 5 ■ **Solving an Equation Having Two Radicals**

$$\sqrt[3]{6x-4} - \sqrt[3]{2-x} = 0$$ Given equation

$$\sqrt[3]{6x-4} = \sqrt[3]{2-x}$$ Isolate radicals

$$6x-4 = 2-x$$ Cube both sides

$$7x = 6$$

$$x = \frac{6}{7}$$ Solution

Ask students why it is necessary to isolate the radicals before cubing both sides in Example 5.

Checking this result is a good test of your skill with fractions. You should end up with the true statement that $\sqrt[3]{\frac{8}{7}} - \sqrt[3]{\frac{8}{7}} = 0$. ■

EXAMPLE 6 ■ **Solving an Equation Having Two Radicals**

$$\sqrt[4]{3x} + \sqrt[4]{2x-5} = 0$$ Given equation

$$\sqrt[4]{3x} = -\sqrt[4]{2x-5}$$ Isolate radicals

$$\left(\sqrt[4]{3x}\right)^4 = \left(-\sqrt[4]{2x-5}\right)^4$$ Raise both sides to fourth power

$$3x = 2x-5$$

$$x = -5$$ Trial solution

Try this simple illustration to show how extraneous roots can be introduced. The equation $x = 2$ has only one solution. Trivially, x is 2. After squaring both sides, $x^2 = 4$ and the resulting equation has two solutions, $x = 2$ and $x = -2$. An "extra" solution was introduced. Stress the checking of solutions in the *original* equation.

A check will show that -5 is not a solution because it yields fourth roots of negative radicands. Therefore, this equation has no solution. ■

The next two examples illustrate how squaring both sides of an equation can yield a *nonlinear* equation.

EXAMPLE 7 ■   A Radical Equation That Converts to a Nonlinear Equation

Encourage students to become profi-
cient at squaring a binomial quickly
and accurately.

$$\sqrt{x} + 2 = x \qquad\qquad \text{Given equation}$$

$$\sqrt{x} = x - 2 \qquad\qquad \text{Isolate radical}$$

$$(\sqrt{x})^2 = (x-2)^2 \qquad\qquad \text{Square both sides}$$

$$x = x^2 - 4x + 4$$

$$-x^2 + 5x - 4 = 0 \qquad\qquad \text{Standard form}$$

$$(-1)(x-4)(x-1) = 0 \qquad\qquad \text{Factor}$$

$$x - 4 = 0 \implies x = 4 \qquad \text{Set first factor equal to 0}$$

$$x - 1 = 0 \implies x = 1 \qquad \text{Set second factor equal to 0}$$

Try checking each of these solutions. When you do, you will find that $x = 4$ is a valid solution, but that $x = 1$ is extraneous. Therefore, the equation has only one solution: $x = 4$. ■

When an equation contains two radicals, it may not be possible to isolate both. In such cases, you may have to raise both sides of the equation to a power at *two* different stages in the solution. This procedure is demonstrated in the next two examples.

EXAMPLE 8 ■   Repeatedly Squaring Both Sides of an Equation

Stress the importance of simplifying
the equation before squaring both
sides a second time.

$$\sqrt{3t+1} = 2 - \sqrt{3t} \qquad\qquad \text{Given equation}$$

$$(\sqrt{3t+1})^2 = (2 - \sqrt{3t})^2 \qquad\qquad \text{Square both sides (first time)}$$

$$3t + 1 = 4 - 4\sqrt{3t} + 3t$$

$$-3 = -4\sqrt{3t} \qquad\qquad \text{Isolate radical}$$

$$(-3)^2 = (-4\sqrt{3t})^2 \qquad\qquad \text{Square both sides (second time)}$$

$$9 = 16(3t)$$

$$\frac{9}{48} = t$$

$$\frac{3}{16} = t$$

A check will show that $t = \frac{3}{16}$ is a valid solution. Therefore, this equation has one solution: $t = \frac{3}{16}$. ■

EXAMPLE 9   ■   Repeatedly Squaring Both Sides of an Equation

$$\sqrt{x^2 + 11} - \sqrt{x^2 - 9} = 2 \qquad \text{Given equation}$$
$$\sqrt{x^2 + 11} = 2 + \sqrt{x^2 - 9} \qquad \text{Isolate one of the radicals}$$
$$\left(\sqrt{x^2 + 11}\right)^2 = \left(2 + \sqrt{x^2 - 9}\right)^2 \qquad \text{Square both sides (first time)}$$
$$x^2 + 11 = 4 + 4\sqrt{x^2 - 9} + (x^2 - 9)$$
$$16 = 4\sqrt{x^2 - 9}$$
$$4 = \sqrt{x^2 - 9} \qquad \text{Isolate radical}$$
$$4^2 = \left(\sqrt{x^2 - 9}\right)^2 \qquad \text{Square both sides (second time)}$$
$$16 = x^2 - 9$$
$$0 = x^2 - 25$$
$$0 = (x + 5)(x - 5)$$

$$x + 5 = 0 \implies x = -5 \qquad \text{Set first factor equal to 0}$$
$$x - 5 = 0 \implies x = 5 \qquad \text{Set second factor equal to 0}$$

Additional problem: You may want to have students solve $4 - \sqrt{x - 1} = \sqrt{3x + 3}$ in class. After they have finished, note that there is an extraneous solution. Answer: $x = 2$ ($x = 26$ is extraneous)

A check will show that both 5 and -5 are solutions. Therefore, this equation has two solutions: 5 and -5.   ■

Applications of Radicals

A common use of radicals occurs in applications involving right triangles.

EXAMPLE 10   ■   An Application Involving a Right Triangle

A softball diamond has the shape of a square with 60-foot sides (see Figure 6.7). The catcher is 5 feet behind home plate. How far does the catcher have to throw the ball to reach second base?

Remind students that a diagram or sketch is often helpful when solving applications. Also, they should ask themselves if the answers they get are reasonable.

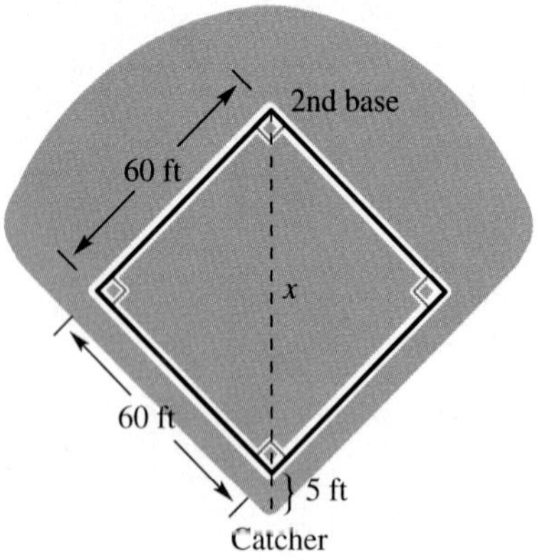

Solution

In Figure 6.7, let x be the hypotenuse of a right triangle with 60-foot legs. Thus, by the Pythagorean Theorem, you have the following equation.

$$x = \sqrt{60^2 + 60^2} \qquad \text{Pythagorean Theorem}$$
$$x = \sqrt{7200}$$
$$x \approx 84.9 \text{ feet}$$

FIGURE 6.7

Thus, the distance from home plate to second base is approximately 84.9 feet. Because the catcher is 5 feet behind home plate, the catcher must make a throw of

$$x + 5 \approx 84.9 + 5 = 89.9 \text{ feet.}$$

Check this solution in the original equation.

■

EXAMPLE 11 ■ An Application Involving the Use of Electricity

The amount of power consumed by an electrical appliance is

$$I = \sqrt{\frac{P}{R}},$$

where I is the current measured in amps, R is the resistance measured in ohms, and P is the power measured in watts. Find the power used by an electric heater, for which $I = 10$ amps and $R = 16$ ohms.

Historical Note: Pythagoras was a 6th century B.C. philosopher and mathematician. He taught the Pythagorean Theorem, but he didn't discover it. The Theorem was known by the Sumerians as early as 2000 B.C. Note the use of the Pythagorean Theorem in Example 10.

Solution

$$I = \sqrt{\frac{P}{R}} \qquad \text{Given equation}$$

$$10 = \sqrt{\frac{P}{16}} \qquad \text{Substitute known values}$$

$$(10)^2 = \left(\sqrt{\frac{P}{16}}\right)^2 \qquad \text{Square both sides}$$

$$100 = \frac{P}{16}$$

$$1600 = P \qquad \text{Solve for } P$$

Therefore, the electric heater uses 1600 watts of power. Check this solution in the original equation. ■

EXAMPLE 12 ■ An Application Involving the Velocity of a Falling Object

The velocity of a free-falling object can be determined from the equation

$$v = \sqrt{2gh},$$

where v is the velocity measured in feet per second, $g = 32$ feet per second, and h is the distance (in feet) the object has fallen. Find the height from which a rock has been dropped if it strikes the ground with a velocity of 50 feet per second.

FIGURE 6.8

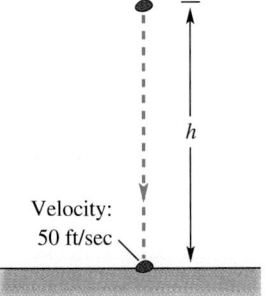

Velocity:
50 ft/sec

Solution

$$v = \sqrt{2gh} \qquad \text{Given equation}$$

$$50 = \sqrt{2(32)h} \qquad \text{Substitute known values}$$

$$(50)^2 = \left(\sqrt{64h}\right)^2 \qquad \text{Square both sides}$$

$$2500 = 64h$$

$$39 \approx h \qquad \text{Solve for } h$$

Thus, the rock has fallen approximately 39 feet when it hits the ground, as shown in Figure 6.8. Check this solution in the original equation. ■

DISCUSSION PROBLEM ■ **Constructing Lengths to Represent Radicals**

FIGURE 6.9

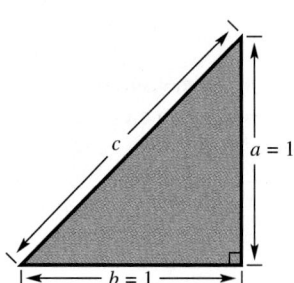

Consider a right triangle that has two sides of length 1, as shown in Figure 6.9. From the Pythagorean Theorem, you can conclude that the length of the hypotenuse is

$$c = \sqrt{a^2 + b^2} = \sqrt{1^2 + 1^2} = \sqrt{2}.$$

Can you construct a right triangle whose three sides have lengths of 1, 2, and $\sqrt{3}$?

■

Warm-Up

The following warm-up exercises involve skills that were covered in earlier sections. You will use these skills in the exercise set for this section.

In Exercises 1–6, solve the equation.

1. $13x + 6 = 32$

2. $7x = 18 + 3x$

3. $17 - 5(x - 2) = 0$

4. $32 + 3(2x - 11) = 0$

5. $3(x - 2)(x + 25) = 0$

6. $-2(12 - x)(114 + x) = 0$

In Exercises 7–10, perform the indicated operations and simplify. (Assume that all variables are positive.)

7. $(\sqrt{x} + 3)(\sqrt{x} - 3)$

8. $\sqrt{u}(\sqrt{20} - \sqrt{5})$

9. $(2\sqrt{t} + 3)^2$

10. $\dfrac{50x}{\sqrt{2}}$

6.5	**EXERCISES**

⊞ means that a graphing utility can help you solve the exercise or check your solution.

In Exercises 1–6, determine whether the value of x is a solution of the equation.

1. $\sqrt{x} - 4 = 4, \quad x = 64$

2. $2\sqrt{x - 3} = 12, \quad x = 39$

3. $(x + 4)^{3/2} - 60 = 4, \quad x = 12$

4. $4(2x + 1)^{2/3} = 36, \quad x = 13$

5. $\sqrt[3]{4x} + 9 = 3, \quad x = -54$

6. $2\sqrt[4]{2x + 1} + 2 = 0, \quad x = 0$

In Exercises 7–40, solve the equation. (Some of the equations have no solutions.)

7. $\sqrt{x} = 20$

8. $\sqrt{x} = 5$

9. $\sqrt{y} - 7 = 0$

10. $\sqrt{t} - 13 = 0$

11. $\sqrt{u} + 13 = 0$

12. $\sqrt{y} + 15 = 0$

13. $\sqrt{a + 100} = 25$

14. $\sqrt{b + 12} = 13$

15. $\sqrt{10x} = 30$

16. $\sqrt{8x} = 6$

17. $\sqrt{3y + 5} - 3 = 4$

18. $\sqrt{5z - 2} + 7 = 10$

19. $5\sqrt{x + 2} = 8$

20. $2\sqrt{x + 4} = 7$

21. $\sqrt{x^2 + 5} = x + 3$

22. $\sqrt{x^2 - 4} = x - 2$

23. $\sqrt{x + 3} = \sqrt{2x - 1}$

24. $\sqrt{3t + 1} = \sqrt{t + 15}$

25. $\sqrt{3y - 5} - 3\sqrt{y} = 0$

26. $\sqrt{2u + 10} - 2\sqrt{u} = 0$

27. $\sqrt[3]{3x - 4} = \sqrt[3]{x + 10}$

28. $2\sqrt[3]{10 - 3x} = \sqrt[3]{2 - x}$

29. $\sqrt[3]{2x + 15} - \sqrt[3]{x} = 0$

30. $\sqrt[4]{2x} + \sqrt[4]{x + 3} = 0$

31. $\sqrt{2x} = x - 4$

32. $\sqrt{x} = x - 6$

33. $\sqrt{8x + 1} = x + 2$

34. $\sqrt{3x + 7} = x + 3$

35. $\sqrt{2t + 3} = 3 - \sqrt{2t}$

36. $\sqrt{x + 3} - \sqrt{x - 1} = 1$

37. $\sqrt{5x - 4} = 2 - \sqrt{5x}$

38. $\sqrt{1 + 4a} = 2\sqrt{a} - 3$

39. $\sqrt{x^2 + 9} + \sqrt{x^2 - 7} = 8$

40. $1 + \sqrt{5 - x^2} = \sqrt{10 - x^2}$

Free-Falling Object In Exercises 41–44, use the equation for the velocity of a free-falling object ($v = \sqrt{2gh}$) as described in Example 12.

41. An object is dropped from a height of 50 feet. Find the velocity of the object when it strikes the ground.

42. An object is dropped from a height of 200 feet. Find the velocity of the object when it strikes the ground.

43. An object that was dropped strikes the ground at a velocity of 60 feet per second. Find the height from which the object was dropped.

44. An object that was dropped strikes the ground at a velocity of 120 feet per second. Find the height from which the object was dropped.

Height of a Building In Exercises 45 and 46, use the equation for time t in seconds for a free-falling object to fall d feet. The equation is

$$t = \sqrt{\frac{d}{16}}.$$

45. A construction worker drops a nail from the top of a building and observes it strike a water puddle after approximately 2 seconds. Estimate the building's height.

46. A construction worker drops a nail from the top of a building and observes it strike a water puddle after approximately 3 seconds. Estimate the building's height.

Length of a Pendulum In Exercises 47 and 48, use the equation for the time t in seconds for a pendulum of length L feet to go through one complete cycle (its period). The equation is

$$t = 2\pi \sqrt{\frac{L}{32}}.$$

47. How long is the pendulum of a grandfather clock with a period of 1.5 seconds (see figure)?

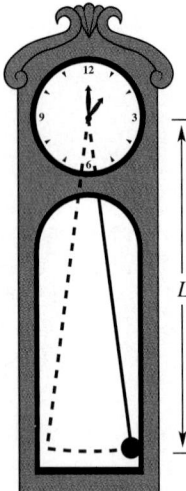

L

48. How long is the pendulum of a mantle clock with a period of 0.75 second?

 49. *Demand for a Product* The demand equation for a product is

$$p = 50 - \sqrt{0.8(x - 1)},$$

where x is the number of units demanded per day and p is the price per unit. Find the demand if the price is $30.02.

50. *Airline Passengers* An airline offers daily flights between Chicago and Denver. The total monthly cost of the flights is

$$C = \sqrt{0.2x + 1}, \qquad 0 \le x,$$

where C is measured in millions of dollars and x is measured in thousands of passengers (see figure). The total cost of the flights for a certain month is 2.5 million dollars. Approximately how many passengers flew that month?

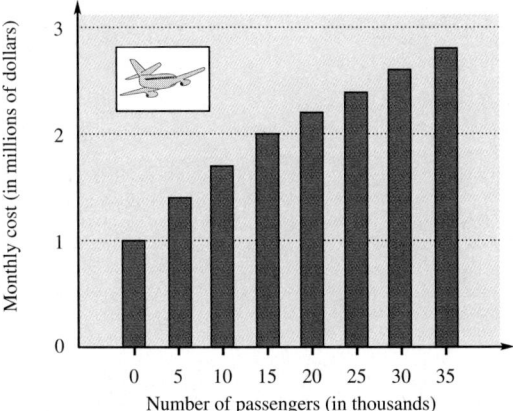

Monthly cost (in millions of dollars)

Number of passengers (in thousands)

51. *Surface Area of a Cone* The surface area of a cone (see figure) is

$$S = \pi r \sqrt{r^2 + h^2}.$$

Solve this equation for h^2.

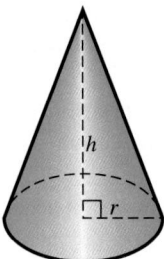

h

r

Using a Graphing Calculator

A graphing calculator (or another type of graphing utility) can be used to investigate the graphs of functions of the form

$$y = x^n, \qquad x \geq 0, \quad n > 0.$$

After some experimentation, you can see that graphs of functions of this form fall into three categories.

1. If $n = 1$, the graph of $y = x^n$ is a straight line.

2. If $n > 1$, the graph of $y = x^n$ curves upward (this is called *concave up*).

3. If $n < 1$, the graph of $y = x^n$ curves downward (this is called *concave down*).

Several examples are shown in the calculator screens below. (Try confirming this observation by sketching graphs of other functions of the form $y = x^n$.)

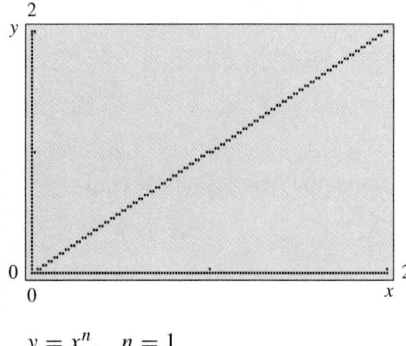

$y = x^n, \quad n = 1$

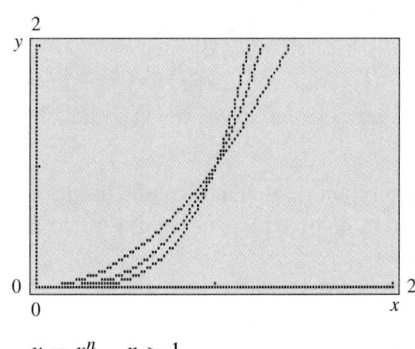

$y = x^n, \quad n > 1$

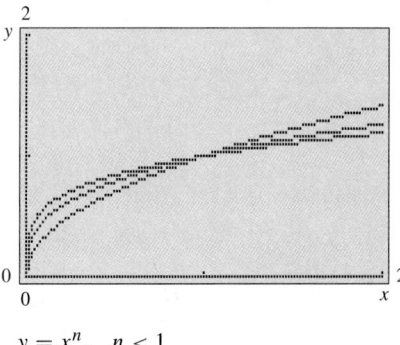

$y = x^n, \quad n < 1$

EXAMPLE 1 ■ Sketching the Graph of a Function

The shoulder height h (in feet) of a male African elephant can be modeled by the equation

$$h = \sqrt[3]{13t} + 3, \qquad 0 \leq t \leq 40,$$

where t is the age of the elephant in years. (By age 40, these elephants are fully grown.) Sketch the graph of this model and use the graph to describe the growth pattern of male African elephants.

Solution

You can enter the equation into a graphing utility as

$$h = (13t)^{1/3} + 3.$$

The graph of the equation is shown below. From the graph, you can see that a newborn male African elephant measures about 3 feet at the shoulders. A fully grown male (40 years old) measures about 11 feet high at the shoulders. From the graph, you can also see that a male African elephant's shoulder height increases most rapidly during its first year. With each succeeding year, the rate of growth becomes less and less.

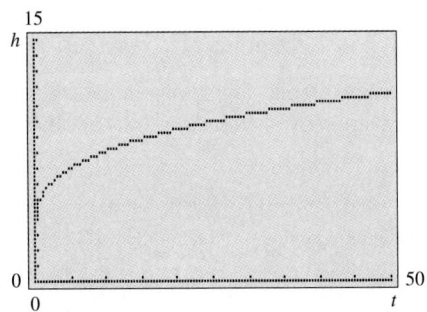

A graphing calculator can be used for visual estimation of solutions of equations that involve radicals, as illustrated in Example 2.

EXAMPLE 2 ■ Estimating a Solution from a Graph

Use the model given in Example 1 to estimate the age of a male African elephant that has a shoulder height of 10 feet.

Solution

You could solve this problem algebraically, as follows.

$h = \sqrt[3]{13t} + 3$	Given model
$10 = \sqrt[3]{13t} + 3$	Substitute 10 for h
$7 = \sqrt[3]{13t}$	Subtract 3 from both sides
$343 = 13t$	Cube both sides
$\dfrac{343}{13} = t$	Divide both sides by 13
$26.4 \approx t$	Use a calculator

From the algebraic solution, you can estimate the elephant's age to be 26 years. Another way to obtain this estimate is to graph the model

$$h = \sqrt[3]{13t} + 3$$

with a graphing calculator (as shown below) and use the *trace* feature of the calculator to estimate the value of t that corresponds to a height of 10 feet.

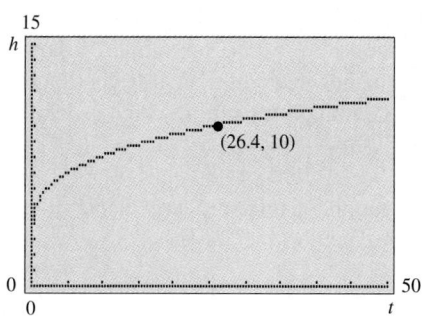

EXERCISES

In Exercises 1–8, sketch the first-quadrant portion of the graph of the function. State whether the graph is concave up or concave down.

1. $y = x^5$

2. $y = x^4$

3. $y = x^{1/5}$

4. $y = x^{1/4}$

5. $y = x^{4/3}$

6. $y = x^{3/4}$

7. $y = x^{5/6}$

8. $y = x^{6/5}$

9. The chest circumferences C (in inches) of five different species of adult primates (tamarins, squirrel monkeys, vervet monkeys, macaques, and baboons) were compared with their weights W (in pounds). The relationship between C and W can be modeled by

$$C = 5W^{37/100}, \qquad 0 < W \leq 40.$$

Sketch the graph of this equation. Then use the graph to describe the relationship between chest circumference and weight for the five species of primates.

10. An adult squirrel monkey has a chest circumference of 16 inches. Estimate the weight of the monkey.

Complex Numbers

The Imaginary Unit i • *Complex Numbers* • *Operations with Complex Numbers* • *Complex Conjugates*

The Imaginary Unit i

In Section 6.2, you learned that a negative number has no *real* square root. For instance, $\sqrt{-1}$ is not real because there is no real number x such that $x^2 = -1$. Thus, as long as you are dealing only with real numbers, you can say that the equation $x^2 = -1$ has no solution. When you are dealing with the *complex* number system, however, this equation does have solutions. To denote these solutions, we use the **imaginary unit i,** which is defined as

$$i = \sqrt{-1}. \qquad \text{Imaginary unit}$$

This number has the property that $i^2 = -1$. Thus, the imaginary unit i is a solution of the equation $x^2 = -1$.

The Square Root of a Negative Number

Let c be a positive real number. Then the square root of $-c$ is given by

$$\sqrt{-c} = \sqrt{c(-1)} = \sqrt{c}\sqrt{-1} = \sqrt{c}\,i.$$

When you write $\sqrt{-c}$ in the form $\sqrt{c}\,i$, you are writing the number in **i-form.**

EXAMPLE 1 ■ Writing Square Roots of Negative Numbers in i-Form

(a) $\sqrt{-36} = \sqrt{36(-1)} = \sqrt{36}\sqrt{-1} = 6i$

(b) $\sqrt{-\frac{16}{25}} = \sqrt{\frac{16}{25}(-1)} = \sqrt{\frac{16}{25}}\sqrt{-1} = \frac{4}{5}\,i$

(c) $\sqrt{-5} = \sqrt{5(-1)} = \sqrt{5}\sqrt{-1} = \sqrt{5}\,i$

(d) $\sqrt{-54} = \sqrt{54(-1)} = \sqrt{54}\sqrt{-1} = 3\sqrt{6}\,i$ ■

To perform operations with square roots of negative numbers, you must first write the numbers in i-form. Once the numbers have been written in i-form, you can add, subtract, and multiply, as follows.

$$ai + bi = (a + b)i \qquad \text{Addition}$$

$$ai - bi = (a - b)i \qquad \text{Subtraction}$$

$$(ai)(bi) = ab(i^2) = ab(-1) = -ab \qquad \text{Multiplication}$$

EXAMPLE 2   ■   Adding Square Roots of Negative Numbers

$$\sqrt{-9} + \sqrt{-49} = \sqrt{9}\sqrt{-1} + \sqrt{49}\sqrt{-1}$$

$$= 3i + 7i \qquad\qquad \text{Write in } i\text{-form}$$

$$= (3 + 7)i \qquad\qquad \text{Distributive Property}$$

$$= 10i \qquad\qquad \text{Simplified form} \qquad ■$$

When performing operations with numbers in i-form, you sometimes need to evaluate powers of the imaginary unit i. The first several powers of i are as follows.

$$i^1 = i$$

$$i^2 = -1$$

$$i^3 = i(i^2) = i(-1) = -i$$

$$i^4 = (i^2)(i^2) = (-1)(-1) = 1$$

$$i^5 = i(i^4) = i(1) = i$$

$$i^6 = (i^2)(i^4) = (-1)(1) = -1$$

$$i^7 = (i^3)(i^4) = (-i)(1) = -i$$

$$i^8 = (i^4)(i^4) = (1)(1) = 1$$

Note how the pattern of values i, -1, $-i$, and 1 repeats for powers greater than 4.

EXAMPLE 3   ■   Operations with Square Roots of Negative Numbers

(a) $\sqrt{-15}\sqrt{-15} = \left(\sqrt{15}\,i\right)\left(\sqrt{15}\,i\right)$ Write in standard form

$$= \left(\sqrt{15}\right)^2 i^2$$

$$= 15(-1)$$

$$= -15 \qquad\qquad \text{Simplified form}$$

(b) $\sqrt{-5}\left(\sqrt{-45} - \sqrt{-4}\right) = \sqrt{5}\,i\left(3\sqrt{5}\,i - 2i\right)$ Write in standard form

$$= \left(\sqrt{5}\,i\right)\left(3\sqrt{5}\,i\right) - \left(\sqrt{5}\,i\right)(2i) \quad \text{Distributive Property}$$

$$= 3(5)(-1) - 2\sqrt{5}(-1)$$

$$= -15 + 2\sqrt{5} \qquad\qquad \text{Simplified form} \qquad ■$$

When multiplying square roots of negative numbers, be sure to write them in i-form *before multiplying*. If you don't, you may obtain incorrect answers. For instance, in Example 3(a), be sure you see the following.

$$\sqrt{-15}\sqrt{-15} \neq \sqrt{(-15)(-15)} = \sqrt{225} = 15$$

Complex Numbers

A number of the form $a + bi$, where a and b are real numbers, is called a **complex number.**

■ **Definition of a Complex Number**	If a and b are real numbers, then the number $$a + bi$$ is called a **complex number,** written in **standard form.** If $b = 0$, then the number $a + bi = a$ is a real number. If $b \neq 0$, then the number $a + bi$ is called an **imaginary number.**

A number cannot be both real and imaginary. For instance, the numbers -2, 0, 1, $\frac{1}{2}$, and $\sqrt{2}$ are real numbers (but they are *not* imaginary numbers), and the numbers $-3i$, $2 + 4i$, and $-1 + i$ are imaginary numbers (but they are *not* real numbers). The diagram shown in Figure 6.10 further illustrates the relationships among real, complex, and imaginary numbers.

FIGURE 6.10

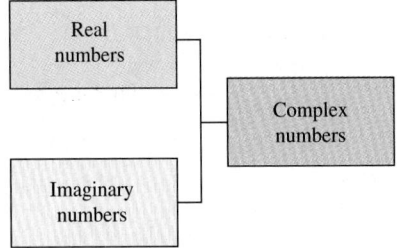

Two complex numbers $a + bi$ and $c + di$, in standard form, are equal if and only if $a = c$ and $b = d$.

EXAMPLE 4   ■   Equality of Two Complex Numbers

(a) Determine whether the complex numbers $\sqrt{9} + \sqrt{-48}$ and $3 - 4\sqrt{3}\,i$ are equal.

(b) Find x and y so that the following equation is valid.

$$3x - \sqrt{-25} = -6 + 3yi$$

Solution

(a) Writing the first number in standard form produces the following.

$$\sqrt{9} + \sqrt{-48} = \sqrt{3^2} + \sqrt{4^2(3)(-1)} = 3 + 4\sqrt{3}\,i$$

This complex number is not equal to $3 - 4\sqrt{3}\,i$, because the two numbers have imaginary parts that differ in sign.

(b) By writing the left side of the equation in standard form, you have the following.

$$3x - \sqrt{-25} = -6 + 3yi \qquad \text{Given equation}$$
$$3x - 5i = -6 + 3yi \qquad \text{Both sides in standard form}$$

For these two numbers to be equal, their real parts must be equal to each other and their imaginary parts must be equal to each other, as follows.

$$3x = -6 \quad \text{and} \quad -5 = 3y$$
$$x = -2 \qquad\qquad y = -\frac{5}{3}$$

■

Operations with Complex Numbers

The real number a is called the **real part** of the complex number $a + bi$, and the real number b is called the **imaginary part** of the complex number. To add or subtract two complex numbers, you add (or subtract) the real and imaginary parts separately. This is similar to combining like terms of a polynomial.

$$(a + bi) + (c + di) = (a + c) + (b + d)i \qquad \text{Addition of complex numbers}$$
$$(a + bi) - (c + di) = (a - c) + (b - d)i \qquad \text{Subtraction of complex numbers}$$

EXAMPLE 5 ■ Adding and Subtracting Complex Numbers

(a) $(3 - i) + (-2 + 4i) = (3 - 2) + (-1 + 4)i = 1 + 3i$

(b) $3i + (5 - 3i) = 5 + (3 - 3)i = 5$

(c) $4 - (-1 + 5i) + (7 + 2i) = [4 - (-1) + 7] + (-5 + 2)i = 12 - 3i$

Note in part (b) that the sum of two imaginary numbers can be a real number. ■

The Commutative, Associative, and Distributive Properties of real numbers are also valid for complex numbers. Note how these properties are used to find the product of two complex numbers.

EXAMPLE 6 ■ Multiplying Complex Numbers

Find the following products, and write your answers in standard form.

(a) $(1 - i)\sqrt{-9}$ (b) $(2 - i)(4 + 3i)$

Solution

(a) $(1 - i)\sqrt{-9} = (1 - i)(3i)$ Write in standard form

$\qquad\qquad\qquad = (1)(3i) - (i)(3i)$ Distributive Property

$\qquad\qquad\qquad = 3i - 3(i^2)$

$\qquad\qquad\qquad = 3i - 3(-1)$

$\qquad\qquad\qquad = 3 + 3i$ Standard form

$$\begin{array}{ccccc} & \overset{F}{} & \overset{O}{} & \overset{I}{} & \overset{L}{} \end{array}$$

(b) $(2 - i)(4 + 3i) = \quad 8 \quad + \quad 6i \quad - \quad 4i \quad - \quad 3i^2$ FOIL Method

$\qquad\qquad\qquad\quad = 8 - 3(-1) + 6i - 4i$ Collect like terms

$\qquad\qquad\qquad\quad = 11 + 2i$ Standard form ■

Stress to students that their answers are not in standard form until they have simplified terms involving i raised to powers greater than 1.

EXAMPLE 7 ■ Multiplying Complex Numbers

Find the following products, and write your answers in standard form.

(a) $(3 + 2i)(3 - 2i)$ (b) $(1 + 4i)^2$

Solution

(a) $(3 + 2i)(3 - 2i) = 3^2 - (2i)^2$ Special product

$\qquad\qquad\qquad\quad = 9 - 2^2 i^2$

$\qquad\qquad\qquad\quad = 9 - 4(-1)$

$\qquad\qquad\qquad\quad = 9 + 4$

$\qquad\qquad\qquad\quad = 13$ Standard form

(b) $(1 + 4i)^2 = 1 + 2(4i) + (4i)^2$ Special product

$\qquad\qquad\quad = 1 + 8i + 4^2 i^2$

$\qquad\qquad\quad = 1 + 8i + 16(-1)$

$\qquad\qquad\quad = -15 + 8i$ Standard form ■

Complex Conjugates

In Example 7(a), note that the product of two imaginary numbers can be a real number. This occurs with pairs of complex numbers of the form $a + bi$ and $a - bi$, called **complex conjugates.** In general, the product of complex conjugates has the following form.

$$(a + bi)(a - bi) = a^2 - (bi)^2$$
$$= a^2 - b^2i^2$$
$$= a^2 - b^2(-1)$$
$$= a^2 + b^2$$

NOTE The $a^2 + b^2$ form of the product of the pair of complex conjugates, $a + bi$ and $a - bi$, is the same result you would obtain by using the **FOIL Method** to find the product. For example,

$$(6 + 7i)(6 - 7i) = 6^2 + 7^2 = 85$$
$$(6 + 7i)(6 - 7i) = 36 - 42i + 42i - 49i^2 = 36 - 49(-1) = 85.$$

Here are some examples.

Complex Number	Complex Conjugate	Product
$4 - 5i$	$4 + 5i$	$4^2 + 5^2 = 41$
$3 + 2i$	$3 - 2i$	$3^2 + 2^2 = 13$
$-2 = -2 + 0i$	$-2 = -2 - 0i$	$(-2)^2 + 0^2 = 4$
$i = 0 + 1i$	$-i = 0 - 1i$	$0^2 + 1^2 = 1$

Complex conjugates are used to divide one complex number by another. To do this, multiply the numerator and denominator by the complex conjugate of the denominator, as shown in Example 8.

EXAMPLE 8   ■   Division of Complex Numbers

Perform the following division, and write your answer in standard form.

$$5 \div (3 - 2i)$$

Solution

$$\frac{5}{3 - 2i} = \frac{5}{3 - 2i} \cdot \frac{3 + 2i}{3 + 2i} \qquad \text{Multiply by complex conjugate}$$

$$= \frac{5(3 + 2i)}{(3 - 2i)(3 + 2i)}$$

$$= \frac{5(3 + 2i)}{3^2 + 2^2} \qquad (3 - 2i)(3 + 2i) = 9 - 4i^2 = 9 + 4$$

$$= \frac{15 + 10i}{13}$$

$$= \frac{15}{13} + \frac{10}{13}i \qquad \text{Standard form} \qquad ■$$

EXAMPLE 9 ■ Division of Complex Numbers

Write the following fraction in standard form.

$$\frac{1}{1+i}$$

Solution

$$\frac{1}{1+i} = \frac{1}{1+i} \cdot \frac{1-i}{1-i} \qquad \text{Multiply by complex conjugate}$$

$$= \frac{1-i}{1^2 + 1^2} \qquad (1+i)(1-i) = 1 - i^2 = 1 + 1$$

$$= \frac{1-i}{2}$$

$$= \frac{1}{2} - \frac{1}{2}i \qquad \text{Standard form} \qquad ■$$

EXAMPLE 10 ■ Division of Complex Numbers

(a) $$\frac{2+3i}{4-2i} = \frac{2+3i}{4-2i} \cdot \frac{4+2i}{4+2i} \qquad \text{Multiply by complex conjugate}$$

$$= \frac{8 + 16i + 6i^2}{4^2 + 2^2} \qquad (4-2i)(4+2i) = 16 - 4i^2 = 16 + 4$$

$$= \frac{8 + 16i + 6(-1)}{20}$$

$$= \frac{2 + 16i}{20}$$

Some students incorrectly rewrite
$(2 + 16i)/(20)$ as $(1 + 16i)/(10)$ or as
$(2 + 4i)/5$.

$$= \frac{1}{10} + \frac{4}{5}i \qquad \text{Standard form}$$

(b) $$\frac{7-i}{5i} = \frac{7-i}{5i} \cdot \frac{-5i}{-5i} \qquad \text{Multiply by complex conjugate}$$

$$= \frac{(7-i)(-5i)}{(5i)(-5i)}$$

$$= \frac{-35i + 5i^2}{-25i^2}$$

$$= \frac{-35i + 5(-1)}{-25(-1)}$$

$$= \frac{-35i - 5}{25}$$

$$= -\frac{1}{5} - \frac{7}{5}i \qquad \text{Standard form} \qquad ■$$

DISCUSSION PROBLEM ■ Factoring the Sum of Two Squares

You know from Section 1.5 that the *difference* of two squares, such as $x^2 - 9$, factors as $(x + 3)(x - 3)$. You also know from that section that the *sum* of two squares, such as $x^2 + 9$, is irreducible over the real numbers.

(a) Using complex numbers, it *is* possible to factor the polynomial $x^2 + 9$. Write a short paragraph showing how this can be done.

(b) As long as you are dealing only with real numbers, the equation $x^2 + 9 = 0$ has no solution. When complex numbers are used, does this equation have a solution? If so, what is the solution? ■

Warm-Up

The following warm-up exercises involve skills that were covered in earlier sections. You will use these skills in the exercise set for this section.

In Exercises 1–6, simplify the expression.

1. $\sqrt{48}$ **2.** $\sqrt{108}$ **3.** $\sqrt{128} + 3\sqrt{50}$

4. $3\sqrt{5}\sqrt{500}$ **5.** $\dfrac{8}{\sqrt{10}}$ **6.** $\dfrac{5}{\sqrt{12} - 2}$

In Exercises 7–10, solve the equation.

7. $x(2x - 35) = 0$ **8.** $(x + 4)\left(x - \frac{5}{4}\right) = 0$

9. $x^2 - 7x - 18 = 0$ **10.** $2x^2 - x - 3 = 0$

6.6 EXERCISES

In Exercises 1–12, write the number in i-form.

1. $\sqrt{-4}$ **2.** $\sqrt{-9}$ **3.** $-\sqrt{-144}$ **4.** $\sqrt{-49}$ **5.** $\sqrt{-\frac{4}{25}}$ **6.** $-\sqrt{-\frac{36}{121}}$

7. $\sqrt{-0.09}$ **8.** $\sqrt{-0.0004}$ **9.** $\sqrt{-8}$ **10.** $\sqrt{-75}$ **11.** $\sqrt{-7}$ **12.** $\sqrt{-15}$

In Exercises 13–16, perform the indicated addition or subtraction and write your answer in standard form.

13. $\sqrt{-16} + 6i$ **14.** $\sqrt{-\frac{1}{4}} - \frac{3}{2}i$ **15.** $\sqrt{-50} - \sqrt{-8}$ **16.** $\sqrt{-500} + \sqrt{-45}$

In Exercises 17–28, perform the indicated operations and write your answer in standard form.

17. $\sqrt{-8}\sqrt{-2}$

18. $\sqrt{-25}\sqrt{-6}$

19. $\sqrt{-18}\sqrt{-3}$

20. $\sqrt{-0.16}\sqrt{-1.21}$

21. $\sqrt{-3}(\sqrt{-3}+\sqrt{-4})$

22. $\sqrt{-12}(\sqrt{-3}-\sqrt{-12})$

23. $\sqrt{-5}(\sqrt{-16}-\sqrt{-10})$

24. $\sqrt{-24}\left(\sqrt{-\frac{4}{9}}+\sqrt{-\frac{1}{4}}\right)$

25. $\left(\sqrt{-6}\right)^2$

26. $\left(\sqrt{-2}\right)^2$

27. $\left(\sqrt{-16}\right)^2$

28. $\left(\sqrt{-25}\right)^2$

In Exercises 29–34, determine a and b.

29. $3-4i = a+bi$

30. $-8+6i = a+bi$

31. $-4-\sqrt{-8} = a+bi$

32. $\sqrt{-36}-3 = a+bi$

33. $(a+5)+(b-1)i = 7-3i$

34. $(2a+1)+(2b+3)i = 5+12i$

In Exercises 35–66, perform the specified operations and write your answer in standard form.

35. $(4-3i)+(6+7i)$

36. $(-10+2i)+(4-7i)$

37. $(-4-7i)+(-10-33i)$

38. $(4+6i)+(15+24i)-(1-i)$

39. $(15+10i)-(2+10i)$

40. $(-21-50i)+(21-20i)$

41. $13i-(14-7i)$

42. $22+(-5+8i)+10i$

43. $\left(\frac{4}{3}+\frac{1}{3}i\right)+\left(\frac{5}{6}+\frac{7}{6}i\right)$

44. $(0.05+2.50i)-(6.2+11.8i)$

45. $15i-(3-25i)+\sqrt{-81}$

46. $(-1+i)-\sqrt{2}-\sqrt{-2}$

47. $(3i)(12i)$

48. $(-5i)(4i)$

49. $(-6i)(-i)(6i)$

50. $\frac{1}{2}(10i)(12i)(-3i)$

51. $(-3i)^3$

52. $(8i)^2$

53. $-5(13+2i)$

54. $10(8-6i)$

55. $4i(-3-5i)$

56. $-3i(10-15i)$

57. $(4+3i)(-7+4i)$

58. $(3+5i)(2+15i)$

59. $(3-4i)^2$

60. $(7+i)^2$

61. $(2+5i)^2$

62. $(8-3i)^2$

63. $\left(-3-\sqrt{-12}\right)\left(4-\sqrt{-12}\right)$

64. $\sqrt{-9}\left(1+\sqrt{-16}\right)$

65. $(2+i)^3$

66. $(3-2i)^3$

In Exercises 67–78, find the conjugate of the complex number. Then find the product of the number and its conjugate.

67. $2+i$

68. $3+2i$

69. $5-\sqrt{6}i$

70. $10-3i$

71. $-2-8i$

72. $-4+\sqrt{2}i$

73. $10i$

74. 20

75. $1 + \sqrt{-3}$　　　　　　**76.** $-3 - \sqrt{-5}$　　　　　**77.** $1.5 + \sqrt{-0.25}$　　　　**78.** $3.2 - \sqrt{-0.04}$

In Exercises 79–90, perform the indicated division and write your answer in standard form.

79. $\dfrac{4}{1-i}$　　　　**80.** $\dfrac{20}{3+i}$　　　　**81.** $\dfrac{-12}{2+7i}$　　　　**82.** $\dfrac{15}{2(1-i)}$　　　　**83.** $\dfrac{4i}{1-3i}$　　　　**84.** $\dfrac{17i}{5+3i}$

85. $\dfrac{20}{2i}$　　　　**86.** $\dfrac{1+i}{3i}$　　　　**87.** $\dfrac{3-4i}{8}$　　　　**88.** $\dfrac{-16+22i}{24}$　　　　**89.** $\dfrac{2+3i}{1+2i}$　　　　**90.** $\dfrac{4-5i}{4+5i}$

In Exercises 91–94, find the indicated sum or difference and write your answer in standard form.

91. $\dfrac{1}{1-2i} + \dfrac{4}{1+2i}$　　　　**92.** $\dfrac{3i}{1+i} + \dfrac{2}{2+3i}$　　　　**93.** $\dfrac{i}{4-3i} - \dfrac{5}{2+i}$　　　　**94.** $\dfrac{1+i}{i} - \dfrac{3}{5-2i}$

In Exercises 95–98, perform the indicated operation on the complex number $a + bi$ and its conjugate $a - bi$.

95. $(a + bi) + (a - bi)$　　　**96.** $(a + bi)(a - bi)$　　　**97.** $(a + bi) - (a - bi)$　　　**98.** $(a + bi)^2 + (a - bi)^2$

In Exercises 99 and 100, show that the value of x is a solution of the equation.

99. $x^2 + 2x + 5 = 0$,　　$x = -1 + 2i$　　　　　　　　**100.** $x^2 - 4x + 13 = 0$,　　$x = 2 - 3i$

6 REVIEW EXERCISES

In Exercises 1–6, evaluate the expression. (Do not use a calculator.)

1. $(2^3 \cdot 3^2)^{-1}$

2. $(2^{-2} \cdot 5^2)^{-2}$

3. $\left(\frac{2}{5}\right)^{-3}$

4. $\left(\frac{1}{3^{-2}}\right)^2$

5. $(6 \times 10^3)^2$

6. $(3 \times 10^{-3})(8 \times 10^7)$

In Exercises 7–14, write the expression without negative or zero exponents and simplify. (Assume that none of the variables is zero.)

7. $\frac{(4x)^2}{2x}$

8. $4(-3x)^3$

9. $(x^3y^{-4})^2$

10. $5yx^0$

11. $\frac{t^{-5}}{t^{-2}}$

12. $\frac{a^5 \cdot a^{-3}}{a^{-2}}$

13. $\left(\frac{y}{3}\right)^{-3}$

14. $(2x^2y^4)^4(2x^2y^4)^{-4}$

In Exercises 15–20, evaluate the expression. (Do not use a calculator.)

15. $\sqrt{1.44}$

16. $\sqrt{0.16}$

17. $\sqrt{\frac{25}{36}}$

18. $-\sqrt{\frac{64}{225}}$

19. $\sqrt{169 - 25}$

20. $\sqrt{16 + 9}$

In Exercises 21–24, use a calculator to approximate the expression. (Round your answer to two decimal places.)

21. $1800(1 + 0.08)^{24}$

22. $0.0024(7{,}658{,}400)$

23. $\sqrt{13^2 - 4(2)(7)}$

24. $\frac{-3.7 + \sqrt{15.8}}{2(2.3)}$

In Exercises 25 and 26, fill in the missing description.

Radical Form	Rational Exponent Form
25. $\sqrt{16} = 4$	
26.	$16^{1/4} = 2$

In Exercises 27 and 28, evaluate the expression. (Do not use a calculator.)

27. $27^{4/3}$

28. $243^{-2/5}$

In Exercises 29 and 30, use a calculator to approximate the expression. (Round your answer to two decimal places.)

29. $75^{-3/4}$

30. $510^{5/3}$

In Exercises 31–34, simplify the expression. (Assume that all variables are positive.)

31. $x^{3/4} \cdot x^{-1/6}$

32. $(2y^2)^{3/2}(2y^{-4})^{1/2}$

33. $\frac{15x^{1/4}y^{3/5}}{5x^{1/2}y}$

34. $\frac{48a^2b^{5/2}}{14a^{-3}b^{-1/2}}$

In Exercises 35–40, simplify the expression. (Assume that all variables are positive.)

35. $\sqrt{360}$

36. $\sqrt{\frac{50}{9}}$

37. $\sqrt{0.25x^4y}$

38. $\sqrt{0.16s^6t^3}$

39. $\sqrt[3]{48a^3b^4}$

40. $\sqrt[4]{32u^4v^5}$

In Exercises 41–46, rationalize the denominator and simplify further when possible. (Assume that all variables are positive.)

41. $\sqrt{\frac{5}{6}}$

42. $\sqrt{\frac{3}{20}}$

43. $\frac{3}{\sqrt{12x}}$

44. $\frac{4y}{\sqrt{10z}}$

45. $\frac{2}{\sqrt[3]{2x}}$

46. $\sqrt[3]{\frac{16t}{s^2}}$

In Exercises 47–56, perform the indicated operations and simplify. (Assume that all variables are positive.)

47. $3\sqrt{40} - 10\sqrt{90}$

48. $9\sqrt{50} - 5\sqrt{8} + \sqrt{48}$

49. $10\sqrt[4]{y+3} - 3\sqrt[4]{y+3}$

50. $\sqrt{25x} + \sqrt{49x} - \sqrt[3]{8x}$

51. $\left(\sqrt{5} + 6\right)^2$

52. $\left(\sqrt{3} - \sqrt{x}\right)\left(\sqrt{3} + \sqrt{x}\right)$

53. $(2\sqrt{3}+7)(\sqrt{6}-2)$

54. $\dfrac{15}{\sqrt{x}+3}$

55. $(\sqrt{x}+10) \div (\sqrt{x}-10)$

56. $(3\sqrt{s}+4) \div (\sqrt{s}+2)$

 In Exercises 57–66, solve the equation.

57. $\sqrt{y}=15$

58. $\sqrt{3x}+9=0$

59. $\sqrt{2(a-7)}=14$

60. $\sqrt{3(2x+3)}=\sqrt{x+15}$

61. $\sqrt{2(x+5)}=x+5$

62. $\sqrt{5t}=1+\sqrt{5(t-1)}$

63. $\sqrt[3]{5x+2}-\sqrt[3]{7x-8}=0$

64. $\sqrt[4]{9x-2}-\sqrt[4]{8x}=0$

65. $\sqrt{1+6x}=2-\sqrt{6x}$

66. $\sqrt{2+9b}+1=3\sqrt{b}$

In Exercises 67–70, write the complex number in standard form.

67. $\sqrt{-48}$

68. $3+2\sqrt{-500}$

69. $\frac{3}{4}-5\sqrt{-\frac{3}{25}}$

70. $-0.5+3\sqrt{-1.21}$

In Exercises 71–80, perform the indicated operation and write your answer in standard form.

71. $(-4+5i)-(-12+8i)$

72. $5(3-8i)+(5+12i)$

73. $(-2)(15i)(-3i)$

74. $-10i(4-7i)$

75. $(4-3i)(4+3i)$

76. $(6-5i)^2$

77. $(12-5i)(2+7i)$

78. $\dfrac{4}{5i}$

79. $\dfrac{5i}{2+9i}$

80. $\dfrac{2+i}{1-9i}$

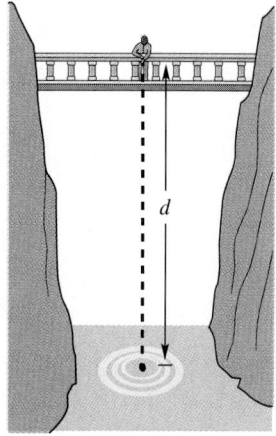

81. *Height of a Bridge* The time t in seconds for a free-falling object to fall d feet is

$$t=\sqrt{\frac{d}{16}}.$$

A child drops a pebble from a bridge and observes it strike the water after approximately 4 seconds (see figure). Estimate the height of the bridge.

82. *Length of a Pendulum* The time t in seconds for a pendulum of length L in feet to go through one complete cycle (its period) is

$$t=2\pi\sqrt{\frac{L}{32}}.$$

How long is the pendulum of a grandfather clock that has period of 1.3 seconds?

6

Chapter Test

Take this test as you would take a test in class. After you are done, check your work against the answers in the back of the book.

In Exercises 1–4, evaluate the expression without using a calculator.

1. $2^{-2} + 2^{-3}$ **2.** $\dfrac{6.3 \times 10^{-3}}{2.1 \times 10^2}$ **3.** $27^{-2/3}$ **4.** $\sqrt{2}\sqrt{18}$

5. Write the real number 0.000032 in scientific notation.

6. Write the real number 3.04×10^7 in decimal notation.

In Exercises 7–12, simplify the expression.

7. $\dfrac{12t^{-2}}{20t^{-1}}$ **8.** $(x + y^{-2})^{-1}$ **9.** $\left(\dfrac{x^{1/2}}{x^{1/3}}\right)^2$

10. $5^{1/4} \cdot 5^{7/4}$ **11.** $\sqrt{\dfrac{32}{9}}$ **12.** $\sqrt[3]{24}$

In Exercises 13 and 14, rationalize the denominator.

13. $\dfrac{3}{\sqrt{6} - 9}$ **14.** $\sqrt[3]{\dfrac{1}{4}}$

15. Combine: $5\sqrt{3x} - 3\sqrt{300x}$

16. Multiply and simplify (assume that x is positive): $\sqrt{5}(\sqrt{15x} + 3)$

17. Expand (assume that x is positive): $\left(4 - \sqrt{2x}\right)^2$

18. Factor: $7\sqrt{27} + 14y\sqrt{12} = 7\sqrt{3}\,(\;\rule{2cm}{0pt}\;)$

In Exercises 19 and 20, solve the equation.

19. $\sqrt{x^2 - 1} = x - 2$ **20.** $\sqrt{x} - x + 6 = 0$

In Exercises 21–26, perform the specified operation and simplify the result.

21. $(2 + 3i) - \sqrt{-25}$ **22.** $(2 - 3i)^2$ **23.** $\sqrt{-16}(1 + \sqrt{-4})$

24. $(3 - 2i)(1 + 5i)$ **25.** $\dfrac{5 - 2i}{i}$ **26.** $\dfrac{2i}{5 - 4i}$

27. The velocity v (in feet per second) of a free-falling object is given by $v = \sqrt{2gh}$, where $g = 32$ feet per second per second, and h is the distance (in feet) the object has fallen. Find the height from which a rock has been dropped if it strikes the ground with a velocity of 80 feet per second.

514

Cumulative Test

Take this test as you would take a test in class. After you are done, check your work against the answers in the back of the book.

In Exercises 1–4, solve the system of equations by the specified method.

1. Graphically

$$\begin{cases} x - y = 1 \\ 2x + y = 5 \end{cases}$$

2. Method of Substitution

$$\begin{cases} 4x + 2y = 8 \\ x - 5y = 13 \end{cases}$$

3. Method of Elimination

$$\begin{cases} 4x - 3y = 8 \\ -2x + y = -6 \end{cases}$$

4. Cramer's Rule

$$\begin{cases} 2x - y = 4 \\ 4y - 5z = -23 \\ 3x + y - 2z = -5 \end{cases}$$

5. Use a determinant to find the area of the triangle with vertices $(-5, 0)$, $(2, 6)$, and $(4, 2)$.

6. The sum of three positive numbers is 44. The second number is four greater than the first, and the third number is three times the first. Find the three numbers.

7. Graph the inequality $5x + 2y > 10$.

8. Find the minimum and maximum values of the objective function $C = 5x + 2y$ subject to the constraints.

$$\begin{cases} x \geq 3 \\ y \geq 2 \\ 3x + 2y \leq 22 \end{cases}$$

9. Multiply and simplify: $\dfrac{x^2 + 8x + 16}{18x^2} \cdot \dfrac{2x^4 + 4x^3}{x^2 - 16}$

10. Perform the indicated operations and simplify: $\dfrac{2}{x} - \dfrac{x}{x^3 + 3x^2} + \dfrac{1}{x + 3}$

11. Simplify: $\dfrac{\left(\dfrac{x}{y} - \dfrac{y}{x}\right)}{\left(\dfrac{x - y}{xy}\right)}$

12. Divide (use synthetic division): $\dfrac{x^3 + 27}{x + 3}$

13. Sketch the graph of the function $f(x) = \dfrac{3(x - 2)}{x + 1}$

14. Solve: $x + \dfrac{4}{x} = 4$

15. Simplify: $\dfrac{-4x^{-3}y^4}{6xy^{-2}}$

16. Simplify: $\left(\dfrac{t^{1/2}}{t^{1/4}}\right)^2$

17. Evaluate without the aid of a calculator: $(4 \times 10^3)^2$

18. Combine the radicals: $10\sqrt{20x} + 3\sqrt{125x}$

19. Multiply: $\left(\sqrt{2x} - 3\right)^2$

20. Rationalize the denominator and simplify: $\dfrac{6}{\sqrt{10} - 2}$

21. Perform the indicated operations and simplify: $\sqrt{-2}\left(\sqrt{-8} + 3\right)$

22. Solve: $\sqrt{x + 10} = x - 2$

23. Working together, two people can complete a task in 12 hours. Working alone, how long would it take each to do the task if one person takes 10 hours longer than the other?

Quadratic Functions, Equations, and Inequalities

7.1   The Factoring and Square Root Methods

7.2   Completing the Square

7.3   The Quadratic Formula and the Discriminant

7.4   Applications of Quadratic Equations

7.5   Graphing Quadratic Functions

7.6   Nonlinear Inequalities in One Variable

Galileo Galilei (1564–1642) discovered a relationship between the height h (in feet) and the falling time t (in seconds) of a free-falling object. The model is $h = -16t^2 + h_0$, where h_0 is the initial height (in feet).

In the balloon race called **Hare and Hound**, balloon operators drop a marker to try to hit a target on the ground. According to this model (which neglects air resistance), how long does it take a marker dropped from 1600 feet to hit the ground?

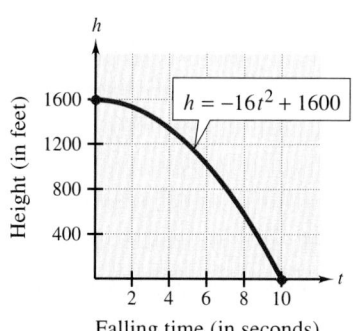

$h = -16t^2 + 1600$

Height (in feet)

Falling time (in seconds)

SECTION	The Factoring and Square Root Methods
7.1	*Solving Quadratic Equations by Factoring* • *Solving Quadratic Equations by Extracting Square Roots* • *Quadratic Equations with Imaginary Solutions* • *Equations of Quadratic Type*

Solving Quadratic Equations by Factoring

In Section 2.1, you saw how to solve a quadratic equation by factoring. In this chapter, you will review that method and look at other ways of solving quadratic equations.

Remember from Section 2.1 that the first step in solving a quadratic equation by factoring is to write the equation in standard form (with the polynomial on the left side of the equation and zero on the right side). Next, you attempt to factor the left side. Finally, you set each factor equal to zero and solve for x.

EXAMPLE 1   ■   Solving Quadratic Equations by Factoring

Solve the following quadratic equations.

(a) $x^2 + 5x = 24$

(b) $3x^2 = 4 - 11x$

Solution

(a)

$$x^2 + 5x = 24 \qquad \text{Given equation}$$
$$x^2 + 5x - 24 = 0 \qquad \text{Standard form}$$
$$(x + 8)(x - 3) = 0 \qquad \text{Factor}$$
$$x + 8 = 0 \implies x = -8 \qquad \text{Set first factor equal to 0}$$
$$x - 3 = 0 \implies x = 3 \qquad \text{Set second factor equal to 0}$$

Thus, the quadratic equation has two solutions: -8 and 3. Check these solutions in the original equation.

(b)

$$3x^2 = 4 - 11x \qquad \text{Given equation}$$
$$3x^2 + 11x - 4 = 0 \qquad \text{Standard form}$$
$$(3x - 1)(x + 4) = 0 \qquad \text{Factor}$$
$$3x - 1 = 0 \implies x = \tfrac{1}{3} \qquad \text{Set first factor equal to 0}$$
$$x + 4 = 0 \implies x = -4 \qquad \text{Set second factor equal to 0}$$

Thus, the quadratic equation has two solutions: $\tfrac{1}{3}$ and -4. Check these solutions in the original equation.   ■

If the two factors of a quadratic equation in standard form are identical, then the corresponding solution is called a **double** or **repeated solution**. This occurs in the next example.

EXAMPLE 2 ■ A Quadratic Equation with a Repeated Solution

$$9x^2 + 12 = 3 + 12x + 5x^2 \qquad \text{Given equation}$$
$$4x^2 - 12x + 9 = 0 \qquad \text{Standard form}$$
$$(2x - 3)(2x - 3) = 0 \qquad \text{Factor}$$
$$2x - 3 = 0 \qquad \text{Set factor equal to 0}$$
$$x = \frac{3}{2} \qquad \text{Repeated solution}$$

Thus, this quadratic equation has only one (repeated) solution: $x = \frac{3}{2}$. Check this solution in the original equation. ■

EXAMPLE 3 ■ Solving Quadratic Equations by Factoring

Solve the following equations by factoring.

(a) $(x + 1)(x - 2) = 18$ (b) $5x^2 = 45$

Solution

Stress to students the need to have the equation set equal to 0 before factoring and setting the factors equal to 0.

(a) Don't be tricked by this quadratic equation. Even though the left side is factored, the right side is not zero. Thus, you must first multiply the left side, then rewrite the equation in standard form, and finally refactor, as follows.

$$(x + 1)(x - 2) = 18 \qquad \text{Given equation}$$
$$x^2 - x - 2 = 18 \qquad \text{Multiply}$$
$$x^2 - x - 20 = 0 \qquad \text{Standard form}$$
$$(x - 5)(x + 4) = 0 \qquad \text{Factor}$$
$$x - 5 = 0 \quad \Longrightarrow \quad x = 5 \qquad \text{Set first factor equal to 0}$$
$$x + 4 = 0 \quad \Longrightarrow \quad x = -4 \qquad \text{Set second factor equal to 0}$$

Thus, this equation has two solutions: 5 and −4. Check these solutions in the original equation.

(b)
$$5x^2 = 45 \qquad \text{Given equation}$$
$$5x^2 - 45 = 0 \qquad \text{Standard form}$$
$$5(x^2 - 9) = 0 \qquad \text{Common monomial factor}$$
$$5(x + 3)(x - 3) = 0 \qquad \text{Factor as difference of squares}$$
$$x + 3 = 0 \quad \Longrightarrow \quad x = -3 \qquad \text{Set first factor equal to 0}$$
$$x - 3 = 0 \quad \Longrightarrow \quad x = 3 \qquad \text{Set second factor equal to 0}$$

Thus, this equation has two solutions: −3 and 3. Check these solutions in the original equation. ■

Solving Quadratic Equations by Extracting Square Roots

There is a nice shortcut for solving quadratic equations of the form

$$u^2 = d,$$

where $d > 0$ and u is an algebraic expression. By factoring, you can see that this equation has two solutions.

$u^2 = d$	Given equation
$u^2 - d = 0$	Standard form
$\left(u + \sqrt{d}\right)\left(u - \sqrt{d}\right) = 0$	Factor
$u + \sqrt{d} = 0 \implies u = -\sqrt{d}$	Set first factor equal to 0
$u - \sqrt{d} = 0 \implies u = \sqrt{d}$	Set second factor equal to 0

Thus, the equation $u^2 = d$ (where $d > 0$) has two solutions. Because the solutions differ only in sign, you can write both of the solutions together, using a "plus or minus sign," as follows.

$$u = \pm\sqrt{d}$$

Note the introduction of the $\pm$ notation.

This form of the solution is read as "u is equal to plus or minus the square root of d." When you solve an equation of the form $u^2 = d$ without going through the steps of factoring, you are **extracting square roots.**

Extracting Square Roots

The equation $u^2 = d$, where $d > 0$, has exactly two solutions:

$$u = \sqrt{d} \quad \text{and} \quad u = -\sqrt{d}.$$

These solutions can also be written as

$$u = \pm\sqrt{d}.$$

EXAMPLE 4 ■ Solving a Quadratic Equation by Extracting Square Roots

Remind students that they are to find *all* solutions when solving equations. Hence, $\pm$ in front of $\sqrt{5}$ in Example 4 denotes both positive and negative square roots of 5.

$3x^2 = 15$	Given equation
$x^2 = 5$	Divide both sides by 3
$x = \pm\sqrt{5}$	Extract square roots

Thus, the equation has two solutions: $\sqrt{5}$ and $-\sqrt{5}$. Note that $x^2 - 5 = 0$ factors as $\left(x + \sqrt{5}\right)\left(x - \sqrt{5}\right) = 0$, which gives the same two solutions. Check these solutions in the original equation. ■

EXAMPLE 5 ■ Solving a Quadratic Equation by Extracting Square Roots

Solve the following quadratic equation by extracting square roots.

$$(x - 2)^2 = 10$$

Solution

In this case, an extra step is needed after extracting the square roots.

$$(x-2)^2 = 10 \qquad \text{Given equation}$$
$$x - 2 = \pm\sqrt{10} \qquad \text{Extract square roots}$$
$$x = 2 \pm \sqrt{10} \qquad \text{Add 2 to both sides}$$

Thus, the equation has two solutions: $2 + \sqrt{10}$ and $2 - \sqrt{10}$. Check these solutions in the original equation. ∎

EXAMPLE 6 ■ **Solving a Quadratic Equation by Extracting Square Roots**

Solve the quadratic equation

$$3(5x+4)^2 - 81 = 0$$

by extracting square roots.

Solution

In this case, some preliminary steps are needed before taking the square roots.

$$3(5x+4)^2 - 81 = 0 \qquad \text{Given equation}$$
$$3(5x+4)^2 = 81$$
$$(5x+4)^2 = 27 \qquad \text{Divide both sides by 3}$$
$$5x + 4 = \pm\sqrt{27} \qquad \text{Extract square roots}$$
$$5x = -4 \pm 3\sqrt{3} \qquad \text{Subtract 4 from both sides}$$
$$x = \frac{-4 \pm 3\sqrt{3}}{5} \qquad \text{Divide both sides by 5}$$

Thus, the solutions are

$$\frac{-4 + 3\sqrt{3}}{5} \approx 0.24 \quad \text{and} \quad \frac{-4 - 3\sqrt{3}}{5} \approx -1.84.$$

FIGURE 7.1

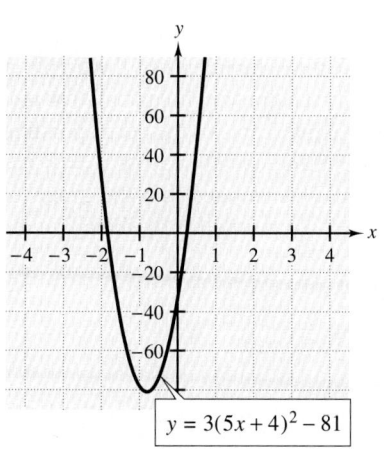

$$y = 3(5x+4)^2 - 81$$

For most equations, we have been emphasizing an algebraic check (by substituting the solutions into the original equation). If you have access to a graphing utility, you can also use a graphic check. To do this, sketch the graph of $y = 3(5x+4)^2 - 81$ and locate the x-intercepts of the graph. From the graph shown in Figure 7.1, you can see that the x-intercepts are $x \approx -1.84$ and $x \approx 0.24$. This graphically confirms the two solutions.

If using graphing calculators, students can use the ZOOM and TRACE features to approximate values of the x-intercepts.

■

Quadratic Equations with Imaginary Solutions

So far in the text, the only solutions you have been finding have been real numbers. But now, having been introduced to imaginary numbers (in Section 6.6), you can find other types of solutions. For instance, although the quadratic equation $x^2 + 1 = 0$ has no solutions that are real numbers, it does have two solutions that are imaginary numbers: i and $-i$. To check this, replace x by i and $-i$, as follows.

$$i^2 + 1 = (-1) + 1 = 0 \quad \Longrightarrow \quad i \text{ is a solution}$$
$$(-i)^2 + 1 = (-1) + 1 = 0 \quad \Longrightarrow \quad -i \text{ is a solution}$$

One way to find imaginary solutions of a quadratic equation is to extend the *extraction of square roots* technique to cover the case where d is a negative number.

Extracting Imaginary Square Roots

The equation $u^2 = d$, where $d < 0$, has exactly two solutions:
$$u = \sqrt{d}\, i \quad \text{and} \quad u = -\sqrt{d}\, i.$$

These solutions can also be written as
$$u = \pm\sqrt{d}\, i.$$

EXAMPLE 7 ■ Extracting Imaginary Square Roots

$x^2 + 8 = 0$	Given equation
$x^2 = -8$	Write in standard form
$x = \pm\sqrt{8}\, i$	Extract imaginary square roots
$x = \pm 2\sqrt{2}\, i$	Simplify

Thus, the given equation has two solutions: $2\sqrt{2}\, i$ and $-2\sqrt{2}\, i$. Check these solutions in the original equation. ■

At several points in the text, we have emphasized the relationship between the real solutions of an equation in one variable and the x-intercepts of the graph of the corresponding equation in two variables. This relationship holds only for *real* solutions. *Imaginary* solutions do not appear as x-intercepts of the corresponding graph. For instance, in Example 7 you found that the equation $x^2 + 8 = 0$ has two imaginary solutions, but no real solutions. This implies that the graph of $y = x^2 + 8$ does not have any x-intercepts. You can confirm this by sketching the graph, as shown in Figure 7.2.

FIGURE 7.2

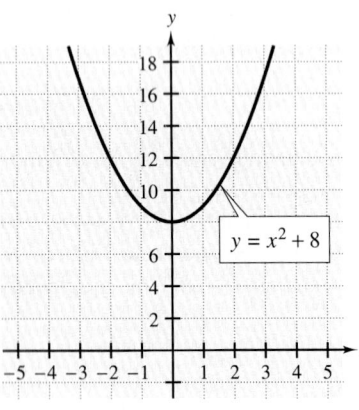

$y = x^2 + 8$

Equations of Quadratic Type

Both the factoring method and the extraction of square roots method can be applied to nonquadratic equations that are of **quadratic type.** An equation is said to be of

Discuss the identifying characteristics of a quadratic-type equation: When written in standard form, an equation has a trinomial of quadratic type if the degree of the leading term is twice the degree of the middle term.

quadratic type if it has the form

$$au^2 + bu + c = 0,$$

where u is an algebraic expression. Here are some examples.

Equation	Written in Quadratic Form	Substitution
$x^4 + 5x^2 + 4 = 0$	$(x^2)^2 + 5(x^2) + 4 = 0$	$u = x^2$
$2x^{2/3} + 3x^{1/3} - 9 = 0$	$2(x^{1/3})^2 + 3(x^{1/3}) - 9 = 0$	$u = x^{1/3}$
$x - 5\sqrt{x} + 6 = 0$	$(\sqrt{x})^2 - 5\sqrt{x} + 6 = 0$	$u = \sqrt{x}$

To solve an equation of quadratic type, rewrite the equation in terms of u to obtain $au^2 + bu + c = 0$, and solve for u. Then replace u with its original value and solve for the original variable. This procedure is demonstrated in Examples 8, 9, and 10.

EXAMPLE 8   ■   Solving an Equation of Quadratic Type

Solve the following equation for x.

$$x^4 - 13x^2 + 36 = 0$$

Solution

This equation is of quadratic type, because it can be written in the form

$$(x^2)^2 - 13(x^2) + 36 = 0.$$

By letting $u = x^2$, you can solve the equation as follows.

$x^4 - 13x^2 + 36 = 0$	Given equation
$(x^2)^2 - 13(x^2) + 36 = 0$	Write in quadratic form
$u^2 - 13u + 36 = 0$	Replace x^2 by u
$(u - 4)(u - 9) = 0$	Factor
$u - 4 = 0 \implies u = 4$	Set first factor equal to 0
$u - 9 = 0 \implies u = 9$	Set second factor equal to 0

At this point, you have found the "u-solutions." To find the "x-solutions," replace u by x^2, as follows.

$$u = 4 \implies x^2 = 4 \implies x = \pm 2$$
$$u = 9 \implies x^2 = 9 \implies x = \pm 3$$

Thus, the original equation has *four* solutions: 2, −2, 3, and −3. Check these solutions in the original equation.   ■

Be sure you see in Example 8 that the u-solutions of 4 and 9 represent only a temporary step in the solution. They are not (in general) solutions of the original equation.

EXAMPLE 9 ■ Solving an Equation of Quadratic Type

Solve the equation $x - 5\sqrt{x} + 6 = 0$.

Solution

This equation is of quadratic type, with $u = \sqrt{x}$.

$x - 5\sqrt{x} + 6 = 0$	Given equation
$(\sqrt{x})^2 - 5(\sqrt{x}) + 6 = 0$	Write in quadratic form
$u^2 - 5u + 6 = 0$	Replace $\sqrt{x}$ by u
$(u - 2)(u - 3) = 0$	Factor
$u - 2 = 0 \implies u = 2$	Set first factor equal to 0
$u - 3 = 0 \implies u = 3$	Set second factor equal to 0

Additional example: Solve the equation $x - 13\sqrt{x} + 36 = 0$.

Solution: The u-solutions are $u = 9$ and $u = 4$. Replacing u with $\sqrt{x}$, we obtain the x-solutions $x = 81$ and $x = 16$. (Note that some students may incorrectly conclude $x = 3$ and $x = 2$.)

Now, using the u-solutions of 2 and 3, you obtain the following x-solutions.

$$u = 2 \implies \sqrt{x} = 2 \implies x = 4$$
$$u = 3 \implies \sqrt{x} = 3 \implies x = 9$$

Thus, the original equation has two solutions: 4 and 9. Check these solutions in the original equation. ■

EXAMPLE 10 ■ Solving an Equation of Quadratic Type

Solve the following equation for x.

$$(x^2 - 4)^2 - 25 = 0$$

Solution

This equation is of quadratic type, with $u = x^2 - 4$.

$(x^2 - 4)^2 - 25 = 0$	Given equation
$u^2 - 25 = 0$	Replace $x^2 - 4$ by u
$(u + 5)(u - 5) = 0$	Factor
$u + 5 = 0 \implies u = -5$	Set first factor equal to 0
$u - 5 = 0 \implies u = 5$	Set second factor equal to 0

Now, using the u-solutions of -5 and 5, you obtain the following x-solutions.

$$u = -5 \quad \text{and} \quad u = 5$$
$$x^2 - 4 = -5 \quad\quad x^2 - 4 = 5$$
$$x^2 = -1 \quad\quad x^2 = 9$$
$$x = \pm i \quad\quad x = \pm 3$$

Thus, the original equation has four solutions: i, $-i$, 3, and -3. Note that two of these solutions are real numbers and two are imaginary numbers. Check these solutions in the original equation. ■

EXAMPLE 11   ■   The Dimensions of a Room

You are working on some house plans. You want the living room in the house to have 200 square feet of floor space.

(a) What dimensions should the living room have if you want it to be square?

(b) What dimensions should the living room have if you want it to be a rectangle whose width is two-thirds of its length?

Solution

(a) If the room is to be square (Figure 7.3a), you can let x represent the length of each side. You can find the value of x by solving the following equation.

$$x^2 = 200 \qquad \text{Formula for area}$$
$$x = \pm\sqrt{200} \qquad \text{Extract square roots}$$
$$x \approx \pm 14.14 \qquad \text{Use a calculator}$$

The negative solution makes no sense in this problem, so you can conclude that the living room should have sides of about 14.14 feet.

(b) If the room is to be rectangular (Figure 7.3b), you can let x represent the length. Because the width is to be two-thirds of the length, you can represent the width by $\frac{2}{3}x$.

$$x\left(\frac{2}{3}x\right) = 200 \qquad \text{Formula for area}$$

$$\frac{2}{3}x^2 = 200 \qquad \text{Simplify left side}$$

$$x^2 = 300 \qquad \text{Divide both sides by } \frac{2}{3}$$
$$x = \pm\sqrt{300} \qquad \text{Extract square roots}$$
$$x \approx \pm 17.32 \qquad \text{Use a calculator}$$

From this you can conclude that the length of the room is $x \approx 17.32$ feet and the width of the room is $\frac{2}{3}x \approx 11.55$ feet. You can check this solution by multiplying 17.32 by 11.55 to see that you obtain approximately 200 square feet of floor space.

FIGURE 7.3   (a)   (b)

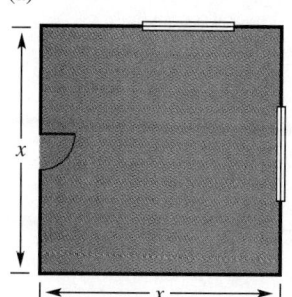
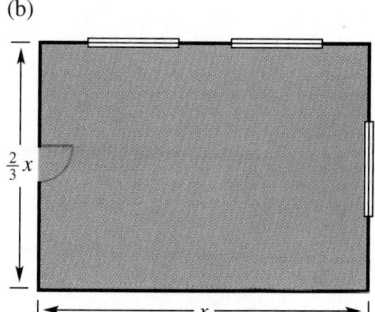

■

DISCUSSION PROBLEM ■ **Constructing a Quadratic Equation from Its Solutions**

Suppose that you are tutoring a student in algebra and want to make up several practice equations for the student to solve by factoring. You might want to start by writing several simple equations with easy solutions such as 1 and 2. Then you might want to put in some tougher problems that have solutions such as 13 and $-\frac{1}{2}$.

(a) Find a quadratic equation that has $x = 1$ and $x = 2$ as solutions.

(b) Find a quadratic equation that has $x = 13$ and $x = -\frac{1}{2}$ as solutions.

(c) Write a short paragraph describing a general procedure for creating a quadratic equation that has two given numbers as solutions. ■

Warm-Up

The following warm-up exercises involve skills that were covered in earlier sections. You will use these skills in the exercise set for this section.

In Exercises 1–6, completely factor the expression.

1. $16x^2 - 121$ **2.** $9t^2 - 24t + 16$

3. $5x^2 - 13x - 6$ **4.** $x(x - 10) - 4(x - 10)$

5. $4y^3 + 4y^2 - 3y$ **6.** $4x^3 - 12x^2 + 16x$

In Exercises 7–10, solve the equation.

7. $5y + 6 = 0$ **8.** $2s - 7 = 0$

9. $(x + 9)(2x - 15) = 0$ **10.** $(5x - 8)(x - 12) = 0$

7.1 EXERCISES

■ means that a graphing utility can help you solve the exercise or check your solution.

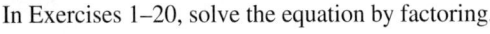

In Exercises 1–20, solve the equation by factoring.

1. $4x^2 - 12x = 0$

2. $25y^2 - 75y = 0$

3. $u(u - 9) - 12(u - 9) = 0$

4. $16x(x - 8) - 12(x - 8) = 0$

5. $4x^2 - 25 = 0$

6. $16y^2 - 121 = 0$

7. $x^2 - 12x + 36 = 0$

8. $9x^2 + 24x + 16 = 0$

9. $x^2 + 60x + 900 = 0$

10. $8x^2 - 10x + 3 = 0$

11. $(y - 4)(y - 3) = 6$

12. $(6 + u)(1 - u) = 10$

13. $2x(3x + 2) = 5 - 6x^2$

14. $(2z + 1)(2z - 1) = -4z^2 - 5z + 2$

15. $3x(x - 6) - 5(x - 6) = 0$

16. $3(4 - x) - 2x(4 - x) = 0$

17. $6x^2 = 54$

18. $5t^2 = 125$

19. $\dfrac{y^2}{2} = 32$

20. $\dfrac{x^2}{6} = 24$

In Exercises 21–36, solve the quadratic equation by extracting square roots.

21. $x^2 = 64$

22. $z^2 = 169$

23. $25x^2 = 16$

24. $9z^2 = 121$

25. $4u^2 - 225 = 0$

26. $16x^2 - 1 = 0$

27. $(x + 4)^2 = 169$

28. $(y - 20)^2 = 625$

29. $(x - 3)^2 = 0.25$

30. $(x + 2)^2 = 0.81$

31. $(x - 2)^2 = 7$

32. $(x + 8)^2 = 28$

33. $(2x + 1)^2 = 50$

34. $(3x - 5)^2 = 48$

35. $(4x - 3)^2 - 98 = 0$

36. $(5x + 11)^2 - 300 = 0$

In Exercises 37–54, solve the equation by extracting imaginary square roots.

37. $z^2 = -36$

38. $x^2 = -9$

39. $x^2 + 4 = 0$

40. $y^2 + 16 = 0$

41. $(t - 3)^2 = -25$

42. $(x + 5)^2 + 81 = 0$

43. $(3z + 4)^2 + 144 = 0$

44. $(2y - 3)^2 + 25 = 0$

45. $9(x + 6)^2 = -121$

46. $4(x - 4)^2 = -169$

47. $(x - 1)^2 = -27$

48. $(2x + 3)^2 = -54$

49. $(x + 1)^2 + 0.04 = 0$

50. $(x - 3)^2 + 2.25 = 0$

51. $\left(c - \frac{2}{3}\right)^2 + \frac{1}{9} = 0$

52. $\left(u + \frac{5}{8}\right)^2 + \frac{49}{16} = 0$

53. $\left(x + \frac{7}{3}\right)^2 = -\frac{38}{9}$

54. $\left(y - \frac{5}{6}\right)^2 = -\frac{4}{5}$

In Exercises 55–60, find all the real solutions of the given equation.

55. $x^4 - 20x^2 + 64 = 0$

56. $x^6 + 7x^3 - 8 = 0$

57. $3x^{2/3} + 8x^{1/3} - 3 = 0$

58. $2x - 9\sqrt{x} + 10 = 0$

59. $(x^2 - 2)^2 - 36 = 0$

60. $(u^2 + 4)^2 - 64 = 0$

In Exercises 61–72, solve the equation. (Find all the real *and* imaginary solutions.)

61. $x^2 - 100 = 0$

62. $y^2 - 400 = 0$

63. $(x - 5)^2 - 100 = 0$

64. $(y + 12)^2 - 400 = 0$

65. $(x - 5)^2 + 100 = 0$

66. $(y + 12)^2 + 400 = 0$

67. $(x^2 - 5)^2 - 100 = 0$

68. $(y^2 + 12)^2 - 400 = 0$

69. $x^4 - 16 = 0$

70. $y^4 - 81 = 0$

71. $x^4 - 4x^2 + 3 = 0$

72. $x^4 + 7x^2 - 8 = 0$

In Exercises 73–76, find a quadratic equation having the given solutions. [*Hint:* An equation with solutions r_1 and r_2 is $(x - r_1)(x - r_2) = 0$.)]

73. $5, -2$

74. $-2, \frac{1}{3}$

75. $1 + \sqrt{2}, 1 - \sqrt{2}$

76. $1 + \sqrt{2}i, 1 - \sqrt{2}i$

77. *Geometry* The surface area of a cube is 326 square meters. How long is each edge (see figure)?

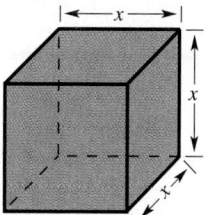

78. *Geometry* Use the Pythagorean Theorem to find x in the right triangle.

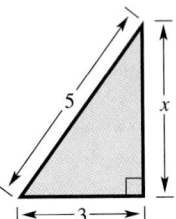

In Exercises 79–84, find the length x of the unknown side of the right triangle. (Round your answer to two decimal places.)

79.

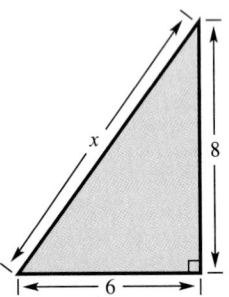

80.

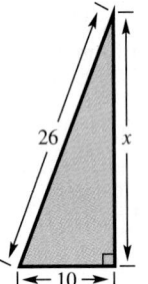

81.

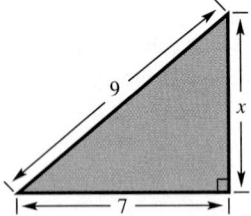

82.

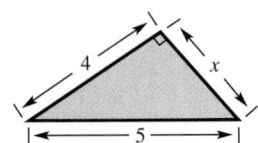

83.

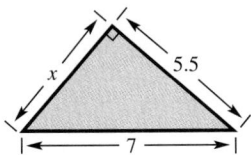

84.

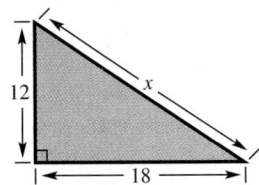

85. *Height of a Ladder* A ladder is 20 feet long and the bottom of the ladder is 4 feet from the wall of a house (see figure). Determine the height at which the top of the ladder rests against the wall.

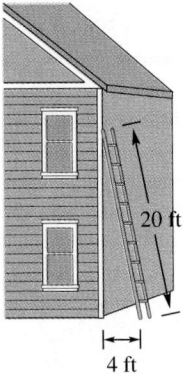

20 ft

4 ft

86. *Length of a Wire* A guy wire on a radio tower is attached to the top of the tower and to an anchor 50 feet from the base of the tower (see figure on next page). Determine the length of the guy wire if the tower is 100 feet high.

Figure for 86

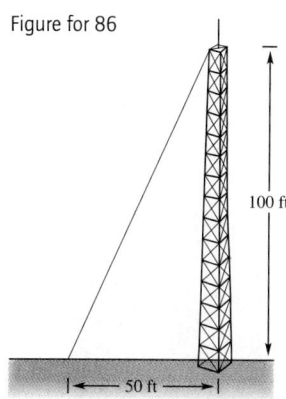

Wisconsin In Exercises 87 and 88, consider the map of Wisconsin (see figure) with a grid superimposed on it. Each unit of the grid represents 10.5 miles.

Figure for 87 and 88

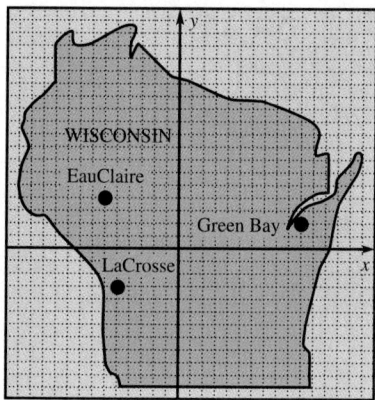

87. Approximate the distance in miles between La Crosse and Green Bay. How long would a flight from La Crosse to Green Bay take at a speed of 225 miles per hour?

88. Approximate the distance in miles between Eau Claire and La Crosse. What is the minimum time necessary to walk from Eau Claire to La Crosse at a rate of 5 miles per hour?

89. *Taking a Trip* You are driving from Deerfield to Campbellsport in Ohio (see figure at top of page). Highway 14 is under construction, so you drive 5.5 miles north on Route 225, then 5.8 miles west on Interstate 76. How many miles longer is this route than taking Highway 14 to its intersection with Interstate 76?

Figure for 89

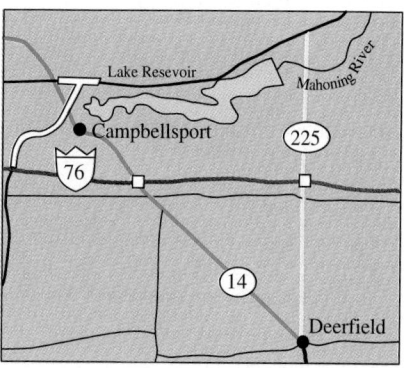

90. *Period of a Pendulum* The *period* of a pendulum is the time the pendulum takes to swing back and forth. The period t (in seconds) of a pendulum of length d (in inches) is given by

$$d = 9.78t^2.$$

The longest pendulum in the world is part of a clock on a building in Tokyo, Japan. This pendulum is about 74 feet long. What is the period of this pendulum? (*Source: Guinness Book of World Records*)

91. *Free-Falling Object* The height above the ground h (in feet) of an object propelled upward from a tower 144 feet high is given by

$$h = 144 + 128t - 16t^2,$$

where t is the time in seconds from the moment the object is released. How long does it take for the object to reach the ground?

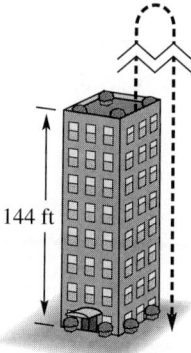

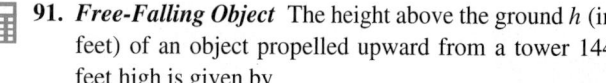

92. *Revenue* The revenue R (in dollars) from the sale of x units of a product is given by

$$R = x\left(120 - \frac{1}{2}x\right).$$

Determine the number of units that must be sold to produce a revenue of $7000.

Fire Loss In Exercises 93 and 94, use the following model, which gives the approximate amount of fire loss to private property in the United States during the years 1970 through 1988.

$$y = 1007.775 + 10.675t^2$$

In this model, y represents the value of private property lost to fire (in millions of dollars) and t represents the year, with $t = 0$ corresponding to 1970 (see figure). (*Source:* Insurance Information Institute)

Figure for 93 and 94

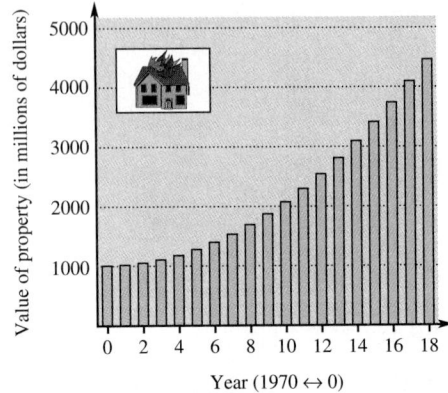

93. Which year had a private property loss of approximately $2075 million due to fire?

94. Which year had a private property loss of approximately $3740 million due to fire?

Completing the Square

Constructing Perfect Square Trinomials • *Solving Quadratic Equations by Completing the Square*

Constructing Perfect Square Trinomials

Consider the quadratic equation

$$(x - 2)^2 = 10. \qquad \text{Completed square form}$$

You know from Example 5 in the previous section that this equation has two solutions: $2 + \sqrt{10}$ and $2 - \sqrt{10}$. Now, let's suppose that you were given the equation in its standard form,

$$x^2 - 4x - 6 = 0. \qquad \text{Standard form}$$

How would you solve this equation if you were given only the standard form? You could try factoring, but after attempting to do this you would find that the left side of the equation is not factorable (using integer coefficients).

In this section, you will study a technique for rewriting an equation so that the equation is in completed square form. This process is called **completing the** square. To complete the square, you use the fact that all perfect square trinomials have a similar form. For instance, consider the following perfect square trinomials.

$$x^2 + 6x + 9 = (x+3)^2$$

$$x^2 - 12x + 36 = (x-6)^2$$

$$x^2 + 5x + \frac{25}{4} = \left(x + \frac{5}{2}\right)^2$$

Consider adding to a pattern such as this

$$x^2 + 10x + \quad = (x + \quad)^2$$
$$x^2 + 11x + \quad = (x + \quad)^2$$
$$x^2 - 12x + \quad = (x - \quad)^2$$

by asking what should be added to create a perfect square trinomial. (See Examples 1 and 2.)

In each case, note that the constant term of the perfect square trinomial is the square of half the coefficient of the x-term. That is, a perfect square trinomial has the following form.

Perfect Square Trinomial   *Square of Binomial*

$$x^2 + bx + \left(\frac{b}{2}\right)^2 = \left(x + \frac{b}{2}\right)^2$$

$$\underbrace{\qquad\qquad}_{(\text{half})^2}$$

Thus, to complete the square for an expression of the form $x^2 + bx$, you must add $(b/2)^2$ to the expression.

Completing the Square

To **complete the square** for the expression

$$x^2 + bx,$$

add $(b/2)^2$, which is the square of half the coefficient of x. Consequently,

$$x^2 + bx + \left(\frac{b}{2}\right)^2 = \left(x + \frac{b}{2}\right)^2.$$

EXAMPLE 1 ■ Constructing a Perfect Square Trinomial

What term should be added to the expression $x^2 - 8x$ so that it becomes a perfect square trinomial?

Solution

Stress that the coefficient of the second-degree term must be 1 before completing the square.

For this expression, the coefficient of the x-term is -8. By taking half of this and squaring the result, you can see that $(-4)^2 = 16$ should be added to the expression to make it a perfect square trinomial. That is,

$$x^2 - 8x + 16 = x^2 - 2(4)x + (-4)^2 = (x-4)^2.$$ ■

EXAMPLE 2 ■ Constructing Perfect Square Trinomials

What terms should be added to the following expressions so that each becomes a perfect square trinomial?

(a) $x^2 + 14x$ (b) $x^2 - 9x$

Solution

(a) For this expression, the coefficient of the x-term is 14. By taking half of this and squaring the result, you can see that $7^2 = 49$ should be added to the expression to make it a perfect square trinomial. That is,

$$x^2 + 14x + 49 = x^2 + 2(7)x + 7^2 = (x + 7)^2.$$

(b) For this expression, the coefficient of the x-term is -9. By taking half of this and squaring the result, you can see that $\left(-\frac{9}{2}\right)^2 = \frac{81}{4}$ should be added to the expression to make it a perfect square trinomial. That is,

$$x^2 - 9x + \frac{81}{4} = x^2 - 2\left(\frac{9}{2}\right)x + \left(-\frac{9}{2}\right)^2 = \left(x - \frac{9}{2}\right)^2. \qquad \blacksquare$$

Solving Quadratic Equations by Completing the Square

You will now see how completing the square can be used to solve a quadratic equation. When using this procedure with an equation, remember that it is essential to *preserve the equality*. Thus, when you add a constant term to one side of the equation, you must be sure to add the same constant to the other side of the equation.

EXAMPLE 3   ■   Completing the Square: Leading Coefficient Is 1

$$x^2 + 12x = 0 \qquad \text{Given equation}$$

$$x^2 + 12x + (6)^2 = 36 \qquad \text{Add } \left(\frac{12}{2}\right)^2 = 36 \text{ to both sides}$$

(half)^2

$$(x + 6)^2 = 36 \qquad \text{Binomial squared}$$

$$x + 6 = \pm\sqrt{36} \qquad \text{Extract square roots}$$

$$x = -6 \pm 6 \qquad \text{Solve for } x$$

$$x = 0 \quad \text{or} \quad x = -12 \qquad \text{Solutions}$$

Thus, the equation has two solutions: 0 and -12. Check these solutions in the original equation.   ■

In Example 3, you used completing the square to solve the quadratic equation simply for the sake of illustration. This particular equation would be easier to solve by factoring. Try reworking the problem by factoring to see that you obtain the same two solutions. The next example involves an equation that cannot be solved by factoring (using integer coefficients).

EXAMPLE 4   ■   Completing the Square: Leading Coefficient Is 1

Solve the following equation by completing the square.

$$x^2 - 6x + 7 = 0$$

Have students check the solutions $3 \pm \sqrt{2}$ by (a) substituting directly and (b) using a calculator, estimating the solution to two decimal places.

Solution

$$x^2 - 6x + 7 = 0 \qquad\qquad \text{Given equation}$$

$$x^2 - 6x = -7 \qquad\qquad \text{Subtract 7 from both sides}$$

$$x^2 - 6x + (-3)^2 = -7 + 9 \qquad\qquad \text{Add } (-3)^2 = 9 \text{ to both sides}$$

$$\underset{(\text{half})^2}{\underbrace{\qquad\qquad}}$$

$$(x - 3)^2 = 2 \qquad\qquad \text{Binomial squared}$$

$$x - 3 = \pm\sqrt{2} \qquad\qquad \text{Extract square roots}$$

$$x = 3 \pm \sqrt{2} \qquad\qquad \text{Solve for } x$$

Thus, the equation has two solutions: $3 + \sqrt{2}$ and $3 - \sqrt{2}$. Check these solutions in the original equation.   ■

If the leading coefficient of a quadratic expression is not 1, you must divide both sides of the equation by this coefficient *before* completing the square, as shown in the following example.

EXAMPLE 5   ■   Completing the Square: Leading Coefficient Is Not 1

Solve the following equation by completing the square.

$$3x^2 + 5x = 2$$

Point out to students that the procedure for completing the square is the same, even when the coefficients are fractional.

Solution

$$3x^2 + 5x = 2 \qquad\qquad \text{Given equation}$$

$$x^2 + \frac{5}{3}x = \frac{2}{3} \qquad\qquad \text{Divide both sides by 3}$$

$$x^2 + \frac{5}{3}x + \left(\frac{5}{6}\right)^2 = \frac{2}{3} + \frac{25}{36} \qquad\qquad \text{Add } \left(\frac{5}{6}\right)^2 = \frac{25}{36} \text{ to both sides}$$

$$\underset{(\text{half})^2}{\underbrace{\qquad\qquad}}$$

$$\left(x + \frac{5}{6}\right)^2 = \frac{49}{36} \qquad\qquad \text{Binomial squared}$$

$$x + \frac{5}{6} = \pm\frac{7}{6} \qquad\qquad \text{Extract square roots}$$

$$x = -\frac{5}{6} \pm \frac{7}{6} \qquad\qquad \text{Subtract } \tfrac{5}{6} \text{ from both sides}$$

$$x = \frac{1}{3} \quad \text{or} \quad x = -2 \qquad\qquad \text{Solutions}$$

Thus, the equation has two solutions: $\frac{1}{3}$ and -2. Check these solutions in the original equation.   ■

NOTE If you solve a quadratic equation by completing the square and obtain solutions that do not involve radicals, then you could have solved the equation by factoring. For instance, verify that the equation in Example 5 could have been factored as

$$(3x - 1)(x + 2) = 0.$$

EXAMPLE 6 ■ Completing the Square: Leading Coefficient Is Not 1

$$2x^2 - x - 2 = 0 \qquad \text{Given equation}$$

$$2x^2 - x = 2 \qquad \text{Add 2 to both sides}$$

$$x^2 - \frac{1}{2}x = 1 \qquad \text{Divide both sides by 2}$$

$$\underbrace{x^2 - \frac{1}{2}x + \left(-\frac{1}{4}\right)^2}_{(\text{half})^2} = 1 + \frac{1}{16} \qquad \text{Add } \left(-\tfrac{1}{4}\right)^2 = \tfrac{1}{16} \text{ to both sides}$$

$$\left(x - \frac{1}{4}\right)^2 = \frac{17}{16} \qquad \text{Binomial squared}$$

$$x - \frac{1}{4} = \pm\frac{\sqrt{17}}{4} \qquad \text{Extract square roots}$$

$$x = \frac{1}{4} \pm \frac{\sqrt{17}}{4} \qquad \text{Solve for } x$$

Thus, this equation has two solutions:

$$x = \frac{1}{4} + \frac{\sqrt{17}}{4} = \frac{1 + \sqrt{17}}{4}$$

and

$$x = \frac{1}{4} - \frac{\sqrt{17}}{4} = \frac{1 - \sqrt{17}}{4}.$$

Check these solutions in the original equation. ■

EXAMPLE 7   ■   Completing the Square

Solve the following quadratic equation and use a calculator to estimate the solutions to two decimal places.

$$2x^2 - 7x - 5 = 0$$

Solution

$$2x^2 - 7x - 5 = 0 \qquad \text{Given equation}$$

$$2x^2 - 7x = 5 \qquad \text{Add 5 to both sides}$$

$$x^2 - \frac{7}{2}x = \frac{5}{2} \qquad \text{Divide both sides by 2}$$

$$x^2 - \frac{7}{2}x + \left(-\frac{7}{4}\right)^2 = \frac{5}{2} + \frac{49}{16} \qquad \text{Add } \left(-\frac{7}{4}\right)^2 = \frac{49}{16} \text{ to both sides}$$

$$\underbrace{\qquad\qquad}_{(\text{half})^2}$$

$$\left(x - \frac{7}{4}\right)^2 = \frac{89}{16} \qquad \text{Binomial squared}$$

$$x - \frac{7}{4} = \pm\frac{\sqrt{89}}{4} \qquad \text{Extract square roots}$$

$$x = \frac{7}{4} \pm \frac{\sqrt{89}}{4} \qquad \text{Solve for } x$$

Thus, the equation has two solutions:

$$x = \frac{7}{4} + \frac{\sqrt{89}}{4} = \frac{7 + \sqrt{89}}{4} \approx 4.11$$

and

$$x = \frac{7}{4} - \frac{\sqrt{89}}{4} = \frac{7 - \sqrt{89}}{4} \approx -0.61.$$

Check these solutions in the original equation.   ■

The method of completing the square can be used to solve *any* quadratic equation, including equations that have imaginary solutions, as demonstrated in the next example.

EXAMPLE 8   ■   **A Quadratic Equation with Imaginary Solutions**

Use completing the square to solve the equation

$$x^2 - 4x + 8 = 0.$$

Solution

Begin by writing the equation in completed square form, as follows.

$x^2 - 4x + 8 = 0$	Given equation
$x^2 - 4x = -8$	Subtract 8 from both sides
$x^2 - 4x + (-2)^2 = -8 + 4$	Add $(-2)^2 = 4$ to both sides
$\underbrace{\qquad}_{(\text{half})^2}$	
$(x - 2)^2 = -4$	Binomial squared
$x - 2 = \pm 2i$	Extract imaginary square roots
$x = 2 \pm 2i$	Solve for x

Thus, the given equation has two solutions: $2 + 2i$ and $2 - 2i$.

Check

$x^2 - 4x + 8 = 0$	Given equation
$(2 + 2i)^2 - 4(2 + 2i) + 8 \stackrel{?}{=} 0$	Replace x by $2 + 2i$
$4 + 8i - 4 - 8 - 8i + 8 \stackrel{?}{=} 0$	
$0 = 0$	Solution checks

Check the other solution in the original equation.   ■

DISCUSSION PROBLEM   ■   **Solving a Quadratic Equation in Completed Square Form**

Consider the following quadratic equation.

$$(x - 1)^2 = d$$

(a) What value or values of d will produce a quadratic equation that has exactly one (repeated) solution?

(b) Describe the values of d that will produce two different solutions, both of which are *rational* numbers.

(c) Describe the values of d that will produce two different solutions, both of which are *irrational* numbers.

(d) Describe the values of d that will produce two different solutions, both of which are *imaginary* numbers.   ■

■

Warm-Up	*The following warm-up exercises involve skills that were covered in earlier sections. You will use these skills in the exercise set for this section.*

In Exercises 1–4, expand and simplify the expression.

1. $(x + 3)^2 - 1$ **2.** $(x + 10)^2 + 50$

3. $\left(u - \frac{1}{2}\right)^2$ **4.** $(v - 12)^2 - 100$

In Exercises 5–10, solve the equation. (Find all real *and* imaginary solutions.)

5. $x^2 = \frac{4}{49}$ **6.** $y^2 = -\frac{9}{16}$

7. $(y - 4)^2 = -36$ **8.** $(z + 2)^2 = 20$

9. $(3x - 2)^2 = 5$ **10.** $3(12 - x) - 8x(12 - x) = 0$

7.2   EXERCISES

⊞ means that a graphing utility can help you solve the exercise or check your solution.

In Exercises 1–12, determine the constant that must be added to the expression so that it becomes a perfect square trinomial.

1. $x^2 + 8x +$ ▨   **2.** $x^2 + 12x +$ ▨   **3.** $y^2 - 20y +$ ▨   **4.** $y^2 - 2y +$ ▨

5. $t^2 + 5t +$ ▨   **6.** $u^2 + 7u +$ ▨   **7.** $x^2 - \frac{6}{5}x +$ ▨   **8.** $y^2 + \frac{4}{3}y +$ ▨

9. $y^2 - \frac{3}{5}y +$ ▨   **10.** $a^2 - \frac{1}{3}a +$ ▨   **11.** $r^2 - 0.4r +$ ▨   **12.** $s^2 + 4.5s +$ ▨

⊞ In Exercises 13–24, solve the quadratic equation (a) by completing the square and (b) by factoring.

13. $x^2 - 25x = 0$ **14.** $x^2 + 32x = 0$ **15.** $t^2 - 8t + 7 = 0$ **16.** $y^2 - 8y + 12 = 0$

17. $x^2 + 2x - 24 = 0$ **18.** $x^2 + 12x + 27 = 0$ **19.** $x^2 + 7x + 12 = 0$ **20.** $z^2 + 3z - 10 = 0$

21. $x^2 - 3x - 18 = 0$ **22.** $t^2 - 5t - 36 = 0$ **23.** $2x^2 - 11x + 12 = 0$ **24.** $3x^2 - 5x - 2 = 0$

In Exercises 25–54, solve the quadratic equation by the method of completing the square. Give the solutions in exact form and in decimal form rounded to two decimal places. (The solutions to some equations are imaginary numbers.)

25. $x^2 - 4x - 3 = 0$ **26.** $x^2 - 6x + 7 = 0$ **27.** $x^2 + 4x - 3 = 0$ **28.** $x^2 + 6x + 7 = 0$

29. $u^2 - 4u + 1 = 0$ **30.** $a^2 - 10a - 15 = 0$ **31.** $x^2 + 2x + 3 = 0$ **32.** $x^2 - 6x + 12 = 0$

33. $x^2 - 10x - 2 = 0$ **34.** $x^2 + 8x - 4 = 0$ **35.** $y^2 + 20y + 10 = 0$ **36.** $y^2 + 6y - 24 = 0$

37. $x^2 - \frac{2}{3}x - 3 = 0$ **38.** $x^2 + \frac{4}{5}x - 1 = 0$ **39.** $t^2 + 5t + 3 = 0$ **40.** $u^2 - 9u - 1 = 0$

41. $v^2 + 3v - 2 = 0$

42. $z^2 - 7z + 9 = 0$

43. $-x^2 + x - 1 = 0$

44. $1 - x - x^2 = 0$

45. $2x^2 + 8x + 3 = 0$

46. $3x^2 - 4x - 1 = 0$

47. $3x^2 + 9x + 5 = 0$

48. $5x^2 - 15x + 7 = 0$

49. $4y^2 + 4y - 9 = 0$

50. $4z^2 - 3z + 2 = 0$

51. $x(x - 7) = 2$

52. $2x \left(x + \frac{4}{3} \right) = 5$

53. $0.1x^2 + 0.5x + 0.2 = 0$

54. $0.02x^2 + 0.10x - 0.05 = 0$

In Exercises 55–58, solve the equation, and give the solutions in decimal form rounded to two decimal places. Check your rounded solutions in the original equation.

55. $\dfrac{x}{2} - \dfrac{1}{x} = 1$

56. $\dfrac{x}{2} + \dfrac{5}{x} = 4$

57. $\sqrt{2x + 1} = x - 3$

58. $\sqrt{3x - 2} = x - 2$

59. *Completing the Square*

(a) Find the total area of the two adjoining rectangles and square in the figure.

(b) Find the area of the small square region that is missing in the lower right-hand corner of the figure and add it to the area found in part (a).

(c) Find the dimensions and the area of the entire figure after adjoining the small square in the lower right-hand corner. Note that you have demonstrated how to complete the square geometrically.

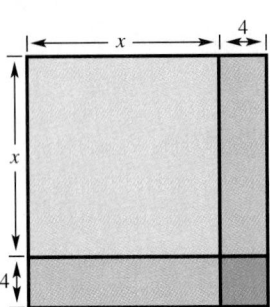

60. *Completing the Square* Use the model given in Exercise 59 to represent geometrically the method of completing the square for $x^2 + 6x$.

61. *Geometry* Find the dimensions of the triangle in the figure if its area is 12 square centimeters.

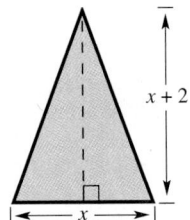

62. *Cutting Across a Lawn* On the sidewalk, the distance from the dormitory to the cafeteria is 400 meters (see figure). By cutting across the lawn, the walking distance is shortened to 300 meters. How long is each part of the L-shaped sidewalk?

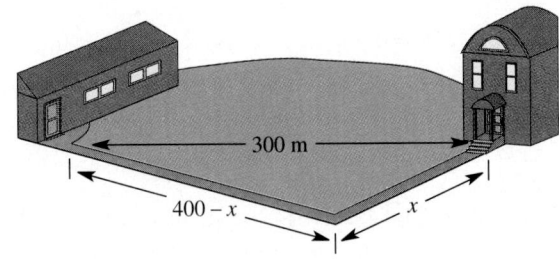

63. *Pulling a Boat in to a Dock* A windlass is used to pull a boat in to the dock (see figure). The rope is attached to the boat at a point 15 feet below the level of the windlass. Find the distance from the boat to the dock when the length of rope is 75 feet.

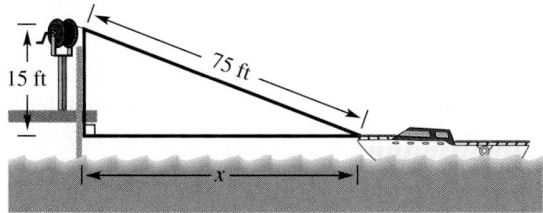

64. *Fencing in a Corral* You have 200 feet of fencing to enclose two adjacent rectangular corrals. The total area of the enclosed region is 1400 square feet. What are the dimensions of each corral? (The corrals are the same size.)

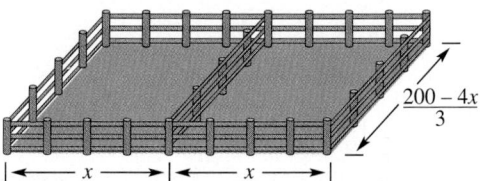

65. *Revenue* The revenue R from selling x units of a certain product is

$$R = x\left(50 - \frac{1}{2}x\right).$$

Find the number of units that must be sold to produce a revenue of $1218.

66. *Revenue* The revenue R from selling x units of a certain product is

$$R = x\left(100 - \frac{1}{10}x\right).$$

Find the number of units that must be sold to produce a revenue of $12,000.

The Quadratic Formula

Before completing the square for the *general* quadratic equation, consider solving a *specific* quadratic equation by completing the square and leaving it on the board for easy reference. Point out that the derivation of the Quadratic Formula is just a generalization of the technique presented in Section 7.2. Both the completing-the-square method and the Quadratic Formula can be used to solve *any* quadratic equation.

In Section 7.2, you solved quadratic equations by completing the square for *each problem separately.* Perhaps you have thought of the possibility of completing the square *once* for the general quadratic equation $ax^2 + bx + c = 0$, and producing a solution model for solving any quadratic equation. This is precisely what is done in this section to develop what is called the **Quadratic Formula.**

$$ax^2 + bx + c = 0 \qquad \text{Standard form, } a \neq 0$$

$$ax^2 + bx = -c \qquad \text{Subtract } c \text{ from both sides}$$

$$x^2 + \frac{b}{a}x = -\frac{c}{a} \qquad \text{Divide both sides by } a$$

$$x^2 + \frac{b}{a}x + \left(\frac{b}{2a}\right)^2 = -\frac{c}{a} + \left(\frac{b}{2a}\right)^2 \qquad \text{Complete the square}$$

$$\underset{\text{(half)}^2}{\underbrace{\qquad\qquad}}$$

$$\left(x + \frac{b}{2a}\right)^2 = \frac{b^2 - 4ac}{4a^2} \qquad \text{Simplify}$$

$$x + \frac{b}{2a} = \pm\sqrt{\frac{b^2 - 4ac}{4a^2}} \qquad \text{Extract square roots}$$

$$x = -\frac{b}{2a} \pm \frac{\sqrt{b^2 - 4ac}}{2\,|a|}$$

$$x = \frac{-b \pm \sqrt{b^2 - 4ac}}{2a} \qquad \text{Solutions}$$

Note that, because $\pm 2\,|a|$ represents the same numbers as $\pm 2a$, you can omit the absolute value bars.

The Quadratic Formula

The solutions of a quadratic equation in the standard form

$$ax^2 + bx + c = 0, \qquad a \neq 0$$

are given by the **Quadratic Formula**

$$x = \frac{-b \pm \sqrt{b^2 - 4ac}}{2a}.$$

Suggest that for each homework problem students write the Quadratic Formula before substituting specific values. Repeatedly writing the formula should help students learn it.

The Quadratic Formula is one of the most important formulas in algebra, and you should memorize it. It helps to try to memorize a verbal statement of the rule. For instance, you might try to remember the following verbal statement of the Quadratic Formula: "Minus b, plus or minus the square root of b squared minus $4ac$, all divided by $2a$."

Solving Quadratic Equations by the Quadratic Formula

When using the Quadratic Formula, remember that *before* the formula can be applied, you must first write the quadratic equation in standard form.

EXAMPLE 1   ■   Using the Quadratic Formula: Two Distinct Solutions

Use the Quadratic Formula to solve the equation

$$x^2 + 6x = 16.$$

Remind students that a quadratic equation should be written in standard form to identify a, b, c easily and correctly.

Solution

To begin, write the equation in standard form, $ax^2 + bx + c = 0$. Then determine the values of a, b, and c. Finally, substitute these values into the Quadratic Formula to obtain the solutions.

$$x^2 + 6x = 16 \qquad \text{Given equation}$$

$$x^2 + 6x - 16 = 0 \qquad \text{Standard form with } a = 1, \ b = 6, \ c = -16$$

$$x = \frac{-b \pm \sqrt{b^2 - 4ac}}{2a} \qquad \text{Quadratic Formula}$$

$$x = \frac{-6 \pm \sqrt{6^2 - 4(1)(-16)}}{2(1)} \qquad \text{Substitute}$$

$$x = \frac{-6 \pm \sqrt{36 + 64}}{2}$$

$$x = \frac{-6 \pm \sqrt{100}}{2}$$

$$x = \frac{-6 \pm 10}{2}$$

$$x = 2 \quad \text{or} \quad x = -8 \qquad \text{Solutions}$$

Therefore, the equation has two solutions: 2 and -8. Check these solutions in the original equation. (Because the simplified versions of the solutions have no radicals, you know that the equation could have been solved by factoring. Try solving the equation that way.)   ■

EXAMPLE 2   ■   **Using the Quadratic Formula: Two Distinct Solutions**

Use the Quadratic Formula to solve the equation

$$-x^2 - 4x + 8 = 0.$$

Solution

This equation is already written in standard form, $ax^2 + bx + c = 0$. However, the leading coefficient is negative. Because it is cumbersome to have negative leading coefficients, multiply both sides of the equation by -1 to produce a positive leading coefficient.

$$-x^2 - 4x + 8 = 0 \qquad\qquad \text{Leading coefficient is negative}$$

$$x^2 + 4x - 8 = 0 \qquad\qquad \text{Standard form with } a = 1, \ b = 4, \ c = -8$$

$$x = \frac{-b \pm \sqrt{b^2 - 4ac}}{2a} \qquad\qquad \text{Quadratic Formula}$$

$$x = \frac{-4 \pm \sqrt{4^2 - 4(1)(-8)}}{2(1)} \qquad\qquad \text{Substitute}$$

$$x = \frac{-4 \pm \sqrt{16 + 32}}{2}$$

$$x = \frac{-4 \pm \sqrt{48}}{2}$$

$$x = \frac{-4 \pm 4\sqrt{3}}{2} \qquad\qquad \text{Simplify radical}$$

$$x = \frac{\cancel{2}(-2 \pm 2\sqrt{3})}{\cancel{2}} \qquad\qquad \text{Divide out common factor}$$

$$x = -2 \pm 2\sqrt{3} \qquad\qquad \text{Solutions}$$

Some students may be tempted to reduce $(-4 \pm 4\sqrt{3})/2$ to $-2 \pm 4\sqrt{3}$. Ask them what is wrong with this reasoning.

Therefore, the equation has two solutions: $-2 + 2\sqrt{3}$ and $-2 - 2\sqrt{3}$. Check these solutions in the original equation.   ■

EXAMPLE 3   ■   **Using the Quadratic Formula: One Repeated Solution**

Use the Quadratic Formula to solve the equation

$$18x^2 - 24x + 8 = 0.$$

Solution

This equation has a common factor of 2. To simplify things, you can first divide both sides of the equation by 2.

$$18x^2 - 24x + 8 = 0 \qquad\qquad \text{Common factor of 2}$$

$$9x^2 - 12x + 4 = 0 \qquad\qquad \text{Standard form with } a = 9, \ b = -12, \ c = 4$$

$$x = \frac{-b \pm \sqrt{b^2 - 4ac}}{2a} \qquad \text{Quadratic Formula}$$

$$x = \frac{-(-12) \pm \sqrt{(-12)^2 - 4(9)(4)}}{2(9)} \qquad \text{Substitute}$$

$$x = \frac{12 \pm \sqrt{144 - 144}}{18}$$

$$x = \frac{12 \pm \sqrt{0}}{18} = \frac{2}{3} \qquad \text{Repeated solution}$$

Therefore, this quadratic equation has only one solution: $\frac{2}{3}$. Check this solution in the original equation. ■

Note in the next example how the Quadratic Formula can be used to solve a quadratic equation that has imaginary solutions.

EXAMPLE 4 ■ **Using the Quadratic Formula: Imaginary Solutions**

Use the Quadratic Formula to solve the equation

$$2x^2 - 4x + 5 = 0.$$

Solution

$$2x^2 - 4x + 5 = 0 \qquad \text{Standard form with } a = 2, \ b = -4, \ c = 5$$

$$x = \frac{-b \pm \sqrt{b^2 - 4ac}}{2a} \qquad \text{Quadratic Formula}$$

$$x = \frac{-(-4) \pm \sqrt{(-4)^2 - 4(2)(5)}}{2(2)} \qquad \text{Substitute}$$

$$x = \frac{4 \pm \sqrt{16 - 40}}{4}$$

$$x = \frac{4 \pm \sqrt{-24}}{4}$$

$$x = \frac{4 \pm 2\sqrt{6}\,i}{4}$$

$$x = \frac{2(2 \pm \sqrt{6}\,i)}{2 \cdot 2} \qquad \text{Divide out common factor of 2}$$

$$x = \frac{2 \pm \sqrt{6}\,i}{2} \quad \text{or} \quad x = 1 \pm \frac{\sqrt{6}}{2}i \qquad \text{Solutions}$$

Thus, the given equation has two imaginary solutions: $(2 + \sqrt{6}\,i)/2$ and $(2 - \sqrt{6}\,i)/2$. Check these solutions in the original equation. ■

The Discriminant

By looking back at the radicand, $b^2 - 4ac$, in the solutions of the equations in Examples 1–4, you can see that its value determines whether the solutions are rational, irrational, repeated, or imaginary. For this reason, the radicand $b^2 - 4ac$ in the Quadratic Formula is called the **discriminant.** The use of the discriminant is summarized as follows.

Using the Discriminant

Let a, b, and c be rational numbers such that $a \neq 0$. The **discriminant** of the quadratic equation $ax^2 + bx + c = 0$ is given by $b^2 - 4ac$, and can be used to classify the solutions of the equation as follows.

Discriminant	Solution Type
1. $b^2 - 4ac =$ perfect square	Two distinct rational solutions (Example 1)
2. $b^2 - 4ac =$ positive nonperfect square	Two distinct irrational solutions (Example 2)
3. $b^2 - 4ac = 0$	One repeated rational solution (Example 3)
4. $b^2 - 4ac =$ negative number	Two distinct imaginary solutions (Example 4)

Look again at Examples 1–4 to see that the equations with rational or repeated solutions could have been solved by *factoring*. In general, quadratic equations (with integer coefficients) for which the discriminant is either zero or a perfect square are factorable (using integer coefficients). Consequently, a quick test of the discriminant can be used to help you decide which solution method to use to solve a quadratic equation.

Figure 7.4 gives a graphical interpretation of the number of real solutions of a quadratic equation.

FIGURE 7.4

If the graph of $y = ax^2 + bx + c$ has two x-intercepts, then the equation $ax^2 + bx + c = 0$ has two real solutions.

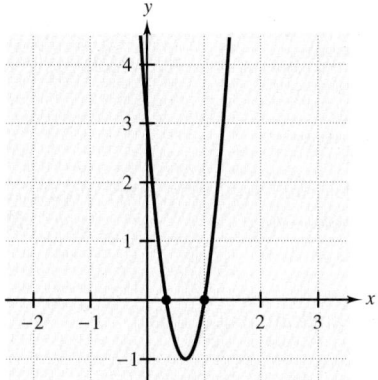

If the graph of $y = ax^2 + bx + c$ has only one x-intercept, then the equation $ax^2 + bx + c = 0$ has only one (repeated) solution.

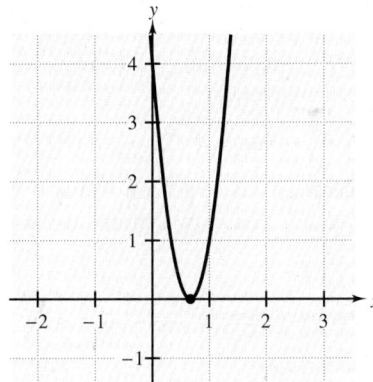

If the graph of $y = ax^2 + bx + c$ has no x-intercepts, then the equation $ax^2 + bx + c = 0$ has no real solution.

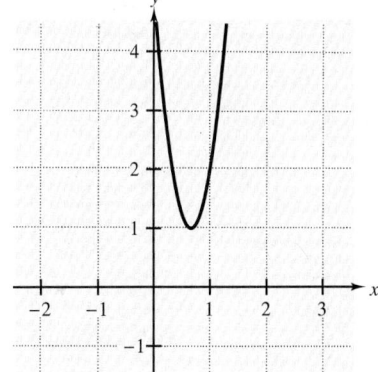

EXAMPLE 5   ■   **Using the Discriminant**

Use the discriminant to determine the type of solutions for each of the following equations. Which of these equations can be solved by factoring using integer coefficients?

(a) $x^2 - x + 2 = 0$ (b) $2x^2 - 3x - 2 = 0$

(c) $x^2 - 2x + 1 = 0$ (d) $x^2 - 2x - 1 = 9$

Solution

(a) For this equation, you have $a = 1$, $b = -1$, and $c = 2$, which yield a discriminant of

$$b^2 - 4ac = (-1)^2 - 4(1)(2) = 1 - 8 = -7.$$

Therefore, the equation has two distinct imaginary solutions. This quadratic equation cannot be factored using integer coefficients.

(b) For this equation, you have $a = 2$, $b = -3$, and $c = -2$, which yield a discriminant of

$$b^2 - 4ac = (-3)^2 - 4(2)(-2) = 9 + 16 = 25.$$

Therefore, the equation has two distinct rational solutions. This quadratic equation can be factored as $(x - 2)(2x + 1)$.

(c) For this equation, you have $a = 1$, $b = -2$, and $c = 1$, which yield a discriminant of

$$b^2 - 4ac = (-2)^2 - 4(1)(1) = 4 - 4 = 0.$$

Therefore, the equation has one repeated rational solution. This quadratic equation can be factored as $(x - 1)^2$.

(d) For this equation, you have $a = 1$, $b = -2$, and $c = -1$, which yield a discriminant of

$$b^2 - 4ac = (-2)^2 - 4(1)(-1) = 4 + 4 = 8.$$

Therefore, the equation has two distinct irrational solutions. This quadratic equation cannot be factored using integer coefficients. ■

You have now looked at five ways to solve quadratic equations.

1. Factoring

2. Extracting square roots

3. Completing the square

4. The Quadratic Formula

5. Graphing

Note this summary.

When choosing one of these five methods, you could first check to see whether the equation is in a form in which you can extract square roots. For instance, it is easy to see that the solutions of $x^2 = 10$ are $x = \pm\sqrt{10}$. Next, you could try factoring. For instance, by rewriting the equation $2x^2 - x - 1 = 0$ in the factored form $(2x + 1)(x - 1) = 0$, you can see that it has two solutions: $-\frac{1}{2}$ and 1. If neither of these two methods works, use the Quadratic Formula (or completing the square); they will work for any quadratic equation. For *real* solutions, remember that you can use a graph to approximate the solutions.

One type of quadratic equation for which the Quadratic Formula is useful is an equation with decimal coefficients. The Quadratic Formula is used for this type of equation in Example 6.

EXAMPLE 6 ■ Using a Calculator with the Quadratic Formula

Use the Quadratic Formula and a calculator to solve the equation

$$1.2x^2 - 17.8x + 8.05 = 0.$$

Solution

This quadratic equation is already in standard form. From this form, you can see that $a = 1.2$, $b = -17.8$, and $c = 8.05$. Therefore, you can use the Quadratic

Formula to solve the equation, as follows.

$$1.2x^2 - 17.8x + 8.05 = 0$$

$$x = \frac{-b \pm \sqrt{b^2 - 4ac}}{2a}$$

$$x = \frac{-(-17.8) \pm \sqrt{(-17.8)^2 - 4(1.2)(8.05)}}{2(1.2)}$$

When using their calculators, some students err in squaring negative numbers. Be sure students know the difference between $(-17.8)^2$ and -17.8^2.

To evaluate these solutions, begin by calculating the square root. Then, *store* this value for later use.*

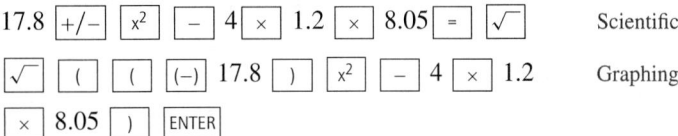

The display for either of these keystroke sequences should be 16.67932852. Storing this result and using the recall key, you obtain the following two solutions.

$$x \approx \frac{17.8 + 16.67932852}{2.4} \approx 14.366 \qquad \text{Add stored value}$$

$$x \approx \frac{17.8 - 16.67932852}{2.4} \approx 0.467 \qquad \text{Subtract stored value} \qquad \blacksquare$$

DISCUSSION PROBLEM   ■   Solutions of a Quadratic Equation

Solve the given quadratic equations and fill in the boxes.

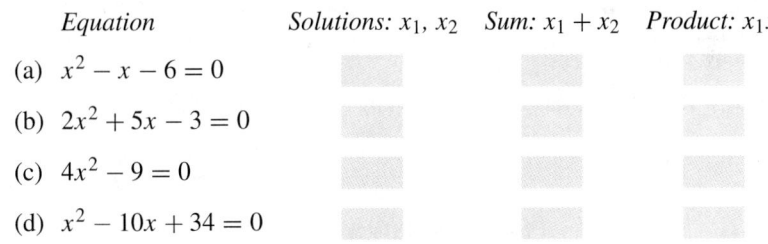

Equation	Solutions: x_1, x_2	Sum: $x_1 + x_2$	Product: $x_1 x_2$
(a) $x^2 - x - 6 = 0$			
(b) $2x^2 + 5x - 3 = 0$			
(c) $4x^2 - 9 = 0$			
(d) $x^2 - 10x + 34 = 0$			

Consider a general quadratic equation $ax^2 + bx + c = 0$ whose solutions are x_1 and x_2. Can you determine a relationship among the coefficients a, b, and c and the sum $(x_1 + x_2)$ and product $(x_1 x_2)$ of the solutions?   ■

*The graphing calculator keystrokes in this text correspond to the TI-81 and TI-82 graphing calculators from Texas Instruments. For other calculators, the keystrokes may differ.

Warm-Up

The following warm-up exercises involve skills that were covered in earlier sections. You will use these skills in the exercise set for this section.

In Exercises 1–4, solve the quadratic equation by factoring.

1. $x^2 - 10x + 21 = 0$ **2.** $x^2 + 14x - 32 = 0$

3. $3x^2 + x - 30 = 0$ **4.** $5x^2 - 18x + 9 = 0$

In Exercises 5 and 6, solve the quadratic equation by completing the square. (Find all real *and* imaginary solutions.)

5. $x^2 + 3x - 2 = 0$ **6.** $3x^2 + 6x + 4 = 0$

In Exercises 7–10, simplify the radical.

7. $\sqrt{25 - 4(2)(2)}$ **8.** $\sqrt{9 - 4(1)(5)}$

9. $\sqrt{64 - 4(2)(-4)}$ **10.** $\sqrt{81 - 4(3)(-6)}$

7.3   EXERCISES

means that a graphing utility can help you solve the exercise or check your solution.

In Exercises 1–4, write the quadratic equation in standard form.

1. $2x^2 = 7 - 2x$ **2.** $7x^2 + 15x = 5$ **3.** $x(10 - x) = 5$ **4.** $x(3x + 8) = 15$

 In Exercises 5–16, solve the quadratic equation by (a) using the Quadratic Formula and (b) factoring.

5. $x^2 - 11x + 28 = 0$ **6.** $x^2 - 12x + 27 = 0$ **7.** $x^2 + 6x + 8 = 0$ **8.** $x^2 + 9x + 14 = 0$

9. $4x^2 + 4x + 1 = 0$ **10.** $9x^2 + 12x + 4 = 0$ **11.** $4x^2 + 12x + 9 = 0$ **12.** $9x^2 - 30x + 25 = 0$

13. $6x^2 - x - 2 = 0$ **14.** $10x^2 - 11x + 3 = 0$ **15.** $x^2 - 5x - 300 = 0$ **16.** $x^2 + 20x - 300 = 0$

In Exercises 17–24, use the discriminant to determine the type of solutions of the quadratic equation.

17. $x^2 + x + 1 = 0$ **18.** $x^2 + x - 1 = 0$ **19.** $2x^2 - 5x - 4 = 0$ **20.** $10x^2 + 5x + 1 = 0$

21. $x^2 + 7x + 15 = 0$ **22.** $3x^2 - 2x - 5 = 0$ **23.** $4x^2 - 12x + 9 = 0$ **24.** $2x^2 + 10x + 6 = 0$

In Exercises 25–44, solve the quadratic equation by using the Quadratic Formula. (Find all real *and* imaginary solutions.)

25. $x^2 - 2x - 4 = 0$

26. $x^2 - 2x - 6 = 0$

27. $t^2 + 4t + 1 = 0$

28. $y^2 + 6y + 4 = 0$

29. $x^2 + 6x - 3 = 0$

30. $x^2 + 8x - 4 = 0$

31. $x^2 - 10x + 23 = 0$

32. $u^2 - 12u + 29 = 0$

33. $x^2 + 3x + 3 = 0$

34. $2x^2 - x + 1 = 0$

35. $2v^2 - 2v - 1 = 0$

36. $4x^2 + 6x + 1 = 0$

37. $2x^2 + 4x - 3 = 0$

38. $2x^2 + 3x + 3 = 0$

39. $9z^2 + 6z - 4 = 0$

40. $8y^2 - 8y - 1 = 0$

41. $x^2 - 0.4x - 0.16 = 0$

42. $x^2 + 0.6x - 0.41 = 0$

43. $2.5x^2 + x - 0.9 = 0$

44. $0.09x^2 - 0.12x - 0.26 = 0$

 In Exercises 45–58, solve the quadratic equation by the most convenient method. (Find all real *and* imaginary solutions.)

45. $z^2 - 169 = 0$

46. $t^2 = 150$

47. $y^2 + 15y = 0$

48. $2y(y - 18) + 3(y - 18) = 0$

49. $25(x - 3)^2 - 36 = 0$

50. $9(x + 2)^2 - 25 = 0$

51. $(x + 4)^2 + 16 = 0$

52. $4u^2 + 49 = 0$

53. $18x^2 + 15x - 50 = 0$

54. $2x^2 - 15x + 225 = 0$

55. $x^2 - 24x + 128 = 0$

56. $1.2x^2 - 0.8x - 5.5 = 0$

57. $x^2 + 8x + 25 = 0$

58. $2x^2 + 8x + 4.5 = 0$

In Exercises 59–62, use a calculator to solve the equation. (Round your answer to three decimal places.)

59. $5x^2 - 18x + 6 = 0$

60. $15x^2 + 3x - 105 = 0$

61. $-0.04x^2 + 4x - 0.8 = 0$

62. $3.7x^2 - 10.2x + 3.2 = 0$

In Exercises 63–66, solve the equation and give the solutions in decimal form rounded to two decimal places. Check your rounded solutions in the original equation.

63. $\dfrac{2x^2}{5} - \dfrac{x}{2} = 1$

64. $\dfrac{x}{3} + \dfrac{1}{x} = 4$

65. $\sqrt{x + 3} = x - 1$

66. $\sqrt{2x - 3} = x - 2$

Exploratory Exercises In Exercises 67–70, describe the values of c such that the equation has (a) two real-number solutions, (b) one real-number solution, and (c) two imaginary-number solutions.

67. $x^2 - 6x + c = 0$

68. $x^2 - 12x + c = 0$

69. $x^2 + 8x + c = 0$

70. $x^2 + 2x + c = 0$

71. *Free-Falling Object* A ball is thrown upward at a velocity of 40 feet per second from a height of 50 feet above the surface of the water below (see figure). The height h (in feet) of the ball at time t seconds after it is thrown is

$$h = -16t^2 + 40t + 50.$$

(a) Find the time at which the ball is again 50 feet above the water.

(b) Find the time at which the ball strikes the water.

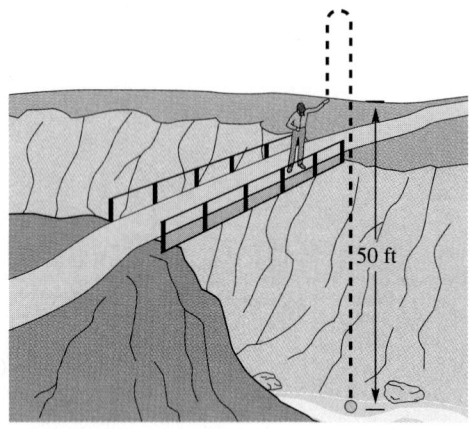

50 ft

72. *Number Problem* Find two consecutive positive even integers whose product is 288.

Practicing Lawyers In Exercises 73 and 74, use the following equation, which gives the approximate number of practicing lawyers in the United States in any year from 1960 through 1985 (see figure).

$$y = 273{,}063 + 2225t + 471t^2$$

In this model, y represents the number of practicing lawyers and t represents the year, with $t = 0$ corresponding to 1960. (*Source:* American Bar Foundation)

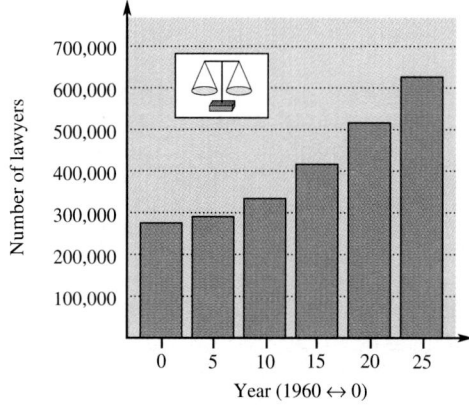

73. Find the year in which there were approximately 321,000 practicing lawyers in the United States.

74. Find the year in which there were approximately 527,500 practicing lawyers in the United States.

Mid-Chapter Quiz

Take this quiz as you would take a quiz in class. After you are done, check your work against the answers in the back of the book.

In Exercises 1–4, solve the equation by factoring.

1. $3y^2 - 27y = 0$

2. $4z^2 - 25 = 0$

3. $x^2 + x - 20 = 0$

4. $12x^2 - x - 6 = 0$

In Exercises 5 and 6, solve the equation by extracting square roots.

5. $2x^2 = 16$

6. $(x - 10)^2 = 81$

In Exercises 7 and 8, solve the equation by completing the square.

7. $x^2 + 6x - 4 = 0$

8. $2x^2 - 6x - 5 = 0$

In Exercises 9 and 10, use the Quadratic Formula to solve the equation.

9. $4x^2 - 6x + 2 = 0$

10. $3x^2 - x + 1 = 0$

In Exercises 11–14, solve the equation.

11. $(y + 3)^2 + 9 = 0$

12. $x^4 + 3x^2 - 10 = 0$

13. $\sqrt{10x + 31} = x + 4$

14. $\dfrac{5}{x} = x - 4$

15. Describe the values of c such that the equation $x^2 + 7x + c = 0$ has (a) one real-number solution, (b) two real-number solutions, and (c) two imaginary-number solutions.

16. The cost for a manufacturer to produce x units of a product is

$$C = 0.125x^2 + 20x + 5000.$$

Find the number of units that can be produced when $C = \$14{,}000$.

17. Find the dimensions of the rectangle in the figure if its area is 160 square feet.

Using a Graphing Calculator

You can use the *zoom* feature of a graphing calculator to approximate the solutions of an equation to any desired accuracy. The four graphs below show four steps in approximating the positive solution of

$$x^2 - 5 = 0$$

to be $x \approx 2.236$. (The exact solution is $x = \sqrt{5}$.) By repeated zooming, you can obtain whatever accuracy you need. *

1. Sketch the graph of $y = x^2 - 5$.

   RANGE Xmin=-10   Ymin=-10
       Xmax=10   Ymax=10
       Xscl=1   Yscl=1

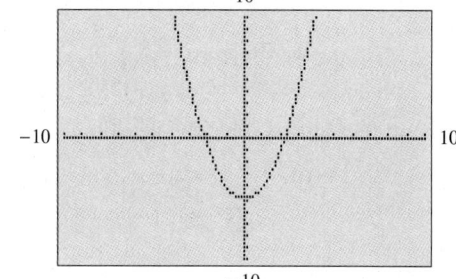

2. Zoom once to get a clearer view of the positive x-intercept. (The viewing window will change automatically.

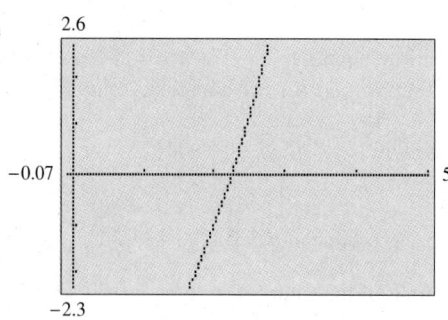

*The graphing calculator keystrokes in this text correspond to the TI-81 and TI-82 graphing calculators from Texas Instruments. For other calculators, the keystrokes may differ.

3. Zoom a second time to get an even clearer view. Use the trace function and the cursor keys to determine that the x-intercept is a little more than 2.2.

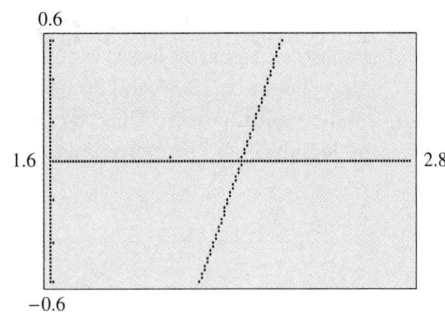

4. Set x-values to range from 2.2 to 2.3, with an x-scale of 0.01. From the graph, you can approximate the x-intercept to be 2.236.

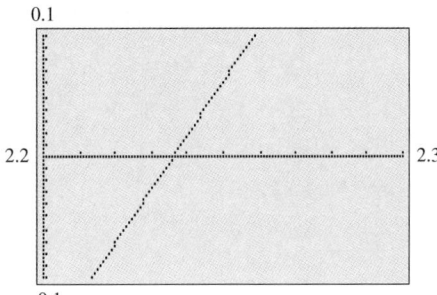

Xmin=2.2	Ymin=-0.1
Xmax=2.3	Ymax=0.1
Xscl=0.01	Yscl=0.01

EXAMPLE 1 ■ Using a Graphing Calculator to Approximate a Solution

On the kite shown at left, the length of $\overline{AB}$ is 39 inches. The combined length of $\overline{AC}$ and $\overline{CB}$ is 50 inches. How long are $\overline{AC}$ and $\overline{CB}$?

Solution

Let x represent the length of $\overline{CB}$. Because the combined length of $\overline{AC}$ and $\overline{CB}$ is 50 inches, it follows that the length of $\overline{AC}$ is $50 - x$. Using the Pythagorean Theorem, you can write the following equation.

$$x^2 + (50 - x)^2 = 39^2$$
$$x^2 + 2500 - 100x + x^2 = 1521$$
$$2x^2 - 100x + 979 = 0$$

In order to solve this equation with a graphing calculator, you must sketch the graph of $y = 2x^2 - 100x + 979$ and approximate the x-intercepts of the graph (see figure below). Using the zoom feature, you can approximate the x-intercepts to be 13.36 and 36.64. From the drawing of the kite, it follows that the smaller solution is x. Thus, the length of $\overline{CB}$ is approximately 13.36 inches and the length of $\overline{AC}$ is approximately 36.64 inches.

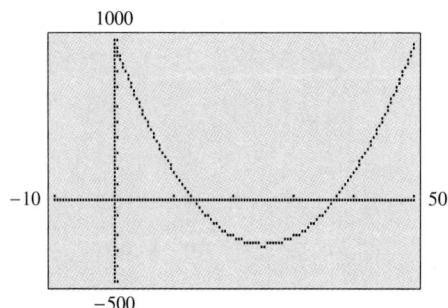

EXERCISES

In Exercises 1–6, use a graphing calculator to approximate the solutions of the equation to one decimal place.

1. $3x^2 - 5x - 7 = 0$

2. $-x^2 - 6x + 12 = 0$

3. $\frac{1}{2}x^2 + 12x + 5 = 0$

4. $\frac{4}{5}x^2 - 7x - 9 = 0$

5. $-0.3x^2 - 4.2x + 8.1 = 0$

6. $4.2x^2 + 5.6x + 1.3 = 0$

In Exercises 7 and 8, solve for x.

7.

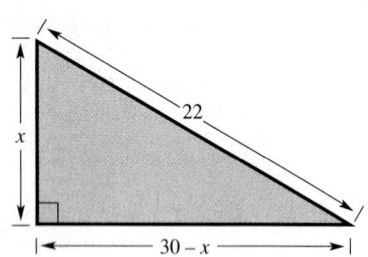

8.

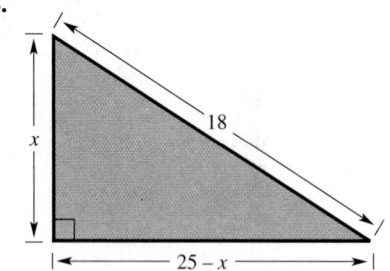

SECTION 7.4	Applications of Quadratic Equations
	Review of Problem-Solving Strategy ● *Applications of Quadratic Equations*

Review of Problem-Solving Strategy

You have completed your study of the technical parts of the algebra of quadratic equations. You have also seen some uses of quadratic equations in solving real-life problems. In this section, you will study other real-life problems involving quadratic equations. To construct these equations from verbal problems, you should use the problem-solving strategy developed in Chapter 3. For convenience, the steps involved in this strategy are reviewed here.

Strategy for Solving Word Problems

1. *Search* for the hidden equality—two expressions said to be equal or known to be equal. A sketch may be helpful.

2. *Write* a verbal model that equates these two expressions.

3. *Assign* numbers to the known quantities, and letters (or algebraic expressions) to the variable quantities.

4. *Rewrite* the verbal model as an algebraic equation using the assigned labels.

5. *Solve* the resulting algebraic equation.

6. *Check* to see that the solution satisfies the word problem as stated.

For word problems that involve quadratic equations, the verbal model is often a *product* of two variable quantities. On other occasions, the model may be a previously known formula that describes the situation. Watch for these variations in the examples that follow.

Applications of Quadratic Equations

EXAMPLE 1   ■   An Application Involving Area

The height of a painting is 4 inches less than twice its width (see Figure 7.5 on the following page). The area of the painting is 240 square inches. Find the dimensions of the painting.

FIGURE 7.5

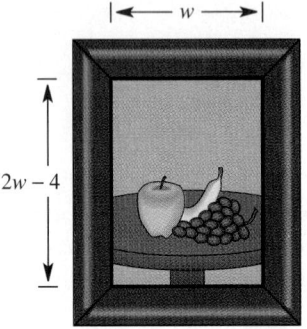

Point out that the equation has two algebraic solutions, $w = 12$ and $w = -10$. However, $w = -10$ is not a feasible solution to the application. The context of the problem will often dictate the domain.

Solution

Verbal model: $\boxed{\text{Area of painting}} = \boxed{\text{Width}} \cdot \boxed{\text{Height}}$

Labels: Picture width $= w$ (inches)
Picture height $= 2w - 4$ (inches)
Area $= 240$ (square inches)

Equation:
$$240 = w(2w - 4)$$
$$0 = 2w^2 - 4w - 240$$
$$0 = 2(w^2 - 2w - 120)$$
$$0 = 2(w - 12)(w + 10)$$
$$w - 12 = 0 \implies w = 12$$
$$w + 10 = 0 \implies w = -10$$

Of the two possible solutions, choose the positive value of w and conclude that $w = 12$. Finally, because the height of the painting is 4 inches less than twice its width, you can see that the height of the painting is $2(12) - 4 = 20$ inches. Check these dimensions in the original problem. ∎

EXAMPLE 2 ∎ **An Interest Problem**

The formula

$$A = P(1 + r)^2$$

represents the amount A of money in an account in which P dollars was deposited for 2 years, at an annual percentage rate of r (in decimal form). Find the interest rate if a deposit of $6000 increased to an amount of $6933.75 over a period of 2 years.

Solution

$A = P(1 + r)^2$	Given formula
$6933.75 = 6000(1 + r)^2$	Substitute known values
$1.155625 = (1 + r)^2$	Divide both sides by 6000
$\pm 1.075 = 1 + r$	Extract square roots
$0.075 = r$	Choose positive solution

Thus, the annual percentage rate is $r = 0.075 = 7.5\%$. Check this solution in the original problem. ∎

EXAMPLE 3   ■   An Investment Problem

A car dealer bought a fleet of cars from a car rental agency for a total of $90,000. By the time the dealer had sold all but six of the cars, at an average profit of $2500 each, the original investment of $90,000 had been regained. How many cars did the dealer sell, and what was the average price per car?

Solution

Although this problem is stated in terms of average prices and average profits per car, you can use a model that assumes that each car sold for the same price.

Verbal model: $\boxed{\dfrac{\text{Selling price}}{\text{per car}}} = \boxed{\dfrac{\text{Cost}}{\text{per car}}} + \boxed{\dfrac{\text{Profit}}{\text{per car}}}$

Remind students that they studied equations involving rational expressions in Section 5.6. Stress the importance of multiplying by the least common denominator to clear fractions.

Labels: Number sold $= x$ (cars)

Total sale $= 90,000$ (dollars)

Selling price $= \dfrac{90,000}{x}$ (dollars per car)

Number bought $= x + 6$ (cars)

Cost $= \dfrac{90,000}{x + 6}$ (dollars per car)

Profit $= 2500$ (dollars per car)

Equation: $\dfrac{90,000}{x} = \dfrac{90,000}{x + 6} + 2500$

$90,000(x + 6) = 90,000x + 2500x(x + 6), \qquad x \neq 0, \; x \neq -6$

$90,000x + 540,000 = 90,000x + 2500x^2 + 15,000x$

$0 = 2500x^2 + 15,000x - 540,000$

$0 = x^2 + 6x - 216$

$0 = (x - 12)(x + 18)$

$x - 12 = 0 \implies x = 12$

$x + 18 = 0 \implies x = -18$

Choosing the positive value, you can conclude that the dealer sold 12 cars at a price of

$\dfrac{90,000}{12} = \$7500$ per car.

Check this solution in the original problem.   ■

EXAMPLE 4   ■   Reduced Rates

A ski club chartered a bus for a ski trip at a cost of $520. In an attempt to lower the bus fare per skier, the club invited nonmembers to go along. When five nonmembers joined the trip, the fare per skier decreased by $5.20. How many club members are going on the bus?

Solution

Verbal model: | Original cost per member | $-$ | $5.20 | $=$ | New cost per skier |

Labels: Number of members $= x$ (people)
Number of skiers $= x + 5$ (people)
Original cost $= \dfrac{520}{x}$ (dollars per person)
New cost $= \dfrac{520}{x + 5}$ (dollars per person)

Equation: $\dfrac{520}{x} - \$5.20 = \dfrac{520}{x + 5}$

$$x(x + 5)\left(\dfrac{520}{x} - 5.2\right) = \left(\dfrac{520}{x + 5}\right) x(x + 5)$$

$$520(x + 5) - 5.2x(x + 5) = 520x$$

$$520x + 2600 - 5.2x^2 - 26x = 520x$$

$$-5.2x^2 - 26x + 2600 = 0$$

$$x^2 + 5x - 500 = 0$$

$$(x + 25)(x - 20) = 0$$

$$x + 25 = 0 \implies x = -25$$

$$x - 20 = 0 \implies x = 20$$

Choosing the positive value of x, you have

$x = 20$ ski club members.

Check this solution in the original problem.   ■

EXAMPLE 5   ■   An Application Involving the Pythagorean Theorem

An L-shaped sidewalk from building A to building B on a college campus is 200 meters long, as shown in Figure 7.6. By cutting diagonally across the grass, students shorten the walking distance to 150 meters. What are the lengths of the two legs of the sidewalk?

FIGURE 7.6

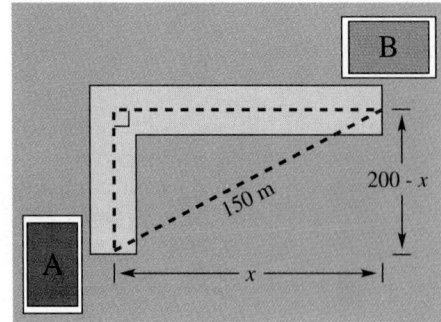

Solution

Common formula: $a^2 + b^2 = c^2$ Pythagorean Theorem

Labels: Length of one leg $(a) = x$ (meters)
Length of other leg $(b) = 200 - x$ (meters)
Length of diagonal $(c) = 150$ (meters)

Equation: $x^2 + (200 - x)^2 = (150)^2$

$$2x^2 - 400x + 40{,}000 = 22{,}500$$
$$2x^2 - 400x + 17{,}500 = 0$$
$$x^2 - 200x + 8750 = 0$$

By using the Quadratic Formula, you can find the solutions of this equation as follows.

$$x = \frac{200 \pm \sqrt{(-200)^2 - 4(1)(8750)}}{2(1)}$$

$$= \frac{200 \pm \sqrt{5000}}{2}$$

$$= \frac{200 \pm 50\sqrt{2}}{2}$$

$$= 100 \pm 25\sqrt{2}$$

Both solutions are positive, and it does not matter which one you choose. If you let

$$x = 100 + 25\sqrt{2} \approx 135.4 \text{ meters,}$$

then the length of the other leg is

$$200 - x \approx 200 - 135.4 \approx 64.6 \text{ meters.}$$

Try choosing the other value of x to see that the same two lengths result.   ■

EXAMPLE 6   ■   Work Problem

An office has two copy machines. Machine B is known to take 12 minutes longer than machine A to copy the company's monthly report. The two machines working together take 8 minutes to reproduce the report. How long would it take each machine alone to reproduce the report?

Additional problem: You may want to have students complete this problem as a class exercise: Working together, Carol and Denise can complete a project in 2 hours. Working alone, how long would it take Denise to complete the same project if she takes 3 hours longer than Carol alone?

Solution

Verbal model:

$$\boxed{\text{Work done by machine A}} + \boxed{\text{Work done by machine B}} = \boxed{\text{1 complete job}}$$

$$\left(\boxed{\text{Rate for A}} \cdot \boxed{\text{Time for both}} \right) + \left(\boxed{\text{Rate for B}} \cdot \boxed{\text{Time for both}} \right) = \boxed{1}$$

Labels: Both machines: Time = 8 minutes, Rate = $\frac{1}{8}$ (job per minute)

Machine A: Time = t minutes, Rate = $\dfrac{1}{t}$ (job per minute)

Machine B: Time = $t + 12$ minutes, Rate = $\dfrac{1}{t + 12}$ (job per minute)

Equation: $\dfrac{1}{t}(8) + \dfrac{1}{t + 12}(8) = 1$

$$\frac{8}{t} + \frac{8}{t + 12} = 1$$

$$t(t + 12)\left(\frac{8}{t} + \frac{8}{t + 12} \right) = t(t + 12)$$

$$8(t + 12) + 8t = t^2 + 12t$$

$$8t + 96 + 8t = t^2 + 12t$$

$$0 = t^2 - 4t - 96$$

$$0 = (t - 12)(t + 8)$$

$$t - 12 = 0 \implies t = 12$$

$$t + 8 = 0 \implies t = -8$$

You should choose the positive value for t and find that

Time for machine A = t = 12 minutes

Time for machine B = $t + 12$ = 24 minutes.

Check these solutions in the original problem. ■

EXAMPLE 7 ■   The Height of a Model Rocket

To help develop intuition about the results produced when numbers are substituted for t in the height equation, ask students to determine the height of the model rocket at $t = 1$ and $t = 4$.

A model rocket is projected upward from ground level according to the height equation

$$h = -16t^2 + 192t, \qquad t \geq 0,$$

where h is height in feet and t is time in seconds.

(a) After how many seconds will the height be 432 feet?

(b) When will the rocket hit the ground?

Solution

(a)

$$h = -16t^2 + 192t \qquad \text{Given equation}$$

$$432 = -16t^2 + 192t \qquad \text{Replace } h \text{ by 432 feet}$$

$$16t^2 - 192t + 432 = 0 \qquad \text{Standard form}$$

$$t^2 - 12t + 27 = 0 \qquad \text{Divide both sides by 16}$$

$$(t - 3)(t - 9) = 0 \qquad \text{Factor}$$

$$t - 3 = 0 \quad \Longrightarrow \quad t = 3 \qquad \text{Set first factor equal to 0}$$

$$t - 9 = 0 \quad \Longrightarrow \quad t = 9 \qquad \text{Set second factor equal to 0}$$

FIGURE 7.7

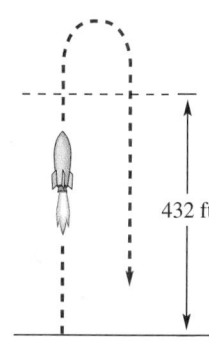

432 ft

So, the rocket will have obtained a height of 432 feet at two different times—once (going up) after 3 seconds, and again (coming down) after 9 seconds (see Figure 7.7).

(b) To find the time it takes for the rocket to hit the ground, let the height be 0.

$$0 = -16t^2 + 192t$$

$$0 = t^2 - 12t$$

$$0 = t(t - 12)$$

$$t = 0 \quad \text{or} \quad t = 12$$

Thus, the rocket will hit the ground after 12 seconds. (Note that the time of $t = 0$ seconds corresponds to the time of lift-off.) ■

DISCUSSION PROBLEM   ■   **The Rocket Problem Revisited**

At what time will the rocket described in Example 7 reach its maximum height? What will the maximum height be? Write a short paragraph describing the procedure you used to determine your answers. ■

Warm-Up

The following warm-up exercises involve skills that were covered in earlier sections. You will use these skills in the exercise set for this section.

In Exercises 1–10, solve the equation. (Find all real *and* imaginary solutions.)

1. $3(x + 7) = 0$

2. $3x(15x - 2) = 0$

3. $2n(n - 2) + (n - 2) = 3$

4. $2n(n + 2) = 240$

5. $3(x + 8)^2 = 243$

6. $t + 3\sqrt{t} - 4 = 0$

7. $x^2 + 3x + 5 = 0$

8. $u^2 + 3u - 5 = 0$

9. $t - \dfrac{6}{t} = 5$

10. $\sqrt{2s + 3} = s$

7.4 **EXERCISES**

 means that a graphing utility can help you solve the exercise or check your solution.

Number Problems In Exercises 1–6, find two positive integers that satisfy the given requirement.

1. The product of two consecutive integers is 240.

2. The product of two consecutive integers is 1122.

3. The product of two consecutive *even* integers is 224.

4. The product of two consecutive *odd* integers is 255.

5. The sum of the squares of two consecutive integers is 313.

6. The sum of the squares of two consecutive integers is 421.

Dimensions of a Rectangle In Exercises 7–16, find the dimensions of the rectangle (l and w represent the length and width of the rectangle, respectively), and fill in the missing value for perimeter or area.

	Width	Length	Perimeter	Area
7.	$0.75l$	l	42 in.	
8.	w	$1.5w$	40 m	
9.	w	$2.5w$		250 ft²
10.	w	$1.5w$		216 cm²
11.	$\frac{1}{3}l$	l		192 in.²
12.	$\frac{3}{4}l$	l		2700 in.²
13.	w	$w+3$	54 km	
14.	$l-6$	l	108 ft	
15.	$l-20$	l		12,000 m²
16.	w	$w+5$		500 ft²

 17. *Lumber Storage Area* A retail lumberyard stores lumber in a rectangular region adjoining the sales office (see figure). The region will be fenced on three sides and the fourth side will be bounded by the wall of the building. Find the dimensions of the region if 350 feet of fencing is available and the area of the region is 12,500 square feet.

Figure for 17

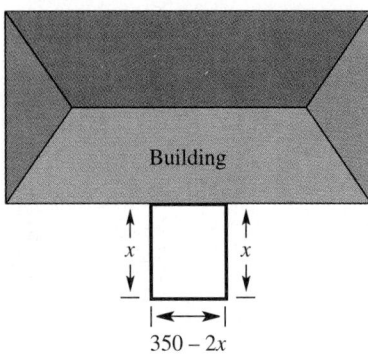

18. *Open Conduit* An open topped rectangular conduit for carrying water in a manufacturing process is made by folding up the edges of a sheet of aluminum 48 inches wide (see figure). A cross section of the conduit must have an area of 288 square inches. Find the width and height of the conduit.

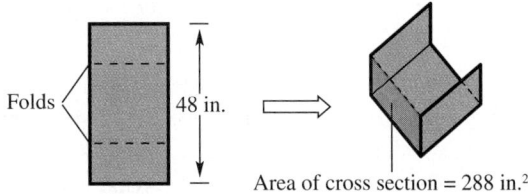

19. *Dimensions of a Triangle* The height of a triangle is two-thirds of its base. The area of the triangle is 192 square inches. Find the dimensions of the triangle.

20. *Dimensions of a Triangle* The height of a triangle is twice its base. The area of the triangle is 625 square inches. Find the dimensions of the triangle.

21. *Fencing a Yard* A friend's family has built a fence around three sides of their property (see figure at top of next page). In total, they used 550 feet of fencing. By their calculations, the area of the lot is 1 acre (43,560 square feet). Is this correct? Explain your answer.

22. *More Fencing* You have 100 feet of fencing. Do you have enough to enclose a rectangular region whose area is 630 square feet? How about a circular region whose area is 630 square feet? Explain.

Figure for 21

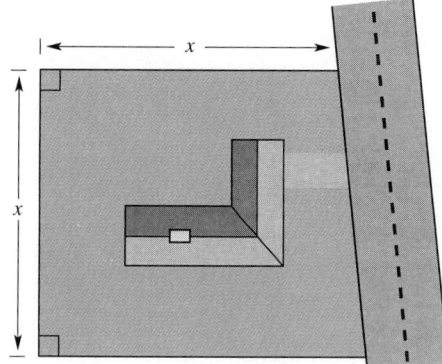

Investment In Exercises 23 and 24, use the compound interest formula $A = P(1 + r)^2$, where A represents the amount after 2 years when a principal of P dollars is invested at an annual percentage rate r (in decimal form). Find the interest rate needed to obtain the amount A when P dollars are invested for 2 years.

23. $P = \$3000$, $A = \$3499.20$

24. $P = \$10,000$, $A = \$11,990.25$

25. *Selling Price* A store owner bought a case of grade A large eggs for $21.60. By the time all but six dozen of the eggs had been sold at a profit of $0.30 per dozen, the original investment of $21.60 had been regained. How many dozen eggs did the owner sell, and what was the selling price per dozen?

26. *Selling Price* A manager of a computer store bought some computers of the same model for $27,000. When all but three of the computers had been sold at a profit of $750 per computer, the original investment of $27,000 had been regained. How many computers were sold, and what was the selling price of each computer?

27. *Reduced Ticket Price* A service organization obtained a block of tickets for a ball game for $240. The block contained eight more tickets than the organization needed for its members. By inviting eight more people to attend (and share the cost), the organization lowered the price per ticket by $1. How many people are going to the game?

28. *Reduced Fare* A science club charters a bus to take the members to a science fair at a cost of $480. In an attempt to lower the bus fare per person, the club invites nonmembers to go along. When two nonmembers join the trip, the fare per person is decreased by $1. How many people are going on the excursion?

29. *Delivery Route* Suppose that you are asked to deliver pizza to offices B and C in your city (see figure), and you are required to keep a log of all the mileages between stops. You forget to look at the odometer at stop B, but after getting to stop C you record the total distance traveled from the pizza shop as 18 miles. The return distance from C to A is 16 miles. The route forms a right triangle, and the distance from A to B is greater than the distance from B to C. Find the distance from A to B.

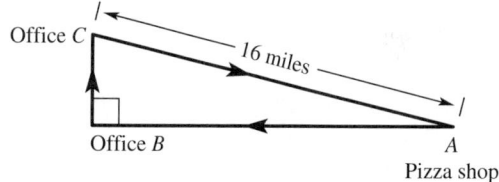

30. *Dimensions of a Rectangle* The perimeter of a rectangle is 102 inches, and the length of its diagonal is 39 inches. Find the dimensions of the rectangle.

31. *Airspeed* An airline runs a commuter flight between two cities that are 720 miles apart. If the average speed of the planes could be increased by 40 miles per hour, the travel time could be decreased by 12 minutes. What airspeed is required to obtain this decrease in travel time?

32. *Average Speed* A truck traveled the first 100 miles of a trip at one speed and the last 135 miles at an average speed of 5 miles per hour less. If the entire trip took 5 hours, what was the average speed for the first part of the trip?

33. *Work Rate Problem* Working together, two people can complete a task in 5 hours. Working alone, how long would it take each to do the task if one person takes 2 hours longer than the other?

34. *Work Rate Problem* An office has two printers. Machine B is known to take 3 minutes longer than machine A to produce the company's monthly financial report. With both machines working together, it takes 6 minutes to produce the report. How long would it take each machine alone to produce the report?

Free-Falling Object In Exercises 35–38, find the time necessary for an object to fall to ground level from an initial height of h_0 feet if its height h at any time t (in seconds) is

$$h = h_0 - 16t^2.$$

35. $h_0 = 144$

36. $h_0 = 625$

37. $h_0 = 1454$ (height of Sears Tower)

38. $h_0 = 984$ (height of Eiffel Tower)

39. *Height of a Baseball* The height h in feet of a baseball that is hit 3 feet above the ground is

$$h = 3 + 75t - 16t^2,$$

where t is time in seconds. Find the time when the ball hits the ground in the outfield.

40. *Hitting Baseballs* Suppose you are hitting baseballs. When you toss the ball into the air, your hand releases the ball at a point 5 feet above the ground. You hit the ball when it falls back to a height of 4 feet. If you toss the ball with an initial velocity of 25 feet per second, the height h of the ball t seconds after leaving your hand is

$$h = 5 + 25t - 16t^2.$$

How much time will pass before you hit the ball?

41. *Cellular Phones* For 1985 through 1990, the number of cellular phone subscribers s (in millions) in the United States can be approximated by the model

$$s = 0.2t^2 - 2t + 5.3, \quad 5 \le t \le 10$$

where $t = 5$ represents 1985 (see figure). In which year did cellular phone companies have 2.1 million subscribers? (*Source:* Cellular Telecommunications Industry Association)

Figure for 41

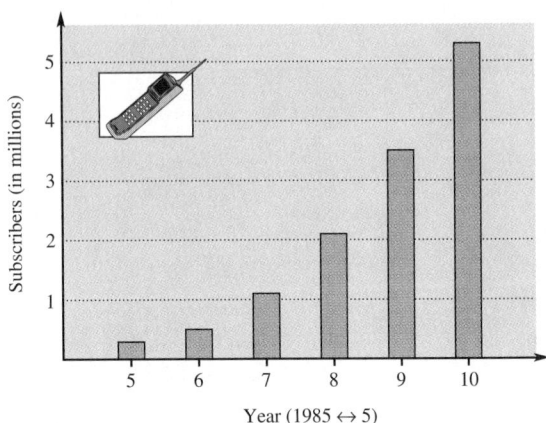

Year (1985 ↔ 5)

42. *Mountain Bike Owners* For the years 1983 through 1990, the number of mountain bike owners m (in millions) in the United States can be approximated by the model

$$m = 0.337t^2 - 2.265t + 3.962, \quad 3 \le t \le 10$$

where $t = 3$ represents 1983 (see figure). In which year did 2.5 million people own mountain bikes? (*Source:* Bicycle Institute of America)

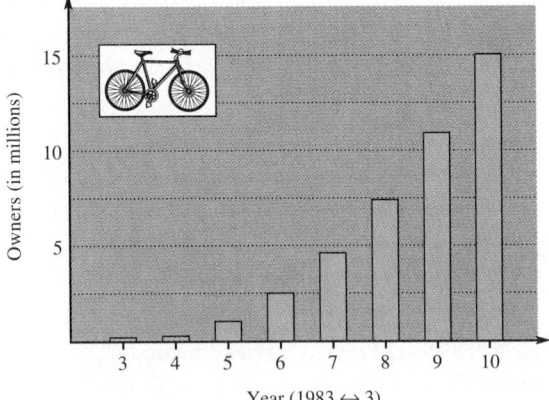

Year (1983 ↔ 3)

Graphing Quadratic Functions

Graphs of Quadratic Functions   •   *Sketching a Parabola*   •   *Finding the Equation of a Parabola*   •
Applications

Graphs of Quadratic Functions

Historical Note: The Greek mathematician Apollonius of Perga (third
century B.C.) was also called the
"Great Geometer." He is credited with
coining the term *parabola*.

In this section, you will study the graphs of quadratic functions of the form

$$f(x) = ax^2 + bx + c. \qquad \text{Quadratic function}$$

Figure 7.8 shows the graph of a simple quadratic function, $y = x^2$. This graph is
an example of a **parabola.** The lowest point on this graph, $(0, 0)$, is the **vertex** of
the parabola, and the vertical line that passes through the vertex (the y-axis, in this
case) is the **axis** of the parabola. Every parabola is *symmetric* about its axis, which
means that if folded along its axis, the two parts of the parabola would match. In
general, the graph of an equation of the form $y = ax^2 + bx + c$, $a \neq 0$, is a
parabola.

FIGURE 7.8

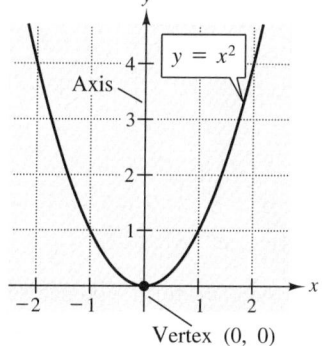

Vertex $(0, 0)$

Equations of Parabolas

The graph of $y = ax^2 + bx + c$, $a \neq 0$, is a **parabola.** The completed-square
form of the equation

$$y = a(x - h)^2 + k \qquad \text{Standard form}$$

is called the **standard form** of the equation. The **vertex** of the parabola occurs
at the point (h, k), and the vertical line passing through the vertex is the **axis**
of the parabola.

You might ask students if they know
of objects with this shape. Supplement their answers with radar and
satellite dishes, car headlights, and
bridge cables.

Parabolas that correspond to equations of the form $y = ax^2 + bx + c$ open
upward if the leading coefficient a is positive, and open *downward* if the leading
coefficient is negative, as shown in Figure 7.9 on the following page.

FIGURE 7.9

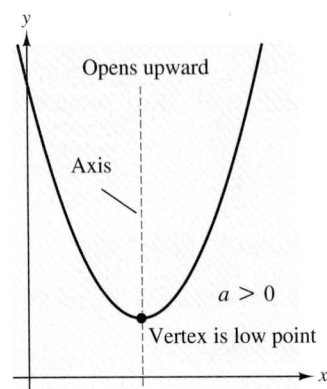

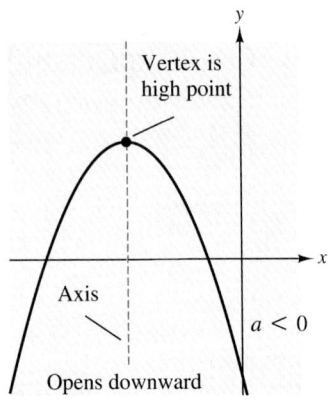

EXAMPLE 1 ■ Finding the Vertex of a Parabola

Find the vertex of the parabola given by the equation $y = x^2 - 6x + 5$.

FIGURE 7.10

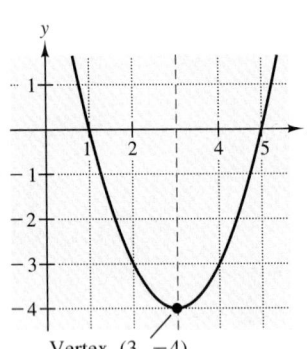

Vertex (3, −4)

Solution

Begin by writing the equation in standard form.

$$y = x^2 - 6x + 5 \qquad \text{Given equation}$$

$$y = x^2 - 6x + (-3)^2 - (-3)^2 + 5 \qquad \text{Add and subtract } (-3)^2$$

half of −6

$$y = (x^2 - 6x + 3^2) - 9 + 5 \qquad \text{Regroup terms}$$

$$y = (x - 3)^2 - 4 \qquad \text{Standard form}$$

Now, from the standard form, you can see that the vertex of the parabola occurs at the point (3, −4), as shown in Figure 7.10. ■

EXAMPLE 2 ■ Finding the Vertex of a Parabola

Discuss standard form and the signs of the coordinates of the vertex.

Find the vertex of the parabola given by the equation $f(x) = x^2 + x$.

Solution

Begin by writing the equation in standard form.

$$f(x) = x^2 + x \qquad \text{Given equation}$$

$$f(x) = x^2 + x + \left(\frac{1}{2}\right)^2 - \left(\frac{1}{2}\right)^2 \qquad \text{Add and subtract } \left(\frac{1}{2}\right)^2$$

half of 1

$$f(x) = \left(x^2 + x + \frac{1}{4}\right) - \frac{1}{4} \qquad \text{Regroup terms}$$

$$f(x) = \left(x + \frac{1}{2}\right)^2 - \frac{1}{4} \qquad \text{Standard form}$$

FIGURE 7.11

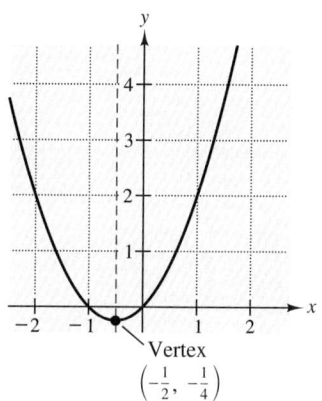

Vertex
$\left(-\frac{1}{2}, -\frac{1}{4}\right)$

Now, from the standard form, you can see that the vertex of the parabola occurs at the point $\left(-\frac{1}{2}, -\frac{1}{4}\right)$, as shown in Figure 7.11. Note that the vertex is $\left(-\frac{1}{2}, -\frac{1}{4}\right)$ because the standard form is

$$k = -\frac{1}{4}$$

$$f(x) = a(x - h)^2 + k = 1\left[x - \left(-\frac{1}{2}\right)\right]^2 + \left(-\frac{1}{4}\right).$$

$$h = -\frac{1}{2}$$

NOTE In Figure 7.11, the graph of f can be obtained by shifting the graph of $y = x^2$ to the left $\frac{1}{2}$ unit and down $\frac{1}{4}$ unit, as discussed in Section 2.6.

EXAMPLE 3 ■ Finding the Vertex of a Parabola

Find the vertex of the parabola given by the equation $y = 2x^2 - 4x$.

Solution

Begin by writing the equation in standard form.

$$y = 2x^2 - 4x \qquad\qquad \text{Given equation}$$
$$y = 2(x^2 - 2x) \qquad\qquad \text{Factor out leading coefficient}$$
$$y = 2\left[x^2 - 2x + (-1)^2 - (-1)^2\right] \qquad\qquad \text{Complete the square}$$

half of -2

$$y = 2(x^2 - 2x + 1^2) - 2(1^2)$$
$$y = 2(x - 1)^2 - 2 \qquad\qquad \text{Standard form}$$

FIGURE 7.12

Now, from the standard form, you can see that the vertex of the parabola occurs at the point $(1, -2)$, as shown in Figure 7.12.

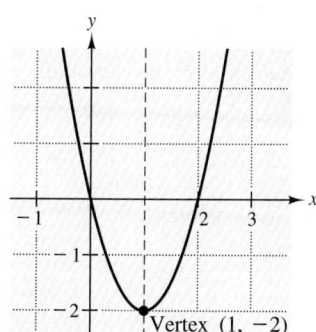

Vertex $(1, -2)$

EXAMPLE 4   ■   Finding the Vertex of a Parabola

Find the vertex of the parabola given by the equation $y = -x^2 - 4x - 3$.

Solution

Begin by writing the equation in standard form.

FIGURE 7.13

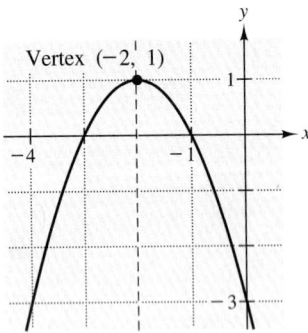

Vertex $(-2, 1)$

$$y = -x^2 - 4x - 3 \qquad \text{Given equation}$$
$$y = -1(x^2 + 4x) - 3 \qquad \text{Factor out leading coefficient}$$
$$y = -(x^2 + 4x + 2^2 - 2^2) - 3 \qquad \text{Complete the square}$$

$$\underbrace{\qquad\qquad}_{\text{half of 4}}$$

$$y = -(x^2 + 4x + 2^2) + 2^2 - 3 \qquad -(-2^2) = 2^2$$
$$y = -(x + 2)^2 + 1 \qquad \text{Standard form}$$

Now, from the standard form, you can see that the vertex of the parabola occurs at the point $(-2, 1)$, as shown in Figure 7.13.   ■

In Examples 1–4, you found the vertex of the given parabola by completing the square *for each* equation. Another technique you can use for this purpose is to complete the square once for the general equation, as follows.

$$y = ax^2 + bx + c$$
$$y = a\left(x^2 + \frac{b}{a}x\right) + c$$
$$y = a\left(x + \frac{b}{2a}\right)^2 + c - \frac{b^2}{4a}$$

From this form, you can see that the vertex occurs when $x = -b/2a$. Try using this technique to find the vertex of each of the parabolas in Examples 1–4.

If you ask students to repeat Examples 1–4 using $x = -(b)/(2a)$, be sure they also find the y-coordinate of the vertex.

Sketching a Parabola

The following steps are helpful in sketching a parabola.

■   **Sketching a Parabola**

To sketch the graph of the parabola given by $y = ax^2 + bx + c$, use the following steps.

1. Write the given equation in standard form and determine the vertex and axis of the parabola.

2. Plot the vertex, axis, and a few additional points on the parabola. (Using the symmetry about the axis can reduce the number of points you need to plot.)

3. Use the fact that the parabola opens *upward* if $a > 0$ and opens *downward* if $a < 0$ to complete the sketch.

EXAMPLE 5 ■ Sketching a Parabola

Sketch the graph of the parabola given by the second-degree equation

$$x^2 - y + 6x + 8 = 0.$$

Solution

Begin by writing the equation in standard form.

$$x^2 - y + 6x + 8 = 0 \qquad \text{Given equation}$$

$$-y = -x^2 - 6x - 8$$

$$y = x^2 + 6x + 8$$

$$y = (x^2 + 6x + 3^2 - 3^2) + 8$$

half of 6

$$y = (x^2 + 6x + 3^2) - 9 + 8$$

$$y = (x + 3)^2 - 1 \qquad \text{Standard form}$$

Therefore, the vertex occurs at the point $(-3, -1)$ and the axis is given by the line $x = -3$. After plotting this information, calculate a few additional points on the parabola, as shown in Table 7.1.

TABLE 7.1

x-value	-5	-4	-3	-2	-1
y-value	3	0	-1	0	3
Solution Points	$(-5, 3)$	$(-4, 0)$	$(-3, -1)$	$(-2, 0)$	$(-1, 3)$

The graph of the parabola is shown in Figure 7.14. Note that the parabola opens upward because the leading coefficient (in the standard form) is positive.

FIGURE 7.14

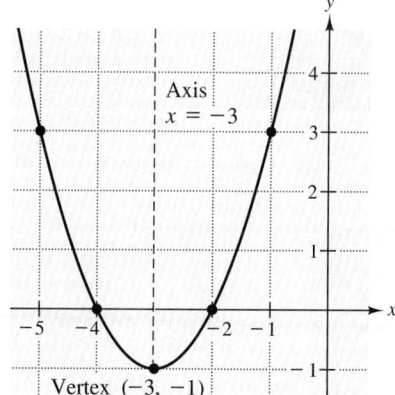

NOTE   In Example 5, two of the additional points chosen for the table, $(-4, 0)$ and $(-2, 0)$, were the x-intercepts for this graph. (Remember that you can find any x-intercepts by setting y equal to zero and solving for x. See Example 6 in Section 2.4.)

Remind students of writing equations of vertical lines in Section 3.6, and relate this to the equation of the axis.

EXAMPLE 6   ■   Sketching a Parabola

Sketch the graph of the parabola given by the second-degree equation

$$x^2 + 2y - 4x + 8 = 0.$$

Solution

Begin by writing the equation in standard form.

$$x^2 + 2y - 4x + 8 = 0 \qquad\qquad \text{Given equation}$$

$$2y = -x^2 + 4x - 8$$

$$y = -\frac{1}{2}x^2 + 2x - 4$$

$$y = -\frac{1}{2}(x^2 - 4x) - 4$$

$$y = -\frac{1}{2}\left[x^2 - 4x + (-2)^2 - (-2)^2\right] - 4$$

half of −4

$$y = -\frac{1}{2}(x^2 - 4x + 2^2) + 2 - 4 \qquad \begin{array}{l}-\frac{1}{2}[-(-2)^2] = \\ -\frac{1}{2}(-4) = 2\end{array}$$

$$y = -\frac{1}{2}(x - 2)^2 - 2 \qquad\qquad \text{Standard form}$$

Therefore, the vertex occurs at the point $(2, -2)$ and the axis is given by the line $x = 2$. After plotting this information, calculate a few additional points on the parabola, as shown in Table 7.2.

TABLE 7.2

x-value	0	1	2	3	4
y-value	-4	$-\frac{5}{2}$	-2	$-\frac{5}{2}$	-4
Solution Points	$(0, -4)$	$\left(1, -\frac{5}{2}\right)$	$(2, -2)$	$\left(3, -\frac{5}{2}\right)$	$(4, -4)$

FIGURE 7.15

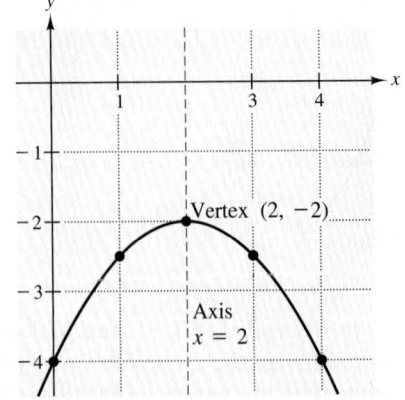

The graph of the parabola is shown in Figure 7.15. Note that the parabola opens downward because the leading coefficient (in the standard form) is negative. ■

Finding the Equation of a Parabola

To write the equation of a parabola with a vertical axis, use the fact that its standard equation has the form

$$y = a(x - h)^2 + k,$$

where (h, k) is the vertex.

EXAMPLE 7 ■ Finding the Equation of a Parabola

Earlier it was noted that ". . . if a point is on the line, then it satisfies the equation of that line . . . " Here a similar idea may help students understand the conceptual steps necessary to build an equation of a parabola from some given information.

Find the equation of the parabola with a vertex of $(-2, 1)$ and a y-intercept of $(0, -3)$, as shown in Figure 7.16.

Solution

Because the vertex occurs at $(h, k) = (-2, 1)$, you can write the following.

$$y = a(x - h)^2 + k \qquad \text{Standard form}$$
$$y = a[x - (-2)]^2 + 1$$
$$y = a(x + 2)^2 + 1$$

Now, to find the value of a, use the fact that the y-intercept is $(0, -3)$.

$$y = a(x + 2)^2 + 1 \qquad \text{Standard form}$$
$$-3 = a(0 + 2)^2 + 1 \qquad \text{Substitute } x = 0 \text{ and } y = -3$$
$$-3 = 4a + 1$$
$$-4 = 4a$$
$$-1 = a \qquad \text{Solve for } a$$

So, the standard form of the equation of the parabola is

$$y = -(x + 2)^2 + 1.$$

FIGURE 7.16

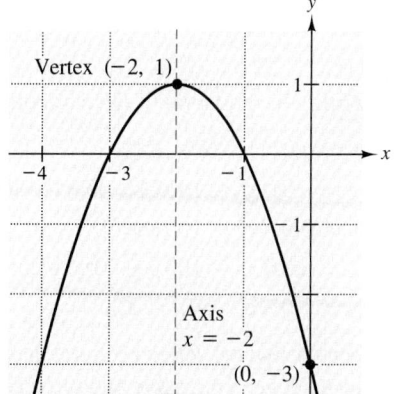

Vertex $(-2, 1)$

Axis
$x = -2$

$(0, -3)$

■

EXAMPLE 8 ■ Finding the Equation of a Parabola

Find the equation of the parabola that has a vertex of $(0, -4)$ and passes through the point $(3, 1)$, as shown in Figure 7.17.

FIGURE 7.17

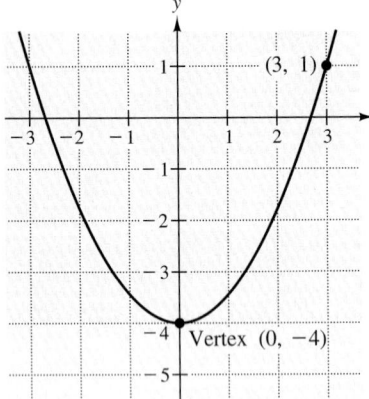

Solution

Because the vertex occurs at $(h, k) = (0, -4)$, you can write the following.

$$y = a(x - h)^2 + k \qquad \text{Standard form}$$
$$y = a(x - 0)^2 - 4$$
$$y = ax^2 - 4$$

To find the value of a, use the fact that the parabola passes through the point $(3, 1)$.

$$y = ax^2 - 4 \qquad \text{Standard form}$$
$$1 = a(3)^2 - 4 \qquad \text{Substitute } x = 3 \text{ and } y = 1$$
$$1 = 9a - 4$$
$$\frac{5}{9} = a \qquad \text{Solve for } a$$

Finally, you can conclude that the standard form of the equation of the parabola is

$$y = \frac{5}{9}x^2 - 4.$$ ■

Applications

EXAMPLE 9 ■ An Application Involving a Parabola

FIGURE 7.18

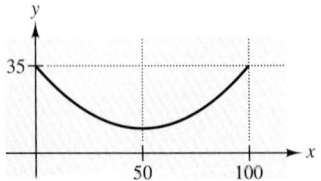

A suspension bridge is 100 feet long, as shown in Figure 7.18. The bridge is supported by cables attached to the top of the tower at each end of the bridge. Each cable hangs in the shape of a parabola given by $y = 0.01x^2 - x + 35$, where x and y are both measured in feet.

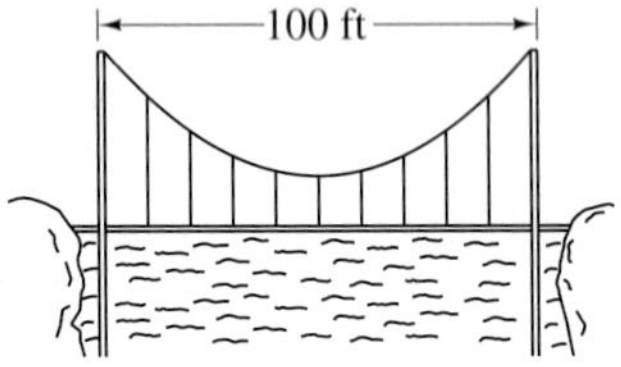

(a) Find the distance between the lowest point of the cable and the roadbed of the bridge.

(b) How tall are the towers?

Solution

(a) By writing the equation of the parabola in standard form,

$$y = 0.01x^2 - x + 35$$
$$y = 0.01(x^2 - 100x) + 35$$
$$y = 0.01(x^2 - 100x + 50^2 - 50^2) + 35$$
$$y = 0.01(x^2 - 100x + 50^2) - 25 + 35$$
$$y = 0.01(x - 50)^2 + 10,$$

you can see that the vertex occurs at the point $(50, 10)$. Therefore, the minimum distance between the cable and the roadbed is 10 feet (see Figure 7.19).

FIGURE 7.19

(b) Because the vertex of the parabola occurs at the midpoint of the bridge, the two towers are located at the points where $x = 0$ and $x = 100$. Substituting an x-value of 0, you can determine that the corresponding y-value is

$$y = 0.01(0^2) - 0 + 35 = 35 \text{ feet.}$$

Therefore, the towers are each 35 feet high. (Try substituting $x = 100$ in the equation to see that you obtain the same y-value.) ∎

DISCUSSION PROBLEM ■ Graphing Calculator Experiment

FIGURE 7.20

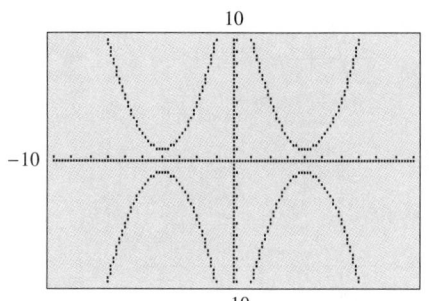

Figure 7.20 shows the display screen of a graphing calculator.* Can you find an equation for each of the four parabolas shown in the figure? If you have access to a graphing calculator, try entering each of the four equations to obtain a display that looks like the one in Figure 7.20.

Warm-Up

The following warm-up exercises involve skills that were covered in earlier sections. You will use these skills in the exercise set for this section.

In Exercises 1–4, sketch the graph of the equation.

1. $y = 3x - 2$ **2.** $y = \frac{1}{2}x + 2$

3. $y = -3x + 5$ **4.** $y = -\frac{1}{2}x + 2$

In Exercises 5–8, solve the equation.

5. $10(x - 2) = 25$ **6.** $30 - 3x = 70 - 5x$

7. $10x(x + 8) = 0$ **8.** $3x(2x + 1) = 15$

In Exercises 9 and 10, expand and simplify the expression.

9. $2(x + 5)^2 - 30$ **10.** $-3(x - 2)^2 - 7$

---

*The graphing calculator screen shown corresponds to the TI-81 and TI-82 graphing calculators from Texas Instruments. However, you can also use other graphing calculators and computer graphing software for these examples and exercises.

7.5	EXERCISES

In Exercises 1–6, match the equation with the correct graph. [The graphs are labeled (a), (b), (c), (d), (e), and (f).]

1. $y = 4 - 2x$ **2.** $y = \frac{1}{2}x - 4$ **3.** $y = x^2 - 3$

4. $y = -x^2 + 3$ **5.** $y = (x - 2)^2$ **6.** $y = 2 - (x - 2)^2$

(a)
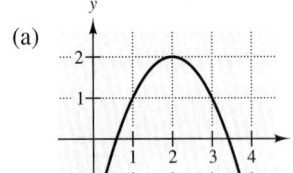

(b)
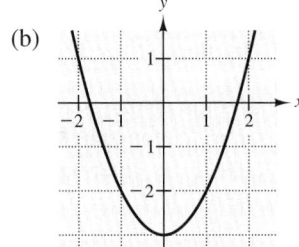

(c)
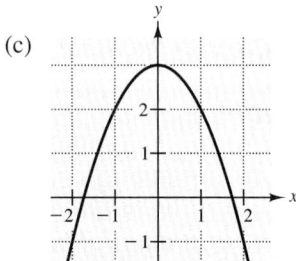

(d)
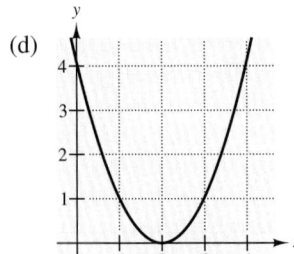

(e)
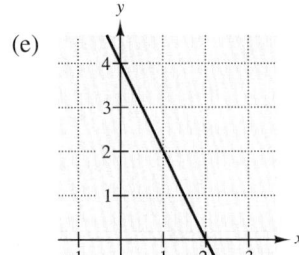

(f)
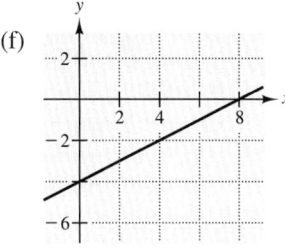

In Exercises 7–14, determine whether the graph of the equation is a parabola that opens up or opens down. Then find the vertex of the parabola.

7. $y = 2(x - 0)^2 + 2$ **8.** $y = -3(x + 5)^2 - 3$ **9.** $y = 4 - (x - 10)^2$ **10.** $y = 2(x - 12)^2 + 3$

11. $y = x^2 - 6$ **12.** $y = -(x + 1)^2$ **13.** $f(x) = 3(x - 3)^2$ **14.** $g(x) = -5(x + 1)^2 + 3$

In Exercises 15–22, find any x-intercepts of the parabola.

15. $y = 25 - x^2$ **16.** $y = x^2 - 49$ **17.** $h(x) = x^2 - 9x$ **18.** $f(x) = x^2 + 4x$

19. $y = 4x^2 - 12x + 9$ **20.** $y = 10 - x - 2x^2$ **21.** $y = x^2 - 3x + 3$ **22.** $y = x^2 - 3x - 10$

In Exercises 23–30, write the equation in standard form and find the vertex of the parabola.

23. $f(x) = x^2 - 4x + 7$ **24.** $g(x) = x^2 + 4x$ **25.** $f(x) = x^2 + 2$ **26.** $y = x^2 + 6x - 5$

27. $y = -x^2 + 2x - 7$ **28.** $y = -x^2 - 10x + 10$ **29.** $y = 2x^2 + 6x + 2$ **30.** $y = 3x^2 - 3x - 9$

In Exercises 31–54, sketch the graph of the quadratic equation. Identify the vertex and any x-intercepts.

31. $y = x^2 - 4$

32. $y = x^2 - 9$

33. $y = -x^2 + 4$

34. $y = -x^2 + 9$

35. $f(x) = x^2 - 3x$

36. $f(x) = x^2 - 4x$

37. $g(x) = -x^2 + 3x$

38. $h(x) = -x^2 + 4x$

39. $y = (x - 4)^2$

40. $y = -(x + 4)^2$

41. $y = x^2 - 8x + 15$

42. $y = x^2 + 4x + 3$

43. $y = -(x^2 + 6x + 5)$

44. $y = -x^2 + 2x + 8$

45. $y = -x^2 + 6x - 7$

46. $y = x^2 + 4x + 2$

47. $y = 3x^2 - 6x + 4$

48. $y = 2(x^2 + 6x + 8)$

49. $y = \frac{1}{2}(x^2 - 2x - 3)$

50. $y = -\frac{1}{2}(x^2 - 6x + 7)$

51. $y = 5 - \dfrac{x^2}{3}$

52. $y = \dfrac{x^2}{3} - 2$

53. $y = \frac{1}{5}(3x^2 - 24x + 38)$

54. $y = \frac{1}{5}(2x^2 - 4x + 7)$

In Exercises 55–58, sketch the graph of the equation and determine the values of x (if any) for which the graph is at the specified height y.

55. $y = -x^2 + 6$, $\quad y = 2$

56. $y = x^2 - 6x + 8$, $\quad y = 3$

57. $y = \frac{1}{2}x^2 - 3x + \frac{13}{2}$, $\quad y = 3$

58. $y = -2x^2 - 4x$, $\quad y = 1$

In Exercises 59–76, find the general form, $y = ax^2 + bx + c$, of the equation of the parabola satisfying the given criteria.

59.
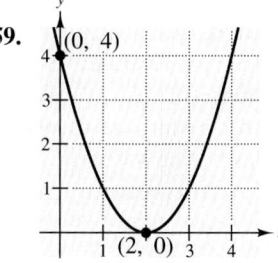

60.
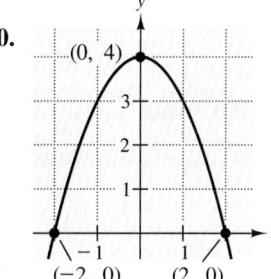

61.
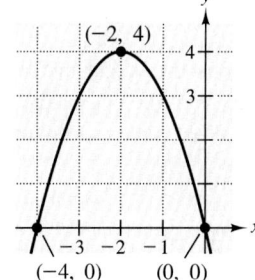

62.
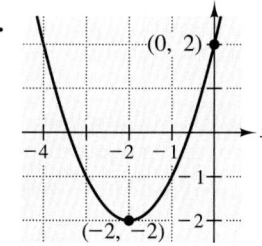

63.
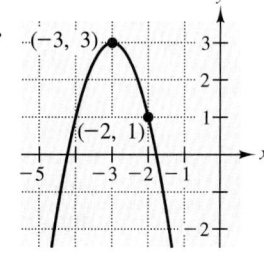

64.
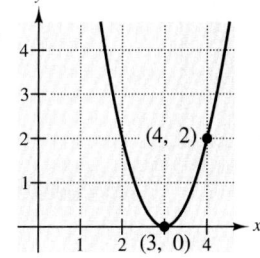

65. Vertex: $(2, 1)$; $a = 1$

66. Vertex: $(-3, -3)$; $a = 1$

67. Vertex: $(-3, 4)$; $a = -1$

68. Vertex: $(3, -2)$; $a = -1$

69. Vertex: $(2, -4)$;   Passes through the point $(0, 0)$

70. Vertex: $(-2, -4)$;   Passes through the point $(0, 0)$

71. Vertex: $(3, 2)$;   Passes through the point $(1, 4)$

72. Vertex: $(-1, -1)$;   Passes through the point $(0, 4)$

73. Vertex: $(-1, 5)$;   Passes through the point $(0, 1)$

74. Vertex: $(5, 10)$;   Passes through the point $(6, 6)$

75. Vertex: $(5, 2)$;   Passes through the point $(10, 3)$

76. Vertex: $(0, 20)$;   Passes through the point $(10, 15)$

77. ***Path of a Ball***   The height y (in feet) of a ball thrown by a child is

$$y = -\frac{1}{12}x^2 + 2x + 4,$$

where x is the horizontal distance (in feet) from where the ball was thrown.
(a) Sketch the path of the ball.
(b) How high was the ball when it left the child's hand?
(c) How high was the ball when it was at its maximum height?
(d) How far from the child did the ball strike the ground?

78. ***Maximum Height of a Diver***   The path of a diver is

$$y = -\frac{4}{9}x^2 + \frac{24}{9}x + 10,$$

where y is the height in feet and x is the horizontal distance from the end of the diving board in feet (see figure). What is the maximum height of the diver?

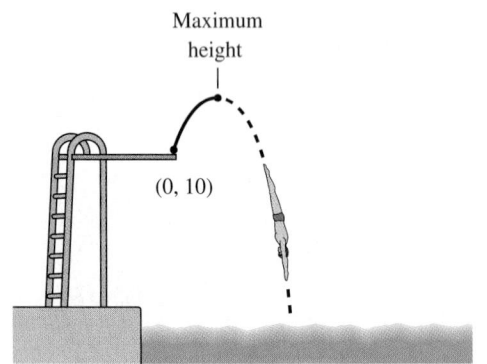

Maximum
height

$(0, 10)$

79. ***Maximum Profit***   A company manufactures radios that cost (the company) $60 each. For buyers who purchase 100 or fewer radios, the purchase price is $90 per radio. To encourage large orders, the company will reduce the price *per radio* for orders over 100, as follows. If 101 radios are purchased, the price is $89.85 per unit. If 102 radios are purchased, the price is $89.70 per unit. If $(100+x)$ radios are purchased, the price is $[90 - x(0.15)]$ dollars per unit.
(a) Show that the profit P is

$$P = (100 + x)[90 - x(0.15)] - (100 + x)60$$
$$= 3000 + 15x - \frac{3}{20}x^2,$$

where x is the number over 100 in the order.
(b) Find the vertex of the profit curve and determine the order size for maximum profit.
(c) Suppose you work for this company and are in charge of setting the price for this model of radio. Would you recommend this pricing scheme with no restrictions? Would you recommend this pricing scheme if price reductions were restricted to orders between 100 and 150 units?

80. ***Maximum Profit***   The profit P (in thousands of dollars) for a certain company is

$$P = 230 + 20s - \frac{1}{2}s^2,$$

where s is the amount (in hundreds of dollars) spent on advertising. Sketch the graph of the profit equation and determine the amount of advertising that yields maximum profit.

81. *Bridge Design* A bridge is to be constructed over a gorge with the main supporting arch being a parabola (see figure). If the equation of the parabola is

$$y = 4\left(100 - \frac{x^2}{2500}\right),$$

where x and y are measured in feet, find the following.
(a) The length of road across the gorge.
(b) The height of the parabolic arch at the center of the span.
(c) The length of the vertical girders at intervals of 100 feet from the center of the bridge.

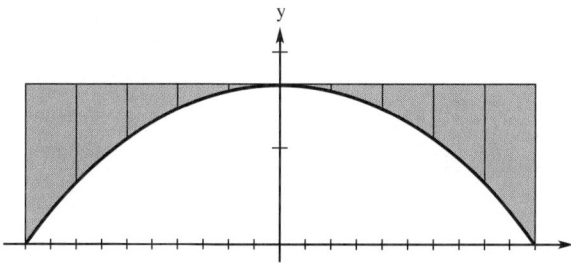

82. *Highway Design* An engineer must design a parabolic arc to create a turn in a freeway around a city. The vertex of the parabola is placed at the origin and must connect with roads represented by the equations $y = -0.4x - 100$ when $x < -500$, and $y = 0.4x - 100$ when $x > 500$ (see figure). Find an equation for the parabolic arc.

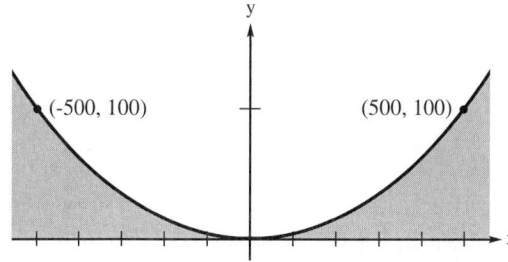

In Exercises 83–86, use a graphing calculator to graph the equation. Use the *trace* feature to approximate the coordinates of the vertex.*

83. $y = \frac{1}{6}(2x^2 - 8x + 11)$

84. $y = -\frac{1}{4}(4x^2 - 20x + 13)$

85. $y = -0.7x^2 - 2.7x + 2.3$

86. $y = 0.75x^2 - 7.50x + 23.00$

87. *Area* The area of a rectangle is given by the equation

$$A = \frac{2}{\pi}(100x - x^2), \qquad 0 < x < 100,$$

where x is the length of the rectangle in feet. Use a graphing calculator to graph this equation and use the *trace* feature to approximate the value of x when A is maximum. Use the following range settings.

Xmin=0
Xmax=100
Xscl=10
Ymin=0
Ymax=1600
Yscl=100

88. *Cost* The cost of producing x units of a product is

$$C = 800 - 10x + \frac{1}{4}x^2, \qquad 0 < x < 40.$$

Use a graphing calculator to graph this equation and use the *trace* feature to approximate the value of x when C is minimum. Use the following range settings.

Xmin=0
Xmax=40
Xscl=5
Ymin=600
Ymax=1000
Yscl=40

* The graphing calculator features shown in this text correspond to the TI-81 and TI-82 graphing calculators from Texas Instruments. However, you can also use other graphing calculators and computer graphing software for these examples and exercises.

Nonlinear Inequalities in One Variable

Test Intervals • *Quadratic Inequalities* • *Rational Inequalities* • *Applications*

Test Intervals

When working with inequalities involving polynomials, it is important to realize that the value of a polynomial can change signs only at its **zeros.** That is, a polynomial can change signs only at the x-values that make the polynomial zero. For instance, the first-degree polynomial $x + 2$ has a zero at -2, and it changes sign at that zero.

$x + 2 < 0$ for $x < -2$		Polynomial is negative
$x + 2 = 0$ for $x = -2$		Zero of polynomial
$x + 2 > 0$ for $x > -2$		Polynomial is positive

You can picture this result on the real number line as shown in Figure 7.21.

FIGURE 7.21

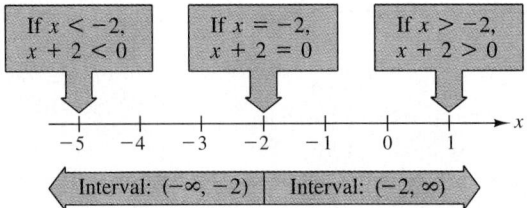

Note in Figure 7.21 that the polynomial $x + 2$ has one zero (given by $x = -2$), and that this zero partitions the real number line into two **test intervals.** The value of the polynomial is negative for *every* x-value in the first test interval $(-\infty, -2)$, and the value of the polynomial is positive for *every* x-value in the second test interval $(-2, \infty)$.

You can use the same basic approach to determine the test intervals for any polynomial.

Finding Test Intervals for a Polynomial

To determine the intervals on which the values of a polynomial are entirely negative or entirely positive, use the following steps.

1. Find all real zeros of the polynomial, and arrange the zeros in increasing order (from smallest to largest). The real zeros of a polynomial are called its **critical numbers.**

2. Use the critical numbers of the polynomial to determine its **test intervals.**

3. Choose one representative x-value in each test interval and evaluate the polynomial at that value. If the value of the polynomial is negative, the polynomial will have negative values for *every* x-value in the interval. If the value of the polynomial is positive, the polynomial will have positive values for *every* x-value in the interval.

EXAMPLE 1 ■ Finding Test Intervals for a Quadratic Polynomial

Determine the intervals on which the following polynomial is entirely negative or entirely positive.

$$x^2 - x - 6$$

Solution

By factoring the given quadratic as

$$x^2 - x - 6 = 0$$

$$(x + 2)(x - 3) = 0,$$

Discuss how to choose test points. For instance, in Example 1, see if students can come up with an alternate set of test points. Ask if this new selection changes the solution to the problem.

you can see that the critical numbers occur at $x = -2$ and $x = 3$. Therefore, the test intervals for the quadratic are

$$(-\infty, -2), \quad (-2, 3), \quad \text{and} \quad (3, \infty). \qquad \text{Test intervals}$$

In each test interval, choose a representative x-value and evaluate the polynomial, as shown in Table 7.3.

TABLE 7.3

Test Interval	Representative x-value	Value of Polynomial	Conclusion
$(-\infty, -2)$	$x = -3$	$(-3)^2 - (-3) - 6 = 6$	Polynomial is positive
$(-2, 3)$	$x = 0$	$(0)^2 - (0) - 6 = -6$	Polynomial is negative
$(3, \infty)$	$x = 4$	$(4)^2 - (4) - 6 = 6$	Polynomial is positive

Therefore, the polynomial has positive values for every x in the intervals $(-\infty, -2)$ and $(3, \infty)$, and negative values for every x in the interval $(-2, 3)$. This result is shown graphically in Figure 7.22.

FIGURE 7.22

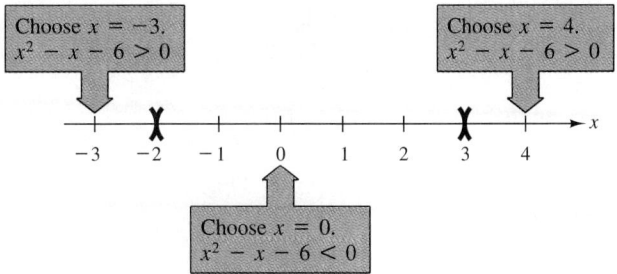

Choose $x = -3$.
$x^2 - x - 6 > 0$

Choose $x = 4$.
$x^2 - x - 6 > 0$

Choose $x = 0$.
$x^2 - x - 6 < 0$

■

FIGURE 7.23

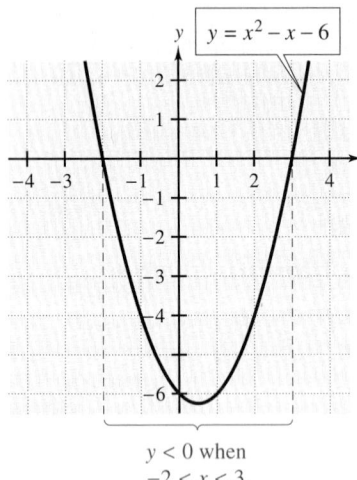

$y < 0$ when
$-2 < x < 3$.

NOTE Another way to determine the intervals on which $x^2 - x - 6$ is entirely negative or entirely positive is to sketch the graph of $y = x^2 - x - 6$, as shown in Figure 7.23. The portions of the graph that lie above the x-axis correspond to $x^2 - x - 6 > 0$ and the portion that lies below the x-axis corresponds to $x^2 - x - 6 < 0$.

Quadratic Inequalities

In Section 3.3, you studied methods for solving linear inequalities of the form

$$ax + b < 0, \quad ax + b \leq 0, \quad ax + b > 0, \quad \text{or} \quad ax + b \geq 0.$$

You will now study methods for solving **quadratic inequalities** of the form

$$ax^2 + bx + c < 0, \quad ax^2 + bx + c \leq 0, \quad ax^2 + bx + c > 0, \quad \text{or}$$

$$ax^2 + bx + c \geq 0.$$

To solve such an inequality, use the guidelines for finding the intervals on which a polynomial is entirely negative or entirely positive. This procedure is demonstrated in Example 2.

EXAMPLE 2 ■ Solving a Quadratic Inequality

Solve the quadratic inequality

$$x^2 - 5x < 0.$$

Solution

$$x^2 - 5x < 0 \qquad\qquad \text{Given inequality}$$

$$x(x - 5) < 0 \qquad\qquad \text{Factor}$$

Critical numbers: $x = 0, x = 5$

Test intervals: $(-\infty, 0), (0, 5), (5, \infty)$

Test: Is $x(x - 5) < 0$?

To test an interval, first choose a convenient number in the interval and then compute the sign of $x(x - 5)$. The results are shown in Figure 7.24.

FIGURE 7.24

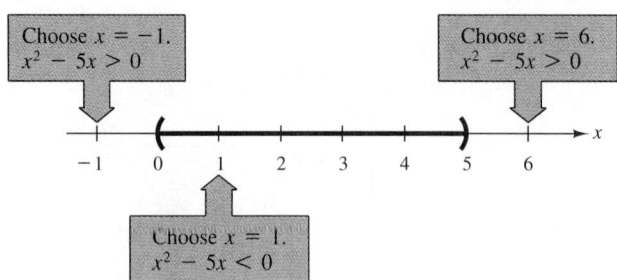

Choose $x = -1$.
$x^2 - 5x > 0$

Choose $x = 6$.
$x^2 - 5x > 0$

Choose $x = 1$.
$x^2 - 5x < 0$

Because the inequality $x^2 - 5x < 0$ is satisfied only by the middle test interval, you can conclude that the solution set of the inequality is the interval

$(0, 5)$.   Solution set   ∎

In Example 2, note that you could have used the same basic procedure if the inequality had been written using the symbols $\leq$, $>$, or $\geq$. For instance, in Figure 7.24 you can see that the solution set of the inequality

$$x^2 - 5x \geq 0$$

consists of the union of the half-open intervals $(-\infty, 0]$ and $[5, \infty)$, which can be written as

$(-\infty, 0] \cup [5, \infty)$.   Solution set

Just as in solving quadratic *equations,* the first step in solving a quadratic *inequality* is to write the inequality in **standard form,** with the polynomial on the left and zero on the right, as demonstrated in Example 3.

EXAMPLE 3   ∎   Solving a Quadratic Inequality

Solve the quadratic inequality

$$2x^2 + 5x > 12.$$

Solution

$2x^2 + 5x > 12$	Given inequality
$2x^2 + 5x - 12 > 0$	Write in standard form
$(x + 4)(2x - 3) > 0$	Factor

Some students may try to solve the quadratic inequality $(x + 4)(2x - 3)$ > 0 by setting each factor greater than 0, as $x + 4 > 0$ and $2x - 3 > 0$, similar to the way they solved quadratic equations by factoring. Point out the error in this reasoning.

Critical numbers: $x = -4,\ x = \dfrac{3}{2}$

Test intervals: $(-\infty, -4),\ \left(-4, \dfrac{3}{2}\right),\ \left(\dfrac{3}{2}, \infty\right)$

Test: Is $(x + 4)(2x - 3) > 0$?

After testing these intervals, as shown in Figure 7.25 on the following page, you can see that the polynomial $2x^2 + 5x - 12$ is positive in the open intervals $(-\infty, -4)$ and $\left(\dfrac{3}{2}, \infty\right)$. Therefore, the solution set of the inequality is

$(-\infty, -4) \cup \left(\dfrac{3}{2}, \infty\right)$.   Solution set

FIGURE 7.25

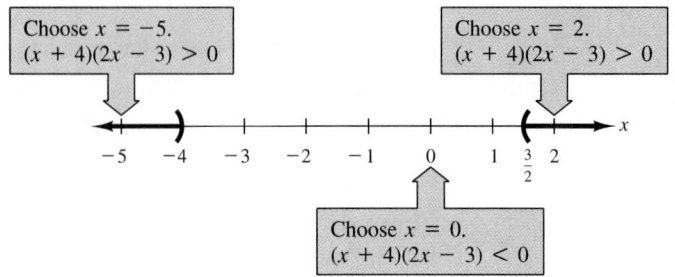

Example 4 shows how the Quadratic Formula can be used to solve a quadratic inequality.

EXAMPLE 4 ■ **Using the Quadratic Formula**

Solve the following quadratic inequality.

$$x^2 - 2x - 1 \le 0$$

Solution

Point out that "does not factor" does not mean "no solution."

Because the quadratic $x^2 - 2x - 1$ does not factor (using integer coefficients), you can use the Quadratic Formula to find its zeros. That is, the solutions of $x^2 - 2x - 1 = 0$ are given by

$$x = \frac{-(-2) \pm \sqrt{(-2)^2 - 4(1)(-1)}}{2(1)} = \frac{2 \pm \sqrt{8}}{2} = \frac{2 \pm 2\sqrt{2}}{2} = 1 \pm \sqrt{2}.$$

Now, having found the zeros of the polynomial $x^2 - 2x - 1$, you can proceed as usual.

Critical numbers: $x = 1 - \sqrt{2}$, $x = 1 + \sqrt{2}$

Test intervals: $\left(-\infty, 1 - \sqrt{2}\right)$, $\left(1 - \sqrt{2}, 1 + \sqrt{2}\right)$, $\left(1 + \sqrt{2}, \infty\right)$

Test: Is $x^2 - 2x - 1 \le 0$?

You may suggest that students convert $1 - \sqrt{2}$ and $1 + \sqrt{2}$ to decimal form to help in choosing test points.

After testing these intervals, as shown in Figure 7.26 on the following page, you can see that the polynomial $x^2 - 2x - 1$ is less than or equal to zero in the *closed* interval $\left[1 - \sqrt{2}, 1 + \sqrt{2}\right]$. Therefore, the solution set of the inequality is

$$\left[1 - \sqrt{2}, 1 + \sqrt{2}\right]. \qquad \text{Solution set}$$

Note that $1 - \sqrt{2} \approx -0.414$ and $1 + \sqrt{2} \approx 2.414$.

FIGURE 7.26

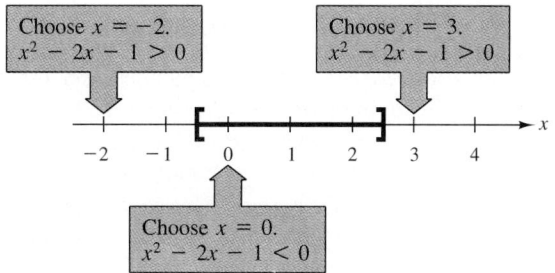

When solving a quadratic inequality, be sure you have accounted for the particular type of inequality symbol given in the inequality. For instance, in Example 4, note that the solution is a closed interval because the original inequality contained a "*less than or equal to*" symbol. If the original inequality had been $x^2 - 2x - 1 < 0$, the solution would have been the *open* interval $\left(1 - \sqrt{2}, 1 + \sqrt{2}\right)$.

So far in this section, each of the quadratic inequalities has had a solution set that consisted of a single interval or the union of two intervals. When solving the exercises for this section, you should be on the watch for some unusual solution sets, as illustrated in Example 5.

EXAMPLE 5 ■ Unusual Solution Sets

Solve the following inequalities to verify that the given solution set is correct.

(a) The solution set of the quadratic inequality

$$x^2 + 2x + 4 > 0$$

consists of the entire set of real numbers, $(-\infty, \infty)$. In other words, the quadratic $x^2 + 2x + 4$ is positive for every real value of x. (Note that this quadratic inequality has *no* critical numbers, because the equation $x^2 + 2x + 4 = 0$ has *no* real solutions. In such a case, there is only one test interval—the entire real line.)

(b) The solution set of the quadratic inequality

$$x^2 + 2x + 1 \leq 0$$

consists of the single real number $\{-1\}$.

(c) The solution set of the quadratic inequality

$$x^2 + 3x + 5 < 0$$

is empty. In other words, the quadratic $x^2 + 3x + 5$ is not less than zero for any value of x.

(d) The solution set of the quadratic inequality

$$x^2 - 4x + 4 > 0$$

consists of all real numbers *except* the number 2. In interval notation, this solution set can be written as $(-\infty, 2) \cup (2, \infty)$. ■

Rational Inequalities

The concepts of critical numbers and test intervals can be extended to inequalities involving rational expressions. To do this, use the fact that the value of a rational expression can change sign only at its *real zeros* (the real x-values for which its numerator is zero) and its *undefined values* (the real x-values for which its denominator is zero). These two types of numbers make up the **critical numbers** of a rational inequality. For instance, the critical numbers of the inequality

$$\frac{x - 2}{(x - 1)(x + 3)} < 0$$

are $x = 2$ (the numerator is zero), and $x = 1$ and $x = -3$ (the denominator is zero). From these three critical numbers, you can see that the given inequality has *four* test intervals:

$$(-\infty, -3), \quad (-3, 1), \quad (1, 2), \quad \text{and} \quad (2, \infty).$$

EXAMPLE 6 ■ Solving a Rational Inequality

Solve the rational inequality

$$\frac{x}{x - 2} > 0.$$

Solution

Critical numbers: $x = 0, x = 2$

Test intervals: $(-\infty, 0), (0, 2), (2, \infty)$

Test: Is $\dfrac{x}{x - 2} > 0$?

After testing these intervals, as shown in Figure 7.27, you can see that the rational expression $x/(x - 2)$ is positive in the open intervals $(-\infty, 0)$ and $(2, \infty)$. Therefore, the solution set of the inequality is

$$(-\infty, 0) \cup (2, \infty). \qquad \text{Solution set}$$

FIGURE 7.27

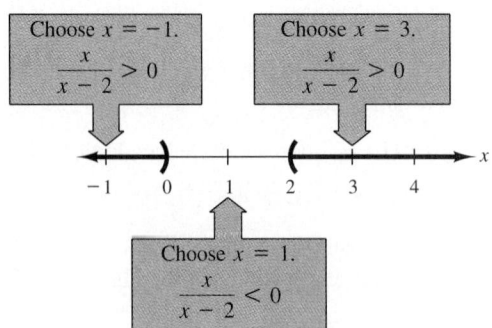

Choose $x = -1$.

$$\frac{x}{x - 2} > 0$$

Choose $x = 3$.

$$\frac{x}{x - 2} > 0$$

Choose $x = 1$.

$$\frac{x}{x - 2} < 0$$

NOTE You can confirm the result in Example 6 by sketching the graph of $y = x/(x-2)$, as shown in Figure 7.28.

FIGURE 7.28

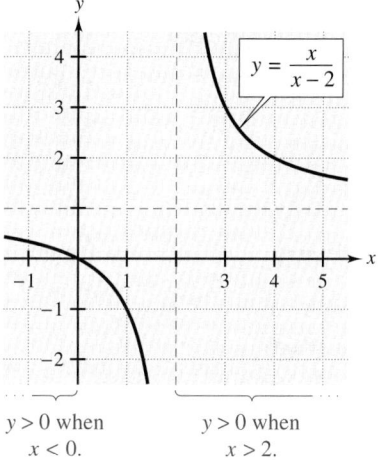

$y > 0$ when $x < 0$.

$y > 0$ when $x > 2$.

In the next example, note that the original inequality is not written in standard form, with the rational expression on the left (written as a *single* fraction) and zero on the right. The first step in solving such an inequality is to write it in standard form.

EXAMPLE 7   ■   Writing in Standard Form First

Solve the rational inequality

$$\frac{2x}{x+3} < 4.$$

Solution

$$\frac{2x}{x+3} < 4 \qquad\qquad \text{Given inequality}$$

$$\frac{2x}{x+3} - 4 < 0 \qquad\qquad \text{Subtract 4 from both sides}$$

$$\frac{2x - 4(x+3)}{x+3} < 0 \qquad\qquad \text{Subtract fractions}$$

$$\frac{-2x - 12}{x+3} < 0 \qquad\qquad \text{Standard form}$$

$$\frac{-2(x+6)}{x+3} < 0 \qquad\qquad \text{Factor}$$

Critical numbers: $x = -6, x = -3$

Test intervals: $(-\infty, -6), (-6, -3), (-3, \infty)$

Test: Is $\dfrac{-2(x+6)}{x+3} < 0$?

After testing these intervals, as shown in Figure 7.29, you can see that the rational expression $-2(x + 6)/(x + 3)$ is negative in the open intervals $(-\infty, -6)$ and $(-3, \infty)$. Therefore, the solution set of the inequality is

$$(-\infty, -6) \cup (-3, \infty). \qquad \text{Solution set}$$

FIGURE 7.29

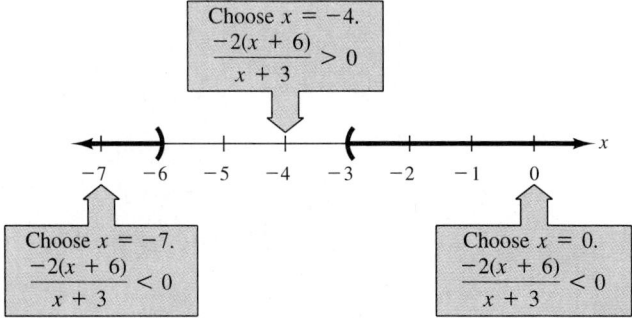

Applications

Examples 8 and 9 describe some applications of quadratic inequalities.

EXAMPLE 8 ■ The Height of a Projectile

A projectile is fired straight up from ground level with an initial velocity of 256 feet per second, so that its height h at any time t is

$$h = -16t^2 + 256t,$$

where h is in feet and t is in seconds. During what interval of time will the height of the projectile exceed 960 feet?

Solution

To solve this problem, you can solve the inequality given by

$$\text{Height} > 960$$
$$-16t^2 + 256t > 960.$$

Using the techniques described in this section, you can solve this inequality as follows.

$$-16t^2 + 256t > 960 \qquad \text{Given inequality}$$
$$-16t^2 + 256t - 960 > 0 \qquad \text{Subtract 960 from both sides}$$
$$t^2 - 16t + 60 < 0 \qquad \text{Divide both sides by } -16 \text{ and reverse inequality}$$
$$(t - 6)(t - 10) < 0 \qquad \text{Factor}$$

FIGURE 7.30

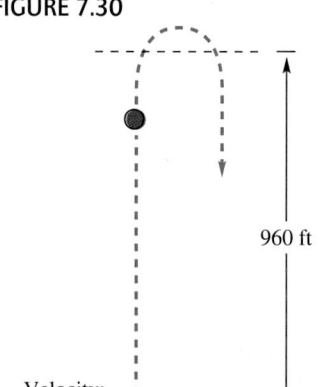

Velocity:
256 ft/sec

960 ft

Thus, the critical numbers are $t = 6$ and $t = 10$. By testing the intervals $(-\infty, 6)$, $(6, 10)$, and $(10, \infty)$, you can determine that the solution interval is $(6, 10)$. Therefore, the height of the object will exceed 960 feet during the time interval given by

$$6 \text{ seconds } < t < 10 \text{ seconds}.$$

(See Figure 7.30.)

To help students review, you may want to ask them to determine the maximum height the projectile attains. The answer, 1024 feet, can be obtained algebraically or graphically.

■

EXAMPLE 9   ■   **Maintaining a Profit Level**

A bakery sells decorated cakes for a profit of

$$P = 5x - \frac{1}{100}x^2,$$

where P is the profit in dollars and x is the number of cakes sold. If the bakery wants its weekly profit from cakes to exceed \$600, how many cakes must it sell each week?

Solution

$600 < P$	Profit must exceed \$600
$600 < 5x - \dfrac{1}{100}x^2$	Profit equation
$\dfrac{1}{100}x^2 - 5x + 600 < 0$	Standard form
$x^2 - 500x + 60{,}000 < 0$	Multiply both sides by 100
$(x - 200)(x - 300) < 0$	Factor

Thus, the critical numbers are $x = 200$ and $x = 300$. By testing the intervals $(-\infty, 200)$, $(200, 300)$, and $(300, \infty)$, you can determine that the solution interval is $(200, 300)$. Therefore, the profit will exceed \$600 provided the production level of cakes falls in the interval

$$200 \text{ cakes } < x < 300 \text{ cakes}. \qquad ■$$

The solution of Example 9 might seem a little unusual, because you might think that the more cakes the bakery sells the more profit it makes. However, in this example the profit actually falls below \$600 when the bakery sells more than

300 cakes. If you take a course in economics, you will see that this type of situation often occurs in business. The reason is that as a business begins to saturate the market, the only way it can increase sales volume is to lower its price per unit.

DISCUSSION PROBLEM   ■   A Graphing Calculator Experiment

FIGURE 7.31

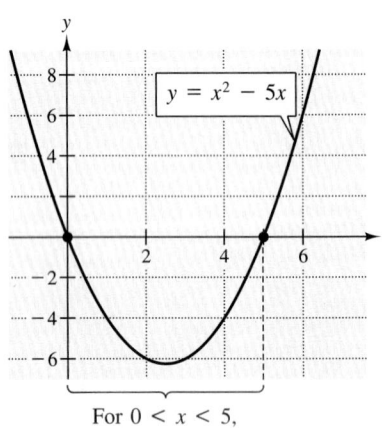

For $0 < x < 5$,
y is negative.

To solve graphically the inequality in Example 2,

$$x^2 - 5x < 0,$$

you can sketch the graph of $y = x^2 - 5x$, as shown in Figure 7.31. In this figure, you can see that the only part of the curve that lies below the x-axis is the portion for which $0 < x < 5$. Thus, the solution of $x^2 - 5x < 0$ is $0 < x < 5$.

Try using this graphing approach to solve some of the other examples in this section.

■

Warm-Up

The following warm-up exercises involve skills that were covered in earlier sections. You will use these skills in the exercise set for this section.

In Exercises 1–6, solve the inequality and sketch the graph of the solution on the real number line.

1. $3x - 15 \leq 0$ **2.** $3x + 5 \geq 0$

3. $7 - 3x > 4 - x$ **4.** $2(x + 6) - 20 < 2$

5. $|x - 3| < 2$ **6.** $|x - 5| > 3$

In Exercises 7–10, determine whether the product is zero, positive, or negative. (It is not necessary to actually find the product.)

7. $(35 - 10)(13 - 25)$ **8.** $(254 - 254)(625 - 500)$

9. $(64 - 82)(12 - 15)$ **10.** $\left(\frac{3}{4} - \frac{3}{5}\right)\left(\frac{1}{6} - \frac{2}{3}\right)$

7.6 EXERCISES

means that a graphing utility can help you solve the exercise or check your solution.

In Exercises 1–6, find the critical numbers of the polynomial and locate them on the real number line.

1. $2x - 6$

2. $4x + 6$

3. $x(2x - 5)$

4. $5x(x - 3)$

5. $x^2 - 4x + 3$

6. $3x^2 - 2x - 8$

In Exercises 7–12, determine the intervals on which the polynomial is entirely negative and entirely positive. Show the intervals graphically on the real number line.

7. $x - 4$

8. $3 - x$

9. $2x(x - 4)$

10. $7x(3 - x)$

11. $x^2 - 4x - 5$

12. $2x^2 - 4x - 3$

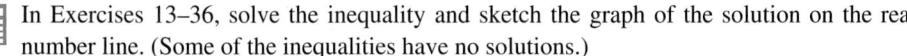 In Exercises 13–36, solve the inequality and sketch the graph of the solution on the real number line. (Some of the inequalities have no solutions.)

13. $2x + 6 \geq 0$

14. $5x - 20 < 0$

15. $3x(x - 2) < 0$

16. $2x(x - 6) > 0$

17. $3x(2 - x) < 0$

18. $2x(6 - x) > 0$

19. $x^2 - 4 \geq 0$

20. $z^2 \leq 9$

21. $x^2 + 3x \leq 10$

22. $t^2 - 15t + 50 < 0$

23. $-2u^2 + 7u + 4 < 0$

24. $-3x^2 - 4x + 4 \leq 0$

25. $x^2 + 4x + 5 < 0$

26. $x^2 + 6x + 10 > 0$

27. $x^2 + 2x + 1 \geq 0$

28. $25 \geq (x - 3)^2$

29. $x^2 - 4x + 2 > 0$

30. $-x^2 + 8x - 11 \leq 0$

31. $(x - 5)^2 < 0$

32. $(y + 3)^2 \geq 0$

33. $6 - (x - 5)^2 < 0$

34. $(y + 3)^2 - 6 \geq 0$

35. $x(x - 2)(x + 2) > 0$

36. $x^2(x - 2) \leq 0$

In Exercises 37–40, determine the critical numbers of the rational expression and locate them on the real number line.

37. $\dfrac{5}{x - 3}$

38. $\dfrac{-6}{x + 2}$

39. $\dfrac{2x}{x + 5}$

40. $\dfrac{x - 2}{x - 10}$

In Exercises 41–52, solve the rational inequality and sketch the graph of the solution on the real number line.

41. $\dfrac{5}{x - 3} > 0$

42. $\dfrac{3}{4 - x} > 0$

43. $\dfrac{-5}{x - 3} > 0$

44. $\dfrac{-3}{4 - x} > 0$

45. $\dfrac{x}{x - 4} \leq 0$

46. $\dfrac{z - 1}{z + 3} < 0$

47. $\dfrac{y - 3}{2y - 11} \geq 0$

48. $\dfrac{x + 5}{3x + 2} \geq 0$

49. $\dfrac{6}{x-4} > 2$

50. $\dfrac{1}{x+2} > -3$

51. $\dfrac{x}{x-3} \le 2$

52. $\dfrac{x+4}{x-5} \ge 10$

53. *Height of a Projectile* A projectile is fired straight up from ground level with an initial velocity of 128 feet per second, so that its height h at any time t is

$$h = -16t^2 + 128t,$$

where h is in feet and t is in seconds. During what interval of time will the height of the projectile exceed 240 feet?

54. *Height of a Projectile* A projectile is fired straight up from ground level with an initial velocity of 88 feet per second, so that its height h at any time t is

$$h = -16t^2 + 88t,$$

where h is in feet and t is in seconds. During what interval of time will the height of the projectile exceed 50 feet?

55. *Length of Fencing* Suppose you have 64 feet of fencing to enclose a rectangular region. Determine the interval for the length such that the area will exceed 240 square feet. [*Hint:* The area is given by $A = l(32 - l)$, where l is the length of the rectangle.]

56. *Length of a Field* A rectangular playing field (see figure) with a perimeter of 100 meters is to have an area of at least 500 square meters. Within what bounds must the length of the field lie?

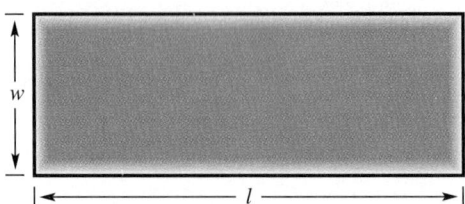

57. *Annual Interest Rate* You plan to invest $1000 in a certificate of deposit for 2 years, and you want the interest for that time period to exceed $150. Find the annual interest rate that must be exceeded if the interest is compounded annually. [*Hint:* You must solve the inequality $1000(1 + r)^2 > 1150$.]

58. *Company Profits* The revenue and cost equations for a product are

$$R = x(50 - 0.0002x)$$
$$C = 12x + 150{,}000,$$

where R and C are in dollars and x represents the number of units sold (see figure). How many units must be sold to obtain a profit of at least $1,650,000? (*Hint:* $P = R - C$)

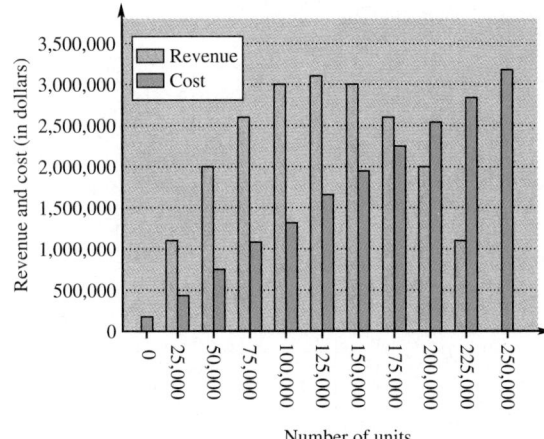

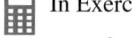

7   REVIEW EXERCISES

In Exercises 1–12, solve the quadratic equation by factoring.

1. $x^2 + 12x = 0$

2. $u^2 - 18u = 0$

3. $3z(z + 10) - 8(z + 10) = 0$

4. $7x(2x - 9) + 4(2x - 9) = 0$

5. $4y^2 - 1 = 0$

6. $2z^2 - 72 = 0$

7. $4y^2 + 20y + 25 = 0$

8. $x^2 + \frac{8}{3}x + \frac{16}{9} = 0$

9. $t^2 - t - 20 = 0$

10. $z^2 + \frac{2}{3}z - \frac{8}{9} = 0$

11. $2x^2 - 2x - 180 = 0$

12. $15x^2 - 30x - 45 = 0$

In Exercises 13–18, solve the quadratic equation by extracting square roots.

13. $x^2 = 10{,}000$

14. $x^2 = 98$

15. $y^2 - 2.25 = 0$

16. $y^2 - 8 = 0$

17. $(x - 16)^2 = 400$

18. $(x + 3)^2 = 0.04$

In Exercises 19–24, solve the quadratic equation by completing the square. (Find all real *and* imaginary solutions.)

19. $x^2 - 6x - 3 = 0$

20. $x^2 + 12x + 6 = 0$

21. $x^2 - 3x + 3 = 0$

22. $t^2 + \frac{1}{2}t - 1 = 0$

23. $2y^2 + 10y + 3 = 0$

24. $3x^2 - 2x + 2 = 0$

In Exercises 25–30, solve the quadratic equation by using the Quadratic Formula. (Find all real *and* imaginary solutions.)

25. $y^2 + y - 30 = 0$

26. $x^2 - x - 72 = 0$

27. $2y^2 + y - 21 = 0$

28. $2x^2 - 3x - 20 = 0$

29. $0.3t^2 - 2t + 5 = 0$

30. $-u^2 + 2.5u + 3 = 0$

In Exercises 31–42, solve the equation by the method of your choice. (Find all real *and* imaginary solutions.)

31. $(v - 3)^2 = 250$

32. $x^2 - 36x = 0$

33. $-x^2 + 5x + 84 = 0$

34. $9x^2 + 6x + 1 = 0$

35. $(x - 9)^2 - 121 = 0$

36. $60 - (x - 6)^2 = 0$

37. $z^2 - 6z + 10 = 0$

38. $z^2 - 14z + 5 = 0$

39. $2y^2 + 3y + 1 = 0$

40. $0.25y^2 + 0.35y - 0.50 = 0$

41. $\dfrac{1}{x} + \dfrac{1}{x + 1} = \dfrac{1}{2}$

42. $x - 5 = \sqrt{x - 2}$

In Exercises 43–46, sketch the graph of the quadratic function. Identify the vertex and any *x*-intercepts.

43. $y = 9 - (x - 3)^2$

44. $y = (x - 2)^2 + 4$

45. $y = x^2 - 8x + 12$

46. $y = x^2 + 4x + 2$

In Exercises 47–50, find the general form, $y = ax^2 + bx + c$, of the equation of the parabola satisfying the given criteria.

47. Vertex: $(3, 5)$; $a = -2$

48. Vertex: $(-2, 3)$; $a = 3$

49. Vertex: $(5, 0)$; Passes through the point $(1, 1)$

50. Vertex: $(-2, 5)$; Passes through the point $(0, 1)$

In Exercises 51–60, solve the inequality and sketch the graph of the solution on the real number line. (Some of the inequalities have no solutions.)

51. $4x - 12 < 0$

52. $3(x + 2) > 0$

53. $5x(7 - x) > 0$

54. $-2x(x - 10) \leq 0$

55. $16 - (x - 2)^2 \leq 0$

56. $(x - 5)^2 - 36 > 0$

57. $2x^2 + 3x - 20 < 0$

58. $3x^2 - 2x - 8 > 0$

59. $\dfrac{x}{2x - 7} \geq 0$

60. $\dfrac{2x - 9}{x - 1} \leq 0$

61. *Number Problem* Find two consecutive positive integers such that the sum of their squares is 265.

62. *Number Problem* Find two consecutive positive integers with a product of 156.

63. *Falling Time* The height h (in feet) of an object above the ground is

$$h = 200 - 16t^2, \quad t > 0,$$

where t is time in seconds. Find the time at which the object strikes the ground.

64. *Falling Time* The height h (in feet) of an object above the ground is

$$h = -16t^2 + 64t + 192, \quad t > 0,$$

where t is time in seconds. Find the time at which the object strikes the ground.

65. *Exploratory Exercise* The perimeter of a rectangle of length l and width w is 48 feet (see figure).

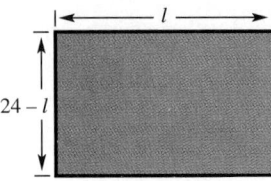

(a) Show that $w = 24 - l$.
(b) Show that the area A is given by $A = lw = l(24 - l)$.
(c) Complete the table by using the equation in part (b).

l	2	4	6	8	10	12	14	16	18
A									

(d) Describe the rectangle that has the largest area.

66. *Dimensions of a Triangle* Find the dimensions of a triangle if its height is $1\frac{2}{3}$ times its base and its area is 3000 square inches.

67. *Decreased Price* A Little League baseball team obtained a block of tickets for a ball game for $96. The block contained three more tickets than the team needed for its members. By inviting three more people to attend (and share the cost), the team lowered the price per ticket by $1.60. How many people are going to the game?

68. *Average Speed* A train travels the first 220 miles of a trip at one speed and the last 180 miles at an average speed of 5 miles per hour faster. If the entire trip takes 7 hours, find the higher of the two speeds.

69. *Work Rate Problem* Working together, two people can complete a task in 6 hours. Working alone, how long would it take each to do the task if one person takes 3 hours longer than the other?

70. *Path of a Projectile* The height y (in feet) of a projectile is

$$y = -\frac{1}{16}x^2 + 5x,$$

where x is the horizontal distance (in feet) from where the projectile was launched.

(a) Sketch the path of the projectile.

(b) How high is the projectile when it is at its maximum height?

(c) How far from the launch point will the projectile strike the ground?

 71. *Percentage of College Graduates* The percentage of the American population that graduated from college between 1940 and 1987 is approximated by the model

Percentage of graduates $= 5.136 + 0.0069t^2,\ 0 \le t,$

where the time t represents the calendar year, with $t = 0$ corresponding to 1940 (see figure). According to this model, when did the percentage of college graduates exceed 25% of the population? (*Source:* U.S. Bureau of Census)

72. Find a quadratic equation with the given solutions.

(a) $x = -3,\ x = 3$

(b) $x = -\frac{1}{2},\ x = 2$

(c) $x = -5,\ x = -1$

(d) $x = -10,\ x = 8$

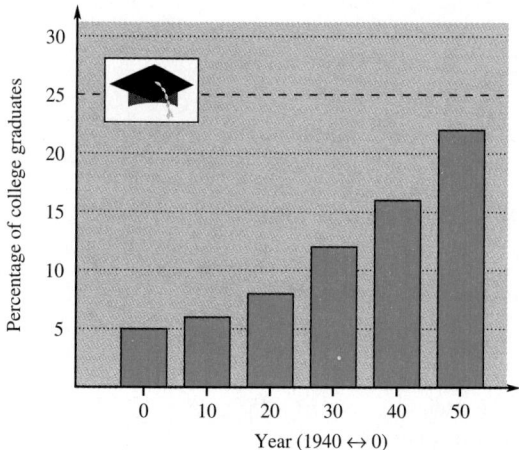

Take this test as you would take a test in class. After you are done, check your work against the answers in the back of the book.

1. Solve by factoring: $x(x+5) - 10(x+5) = 0$

2. Solve by factoring: $8x^2 - 21x - 9 = 0$

3. Solve by extracting square roots: $(x-2)^2 = 0.09$

4. Solve by extracting square roots: $(x+3)^2 + 81 = 0$

5. Find the real number c such that $x^2 - 3x + c$ is a perfect square trinomial.

6. Solve by completing the square: $2x^2 - 6x + 3 = 0$

7. Find the discriminant and use it to determine the type of the solutions of the quadratic equation $5x^2 - 12x + 10 = 0$.

8. Solve by using the Quadratic Formula: $3x^2 - 8x + 3 = 0$

9. Solve and round the result to two decimal places: $\sqrt{3x+7} - \sqrt{x+12} = 1$

10. Find a quadratic equation having the solutions -4 and 5.

11. Sketch the graph of the parabola given by the equation $y = -2(x-2)^2 + 8$. Identify the coordinates of the vertex and any x-intercepts of the graph.

12. Find the general form of the equation of the parabola that has its vertex at $(3, -2)$ and that passes through the point $(0, 4)$.

In Exercises 13–15, solve and sketch the solution on the real number line.

13. $2x(x-3) < 0$

14. $16 \le (x-2)^2$

15. $\dfrac{3}{x-2} > 4$

16. Find two consecutive positive integers whose product is 210.

17. The width of a rectangle is 8 feet less than the length. The area of the rectangle is 240 square feet. Find the dimensions of the rectangle.

18. A train traveled the first 125 miles of a trip at one speed and the last 180 miles at an average speed of 5 miles per hour slower. If the total time for the trip was $6\frac{1}{2}$ hours, what was the average speed for the first part of the trip?

19. The height of an object dropped from an initial height of 75 feet is Height $= -16t^2 + 75$, where the height is measured in feet and the time t is measured in seconds. How long will it take the object to fall to a height of 35 feet?

20. The revenue R for a charter bus company is

$$R = -\frac{1}{20}(n^2 - 240n), \quad 80 \le n \le 160,$$

where n is the number of passengers that charter buses for a particular trip. Find the number of passengers that will give the bus company maximum revenue.

Additional Functions and Relations

8.1   Combinations of Functions

8.2   Inverse Functions

8.3   Variation and Mathematical Models

8.4   Polynomial Functions and Their Graphs

8.5   Circles, Ellipses, and Hyperbolas

8.6   Nonlinear Systems of Equations

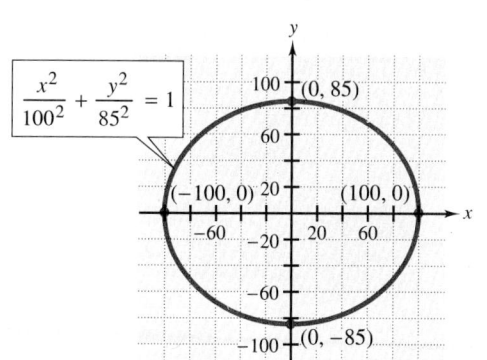

*A*ustralian football (or rugby) is played on elliptical fields. The field can be a maximum of 170 yards wide and 200 yards long. If the origin is the center of the field, then the equation of the elliptical boundary of a field of maximum dimensions is

$$\frac{x^2}{100^2} + \frac{y^2}{85^2} = 1.$$

The graph of this equation is shown at left. Consider a different playing field whose boundary is given by

$$\frac{x^2}{90^2} + \frac{y^2}{75^2} = 1.$$

How wide is this field? How long is it?

SECTION	**Combinations of Functions**
8.1	*Arithmetic Combinations of Functions* • *Composition of Functions*

Arithmetic Combinations of Functions

In Section 2.5, you were introduced to the concept of a function. In this and the next section, you will learn more about functions.

Just as two real numbers can be combined by the operations of addition, subtraction, multiplication, and division to form other real numbers, two functions can be combined to form new functions. For example, if

$$f(x) = 3x - 1 \quad \text{and} \quad g(x) = x^2 - 4,$$

you can form the sum, difference, product, and quotient of f and g as follows.

$$f(x) + g(x) = (3x - 1) + (x^2 - 4) = x^2 + 3x - 5 \qquad \text{Sum}$$
$$f(x) - g(x) = (3x - 1) - (x^2 - 4) = -x^2 + 3x + 3 \qquad \text{Difference}$$
$$f(x)g(x) = (3x - 1)(x^2 - 4) = 3x^3 - x^2 - 12x + 4 \qquad \text{Product}$$
$$\frac{f(x)}{g(x)} = \frac{3x - 1}{x^2 - 4}, \qquad x \neq \pm 2 \qquad \text{Quotient}$$

Students often think $(fg)(x)$ means f times g times x. Make sure they understand this notation.

The domain of an arithmetic combination of functions f and g consists of all real numbers that are common to the domains of f and g. In the case of the quotient $f(x)/g(x)$, there is the further restriction that $g(x) \neq 0$.

Definition of Sum, Difference, Product, and Quotient of Functions

Let f and g be two functions with overlapping domains. Then for all x common to both domains, the **sum, difference, product,** and **quotient** of f and g are defined as follows.

1. *Sum:* $(f + g)(x) = f(x) + g(x)$

2. *Difference:* $(f - g)(x) = f(x) - g(x)$

3. *Product:* $(fg)(x) = f(x) \cdot g(x)$

4. *Quotient:* $\left(\dfrac{f}{g}\right)(x) = \dfrac{f(x)}{g(x)}, \qquad g(x) \neq 0$

EXAMPLE 1 ■ Finding the Sum and Difference of Two Functions

Given $f(x) = 4x - 3$ and $g(x) = 2x^2 + x$, find the following.

(a) $(f + g)(x)$ (b) $(f - g)(x)$

Then evaluate each of these combinations when $x = 3$.

Have students evaluate $f(3)$ and $g(3)$ and then calculate $f(3) + g(3)$. Have them compare their results with $(f + g)(3)$ in Example 1a.

Solution

(a) The sum of the functions f and g is given by

$$(f + g)(x) = f(x) + g(x)$$
$$= (4x - 3) + (2x^2 + x)$$
$$= 2x^2 + 5x - 3.$$

When $x = 3$, the value of this sum is

$$(f + g)(3) = 2\left(3^2\right) + 5(3) - 3 = 30.$$

(b) The difference of the functions f and g is given by

$$(f - g)(x) = f(x) - g(x)$$
$$= (4x - 3) - (2x^2 + x)$$
$$= -2x^2 + 3x - 3.$$

When $x = 3$, the value of this difference is

$$(f - g)(3) = -2(3^2) + 3(3) - 3 = -12.$$ ■

NOTE In Example 1, $(f + g)(3)$ also could have been evaluated as

$$(f + g)(3) = f(3) + g(3)$$
$$= [4(3) - 3] + [2(3)^2 + 3]$$
$$= 9 + 21$$
$$= 30.$$

Similarly, $(f - g)(3)$ could have been evaluated as

$$(f - g)(3) = f(3) - g(3) = 9 - 21 = -12.$$

In Example 1, both f and g have domains that consist of all real numbers. Thus, the domains of both $(f + g)$ and $(f - g)$ are also the set of all real numbers. In the next example, the domain of the sum of f and g is smaller than the domains of f and g.

EXAMPLE 2 ■ Finding the Domain of the Sum of Two Functions

Given the functions $f(x) = \sqrt{x + 1}$ and $g(x) = \sqrt{1 - x}$, find the sum of f and g. Then find the domain of $(f + g)$.

Solution

The sum of f and g is given by

$$(f + g)(x) = f(x) + g(x)$$
$$= \sqrt{x + 1} + \sqrt{1 - x}.$$

Now, because f is defined for $x + 1 \geq 0$ and g is defined for $1 - x \geq 0$, it follows that the domains of f and g are as follows.

$$(\text{Domain of } f) = [-1, \infty)$$
$$(\text{Domain of } g) = (-\infty, 1]$$

The set that is common to both of these domains is the closed interval $[-1, 1]$, as shown in Figure 8.1. Therefore, the domain of $(f + g)$ is

$$(\text{Domain of } f + g) = [-1, 1].$$ ■

FIGURE 8.1

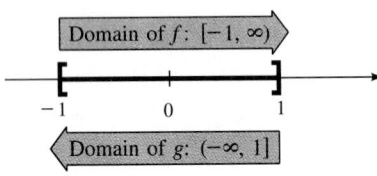

Domain of f: $[-1, \infty)$

-1 0 1

Domain of g: $(-\infty, 1]$

EXAMPLE 3   ■   **Finding the Product of Two Functions**

Given $f(x) = x^2$ and $g(x) = x - 3$, find the product of f and g. Then evaluate the product when $x = 4$.

Solution

The product of f and g is given by

$$\begin{aligned}
(fg)(x) &= f(x)g(x) \\
&= (x^2)(x - 3) \\
&= x^3 - 3x^2.
\end{aligned}$$

The value of this product when $x = 4$ is

$$(fg)(4) = 4^3 - 3(4^2) = 16.$$ ■

EXAMPLE 4   ■   **Finding the Quotient of Two Functions**

Given $f(x) = x^2$ and $g(x) = x - 3$, find the quotient of f and g. Then evaluate the quotient when $x = 5$.

Solution

Using the functions f and g from Example 4, have students compare the domains of these functions to the domain of $(f/g)(x)$. Ask students if they can think of functions $f(x)$ and $g(x)$ in which the domain of $(f/g)(x)$ doesn't have to be restricted.

The quotient of f and g is given by

$$\begin{aligned}
\left(\frac{f}{g}\right)(x) &= \frac{f(x)}{g(x)} \\
&= \frac{x^2}{x - 3}, \qquad x \neq 3.
\end{aligned}$$

(Note that the domain of the quotient has to be restricted to exclude the number 3.) The value of this quotient when $x = 5$ is

$$\left(\frac{f}{g}\right)(5) = \frac{5^2}{5 - 3} = \frac{25}{2}.$$ ■

Composition of Functions

Another way of combining two functions is to form the **composition** of one with the other. For instance, if $f(x) = 2x^2$ and $g(x) = x - 1$, then the composition of f with g is given by

$$f(g(x)) = f(x - 1) = 2(x - 1)^2.$$

This composition is denoted as $f \circ g$.

Definition of Composition of Two Functions

The **composition** of the functions f and g is given by

$$(f \circ g)(x) = f(g(x)).$$

The domain of $(f \circ g)$ is the set of all x in the domain of g such that $g(x)$ is in the domain of f (see Figure 8.2).

FIGURE 8.2

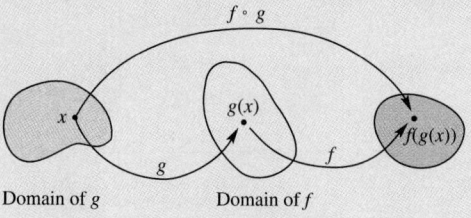

Domain of g Domain of f

EXAMPLE 5 ■ Forming the Composition of Two Functions

Given $f(x) = 2x + 4$ and $g(x) = 3x - 1$, find the composition of f with g. Then find the value of the composition when $x = 1$ in two ways.

Remind students that composition can be described as the substitution of one function for the variable position of another function. Have students complete a pattern such as the following. Add additional substitutions if necessary.

$$f(x) = 4x - 1$$
$$f(\ ) = 4(\ ) - 1$$
$$f(2) = 4(2) - 1 = 7$$
$$f(a) = 4(a) - 1 = 4a - 1$$
$$f(x + 2) = 4(x + 2) - 1 = 4x + 7$$

Note that $f(x + 2)$ is the same as $(f \circ g)(x)$ where $g(x) = x + 2$.

Solution

The composition of f with g is given by

$$
\begin{aligned}
(f \circ g)(x) &= f(g(x)) \\
&= f(3x - 1) \\
&= 2(3x - 1) + 4 \\
&= 6x - 2 + 4 \\
&= 6x + 2.
\end{aligned}
$$

To evaluate this composition when $x = 1$, you can use either of the following two methods. Note that you obtain the same value with either method.

(a) $(f \circ g)(1) = 6(1) + 2 = 8$

(b) $(f \circ g)(1) = f(3(1) - 1) = f(2) = 2(2) + 4 = 8$ ■

The composition of f with g is generally *not* the same as the composition of g with f. This is illustrated in Example 6.

EXAMPLE 6   ■   Comparing the Compositions of Functions

Given $f(x) = 2x - 3$ and $g(x) = x^2 + 1$, find the following.

(a) $(f \circ g)(x)$ (b) $(g \circ f)(x)$

Solution

(a) The composition of f with g is as follows.

$$
\begin{aligned}
(f \circ g)(x) &= f(g(x)) && \text{Definition of } f \circ g \\
&= f(x^2 + 1) && \text{Definition of } g(x) \\
&= 2(x^2 + 1) - 3 && \text{Definition of } f(x) \\
&= 2x^2 + 2 - 3 \\
&= 2x^2 - 1
\end{aligned}
$$

(b) The composition of g with f is as follows.

$$
\begin{aligned}
(g \circ f)(x) &= g(f(x)) && \text{Definition of } g \circ f \\
&= g(2x - 3) && \text{Definition of } f(x) \\
&= (2x - 3)^2 + 1 && \text{Definition of } g(x) \\
&= 4x^2 - 12x + 9 + 1 \\
&= 4x^2 - 12x + 10
\end{aligned}
$$

Note that $(f \circ g)(x) \neq (g \circ f)(x)$.   ■

EXAMPLE 7   ■   Finding the Domain of the Composition of Two Functions

Given $f(x) = x^2$ and $g(x) = \sqrt{x}$, find the composition of f with g, and find the domain of the composition.

Solution

The composition of f with g is given by

$$
\begin{aligned}
(f \circ g)(x) &= f(g(x)) \\
&= f(\sqrt{x}) \\
&= (\sqrt{x})^2 \\
&= x, \quad x \geq 0.
\end{aligned}
$$

To find the domain of $(f \circ g)$, note that the domain of g consists of all nonnegative real numbers, $[0, \infty)$. Thus, you must restrict the composition of f with g to this same set, and conclude that

(Domain of $f \circ g$) $= [0, \infty)$.   ■

EXAMPLE 8   ■   Forming Composite Functions in Different Orders

The regular price of a certain new car is $15,800. The dealership advertised a factory rebate of $1500 *and* a 12% discount. Compare the sale price obtained by subtracting the rebate first and then taking the discount with the sale price obtained by taking the discount first and then subtracting the rebate.

Solution

Using function notation, the price after rebate, $f(x)$, and the price after discount, $g(x)$, can be represented by

$$f(x) = x - 1500 \qquad\qquad \text{Rebate of \$1500}$$
$$g(x) = 0.88x, \qquad\qquad \text{Discount of 12\%}$$

where x is the price of the car.

Rebate First: If you subtract the rebate first, the sale price is given by the composition of g with f.

$$
\begin{aligned}
g(f(15,800)) &= g(15,800 - 1500) & \text{Subtract rebate first}\\
&= g(14,300) & \text{Simplify}\\
&= 0.88 \cdot 14,300 & \text{Take discount}\\
&= \$12,584 & \text{Simplify}
\end{aligned}
$$

Discount First: If you take the discount first, the sale price is given by the composition of f with g.

$$
\begin{aligned}
f(g(15,800)) &= f(0.88 \cdot 15,800) & \text{Take discount first}\\
&= f(13,904) & \text{Simplify}\\
&= 13,904 - 1500 & \text{Subtract rebate}\\
&= \$12,404 & \text{Simplify}
\end{aligned}
$$

Thus, given an option, you should take the discount first.   ■

Car dealers often offer two types of incentives to buyers. One is a factory *rebate*, which is a dollar amount that is subtracted from the price of the car. The other is a percentage discount that is taken off the price of the car.

DISCUSSION PROBLEM   ■   Comparing the Compositions of Functions

You have seen that the composition of f with g is generally not the same as the composition of g with f. There are, however, special cases in which these two compositions are the same. For instance, let $f(x) = 2x$ and $g(x) = \frac{1}{2}x$. Show that for these two functions, the compositions $(f \circ g)$ and $(g \circ f)$ are the same. Can you find other examples of functions f and g such that the compositions $(f \circ g)$ and $(g \circ f)$ are the same?   ■

Warm-Up

The following warm-up exercises involve skills that were covered in earlier sections. You will use these skills in the exercise set for this section.

In Exercises 1–4, find the domain of the function.

1. $f(x) = x^3 + 3x^2 - 2$

2. $A(x) = \sqrt{10 - x}$

3. $H(x) = \dfrac{12}{x - 5}$

4. $g(x) = \dfrac{25}{x^2 - 25}$

In Exercises 5–10, sketch the graph of the function.

5. $f(x) = \frac{1}{2}x$

6. $f(x) = -2x + 3$

7. $f(x) = x^2 - 4$

8. $f(x) = 2x^2 - x$

9. $f(x) = |x| + 1$

10. $f(x) = x^3 + 3$

8.1 EXERCISES

means that a graphing utility can help you solve the exercise or check your solution.

In Exercises 1–8, use the given functions f and g to find the combinations $(f + g)(x)$, $(f - g)(x)$, $(fg)(x)$, and $(f/g)(x)$. Find the domain of each combination.

1. $f(x) = 2x$
$g(x) = x^2$

2. $f(x) = x + 1$
$g(x) = \sqrt{x}$

3. $f(x) = 4x - 3$
$g(x) = x^2 - 9$

4. $f(x) = x^3$
$g(x) = x^2 + 1$

5. $f(x) = \dfrac{1}{x}$
$g(x) = 5x$

6. $f(x) = \dfrac{1}{x + 2}$
$g(x) = \dfrac{1}{x - 2}$

7. $f(x) = \sqrt{x + 4}$
$g(x) = \sqrt{4 - x}$

8. $f(x) = \sqrt{x}$
$g(x) = \sqrt{x^2 - 16}$

In Exercises 9–16, evaluate the indicated combination of the functions f and g at the given values of the independent variable. (If not possible, state the reason.)

9. $f(x) = x^2$, $g(x) = 2x + 3$
  (a) $(f + g)(2)$ (b) $(f - g)(3)$

10. $f(x) = x^3$, $g(x) = x - 5$
  (a) $(f + g)(-2)$ (b) $(f - g)(5)$

11. $f(x) = \sqrt{x}$, $g(x) = x^2 + 1$
  (a) $(f + g)(4)$ (b) $(f - g)(9)$

12. $f(x) = \sqrt{x - 3}$, $g(x) = x^2 - 4$
  (a) $(f + g)(7)$ (b) $(f - g)(1)$

13. $f(x) = |x|$, $g(x) = 5$
  (a) $(fg)(-2)$ (b) $\left(\dfrac{f}{g}\right)(-7)$

14. $f(x) = |x - 3|$, $g(x) = (x - 2)^3$
  (a) $(fg)(3)$ (b) $\left(\dfrac{f}{g}\right)\left(\dfrac{5}{2}\right)$

15. $f(x) = \dfrac{1}{x-4}, \quad g(x) = x$

    (a) $(fg)(-2)$ (b) $\left(\dfrac{f}{g}\right)(4)$

16. $f(x) = \dfrac{1}{x^2}, \quad g(x) = \dfrac{1}{x+1}$

    (a) $(fg)\left(\dfrac{1}{2}\right)$ (b) $\left(\dfrac{f}{g}\right)(0)$

In Exercises 17–24, find the indicated composite functions.

17. $f(x) = x - 3, \quad g(x) = x^2$ (a) $(f \circ g)(x)$ (b) $(g \circ f)(x)$ (c) $(f \circ g)(4)$ (d) $(g \circ f)(7)$

18. $f(x) = x + 5, \quad g(x) = x^3$ (a) $(f \circ g)(x)$ (b) $(g \circ f)(x)$ (c) $(f \circ g)(2)$ (d) $(g \circ f)(-3)$

19. $f(x) = |x - 3|, \quad g(x) = 3x$ (a) $(f \circ g)(x)$ (b) $(g \circ f)(x)$ (c) $(f \circ g)(1)$ (d) $(g \circ f)(2)$

20. $f(x) = |x|, \quad g(x) = 2x + 5$ (a) $(f \circ g)(x)$ (b) $(g \circ f)(x)$ (c) $(f \circ g)(-2)$ (d) $(g \circ f)(-4)$

21. $f(x) = \sqrt{x}, \quad g(x) = x + 5$ (a) $(f \circ g)(x)$ (b) $(g \circ f)(x)$ (c) $(f \circ g)(4)$ (d) $(g \circ f)(9)$

22. $f(x) = \sqrt{x + 6}, \quad g(x) = 2x - 3$ (a) $(f \circ g)(x)$ (b) $(g \circ f)(x)$ (c) $(f \circ g)(3)$ (d) $(g \circ f)(-2)$

23. $f(x) = \dfrac{1}{x-3}, \quad g(x) = \sqrt{x}$ (a) $(f \circ g)(x)$ (b) $(g \circ f)(x)$ (c) $(f \circ g)(49)$ (d) $(g \circ f)(12)$

24. $f(x) = \dfrac{4}{x^2 - 4}, \quad g(x) = \dfrac{1}{x}$ (a) $(f \circ g)(x)$ (b) $(g \circ f)(x)$ (c) $(f \circ g)(-2)$ (d) $(g \circ f)(1)$

In Exercises 25–30, find the domains of the compositions (a) $f \circ g$ and (b) $g \circ f$.

25. $f(x) = x^2 + 1$
    $g(x) = 2x$

26. $f(x) = 2 - 3x$
    $g(x) = 5x + 3$

27. $f(x) = \sqrt{x}$
    $g(x) = x - 2$

28. $f(x) = \sqrt{x - 5}$
    $g(x) = x^2$

29. $f(x) = \dfrac{9}{x + 9}$
    $g(x) = x^2$

30. $f(x) = \dfrac{x}{x - 4}$
    $g(x) = \sqrt{x}$

In Exercises 31–40, find the indicated combination of the functions $f(x) = x^2 - 3x$ and $g(x) = 5x + 3$.

31. $(f - g)(t)$ **32.** $(f/g)(2)$

33. $(f + g)(t - 2)$ **34.** $(fg)(z)$

35. $\dfrac{g(x + h) - g(x)}{h}$ **36.** $\dfrac{f(x + h) - f(x)}{h}$

37. $(f \circ g)(-1)$ **38.** $(g \circ f)(3)$

39. $(g \circ f)(y)$ **40.** $(f \circ g)(z)$

In Exercises 41–46, use the functions f and g, which are given as sets of ordered pairs.

    f: $\{(-2, 3), (-1, 1), (0, 0), (1, -1), (2, -3)\}$

    g: $\{(-3, 1), (-1, -2), (0, 2), (2, 2), (3, 1)\}$

41. (a) Find $f(1)$.
    (b) Find $g(-1)$.
    (c) Find $(g \circ f)(1)$.

42. (a) Find $g(0)$.
    (b) Find $f(2)$.
    (c) Find $(f \circ g)(0)$.

43. Find $(f \circ g)(-3)$.   **44.** Find $(g \circ f)(-2)$.

45. Find $(f \circ g)(2)$.   **46.** Find $(g \circ f)(2)$.

In Exercises 47–50, sketch the graphs of the functions f, g, and $f + g$ on the same set of coordinate axes.

47. $f(x) = \frac{1}{2}x$
    $g(x) = x - 1$

48. $f(x) = \frac{1}{3}x$
    $g(x) = -x + 4$

49. $f(x) = x^2$
    $g(x) = -2x$

50. $f(x) = 4 - x^2$
    $g(x) = x$

51. *Stopping Time* While traveling in a car at x miles per hour, you are required to make a panic stop in order to avoid an accident. The distance (in feet) the car travels during your reaction time is $R(x) = \frac{3}{4}x$. The distance traveled while braking is $B(x) = \frac{1}{15}x^2$. Find the function giving the total stopping distance T (in feet). Graph the functions R, B, and T on the same set of coordinate axes for $0 \le x \le 60$.

52. *Comparing Sales* Suppose you own two fast-food restaurants. From 1985 to 1990, sales for the first restaurant have been decreasing according to the function

$$R_1 = 500 - 0.8t^2, \qquad t = 5, 6, 7, 8, 9, 10,$$

where R_1 represents the sales for the first restaurant (in thousands of dollars) and t represents the calendar year, with $t = 5$ corresponding to 1985. During the same 6-year period, sales for the second restaurant have been increasing according to the function

$$R_2 = 250 + 0.78t, \qquad t = 5, 6, 7, 8, 9, 10.$$

(a) Write a function that represents the total sales for the two restaurants over the 6-year period.

(b) Use the *stacked bar graph* in the figure, which represents the total sales during this 6-year period, to determine whether the total sales have been increasing or decreasing.

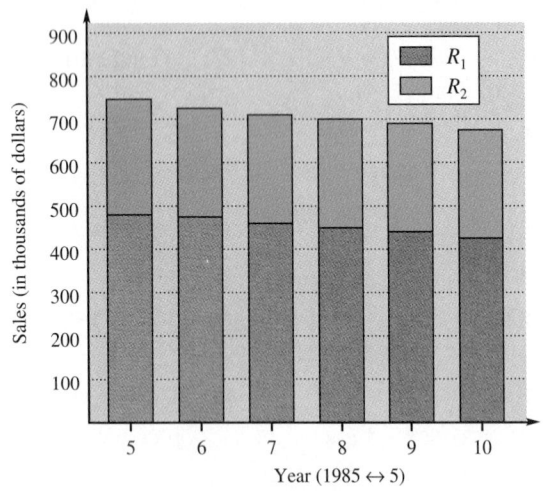

Year (1985 ↔ 5)

53. *Sales Bonus* Suppose you are a sales representative for a clothing manufacturer. You are paid an annual salary plus a bonus of 2% of your sales *over* $200,000. Consider the two functions $f(x) = x - 200{,}000$ and $g(x) = 0.02x$. If x is greater than $200,000, which of the following represents your bonus? Explain.

(a) $f(g(x))$ (b) $g(f(x))$

54. *A Bridge over Troubled Waters* You are standing on a bridge over a calm pond and drop a pebble, causing ripples of concentric circles in the water. The radius (in feet) of the outer ripple is given by $r(t) = 0.6t$, where t is time in seconds after the pebble hits the water. The area of the circle is given by the function $A(r) = \pi r^2$. Find an equation for the composite function $A(r(t))$.

Because of surface tension, a drop of water assumes the shape that has the smallest surface area—a sphere.

In Exercises 55 and 56, use a graphing calculator to graph the functions f, g, and $f + g$ on the same set of coordinate axes.* Observe from the graphs which of the two functions is more significant in determining the magnitude of the sum for different values of x. Use the given range settings.

55. $f(x) = x^3$
$g(x) = 3(1 - x^2)$

| Xmin=-2 |
| Xmax=3 |
| Xscl=1 |
| Ymin=-3 |
| Ymax=3 |
| Yscl=1 |

56. $f(x) = x^3$
$g(x) = -3x$

| Xmin=-3 |
| Xmax=3 |
| Xscl=1 |
| Ymin=-4 |
| Ymax=4 |
| Yscl=1 |

* The graphing calculator features shown in the examples and exercises in this text correspond to the TI-81 and TI-82 graphing calculators from Texas Instruments. However, you can also use other graphing calculators and computer graphing software for these examples and exercises.

Inverse Functions

The Inverse of a Function • *Finding the Inverse of a Function* • *The Graph of the Inverse of a Function*

The Inverse of a Function

When functions were introduced in Section 2.5, you saw that one way to represent a function is by a set of ordered pairs. For instance, the function $f(x) = x+2$ from the set $A = \{1, 2, 3, 4\}$ to the set $B = \{3, 4, 5, 6\}$ can be written as follows.

$$f(x) = x + 2: \quad \{(1,3), (2,4), (3,5), (4,6)\}$$

By interchanging the first and second coordinates in each of these ordered pairs, you can form another function that is called the **inverse function** of f. This function is denoted by f^{-1}. It is a function from the set B to the set A, and can be written as follows.

$$f^{-1}(x) = x - 2: \quad \{(3,1), (4,2), (5,3), (6,4)\}$$

FIGURE 8.3

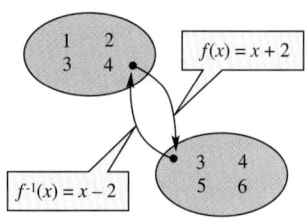

Note that the domain of f is equal to the range of f^{-1}, and vice versa, as shown in Figure 8.3. Also note that the functions f and f^{-1} have the effect of "undoing" each other. In other words, when you form the composition of f with f^{-1}, or the composition of f^{-1} with f, you obtain the identity function, as follows.

$$f(f^{-1}(x)) = f(x - 2) = (x - 2) + 2 = x$$
$$f^{-1}(f(x)) = f^{-1}(x + 2) = (x + 2) - 2 = x$$

NOTE Don't be confused by the use of -1 to denote the inverse function f^{-1}. Whenever we write f^{-1}, we will *always* be referring to the inverse of the function f and *not* to the reciprocal of $f(x)$.

EXAMPLE 1 ■ Determining Inverse Functions Informally

Find the inverse of the function $f(x) = 3x$ and verify that both $f(f^{-1}(x))$ and $f^{-1}(f(x))$ are equal to the identity function.

Solution

The given function is a function that *multiplies* each input by 3. To "undo" this function, you need to *divide* each input by 3. Thus, the inverse function of $f(x) = 3x$ is given by

$$f^{-1}(x) = \frac{x}{3}.$$

You can verify that both $f(f^{-1}(x))$ and $f^{-1}(f(x))$ are equal to the identity function, as follows.

$$f(f^{-1}(x)) = f\left(\frac{x}{3}\right) = 3\left(\frac{x}{3}\right) = x$$

$$f^{-1}(f(x)) = f^{-1}(3x) = \frac{3x}{3} = x$$ ■

EXAMPLE 2 ■ Determining Inverse Functions Informally

Find the inverse of the function $f(x) = x - 4$ and verify that both $f(f^{-1}(x))$ and $f^{-1}(f(x))$ are equal to the identity function.

Solution

The given function is a function that *subtracts* 4 from each input. To "undo" this function, you need to *add* 4 to each input. Thus, the inverse function of $f(x) = x - 4$ is given by

$$f^{-1}(x) = x + 4.$$

You can verify that both $f(f^{-1}(x))$ and $f^{-1}(f(x))$ are equal to the identity function, as follows.

$$f(f^{-1}(x)) = f(x + 4) = (x + 4) - 4 = x$$
$$f^{-1}(f(x)) = f^{-1}(x - 4) = (x - 4) + 4 = x$$ ■

The formal definition of the inverse of a function is given as follows.

Definition of the Inverse of a Function

Let f and g be two functions such that

$$f(g(x)) = x \quad \text{for every } x \text{ in the domain of } g$$

and

$$g(f(x)) = x \quad \text{for every } x \text{ in the domain of } f.$$

Then the function g is called the **inverse** of the function f, and is denoted by f^{-1} (read "f-inverse"). Thus, $f(f^{-1}(x)) = x$ and $f^{-1}(f(x)) = x$. The domain of f must be equal to the range of f^{-1}, and vice versa.

Note from this definition that if the function g is the inverse of the function f, then it must also be true that the function f is the inverse of the function g. For this reason, the functions f and g are sometimes called *inverses of each other*.

EXAMPLE 3   ■   Verifying Inverse Functions

Show that the functions $f(x) = x^3 + 1$ and $g(x) = \sqrt[3]{x - 1}$ are inverses of each other.

Solution

Begin by noting that the domain and range of both functions is the entire set of real numbers. To show that f and g are inverses of each other, you need to show that $f(g(x)) = x$ and $g(f(x)) = x$, as follows.

$$\begin{aligned} f(g(x)) &= f\left(\sqrt[3]{x - 1}\right) \\ &= \left(\sqrt[3]{x - 1}\right)^3 + 1 \\ &= (x - 1) + 1 \\ &= x \end{aligned}$$

$$\begin{aligned} g(f(x)) &= g(x^3 + 1) \\ &= \sqrt[3]{(x^3 + 1) - 1} \\ &= \sqrt[3]{x^3} \\ &= x \end{aligned}$$

Note that the two functions f and g "undo" each other in the following verbal sense. The function f first cubes the input x and then adds 1, whereas the function g first subtracts 1, and then takes the cube root of the result.   ■

EXAMPLE 4   ■   Verifying Inverse Functions

Which of the following functions is the inverse of the function $f(x) = 1/(5x)$?

$$g(x) = \frac{5}{x} \quad \text{or} \quad h(x) = \frac{1}{5x}$$

Solution

By forming the composition of f with g, you find that

$$\begin{aligned} f(g(x)) &= f\left(\frac{5}{x}\right) \\ &= \frac{1}{5\left(\dfrac{5}{x}\right)} \\ &= \frac{1}{\left(\dfrac{25}{x}\right)} \\ &= \frac{x}{25}. \end{aligned}$$

Because this composition does not yield x, you conclude that g *is not* the inverse of f. By forming the composition of f with h, you find that

$$f(h(x)) = f\left(\frac{1}{5x}\right)$$

$$= \frac{1}{5\left(\dfrac{1}{5x}\right)}$$

$$= \frac{1}{\left(\dfrac{1}{x}\right)}$$

$$= x.$$

Emphasize that *both* conditions, $f(g(x)) = x$ and $g(f(x)) = x$, must be satisfied.

Thus, it appears that h is the inverse of f. You can confirm this by showing that the composition of h with f is also equal to the identity function. (Try doing this.) ■

Finding the Inverse of a Function

For simple functions (such as the ones in Examples 1 and 2), you can find inverse functions by inspection. For instance, the inverse of $f(x) = 10x$ is equal to $f^{-1}(x) = x/10$. For more complicated functions, however, it is best to use the following steps for finding the inverses. The key step in these guidelines is switching the roles of x and y. This step corresponds to the fact that inverse functions have ordered pairs with the coordinates reversed.

Finding the Inverse of a Function

To find the inverse of a function f, use the following steps.

1. In the equation for $f(x)$, replace $f(x)$ by y.

2. Interchange the roles of x and y.

3. If the new equation does not represent y as a function of x, then the function f does not have an inverse function. If the new equation does represent y as a function of x, then solve the new equation for y.

4. Replace y by $f^{-1}(x)$.

5. Verify that f and f^{-1} are inverses of each other by showing that $f(f^{-1}(x)) = x = f^{-1}(f(x))$.

EXAMPLE 5　■　Finding the Inverse of a Function

Determine whether the function $f(x) = 2x + 3$ has an inverse. If it does, find its inverse.

Solution

$$f(x) = 2x + 3 \qquad \text{Given function}$$

$$y = 2x + 3 \qquad \text{Replace } f(x) \text{ by } y$$

$$x = 2y + 3 \qquad \text{Interchange } x \text{ and } y$$

$$y = \frac{x - 3}{2} \qquad \text{Solve for } y$$

$$f^{-1}(x) = \frac{x - 3}{2} \qquad \text{Replace } y \text{ by } f^{-1}(x)$$

Thus, the inverse of $f(x) = 2x + 3$ is

$$f^{-1}(x) = \frac{x - 3}{2}.$$

You can verify that $f(f^{-1}(x)) = x$ and $f^{-1}(f(x)) = x$.　■

Note in step 3 of the guidelines for finding the inverse of a function that it is possible that a function has no inverse. This possibility is illustrated in Example 6.

EXAMPLE 6　■　A Function That Has No Inverse

Determine whether the function $f(x) = x^2$ has an inverse. If it does, find its inverse.

Solution

$$f(x) = x^2 \qquad \text{Given function}$$

$$y = x^2 \qquad \text{Replace } f(x) \text{ by } y$$

$$x = y^2 \qquad \text{Interchange } x \text{ and } y$$

Because the equation $x = y^2$ does not represent y as a function of x, you can conclude that the original function f does not have an inverse.　■

NOTE　The equation $x = y^2$ does not represent y as a function of x because you can find two different y-values that correspond to the same x-value. For example, when $x = 9$, y can be 3 or -3.

The Graph of the Inverse of a Function

The graphs of a function f and its inverse f^{-1} are related to each other in the following way. If the point (a, b) lies on the graph of f, then the point (b, a) must lie on the graph of f^{-1}, and vice versa. This means that the graph of f^{-1} is a reflection of the graph of f in the line $y = x$, as shown in Figure 8.4.

FIGURE 8.4

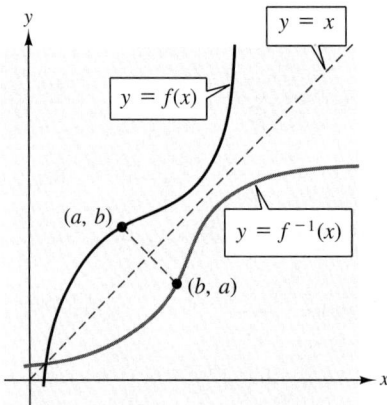

The graph of f^{-1} is a reflection of the graph of f in the line $y = x$.

This "reflective property" of the graphs of f and f^{-1} is illustrated in Examples 7 and 8.

EXAMPLE 7 ■ The Graphs of f and f^{-1}

Sketch the graphs of the inverse functions $f(x) = 2x - 3$ and $f^{-1}(x) = \frac{1}{2}(x+3)$ on the same rectangular coordinate system, and show that the graphs are reflections of each other in the line $y = x$.

Solution

The graphs of f and f^{-1} are shown in Figure 8.5 on the next page. Visually, it appears that the graphs are reflections of each other in the line $y = x$. You can further verify this reflective property by testing a few points on each graph. Note in the following list that if the point (a, b) is on the graph of f, then the point (b, a) is on the graph of f^{-1}.

$f(x) = 2x - 3$	$f^{-1}(x) = \frac{1}{2}(x + 3)$
$(-1, -5)$	$(-5, -1)$
$(0, -3)$	$(-3, 0)$
$(1, -1)$	$(-1, 1)$
$(2, 1)$	$(1, 2)$
$(3, 3)$	$(3, 3)$

FIGURE 8.5

Enhancement activity: If you are using graphing calculators, you may show or have students try the following. Set the calculator in parametric mode and the viewing rectangle to show between -10 and 10 for T, X, and Y. Then graph together on the same grid these expressions:

$$X_{1T} = T$$
$$Y_{1T} = 2T - 3$$
$$X_{2T} = 2T - 3$$
$$Y_{2T} = T$$
$$X_{3T} = T$$
$$Y_{3T} = T$$

Have students compare the display to Figure 8.5.

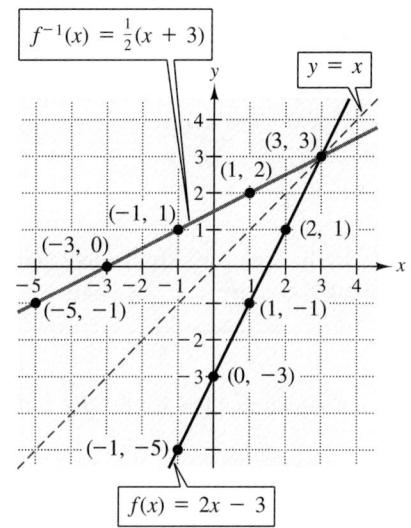

$f^{-1}(x) = \frac{1}{2}(x + 3)$

$y = x$

$f(x) = 2x - 3$

In Example 6, you saw that the function $f(x) = x^2$ has no inverse. A more complete way of saying this is *"assuming that the domain of f is the entire real line,* the function $f(x) = x^2$ has no inverse." If, however, you restrict the domain of f to the nonnegative real numbers, then f does have an inverse, as demonstrated in Example 8.

EXAMPLE 8 ■ The Graphs of f and f^{-1}

Sketch the graphs of the inverse functions

$$f(x) = x^2, \quad x \geq 0 \quad \text{and} \quad f^{-1}(x) = \sqrt{x}$$

on the same rectangular coordinate system and show that the graphs are reflections of each other in the line $y = x$.

Solution

The graphs of f and f^{-1} are shown in Figure 8.6. Visually, it appears that the graphs are reflections of each other in the line $y = x$. You can further verify this reflective property by testing a few points on each graph. Note in the following list that if the point (a, b) is on the graph of f, then the point (b, a) is on the graph of f^{-1}.

$f(x) = x^2, \quad x \geq 0$	$f^{-1}(x) = \sqrt{x}$
$(0, 0)$	$(0, 0)$
$(1, 1)$	$(1, 1)$
$(2, 4)$	$(4, 2)$
$(3, 9)$	$(9, 3)$

FIGURE 8.6

$f(x) = x^2$
$x \geq 0$

$y = x$

$f^{-1}(x) = \sqrt{x}$

In the guidelines for finding the inverse of a function, we included an *algebraic* test for determining whether a function has an inverse. The reflective property of the graphs of inverse functions gives you a nice *geometric* test for determining whether a function has an inverse. This is called the **horizontal line test** for inverse functions.

Horizontal Line Test for Inverse Functions	A function f has an inverse function if and only if no *horizontal* line intersects the graph of f at more than one point.

The horizontal line test for inverse functions is demonstrated in Example 9.

EXAMPLE 9 ■ Applying the Horizontal Line Test for Inverse Functions

Historical Note: Sonya Kovalevsky was a brilliant Russian mathematician and mathematical physicist who lived from 1850 until 1891. She worked extensively with inverse functions in higher mathematics.

(a) The graph of the function $f(x) = x^3 - 1$ is shown in Figure 8.7. Because no horizontal line intersects the graph of f at more than one point, you can conclude that f *does* possess an inverse function.

(b) The graph of the function $f(x) = x^2 - 1$ is shown in Figure 8.8. Because it is possible to find a horizontal line that intersects the graph of f at more than one point, you can conclude that f *does not* possess an inverse function.

FIGURE 8.7

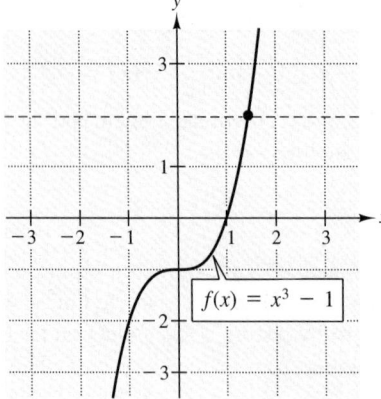

$f(x) = x^3 - 1$

FIGURE 8.8

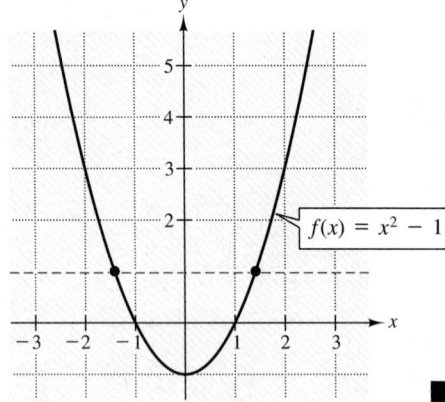

$f(x) = x^2 - 1$

DISCUSSION PROBLEM ■ The Existence of an Inverse Function

Write a short paragraph describing why the following functions do or do not possess inverse functions.

(a) Let x represent the retail price of an item (in dollars), and let $f(x)$ represent the sales tax on the item. Assume that the sales tax is 6% of the retail price *and* that the sales tax is rounded to the nearest cent. Does this function possess an inverse? (*Hint*: Can you undo this function? For instance, if you know that the sales tax is $0.12, can you determine *exactly* what the retail price is?)

(b) Let x represent the temperature in degrees Celsius, and $f(x)$ represent the temperature in degrees Fahrenheit. Does this function possess an inverse? (*Hint*: The formula for converting from degrees Celsius to degrees Fahrenheit is $F = \frac{9}{5}C + 32$.) ■

Warm-Up

The following warm-up exercises involve skills that were covered in earlier sections. You will use these skills in the exercise set for this section.

In Exercises 1–4, find the domain of the function.

1. $f(x) = x^3 - 2x$ **2.** $g(x) = \sqrt[3]{x}$

3. $h(x) = \sqrt{16 - x^2}$ **4.** $A(x) = \dfrac{3}{36 - x^2}$

In Exercises 5–10, use the functions f and g to find the combinations $(f + g)(x)$, $(f - g)(x)$, $(fg)(x)$, $(f/g)(x)$, $(f \circ g)(x)$, and $(g \circ f)(x)$. Find the domain of each combination.

5. $f(x) = x^2 - 9$ **6.** $f(x) = 2x - 3$ **7.** $f(x) = \sqrt[3]{x - 2}$

   $g(x) = 3x$ $g(x) = \frac{1}{2}x^2$ $g(x) = x^2$

8. $f(x) = \sqrt{x}$ **9.** $f(x) = \dfrac{1}{x}$ **10.** $f(x) = x + 2$

   $g(x) = x^2 - 4$ $g(x) = \dfrac{1}{x + 2}$

                              $g(x) = \dfrac{1}{x - 2}$

8.2 **EXERCISES**

▦ means that a graphing utility can help you solve the exercise or check your solution.

In Exercises 1–8, find the inverse of the function f informally. Verify that $f(f^{-1}(x))$ and $f^{-1}(f(x))$ are equal to the identity function.

1. $f(x) = 5x$

2. $f(x) = \frac{1}{3}x$

3. $f(x) = x + 10$

4. $f(x) = x - 5$

5. $f(x) = x^7$

6. $f(x) = x^5$

7. $f(x) = \sqrt[3]{x}$

8. $f(x) = x^{1/5}$

In Exercises 9–20, verify algebraically that the functions f and g are inverses of each other.

9. $f(x) = 10x$

$g(x) = \frac{1}{10}x$

10. $f(x) = \frac{2x}{3}$

$g(x) = \frac{3x}{2}$

11. $f(x) = x + 15$

$g(x) = x - 15$

12. $f(x) = 3 - x$

$g(x) = 3 - x$

13. $f(x) = 1 - 2x$

$g(x) = \frac{1 - x}{2}$

14. $f(x) = 2x - 1$

$g(x) = \frac{1}{2}(x + 1)$

15. $f(x) = 2 - 3x$

$g(x) = \frac{1}{3}(2 - x)$

16. $f(x) = -\frac{1}{4}x + 3$

$g(x) = -4(x - 3)$

17. $f(x) = \sqrt[3]{x + 1}$

$g(x) = x^3 - 1$

18. $f(x) = x^5$

$g(x) = \sqrt[5]{x}$

19. $f(x) = \frac{1}{x}$

$g(x) = \frac{1}{x}$

20. $f(x) = \frac{1}{x - 3}$

$g(x) = 3 + \frac{1}{x}$

In Exercises 21–24, match the graph with the graph of its inverse. [The graphs of the inverse functions are labeled (a), (b), (c), and (d).]

21.

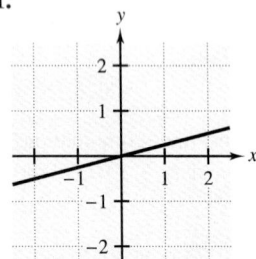

22.

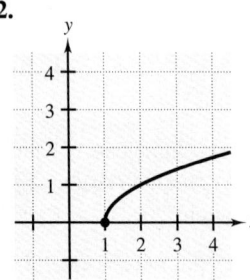

23.

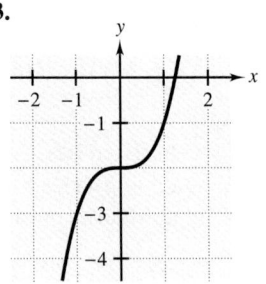

24.

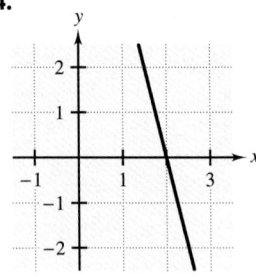

(a)

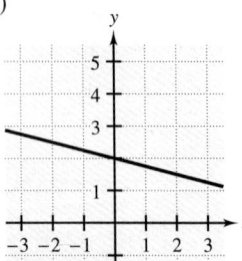

(b)

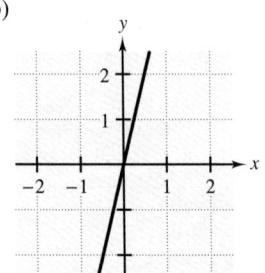

(c)

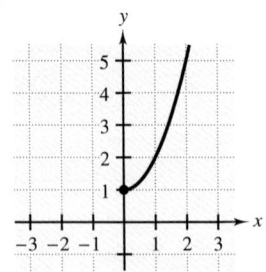

(d)

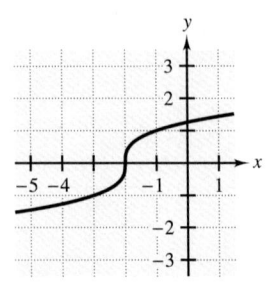

In Exercises 25–30, verify geometrically (as in Example 7) that the functions f and g are inverses of each other.

25. $f(x) = \frac{1}{3}x$

$g(x) = 3x$

26. $f(x) = \frac{1}{5}x - 1$

$g(x) = 5x + 5$

27. $f(x) = \sqrt{x+1}$

$g(x) = x^2 - 1, \quad x \geq 0$

28. $f(x) = \sqrt{4-x}$

$g(x) = 4 - x^2, \quad x \geq 0$

29. $f(x) = \frac{1}{8}x^3$

$g(x) = 2\sqrt[3]{x}$

30. $f(x) = \sqrt[3]{x+2}$

$g(x) = x^3 - 2$

 In Exercises 31–42, find the inverse of the function.

31. $f(x) = 8x$

32. $f(x) = \frac{x}{10}$

33. $g(x) = x + 25$

34. $f(x) = 7 - x$

35. $g(x) = 3 - 4x$

36. $g(x) = 6x + 1$

37. $h(x) = \sqrt{x}$

38. $h(x) = \sqrt{x+5}$

39. $f(t) = t^3 - 1$

40. $h(t) = t^5$

41. $g(s) = \frac{5}{s}$

42. $f(s) = \frac{2}{3-s}$

In Exercises 43–56, determine whether the function has an inverse.

43. $f(x) = 9 - x^2$

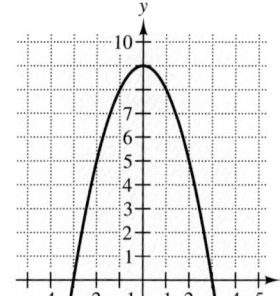

44. $f(x) = \frac{1}{5}x$

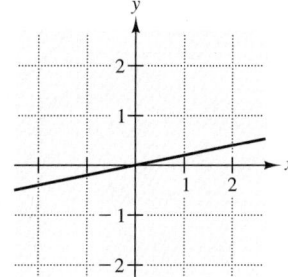

45. $f(x) = \frac{1}{4}x^3$

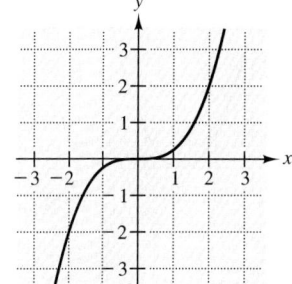

46. $f(x) = x^2 - 2$

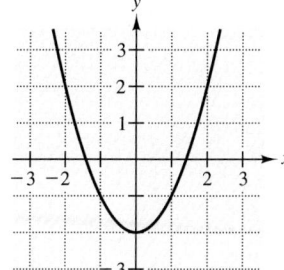

47. $g(x) = \sqrt{25 - x^2}$

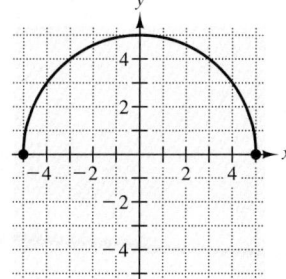

48. $g(x) = |x - 4|$

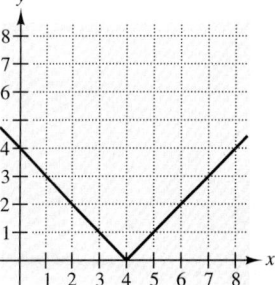

49. $f(x) = \sqrt[3]{5 - x}$

50. $h(x) = 4 - \sqrt[3]{x}$

51. $g(x) = x^4$

52. $f(x) = (x+2)^5$

53. $h(t) = \frac{5}{t}$

54. $g(t) = \frac{5}{t^2}$

55. $f(s) = \frac{4}{s^2 + 1}$

56. $f(x) = \frac{1}{x - 2}$

In Exercises 57–60, place a restriction on the domain of the function so that the graph of the restricted function satisfies the horizontal line test for inverses. Find the inverse of the restricted function. (*Note:* There is more than one correct answer to each problem.)

57. $f(x) = x^4$

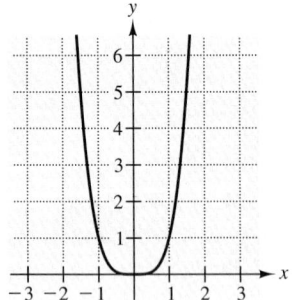

58. $f(x) = 9 - x^2$

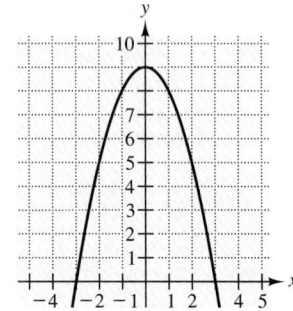

59. $f(x) = (x - 2)^2$

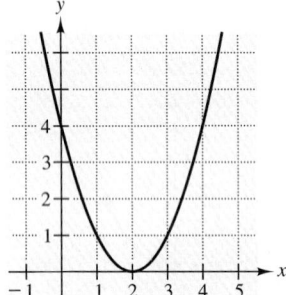

60. $f(x) = |x - 2|$

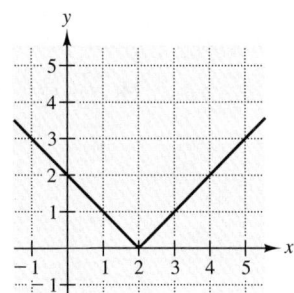

In Exercises 61–64, use the graph of f to complete the table and sketch the graph of f^{-1}.

61.

x	0	1	3	4
f^{-1}				

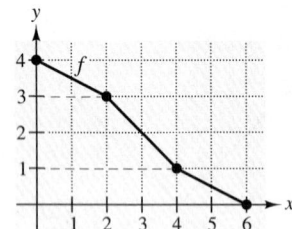

62.

x	−1	0	1	5
f^{-1}				

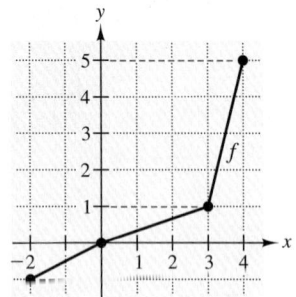

63.

x	−4	−2	2	3
f^{-1}				

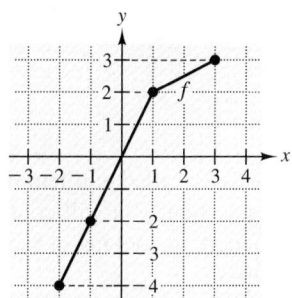

64.

x	−3	−2	0	6
f^{-1}				

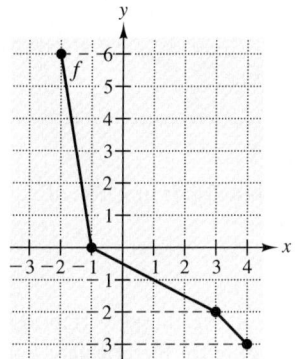

65. Consider the function $f(x) = 3 - 2x$.

(a) Find $f^{-1}(x)$.

(b) Find $(f^{-1})^{-1}(x)$.

66. *Exploratory Exercise* Consider the functions $f(x) = 4x$ and $g(x) = x + 6$.

(a) Find $(f \circ g)(x)$.

(b) Find $(f \circ g)^{-1}(x)$.

(c) Find $f^{-1}(x)$ and $g^{-1}(x)$.

(d) Find $(g^{-1} \circ f^{-1})(x)$ and compare the result with that in part (b).

In Exercises 67 and 68, find the inverse of the function f. Use a graphing calculator to graph f and f^{-1} on the same screen using the given range settings.*

67. $f(x) = x^3 + 1$

Xmin=-2
Xmax=4
Xscl=1
Ymin=-2
Ymax=4
Yscl=1

68. $f(x) = \sqrt{x^2 - 4}, \quad x \geq 2$

Xmin=0
Xmax=6
Xscl=1
Ymin=0
Ymax=6
Yscl=1

69. *Hourly Wage* Your wage is $9.00 per hour plus $0.65 for each unit produced per hour. Thus, your hourly wage y in terms of the number of units produced is given by $y = 9 + 0.65x$.

(a) Determine the inverse of the function.

(b) What does each variable represent in the inverse function?

(c) Determine the number of units produced when your hourly wage averages $14.20.

*The graphing calculator features given in the examples and exercises in this text correspond to the TI-81 and TI-82 graphing calculators from Texas Instruments. However, you can also use other graphing calculators and computer graphing software for these examples and exercises.

Direct Variation

One of the goals of applied mathematics is to find equations, called **mathematical models,** that describe real-world phenomena. In this section, you will continue your study of methods for creating mathematical models by looking at models that are related to the concept of **variation.**

Direct Variation

The following statements are equivalent.

1. y **varies directly** as x.

2. y is **directly proportional** to x.

3. $y = kx$ for some constant k.

The number k is called the **constant of proportionality.**

In the mathematical model for direct variation, y is a *linear* function of x. That is,

$$y = kx.$$

To use this mathematical model in applications involving direct variation, you need to use the given values of x and y to find the value for the constant k. This procedure is demonstrated in Examples 1 and 2.

EXAMPLE 1   ■   Direct Variation

Assume that the total revenue R (in dollars) obtained from selling x units of a given product is directly proportional to the number of units sold. When 10,000 units are sold, the total revenue is $142,500.

(a) Find a mathematical model that relates the total revenue R to the number of units sold x.

(b) Find the total revenue obtained from selling 12,000 units.

Solution

(a) Because the total revenue is directly proportional to the number of units sold, you have the model

$$R = kx.$$

To find the value of the constant k, use the fact that $R = 142,500$ when $x = 10,000$. Substituting these values into the model produces

$$
\begin{array}{cc}
R & x \\
\downarrow & \downarrow
\end{array}
$$
$$142,500 = k(10,000)$$

which implies that $k = 142,500/10,000 = 14.25$. Thus, the equation relating the total revenue to the total number of units sold is

$$R = 14.25x.$$

(b) When $x = 12,000$, the total revenue is

$$R = 14.25(12,000) = \$171,000.$$ ∎

EXAMPLE 2 ∎ **Direct Variation**

Hooke's Law for springs states that the distance a spring is stretched (or compressed) is proportional to the force on the spring. A force of 20 pounds stretches a certain spring 5 inches.

(a) Find a mathematical model that relates the distance the spring is stretched to the force applied to the spring.

(b) How far will a force of 30 pounds stretch the spring?

Solution

(a) For this problem, let d represent the distance (in inches) that the spring is stretched and let F represent the force (in pounds) that is applied to the spring. Because the distance d is proportional to the force F, you have the model

$$d = kF.$$

To find the value of the constant k, use the fact that $d = 5$ when $F = 20$. Substituting these values into the model produces

$$
\begin{array}{cc}
d & F \\
\downarrow & \downarrow
\end{array}
$$
$$5 = k(20),$$

which implies that $k = \frac{5}{20} = \frac{1}{4}$. Thus, the equation relating distance and force is

$$d = \frac{1}{4}F.$$

(b) When $F = 30$, the distance is

$$d = \frac{1}{4}(30) = 7.5 \text{ inches.}$$

(See Figure 8.9.) ∎

FIGURE 8.9

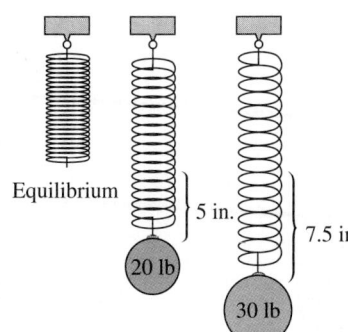

Equilibrium

5 in.

7.5 in.

20 lb

30 lb

In Examples 1 and 2, the direct variations were such that an *increase* in one variable corresponded to an *increase* in the other variable. For instance, in the model $d = \frac{1}{4}F$, if the force F increases, then the distance d also increases. There are, however, other applications of direct variation in which an *increase* in one variable corresponds to a *decrease* in the other variable. For instance, in the model $y = -2x$, an increase in x will yield a decrease in y.

Another type of direct variation relates one variable to a *power* of another variable.

Direct Variation as *n*th Power

The following statements are equivalent.

1. y **varies directly as the *n*th power** of x.

2. y is **directly proportional to the *n*th power** of x.

3. $y = kx^n$ for some constant k.

EXAMPLE 3 ■ Direct Variation as a Power

Additional example using the perimeter and area of a square: As the length of its side increases, the area of a square increases at a much faster rate than does its perimeter.

s	$P = 4s$	$A = s^2$
8	32	64
9	36	81
10	40	100
11	44	121
12	48	144

The distance a ball rolls down an inclined plane is directly proportional to the square of the time it rolls. Assume a case in which, during the first second, the ball rolls 6 feet.

(a) Find a mathematical model that relates the distance traveled to the time.

(b) How far will the ball roll during the first 2 seconds?

Solution

(a) Letting d be the distance (in feet) that the ball rolls and letting t be the time (in seconds), you obtain the model

$$d = kt^2.$$

Because $d = 6$ when $t = 1$, it follows that $k = 6$. Therefore, the equation relating distance to time is

$$d = 6t^2.$$

(b) When $t = 2$, the distance traveled is

$$d = 6(2^2) = 6(4) = 24 \text{ feet.}$$

(See Figure 8.10.)

FIGURE 8.10

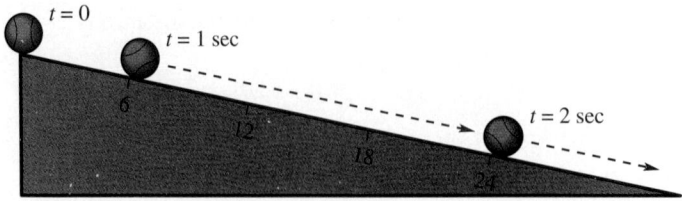

Inverse Variation

A second type of variation is called **inverse variation.** With this type of variation, one of the variables is said to be inversely proportional to the other variable.

Inverse Variation	The following statements are equivalent.
	1. y **varies inversely** as x.
	2. y is **inversely proportional** to x.
	3. $y = \dfrac{k}{x}$ for some constant k.

Remind students that direct variation is translated as $y = kx$. With inverse variation, $y = k/x$ can be written as $y = k(1/x)$, for some constant k.

If x and y are related by an equation of the form $y = k/x^n$, then it is said that y varies inversely as the nth power of x (or that y is inversely proportional to the nth power of x).

EXAMPLE 4   ■   Inverse Variation

The marketing department of a large company has found that the demand for one of its products varies inversely with the price of the product. (When the price is low, more people are willing to buy the product than when the price is high.) When the price of the product is $7.50, the monthly demand is 50,000 units. Approximate the monthly demand if the price is reduced to $6.00.

Solution

Let x represent the number of units that are sold each month (the demand), and let p represent the price per unit (in dollars). Because the demand is inversely proportional to the price, you obtain the model

$$x = \frac{k}{p}.$$

By substituting $x = 50{,}000$ when $p = 7.50$, you can determine that $k = (7.5)(50{,}000) = 375{,}000$. Thus, the model is

$$x = \frac{375{,}000}{p}.$$

To find the demand that corresponds to a price of $6.00, substitute 6 for p in the equation and obtain a demand of

$$x = \frac{375{,}000}{6} = 62{,}500 \text{ units.}$$

Thus, if the price were lowered from $7.50 per unit to $6.00 per unit, you could expect the monthly demand to increase from 50,000 units to 62,500 units.   ■

Some applications of variation involve problems with *both* direct and inverse variation in the same model. These types of models are said to have **combined variation.** For instance, in the model

$$z = k \left(\frac{x}{y} \right),$$

z is directly proportional to x *and* inversely proportional to y. This type of model is shown in Example 5.

EXAMPLE 5 ■ Direct and Inverse Variation

A company determines that the demand for one of its products is directly proportional to the amount spent on advertising and inversely proportional to the price of the product. When \$40,000 is spent on advertising and the price per unit is \$20, the monthly demand is 10,000 units.

(a) If the amount spent on advertising were increased to \$50,000, how much could the price be increased to maintain a monthly demand of 10,000 units?

(b) If you were in charge of the advertising department, would you recommend this increased expenditure for advertising?

Solution

Additional problem: Have students write a mathematical model for the statement "N varies directly as the square of r and inversely as the cube of s." Answer: $N = (kr^2)/(s^3)$

(a) Let x represent the number of units sold each month (the demand), let a represent the amount spent on advertising (in dollars), and let p represent the price per unit (in dollars). Because the demand is directly proportional to the advertising expenditure and inversely proportional to the price, you obtain the model

$$x = \frac{ka}{p}.$$

By substituting 10,000 for x when $a = 40{,}000$ and $p = 20$, you can determine that $k = (10{,}000)(20)/(40{,}000) = 5$. Thus, the model is

$$x = \frac{5a}{p}.$$

To find the price that corresponds to a demand of 10,000 and an advertising expenditure of \$50,000, substitute 10,000 for x and 50,000 for a in the model and solve for p.

$$10{,}000 = \frac{5(50{,}000)}{p} \qquad \Longrightarrow \qquad p = \frac{5(50{,}000)}{10{,}000} = \$25$$

(b) The total revenue for selling 10,000 units at \$20 is \$200,000, and the total revenue for selling 10,000 units at \$25 is \$250,000. Thus, by increasing the advertising expenditure from \$40,000 to \$50,000, the company can increase its revenue by \$50,000. This implies that you should recommend the increased expenditure for advertising.

Amount of Advertising	Price	Revenue
$40,000	$20	$10,000 \times 20 = $200,000
$50,000	$25	$10,000 \times 25 = $250,000

The graph in Figure 8.11 shows the "banking angles" for a bicycle at a speed of 15 miles per hour. The banking angle and the radius of the turn vary inversely. As the radius of the turn gets smaller, the bicyclist must lean at greater angles to avoid falling over.

FIGURE 8.11

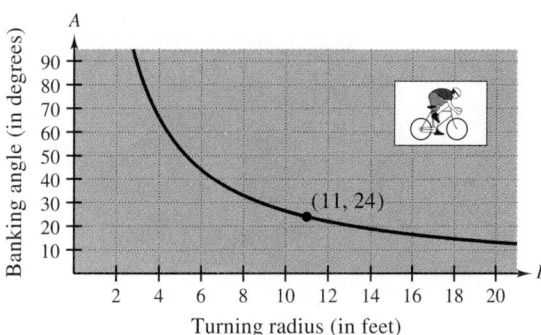

EXAMPLE 6   ■   Finding a Model for "Banking Angles"

Find a model that relates the banking angle A with the turning radius R for a bicyclist at a speed of 15 miles per hour. Use the model to find the banking angle for a turning radius of 8 feet.

Solution

From the graph, you can see that A is $24°$ when the turning radius R is 11 feet.

$$A = \frac{k}{R} \qquad \text{Model for inverse variation}$$

$$24 = \frac{k}{11} \qquad \text{Substitute 24 for } A \text{ and 11 for } R$$

$$264 = k \qquad \text{Solve for } k$$

The model is given by

$$A = \frac{264}{R},$$

where A is measured in degrees and R is measured in feet. When the turning radius is 8 feet, the banking angle is

$$A = \frac{264}{8} = 33°.$$

You can use the graph to confirm this result.   ■

The most popular bicycle road race is the Tour de France, lasting 24 days and covering 2500 miles in Europe. Greg LeMond was the first American winner.

Joint Variation

The model used in Example 5 involved both direct and inverse variation, and the word *and* was used to couple the two types of variation together. To describe two different *direct* variations in the same statement, the word *jointly* is used. For instance, the model

$$z = kxy$$

can be described by saying that z is *jointly* proportional to x and to y.

Joint Variation

The following statements are equivalent.

1. z **varies jointly** as x and y.

2. z is **jointly proportional** to x and y.

3. $z = kxy$ for some constant k.

If x, y, and z are related by the equation $z = kx^n y^m$, then z is said to vary jointly as the nth power of x and the mth power of y.

EXAMPLE 7 ■ Joint Variation

The *simple interest* for a certain savings account is jointly proportional to the time and the principal. After one quarter (three months), the interest for a principal of $6000 is $120. How much interest would a principal of $7500 earn in 5 months?

Solution

To begin, let I represent the interest earned (in dollars), let P represent the principal (in dollars), and let t represent the time (in years). Then, because the interest is jointly proportional to the time and the principal, you have the following model.

$$I = ktP$$

Because $I = 120$ when $P = 6000$ and $t = \frac{1}{4}$, you can determine that $k = 120/\left(6000 \cdot \frac{1}{4}\right) = 0.08$. Therefore, the model that relates the interest to the time and principal is

$$I = 0.08t P.$$

To find the interest earned for a principal of $7500 over a 5-month period of time, substitute 7500 for P and $\frac{5}{12}$ for t in the model and obtain an interest of

$$I = 0.08 \left(\frac{5}{12} \right)(7500) = \$250.$$

■

DISCUSSION PROBLEM ■ You Be the Instructor

Suppose you are teaching an algebra class and are writing a test that covers the material in this section. Write two test questions that you think give a fair representation of this material. (Assume that your students can spend 10 minutes on each question.) ■

Warm-Up

The following warm-up exercises involve skills that were covered in earlier sections. You will use these skills in the exercise set for this section.

In Exercises 1–6, solve the equation for k.

1. $4 = k(12)$ **2.** $250 = k(75)$ **3.** $46 = \dfrac{k}{0.02}$

4. $19 = \dfrac{k}{27}$ **5.** $360 = \dfrac{k(72)}{8}$ **6.** $700 = \dfrac{k(49)}{2^2}$

In Exercises 7–10, sketch the graph of $y = kx^2$ for the value of k.

7. $k = \frac{1}{4}$ **8.** $k = 2$ **9.** $k = -4$ **10.** $k = -\frac{1}{2}$

8.3 EXERCISES

In Exercises 1–12, write a mathematical model for the statement.

1. I varies directly as V.

2. C varies directly as r.

3. u is directly proportional to the square of v.

4. V varies directly as the cube of x.

5. p varies inversely as d.

6. S is inversely proportional to the square of v.

7. P is inversely proportional to the square root of $1 + r$.

8. A varies inversely as the fourth power of t.

9. A varies jointly as l and w.

10. V varies jointly as h and the square of r.

11. *Boyle's Law* If the temperature of a gas is not allowed to change, its absolute pressure P is inversely proportional to its volume V.

12. *Newton's Law of Universal Gravitation* The gravitational attraction F between two particles of masses m_1 and m_2 is directly proportional to the product of the masses and inversely proportional to the square of the distance r between the particles.

In Exercises 13–16, write a sentence using the variation terminology in this section to describe the formula.

13. *Area of a Triangle* $A = \frac{1}{2}bh$

14. *Area of a Circle* $A = \pi r^2$

15. *Volume of a Right Circular Cylinder* $V = \pi r^2 h$

16. *Average Speed* $r = \dfrac{d}{t}$

In Exercises 17–26, find the constant of proportionality and give the equation relating the variables.

17. s varies directly as t, and $s = 20$ when $t = 4$.

18. h is directly proportional to r, and $h = 28$ when $r = 12$.

19. F is directly proportional to the square of x, and $F = 500$ when $x = 40$.

20. v varies directly as the square root of s, and $v = 24$ when $s = 16$.

21. n varies inversely as m, and $n = 32$ when $m = 1.5$.

22. q is inversely proportional to p, and $q = \frac{3}{2}$ when $p = 50$.

23. F varies jointly as x and y, and $F = 500$ when $x = 15$ and $y = 8$.

24. V varies jointly as h and the square of b, and $V = 288$ when $h = 6$ and $b = 12$.

25. d varies directly as the square of x and inversely with r, and $d = 3000$ when $x = 10$ and $r = 4$.

26. z is directly proportional to x and inversely proportional to the square root of y, and $z = 720$ when $x = 48$ and $y = 81$.

27. *Revenue*   The total revenue R is directly proportional to the number of units sold x. When 500 units are sold, the revenue is $3875. Find the revenue when 635 units are sold.

28. *Revenue*   The total revenue R is directly proportional to the number of units sold x. When 25 units are sold, the revenue is $300. Find the revenue when 42 units are sold.

29. *Hooke's Law*   A force of 50 pounds stretches a spring 5 inches (see figure).
 (a) How far will a force of 20 pounds stretch the spring?
 (b) What force is required to stretch the spring 1.5 inches?

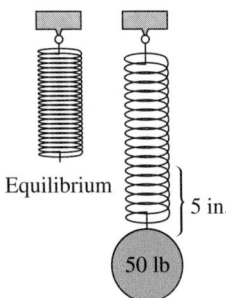

Equilibrium   5 in.

50 lb

30. *Hooke's Law*   A force of 50 pounds stretches a spring 3 inches.
 (a) How far will a force of 20 pounds stretch the spring?
 (b) What force is required to stretch the spring 1.5 inches?

31. *Hooke's Law*   A baby weighing $10\frac{1}{2}$ pounds compresses the spring of a baby scale 7 millimeters (see figure). Determine the weight of a baby that will compress the spring 12 millimeters.

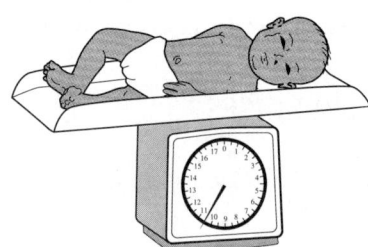

32. *Hooke's Law*   A force of 50 pounds compresses the spring in a scale 1.5 inches. Find the distance the spring will be compressed when a 20-pound object is placed on the scale.

33. *Stopping Distance*   The stopping distance d of an automobile is directly proportional to the square of its speed s. On a certain road surface, a car requires 75 feet to stop when its speed is 30 miles per hour. Estimate the stopping distance if the brakes are applied when the car is traveling at 50 miles per hour under similar road conditions.

34. *Distance* Neglecting air resistance, the distance d that an object falls varies directly as the square of the time t it has been falling. If an object falls 64 feet in 2 seconds, determine the distance it will fall in 6 seconds.

35. *Power Generation* The power P generated by a wind turbine varies directly as the cube of the wind speed w. The turbine generates 750 watts of power in a 25-mile-per-hour wind. Find the power it generates in a 40-mile-per-hour wind.

36. *Velocity of a Stream* The diameter d of a particle moved by a stream is directly proportional to the square of the velocity v of the stream. A stream with a velocity of $\frac{1}{4}$ mile per hour can move coarse sand particles about 0.02 inch in diameter. What must the velocity be to carry particles with a diameter of 0.12 inch?

37. *Weight* A person's weight on the moon varies directly with the person's weight on earth. Neil Armstrong, the first man on the moon, weighed 360 pounds on earth when wearing his heavy equipment. On the moon, he weighed only 60 pounds, with equipment. If the first woman in space, Valentina V. Tereshkova, had landed on the moon and weighed 54 pounds, with equipment, how much would she have weighed on earth, with equipment?

38. *Snowshoes* When a person walks, the pressure P on each boot sole varies inversely with the area A of the sole. Denise is trudging through deep snow, wearing boots that have a sole area of 29 square inches each. The boot-sole pressure is 4 pounds per square inch. If Denise were wearing snowshoes, each with an area 11 times that of her boot soles, what would be the pressure on each snowshoe? The constant of proportionality in this problem is Denise's weight. How much does she weigh?

Snowshoes distribute a person's weight over a large area, preventing the person from sinking into the snow.

39. *Oil Spill* The graph shows the percentage p of oil that remained in Chedabucto Bay, Nova Scotia, after an oil spill. The cleaning of the spill was left primarily to natural actions such as wave motion, evaporation, photochemical decomposition, and bacterial decomposition. After about a year, the percentage that remained varied inversely with time. Find a model that relates p and t, where t is the number of years since the spill. Then use the model to find the percentage of the oil that remained $6\frac{1}{2}$ years after the spill.

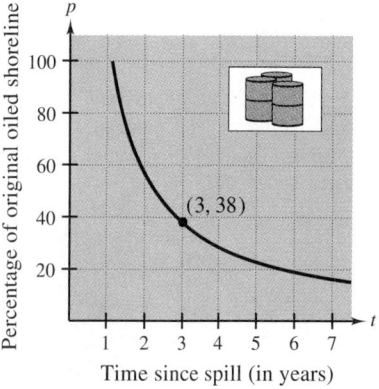

40. *Ocean Temperatures* The graph shows the temperature of the water in the north central Pacific Ocean. At depths greater than 900 meters, the water temperature varies inversely with the water depth. Find a model that relates the temperature T with the depth d. What is the temperature at a depth of 4385 meters?

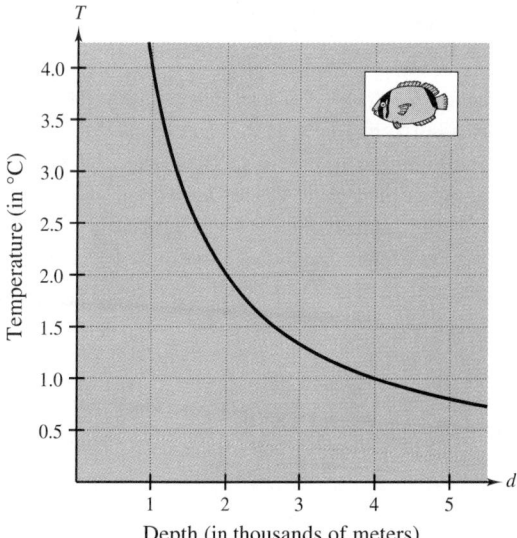

41. Demand Function A company has found that the daily demand x for its product is inversely proportional to the price p. When the price is $5, the demand is 800 units. Approximate the demand if the price is increased to $6.

42. Weight of an Astronaut The gravitational force F with which an object is attracted to the earth is inversely proportional to the square of its distance r from the center of the earth. If an astronaut weighs 190 pounds on the surface of the earth ($r \approx 4000$ miles), what will the astronaut weigh 1000 miles above the earth's surface?

43. Amount of Illumination The illumination I from a light source varies inversely as the square of the distance d from the light source (see figure). If you raise a study lamp from 18 inches to 36 inches over your study desk, the illumination will change by what factor?

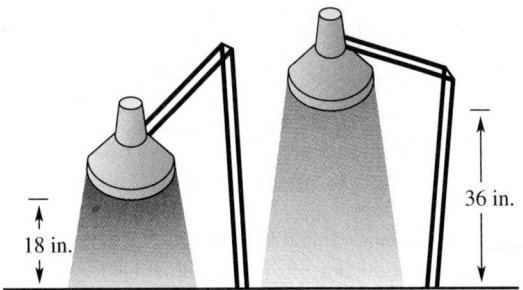

44. Frictional Force The frictional force F between the tires and the road required to keep a car on a curved section of a highway is directly proportional to the square of the speed s of the car. If the speed of the car is doubled, the force will change by what factor?

45. Load of a Beam The load that can be safely supported by a horizontal beam varies jointly as the width of the beam and the square of its depth and inversely as the length of the beam. A beam with width 3 inches, depth 8 inches, and length 10 feet can safely support 2000 pounds. Determine the safe load of a beam made from the same material if its depth is increased to 10 inches.

46. Best Buy The prices of the 9-inch, 12-inch, and 15-inch diameter pizzas at a certain pizza shop are $6.78, $9.78, and $12.18, respectively. One would expect that the price of a certain size pizza would be directly proportional to its surface area. Is that the case for this pizza shop? If not, which size of pizza is the best buy?

Exploratory Exercise In Exercises 47–50, use the value of k to complete the table for the direct variation model $y = kx^2$. Plot the points on the rectangular coordinate system.

x	2	4	6	8	10
$y = kx^2$					

47. $k = 1$ **48.** $k = 2$

49. $k = \frac{1}{2}$ **50.** $k = \frac{1}{4}$

Exploratory Exercise In Exercises 51–54, use the value of k to complete the table for the inverse variation model $y = k/(x^2)$. Plot the points on the rectangular coordinate system.

x	2	4	6	8	10
$y = \dfrac{k}{x^2}$					

51. $k = 2$ **52.** $k = 5$

53. $k = 10$ **54.** $k = 20$

Mid-Chapter Quiz

Take this quiz as you would take a quiz in class. After you are done, check your work against the answers in the back of the book.

In Exercises 1–8, let $f(x) = 2x - 8$ and $g(x) = x^2$. Find the specified combination of the functions and find its domain.

1. $(f - g)(x)$

2. $(f + g)(x)$

3. $(fg)(x)$

4. $(g - f)(x)$

5. $(f/g)(x)$

6. $(g/f)(x)$

7. $(f \circ g)(x)$

8. $(g \circ f)(x)$

9. *Weekly Production Cost* The weekly cost of producing x units in a manufacturing process is given by the function $C(x) = 60x + 750$. The number of units produced in t hours during a week is given by $x(t) = 50t$, $0 \le t \le 40$. Find, simplify, and interpret $C(x(t))$.

10. Verify algebraically that the functions $f(x) = 100 - 15x$ and $g(x) = \frac{1}{15}(100 - x)$ are inverses of each other.

11. Find the inverse of the function $f(x) = x^3 - 8$. Sketch the graph of the function and its inverse.

12. Determine whether the function $g(x) = 1/x^2$ has an inverse. If not, place a restriction on the domain such that the graph of the restricted function satisfies the horizontal line test for inverses.

In Exercises 13–16, find the constant of proportionality and give the equation relating the variables.

13. z varies directly as t, and $z = 12$ when $t = 4$.

14. m varies inversely as n, and $m = 2$ when $n = 7$.

15. S varies jointly as h and the square of r, and $S = 120$ when $h = 6$ and $r = 2$.

16. N varies directly as the square of t and inversely with s, and $N = 300$ when $t = 10$ and $s = 5$.

17. *Mining Danger* One of the dangers of coal mining is the methane gas that can leak out of seams in the rock. Methane forms an explosive mixture with air at a concentration of 5% or greater. Suppose a steady leak of methane begins in a coal mine so that the concentration of methane gas varies directly with time. Twelve minutes after a leak begins, the concentration of methane in the air is 2%. If the leak continues at the same rate, when could an explosion occur?

SECTION	Polynomial Functions and Their Graphs
8.4	*Graphs of Polynomial Functions* • *Sketching Graphs of Polynomial Functions* • *Applications of Polynomial Functions*

Graphs of Polynomial Functions

A function of the form

$$f(x) = a_n x^n + a_{n-1} x^{n-1} + \cdots + a_1 x + a_0, \qquad a_n \neq 0$$

Discuss the general description of a polynomial function. Explain that a polynomial function has real number coefficients and that the exponents of the variables are whole numbers.

is a **polynomial function** of degree n. You already know how to sketch the graph of a polynomial function of degree 0, 1, or 2.

Function	*Degree*	*Graph*
$f(x) = a$	0	Horizontal line
$f(x) = ax + b$	1	Line of slope a
$f(x) = ax^2 + bx + c$	2	Parabola

EXAMPLE 1   ■   Polynomial Functions of Degrees 0, 1, and 2

(a) The graph of $f(x) = 3$ is a horizontal line, as shown in Figure 8.12(a). This polynomial function has degree 0 and is called a *constant* function.

(b) The graph of $f(x) = -\frac{1}{2}x + 3$ is a line with a slope of $-\frac{1}{2}$ and a y-intercept of 3, as shown in Figure 8.12(b). This polynomial function has degree 1 and is called a *linear* function.

(c) The graph of $f(x) = x^2 + 2x - 3$ is a parabola with a vertex of $(-1, -4)$, as shown in Figure 8.12(c). This polynomial function is of degree 2 and is called a *quadratic* function.

FIGURE 8.12

(a)
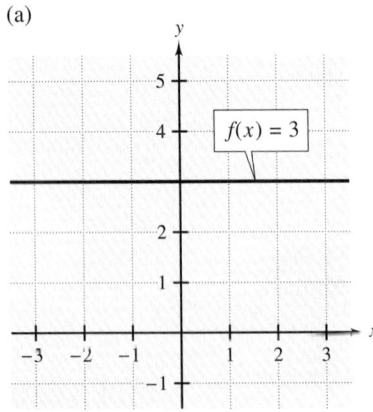

(b)
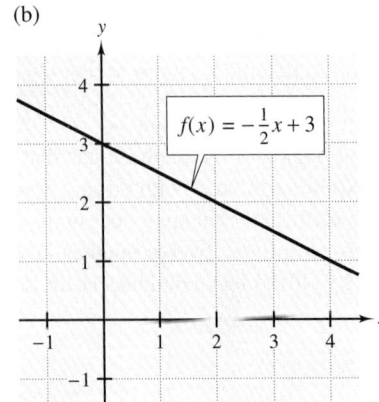

(c)
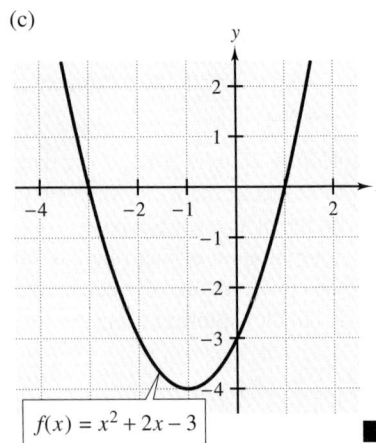

■

FIGURE 8.13

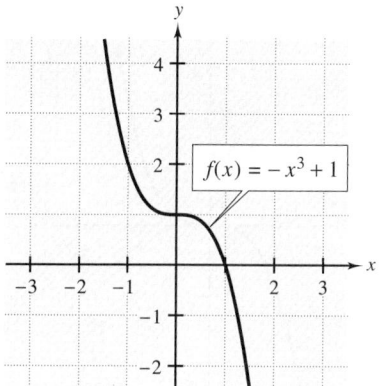
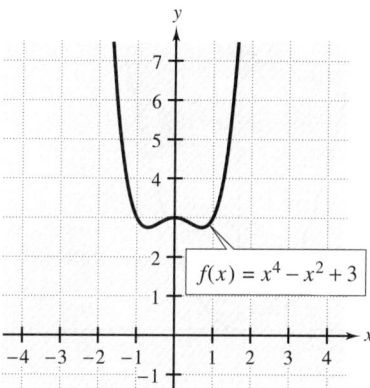
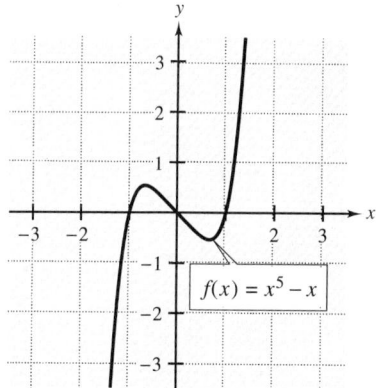

The graphs of polynomial functions of degree greater than 2 are more difficult to sketch, as illustrated in Figure 8.13. In this lesson, however, you will learn to recognize some of their basic features. With these features and point plotting, you will be able to make rough sketches *by hand.* (If you go on to take a class in calculus, you will learn to sketch more accurate graphs of polynomial functions of degree greater than 2.)

Features of Graphs of Polynomial Functions

1. The graph of a polynomial function is **continuous.** This means that the graph has no breaks—you could sketch the graph without lifting your pencil from the paper.

2. The graph of a polynomial function has only smooth turns. A function of degree n has *at most* $n - 1$ turns.

3. If the leading coefficient of the polynomial function is positive, then the graph rises to the right. If the leading coefficient is negative, then the graph falls to the right.

The polynomial functions (of degree 2 or greater) that have the simplest graphs are the monomial functions $f(x) = a_n x^n$. When n is *even*, the graph is similar to the graph of $f(x) = x^2$, as shown in Figure 8.14(a) on the next page. When n is *odd*, the graph is similar to the graph of $f(x) = x^3$, as shown in Figure 8.14(b). Moreover, the greater the value of n, the flatter the graph of a monomial is on the interval $-1 \leq x \leq 1$.

FIGURE 8.14

(a) For n even, the graph of $y = x^n$ is similar to the graph of $y = x^2$.

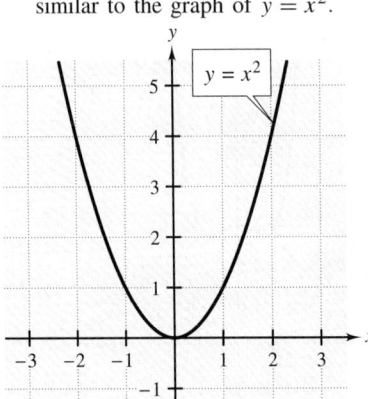

(b) For n odd, the graph of $y = x^n$ is similar to the graph of $y = x^3$.

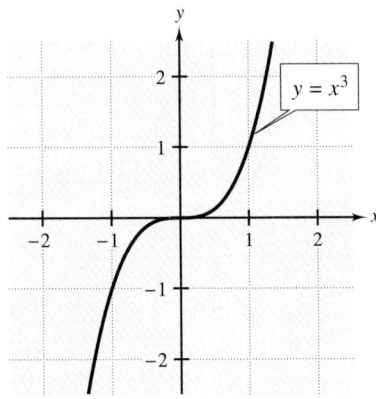

EXAMPLE 2 ■ **Sketching Transformations of Monomial Functions**

As a classroom demonstration or class exercise, graph $f(x) = x^4$, $g(x) = -x^4$, $h(x) = x^4 - 3$, and $r(x) = (x + 5)^4$ on a single grid using a graphing utility. Compare and contrast the graphs. Help students see the value of technology as a tool for investigation.

(a) *Reflection:* To sketch the graph of

$$g(x) = -x^5,$$

reflect the graph of $f(x) = x^5$ in the x-axis, as shown in Figure 8.15(a).

(b) *Vertical Shift:* To sketch the graph of

$$g(x) = x^4 + 1,$$

shift the graph of $f(x) = x^4$ up one unit, as shown in Figure 8.15(b).

(c) *Horizontal Shift:* To sketch the graph of

$$g(x) = (x + 1)^3,$$

shift the graph of $f(x) = x^3$ one unit to the left, as shown in Figure 8.15(c).

FIGURE 8.15

(a)

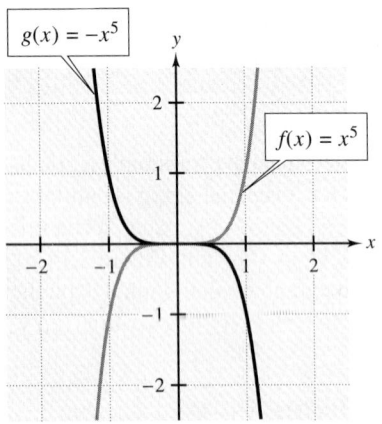

(b)

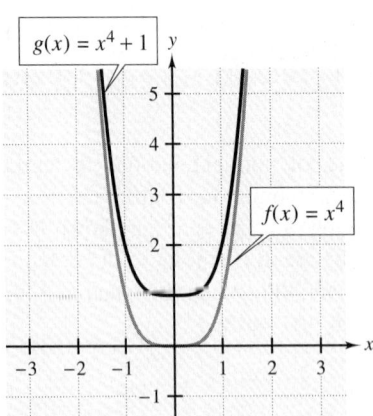

(c)

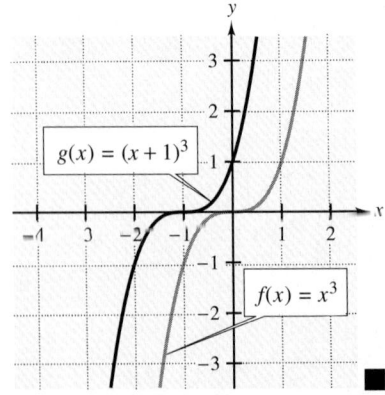

Sketching Graphs of Polynomial Functions

When sketching the graph of a polynomial function, it helps to label the intercepts. The y-intercept of the graph of a polynomial function is easy to find—it is the value of the function when $x = 0$. For instance, the y-intercept of the graph of $f(x) = x^3 - 7x + 6$ is

$$f(0) = 0^3 - 7(0) + 6 = 6, \qquad y\text{-intercept}$$

Have students discuss why a polynomial function has exactly *one* y-intercept.

as shown in Figure 8.16. The graph of a polynomial function always has exactly one y-intercept.

The x-intercepts of the graph of a polynomial function are not as easy to find. Moreover, the graph can have no x-intercepts, one x-intercept, or several x-intercepts. The x-intercepts of the graph of a polynomial function are the x-values for which $f(x) = 0$. You can find these values by solving the equation

$$f(x) = 0. \qquad x\text{-intercepts}$$

These values are also called the **zeros** of the function. If you go on to take a course in college algebra, you will study techniques for solving polynomial equations of degree 3 or greater. You will also learn that the graph of a polynomial function of degree n has *at most n* x-intercepts. For instance, the graph of $f(x) = x^3 - 7x + 6$ has three x-intercepts, as shown in Figure 8.16. Try showing that $f(x) = 0$ when $x = -3$, $x = 1$, and $x = 2$.

FIGURE 8.16

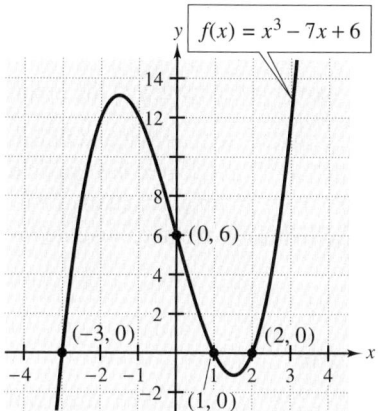

EXAMPLE 3 ■ Sketching the Graph of a Cubic Polynomial Function

Sketch the graph of $f(x) = x^3 - 9x$.

Solution

Because the leading coefficient is positive, you know that the graph rises to the right. Also, because the degree of the polynomial function is 3, you know that the graph can have at most two turns. Using these general observations with the values given in the following table, you can sketch the graph as shown in Figure 8.17. Notice that the graph has three x-intercepts: -3, 0, and 3.

FIGURE 8.17

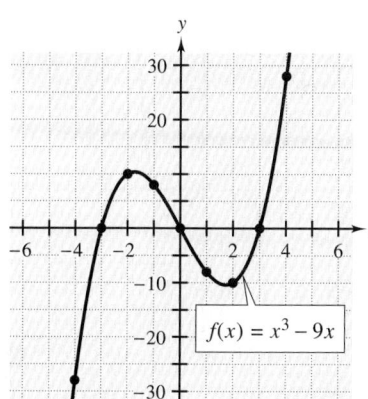

x	-4	-3	-2	-1	0	1	2	3	4
$f(x)$	-28	0	10	8	0	-8	-10	0	28

■

NOTE The x-intercepts of the graphs of some polynomial functions can be found by setting $f(x) = 0$ and factoring. For the equation in Example 3, this is done as follows.

$$x^3 - 9x = 0$$
$$x(x^2 - 9) = 0$$
$$x(x + 3)(x - 3) = 0$$
$$x = 0$$
$$x + 3 = 0 \implies x = -3$$
$$x - 3 = 0 \implies x = 3$$

Thus, the three x-intercepts are 0, -3, and 3, as shown in the table and the graph.

In Figure 8.17, note that the graph of the function rises to the right and falls to the left. In general, the graph of a polynomial function of *odd* degree has opposite behaviors to the right and left, whereas the graph of a polynomial function of *even* degree has the same behavior to the right and left. This is further illustrated in Example 4.

EXAMPLE 4 ■ Identifying the Left and Right Behavior of a Graph

(a) Because the leading coefficient of $f(x) = -x^3 + 4x$ is negative, the graph falls to the right. Because the degree is odd, the graph rises to the left, as shown in Figure 8.18(a).

(b) Because the leading coefficient of $f(x) = x^4 - 5x^2 + 4$ is positive, the graph rises to the right. Because the degree is even, the graph also rises to the left, as shown in Figure 8.18(b).

(c) Because the leading coefficient of $f(x) = x^5 - x$ is positive, the graph rises to the right. Because the degree is odd, the graph falls to the left, as shown in Figure 8.18(c).

FIGURE 8.18

(a)

(b)

(c)

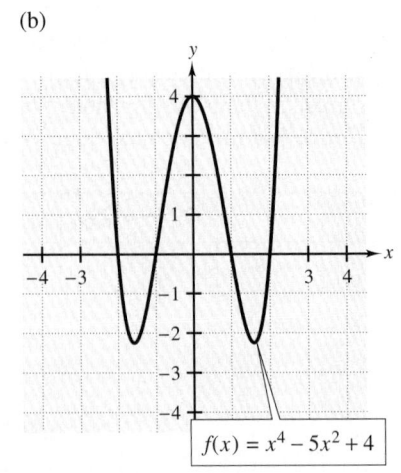

$f(x) = -x^3 + 4x$

$f(x) = x^4 - 5x^2 + 4$

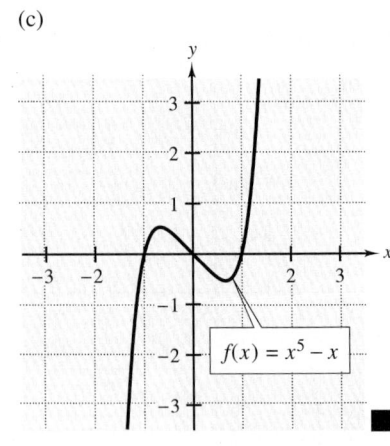

$f(x) = x^5 - x$

Applications of Polynomial Functions

EXAMPLE 5 ■ Charitable Contributions

In 1990, the average percentage P (in decimal form) of income that Americans gave to charitable organizations was related to their income x (in thousands of dollars) by the model

$$P = 0.000014x^2 - 0.001529x + 0.05855, \qquad 5 \leq x \leq 100.$$

Use this model to write a model for the average *amount* given. (*Source: Giving and Volunteering in the United States*)

Solution

The graph of the given model is shown in Figure 8.19. Notice that the vertex of the graph occurs when $x \approx 55$. This means that in 1990 the income level that averaged the smallest percentage contributions to charity was the $55,000 level. To find the average amount A that each income level gave to charity, you can multiply the average percentage-of-income model by the income ($1000x$) to obtain

$$A = 1000x(0.000014x^2 - 0.001529x + 0.05855), \qquad 5 \leq x \leq 100$$
$$= 0.014x^3 - 1.529x^2 + 58.55x.$$

The graph of this third-degree polynomial is shown in Figure 8.20. Note in the graph that the average *amount* given to charity increases as the income level increases.

FIGURE 8.19

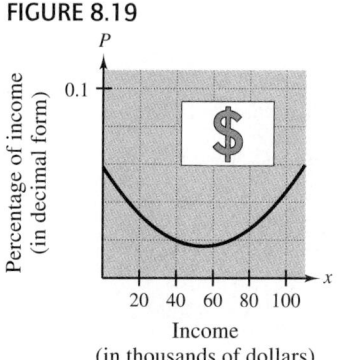

Income
(in thousands of dollars)

FIGURE 8.20

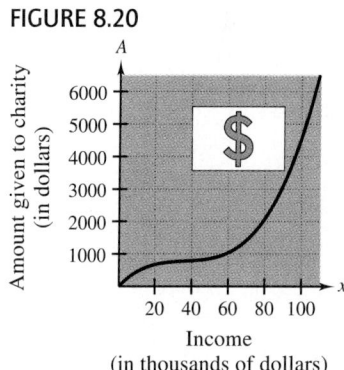

Income
(in thousands of dollars)

DISCUSSION PROBLEM ■ Finding x-Intercepts of Polynomial Functions

In Section 7.1, you learned to solve polynomial equations of quadratic type. Use that technique to find the x-intercepts of

$$f(x) = x^4 - 5x + 4.$$

Warm-Up

The following warm-up exercises involve skills that were covered in earlier sections. You will use these skills in the exercise set for this section.

In Exercises 1–6, factor the expression completely.

1. $12x^2 + 7x - 10$ **2.** $25x^3 - 60x^2 + 36x$ **3.** $12z^4 + 17z^3 + 5z^2$

4. $y^3 + 125$ **5.** $x^3 + 3x^2 - 4x - 12$ **6.** $x^3 + 2x^2 + 3x + 6$

In Exercises 7–10, find all real solutions of the equation.

7. $5x^2 + 8 = 0$ **8.** $x^2 - 6x + 4 = 0$

9. $4x^2 + 4x - 11 = 0$ **10.** $x^4 - 18x^2 + 81 = 0$

8.4   EXERCISES

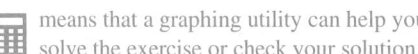 means that a graphing utility can help you solve the exercise or check your solution.

 In Exercises 1–6, identify and sketch the graph of the polynomial function of degree 0, 1, or 2.

1. $g(x) = -2$ **2.** $f(x) = 8$ **3.** $f(x) = 7 - 3x$

4. $h(x) = \frac{1}{2}x + 1$ **5.** $f(x) = 4 - (x - 3)^2$ **6.** $h(t) = t^2 + 2t + 1$

In Exercises 7–12, state the maximum number of turns in the graph of the function.

7. $f(x) = 2x^5 - x + 6$ **8.** $f(x) = \frac{1}{2}x^3 + 1$ **9.** $g(s) = -3s^2 + 2s + 4$

10. $h(u) = (u - 3)^2 - 1$ **11.** $h(x) = x^4 + 9x - 1$ **12.** $f(x) = -2x^6 + 3x + 7$

In Exercises 13–18, describe the transformation of g that would produce f.

13. $f(x) = (x - 5)^3$ **14.** $f(x) = (x + 3)^2$ **15.** $f(x) = x^4 + 3$

     $g(x) = x^3$ $g(x) = x^2$ $g(x) = x^4$

16. $f(x) = x^5 - 2$ **17.** $f(x) = -x^6$ **18.** $f(x) = -x$

     $g(x) = x^5$ $g(x) = x^6$ $g(x) = x$

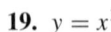 In Exercises 19 and 20, use the graph of the given equation to sketch the graphs of the specified transformations.

19. $y = x^3$

   (a) $f(x) = (x - 2)^3$

   (b) $f(x) = x^3 - 2$

   (c) $f(x) = (x - 2)^3 - 2$

   (d) $f(x) = -x^3$

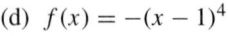

20. $y = x^4$

   (a) $f(x) = (x + 3)^4$

   (b) $f(x) = x^4 - 3$

   (c) $f(x) = 4 - x^4$

   (d) $f(x) = -(x - 1)^4$

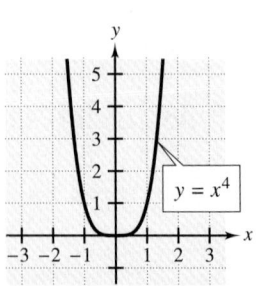

In Exercises 21–26, determine the right and left behavior of the graph of the polynomial function.

21. $f(x) = 2x^2 - 3x + 1$

22. $g(x) = 5 - \frac{7}{2}x - 3x^2$

23. $f(x) = \frac{1}{3}x^3 + 5x$

24. $f(x) = -2.1x^5 + 4x^3 - 2$

25. $f(x) = 6 - 2x + 4x^2 - 5x^3$

26. $h(x) = 1 - x^6$

In Exercises 27–38, find the x- and y-intercepts of the polynomial function.

27. $f(x) = x^2 - 25$

28. $f(x) = 49 - x^2$

29. $h(x) = x^2 - 6x + 9$

30. $f(x) = x^2 + 10x + 25$

31. $f(x) = x^2 + x - 2$

32. $f(x) = 3x^2 - 12x + 3$

33. $f(x) = x^3 - 4x^2 + 4x$

34. $f(x) = x^4 - x^3 - 20x^2$

35. $g(x) = \frac{1}{2}x^4 - \frac{1}{2}$

36. $f(x) = 2x^4 - 2x^2 - 40$

37. $f(x) = 5x^4 + 15x^2 + 10$

38. $f(x) = x^3 - 4x^2 - 25x + 100$

In Exercises 39–46, match the polynomial function with the correct graph. [The graphs are labeled (a), (b), (c), (d), (e), (f), (g), and (h).]

39. $f(x) = -3x + 5$

40. $f(x) = x^2 - 2x$

41. $f(x) = -2x^2 - 8x - 9$

42. $f(x) = 3x^3 - 9x + 1$

43. $f(x) = -\frac{1}{3}x^3 + x - \frac{2}{3}$

44. $f(x) = -\frac{1}{4}x^4 + 2x^2$

45. $f(x) = 3x^4 + 4x^3$

46. $f(x) = x^5 - 5x^3 + 4x$

(a)
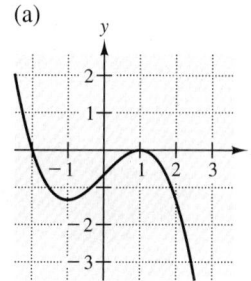

(b)
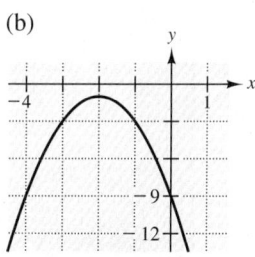

(c)
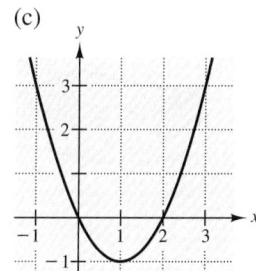

(d)
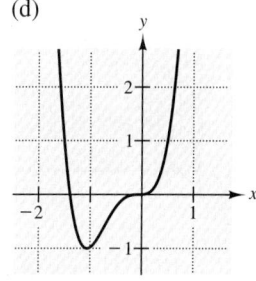

(e)
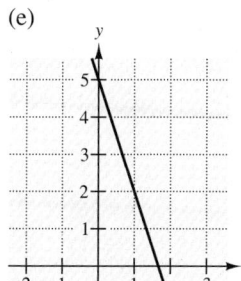

(f)
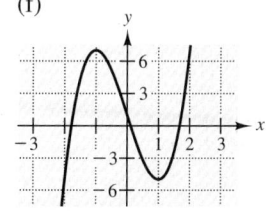

(g)
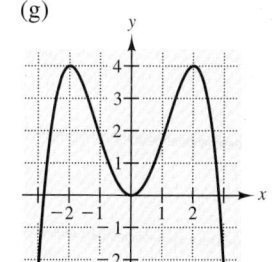

(h)
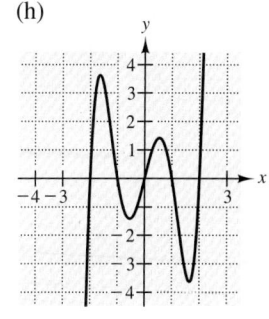

In Exercises 47–58, sketch the graph of the function.

47. $f(x) = -\frac{3}{2}$ **48.** $h(x) = \frac{1}{3}x - 3$ **49.** $f(x) = -x^3$ **50.** $g(x) = -x^4$

51. $f(x) = x^5 + 1$ **52.** $f(x) = x^4 - 2$ **53.** $h(x) = (x + 2)^4$ **54.** $g(x) = (x - 4)^3$

55. $f(x) = 1 - x^6$ **56.** $g(x) = 1 - (x + 1)^6$ **57.** $f(x) = x^3 - 3x^2$ **58.** $f(x) = x^3 - 4x$

 In Exercises 59–64, use a graphing calculator to graph the function.

59. $f(x) = -x^3 + 4x^2$ **60.** $f(x) = \frac{1}{4}x^4 - 2x^2$ **61.** $g(t) = -\frac{1}{4}(t - 2)^2(t + 2)^2$

62. $f(x) = x^2(x - 4)$ **63.** $h(s) = s^5 + s$ **64.** $g(t) = t^5 + 3t^3 - t$

65. *Volume of a Box* An open box is made from a 12-inch-square piece of material by cutting equal squares from each corner and turning up the sides (see figure).

(a) Verify that the volume of the box is

$$V(x) = 4x(6 - x)^2.$$

(b) Determine the domain of the function V.

(c) Sketch the graph of the function and use the graph to estimate the value of x for which $V(x)$ is maximum.

66. *Volume of a Box* An open box with locking tabs is made from a 12-inch-square piece of material. This is done by cutting equal squares from each corner and folding along the dashed lines as shown in the figure.

(a) Verify that the volume of the box is

$$V(x) = 8x(3 - x)(6 - x).$$

(b) Determine the domain of the function V.

(c) Sketch the graph of the function and use the graph to estimate the value of x for which $V(x)$ is maximum.

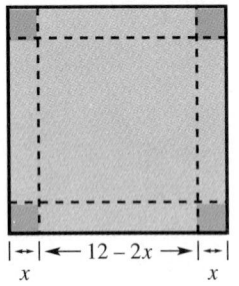
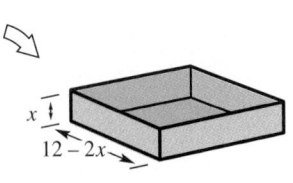
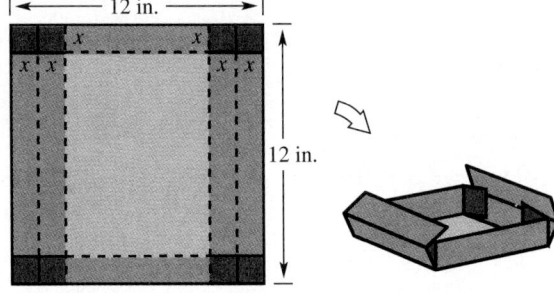

SECTION
8.5

Circles, Ellipses, and Hyperbolas

The Conics   •   Circles   •   Ellipses   •   Hyperbolas   •   Applications of Conics

The Conics

Historical Note: The first notable work of the French mathematician Blaise Pascal (1623–1662) was a paper about the conic sections.

In Section 7.5, you saw that the graph of a second-degree equation $y = ax^2 + bx + c$ is a parabola. A parabola is one of four types of curves that are called **conics** or **conic sections.** The other three types are circles, ellipses, and hyperbolas. As indicated in Figure 8.21, the name "conic" relates to the fact that each of these figures can be obtained by intersecting a plane with a double-napped cone.

FIGURE 8.21

Circle   Parabola   Ellipse   Hyperbola

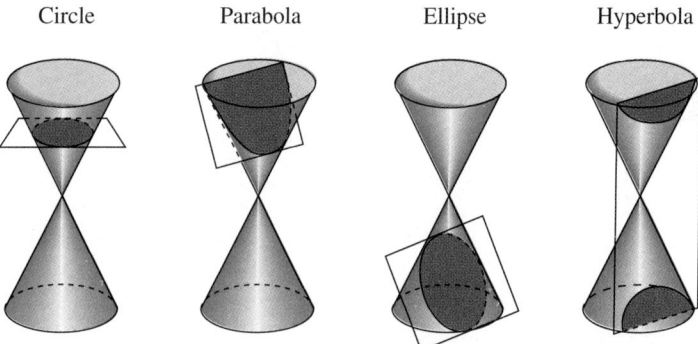

Practical applications of the conic sections are numerous and varied. Encourage students to suggest other applications of parabolas, ellipses, and hyperbolas.

Conic sections occur in many practical applications. Reflective surfaces in satellite dishes, flashlights, and telescopes often have a parabolic shape. The orbits of planets are elliptical, and the orbits of comets are usually either elliptical or hyperbolic. Ellipses and parabolas are also used in the construction of archways and bridges.

Circles

FIGURE 8.22

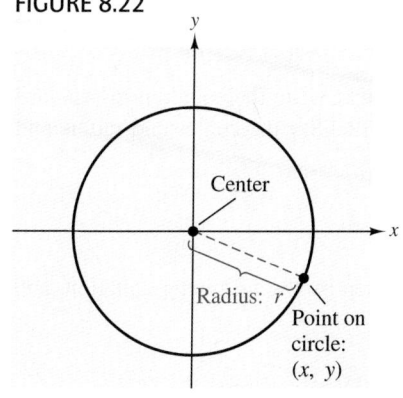

A **circle** in the rectangular coordinate plane consists of all points (x, y) that are a given positive distance r from a fixed point, called the **center** of the circle. The distance r is called the **radius** of the circle. If the center of the circle is the origin, as shown in Figure 8.22, then the relationship between the coordinates of any point (x, y) on the circle and the radius r is given

$$\text{Radius} = r = \sqrt{(x - 0)^2 + (y - 0)^2} \qquad \text{Distance Formula}$$
$$= \sqrt{x^2 + y^2}. \qquad \text{Here is another application of the distance formula.}$$

By squaring both sides of this equation, you obtain the equation given on the next page, which is called the **standard form of the equation of a circle.**

Standard Form of the Equation of a Circle (Center at Origin)	The **standard form of the equation of a circle** centered at the origin is $$x^2 + y^2 = r^2.$$ The positive number r is called the **radius** of the circle.

EXAMPLE 1 ■   Finding an Equation of a Circle

Mention that the major axis is the longer axis.

Find an equation of the circle that is centered at the origin and that has a radius of 2.

Solution

Using the standard form of the equation of a circle (with center at the origin)

$$x^2 + y^2 = r^2$$

and $r = 2$, you obtain

$$x^2 + y^2 = 2^2$$
$$x^2 + y^2 = 4.$$

The circle given by this equation is shown in Figure 8.23.

FIGURE 8.23

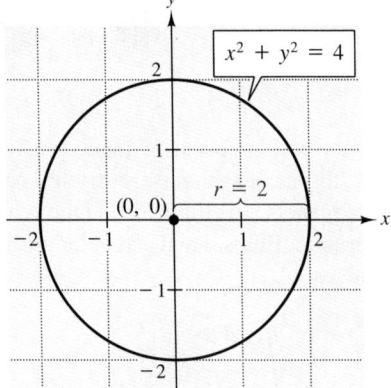

To sketch the circle for a given equation, first write the equation in standard form. Then, from the standard form, you can identify the center and radius and sketch the circle.

EXAMPLE 2 ■   Sketching a Circle

Identify the center and radius of the circle given by the following equation, and sketch the circle.

$$4x^2 + 4y^2 - 25 = 0$$

FIGURE 8.24

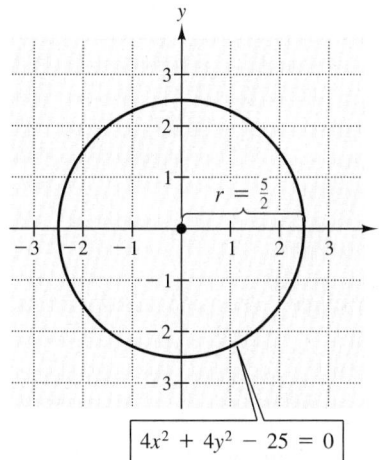

$$4x^2 + 4y^2 - 25 = 0$$

Solution

Begin by writing the given equation in standard form.

$$4x^2 + 4y^2 - 25 = 0 \qquad \text{Given equation}$$

$$4x^2 + 4y^2 = 25 \qquad \text{Add 25 to both sides}$$

$$x^2 + y^2 = \frac{25}{4} \qquad \text{Divide both sides by 4}$$

$$x^2 + y^2 = \left(\frac{5}{2}\right)^2 \qquad \text{Standard form}$$

Now, from this standard form, you can see that the graph of the equation is a circle that is centered at the origin and that has a radius of $\frac{5}{2}$, as shown in Figure 8.24.

■

Ellipses

Another type of conic, called an **ellipse,** is defined as follows. An **ellipse** in the rectangular coordinate system consists of all points (x, y) such that the sum of the distances between (x, y) and two distinct fixed points is a constant, as shown in Figure 8.25. Each of the two fixed points is called a **focus** of the ellipse. (The plural of focus is **foci.**) In this text, the discussion of ellipses is restricted to those that are centered at the origin and that have foci that lie on the x-axis or on the y-axis.

FIGURE 8.25

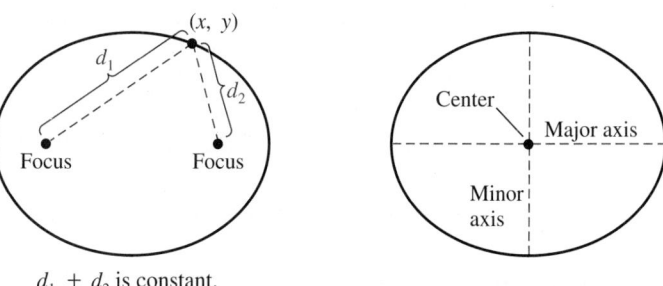

$d_1 + d_2$ is constant.

The line through the foci intersects the ellipse at two points, called the **vertices.** The line segment joining the vertices is called the **major axis,** and its midpoint is called the **center** of the ellipse. The line segment perpendicular to the major axis at the center is called the **minor axis** of the ellipse, and the points at which the minor axis intersect the ellipse are called **co-vertices.**

FIGURE 8.26

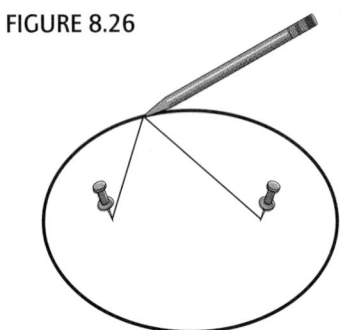

You can visualize the definition of an ellipse by imagining two thumbtacks placed at the foci, as shown in Figure 8.26. If the ends of a fixed length of string are fastened to the thumbtacks and the string is drawn taut with a pencil, the path traced by the pencil will be an ellipse.

The standard form of the equation of an ellipse takes one of two forms, depending on whether the major axis is horizontal or vertical.

Standard Form of the Equation of an Ellipse (Center at Origin)

The **standard form of the equation of an ellipse** with its center at the origin and with major and minor axes of lengths $2a$ and $2b$, respectively, is

$$\frac{x^2}{a^2} + \frac{y^2}{b^2} = 1 \quad \text{or} \quad \frac{x^2}{b^2} + \frac{y^2}{a^2} = 1, \qquad 0 < b < a.$$

The vertices lie on the major axis, a units from the center, and the co-vertices lie on the minor axis, b units from the center, as shown in Figure 8.27.

Major axis is horizontal.   Major axis is vertical.
Minor axis is vertical.   Minor axis is horizontal.

FIGURE 8.27

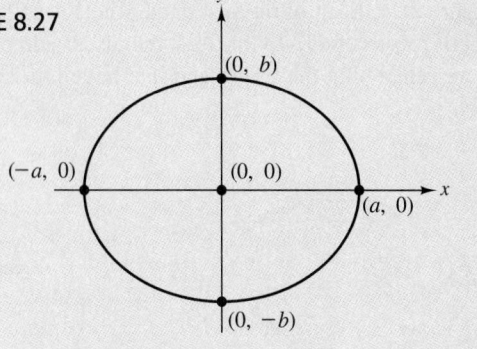

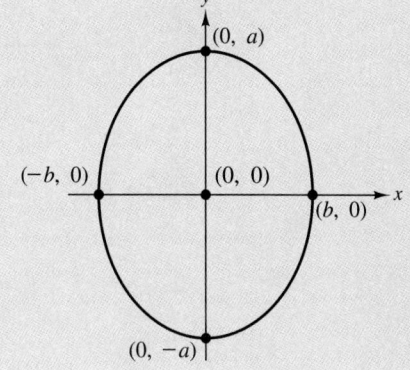

EXAMPLE 3   ■   Finding an Equation of an Ellipse

Find an equation of the ellipse with vertices $(-3, 0)$ and $(3, 0)$ and with co-vertices $(0, -2)$ and $(0, 2)$.

FIGURE 8.32   Transverse axis is horizontal.   Transverse axis is vertical.

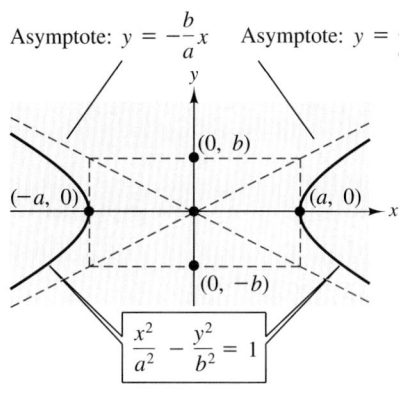

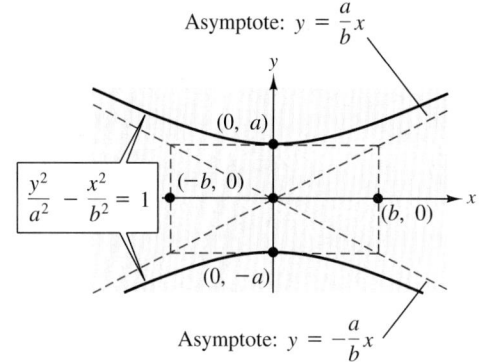

EXAMPLE 5   ■   Sketching a Hyperbola

Identify the vertices of the hyperbola given by the following equation, and sketch the hyperbola.

$$\frac{x^2}{36} - \frac{y^2}{16} = 1$$

Solution

From the standard form of the equation

$$\frac{x^2}{6^2} - \frac{y^2}{4^2} = 1,$$

you can see that the center of the hyperbola is the origin and the transverse axis is horizontal. Therefore, the vertices lie six units to the left and right of the center at the points $(-6, 0)$ and $(6, 0)$. Because $a = 6$ and $b = 4$, you can sketch the hyperbola by first drawing a central rectangle with a width of $2a = 12$ and a height of $2b = 8$, as shown in Figure 8.33(a). Next, draw the asymptotes of the hyperbola through the corners of the central rectangle and plot the vertices. Finally, draw the hyperbola, as shown in Figure 8.33(b).

FIGURE 8.33   (a)

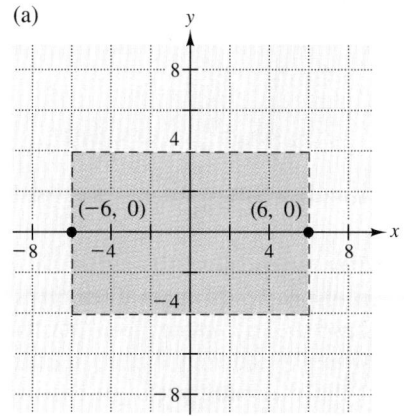

(b)

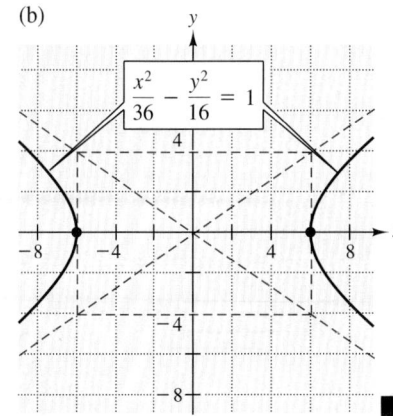

EXAMPLE 6 ■ Sketching a Hyperbola

Identify the vertices of the hyperbola given by the following equation, and sketch the hyperbola.

$$4y^2 - 9x^2 = 36$$

Solution

To begin, write the given equation in standard form.

$$4y^2 - 9x^2 = 36 \qquad \text{Given equation}$$

$$\frac{4y^2}{36} - \frac{9x^2}{36} = \frac{36}{36} \qquad \text{Divide both sides by 36}$$

$$\frac{y^2}{9} - \frac{x^2}{4} = 1 \qquad \text{Simplify}$$

$$\frac{y^2}{3^2} - \frac{x^2}{2^2} = 1 \qquad \text{Standard form}$$

Now, from the standard form, you can see that the center of the hyperbola is the origin and the transverse axis is vertical. Thus, the vertices lie three units above and below the center at $(0, 3)$ and $(0, -3)$. Because $a = 3$ and $b = 2$, you can sketch the hyperbola by first drawing a central rectangle with a width of $2b = 4$ and a height of $2a = 6$, as shown in Figure 8.34(a). Next, draw the asymptotes of the hyperbola through the corners of the central rectangle and plot the vertices. Finally, draw the hyperbola, as shown in Figure 8.34(b).

FIGURE 8.34   (a)

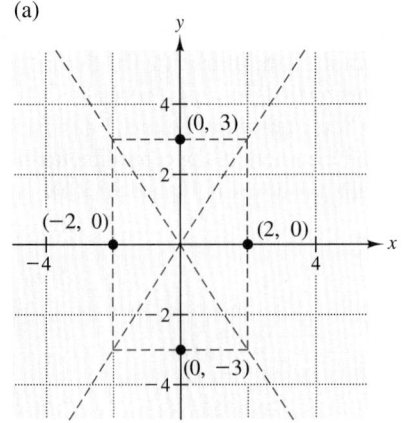

(b)

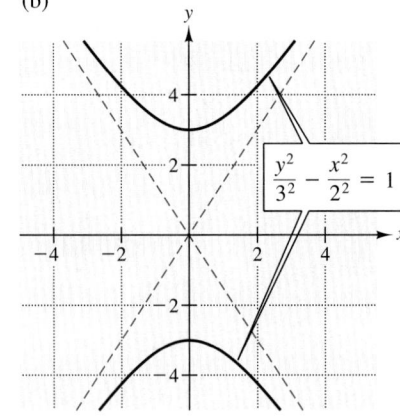

To help students determine whether a transverse axis is horizontal or vertical, have them write the following equations in standard form and then identify the transverse axis:

(a) $3x^2 - 2y^2 = 12$

(b) $8y^2 - 5x^2 = 40$

(c) $6x^2 - 24y^2 = -24$

Answers:

(a) $\dfrac{x^2}{4} - \dfrac{y^2}{6} = 1$ Horizontal

(b) $\dfrac{y^2}{5} - \dfrac{x^2}{8} = 1$ Vertical

(c) $\dfrac{y^2}{1} - \dfrac{x^2}{4} = 1$ Vertical

Finding the equation of a hyperbola is a little more difficult than finding the equation of one of the other three types of conics. However, if you know the vertices and the asymptotes, you can find the values of a and b, which enable you to write the equation. Notice in Example 7 that the key to this procedure is knowing that the central rectangle has a width of $2b$ and a height of $2a$.

EXAMPLE 7   ■   **Finding the Equation of a Hyperbola**

Find an equation of the hyperbola with a vertical transverse axis and with vertices $(0, 3)$ and $(0, -3)$. The asymptotes of the hyperbola are $y = \frac{3}{5}x$ and $y = -\frac{3}{5}x$.

Solution

Remind students that asymptotes are lines and that the students should graph asymptotes just as they graphed lines in Section 2.4.

To begin, sketch the lines that represent the asymptotes, as shown in Figure 8.35(a). Note that these two lines intersect at the origin, which implies that the center of the hyperbola is $(0, 0)$. Next, plot the two vertices at the points $(0, 3)$ and $(0, -3)$. Because you know where the vertices are located, you can sketch the central rectangle of the hyperbola, as shown in Figure 8.35(a). Note that the corners of the central rectangle occur at the points $(-5, 3)$, $(5, 3)$, $(-5, -3)$, and $(5, -3)$. Because the width of the central rectangle is $2b = 10$, it follows that $b = 5$. Similarly, because the height of the central rectangle is $2a = 6$, it follows that $a = 3$. Thus, you can conclude that the standard form of the equation of the hyperbola is

$$\frac{y^2}{3^2} - \frac{x^2}{5^2} = 1.$$

The graph is shown in Figure 8.35(b).

FIGURE 8.35   (a)

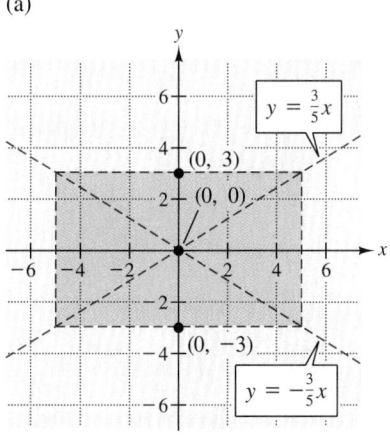

(b)

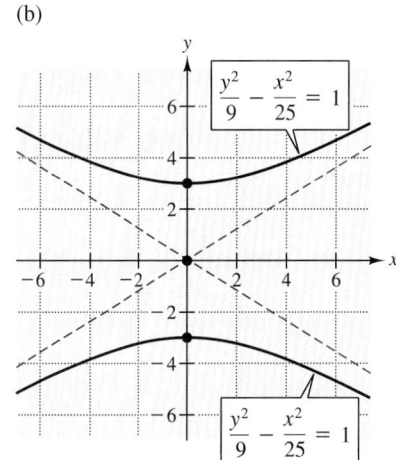

Applications of Conics

EXAMPLE 8   ■   **An Application Involving an Ellipse**

FIGURE 8.36

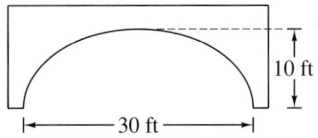

Suppose you are responsible for designing a semielliptical archway, as shown in Figure 8.36. The height of the archway is 10 feet, and the width is 30 feet. Find an equation of the ellipse and use the equation to sketch an accurate drawing of the archway.

Solution

To make the equation simple, place the origin at the center of the ellipse. This means that the standard form of the equation is

$$\frac{x^2}{a^2} + \frac{y^2}{b^2} = 1.$$

Because the major axis is horizontal, you can see that $a = 15$ and $b = 10$, which imply that the equation is

$$\frac{x^2}{15^2} + \frac{y^2}{10^2} = 1.$$

To make an accurate sketch of the ellipse, it is helpful to solve this equation for y, as follows.

$$\frac{x^2}{15^2} + \frac{y^2}{10^2} = 1$$

$$\frac{x^2}{225} + \frac{y^2}{100} = 1$$

$$\frac{y^2}{100} = 1 - \frac{x^2}{225}$$

$$y^2 = 100\left(1 - \frac{x^2}{225}\right)$$

$$y = 10\sqrt{1 - \frac{x^2}{225}}$$

(Note that you are allowed to take the positive square root because the y-values must be positive.) Finally, calculate several y-values for the archway, as shown in Table 8.1.

TABLE 8.1

x-Value	±15	±12.5	±10	±7.5	±5	±2.5	0
y-Value	0	5.53	7.45	8.66	9.43	9.86	10

Using the values shown in the table, you can make a sketch of the archway, as shown in Figure 8.37.

FIGURE 8.37

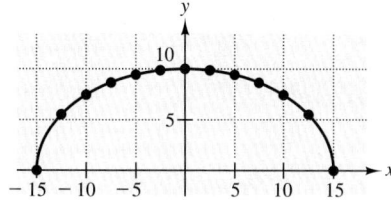

DISCUSSION PROBLEM ■ Sketching a Hyperbola

In this section, the only types of hyperbolas that were discussed were those with vertices that lie on the x-axis or on the y-axis. There are other types of hyperbolas. For instance, the graph of the rational function

$$f(x) = \frac{1}{x}$$

is a hyperbola. Sketch the graph of this hyperbola. Then label both its vertices and asymptotes.

Warm-Up

The following warm-up exercises involve skills that were covered in earlier sections. You will use these skills in the exercise set for this section.

In Exercises 1–4, solve the equation for y.

1. $2x + 3y - 9 = 0$ **2.** $x^2 + y^2 = 36$

3. $2x^2 + 3y^2 - 6 = 0$ **4.** $2x^2 - y^2 - 3 = 0$

In Exercises 5 and 6, simplify the radical.

5. $\sqrt{450}$ **6.** $\sqrt{1350}$

In Exercises 7–10, sketch the graph of the equation.

7. $y = 2x - 3$ **8.** $y = -\frac{3}{4}x + 2$

9. $y = x^2 - 4x + 4$ **10.** $y = 9 - x^2$

8.5 EXERCISES

In Exercises 1–6, match the equation with the correct graph. [The graphs are labeled (a), (b), (c), (d), (e), and (f).]

1. $x^2 + y^2 = 9$

2. $4x^2 + 4y^2 = 9$

3. $\dfrac{x^2}{4} + \dfrac{y^2}{9} = 1$

4. $\dfrac{x^2}{9} + \dfrac{y^2}{4} = 1$

5. $x^2 - y^2 = 4$

6. $x^2 - y^2 = -4$

(a)
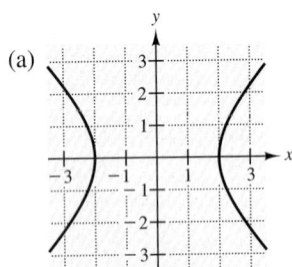

(b)
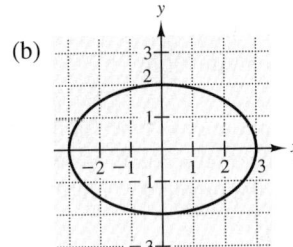

(c)
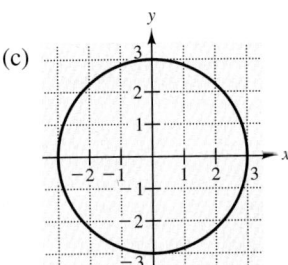

(d)
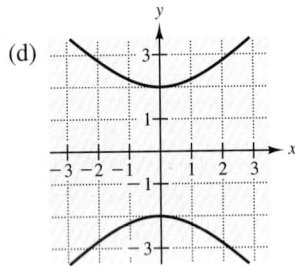

(e)
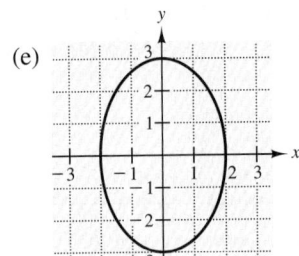

(f)
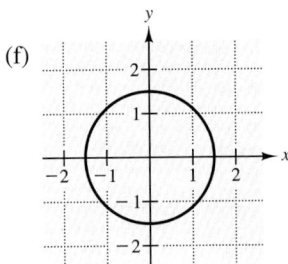

In Exercises 7–14, find an equation of the circle with center $(0, 0)$ that satisfies the given criteria.

7. Radius: 5

8. Radius: 7

9. Radius: $\frac{2}{3}$

10. Radius: $\frac{5}{2}$

11. Passes through the point $(0, 8)$

12. Passes through the point $(-2, 0)$

13. Passes through the point $(5, 2)$

14. Passes through the point $(-1, -4)$

In Exercises 15–20, identify the center and radius of the circle and sketch its graph.

15. $x^2 + y^2 = 16$

16. $x^2 + y^2 = 25$

17. $4x^2 + 4y^2 = 1$

18. $9x^2 + 9y^2 = 64$

19. $25x^2 + 25y^2 - 144 = 0$

20. $\dfrac{x^2}{4} + \dfrac{y^2}{4} - 1 = 0$

In Exercises 21–26, find the standard form of the equation of the ellipse that is centered at the origin and that satisfies the given criteria.

Vertices	Co-vertices		Vertices	Co-vertices

21. $(-4, 0)$, $(4, 0)$ $(0, -3)$, $(0, 3)$ **22.** $(-4, 0)$, $(4, 0)$ $(0, -1)$, $(0, 1)$

23. $(0, -4)$, $(0, 4)$ $(-3, 0)$, $(3, 0)$ **24.** $(0, -5)$, $(0, 5)$ $(-1, 0)$, $(1, 0)$

25. The major axis is horizontal with length 20, and the minor axis has length 12.

26. The minor axis is horizontal with length 30, and the major axis has length 50.

In Exercises 27–34, identify the vertices and co-vertices of the ellipse and sketch the graph of the ellipse.

27. $\dfrac{x^2}{16} + \dfrac{y^2}{4} = 1$ **28.** $\dfrac{x^2}{25} + \dfrac{y^2}{9} = 1$ **29.** $\dfrac{x^2}{4} + \dfrac{y^2}{16} = 1$ **30.** $\dfrac{x^2}{9} + \dfrac{y^2}{25} = 1$

31. $\dfrac{x^2}{\frac{25}{9}} + \dfrac{y^2}{\frac{16}{9}} = 1$ **32.** $\dfrac{x^2}{1} + \dfrac{y^2}{\frac{1}{4}} = 1$ **33.** $4x^2 + y^2 - 4 = 0$ **34.** $4x^2 + 9y^2 - 36 = 0$

In Exercises 35–46, identify the vertices of the hyperbola and sketch its graph. Show the asymptotes and write an equation for each.

35. $x^2 - y^2 = 9$ **36.** $x^2 - y^2 = 1$ **37.** $y^2 - x^2 = 9$ **38.** $y^2 - x^2 = 1$

39. $\dfrac{x^2}{9} - \dfrac{y^2}{25} = 1$ **40.** $\dfrac{x^2}{4} - \dfrac{y^2}{9} = 1$ **41.** $\dfrac{y^2}{9} - \dfrac{x^2}{25} = 1$ **42.** $\dfrac{y^2}{4} - \dfrac{x^2}{9} = 1$

43. $\dfrac{x^2}{1} - \dfrac{y^2}{\frac{9}{4}} = 1$ **44.** $\dfrac{y^2}{\frac{1}{4}} - \dfrac{x^2}{\frac{25}{4}} = 1$ **45.** $4y^2 - x^2 + 16 = 0$ **46.** $4y^2 - 9x^2 - 36 = 0$

In Exercises 47–52, find an equation of the hyperbola that is centered at the origin and that satisfies the given criteria.

Vertices	Asymptotes		Vertices	Asymptotes

47. $(-4, 0)$, $(4, 0)$ $y = 2x$, $y = -2x$ **48.** $(-2, 0)$, $(2, 0)$ $y = \frac{1}{3}x$, $y = -\frac{1}{3}x$

49. $(0, -4)$, $(0, 4)$ $y = \frac{1}{2}x$, $y = -\frac{1}{2}x$ **50.** $(0, -2)$, $(0, 2)$ $y = 3x$, $y = -3x$

51. $(-9, 0)$, $(9, 0)$ $y = \frac{2}{3}x$, $y = -\frac{2}{3}x$ **52.** $(0, -5)$, $(0, 5)$ $y = x$, $y = -x$

In Exercises 53–64, identify the graph of the equation as a line, circle, parabola, ellipse, or hyperbola.

53. $y = 2x^2 - 8x + 2$ **54.** $y = 10 - \frac{3}{2}x$ **55.** $4x^2 + 9y^2 - 36 = 0$ **56.** $4x^2 + 4y^2 - 36 = 0$

57. $4x^2 - 9y^2 - 36 = 0$ **58.** $x^2 - 4y + 2x = 0$ **59.** $x^2 + y^2 - 1 = 0$ **60.** $2x^2 + 2y^2 - 9 = 0$

61. $3x + 2 = 0$ **62.** $y - 4 = 0$ **63.** $x^2 - y - 4 = 0$ **64.** $y^2 = x^2 + 2$

65. Satellite Orbit Find an equation of the circular orbit of a satellite 500 miles above the surface of the earth. Place the origin of the rectangular coordinate system at the center of the earth and assume that the radius of the earth is 4000 miles.

66. Architecture The top portion of a stained glass window is in the form of a pointed Gothic arch (see figure). Each side of the arch is an arc of a circle that has a radius 12 feet and a center at the base of the opposite arch. Find an equation of one of the circles and use it to determine the height of the point of the arch above the base.

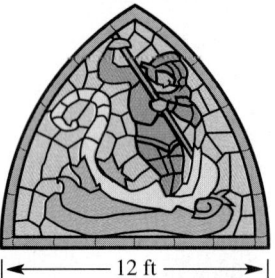

|←———— 12 ft ————→|

67. Geometry A rectangle centered at the origin with sides parallel to the coordinate axes is placed in a circle of radius 25 inches centered at the origin (see figure). If the length of the rectangle is $2x$ inches, show that its width and area are given by $2\sqrt{625 - x^2}$ and $4x\sqrt{625 - x^2}$, respectively.

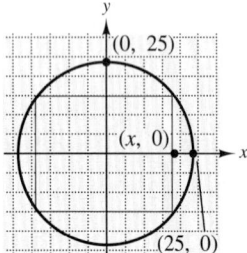

68. Height of an Arch A *semicircular* arch for a tunnel under a river has a diameter of 100 feet (see figure). Determine the height of the arch 5 feet from the edge of the tunnel.

Figure for 68

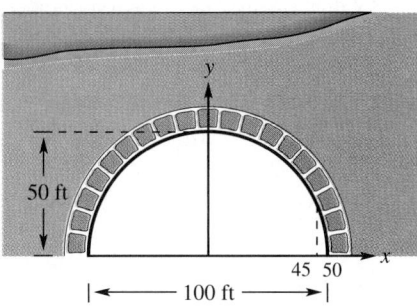

|←———— 100 ft ————→|

69. Height of an Arch A *semielliptical* arch for a tunnel under a river has a width of 100 feet and a height of 40 feet (see figure). Determine the height of the arch 5 feet from the edge of the tunnel.

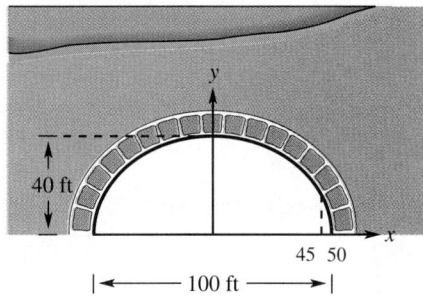

|←———— 100 ft ————→|

70. Bicycle Chainwheel The pedals of a bicycle drive a *chainwheel*, which drives a smaller *sprocket wheel* on the rear axle (see figure). Many chainwheels are circular. Some, however, are slightly elliptical, which tends to make pedaling easier. Find an equation of an elliptical chainwheel that is 8 inches at its widest point and $7\frac{1}{2}$ inches at its narrowest point.

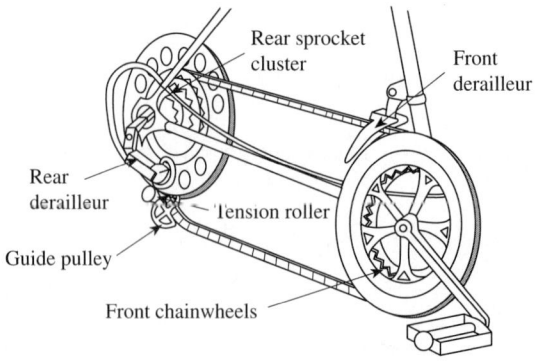

71. Sketch a graph of the ellipse that consists of all points (x, y) such that the sum of the distances between (x, y) and two fixed points is 15 units and the foci are located at the centers of the two sets of concentric circles in the figure.

72. Sketch a graph of the hyperbola that consists of all points (x, y) such that the difference of the distances between (x, y) and two fixed points is eight units and the foci are located at the centers of the two sets of concentric circles in the figure.

Figure for 71 and 72

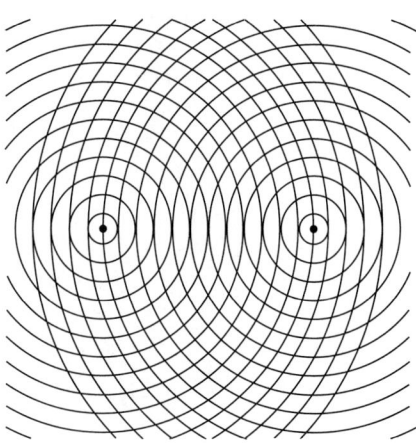

SECTION	**Nonlinear Systems of Equations**
8.6	*Solving Nonlinear Systems of Equations by Graphing* • *Solving Nonlinear Systems of Equations by Substitution* • *Solving Nonlinear Systems of Equations by Elimination* • *Applications*

Solving Nonlinear Systems of Equations by Graphing

In Chapter 4, you studied several methods for solving a system of linear equations. For instance, the linear system

$$\begin{cases} 2x - 3y = 7 \\ x + 4y = -2 \end{cases}$$

has one solution: $(2, -1)$. Graphically, this means that the point $(2, -1)$ is a point of intersection of the two lines whose equations make up the system, as shown in Figure 8.38.

FIGURE 8.38

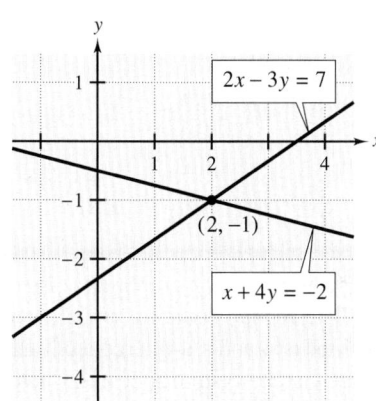

In Chapter 4, you also learned that a linear system can have no solution, exactly one solution, or infinitely many solutions. A **nonlinear system of equations** is a system that contains at least one nonlinear equation. Nonlinear systems of equations can have no solution, one solution, or two or more solutions. For instance, in Figure 8.39(a), the circle and line have no point of intersection, in Figure 8.39(b), the circle and line have one point of intersection, and in Figure 8.39(c), the parabola and line have two points of intersection.

FIGURE 8.39

Review what it means graphically and algebraically for an ordered pair to be a solution to a system of linear equations.

(a) 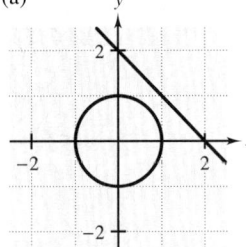 (b) 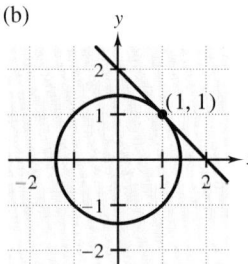 (c)

To solve a nonlinear system of equations graphically, use the following procedure.

Solving a Nonlinear System Graphically

1. Sketch the graph of each equation in the system.

2. Locate the point(s) of intersection of the graphs (if any) and graphically approximate the coordinates of the points.

3. Check the coordinate values by substituting in each equation in the original system. If the coordinate values do not check, then you may have to use an algebraic approach such as one of those discussed later in this section.

EXAMPLE 1   ■   Solving a Nonlinear System Graphically

Find all solutions of the following nonlinear system of equations.

$$\begin{cases} x^2 + y^2 = 25 & \text{Equation 1} \\ x - y = 1 & \text{Equation 2} \end{cases}$$

Solution

Begin by sketching the graph of each equation. The first equation graphs as a circle centered at the origin and having a radius of 5. The second equation, which can be written as $y = x - 1$, graphs as a line with a slope of 1 and a y-intercept of -1. From the graphs shown in Figure 8.40, you can see that the system appears to have two solutions: $(-3, -4)$ and $(4, 3)$. You can check these solutions as follows.

FIGURE 8.40

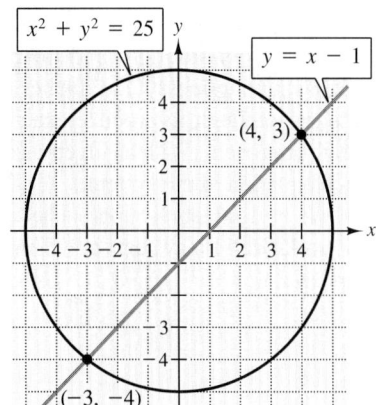

Check $(-3, -4)$: Substitute -3 for x and -4 for y in each equation.

$$(-3)^2 + (-4)^2 = 9 + 16 = 25 \qquad \text{Solution checks in Equation 1}$$
$$(-3) - (-4) = -3 + 4 = 1 \qquad \text{Solution checks in Equation 2}$$

Check $(4, 3)$: Substitute 4 for x and 3 for y in each equation.

$$4^2 + 3^2 = 16 + 9 = 25 \qquad \text{Solution checks in Equation 1}$$
$$4 - 3 = 1 \qquad \text{Solution checks in Equation 2}$$

From these checks, you can conclude that both points are actual solutions of the given system of equations.

■

EXAMPLE 2   ■   Solving a Nonlinear System Graphically

You might demonstrate or have students graph functions such as those from Example 2 with a graphing utility. Use the ZOOM and TRACE features to determine the solutions of the nonlinear system.

Find all solutions of the following nonlinear system of equations.

$$\begin{cases} y = (x - 3)^2 & \text{Equation 1} \\ x + y = 5 & \text{Equation 2} \end{cases}$$

Solution

Begin by sketching the graph of each equation. The first equation graphs as a parabola with vertex at point $(3, 0)$. The second equation, which can be written as $y = -x + 5$, graphs as a line with a slope of -1 and a y-intercept of 5. From the graphs shown in Figure 8.41, you can see that the system appears to have two solutions: $(1, 4)$ and $(4, 1)$. You can check these solutions as follows.

FIGURE 8.41

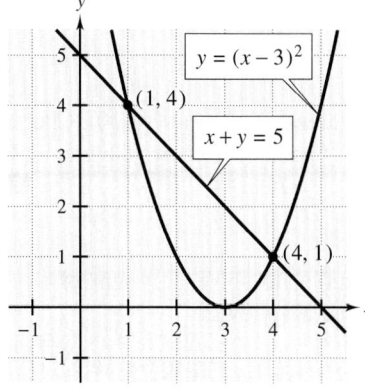

Check $(1, 4)$: Substitute 1 for x and 4 for y in each equation.

$$4 = (1 - 3)^2 = (-2)^2 \qquad \text{Solution checks in Equation 1}$$
$$1 + 4 = 5 \qquad \text{Solution checks in Equation 2}$$

Check $(4, 1)$: Substitute 4 for x and 1 for y in each equation.

$$1 = (4 - 3)^2 = (1)^2 \qquad \text{Solution checks in Equation 1}$$
$$4 + 1 = 5 \qquad \text{Solution checks in Equation 2}$$

From these checks, you can conclude that both points are actual solutions of the given system of equations. ■

Solving Nonlinear Systems of Equations by Substitution

The graphical approach to solving any type of system (linear or nonlinear) in two variables is very useful for helping you see the number of solutions and their approximate coordinates. For systems with solutions having messy coordinates, however, a graphical approach is usually not accurate enough to produce exact solutions. In such cases, you should use an algebraic approach. (With an algebraic approach, you should still sketch the graph of each equation in the system.)

As with systems of *linear* equations, there are two basic algebraic approaches: substitution and elimination. Substitution usually works well for systems in which one of the equations is linear, as shown in Example 3.

EXAMPLE 3   ■   Using Substitution to Solve a Nonlinear System

Solve the following system of equations.

$$\begin{cases} 4x^2 + y^2 = 4 & \text{Equation 1} \\ -2x + y = 2 & \text{Equation 2} \end{cases}$$

Solution

Begin by solving for y in Equation 2.

$$y = 2x + 2 \qquad\qquad \text{Solve for } y \text{ in Equation 2}$$

Next, substitute this expression for y into Equation 1.

$$4x^2 + y^2 = 4 \qquad\qquad \text{Equation 1}$$
$$4x^2 + (2x + 2)^2 = 4 \qquad\qquad \text{Replace } y \text{ by } 2x + 2$$
$$4x^2 + 4x^2 + 8x + 4 = 4$$
$$8x^2 + 8x = 0 \qquad\qquad \text{Write in standard form}$$
$$8x(x + 1) = 0 \qquad\qquad \text{Factor}$$
$$x = 0,\ -1 \qquad\qquad \text{Solve for } x$$

Finally, back-substitute these values of x into the revised Equation 2 to solve for y.

$$\text{For } x = 0: \quad y = 2(0) + 2 = 2$$
$$\text{For } x = -1: \quad y = 2(-1) + 2 = 0$$

Thus, the given system of equations has two solutions: $(0, 2)$ and $(-1, 0)$. Check these solutions in each of the original equations. Figure 8.42 shows the graph of the system.   ■

FIGURE 8.42

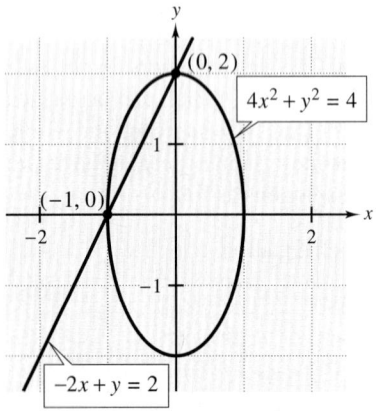

Remind students that *both* solutions must be checked in *both* equations of the system.

The steps for using the method of substitution to solve a system of two equations involving two variables are summarized as follows.

■

Method of Substitution	To solve a system of two equations in two variables, use the following steps.

1. Solve one of the equations for one variable in terms of the other.

2. Substitute the expression found in step 1 into the other equation to obtain an equation of one variable.

3. Solve the equation obtained in step 2.

4. Back-substitute the solution from step 3 into the expression obtained in step 1 to find the value of the other variable.

5. Check the solution to see that it satisfies *each* of the original equations.

EXAMPLE 4   ■   Method of Substitution: No-Solution Case

Solve the following system of equations.

$$\begin{cases} x^2 - y = 0 & \text{Equation 1} \\ x \ \ - y = 1 & \text{Equation 2} \end{cases}$$

FIGURE 8.43

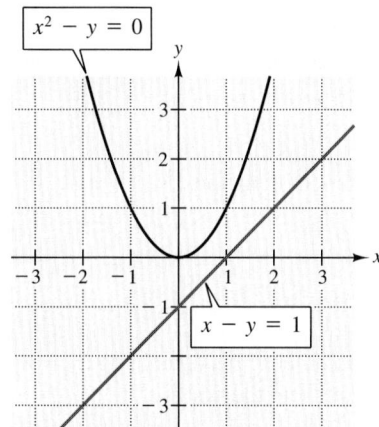

Stress to students that algebraically "no solution" means graphically "no points of intersection."

Solution

Begin by solving for y in Equation 2.

$$y = x - 1 \qquad \text{Solve for } y \text{ in Equation 2}$$

Next, substitute this expression for y into Equation 1.

$$x^2 - y = 0 \qquad \text{Equation 1}$$
$$x^2 - (x - 1) = 0 \qquad \text{Replace } y \text{ by } x - 1$$
$$x^2 - x + 1 = 0$$
$$x = \frac{1 \pm \sqrt{1 - 4}}{2} \qquad \text{Use Quadratic Formula}$$

Now, because the Quadratic Formula yields a negative number inside the square root sign, you can conclude that the equation $x^2 - x + 1$ has no (real) solution. Hence, the given system has no (real) solution. Figure 8.43 shows the graph of the original system. Notice that the parabola and line have no point of intersection.

■

Solving Nonlinear Systems of Equations by Elimination

In Section 4.1, you learned how to use the method of elimination to solve a linear system. This method can also be used with special types of nonlinear systems, as demonstrated in Example 5.

EXAMPLE 5 ■ Using the Method of Elimination

Find all solutions of the following nonlinear system of equations.

$$\begin{cases} 4x^2 + y^2 = 64 & \text{Equation 1} \\ x^2 + y^2 = 52 & \text{Equation 2} \end{cases}$$

Solution

Because both equations have y^2 as a term (and no other terms containing y), you can eliminate y by subtracting the second equation from the first.

$$\begin{aligned} 4x^2 + y^2 &= 64 \\ -x^2 - y^2 &= -52 \\ \hline 3x^2 &= 12 \end{aligned}$$ Subtract Equation 2 from Equation 1

After eliminating y, solve the remaining equation for x.

$$3x^2 = 12$$
$$x^2 = 4$$
$$x = \pm 2$$

When $x = \pm 2$, $y = \pm\sqrt{48} = \pm4\sqrt{3}$. This implies that the system has four solutions:

$$\left(2, 4\sqrt{3}\right), \qquad \left(2, -4\sqrt{3}\right), \qquad \left(-2, 4\sqrt{3}\right), \qquad \left(-2, -4\sqrt{3}\right).$$

Figure 8.44 shows the graph of the system. Notice that the graph of the first equation is an ellipse and the graph of the second equation is a circle. ■

FIGURE 8.44

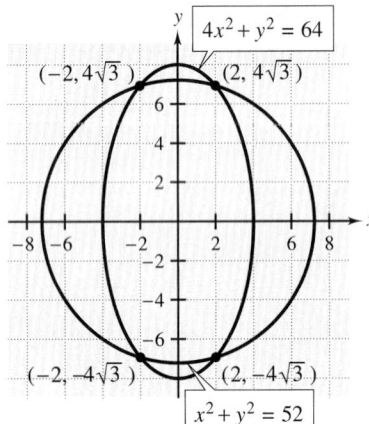

EXAMPLE 6 ■ Using the Method of Elimination

Find all solutions of the following nonlinear system of equations.

$$\begin{cases} x^2 - 2y = 4 & \text{Equation 1} \\ x^2 - y^2 = 1 & \text{Equation 2} \end{cases}$$

Solution

Because both equations have x^2 as a term (and no other terms containing x), you can eliminate x by subtracting the second equation from the first.

$$\begin{aligned} x^2 - 2y &= 4 \\ -x^2 + y^2 &= -1 \\ \hline y^2 - 2y &= 3 \end{aligned}$$ Subtract Equation 2 from Equation 1

After eliminating x, solve the remaining equation for y.

$$y^2 - 2y = 3$$
$$y^2 - 2y - 3 = 0$$
$$(y - 3)(y + 1) = 0$$
$$y = -1, \ 3$$

Ask students what would occur if they tried to eliminate the y terms listed instead of the x terms in Example 6.

FIGURE 8.45

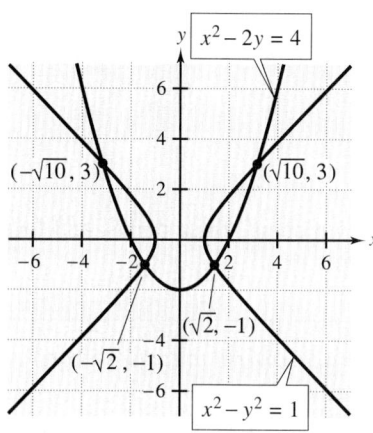

When $y = -1$, $x = \pm\sqrt{2}$, and when $y = 3$, $x = \pm\sqrt{10}$. This implies that the system has four solutions:

$$\left(\sqrt{2}, -1\right), \qquad \left(-\sqrt{2}, -1\right), \qquad \left(\sqrt{10}, 3\right), \qquad \left(-\sqrt{10}, 3\right).$$

Figure 8.45 shows the graph of the system. Notice that the graph of the first equation is a parabola and the graph of the second equation is a hyperbola.

Applications

EXAMPLE 7 ■ Comparing the Revenues of Two Companies

From 1980 through 1995, the revenues R (in millions of dollars) of company A and company B can be modeled by

$$\begin{cases} R = 0.1t + 2.5 & \text{Company A} \\ R = 0.02t^2 - 0.2t + 3.2, & \text{Company B} \end{cases}$$

where t represents the year, with $t = 0$ corresponding to 1980. Sketch the graphs of these two models. During which two years did the companies have approximately equal revenues?

FIGURE 8.46

Solution

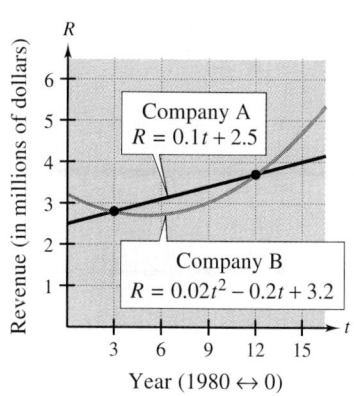

The graphs of the two models are shown in Figure 8.46. From the graph, you can see that company A's revenue followed a linear pattern. It had a revenue of $2.5 million in 1980 and had an increase of $0.1 million each year. Company B's revenue followed a quadratic pattern. Between 1980 and 1985, the company's revenue was decreasing. Then, from 1985 through 1995, the revenue was increasing. From the graph, you can see that the two companies had approximately equal revenues in 1983 (company A had $2.8 million and company B had $2.78 million) and again in 1992 (company A had $3.7 million and company B had $3.68 million).

DISCUSSION PROBLEM ■ **Creating Examples**

Sketch the graph of the circle given by

$$x^2 + y^2 = 4. \qquad \text{Circle}$$

Then find values of C such that the parabola

$$y = x^2 + C \qquad \text{Parabola}$$

intersects the circle in zero points, one point, two points, three points, and four points. ■

Warm-Up

The following warm-up exercises involve skills that were covered in earlier sections. You will use these skills in the exercise set for this section.

In Exercises 1–6, sketch the graph of the equation.

1. $y = \sqrt{x - 3}$ **2.** $x^2 + y^2 = 49$ **3.** $x^2 + 4y^2 = 16$

4. $y^2 - x^2 = 16$ **5.** $y = (x - 2)^2$ **6.** $y = 3 - x^3$

In Exercises 7–10, solve the system of linear equations.

7. $\begin{cases} x - 4y = 0 \\ 3x + 4y = 8 \end{cases}$ **8.** $\begin{cases} 4x + 5y = 6 \\ 3x - y = 14 \end{cases}$

9. $\begin{cases} x + y + z = 0 \\ 4x + y - 3z = -9 \\ 2x - 3y + 2z = 20 \end{cases}$ **10.** $\begin{cases} 2x - 3y + 2z = 20 \\ 3x + y - z = 6 \\ 2x + 3y - z = 2 \end{cases}$

8.6 **EXERCISES**

■ means that a graphing utility can help you solve the exercise or check your solution.

In Exercises 1–6, use the graphs of the equations to determine whether the system has any solutions. If solutions exist, use the graph to identify the solutions. Check your results.

1. $\begin{cases} x + y = 2 \\ x^2 - y = 0 \end{cases}$

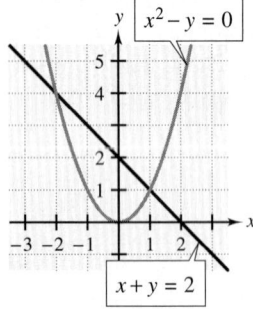

2. $\begin{cases} 2x + y = 10 \\ x^2 + y^2 = 25 \end{cases}$

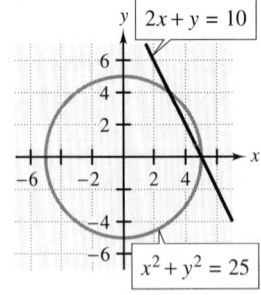

3. $\begin{cases} x - 2y = 4 \\ x^2 - y = 0 \end{cases}$

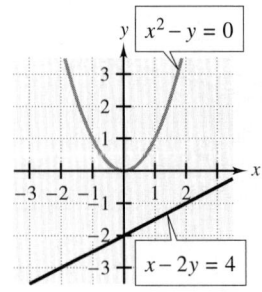

4. $\begin{cases} y = \sqrt{x-2} \\ x - 2y = 1 \end{cases}$

5. $\begin{cases} \dfrac{x^2}{4} + \dfrac{y^2}{9} = 1 \\ 3x - 2y + 6 = 0 \end{cases}$

6. $\begin{cases} \dfrac{x^2}{4} - \dfrac{y^2}{9} = 1 \\ 5x - 2y = 0 \end{cases}$

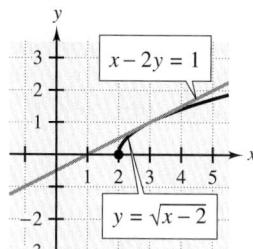

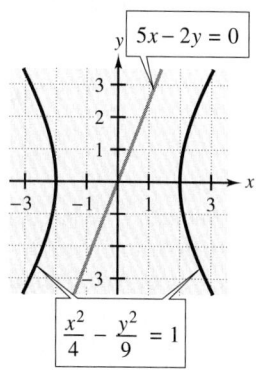

In Exercises 7–12, sketch the graphs of the equations and find any solutions of the system of equations.

7. $\begin{cases} y = x \\ y = x^3 \end{cases}$

8. $\begin{cases} y = x^2 \\ y = x + 2 \end{cases}$

9. $\begin{cases} y = x^2 \\ y = -x^2 + 4x \end{cases}$

10. $\begin{cases} y = 8 - x^2 \\ y = 6 - x \end{cases}$

11. $\begin{cases} \sqrt{x} - y = 0 \\ x - 5y = -6 \end{cases}$

12. $\begin{cases} x^2 - y^2 = 12 \\ x - 2y = 0 \end{cases}$

In Exercises 13–28, solve the system of equations by the method of substitution.

13. $\begin{cases} y = 2x^2 \\ y = -2x + 12 \end{cases}$

14. $\begin{cases} y = 5x^2 \\ y = -15x - 10 \end{cases}$

15. $\begin{cases} x^2 + y = 9 \\ x - y = -3 \end{cases}$

16. $\begin{cases} x - y^2 = 0 \\ x - y = 2 \end{cases}$

17. $\begin{cases} x^2 + 2y = 6 \\ x - y = -4 \end{cases}$

18. $\begin{cases} x^2 + y^2 = 100 \\ x = 12 \end{cases}$

19. $\begin{cases} x^2 + y^2 = 25 \\ 2x - y = -5 \end{cases}$

20. $\begin{cases} x^2 + y^2 = 169 \\ x + y = 7 \end{cases}$

21. $\begin{cases} y = \sqrt{4-x} \\ x + 3y = 6 \end{cases}$

22. $\begin{cases} y = \sqrt[3]{x} \\ y = x \end{cases}$

23. $\begin{cases} 16x^2 + 9y^2 = 144 \\ 4x + 3y = 12 \end{cases}$

24. $\begin{cases} y = 2x^2 \\ y = x^4 - 2x^2 \end{cases}$

25. $\begin{cases} x^2 - y^2 = 9 \\ x^2 + y^2 = 1 \end{cases}$

26. $\begin{cases} x^2 - y^2 = 16 \\ 3x - y = 12 \end{cases}$

27. $\begin{cases} 3x + 2y = 90 \\ xy = 300 \end{cases}$

28. $\begin{cases} x + 2y = 40 \\ xy = 150 \end{cases}$

In Exercises 29–36, solve the system of equations by elimination.

29. $\begin{cases} x^2 + 2y = 1 \\ x^2 + y^2 = 4 \end{cases}$

30. $\begin{cases} x + y^2 = 5 \\ 2x^2 + y^2 = 6 \end{cases}$

31. $\begin{cases} -x + y^2 = 10 \\ x^2 - y^2 = -8 \end{cases}$

32. $\begin{cases} x^2 + y = 9 \\ x^2 - y^2 = 7 \end{cases}$

33. $\begin{cases} x^2 + y^2 = 7 \\ x^2 - y^2 = 1 \end{cases}$

34. $\begin{cases} x^2 + y^2 = 25 \\ y^2 - x^2 = 7 \end{cases}$

35. $\begin{cases} \dfrac{x^2}{4} + y^2 = 1 \\ x^2 + \dfrac{y^2}{4} = 1 \end{cases}$

36. $\begin{cases} x^2 - y^2 = 1 \\ \dfrac{x^2}{2} + y^2 = 1 \end{cases}$

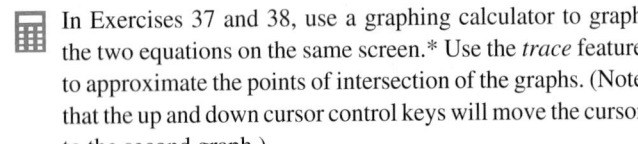 In Exercises 37 and 38, use a graphing calculator to graph the two equations on the same screen.* Use the *trace* feature to approximate the points of intersection of the graphs. (Note that the up and down cursor control keys will move the cursor to the second graph.)

37. $\begin{cases} y = x^3 \\ y = x^3 - 3x^2 + 3x \end{cases}$ **38.** $\begin{cases} y = \frac{1}{5}(24 - x) \\ y = \sqrt{64 - x^2} \end{cases}$

39. *Busing Boundary* To be eligible to ride the school bus to East High School, a student must live at least 1 mile from the school (see figure). Describe the portion of Clark Street from which the residents are *not* eligible to ride the school bus. Use a coordinate system in which the school is at $(0, 0)$ and each unit represents 1 mile.

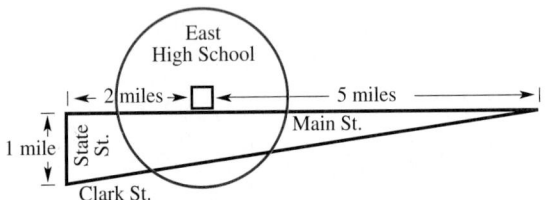

40. *Geometry* A theorem from geometry states that if a triangle is inscribed in a circle so that one side of the triangle is a diameter of the circle, then the triangle is a right triangle (see figure). Show that this theorem is true for the circle $x^2 + y^2 = 100$ and the triangle formed by the lines $y = 0$, $y = \frac{1}{2}x + 5$, and $y = -2x + 20$. (Find the vertices of the triangle and verify that it is a right triangle.)

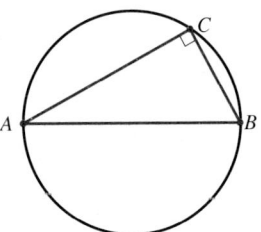

41. *Comparing Populations* From 1982 to 1988, the northeastern part of the United States grew at a slower rate than the western part. (*Source:* U.S. Bureau of Census) Two models that represent the populations of the two regions are

$$\begin{cases} P = 49{,}094.5 + 106.9t + 9.7t^2 & \text{Northeast} \\ P = 43{,}331.9 + 907.8t, & \text{West} \end{cases}$$

where P is the population in thousands and t is the calendar year, with $t = 3$ corresponding to 1983. According to these models, when did the population of the West overtake the population of the Northeast?

*The graphing calculator features given in the examples and exercises in this text correspond to the TI-81 and TI-82 graphing calculators from Texas Instruments. However, you can also use other graphing calculators and computer graphing software for these examples and exercises.

Using a Graphing Calculator

When sketching a conic with a graphing calculator, you must remember two things. First, most graphing calculators are set up to sketch the graphs of functions, not relations—so you must first write the relation as one *or more* functions. Second, to obtain graphs with true perspectives (in which circles look like circles), you must use a "square setting." For a graphing calculator with a display screen height of two-thirds of its width, you can obtain a square setting by using

$$\frac{\text{Ymax} - \text{Ymin}}{\text{Xmax} - \text{Xmin}} = \frac{2}{3}.$$ Square setting

EXAMPLE 1 ■ Sketching a Circle with a Graphing Calculator

Use a graphing calculator to graph $x^2 + y^2 = 4$.

Solution

Begin by solving the given equation for y.

$$x^2 + y^2 = 4$$
$$y^2 = 4 - x^2$$
$$y = \pm\sqrt{4 - x^2}$$

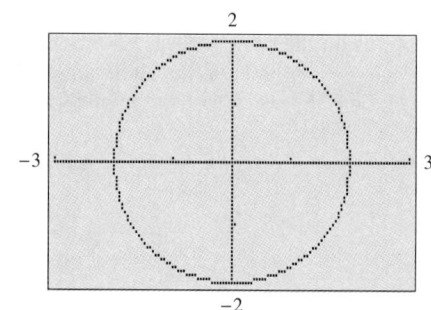

Next, set the calculator's range setting to $-3 \le x \le 3$ and $-2 \le y \le 2$. Finally, enter each of the following equations—one to sketch the top half of the circle and the other to sketch the bottom half of the circle.

$$y = \sqrt{4 - x^2}$$ Top half of circle
$$y = -\sqrt{4 - x^2}$$ Bottom half of circle ■

A graphing calculator can be used to estimate the points of intersection of two graphs. For instance, to find the points of intersection of the line $y = 2x - 2$ and the circle $x^2 + y^2 = 4$, you could show the graph of the line with the graph in Example 1. Using the *trace* feature, you can approximate the points of intersection as $(0, -2)$ and $(1.6, 1.2)$. Both points can be verified algebraically, as follows.

$$x^2 + y^2 = 4$$ Given equation of circle
$$x^2 + (2x - 2)^2 = 4$$ Substitute $2x - 2$ for y.
$$5x^2 - 8x = 0$$ Standard form
$$x(5x - 8) = 0$$ Factor
$$x = 0 \text{ or } x = \frac{8}{5}$$ Zero-Factor Property

663

EXAMPLE 2 ■ Investigating Asymptotic Behavior

Sketch the hyperbola $\dfrac{x^2}{3^2} - \dfrac{y^2}{1^2} = 1$ and its asymptotes on the same display screen.

Solution

To sketch the hyperbola, solve the given equation for y.

$$\frac{x^2}{3^2} - \frac{y^2}{1^2} = 1$$

$$\frac{1}{9}x^2 - 1 = y^2$$

$$\pm\sqrt{\frac{1}{9}x^2 - 1} = y$$

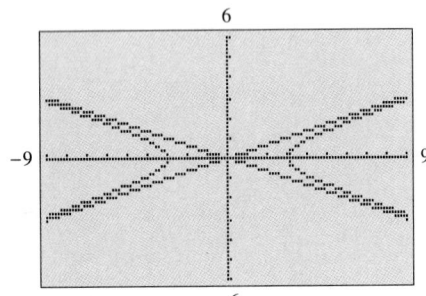

The "plus version" is the equation of the top half of each branch, and the "minus version" is the equation of the bottom half of each branch. The asymptotes of the hyperbola pass through the point $(0, 0)$ with slopes of $\pm(b/a) = \pm\frac{1}{3}$.

$$y = \frac{1}{3}x \quad \text{and} \quad y = -\frac{1}{3}x \qquad \text{Asymptotes}$$

By sketching the hyperbola and the two asymptotes on the same screen, you can see that the hyperbola gets closer to its asymptotes as you move to the right side or left side of the screen. ■

EXERCISES

In Exercises 1–6, use a graphing calculator to graph the equation.

1. $x^2 + y^2 = 16$

2. $\dfrac{x^2}{2^2} + \dfrac{y^2}{5^2} = 1$

3. $\dfrac{y^2}{4^2} - \dfrac{x^2}{6^2} = 1$

4. $x^2 + y^2 = 10$

5. $\dfrac{x^2}{12} + \dfrac{y^2}{9} = 1$

6. $\dfrac{x^2}{16} - \dfrac{y^2}{10} = 1$

In Exercises 7–9, find the points of intersection of the graphs. Use a graphing calculator to verify your result.

7. $\begin{cases} x^2 + y^2 = 10 \\ x + 2y = -5 \end{cases}$

8. $\begin{cases} \dfrac{4x^2}{9} + y^2 = 5 \\ x + y = 2 \end{cases}$

9. $\begin{cases} y^2 = \dfrac{1}{2}x \\ x - 2y = 0 \end{cases}$

10. Find the equations of the asymptotes of the hyperbola. Then use a graphing calculator to graph the hyperbola and the asymptotes on the same display screen.

$$\frac{x^2}{5^2} - \frac{y^2}{4^2} = 1$$

REVIEW EXERCISES

In Exercises 1–6, evaluate the indicated combination of the functions f and g at the given values of the independent variable. (If not possible, state the reason.)

1. $f(x) = x^2$, $g(x) = 4x - 5$ (a) $(f + g)(-5)$ (b) $(f - g)(0)$ (c) $(fg)(2)$ (d) $\left(\dfrac{f}{g}\right)(1)$

2. $f(x) = \dfrac{3}{4}x^3$, $g(x) = x + 1$ (a) $(f + g)(-1)$ (b) $(f - g)(2)$ (c) $(fg)\left(\dfrac{1}{3}\right)$ (d) $\left(\dfrac{f}{g}\right)(2)$

3. $f(x) = \dfrac{2}{3}\sqrt{x}$, $g(x) = -x^2$ (a) $(f + g)(1)$ (b) $(f - g)(9)$ (c) $(fg)\left(\dfrac{1}{4}\right)$ (d) $\left(\dfrac{f}{g}\right)(2)$

4. $f(x) = |x|$, $g(x) = 3$ (a) $(f + g)(-2)$ (b) $(f - g)(3)$ (c) $(fg)(-10)$ (d) $\left(\dfrac{f}{g}\right)(-3)$

5. $f(x) = \dfrac{2}{x - 1}$, $g(x) = x$ (a) $(f + g)\left(\dfrac{1}{3}\right)$ (b) $(f - g)(3)$ (c) $(fg)(-1)$ (d) $\left(\dfrac{f}{g}\right)(5)$

6. $f(x) = \dfrac{1}{x}$, $g(x) = \dfrac{1}{x - 4}$ (a) $(f + g)(2)$ (b) $(f - g)(-2)$ (c) $(fg)\left(\dfrac{7}{2}\right)$ (d) $\left(\dfrac{f}{g}\right)(1)$

In Exercises 7–10, find the indicated composite functions.

7. $f(x) = x + 2$, $g(x) = x^2$ (a) $(f \circ g)(x)$ (b) $(g \circ f)(x)$ (c) $(f \circ g)(2)$ (d) $(g \circ f)(-1)$

8. $f(x) = \sqrt[3]{x}$, $g(x) = x + 2$ (a) $(f \circ g)(x)$ (b) $(g \circ f)(x)$ (c) $(f \circ g)(6)$ (d) $(g \circ f)(64)$

9. $f(x) = \sqrt{x + 1}$, $g(x) = x^2 - 1$ (a) $(f \circ g)(x)$ (b) $(g \circ f)(x)$ (c) $(f \circ g)(5)$ (d) $(g \circ f)(-1)$

10. $f(x) = \dfrac{1}{x - 5}$, $g(x) = \dfrac{5x + 1}{x}$ (a) $(f \circ g)(x)$ (b) $(g \circ f)(x)$ (c) $(f \circ g)(1)$ (d) $(g \circ f)\left(\dfrac{1}{5}\right)$

In Exercises 11 and 12, find the domains of the compositions (a) $f \circ g$ and (b) $g \circ f$.

11. $f(x) = \sqrt{x - 4}$, $g(x) = 2x$

12. $f(x) = \dfrac{2}{x - 4}$, $g(x) = x^2$

In Exercises 13–16, determine whether the function has an inverse.

13. $f(x) = x^2 - 25$

14. $f(x) = \dfrac{1}{4}x^3$

15. $h(x) = 4\sqrt[3]{x}$

16. $g(x) = \sqrt{9 - x^2}$

In Exercises 17–22, find the inverse of the function. (If not possible, state the reason.)

17. $f(x) = \dfrac{1}{4}x$

18. $f(x) = 2x - 3$

19. $h(x) = \sqrt{x}$

20. $g(x) = x^2 + 2$, $x \geq 0$

21. $f(t) = |t + 3|$

22. $h(t) = t$

In Exercises 23–26, find the constant of proportionality and write an equation that relates the variables.

23. y varies directly as the cube root of x, and $y = 12$ when $x = 8$.

24. r varies inversely as s, and $r = 45$ when $s = \frac{3}{5}$.

25. T varies jointly as r and the square of s, and $T = 5000$ when $r = 0.09$ and $s = 1000$.

26. D is directly proportional to the cube of x and inversely proportional to y, and $D = 810$ when $x = 3$ and $y = 25$.

27. *Power Generation* The power generated by a wind turbine is given by the function $P = kw^3$, where P is the number of kilowatts produced at a wind speed of w miles per hour and k is the constant of proportionality.
  (a) Find k if $P = 1000$ when $w = 20$.
  (b) Find the output for a wind speed of 25 miles per hour.

28. *Hooke's Law* A force of 100 pounds stretches a spring 4 inches.
  (a) How far will a force of 200 pounds stretch the spring?
  (b) What force is required to stretch the spring 2.5 inches?

29. *Hooke's Law* A force of 100 pounds stretches a spring 4 inches. Find the force required to stretch the same spring 6 inches.

30. *Stopping Distance* The stopping distance d of an automobile is directly proportional to the square of its speed s. How will the stopping distance be changed by doubling the speed of the car?

31. *Demand Function* A company has found that the daily demand x for its product varies inversely as the square root of the price p. When the price is $25, the demand is approximately 1000 units. Approximate the demand if the price is increased to $28.

32. *Weight of an Astronaut* The gravitational force F with which an object is attracted to the earth is inversely proportional to the square of its distance r from the center of the earth. If an astronaut weighs 200 pounds on the surface of the earth ($r \approx 4000$ miles), what will the astronaut weigh 500 miles above the earth's surface?

In Exercises 33–36, determine the right and left behavior of the graph of the polynomial function.

33. $f(x) = -x^2 + 6x + 9$

34. $f(x) = \frac{1}{2}x^3 + 2x$

35. $g(x) = \frac{3}{4}(x^4 + 3x^2 + 2)$

36. $h(x) = -x^5 - 7x^2 + 10x$

In Exercises 37–42, sketch the graph of the function and identify any intercepts.

37. $f(x) = -(x - 2)^3$ **38.** $f(x) = (x + 1)^3$

39. $g(x) = x^4 - x^3 - 2x^2$ **40.** $f(x) = x^3 - 4x$

41. $f(x) = x(x + 3)^2$ **42.** $f(x) = x^4 - 4x^2$

In Exercises 43–50, identify, and sketch the graph of, the conic represented by the equation.

43. $x^2 - 2y = 0$ **44.** $x^2 + y^2 = 64$

45. $x^2 - y^2 = 64$ **46.** $x^2 + 4y^2 = 64$

47. $\dfrac{x^2}{25} + \dfrac{y^2}{4} = 1$ **48.** $\dfrac{x^2}{25} - \dfrac{y^2}{4} = -1$

49. $4x^2 + 4y^2 - 9 = 0$ **50.** $x^2 + 9y^2 - 9 = 0$

In Exercises 51–56, find the general form of the equation for the conic meeting the given criteria.

51. Hyperbola: Vertices $(-6, 0)$, $(6, 0)$
  Asymptotes: $y = \frac{1}{3}x$, $y = -\frac{1}{3}x$

52. Circle: Center $(0, 0)$
  Passes through the point $(-6, 8)$

53. Ellipse: Vertices $(0, -5)$, $(0, 5)$
  Co-vertices $(-2, 0)$, $(2, 0)$

54. Ellipse: Vertices $(-10, 0)$, $(10, 0)$
  Co-vertices $(0, -6)$, $(0, 6)$

55. Circle: Center $(0, 0)$
  Radius 20

56. Hyperbola: Vertices $(0, -4)$, $(0, 4)$
  Asymptotes $y = 2x$, $y = -2x$

57. *Satellite Orbit* Find an equation of the circular orbit of a satellite 1000 miles above the surface of the earth. Place the origin of the rectangular coordinate system at the center of the earth and assume the radius of the earth to be 4000 miles.

58. *Satellite Orbit* Find an equation of the elliptical orbit of a satellite that varies in altitude between 500 miles and 1000 miles above the surface of the earth. Place the origin of the rectangular coordinate system at the center of the earth and assume the radius of the earth to be 4000 miles.

In Exercises 59–66, find any solutions of the nonlinear system of equations.

59. $\begin{cases} y = 5x^2 \\ y = -15x - 10 \end{cases}$

60. $\begin{cases} y^2 = 16x \\ 4x - y = -24 \end{cases}$

61. $\begin{cases} x^2 + y^2 = 1 \\ x + y = -1 \end{cases}$

62. $\begin{cases} x^2 + y^2 = 100 \\ x + y = 0 \end{cases}$

63. $\begin{cases} \dfrac{x^2}{16} + \dfrac{y^2}{4} = 1 \\ y = x + 2 \end{cases}$

64. $\begin{cases} \dfrac{x^2}{100} + \dfrac{y^2}{25} = 1 \\ y = -x - 5 \end{cases}$

65. $\begin{cases} \dfrac{x^2}{25} + \dfrac{y^2}{9} = 1 \\ \dfrac{x^2}{25} - \dfrac{y^2}{9} = 1 \end{cases}$

66. $\begin{cases} x^2 + y^2 = 16 \\ -x^2 + \dfrac{y^2}{16} = 1 \end{cases}$

Chapter Test

Take this test as you would take a test in class. After you are done, check your work against the answers in the back of the book.

In Exercises 1–4, use the functions $f(x) = \frac{1}{2}x$ and $g(x) = x^2 - 1$ to find the following.

1. $(f - g)(4)$

2. $(fg)(-4)$

3. $\left(\dfrac{f}{g}\right)(2)$

4. $(f \circ g)(-5)$

5. Find the domain of the composite function $(f \circ g)(x)$ if $f(x) = \sqrt{25 - x}$ and $g(x) = x^2$.

6. Find the inverse of the function $f(x) = \frac{1}{7}x - 1$. Verify that $f(f^{-1}(x)) = x$.

7. Write a mathematical model for the statement "S varies directly as the square of x and inversely as y".

8. Find a mathematical model that relates u and v if v varies directly as the square root of u, and $v = \frac{3}{2}$ when $u = 36$.

9. *Boyle's Law* If the temperature of a gas is not allowed to change, its absolute pressure P is inversely proportional to its volume V, according to Boyle's Law. A large balloon is filled with 180 cubic meters of helium at atmospheric pressure (1 atm) at sea level. What is the volume of the helium if the balloon rises to an altitude where the atmospheric pressure is 0.75 atm? Assume that the temperature does not change.

10. Determine the right and left behavior of the polynomial function $f(x) = -2x^3 + 3x^2 - 4$.

11. Sketch the graph of the function $f(x) = 1 - (x - 2)^3$. Identify any intercepts of the graph.

12. Find an equation of the circle that is centered at the origin and that passes through the point $(-3, 4)$.

13. Find the standard form of the equation of the ellipse centered at the origin with vertices $(0, -10)$ and $(0, 10)$ and co-vertices $(-3, 0)$ and $(3, 0)$.

14. Find the standard form of the equation of the hyperbola centered at the origin with vertices $(-3, 0)$ and $(3, 0)$ and asymptotes $y = \frac{1}{2}x$ and $y = -\frac{1}{2}x$.

15. Sketch the graph of each of the equations.

(a) $x^2 + y^2 = 9$ (b) $\dfrac{x^2}{9} + \dfrac{y^2}{16} = 1$ (c) $\dfrac{x^2}{9} - \dfrac{y^2}{16} = 1$ (d) $\dfrac{x}{3} - \dfrac{y}{4} = 1$

16. Find any solutions of the nonlinear system of equations.

(a) $\begin{cases} y = -\frac{1}{2}x^2 \\ y = -4x + 6 \end{cases}$ (b) $\begin{cases} x^2 + y^2 = 100 \\ x + y = 14 \end{cases}$

Exponential and Logarithmic Functions

9.1   Exponential and Logarithmic Expressions

9.2   Properties of Logarithms

9.3   Exponential Functions and Their Graphs

9.4   Logarithmic Functions and Their Graphs

9.5   Solving Exponential and Logarithmic Equations

9.6   Applications of Exponential and Logarithmic Functions

I n 1638, the French mathematician René Descartes developed a mathematical model for the spiral pattern of a chambered nautilus. Each time the nautilus grows out of its old living chamber, it creates a new empty chamber that is 6.3% larger. If

the volume of zeroth chamber is 1, then the volume of the nth chamber is given by the exponential model

$$\text{Volume} = (1.063)^n.$$

Use this model to find the volume of the fourth and fifth chambers. Is the fifth chamber 6.3% larger than the fourth?

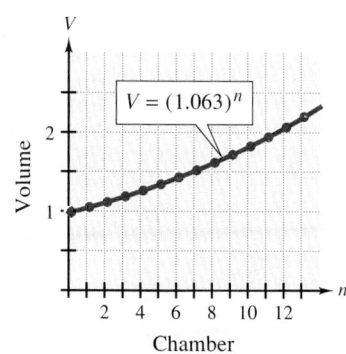

Chamber

669

Exponential and Logarithmic Expressions

Exponential and Logarithmic Expressions • Evaluating Logarithmic Expressions • Evaluating Common Logarithms and Natural Logarithms • Applications

Exponential and Logarithmic Expressions

In this chapter, you will study two important types of expressions: exponential expressions and logarithmic expressions. You are already familiar with exponential expressions. For instance, 2^3 is an exponential expression.

In many applications involving exponential expressions, it is helpful to rewrite the expression in **logarithmic form.** For instance, the equation

$$2^3 = 8 \qquad \text{Exponential form}$$

can be rewritten in logarithmic form as

$$\log_2 8 = 3. \qquad \text{Logarithmic form}$$

This equation is read as "the logarithm of 8 with base 2 is 3."

Definition of Logarithm

Let a and x be positive real numbers such that $a \neq 1$. The **logarithm of x with base a** is denoted by $\log_a x$ and is defined as follows.

$$y = \log_a x \quad \text{if and only if} \quad x = a^y$$

In both forms, note that a is the base. In the second form, note that y is the exponent, which, from the first form, implies that a *logarithm is an exponent.*

Stress that a logarithm is an exponent.

From this definition, you can see that the equations $y = \log_a x$ and $x = a^y$ are equivalent. The first equation is in *logarithmic* form and the second is in *exponential* form.

EXAMPLE 1   ■   Rewriting Exponential Equations in Logarithmic Form

Rewrite the following exponential equations in logarithmic form.

(a) $4^3 = 64$ (b) $2^0 = 1$ (c) $3^{-1} = \frac{1}{3}$ (d) $9^{1/2} = 3$

Solution

(a) For this equation, the base is 4, and the exponent is 3. Using the fact that a logarithm is an exponent, you can write

$$3 = \log_4 64. \qquad \text{Logarithmic form}$$

(b) For this equation, the base is 2, and the exponent is 0. Using the fact that a logarithm is an exponent, you can write

$$0 = \log_2 1. \qquad \text{Logarithmic form}$$

(c) For this equation, the base is 3, and the exponent is -1. Using the fact that a logarithm is an exponent, you can write

$$-1 = \log_3 \frac{1}{3}.$$ Logarithmic form

(d) For this equation, the base is 9, and the exponent is $\frac{1}{2}$. Using the fact that a logarithm is an exponent, you can write

$$\frac{1}{2} = \log_9 3.$$ Logarithmic form ■

EXAMPLE 2 ■ Rewriting Logarithmic Equations in Exponential Form

Rewrite the following logarithmic equations in exponential form.

(a) $\log_2 16 = 4$ (b) $\log_4 2 = \frac{1}{2}$ (c) $\log_{10} \frac{1}{100} = -2$

Solution

(a) For this equation, the base is 2, and the logarithm is 4. Using the fact that a logarithm is an exponent, you can write
$$2^4 = 16.$$ Exponential form

(b) For this equation, the base is 4, and the logarithm is $\frac{1}{2}$. Using the fact that a logarithm is an exponent, you can write
$$4^{1/2} = 2.$$ Exponential form

(c) For this equation, the base is 10, and the logarithm is -2. Using the fact that a logarithm is an exponent, you can write
$$10^{-2} = \frac{1}{100}.$$ Exponential form ■

Evaluating Logarithmic Expressions

When evaluating logarithmic expressions, remember that a *logarithm is an exponent.* For instance, because the exponent in the expression

$$\overset{\text{Exponent}}{2^3} = 8$$

is 3, the value of the logarithm $\log_2 8$ is 3. That is,

$$\log_2 8 = 3.$$

Therefore, to evaluate the logarithmic expression $\log_a x$, you need to answer the question, "To what power must a be raised to obtain x?"

EXAMPLE 3 ■   Evaluating Logarithms

Evaluate the following logarithms.

(a) $\log_2 32$ (b) $\log_3 9$ (c) $\log_5 1$

Solution

(a) The power to which 2 must be raised to obtain 32 is 5. That is,
$$2^5 = 32 \quad \Longrightarrow \quad \log_2 32 = 5.$$

(b) The power to which 3 must be raised to obtain 9 is 2. That is,
$$3^2 = 9 \quad \Longrightarrow \quad \log_3 9 = 2.$$

(c) The power to which 5 must be raised to obtain 1 is 0. That is,
$$5^0 = 1 \quad \Longrightarrow \quad \log_5 1 = 0.$$ ■

EXAMPLE 4 ■   Evaluating Logarithms

Evaluate the following logarithms.

(a) $\log_4 2$ (b) $\log_{10} \frac{1}{10}$ (c) $\log_3(-1)$

Solution

(a) The power to which 4 must be raised to obtain 2 is $\frac{1}{2}$. That is,
$$4^{1/2} = 2 \quad \Longrightarrow \quad \log_4 2 = \frac{1}{2}.$$

(b) The power to which 10 must be raised to obtain $\frac{1}{10}$ is -1. That is,
$$10^{-1} = \frac{1}{10} \quad \Longrightarrow \quad \log_{10} \frac{1}{10} = -1.$$

(c) There is *no power* to which 3 can be raised to obtain -1. The reason for this is that for any value of x, 3^x is a positive number. Therefore, $\log_3(-1)$ is undefined. ■

The following properties of logarithms follow directly from the definition of the logarithm with base a.

Properties of Logarithms   Let a and x be positive real numbers such that $a \neq 1$.

1. $\log_a 1 = 0$ because $a^0 = 1$.

2. $\log_a a = 1$ because $a^1 = a$.

3. $\log_a a^x = x$ because $a^x = a^x$.

EXAMPLE 5   ■   Evaluating Logarithmic Expressions

(a) $\log_4 1 = 0$ (because $4^0 = 1$)

(b) $\log_8 8 = 1$ (because $8^1 = 8$)

(c) $\log_3(3^5) = 5$ (because $\log_a a^x = x$)   ■

Evaluating Common Logarithms and Natural Logarithms

A logarithm with base 10 is called a **common logarithm.**

EXAMPLE 6   ■   Evaluating Common Logarithms

Inform students that $\log_{10} 100$, $\log_{10} 0.01$, and $\log_{10} \frac{1}{1000}$ can be written as $\log 100$, $\log 0.01$, and $\log \frac{1}{1000}$.

Evaluate the following common logarithms.

(a) $\log_{10} 100$ (b) $\log_{10} 0.01$ (c) $\log_{10} \frac{1}{1000}$

Solution

(a) Because $10^2 = 100$, it follows that

$$\log_{10} 100 = 2.$$

(b) Because $10^{-2} = \frac{1}{100} = 0.01$, it follows that

$$\log_{10} 0.01 = -2.$$

(c) Because $10^{-3} = \frac{1}{1000}$, it follows that

$$\log_{10} \frac{1}{1000} = -3.$$ ■

So far, all logarithmic bases have been positive integers. In many applications of logarithms, the most convenient choice for a base is the following irrational number, denoted by the letter e.

$$e \approx 2.71828\ldots$$

This number is called the **natural base,** and a logarithm with base e is called a **natural logarithm.** This logarithm is denoted by the symbol $\ln x$, which is read as "el en of x."

■

Natural Logarithm	The logarithm given by
	$$\log_e x = \ln x,$$
	where $x > 0$, is called the **natural logarithm** of x.

Orally or in writing, have students discuss the difference between a common logarithm and a natural logarithm.

The three properties of logarithms listed earlier in this section are also valid for natural logarithms.

Properties of Natural Logarithms	Let x be a positive real number.
	1. $\ln 1 = 0$ because $e^0 = 1$.
	2. $\ln e = 1$ because $e^1 = e$.
	3. $\ln e^x = x$ because $e^x = e^x$.

These properties are used to evaluate natural logarithms in the following example.

EXAMPLE 7   ■   **Evaluating Natural Logarithms**

Using the property that $\ln e^x = x$, you obtain the following.

(a) $\ln e^2 = 2$

(b) $\ln \dfrac{1}{e} = \ln e^{-1} = -1$ ■

On most calculators, the common logarithm key is denoted by [log] and the natural logarithm key is denoted by [ln] .* Here are some examples of the use of a calculator to evaluate common and natural logarithms.

EXAMPLE 8   ■   **Using a Calculator to Evaluate Common Logarithms**

Scientific

Logarithm	*Keystrokes*	*Display*
$\log 4$	4 [log]	0.60206
$\log 0.5$	.5 [log]	-0.30103
$\log(-1)$	1 [+/−] [log]	ERROR

Graphing

Logarithm	*Keystrokes*	*Display*
$\log 4$	[LOG] 4 [ENTER]	0.60206
$\log 0.5$	[LOG] .5 [ENTER]	-0.30103
$\log(-1)$	[LOG] [(−)] 1 [ENTER]	ERROR

  ■

NOTE Be sure you see that $\log(-1)$ gives an error. This occurs because the domain of $\log x$ is the set of positive real numbers, which means that $\log(-1)$ is undefined.

*The graphing calculator keystrokes given in this text correspond to the TI-81 and TI-82 graphing calculators from Texas Instruments. For other graphing calculators, the keystrokes may differ.

EXAMPLE 9   ■   Using a Calculator to Evaluate Natural Logarithms

Scientific

Logarithm	*Keystrokes*	*Display*
ln 2	2 ln	0.6931472
ln 0.3	.3 ln	−1.2039728

Graphing

Logarithm	*Keystrokes*	*Display*
ln 2	LN 2 ENTER	0.6931472
ln 0.3	LN .3 ENTER	−1.2039728

Applications

EXAMPLE 10   ■   Evaluating a Logarithmic Expression

The slope s of a beach is related to the average diameter d (in millimeters) of the sand particles on the beach by the model

$$s = 0.159 + 0.118 \log_{10} d.$$

Find the slope of a beach if the average diameter of its sand particles is 0.25 millimeter.

Solution

If $d = 0.25$, then the slope of the beach is

$s = 0.159 + 0.118 \log_{10} 0.25$	Substitute 0.25 for d
$\approx 0.159 + 0.118(-0.602)$	Use a calculator
$\approx 0.09.$	

The slope of the beach is about 0.09. This is a gentle slope, which has a rise of only 9 meters for a run of 100 meters.   ■

Diameter (in mm)	Type of Sand Particle
4	Pebble
2	Granule
1	Very coarse sand
0.5	Coarse sand
0.25	Medium sand
0.125	Fine sand
0.0625	Very fine sand

The slope of a beach depends on the size of sand particles. Coarse sand produces a steeper slope than does fine sand. The photo at the left shows a "coral sand" beach in St. Thomas, Virgin Islands. Its gentle slope is the result of its fine sand.

DISCUSSION PROBLEM   ■   What Is a Logarithm?

Write a paragraph explaining in your own words what is meant by the statement, *A logarithm is an exponent.* Illustrate your explanation with examples.   ■

Warm-Up

The following warm-up exercises involve skills that were covered in earlier sections. You will use these skills in the exercise set for this section.

In Exercises 1–6, evaluate the expression.

1. $\left(\frac{3}{4}\right)^3$ **2.** $(3 \cdot 10)^2$ **3.** 2^{-2} **4.** 3^{-4}

5. $25^{1/2}$ **6.** $81^{-1/2}$

In Exercises 7–10, use the properties of exponents to simplify the expression.

7. z^0 **8.** $y^3 \cdot y^2$ **9.** $\dfrac{26x^4}{2x^3}$ **10.** $\left(-\dfrac{2}{y}\right)^3$

9.1   EXERCISES

In Exercises 1–6, write the exponential equation in logarithmic form.

1. $7^2 = 49$ **2.** $5^4 = 625$ **3.** $3^{-2} = \frac{1}{9}$

4. $6^{-3} = \frac{1}{216}$ **5.** $8^{2/3} = 4$ **6.** $81^{3/4} = 27$

In Exercises 7–12, write the logarithmic equation in exponential form.

7. $\log_5 25 = 2$ **8.** $\log_{10} 10{,}000 = 4$ **9.** $\log_4 \frac{1}{16} = -2$

10. $\log_3 \frac{1}{243} = -5$ **11.** $\log_{36} 6 = \frac{1}{2}$ **12.** $\log_{32} 4 = \frac{2}{5}$

In Exercises 13–32, evaluate the expression without using a calculator. If not possible, state the reason.

13. $\log_2 8$ **14.** $\log_3 27$ **15.** $\log_2 \frac{1}{4}$ **16.** $\log_3 \frac{1}{9}$

17. $\log_4 1$ **18.** $\log_5(-6)$ **19.** $\log_2(-3)$ **20.** $\log_3 1$

21. $\log_{10} 10$ **22.** $\log_8 8$ **23.** $\log_{10} 1000$ **24.** $\log_{10} \frac{1}{100}$

25. $\log_9 3$ **26.** $\log_{144} 12$ **27.** $\log_{16} 8$ **28.** $\log_{25} 125$

29. $\log_7 7^4$ **30.** $\log_5 5^3$ **31.** $\log_3 3^{-2}$ **32.** $\log_2 0$

In Exercises 33–40, use the properties of natural logarithms to evaluate the expression.

33. $\ln e^2$

34. $\ln e^{-4}$

35. $\ln 1$

36. $\ln 8^0$

37. $\ln e$

38. $\ln \dfrac{1}{e^3}$

39. $\ln \dfrac{e^3}{e^2}$

40. $\ln(e^2 \cdot e^4)$

In Exercises 41–52, use a calculator to evaluate the logarithm. (Round your answer to four decimal places.)

41. $\log_{10} 31$

42. $\log_{10} 5310$

43. $\log_{10} 0.85$

44. $\log_{10} 0.345$

45. $\log_{10}\left(\sqrt{2}+4\right)$

46. $\log_{10} \dfrac{\sqrt{3}}{2}$

47. $\ln 25$

48. $\ln 6.57$

49. $\ln 0.75$

50. $\ln\left(\sqrt{3}-1\right)$

51. $\ln\left(\dfrac{1+\sqrt{5}}{3}\right)$

52. $\ln\left(1+\dfrac{0.10}{12}\right)$

53. *Doubling Time* The time t (in years) for an investment to double in value when compounded continuously at the rate r is

$$t = \frac{\ln 2}{r}.$$

Complete the table, which shows the "doubling times" for several annual percentage rates.

r	0.07	0.08	0.09	0.10	0.11	0.12
t						

54. *Intensity of Sound* The relationship between the number of decibels B and the intensity of a sound I (in watts per meter squared) is

$$B = 10 \log_{10}\left(\frac{I}{10^{-16}}\right).$$

Determine the number of decibels of a sound with an intensity of 10^{-4} watts per meter squared.

55. *American Elk* The antler spread a (in inches) and shoulder height h (in inches) of an adult male American elk are related by the model

$$h = 116 \log_{10}(a + 40) - 176.$$

Approximate the shoulder height of a male American elk with an antler spread of 55 inches.

56. *Tornadoes* Most tornadoes last less than an hour and travel less than 20 miles. The speed of the wind S (in miles per hour) near the center of the tornado is related to the distance the tornado travels d (in miles) by the model

$$S = 93 \log_{10} d + 65.$$

On March 18, 1925, a large tornado struck portions of Missouri, Illinois, and Indiana, covering a distance of 220 miles. Approximate the speed of the wind near the center of this tornado.

Male American elks grow antlers with a spread of about 5 feet.

Properties of Logarithms

Properties of Logarithms   •   *Rewriting Logarithmic Expressions*   •   *Applications*

Properties of Logarithms

You know from the previous section that an exponential equation can be rewritten as an equivalent logarithmic equation. Thus, it makes sense that each property of exponents should have a corresponding property of logarithms. For instance, the exponential property $a^0 = 1$ has the corresponding logarithmic property $\log_a 1 = 0$. In this section you will study the logarithmic properties that correspond to the following three exponential properties.

1. $a^n a^m = a^{n+m}$

2. $\dfrac{a^n}{a^m} = a^{n-m}$

3. $(a^n)^m = a^{nm}$

Properties of Logarithms

Let a be a positive real number such that $a \neq 1$, let n be a real number, and let u and v be real numbers, variables, or algebraic expressions such that $u > 0$ and $v > 0$.

	Logarithm with Base a	*Natural Logarithm*
1.	$\log_a(uv) = \log_a u + \log_a v$	$\ln(uv) = \ln u + \ln v$
2.	$\log_a \dfrac{u}{v} = \log_a u - \log_a v$	$\ln \dfrac{u}{v} = \ln u - \ln v$
3.	$\log_a u^n = n \log_a u$	$\ln u^n = n \ln u$

Ask students to use their calculators to evaluate $\log(2+3)$ and $\log 2 + \log 3$ and then to compare the results.

NOTE There is no general property of logarithms that can be used to simplify $\log_a(u + v)$. Specifically, $\log_a(u + v)$ *does not equal* $\log_a u + \log_a v$.

EXAMPLE 1   ■   Using Properties of Logarithms

Use the facts that $\ln 2 \approx 0.693$, $\ln 3 \approx 1.099$, and $\ln 5 \approx 1.609$ to approximate the following.

(a) $\ln \frac{2}{3}$ (b) $\ln 10$ (c) $\ln 30$

Solution

(a) $\ln \frac{2}{3} = \ln 2 - \ln 3$ Property 2

   $\approx 0.693 - 1.099$

   $= -0.406$

(b) $\ln 10 = \ln(2 \cdot 5)$ Property 1

   $= \ln 2 + \ln 5$

   $\approx 0.693 + 1.609$

   $= 2.302$

Point out that Logarithm Property 1 applies even if there are more than two factors.

(c) $\ln 30 = \ln(2 \cdot 3 \cdot 5)$ Property 1

   $= \ln 2 + \ln 3 + \ln 5$

   $\approx 0.693 + 1.099 + 1.609$

   $= 3.401$

Check these results on your calculator.   ■

Have students verbalize Logarithm Property 3.

When you are using the properties of logarithms, it helps to state the properties *verbally*. For instance, the verbal form of the property $\ln(uv) = \ln u + \ln v$ is *the log of a product is the sum of the logs of the factors.* Similarly, the verbal form of the property $\ln(u/v) = \ln u - \ln v$ is *the log of a quotient is the difference of the logs of the numerator and denominator.*

EXAMPLE 2   ■   Using Properties of Logarithms

Use the properties of logarithms to verify that

$$-\ln 2 = \ln \frac{1}{2}.$$

Solution

Using Property 3 of logarithms, you can write the following.

$$-\ln 2 = (-1)\ln 2 = \ln 2^{-1} = \ln \frac{1}{2}$$

Try checking this result on your calculator.   ■

Rewriting Logarithmic Expressions

In Examples 1 and 2, the properties of logarithms were used to rewrite logarithmic expressions involving the log of a *constant*. A more common use of these properties is to rewrite the log of a *variable expression*. This is illustrated in the next several examples.

EXAMPLE 3 ■ Rewriting the Logarithm of a Product

Use the properties of logarithms to rewrite the following logarithmic expression.

$$\log_{10} 7x^3$$

Solution

$$\log_{10} 7x^3 = \log_{10} 7 + \log_{10} x^3 \qquad \text{Property 1}$$
$$= \log_{10} 7 + 3 \log_{10} x \qquad \text{Property 3} \qquad ■$$

When you rewrite a logarithmic expression as in Example 3, you are **expanding** the expression. The reverse procedure, demonstrated in Example 4, is called **condensing** a logarithmic expression.

EXAMPLE 4 ■ Condensing a Logarithmic Expression

Use the properties of logarithms to condense the following logarithmic expression.

$$\log_3 x - \log_3 5$$

Solution

$$\log_3 x - \log_3 5 = \log_3 \frac{x}{5} \qquad \text{Property 2} \qquad ■$$

Sometimes, expanding or condensing logarithmic expressions involves several steps. In the next three examples, be sure you can justify each step in the solution.

EXAMPLE 5 ■ Expanding a Logarithmic Expression

Expand the following logarithmic expression.

$$\log_2 \frac{3x}{y^2}, \qquad y > 0$$

Solution

$$\log_2 \frac{3x}{y^2} = \log_2 3 + \log_2 x - \log_2 y^2 \qquad \text{Properties 1 and 2}$$
$$= \log_2 3 + \log_2 x - 2 \log_2 y \qquad \text{Property 3} \qquad ■$$

EXAMPLE 6 ■   Expanding a Logarithmic Expression

Expand the following logarithmic expression.

$$\ln \sqrt{x^2 - 1}, \qquad x > 1$$

Solution

$$\ln \sqrt{x^2 - 1} = \ln(x^2 - 1)^{1/2} \qquad\qquad \text{Rewrite using fractional exponent}$$

$$= \frac{1}{2} \ln(x^2 - 1) \qquad\qquad \text{Property 3}$$

$$= \frac{1}{2} \ln(x - 1)(x + 1) \qquad\qquad \text{Factor}$$

$$= \frac{1}{2}[\ln(x - 1) + \ln(x + 1)] \qquad \text{Property 1}$$

Stress the importance of grouping
symbols, and point out that $\frac{1}{2}$ was
distributed to both terms.

$$= \frac{1}{2} \ln(x - 1) + \frac{1}{2} \ln(x + 1) \qquad \text{Distributive Property} \qquad ■$$

EXAMPLE 7 ■   Condensing a Logarithmic Expression

Use the properties of logarithms to condense the following expression.

$$\ln 2 - 2 \ln x$$

Solution

$$\ln 2 - 2 \ln x = \ln 2 - \ln x^2, \qquad x > 0 \qquad \text{Property 3}$$

$$= \ln \frac{2}{x^2}, \qquad x > 0 \qquad \text{Property 2} \qquad ■$$

Note that when you expand or condense a logarithmic expression, it is possible
to change the domain of the expression. For instance, the domain of the expression

$$2 \ln x$$

is the set of positive real numbers, whereas the domain of

$$\ln x^2$$

is the set of nonzero real numbers. Thus, when you expand or condense a logarithmic expression, you should check to see whether the rewriting changed the
domain of the expression. In such cases, you should qualify the condensing or
expanding by restricting the domain. For instance, you could write

$$f(x) = 2 \ln x = \ln x^2, \qquad x > 0.$$

Applications

EXAMPLE 8 ■ An Application: Human Memory Model

Students participating in a psychological experiment attended several lectures on a subject. Every month for a year after that, the students were tested to see how much of the material they remembered. The average scores s for the group were given by the **human memory model**

$$s = 80 - \ln(t + 1)^9, \qquad 0 \le t \le 12,$$

where t is the time in months. Find the average scores for the group after 2 months and after 8 months.

Solution

Common error: Be aware that some students may incorrectly simplify $80 - 9\ln 9$ as $71\ln 9$. Review the order of operations.

To make the calculations easier, you can begin by rewriting the model as

$$s = 80 - 9\ln(t + 1), \qquad 0 \le t \le 12.$$

After 2 months, the average score will be

$$s = 80 - 9\ln 3 \approx 70.1,$$ Average score after 2 months

and after 8 months, the average score will be

$$s = 80 - 9\ln 9 \approx 60.2.$$ Average score after 8 months ■

EXAMPLE 9 ■ Estimating the Time of Death

At 8:30 A.M., a coroner was called to the home of a person who had died during the night. To estimate the time of death, the coroner took the person's body temperature twice. At 9:00 A.M. the temperature was 85.7° F, and at 9:30 A.M. the temperature was 82.8° F. The room temperature was 70° F. When did the person die?

Solution

To estimate the time of death, the coroner used Newton's Law of Cooling,

$$kt = \ln \frac{T - S}{T_0 - S},$$ Newton's Law of Cooling

where k is a constant, S is the temperature of the surrounding air, and t is the time it takes for the body temperature to cool from $T_0 = 98.6$ to T. From the first temperature reading, the coroner could obtain

$$kt = \ln \frac{85.7 - 70}{98.6 - 70}$$

$$\approx \ln 0.54895$$

$$\approx -0.6.$$

Forensic scientists work in crime laboratories of police departments. Modern crime laboratories use computers and science to examine clues.

From the second temperature reading, the coroner could obtain

$$k\left(t + \frac{1}{2}\right) = \ln\frac{82.8 - 70}{98.6 - 70}$$
$$\approx \ln 0.44755$$
$$\approx -0.8.$$

Solving the two equations

$$kt = -0.6$$
$$k\left(t + \frac{1}{2}\right) = -0.8$$

for k and t produces $k = -0.4$ and $t = 1.5$. Therefore, the coroner could approximate that the first temperature reading took place 1.5 hours after death, which implies that the death occurred at about 7:30 A.M.   ■

DISCUSSION PROBLEM   ■   Historical Use of Logarithms

The English mathematician Henry Briggs (1561-1630) was also a founder of calculations by logarithms. He worked with Napier, and they both published papers on the advantage of having tables of logarithms using the base 10.

Logarithms were invented in 1594 by John Napier, a Scottish mathematician. Using the reference books in the library, write a short paragraph describing the use of logarithms in calculating products, quotients, and powers of large numbers. For instance, how could a table of logarithms be used to approximate the number $(1.34)^{2.7}$? (Remember that electronic calculators and computers are 20th-century inventions.)   ■

Warm-Up

The following warm-up exercises involve skills that were covered in earlier sections. You will use these skills in the exercise set for this section.

In Exercises 1–6, use the properties of exponents to simplify the expression.

1. $(x^2 \cdot x^3)^4$ **2.** $4^{-2} \cdot x^2$ **3.** $\dfrac{15y^{-3}}{10y^2}$

4. $\left(\dfrac{3x^2}{2y}\right)^{-2}$ **5.** $\dfrac{3x^2 y^3}{18x^{-1}y^2}$ **6.** $(x^2 + 1)^0$

In Exercises 7–10, rewrite the equation in exponential form.

7. $\log_4 64 = 3$ **8.** $\log_3 \frac{1}{81} = -4$

9. $\ln e^{2x} = 2x$ **10.** $\ln 1 = 0$

9.2 EXERCISES

In Exercises 1–8, use the properties of logarithms to evaluate the expression (without using a calculator). If not possible, state the reason.

1. $\log_5 5^2$

2. $\ln e^4$

3. $\log_3 36 - \log_3 12$

4. $\log_6 2 + \log_6 3$

5. $\ln 1$

6. $\log_{10} 1$

7. $\log_2 \frac{1}{8}$

8. $\log_3(-3)$

In Exercises 9–20, use the facts that $\log_5 2 \approx 0.431$ and $\log_5 3 \approx 0.683$ to approximate the logarithm. Do not use a calculator.

9. $\log_5 4$

10. $\log_5 8$

11. $\log_5 6$

12. $\log_5 24$

13. $\log_5 \frac{3}{2}$

14. $\log_5 \frac{9}{2}$

15. $\log_5 \sqrt{2}$

16. $\log_5 \sqrt[3]{9}$

17. $\log_5(3 \cdot 2^4)$

18. $\log_5 \sqrt{3 \cdot 2^5}$

19. $\log_5 3^0$

20. $\log_5 4^3$

In Exercises 21–36, use the properties of logarithms to expand the expression as a sum, difference, or multiple of logarithms.

21. $\log_3 11x$

22. $\ln 5x$

23. $\ln y^3$

24. $\log_7 x^2$

25. $\log_2 \frac{z}{17}$

26. $\log_{10} \frac{7}{y}$

27. $\log_5 x^{-2}$

28. $\log_2 \sqrt{s}$

29. $\log_3 \sqrt[3]{x+1}$

30. $\log_4 \frac{1}{\sqrt{t}}$

31. $\ln 3x^2 y$

32. $\ln y(y-1)^2$

33. $\log_2 \frac{x^2}{x-3}$

34. $\log_5 \sqrt{\frac{x}{y}}$

35. $\ln \sqrt[3]{x(x+5)}$

36. $\ln(3x(x-5))^2$

In Exercises 37–52, use the properties of logarithms to condense the expression as a logarithm of a single quantity.

37. $\log_2 3 + \log_2 x$

38. $\log_5 2x + \log_5 3y$

39. $\log_{10} 4 - \log_{10} x$

40. $\ln 10x - \ln z$

41. $4 \ln b$

42. $10 \log_4 z$

43. $-2 \log_5 2x$

44. $-5 \ln(x+3)$

45. $\frac{1}{3} \ln(2x+1)$

46. $-\frac{1}{2} \log_3 5y$

47. $\log_3 2 + \frac{1}{2} \log_3 y$

48. $\ln 6 - 3 \ln z$

49. $2 \ln x + 3 \ln y - \ln z$

50. $4 \ln 3 - 2 \ln x - \ln y$

51. $4(\ln x + \ln y)$

52. $2[\ln x - \ln(x+1)]$

In Exercises 53–58, determine whether the equation is true or false.

53. $\log_2 8x = 3 + \log_2 x$

54. $\log_8 4 + \log_8 16 = 2$

55. $\log_3(u+v) = \log_3 u + \log_3 v$

56. $\log_3(u+v) = \log_3 u \cdot \log_3 v$

57. $\frac{\log_6 10}{\log_6 3} = \log_6 10 - \log_6 3$

58. $\ln e^{2-x} = 2 - x$

In Exercises 59–64, simplify the logarithmic expression.

59. $\log_4 \frac{4}{x}$

60. $\log_3(3^2 \cdot 4)$

61. $\log_5 \sqrt{50}$

62. $\log_2 \sqrt{22}$

63. $\ln 3e^2$

64. $\ln \frac{6}{e^5}$

65. *Exploratory Exercise* Approximate the natural logarithms of as many integers as possible between 1 and 20 using $\ln 2 \approx 0.6931$, $\ln 3 \approx 1.0986$, and $\ln 5 \approx 1.6094$. (Do not use a calculator.)

66. Use a calculator to demonstrate that

$$\frac{\ln x}{\ln y} \neq \ln \frac{x}{y} = \ln x - \ln y$$

by evaluating each expression when $x = 3$ and $y = 5$.

67. *Intensity of Sound* The relationship between the number of decibels B and the intensity of a sound I (in watts per meter squared) is

$$B = 10 \log_{10} \left(\frac{I}{10^{-16}} \right).$$

Use the properties of logarithms to write the formula in simpler form, and determine the number of decibels of a sound with an intensity of 10^{-10} watts per meter squared.

68. *Human Memory Model* Students participating in a psychological experiment attended several lectures on a subject. Every month for a year after that, the students were tested to see how much of the material they remembered. The average scores for the group were given by the human memory model

$$f(t) = 80 - \log_{10}(t + 1)^{12}, \qquad 0 \le t \le 12,$$

where t is the time in months. Use the properties of logarithms to write the formula in simpler form, and find the average scores for the group after 2 months and after 8 months.

Molecular Transport In Exercises 69 and 70, use the following information. The energy E (in kilocalories per gram molecule) required to transport a substance from outside to inside a living cell is

$$E = 1.4(\log_{10} C_2 - \log_{10} C_1),$$

where C_1 is the concentration of the substance outside the cell and C_2 is the concentration inside the cell.

69. Condense the equation.

70. The concentration of a particular substance inside a cell is twice the concentration outside the cell. How much energy is required to transport the substance from outside the cell to inside the cell?

SECTION	Exponential Functions and Their Graphs
9.3	*Exponential Functions* • *Graphs of Exponential Functions* • *The Natural Exponential Function* • *Compound Interest*

Exponential Functions

Whereas polynomial and rational functions have terms with variable bases and constant exponents, exponential functions have terms with *constant bases* and *variable exponents*. Here are some examples.

Polynomial or Rational Function   *Exponential Function*

Constant exponent   Variable exponent

$$f(x) = x^2$$ $$f(x) = 2^x$$

Variable base   Constant base

Constant exponent   Variable exponent

$$f(x) = x^{-3}$$ $$f(x) = 3^{-x}$$

Variable base   Constant base

The definition of the **exponential function with base *a*** follows on page 686.

Definition of Exponential Function	The **exponential function f with base a** is denoted by $$f(x) = a^x,$$ where $a > 0$, $a \neq 1$, and x is any real number.

Have students determine the domain and range of the exponential function $f(x) = a^x$.

NOTE The base $a = 1$ is excluded because $f(x) = 1^x = 1$ is a constant function, *not* an exponential function.

In Chapter 6, you learned to evaluate a^x for integer and rational values of x. For example, you know that

$$8^3 = 8 \cdot 8 \cdot 8 = 512 \quad \text{and} \quad 8^{2/3} = \left(\sqrt[3]{8}\right)^2 = 2^2 = 4.$$

To evaluate 8^x for *any* real number x, you need to interpret forms with *irrational* exponents, such as $8^{\sqrt{2}}$. For the purpose of this text, it is sufficient to think of

$$a^{\sqrt{2}},$$

where $\sqrt{2} \approx 1.414214$, as the number having successively closer approximations

$$a^{1.4}, \quad a^{1.41}, \quad a^{1.414}, \quad a^{1.4142}, \quad a^{1.41421}, \quad a^{1.414214}, \quad \ldots .$$

The properties of exponents that were discussed in Section 6.1 can be extended to cover exponential functions, as follows.

Properties of Exponential Functions	Let a be a positive real number, and let x and y be real numbers, variables, or algebraic expressions.

1. $a^x \cdot a^y = a^{x+y}$ 3. $\dfrac{a^x}{a^y} = a^{x-y}$

2. $(a^x)^y = a^{xy}$ 4. $a^{-x} = \dfrac{1}{a^x} = \left(\dfrac{1}{a}\right)^x$

To evaluate exponential functions with a calculator, you can use the exponential key $\boxed{y^x}$ (where y is the base and x is the exponent) or $\boxed{\wedge}$. The base is entered first, then the exponent.* For example, to evaluate $3^{-1.3}$, you can use the following keystrokes.

Keystrokes	*Display*	
3 $\boxed{y^x}$ 1.3 $\boxed{+/-}$ $\boxed{=}$	0.239741	Scientific
3 $\boxed{\wedge}$ $\boxed{(-)}$ 1.3 $\boxed{\text{ENTER}}$	0.239741	Graphing

*The graphing calculator keystrokes given in this text correspond to the TI-81 and TI-82 graphing calculators from Texas Instruments. For other graphing calculators, the keystrokes may differ.

EXAMPLE 1   ■   **Evaluating Exponential Functions**

Evaluate the following functions at the indicated values of x. Use a calculator only if it is necessary or more efficient.

Function	*Values*
(a) $f(x) = 2^x$	$x = 3, \ x = -4, \ x = \pi$
(b) $g(x) = 12^x$	$x = 3, \ x = -0.1, \ x = \frac{5}{7}$
(c) $h(x) = (1.085)^x$	$x = 0, \ x = -3$

Solution

(a) $f(3) = 2^3 = 8$ Calculator is not necessary.

$f(-4) = 2^{-4} = \dfrac{1}{2^4} = \dfrac{1}{16}$ Calculator is not necessary.

$f(\pi) = 2^\pi \approx 8.825$ Calculator is necessary.

(b) $g(3) = 12^3 = 1728$ Calculator is more efficient.

$g(-0.1) = 12^{-0.1} \approx 0.7800$ Calculator is necessary.

$g\left(\frac{5}{7}\right) = 12^{5/7} \approx 5.900$ Calculator is necessary.

(c) $h(0) = (1.085)^0 = 1$ Calculator is not necessary.

$h(-3) = (1.085)^{-3} \approx 0.7829$ Calculator is more efficient.   ■

When using their calculators to evaluate $12^{5/7}$, some students may incorrectly enter 12 $\boxed{\wedge}$ 5 $\boxed{\div}$ 7. Stress the importance of using grouping symbols—even with calculators.

Graphs of Exponential Functions

The basic nature of the graph of an exponential function can be determined by plotting several points. (A computer or calculator with graphing capabilities also helps.)

EXAMPLE 2   ■   **Sketching the Graph of an Exponential Function ($y = a^x$)**

On the same coordinate plane, sketch the graphs of the following functions.

(a) $f(x) = 2^x$ (b) $g(x) = 4^x$

Determine the domain and range of each function.

Solution

On the next page, Table 9.1 lists some values of each function, and Figure 9.1 shows their graphs. From the graphs, you can see that the domain and range of each function is as follows.

Domain: all real numbers

Range: all positive real numbers

FIGURE 9.1

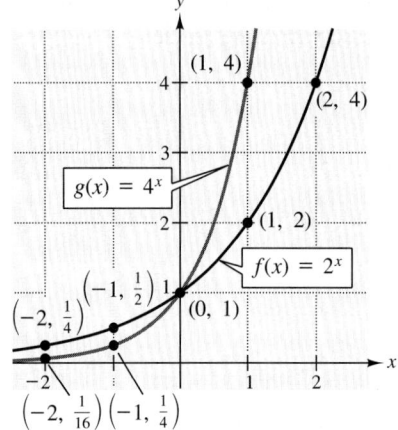

TABLE 9.1

x	-2	-1	0	1	2	3
2^x	$\frac{1}{4}$	$\frac{1}{2}$	1	2	4	8
4^x	$\frac{1}{16}$	$\frac{1}{4}$	1	4	16	64

■

You know from your study of the graphs of functions (see Section 2.6) that the graph of $h(x) = f(-x) = 2^{-x}$ is a reflection of the graph of $f(x) = 2^x$ in the y-axis. This is reinforced in the following example.

EXAMPLE 3 ■ Sketching the Graph of an Exponential Function $(y = a^{-x})$

On the same coordinate plane, sketch the graphs of the following functions.

(a) $y = 2^{-x}$

(b) $y = 4^{-x}$

Solution

Table 9.2 lists some values of each function, and Figure 9.2 shows their graphs.

FIGURE 9.2

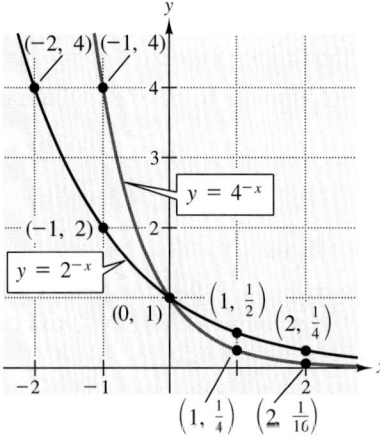

TABLE 9.2

x	-3	-2	-1	0	1	2
2^{-x}	8	4	2	1	$\frac{1}{2}$	$\frac{1}{4}$
4^{-x}	64	16	4	1	$\frac{1}{4}$	$\frac{1}{16}$

■

Examples 2 and 3 suggest that for $a > 1$, the graph of $y = a^x$ increases and the graph of $y = a^{-x}$ decreases. The graphs shown in Figure 9.3 are typical of the graphs of these two types of exponential functions. That is, each has one y-intercept at $(0, 1)$ and one horizontal asymptote (the x-axis).

FIGURE 9.3

Graph of $y = a^x$, $a > 1$
- Domain: $(-\infty, \infty)$
- Range: $(0, \infty)$
- Intercept: $(0, 1)$
- Increasing
- Asymptote: x-axis

Graph of $y = a^{-x}$, $a > 1$
- Domain: $(-\infty, \infty)$
- Range: $(0, \infty)$
- Intercept: $(0, 1)$
- Decreasing
- Asymptote: x-axis

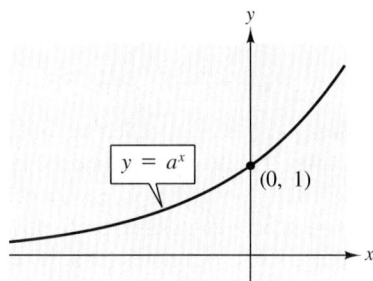

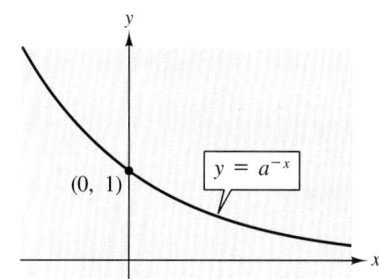

EXAMPLE 4 ■ An Application: Radioactive Decay

Let y represent the mass of a particular radioactive element whose half-life is 25 years. The initial mass is 10 grams. After t years, the mass (in grams) is given by

$$y = 10\left(\frac{1}{2}\right)^{t/25}, \quad t \ge 0.$$

How much of the initial mass is present after 120 years?

Solution

When $t = 120$, the mass is given by

$$y = 10\left(\frac{1}{2}\right)^{120/25} = 10\left(\frac{1}{2}\right)^{4.8} \approx 0.359 \text{ gram}.$$

Thus, after 120 years, the mass has decayed from an initial amount of 10 grams to only 0.359 gram. Note in Figure 9.4 on the next page that the graph of the function shows the 25-year half-life. That is, after 25 years the mass is 5 grams (half of the original). After another 25 years, the mass is 2.5 grams, and so on.

FIGURE 9.4

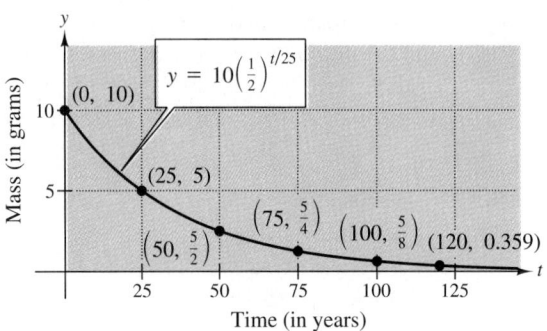

The Natural Exponential Function

FIGURE 9.5

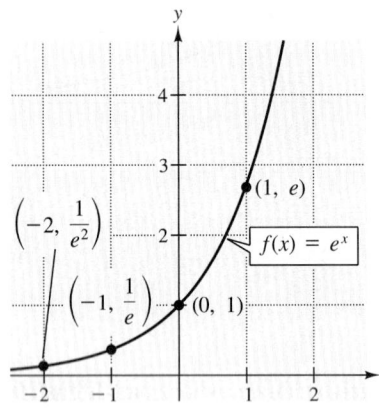

The function $f(x) = e^x$ is called the **natural exponential function.** Its graph is shown in Figure 9.5. Be sure you understand that, for this function, e is the constant number $2.71828\ldots$ and x is a variable. To evaluate the natural exponential function, you need a calculator, preferably one having a natural exponential key $\boxed{e^x}$. Here are some examples of how to use such a calculator to evaluate the natural exponential function.

Scientific Calculator

Function Value	*Keystrokes*	*Calculator Display*
$f(2) = e^2$	2 $\boxed{e^x}$	7.3890561
$f(-3) = e^{-3}$	3 $\boxed{+/-}$ $\boxed{e^x}$	0.049787
$f(0.32) = e^{0.32}$	.32 $\boxed{e^x}$	1.3771278

Graphing Calculator

Function Value	*Keystrokes*	*Calculator Display*
$f(2) = e^2$	$\boxed{e^x}$ 2 $\boxed{\text{ENTER}}$	7.3890561
$f(-3) = e^{-3}$	$\boxed{e^x}$ $\boxed{(-)}$ 3 $\boxed{\text{ENTER}}$	0.049787
$f(0.32) = e^{0.32}$	$\boxed{e^x}$ .32 $\boxed{\text{ENTER}}$	1.3771278

NOTE Some calculators do not have a key labeled $\boxed{e^x}$. If you have a calculator that does not have this key, but does have a key labeled $\boxed{\ln x}$, you will have to use a two-keystroke sequence such as $\boxed{\text{INV}}$ $\boxed{\ln x}$ in place of $\boxed{e^x}$.

Compound Interest

One of the most familiar uses of exponential functions involves **compound interest.** Suppose that a principal P is invested at an annual percentage rate r (in decimal form), compounded once a year. If the interest is added to the principal at the end of the year, then the balance is

$$A = P + Pr = P(1+r).$$

This pattern of multiplying the previous principal by $(1+r)$ is then repeated each successive year, as shown in Table 9.3.

TABLE 9.3

Time in Years	Balance After Each Compounding
0	$A = P$
1	$A = P(1+r)$
2	$A = P(1+r)(1+r) = P(1+r)^2$
3	$A = P(1+r)^2(1+r) = P(1+r)^3$
$\vdots$	$\vdots$
t	$A = P(1+r)^t$

To account for more frequent compounding of interest (such as quarterly or monthly compounding), let n be the number of compoundings per year and let t be the number of years. Then the rate per compounding is r/n and the account balance after t years is

$$A = P\left(1 + \frac{r}{n}\right)^{nt}.$$

EXAMPLE 5 ■ Finding the Balance for Compound Interest

A sum of $10,000 is invested at an annual percentage rate of 7.5%, compounded monthly. Find the balance in the account after 10 years.

Solution

Using the formula

$$A = P\left(1 + \frac{r}{n}\right)^{nt},$$

with $P = 10{,}000$, $r = 0.075$, $n = 12$ (for monthly compounding), and $t = 10$, you obtain the following balance.

$$A = 10{,}000\left(1 + \frac{0.075}{12}\right)^{12(10)} \approx \$21{,}120.65$$ ■

A second method used to compute interest is called **continuous compounding.** The formula for the balance for this type of compounding is

$$A = Pe^{rt}.$$

The formulas for both types of compounding are summarized as follows.

Formulas for Compound Interest

Let A be the balance in an account after t years with principal P and annual percentage rate r (in decimal form).

1. For n compoundings per year: $A = P\left(1 + \dfrac{r}{n}\right)^{nt}$

2. For continuous compounding: $A = Pe^{rt}$

EXAMPLE 6 ■ Comparing Two Types of Compounding

A total of $15,000 is invested at an annual percentage rate of 8%. Find the balance after 6 years if the interest is compounded (a) quarterly and (b) continuously.

Additional problem: $1000 is deposited for 9 months at 12%. What is the total amount in the account (a) at simple interest, (b) if interest is compounded quarterly, and (c) if interest is compounded continuously?

Answers: (a) $1088.71 (b) $1092.73 (c) $1094.17

Solution

(a) Letting $P = 15{,}000$, $r = 0.08$, $n = 4$, and $t = 6$, you can find the balance after 6 years at quarterly compounding to be

$$A = 15{,}000\left(1 + \frac{0.08}{4}\right)^{4(6)} = \$24{,}126.56.$$

(b) Letting $P = 15{,}000$, $r = 0.08$, and $t = 6$, you can find the balance after 6 years at continuous compounding to be

$$A = 15{,}000e^{0.08(6)} = \$24{,}241.12.$$

Note that the balance is greater with continuous compounding than with quarterly compounding. ■

NOTE Example 6 illustrates the following general rule. For a given principal, interest rate, and time, the more often the interest is compounded, the greater the balance will be. Moreover, the balance obtained by continuous compounding is larger than the balance obtained by compounding n times per year.

DISCUSSION PROBLEM   ■   A Calculator Experiment

The natural number e arises in several fascinating ways in mathematics. The following steps show one way that this number can be approximated with a calculator.

(a) Use a calculator to complete the following table for the indicated values of x.

x	0	10	100	1000	10,000
$\left(1+\dfrac{1}{x}\right)^x$					

(b) Use the information in the table to sketch the graph of the function

$$y = \left(1 + \frac{1}{x}\right)^x.$$

Does this graph appear to be approaching a horizontal asymptote?

(c) From parts (a) and (b), what conclusions can you draw about the value of

$$\left(1 + \frac{1}{x}\right)^x$$

as x gets larger and larger?   ■

Warm-Up

The following warm-up exercises involve skills that were covered in earlier sections. You will use these skills in the exercise set for this section.

In Exercises 1–10, use the properties of exponents to simplify the expression.

1. $x^5 \cdot x^3$

2. $(x + 10)^0, \quad x \neq -10$

3. $\dfrac{z^7}{z^4}, \quad z \neq 0$

4. $\dfrac{18(x - 3)^5}{24(x - 3)}, \quad x \neq 3$

5. $(x^2 y)^4$

6. $\left(\dfrac{x}{2y}\right)^5, \quad y \neq 0$

7. $x^4 y^{-3}, \quad y \neq 0$

8. $(a^2)^{-4}, \quad a \neq 0$

9. $(b^6)^{1/2}$

10. $(8x^3)^{1/3}$

9.3 **EXERCISES**

means that a graphing utility can help you solve the exercise or check your solution.

In Exercises 1–12, evaluate the function at the given values. Use a calculator only if it is necessary or more efficient. (Round your answer to three decimal places.)

1. $f(x) = 3^x$

   (a) $x = -2$

   (b) $x = 0$

   (c) $x = 1$

2. $g(x) = 5^x$

   (a) $x = -1$

   (b) $x = 1$

   (c) $x = 3$

3. $F(x) = 3^{-x}$

   (a) $x = -2$

   (b) $x = 0$

   (c) $x = 1$

4. $G(x) = 5^{-x}$

   (a) $x = -1$

   (b) $x = 1$

   (c) $x = \sqrt{3}$

5. $f(t) = 500\left(\dfrac{1}{2}\right)^t$

   (a) $t = 0$

   (b) $t = 1$

   (c) $t = \pi$

6. $g(s) = 1200\left(\dfrac{2}{3}\right)^s$

   (a) $s = 0$

   (b) $s = 2$

   (c) $s = 4$

7. $f(x) = 1000(1.05)^{2x}$

   (a) $x = 0$

   (b) $x = 5$

   (c) $x = 10$

8. $P(t) = \dfrac{10{,}000}{(1.01)^{12t}}$

   (a) $t = 2$

   (b) $t = 10$

   (c) $t = 20$

9. $f(x) = e^x$

   (a) $x = -1$

   (b) $x = 0$

   (c) $x = \frac{1}{2}$

10. $A(t) = 200e^{0.1t}$

   (a) $t = 10$

   (b) $t = 20$

   (c) $t = 40$

11. $g(x) = 10e^{-0.5x}$

   (a) $x = -4$

   (b) $x = 4$

   (c) $x = 8$

12. $f(z) = \dfrac{100}{1 + e^{-0.05z}}$

   (a) $z = 0$

   (b) $z = 10$

   (c) $z = 20$

In Exercises 13–18, match the exponential function with its graph. [The graphs are labeled (a), (b), (c), (d), (e), and (f).]

13. $f(x) = 2^x$

14. $f(x) = -2^x$

15. $f(x) = 2^{-x}$

16. $f(x) = 2^x - 1$

17. $f(x) = 2^{x-1}$

18. $f(x) = 2^{x+1}$

(a)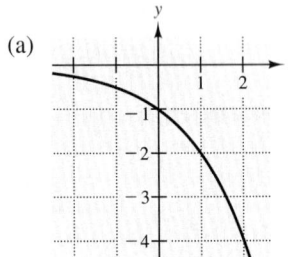

(b)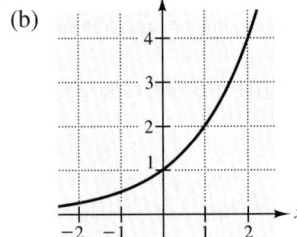

(c)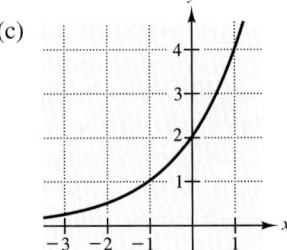

(d)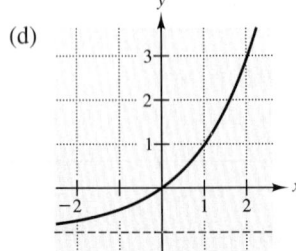

(e)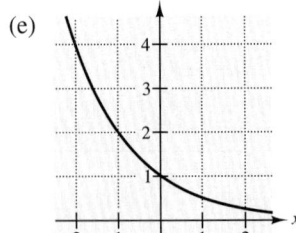

(f)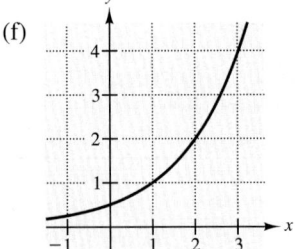

In Exercises 19–34, sketch the graph of the function.

19. $f(x) = 3^x$

20. $f(x) = 3^{-x} = \left(\frac{1}{3}\right)^x$

21. $g(x) = 3^x - 2$

22. $g(x) = 3^x + 1$

23. $h(x) = \frac{1}{2}(3^x)$

24. $h(x) = \frac{1}{2}(3^{-x})$

25. $f(t) = 2^{-t^2}$

26. $f(t) = 2^{t^2}$

27. $h(x) = 2^{0.5x}$

28. $g(t) = 2^{-0.5t}$

29. $f(x) = -2^{0.5x}$

30. $h(t) = -2^{-0.5t}$

31. $f(x) = e^{0.2x}$

32. $f(x) = e^{-0.2x}$

33. $P(t) = 100e^{-0.1t}$

34. $A(t) = 1000e^{0.08t}$

35. *Population Growth* The population of the United States (in recent years) can be approximated by the exponential function

$$P(t) = 203(1.0118)^{t-1970},$$

where t is the year and P is the population in millions. Use this model to approximate the population in the years (a) 1995 and (b) 2000.

36. *Property Value* Suppose that the value of a piece of property doubles every 15 years. If you buy the property for $64,000, then its value t years after the date of purchase should be

$$V(t) = 64{,}000(2)^{t/15}.$$

Use this model to approximate the value of the property (a) 5 years and (b) 20 years after the date of purchase.

37. *Depreciation* After t years, the value of a car that cost $16,000 is

$$V(t) = 16{,}000\left(\frac{3}{4}\right)^t.$$

Sketch a graph of the function and determine the value of the car 2 years after it was purchased.

38. *Depreciation* Suppose that straight-line depreciation is used to determine the value of the car in Exercise 37. If the car depreciates $3000 per year, the model for its value after t years is

$$V(t) = 16{,}000 - 3000t.$$

Sketch the graph of this line on the same set of coordinate axes used for the graph in Exercise 37. If you were selling the car after having it for 2 years, which depreciation model would you prefer? If you sell it after 4 years, which model would be to your advantage?

39. *Inflation Rate* Suppose that the annual rate of inflation averages 5% over the next 10 years. With this rate of inflation, the approximate cost C of goods or services during any year in that decade will be

$$C(t) = P(1.05)^t, \qquad 0 \le t \le 10,$$

where t is time in years and P is the present cost. If the price of an oil change for your car is presently $19.95, estimate the price 10 years from now.

40. *Price and Demand* The daily demand x for a certain product is given by

$$p = 25 - 0.4e^{0.02x},$$

where p is the price of the product. Find the price if the demand is (a) $x = 100$ units and (b) $x = 125$ units.

In Exercises 41–46, complete the table to determine the balance A for P dollars invested at rate r for t years and compounded n times per year.

n	1	4	12	365	Continuous compounding
A					

	Principal	Rate	Time
41.	$100	8%	20 years
42.	$2000	9%	10 years
43.	$5000	10%	40 years
44.	$1500	7%	2 years
45.	$10,000	9.5%	30 years
46.	$400	8%	50 years

In Exercises 47–50, complete the table to determine the principal P that will yield a balance of A dollars when invested at rate r for t years and compounded n times per year.

n	1	4	12	365	Continuous compounding
P					

	Balance	*Rate*	*Time*
47.	$5000	7%	10 years
48.	$100,000	9%	20 years
49.	$1,000,000	10.5%	40 years
50.	$2500	7.5%	2 years

 51. On the same set of coordinate axes, sketch the graphs of the following functions.

    (a) $f(x) = 2x$ (b) $f(x) = 2x^2$
    (c) $f(x) = 2^x$ (d) $f(x) = 2^{-x}$

52. *Savings Plan* Suppose you decide to start saving pennies according to the following pattern. You save one penny the first day, two pennies the second day, four the third day, etc. Each day you save twice the number of pennies as you did on the previous day. Which function in Exercise 51 models this problem, and how many pennies do you save on the 30th day? (In the next chapter, you will learn how to find the total number saved.)

53. *Parachute Drop* A parachutist jumps from a plane and opens the parachute at a height of 2000 feet (see figure). The height of the parachutist is then

$$h = 1950 + 50e^{-1.6t} - 20t,$$

where h is the height in feet and t is the time in seconds. (The time $t = 0$ corresponds to the time when the parachute is opened.) Find the height of the parachutist when $t = 0$, 25, 50, and 75 seconds. When do you think the parachutist will reach the ground?

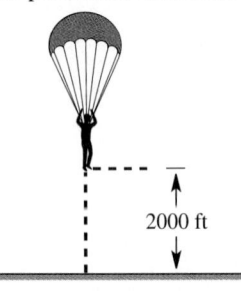

2000 ft

54. *Parachute Drop* A parachutist jumps from a plane and opens the parachute at a height of 3000 feet. The height of the parachutist is then

$$h = 2940 + 60e^{-1.7t} - 22t,$$

where h is the height in feet and t is the time in seconds. (The time $t = 0$ corresponds to the time when the parachute is opened.) Find the height of the parachutist when $t = 0$, 50, and 100 seconds. When do you think the parachutist will reach the ground?

55. *8-Track Cartridges* Eight-track cartridges increased in popularity until about 1980. For the years 1980 through 1985, however, the number n (in millions) of 8-track cartridges sold each year can be modeled by

$$n = 88\left(\frac{10}{21}\right)^t,$$

where $t = 0$ corresponds to 1980 (see figure). How many were sold in 1985? (*Source:* Recording Industry Association of America)

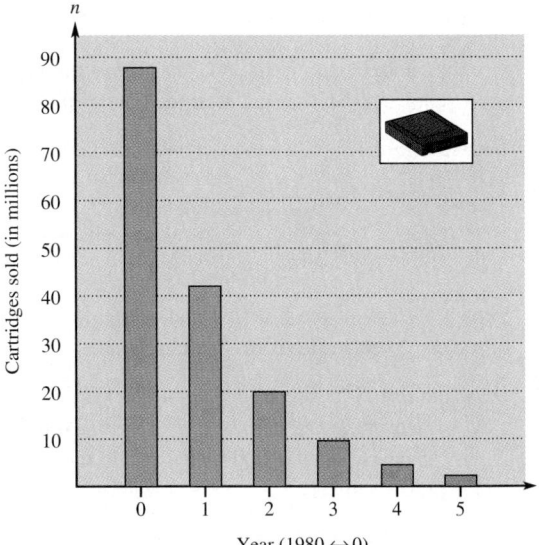

Year (1980 ↔ 0)

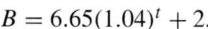 **56.** *Recreational Boats* From 1970 through 1990, the number of recreational boats B (in millions) owned in the United States can be modeled by

$$B = 6.65(1.04)^t + 2,$$

where $t = 0$ represents 1970. Sketch the graph of this function. How many recreational boats were owned in the United States in 1984? (*Source:* National Marine Manufacturers Association)

 In Exercises 57–62, use a graphing calculator to graph the function, using the following viewing window settings.*

> Xmin=-4
> Xmax=4
> Xscl=1
> Ymin=0
> Ymax=8
> Yscl=1

57. $y = 5^{x/3}$ **58.** $y = 5^{(x-2)/3}$

59. $y = 5^{-x/3}$ **60.** $y = 5^{-x/3} + 2$

61. $y = \frac{3}{2}e^{x/3}$ **62.** $y = 6e^{-x^2/3}$

63. *Compound Interest* Compare the rate of increase of an investment of $500 in two different accounts with respective interest rates of 6% and 8% compounded continuously. Perform the comparison by using a graphing calculator to graph the equations $y = 500e^{0.06x}$ and $y = 500e^{0.08x}$ on the same screen, using the following viewing window settings.

> Xmin=0
> Xmax=40
> Xscl=5
> Ymin=0
> Ymax=6000
> Yscl=500

* The graphing calculator features given in the examples and exercises in this text correspond to the TI-81 and TI-82 graphing calculators from Texas Instruments. However, you can also use other graphing calculators and computer graphing software for these examples and exercises.

Mid-Chapter Quiz

Take this quiz as you would take a quiz in class. After you are done, check your work against the answers in the back of the book.

1. Write the exponential equation in logarithmic form.
   (a) $5^3 = 125$
   (b) $e^0 = 1$

2. Write the logarithmic equation in exponential form.
   (a) $\log_{16} 4 = \frac{1}{2}$
   (b) $\log_5 \frac{1}{25} = -2$

3. Evaluate $\log_9 81$ without the aid of a calculator.

4. Evaluate $\ln \dfrac{1}{e^2}$ without the aid of a calculator.

5. Use a calculator to evaluate the following. (Round your answer to four decimal places.)
   (a) $\log_{10} 516$
   (b) $\ln 2115$

6. Use the properties of logarithms to expand the expression
   $$\ln\left(\frac{x}{x+3}\right)^3$$
   as a sum, difference, or multiple of logarithms.

7. Use the properties of logarithms to condense the expression
   $$2\log_3 x + 4\log_3(x+5)$$
   as a logarithm of a single quantity.

8. Simplify the logarithmic expression $\log_5 \dfrac{5^3}{x}$.

9. Find $f(4)$ if $f(x) = 1000(1 - e^{-0.5x})$. (Round your answer to three decimal places.)

10. Sketch the graph of each function.
    (a) $f(x) = 4^{x-1}$
    (b) $g(x) = 800(2^{-x})$

11. ***Compound Interest*** Find the balance after 8 years if $1250 is deposited in an account that pays 6.5% compounded (a) monthly and (b) continuously.

12. ***Air Pressure*** The air pressure P at sea level is approximately 14.7 pounds per square inch. As the altitude h (in feet above sea level) increases, the air pressure decreases. The relationship between air pressure and altitude can by modeled by
    $$P = 14.7e^{-0.00004h}.$$
    Mount Everest in Tibet and Nepal rises to a height of 29,108 feet above sea level. What is the air pressure at the peak of Mount Everest?

<table>
<tr><td>SECTION
9.4</td><td>**Logarithmic Functions and Their Graphs**
Graphs of Logarithmic Functions • *The Natural Logarithmic Function* • *Change of Base*</td></tr>
</table>

Graphs of Logarithmic Functions

In Section 8.2, you were introduced to the concept of the inverse of a function. There, you saw that if a function has the property that no horizontal line intersects the graph of a function more than once, then the function must have an inverse. By looking back at the graphs of the exponential functions introduced in Section 9.3, you will see that every function of the form $f(x) = a^x$ passes the "horizontal line test," and therefore must have an inverse. This inverse function is called the **logarithmic function with base a.**

Definition of Logarithmic Function	Let a and x be positive real numbers, with $a \neq 1$. The function given by $$f(x) = \log_a x$$ is called the **logarithmic function with base a.**

To sketch the graph of $y = \log_a x$, you can use the fact that the graphs of inverse functions are reflections of each other in the line $y = x$, as demonstrated in Example 1.

EXAMPLE 1   ■   Graphs of Exponential and Logarithmic Functions

Sketch the graphs of the following functions on the same coordinate system.

(a) $f(x) = 2^x$ (b) $g(x) = \log_2 x$

FIGURE 9.6

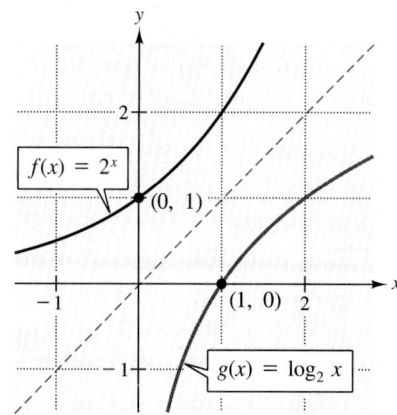

Solution

(a) Begin by constructing a table of values for $f(x) = 2^x$.

x	-2	-1	0	1	2	3
$f(x) = 2^x$	$\frac{1}{4}$	$\frac{1}{2}$	1	2	4	8

By plotting these points and connecting them with a smooth curve, you obtain the graph shown in Figure 9.6.

(b) Because $g(x) = \log_2 x$ is the inverse function of $f(x) = 2^x$, the graph of g is obtained by reflecting the graph of f in the line $y = x$, as shown in Figure 9.6.   ■

Notice from the graph of $g(x) = \log_2 x$ shown in Figure 9.6 that the domain of the function is the set of positive numbers and the range is the set of all real numbers. The basic characteristics of the graph of a logarithmic function are summarized in Figure 9.7. In this figure, note that the graph has one x-intercept, at $(1, 0)$. Also note that the y-axis is a vertical asymptote of the graph.

FIGURE 9.7

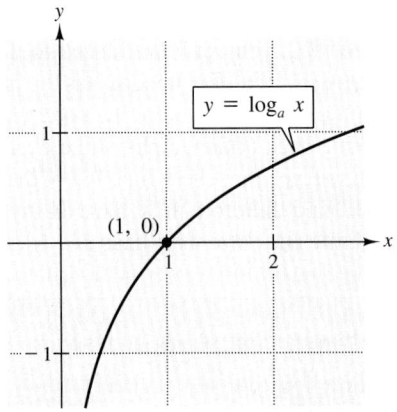

Graph of $y = \log_a x$, $a > 1$

- Domain: $(0, \infty)$
- Range: $(-\infty, \infty)$
- Intercept: $(1, 0)$
- Increasing
- Asymptote: y-axis

In Example 1, the inverse property of logarithmic functions was used to sketch the graph of $g(x) = \log_2 x$. You can also use a standard point-plotting approach, as demonstrated in Example 2.

EXAMPLE 2 ■ Sketching the Graph of a Logarithmic Function

Sketch the graph of the common logarithmic function

$$f(x) = \log_{10} x.$$

Solution

Begin by making a table of values. Note that some of the values can be obtained without a calculator, whereas others require a calculator. Using the points listed in the table below, sketch the graph as shown in Figure 9.8. Notice how slowly the graph rises for $x > 1$. In Figure 9.8, you would need to move out to $x = 1000$ before the graph would rise to $y = 3$.

FIGURE 9.8

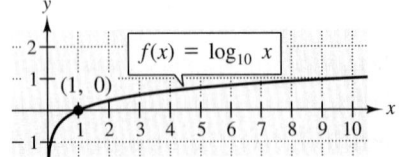

	Without Calculator				With Calculator		
x	$\frac{1}{100}$	$\frac{1}{10}$	1	10	2	5	8
$\log_{10} x$	-2	-1	0	1	0.301	0.699	0.903

■

EXAMPLE 3 ■ Sketching the Graph of a Logarithmic Function

Sketch the graph of $g(x) = 2 + \log_{10} x$.

Solution

FIGURE 9.9

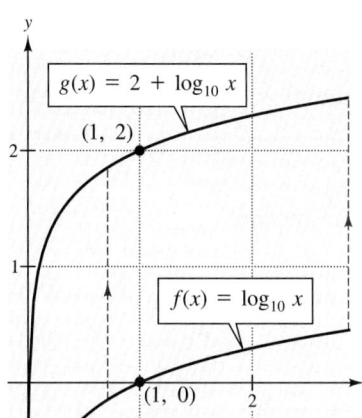

You can sketch the graph of g by shifting the graph of $f(x) = \log_{10} x$ two units up, as shown in Figure 9.9. Once you have sketched the graph, you should check a few points to make sure that you have shifted the graph properly. For instance, when $x = 1$, you have

$$g(1) = 2 + \log_{10} 1 = 2 + 0 = 2.$$

This implies that $(1, 2)$ is a point on the graph of g, as shown in Figure 9.9.

■

EXAMPLE 4 ■ Sketching the Graph of a Logarithmic Function

Sketch the graph of $g(x) = \log_{10}(x - 1)$.

Solution

FIGURE 9.10

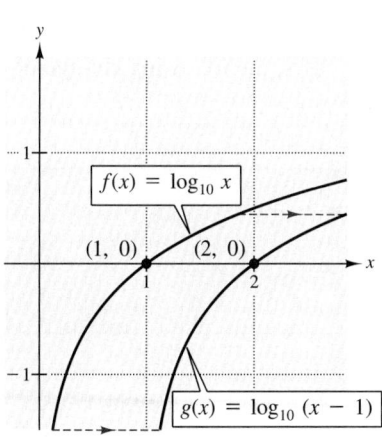

You can sketch the graph of g by shifting the graph of $f(x) = \log_{10} x$ one unit to the right, as shown in Figure 9.10. After sketching the graph of g, try checking some points to make sure that you have shifted the graph correctly. For instance, when $x = 2$, you have

$$g(2) = \log_{10}(2 - 1) = \log_{10} 1 = 0.$$

This implies that $(2, 0)$ is a point on the graph of g, as shown in Figure 9.10.

■

The Natural Logarithmic Function

As with exponential functions, the most widely used base for logarithmic functions is the number e.

The Natural Logarithmic Function	The function defined by $$f(x) = \log_e x = \ln x,$$ where $x > 0$, is called the **natural logarithmic function.**

EXAMPLE 5   ■   Sketching the Graph of the Natural Logarithmic Function

Sketch the graph of $f(x) = \ln x$.

Alternatively, you might discuss the inverse of $f(x) = \ln x$ and how to use its graph to generate the graph of $f(x) = \ln x$.

FIGURE 9.11

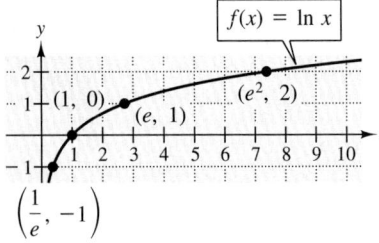

Solution

Begin by constructing a table of values.

x	$\dfrac{1}{e^2} \approx 0.14$	$\dfrac{1}{e} \approx 0.37$	1	$e \approx 2.72$	$e^2 \approx 7.39$
$f(x) = \ln x$	-2	-1	0	1	2

By plotting these points and connecting them with a smooth curve, you obtain the graph shown in Figure 9.11.   ■

Change of Base

Although 10 and e are the most frequently used bases, you may occasionally need to evaluate logarithms with other bases. In such cases, the following **change-of-base formula** is useful.

Change-of-Base Formula	Let a, b, and x be positive real numbers, with $a \neq 1$ and $b \neq 1$. $$\log_a x = \frac{\log_b x}{\log_b a} \quad \text{or} \quad \log_a x = \frac{\ln x}{\ln a}$$

Be aware that some students confuse the property of logarithms that states $\log_b u - \log_b v = \log_b \dfrac{u}{v}$ with $\log_v u = \dfrac{\log_b u}{\log_b v}$. Hence, they may be tempted to rewrite incorrectly $\log_a x = \dfrac{\log_b x}{\log_b a}$ as $\log_b x - \log_b a$.

The usefulness of this change-of-base formula is that you can use a calculator that has only the common logarithm key $\boxed{\text{log}}$ and the natural logarithm key $\boxed{\text{ln}}$ to evaluate logarithms with any base.* This is demonstrated in Examples 6 and 7.

---

* The graphing calculator keystrokes given in this text correspond to the TI-81 and TI-82 graphing calculators from Texas Instruments. For other graphing calculators, the keystrokes may differ.

EXAMPLE 6 ■ Changing Base to Convert to Common Logarithms

Use *common* logarithms to evaluate $\log_3 5$.

Solution

Using the change-of-base formula with $a = 3$, $b = 10$, and $x = 5$, you can convert to common logarithms by writing

$$\log_3 5 = \frac{\log_{10} 5}{\log_{10} 3}.$$

Now, using the following keystrokes

Keystrokes	*Display*	
5 ☐log ☐÷ 3 ☐log ☐=	1.4649735	Scientific
☐LOG 5 ☐÷ ☐LOG 3 ☐ENTER	1.4649735	Graphing

you find that the value of $\log_3 5$ is

$$\log_3 5 \approx 1.465.$$ ■

The evaluation in Example 6 used common logarithms, but you could just as easily have used natural logarithms, as demonstrated in Example 7.

EXAMPLE 7 ■ Changing Base to Convert to Natural Logarithms

Use *natural* logarithms to evaluate $\log_6 2$.

Solution

Using the change-of-base formula with $a = 6$ and $x = 2$, you can convert to natural logarithms by writing

$$\log_6 2 = \frac{\ln 2}{\ln 6}.$$

Now, using the following keystrokes

Keystrokes	*Display*	
2 ☐ln ☐÷ 6 ☐ln ☐=	0.3868528	Scientific
☐LN 2 ☐÷ ☐LN 6 ☐ENTER	0.3868528	Graphing

you find that the value of $\log_6 2$ is

$$\log_6 2 \approx 0.387.$$

(Try this problem with common logarithms to see that you obtain the same result.)

■

DISCUSSION PROBLEM ■ Identifying Graphs of Common Functions

Classify each of the following graphs as *linear, quadratic, exponential,* or *logarithmic.* In each case, write a short paragraph to support your answer. Include statements regarding the domain, range, intercepts, and asymptotes (if any) of each.

(a)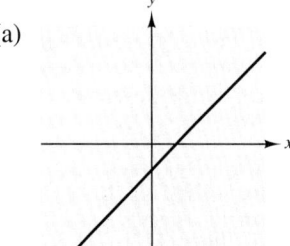

(b)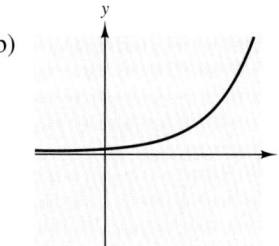

(c)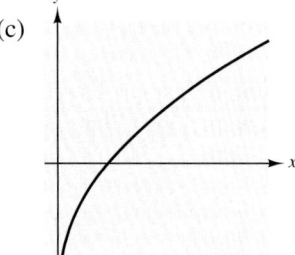

(d)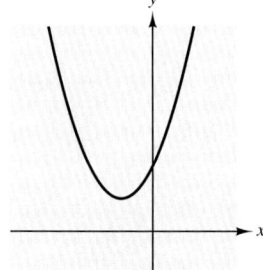

Warm-Up

The following warm-up exercises involve skills that were covered in earlier sections. You will use these skills in the exercise set for this section.

In Exercises 1–6, evaluate the expression. Use a calculator only if it is necessary or more efficient.

1. $\log_{10} 10{,}000$ **2.** $\log_4 64$ **3.** $\log_3 \frac{1}{81}$

4. $\ln e^3$ **5.** e^{-1} **6.** $e^{1/2}$

In Exercises 7–10, describe the relationship between the graphs of f and g.

7. $g(x) = f(x) - 4$ **8.** $g(x) = f(x - 4)$

9. $g(x) = -f(x)$ **10.** $g(x) = f(-x)$

9.4   **EXERCISES**   means that a graphing utility can help you solve the exercise or check your solution.

In Exercises 1–4, sketch the graphs of f and g on the same set of coordinate axes. What can you conclude about the relationship between f and g?

1. $f(x) = \log_3 x$
$g(x) = 3^x$

2. $f(x) = \log_4 x$
$g(x) = 4^x$

3. $f(x) = \log_6 x$
$g(x) = 6^x$

4. $f(x) = \ln x$
$g(x) = e^x$

In Exercises 5–10, use the graph of $y = \log_3 x$ (see Exercise 1) to match the function with its graph. [The graphs are labeled (a), (b), (c), (d), (e), and (f).]

5. $f(x) = 4 + \log_3 x$

6. $f(x) = -2 + \log_3 x$

7. $f(x) = -\log_3 x$

8. $f(x) = \log_3(-x)$

9. $f(x) = \log_3(x - 4)$

10. $f(x) = \log_3(x + 2)$

(a)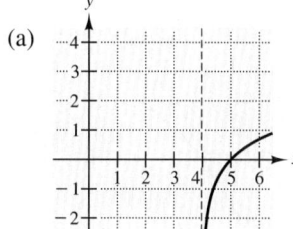

(b)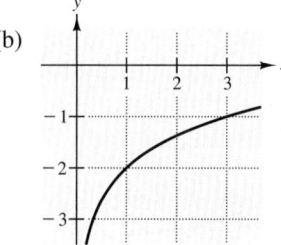

(c)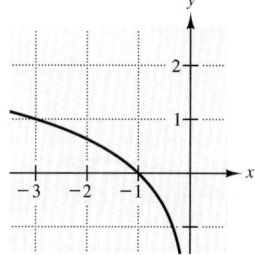

(d)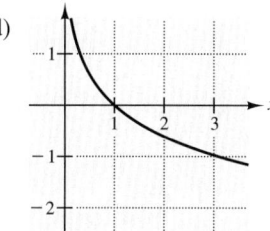

(e)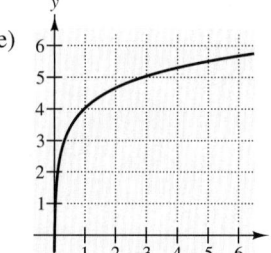

(f)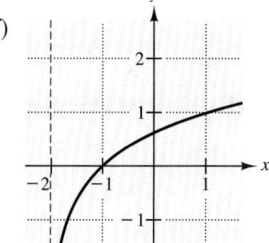

In Exercises 11–20, sketch the graph of the function.

11. $f(x) = \log_5 x$

12. $g(x) = \log_8 x$

13. $g(t) = -\log_2 t$

14. $h(s) = -2\log_3 s$

15. $f(x) = 3 + \log_2 x$

16. $f(x) = -2 + \log_3 x$

17. $g(x) = \log_2(x - 3)$

18. $h(x) = \log_3(x + 1)$

19. $f(x) = \log_{10}(10x)$

20. $g(x) = \log_4(4x)$

In Exercises 21–26, use the graph of $y = \ln x$ to match the function with its graph. [The graphs are labeled (a), (b), (c), (d), (e), and (f).]

21. $f(x) = \ln(x + 1)$

22. $f(x) = 4 - \ln(x + 4)$

23. $f(x) = \ln\left(x - \frac{3}{2}\right)$

24. $f(x) = -\frac{3}{2}\ln x$

25. $f(x) = 10\ln x$

26. $f(x) = \ln(-x)$

(a)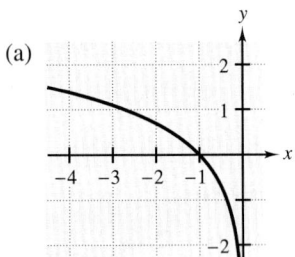

(b)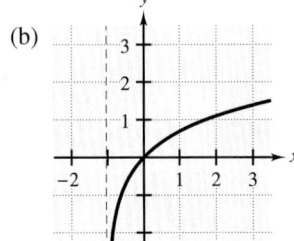

(c)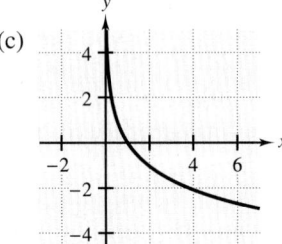

(d)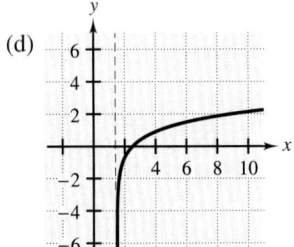

(e)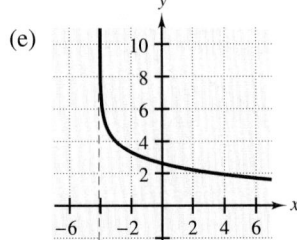

(f)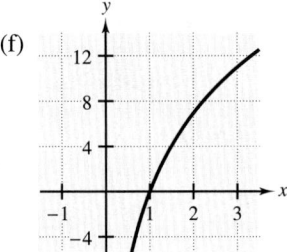

In Exercises 27–38, sketch the graph of the function.

27. $f(x) = -\ln x$

28. $f(x) = -2\ln x$

29. $f(x) = 3\ln x$

30. $h(t) = 4\ln t$

31. $f(x) = 1 + \ln x$

32. $h(x) = 2 + \ln x$

33. $g(x) = \ln(x + 6)$

34. $f(x) = -\ln(x - 2)$

35. $h(x) = -3 - \ln(x + 4)$

36. $g(x) = \ln(x + 2) - 1$

37. $f(x) = 2\ln(x - 1)$

38. $g(x) = 3\ln(x + 2)$

In Exercises 39–50, evaluate the logarithm by means of the change-of-base formula. Use both (a) the common logarithm key and (b) the natural logarithm key on a calculator. (Round your answer to four decimal places.)

39. $\log_8 132$

40. $\log_5 510$

41. $\log_3 7$

42. $\log_7 4$

43. $\log_2 0.72$

44. $\log_{12} 0.6$

45. $\log_{15} 1250$

46. $\log_{20} 125$

47. $\log_{1/2} 4$

48. $\log_{1/3}(0.015)$

49. $\log_4 \sqrt{42}$

50. $\log_3\left(1 + e^2\right)$

In Exercises 51–56, answer the question, or find the required element, for the function $f(x) = \log_{10} x$. (Do not use a calculator.)

51. What is the domain of f?

52. Find the inverse function of f.

53. Find the interval in which $f(x)$ lies, given that $1000 \le x \le 10{,}000$.

54. Find the interval in which x lies, given that $f(x)$ is negative.

55. By what amount will x increase, given that $f(x)$ is increased by one unit?

56. Find the ratio of a to b, given that $f(a) = 3 + f(b)$.

In Exercises 57–62, use a graphing calculator to graph the function, using the following viewing window settings.*

```
Xmin=0
Xmax=10
Xscl=1
Ymin=-3
Ymax=7
Yscl=1
```

57. $y = 5 \log_{10} x$

58. $y = 5 \log_{10}(x - 3)$

59. $y = -3 + 5 \log_{10} x$

60. $y = 5 \log_{10}(3x)$

61. $y = 2 \ln x$

62. $y = -2 \ln x$

*The graphing calculator features given in the examples and exercises in this text correspond to the TI-81 and TI-82 graphing calculators from Texas Instruments. However, you can also use other graphing calculators and computer graphing software for these examples and exercises.

Using a Graphing Calculator

A graphing calculator can be used to fit exponential and logarithmic models of the form

$$y = a(b^x) \quad \text{and} \quad y = a + b \ln x$$

to data. For details on the actual keystrokes that must be used, consult the user's manual that accompanies your graphing calculator.

EXAMPLE 1 ■ Finding an Exponential Model

For the years 1950 through 1990, the total amount y (in billions of dollars) of life insurance held in the United States is shown in the table. Use a graphing calculator to find an exponential model that approximates these data. (*Source:* American Council of Life Insurance)

Year	Amount
1950	234
1955	373
1960	586
1965	901
1970	1402
1975	2314
1980	3541
1985	6053
1990	9198

Solution

Let $x = 0$ represent 1950. Then enter the following coordinates into the statistical data bank of the calculator.

$$(0, 234), \quad (5, 373), \quad (10, 586), \quad (15, 901), \quad (20, 1402),$$
$$(25, 2314), \quad (30, 3541), \quad (35, 6053), \quad (40, 9198)$$

After selecting the "exponential regression" program, the display should read $a \approx 231.5$ and $b \approx 1.0963$. (The correlation of $r \approx 0.9998$ tells you that the fit is very good.) Thus, a model for the data is

$$y = 231.5(1.0963)^x.$$

The graph below compares the actual data with the model.

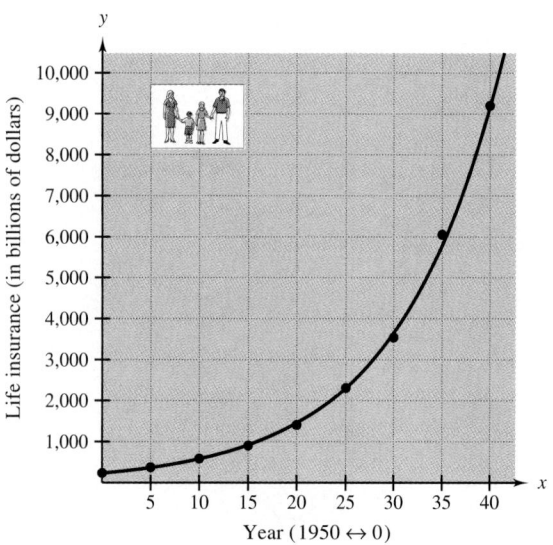

EXAMPLE 2 ■ Finding a Logarithmic Model

A savings account contains $1000, which earns 5% annual interest, compounded continuously. The number of years y the money must be left in the account for it to grow to a balance of x dollars is given in the table. Find a logarithmic model for these data.

Balance, x	1000	2000	3000	4000	5000	6000	7000
Years, y	0	13.86	21.97	27.73	32.19	35.84	38.92

Solution

To find a logarithmic model for these data, enter the seven ordered pairs from the table into the statistical data bank of the calculator. Then select the "logarithmic regression" program. The display should read $a \approx -138.2$ and $b \approx 20$. (The correlation of $r \approx 1$ tells you that the fit is nearly perfect.) Thus, the model for the data is

$$y = -138.2 + 20 \ln x.$$

The graph below compares the actual data with the model.

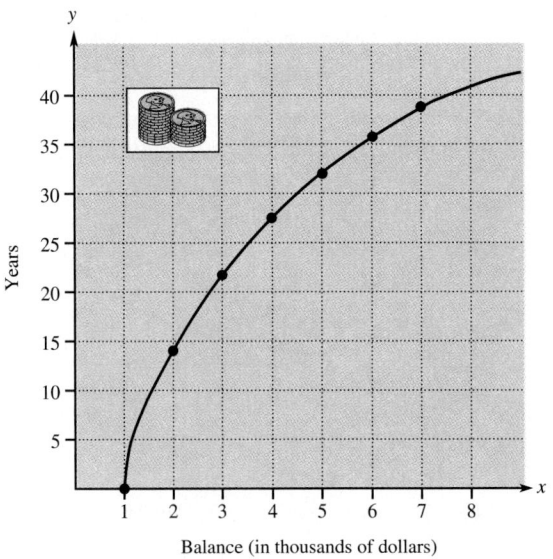

Balance (in thousands of dollars)

EXERCISES

In Exercises 1–4, plot the data. Then find an exponential or logarithmic model that approximates the data. (Round your answer to four decimal places.)

1.

x	1	2	3	4	5	6	7	8	9	10
y	0.58	0.66	0.76	0.87	1.01	1.16	1.33	1.53	1.76	2.02

2.

x	1	2	3	4	5	6	7	8	9	10
y	2.42	2.93	3.54	4.29	5.19	6.28	7.59	9.19	11.12	13.45

3.

x	1	2	3	4	5	6	7	8	9	10
y	2.00	2.35	2.55	2.69	2.80	2.90	2.97	3.04	3.10	3.15

4.

x	1	2	3	4	5	6	7	8	9	10
y	5.00	4.58	4.34	4.17	4.03	3.92	3.83	3.75	3.68	3.62

5. *Liz Claiborne, Inc.* The table shows the annual sales y (in millions of dollars) for Liz Claiborne, Inc. for the years 1981 through 1990. Plot the data, letting $x = 1$ represent 1981. Then find an exponential or logarithmic model that approximates the data. Use your model to estimate the 1991 sales. (*Source:* Liz Claiborne, Inc.)

Year, x	1981	1982	1983	1984	1985	1986	1987	1988	1989	1990
Sales, y	116.8	165.7	228.7	391.3	556.6	813.5	1053.3	1184.2	1410.7	1728.9

Solving Exponential and Logarithmic Equations

Properties of Exponential and Logarithmic Equations • *Solving Exponential Equations* • *Solving Logarithmic Equations* • *Applications*

Properties of Exponential and Logarithmic Equations

So far in this chapter, you have focused on the definitions, graphs, and properties of exponential and logarithmic functions. In this section, you will study procedures for *solving equations* that involve exponential or logarithmic expressions. As a simple example, consider the exponential equation

$$2^x = 16.$$

By rewriting this equation in the form

$$2^x = 2^4,$$

you can see that the solution is $x = 4$. To solve this equation, you can use one of the following properties.

Properties of Exponential and Logarithmic Equations

Let a be a positive real number such that $a \neq 1$, and let x and y be real numbers.

1. $a^x = a^y$ if and only if $x = y$

2. $\log_a x = \log_a y$ if and only if $x = y$ $(x > 0, \ y > 0)$

EXAMPLE 1 ■ Solving Exponential and Logarithmic Equations

Solve the following equations.

(a) $4^{x+2} = 4^5$ (b) $\ln(2x - 3) = \ln 11$

Solution

Stress that the bases must be the same before exponents can be equated.

(a)
$4^{x+2} = 4^5$	Given equation
$x + 2 = 5$	Property 1
$x = 3$	Subtract 2 from both sides

Thus, the solution is 3. Check this solution in the original equation.

(b)
$\ln(2x - 3) = \ln 11$	Given equation
$2x - 3 = 11$	Property 2
$2x = 14$	Add 3 to both sides
$x = 7$	Divide both sides by 2

Thus, the solution is 7. Check this solution in the original equation. ■

Solving Exponential Equations

In Example 1(a), you were able to solve the given equation because both sides of the equation were written in exponential form (with the same base). However, if only one side of the equation is written in exponential form, it is more difficult to solve the equation. For example, how would you solve the following equation?

$$2^x = 7$$

To solve this equation, you must find the power to which 2 can be raised to obtain 7. To do this, you can rewrite the equation in logarithmic form. Recall that

$$\underbrace{x = a^y}_{\substack{\text{Exponential} \\ \text{form}}} \qquad \text{if and only if} \qquad \underbrace{\log_a x = y}_{\substack{\text{Logarithmic} \\ \text{form}}}.$$

EXAMPLE 2   ■   **Solving an Exponential Equation**

Solve the equation $2^x = 7$.

Solution

$2^x = 7$	Given equation
$x = \log_2 7$	Rewrite equation in logarithmic form

Thus, the solution of the equation is $x = \log_2 7 \approx 2.807$. Try checking this solution with your calculator by evaluating $2^{2.807}$. The result should be $2^{2.807} \approx 7$.   ■

NOTE   In Example 2, you can use the change-of-base formula to evaluate $\log_2 7$.

$$\log_2 7 = \frac{\log_{10} 7}{\log_{10} 2} = \frac{\ln 7}{\ln 2} \approx 2.807$$

EXAMPLE 3   ■   **Solving an Exponential Equation**

Solve the equation $2e^x = 10$.

Solution

$2e^x = 10$	Given equation
$e^x = 5$	Divide both sides by 2
$x = \log_e 5$	Rewrite equation in logarithmic form
$x = \ln 5$	

Thus, the solution of the equation is $x = \ln 5 \approx 1.609$. Try checking this solution in the original equation.   ■

EXAMPLE 4   ■   Solving an Exponential Equation

Solve the equation $5 + e^{x+1} = 20$.

Point out the need to isolate the exponential term before rewriting the equation in logarithmic form.

Solution

$5 + e^{x+1} = 20$	Given equation
$e^{x+1} = 15$	Subtract 5 from both sides
$x + 1 = \ln 15$	Rewrite equation in logarithmic form
$x = -1 + \ln 15$	Subtract 1 from both sides

Thus, the solution of the equation is $x = -1 + \ln 15 \approx 1.708$. Try checking this solution in the original equation.   ■

Solving Logarithmic Equations

You know how to solve an exponential equation by writing it in logarithmic form. To solve a logarithmic equation, you can write it in exponential form. For instance, to solve a logarithmic equation such as

$$\ln x = 2,$$

you would rewrite the equation in exponential form, as follows.

$\ln x = 2$	Given equation
$x = e^2$	Rewrite equation in exponential form

This procedure is demonstrated in the next four examples.

Guidelines for solving exponential and logarithmic equations are listed below.

Guidelines for Solving Exponential and Logarithmic Equations

1. To solve an exponential equation, first isolate the exponential expression, then rewrite the equation in logarithmic form and solve for the variable.

2. To solve a logarithmic equation, first isolate the logarithmic expression, then rewrite the equation in exponential form and solve for the variable.

EXAMPLE 5   ■   Solving a Logarithmic Equation

Solve the equation $2 \ln x = 5$.

Solution

Begin by isolating the logarithmic expression, as follows.

$2 \ln x = 5$	Given equation
$\ln x = \dfrac{5}{2}$	Divide both sides by 2
$x = e^{5/2}$	Rewrite equation in exponential form

Thus, the solution is $x = e^{5/2} \approx 12.182$. Try checking this solution in the original equation. ∎

EXAMPLE 6 ■ **Solving a Logarithmic Equation**

Solve the equation $3 \log_{10} x = 6$.

Solution

$3 \log_{10} x = 6$	Given equation
$\log_{10} x = 2$	Divide both sides by 3
$x = 10^2$	Rewrite equation in exponential form

Thus, the solution is $x = 10^2 = 100$. Try checking this solution in the original equation. ∎

EXAMPLE 7 ■ **Solving a Logarithmic Equation**

Solve the equation $20 \ln 0.2x = 30$.

Solution

$20 \ln 0.2x = 30$	Given equation
$\ln 0.2x = 1.5$	Divide both sides by 20
$0.2x = e^{1.5}$	Rewrite equation in exponential form
$x = 5e^{1.5}$	Divide both sides by 0.2

Thus, the solution is $x = 5e^{1.5} \approx 22.408$. Try checking this solution in the original equation. ∎

EXAMPLE 8 ■ **Solving a Logarithmic Equation**

Solve the equation $\log_{10} 2x - \log_{10}(x - 3) = 1$.

Review the three basic properties of logarithms.

Solution

$\log_{10} 2x - \log_{10}(x - 3) = 1$	Given equation
$\log_{10} \dfrac{2x}{x - 3} = 1$	Condense the left side
$\dfrac{2x}{x - 3} = 10^1$	Rewrite equation in exponential form
$2x = 10x - 30$	Multiply both sides by $x - 3$
$-8x = -30$	Subtract $10x$ from both sides
$x = \dfrac{15}{4}$	Solution

Thus, the solution is $x = \frac{15}{4}$. Try checking this solution in the original equation. ∎

Applications

The following application uses the formula for continuously compounded interest, $A = Pe^{rt}$.

EXAMPLE 9   ■   **An Application: Compound Interest**

A deposit of $5000 is placed in a savings account for 2 years. The interest for the account is compounded continuously. At the end of 2 years, the balance in the account is $5867.56. What is the annual percentage rate for this account?

Solution

Using the formula for continuously compounded interest, $A = Pe^{rt}$, you have the following solution.

Common formula: $A = Pe^{rt}$

Labels: Principal $= P = 5000$ (dollars)
Amount $= A = 5867.56$ (dollars)
Time $= t = 2$ (years)
Annual percentage rate $= r$ (in decimal form)

Equation: $5867.56 = 5000e^{2r}$

$$\frac{5867.56}{5000} = e^{2r}$$

$$1.1735 \approx e^{2r}$$

$$\ln 1.1735 \approx 2r$$

$$0.16 \approx 2r$$

$$0.08 \approx r$$

Therefore, the annual percentage rate is 8%. Try checking this solution in the original statement of the problem.   ■

DISCUSSION PROBLEM   ■   **Solving an Exponential Equation**

The *polynomial* equation

$$x^2 - 3x + 2 = 0$$

can be solved by factoring the left side and setting each factor equal to zero, as follows.

$$(x - 2)(x - 1) = 0$$

$$x - 2 = 0 \implies x = 2$$

$$x - 1 = 0 \implies x = 1$$

Use the same procedure to solve the equation

$$e^{2x} - 3e^x + 2 = 0.$$ ■

Warm-Up

The following warm-up exercises involve skills that were covered in earlier sections. You will use these skills in the exercise set for this section.

In Exercises 1–6, solve for x.

1. $10x - 3 = 22$ **2.** $3 - 2(x - 1) = 12$ **3.** $(x - 4)^2 - 36 = 0$

4. $x^2 - 5x + 6 = 0$ **5.** $-28 + x \ln 5 = 0$ **6.** $5xe^2 = 0$

In Exercises 7–10, simplify the expression.

7. $\log_{10} 10^3$ **8.** $\log_5 5^6$ **9.** $\log_2 1$ **10.** $\ln e^4$

9.5   EXERCISES

In Exercises 1–16, solve the equation for x. (Do not use a calculator.)

1. $2^x = 2^5$ **2.** $5^x = 5^3$ **3.** $3^{x+4} = 3^{12}$ **4.** $10^{1-x} = 10^4$

5. $4^{x-1} = 16$ **6.** $3^{2x} = 81$ **7.** $5^x = \frac{1}{125}$ **8.** $4^{x+1} = \frac{1}{64}$

9. $\log_2(x + 3) = \log_2 7$ **10.** $\log_5 2x = \log_5 36$ **11.** $\ln 5x = \ln 22$ **12.** $\ln(2x - 3) = \ln 17$

13. $\log_3 x = 4$ **14.** $\log_5 x = 3$ **15.** $\log_{10} 2x = 6$ **16.** $\log_2(3x - 1) = 5$

 In Exercises 17–36, solve the exponential equation. (Round your answer to two decimal places.)

17. $2^x = 45$ **18.** $5^x = 212$ **19.** $10^{2y} = 52$ **20.** $12^{x-1} = 1500$

21. $\frac{1}{5}(4^{x+2}) = 300$ **22.** $3(2^{t+4}) = 350$ **23.** $4 + e^{2x} = 150$ **24.** $500 - e^{x/2} = 35$

25. $23 - 5e^{x+1} = 3$ **26.** $2e^x + 5 = 115$ **27.** $300e^{x/2} = 9000$ **28.** $1000^{0.12x} = 25{,}000$

29. $6000e^{-2t} = 1200$ **30.** $10{,}000e^{-0.1t} = 4000$ **31.** $250(1.04)^x = 1000$ **32.** $32(1.5)^x = 640$

33. $\dfrac{1600}{(1.1)^x} = 200$ **34.** $\dfrac{5000}{(1.05)^x} = 250$ **35.** $4(1 + e^{x/3}) = 84$ **36.** $50(3 - e^{2x}) = 125$

 In Exercises 37–52, solve the logarithmic equation. (Round your answer to two decimal places.)

37. $\log_{10} x = 0$ **38.** $\ln x = 1$ **39.** $\log_{10} x = 3$ **40.** $\log_{10} x = -2$

41. $\log_{10} 4x = \frac{3}{2}$ **42.** $\log_{10}(x + 3) = \frac{5}{3}$ **43.** $\ln x = 2.1$ **44.** $\ln 2x = 3$

45. $\frac{2}{3} \ln(x + 1) = -1$ **46.** $8 \ln(3x - 2) = 1.5$ **47.** $2 \log_{10}(x + 5) = 15$ **48.** $-1 + 3 \log_{10} \dfrac{x}{2} = 8$

49. $\ln x^2 = 6$ **50.** $\ln \sqrt{x} = 6.5$ **51.** $\log_{10} x(x - 3) = 1$ **52.** $\log_{10} x(x + 1) = 0$

53. ***Doubling Time*** Solve the exponential equation

$$5000 = 2500e^{0.09t}$$

for t to determine the number of years required for an investment of \$2500 to double in value when compounded continuously at the rate of 9%.

 54. ***Doubling Time*** Solve the exponential equation

$$10,000 = 5000e^{10r}$$

for r to determine the interest rate required for an investment of \$5000 to double in value when compounded continuously for 10 years.

 55. ***Intensity of Sound*** The relationship between the number of decibels B and the intensity of a sound I (in watts per meter squared) is

$$B = 10 \log_{10}\left(\frac{I}{10^{-16}}\right).$$

Determine the intensity of a sound I if it registers 75 decibels on an intensity meter.

 56. ***Human Memory Model*** The average score A for a group of students who took a test t months after the completion of a course is given by the memory model

$$A = 80 - \log_{10}(t + 1)^{12}.$$

Determine how long it would take after the completion of the course for the average score to fall to $A = 72$.

 57. ***Oceanography*** Oceanographers use the density d (in grams per cubic centimeter) of seawater to obtain information about the circulation of water masses and the rates at which waters of different densities mix. For water with a salinity of 30%, the water temperature T (in degrees Celsius) is related to the density by

$$T = 7.9 \ln(1.0245 - d) + 61.84.$$

Find the densities of the subantarctic water and the antarctic bottom water shown in the figure at the top of the next column.

Figure for 57

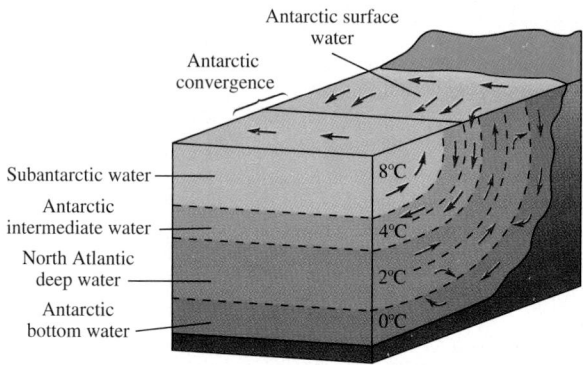

This cross section shows complex currents at various depths in the South Atlantic Ocean off Antarctica.

 58. ***Friction*** To restrain an untrained horse, a person partially wraps the horse's rope around a cylindrical post in the corral (see figure). If the horse is pulling on the rope with a force of 200 pounds, the force F (in pounds) that must be exerted by the person is

$$F = 200e^{-0.2\pi\theta/180},$$

where θ is the angle of wrap in degrees. Find the smallest value of θ if F cannot exceed 80 pounds.

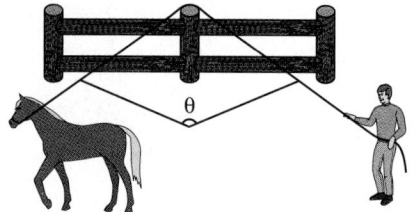

Applications of Exponential and Logarithmic Functions
Compound Interest • *Effective Yield* • *Growth and Decay* • *Intensity Models*

Notice that the formulas for periodic compounding and continuous compounding have five variables and four variables, respectively. Using basic algebraic skills and the properties of exponents and logarithms, we are now able to solve for A, P, r, or t in either formula, given the value of all of the other variables in the formula.

Compound Interest

In Section 9.3, the following two formulas for compound interest were introduced. In these formulas, A is the balance, P is the principal, r is the annual percentage rate (in decimal form), and t is the time in years.

n Compoundings per Year

$$A = P\left(1 + \frac{r}{n}\right)^{nt}$$

Continuous Compounding

$$A = Pe^{rt}$$

Example 9 in the previous section demonstrated a procedure for finding the annual percentage rate for continuously compounded interest. Example 1 below shows how to find the annual percentage rate for interest that is compounded n times per year.

EXAMPLE 1 ■ Finding the Annual Percentage Rate

An investment of $50,000 is made in an account that compounds interest quarterly. After 4 years, the balance in the account is $71,381.07. What is the annual percentage rate for this account?

Solution

Common formula: $\quad A = P\left(1 + \dfrac{r}{n}\right)^{nt}$

Labels: Principal $= P = 50,000$ (dollars)
            Amount $= A = 71,381.07$ (dollars)
            Time $= t = 4$ (years)
            Number of compoundings per year $= n = 4$ (number per year)
            Annual percentage rate $= r$ (in decimal form)

Equation:

$$71{,}381.07 = 50{,}000\left(1 + \frac{r}{4}\right)^{(4)(4)}$$

$$\frac{71{,}381.07}{50{,}000} = \left(1 + \frac{r}{4}\right)^{16}$$

$$1.42762 \approx \left(1 + \frac{r}{4}\right)^{16}$$

$$(1.42762)^{1/16} \approx 1 + \frac{r}{4} \qquad \text{Raise both sides to } \tfrac{1}{16} \text{ power}$$

$$1.0225 \approx 1 + \frac{r}{4}$$

$$0.0225 \approx \frac{r}{4}$$

$$0.09 \approx r$$

Therefore, the annual percentage rate is 9%. Check this solution in the original statement of the problem. ∎

NOTE In Example 1, to "get rid of" the 16th power on the right side of the equation, you needed to raise both sides of the equation to the $\frac{1}{16}$ power.

EXAMPLE 2 ■ **Doubling Time for Continuous Compounding**

An investment is made in a trust fund at an annual percentage rate of 8.75%, compounded continuously. How long will it take for the investment to double?

Solution

Common formula: $\quad A = Pe^{rt}$

Labels: Principal $= P$ (dollars)
Amount $= A = 2P$ (dollars)
Time $= t$ (years)
Annual percentage rate $= r = 0.0875$ (in decimal form)

Equation:

$$2P = Pe^{0.0875t}$$

$$2 = e^{0.0875t}$$

$$\ln 2 = 0.0875t \qquad \text{Rewrite equation in logarithmic form}$$

$$\frac{\ln 2}{0.0875} = t$$

$$7.92 \approx t$$

Interest rates paid by banks and other savings institutions vary with types of savings accounts. Passbook accounts typically pay lower interest rates than certificates of deposit (which require commitment for a specific period of time).

Therefore, it will take approximately 7.92 years for the investment to double. Check this solution in the original statement of the problem. (Note that you do not have to know a dollar amount for the principal in order to answer this question.) ∎

EXAMPLE 3 ■ Finding the Type of Compounding

Suppose you deposit $1000 in a savings account. At the end of 1 year, your balance is $1077.63. If the bank tells you that the annual percentage rate for the account is 7.5%, how was the interest compounded?

Solution

If the interest had been compounded continuously at 7.5%, the balance would have been $A = 1000e^{(0.075)(1)} = \$1077.88$. Because the actual balance is slightly less than this, you can use the formula for interest that is compounded n times per year.

Common formula: $A = P\left(1 + \dfrac{r}{n}\right)^{nt}$

Labels: Principal $= P = 1000$ (dollars)
Amount $= A = 1077.63$ (dollars)
Time $= t = 1$ (year)
Annual percentage rate $= r = 0.075$ (in decimal form)
Number of compoundings per year $= n$ (number per year)

Equation: $1077.63 = 1000\left(1 + \dfrac{0.075}{n}\right)^{n}$

$$1.07763 = \left(1 + \dfrac{0.075}{n}\right)^{n}$$

At this point, it is not clear what you should do to try to solve the equation for n. However, you know that the more common types of compoundings are annual ($n = 1$), quarterly ($n = 4$), monthly ($n = 12$), and daily ($n = 365$). By testing these four values of n, you can find that the solution is $n = 12$.

n	1	4	12	365
$\left(1 + \dfrac{0.075}{n}\right)^{n}$	1.075	1.07714	1.07763	1.07788

Therefore, the interest was compounded monthly. ■

Effective Yield

In Example 3, notice that an investment of $1000 at an interest rate of 7.5% compounded monthly produced a balance of $1077.63 at the end of 1 year. Because $77.63 of this amount is interest, the **effective yield** for this investment is

$$\text{Effective yield} = \frac{\text{year's interest}}{\text{amount invested}} = \frac{77.63}{1000} = 0.07763 = 7.763\%.$$

In other words, the effective yield for an investment collecting compound interest is the *simple interest rate* that would yield the same balance at the end of 1 year.

EXAMPLE 4 ■ Finding the Effective Yield

An investment is made in an account that pays 6.75% interest, compounded continuously. What is the effective yield for this investment?

Solution

Notice that you do not have to know the principal or the time that the money will be left in the account. Instead, you can choose an arbitrary principal, such as $1000.00. Then, because effective yield is based on the balance at the end of 1 year, you can use the following formula.

$A = Pe^{rt}$ Common formula

$A = 1000e^{0.0675(1)}$ Substitute known values

$A = 1069.83$

Now, because the account would earn $69.83 in interest after 1 year for a principal of $1000, you can conclude that the effective yield is

$$\text{Effective yield} = \frac{69.83}{1000} = 0.06983 = 6.983\%.$$ ■

Growth and Decay

The balance in an account earning *continuously* compounded interest is one example of a quantity that increases over time according to the **exponential growth model**

$$y = Ce^{kt}.$$

Exponential Growth and Decay

The mathematical model for exponential growth or decay is given by

$$y = Ce^{kt}.$$

For this model, t is the time, C is the original size (or amount) of the quantity, and y is the size (or amount) after time t. The number k is a constant that is determined by the rate of growth. If $k > 0$, the model represents **exponential growth,** and if $k < 0$, the model represents **exponential decay.**

One common application of exponential growth is in modeling the growth of a population, as shown in Example 5.

EXAMPLE 5 ■ Population Growth

Suppose you are doing research about a particular country and discover that the country's population is two million in 1980 and three million in 1990. What would you predict the population of the country to be in the year 2000?

Solution

If you assumed a *linear growth model,* you would simply predict the population in the year 2000 to be four million. (Try verifying this.) However, social scientists and demographers have discovered that exponential growth models are better than linear growth models for representing population growth. Thus, you can use the exponential growth model

$$y = Ce^{kt}.$$

In this model, let $t = 0$ represent the year 1980. Then, the given information about the population can be described by the following table.

t (years)	0	10	20
Ce^{kt} (million)	$Ce^{k(0)} = 2$	$Ce^{k(10)} = 3$	$Ce^{k(20)} = ?$

To find the population when $t = 20$, first find the values of C and k. From the table, use the fact that $Ce^{k(0)} = Ce^0 = 2$ to conclude that $C = 2$. Then, using this value of C, you can solve for k as follows.

$Ce^{k(10)} = 3$	From table
$2e^{10k} = 3$	Substitute value of C
$e^{10k} = \dfrac{3}{2}$	Divide both sides by 2
$10k = \ln \dfrac{3}{2}$	Rewrite equation in logarithmic form
$k = \dfrac{1}{10} \ln \dfrac{3}{2}$	Divide both sides by 10
$k \approx 0.0405$	

Finally, you can use this value of k to conclude that the population in the year 2000 is given by

$$2e^{0.0405(20)} \approx 2(2.25) = 4.5 \text{ million.}$$

Figure 9.12 graphically compares the exponential growth model with a linear growth model.

FIGURE 9.12

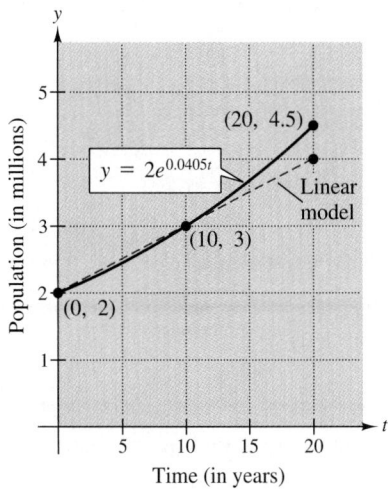

EXAMPLE 6 ■ Radioactive Decay

Radioactive iodine is a by-product of some types of nuclear reactors. Its **half-life** is 60 days. That is, after 60 days, a given amount of radioactive iodine will have decayed to half of that amount. Suppose a nuclear accident occurs and releases 20 grams of radioactive iodine. How long will it take for the radioactive iodine to decay to an amount of 1 gram?

Solution

To solve this problem, use the model for exponential decay.

$$y = Ce^{kt}$$

Next, use the information given in the problem to set up the following table.

t (days)	0	60	?
Ce^{kt} (grams)	$Ce^{k(0)} = 20$	$Ce^{k(60)} = 10$	$Ce^{k(t)} = 1$

Because $Ce^{k(0)} = Ce^0 = 20$, you can conclude that $C = 20$. Then, using this value of C, you can solve for k as follows.

Discussion project: Have students research and bring in articles relating to the Chernobyl disaster. Note the numerous references to radioactive exposure, the half-life of uranium, and the length of time that the area will be adversely affected.

$Ce^{k(60)} = 10$	From table
$20e^{60k} = 10$	Substitute value of C
$e^{60k} = \dfrac{1}{2}$	Divide both sides by 20
$60k = \ln \dfrac{1}{2}$	Rewrite equation in logarithmic form
$k = \dfrac{1}{60} \ln \dfrac{1}{2} \approx -0.01155$	Solve for k

FIGURE 9.13

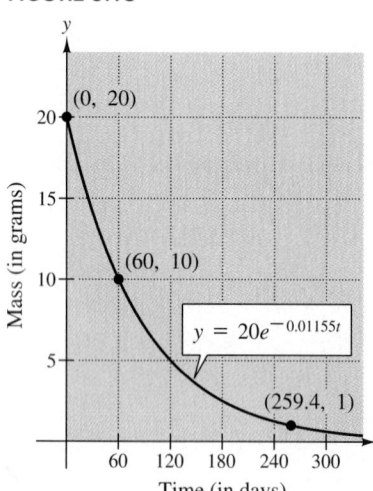

Finally, you can use this value of k to find the time when the amount is 1 gram as follows.

$Ce^{kt} = 1$	From table
$20e^{-0.01155t} = 1$	Substitute values of C and k
$e^{-0.01155t} = \dfrac{1}{20}$	Divide both sides by 20
$-0.01155t = \ln \dfrac{1}{20}$	Rewrite equation in logarithmic form
$t = \dfrac{1}{-0.01155} \ln \dfrac{1}{20}$	Solve for t
≈ 259.4 days	

Thus, 20 grams of radioactive iodine will have decayed to 1 gram after about 259.4 days. This solution is shown graphically in Figure 9.13. ■

Intensity Models

The magnitude of an earthquake is measured on the *Richter scale.* The magnitude R of an earthquake of energy E is given by $R = \frac{2}{3}(\log_{10} E - 11.4)$. Measuring 7.1 on the Richter scale, the San Francisco Earthquake of 1989 caused much devastation.

EXAMPLE 7   ■   Using Properties of Logarithms

Find the energy levels of the following earthquakes.

(a) Great San Francisco in 1906, $R = 8.3$.

(b) San Francisco Bay Area in 1989, $R = 7.1$.

Solution

(a) Using the formula $R = \frac{2}{3}(\log_{10} E - 11.4)$ and $R = 8.3$, you can write $8.3 = \frac{2}{3}(\log_{10} E - 11.4)$ or

$$E = 10^{23.85} \approx 7.08 \times 10^{23}.$$

(b) Using $R = 7.1$, you can write $7.1 = \frac{2}{3}(\log_{10} E - 11.4)$ or
$$E = 10^{22.05} \approx 1.12 \times 10^{22}.$$

Note that an increase of 1.2 units on the Richter Scale (from 7.1 to 8.3) represents an energy change of a factor of

$$\frac{7.08 \times 10^{23}}{1.12 \times 10^{22}} \approx 63.$$

In other words, the "Great San Francisco Earthquake" of 1906 was about 63 times greater in energy than the one in 1989.   ■

DISCUSSION PROBLEM   ■   Creating Word Problems

We spent a lot of time looking for real-life applications that could be used in the examples and exercises. If you can think of any good applications that we could use in future editions, we welcome your suggestions. Please send your applications to our publisher (D. C. Heath and Company, c/o Mathematics, 125 Spring Street, Lexington, MA 02173), who will forward them to us.   ■

Warm-Up

The following warm-up exercises involve skills that were covered in earlier sections. You will use these skills in the exercise set for this section.

In Exercises 1–10, solve the equation. (Round your answer to two decimal places.)

1. $6 + 3(x - 5) = 27 - 5(x - 6)$ **2.** $5x^2 + 2x - 3 = 0$

3. $\sqrt{x - 5} = 6$ **4.** $\dfrac{4}{t} + \dfrac{3}{2t} = 1$

5. $\log_6 2x = 2$ **6.** $\log_2(x - 5) = 6$ **7.** $e^{x/2} = 8$

8. $e^{x-5} = 550$ **9.** $300(1.01)^t = 1200$ **10.** $100(1 - e^t) = 75$

9.6 EXERCISES

means that a graphing utility can help you solve the exercise or check your solution.

Annual Interest Rate In Exercises 1–6, find the annual interest rate, given the principal P, the balance A, the time t, and the type of compounding.

Principal	Balance	Time	Compounding
1. $500	$1004.83	10 years	Monthly
2. $3000	$21,628.70	20 years	Quarterly
3. $1000	$36,581.00	40 years	Daily
4. $200	$314.85	5 years	Yearly
5. $750	$8267.38	30 years	Continuous
6. $2000	$4234.00	10 years	Continuous

Doubling Time In Exercises 7–12, find the time t for an investment to double in value, given the principal P, the annual interest rate r, and the type of compounding.

Principal	Rate	Compounding
7. $6000	8%	Quarterly
8. $500	$5\frac{1}{4}$%	Monthly
9. $2000	10.5%	Daily
10. $10,000	9.5%	Yearly
11. $1500	7.5%	Continuous
12. $100	6%	Continuous

Type of Compounding In Exercises 13–16, determine the type of compounding, given the principal P, the balance A, the time t, and the interest rate r. Solve the problem by trying the more common types of compounding.

Principal	Balance	Time	Rate
13. $750	$1587.75	10 years	7.5%
14. $10,000	$73,890.56	20 years	10%
15. $100	$141.48	5 years	7%
16. $4000	$4788.76	2 years	9%

Effective Yield In Exercises 17–22, find the effective yield, given the annual interest rate and the type of compounding.

Rate	Compounding
17. 8%	Continuous
18. 9.5%	Daily
19. 7%	Monthly
20. 8%	Yearly
21. $5\frac{1}{4}$%	Daily
22. 9%	Quarterly

Principal In Exercises 23–26, determine the principal P that must be deposited in an account to obtain the balance A for the given interest rate r and type of compounding.

	Balance	Rate	Time	Compounding
23.	$10,000	9%	20 years	Continuous
24.	$5000	8%	5 years	Continuous
25.	$750	6%	3 years	Daily
26.	$3000	7%	10 years	Monthly

Balance After Monthly Deposits In Exercises 27–30, suppose you make monthly deposits of P dollars into a savings account at an annual interest rate r, compounded continuously. Find the balance A after t years given that

$$A = \frac{P(e^{rt} - 1)}{e^{r/12} - 1}.$$

	Principal	Rate	Time
27.	$30	8%	10 years
28.	$100	9%	30 years
29.	$50	10%	40 years
30.	$20	7%	20 years

Balance After Monthly Deposits In Exercises 31 and 32, suppose you make monthly deposits of $30 in a savings account at an annual interest rate of 8%, compounded continuously (see figure).

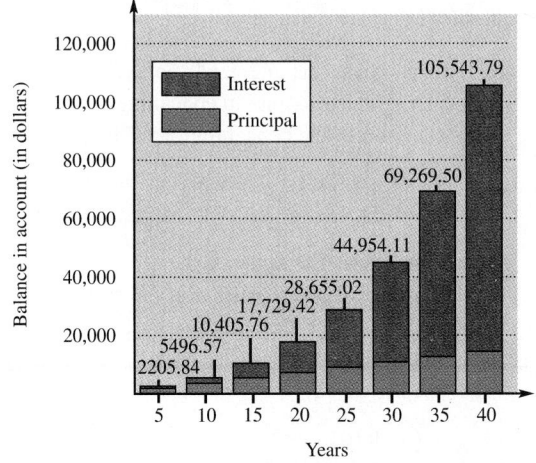

31. Find the total amount that has been deposited in the account in the first 20 years and the total interest earned.

32. Find the total amount that has been deposited in the account in 40 years and the total interest earned.

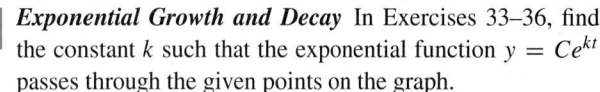

Exponential Growth and Decay In Exercises 33–36, find the constant k such that the exponential function $y = Ce^{kt}$ passes through the given points on the graph.

33.
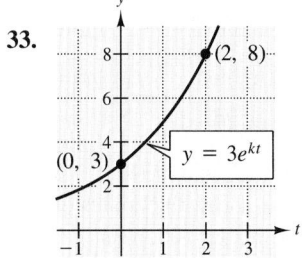

34.
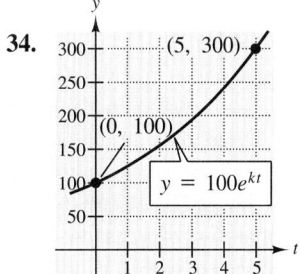

35.
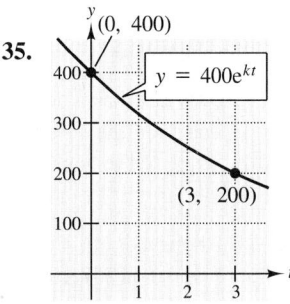

36.
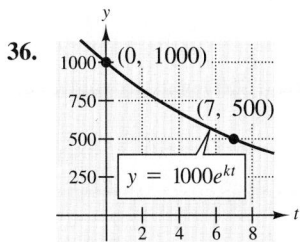

Population of a City In Exercises 37–40, the population of a city is given for the year 1985, along with the predicted population for the year 2000. Find the constants C and k to obtain the exponential growth model $y = Ce^{kt}$ for the population growth. (Let $t = 0$ correspond to the year 1985.) Use the model to predict the population of the city in the year 2005.

City	1985	2000
37. Los Angeles	9.6 million	10.7 million
38. Houston	2.1 million	2.6 million
39. Dhaka, Bangladesh	3.3 million	6.5 million
40. Lagos, Nigeria	6.1 million	12.5 million

41. Radioactive Decay Radioactive radium (Ra^{226}) has a half-life of 1620 years. If you start with 5 grams of this isotope, how much will remain after 1000 years?

42. Radioactive Decay The isotope Pu^{230} has a half-life of 24,360 years. If you start with 10 grams of this isotope, how much will remain after 10,000 years?

43. Carbon 14 Dating C^{14} dating assumes that the carbon dioxide on earth today has the same radioactive content as it did centuries ago. If this is true, then the amount of C^{14} absorbed by a tree that grew several centuries ago should be the same as the amount of C^{14} absorbed by a tree growing today. A piece of ancient charcoal contains only 15% as much of the radioactive carbon as a piece of modern charcoal. How long ago did the tree burn to make the ancient charcoal if the half-life of C^{14} is 5730 years? (Round your answer to the nearest 100 years.)

44. Depreciation A car that cost $22,000 new has a depreciated value of $16,500 after 1 year. Find the value of the car when it is 3 years old by using the exponential model $y = Ce^{kt}$.

45. Advertising Effect The sales S (in thousands of units) of a product after x hundred dollars is spent on advertising is

$$S = 10(1 - e^{kx}).$$

(a) Find S as a function of x if 2500 units are sold when $500 is spent on advertising.

(b) How many units will be sold if advertising expenditures are raised to $700?

46. Population Growth The population p of a certain species t years after it is introduced into a new habitat is

$$p(t) = \frac{5000}{1 + 4e^{-t/6}}.$$

(a) Determine the size of the population that was introduced into the habitat.

(b) Determine the size of the population after 9 years.

(c) After how many years will the population be 2000?

47. Earthquake Intensity On March 27, 1964, Alaska experienced an earthquake that measured 8.4 on the Richter scale. On February 9, 1971, an earthquake in the San Fernando Valley in California measured 6.6 on the Richter scale. How much greater in intensity was the Alaskan earthquake than the one in California?

48. Earthquake Intensity On March 10, 1933, Long Beach, California experienced an earthquake that measured 6.2 on the Richter scale. On December 3, 1988, an earthquake in Pasadena, California measured 5.0 on the Richter scale. How much greater in intensity was the Long Beach earthquake than the one in Pasadena?

Acidity Model In Exercises 49–52, use the acidity model $\text{pH} = -\log_{10} [H^+]$, where acidity (pH) is a measure of the hydrogen ion concentration $[H^+]$ (measured in moles of hydrogen per liter) of solution.

49. Find the pH of a solution that has a hydrogen ion concentration of 9.2×10^{-8}.

50. Compute the hydrogen ion concentration if the pH of a solution is 4.7.

51. A certain fruit has a pH of 2.5, and an antacid tablet has a pH of 9.5. The hydrogen ion concentration of the fruit is how many times the concentration of the tablet?

52. If the pH of a solution is decreased by one unit, the hydrogen ion concentration is increased by what factor?

53. World Population The figure on the next page shows the population P (in billions) of the world as projected by the Population Reference Bureau. The bureau's projection can be modeled by

$$P = \frac{11.14}{1 + 1.101e^{-0.051t}},$$

where $t = 0$ represents 1990. Use the model to estimate the population in 2020.

Figure for 53

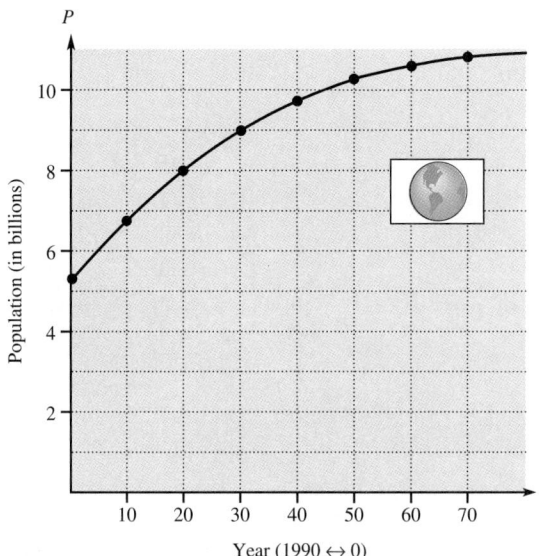

Population (in billions)

Year (1990 ↔ 0)

55. *Sales Growth*   Annual sales y of a product x years after it is introduced is approximated by

$$y = \frac{2000}{1 + 4e^{-x/2}}.$$

(a) Use a graphing calculator to graph the equation. Use the following viewing window settings.

Xmin=0
Xmax=10
Xscl=1
Ymin=0
Ymax=2000
Yscl=100

(b) Use the *trace* feature to approximate annual sales when $x = 4$.

(c) Use the *trace* feature to approximate the time when annual sales are $y = 1100$ units.

(d) According to this model, the annual sales will approach what maximum level?

54. *Depreciation*   After x years, the value y of a truck that cost \$32,000 is

$$y = 32{,}000(0.8)^x.$$

(a) Use a graphing calculator to graph the equation. Use the following viewing window settings.*

Xmin=0
Xmax=8
Xscl=1
Ymin=0
Ymax=32000
Yscl=4000

(b) Use the *trace* feature to approximate the value of the truck after 1 year.

(c) Use the *trace* feature to approximate the number of years after which the value of the truck will be \$16,000.

* The graphing calculator features given in the examples and exercises in this text correspond to the TI-81 and TI-82 graphing calculators from Texas Instruments. However, you can also use other graphing calculators and computer graphing software for these examples and exercises.

REVIEW EXERCISES

In Exercises 1 and 2, write the exponential equation in logarithmic form.

1. $4^3 = 64$

2. $25^{3/2} = 125$

In Exercises 3–10, evaluate the expression without using a calculator.

3. $\log_{10} 1000$

4. $\log_9 3$

5. $\log_3 \frac{1}{9}$

6. $\log_4 \frac{1}{16}$

7. $\ln e^7$

8. $\log_a \frac{1}{a}$

9. $\ln 1$

10. $\ln e^{-3}$

In Exercises 11–16, use the properties of logarithms to expand the expression as a sum, difference, or multiple of logarithms.

11. $\log_4 6x^4$

12. $\log_{10} 2x^{-3}$

13. $\log_5 \sqrt{x+2}$

14. $\ln \sqrt[3]{\dfrac{x}{5}}$

15. $\ln \dfrac{x+2}{x-2}$

16. $\ln x(x-3)^2$

In Exercises 17–22, use the properties of logarithms to condense the expression into a logarithm of a single quantity.

17. $\log_4 x - \log_4 10$

18. $5 \log_2 y$

19. $\log_8 16x + \log_8 2x^2$

20. $4(1 + \ln x + \ln x)$

21. $-2(\ln 2x - \ln 3)$

22. $-\frac{2}{3} \ln 3y$

In Exercises 23–28, determine whether the equation is true or false.

23. $\log_4 \dfrac{16}{x} = 2 - \log_4 x$

24. $\log_2 4x = 2 \log_2 x$

25. $\dfrac{\ln 5x}{\ln 10x} = \ln \dfrac{1}{2}$

26. $\log_{10} 10^{2x} = 2x$

27. $e^{\ln t} = t$

28. $e^{2x} - 1 = (e^x + 1)(e^x - 1)$

In Exercises 29–34, approximate the logarithm, given that $\log_5 2 \approx 0.4307$ and $\log_5 3 \approx 0.6826$. (Do not use a calculator.)

29. $\log_5 18$

30. $\log_5 \sqrt{6}$

31. $\log_5 \frac{1}{2}$

32. $\log_5 \frac{2}{3}$

33. $\log_5 (12)^{2/3}$

34. $\log_5 (5^2 \cdot 6)$

In Exercises 35–44, evaluate the function at the given values of the independent variable.
Use a calculator only if it is necessary or more efficient. (Round your answer to three decimal places.)

35. $f(x) = 2^x$
    (a) $x = -3$ (b) $x = 1$ (c) $x = 2$

36. $g(x) = 2^{-x}$
    (a) $x = -2$ (b) $x = 0$ (c) $x = 2$

37. $g(t) = e^{-t/3}$
    (a) $t = -3$ (b) $t = \pi$ (c) $t = 6$

38. $h(s) = 1 - e^{0.2s}$
    (a) $s = 0$ (b) $s = 2$ (c) $s = \sqrt{10}$

39. $f(x) = \log_3 x$
    (a) $x = 1$ (b) $x = 27$ (c) $x = 0.5$

40. $g(x) = \log_{10} x$
    (a) $x = 0.01$ (b) $x = 0.1$ (c) $x = 30$

41. $f(x) = \ln x$
    (a) $x = e$ (b) $x = \frac{1}{3}$ (c) $x = 10$

42. $h(x) = \ln x$
    (a) $x = e^2$ (b) $x = \frac{5}{4}$ (c) $x = 1200$

43. $g(x) = \ln e^{3x}$
    (a) $x = -2$ (b) $x = 0$ (c) $x = 7.5$

44. $f(x) = \log_2 \sqrt{x}$
    (a) $x = 4$ (b) $x = 64$ (c) $x = 5.2$

In Exercises 45–50, match the function with the sketch of its graph. [The graphs are labeled
(a), (b), (c), (d), (e), and (f).]

45. $f(x) = 2^x$

46. $f(x) = 2^{-x}$

47. $f(x) = -2^x$

48. $f(x) = 2^x + 1$

49. $f(x) = \log_2 x$

50. $f(x) = \log_2(x - 1)$

(a)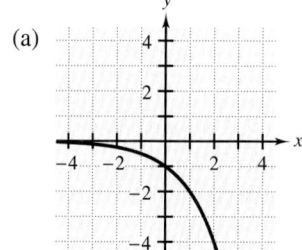

(b)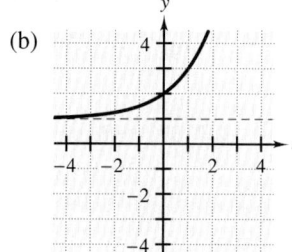

(c)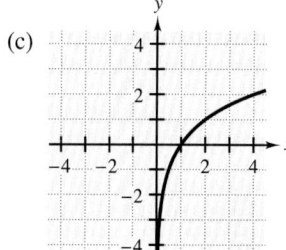

(d)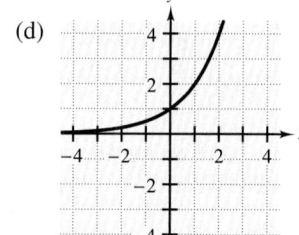

(e)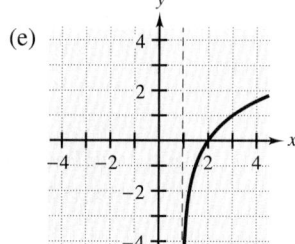

(f)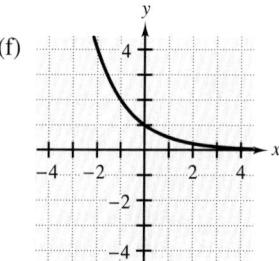

In Exercises 51–62, sketch the graph of the function.

51. $y = 3^{x/2}$

52. $y = 3^{x/2} - 2$

53. $f(x) = 3^{-x/2}$

54. $f(x) = -3^{-x/2}$

55. $f(x) = 3^{-x^2}$

56. $g(t) = 3^{|t|}$

57. $f(x) = -2 + \log_3 x$

58. $f(x) = 2 + \log_3 x$

59. $g(x) = \ln 2x$

60. $f(x) = 3 + \ln x$

61. $f(x) = e^{x+2}$

62. $g(x) = \ln(x - 5)$

In Exercises 63–66, evaluate the logarithm using the change-of-base formula. Do each problem twice, once with common logarithms and once with natural logarithms. (Round your answer to three decimal places.)

63. $\log_4 9$

64. $\log_{1/2} 5$

65. $\log_{12} 200$

66. $\log_3 0.28$

In Exercises 67–72, solve the equation without the aid of a calculator.

67. $2^x = 64$

68. $3^{x-2} = 81$

69. $4^{x-3} = \frac{1}{16}$

70. $\log_2 2x = \log_2 100$

71. $\log_3 x = 5$

72. $\log_5(x - 10) = 2$

In Exercises 73–84, solve the equation. (Round your answer to two decimal places.)

73. $3^x = 500$

74. $8^x = 1000$

75. $2e^{x/2} = 45$

76. $100e^{-0.6x} = 20$

77. $\dfrac{500}{(1.05)^x} = 100$

78. $25(1 - e^t) = 12$

79. $\log_{10} 2x = 1.5$

80. $\frac{1}{3}\log_2 x + 5 = 7$

81. $\ln x = 7.25$

82. $\ln x = -0.5$

83. $\log_2 2x = -0.65$

84. $\log_5 x + 1 = 4.8$

Balance in an Account In Exercises 85–88, complete the table to determine the balance A for P dollars invested at rate r for t years and compounded n times per year.

n	1	4	12	365	Continuous compounding
A					

85. $P = \$500$ $r = 7\%$ $t = 30$ years

86. $P = \$100$ $r = 5\frac{1}{4}\%$ $t = 60$ years

87. $P = \$10,000$ $r = 10\%$ $t = 20$ years

88. $P = \$2500$ $r = 8\%$ $t = 1$ year

Principal In Exercises 89 and 90, complete the table to determine the principal P that will yield a balance of A dollars when invested at rate r for t years and compounded n times per year.

n	1	4	12	365	Continuous compounding
P					

89. $A = \$50,000$ $r = 8\%$ $t = 40$ years

90. $A = \$1000$ $r = 6\%$ $t = 1$ year

91. ***Inflation Rate*** If the annual rate of inflation averages 5% over the next 10 years, then the approximate cost C of goods and services during any year in that decade will be

$$C(t) = P(1.05)^t, \qquad 0 \le t \le 10,$$

where t is time in years and P is the present cost. If the price of an oil change for your car is presently $19.95, estimate the number of years until it will cost $25.00.

92. ***Doubling Time*** Find the time required for an investment of $1000 to double in value when invested at 8% compounded monthly.

93. ***Product Demand*** The price p of a certain product is

$$p = 25 - 0.4e^{0.02x},$$

where x is the demand for the product. Approximate the demand if the price is $16.97.

Solution

$$a_1 = \frac{(-1)^1}{2(1)-1} = -\frac{1}{1} \qquad a_2 = \frac{(-1)^2}{2(2)-1} = \frac{1}{3}$$

$$a_3 = \frac{(-1)^3}{2(3)-1} = -\frac{1}{5} \qquad a_4 = \frac{(-1)^4}{2(4)-1} = \frac{1}{7}$$

$$a_5 = \frac{(-1)^5}{2(5)-1} = -\frac{1}{9} \qquad a_6 = \frac{(-1)^6}{2(6)-1} = \frac{1}{11}$$

The entire sequence can be written as follows.

$$-1, \quad \frac{1}{3}, \quad -\frac{1}{5}, \quad \frac{1}{7}, \quad -\frac{1}{9}, \quad \frac{1}{11}, \quad \cdots, \quad \frac{(-1)^n}{2n-1}, \quad \cdots \qquad \blacksquare$$

Factorial Notation

Some very important sequences in mathematics involve terms that are defined with special types of products called **factorials.**

Definition of Factorial	If n is a positive integer, then n **factorial** is defined as
	$$n! = 1 \cdot 2 \cdot 3 \cdot 4 \cdots (n-1) \cdot n.$$
	As a special case, zero factorial is defined as $0! = 1$.

An equivalent expression for n factorial is $n! = n(n-1)(n-2)\cdots 3 \cdot 2 \cdot 1$. The first several factorial values are as follows.

$0! = 1$	$1! = 1$
$2! = 1 \cdot 2 = 2$	$3! = 1 \cdot 2 \cdot 3 = 6$
$4! = 1 \cdot 2 \cdot 3 \cdot 4 = 24$	$5! = 1 \cdot 2 \cdot 3 \cdot 4 \cdot 5 = 120$

Most scientific calculators have a factorial key, denoted by $\boxed{n!}$. If your calculator has such a key, try using it to evaluate $n!$ for several values of n. You will see that the value of n does not have to be very large in order for the value of $n!$ to be huge. For instance, $10! = 3,628,800$.

EXAMPLE 5 ■ Finding the Terms of a Sequence Involving Factorials

Write the first six terms of the sequence having an nth term of

Remind students that $0! = 1$, not 0; hence, it is acceptable to have $0!$ in a denominator.

$$a_n = \frac{1}{n!}.$$

(Assume that the sequence begins with $n = 0$.)

Solution

$$a_0 = \frac{1}{0!} = \frac{1}{1} = 1 \qquad\qquad a_1 = \frac{1}{1!} = \frac{1}{1} = 1$$

$$a_2 = \frac{1}{2!} = \frac{1}{2} \qquad\qquad a_3 = \frac{1}{3!} = \frac{1}{1 \cdot 2 \cdot 3} = \frac{1}{6}$$

$$a_4 = \frac{1}{4!} = \frac{1}{1 \cdot 2 \cdot 3 \cdot 4} = \frac{1}{24} \qquad\qquad a_5 = \frac{1}{5!} = \frac{1}{1 \cdot 2 \cdot 3 \cdot 4 \cdot 5} = \frac{1}{120}$$

The entire sequence can be written as follows.

$$1, \quad 1, \quad \frac{1}{2}, \quad \frac{1}{6}, \quad \frac{1}{24}, \quad \frac{1}{120}, \quad \cdots \quad \frac{1}{n!}, \quad \cdots \qquad\blacksquare$$

Factorials follow the same conventions for order of operation as do exponents. For instance, $2n!$ means $2(n!)$. Notice how these conventions are used in the next example.

EXAMPLE 6 ■ Finding the Terms of a Sequence Involving Factorials

Write the first six terms of the sequence having an nth term of

$$a_n = \frac{2n!}{(2n)!}.$$

(Assume that the sequence begins with $n = 0$.)

Solution

$$a_0 = \frac{2(0!)}{0!} = \frac{2}{1} = 2 \qquad\qquad a_1 = \frac{2(1!)}{2!} = \frac{2}{2} = 1$$

$$a_2 = \frac{2(2!)}{4!} = \frac{4}{24} = \frac{1}{6} \qquad\qquad a_3 = \frac{2(3!)}{6!} = \frac{12}{720} = \frac{1}{60}$$

$$a_4 = \frac{2(4!)}{8!} = \frac{48}{40,320} = \frac{1}{840} \qquad\qquad a_5 = \frac{2(5!)}{10!} = \frac{240}{3,628,800} = \frac{1}{15,120}$$

The entire sequence can be written as follows.

$$2, \quad 1, \quad \frac{1}{6}, \quad \frac{1}{60}, \quad \frac{1}{840}, \quad \frac{1}{15,120}, \quad \cdots \quad \frac{2n!}{(2n)!}, \quad \cdots \qquad\blacksquare$$

In Example 6, we multiplied the numerators and denominators before reducing the fractions. When you are finding the terms of a sequence, reducing is often easier if you leave the numerator and denominator in factored form. For instance, notice how the factors divide out in the following fraction.

$$a_5 = \frac{2(5!)}{10!} = \frac{2 \cdot 1 \cdot 2 \cdot 3 \cdot 4 \cdot 5}{1 \cdot 2 \cdot 3 \cdot 4 \cdot 5 \cdot 6 \cdot 7 \cdot 8 \cdot 9 \cdot 10} = \frac{1}{3 \cdot 7 \cdot 8 \cdot 9 \cdot 10} = \frac{1}{15,120}$$

Sigma Notation

Many applications involve finding the sum of the first n terms of a sequence. A convenient shorthand notation for such a sum is **sigma notation.** This name comes from the use of the uppercase Greek letter sigma, written as Σ.

Definition of Sigma Notation	The sum of the first n terms of the sequence having an nth term of a_n is

$$\sum_{i=1}^{n} a_i = a_1 + a_2 + a_3 + a_4 + \cdots + a_n,$$

where i is called the **index of summation,** n is the **upper limit of summation,** and 1 is the **lower limit of summation.**

EXAMPLE 7 ■ Sigma Notation for Sums

Evaluate the sum $\displaystyle\sum_{i=1}^{6} 2i$.

Solution

By letting i take on successive integer values from 1 to 6, you obtain the following.

$$\sum_{i=1}^{6} 2i = 2(1) + 2(2) + 2(3) + 2(4) + 2(5) + 2(6)$$
$$= 2 + 4 + 6 + 8 + 10 + 12$$
$$= 42 \qquad ■$$

Sigma notation expresses a broad idea more concisely. Use several different examples to help students become more comfortable with it. As a first example, consider placing a_1, a_2, a_3, . . . , a_n above each of the corresponding values of the sequence to help reinforce the meaning of the notation.

    In Example 7, the index of summation was i and the summation began with $i = 1$. Any letter can be used as the index of summation, and the summation can begin with any integer. For instance, the index of summation in the next example is k and the summation begins with $k = 0$.

EXAMPLE 8 ■ Sigma Notation for Sums

Evaluate the sum $\displaystyle\sum_{k=0}^{8} \frac{1}{k!}$.

Solution

By letting k take on successive integer values from 0 to 8, you obtain the following.

$$\sum_{k=0}^{8} \frac{1}{k!} = \frac{1}{0!} + \frac{1}{1!} + \frac{1}{2!} + \frac{1}{3!} + \frac{1}{4!} + \frac{1}{5!} + \frac{1}{6!} + \frac{1}{7!} + \frac{1}{8!}$$

$$= 1 + 1 + \frac{1}{2} + \frac{1}{6} + \frac{1}{24} + \frac{1}{120} + \frac{1}{720} + \frac{1}{5040} + \frac{1}{40,320}$$

$$\approx 2.71828$$

Note that this sum is approximately $e = 2.71828$. . . . ■

EXAMPLE 9 ■ Sigma Notation for Sums

Evaluate the sum $\displaystyle\sum_{i=1}^{4} 5$.

Solution

By letting i take on successive integer values from 1 to 4, you obtain the following.

$$\sum_{i=1}^{4} 5 = 5 + 5 + 5 + 5 = 20$$

Note that each of the terms of this sum is constant because $a_i = 5$ for all values of i. ■

EXAMPLE 10 ■ Writing a Sum in Sigma Notation

Write the following sum using sigma notation.

$$\frac{2}{2} + \frac{2}{3} + \frac{2}{4} + \frac{2}{5} + \frac{2}{6}$$

Remind students that the process of going from terms of a series to sigma notation requires some trial-and-error, observation, and conjecture. What is the pattern? squares? multiples? cubes less one?

Solution

To write this sum using sigma notation, you must find a pattern for the terms. After examining the terms, you can see that each has a numerator of 2, and the denominators are the successive integers from 2 to 6. Thus, one possible sigma notation for this sum is as follows.

$$\sum_{i=1}^{5} \frac{2}{i+1} = \frac{2}{2} + \frac{2}{3} + \frac{2}{4} + \frac{2}{5} + \frac{2}{6}$$ ■

DISCUSSION PROBLEM ■ Writing Sigma Notation in Different Ways

There is more than one way to write a given sum in sigma notation. For instance, the sum

$$1 + 2 + 3 + 4 + 5 + 6$$

could be written as

$$\sum_{i=1}^{6} i \quad \text{or} \quad \sum_{i=0}^{5} (i + 1).$$

Write each of the following sums in two ways and determine which is the simpler.

(a) $1 + 4 + 9 + 16 + 25 + 36$ (b) $3 + 5 + 7 + 9 + 11 + 13$ ■

Warm-Up

The following warm-up exercises involve skills that were covered in earlier sections. You will use these skills in the exercise set for this section.

In Exercises 1–10, perform the indicated operations and simplify.

1. $\dfrac{1}{3} \cdot \dfrac{3}{5} \cdot \dfrac{5}{7}$

2. $\dfrac{3 \cdot 6 \cdot 9 \cdot 12}{3^4}$

3. $\dfrac{1}{1 \cdot 2} + \dfrac{1}{2 \cdot 3} + \dfrac{1}{3 \cdot 4}$

4. $\dfrac{1}{2} + \dfrac{1}{3} + \dfrac{1}{4}$

5. $\dfrac{n(n-1)(n-2)}{(n-1)(n-2)}$

6. $\dfrac{n^2-1}{n+1}$

7. $\dfrac{2n^2 + 7n - 15}{n^2 - 25}$

8. $\dfrac{n^2 + 3n + 2}{n^2 + 2n}$

9. $\dfrac{1}{n} + \dfrac{2}{n^2}$

10. $\dfrac{1}{n-1} + \dfrac{1}{n+1}$

10.1   EXERCISES

In Exercises 1–20, write the first five terms of the infinite sequence. (Assume that the sequence begins with $n = 1$.)

1. $a_n = 2n$

2. $a_n = 3n$

3. $a_n = (-1)^n 2n$

4. $a_n = (-1)^{n+1} 3n$

5. $a_n = \left(\dfrac{1}{2}\right)^n$

6. $a_n = \left(\dfrac{1}{3}\right)^n$

7. $a_n = \left(-\dfrac{1}{2}\right)^{n+1}$

8. $a_n = \left(-\dfrac{1}{3}\right)^n$

9. $a_n = \dfrac{1}{n+1}$

10. $a_n = \dfrac{3}{2n+1}$

11. $a_n = \dfrac{2n}{3n+2}$

12. $a_n = \dfrac{5n}{4n+3}$

13. $a_n = \dfrac{(-1)^n}{n^2}$

14. $a_n = \dfrac{1}{\sqrt{n}}$

15. $a_n = \dfrac{2^n}{n!}$

16. $a_n = \dfrac{n!}{(n-1)!}$

17. $a_n = 5 - \dfrac{1}{2^n}$

18. $a_n = 7 + \dfrac{1}{3^n}$

19. $a_n = 2 + (-2)^n$

20. $a_n = \dfrac{1 + (-1)^n}{n^2}$

In Exercises 21–28, simplify the ratio of factorials.

21. $\dfrac{5!}{4!}$

22. $\dfrac{18!}{17!}$

23. $\dfrac{10!}{12!}$

24. $\dfrac{5!}{8!}$

25. $\dfrac{n!}{(n+1)!}$

26. $\dfrac{(n+2)!}{n!}$

27. $\dfrac{(2n)!}{(2n-1)!}$

28. $\dfrac{(2n+2)!}{(2n)!}$

In Exercises 29–46, find the specified sum.

29. $\displaystyle\sum_{k=1}^{6} 3k$

30. $\displaystyle\sum_{k=1}^{4} 5k$

31. $\displaystyle\sum_{i=0}^{4} (2i + 3)$

32. $\displaystyle\sum_{i=2}^{7} (4i - 1)$

33. $\displaystyle\sum_{j=1}^{5} \frac{(-1)^{j+1}}{j}$

34. $\displaystyle\sum_{j=0}^{3} \frac{1}{j^2 + 1}$

35. $\displaystyle\sum_{k=1}^{6} (-8)$

36. $\displaystyle\sum_{n=3}^{12} 10$

37. $\displaystyle\sum_{i=1}^{8} \left(\frac{1}{i} - \frac{1}{i+1}\right)$

38. $\displaystyle\sum_{k=1}^{5} \left(\frac{2}{k} - \frac{2}{k+2}\right)$

39. $\displaystyle\sum_{n=0}^{5} \left(-\frac{1}{3}\right)^n$

40. $\displaystyle\sum_{n=0}^{6} \left(\frac{3}{2}\right)^n$

41. $\displaystyle\sum_{n=1}^{6} n(n + 1)$

42. $\displaystyle\sum_{n=0}^{5} 2n^2$

43. $\displaystyle\sum_{j=2}^{6} (j! - j)$

44. $\displaystyle\sum_{j=0}^{4} \frac{6}{j!}$

45. $\displaystyle\sum_{k=1}^{6} \ln k$

46. $\displaystyle\sum_{k=2}^{4} \frac{k}{\ln k}$

In Exercises 47–60, write the sum using sigma notation. (Use k as the index of summation and begin with $k = 0$ or $k = 1$.)

47. $1 + 2 + 3 + 4 + 5$

48. $8 + 9 + 10 + 11 + 12 + 13$

49. $2 + 4 + 6 + 8 + 10$

50. $24 + 30 + 36 + 42$

51. $\dfrac{1}{2(1)} + \dfrac{1}{2(2)} + \dfrac{1}{2(3)} + \dfrac{1}{2(4)} + \cdots + \dfrac{1}{2(10)}$

52. $\dfrac{3}{1+1} + \dfrac{3}{1+2} + \dfrac{3}{1+3} + \dfrac{3}{1+4} + \cdots + \dfrac{3}{1+50}$

53. $\dfrac{1}{1^2} + \dfrac{1}{2^2} + \dfrac{1}{3^2} + \dfrac{1}{4^2} + \cdots + \dfrac{1}{20^2}$

54. $\dfrac{1}{2^0} + \dfrac{1}{2^1} + \dfrac{1}{2^2} + \dfrac{1}{2^3} + \cdots + \dfrac{1}{2^{12}}$

55. $\dfrac{1}{3^0} - \dfrac{1}{3^1} + \dfrac{1}{3^2} - \dfrac{1}{3^3} + \cdots - \dfrac{1}{3^9}$

56. $\left(-\dfrac{2}{3}\right)^0 + \left(-\dfrac{2}{3}\right)^1 + \left(-\dfrac{2}{3}\right)^2 + \cdots + \left(-\dfrac{2}{3}\right)^{20}$

57. $\frac{1}{2} + \frac{2}{3} + \frac{3}{4} + \frac{4}{5} + \frac{5}{6} + \cdots + \frac{11}{12}$

58. $\frac{2}{4} + \frac{4}{7} + \frac{6}{10} + \frac{8}{13} + \frac{10}{16} + \cdots + \frac{20}{31}$

59. $1 + 1 + 2 + 6 + 24 + 120 + 720$

60. $1 + 1 + \frac{1}{2} + \frac{1}{6} + \frac{1}{24} + \frac{1}{120} + \frac{1}{720}$

Arithmetic Mean In Exercises 61–64, use the definition of the *arithmetic mean* $\bar{x}$ of a set of n measurements $x_1, x_2, x_3,$ $\ldots, x_n$ to find the mean (average) of the given measurements.

$$\bar{x} = \frac{x_1 + x_2 + x_3 + \cdots + x_n}{n}$$

61. $3, 7, 2, 1, 5$

62. $84, 69, 66, 96$

63. $0.5, 0.8, 1.1, 0.8, 0.7, 0.7, 1.0$

64. $-1.0, 4.2, 5.4, -3.2, 3.6$

65. *Compound Interest* A deposit of $500 is made in an account that earns 7% interest compounded monthly. The balance in the account after N months is

$$A_N = 500\left(1 + \frac{0.07}{12}\right)^N, \qquad N = 1,\ 2,\ 3,\ \ldots.$$

(a) Compute the first eight terms of this sequence.

(b) Find the balance in this account after 20 years by computing A_{240}.

66. *Depreciation* At the end of each year, the value of a car with an initial cost of $16,000 is three-fourths what it was at the beginning of the year. Thus, after n years its value is

$$a_n = 16{,}000\left(\frac{3}{4}\right)^n, \qquad n = 1,\ 2,\ 3,\ \ldots .$$

Find the value of the car 5 years after it was purchased by computing a_5.

67. *Soccer Ball* The number of degrees a_n in each angle of a regular n-sided polygon is

$$a_n = \frac{180(n-2)}{n}, \qquad n \geq 3.$$

The surface of a soccer ball is made of regular hexagons and pentagons. If a soccer ball is taken apart and flattened, as shown in the figure, the sides of the hexagons don't meet each other. Use the terms a_5 and a_6 to explain why there are gaps between adjacent hexagons.

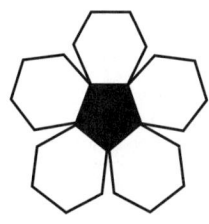

68. *Stars* The stars in Example 3 were formed by placing n equally spaced points on a circle and connecting each point with the second point from it on the circle. The stars in the accompanying figure are formed in a similar way except each point is connected with the third point from it. For these stars, the number of degrees in a tip is

$$d_n = \frac{180(n-6)}{n},$$
$$n \geq 7.$$

Write the first five terms of this sequence.

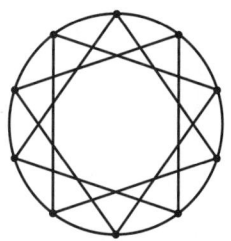
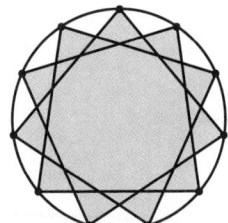

Arithmetic Sequences

Arithmetic Sequences • *The Sum of an Arithmetic Sequence* • *Applications*

Arithmetic Sequences

A sequence with consecutive terms that have a common difference is called an **arithmetic sequence.**

Definition of an Arithmetic Sequence	A sequence is called **arithmetic** if the differences between consecutive terms are the same. Thus, the sequence $$a_1, \quad a_2, \quad a_3, \quad a_4, \quad \ldots, \quad a_n, \quad \ldots$$ is arithmetic if there is a number d such that $$a_2 - a_1 = d, \qquad a_3 - a_2 = d, \qquad a_4 - a_3 = d,$$ and so on. The number d is called the **common difference** of the arithmetic sequence.

EXAMPLE 1 ■

Examples of Arithmetic Sequences

Additional problem: Have students decide whether the sequence 1, 7, 13, 19, 26, 33, 39, . . . is an arithmetic sequence. Answer: No

(a) The sequence having an nth term of $3n + 2$ is arithmetic. For this sequence, the common difference between consecutive terms is 3.

$$\underbrace{5, \quad 8,}_{8-5=3} \quad 11, \quad 14, \quad \ldots, \quad 3n + 2, \quad \ldots$$

(b) The sequence having an nth term of $7 - 5n$ is arithmetic. For this sequence, the common difference between consecutive terms is -5.

$$\underbrace{2, \quad -3,}_{-3-2=-5} \quad -8, \quad -13, \quad \ldots, \quad 7 - 5n, \quad \ldots$$

(c) The sequence having an nth term of $\frac{1}{4}(n + 3)$ is arithmetic. For this sequence, the common difference between consecutive terms is $\frac{1}{4}$.

$$\underbrace{1, \quad \frac{5}{4},}_{\frac{5}{4}-1=\frac{1}{4}} \quad \frac{3}{2}, \quad \frac{7}{4}, \quad \ldots, \quad \frac{n+3}{4}, \quad \ldots$$

■

Point out that the common difference is d, not c.

In Example 1, notice that each of the arithmetic sequences has an nth term that is of the form $dn + c$, where the common difference of the sequence is d. We summarize this result as follows.

The nth Term of an Arithmetic Sequence	The nth term of an arithmetic sequence has the form $$a_n = a_1 + (n-1)d,$$ where d is the common difference between the terms of the sequence, and a_1 is the first term.

EXAMPLE 2 ■ Finding the nth Term of an Arithmetic Sequence

Find a formula for the nth term of the arithmetic sequence with a common difference of 2 and a first term of 5.

Solution

You know that the formula for the nth term is of the form $a_n = a_1 + (n-1)d$. Moreover, because the common difference is $d = 2$, and the first term is $a_1 = 5$, the formula must have the form

$$a_n = 5 + 2(n-1).$$

Thus, the formula for the nth term is

$$a_n = 2n + 3.$$

The sequence, therefore, has the following form.

$$5, \quad 7, \quad 9, \quad 11, \quad 13, \quad \ldots, \quad 2n+3, \quad \ldots \qquad ■$$

EXAMPLE 3 ■ Finding the nth Term of an Arithmetic Sequence

Find a formula for the nth term of the arithmetic sequence with a common difference of 6 and a *second* term of 11. What is the 19th term of this sequence?

Given any two variables, we can easily solve for the third variable by using substitution.

Solution

You know that the formula for the nth term is of the form $a_n = a_1 + (n-1)d$. Moreover, because the common difference is $d = 6$, the formula must have the form

$$a_n = a_1 + 6(n-1).$$

Owing to the fact that the second term is

$$a_2 = 11 = a_1 + 6(2-1) = a_1 + 6,$$

it follows that $a_1 = 5$.

$$11 = a_1 + 6$$
$$5 = a_1$$

Thus, the formula for the nth term is

$$a_n = 5 + 6(n - 1) = 6n - 1.$$

The sequence, therefore, has the following form.

$$5, \quad 11, \quad 17, \quad 23, \quad 29, \quad \ldots, \quad 6n - 1, \quad \ldots$$

The 19th term of the sequence is

$$a_{19} = 6(19) - 1 = 113.$$ ■

If you know the nth term of an arithmetic sequence *and* you know the common difference of the sequence, then you can find the $(n + 1)$th term by using the following **recursive formula.**

$$a_{n+1} = a_n + d$$

EXAMPLE 4   ■   Using a Recursive Formula

The 12th term of an arithmetic sequence is 52 and the common difference is 3. What is the 13th term of the sequence?

Solution

$$a_{13} = a_{12} + 3 = 52 + 3 = 55$$ ■

The Sum of an Arithmetic Sequence

It may be helpful to ask students to find several partial sums of the sequence whose nth term is $3n + 4$.

The sum of the first n terms of an arithmetic sequence is called the **nth partial sum** of the sequence. For instance, the fifth partial sum of the arithmetic sequence whose nth term is $3n + 4$ is

$$\sum_{i=1}^{5} (3i + 4) = 7 + 10 + 13 + 16 + 19 = 65.$$

There is a nice shortcut for finding the nth partial sum of an arithmetic sequence.

The nth Partial Sum of an Arithmetic Sequence

The nth partial sum of the arithmetic sequence having an nth term of a_n is

$$\sum_{i=1}^{n} a_i = a_1 + a_2 + a_3 + a_4 + \cdots + a_n = \frac{n}{2}(a_1 + a_n).$$

In other words, to find the sum of the first n terms of an arithmetic sequence, find the average of the first term and the nth term, and multiply by n.

EXAMPLE 5   ■   Finding the nth Partial Sum of an Arithmetic Sequence

Find the sum of the first 20 terms of the arithmetic sequence having an nth term of $4n + 1$.

Solution

The first term of this sequence is $a_1 = 4(1) + 1 = 5$, and the 20th term is $a_{20} = 4(20) + 1 = 81$. Therefore, the sum of the first 20 terms is given by

$$\sum_{i=1}^{n} a_i = \frac{n}{2}(a_1 + a_n)$$

$$\sum_{i=1}^{20} (4i + 1) = \frac{20}{2}(a_1 + a_{20})$$

$$= 10(5 + 81)$$
$$= 10(86)$$
$$= 860. \qquad \blacksquare$$

EXAMPLE 6 ■ Finding the nth Partial Sum of an Arithmetic Sequence

Find the following sum.

$$7 + 10 + 13 + 16 + 19 + 22 + 25 + 28 + 31 + 34 + 37 + 40 + 43$$

Solution

One way to find this sum is to simply add all of the numbers. However, by recognizing that the numbers form an arithmetic sequence that has 13 terms, you can find the sum using the formula for the nth partial sum of an arithmetic sequence.

$$\text{Sum} = \frac{13}{2}(7 + 43) = \frac{13}{2}(50) = 13(25) = 325$$

Check this result on your calculator by actually adding the 13 terms. ■

EXAMPLE 7 ■ Finding the nth Partial Sum of an Arithmetic Sequence

Find the sum of the integers from 1 to 100.

Historical Note: Karl Friedrich Gauss's (1777-1855) teacher asked him to add all the integers 1 to 100. When Gauss returned with the correct answer after only a few moments, the teacher could only look at him in astounded silence. This is what Gauss did:

$$1 + \quad 2 + \quad 3 + \cdots + 100$$
$$100 + \ 99 + \ 98 + \cdots + \quad 1$$
$$\overline{101 + 101 + 101 + \cdots + 101}$$

$$\frac{101 \times 100}{2} = 5050$$

Solution

Because the integers

$$1, \quad 2, \quad 3, \quad 4, \quad \ldots, \quad 100$$

form an arithmetic sequence, you can find the sum as follows.

$$\sum_{i=1}^{100} i = \frac{100}{2}(1 + 100)$$

$$= 50(101)$$
$$= 5050 \qquad \blacksquare$$

Applications

The next two examples give you some idea of how arithmetic sequences can be used to solve problems in business and science.

EXAMPLE 8   ■   An Application: Total Sales

A small business sells $100,000 worth of products during its first year. The owner of the business has set a goal of increasing annual sales by $25,000 each year for 9 years. Assuming that this goal is met, find the total sales during the first 10 years the business is in operation.

Solution

The annual sales during the first 10 years form the following arithmetic sequence.

$100,000, $125,000, $150,000, $175,000, $200,000,

$225,000, $250,000, $275,000, $300,000, $325,000

Using the formula for the nth partial sum of an arithmetic sequence, you can find the total sales during the first 10 years as follows.

$$\text{Total sales} = \frac{10}{2}(100,000 + 325,000) = 5(425,000) = \$2,125,000$$

From the bar graph shown in Figure 10.2, notice that the annual sales for this company follow a *linear growth* pattern. In other words, saying that a quantity increases arithmetically is the same as saying that it increases linearly.

FIGURE 10.2

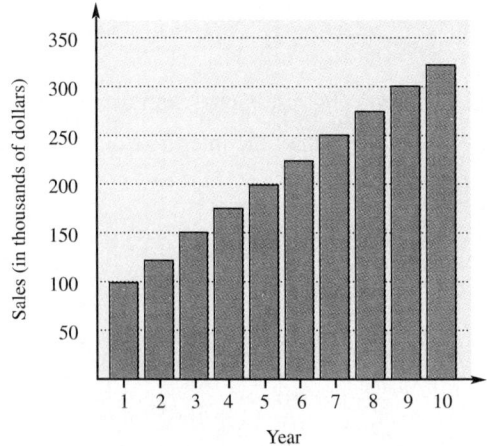

Year

Red Rocks Park is an open-air amphitheater carved out of rock near Denver, Colorado. The amphitheater has 69 rows of seats. Rows 46 through 69 have seats for a total of 3318 people. The number of seats in the first 45 rows can be modeled by the arithmetic sequence whose nth term is $87\frac{1}{2} + \frac{3}{2}n$.

EXAMPLE 9 ■ Seating Capacity

You are organizing a concert at Red Rocks Park. How much should you charge per ticket in order to receive $50,000 for the ticket sales of a performance that is sold out?

Solution

From the information given in the figure caption above, you know that there are 3318 seats in the last 24 rows (see Figure 10.3). To approximate the number of seats in the first 45 rows, you can use the formula for the sum of an arithmetic series.

$$\sum_{n=1}^{45}\left(87\frac{1}{2} + \frac{3}{2}n\right) = 89 + 90\frac{1}{2} + 92 + 93\frac{1}{2} + \cdots + 155$$

$$= \frac{45}{2}(89 + 155)$$

$$= 5490$$

Thus, the total number of seats is about $3318 + 5490 = 8808$. To bring in $50,000, you should charge about

$$\frac{1}{8808}(50,000) \approx \$5.68 \text{ per ticket.}$$

FIGURE 10.3

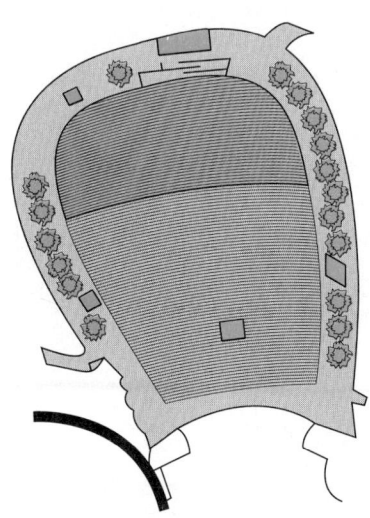

DISCUSSION PROBLEM ■ An Experiment Involving an Arithmetic Sequence

The following sequence of perfect squares is *not* arithmetic.

$$1, \quad 4, \quad 9, \quad 16, \quad 25, \quad 36, \quad 49, \quad 64, \quad 81, \quad \ldots$$

However, you can form a related sequence that is arithmetic by finding the differences of consecutive terms as follows.

$$
\begin{array}{ccccccccc}
 & \overbrace{}^{5} & & \overbrace{}^{9} & & \overbrace{}^{13} & & \overbrace{}^{17} & \\
1 & 4 & 9 & 16 & 25 & 36 & 49 & 64 & 81 \\
 & \underbrace{}_{3} & & \underbrace{}_{7} & & \underbrace{}_{11} & & \underbrace{}_{15} &
\end{array}
$$

Thus, the related arithmetic sequence is

$$3, \quad 5, \quad 7, \quad 9, \quad 11, \quad 13, \quad 15, \quad 17.$$

Can you think of a way to find an arithmetic sequence that is related to the following sequence of perfect cubes?

$$1, \quad 8, \quad 27, \quad 64, \quad 125, \quad 216, \quad 343, \quad 512, \quad 729, \quad \ldots$$ ■

Warm-Up

The following warm-up exercises involve skills that were covered in earlier sections. You will use these skills in the exercise set for this section.

In Exercises 1–4, find the sum.

1. $\displaystyle\sum_{i=1}^{6} \frac{1}{i}$

2. $\displaystyle\sum_{i=0}^{6} \frac{5}{i+1} - \frac{5}{i+3}$

3. $\displaystyle\sum_{n=1}^{4} 3^n$

4. $\displaystyle\sum_{n=1}^{5} (3n - 1)$

In Exercises 5–10, evaluate the expression.

5. $15\left(\dfrac{5+33}{2}\right)$

6. $35\left(\dfrac{5+107}{2}\right)$

7. $11\left(\dfrac{\frac{16}{3} + \frac{26}{3}}{2}\right)$

8. $20\left(\dfrac{\frac{7}{4} + 16}{2}\right)$

9. $\frac{7}{2}[2(-3) + 6(4)]$

10. $\frac{9}{2}[2(4) + 8(3)]$

10.2 EXERCISES

In Exercises 1–6, find the common difference of the arithmetic sequence.

1. 2, 5, 8, 11, . . .

2. −8, 0, 8, 16, . . .

3. 100, 94, 88, 82, . . .

4. 3200, 2800, 2400, 2000, . . .

5. 1, $\frac{5}{3}$, $\frac{7}{3}$, 3, . . .

6. $\frac{1}{2}$, $\frac{5}{4}$, 2, $\frac{11}{4}$, . . .

In Exercises 7–18, determine whether the sequence is arithmetic. If it is, find the common difference.

7. 2, 4, 6, 8, . . .

8. 2, 6, 10, 14, . . .

9. 2, $\frac{7}{2}$, 5, $\frac{13}{2}$, . . .

10. 5, 13, 21, 29, 37, . . .

11. 32, 16, 0, −16, . . .

12. 32, 16, 8, 4, . . .

13. $\frac{1}{3}$, $\frac{2}{3}$, $\frac{4}{3}$, $\frac{8}{3}$, $\frac{16}{3}$, . . .

14. $\frac{1}{3}$, $\frac{1}{2}$, $\frac{2}{3}$, $\frac{5}{6}$, 1, . . .

15. 3.2, 4, 4.8, 5.6, . . .

16. 8, 4, 2, 1, 0.5, 0.25, . . .

17. ln 4, ln 8, ln 12, ln 16, . . .

18. e, e^2, e^3, e^4, . . .

In Exercises 19–26, write the first five terms of the arithmetic sequence. (Assume that n begins with 1.)

19. $a_n = 3n + 4$

20. $a_n = 5n - 4$

21. $a_n = -2n + 8$

22. $a_n = -10n + 100$

23. $a_n = \frac{5}{2}n - 1$

24. $a_n = \frac{2}{3}n + 2$

25. $a_n = -\frac{1}{4}(n - 1) + 4$

26. $a_n = 4(n + 2) + 24$

In Exercises 27–30, match the arithmetic sequence with its graph. [The graphs are labeled (a), (b), (c), and (d).]

27. $a_n = -\frac{1}{3}n + 2$

28. $a_n = 5n - 4$

29. $a_n = 2n - 3$

30. $a_n = -\frac{1}{2}n + 4$

(a)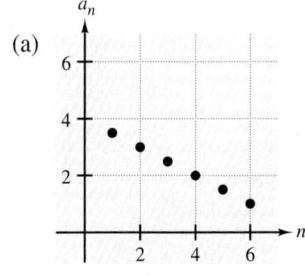

(b)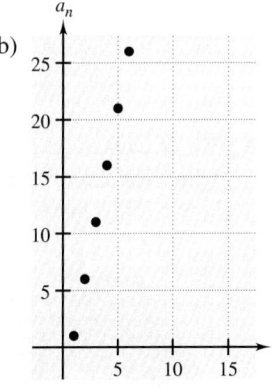

(c)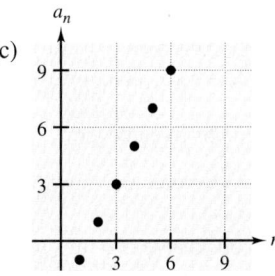

(d)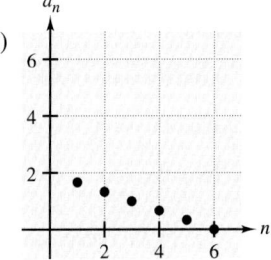

In Exercises 31–38, find a formula for the nth term of the arithmetic sequence.

31. $a_1 = 3$, $d = \frac{1}{2}$

32. $a_1 = -1$, $d = 1.2$

33. $a_1 = 1000$, $d = -25$

34. $a_1 = 64$, $d = -8$

35. $a_1 = 20$, $a_2 = 24$

36. $a_1 = 16$, $a_2 = 22$

37. $a_1 = 50$, $a_3 = 30$

38. $a_1 = 10$, $a_3 = 12$

In Exercises 39–44, write the first five terms of the arithmetic sequence.

39. $a_1 = 5, \quad a_{k+1} = a_k + 3$

40. $a_1 = 8, \quad a_{k+1} = a_k + 7$

41. $a_1 = 9, \quad a_{k+1} = a_k - 3$

42. $a_1 = 12, \quad a_{k+1} = a_k - 6$

43. $a_1 = -10, \quad a_{k+1} = a_k + 6$

44. $a_1 = -20, \quad a_{k+1} = a_k - 4$

In Exercises 45–52, find the sum.

45. $\displaystyle\sum_{k=1}^{20} k$

46. $\displaystyle\sum_{k=1}^{30} 4k$

47. $\displaystyle\sum_{k=1}^{50} (2k + 3)$

48. $\displaystyle\sum_{k=1}^{100} (4k - 1)$

49. $\displaystyle\sum_{n=1}^{40} (1000 - 25n)$

50. $\displaystyle\sum_{n=1}^{20} (500 - 10n)$

51. $\displaystyle\sum_{n=1}^{500} \frac{n}{2}$

52. $\displaystyle\sum_{n=1}^{600} \frac{2n}{3}$

In Exercises 53–60, find the nth partial sum of the arithmetic sequence.

53. 5, 12, 19, 26, 33, . . . , $n = 12$

54. 2, 12, 22, 32, 42, . . . , $n = 20$

55. 200, 175, 150, 125, 100, . . . , $n = 8$

56. 800, 785, 770, 755, 740, . . . , $n = 25$

57. $-50, -38, -26, -14, -2, . . . ,$ $n = 50$

58. $-16, -8, 0, 8, 16, . . . ,$ $n = 30$

59. 1, 4.5, 8, 11.5, 15, . . . , $n = 12$

60. 2.2, 2.8, 3.4, 4.0, 4.6, . . . , $n = 12$

61. Find the sum of the first 75 positive integers.

62. Find the sum of the integers from 35 to 100 inclusive.

63. Find the sum of the first 50 even positive integers.

64. Find the sum of the first 100 positive odd integers.

65. *Salary Increases* In your new job you are told that your starting salary will be $36,000 with an increase of $2000 at the end of each of the first 5 years. How much will you be paid through the end of your first 6 years of employment with the company?

66. *Would You Accept This Job?* Suppose that you receive 25¢ on the first day of the month, 50¢ the second day, 75¢ the third day, and so on. Determine the total amount that you will receive during a 30-day month.

67. *Ticket Prices* There are 20 rows of seats on the main floor of a concert hall: 20 seats in the first row, 21 seats in the second row, 22 seats in the third row, and so on (see figure). How much should you charge per ticket in order to obtain $15,000 for the sale of all of the seats on the main floor?

68. *Pile of Logs* Logs are stacked in a pile as shown in the figure. The top row has 15 logs and the bottom row has 21 logs. How many logs are in the stack?

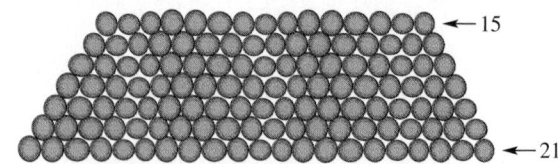

69. *Baling Hay* In the first two trips baling hay around a large field (see figure), a farmer obtains 93 bales and 89 bales, and the farmer estimates that the same pattern will continue. Estimate the total number of bales if the farmer makes another six trips around the field.

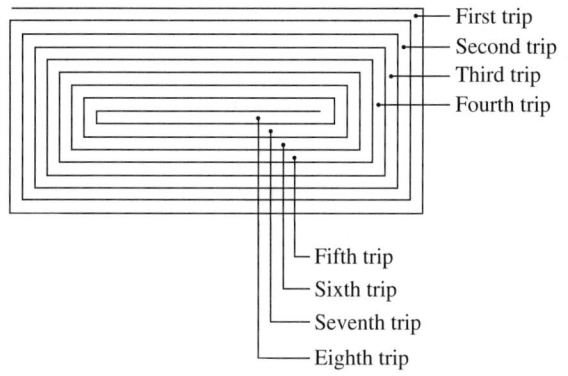

— First trip
— Second trip
— Third trip
— Fourth trip

Fifth trip
Sixth trip
Seventh trip
Eighth trip

70. *Clock Chimes* A clock chimes once at 1:00, twice at 2:00, three times at 3:00, and so on. The clock also chimes once at 15-minute intervals that are not on the hour. How many times does the clock chime in a 12-hour period?

71. *Free-Falling Object* A free-falling object will fall 16 feet during the first second, 48 more feet during the next second, 80 more during the third second, and so on. What is the total distance the object will fall in 8 seconds if this pattern continues?

Geometric Sequences

Geometric Sequences • Sum of a Geometric Sequence • Applications

---

Students often confuse arithmetic and geometric sequences. Remind them that arithmetic sequences have a common *difference* and geometric sequences have a common *ratio* (sometimes called a factor or constant multiplier).

Geometric Sequences

In Section 10.2, you saw that a sequence with consecutive terms that have a common *difference* is an arithmetic sequence. In this section, you will study another important type of sequence called a **geometric sequence.** Consecutive terms of a geometric sequence have a common *ratio,* as indicated in the following definition.

Definition of a Geometric Sequence

A sequence is called **geometric** if the ratios of consecutive terms are the same. Thus, the sequence

$$a_1, \quad a_2, \quad a_3, \quad a_4, \quad \ldots, \quad a_n, \quad \ldots$$

is geometric if there is a number r, $r \neq 0$, such that

$$\frac{a_2}{a_1} = r, \qquad \frac{a_3}{a_2} = r, \qquad \frac{a_4}{a_3} = r,$$

and so on. The number r is called the **common ratio** of the geometric sequence.

EXAMPLE 1   ■   Examples of Geometric Sequences

(a) The sequence having an nth term of 2^n is geometric. For this sequence, the common ratio between consecutive terms is 2.

$$2, \quad 4, \quad 8, \quad 16, \quad \ldots$$

$$\tfrac{4}{2} = 2$$

(b) The sequence having an nth term of $4(3^n)$ is geometric. For this sequence, the common ratio between consecutive terms is 3.

$$12, \quad 36, \quad 108, \quad 324, \quad \ldots$$

$$\tfrac{36}{12} = 3$$

Additional problem: Have students determine if the sequence 3, −6, 12, −24, −48, 96, . . . is geometric. Answer: No

(c) The sequence having an nth term of $\left(-\frac{1}{3}\right)^n$ is geometric. For this sequence, the common ratio between consecutive terms is $-\frac{1}{3}$.

$$-\frac{1}{3}, \quad \frac{1}{9}, \quad -\frac{1}{27}, \quad \frac{1}{81}, \quad \ldots$$

$$\frac{\frac{1}{9}}{-\frac{1}{3}} = -\frac{1}{3}$$

■

In Example 1, notice that each of the geometric sequences has an nth term that is of the form $a_1 r^{n-1}$, where the common ratio of the sequence is r. The result is summarized as follows.

The nth Term of a Geometric Sequence

The nth term of a geometric sequence has the form

$$a_n = a_1 r^{n-1},$$

where r is the common ratio of consecutive terms of the sequence. Thus, every geometric sequence can be written in the following form.

$$a_1, \quad a_2, \quad a_3, \quad a_4, \quad a_5, \quad \ldots, \quad a_n, \quad \ldots$$

$$\downarrow \quad \downarrow \quad \downarrow \quad \downarrow \quad \downarrow \quad \ldots, \quad \downarrow \quad \ldots$$

$$a_1, \quad a_1 r, \quad a_1 r^2, \quad a_1 r^3, \quad a_1 r^4, \quad \ldots, \quad a_1 r^{n-1}$$

Point out that the factor a_1 remains constant, while the exponent of r increases by 1 in each successive term.

NOTE   If you know the nth term of a geometric sequence, then the $(n + 1)$th term can be found by multiplying by r. That is,

$$a_{n+1} = r a_n.$$

EXAMPLE 2   ■   Finding the nth Term of a Geometric Sequence

Find a formula for the nth term of the geometric sequence with a common ratio of 3 and a first term of 1. What is the eighth term of this sequence?

Solution

You know that the formula for the nth term is of the form $a_n = a_1 r^{n-1}$. Moreover, because the common ratio is $r = 3$, the formula must have the form

$$a_n = a_1 r^{n-1} = (1)(3^{n-1}) = 3^{n-1}.$$

The sequence, therefore, has the following form.

$$1, \quad 3, \quad 9, \quad 27, \quad 81, \quad \ldots, \quad 3^{n-1}, \quad \ldots$$

The eighth term of the sequence is

$$a_8 = 3^{8-1} = 3^7 = 2187.$$ ■

EXAMPLE 3   ■   **Finding the nth Term of a Geometric Sequence**

Find a formula for the nth term of the geometric sequence of which the first two terms are 4 and 2. What is the 10th term of this sequence?

Solution

Have students compare Examples 2 and 3. Note that in Example 2 $r > 1$ and the sequence increases; in Example 3, $r < 1$ and the sequence decreases. Ask students to explain why.

You know that the formula for the nth term is of the form $a_n = a_1 r^{n-1}$. Moreover, because the common ratio is

$$\frac{a_2}{a_1} = \frac{2}{4} = \frac{1}{2},$$

the formula for the nth term must have the form

$$a_n = a_1 r^{n-1} = 4\left(\frac{1}{2}\right)^{n-1}.$$

The sequence, therefore, has the following form.

$$4, \quad 2, \quad 1, \quad \frac{1}{2}, \quad \frac{1}{4}, \quad \ldots, \quad 4\left(\frac{1}{2}\right)^{n-1}, \quad \ldots$$

The 10th term of the sequence is

$$a_{10} = 4\left(\frac{1}{2}\right)^{10-1} = 2^2\left(\frac{1}{2^9}\right) = \frac{1}{2^7} = \frac{1}{128}.$$ ■

NOTE   Another way of solving Example 3 is simply to multiply 4 by $r = \frac{1}{2}$ several times to obtain the first 10 terms in the sequence.

1	2	3	4	5	6	7	8	9	10
↓	↓	↓	↓	↓	↓	↓	↓	↓	↓
4,	2,	1,	$\frac{1}{2}$,	$\frac{1}{4}$,	$\frac{1}{8}$,	$\frac{1}{16}$,	$\frac{1}{32}$,	$\frac{1}{64}$,	$\frac{1}{128}$

Sum of a Geometric Sequence

In Section 10.2, you saw that there is a simple formula for finding the sum of the first n terms of an arithmetic sequence. There is also a formula for finding the sum of the first n terms of a geometric sequence.

The nth Partial Sum of a Geometric Sequence	The nth partial sum of the geometric sequence having an nth term of $a_n = a_1 r^{n-1}$ is $$\sum_{i=1}^{n} a_1 r^{i-1} = a_1 + a_1 r + a_1 r^2 + a_1 r^3 + \cdots + a_1 r^{n-1} = a_1 \left(\frac{r^n - 1}{r - 1} \right).$$

EXAMPLE 4 ■ Finding the nth Partial Sum of a Geometric Sequence

Find the following sum.

$$1 + 2 + 4 + 8 + 16 + 32 + 64 + 128$$

Solution

Because the terms of this sum form a geometric sequence with the common ratio of $r = 2$, you can use the formula for the nth partial sum of a geometric sequence. Moreover, because the first term of the sequence is $a_1 = 1$, it follows that the sum is given by

$$\sum_{i=1}^{n} a_1 r^{i-1} = a_1 \left(\frac{r^n - 1}{r - 1} \right)$$

$$\sum_{i=1}^{8} 2^{i-1} = (1) \left(\frac{2^8 - 1}{2 - 1} \right)$$

$$= \frac{256 - 1}{2 - 1}$$

$$= 255.$$

EXAMPLE 5 ■ Finding the nth Partial Sum of a Geometric Sequence

Be sure to have students distinguish between r^{n-1} and $r^n - 1$.

Find the sum of the first five terms of the geometric sequence having an nth term of $a_n = \left(\frac{2}{3} \right)^n$.

Solution

For this geometric sequence, the first term is $a_1 = \frac{2}{3}$, and the common ratio is $r = \frac{2}{3}$. Therefore, the sum of the first five terms is as follows on the next page.

$$\sum_{i=1}^{n} a_1 r^{i-1} = a_1 \left(\frac{r^n - 1}{r - 1} \right)$$

$$\sum_{i=1}^{5} \left(\frac{2}{3} \right)^i = \frac{2}{3} \left[\frac{\left(\frac{2}{3} \right)^5 - 1}{\frac{2}{3} - 1} \right]$$

$$= \frac{2}{3} \left(\frac{\frac{32}{243} - 1}{-\frac{1}{3}} \right)$$

$$= \frac{2}{3} \left(-\frac{211}{243} \right) (-3)$$

$$= \frac{422}{243}$$

$$\approx 1.737$$ ■

Remind students that they can invert and multiply to simplify a compound fraction.

EXAMPLE 6   ■   Finding the *n*th Partial Sum of a Geometric Sequence

Find the following sum.

$$\sum_{i=1}^{7} \frac{3^{i-1}}{2}$$

Solution

By writing out a few terms of the sum,

$$\sum_{i=1}^{7} \frac{3^{i-1}}{2} = \sum_{i=1}^{7} \frac{1}{2} (3^{i-1}) = \frac{1}{2} + \frac{1}{2}(3) + \frac{1}{2}(3^2) + \cdots + \frac{1}{2}(3^6),$$

you can see that $a_1 = \frac{1}{2}$ and $r = 3$. Thus, the sum is as follows.

$$\sum_{i=1}^{n} a_1 r^{i-1} = a_1 \left(\frac{r^n - 1}{r - 1} \right)$$

$$\sum_{i=1}^{7} \frac{1}{2} (3^{i-1}) = \frac{1}{2} \left(\frac{3^7 - 1}{3 - 1} \right)$$

$$= \frac{1}{2} \left(\frac{2187 - 1}{2} \right)$$

$$= \frac{2186}{4}$$

$$= 546.5$$ ■

Applications

This section closes with two applications of geometric sequences.

EXAMPLE 7 ■ An Application: A Lifetime Salary

Suppose you accept a job that pays a salary of $28,000 the first year. During the next 39 years, suppose you receive a 6% raise each year. What would your total salary be over the 40-year period?

Solution

Using a geometric sequence, your salary during the first year would be

$$a_1 = 28,000,$$

and, with a 6% raise each year, your salary during the second and third years would be as follows.

$$a_2 = 28,000 + 28,000(0.06) = 28,000(1.06)^1$$
$$a_3 = 28,000(1.06) + 28,000(1.06)(0.06) = 28,000(1.06)^2$$

From this pattern, you can see that the common ratio of the geometric sequence is $r = 1.06$. Using the formula for the nth partial sum of a geometric sequence, you can determine that the total salary over the 40-year period would be as follows.

$$\text{Total salary} = a_1\left(\frac{r^n - 1}{r - 1}\right)$$
$$= 28,000\left[\frac{(1.06)^{40} - 1}{1.06 - 1}\right]$$
$$= 28,000\left[\frac{(1.06)^{40} - 1}{0.06}\right]$$
$$= \$4,333,335$$

The bar graph in Figure 10.4 graphically illustrates your salary during the 40-year period.

FIGURE 10.4

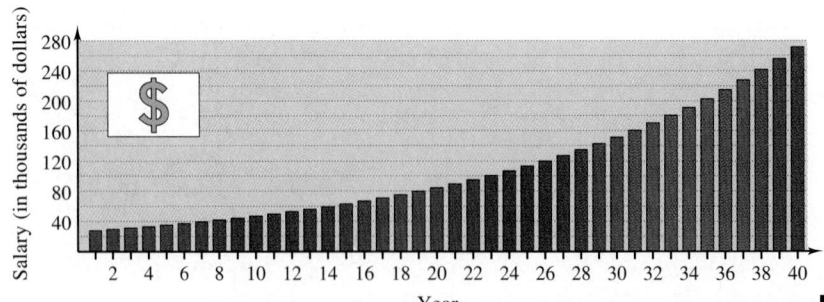

EXAMPLE 8 ■   An Application: Increasing Annuity

Suppose you deposit $100 in an account each month for 2 years. The account pays an annual interest rate of 9%, compounded monthly. What is your balance at the end of 2 years? (This type of savings plan is called an **increasing annuity.**)

Solution

Using the formula for compound interest, you can see that the first deposit would earn interest for the full 24 months. Thus, by the end of 24 months, the first deposit will have grown to a balance of

Encourage students to write out the first few terms of the sequence to see "what's going on" and to help determine a_1 and r.

$$a_{24} = 100\left(1 + \frac{0.09}{12}\right)^{24} = 100(1.0075)^{24}.$$

Because the second deposit will have earned interest for only 23 months, its balance will be

$$a_{23} = 100\left(1 + \frac{0.09}{12}\right)^{23} = 100(1.0075)^{23}.$$

Thus, the total of the 24 deposits will be

$$\text{Total} = 100(1.0075)^1 + 100(1.0075)^2 + \cdots + 100(1.0075)^{24}$$

$$= 100(1.0075)\left(\frac{1.0075^{24} - 1}{1.0075 - 1}\right) \qquad a_1\left(\frac{r^n - 1}{r - 1}\right)$$

$$= \$2638.49. \qquad\qquad\qquad\qquad\qquad\qquad\qquad ■$$

DISCUSSION PROBLEM ■   Annual Revenue

The bar graphs shown in Figure 10.5 show the annual revenue R (in millions of dollars) for two companies over a 10-year period. One company's revenue growth followed an arithmetic pattern; the other company's growth followed a geometric pattern. Find a model for each company's revenue and find the total revenue over the 10-year period. Which company earned the most during the 10-year period? Which company would you rather own? Explain your reasoning.

FIGURE 10.5

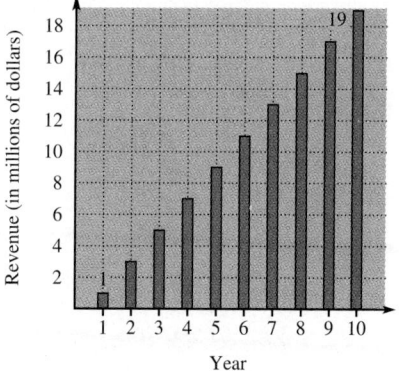
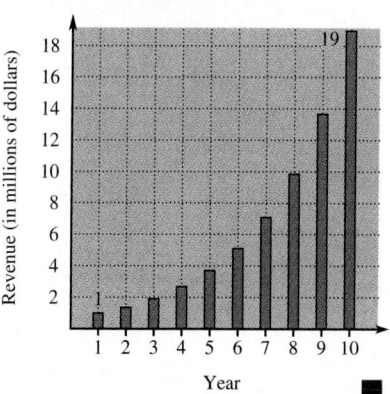

■

Warm-Up

The following warm-up exercises involve skills that were covered in earlier sections. You will use these skills in the exercise set for this section.

In Exercises 1–4, evaluate the quantity.

1. $\left(\dfrac{2}{3}\right)^4$ **2.** $\left(\dfrac{3}{5}\right)^3$ **3.** $\dfrac{2^{-3}}{3^2}$ **4.** $\dfrac{8^2}{4^3}$

In Exercises 5–10, simplify the expression.

5. $\dfrac{n!}{(n-1)!}$ **6.** $\dfrac{(n+1)!}{n!}$ **7.** $(3n)(4n^2)(5n^3)$

8. $n(2n)^4$ **9.** $\dfrac{(3n)^2}{18n}$ **10.** $\dfrac{18n}{10n^{-1}}$

10.3 EXERCISES

In Exercises 1–8, find the common ratio for the geometric sequence.

1. 2, 6, 18, 54, . . . **2.** 5, −10, 20, −40, . . . **3.** 12, −6, 3, $-\frac{3}{2}$, . . .

4. 54, 18, 6, 2, . . . **5.** 1, $-\frac{3}{2}$, $\frac{9}{4}$, $-\frac{27}{8}$, . . . **6.** 9, 6, 4, $\frac{8}{3}$, . . .

7. e, e^2, e^3, e^4, . . . **8.** 1.1, $(1.1)^2$, $(1.1)^3$, $(1.1)^4$, . . .

In Exercises 9–16, determine whether the sequence is geometric. If it is, find the common ratio.

9. 10, 15, 20, 25, . . . **10.** 10, 20, 40, 80, . . . **11.** 64, 32, 16, 8, . . .

12. 64, 32, 0, −32, . . . **13.** 1, $-\frac{2}{3}$, $\frac{4}{9}$, $-\frac{8}{27}$, . . . **14.** $\frac{1}{3}$, $-\frac{2}{3}$, $\frac{4}{3}$, $-\frac{8}{3}$, . . .

15. $10(1+0.02)$, $10(1+0.02)^2$, $10(1+0.02)^3$, $10(1+0.02)^4$, . . . **16.** 1, 0.2, 0.04, 0.008, . . .

In Exercises 17–24, write the first five terms of the geometric sequence.

17. $a_1 = 4$, $r = 2$ **18.** $a_1 = 3$, $r = 4$ **19.** $a_1 = 6$, $r = \frac{1}{3}$ **20.** $a_1 = 4$, $r = \frac{1}{2}$

21. $a_1 = 1$, $r = -\frac{1}{2}$ **22.** $a_1 = 32$, $r = -\frac{3}{4}$ **23.** $a_1 = 1000$, $r = 1.01$ **24.** $a_1 = 4000$, $r = \frac{1}{1.01}$

In Exercises 25–32, find the formula for the nth term of the geometric sequence. (Assume that n begins with 1.)

25. $a_1 = 2$, $r = 3$ **26.** $a_1 = 5$, $r = 4$ **27.** $a_1 = 1$, $r = 2$ **28.** $a_1 = 1$, $r = -5$

29. $a_1 = 4$, $r = -\frac{1}{2}$ **30.** $a_1 = 9$, $r = \frac{2}{3}$ **31.** $a_1 = 8$, $a_2 = 2$ **32.** $a_1 = 18$, $a_2 = 8$

In Exercises 33–40, find the nth term of the geometric sequence.

33. $a_1 = 6$, $r = \frac{1}{2}$, $n = 10$ **34.** $a_1 = 8$, $r = \frac{3}{4}$, $n = 8$ **35.** $a_1 = 3$, $r = \sqrt{2}$, $n = 10$

36. $a_1 = 500$, $r = 1.06$, $n = 40$ **37.** $a_1 = 4$, $a_2 = 3$, $n = 5$ **38.** $a_1 = 1$, $a_2 = 9$, $n = 7$

39. $a_1 = 1$, $a_3 = \frac{9}{4}$, $n = 6$ **40.** $a_1 = 1$, $a_3 = 16$, $n = 4$

In Exercises 41–48, find the sum.

41. $\displaystyle\sum_{i=1}^{10} 2^{i-1}$ **42.** $\displaystyle\sum_{i=1}^{6} 3^{i-1}$ **43.** $\displaystyle\sum_{i=1}^{12} 3\left(\frac{3}{2}\right)^{i-1}$ **44.** $\displaystyle\sum_{i=1}^{20} 12\left(\frac{2}{3}\right)^{i-1}$

45. $\displaystyle\sum_{i=1}^{15} 3\left(-\frac{1}{3}\right)^{i-1}$ **46.** $\displaystyle\sum_{i=1}^{8} 8\left(-\frac{1}{4}\right)^{i-1}$ **47.** $\displaystyle\sum_{i=1}^{20} 100(1.1)^{i-1}$ **48.** $\displaystyle\sum_{i=1}^{40} 50(1.07)^{i-1}$

In Exercises 49–56, find the nth partial sum of the geometric sequence.

49. $1, -3, 9, -27, 81, \ldots$, $n = 10$ **50.** $3, -6, 12, -24, 48, \ldots$, $n = 12$

51. $8, 4, 2, 1, \frac{1}{2}, \ldots$, $n = 15$ **52.** $9, 6, 4, \frac{8}{3}, \frac{16}{9}, \ldots$, $n = 10$

53. $1, \sqrt{2}, 2, 2\sqrt{2}, 4, \ldots$, $n = 12$ **54.** $40, -10, \frac{5}{2}, -\frac{5}{8}, \frac{5}{32}, \ldots$, $n = 10$

55. $30, 30(1.06), 30(1.06)^2, 30(1.06)^3, 30(1.06)^4, \ldots$, $n = 20$

56. $100, 100(1.08), 100(1.08)^2, 100(1.08)^3, 100(1.08)^4, \ldots$, $n = 40$

57. *Depreciation* A company pays $250,000 for a machine. During the next 5 years, the machine depreciates at the rate of 25% per year. (That is, at the end of each year, the depreciated value is 75% of what it was at the beginning of the year.)
(a) Find a formula for the nth term of the geometric sequence that gives the value of the machine n full years after it was purchased.
(b) Find the depreciated value of the machine at the end of 5 full years.

58. *Population Increase* A city of 500,000 people is growing at the rate of 1% per year. (That is, at the end of each year the population is 1.01 times the population at the beginning of the year.)
(a) Find a formula for the nth term of the geometric sequence that gives the population n years from now.
(b) Estimate the population 20 years from now.

59. *Salary Increases* Suppose that you accept a job that pays a salary of $30,000 the first year. During the next 39 years, suppose you receive a 5% raise each year. What would your *total* salary be over the 40-year period?

60. *Salary Increases* Suppose you accept a job that pays a salary of $30,000 the first year. During the next 39 years, suppose you receive a 5.5% raise each year. What would your *total* salary be over the 40-year period?

Increasing Annuity In Exercises 61–64, find the balance in an increasing annuity (see Example 8) when a principal of P dollars is invested at the beginning of each month for t years. Assume that the amount in the fund is compounded monthly at rate r.

61. $P = \$100$ $t = 10$ years $r = 9\%$

62. $P = \$50$ $t = 5$ years $r = 7\%$

63. $P = \$30$ $t = 40$ years $r = 8\%$

64. $P = \$200$ $t = 30$ years $r = 10\%$

65. *Would You Accept This Job?* Suppose you begin working at a company that pays $0.01 for the first day, $0.02 for the second day, $0.04 for the third day, and so on. If the daily wage keeps doubling, what will your total income be for working (a) 29 days and (b) 30 days?

66. *Would You Accept This Job?* Suppose you begin working at a company that pays $0.01 for the first day, $0.03 for the second day, $0.09 for the third day, and so on. If the daily wage keeps tripling, what will your total income be for working (a) 25 days and (b) 26 days?

67. *Area* The sides of a square are 12 inches. A new square is formed by connecting the midpoints of the sides of the original square, and two of the resulting triangles are shaded (see figure). If this process is repeated five more times, what will be the total area of shaded region?

 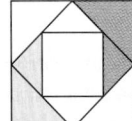 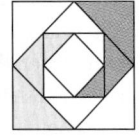 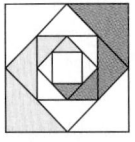

68. *Area* The sides of a square are 12 inches. The square is divided into nine smaller squares, and the center square is shaded (see figure). Each of the eight unshaded squares is then divided into nine smaller squares of which the center square is shaded. If this process is repeated four more times, what will be the total area of shaded region?

Figure for 68

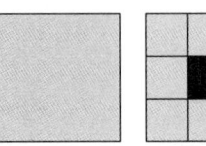 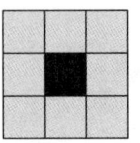 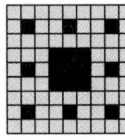 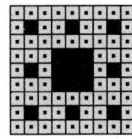

69. *Number of Ancestors* The number of direct ancestors a person has had is as follows.

$$2 + 2^2 + 2^3 + 2^4 + \cdots + 2^n + \cdots$$

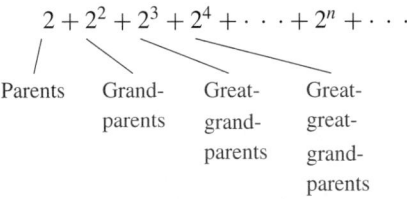

Parents   Grand-parents   Great-grand-parents   Great-great-grand-parents

This formula is valid provided the person has no common ancestors. (A common ancestor is one to whom you are related in more than one way. For example, one of your great-grandmothers on your father's side might also be one of your great-grandmothers on your mother's side; see figure.) How many direct ancestors have you had who have lived since the year A.D. 1? Assume that the average time between generations has been 30 years (resulting in 66 generations), so that the total is

$$2 + 2^2 + 2^3 + 2^4 + \cdots + 2^{66}.$$

Considering the total, is it reasonable to assume that you have had no common ancestors in the past 2000 years?

Have you had common ancestors?

Great-great-grandparents

Great-grandparents

Grandparents

Parents

Take this quiz as you would take a quiz in class. After you are done, check your work against the answers in the back of the book.

1. Write the first five terms of the sequence $a_n = 30 - 5n$. Begin with $n = 1$.

2. Write the first five terms of the sequence $a_n = \left(-\frac{1}{4}\right)^n$. Begin with $n = 0$.

3. Write the sum $1 + 7 + 13 + 19 + 25 + 31 + 37$ using sigma notation.

4. Write the sum $\left(\frac{4}{3}\right)^0 + \left(\frac{4}{3}\right)^1 + \left(\frac{4}{3}\right)^2 + \left(\frac{4}{3}\right)^3 + \left(\frac{4}{3}\right)^4$ using sigma notation.

5. Simplify: $\dfrac{12!}{10!}$ **6.** Simplify: $\dfrac{n!}{(n-2)!}$

In Exercises 7 and 8, determine whether the sequence is arithmetic or geometric. If it is arithmetic, state its common difference. If it is geometric, state its common ratio.

7. $\frac{1}{2}$, 2, 8, 32, 128, . . .   **8.** 1, $\frac{1}{2}$, 0, $-\frac{1}{2}$, -1, . . .

9. Find a formula for the nth term of the arithmetic sequence in which $a_1 = 3$ and $a_2 = 12$.

10. Find a formula for the nth term of the geometric sequence in which $a_1 = 5$ and $r = -0.5$.

In Exercises 11–14, find the sum of the series.

11. $\displaystyle\sum_{k=1}^{6} (2k - 1)$ **12.** $\displaystyle\sum_{j=1}^{4} (32 - 4j)$ **13.** $\displaystyle\sum_{i=1}^{20} 16(4)^{i-1}$ **14.** $\displaystyle\sum_{n=1}^{4} 6\left(-\frac{1}{4}\right)^{n-1}$

15. A deposit of \$10,000 is made in an account that earns 8% interest compounded monthly. The balance in the account after n months is

$$A_n = 10,000\left(1 + \frac{0.08}{12}\right)^n, \qquad n = 1, \quad 2, \quad 3, \quad . \, . \, . \, .$$

(a) Compute the first five terms of the sequence.
(b) Find the balance in this account after 10 years by computing A_{120}.

16. A well-drilling company charges \$15 for drilling the first foot of a well, \$15.25 for the second foot, \$15.50 for the third foot, and so on. How much would it cost to have this company drill a 100-foot well?

17. For the years 1982 through 1990, the amount a_n (in millions of dollars) of guaranteed student loans can be modeled by

$$a_n = 5650.2(1.2)^{n-1},$$

where $n = 2$ corresponds to 1982. Use this model to approximate the total amount of guaranteed student loans for the specified years. (*Source:* Student Loan Marketing Assn.)

Binomial Coefficients

Recall that a **binomial** is a polynomial that has two terms. In this section, you will learn a formula that gives a quick method of raising a binomial to a power. To begin, let's look at the expansion of $(x + y)^n$ for several values of n.

Remind students to use the distributive property to expand $(x + y)^3$, $(x + y)^4$, $(x + y)^5$, etc.

$$(x + y)^0 = 1$$
$$(x + y)^1 = x + y$$
$$(x + y)^2 = x^2 + 2xy + y^2$$
$$(x + y)^3 = x^3 + 3x^2y + 3xy^2 + y^3$$
$$(x + y)^4 = x^4 + 4x^3y + 6x^2y^2 + 4xy^3 + y^4$$
$$(x + y)^5 = x^5 + 5x^4y + 10x^3y^2 + 10x^2y^3 + 5xy^4 + y^5$$

One strategy is to let students make these observations for the other binomials expanded above. Have students look for the symmetric pattern alluded to in item 5 and see if they can come up with Pascal's Triangle *before* it is introduced.

There are several characteristics to observe about these expansions of $(x + y)^n$.

1. In each expansion, there are $n + 1$ terms.

2. In each expansion, x and y have symmetrical roles. The powers of x decrease by 1 in successive terms, whereas the powers of y increase by 1.

3. The sum of the powers of each term in a binomial expansion is n. For example, in the expansion of $(x + y)^5$, the sum of the powers of each term is 5, as follows.

$$\overbrace{4 + 1 = 5} \qquad \overbrace{3 + 2 = 5}$$
$$(x + y)^5 = x^5 + 5\,x^4 y^1 + 10\,x^3 y^2 + 10x^2 y^3 + 5xy^4 + y^5$$

4. The first term is x^n, the last term is y^n, and each of these terms has a coefficient of 1.

5. The coefficients increase and then decrease in a symmetrical pattern. For $(x + y)^5$, the pattern is

$$1 \qquad 5 \qquad 10 \qquad 10 \qquad 5 \qquad 1.$$

The most difficult part of a binomial expansion is finding the coefficients of the interior terms. To find these **binomial coefficients,** you can use a well-known theorem called the **Binomial Theorem.**

The Binomial Theorem

In the expansion of $(x + y)^n$,

$$(x + y)^n = x^n + nx^{n-1}y + \cdots + {}_nC_m x^{n-m}y^m + \cdots + nxy^{n-1} + y^n,$$

the coefficient of $x^{n-m}y^m$ is given by

$${}_nC_m = \frac{n!}{(n-m)!m!}.$$

NOTE Other notations that are commonly used for ${}_nC_m$ are $\binom{n}{m}$ and $C(n, m)$. The notation in this text is similar to that used by the TI-81 and TI-82 graphing calculators.

EXAMPLE 1 ■ Finding Binomial Coefficients

Find the following binomial coefficients.

(a) ${}_8C_2$ (b) ${}_{10}C_3$ (c) ${}_7C_0$

Some students may have calculators that will evaluate ${}_nC_m$. Students can learn how to do this on their calculators, verifying computations. For instance, on the TI-81 or TI-82 graphing calculator, ${}_8C_2$ would be evaluated using the key sequence

8 [MATH] [◁] 3 2 [ENTER] .

Solution

Note in parts (a) and (b) how you can factor the numerator factorial in order to divide out one of the denominator factorials.

(a) ${}_8C_2 = \dfrac{8!}{6!2!} = \dfrac{(8 \cdot 7) \cdot \cancel{6!}}{\cancel{6!} \cdot 2!} = \dfrac{8 \cdot 7}{2 \cdot 1} = 28$

(b) ${}_{10}C_3 = \dfrac{10!}{7!3!} = \dfrac{(10 \cdot 9 \cdot 8) \cdot \cancel{7!}}{\cancel{7!} \cdot 3!} = \dfrac{10 \cdot 9 \cdot 8}{3 \cdot 2 \cdot 1} = 120$

(c) ${}_7C_0 = \dfrac{7!}{7!0!} = 1$ ■

NOTE For the cases when $m \neq 0$ or $m \neq n$, as in parts (a) and (b) above, there is a simple pattern for evaluating binomial coefficients.

$$
{}_8C_2 = \frac{\overbrace{8 \cdot 7}^{2 \text{ factors}}}{\underbrace{2 \cdot 1}_{2 \text{ factorial}}} \quad \text{and} \quad {}_{10}C_3 = \frac{\overbrace{10 \cdot 9 \cdot 8}^{3 \text{ factors}}}{\underbrace{3 \cdot 2 \cdot 1}_{3 \text{ factorial}}}
$$

In general, we have the following.

$$
{}_nC_m = \frac{\overbrace{n(n-1)(n-2) \cdots}^{m \text{ factors}}}{m!}, \qquad a < m < n
$$

EXAMPLE 2 ■ Finding Binomial Coefficients

Find the following binomial coefficients.

(a) $_7C_3$ (b) $_7C_4$ (c) $_{12}C_1$ (d) $_{12}C_{11}$

Solution

(a) $_7C_3 = \dfrac{7 \cdot 6 \cdot 5}{3 \cdot 2 \cdot 1} = 35$

(b) $_7C_4 = \dfrac{7 \cdot 6 \cdot 5 \cdot 4}{4 \cdot 3 \cdot 2 \cdot 1} = 35$

(c) $_{12}C_1 = \dfrac{12}{1} = 12$

(d) $_{12}C_{11} = \dfrac{12!}{1!11!} = \dfrac{(12) \cdot \cancel{11!}}{1! \cdot \cancel{11!}} = \dfrac{12}{1} = 12$ ■

The observation $_nC_m = {_nC_{n \cdot m}}$ should be emphasized. It can greatly simplify some computations.

It is not a coincidence that the results in parts (a) and (b), and similarly in parts (c) and (d), are the same. In general, it is true that $_nC_m = {_nC_{n-m}}$. For instance,

$$_6C_0 = {_6C_6} = 1, \quad {_6C_1} = {_6C_5} = 6, \quad \text{and} \quad {_6C_2} = {_6C_4} = 15.$$

This shows the symmetric property of binomial coefficients that was identified earlier. Notice how this property is used in Example 3.

EXAMPLE 3 ■ Finding a Binomial Coefficient

Find the binomial coefficient $_{12}C_{10}$.

Solution

Rather than calculating the binomial coefficient

$$_{12}C_{10} = \frac{12 \cdot 11 \cdot 10 \cdot 9 \cdot 8 \cdot 7 \cdot 6 \cdot 5 \cdot 4 \cdot 3}{10 \cdot 9 \cdot 8 \cdot 7 \cdot 6 \cdot 5 \cdot 4 \cdot 3 \cdot 2 \cdot 1},$$

you can use the fact that $_{12}C_{10} = {_{12}C_2}$, as follows.

$$_{12}C_{10} = {_{12}C_2} = \frac{12 \cdot 11}{2 \cdot 1} = 66$$ ■

Pascal's Triangle

There is a convenient way to remember a pattern for binomial coefficients. Arranging the coefficients in a triangular pattern produces the array on the following page, which is called **Pascal's Triangle.** This triangle is named after the famous French mathematician Blaise Pascal (1623–1662).

$$
\begin{array}{ccccccccccccccc}
 & & & & & & & 1 & & & & & & & \\
 & & & & & & 1 & & 1 & & & & & & \\
 & & & & & 1 & & 2 & & 1 & & & & & \\
 & & & & 1 & & 3 & & 3 & & 1 & & & & \\
 & & & 1 & & 4 & & 6 & & 4 & & 1 & & & \\
 & & 1 & & 5 & & 10 & & 10 & & 5 & & 1 & & \\
 & 1 & & 6 & & 15 & & 20 & & 15 & & 6 & & 1 & \\
1 & & 7 & & 21 & & 35 & & 35 & & 21 & & 7 & & 1
\end{array}
$$

The first and last number in each row of Pascal's Triangle is 1. Every remaining number in each row is the sum of the two numbers that lie diagonally above the number. For example, the two numbers above 35 are 15 and 20.

$$15 \quad\quad 20$$
$$\diagdown \quad \diagup$$
$$35$$

$$15 + 20 = 35$$

Pascal noticed that numbers in this triangle are precisely the same numbers that are the coefficients of binomial expansions, as follows.

$$(x + y)^0 = 1$$
$$(x + y)^1 = 1x + 1y$$
$$(x + y)^2 = 1x^2 + 2xy + 1y^2$$
$$(x + y)^3 = 1x^3 + 3x^2y + 3xy^2 + 1y^3$$
$$(x + y)^4 = 1x^4 + 4x^3y + 6x^2y^2 + 4xy^3 + 1y^4$$
$$(x + y)^5 = 1x^5 + 5x^4y + 10x^3y^2 + 10x^2y^3 + 5xy^4 + 1y^5$$
$$(x + y)^6 = 1x^6 + 6x^5y + 15x^4y^2 + 20x^3y^3 + 15x^2y^4 + 6xy^5 + 1y^6$$
$$(x + y)^7 = 1x^7 + 7x^6y + 21x^5y^2 + 35x^4y^3 + 35x^3y^4 + 21x^2y^5 + 7xy^6 + 1y^6$$

Because the top row in Pascal's Triangle corresponds to the binomial expansion $(x + y)^0 = 1$, it is called the **zero row.** Similarly, the next row corresponds to the binomial expansion $(x + y)^1 = 1(x) + 1(y)$, and it is called the **first row.** In general, the ***n*th row** in Pascal's Triangle gives the coefficients of $(x + y)^n$.

EXAMPLE 4   ■   **Using Pascal's Triangle**

Use Pascal's Triangle to find the following binomial coefficients.

$$_8C_0, \quad _8C_1, \quad _8C_2, \quad _8C_3, \quad _8C_4, \quad _8C_5, \quad _8C_6, \quad _8C_7, \quad _8C_8$$

Solution

These nine binomial coefficients represent the eighth row of Pascal's Triangle. Thus, using the seventh row of the triangle, you can calculate the numbers in the eighth row, as follows.

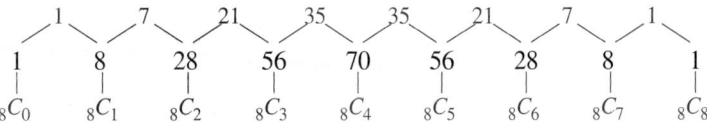

■

Binomial Expansions

As mentioned at the beginning of this section, when you write out the terms of a binomial that is raised to a power, you are **expanding a binomial.** The Binomial Theorem and Pascal's Triangle give you an easy way to expand binomials, as demonstrated in the next four examples.

EXAMPLE 5 ■ Expanding a Binomial

Write the expansion for the expression $(x + 1)^3$.

Solution

The binomial coefficients from the third row of Pascal's Triangle are 1, 3, 3, 1. Therefore, the expansion is as follows.

$$(x + 1)^3 = (1)x^3 + (3)x^2(1) + (3)x(1^2) + (1)(1^3)$$
$$= x^3 + 3x^2 + 3x + 1$$ ■

Stress that when binomials representing differences are expanded, the signs alternate.

To expand binomials representing *differences*, rather than sums, alternate the signs, as follows.

$$(x - 1)^3 = x^3 - 3x^2 + 3x - 1$$
$$(x - 1)^4 = x^4 - 4x^3 + 6x^2 - 4x + 1$$

EXAMPLE 6 ■ Expanding a Binomial

Write the expansion for the expression $(x - 2)^3$.

Solution

The binomial coefficients from the third row of Pascal's Triangle are 1, 3, 3, 1. Therefore, the expansion is as follows.

$$(x - 2)^3 = (1)x^3 - (3)x^2(2) + (3)x(2^2) - (1)(2^3)$$
$$= x^3 - 6x^2 + 12x - 8$$ ■

EXAMPLE 7 ■ Expanding a Binomial

Write the expansion for the expression $(x + 3)^4$.

Solution

The binomial coefficients from the fourth row of Pascal's Triangle are 1, 4, 6, 4, 1. Therefore, the expansion is as follows.

$$(x + 3)^4 = (1)x^4 + (4)x^3(3) + (6)x^2(3^2) + (4)x(3^3) + (1)(3^4)$$
$$= x^4 + 12x^3 + 54x^2 + 108x + 81$$ ■

EXAMPLE 8   ■   Expanding a Binomial

Write the expansion for the expression $(x - 2y)^4$.

Solution

Remind students that $(2y)^m = 2^m y^m$.

The binomial coefficients from the fourth row of Pascal's Triangle are 1, 4, 6, 4, 1. Therefore, the expansion is as follows.

$$(x - 2y)^4 = (1)x^4 - (4)x^3(2y) + (6)x^2(2y)^2 - (4)x(2y)^3 + (1)(2y)^4$$
$$= x^4 - 8x^3y + 24x^2y^2 - 32xy^3 + 16y^4 \qquad ■$$

DISCUSSION PROBLEM   ■   The Rows of Pascal's Triangle

By adding the terms in each of the rows of Pascal's Triangle, we obtain the following.

Row 0: $1 = 1$
Row 1: $1 + 1 = 2$
Row 2: $1 + 2 + 1 = 4$
Row 3: $1 + 3 + 3 + 1 = 8$
Row 4: $1 + 4 + 6 + 4 + 1 = 16$

Can you find a pattern for this sequence? Use this pattern to find the sum of the terms in the 10th row of Pascal's Triangle. Then check your answer by actually adding the terms of the 10th row.   ■

Warm-Up

The following warm-up exercises involve skills that were covered in earlier sections. You will use these skills in the exercise set for this section.

In Exercises 1–6, perform the indicated operations and simplify.

1. $-3x(x - 2)$ **2.** $(x + 3)(x - 3)$ **3.** $(x - 2)^2$

4. $(2x + 3)^2$ **5.** $x^2y(5xy^{-3})$ **6.** $(-2x^2y)^3$

In Exercises 7–10, evaluate the quantity.

7. $4!$ **8.** $\dfrac{8!}{7!}$ **9.** $\dfrac{12!}{9!}$ **10.** $\dfrac{10!}{2!\,8!}$

10.4   **EXERCISES**   ▦ means that a graphing utility can help you solve the exercise or check your solution.

In Exercises 1–10, evaluate $_nC_m$.

1. $_6C_4$ **2.** $_7C_3$

3. $_{10}C_5$ **4.** $_{12}C_9$

5. $_{20}C_{20}$ **6.** $_{15}C_0$

7. $_{50}C_{48}$ **8.** $_{200}C_1$

9. $_{25}C_4$ **10.** $_{18}C_5$

In Exercises 11–14, evaluate the binomial coefficient $_nC_m$. Also, evaluate its symmetric coefficient $_nC_{n-m}$ to demonstrate that it is equal to $_nC_m$.

11. $_{15}C_3$ **12.** $_9C_4$

13. $_{25}C_5$ **14.** $_{30}C_3$

15. Find the ninth row of Pascal's Triangle.

16. Find the 10th row of Pascal's Triangle.

17. Use Pascal's Triangle to evaluate the coefficients.
    (a) $_7C_3$ (b) $_9C_5$

18. Use Pascal's Triangle to evaluate the coefficients.
    (a) $_6C_2$ (b) $_9C_3$

In Exercises 19–34, use the Binomial Theorem and Pascal's Triangle to expand the expression.

19. $(x + y)^4$ **20.** $(u + v)^6$

21. $(x + 1)^5$ **22.** $(x + 2)^5$

23. $(x + 3)^6$ **24.** $(x + 4)^4$

25. $(x - y)^5$ **26.** $(x - 5)^4$

27. $(u - 2v)^3$ **28.** $(2x + y)^5$

29. $(3a + 2b)^4$ **30.** $(4u - 3v)^3$

31. $(x + y)^8$ **32.** $(r - s)^7$

33. $(x - 2)^6$ **34.** $(2x + 3)^5$

In Exercises 35–40, find the coefficient of the term of the expression.

35. $(x + 1)^{10}$, x^7 **36.** $(x + 3)^{12}$, x^9

37. $(x - y)^{15}$, x^4y^{11} **38.** $(x + y)^{10}$, x^7y^3

39. $(2x + y)^{12}$, x^3y^9 **40.** $(x - 3y)^{14}$, x^3y^{11}

In Exercises 41–44, use the Binomial Theorem to expand the imaginary number. Simplify your answer by using the fact that $i^2 = -1$.

41. $(1 + i)^4$ **42.** $(2 - i)^5$

43. $(2 - 3i)^6$ **44.** $\left(5 - \sqrt{-9}\right)^3$

Probability In the study of probability, it is sometimes necessary to use the expansion $(p + q)^n$, where $p + q = 1$. In Exercises 45–48, use the Binomial Theorem to expand the expression.

45. $\left(\frac{1}{2} + \frac{1}{2}\right)^5$ **46.** $\left(\frac{2}{3} + \frac{1}{3}\right)^4$

47. $\left(\frac{1}{4} + \frac{3}{4}\right)^4$ **48.** $\left(\frac{2}{5} + \frac{3}{5}\right)^3$

In Exercises 49–52, use the Binomial Theorem to approximate the quantity accurate to three decimal places. For example,

$$(1.02)^{10} = (1 + 0.02)^{10}$$
$$= 1 + 10(0.02) + 45(0.02)^2 + \ldots\ .$$

49. $(1.02)^8$ **50.** $(2.005)^{10}$

51. $(2.99)^{12}$ **52.** $(1.98)^9$

53. Decide whether the formula

$$_nC_m = {}_nC_{n-m}$$

is true or false for all integers m and n such that $1 \le m \le n$. Explain your reasoning.

54. *Patterns in Pascal's Triangle* Use each encircled group of numbers in the figure to form a 2×2 matrix. Find the determinant of each matrix. Describe the pattern.

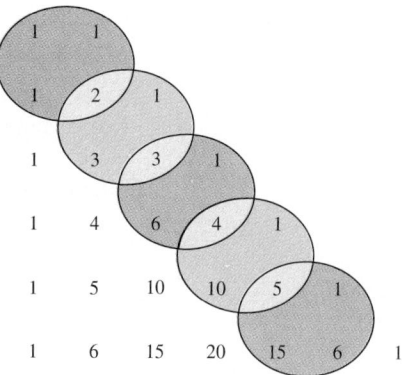

Simple Counting Problems

The last two sections of this chapter provide a brief introduction to some of the basic counting principles and their application to probability. In the next section, you will see that much of probability has to do with counting the number of ways an event can occur. Examples 1, 2, and 3 below describe some simple cases.

EXAMPLE 1 ■ A Random Number Generator

A random number generator (on a computer) selects an integer from 1 to 30. Find the number of ways the following events can occur.
(a) An even integer is selected.
(b) A number that is less than 12 is selected.
(c) A prime number is selected.

Solution

(a) Because half of the numbers from 1 to 30 are even, this event can occur in 15 different ways.

(b) The integers from 1 to 30 that are less than 12 are listed in the set

$$\{1, \ 2, \ 3, \ 4, \ 5, \ 6, \ 7, \ 8, \ 9, \ 10, \ 11\}.$$

Because this set has 11 members, you can conclude that there are 11 different ways this event can happen.

(c) The prime numbers in the set of integers from 1 to 30 are as follows.

$$\{2, \ 3, \ 5, \ 7, \ 11, \ 13, \ 17, \ 19, \ 23, \ 29\}$$

Because this set has 10 members, you can conclude that there are 10 different ways this event can happen. ■

EXAMPLE 2 ■ Selecting Pairs of Numbers at Random

Eight pieces of paper are numbered from 1 to 8 and placed in a box. One piece of paper is drawn from the box, its number is written down, and the piece of paper is replaced in the box. Then, a second piece of paper is drawn from the box, and its number is written down. Finally, the two numbers are added together. How many different ways can a total of 12 be obtained?

Solution

To solve this problem, count the different ways that a total of 12 can be obtained using two numbers between 1 and 8.

$$\boxed{\text{First number}} \ + \ \boxed{\text{Second number}} \ = \ \boxed{12}$$

After considering the various possibilities, you can see that this equation can be solved in the following five ways.

First Number	Second Number
4	8
5	7
6	6
7	5
8	4

Thus, a total of 12 can be obtained in five different ways. ■

Solving counting problems can be tricky. Often, seemingly minor changes in the statement of a problem can affect the answer. For instance, compare the counting problem in the next example with that given in Example 2.

EXAMPLE 3 ■ Selecting Pairs of Numbers at Random

Eight pieces of paper are numbered from 1 to 8 and placed in a box. Two pieces of paper are drawn from the box, and the numbers on each piece of paper are written down and totaled. How many different ways can a total of 12 be obtained?

Solution

To solve this problem, count the different ways that a total of 12 can be obtained *using two different numbers* between 1 and 8.

First Number	Second Number
4	8
5	7
7	5
8	4

Thus, a total of 12 can be obtained in four different ways. ■

NOTE The difference between the counting problems in Examples 2 and 3 can be distinguished by saying that the random selection in Example 2 occurs **with replacement,** whereas the random selection in Example 3 occurs **without replacement,** which eliminates the possibility of choosing two 6's.

Counting Principles

The first three examples in this section discussed simple counting problems in which you could *list* each possible way that an event can occur. This is always the best way to solve a counting problem when it can be done. However, some events can occur in so many different ways that it is not feasible to write out the entire list. In such cases, you must rely on formulas and counting principles. The most important of these is called the **Fundamental Counting Principle.**

Fundamental Counting Principle	Let E_1 and E_2 be two events. The first event E_1 can occur in m_1 different ways. After E_1 has occurred, E_2 can occur in m_2 different ways. The number of ways that the two events can occur is $$m_1 \cdot m_2.$$

NOTE The Fundamental Counting Principle can be extended to three or more events. For instance, the number of ways that three events E_1, E_2, and E_3 can occur is $m_1 \cdot m_2 \cdot m_3$.

EXAMPLE 4 ■ Applying the Fundamental Counting Principle

How many different pairs of letters from the English alphabet are possible? (Disregard the difference between uppercase and lowercase letters.)

Solution

Additional problem: After completing Example 4, ask students to determine how many different pairs of letters from the English alphabet are possible if the two letters must be different. Answer: 650

This experiment has two events. The first event is the choice of the first letter, and the second event is the choice of the second letter. Because the English alphabet contains 26 letters, it follows that each event can occur in 26 ways.

Two-letter "words"

26   26

Thus, using the Fundamental Counting Principle, it follows that the number of two-letter words is

$26 \cdot 26 = 676.$ ■

EXAMPLE 5 ■ Applying the Fundamental Counting Principle

Telephone numbers in the United States have 10 digits. The first three make up the *area code,* and the next seven the *local telephone number.* How many different telephone numbers are possible within each of the regions shown in blue on the map in Figure 10.6? (A local telephone number cannot have 0 or 1 as its first or second digit.)

FIGURE 10.6

In 1990, each of the 24 regions (22 states, Puerto Rico, and the District of Columbia) shown in blue on the map had a single area code for all telephone numbers in the region.

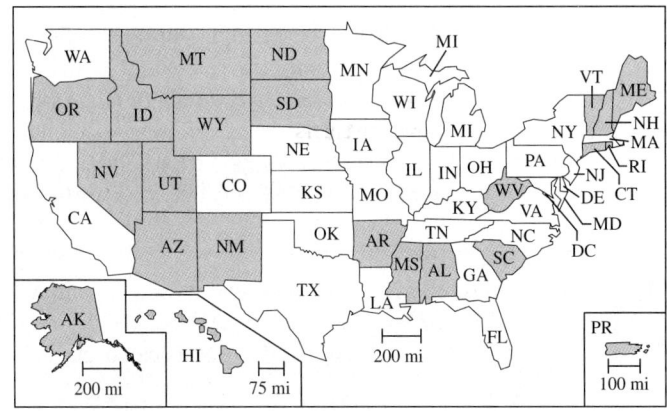

Solution

There are only eight choices for the first two digits, because neither can be 0 or 1. For each of the other digits, there are 10 choices.

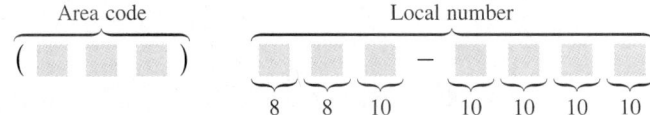

Thus, by the Fundamental Counting Principle, the number of local telephone numbers that are possible within each area code is

$$8 \cdot 8 \cdot 10 \cdot 10 \cdot 10 \cdot 10 \cdot 10 = 6{,}400{,}000.$$

Thus, each of the regions had, at most, 6.4 million phone numbers in 1990. ■

EXAMPLE 6 ■ Applying the Fundamental Counting Principle

Some states allow people to order special license plates for their cars. Suppose that the state allows no more than six letters for the license plate. How many different six-letter license plates are possible?

Additional problem: John has some ham, bologna, and salami to make a sandwich. There is some mayonnaise and mustard in the refrigerator and both white and brown bread in the pantry. If he chooses one from each group, how many different sandwiches can he make? Answer: 12

Solution

There are 26 letters that can be used for each of the six letters on the plate.

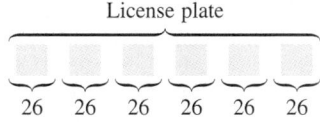

License plate

26   26   26   26   26   26

Thus, the number of different plates that are possible is

$$26 \cdot 26 \cdot 26 \cdot 26 \cdot 26 \cdot 26 = 26^6 = 308{,}915{,}776.$$ ■

Permutations

One important application of the Fundamental Counting Principle is determining the number of ways that n elements can be arranged (in order). An ordering of n elements is called a **permutation** of the elements.

Definition of Permutation	A **permutation** of n different elements is an ordering of the elements such that one element is first, one is second, one is third, and so on.

EXAMPLE 7   ■   Listing Permutations

Write the different permutations of the letters A, B, and C.

Solution

These three letters can be arranged in the following six different ways.

A, B, C   B, A, C   C, A, B
A, C, B   B, C, A   C, B, A

Thus, the three letters have six different permutations.   ■

In Example 7, you could list the different permutations of three letters. However, you could also have used the Fundamental Counting Principle. To do this, you could reason that there are three choices for the first letter, two choices for the second, and only one choice for the third, as follows.

Permutations of three letters

3   2   1

Thus, the number of permutations of three letters is

$$3 \cdot 2 \cdot 1 = 3! = 6.$$

EXAMPLE 8 ■ Finding the Number of Permutations of *n* Elements

How many different permutations are possible for the letters A, B, C, D, E, and F?

Solution

There are too many different permutations to list, so you can use the following reasoning.

First position: Any of the *six* letters.

Second position: Any of the remaining *five* letters.

Third position: Any of the remaining *four* letters.

Fourth position: Any of the remaining *three* letters.

Fifth position: Either of the remaining *two* letters.

Sixth position: The *one* remaining letter.

Thus, the number of choices for the six positions is as follows.

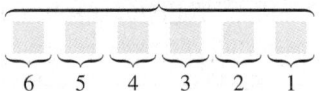

Permutations of six letters

6   5   4   3   2   1

Using the Fundamental Counting Principle, you can find that the total number of permutations of the six letters is

$$6 \cdot 5 \cdot 4 \cdot 3 \cdot 2 \cdot 1 = 6! = 720.$$ ■

The results obtained in Examples 7 and 8 can be generalized to conclude that the number of permutations of *n* different elements is *n*!.

Number of Permutations of *n* Elements

The number of permutations of *n* elements is

$$n \cdot (n - 1) \cdots 4 \cdot 3 \cdot 2 \cdot 1 = n!.$$

In other words, there are *n*! different ways that *n* elements can be ordered.

EXAMPLE 9 ■ Finding the Number of Permutations

Suppose that you are a supervisor for 11 employees. One of your responsibilities is to perform an annual evaluation for each employee, and then rank the 11 different performances. (In other words, one employee must be ranked as having the best performance, one must be ranked second, and so on.) How many different rankings are possible?

Solution

Because there are 11 different employees, you have 11 choices for first ranking. After choosing the first ranking, you can choose any of the remaining 10 for second ranking, and so on.

Rankings of 11 employees

11   10   9   8   7   6   5   4   3   2   1

Thus, the number of different rankings is

$$11! = 39,916,800.$$ ■

Combinations

Students often confuse a permutation with a combination. Help them see the difference using comparison and contrast. $\{A, B, C\}$ and $\{B, A, C\}$ are distinct as a permutation, but identical as a combination.

When counting the number of possible permutations of a set of elements, order is important. The final topic in this section describes a method of selecting subsets of a larger set in which order is *not important*. Such subsets are called **combinations of n elements taken m at a time.** For instance, the combinations

$$\{A, B, C\} \quad \text{and} \quad \{B, A, C\}$$

are equivalent because both sets contain the same three elements, and the order in which the elements are listed is *not important*. Hence, you would count only one of the two sets. A common example of how a combination occurs is a card game in which the player is free to reorder the cards after they have been dealt.

EXAMPLE 10   ■   Combination of n Elements Taken m at a Time

In how many different ways can three letters be chosen from the letters A, B, C, D, and E? (The order of the three letters is not important.)

Solution

The following subsets represent the different combinations of three letters that can be chosen from five letters.

$\{A, B, C\}$	$\{A, B, D\}$
$\{A, B, E\}$	$\{A, C, D\}$
$\{A, C, E\}$	$\{A, D, E\}$
$\{B, C, D\}$	$\{B, C, E\}$
$\{B, D, E\}$	$\{C, D, E\}$

From this list, you can conclude that there are 10 different ways that three letters can be chosen from five letters. ■

The formula for the number of combinations of n elements taken m at a time is as follows.

Number of Combinations of n Elements Taken m at a Time	The number of combinations of n elements taken m at a time is $$_nC_m = \frac{n!}{(n-m)!\,m!}.$$

Note that the formula for $_nC_m$ is the same one given for binomial coefficients. Suppose you wanted to use this formula to solve the counting problem in Example 10, in which you were asked to find the number of combinations of 5 elements taken 3 at a time. Thus, $n = 5$, $m = 3$, and the number of combinations is

Remind students of how some calculators will evaluate $_nC_m$.

$$_5C_3 = \frac{5!}{2!3!} = \frac{5 \cdot 4 \cdot 3}{3 \cdot 2 \cdot 1} = 10,$$

which is the same as the answer obtained in Example 10.

EXAMPLE 11   ■   Combinations of n Elements Taken m at a Time

A standard poker hand consists of five cards dealt from a deck of 52. How many different poker hands are possible? (After the cards are dealt, the player may reorder them, and therefore order is not important.)

Solution

Use the formula for the number of combinations of 52 elements taken five at a time, as follows.

$$_{52}C_5 = \frac{52!}{47!5!} = \frac{52 \cdot 51 \cdot 50 \cdot 49 \cdot 48}{5 \cdot 4 \cdot 3 \cdot 2 \cdot 1} = 2{,}598{,}960 \text{ different hands} \quad ■$$

DISCUSSION PROBLEM   ■   You Be the Instructor

Suppose you are teaching an algebra class and are writing a test for this chapter. Create two word problems that you think are appropriate for the test. One of the word problems should deal with permutations and the other should deal with combinations. (Assume that your students will have only 5 minutes to solve each problem.)   ■

Warm-Up

The following warm-up exercises involve skills that were covered in earlier sections. You will use these skills in the exercise set for this section.

In Exercises 1–4, evaluate $_nC_m$.

1. $_5C_2$ **2.** $_{10}C_3$ **3.** $_9C_6$ **4.** $_{20}C_{20}$

In Exercises 5–10, evaluate the quantity.

5. $8 \cdot 2^3 \cdot 3^2$ **6.** $6^2 \cdot 5 \cdot 4$ **7.** $\dfrac{30!}{28!}$

8. $\dfrac{9!}{2!(4!)(3!)}$ **9.** $\dfrac{20!}{16!\,4!}$ **10.** $\dfrac{2 \cdot 4 \cdot 6 \cdot 8}{2^4}$

10.5   EXERCISES

Random Selection In Exercises 1–10, determine the number of ways the specified event can occur when one or more marbles are selected from a bowl containing 20 marbles numbered 1 through 20.

1. An odd number is obtained when one marble is drawn.

2. An even number is obtained when one marble is drawn.

3. A prime number is obtained when one marble is drawn.

4. A number greater than 12 is obtained when one marble is drawn.

5. A number divisible by 3 is obtained when one marble is drawn.

6. A number divisible by 6 is obtained when one marble is drawn.

7. Two marbles are drawn one after the other, with the first being replaced before the second is drawn. The sum of the numbers is 8.

8. Two marbles are drawn one after the other, with the first being replaced before the second is drawn. The sum of the numbers is 15.

9. Two marbles are drawn one after the other without replacement of the first one drawn. The sum of the numbers is 8.

10. Two marbles are drawn one after the other without replacement of the first one drawn. The sum of the numbers is 15.

11. *Identification Numbers* In a statistical study, each participant was given an identification label consisting of a letter of the alphabet followed by a single digit. How many distinct identification labels can be made in this way?

12. *Identification Numbers* How many identification labels (see Exercise 11) can be made by one letter of the alphabet followed by a two-digit number?

13. *License Plates* How many distinct automobile license plates can be formed by using a four-digit number followed by two letters?

14. *Three-Digit Numbers* How many three-digit numbers can be formed in each situation? (Leading zeros are allowed.)
  (a) The hundreds digit cannot be 0.
  (b) No repetition of digits is allowed.
  (c) The number must be less than 400.

15. *Staffing Choices* A small grocery store needs to open another checkout line. Three people who can run the cash register are available and two people are available to bag groceries. How many different ways can the additional checkout line be staffed?

16. *Computer System* You are in the process of purchasing a new computer system. You must choose one of three monitors, one of two computers, and one of two keyboards. How many different configurations of the system are available to you?

17. *Toboggan Ride* Five people line up on a toboggan at the top of the hill. In how many ways can they be seated if only two of the five are willing to sit in the front seat?

18. *Taking a Trip* Four people are taking a long trip in a car. Two sit in the front seat and two in the back seat. Three of the people agree to share the driving. In how many different arrangements can the four people sit?

19. *Morse Code* In Morse Code, all characters are transmitted using a sequence of *dots* and *dashes*. How many different characters can be formed with a sequence of four symbols, each of which is a dot or a dash?

20. *Task Assignment* Four people are assigned to four different tasks. In how many ways can the assignments be made if one of the four is not qualified for the first task?

21. *Permutations* List all the permutations of the letters X, Y, and Z .

22. *Permutations* List all the permutations of the letters A, B, C, and D.

23. *Seating Arrangement* In how many ways can five children be seated in a single row of chairs?

24. *Seating Arrangement* In how many ways can six people be seated in a six-passenger car?

25. *Time Management Study* Eight steps must be performed to accomplish a certain task, and these steps can be performed in any order. Management wants to test each possible order to determine which is least time consuming. How many different orders will have to be tested?

26. *Time Management Study* Repeat Exercise 25 if one step must be done first.

27. *Number of Subsets* List all the subsets with two elements that can be formed from the set of letters {A, B, C, D, E, F}.

28. *Number of Subsets* List all the subsets with three elements that can be formed from the set of letters {A, B, C, D, E, F}.

29. *Committee Selection* Three students are selected from a class of 20 to form a fund-raising committee. In how many ways can the committee be formed?

30. *Committee Selection* In how many ways can a committee of five be formed from a group of 30 people?

31. *Menu Selection* A group of four people goes out to dinner at a restaurant. There are nine entrees on the menu and the four people decide that no two will order the same thing. How many ways can the four order from the nine entrees?

32. *Test Questions* A student is required to answer any nine questions from the 12 questions on an exam. In how many ways can the student select the nine questions?

33. *Relationships* The number of interpersonal relationships increases dramatically as the size of a group increases (see figure). Determine the number of different two-person relationships for each of the following group sizes.

  (a) 3   (b) 4   (c) 6   (d) 8   (e) 10   (f) 12

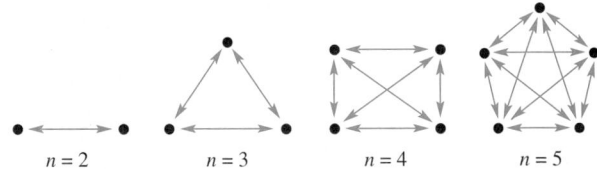

34. *Group Selection* Four people are to be selected from four couples. In how many ways can this be done if (a) there are no restrictions and (b) one person from each couple must be selected?

35. *Softball League* Six churches form a softball league. If each team must play every other team twice during the season, what is the total number of league games that will be played?

36. *Basketball Lineup* A high school basketball team has 15 players. In how many ways can the coach choose the starting lineup? (Assume that every player can play each position.)

37. *Pizza Toppings* A pizza shop offers nine toppings. How many different "three-topping pizzas" can be formed with the nine toppings? (Assume that no topping is used twice.)

38. *Geometry* Three points that are not on a line determine three lines. How many lines are determined by seven points, no three of which are on a line?

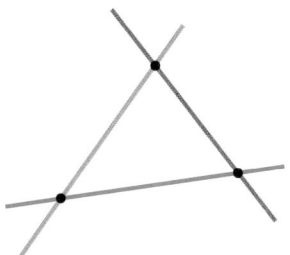

Diagonals of a Polygon In Exercises 39–42, find the number of diagonals of the polygon. (A line segment connecting any two nonadjacent vertices of a polygon is called a *diagonal* of the polygon.)

39. Pentagon

40. Hexagon

41. Octagon

42. Decagon (10 sides)

<table>
<tr><td>SECTION
10.6</td><td>**Probability**
The Probability of an Event   •   Using Counting Methods to Find Probabilities</td></tr>
</table>

The Probability of an Event

The **probability of an event** is a number from 0 to 1 that indicates the likelihood that the event will occur. An event that is certain to occur has a probability of 1. An event that *cannot* occur has a probability of 0. An event that is *equally likely* to occur or not occur has a probability of $\frac{1}{2}$, or 0.5. For instance, the probability that a phone number selected at random ends in an even digit is 0.5.

Stress that the probability of an event must be a number between 0 and 1, inclusive.

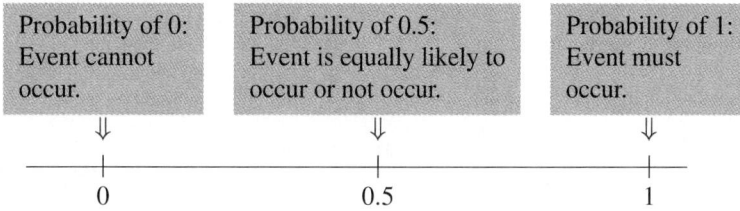

Probability of 0: Event cannot occur.	Probability of 0.5: Event is equally likely to occur or not occur.	Probability of 1: Event must occur.
⇓	⇓	⇓
0	0.5	1

| **The Probability of an Event** | Consider a set S, called a **sample space,** that is composed of a finite number of outcomes, each of which is equally likely to occur. A subset E of the sample space is an **event.** The probability P that an outcome in E will occur is the ratio of the number of outcomes in E to the number of outcomes in S.

$$P = \frac{\text{number of outcomes in event}}{\text{number of outcomes in sample space}}$$ |
|---|---|

EXAMPLE 1 ■ Finding the Probability of an Event

(a) You are dialing a friend's phone number, but can't remember the last digit. If you choose a digit at random, what is the probability that you will dial the correct number?

$$P = \frac{\text{number of correct digits}}{\text{number of possible digits}} = \frac{1}{10}$$

(b) On a multiple-choice test, you know that the answer to question 8 is not **a** or **d,** but you are not sure about **b, c,** and **e.** If you guess, what is the probability that you are *wrong*?

$$P = \frac{\text{number of wrong answers}}{\text{number of possible answers}} = \frac{2}{3}$$ ■

EXAMPLE 2 ■ Conducting a Poll

In 1990, the Centers for Disease Control took a survey of 11,631 high school students. The students were asked whether they considered themselves to be at a good weight, underweight, or overweight. The results of the survey are shown in Figure 10.7.

FIGURE 10.7

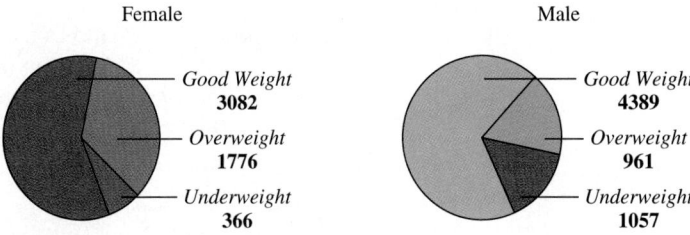

Female

Good Weight
3082

Overweight
1776

Underweight
366

Male

Good Weight
4389

Overweight
961

Underweight
1057

(a) If you choose a female at random from those surveyed, the probability that she said she was underweight is

$$P = \frac{\text{number of females who answered "underweight"}}{\text{number of females in survey}}$$

$$= \frac{366}{3082 + 366 + 1776}$$

$$= \frac{366}{5224}$$

$$\approx 0.07.$$

(b) If you choose a person who answered "underweight" from those surveyed, the probability that the person is female is

$$P = \frac{\text{number of females who answered "underweight"}}{\text{number in survey who answered "underweight"}}$$

$$= \frac{366}{366 + 1057}$$

$$= \frac{366}{1423}$$

$$\approx 0.26.$$ ∎

Polls such as the one described in Example 2 are often used to make inferences about a population that is larger than the sample. For instance, from Example 2 you might infer that 7% of *all* high school girls consider themselves to be underweight. When you make such an inference, it is important that those surveyed are representative of the entire population.

EXAMPLE 3 ∎ Using Area to Find Probability

You have just stepped into the tub to take a shower when one of your contact lenses falls out. (You have not yet turned on the shower.) Assuming that the lens is equally likely to land anywhere on the bottom of the tub, what is the probability that it lands in the drain? Use the dimensions shown in Figure 10.8 to answer the question.

Solution

Because the area of the tub bottom is $(26)(50) = 1300$ square inches and the area of the drain is $\pi(1^2) = \pi$ square inches, the probability that the lens lands in the drain is about

$$P = \frac{\pi}{1300} \approx 0.0024.$$

FIGURE 10.8

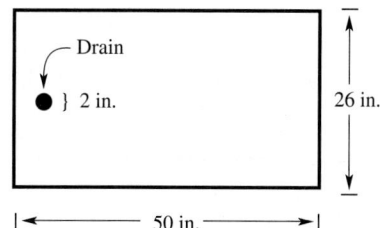

∎

EXAMPLE 4   ■   The Probability of Inheriting Certain Genes

Common parakeets have genes that can produce any one of four feather colors: green (BBCC, BBCc, BbCC, or BbCc), blue (BBcc or Bbcc), yellow (bbCC or bbCc), or white (bbcc). Use the *Punnett square* in Figure 10.9 to find the probability that an offspring of two green parents (both with BbCc feather genes) will be yellow. Note that each parent passes along a B or b gene and a C or c gene.

FIGURE 10.9

	BC	Bc	bC	bc
BC	BBCC	BBCc	BbCC	BbCc
Bc	BBCc	BBcc	BbCc	Bbcc
bC	BbCC	BbCc	bbCC	bbCc
bc	BbCc	Bbcc	bbCc	bbcc

Solution

The probability that an offspring will be yellow is

$$P = \frac{\text{number of yellow possibilities}}{\text{number of possibilities}}$$

$$= \frac{3}{16}.$$

Using Counting Methods to Find Probabilities

EXAMPLE 5   ■   The Probability of a Royal Flush

Five cards are dealt at random from a standard deck of 52 playing cards (see Figure 10.10). What is the probability that the cards are 10-J-Q-K-A of the same suit?

Solution

The number of five-card hands possible from a deck of 52 cards is given by the number of combinations of 52 elements taken 5 at a time.

$$_{52}C_5 = \frac{52!}{47! \cdot 5!}$$

$$= \frac{52 \cdot 51 \cdot 50 \cdot 49 \cdot 48}{5 \cdot 4 \cdot 3 \cdot 2 \cdot 1}$$

$$= 2,598,960$$

FIGURE 10.10

Standard 52-Card Deck

A ♠	A ♥	A ♦	A ♣
K ♠	K ♥	K ♦	K ♣
Q ♠	Q ♥	Q ♦	Q ♣
J ♠	J ♥	J ♦	J ♣
10 ♠	10 ♥	10 ♦	10 ♣
9 ♠	9 ♥	9 ♦	9 ♣
8 ♠	8 ♥	8 ♦	8 ♣
7 ♠	7 ♥	7 ♦	7 ♣
6 ♠	6 ♥	6 ♦	6 ♣
5 ♠	5 ♥	5 ♦	5 ♣
4 ♠	4 ♥	4 ♦	4 ♣
3 ♠	3 ♥	3 ♦	3 ♣
2 ♠	2 ♥	2 ♦	2 ♣

Because only four of these five-card hands are 10-J-Q-K-A of the same suit, the probability that the hand contains these cards is

$$P = \frac{4}{2,598,960}$$

$$= \frac{1}{649,740}.$$

EXAMPLE 6 ■ Conducting a Survey

In 1990, a survey was conducted of 500 adults who had worn Halloween costumes. Each person was asked how he or she acquired a Halloween costume: created it, rented it, bought it, or borrowed it. The results are shown in Figure 10.11. What is the probability that the first four people who were polled all created their costumes?

FIGURE 10.11

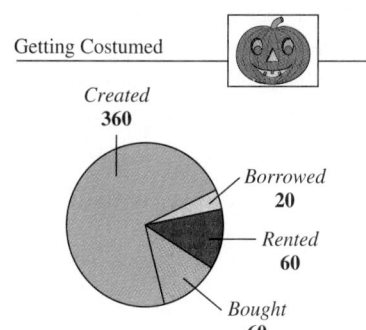

Getting Costumed

Created
360

Borrowed
20

Rented
60

Bought
60

Solution

To answer this question, you need to use the formula for the number of combinations *twice*. First, find the number of ways to choose four people from the 360 who created their own costumes.

$$_{360}C_4 = \frac{360 \cdot 359 \cdot 358 \cdot 357}{4 \cdot 3 \cdot 2 \cdot 1} = 688,235,310$$

Next, find the number of ways to choose four people from the 500 who were surveyed.

$$_{500}C_4 = \frac{500 \cdot 499 \cdot 498 \cdot 497}{4 \cdot 3 \cdot 2 \cdot 1} = 2,573,031,125$$

The probability that all of the first four people surveyed created their own costumes is the ratio of these two numbers.

$$P = \frac{\text{number of ways to choose 4 from 360}}{\text{number of ways to choose 4 from 500}}$$

$$= \frac{688,235,310}{2,573,031,125}$$

$$\approx 0.267$$

EXAMPLE 7 ■ Forming a Committee

To obtain input from 200 company employees, the management of a company selected a committee of five. Of the 200 employees, 56 were from minority groups. None of the 56, however, was selected to be on the committee. Does this indicate that the management's selection was biased?

Solution

Part of the solution is similar to that of Example 6. If the five committee members were selected at random, the probability that all five would be nonminority is

$$P = \frac{\text{number of ways to choose 5 from 144 nonminority employees}}{\text{number of ways to choose 5 from 200 employees}}$$

$$= \frac{_{144}C_5}{_{200}C_5}$$

$$= \frac{481{,}008{,}528}{2{,}535{,}650{,}040}$$

$$\approx 0.19.$$

Thus, if the committee were chosen at random (that is, without bias), the likelihood that it would have no minority members is about 0.19. Although this does not *prove* that there was bias, it does suggest a bias. ■

EXAMPLE 8 ■ Probability of Guessing Correctly

You are taking a chemistry test and are asked to arrange the first 10 elements *in the order* that they appear on the periodic table of elements. Suppose that you have no idea of the correct order and simply guess. What is the probability that you guess correctly?

Solution

You have 10 choices for the first element, nine choices for the second, eight choices for the third, and so on.

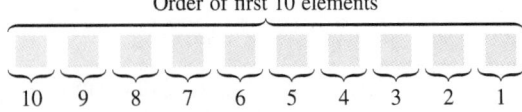

Order of first 10 elements

10 9 8 7 6 5 4 3 2 1

The number of different orders is $10! = 3{,}628{,}800$, which means that your probability of guessing correctly is

$$P = \frac{1}{3{,}628{,}800}.$$

■

DISCUSSION PROBLEM   ■   A Computer Experiment

A well-known experiment in probability is called the "Birthday Problem." This problem concerns the probability that in a group of n people at least two of the people will have the same birthday. The following BASIC program is a computer simulation of the birthday problem. Try running the program a few times with a computer that has the BASIC language. How many selections did the computer make each time before it found two people with the same birthday?

```
BASIC Program
10   RANDOMIZE 10*VAL(MID$(TIME$,4,2)) + VAL(RIGHT$(TIME$,2))
20   DIM DAYS (365)
30   DONE = 0: SELECT = 0
40   WHILE (DONE = 0)
50     BIRTHDAY = INT(365*RND) + 1
60     IF DAYS(BIRTHDAY)=1 THEN DONE=1 ELSE DAYS(BIRTHDAY)=1
70     IF DONE=0 THEN PRINT BIRTHDAY;: SELECT = SELECT + 1
80   WEND
90   PRINT BIRTHDAY
100  PRINT "SELECTION NUMBER" ;SELECT + 1; "IS A REPEAT."
110  END
```

Warm-Up

The following warm-up exercises involve skills that were covered in earlier sections. You will use these skills in the exercise set for this section.

In Exercises 1–10, evaluate the quantity.

1. $\frac{1}{6} + \frac{3}{4}$ **2.** $\frac{3}{8} + \frac{1}{4}$ **3.** $\frac{3}{5} + \frac{7}{10} - \frac{1}{15}$ **4.** $\frac{5}{9} + \frac{5}{12} - \frac{5}{16}$

5. $\dfrac{6 \cdot 5}{6!}$ **6.** $\dfrac{8 \cdot 7 \cdot 6}{8!}$ **7.** $\dfrac{_4C_2}{_8C_2}$ **8.** $\dfrac{_6C_3}{_{10}C_3}$

9. $\frac{7}{8} \cdot \frac{6}{7} \cdot \frac{5}{6}$ **10.** $1 - \left(\frac{2}{5}\right)^3$

10.6 EXERCISES

Sample Space In Exercises 1–4, determine the sample space for the experiment.

1. One letter from the alphabet is chosen.

2. A six-sided die is tossed twice and the sum is recorded.

3. Two county supervisors are selected from five supervisors, A, B, C, D, and E, to study a recycling plan.

4. A salesperson makes a presentation about a product in three homes per day. In each home, there may be a sale (denote by Y) or there may be no sale (denote by N).

Coin Tossing In Exercises 5–8, a coin is tossed three times. Find the probability of the specified event. Use the sample space

$$S = \{HHH, HHT, HTH, HTT, THH, THT, TTH, TTT\}.$$

 5. The event of getting 2 heads.

 6. The event of getting a tail on the second toss.

 7. The event of getting at least one head.

 8. The event of getting no more than two heads.

Playing Cards In Exercises 9–12, a card is drawn from a standard deck of playing cards. Find the probability of the specified event.

 9. The event of getting a red card.

10. The event of getting a queen.

11. The event of getting a face card.

12. The event of getting a black face card.

Tossing a Die In Exercises 13–16, a six-sided die is tossed. Find the probability of the specified event.

13. The event that the number that turns up is 5.

14. The event that the number that turns up is 7.

15. The event that the number that turns up is no more than 5.

16. The event that the number that turns up is at least 1.

17. *Multiple-Choice Test* A student takes a multiple-choice test in which there are five choices for each question. Find the probability that the first question is answered correctly given the following conditions.
    (a) The student has no idea of the answer and guesses at random.
    (b) The student can eliminate two of the choices and guesses from the remaining choices.
    (c) The student knows the answer.

18. *Multiple-Choice Test* A student takes a multiple-choice test in which there are four choices for each question. Find the probability that the first question is answered correctly given the following conditions.
    (a) The student has no idea of the answer and guesses at random.

    (b) The student can eliminate two of the choices and guesses from the remaining choices.
    (c) The student knows the answer.

19. *Random Selection* Twenty marbles numbered 1 through 20 are placed in a bag, and one is selected. Find the probability that the number selected is the following.
    (a) The number 12
    (b) A prime number
    (c) An odd number
    (d) A number less than 6

20. *Class Election* Three people are running for class president. From a small opinion poll, it is estimated that the probability that candidate A will win is 0.5 and the probability that candidate B will win is 0.3. What is the probability that the third candidate will win?

21. *Continuing Education* In a high school graduating class of 325 students, 255 are going to continue their education. What is the probability that a student selected at random from the class will not be furthering his or her education?

22. *Study Questions* An instructor gives the class a list of four study questions for the next exam. Two of the four study questions will be on the exam. Find the probability that a student who knows the material relating to three of the four questions will be able to answer both questions selected for the exam.

United States Blood Types In Exercises 23 and 24, use the circle graph in the figure which shows the number of people in the United States in 1990 with each blood type. (*Source:* American Association of Blood Banks)

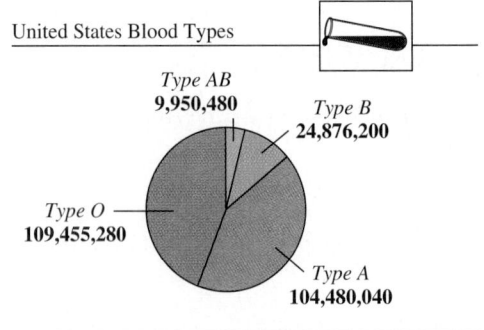

United States Blood Types

Type AB
9,950,480

Type B
24,876,200

Type O
109,455,280

Type A
104,480,040

23. A person is selected at random from the United States population. What is the probability that the person does *not* have blood type B?

24. What is the probability that a person selected at random from the United States population *does* have blood type B? How is this probability related to the probability found in Exercise 23?

25. *Meeting Time* You and a friend agree to meet at your favorite fast food restaurant between 5:00 and 6:00 P.M. The one who arrives first will wait 15 minutes for the other, after which that person will leave (see figure). What is the probability that the two of you actually meet, assuming that your arrival times are random within the hour?

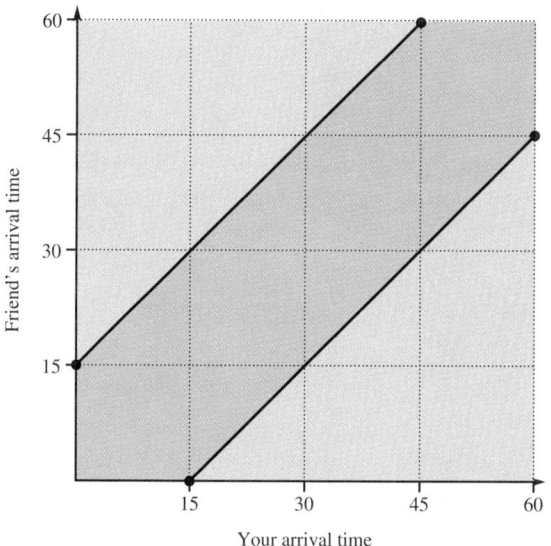

26. *Exploratory Exercise Estimating* π A coin of diameter d is dropped onto a paper that contains a grid of squares d units on a side (see figure).
 (a) Find the probability that the coin covers a vertex of one of the squares in the grid.
 (b) Repeat the experiment 100 times and use the results to approximate π.
 (It is best to lay the grid on a magazine so that the coin does not bounce when it is dropped from a height of about 3 feet.)

Figure for 26

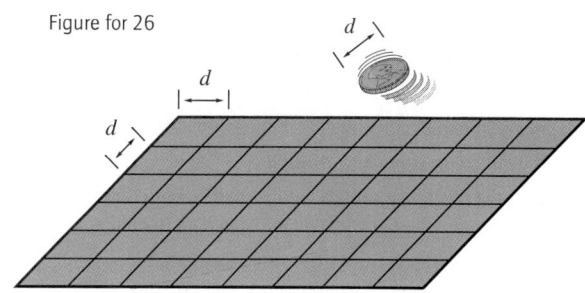

27. *Girl or Boy?* The genes that determine the sex of humans are denoted by XX (female) and XY (male). Complete the Punnett square below. Then use the result to explain why it is equally likely that a newborn baby will be a boy or a girl.

	Female	
	X	X
Male X	?	?
Male Y	?	?

28. *Blood Types* There are four basic human blood types: A (AA or Ao), B (BB or Bo), AB (AB), and O (oo). Complete the Punnett square below. What is the blood type of each parent? What is the probability that their offspring will have blood type A? B? AB? O?

	A	o
B	?	?
o	?	?

In Exercises 29–34, the sample spaces are large and it will be necessary to use the counting principles discussed in Section 10.5.

29. *Game Show* On a game show, you are given five digits to arrange in the proper order to give the price of a car. If you arrange them correctly, you win the car. Find the probability of winning if you know the correct position of only one digit and must guess at the digits in all the other positions.

30. *Committee Selection* A committee of three students is to be selected from a group of three women and five men. Find the probability that the committee will be composed entirely of women.

31. *Defective Units* A shipment of 10 food processors to a certain store contained two defective units. If you purchase two of these 10 food processors as birthday gifts for friends, determine the probability that you get both defective units.

32. *Book Selection* Four books are selected at random from a shelf containing six novels and four autobiographies. Find the probability that the four autobiographies are selected.

33. *Card Selection* Five cards are selected from a standard deck of 52 cards. Find the probability that you obtain the four aces.

34. *Card Selection* Five cards are selected from a standard deck of 52 cards. Find the probability of getting all hearts.

35. If the probability of an event occurring is p, what is the probability that the event does not occur?

36. What is the sum of the probabilities of all elements in a sample space?

Using a Random Number Generator

Computer programming languages (such as BASIC) and graphing calculators are capable of generating collections of random numbers that can be used to perform probability experiments.

EXAMPLE 1 ■ Using a Random Number Generator

The following BASIC program will generate 10 randomly chosen integers from 0 to 9, inclusive.

```
10 RANDOMIZE
20 FOR I=1 TO 10
30 N=INT(10*RND)
40 PRINT N
50 NEXT
60 END
```

When we ran this program, the output was as follows. (If you run the program, the output should differ.)

0, 0, 3, 4, 8, 6, 4, 6, 2, 4

Each time the command "RND" is called by the program, it randomly chooses a number RND such that

$0 \leq \text{RND} < 1.$

By multiplying RND by 10, you obtain numbers such that

$0 \leq (10 * \text{RND}) < 10.$

Finally, the INT command converts the random number to an integer by dropping the decimal part of the number. For instance, $\text{INT}(2.367) = 2$. ■

Example 1 illustrates a very important feature of collections of random numbers. That is, a collection of randomly chosen numbers will not necessarily contain an "even distribution" of the numbers in the sample space. For instance, of the 10 integers chosen in Example 1, the number 4 was chosen three times and the numbers 1, 5, 7, and 9 were not chosen.

Thus, *over the short run,* a collection of randomly chosen numbers can differ quite substantially from those that would be predicted by probability. *Over the long run,* the collection should agree closely with those predicted by probability.

It is precisely this feature of randomly chosen numbers that is the basis of gambling. Because the "house" is dealing with large collections of numbers, it can predict its outcome with a fair degree of certainty. A "player's" outcome is much less certain because he or she is generally dealing with relatively small collections of random numbers.

EXAMPLE 2 ■ Using a Random Number Generator

The following BASIC program will generate and tally M randomly chosen integers between 0 and 9, inclusive.

```
10 RANDOMIZE
20 INPUT; M
30 DIM C(10)
40 FOR I=1 TO M
50 N=INT(10*RND)
60 C(N+1)=C(N+1)+1
70 NEXT
80 FOR N=0 TO 9
90 PRINT ``THE NUMBER'';N;``WAS CHOSEN'';C(N+1);``TIMES.''
100 NEXT
110 END
```

When we ran this program for $M = 100$, 1000, and 10,000, we obtained the results shown in the table below.

Number	0	1	2	3	4	5	6	7	8	9
Times Chosen (M=100)	11	6	9	12	14	9	14	9	10	6
Times Chosen (M=1000)	101	99	105	79	97	97	113	109	103	97
Times Chosen (M=10,000)	985	972	1018	1013	978	994	978	1022	1055	985

■

EXERCISES

1. In Example 2, how many times would you expect the number 1 to be chosen when $M = 100$? By what percentage did the actual number of times 1 was chosen differ from the expected number of times?

2. In Example 2, how many times would you expect the number 3 to be chosen when $M = 1000$? By what percentage did the actual number of times 3 was chosen differ from the expected number of times?

3. In Example 2, how many times would you expect the number 8 to be chosen when $M = 10,000$? By what percentage did the actual number of times 8 was chosen differ from the expected number of times?

4. What can you conclude from the results of Exercises 1, 2, and 3?

5. Enter the program shown in Example 2 in a computer. Then run the program with $M = 1000$. Describe the results.

6. If you have access to a graphing calculator, use it to write a random number generating program that is similar to that in Example 1. Then run the program to generate 20 random integers from 0 to 9, inclusive. Describe your results. (The following program is for the TI-81 and TI-82. Programs for other graphing calculators are similar.)

```
:Lbl 1
:IPart (10*Rand)→X
:Disp X
:Pause
:Goto 1
```

10 REVIEW EXERCISES

In Exercises 1–4, use sigma notation to write the sum. (Use k as the index of summation and begin with $k = 0$ or $k = 1$.)

1. $[5(1) - 3] + [5(2) - 3] + [5(3) - 3] + [5(4) - 3]$

2. $[90 - 10(1)] + [90 - 10(2)] + [90 - 10(3)] + [90 - 10(4)] + [90 - 10(5)]$

3. $\dfrac{1}{3(1)} + \dfrac{1}{3(2)} + \dfrac{1}{3(3)} + \dfrac{1}{3(4)} + \dfrac{1}{3(5)} + \dfrac{1}{3(6)}$

4. $\left(-\dfrac{1}{3}\right)^0 + \left(-\dfrac{1}{3}\right)^1 + \left(-\dfrac{1}{3}\right)^2 + \left(-\dfrac{1}{3}\right)^3 + \left(-\dfrac{1}{3}\right)^4$

In Exercises 5–8, simplify the ratio of factorials.

5. $\dfrac{20!}{18!}$

6. $\dfrac{50!}{53!}$

7. $\dfrac{n!}{(n-3)!}$

8. $\dfrac{(n-1)!}{(n+1)!}$

In Exercises 9–12, write the first five terms of the arithmetic sequence. (Assume that the sequence begins with $n = 1$.)

9. $a_n = 132 - 5n$

10. $a_n = 2n + 3$

11. $a_n = \frac{3}{4}n + \frac{1}{2}$

12. $a_n = -\frac{3}{5}n + 1$

In Exercises 13–16, write the first five terms of the geometric sequence.

13. $a_1 = 10, \quad r = 3$

14. $a_1 = 2, \quad r = -5$

15. $a_1 = 100, \quad r = -\frac{1}{2}$

16. $a_1 = 12, \quad r = \frac{1}{6}$

In Exercises 17–26, find a formula for the nth term of the sequence.

17. Arithmetic sequence: $a_1 = 10, \quad d = 4$

18. Arithmetic sequence: $a_1 = 32, \quad d = -2$

19. Arithmetic sequence: $a_1 = 1000, \quad a_2 = 950$

20. Arithmetic sequence: $a_1 = 12, \quad a_2 = 20$

21. Geometric sequence: $a_1 = 1, \quad r = -\frac{2}{3}$

22. Geometric sequence: $a_1 = 100, \quad r = 1.07$

23. Geometric sequence: $a_1 = 24, \quad a_2 = 48$

24. Geometric sequence: $a_1 = 16, \quad a_2 = -4$

25. Geometric sequence: $a_1 = 12, \quad a_4 = -\frac{3}{2}$

26. Geometric sequence: $a_2 = 1, \quad a_3 = \frac{1}{3}$

In Exercises 27–42, find the sum.

27. $\displaystyle\sum_{k=1}^{4} 7$

28. $\displaystyle\sum_{k=1}^{4} \dfrac{(-1)^k}{k}$

29. $\displaystyle\sum_{n=1}^{4} \left(\dfrac{1}{n} - \dfrac{1}{n+1}\right)$

30. $\displaystyle\sum_{n=1}^{4} \left(\dfrac{1}{n} - \dfrac{1}{n+2}\right)$

31. $\displaystyle\sum_{k=1}^{12} (7k - 5)$

32. $\displaystyle\sum_{k=1}^{10} (100 - 10k)$

33. $\displaystyle\sum_{j=1}^{100} \dfrac{j}{4}$

34. $\displaystyle\sum_{j=1}^{50} \dfrac{3j}{2}$

35. $\displaystyle\sum_{n=1}^{12} 2^n$

36. $\displaystyle\sum_{n=1}^{12} (-2)^n$

37. $\displaystyle\sum_{k=1}^{8} 5\left(-\dfrac{3}{4}\right)^{k-1}$

38. $\displaystyle\sum_{k=1}^{10} 4\left(\dfrac{3}{2}\right)^{k-1}$

39. $\displaystyle\sum_{i=1}^{8} (1.25)^{i-1}$

40. $\displaystyle\sum_{i=1}^{8} (-1.25)^{i-1}$

41. $\displaystyle\sum_{n=1}^{120} 500(1.01)^{n-1}$

42. $\displaystyle\sum_{n=1}^{40} 1000(1.1)^{n-1}$

43. Find the sum of the first 50 positive integers that are multiples of 4.

44. Find the sum of the integers from 225 to 300.

45. *Auditorium Seating* Each row in a small auditorium has three more seats than the preceding row. Find the seating capacity of the auditorium if the front row seats 22 people and there are 12 rows of seats.

46. *Depreciation* A company pays $120,000 for a machine. During the next 5 years, the machine depreciates at the rate of 30% per year. (That is, at the end of each year, the depreciated value is 70% of what it was at the beginning of the year.)
(a) Find a formula for the nth term of the geometric sequence that gives the value of the machine n full years after it was purchased.
(b) Find the depreciated value of the machine at the end of 5 full years.

47. *Population Increase* A city of 85,000 people is growing at the rate of 1.2% per year. (That is, at the end of each year, the population is 1.012 times the population at the beginning of the year.)
(a) Find a formula for the nth term of the geometric sequence that gives the population n years from now.
(b) Estimate the population 50 years from now.

48. *Salary Increase* Suppose that you accept a job that pays a salary of $32,000 the first year. During the next 39 years, suppose you receive a 5.5% raise each year. What would your total salary be over the 40-year period?

In Exercises 49–54, use the Binomial Theorem and Pascal's Triangle to expand the expression. Simplify your answer.

49. $(x + 1)^{10}$ **50.** $(u - v)^9$

51. $(y - 2)^6$ **52.** $(x + 3)^5$

53. $\left(\frac{1}{2} - x\right)^8$ **54.** $(3x - 2y)^4$

55. *Morse Code* In Morse Code, all characters are transmitted using a sequence of *dots* and *dashes*. How many different characters can be formed by using a sequence of three dots and dashes? (These can be repeated. For example, dash-dot-dot represents the letter d.)

56. *Forming Line Segments* How many straight line segments can be formed by seven points such that no three are collinear?

57. *Committee Selection* Determine the number of ways a committee of five people can be formed from a group of 15 people.

58. *Program Listing* There are seven participants in a piano recital. In how many orders can their names be listed in the program?

59. *Rolling a Die* Find the probability of obtaining a number greater than 4 when a single six-sided die is rolled.

60. *Coin Tossing* Find the probability of obtaining at least one head when a coin is tossed four times.

61. *Book Selection* A child who does not know how to read carries a four-volume set of books to a bookshelf. Find the probability that the child will put the books on the shelf in the correct order.

62. *Rolling a Die* Are the chances of rolling a 3 with one six-sided die the same as those of rolling a total of 6 with two six-sided dice? If not, which has the greater probability of occurring?

63. *Hospital Inspection* As part of a monthly inspection at a hospital, the inspection team randomly selects reports from eight of the 84 nurses who are on duty. What is the probability that none of the reports selected will be from the 10 most experienced nurses on duty?

64. *Target Shooting* An archer shoots an arrow at the target shown in the figure. Suppose that the arrow is equally likely to hit any point on the target. What is the probability that the arrow hits the bull's eye? What is the probability that the arrow hits the blue ring?

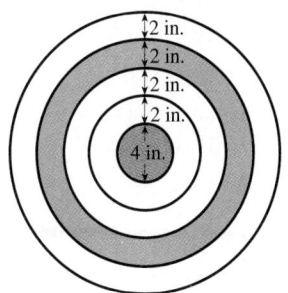

Chapter Test

Take this test as you would take a test in class. After you are done, check your work against the answers in the back of the book.

1. Write the first five terms of the infinite sequence $a_n = \left(-\frac{2}{3}\right)^{n-1}$. (Assume that the sequence starts with $n = 1$.)

2. Evaluate: $\displaystyle\sum_{j=0}^{4} (3j + 1)$

3. Use sigma notation to write the sum. (Use k as the index of summation and begin with $k = 0$ or $k = 1$.)
$$\frac{2}{3(1) + 1} + \frac{2}{3(2) + 1} + \cdots + \frac{2}{3(12) + 1}$$

4. Write the first five terms of the arithmetic sequence with first term $a_1 = 12$ and common difference $d = 4$.

5. Find a formula for the nth term of the arithmetic sequence with first term $a_1 = 5000$ and common difference $d = -100$.

6. Find the sum of the first 50 positive integers that are multiples of 3.

7. Find the common ratio of the geometric sequence $2, \ -3, \ \frac{9}{2}, \ -\frac{27}{4}, \ \frac{81}{8}, \ \ldots \ .$

8. Find a formula for the nth term of the geometric sequence with first term $a_1 = 4$ and common ratio $r = \frac{1}{2}$.

9. Find the sum of the first eight terms of the geometric sequence in which $a_1 = 3$ and $a_2 = 6$.

10. Find the balance in an increasing annuity when a principal of $50 is deposited at the beginning of each month for 25 years. Assume the money is compounded monthly at 8%.

11. Evaluate: $\ _{20}C_3$

12. Use the Binomial Theorem and Pascal's Triangle to expand $(x - 2)^5$.

13. Find the coefficient of the term $x^3 y^5$ in the expansion of $(x + y)^8$.

14. How many distinct license plates can be issued with one letter followed by a three-digit number?

15. Four students are randomly selected from a class of 25 to answer questions from a reading assignment. In how many ways can four students be selected? (Assume that the order in which the students are selected is not important.)

16. The weather report indicates that the probability of snow tomorrow is 0.75. What is the probability that it will not snow?

17. A card is drawn from a standard deck of 52 cards. Find the probability that it is a red face card.

18. Suppose that two spark plugs require replacement in a four-cylinder engine. If the mechanic randomly removes two plugs, find the probability that they are the two defective plugs.

Cumulative Test

Take this test as you would take a test in class. After you are done, check your work against the answers in the back of the book.

1. Solve by extracting square roots: $(x-5)^2 + 50 = 0$

2. Solve by completing the square: $3x^2 + 6x + 2 = 0$

3. ***Charitable Giving*** A group of n people decides to buy a \$36,000 minibus for a charitable organization. Each person will pay an equal share of the cost. If three additional people were to join the group, the cost per person would decrease by \$1000. Find n.

4. Sketch the graph of each of the equations.

   (a) $y = \dfrac{1}{2}(x^2 - 2x - 3)$ (b) $x^2 + y^2 = 8$ (c) $\dfrac{x^2}{1} + \dfrac{y^2}{4} = 1$ (d) $\dfrac{x^2}{1} - \dfrac{y^2}{4} = 1$

5. Find an equation of the parabola that has its vertex at $(3, -2)$ and that passes through the point $(0, 4)$.

6. ***Architecture*** A semicircular arch is positioned over the roadway onto the grounds of an estate. The roadway is 10 feet wide, and the arch is sitting on pillars that are 8 feet tall. Find the maximum height of a truck that can be driven onto the estate if its width is 8 feet.

7. Given the functions $f(x) = \sqrt{x}$ and $g(x) = 2x - 1$, find

   (a) $(f - g)(4)$ (b) $(fg)(9)$ (c) $\left(\dfrac{f}{g}\right)(3)$ (d) $(f \circ g)(5)$.

8. Find the inverse f^{-1} of the function $f(x) = 3x - 2$. Sketch the graphs of f and f^{-1} on the same set of coordinate axes.

9. ***Stopping Distance*** The stopping distance d of a car is directly proportional to the square of its speed s. On a certain type of pavement, a car required 50 feet to stop when its speed was 25 miles per hour. Estimate the stopping distance when the speed of the car is 40 miles per hour.

10. Solve the system of nonlinear equations and sketch a graph of the solution.

$$\begin{cases} x^2 - 4y = -1 \\ x^2 + y = 4 \end{cases}$$

11. Evaluate without using a calculator: $\log_4 \frac{1}{16}$

12. Sketch the graph of the functions $f(x) = e^x$ and $g(x) = \ln x$. What relationship exists between the functions f and g?

13. Use the properties of logarithms to condense $3(\log_2 x + \log_2 y) - \log_2 z$ into a logarithm of a single quantity.

14. Solve each of the equations. (Round your answer to three decimal places.)

(a) $\log_x \left(\frac{1}{9}\right) = -2$ (b) $4\ln x = 10$ (c) $500(1.08)^t = 2000$ (d) $3(1 + e^{2x}) = 20$

15. *Effective Yield* Determine the effective yield of an 8% interest rate compounded continuously.

16. *Investment Growth* Determine the length of time required for an investment of $1000 to quadruple in value if the investment earns 9% compounded continuously.

17. Find the sum of the first 20 terms of an arithmetic sequence if the first and second terms are $a_1 = 2$ and $a_2 = 6$.

18. Find the seventh term of a geometric sequence if $a_1 = 36$ and the common ratio is $r = -\frac{1}{3}$.

19. Find the coefficient of the term $x^4 y^3$ in the expansion of $(x - y)^7$.

20. *Task Assignment* Five people are assigned to five different tasks in an assembly plant. How many ways can the assignments be made if only two of the five people are qualified to do the first task?

21. *Committee Assignment* In how many ways can a committee of three be selected from a group of 10 people?

22. *Political Poll* Two people are running for political office. The news media report that the latest polls indicate that one candidate is leading the other by seven percentage points. What is the probability of each candidate winning if the poll is correct?

23. *Tossing a Die* Find the probability of obtaining a sum of at least nine when a die is tossed twice.

Introduction to Logic

A.1   Introduction to Logic
A.2   Implications, Quantifiers, and Venn Diagrams
A.3   Logical Arguments

Introduction to Logic
Statements • Truth Tables

Statements

In everyday speech and in mathematics we make inferences that adhere to common **laws of logic**. These laws (or methods of reasoning) allow us to build an algebra of statements by using logical operations to form compound statements from simpler ones. One of the primary goals of logic is to determine the truth value (true or false) of a compound statement knowing the truth value of its simpler component statements. For instance, we will learn that the compound statement "The temperature is below freezing and it is snowing" is true only if both component statements are true.

Definition of a Statement	1. A **statement** is a sentence to which only one truth value (either true or false) can be meaningfully assigned.
	2. An **open statement** is a sentence that contains one or more variables and becomes a statement when each variable is replaced by a specific item from a designated set.

NOTE: In the above definition, the word *statement* can be replaced by the word *proposition*.

EXAMPLE 1 ■ Statements, Nonstatements, and Open Statements

Statement	*Truth Value*
A square is a rectangle.	T
-3 is less than -5.	F

Nonstatement	*Truth Value*
Do your homework.	No truth value can be meaningfully assigned.
Did you call the police?	No truth value can be meaningfully assigned.

Open Statement	*Truth Value*
x is an irrational number.	We need a value of x.
She is a computer science major.	We need a specific person. ■

Symbolically, we represent statements by lowercase letters p, q, r, and so on. Statements can be changed or combined to form **compound statements** by means of the three logical operations **and**, **or**, and **not**, which we represent by $\wedge$ (and), $\vee$ (or), and $\sim$ (not). In logic we use the word *or* in the *inclusive* sense (meaning "and/or" in everyday language). That is, the statement "p or q" is true if p is true, q is true, or both p and q are true. The following list summarizes the terms and symbols used with these three operations of logic.

Operations of Logic	Operation	Verbal Statement	Symbolic Form	Name of Operation
	$\sim$	not p	$\sim p$	**Negation**
	$\wedge$	p and q	$p \wedge q$	**Conjunction**
	$\vee$	p or q	$p \vee q$	**Disjunction**

Compound statements can be formed using more than one logical operation, as demonstrated in Example 2.

EXAMPLE 2   ■   Forming Negations and Compound Statements

The statements p and q are as follows.

   p: The temperature is below freezing.
   q: It is snowing.

Write the verbal form for each of the following.

(a) $p \wedge q$ (b) $\sim p$

(c) $\sim(p \vee q)$ (d) $\sim p \wedge \sim q$

Solution

(a) The temperature is below freezing and it is snowing.

(b) The temperature is not below freezing.

(c) It is not true that the temperature is below freezing or it is snowing.

(d) The temperature is not below freezing and it is not snowing.   ■

EXAMPLE 3   ■   Forming Compound Statements

The statements p and q are as follows.

   p: The temperature is below freezing.
   q: It is snowing.

(a) Write the symbolic form for: *The temperature is not below freezing or it is not snowing.*

(b) Write the symbolic form for: *It is not true that the temperature is below freezing and it is snowing.*

Solution

(a) The symbolic form is: $\sim p \vee \sim q$

(b) The symbolic form is: $\sim(p \wedge q)$ ■

Truth Tables

To determine the truth value of a compound statement, we use charts called **truth tables**. The following tables represent the three basic logical operations.

TABLE 1 Negation

p	q	$\sim p$	$\sim q$
T	T	F	F
T	F	F	T
F	T	T	F
F	F	T	T

TABLE 2 Conjunction

p	q	$p \wedge q$
T	T	T
T	F	F
F	T	F
F	F	F

TABLE 3 Disjunction

p	q	$p \vee q$
T	T	T
T	F	T
F	T	T
F	F	F

For the sake of uniformity, all truth tables with two component statements will have T and F values for p and q assigned in the order shown in the first column of these three tables. Truth tables for several operations can be combined into one chart by using the same two first columns. For each operation a new column is added. Such an arrangement is especially useful with compound statements that involve more than one logical operation and for showing that two statements are logically equivalent.

Logical Equivalence

Two compound statements are **logically equivalent** if they have identical truth tables. Symbolically, we denote the equivalence of the statements p and q by writing $p \equiv q$.

EXAMPLE 4 ■ Logical Equivalence

Use a truth table to show the logical equivalence of the statements $\sim p \wedge \sim q$ and $\sim(p \vee q)$.

Solution

TABLE 4

p	q	$\sim p$	$\sim q$	$\sim p \wedge \sim q$	$p \vee q$	$\sim(p \vee q)$
T	T	F	F	F	T	F
T	F	F	T	F	T	F
F	T	T	F	F	T	F
F	F	T	T	T	F	T

⌐——— Identical ———⌐

Since the fifth and seventh columns in Table 4 are identical, the two given statements are logically equivalent. ■

The equivalence established in Example 4 is one of two well-known rules in logic called **DeMorgan's Laws**. Verification of the second of DeMorgan's Laws is left as an exercise.

| **DeMorgan's Laws** | 1. $\sim(p \vee q) \equiv \sim p \wedge \sim q$ |
| | 2. $\sim(p \wedge q) \equiv \sim p \vee \sim q$ |

Compound statements that are true, no matter what the truth values of component statements, are called **tautologies**. One simple example is the statement "p or not p," as shown in Table 5.

TABLE 5 $p \vee \sim p$ **is a tautology**

p	$\sim p$	$p \vee \sim p$
T	F	T
F	T	T

A.1   EXERCISES   See also Answers to Exercises.

In Exercises 1–12, classify the sentence as a statement, a nonstatement, or an open statement.

1. All dogs are brown.   Statement

2. Can I help you?   Nonstatement

3. That figure is a circle.   Open statement

4. Substitute 4 for x.   Nonstatement

5. x is larger than 4.   Open statement

6. 8 is larger than 4.   Statement

7. $x + y = 10$ Open statement

8. $12 + 3 = 14$ Statement

9. Hockey is fun to watch.   Nonstatement

10. One mile is greater than one kilometer.   Statement

11. It is more than one mile to the school.   Open statement

12. Come to the party.   Nonstatement

In Exercises 13–20, determine whether the open statement is true for the given values of x.

Open Statement   *Values of x*

13. $x^2 - 5x + 6 = 0$ (a) $x = 2$ True (b) $x = -2$ False

14. $x^2 - x - 6 = 0$ (a) $x = 2$ False (b) $x = -2$ True

15. $x^2 \le 4$ (a) $x = -2$ (b) $x = 0$ True
                          True

16. $|x - 3| = 4$ (a) $x = -1$ (b) $x = 7$ True
                          True

17. $4 - |x| = 2$ (a) $x = 0$ False (b) $x = 1$ False

18. $\sqrt{x^2} = x$ (a) $x = 3$ True (b) $x = -3$ False

19. $\dfrac{x}{x} = 1$ (a) $x = -4$ (b) $x = 0$ False
                              True

20. $\sqrt[3]{x} = -2$ (a) $x = 8$ False (b) $x = -8$ True

In Exercises 21–24, write the verbal form for each of the following.

(a) $\sim p$ (b) $\sim q$ (c) $p \wedge q$ (d) $p \vee q$

21. *p:* The sun is shining.   **22.** *p:* The car has a radio.
   q: It is hot.   *q:* The car is red.

23. *p:* Lions are mammals.   **24.** *p:* Twelve is less than fifteen.
   q: Lions are carnivorous.   *q:* Seven is a prime number.

In Exercises 25–28, write the verbal form for each of the following.

(a) $\sim p \wedge q$ (b) $\sim p \vee q$ (c) $p \wedge \sim q$ (d) $p \vee \sim q$

25. *p:* The sun is shining.   **26.** *p:* The car has a radio.
   q: It is hot.   *q:* The car is red.

27. *p:* Lions are mammals.   **28.** *p:* Twelve is less than fifteen.
   q: Lions are carnivorous.   *q:* Seven is a prime number.

In Exercises 29–32, write the symbolic form of the given compound statement. In each case let p represent the statement "It is four o'clock," and let q represent the statement "It is time to go home."

29. It is four o'clock and it is not time to go home. $p \wedge \sim q$ **30.** It is not four o'clock or it is not time to go home.
$\sim p \vee \sim q$

31. It is not four o'clock or it is time to go home. $\sim p \vee q$ **32.** It is four o'clock and it is time to go home.
$p \wedge q$

In Exercises 33–36, write the symbolic form of the given compound statement. In each case let p represent the statement "The dog has fleas," and let q represent the statement "The dog is scratching."

$\sim p \vee \sim q$

33. The dog does not have fleas or the dog is not scratching.

34. The dog has fleas and the dog is scratching. $p \wedge q$

35. The dog does not have fleas and the dog is scratching.

$\sim p \wedge q$

36. The dog has fleas or the dog is not scratching.

$p \vee \sim q$

In Exercises 37–42, write the negation of the given statement.

37. The bus is not blue.
  The bus is blue.

38. Frank is not six feet tall. Frank is six feet tall.

39. x is equal to 4.
  x is not equal to 4.

40. x is not equal to 4.
  x is equal to 4.

41. The Earth is not flat.
  The Earth is flat.

42. The Earth is flat.
  The Earth is not flat.

In Exercises 43–48, construct a truth table for the given compound statement.

43. $\sim p \wedge q$

44. $\sim p \vee q$

45. $\sim p \vee \sim q$

46. $\sim p \wedge \sim q$

47. $p \vee \sim q$

48. $p \wedge \sim q$

In Exercises 49–54, use a truth table to determine whether the given statements are logically equivalent.

49. $\sim p \wedge q, \quad p \vee \sim q$

50. $\sim (p \wedge \sim q), \quad \sim p \vee q$

51. $\sim (p \vee \sim q), \quad \sim p \wedge q$

52. $\sim (p \vee q), \quad \sim p \vee \sim q$

53. $p \wedge \sim q, \quad \sim (\sim p \vee q)$

54. $p \wedge \sim q, \quad \sim (\sim p \wedge q)$

43.

p	q	$\sim p$	$\sim p \wedge q$
T	T	F	F
T	F	F	F
F	T	T	T
F	F	T	F

44.

p	q	$\sim p$	$\sim p \vee q$
T	T	F	T
T	F	F	F
F	T	T	T
F	F	T	T

49. Not logically equivalent
51. Logically equivalent
53. Logically equivalent
50. Logically equivalent
52. Not logically equivalent
54. Not logically equivalent

In Exercises 55–58, determine whether the statements are logically equivalent.

55. (a) The house is red and it is not made of wood.
  (b) The house is red or it is not made of wood.

56. (a) It is not true that the tree is not green.
  (b) The tree is green.

57. (a) The statement that the house is white or blue is not true.
  (b) The house is not white and it is not blue.

58. (a) I am not twenty-five years old and I am not applying for this job.
  (b) The statement that I am twenty-five years old and applying for this job is not true.

In Exercises 59–62, use a truth table to determine whether the given statement is a tautology.

59. $\sim p \wedge p$

60. $\sim p \vee p$

61. $\sim (\sim p) \vee \sim p$

62. $\sim (\sim p) \wedge \sim p$

63. Use a truth table to verify the second of DeMorgan's Laws:

$$\sim (p \wedge q) \equiv \sim p \vee \sim q$$

55. Not logically equivalent
57. Logically equivalent
59. Not a tautology
61. A tautology
56. Logically equivalent
58. Not logically equivalent
60. A tautology
62. Not a tautology

63.
┌─Identical─┐

p	q	$\sim p$	$\sim q$	$p \wedge q$	$\sim (p \wedge q)$	$\sim p \vee \sim q$
T	T	F	F	T	F	F
T	F	F	T	F	T	T
F	T	T	F	F	T	T
F	F	T	T	F	T	T

45.

p	q	$\sim p$	$\sim q$	$\sim p \vee \sim q$
T	T	F	F	F
T	F	F	T	T
F	T	T	F	T
F	F	T	T	T

Implications, Quantifiers, and Venn Diagrams

Implications • Logical Quantifiers • Venn Diagrams

Implications

A statement of the form "If p, then q," is called an **implication** (or a conditional statement) and is denoted by

$$p \rightarrow q.$$

We call p the **hypothesis** and q the **conclusion**. There are many different ways to express the implication $p \rightarrow q$, as shown in the following list.

Different Ways of Stating Implications	The implication $p \rightarrow q$ has the following equivalent verbal forms.
	1. If p, then q. 4. q follows from p.
	2. p implies q. 5. q is necessary for p.
	3. p, only if q. 6. p is sufficient for q.

Normally, we think of the implication $p \rightarrow q$ as having a cause-and-effect relationship between the hypothesis p and the conclusion q. However, you should be careful not to confuse the truth value of the component statements with the truth value of the implication. The following truth table should help you keep this distinction in mind.

TABLE 6 Implication

p	q	$p \rightarrow q$
T	T	T
T	F	F
F	T	T
F	F	T

Note in Table 6 that the implication $p \rightarrow q$ is false only when p is true and q is false. This is like a promise. Suppose you promise a friend that "If the sun shines, I will take you fishing." The only way you can break your promise is for the sun to shine (p is true) and you do not take your friend fishing (q is false). If the sun doesn't shine (p is false), you have no obligation to go fishing, and hence, the promise cannot be broken.

EXAMPLE 1   ■   **Finding Truth Values of Implications**

Give the truth value of each implication.

(a) If 3 is odd, then 9 is odd.

(b) If 3 is odd, then 9 is even.

(c) If 3 is even, then 9 is odd.

(d) If 3 is even, then 9 is even.

Solution

	Hypothesis	*Conclusion*	*Implication*
(a)	T	T	T
(b)	T	F	F
(c)	F	T	T
(d)	F	F	T

■

The next example shows how to write an implication as a disjunction.

EXAMPLE 2   ■   **Identifying Equivalent Statements**

Use a truth table to show the logical equivalence of the following statements.

(a) If I get a raise, I will take my family on a vacation.

(b) I will not get a raise *or* I will take my family on a vacation.

Solution

We let p represent the statement "I will get a raise," and let q represent the statement "I will take my family on a vacation." Then, we can represent the statement in part (a) as $p \rightarrow q$ and the statement in part (b) as $\sim p \lor q$. The logical equivalence of these two statements is shown in the following truth table.

TABLE 7 $p \rightarrow q \equiv \sim p \lor q$

p	q	$\sim p$	$\sim p \lor q$	$p \rightarrow q$
T	T	F	T	T
T	F	F	F	F
F	T	T	T	T
F	F	T	T	T

└─ Identical ─┘

Because the fourth and fifth columns of the truth table are identical, we can conclude that the two statements $p \rightarrow q$ and $\sim p \lor q$ are equivalent.   ■

From Table 7 and the fact that $\sim(\sim p) \equiv p$, we can write the **negation of an implication**. That is, since $p \rightarrow q$ is equivalent to $\sim p \vee q$, it follows that the negation of $p \rightarrow q$ must be $\sim(\sim p \vee q)$, which by DeMorgan's Laws can be written as follows.

$$\sim(p \rightarrow q) \equiv p \wedge \sim q$$

For the implication $p \rightarrow q$, there are three important associated implications.

1. The **converse** of $p \rightarrow q$: $q \rightarrow p$

2. The **inverse** of $p \rightarrow q$: $\sim p \rightarrow \sim q$

3. The **contrapositive** of $p \rightarrow q$: $\sim q \rightarrow \sim p$

From Table 8 you can see that these four statements yield two pairs of logically equivalent implications.

TABLE 8

p	q	$\sim p$	$\sim q$	$p \rightarrow q$	$\sim q \rightarrow \sim p$	$q \rightarrow p$	$\sim p \rightarrow \sim q$
T	T	F	F	T	T	T	T
T	F	F	T	F	F	T	T
F	T	T	F	T	T	F	F
F	F	T	T	T	T	T	T

└─ Identical ─┘   └─ Identical ─┘

NOTE: The connective "$\rightarrow$" is used to determine the truth values in the last three columns of Table 8.

EXAMPLE 3   ■   Writing the Converse, Inverse, and Contrapositive

Write the converse, inverse, and contrapositive for the implication "If I get a B on my test, then I will pass the course."

Solution

(a) *Converse:* If I pass the course, then I got a B on my test.

(b) *Inverse:* If I do not get a B on my test, then I will not pass the course.

(c) *Contrapositive:* If I do not pass the course, then I did not get a B on my test. ■

NOTE: In Example 3 be sure you see that neither the converse nor the inverse are logically equivalent to the original implication. To see this, consider that the original implication simply states that if you get a B on your test, then you will pass the course. The converse is not true because knowing that you passed the course does not imply that you got a B on the test. After all, you might have gotten an A on the test!

A **biconditional statement**, denoted by $p \leftrightarrow q$, is the conjunction of the implications $p \rightarrow q$ and $q \rightarrow p$. We often write a biconditional statement as "p if and only if q," or in shorter form as "p iff q." A biconditional statement is true when both components are true and when both components are false, as shown in the following truth table.

TABLE 9   Biconditional Statement: p if and only if q

p	q	$p \rightarrow q$	$q \rightarrow p$	$p \leftrightarrow q$	$(p \rightarrow q) \wedge (q \rightarrow p)$
T	T	T	T	T	T
T	F	F	T	F	F
F	T	T	F	F	F
F	F	T	T	T	T

The following list summarizes some of the laws of logic that we have discussed up to this point.

Laws of Logic

1. For every statement p, either p is true or p is false.   Law of Excluded Middle

2. $\sim(\sim p) \equiv p$ Law of Double Negation

3. $\sim(p \vee q) \equiv \sim p \wedge \sim q$ DeMorgan's Law

4. $\sim(p \wedge q) \equiv \sim p \vee \sim q$ DeMorgan's Law

5. $p \rightarrow q \equiv \sim p \vee q$ Law of Implication

6. $p \rightarrow q \equiv \sim q \rightarrow \sim p$ Law of Contraposition

Logical Quantifiers

Logical quantifiers are words such as *some*, *all*, *every*, *each*, *one*, and *none*. Here are some examples of statements with quantifiers.

Some isosceles triangles are right triangles.

Every painting on display is for sale.

Not all corporations have male chief executive officers.

All squares are parallelograms.

Being able to recognize the negation of a statement involving a quantifier is one of the most important skills in logic. For instance, consider the statement "All dogs are brown." In order for this statement to be false, we do not have to show that *all* dogs are not brown, we must simply find at least one dog that is not brown. Thus, the negation of the statement is "Some dogs are brown."

Next we list some of the more common negations involving quantifiers.

Negating Statements with Quantifiers

Statement	*Negation*
1. All *p* are *q*.	Some *p* are not *q*.
2. Some *p* are *q*.	No *p* is *q*.
3. Some *p* are not *q*.	All *p* are *q*.
4. No *p* is *q*.	Some *p* are *q*.

When using logical quantifiers, the word *all* can be replaced by the words *each* or *every*. For instance, the following are equivalent.

All *p* are *q*.   Each *p* is *q*.   Every *p* is *q*.

Similarly, the word *some* can be replaced by the words *at least one*. For instance, the following are equivalent.

Some *p* are *q*.   At least one *p* is *q*.

EXAMPLE 4 ■ Negating Quantifying Statements

Write the negation of each of the following.

(a) All students study.

(b) Not all prime numbers are odd.

(c) At least one mammal can fly.

(d) Some bananas are not yellow.

Solution

(a) Some students do not study.

(b) All prime numbers are odd.

(c) No mammals can fly.

(d) All bananas are yellow.

Venn Diagrams

Venn diagrams are figures that are used to show relationships between two or more sets of objects. They can help us interpret quantifying statements. Study the following Venn diagrams in which the circle marked A represents people over six feet tall and the circle marked B represents the basketball players.

1. All basketball players are over six feet tall.

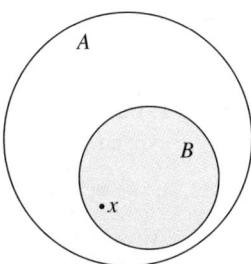

2. Some basketball players are over six feet tall.

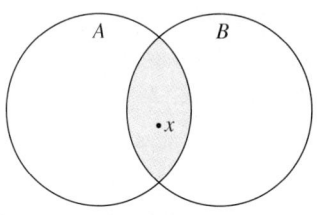

3. Some basketball players are not over six feet tall.

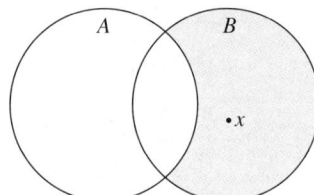

4. No basketball player is over six feet tall.

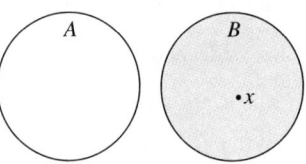

A.2   EXERCISES   See also Answers to Exercises.

In Exercises 1–4, write the verbal form for each of the following.

(a) $p \rightarrow q$ (b) $q \rightarrow p$ (c) $\sim q \rightarrow \sim p$ (d) $p \rightarrow \sim q$

1. p: The engine is running.
   q: The engine is wasting gasoline.

2. p: The student is at school.
   q: It is nine o'clock.

3. p: The integer is even.
   q: It is divisible by two.

4. p: The person is generous.
   q: The person is rich.

1. (a) If the engine is running, then the engine is wasting gasoline.

(b) If the engine is wasting gasoline, then the engine is running.

(c) If the engine is not wasting gasoline, then the engine is not running.

(d) If the engine is running, then the engine is not wasting gasoline.

In Exercises 5–10, write the symbolic form of the compound statement. Let p represent the statement "The economy is expanding," and let q represent the statement "Interest rates are low."

5. If interest rates are low, then the economy is expanding.
$q \rightarrow p$

6. If interest rates are not low, then the economy is not expanding. $\sim q \rightarrow \sim p$

7. An expanding economy implies low interest rates.
$p \rightarrow q$

8. Low interest rates are sufficient for an expanding economy. $q \rightarrow p$

9. Low interest rates are necessary for an expanding economy. $p \rightarrow q$

10. The economy will expand only if interest rates are low.
$p \rightarrow q$

In Exercises 11–20, give the truth value of the implication.

11. If 4 is even, then 12 is even. True

12. If 4 is even, then 2 is odd. False

13. If 4 is odd, then 3 is odd. True

14. If 4 is odd, then 2 is odd. True

15. If $2n$ is even, then $2n + 2$ is odd. False

16. If $2n + 1$ is even, then $2n + 2$ is odd. True

17. $3 + 11 > 16$ only if $2 + 3 = 5$. True

18. $\frac{1}{6} < \frac{2}{3}$ is necessary for $\frac{1}{2} > 0$. True

19. $x = -2$ follows from $2x + 3 = x + 1$. True

20. If $2x = 224$, then $x = 10$. False

In Exercises 21–26, write the converse, inverse, and contrapositive of the statement.

21. If the sky is clear, then you can see the eclipse.

22. If the person is nearsighted, then he is ineligible for the job.

23. If taxes are raised, then the deficit will increase.

24. If wages are raised, then the company's profits will decrease.

25. It is necessary to have a birth certificate to apply for the visa.

26. The number is divisible by three only if the sum of its digits is divisible by three.

In Exercises 27–40, write the negation of the statement.

27. Paul is a junior or senior. Paul is not a junior and not a senior.

28. Jack is a senior and he plays varsity basketball.
Jack is not a senior or he does not play varsity basketball.

29. If the temperature increases, then the metal rod will expand. The temperature increases and the metal rod will not expand.

30. If the test fails, then the project will be halted.
The test fails and the project will not be halted.

31. We will go to the ocean only if the weather forecast is good. We will go to the ocean and the weather forecast is not good.

32. Completing the pass on this play is necessary if we are going to win the game. We are going to win the game and not complete the pass on this play.

33. Some students are in extracurricular activities.
No student is in an extracurricular activity.

34. Some odd integers are not prime numbers.
No odd integer is not a prime number.

35. All contact sports are dangerous.
Some contact sports are not dangerous.

36. All members must pay their dues prior to June 1.
Some members must not pay their dues prior to June 1.

37. No child is allowed at the concert.
Some children are allowed at the concert.

38. No contestant is over the age of twelve.
Some contestants are over the age of twelve.

39. At least one of the $20 bills is counterfeit.
No $20 bills are counterfeit.

40. At least one unit is defective.
No units are defective.

In Exercises 41–48, construct a truth table for the compound statement.

41. $\sim(p \rightarrow \sim q)$

42. $\sim q \rightarrow (p \rightarrow q)$

43. $\sim(q \rightarrow p) \wedge q$

44. $p \rightarrow (\sim p \vee q)$

45. $[(p \vee q) \wedge (\sim p)] \rightarrow q$

46. $[(p \rightarrow q) \wedge (\sim q)] \rightarrow p$

47. $(p \leftrightarrow \sim q) \rightarrow \sim p$

48. $(p \vee \sim q) \leftrightarrow (q \rightarrow \sim p)$

In Exercises 49–56, use a truth table to show the logical equivalence of the two statements.

49. $q \to p$ $\sim p \to \sim q$

50. $\sim p \to q$ $p \vee q$

51. $\sim(p \to q)$ $p \wedge \sim q$

52. $(p \vee q) \to q$ $p \to q$

53. $(p \to q) \vee \sim q$ $p \vee \sim p$

54. $q \to (\sim p \vee q)$ $q \vee \sim q$

55. $p \to (\sim p \wedge q)$ $\sim p$

56. $\sim(p \wedge q) \to \sim q$ $p \vee \sim q$

57. Select the statement that is logically equivalent to the statement "If a number is divisible by six, then it is divisible by two."   (c)
   (a) If a number is divisible by two, then it is divisible by six.
   (b) If a number is not divisible by six, then it is not divisible by two.
   (c) If a number is not divisible by two, then it is not divisible by six.
   (d) Some numbers are divisible by six and not divisible by two.

58. Select the statement that is logically equivalent to the statement "It is not true that Pam is a conservative and a Democrat."   (c)
   (a) Pam is a conservative and a Democrat.
   (b) Pam is not a conservative and not a Democrat.
   (c) Pam is not a conservative or she is not a Democrat.
   (d) If Pam is not a conservative, then she is a Democrat.

59. Select the statement that is *not* logically equivalent to the statement "Every citizen over the age of 18 has the right to vote."   (a)
   (a) Some citizens over the age of 18 have the right to vote.
   (b) Each citizen over the age of 18 has the right to vote.
   (c) All citizens over the age of 18 have the right to vote.
   (d) No citizen over the age of 18 can be restricted from voting.

60. Select the statement that is *not* logically equivalent to the statement "It is necessary to pay the registration fee to take the course."   (c)
   (a) If you take the course, then you must pay the registration fee.
   (b) If you do not pay the registration fee, then you cannot take the course.
   (c) If you pay the registration fee, then you may take the course.
   (d) You may take the course only if you pay the registration fee.

In Exercises 61–70, sketch a Venn diagram and shade the region that illustrates the given statement. Let A be a circle that represents people who are happy, and let B be a circle that represents college students.

61. All college students are happy.

62. All happy people are college students.

63. No college students are happy.

64. No happy people are college students.

65. Some college students are not happy.

66. Some happy people are not college students.

67. At least one college student is happy.

68. At least one happy person is not a college student.

69. Each college student is sad.

70. Each sad person is not a college student.

In Exercises 71–74, state whether the statement follows from the given Venn diagram. Assume that each area shown in the Venn diagram is non-empty. (*Note:* Use only the information given in the diagram. Do not be concerned with whether the statement is actually true or false.)

71. (a) All toads are green.
     (b) Some toads are green.

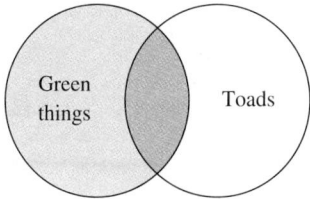

   (a) Statement does not follow.
   (b) Statement follows.

72. (a) All men are company presidents.
(b) Some company presidents are women.

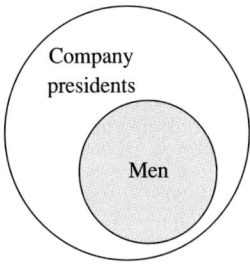

Company presidents

Men

(a) Statement follows.
(b) Statement follows.

73. (a) All blue cars are old.
(b) Some blue cars are not old.

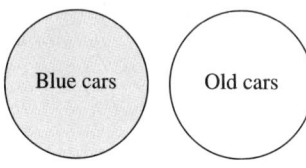

Blue cars   Old cars

(a) Statement does not follow.
(b) Statement does not follow.

74. (a) No football players are over six feet tall.
(b) Every football player is over six feet tall.

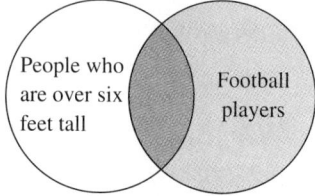

People who are over six feet tall   Football players

(a) Statement does not follow.
(b) Statement does not follow.

SECTION
A.3

Logical Arguments
Arguments • Venn Diagrams and Arguments • Proofs

Arguments

An **argument** is a collection of statements, listed in order. The last statement is called the **conclusion** and the other statements are called the **premises**. An argument is **valid** if the conjunction of all the premises implies the conclusion. The most common type of argument takes the following form.

Premise #1: $p \rightarrow q$
Premise #2: p
Conclusion: q

This form of argument is called the **Law of Detachment** or *Modus Ponens*. It is illustrated in the following example.

EXAMPLE 1   ■   A Valid Argument

Show that the following argument is valid.

Premise #1:   If Sean is a freshman, then he is taking algebra.
Premise #2:   Sean is a freshman.
Conclusion: Therefore, Sean is taking algebra.

Solution

We let p represent the statement "Sean is a freshman," and q represent the statement "Sean is taking algebra." Then the argument fits the Law of Detachment, which can be written as follows.

$$[(p \rightarrow q) \wedge p] \rightarrow q$$

The validity of this argument is shown in the following truth table.

TABLE 10   **Law of Detachment**

p	q	$p \rightarrow q$	$(p \rightarrow q) \wedge p$	$[(p \rightarrow q) \wedge p] \rightarrow q$
T	T	T	T	T
T	F	F	F	T
F	T	T	F	T
F	F	T	F	T

Keep in mind that the validity of an argument has nothing to do with the truthfulness of the premises or conclusion. For instance, the following argument is valid—the fact that it is fanciful does not alter its validity.

Premise #1:	If I snap my fingers, elephants will stay out of my house.
Premise #2:	I am snapping my fingers.
Conclusion:	Therefore, elephants will stay out of my house.

We have discussed the most common form of logical argument. This and three other commonly used forms of valid arguments are summarized in the following list.

Four Types of Valid Arguments	Name	Pattern		
	1. **Law of Detachment** or *Modus Ponens*	Premise #1:	$p \rightarrow q$	
		Premise #2:	p	
		Conclusion:	q	
	2. **Law of Contraposition** or *Modus Tollens*	Premise #1:	$p \rightarrow q$	
		Premise #2:	$\sim q$	
		Conclusion:	$\sim p$	
	3. **Law of Transitivity** or *Syllogism*	Premise #1:	$p \rightarrow q$	
		Premise #2:	$q \rightarrow r$	
		Conclusion:	$p \rightarrow r$	
	4. **Law of Disjunctive Syllogism**	Premise #1:	$p \vee q$	
		Premise #2:	$\sim p$	
		Conclusion:	q	

EXAMPLE 2   ■   An Invalid Argument

Determine whether the following argument is valid.

Premise #1:	If John is elected, the income tax will be increased.
Premise #2:	The income tax was increased.
Conclusion:	Therefore, John was elected.

Solution

This argument has the following form.

	Pattern	*Implication*
Premise #1:	$p \rightarrow q$	$[(p \rightarrow q) \wedge q] \rightarrow p$
Premise #2:	q	
Conclusion:	p	

This is not one of the four valid forms of arguments that we listed. We can construct a truth table to verify that the argument is invalid, as follows.

TABLE 11   **An Invalid Argument**

p	q	$p \rightarrow q$	$(p \rightarrow q) \wedge q$	$[(p \rightarrow q) \wedge q] \rightarrow p$
T	T	T	T	T
T	F	F	F	T
F	T	T	T	F
F	F	T	F	T

An invalid argument, like the one in Example 2, is called a **fallacy**. Other common fallacies are given in the following example.

EXAMPLE 3   ■   Common Fallacies

Each of the following arguments is invalid.

(a) *Arguing from the Converse:* If the football team wins the championship, then students will skip classes. The students skipped classes. Therefore, the football team won the championship.

(b) *Arguing from the Inverse:* If the football team wins the championship, then students will skip classes. The football team did not win the championship. Therefore, the students did not skip classes.

(c) *Arguing from False Authority:* Wheaties are best for you because Joe Montana eats them.

(d) *Arguing from an Example:* Beta Brand products are not reliable because my Beta Brand snowblower does not start in cold weather.

(e) *Arguing from Ambiguity:* If automobile carburetors are modified, the automobile will pollute. Brand X automobiles have modified carburetors. Therefore, Brand X automobiles pollute.

(f) *Arguing by False Association:* Joe was running through the alley when the fire alarm went off. Therefore, Joe started the fire. ■

EXAMPLE 4 ■ A Valid Argument

Determine whether the following argument is valid.

Premise #1:	You like strawberry pie or you like chocolate pie.
Premise #2:	You do not like strawberry pie.
Conclusion:	Therefore, you like chocolate pie.

Solution

This argument has the following form.

Premise #1:	$p \vee q$
Premise #2:	$\sim p$
Conclusion:	q

This argument is a disjunctive syllogism, which is one of the four common types of valid arguments. ■

In a valid argument, the conclusion drawn from the premise is called a **valid conclusion**.

EXAMPLE 5 ■ Making Valid Conclusions

Given the following two premises, which of the conclusions are valid?

Premise #1:	If you like boating, then you like swimming.
Premise #2:	If you like swimming, then you are a scholar.

(a) Conclusion: If you like boating, then you are a scholar.

(b) Conclusion: If you do not like boating, then you are not a scholar.

(c) Conclusion: If you are not a scholar, then you do not like boating.

Solution

(a) This conclusion is valid. It follows from the Law of Transitivity (or syllogism).

(b) This conclusion is invalid. The fallacy stems from arguing from the inverse.

(c) This conclusion is valid. It follows from the Law of Contraposition. ■

FIGURE A.1

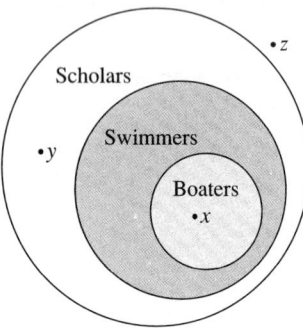

Venn Diagrams and Arguments

Venn diagrams can be used to informally test the validity of an argument. For instance, a Venn diagram for the premises in Example 5 is shown in Figure A.1. In this figure the validity of Conclusion (a) is seen by choosing a boater x in all three sets. Conclusion (b) is seen to be invalid by choosing a person y who is a scholar but does not like boating. Finally, person z indicates the validity of Conclusion (c).

Venn diagrams work well for testing arguments that involve quantifiers, as shown in the next two examples.

EXAMPLE 6 ■ **Using a Venn Diagram to Show that an Argument Is Not Valid**

Use a Venn diagram to test the validity of the following argument.

Premise #1:	Some plants are green.
Premise #2:	All lettuce is green.
Conclusion:	Therefore, lettuce is a plant.

Solution

FIGURE A.2

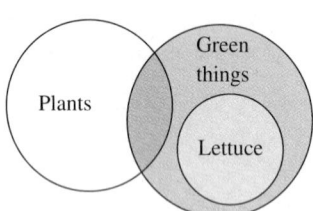

From the Venn diagram shown in Figure A.2, we can see that this is not a valid argument. Remember that even though the conclusion is true (lettuce is a plant), this does not imply that the argument is true. ■

NOTE: When you are using Venn diagrams, you must remember to draw the most general case. For example, in Figure A.2 the circle representing plants is not drawn entirely within the circle representing green things because we are told that only *some* plants are green.

EXAMPLE 7 ■ Using a Venn Diagram to Show that an Argument Is Valid

Use a Venn diagram to test the validity of the following argument.

Premise #1:	All good tennis players are physically fit.
Premise #2:	Some golfers are good tennis players.
Conclusion:	Therefore, some golfers are physically fit.

FIGURE A.3

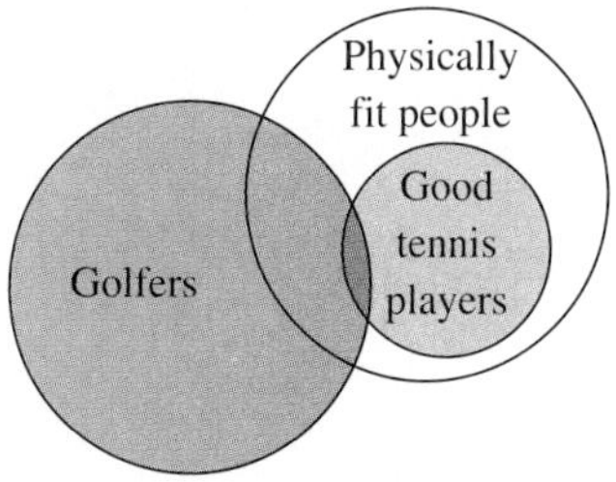

Solution

Since the set of golfers intersects the set of good tennis players, we see from Figure A.3 that the set of golfers must also intersect the set of physically fit people. Therefore, the argument is valid. ∎

Proofs

What does the word *proof* mean to you? In mathematics we use the word *proof* to simply mean a valid argument. Many proofs involve more than two premises and a conclusion. For instance, the proof in Example 8 involves three premises and a conclusion.

EXAMPLE 8   ∎   **A Proof by Contraposition**

Use the following three premises to prove that "It is not snowing today."

> Premise #1:   If it is snowing today, Greg will go skiing.
> Premise #2:   If Greg is skiing today, then he is not studying.
> Premise #3:   Greg is studying today.

Solution

We let p represent the statement "It is snowing today," let q represent "Greg is skiing," and let r represent "Greg is studying today." Thus, the given premises have the following form.

> Premise #1: $p \to q$
> Premise #2: $q \to \sim r$
> Premise #3: r

By noting that $r \equiv \sim(\sim r)$, reordering the premises, and writing the contrapositives of the first and second premises, we can obtain the following valid argument.

> Premise #3: r
> Contrapositive of Premise #2: $r \to \sim q$
> Contrapositive of Premise #1: $\sim q \to \sim p$
> Conclusion: $\sim p$

Thus, we can conclude $\sim p$. That is, "It is not snowing today." ∎

EXERCISES   See also Answers to Exercises.

In Exercises 1–4, use a truth table to show that the given argument is valid.

1. Premise #1: $p \to \sim q$
   Premise #2: q
   Conclusion: $\sim p$

2. Premise #1: $p \leftrightarrow q$
   Premise #2: p
   Conclusion: q

3. Premise #1: $p \lor q$
   Premise #2: $\sim p$
   Conclusion: q

4. Premise #1: $p \land q$
   Premise #2: $\sim p$
   Conclusion: q

In Exercises 5–8, use a truth table to show that the given argument is invalid.

5. Premise #1: $\sim p \to q$
   Premise #2: p
   Conclusion: $\sim q$

6. Premise #1: $p \to q$
   Premise #2: $\sim p$
   Conclusion: $\sim q$

7. Premise #1: $p \lor q$
   Premise #2: q
   Conclusion: p

8. Premise #1: $\sim(p \land q)$
   Premise #2: q
   Conclusion: p

In Exercises 9–22, determine whether the argument is valid or invalid.

9. Premise #1:   If taxes are increased, then businesses will leave the state.
   Premise #2:   Taxes are increased.
   Conclusion: Therefore, businesses will leave the state.   Valid

10. Premise #1:   If a student does the homework, then a good grade is certain.
   Premise #2:   Liza does the homework.
   Conclusion: Therefore, Liza will receive a good grade for the course.   Valid

11. Premise #1:   If taxes are increased, then businesses will leave the state.
   Premise #2:   Businesses are leaving the state.
   Conclusion: Therefore, taxes were increased.   Invalid

12. Premise #1:   If a student does the homework, then a good grade is certain.
   Premise #2:   Liza received a good grade for the course.
   Conclusion: Therefore, Liza did her homework.   Invalid

13. Premise #1:   If the doors are kept locked, then the car will not be stolen.
   Premise #2:   The car was stolen.
   Conclusion: Therefore, the car doors were unlocked.   Valid

14. Premise #1:   If Jan passes the exam, she is eligible for the position.
   Premise #2:   Jan is not eligible for the position.
   Conclusion: Therefore, Jan did not pass the exam.   Valid

15. Premise #1:   All cars manufactured by the Ford Motor Company are reliable.
     Premise #2:   Lincolns are manufactured by Ford.
     Conclusion: Therefore, Lincolns are reliable cars.   Valid

16. Premise #1:   Some cars manufactured by the Ford Motor Company are reliable.
     Premise #2:   Lincolns are manufactured by Ford.
     Conclusion: Therefore, Lincolns are reliable.   Invalid

17. Premise #1:   All federal income tax forms are subject to the Paperwork Reduction Act of 1980.
     Premise #2:   The 1040 Schedule A form is subject to the Paperwork Reduction Act of 1980.
     Conclusion: Therefore, the 1040 Schedule A form is a federal income tax form.   Invalid

18. Premise #1:   All integers divisible by six are divisible by three.
     Premise #2:   Eighteen is divisible by six.
     Conclusion: Therefore, eighteen is divisible by three.   Valid

19. Premise #1:   Eric is at the store or the handball court.
     Premise #2:   He is not at the store.
     Conclusion: Therefore, he must be at the handball court.   Valid

20. Premise #1:   The book must be returned within two weeks or you pay a fine.
     Premise #2:   The book was not returned within two weeks.
     Conclusion: Therefore, you must pay a fine.   Valid

21. Premise #1:   It is not true that it is a diamond and it sparkles in the sunlight.
     Premise #2:   It does sparkle in the sunlight.
     Conclusion: Therefore, it is a diamond.   Invalid

22. Premise #1:   Either I work tonight or I pass the mathematics test.
     Premise #2:   I'm going to work tonight.
     Conclusion: Therefore, I will fail the mathematics test.   Invalid

In Exercises 23–30, determine which conclusion is valid from the given premises.

23. Premise #1:   If seven is a prime number, then seven does not divide evenly into twenty-one.
     Premise #2:   Seven divides evenly into twenty-one.
     (a) Conclusion: Therefore, seven is a prime number.
     (b) Conclusion: Therefore, seven is not a prime number.   (b)
     (c) Conclusion: Therefore, twenty-one divided by seven is three.

24. Premise #1:   If the fuel is shut off, then the fire will be extinguished.
     Premise #2:   The fire continues to burn.
     (a) Conclusion: Therefore, the fuel was not shut off.
     (b) Conclusion: Therefore, the fuel was shut off.
     (c) Conclusion: Therefore, the fire becomes hotter.   (c)

25. Premise #1:   It is necessary that interest rates be lowered for the economy to improve.
    Premise #2:   Interest rates were not lowered.
    (a) Conclusion: Therefore, the economy will improve.
    (b) Conclusion: Therefore, interest rates are irrelevant to the performance of the economy.
    (c) Conclusion: Therefore, the economy will not improve.   (c)

26. Premise #1:   It will snow only if the temperature is below 32° at some level of the atmosphere.
    Premise #2:   It is snowing.
    (a) Conclusion: Therefore, the temperature is below 32° at ground level.
    (b) Conclusion: Therefore, the temperature is above 32° at some level of the atmosphere.
    (c) Conclusion: Therefore, the temperature is below 32° at some level of the atmosphere.   (c)

27. Premise #1:   Smokestack emissions must be reduced or acid rain will continue as an environmental problem.
    Premise #2:   Smokestack emissions have not decreased.
    (a) Conclusion: Therefore, the ozone layer will continue to be depleted.
    (b) Conclusion: Therefore, acid rain will continue as an environmental problem.   (b)
    (c) Conclusion: Therefore, stricter automobile emission standards must be enacted.

28. Premise #1:   The library must upgrade its computer system or service will not improve.
    Premise #2:   Service at the library has improved.
    (a) Conclusion: Therefore, the computer system was upgraded.   (a)
    (b) Conclusion: Therefore, more personnel were hired for the library.
    (c) Conclusion: Therefore, the computer system was not upgraded.

29. Premise #1:   If Rodney studies, then he will make good grades.
    Premise #2:   If he makes good grades, then he will get a good job.
    (a) Conclusion: Therefore, Rodney will get a good job.
    (b) Conclusion: Therefore, if Rodney doesn't study, then he won't get a good job.
    (c) Conclusion: Therefore, if Rodney doesn't get a good job, then he didn't study.   (c)

30. Premise #1:   It is necessary to have a ticket and an ID card to get into the arena.
    Premise #2:   Janice entered the arena.
    (a) Conclusion: Therefore, Janice does not have a ticket.
    (b) Conclusion: Therefore, Janice has a ticket and an ID card.   (b)
    (c) Conclusion: Therefore, Janice has an ID card.

In Exercises 31–34, use a Venn diagram to test the validity of the argument.

31. Premise #1:   All numbers divisible by ten are divisible by five.
    Premise #2:   Fifty is divisible by ten.
    Conclusion: Therefore, fifty is divisible by five.   Valid

32. Premise #1:   All human beings require adequate rest.
    Premise #2:   All infants are human beings.
    Conclusion: Therefore, all infants require adequate rest.   Valid

33. Premise #1:   No person under the age of eighteen is eligible to vote.
     Premise #2:   Some college students are eligible to vote.
     Conclusion: Therefore, some college students are under the age of eighteen.   Invalid

34. Premise #1:   Every amateur radio operator has a radio license.
     Premise #2:   Jackie has a radio license.
     Conclusion: Therefore, Jackie is an amateur radio operator.   Invalid

In Exercises 35–38, use the premises to prove the given conclusion.

35. Premise #1:   If Sue drives to work, then she will stop at the grocery store.
     Premise #2:   If she stops at the grocery store, then she'll buy milk.
     Premise #3:   Sue drove to work today.
     Conclusion: Therefore, Sue will get milk.

36. Premise #1:   If Bill is patient, then he will succeed.
     Premise #2:   Bill will get bonus pay if he succeeds.
     Premise #3:   Bill did not get bonus pay.
     Conclusion: Therefore, Bill is not patient.

37. Premise #1:   If this is a good product, then we should buy it.
     Premise #2:   Either it was made by XYZ Corporation, or we will not buy it.
     Premise #3:   It is not made by XYZ Corporation.
     Conclusion: Therefore, it is not a good product.

38. Premise #1:   If the book is returned within two weeks, then there is no fine.
     Premise #2:   You pay a fine or you may not check out another book.
     Premise #3:   You are allowed to check out another book.
     Conclusion: Therefore, the book was not returned within two weeks.

35. Let p be the statement "Sue drives to work," let q represent "She will stop at the grocery store," and let r represent "She'll buy milk."

First write:

    Premise #1: $p \rightarrow q$
    Premise #2: $q \rightarrow r$
    Premise #3: p

Reorder the premises:

    Premise #3: p
    Premise #1: $p \rightarrow q$
    Premise #2: $q \rightarrow r$
    Conclusion: r

Then we can conclude r. That is, "Sue will get milk."

36. Let p represent "Bill is patient," let q represent "He will succeed," let r represent "Bill will get bonus pay."

First write:

    Premise #1: $p \rightarrow q$
    Premise #2: $q \rightarrow r$
    Premise #3: $\sim r$

Conclusion from Premise #1, Premise #2: $p \rightarrow r$
Conclusion from $p \rightarrow r$ and Premise #3: $\sim p$

That is, "Bill is not patient."

38. Let p represent "The book is returned in two weeks," let q represent "There is a fine," and let r represent "You may not check out another book."

First write:

Premise #1: $p \rightarrow \sim q$
Premise #2: $q \lor r$
Premise #3: $\sim r$

Conclusion from Premise #2, Premise #3: q
Note $q \equiv \sim(\sim q)$
Conclusion from q, Premise #1: $\sim p$
That is, "The book was not returned within two weeks."

Answers to Odd-Numbered Exercises

Chapter 1

Chapter Opener   (page 1)

$1080°$, $135°$

SECTION 1.1   (page 13)

1. (a) 1, 4, 6   (b) -10, 0, 1, 4, 6
   (c) -10, $-\frac{2}{3}$, $-\frac{1}{4}$, 0, $\frac{5}{8}$, 1, 4, 6
   (d) $-\sqrt{5}$, $\sqrt{3}$, 2π

3. $2 < 5$

5. $-7 < -2$

7. $\frac{1}{3} > \frac{1}{4}$

9. $-\frac{5}{8} < \frac{1}{2}$

11. 6   13. 19   15. 8   17. 39   19. 225
21. -16 23. -85 25. -34, 34   27. $\frac{3}{11}$, $\frac{3}{11}$
29.

31.

33. $|3| < |-6|$ 35. $|-4| = |4|$
37. $-|-2| > -|-3|$ 39. $x < 0$ 41. $x \geq 0$
43. $p < \$225$ 45. 45   47. -19 49. -20
51. $\frac{5}{4}$ 53. $\frac{1}{10}$ 55. $\frac{105}{8}$ or $13\frac{1}{8}$ 57. 60
59. -28 61. -30 63. 32.13   65. $\frac{1}{2}$
67. $-\frac{1}{3}$ 69. 6   71. $-\frac{5}{2}$ 73. $\frac{46}{17}$
75. 4.03   77. $(4)(4)(4)$
79. $\left(-\frac{4}{5}\right)\left(-\frac{4}{5}\right)\left(-\frac{4}{5}\right)\left(-\frac{4}{5}\right)\left(-\frac{4}{5}\right)\left(-\frac{4}{5}\right)$ 81. $\left(\frac{5}{8}\right)^4$

83. $(-4)^6$ 85. 16   87. -1 89. -16
91. -64 93. $\frac{49}{64}$ 95. $-\frac{49}{64}$ 97. 9
99. 7   101. 5   103. -2 105. -1
107. 3   109. 36   111. 57   113. 135
115. -7.8 117. $\$2533.56$ 119. $\frac{17}{180}$
121.

Day	Daily Gain or Loss
Tuesday	$+\$5$
Wednesday	$+\$8$
Thursday	$-\$5$
Friday	$+\$16$

123. 15 m^2 125. 36 ft^2 127. 6.125 ft^3
129. True. Every integer p can be written in the form of the rational number $p/1$.
131. False. Given the integer 2, its reciprocal is the rational number $\frac{1}{2}$.
133. True. Any nonzero real number raised to an even power is positive.

USING A GRAPHING CALCULATOR   (page 20)

1. $(-4)^2$ 3. $(1+4)^3$ 5. 397.440
7. 4764.250   9. 15,379.913   11. -0.218
13. The calculator steps correspond to $0.05(1.24) + 2.36$ instead of $0.05(1.24 + 2.36)$.
   Correct Calculator Steps:
   $.05$ [×] [(] 1.24 [+] 2.36 [)] [ENTER]
   Display: .18

15. The calculator steps correspond to $126 + \frac{37}{4}$ instead of $\frac{126 + 37}{4}$.
   Correct Calculator Steps:
   [(] 126 [+] 37 [)] [÷] 4 [ENTER]
   Display: 40.75

SECTION 1.2 *(page 28)*

Warm-Up *(page 28)*

1. 225 **2.** −240 **3.** −150 **4.** 120
5. $\frac{8}{5}$ **6.** $-\frac{1}{4}$ **7.** $\frac{6}{5}$ **8.** $\frac{3}{10}$
9. −27 **10.** $\frac{25}{64}$

1. Commutative Property of Addition
3. Commutative Property of Multiplication
5. Associative Property of Multiplication
7. Multiplicative Identity Property
9. Associative Property of Addition
11. Distributive Property
13. Additive Identity Property
15. Additive Inverse Property
17. Multiplicative Inverse Property
19. Associative Property of Addition
21. Commutative Property of Multiplication
23. Distributive Property
25. $(3 \cdot 6)y$ **27.** $-3(15)$ **29.** $5 \cdot 6 + 5 \cdot z$ or $30 + 5z$
31. $-x + 25$ **33.** $x + 8$ **35.** (a) -10 (b) $\frac{1}{10}$
37. (a) 16 (b) $-\frac{1}{16}$ **39.** (a) $-6z$ (b) $\frac{1}{6z}$
41. (a) $-x - 1$ or $-(x+1)$ (b) $\frac{1}{x+1}$
43. $x + (5 + 3)$ **45.** $(32 + 4) + y$ **47.** $(3 \cdot 4)5$
49. $(6 \cdot 2)y$ **51.** $20 \cdot a + 20 \cdot 5$ or $20a + 100$
53. $5 \cdot 3x + 5 \cdot 4$ or $15x + 20$
55. $x \cdot (-2) + 6 \cdot (-2)$ or $-2x - 12$
57. Given
Addition Property of Equality
Associative Property of Addition
Additive Inverse Property
Additive Identity Property
59. Given
Addition Property of Equality
Associative Property of Addition
Additive Inverse Property
Additive Identity Property
Multiplication Property of Equality
Associative Property of Multiplication
Multiplicative Inverse Property
Multiplicative Identity Property
61. $3x + 15$ **63.** $-2x - 16$ **65.** 0
67. 28 **69.** 434 **71.** 62.82
73. $a(b + c) = ab + ac$ **75.** $9.5 billion

SECTION 1.3 *(page 40)*

Warm-Up *(page 40)*

1. 168 **2.** 20 **3.** $-\frac{21}{2}$ **4.** 7
5. $\frac{72}{5}$ **6.** $-\frac{19}{10}$
7. Multiplicative Inverse Property
8. Associative Property of Addition
9. Distributive Property
10. Additive Identity Property

1. $10x$, 5 **3.** $-3y^2$, $2y$, -8
5. $4x^2$, $-3y^2$, $-5x$, $2y$ **7.** x^2, $-2.5x$, $-\frac{1}{x}$
9. 5 **11.** $-\frac{3}{4}$
13. Commutative Property of Addition
15. Associative Property of Multiplication
17. Multiplicative Inverse Property
19. Distributive Property
21. Multiplicative Identity Property
23. (a) $5x + 5 \cdot 6$ or $5x + 30$ (b) $(x+6)5$
25. (a) $(xy)6$ (b) $(6x)y$
27. (a) 0 (b) $(-4t^2) + 4t^2$
29. $(x \cdot x \cdot x) \cdot (x \cdot x \cdot x \cdot x)$
31. $(-2x)(-2x)(-2x)$ **33.** $\left(\frac{y}{5}\right)\left(\frac{y}{5}\right)\left(\frac{y}{5}\right)\left(\frac{y}{5}\right)\left(\frac{y}{5}\right)$
35. $(5x)^4$ **37.** $x^3 y^3$ **39.** x^{12} **41.** $27y^6$
43. $16x^2$ **45.** $-125z^6$ **47.** $6x^3 y^4$
49. $-54u^5 v^3$ **51.** x^2 **53.** $81x^2$
55. $\frac{4}{3}xy$ **57.** $\frac{16}{3}x^2 y^2$ **59.** $-25y^6$
61. $-3125z^8$ **63.** $16a^4$ **65.** $8xy$ **67.** $\frac{2}{3}a^7 b$
69. $\frac{a^6}{27}$ **71.** $\frac{-4x^8}{25y^2}$ **73.** x **75.** x^{4n}
77. x^{n+4} **79.** rs^{m+3} **81.** $7x$ **83.** $8y$
85. $8x + 18y$ **87.** $\frac{11}{2}z^2 + \frac{3}{2}z + 10$
89. $4u^2 v^2 + uv$ **91.** $5a^2 b^2 - 2ab$
93. $8x^2 + 4x - 12$ **95.** $-18y^2 + 3y + 6$
97. $12x - 35$ **99.** $-3y^2 - 7y - 7$
101. $-2b^2 + 4b - 36$ **103.** $9x^3 - 5$
105. $2y^3 + y^2 + y$ **107.** $-x^2 y^2 - 2xy$
109. $-51a^7$ **111.** $4x - 14$
113. (a) 3 (b) -10 **115.** (a) 7 (b) 7
117. (a) 0 (b) $\frac{3}{10}$ **119.** (a) 13 (b) -36
121. (a) 0 (b) Undefined
123. (a) 210 (b) 140

125. (a) $-7, -5, -3, -1, 1, 3$
    (b) A two-unit increase
    (c) A $\frac{3}{4}$-unit increase
127. (a) \$2759.4009 million   (b) Appear the same
129. (a) No   (b) The bar graph   **131.** $\frac{57}{2}$

CHAPTER 1 MID-CHAPTER QUIZ   *(page 45)*

1. $-\frac{3}{2} > -\frac{5}{2}$ **2.** 50   **3.** -68 **4.** $\frac{1}{24}$
5. 71.03   **6.** $\frac{2}{5}$ **7.** $\frac{27}{8}$ **8.** -2
9. $(5x)^4$ **10.** Distributive Property
11. Multiplicative Identity Property
12. Additive Inverse Property
13. $\dfrac{1}{x+1}$ **14.** $5x^4, -3x^2, 2, \dfrac{-1}{x^2}$ **15.** $4x^5$
16. $16x^4$ **17.** $\dfrac{y^6}{27}$ **18.** $\dfrac{3}{2}xy^2$
19. $-x^2 + 2xy$ **20.** $-6x^3 - 3x$
21. (a) 225   (b) 5   **22.** $6x + 16$
23. Loss of \$498,833.36

SECTION 1.4   *(page 55)*

Warm-Up   *(page 54)*

1. $23y - 69$ **2.** $12 - 30z$ **3.** $-10 + 4x$
4. $-25x + 20$ **5.** $32x, -10y, -7$
6. $-3s, 4t, 1$ **7.** $7x - 15$ **8.** $3y + 36$
9. $-4u$ **10.** $-5v - 18$

---

1. $10x - 4, 1, 10$ **3.** $-3y^4 + 5, 4, -3$
5. $4t^5 - t^2 + 6t + 3, 5, 4$ **7.** $-4, 0, -4$
9. Binomial   **11.** Trinomial
13. Exponent in y^{-3} is not nonnegative.
15. ax^3 where $a \neq 0$ **17.** $8x^2 + 4$
19. (a) 16   (b) 0   (c) -16 (d) 16
21. (a) -27 (b) -16 (c) 0   (d) $\frac{9}{16}$
23. $7x^2 + 3$ **25.** $6x^2 - 7x + 8$ **27.** $-2x^2 + 15$
29. $2x^2 - 3x$ **31.** 4   **33.** $x^2 - 3x + 2$
35. $7x^3 + 2x$ **37.** $4x^3 + x + 4$ **39.** $x^2 - 2x + 5$
41. $-4x^3 - 2x + 13$ **43.** $-2y^4 - 5y + 4$
45. $2x^3 - 2x + 3$ **47.** $q^2 + 15$
49. $7x^3 + 22x^2 + 4$ **51.** $29s + 8$

53. $3t^2 + 29$ **55.** $3v^2 + 78v + 27$ **57.** $16a^3$
59. $10y - 2y^2$ **61.** $8x^5 - 12x^4 + 20x^3$
63. $-10x^2 - 6x^4 + 14x^5$ **65.** $x^2 + 3x - 28$
67. $2a^2 - 11a + 15$ **69.** $6x^2 + 7xy + 2y^2$
71. $48y^2 + 32y - 3$ **73.** $75x^3 + 30x^2$
75. $-a^2 + 19a$ **77.** $x^4 - 2x^3 - 3x^2 + 8x - 4$
79. $2u^3 + 13u^2 + 11u - 20$ **81.** $4x^4 - 4x^3 + 6x - 9$
83. $a^3 + 15a^2 + 75a + 125$ **85.** $x^3 + 27$
87. $t^4 - t^2 + 4t - 4$ **89.** $-6x^3 - 5x^2 + 4x - 3$
91. $x^2 - 4$ **93.** $a^2 - 36c^2$ **95.** $4x^2 - \frac{1}{16}$
97. $0.04t^2 - 0.25$ **99.** $x^2 + 10x + 25$
101. $25x^2 - 20x + 4$ **103.** $4a^2 + 12ab + 9b^2$
105. $x^2 - 2xy + y^2 + 4x - 4y + 4$
107. $u^2 - 2uv + 6u + v^2 - 6v + 9$ **109.** $-8x - 7$
111. $6y^2 - 32y + 36$ **113.** $12t$ **115.** $8x^2 + 26x$
117. (a) $A = (x + a)(x + b) = x^2 + ax + bx + ab$
    (b) Because both expressions represent the area,
        the two must be equal. Therefore,
        $(x + a)(x + b) = x^2 + bx + ax + ab$.
        This statement illustrates the FOIL Method.
119. $1000 + 2000r + 1000r^2$ **121.** Dropped, 100 ft
123. Thrown downward, 50 ft
125. 224 ft, 216 ft, 176 ft   **127.** $4.29x^2$
129. $-7.148a^2 + 15.691a$
131. (a) $y = 31.43 - 0.55t + 0.013t^2$
        1975 — 29.005 gal
        1985 — 26.105 gal
    (b) Decreasing

SECTION 1.5   *(page 67)*

Warm-Up   *(page 67)*

1. $24x^2 - 36x$ **2.** $28y - 21y^2$
3. $-60x + 42x^2$ **4.** $-2y^2 - 2y$ **5.** $2t$
6. $2z^2 + 3z - 35$ **7.** $121 - x^2$
8. $36r^2 - 25s^2$ **9.** $x^3 - 8$ **10.** $x^3 + 1$

---

1. 6   **3.** $3x$ **5.** $6z^2$ **7.** $14b^2$
9. $21(x + 8)^2$ **11.** $x(1 - z)^2$ **13.** $8(z - 1)$
15. $2(2u + 5)$ **17.** $6(4x^2 - 3)$ **19.** $x(2x + 1)$
21. $7u(3u - 2)$ **23.** $1(11u^2 + 9)$
25. $3y(x^2y - 5)$ **27.** $4(7x^2 + 4x - 2)$

29. $x^2(14x^2 + 21x + 9)$ **31.** $-5(x - 2)$
33. $-7(2x - 1)$ **35.** $-2(x^2 - 2x - 4)$
37. $-1(4t^2 - 2t + 15)$ **39.** $(y - 3)(2y + 5)$
41. $(t^2 + 1)(5t - 4)$ **43.** $a(a + 6)(1 - a)$
45. $(y - 6)(y + 2)$ **47.** $(x - 3)(x^2 + 2)$
49. $(x + 25)(x + 1)$ **51.** $(x + 2)(x^2 + 1)$
53. $(a - 4)(a^2 + 2)$ **55.** $(z + 3)(z^3 - 2)$
57. $(c - 3)(d + 3)$ **59.** $(x + 8)(x - 8)$
61. $(4y + 3z)(4y - 3z)$ **63.** $(x + 2y)(x - 2y)$
65. $(a^4 + 6)(a^4 - 6)$ **67.** $(10 - 3y)(10 + 3y)$
69. $(ab - 4)(ab + 4)$ **71.** $(a + 11)(a - 3)$
73. $(4 - z)(14 + z)$ **75.** $(x - 2)(x^2 + 2x + 4)$
77. $(y + 4z)(y^2 - 4yz + 16z^2)$
79. $(2t - 3)(4t^2 + 6t + 9)$ **81.** $2(2 - 5x)(2 + 5x)$
83. $(y + 3x)(y - 3x)(y^2 + 9x^2)$
85. $2(x - 3)(x^2 + 3x + 9)$ **87.** 6399
89. $(2x^n - 5)(2n^n + 5)$
91. $x^2(3x + 4) - (3x + 4) = (x - 1)(x + 1)(3x + 4)$
       $3x(x^2 - 1) + 4(x^2 - 1) = (x - 1)(x + 1)(3x + 4)$
93. $p = 800 - 0.25x$ **95.** $4x$ **97.** $P(1 + rt)$
99. $\pi(R - r)(R + r)$

SECTION 1.6   *(page 80)*

Warm-Up   *(page 80)*

1. $x^2 + 2x + 1$ **2.** $x^2 - 10x + 25$
3. $4 - 4y + y^2$ **4.** $6 + y - 2y^2$
5. $10z^2 - 29z - 21$ **6.** $2t^3 - 5t^2 - 12t$
7. $3x(x - 7)$ **8.** $-(x - 3)(x^2 + 1)$
9. $(2t - 13)(2t + 13)$ **10.** $(y - 4)(y^2 + 4y + 16)$

---

1. $(x + 2)^2$ **3.** $(a - 6)^2$ **5.** $(5y - 1)^2$
7. $(3b + 2)^2$ **9.** $(2x - y)^2$ **11.** $(u + 4v)^2$
13. $x + 1$ **15.** $y - 5$ **17.** $z - 2$
19. $(x + 3)(x + 1)$ **21.** $(x - 3)(x - 2)$
23. $(y + 10)(y - 3)$ **25.** $(t - 7)(t + 3)$
27. $(x - 12)(x - 8)$ **29.** $(x - 7y)(x + 5y)$
31. $5x + 3$ **33.** $5a - 3$ **35.** $2y - 9$
37. $(3x + 1)(x + 1)$ **39.** $(2x + 3)(4x - 3)$
41. $(3b - 1)(2b + 7)$ **43.** $(6x + y)(4x - 3y)$
45. Not factorable   **47.** $(2u - 5)(u + 7v)$
49. $(-1)(2x - 3)(x + 2)$ **51.** $(-1)(4x + 1)(15x - 1)$
53. $(3x + 4)(x + 2)$ **55.** $(2x - 1)(3x + 2)$

57. $(3x - 1)(5x - 2)$ **59.** $3x^3(x - 2)(x + 2)$
61. $2t(5t - 9)(t + 2)$ **63.** $(3x - 2)(7x - 2)$
65. $(3 - z)(9 + z)$ **67.** $2(3x - 1)(9x^2 + 3x + 1)$
69. $3a(3a - b)(3a + b)$ **71.** ± 18 **73.** ± 6
75. ± 12 **77.** 16   **79.** 9   **81.** 25
83. 14   **85.** $\pm 9, \pm 11, \pm 19$
87. $\pm 4, \pm 20$ **89.** $\pm 13, \pm 14, \pm 22, \pm 41$
91. $8, -16$ **93.** $2, -40$ **95.** $-5, -32$
97. 2704   **99.** c   **101.** a
103. $(x + 3)(x + 1)$

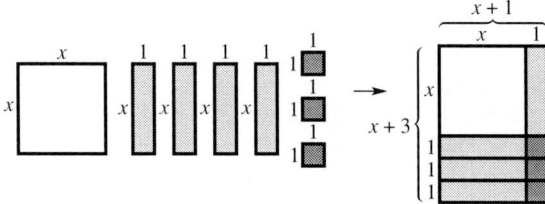

CHAPTER 1 REVIEW EXERCISES   *(page 84)*

1. $3 > -\frac{1}{8}$

3. $-\frac{8}{5} < -\frac{2}{5}$

5. 7.2   **7.** -7.2 **9.** 11   **11.** 230
13. -38 **15.** -4200 **17.** 14   **19.** 0
21. $\frac{11}{21}$ **23.** $\frac{1}{6}$ **25.** $\frac{17}{8}$ **27.** $-\frac{1}{20}$
29. 2   **31.** $\frac{6}{5}$ **33.** 5.25
35. -216 **37.** $\frac{1}{12}$ **39.** 20
41. Additive Inverse Property
43. Distributive Property
45. Associative Property of Addition
47. Commutative Property of Multiplication
49. Multiplicative Inverse Property
51. Associative Property of Addition
53. $4s - 8t$ **55.** $-3y^2 + 10y$ **57.** $u - 3v$
59. $5x - 10$ **61.** $5x - y$ **63.** $20u$
65. $18b - 15a$ **67.** x^6 **69.** $-64a^7$
71. $8u^2v^2$ **73.** $162a^{13}b^6$ **75.** (a) 0   (b) -3
77. (a) 15   (b) 3   **79.** $6x - x^2$
81. $-9x^3 + 9x - 4$ **83.** $6y^2 - 2y + 15$
85. $15x^2 - 11x - 12$ **87.** $4x^3 - 5x + 6$

89. $u^2 - 8u + 7$ **91.** $16x^2 - 56x + 49$
93. $25u^2 - 64$ **95.** $u^2 - v^2 - 6u + 9$
97. $3x^2(2 + 5x)$ **99.** $4a^2b^2(2a - 3b)$
101. $7(x + 5)(x + 9)$ **103.** $(3a - 10)(3a + 10)$
105. $(u - 3)(u + 15)$ **107.** $(x - 20)^2$
109. $(2s + 9t)^2$ **111.** $4a(1 - 4a)(1 + 4a)$
113. $4(2x - 3)(2x - 1)$ **115.** $(2x + 1)(4x^2 - 2x + 1)$
117. $(2u - 7)^2$ **119.** Not factorable
121. $t(t + 2)(4t - 1)$ **123.** ± 6
125. $\pm 1, \pm 7, \pm 13, \pm 29$ **127.** 55,578,932
129. \$644   **131.** 1   **133.** $10p^3 - 20p^4 + 10p^5$

CHAPTER 1 TEST   *(page 87)*

1. $-\frac{1}{2}$ **2.** $\frac{1}{6}$ **3.** $\frac{4}{27}$
4. $-\frac{27}{125}$ **5.** 11   **6.** 15
7. Associative Property of Multiplication
8. Multiplicative Inverse Property
9. $-5x$ **10.** $10x - 15$ **11.** $3x^4 y^3$
12. $-2x^2 + 5x - 1$ **13.** $-2y^2 - 2y$
14. $8x^2 - 4x + 10$ **15.** $11t + 7$ **16.** $\frac{y^3}{8}$
17. $2x^2 + 7xy - 15y^2$ **18.** $6s^3 - 17s^2 + 26s - 21$
19. $16x^2 - 24x + 9$ **20.** $16 - a^2 - 2ab - b^2$
21. $6y(3y - 2)$ **22.** $(x - 2)(5x^2 - 6)$
23. $(3u - 1)^2$ **24.** $2(3x + 2)(x - 5)$
25. 640 ft^3 **26.** $x^2 + 26x$

Chapter 2

Chapter Opener   *(page 88)*

$s \approx 342,119.44$ ft^2

SECTION 2.1   *(page 100)*

Warm-Up   *(page 99)*

1. $-2x$ **2.** $6x + 64$ **3.** $1 - 6x$
4. $-3x + 22$ **5.** $12y$ **6.** $0.02x + 100$
7. $2v(2v + 9)$ **8.** $(4x + 5)(4x - 5)$
9. $(x + 9)(x - 3)$ **10.** $-(4x - 3)(2x - 1)$

1. (a) No   (b) Yes   **3.** (a) No   (b) Yes
5. (a) Yes   (b) No   **7.** 15, 3, -5
9. 5, 7, $-2x$, $-\frac{7}{2}$ **11.** 4   **13.** $-\frac{2}{3}$
15. 2   **17.** $\frac{1}{3}$ **19.** 0
21. Not possible since $7 \neq 0$.   **23.** 11   **25.** -3
27. -3 **29.** $\frac{6}{5}$ **31.** 50   **33.** $\frac{19}{10}$
35. 0   **37.** $-\frac{10}{3}$ **39.** 72   **41.** 23
43. 0, 8   **45.** 3, -10 **47.** $-\frac{5}{2}, -\frac{1}{3}$
49. 5, -2 **51.** 4   **53.** -8
55. $\frac{5}{4}, -5$ **57.** ± 5 **59.** $-2, 6$
61. 0, $-\frac{5}{3}$ **63.** $-10, 2$ **65.** 5, -2
67. $-\frac{7}{2}, 5$ **69.** $-7, 0$ **71.** -2
73. -2 **75.** $\frac{25}{3}$ **77.** -20 **79.** 12
81. ± 4 **83.** $\frac{3}{2}$ **85.** $-\frac{1}{2}, 7$
87. $-6, 5$ **89.** 3.89   **91.** 30.28
93. (a)

t	1	1.5	2
Width	300	240	200
Length	300	360	400
Area	90,000	86,400	80,000

t	3	4	5
Width	150	120	100
Length	450	480	500
Area	67,500	57,600	50,000

(b) In a rectangle of fixed perimeter with length l equal to t times width w and $t \geq 1$, as t increases, w decreases, l increases, and the area A decreases. The maximum area occurs when the length and width are equal (when $t = 1$).

95. 6 hr   **97.** 15 ft $\times$ 22 ft
99. (a) $V = $ length $\cdot$ width $\cdot$ height
$V = x \cdot x \cdot 5 = 5x^2$

(b) 20, 80, 180, 320

(c) 22 in. $\times$ 22 in.

101. A linear equation in x is an equation that can be written in the form $ax + b = 0$, $a \neq 0$. A quadratic equation in x is an equation that can be written in the form
$ax^2 + bx + c = 0$, $a \neq 0$.

SECTION 2.2   *(page 111)*

Warm-Up   *(page 110)*

1. $\frac{22}{5}$ **2.** -14 **3.** 5   **4.** $\frac{4}{3}$

5. 3   **6.** 6   **7.** $-4, 2$ **8.** $-\frac{3}{2}$

9. $-12, 8$ **10.** $-4, 0$

1. (a) $x = \dfrac{6 + 3y}{2}$ (b) $y = \dfrac{2x - 6}{3}$

3. (a) $x = \dfrac{10y - 11}{7}$ (b) $y = \dfrac{7x + 11}{10}$

5. (a) $x = -6y$ (b) $y = \frac{1}{6}x$

7. (a) $x = \dfrac{10 - 2y}{5}$ (b) $y = \dfrac{10 - 5x}{2}$

19. $R = \dfrac{E}{I}$ **21.** $L = \dfrac{S}{1 - r}$

23. $b = \dfrac{2A - ah}{h}$ **25.** $r = \dfrac{A - P}{Pt}$

27. $n = \dfrac{S}{a_1 + a_n}$ **29.** $r = \dfrac{-a_1 + S}{S}$

31. 15 in.   **33.** 81 in.^{2} **35.** 2 ft

37. (a) $p = \dfrac{A}{2\pi w}$ (b) $\dfrac{11}{2\pi}$ cm

39. 7%   **41.** $6340.96   **43.** 2275 mi

45. $\dfrac{100}{11}$ hr   **47.** $\dfrac{2000}{3}$ ft/sec

49. $\frac{5}{17}$ hr   **51.** 12.3 mi/hr

53. 30 days   **55.** $x = 0,\ x = -\dfrac{b}{a}$

SECTION 2.3   *(page 124)*

Warm-Up   *(page 124)*

1. $-6 < 2$
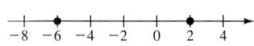

2. $\frac{11}{3} > 2$

3. $-3 > -7$
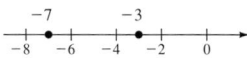

4. $-5 < 0$
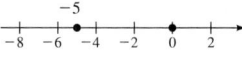

5. $\frac{13}{3} > -\frac{3}{2}$
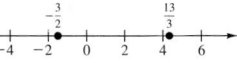

6. $\frac{3}{5} < \frac{13}{16}$
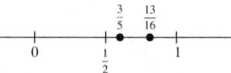

7. 3   **8.** 27   **9.** $\frac{33}{4}$ **10.** 3

1.
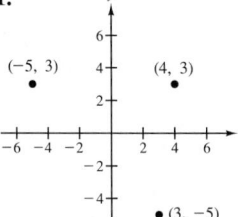

3.
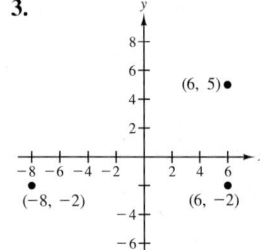

5.
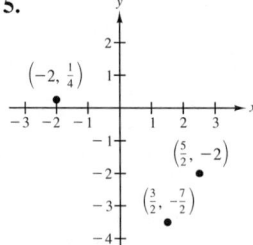

7. (a) Quadrant IV **(b)** Quadrant II

9. A: $(4, -2)$ B: $(-3, -2.5)$ C: $(3, 0.5)$

11.

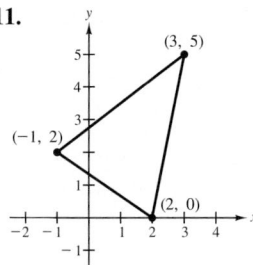

13.

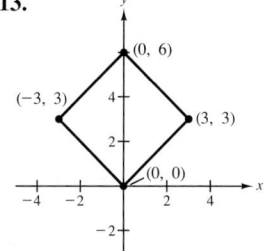

15.

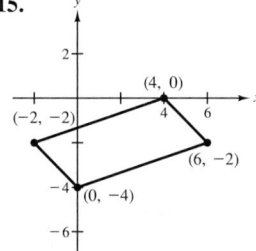

17. (a) Quadrant II, Quadrant III

(b) If $y > 0$, the point would be located to the *left* of the vertical axis and *above* the horizontal axis in the second quadrant. If $y < 0$, the point would be located to the *left* of the vertical axis and *below* the horizontal axis in the third quadrant. Note: If $y = 0$, the point would be located on the x-axis *between* the second and third quadrants.

19. (a) Quadrant I, Quadrant III

(b) The product xy is positive; therefore x and y have the same sign. If x and y are positive, the point would be located to the *right* of the vertical axis and *above* the horizontal axis in the first quadrant. If x and y are negative, the point would be located to the *left* of the vertical axis and *below* the horizontal axis in the third quadrant.

21. $(3, -4)$ **23.** $(-10, -10)$

25.

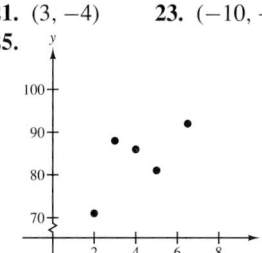

27.

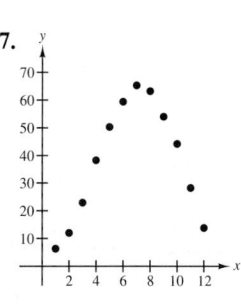

29. (a) Yes   **(b)** No   **(c)** No   **(d)** Yes

31. (a) No   **(b)** Yes   **(c)** Yes   **(d)** No

33. (a) Yes   **(b)** No   **(c)** Yes   **(d)** Yes

35.

x	-2	0	2	4	6
$y = 5x - 1$	-11	-1	9	19	29

37.

x	-2	-1	0	1	2
$y = x^2 - x + 3$	9	5	3	3	5

39.

x	100	150	200
$y = 28x + 3000$	5800	7200	8600

x	250	300
$y = 28x + 3000$	$10,000$	$11,400$

41. 7

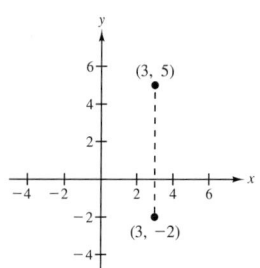

43. 7

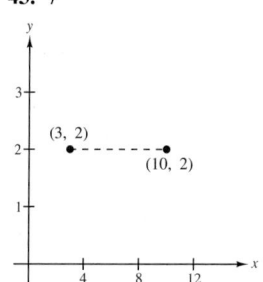

45. 11

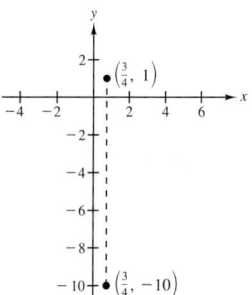

47. (a) $(x, y) = (10, 2)$ **(b)** 6, 8
(c) 10   **(d)** 10
49. (a) $(x, y) = (4, -4)$ **(b)** 8, 7
(c) $\sqrt{113}$ **(d)** $\sqrt{113}$
51. 5   **53.** 15   **55.** 7.81   **57.** 5.39
59. Yes   **61.** No   **63.** $2 + \sqrt{274} \approx 18.55$ ft
65. $(1, 4)$ **67.** $\left(\frac{7}{2}, \frac{9}{2}\right)$

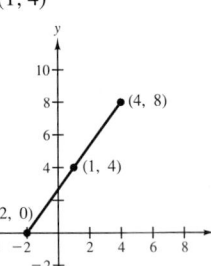

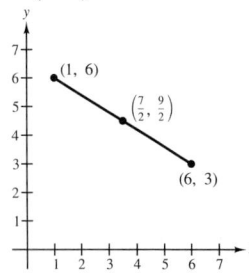

69. $(-1, 1)$, $(3, 2)$, $(0, 4)$
71. Reflection about the y-axis

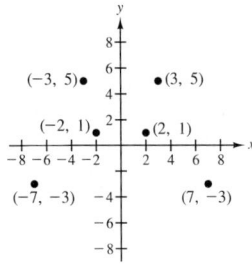

CHAPTER 2 MID-CHAPTER QUIZ   *(page 129)*

1. 8   **2.** 2   **3.** 2   **4.** 12
5. $-15, 16$ **6.** $-12, 6$ **7.** $\frac{1}{2}$
8. $-\frac{2}{3}, 2$ **9.** $C = \dfrac{5F - 160}{9}$ **10.** $l = \dfrac{P - 2w}{2}$

11. $h = \dfrac{S - 2\pi r^2}{2\pi r}$
$h = \dfrac{35 - 6\pi}{2\pi} \approx 2.57$ in.

12. 4 sec   **13.** $\frac{15}{2}$ hr

14. (a) Quadrant II, Quadrant IV
(b) The product xy is negative; therefore, x and y have opposite signs. If x is negative and y is positive, the point would be located to the *left* of the vertical axis and *above* the horizontal axis. Thus, the point would be in the second quadrant. If x is positive and y is negative, the point would be located to the *right* of the vertical axis and *below* the horizontal axis. Thus, the point would be in the fourth quadrant.

15.   **16.** 10.44

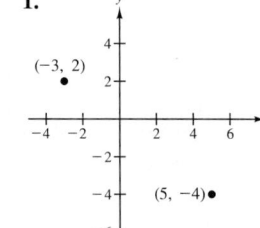

17. (a) Yes   **(b)** No   **(c)** No   **(d)** Yes

SECTION 2.4   *(page 143)*

Warm-Up   *(page 142)*

1.   **2.**

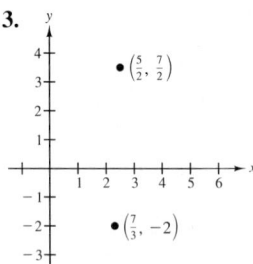

5. $\frac{48}{5}$ **6.** 6   **7.** 13   **8.** $-\frac{11}{3}$
9. $\frac{130}{19} \approx 6.84$ **10.** 7

1.

x	-4	-1	0
y	14	5	2
(x, y)	$(-4, 14)$	$(-1, 5)$	$(0, 2)$

x	2	4
y	-4	-10
(x, y)	$(2, -4)$	$(4, -10)$

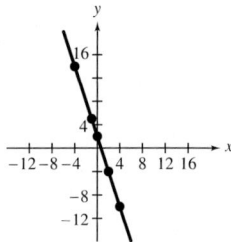

3.

x	± 2	-1	0
y	0	3	4
(x, y)	$(\pm 2, 0)$	$(-1, 3)$	$(0, 4)$

x	2	± 3
y	0	-5
(x, y)	$(2, 0)$	$(\pm 3, -5)$

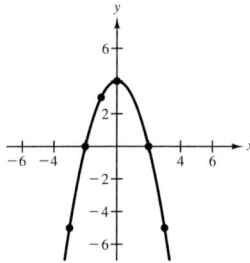

5.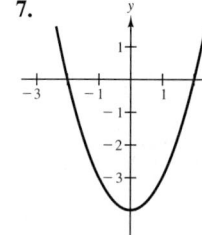

7.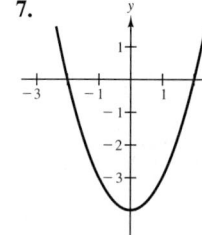

9.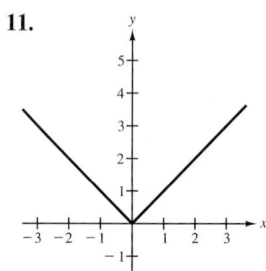

11.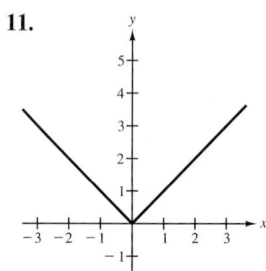

13. $(10, 0)$, $(0, 5)$ **15.** $(-20, 0)$, $(0, 15)$
17. $(\pm 5, 0)$, $(0, -25)$ **19.** $(0, 0)$
21. No x-intercept, $(0, 3)$ **23.** $\left(\frac{2}{3}, 0\right)$, $(-1, 0)$, $(0, -2)$

25.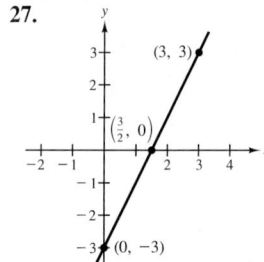

27.

29.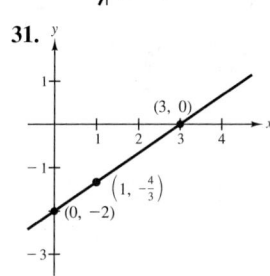

31.

33.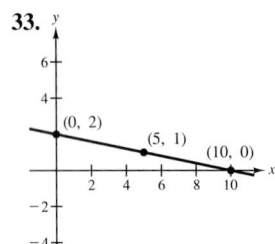

35.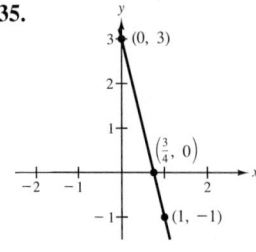

49. $x = \pm 2$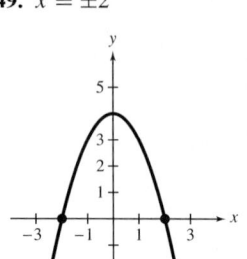

51. $x = 1$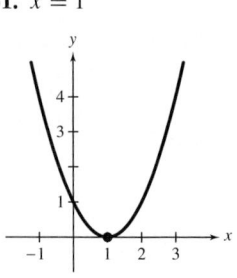

37.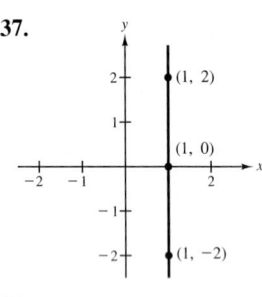

39.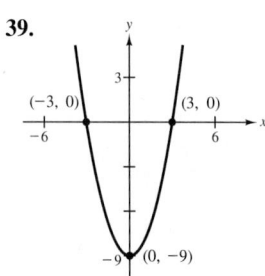

53.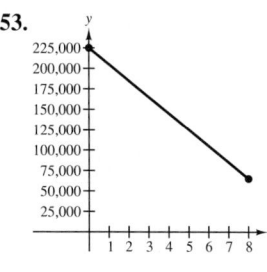

55. (b)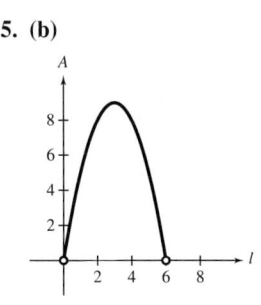

41.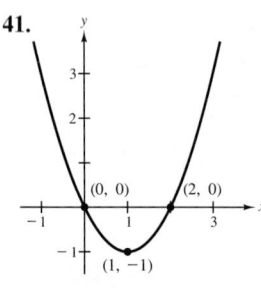

43.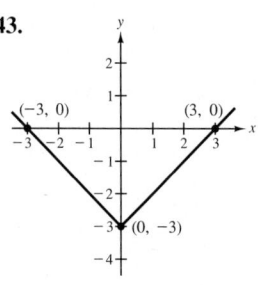

(c) 3 m × 3 m

57. (a)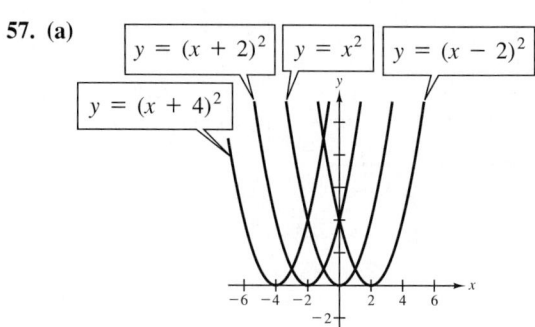

45. $x = 4$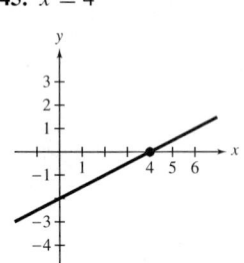

47. $x = \frac{9}{2}$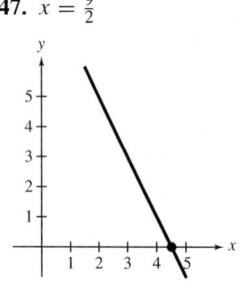

(b) The graph of $y = (x + c)^2$, $c > 0$, is obtained by shifting the graph of $y = x^2$ to the *left* c units.
The graph of $y = (x - c)^2$, $c > 0$, is obtained by shifting the graph of $y = x^2$ to the *right* c units.

USING A GRAPHING CALCULATOR   *(page 149)*

1.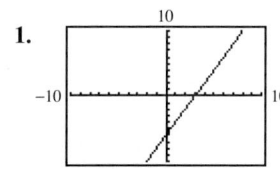

3.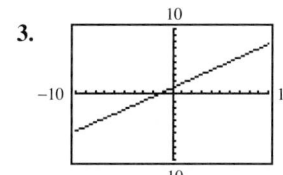

5.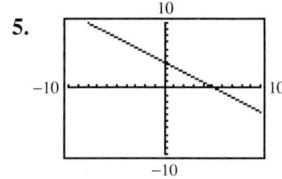

7.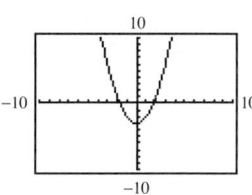

9.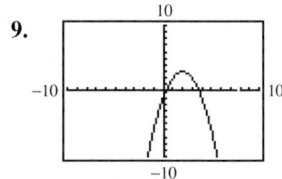

11.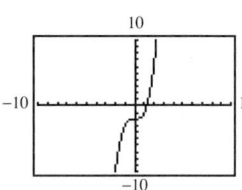

13.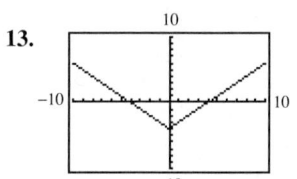

15.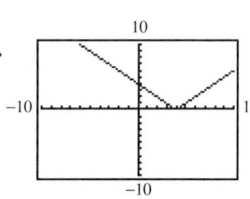

17.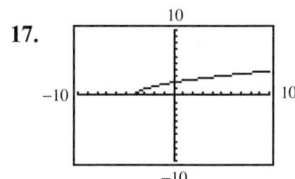

19.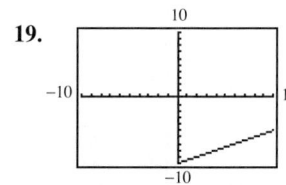

21.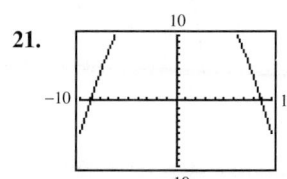

23.
```
Xmin=-8
Xmax=8
Xscl=1
Ymin=-1
Ymax=17
Yscl=1
```

25.
```
Xmin=-2
Xmax=10
Xscl=1
Ymin=-5
Ymax=5
Yscl=1
```

27.
```
Xmin=-5
Xmax=5
Xscl=1
Ymin=-3
Ymax=6
Yscl=1
```

SECTION 2.5   *(page 161)*

Warm-Up   *(page 161)*

1. $2h$ **2.** $3t$ **3.** $2x^2 - 16x + 27$
4. $-3x^2 - 12x - 2$ **5.** $-8x - 10$
6. $-2x^2 + 14x$
7. (a) -4 (b) 8 (c) 8 (d) 71
8. (a) 0 (b) 9 (c) 0 (d) 8.75
9. (a) 0 (b) Undefined   (c) 8 (d) -6
10. (a) 2 (b) -2 (c) $\frac{5}{2}$ (d) $\frac{101}{10}$

1. (a) Function from A to B
(b) Not a function from A to B
(c) Function from A to B
(d) Not a function from A to B

3. $y = 10x + 12$ **5.** $y = \dfrac{-3x + 2}{7}$

7. $y = -x^2 + 2x + 3$ **9.** $y = \dfrac{-|x| + 4}{2}$

11. There are two values of y associated with one value of x.

13. There are two values of y associated with one value of x.

15. (a) 2 (b) -2 (c) k (d) $k + 1$

17. (a) $3, 3$ (b) $-4, -4$
(c) s, s (d) $s - 2, s - 2$

19. (a) 29 (b) 11 (c) $12a - 2$ (d) $12a + 5$

21. (a) 8 (b) $\frac{2}{9}$ (c) $\frac{1}{2}a^2 + 4a + 8$ (d) $8 + \frac{1}{2}a^2$

23. (a) 2 (b) 3 (c) $\sqrt{21}$ (d) $\sqrt{5z + 5}$

25. (a) 0 (b) $-\dfrac{3}{2}$ (c) $-\dfrac{5}{2}$ (d) $\dfrac{3x + 12}{x - 1}$

27. (a) 2 (b) -2 (c) 10 (d) -8

29. (a) 0 (b) $\frac{7}{4}$ (c) 3 (d) 0

31. (a) 2 (b) $\dfrac{2(x - 6)}{x}$

33. (a) $-x^2 - 4x - 3$ (b) $-2x - 2$

35. Domain: {0, 2, 4, 6}
   Range: {0, 1, 8, 27}
37. Domain: All real numbers r such that $r > 0$
   Range: All real numbers C such that $C > 0$
39. All real numbers x such that $x \neq 3$
41. All real numbers x such that $x \neq 2, 1$
43. All real numbers t such that $t \neq 0, -2$
45. All real numbers x
47. All real numbers x such that $x \geq 2$
49. All real numbers x such that $x \geq 0$
51. All real numbers x **53.** All real numbers t
55. $\left\{ (3, 150), (2, 100), (8, 400), (6, 300), \left(\frac{1}{2}, 25 \right) \right\}$
57. {(1986, New York), (1987, Minnesota),
   (1988, Los Angeles), (1989, Oakland),
   (1990, Cincinnati), (1991, Minnesota),
   (1992, Toronto)}
59. (a) 80.6   **(b)** 89.3   **(c)** 74.8   **(d)** 63.9
61. $P(x) = 4x$ **63.** $V(x) = x^3$
65. $A(x) = (32 - x)^2$
67. (a) \$240   **(b)** \$320   **(c)** \$380   **(d)** \$440
69. Function   **71.** Not a function

5.
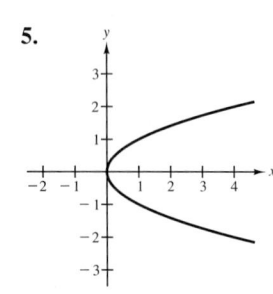
6.
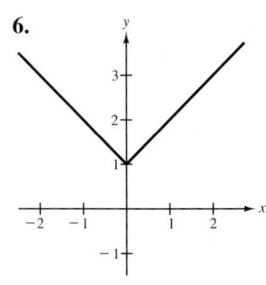

7. (a) 12   **(b)** $\frac{3}{16}$ **8. (a)** -7 **(b)** $-2x$
9. (a) $\dfrac{1}{3}$ **(b)** $\dfrac{c-6}{c+4}$
10. (a) $2\sqrt{3}$ **(b)** $\sqrt{t-1}$

1. c   **2.** b   **3.** e   **4.** a   **5.** f   **6.** d
7.
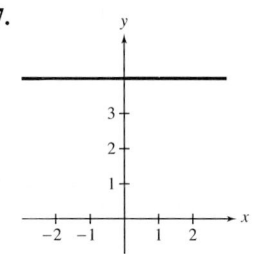
9.
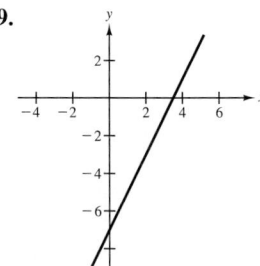

SECTION 2.6   *(page 176)*

Warm-Up   *(page 176)*

1.
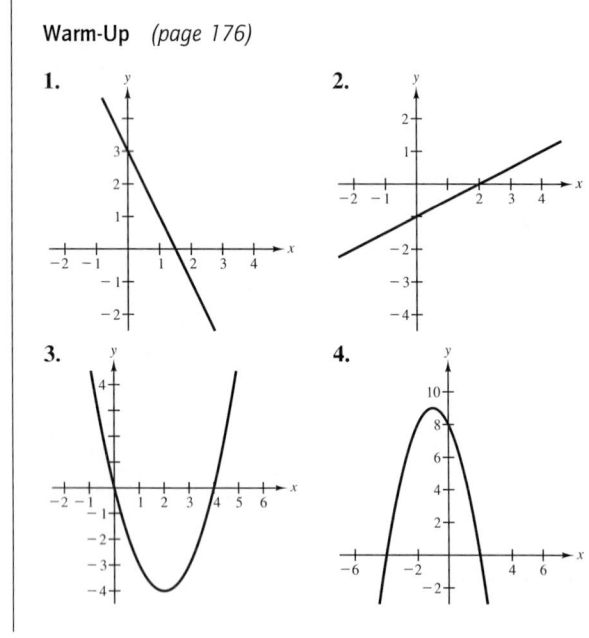
2.

3.

4.

11.
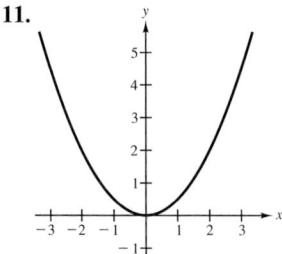
13.
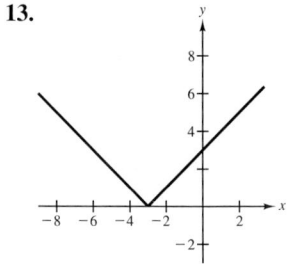

15.
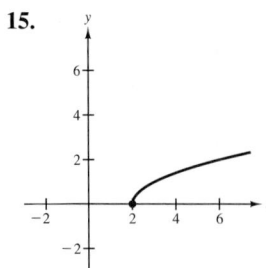
17.
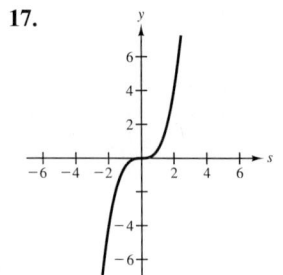

19.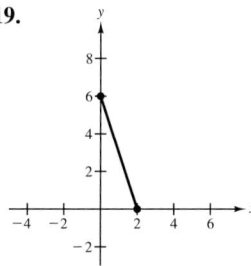

21.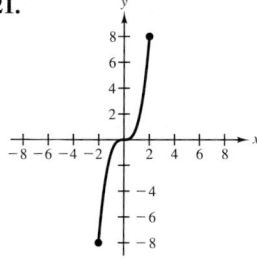

37. Horizontal shift two units to the left

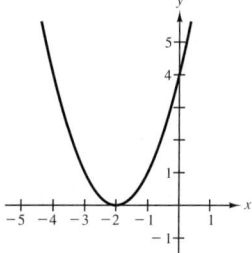

23. y is a function of x.   **25.** y is not a function of x.

27. y is a function of x.   **29.** y is a function of x.

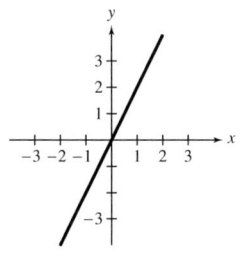   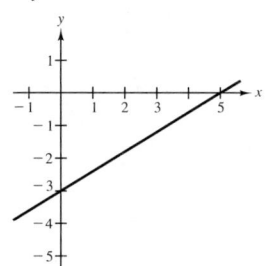

39. Reflection in the x-axis

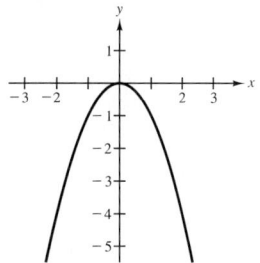

31. y is not a function of x.   **33.** y is a function of x.

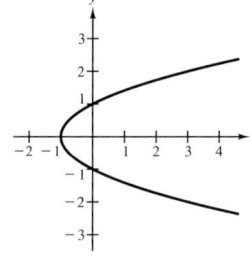   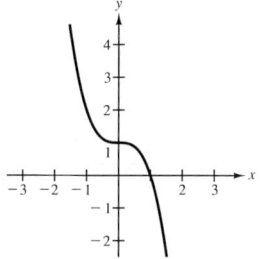

41. Horizontal shift three units to the right and a reflection in the x-axis

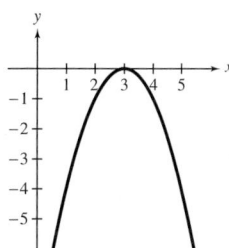

35. Vertical shift two units upward

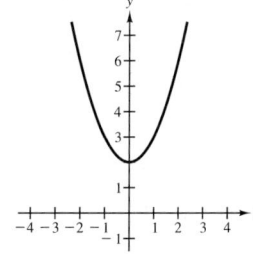

43. Vertical shift three units upward

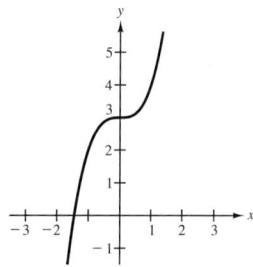

45. Horizontal shift three units to the right

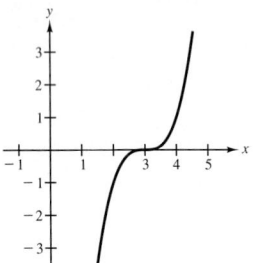

47. Reflection in the y-axis

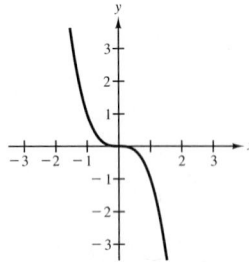

49. Reflection in the x-axis followed by a horizontal shift one unit to the right followed by a vertical shift two units upward

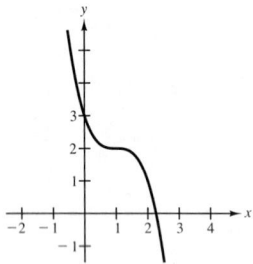

51. Horizontal shift five units to the right

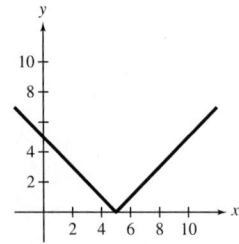

53. Reflection in the x-axis

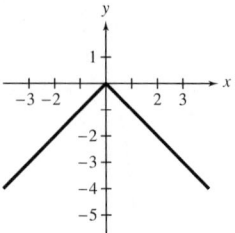

55. Vertical shift two units upward

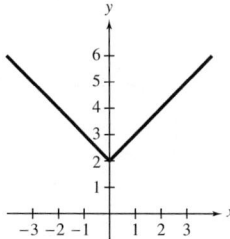

57. Horizontal shift three units to the left and a vertical shift one unit downward

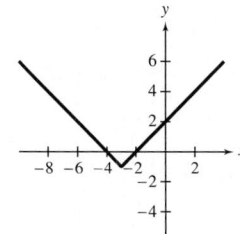

59. $y = -\sqrt{x}$ **61.** $y = \sqrt{x + 2}$
63. $y = \sqrt{-x} + 2$

65.

67.

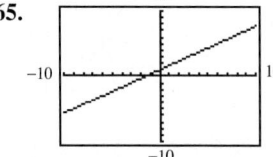

69.

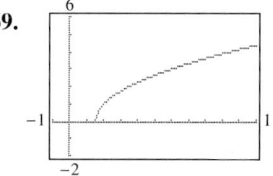

71. The graph of $f(x) = \sqrt{x - 3}$ indicates that only those x-values that are greater than or equal to 3 are included.
Domain: All real numbers $x \geq 3$.

73. (a) $0 < x < 4$
(b)

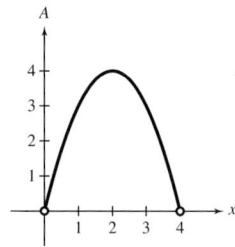

$x = 2$

CHAPTER 2 REVIEW EXERCISES *(page 180)*

1. (a) No   (b) Yes   **3.** (a) Yes   (b) No
5. 2   **7.** 14   **9.** -8.2
11. $\frac{16}{3}$ **13.** 2   **15.** $\frac{20}{17}$
17. 0, 3   **19.** 10, -10 **21.** 15, 10
23. $-\frac{4}{3}$, 2   **25.** 3.38   **27.** 33.32
29. $x = \frac{1}{2}(7y - 4)$ **31.** $h = \dfrac{V}{\pi r^2}$

33.

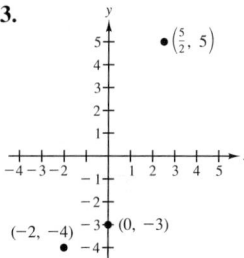

35. $d = 5$

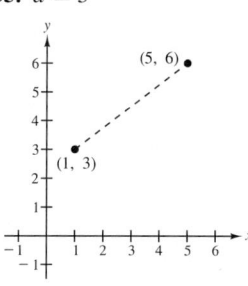

37. $d \approx 7.62$

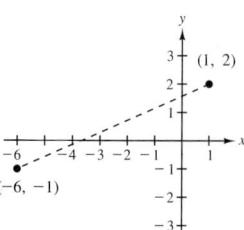

39.

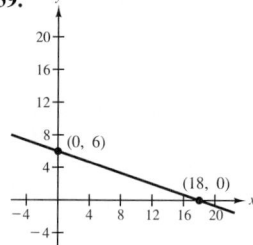

41.

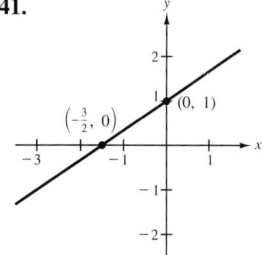

43.

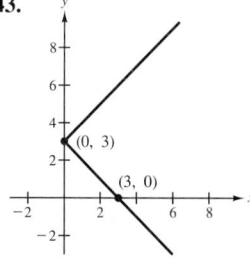

45. y is not a function of x.   **47.** y is a function of x.

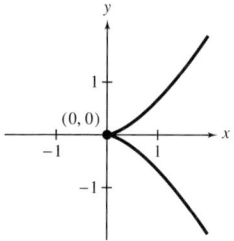

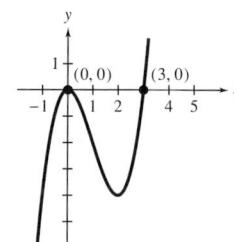

49. $(9, 0)$, $(6, 0)$ **51.** No x-intercept, $(0, -1)$, $(0, 1)$
53. $(8, 0)$, $(-8, 0)$, $(0, 8)$ **55.** $(2, 0)$, $(0, 4)$
57. (a) 29   (b) 3   (c) $\dfrac{36 - 5t}{2}$ (d) $4 - \dfrac{5}{2}(x + h)$
59. (a) 3   (b) 0   (c) $\sqrt{2}$ (d) $\sqrt{5 - 5x}$
61. (a) -3 (b) 2   (c) 0   (d) -7
63. (a) -2 (b) $\dfrac{2(6 - x)}{x}$
65. (a) $(x + h)^2 = x^2 + 2xh + h^2$ (b) $2xh + h^2$
67. All real numbers x
69. All real numbers x such that $x \neq 5$

71. (a) Reflection of the graph of $f(x) = x^2$ in the x-axis
with a left shift of one unit
(b)

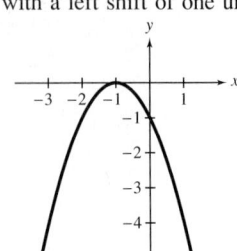

73. (a) The graph of $f(t) = |t|$ shifted two units to the
right and one unit downward
(b)

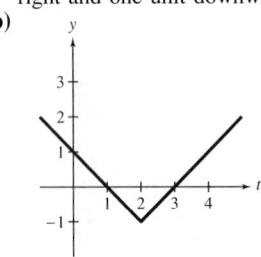

75. (a) The graph of $f(x) = x^3$ shifted downward two units.
(b)

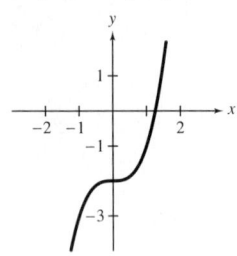

77. 7.96   **79.** 6 in. × 8 in.

CHAPTER 2 TEST   *(page 183)*

1. (a) No   **(b)** Yes   **2.** $r = \frac{1}{3}(2s - 1)$
3. 4   **4.** 4   **5.** 24
6. $\frac{19}{2}$ **7.** −5, 1   **8.** $-\frac{4}{3}$, 3
9. (a) Quadrant I, Quadrant III
  (b) The product xy is positive, and therefore x and y
have the same sign. If x and y are both positive, the
point is located to the *right* of the vertical axis and
above the horizontal axis in the first quadrant. If x
and y are both negative, the point is located to the
left of the vertical axis and *below* the horizontal axis
in the third quadrant.

10. $\sqrt{73} \approx 8.54$ **11.** (3, 0), (−1, 0), (0, −3)
12.

13. y is not a function of x.
14. (a) 16   **(b)** $3h$ **(c)** $3t + 1$ **(d)** $s − 2$
15.

16. (a) $y = |x − 2|$ **(b)** $y = |x| − 2$
   (c) $y = 2 − |x|$

17. 9 cm × 6 cm   **18.** $2000

Chapter 3

Chapter Opener *(page 185)*

1120 miles, 30 days

SECTION 3.1 *(page 198)*

Warm-Up *(page 197)*

1. $43 - 6x$ **2.** $u^2 - 16$ **3.** $2n - 3$
4. $16n + 6$ **5.** -4 **6.** 27   **7.** 14
8. -19 **9.** $\frac{940}{3}$ **10.** $\frac{1}{10}$

---

1. $8 + n$ **3.** $15 - 3n$ **5.** $\frac{1}{3}n$ **7.** $0.30L$
9. $\frac{x}{6}$ **11.** $\frac{3 + 4x}{8}$ **13.** $|n - 5|$
15. The sum of three times a number and 2
17. Eight times the difference of a number and 5
19. The ratio of a number to 8
21. The sum of a number and 10, all divided by 3
23. $0.25n$ **25.** $55t$ **27.** $\frac{100}{r}$ **29.** $0.45y$
31. $0.0125I$ **33.** $0.80L$ **35.** $8.25 + 0.60q$
37. s^2 **39.** $3.24l$ **41.** $4n + 2$
43. $6w^2$ ft^2 **45.** 44, 46, 48   **47.** 10

49. | Number | + | 30 | = | 82 |

$x + 30 = 82$
$x = 52$

51. | 3 | · | Side | = | Perimeter |

$3s = 129$
$s = 43$

53. | Cost for parts | + | Labor cost per hour | · |

| Number of hours | = | Total cost |

$275 + 35x = 380$
$x = 3$

55. | Number of people laid off | = |

| Percentage of work force | · |

| Number of employees |

$25 = p(160)$
$0.15625 = p$ or 15.625%

57. | Number of defective parts in sample | = |

| Percentage of defective parts in sample | · |

| Size of sample |

$3 = 0.015x$
$200 = x$

59. $0.80 \cdot$ | Possible points | $- 372 = 8$

$0.80x - 372 = 8$
$x = 475$

61. 2004   **63.** 7315   **65.** $\frac{3}{4}$ **67.** $\frac{1}{25}$
69. $\frac{85}{4}$ **71.** 4   **73.** $\frac{15}{12}$ **75.** 16
77. 20 pt   **79.** \$1800   **81.** 350 mi
83. 3   **85.** 46.9 ft

SECTION 3.2 *(page 210)*

Warm-Up *(page 210)*

1. $\frac{11}{4}$ **2.** 5   **3.** $\frac{5}{9}$ **4.** $\frac{10}{3}$
5. $\frac{2}{5}$ **6.** 20   **7.** -4 **8.** $-\frac{6}{5}$
9. 9   **10.** 50

---

1. \$18.38, 40%   **3.** \$152.00, 65%
5. \$22,250.00, 21%   **7.** \$416.70, \$191.70
9. \$23.76, 48%   **11.** \$111.00, 63%
13. \$33.25, \$61.75   **15.** \$1145.00, 22%
17. \$54.15   **19.** Department store   **21.** 2.5 hr
23. 9 min, \$2.06   **25.** 18.3%   **27.** 9%
29. Tax = \$267;   Total bill = \$4717;
Amount financed = \$3717
31. 30 \$0.15 stamps; 40 \$0.30 stamps   **33.** 700
35. 500–\$3.00, 400–\$2.74, 300–\$2.48, 200–\$2.22,
100–\$1.96, 0–\$1.70
(a) Decreases   **(b)** Decreases
(c) The price of the mixture would be equal to the
average price of the oats and the corn per bushel.
37. 4 roses   **39.** \$30,000 at 8.5%   \$10,000 at 10%
41. 50 gal at 20%   50 gal at 60%
43. 8 qt at 15%   16 qt at 60%   **45.** $\frac{5}{6}$ gal
47. 1440 mi   **49.** $\frac{1}{3}$ hr   **51.** 43.6 mi/hr
53. $83\frac{1}{3}$ mi/hr   **55.** $\frac{8}{15}t$ **57.** 10 min
59. $\frac{1}{5}$, $\frac{1}{8}$, $\frac{40}{13}$ or $3\frac{1}{13}$ hr   **61.** 1985, \$0.286   **63.** 8.3%

SECTION 3.3   *(page 226)*

Warm-Up *(page 226)*

1. $-\frac{3}{4} > -5$ **2.** $-\frac{1}{5} > -\frac{1}{3}$ **3.** $\pi > -3$

4. $6 < \frac{13}{2}$ **5.** $\frac{7}{2}$ **6.** $\frac{48}{7}$ **7.** 40

8. 153   **9.** -12 **10.** 24

---

1. (a) Yes   **(b)** No   **(c)** Yes   **(d)** No
3. (a) No   **(b)** Yes   **(c)** Yes   **(d)** No
5. f   **6.** c   **7.** a   **8.** e   **9.** d   **10.** b
11. $(-5, 3]$

13. $\left(0, \frac{3}{2}\right]$

15. $\left(-\infty, -\frac{5}{2}\right)$

17. $x \le 2$

19. $x < \frac{11}{2}$

21. $x \le -4$

23. $x > 8$

25. $x > 7$

27. $x \ge 7$

29. $x > -\frac{2}{3}$

31. $x > \frac{9}{2}$

33. $2 < y$

35. $-10 \ge y$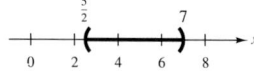

37. $x \ge -12$

39. $\frac{5}{2} < x < 7$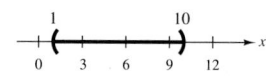

41. $-\frac{3}{2} < x < \frac{9}{2}$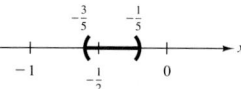

43. $1 < x < 10$

45. $-\frac{3}{5} < x < -\frac{1}{5}$

47. $x \ge 0$ **49.** $z \ge 2$ **51.** x is at least $\frac{5}{2}$.
53. z is greater than 0 and no more than π.
55. \$2600
57. The average temperature in Miami is greater than the average temperature in New York.
59. $m < 25{,}357.14286$ or $m \le 25{,}357$ **61.** $x \ge 31$
63. The call must be less than 6.38 minutes. If a portion of a minute is billed as a full minute, then the call must be less than or equal to 6 minutes.
65. $2 \le d \le 8$ **67.** $-8 < -t \le 5$
69. $3 \le n \le \frac{15}{2}$ **71.** $n > 21$ **73.** 1970–1974
75. 1993   **77.** $10{,}000 \le b \le 120{,}000$

CHAPTER 3 MID-CHAPTER QUIZ   *(page 230)*

1. $5m + 10n$ **2.** $L + 0.06L = 1.06L$ **3.** $\frac{4}{3}$
4. 262, 263   **5.** 2 in. × 24 in.   **6.** \$563,952
7. Taxes: 7.12%
Employee benefits: 10.98%
Miscellaneous: 12.46%
Insurance: 2.37%
Supplies: 2.67%
Utilities: 1.19%
Rent: 3.86%
Wages: 59.35%

8. $\frac{1}{50}$ **9.** 17.1 gal **10.** $\frac{10}{3}$ **11.** $188

12. 10 hr **13.** 6 ml **14.** $\frac{18}{7}$ or $2\frac{4}{7}$ hr

15. $x \geq 4$

16. $t > \frac{1}{3}$

17. $-6 \leq x < 4$

18. $x \geq \frac{6}{5}$

19. $x > 33$

SECTION 3.4 *(page 240)*

Warm-Up *(page 240)*

1. 15 **2.** 3.2 **3.** -72 **4.** -4
5. 18 **6.** -19.8 **7.** $-\frac{1}{2} < x < \frac{1}{2}$
8. $-\frac{1}{2} \leq x \leq \frac{3}{2}$ **9.** $-6 \leq x \leq 6$
10. $20 < x < 30$

1. 45, -45 **3.** 0 **5.** 21, 11 **7.** 11, -14
9. $\frac{16}{3}$, 16 **11.** No solution **13.** $-\frac{11}{5}$, $\frac{17}{5}$
15. $-\frac{39}{2}$, $\frac{15}{2}$ **17.** 18.75, -6.25 **19.** -3, 7
21. 11, 13 **23.** $-\frac{15}{4}$
25. (a) Yes (b) No (c) No (d) Yes
27. (a) No (b) Yes (c) Yes (d) No
29. (a) Yes (b) No (c) No (d) Yes
31. (a) No (b) Yes (c) Yes (d) No
33.

35.
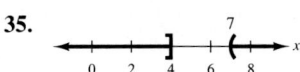

37. $-4 < y < 4$

39. $-2 \leq y \leq 6$

41. $-6 < y < 2$

43. $y \leq -4$ or $y \geq 4$

45. $y < -2$ or $y > 6$

47. $y \leq -6$ or $y \geq 2$

49. $-7 < x < 7$

51. $-9 \leq y \leq 9$

53. $x < -2$ or $x > \frac{2}{3}$

55. $x < 3$ or $x > 7$

57. $-5 < x < 35$ **59.** $-82 \leq x \leq 78$
61. $t \leq -\frac{15}{2}$ or $t \geq \frac{5}{2}$ **63.** $s > 23$ or $s < -17$
65. $z < -50$ or $z > 110$
67.

69. $|x| \leq 2$ **71.** $|x - 10| < 3$ **73.** $|x - 19| < 3$
75. $|x| < 3$ **77.** $|x - 5| > 6$
79. $|N - 222.5| \leq 102.5$

81. $-1 < x < 7$

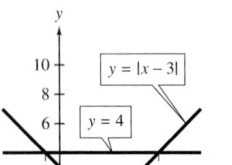

$|x - 3| < 4$

83. $x < -6$ or $x > 2$

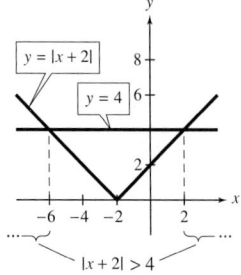

$|x + 2| > 4$

19. $\frac{5}{7}$, rises

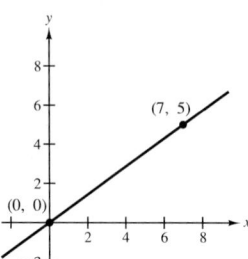

21. $\frac{1}{2}$, rises

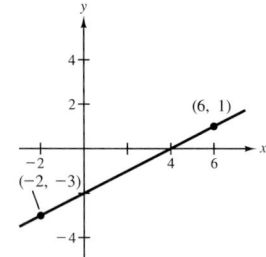

23. $-\frac{3}{2}$, falls

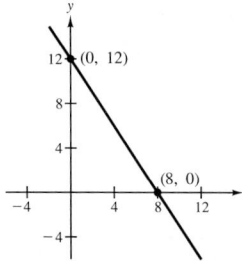

25. Undefined, vertical

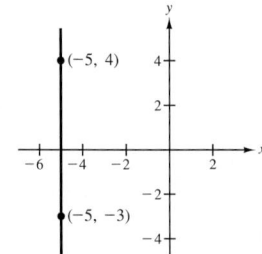

SECTION 3.5   *(page 255)*

Warm-Up   *(page 255)*

1. $\frac{1}{2}$ **2.** $\frac{1}{2}$ **3.** $-\frac{29}{2}$ **4.** $-\frac{1}{2}$
5. $\frac{16}{5}$ **6.** $-\frac{7}{3}$ **7.** $y = \frac{5}{8}(x - 1)$
8. $y = -\frac{9}{2}x$ **9.** $y = 2x - 1$
10. $y = \frac{1}{4}(3x + 20)$

1. $f(x) = 3x + 10$ **3.** $f(x) = -\frac{1}{4}x + \frac{45}{8}$
5. $f(x) = 4$
7. $(0, -6), (3, 0)$ **9.** $(0, 1), \left(\frac{4}{3}, 0\right)$

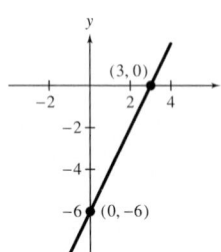

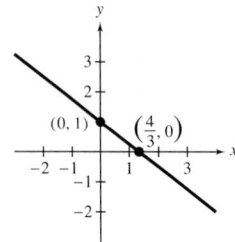

11. $\frac{2}{3}$ **13.** -2 **15.** Undefined
17. (a) L_3 **(b)** L_2 **(c)** L_1

27. 0, horizontal

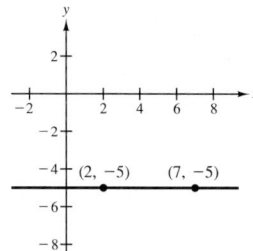

29. $-\frac{18}{17}$, falls

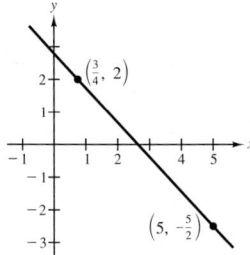

31. $-\frac{5}{6}$, falls

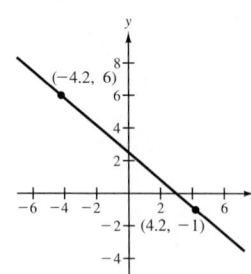

33. $\frac{29}{9}$, rises

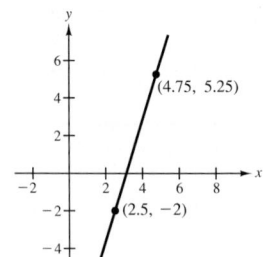

35. $\frac{5}{3}$, rises

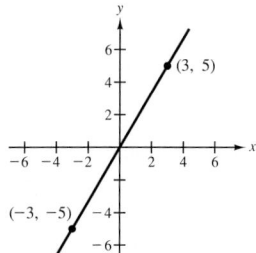

37. $x = 1$ **39.** $y = -15$

41.

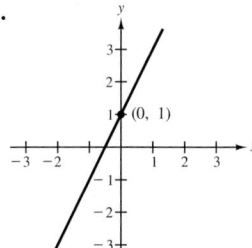

43.

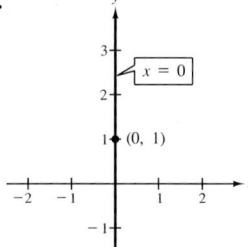

45.

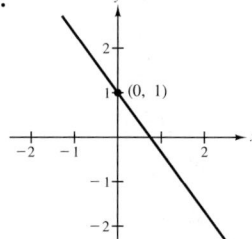

47.

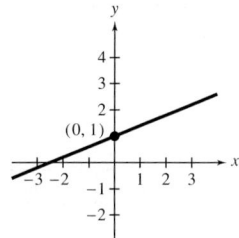

49.

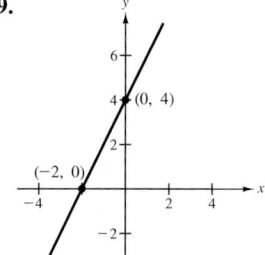

51.

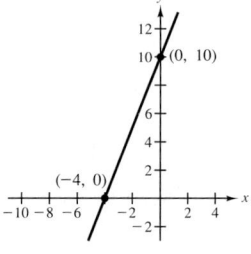

53. 3, $(0, -2)$ **55.** -1, $(0, 0)$

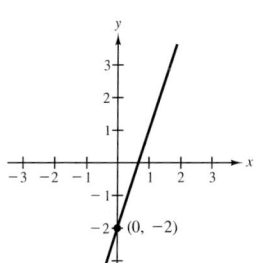

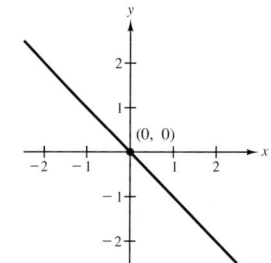

57. $-\frac{3}{2}$, $(0, 1)$ **59.** $\frac{1}{4}$, $\left(0, \frac{1}{2}\right)$

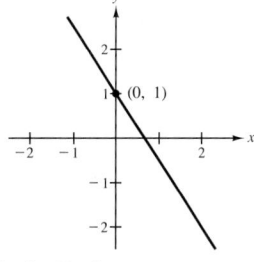

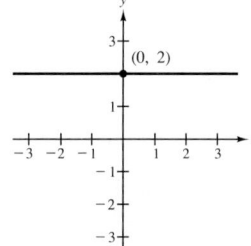

61. 0, $(0, 2)$ **63.** 5, $(0, -5)$

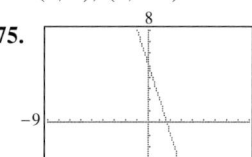

65. $(6, 2)$, $(10, 2)$ **67.** $(4, -1)$, $(5, 2)$
69. $(1, 2)$, $(2, 1)$ **71.** $(-2, 4)$, $(1, 8)$
73. $(4, 0)$, $(4, -1)$

75.

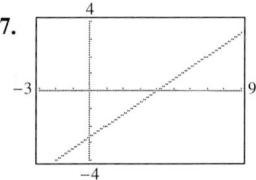

77.

79. Perpendicular   **81.** Parallel

83. Parallel   **85.** Perpendicular

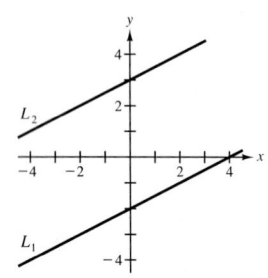 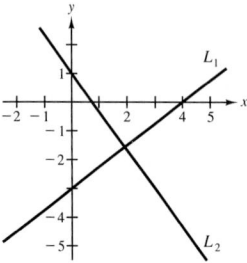

87. (a) 1988 and 1989   **(b)** 1985 and 1986

89. (a)

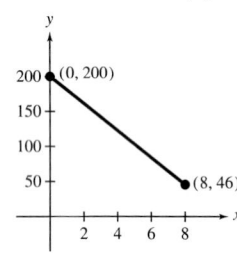

(b) $19.25/day

91. 16,667 ft

USING A GRAPHING CALCULATOR   *(page 262)*

1. 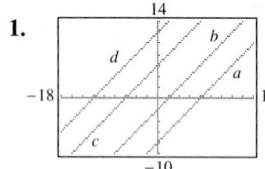   **3.**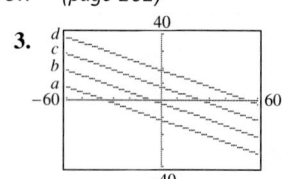

5.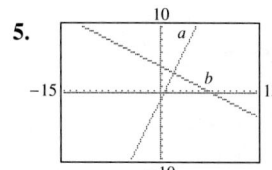

The lines are perpendicular because their slopes are negative reciprocals. Use a square setting for the viewing rectangle.

7.

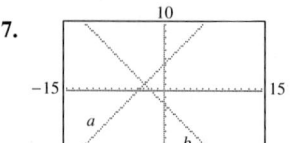

The lines are perpendicular because their slopes are negative reciprocals. Use a square setting for the viewing rectangle.

9. No. Parallel lines appear parallel with square *or* nonsquare settings for the viewing rectangle.

11. No. Intersecting lines appear to intersect with square *or* nonsquare settings for the viewing rectangle.

SECTION 3.6   *(page 273)*

Warm-Up   *(page 273)*

1. $-\frac{1}{4}$ **2.** 1   **3.** Undefined   **4.** $-\frac{9}{10}$

5. 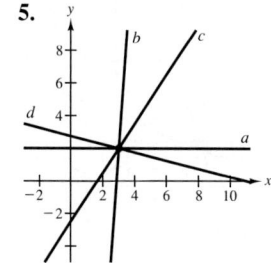   **6.**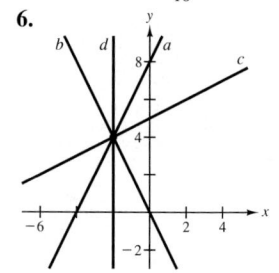

7. $y = -\frac{3}{5}x + \frac{21}{5}$ **8.** $y = \frac{4}{3}x - 9$

9. $y = -4x + 18$ **10.** $y = x + 9$

1. $m = 6, (-4, 3)$ **3.** $m = \frac{3}{4}, \left(-2, -\frac{5}{8}\right)$

5. $m = \frac{2}{3}, (0, -2)$ **7.** $m = \frac{3}{2}, (0, 0)$

9. $m = 0, (0, 3)$ **11.** $m = \frac{5}{2}, (0, 12)$

13. $x - 2y - 12 = 0$ **15.** $4x + 2y - 3 = 0$

17. $2x + y - 4 = 0$ **19.** $2x + 3y - 10 = 0$

21. $x - 2 = 0$ **23.** $8x - 6y + 9 = 0$

25. $3x - 2y = 0$

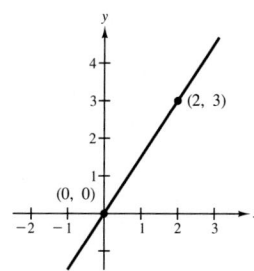

27. $3x + 7y - 15 = 0$

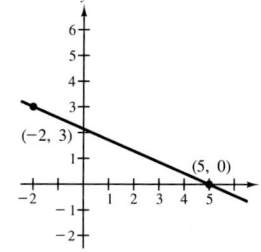

29. $x - 2y + 6 = 0$

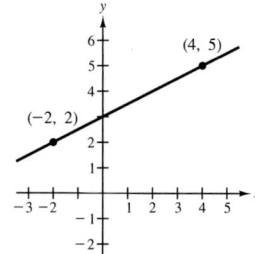

31. $y - 3 = 0$

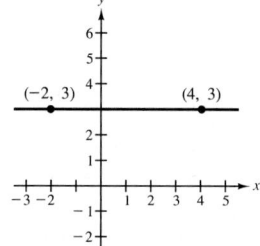

33. $14x + 6y - 39 = 0$

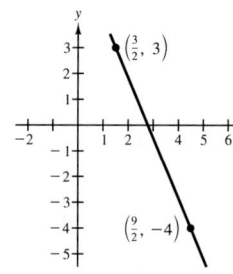

35. $10x + 90y - 27 = 0$

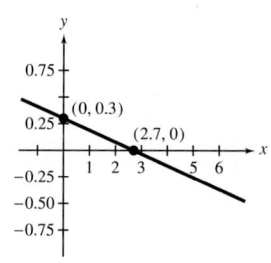

37. $f(x) = -\frac{2}{5}x$ **39.** $f(x) = \frac{5}{2}x - 2$

41. $f(x) = \frac{7}{6}x + \frac{1}{4}$

43. (a) $3x - y - 5 = 0$ (b) $x + 3y - 5 = 0$

45. (a) $5x + 4y + 9 = 0$ (b) $4x - 5y + 40 = 0$

47. (a) $y - 2 = 0$ (b) $x + 1 = 0$

49.

x	0	50	100	500	1000
C	5000	6000	7000	15,000	25,000

51. $S = 1500 + 0.03M$ **53.** $S = 0.70L$

55. $y = 72 + 5t$

57. (a) $x = -\frac{1}{15}p + 80$ (b) 42   (c) 48

59. $C = 0.32t + 2.3$, \$7.1 billion

61. $s = 0.6t + 4.9$

63. (a) and (b)

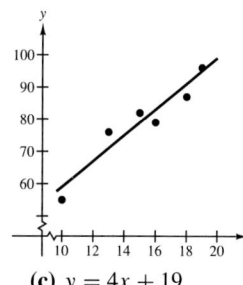

65. Parallel

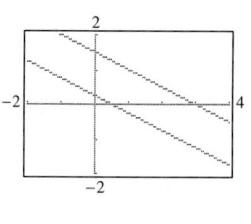

(c) $y = 4x + 19$

(d) 87

CHAPTER 3 REVIEW EXERCISES   *(page 277)*

1. $200 - 3n$ **3.** $n^2 + 49$

5. The sum of twice a number and 7

7. The difference of a number and 5, all divided by 4

9. $0.18I$ **11.** $l(l - 5) = l^2 - 5l$ **13.** \$350.93

15. \$194.25   **17.** 15.5%

19. 25 dimes, 35 quarters   **21.** 6.35 hr

23. 2800 mi   **25.** $\frac{15}{4}$ or $3\frac{3}{4}$ hr

27. \$20,000 at 8.5%; \$30,000 at 10%   **29.** $\frac{4}{3}$

31. $\frac{3}{20}$ **33.** $y = \frac{7}{2}$ **35.** $b = \frac{25}{2}$

37. \$1856.25   **39.** 487.5 mi   **41.** 3

43. $z \le 10$ **45.** $7 \le y < 14$

47. $x > 3$

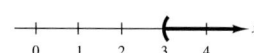

49. $y > -\frac{70}{3}$

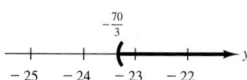

51. $-20 < x \le 20$

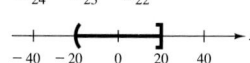

53. $-16 < x < -1$

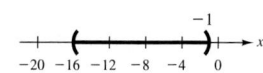

55. $-4 < x < 11$

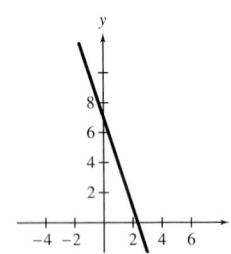

57. $x > 7$ or $x < 1$

59. $b > 5$ or $b < -9$

61. $-4, 8$ **63.** $\frac{1}{2}, 3$ **65.** $\frac{2}{7}$ **67.** 0
69. $-3, (0, 7)$ **71.** $\frac{5}{2}, (0, -5)$

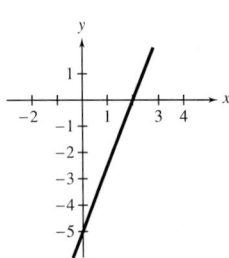

73. $2x - y - 6 = 0$ **75.** $4x + y = 0$
77. $2x + 3y - 17 = 0$ **79.** $x - 7 = 0$
81. $x + 2y + 6 = 0$ **83.** $y - 10 = 0$
85. $55x + 32y - 176 = 0$ **87.** $2x - 5 = 0$
89. (a) $x + 2y - 9 = 0$ (b) $2x - y + 7 = 0$
91. (a) $8x - 6y + 15 = 0$ (b) $24x + 32y - 105 = 0$
93. (a) $C = 8.55x + 25,000$ (b) $P = 4.05x - 25,000$

CHAPTER 3 TEST *(page 280)*

1. $5n - 8$ **2.** $0.6l^2$ **3.** $\$8000$
4. $\$1466.67$ **5.** $2\frac{1}{2}$ hr
6. $33\frac{1}{3}$ liters at 10%; $66\frac{2}{3}$ liters at 40% **7.** $\frac{2}{3}$ hr
8. $x > 2$

9. $-7 < x \leq 1$

10. $1 \leq x \leq 5$

11. $x > -3$ or $x < -5$

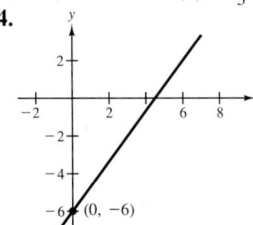

12. $-8, 2$ **13.** (a) $-\frac{2}{3}$ (b) Undefined
14. **15.**

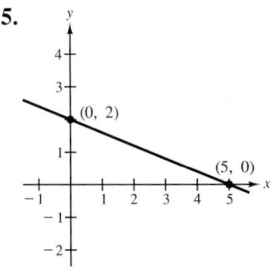

16. $y = -\frac{3}{2}x + 4$ **17.** $x - 2y - 55 = 0$
18. $\frac{3}{5}$ **19.** $V = -2250t + 16,000, 2\frac{2}{3}$ yrs

CUMULATIVE TEST: CHAPTERS 1-3 *(page 281)*

1. 23 **2.** $-\frac{10}{27}$ **3.** $8a^8 b^7$
4. $2x^3 y - 11xy$ **5.** $t^2 - 9t$
6. $x^2 - 2xy + y^2 + 4x + 4$ **7.** $\frac{3}{2}$
8. $-\frac{3}{2}$ **9.** ± 8 **10.** $-\frac{1}{2}, 3$
11. y is a function of x.
12. All real numbers x such that $x \geq 2$
13. (a) 4 (b) $c^2 + 3c$ **14.** $3n - 8$
15. 10 and 11 or -9 and -8 **16.** $\$1408.75$
17. $\frac{14}{3}$ **18.** 6.5
19. $x \leq -1$ or $x \geq 5$

20. $x \geq 103$ **21.** (a) 10 (b) $3x - 4y + 12 = 0$
22.

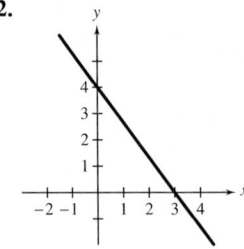

Chapter 4

Chapter Opener *(page 282)*

Increasing; Compact disc sales
Decreasing; Long playing record sales
1986

SECTION 4.1 *(page 294)*

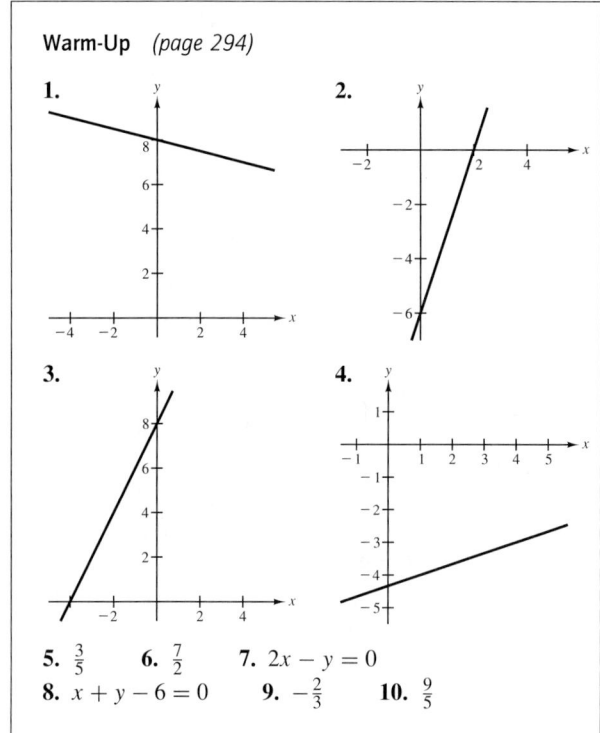

Warm-Up *(page 294)*

1.

2.

3.

4.

5. $\frac{3}{5}$ **6.** $\frac{7}{2}$ **7.** $2x - y = 0$
8. $x + y - 6 = 0$ **9.** $-\frac{2}{3}$ **10.** $\frac{9}{5}$

1. (a) Yes **(b)** No **3. (a)** No **(b)** Yes
5. No solution **7.** $\left(1, \frac{1}{3}\right)$
9. Infinite number of solutions

11. $(-1, -1)$

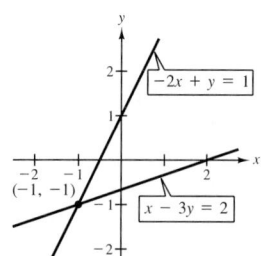

13. $(10, 0)$

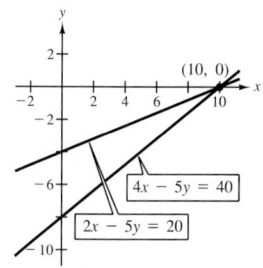

15. $(9, 12)$

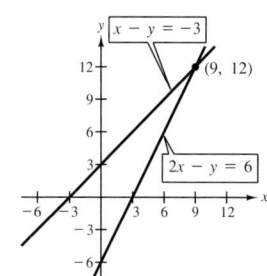

17. $\left(-\frac{3}{2}, \frac{5}{2}\right)$

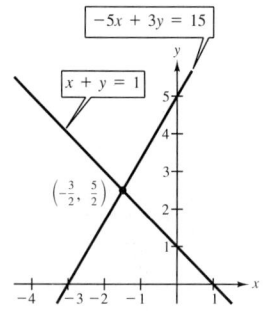

19. Infinite number of solutions **21.** One solution
23. No solution **25.** $\left(2992, \frac{798}{25}\right)$ **27.** $(2, 1)$
29. $(4, 3)$ **31.** $\left(4, -\frac{1}{2}\right)$ **33.** $(1, 2)$
35. $(4, -2)$ **37.** $(7, 2)$ **39.** $(-2, -1)$
41. No solution **43.** Infinite number of solutions
45. $(5, 3)$ **47.** $(2, -2)$

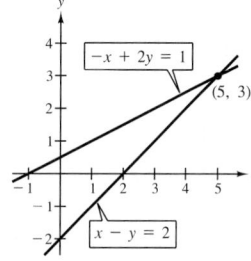

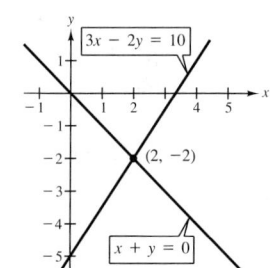

49. No solution

51. Infinite number of solutions

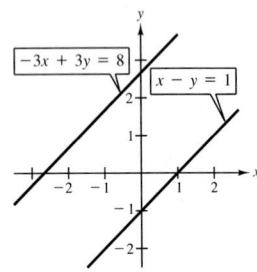

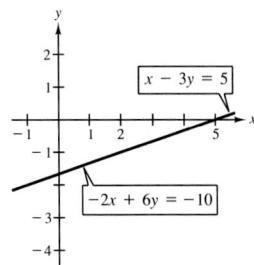

53. $(3, 2)$ **55.** $(-2, 5)$ **57.** $(-1, -1)$

59. $(7, -2)$ **61.** No solution

63. $\left(\frac{3}{2}, 1\right)$ **65.** $(-2, -1)$

67. Infinite number of solutions **69.** $\left(\frac{25}{2}, -\frac{1}{5}\right)$

71. $(3000, -2000)$ **73.** $(4, 3)$ **75.** $(2, 7)$

77. $(15, 10)$ **79.** $\left(\frac{33}{4}, \frac{25}{2}\right)$

81. 5 dimes, 10 quarters **83.** 11 nickels, 14 dimes

85. Regular: \$1.11; Premium: \$1.22

87. \$5.65 variety: 6 lb; \$8.95 variety: 4 lb

89. 40% solution: 12 l; 65% solution: 8 l

91. \$15,000 at 8%, \$5000 at 9.5%

93. 10,000 units **95.** 63

97. (a) $y = \frac{3}{2}x - \frac{1}{6}$

(b)

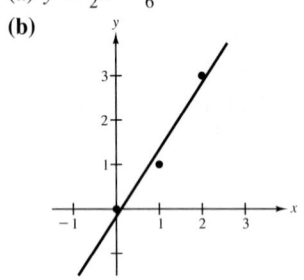

99. Depth: 10 ft; Length: 122 ft, 125 ft

101. $\begin{cases} 3x + 2y = 22 \\ 2x - y = 3 \end{cases}$

103. $\begin{cases} x + y = -3 \\ 7x - 3y = -1 \end{cases}$

USING A GRAPHING CALCULATOR *(page 303)*

1.

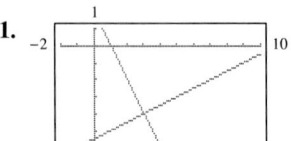

The solution appears to be approximately $(3, -4)$.

Check:

$$y = -2x + 2 \qquad\qquad y = \frac{1}{2}x - \frac{11}{2}$$
$$-4 \overset{?}{=} -2(3) + 2 \qquad -4 \geq \frac{1}{2}(3) - \frac{11}{2}$$
$$-4 = -4 \qquad\qquad\qquad -4 = -\frac{8}{2}$$

The solution is $(3, -4)$.

3.

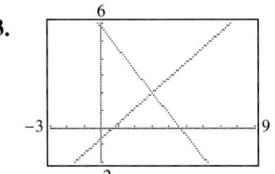

Rewrite: $\begin{cases} y = \frac{7}{8}x - \frac{5}{8} \\ y = -\frac{5}{4}x + \frac{23}{4} \end{cases}$

The solution appears to be approximately $(3, 2)$.

Check:

$$7x - 8y = 5 \qquad\qquad 5x + 4y = 23$$
$$7(3) - 8(2) \overset{?}{=} 5 \qquad 5(3) + 4(2) \overset{?}{=} 23$$
$$21 - 16 = 5 \qquad\qquad 15 + 8 = 23$$

The solution is $(3, 2)$.

5. 1984

SECTION 4.2 *(page 314)*

Warm-Up *(page 313)*

1. $(5, 4)$ **2.** $(-1, 1)$ **3.** Inconsistent

4. Infinite number of solutions **5.** $(3, -2)$

6. $(-2, -2)$ **7.** $5x - z$ **8.** $34y - 5x$

9. $35z - 17y$ **10.** $7x + 26z$

1. $(22, -1, -5)$ **3.** $(14, 3, -1)$
5. $(1, 2, 3)$ **7.** $(1, 2, 3)$ **9.** $(2, -3, -2)$
11. No solution   **13.** $(-4, 8, 5)$
15. $(3a + 1, 3a - 1, a)$, where a is any real number
17. $(1, 0, -2)$ **19.** $(-4, 2, 3)$ **21.** $(1, -1, 2)$
23. $(-2a - 2, a + 1, a)$, where a is any real number
25. $\left(\frac{7}{5}a + 1, \frac{1}{5}a - a, a\right)$, where a is any real number
27. $\left(-\frac{1}{2}a + \frac{1}{4}, \frac{1}{2}a + \frac{5}{4}, a\right)$, where a is any real number
29. $y = 2x^2 + 3x - 4$ **31.** $y = x^2 - 4x + 3$
33. $x^2 + y^2 - 4x = 0$ **35.** $x^2 + y^2 - 6x - 8y = 0$
37. $s = -16t^2 + 144$ **39.** $s = -16t^2 + 48t$
41. 20 gal of spray X;
   18 gal of spray Y;
   16 gal of spray Z
43. Strings: 50; Wind: 20; Percussion: 8
45. $\begin{cases} x + y + z = 3 \\ x - y - 3z = 1 \\ 2x + 3y - z = -3 \end{cases}$

SECTION 4.3   *(page 327)*

Warm-Up   *(page 327)*

1. $(2, 1)$ **2.** $(3, -2)$ **3.** $(5, 4)$
4. $\left(\frac{1}{2}, \frac{1}{3}\right)$ **5.** $(1, 2, 1)$ **6.** $(4, -1, 3)$
7. 9 **8.** $\frac{14}{5}$ **9.** $-\frac{19}{4}$ **10.** $-\frac{11}{6}$

---

1. 4×2 **3.** 3×4
5. $\begin{bmatrix} 4 & -5 & \vdots & -2 \\ -1 & 8 & \vdots & 10 \end{bmatrix}$ **7.** $\begin{bmatrix} 1 & 10 & -3 & \vdots & 2 \\ 5 & -3 & 4 & \vdots & 0 \\ 2 & 4 & 0 & \vdots & 6 \end{bmatrix}$
9. $\begin{cases} 4x + 3y = 8 \\ x - 2y = 3 \end{cases}$ **11.** $\begin{cases} x + 2z = -10 \\ 3y - z = 5 \\ 4x + 2y = 3 \end{cases}$
13. $\begin{bmatrix} 1 & 4 & 3 \\ 0 & 2 & -1 \end{bmatrix}$
15. $\begin{bmatrix} 1 & 1 & 4 & -1 \\ 0 & 5 & -2 & 6 \\ 0 & 3 & 20 & 4 \end{bmatrix}$ $\begin{bmatrix} 1 & 1 & 4 & -1 \\ 0 & 1 & -\frac{2}{5} & \frac{6}{5} \\ 0 & 3 & 20 & 4 \end{bmatrix}$

17. $\begin{bmatrix} 1 & 2 & 3 \\ 0 & 1 & 2 \end{bmatrix}$ **19.** $\begin{bmatrix} 1 & -1 & -2 \\ 0 & 1 & \frac{9}{10} \end{bmatrix}$
21. $\begin{bmatrix} 1 & 1 & 0 & 5 \\ 0 & 1 & 2 & 0 \\ 0 & 0 & 1 & -1 \end{bmatrix}$ **23.** $\begin{bmatrix} 1 & -1 & -1 & 1 \\ 0 & 1 & 6 & 3 \\ 0 & 0 & 1 & \frac{4}{5} \end{bmatrix}$
25. $\begin{cases} x - 2y = 4 \\ y = -3 \end{cases}$
   $(-2, -3)$
27. $\begin{cases} x - y + 2z = 4 \\ y - z = 2 \\ z = -2 \end{cases}$
   $(8, 0, -2)$
29. $(3, 2)$ **31.** $(1, 1)$ **33.** No solution
35. $(4, -3, 2)$ **37.** $(1, -1, 2)$ **39.** $(1, 2, -1)$
41. $(2a + 1, 3a + 2, a)$ **43.** No solution
45. $\left(2, 5, \frac{5}{2}\right)$
47. 8%: \$800,000; 9%: \$500,000; 12%: \$200,000
49. There are infinitely many solutions.
   One possible allocation is: \$250,000 in certificates of deposit, \$125,000 in municipal bonds, and \$125,000 in blue-chip stocks.
51. $y = x^2 + 2x + 4$ **53.** $x^2 + y^2 - 5x - 3y + 6 = 0$

CHAPTER 4 MID-CHAPTER QUIZ   *(page 330)*

1. $(2, 1)$ **2.** No solution   **3.** $(1, 3)$
4. $(5, 3)$ **5.** $(9, 4)$
6. Infinite number of solutions
7. $(-18, 4, 2)$ **8.** $(-1, 5, 5)$
9. Their associated systems of equations have the same solution.
10. $(16, -5, 2)$ **11.** $\begin{cases} 2l + 2w = 90 \\ 2l - 3w = 0 \end{cases}$
12. $\begin{cases} x + y = 300 \\ x - 3y = 0 \end{cases}$
13. $\begin{cases} 0.94x + 0.92y + 0.80z = 20,144 \\ 0.04x + 0.02y + 0.04z = 766 \\ 0.02x + 0.06y + 0.16z = 1,990 \end{cases}$
14. $\begin{cases} 2x - y + z = 8 \\ x + 2y + 5z = 1 \\ x + z = 3 \end{cases}$

SECTION 4.4 *(page 341)*

Warm-Up *(page 341)*

1. 10   **2.** −56   **3.** −82   **4.** 153
5. 32   **6.** $-\frac{65}{3}$ **7.** −2
8. $\frac{9}{11}$ **9.** $\frac{6}{5}$ **10.** $-\frac{11}{3}$

---

1. 5   **3.** 27   **5.** 0   **7.** 6
9. −24   **11.** −0.16   **13.** −24   **15.** −2
17. −30   **19.** 3   **21.** 0   **23.** −75
25. −58   **27.** 0   **29.** −0.22   **31.** $x - 5y + 2$
33. (1, 2)   **35.** (2, −2)   **37.** $\left(\frac{3}{4}, -\frac{1}{2}\right)$
39. $D = 0$, cannot use Cramer's Rule   **41.** $\left(\frac{2}{3}, \frac{1}{2}\right)$
43. (−1, 3, 2)   **45.** $\left(1, \frac{1}{2}, \frac{3}{2}\right)$ **47.** (1, −2, 1)
49. $D = 0$, cannot use Cramer's Rule
51. 7   **53.** $\frac{31}{2}$ **55.** $\frac{53}{2}$
57. 250 mi^2 **59.** Collinear
61. Not collinear   **63.** $3x - 5y = 0$
65. $7x - 6y - 28 = 0$ **67.** $y = 2x^2 - 6x + 1$
69. $y = -3x^2 + 2x$ **71.** 248   **73.** $\left(\frac{51}{16}, -\frac{7}{16}, -\frac{13}{16}\right)$
75. (a) $\left(\frac{4k - 3}{2k - 1}, \frac{4k - 1}{2k - 1}\right)$ **(b)** $\frac{1}{2}$

SECTION 4.5 *(page 349)*

Warm-Up *(page 349)*

1.

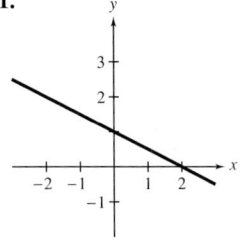

2.

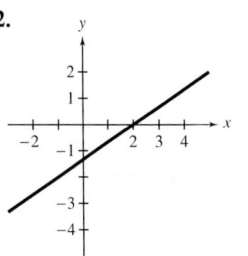

3.

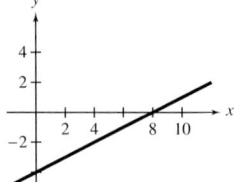

4.

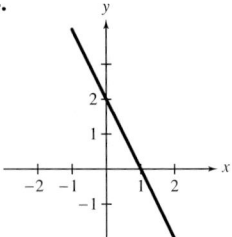

5.

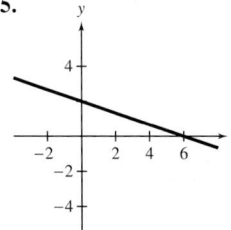

6.

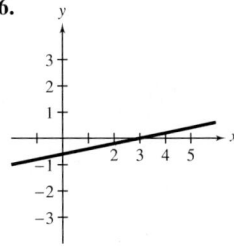

7.

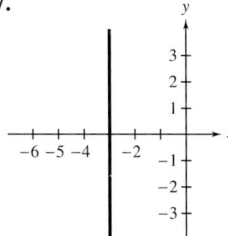

8.

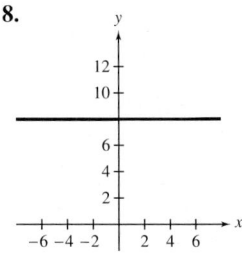

9.

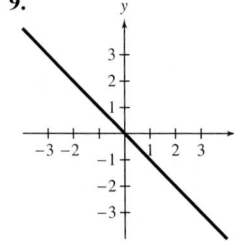

10.

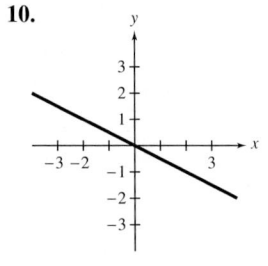

1. b   **2.** a   **3.** d   **4.** e   **5.** f   **6.** c
7. (a) Yes   **(b)** No   **(c)** Yes   **(d)** Yes
9. (a) No   **(b)** No   **(c)** Yes   **(d)** Yes
11. (a) Yes   **(b)** No   **(c)** No   **(d)** Yes
13. (a) No   **(b)** Yes   **(c)** No   **(d)** Yes

15.
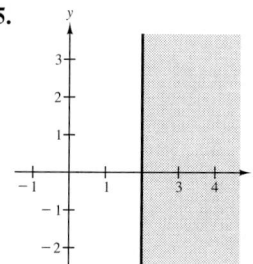

17.
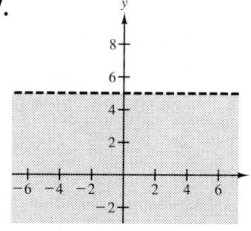

39. (a) and **(b)**
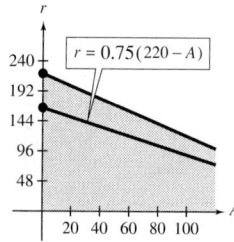

19.
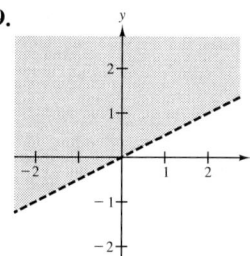

21.
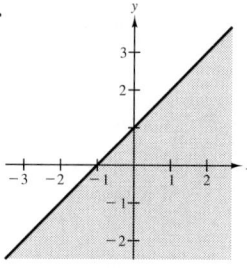

SECTION 4.6 *(page 361)*

Warm-Up *(page 360)*

1. $-11x$ **2.** $-41v$ **3.** $16x - 3$
4. $4y - 2$ **5.** 4 **6.** 11
7. Perpendicular **8.** Parallel
9. Parallel **10.** Neither

23.
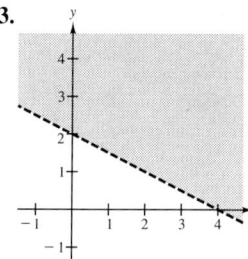

25.
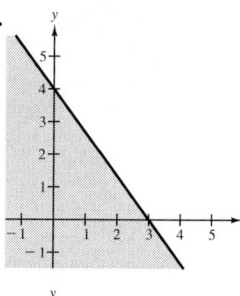

1. c **2.** b **3.** f **4.** e **5.** a **6.** d

7.
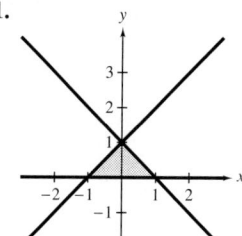

9.

27.
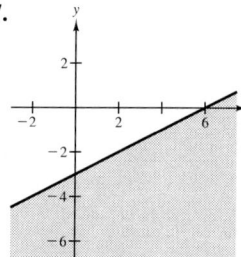

29.
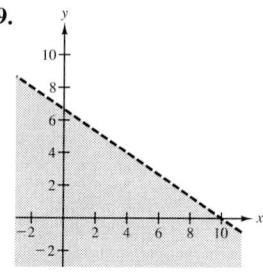

31. $3x + 4y > 17$ **33.** $y < 2$ **35.** $x - 2y < 0$
37. $2x + 2y \le 500$ or $y \le -x + 250,\ x \ge 0,\ y \ge 0$

11.
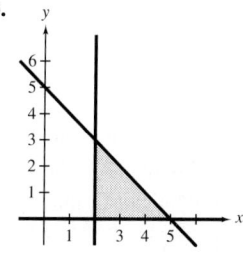

13.

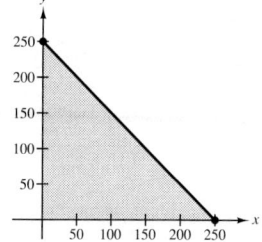

15.

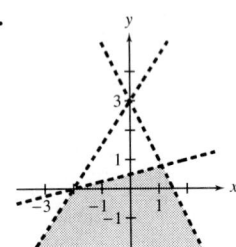

17.

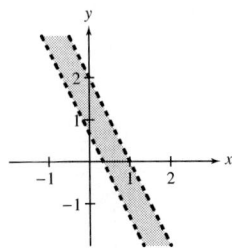

19.

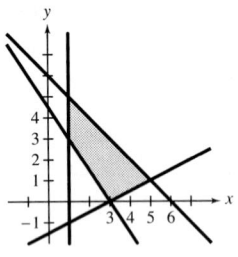

21. $x \geq 1$
$x \leq 8$
$y \geq -5$
$y \leq 3$

23. $y \leq \frac{9}{10}x + \frac{42}{5}$
$y \geq 3x$
$y \geq \frac{2}{3}x + 7$

25. $x \leq 90$
$y \leq 0$
$y \geq -10$
$y \geq -\frac{1}{7}x$

27. $x + y \leq 20{,}000$
$x \geq 5000$
$y \geq 5000$
$y \geq 2x$

29. $x + y \geq 15{,}000$
$15x + 25y \geq 275{,}000$
$x \geq 8000$
$y \geq 4000$

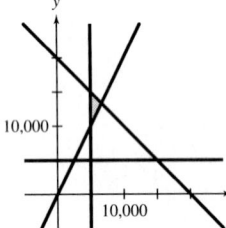

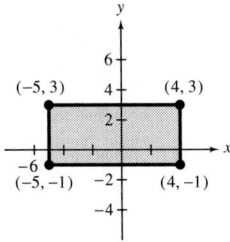

31. Minimum value at (0, 0): 0
Maximum value at (0, 6): 30

33. Minimum value at (0, 0): 0
Maximum value at (6, 0): 60

35. Minimum value at (0, 0): 0
Maximum value at (3, 4): 17

37. Minimum value at (0, 0): 0
Maximum value at (4, 0): 20

39. Minimum value at (0, 0): 0
Maximum value at (60, 20): 740

41. Minimum value at (0, 0): 0
Maximum value at (30, 45): 2325

43. Minimum value at (4, −1): −9
Maximum value at (−5, 3): 13

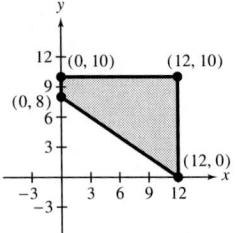

45. Minimum value at (12, 0): 12
Maximum value at (12, 10): 52

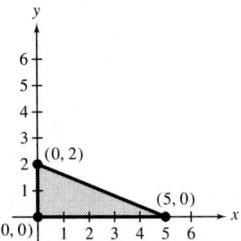

47. Minimum value at (0, 0): 0
Maximum value at (5, 0): 30

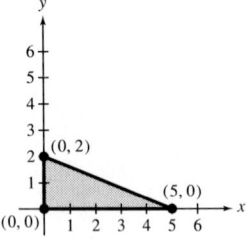

49. Minimum value at (0, 0): 0
Maximum value at (5, 0): 45

51. Minimum value at (5, 3): 35
No maximum value

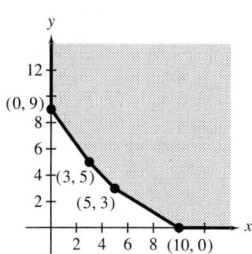

53. Minimum value at (10, 0): 20
No maximum value

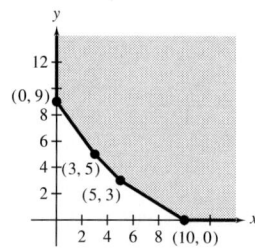

55. 750 units of Model A; 1000 units of Model B
57. 8 audits; 8 tax returns
59. 3 bags of Brand X; 6 bags of Brand Y

CHAPTER 4 REVIEW EXERCISES *(page 366)*

1. (1, 1)

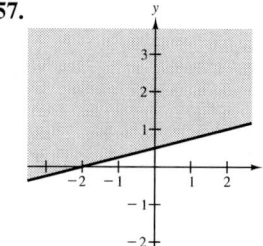

3. No solution

5. (4, 8)

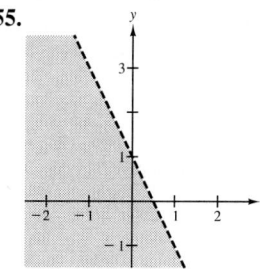

7. (2, −1)   **9.** No solution   **11.** (−10, −5)
13. (0, 0)   **15.** $\left(\frac{5}{2}, 3\right)$ **17.** (2, −3, 3)
19. (10, −12)   **21.** $\left(\frac{3}{5}, \frac{1}{2}\right)$
23. $\left(\frac{24}{5}, \frac{22}{5}, -\frac{8}{5}\right)$ **25.** (−3, 7)
27. $D = 0$, cannot use Cramer's Rule   **29.** (2, −3, 3)
31. 16   **33.** 7   **35.** $x - 2y + 4 = 0$
37. $2x + 6y - 13 = 0$
39. $\begin{cases} 3x + y = -2 \\ 6x + y = 0 \end{cases}$

41. 16,667 units
43. 75% solution: 40 gal; 50% solution: 60 gal
45. \$9.95 tapes: 400; \$14.95 tapes: 250
47. \$8000 at 7%; \$5000 at 9%; \$7000 at 11%
49. $y = 2x^2 + x - 6$ **51.** $x^2 + y^2 - 4x + 2y - 4 = 0$
53.

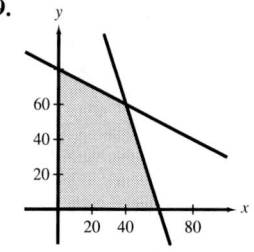

55.

57.

59.

61.

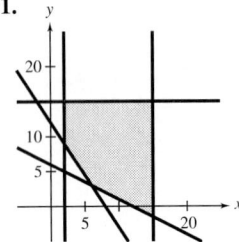

63. $\begin{cases} 2x + y \geq 7 \\ x - y \leq 2 \\ 2x + y \leq 22 \\ x - y \geq -4 \end{cases}$

65. $\begin{cases} x + y \leq 1500 \\ x \qquad \geq 400 \\ \qquad y \geq 600 \end{cases}$

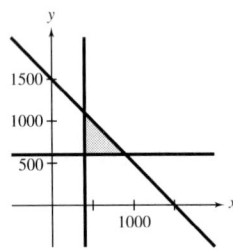

67. Maximum at (5, 8): 47
69. Minimum at (15, 0): 26.25
71. 5 units of Product A; 2 units of Product B

CHAPTER 4 TEST *(page 371)*

1. (b)   **2.** (3, 2)   **3.** (2, 4)
4. (−2, 2)   **5.** $\left(\frac{1}{4}, \frac{1}{3}\right)$ **6.** (2, 2a − 1, a)
7. (−1, 3, 3)   **8.** $\left(4, \frac{1}{7}\right)$ **9.** −62
10. $\begin{cases} x + 2y = -1 \\ x + y = 2 \end{cases}$ **11.** $\begin{cases} x + y = 200 \\ x = 4y \end{cases}$
                                    40 mi, 160 mi
12. $y = 2x^2 - 3x + 4$ **13.** 12

14.

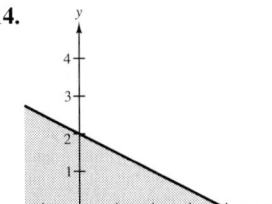

15.

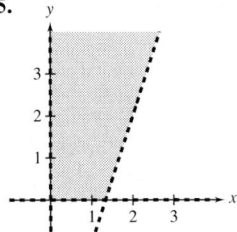

16.

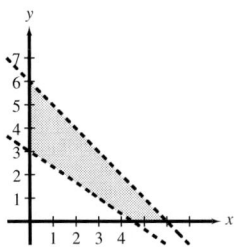

17. Minimum at (0, 0): 0
    Maximum at (3, 3): 48

18. Let x and y be the numbers of units of Model A and Model B, respectively. Then

Objective function: $C = 100x + 150y$

Constraints: $3.5x + 8y \leq 5600$
                    $2.5x + 2y \leq 2000$
                    $1.3x + 0.9y \leq 900$
                    $x \qquad \geq 0$
                    $\qquad y \geq 0$

Chapter 5

Chapter Opener *(page 373)*

Hummingbird ≈ 62 times per second

California gull ≈ 4 times per second

SECTION 5.1 *(page 382)*

Warm-Up *(page 382)*

1. -20 **2.** 70 **3.** -21 **4.** -27

5. $\frac{15}{26}$ **6.** $\frac{11}{23}$ **7.** $\frac{5}{8}$ **8.** $\frac{7}{12}$

9. $20a(x^2 + 3x + 4)$ **10.** $y^2(x-8)(2x+7)$
---

1. $(-\infty, 8) \cup (8, \infty)$ **3.** $(-\infty, -4) \cup (-4, \infty)$
5. $(-\infty, \infty)$ **7.** $(-\infty, -4) \cup (-4, 4) \cup (4, \infty)$
9. $(-\infty, -1) \cup (-1, 5) \cup (5, \infty)$
11. $(-\infty, 2) \cup (2, 4) \cup (4, \infty)$
13. $\{1, 2, 3, 4, \ldots\}$ **15.** $\dfrac{x}{5}$

17. $6y,\ y \neq 0$ **19.** $\dfrac{6x}{5y^3},\ x \neq 0$

21. $x,\ x \neq 8,\ x \neq 0$ **23.** $\frac{1}{2},\ x \neq \frac{3}{2}$

25. $-\dfrac{1}{3},\ x \neq 5$ **27.** $\dfrac{3y^2}{y^2 + 1},\ x \neq 0$

29. $\dfrac{y-8}{15},\ y \neq -8$ **31.** $\dfrac{1}{a+3}$

33. $\dfrac{x}{x-7}$ **35.** $\dfrac{y(y+2)}{y+6},\ y \neq 2$

37. $\dfrac{-1}{2x+3},\ x \neq 3$ **39.** $\dfrac{5x+4}{5x+2},\ x \neq \frac{1}{3}$

41. $\dfrac{5+3xy}{y^2},\ x \neq 0$ **43.** $\dfrac{3(m-2n)}{m+2n}$

45. $\dfrac{x^2 + 2xy + 4y^2}{x+2y},\ x \neq 2y$

47. $y + 5z,\ y^2 - 5yz + 25z^2$

49. $\dfrac{u-4}{v-2},\ v \neq 3$ **51.** $\frac{3}{2} \neq 9$

53. $\frac{16}{21} \neq \frac{3}{4}$ **55.** $x^n - 3,\ x \neq 0$

57. $x^n - 2,\ x^n \neq -2$

59.

-1	0	1	2	Undefined	4	5
-1	0	1	2	3	4	5

The two expressions are equivalent for all x-values except $x = 2$.

61. (a) 0 (b) $\dfrac{x-10}{4x}$ is undefined for $x = 0$

(c) $\frac{3}{2}$ (d) $\frac{1}{24}$

63. (a) $\dfrac{2500 + 9.25x}{x}$ (b) $\{1, 2, 3, 4, \ldots\}$ (c) \$34.25

65. $\frac{1}{9},\ x \neq 0$ **67.** π

69. $\dfrac{107.1 + 12.64t + 0.54t^2}{31.6 + 0.51t - 0.14t^2} \times 1000$

SECTION 5.2 *(page 392)*

Warm-Up *(page 392)*

1. $\frac{3}{10}$ **2.** $\frac{1}{20}$ **3.** $\frac{5}{56}$

4. $-\frac{15}{112}$ **5.** 5 **6.** $-\frac{4}{35}$

7. $(x-1)(x-2)$ **8.** $(x-1)(2x+7)$

9. $(x+1)(11x-5)$ **10.** $(2x-7)^2$
---

1. $\dfrac{3x}{2},\ x \neq 0$ **3.** $\dfrac{s^3}{6},\ s \neq 0$

5. $24u^2,\ u \neq 0$ **7.** 24

9. $\dfrac{2uv(u+v)}{3(3u+v)},\ u \neq 0$ **11.** $-1,\ r \neq 12$

13. $-\dfrac{x+8}{x^2},\ x \neq \frac{3}{2}$ **15.** $\dfrac{2(r+2)}{11r},\ r \neq 2$

17. $2t + 5,\ t \neq 3,\ t \neq -2$ **19.** $\dfrac{xy(x+2y)}{(x-2y)}$

21. $\dfrac{(x-y)^2}{x+y},\ x \neq -3y$

23. $\dfrac{(u+v)^2}{(u-v)^2},\ x \neq y,\ x \neq -y$

25. $\dfrac{(x-1)(2x+1)}{(3x-2)(x+2)},\ x \neq 5,\ x \neq -5,\ x \neq -1$

27. $\dfrac{3y^2}{2ux^2},\ v \neq 0$ **29.** $\dfrac{3}{2(a+b)}$

31. $x^4 y(x + 2y)$, $x \neq 0$, $y \neq 0$, $x \neq -2y$

33. $\dfrac{3x}{10}$, $x \neq 0$ **35.** $-\dfrac{5x}{2}$, $x \neq 0$, $x \neq 5$

37. $\dfrac{(x + 3)(4x + 1)}{(3x - 1)(x - 1)}$, $x \neq -3$, $x \neq -\dfrac{1}{4}$

39. $\dfrac{(x + 2)(x + 3)}{x}$, $x \neq -2$, $x \neq 2$

41. $\dfrac{(x + 1)(2x - 5)}{x}$, $x \neq -1$, $x \neq -5$, $x \neq -\dfrac{2}{3}$

43. $\dfrac{x^4}{(x^n + 1)^2}$, $x^n \neq -3$, $x^n \neq 3$, $x \neq 0$

45. $\dfrac{2w^2 + 3w}{6}$ **47.** $\dfrac{x}{4(2x + 1)}$ **49.** $\dfrac{\pi x}{4(2x + 1)}$

SECTION 5.3   *(page 403)*

Warm-Up   *(page 403)*

1. $\dfrac{7}{9}$ **2.** $\dfrac{1}{2}$ **3.** $\dfrac{3}{16}$ **4.** $-\dfrac{3}{5}$ **5.** $\dfrac{37}{36}$

6. $\dfrac{37}{48}$ **7.** $-\dfrac{146}{225}$ **8.** $\dfrac{181}{21}$ **9.** 26   **10.** $\dfrac{5}{64}$

1. $-\dfrac{2}{9}$ **3.** $\dfrac{2 + y}{2}$ **5.** $-\dfrac{3}{a}$

7. $\dfrac{x + 6}{3}$ **9.** $-\dfrac{4}{3}$ **11.** $\dfrac{1}{x - 3}$, $x \neq 0$

13. 1, $y \neq 6$ **15.** $20x^3$ **17.** $15x^2(x + 5)$

19. $6x(x + 2)(x - 2)$ **21.** x^2 **23.** $(u + 1)$

25. $-(x + 2)$ **27.** $\dfrac{2n^2(n + 8)}{6n^2(n - 4)}$, $\dfrac{10(n - 4)}{6n^2(n - 4)}$

29. $\dfrac{2(x + 3)}{x^2(x + 3)(x - 3)}$, $\dfrac{5x(x - 3)}{x^2(x + 3)(x - 3)}$

31. $\dfrac{(x - 8)(x - 5)}{(x + 5)(x - 5)^2}$, $\dfrac{9x(x + 5)}{(x + 5)(x - 5)^2}$

33. $\dfrac{25 - 12x}{20x}$ **35.** $\dfrac{7(a + 2)}{a^2}$ **37.** 0, $x \neq 4$

39. $\dfrac{3(x + 2)}{x - 8}$ **41.** 1, $x \neq \dfrac{2}{3}$

43. $\dfrac{5(5x + 22)}{x + 4}$ **45.** $\dfrac{1}{2x(x - 3)}$

47. $\dfrac{x^2 - 7x - 15}{(x + 3)(x - 2)}$ **49.** $\dfrac{x - 2}{x(x + 1)}$

51. $\dfrac{5(x + y)}{(x + 5y)(x - 5y)}$ **53.** $\dfrac{4}{x^2(x^2 + 1)}$

55. $\dfrac{x^2 + 3x + 9}{x(x - 3)(x + 3)}$ **57.** $\dfrac{4x}{(x - 4)^2}$

59. $\dfrac{2(4x^2 + 5x - 3)}{x^2(x + 3)}$ **61.** $\dfrac{x}{x - 1}$, $x \neq -6$

63. $\dfrac{3x + 2y + 3}{(3x + 2y)(2x - y)}$ **65.** $\dfrac{10x - 3}{(2x - 1)(x - 1)}$

67. $-\dfrac{5x + 3}{3x(x - 1)(3x + 1)}$ **69.** $\dfrac{y - x}{xy}$, $x \neq -y$

71. $\dfrac{5u - 2v}{(u - v)^2}$ **73.** $\dfrac{x}{2(3x + 1)}$

75. $\dfrac{4 + 3x}{4 - 3x}$, $x \neq 0$ **77.** $-4x - 1$, $x \neq 0$, $x \neq \dfrac{1}{4}$

79. $\dfrac{3}{4}$, $x \neq 0$, $x \neq 3$ **81.** $\dfrac{y + 1}{y - 3}$, $y \neq 0$, $y \neq 1$

83. $y - x$, $x \neq 0$, $y \neq 0$, $x \neq -y$ **85.** $\dfrac{5t}{12}$

87. $\dfrac{5x}{24}$ **89.** $\dfrac{x}{4}$, $\dfrac{x}{3}$, $\dfrac{5x}{12}$

91. (a) 12.7%   **(b)** $r = \dfrac{288(NM - P)}{N(12P + NM)}$, 12.7%

93. $\dfrac{R_1 R_2}{R_1 + R_2}$

95. $(-4, 2, 2)$

$$\dfrac{4}{x^3 - x} = -\dfrac{4}{x} + \dfrac{2}{x + 1} + \dfrac{2}{x - 1}$$

CHAPTER 5 MID-CHAPTER QUIZ   *(page 407)*

1. $\dfrac{5x}{2y^3}$, $x \neq 0$ **2.** $\dfrac{-1}{2x + 1}$, $x \neq 4$

3. $36x^2(x + 2)^2$ **4.** $5x(x + 1)(x - 1)$

5. $\dfrac{3(x + 4)}{x^2}$, $x \neq 4$ **6.** $\dfrac{7(x + 1)}{xy(x - 2)}$

7. $\dfrac{x + 5}{2x + 3}$, $x \neq \dfrac{3}{2}$, $x \neq -5$ **8.** $\dfrac{x + 6}{3x^2}$

9. $\dfrac{-5(4x + 5)}{(x - 6)(x + 5)}$ **10.** $\dfrac{49x^2 + 24x - 5}{6x(x + 1)(x - 1)}$

11. $\dfrac{4(x + 2)}{1 - 16x}$, $x \neq 0$ **12.** $\dfrac{x + 5}{x}$, $x \neq -3$, $x \neq 3$

13. $\dfrac{7}{6x(x + 5)}$ **14.** $\dfrac{11t}{15}$

SECTION 5.4 *(page 417)*

Warm-Up *(page 417)*

1. $\dfrac{1}{4}$ **2.** $\dfrac{29}{90}$ **3.** $-\dfrac{4x}{3}$

4. $\dfrac{t^2}{9}$ **5.** $\dfrac{2u}{9v^2}$ **6.** $-\dfrac{r^3}{7}$

7. $-15x^4 + 12x^3$ **8.** $70y^3 - 20y$

9. $2x^3 + 7x^2 - 3x - 4$ **10.** $8x^2 + 10x + 3$

1. $3z + 5$ **3.** $\frac{5}{2}z^2 + z - 3$ **5.** $7x^2 - 2x,\ x \neq 0$

7. $-10z^2 - 6,\ z \neq 0$ **9.** $4z^2 + \frac{3}{2}z - 1,\ z \neq 0$

11. $m^3 + 2m - \dfrac{7}{m}$ **13.** $\dfrac{5}{2}x - 4 + \dfrac{7}{2}y,\ x \neq 0,\ y \neq 0$

15. $x + 1,\ x \neq -5$ **17.** $x - 5,\ x \neq 3$

19. $x + 10,\ x \neq -5$

21. $x + 7,\ x \neq 3$ **23.** $y + 3,\ y \neq -\frac{1}{2}$

25. $6t - 5,\ t \neq \frac{5}{2}$ **27.** $4x - 1,\ x \neq -\frac{1}{4}$

29. $x^2 - 5x + 25,\ x \neq -5$ **31.** $x^2 + 4,\ x \neq 2$

33. $2 + \dfrac{5}{x + 2}$ **35.** $x - 4 + \dfrac{32}{x + 4}$

37. $5x - 8 + \dfrac{19}{x + 2}$ **39.** $4x + 3 - \dfrac{11}{3x + 2}$

41. $\dfrac{6}{5}z + \dfrac{41}{25} + \dfrac{41}{25(5z - 1)}$

43. $2x^2 + x + 4 + \dfrac{6}{x - 3}$

45. $x^5 + x^4 + x^3 + x^2 + x + 1,\ x \neq 1$ **47.** $x + 2$

49. $x + 4,\ x^2 + 3x - 2 \neq 0$

51. $x^2 - 3x - 5,\ x^2 + x - 3 \neq 0$

53. $x^{2n} + x^n + 4,\ x^n \neq -2$ **55.** $(x - 7)(x - 8)$

57. $(2a - 5)(a + 9)$ **59.** $(15x + 10)\left(x - \frac{4}{5}\right)$

61. $(y^2 + 3y + 2)(y - 2)$ **63.** $(2t^2 + 5t - 6)(t + 5)$

65. $(x^3 - 3x + 1)(x - 1)$ **67.** $x^2 - x + 4 - \dfrac{17}{x + 4}$

69. $x^3 - 2x^2 - 4x - 7 - \dfrac{4}{x - 2}$

71. $5x^2 - 25x + 125 - \dfrac{613}{x + 5}$

73. $5x^2 + 14x + 56 + \dfrac{232}{x - 4}$

75. $10x^3 + 10x^2 + 60x + 360 + \dfrac{1360}{x - 6}$

77. $0.1x + 0.82 + \dfrac{1.164}{x - 0.2}$ **79.** -8

81. $2x,\ x \neq 0$ **83.** $7uv,\ u \neq 0,\ v \neq 0$

85. Invalid. 5's are terms, not factors.   **87.** Valid

89. $x^3 - 5x^2 - 5x - 10$ **91.** $x^2 - 3$ **93.** $2x + 8$

95. (a)

36	$\frac{51}{8}$	-12	16	148

(b)

36	$\frac{51}{8}$	-12	16	148

(c) The value of $f(c)$ is equal to the remainder when the polynomial is divided by $x - c$.

$f(4) = 66$

SECTION 5.5 *(page 428)*

Warm-Up *(page 428)*

1. $(x - 5)(x + 2)$ **2.** $(x - 5)(x - 2)$

3. $x(x + 1)(x + 3)$ **4.** $(x^2 - 2)(x - 4)$

5. 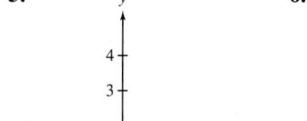   **6.**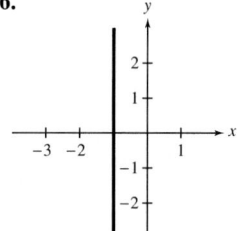

7. 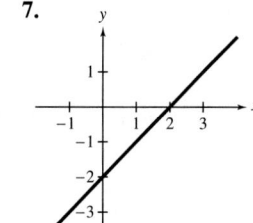   **8.**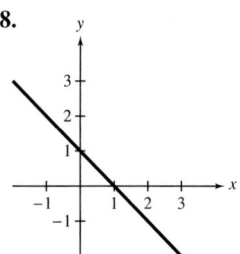

9. $5 + \dfrac{26}{x - 4}$ **10.** $-2 + \dfrac{7}{x + 2}$

1. e   **2.** f   **3.** a   **4.** b   **5.** c   **6.** d

7. Domain: all real numbers x such that $x \neq 5$
Vertical asymptote: $x = 5$
Horizontal asymptote: $y = 0$

9. Domain: all real numbers x such that $x \neq 0$
Vertical asymptote: $x = 0$
Horizontal asymptote: $y = 0$

11. Domain: all real numbers x such that $x \neq \pm 3$
Vertical asymptotes: $x = -3$, $x = 3$
Horizontal asymptote: $y = 0$

13. Domain: all real numbers x such that $x \neq 2$
Vertical asymptote: $x = 2$
Horizontal asymptote: $y = -1$

15. Domain: all real numbers x
Vertical asymptote: none
Horizontal asymptote: $y = 3$

17. Domain: all real numbers x such that $x \neq \pm 1$
Vertical asymptotes: $x = -1$, $x = 1$
Horizontal asymptote: none

19.
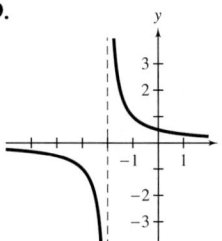

21.
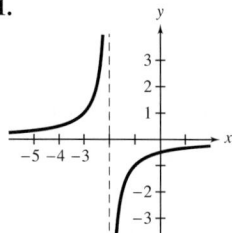

23.
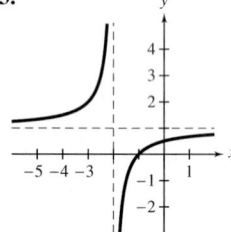

25.
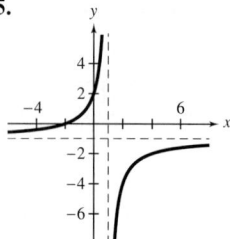

27.
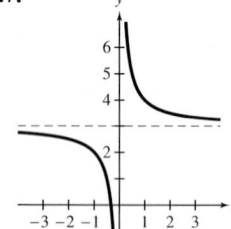

29.
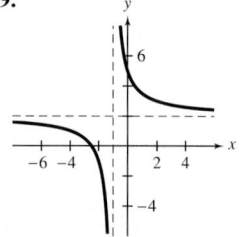

31.
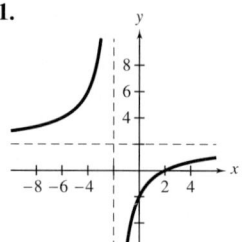

33.
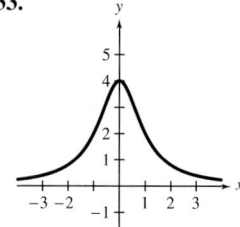

35.
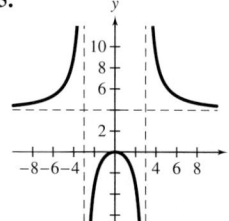

37.
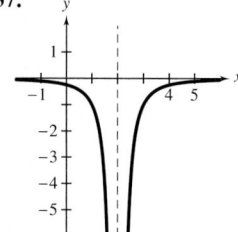

39.
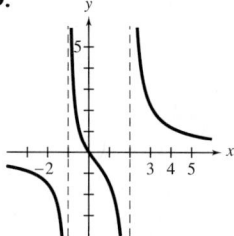

41. **(a)** 28.33 million dollars
(b) 170 million dollars
(c) 765 million dollars
(d) No

43. **(a)** 167, 250, 400
(b) 750; The limiting size of the population
(c)
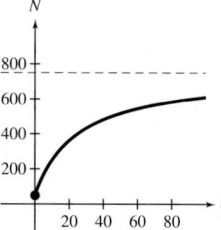

45.

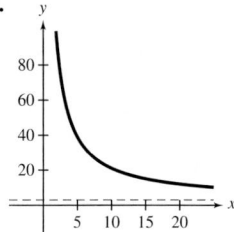

$$\overline{C} = \frac{3x + 180}{x}, \ x > 0, \ \overline{C} = 3$$

As the number of movies rented increases, the average cost per video approaches the rental price.

USING A GRAPHING CALCULATOR *(page 432)*

1.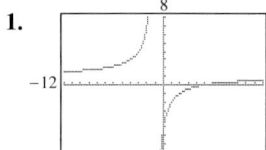

3.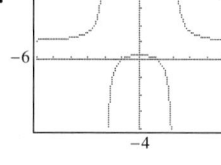

5.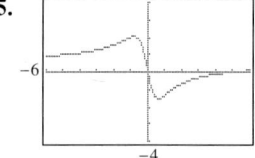

7. $g(x) = x + \dfrac{1}{x + 1}$

For large values of $|x|$, $g(x)$ is close to $f(x) = x$ because $\dfrac{1}{|x| + 1}$ is very small.

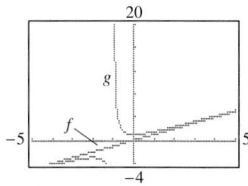

SECTION 5.6 *(page 441)*

Warm-Up *(page 441)*

1. $\frac{16}{3}$ **2.** -19 **3.** 3 **4.** $\frac{5}{2}$

5. $\frac{2}{3}, -8$ **6.** $16, -\frac{3}{2}$ **7.** $0, \frac{21}{2}$

8. 5 **9.** $6, -7$ **10.** $0, 8$

---

1. (a) No **(b)** No **(c)** No **(d)** Yes
3. (a) No **(b)** Yes **(c)** Yes **(d)** No
5. $\frac{3}{2}$ **7.** $\frac{1}{4}$ **9.** $-\frac{8}{3}$ **11.** 10 **13.** $\frac{7}{4}$
15. $\frac{1}{3}$ **17.** 61 **19.** $\frac{18}{5}$ **21.** 3 **23.** 3
25. $-\frac{11}{5}$ **27.** $\frac{4}{3}$ **29.** $-\frac{4}{15}$ **31.** No solution
33. No solution   **35.** $1, \frac{3}{2}$ **37.** 3 **39.** 1
41. ± 6 **43.** ± 4 **45.** $8, -9$ **47.** $3, 13$
49. $0, 2$ **51.** $8, -3$ **53.** $2, 3$ **55.** $8, \frac{1}{8}$
57. 40 mph   **59.** 15 **61.** 85%
63. 3 hr; $1\frac{7}{8}$ min; $1\frac{2}{3}$ hr
65. (a) $\dfrac{1}{20}$ min   **(b)** $\dfrac{x}{20}$ min   **(c)** $1\dfrac{3}{4}$ min
67. 15 hr; $22\frac{1}{2}$ hr   **69.** $11\frac{1}{4}$ hr
71. 1986 **73.** 1984

CHAPTER 5 REVIEW EXERCISES *(page 445)*

1. $(-\infty, 8) \cup (8, \infty)$
3. $(-\infty, 1) \cup (1, 6) \cup (6, \infty)$
5. $\dfrac{2x^3}{5}, \ x \neq 0, \ y \neq 0$
7. $\dfrac{b - 3}{6(b - 4)}$ **9.** $-9, \ x \neq y$
11. $\dfrac{y}{8x}, \ y \neq 0$ **13.** $12z(z - 6), \ z \neq -6$
15. $-\frac{1}{4}, \ u \neq 0, \ v \neq 3$ **17.** $3x^2, \ x \neq 0$
19. $\dfrac{125y}{x}, \ y \neq 0$ **21.** $\dfrac{x(x - 1)}{x - 7}, \ x \neq -1, \ x \neq 1$
23. $-\dfrac{7}{9}a$ **25.** $\dfrac{-48x + 35}{48x}$
27. $\dfrac{4x + 3}{(x + 5)(x - 12)}$ **29.** $\dfrac{5x^3 - 5x^2 - 31x + 13}{(x + 2)(x - 3)}$
31. $\dfrac{x + 24}{x(x^2 + 4)}$ **33.** $\dfrac{6(x - 9)}{(x + 3)^2(x - 3)}$

35. $\dfrac{6(x+5)}{x(x+7)}$, $x \neq 5$, $x \neq -5$ **37.** $\dfrac{3t^2}{5t-2}$, $t \neq 0$

39. $\dfrac{-a^2 + a + 16}{(4a^2 + 16a + 1)(a-4)}$ **41.** $2x^2 - \dfrac{1}{2}$, $x \neq 0$

43. $2x^2 + x - 1 + \dfrac{1}{3x-1}$

45. $x^2 - 2$, $x \neq 1$, $x \neq -1$ **47.** $x^2 + 5x - 7$

49. $x^3 + 3x^2 + 6x + 18 + \dfrac{29}{x-3}$

51. d   **53.** c

55.

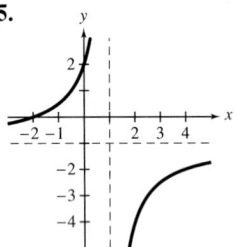

57.

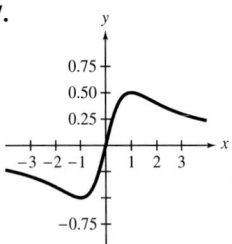

59.

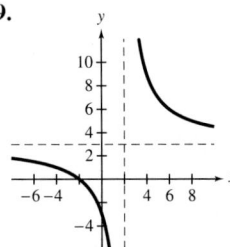

61.

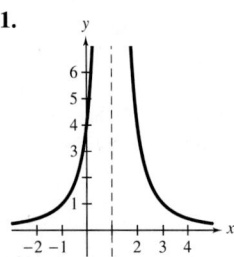

63.

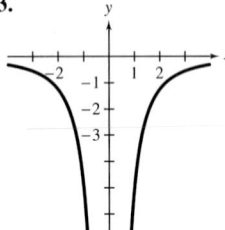

65.

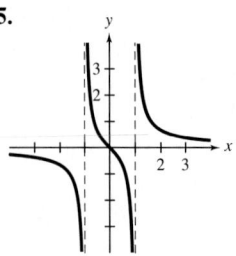

67.

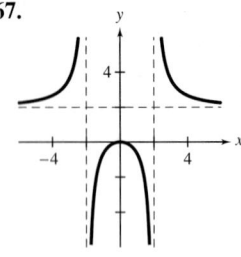

69. Horizontal asymptote: $y = \frac{1}{2}$
As x gets larger, the average cost per unit approaches $0.50.

71. (a) 176 million dollars   **(b)** 528 million dollars
(c) 1584 million dollars   **(d)** No

73. -40 **75.** $\frac{3}{2}$ **77.** 5   **79.** 6, -4

81. 2, -6 **83.** -2, 2   **85.** $-\frac{9}{5}$, 3   **87.** 56 mph

89. 4   **91.** $6\frac{2}{3}$ min

CHAPTER 5 TEST   *(page 449)*

1. $(\infty, -5) \cup (-5, 5) \cup (5, \infty)$

2. $x^3(x+3)^2(x-3)$ **3.** $-\frac{1}{3}$, $x \neq 2$

4. $\dfrac{2a+3}{5}$, $a \neq 4$ **5.** $\dfrac{5z}{3}$, $z \neq 0$

6. $\dfrac{4}{y+4}$, $y \neq 2$ **7.** $\dfrac{(2x+3)^2}{x+1}$, $x \neq \dfrac{3}{2}$

8. $\dfrac{14y^2}{15}$, $x \neq 0$ **9.** $\dfrac{x^3}{4}$, $x \neq -2$

10. 4, $x \neq -1$ **11.** $-\dfrac{2x^2 - 2x - 1}{x+1}$

12. $\dfrac{5x^2 - 15x - 2}{(x+2)(x-3)}$ **13.** $\dfrac{5x^3 + x^2 - 7x - 5}{x^2(x+1)^2}$

14. $-(3x+1)$ **15.** $t^2 + 3 - \dfrac{6t-6}{t^2-2}$

16. $2x^3 + 6x^2 + 3x + 9 + \dfrac{20}{x-3}$

17.

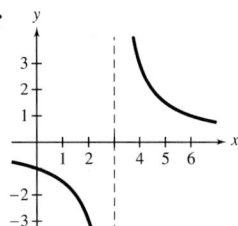

18.

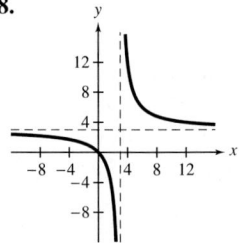

19. 22   **20.** -1, $-\frac{15}{2}$

21. No solution   **22.** $6\frac{2}{3}$ hr, 10 hr

Chapter 6

Chapter Opener   *(page 450)*

≈ 29.47 yr

SECTION 6.1   *(page 459)*

Warm-Up   *(page 459)*

1. a^{10} **2.** $x^5 y^4$ **3.** $9y^3$ **4.** $4rs^2$

5. $-t^6$ **6.** $16z^2$ **7.** $36x^5 y^8$

8. $-128u^5 v^7$ **9.** $\dfrac{16a^8}{81b^4}$ **10.** $\dfrac{9x^2}{16y^6}$

---

1. $\frac{1}{25}$ **3.** $-\frac{1}{1000}$ **5.** $-\frac{1}{243}$ **7.** 64 **9.** -32

11. $\frac{3}{2}$ **13.** 1 **15.** 1 **17.** 729 **19.** $100{,}000$

21. $\frac{1}{64}$ **23.** $\frac{1}{16}$ **25.** $\frac{3}{16}$ **27.** $\frac{64}{121}$ **29.** $\frac{16}{15}$

31. y^2 **33.** x^6 **35.** x^6 **37.** a **39.** t^2

41. $-\dfrac{12}{xy^3}$ **43.** $\dfrac{y^4}{9x^4}$ **45.** $\dfrac{3x^5}{y^4}$ **47.** $\dfrac{81v^8}{u^6}$

49. $\dfrac{x^{12} y^8}{16}$ **51.** $\dfrac{1}{4x^4}$ **53.** $\dfrac{z^2}{16}$ **55.** $\dfrac{125x^9}{y^{12}}$

57. 1 **59.** $\dfrac{10}{x}$ **61.** $\dfrac{b^5}{a^5}$ **63.** $\dfrac{t^{32}}{s^{32}}$

65. $4096\, m^6$ **67.** $\dfrac{1}{x^8 y^{12}}$ **69.** $\dfrac{1}{a^4 b^4}$

71. $\dfrac{v^2}{uv^2 + 1}$ **73.** $\dfrac{ab}{b - a}$ **75.** 5.75×10^7

77. 9.461×10^{15} **79.** 8.99×10^{-5}

81. $524{,}000{,}000$ **83.** $13{,}000{,}000$

85. 0.00000000048 **87.** $680{,}000$ **89.** 4000

91. $9{,}000{,}000{,}000{,}000{,}000$

93. 4.703×10^{11} **95.** 3.463×10^{10} **97.** $4x^8$

99. $80{,}000{,}000{,}000{,}000{,}000{,}000{,}000$

101. 1.58×10^{-5} yr ≈ 8.3 min

103. 0.333×10^6 or $333{,}000$

105. $\$11{,}587.63$

SECTION 6.2   *(page 471)*

Warm-Up   *(page 470)*

1. 169 **2.** -169 **3.** $\frac{8}{27}$ **4.** $\frac{3}{10}$

5. 48 **6.** $\dfrac{27}{8}$ **7.** $\dfrac{1}{x^3}$ **8.** $-\dfrac{15y^4}{x^2}$

9. $\dfrac{9y^2}{4x^2}$ **10.** $\dfrac{v^4}{u^5}$

---

1. 7 **3.** 4.2 **5.** Square root   **7.** 9

9. $\frac{3}{4}$ **11.** 8 **13.** Not a real number

15. $-\frac{2}{3}$ **17.** 0.4 **19.** 5 **21.** $-\frac{1}{4}$ **23.** 3

25. Not a real number   **27.** -0.3 **29.** 13

31. 14 **33.** 13 **35.** -20 **37.** 16

39. $16^{1/2} = 4$ **41.** $27^{2/3} = 9$ **43.** $\sqrt[4]{256^3} = 64$

45. 5 **47.** 8 **49.** $\frac{1}{4}$ **51.** $\frac{4}{9}$ **53.** $\frac{3}{11}$

55. 1 **57.** Irrational   **59.** Rational

61. 8.5440 **63.** 1.0420 **65.** 0.0038 **67.** 4.3004

69. 66.7213 **71.** $|t|$ **73.** y^3 **75.** x^3

77. $\dfrac{9y^{3/2}}{x^{2/3}}$ **79.** $\dfrac{3y^2}{4z^{4/3}}$ **81.** $x^{1/4}$ **83.** $c^{1/2}$

85. $\sqrt[8]{y}$ **87.** $2x^{3/2} - 3x^{1/2}$ **89.** $1 + 5y$

91. All real numbers $x \geq 0$ **93.** All real numbers x

95. All real numbers $x > 0$ **97.** 12.8%

99. 13 in. $\times$ 13 in. $\times$ 13 in.   **101.** 0.026 in.

SECTION 6.3   *(page 480)*

Warm-Up   *(page 480)*

1. 100 **2.** -10 **3.** $-\frac{3}{5}$ **4.** $\frac{2}{5}$

5. $x^{11/6}$ **6.** $a^{1/4}$ **7.** $16x^4$ **8.** $4096x^2 y^8$

9. $\dfrac{16}{x^{1/4} y^{1/4}}$ **10.** $\dfrac{8y^2}{z^{5/2}}$

---

1. $\sqrt{30}$ **3.** $\sqrt[3]{110}$ **5.** $\sqrt{\frac{15}{31}}$ **7.** $\sqrt[5]{\frac{152}{3}}$

9. $3\sqrt{35}$ **11.** $3\sqrt[4]{11}$ **13.** $\dfrac{\sqrt{35}}{3}$ **15.** $\dfrac{\sqrt[3]{11}}{10}$

17. $2\sqrt{5}$ **19.** $3\sqrt{3}$ **21.** 0.2 **23.** $2\sqrt[3]{3}$

25. $10\sqrt[4]{3}$ **27.** $\dfrac{\sqrt{15}}{2}$ **29.** $\dfrac{\sqrt[3]{35}}{4}$ **31.** $\dfrac{\sqrt[5]{15}}{3}$

33. 0.3 **35.** $3x^2\sqrt{x}$ **37.** $xy\sqrt[3]{x}$ **39.** $2xy\sqrt[5]{y}$

41. $\dfrac{\sqrt{13}}{5}$ **43.** $\dfrac{2\sqrt[5]{x^2}}{y}$ **45.** $\dfrac{3a\sqrt[3]{2a}}{b^3}$

47. $4y^2\sqrt{3x}$ **49.** $\dfrac{4a^2\sqrt{2}}{|b|}$ **51.** $3x^2$

53. $2x\sqrt[5]{3}$ **55.** $\dfrac{2\sqrt[3]{2}}{3}$ **57.** $\dfrac{\sqrt{3}}{3}$

59. $4\sqrt{3}$ **61.** $\dfrac{\sqrt[4]{20}}{2}$ **63.** $\dfrac{3\sqrt[3]{2}}{2}$

65. $\dfrac{\sqrt{y}}{y}$ **67.** $\dfrac{\sqrt{3x}}{x}$ **69.** $\dfrac{2\sqrt{x}}{x^2}$

71. $\dfrac{\sqrt[3]{18xy^2}}{3y}$ **73.** $\dfrac{a^2\sqrt[3]{a^2b}}{b}$ **75.** $\dfrac{2\sqrt{3b}}{b^2}$

77. $\dfrac{b\sqrt[4]{ab}}{a}$ **79.** $\dfrac{\sqrt[3]{2178x^2z^2}}{11z^2}$ **81.** $2\sqrt{2}$

83. $-5\sqrt[4]{7}-11\sqrt[4]{3}$ **85.** $24\sqrt{2}-6$ **87.** $30\sqrt[3]{2}$

89. These radical expressions cannot be combined.

91. $13\sqrt{y}$ **93.** $(10-z)\sqrt[3]{z}$ **95.** $\dfrac{2\sqrt{5}}{5}$

97. $\dfrac{9\sqrt{5}}{5}$ **99.** 5.667 **101.** 1.993

103. $\sqrt{7}+\sqrt{18}>\sqrt{7+18}$ **105.** $5>\sqrt{3^2+2^2}$

107. 89.44 cycles/sec

109. ≈ 175 vibrations/sec ≈ 932 vibrations/sec

111. 1

CHAPTER 6 MID-CHAPTER QUIZ *(page 483)*

1. $\dfrac{1}{8}$ **2.** $\dfrac{1}{64}$ **3.** $\dfrac{16}{9}$ **4.** $\dfrac{1}{6}$ **5.** 3

6. 8 **7.** $-125x^3$ **8.** $12x^4$ **9.** $\dfrac{y^2}{2}$

10. $\dfrac{7x^{1/6}}{6}$ **11.** $6\sqrt{2}$ **12.** $\dfrac{2c\sqrt{10c}}{5d^3}$

13. $2x\sqrt[3]{2xy}$ **14.** $14\sqrt{2x}-3\sqrt{x}$

15. 1.3837×10^{-2} **16.** 1.573×10^{10}

17. $\dfrac{\sqrt{15y}}{5y}$ **18.** $\dfrac{\sqrt[3]{6t^2}}{2t}$

19. Factor of 4 **20.** 24 in. × 24 in. × 24 in.

SECTION 6.4 *(page 488)*

Warm-Up *(page 488)*

1. $34-5x$ **2.** $18x-28$ **3.** $t^2+\frac{2}{3}t+\frac{1}{9}$
4. $4x^2-20x+25$ **5.** $25x^2-16$
6. $2-\frac{23}{2}x-3x^2$ **7.** $2xy\sqrt[3]{2x}$
8. $6xy\sqrt{2y}$ **9.** $(x-4)\sqrt{3}$ **10.** $2\sqrt{7x}$

1. 4 **3.** $2\sqrt{5}-\sqrt{15}$ **5.** $2\sqrt{10}+8\sqrt{2}$ **7.** -1
9. 4 **11.** $8\sqrt{5}+24$ **13.** $\sqrt{15}+3\sqrt{3}-5\sqrt{5}-15$
15. $y+4\sqrt{y}$ **17.** $9x+25$ **19.** $2x+20\sqrt{2x}+100$
21. $45x-17\sqrt{x}-6$ **23.** $x-y$
25. $2-7\sqrt[3]{4}$ **27.** $\sqrt[3]{4x^2}+10\sqrt[3]{2x}+25$
29. $2y-10\sqrt[3]{2y}+10\sqrt[3]{4y^2}-100$ **31.** $(x+3)$
33. $(4-3x)$ **35.** $(2u+\sqrt{2u})$
37. $\dfrac{1-2\sqrt{x}}{3}$ **39.** $\dfrac{-1+\sqrt{3y}}{4}$
41. $2-\sqrt{5},\ -1$ **43.** $\sqrt{11}+\sqrt{3},\ 8$
45. $\sqrt{x}+3,\ x-9$ **47.** $\sqrt{2u}+\sqrt{3},\ 2u-3$
49. $\dfrac{\sqrt{22}+2}{3}$ **51.** $-4\sqrt{7}+12$
53. $\dfrac{4\sqrt{7}+11}{3}$ **55.** $\dfrac{x\sqrt{15}+x\sqrt{3}}{4}$
57. $\dfrac{t\sqrt{5t}+t\sqrt{t}}{2}$ **59.** $\dfrac{2x-9\sqrt{x}-5}{4x-1}$
61. $192\sqrt{2}$ in.^{2}

SECTION 6.5 *(page 496)*

Warm-Up *(page 496)*

1. 2 **2.** $\frac{9}{2}$ **3.** $\frac{27}{5}$ **4.** $\frac{1}{6}$
5. $2,-25$ **6.** $12,-114$ **7.** $x-9$
8. $\sqrt{5u}$ **9.** $4t+12\sqrt{t}+9$ **10.** $25\sqrt{2}x$

1. Yes **3.** Yes **5.** Yes **7.** 400 **9.** 49
11. No solution **13.** 525 **15.** 90 **17.** $\frac{44}{3}$
19. $\frac{14}{25}$ **21.** $-\frac{2}{3}$ **23.** 4 **25.** No solution
27. 7 **29.** -15 **31.** 8 **33.** $1,3$ **35.** $\frac{1}{2}$

37. $\frac{4}{5}$ **39.** ± 4 **41.** 56.57 ft/sec
43. 56.25 ft **45.** 64 ft **47.** 1.82 ft
49. 500 units **51.** $h^2 = \dfrac{s^2 - \pi^2 r^4}{\pi^2 r^2}$

USING A GRAPHING CALCULATOR   *(page 501)*

1.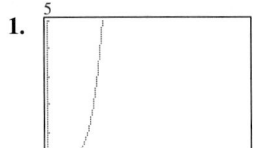

3.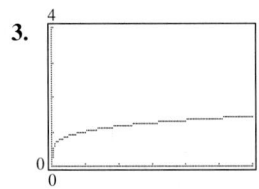

5.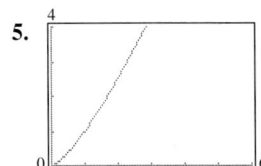

7.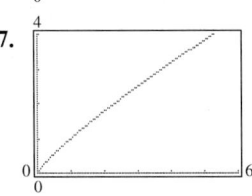

9.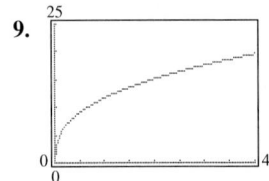

As the weight of the primate increases, the chest circumference increases. The *rate* of increase of the chest circumference is highest at lower weights, and the *rate* of growth of the chest circumference becomes lower as the weight continues to increase.

SECTION 6.6   *(page 509)*

Warm-Up *(page 509)*

1. $4\sqrt{3}$ **2.** $6\sqrt{3}$ **3.** $23\sqrt{2}$ **4.** 150
5. $\dfrac{4\sqrt{10}}{5}$ **6.** $\dfrac{5(\sqrt{3}+1)}{4}$ **7.** $0, \dfrac{35}{2}$
8. $-4, \frac{5}{4}$ **9.** $-2, 9$ **10.** $-1, \frac{3}{2}$

1. $2i$ **3.** $-12i$ **5.** $\frac{2}{5}i$ **7.** $0.3i$
9. $2\sqrt{2}i$ **11.** $\sqrt{7}i$ **13.** $10i$ **15.** $3\sqrt{2}i$
17. -4 **19.** $-3\sqrt{6}$ **21.** $-2\sqrt{3}-3$
23. $5\sqrt{2}-4\sqrt{5}$ **25.** -6 **27.** -16
29. $a=3, b=-4$ **31.** $a=-4, b=-2\sqrt{2}$
33. $a=2, b=-2$ **35.** $10+4i$ **37.** $-14-40i$
39. 13 **41.** $-14+20i$ **43.** $\frac{13}{6}+\frac{3}{2}i$
45. $-3+49i$ **47.** -36 **49.** $-36i$ **51.** $27i$
53. $-65-10i$ **55.** $20-12i$ **57.** $-40-5i$
59. $-7-24i$ **61.** $-21+20i$ **63.** $-24-2\sqrt{3}i$
65. $2+11i$ **67.** $2-i, 5$ **69.** $5+\sqrt{6}i, 31$
71. $-2+8i, 68$ **73.** $-10i, 100$ **75.** $1-\sqrt{3}i, 4$
77. $1.5-\sqrt{-0.25}, 2.5$ **79.** $2+2i$
81. $-\frac{24}{53}+\frac{84}{53}i$ **83.** $-\frac{6}{5}+\frac{2}{5}i$ **85.** $0-10i$
87. $\frac{3}{8}-\frac{1}{2}i$ **89.** $\frac{8}{5}-\frac{1}{5}i$ **91.** $1-\frac{6}{5}i$
93. $-\frac{53}{25}+\frac{29}{25}i$ **95.** $2a+0i$ **97.** $0+2bi$
99. $(-1+2i)^2 + 2(-1+2i) + 5$
$$= 1 - 4i + 4i^2 - 2 + 4i + 5$$
$$= (-4+4)i + (1-4-2+5)$$
$$= 0$$

CHAPTER 6 REVIEW EXERCISES   *(page 512)*

1. $\frac{1}{72}$ **3.** $\frac{125}{8}$ **5.** 3.6×10^7 **7.** $8x$
9. $\dfrac{x^6}{y^8}$ **11.** $\dfrac{1}{t^3}$ **13.** $\dfrac{27}{y^3}$ **15.** 1.2
17. $\frac{5}{6}$ **19.** 12 **21.** $11,414.13$ **23.** 10.63
25. $16^{1/2}=4$ **27.** 81 **29.** 0.04 **31.** $x^{7/12}$
33. $\dfrac{3}{x^{1/4}y^{2/5}}$ **35.** $6\sqrt{10}$ **37.** $0.5x^2\sqrt{y}$
39. $2ab\sqrt[3]{6b}$ **41.** $\dfrac{\sqrt{30}}{6}$ **43.** $\dfrac{\sqrt{3x}}{2x}$
45. $\dfrac{\sqrt[3]{4x^2}}{x}$ **47.** $-24\sqrt{10}$ **49.** $7\sqrt[4]{y+3}$
51. $12\sqrt{5}+41$ **53.** $6\sqrt{2}+7\sqrt{6}-4\sqrt{3}-14$
55. $\dfrac{x+20\sqrt{x}+100}{x-100}$ **57.** 225 **59.** 105
61. $-3, -5$ **63.** 5 **65.** $\frac{3}{32}$ **67.** $4\sqrt{3}i$
69. $\frac{3}{4}-\sqrt{3}i$ **71.** $8-3i$ **73.** -90 **75.** 25
77. $59+74i$ **79.** $\frac{9}{17}+\frac{2}{17}i$ **81.** 256 ft

CHAPTER 6 TEST   *(page 514)*

1. $\frac{3}{8}$ **2.** 3×10^{-5} **3.** $\frac{1}{9}$ **4.** 6
5. 3.2×10^{-5} **6.** 30,400,000

7. $\frac{3}{5t}$ **8.** $\frac{y^2}{xy^2 + 1}$ **9.** $x^{1/3}$ **10.** 25

11. $\frac{4}{3}\sqrt{2}$ **12.** $2\sqrt[3]{3}$ **13.** $-\frac{\sqrt{6} + 9}{25}$

14. $\frac{\sqrt[3]{2}}{2}$ **15.** $-25\sqrt{3x}$ **16.** $5\sqrt{3x} + 3\sqrt{5}$
17. $16 - 8\sqrt{2x} + 2x$ **18.** $(4y + 3)$
19. No solution   **20.** 9   **21.** $2 - 2i$
22. $-5 - 12i$ **23.** $-8 + 4i$ **24.** $13 + 13i$
25. $-2 - 5i$ **26.** $-\frac{8}{41} + \frac{10}{41}i$ **27.** 100 ft

CUMULATIVE TEST: CHAPTERS 4-6   *(page 515)*

1. $(2, 1)$ **2.** $(3, -2)$ **3.** $(5, 4)$
4. $(1, -2, 3)$ **5.** 20   **6.** 8, 12, 24
7.

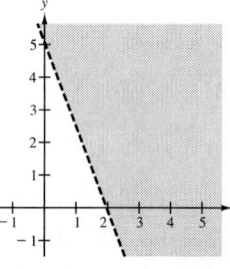

8. Minimum at $(3, 2)$: 19, Maximum at $(6, 2)$: 34
9. $\frac{x(x + 2)(x + 4)}{9(x - 4)}$, $x \neq 0$, $x \neq -4$

10. $\frac{3x + 5}{x(x + 3)}$ **11.** $x + y$, $x \neq y$, $x \neq 0$, $y \neq 0$
12. $x^2 - 3x + 9$
13.

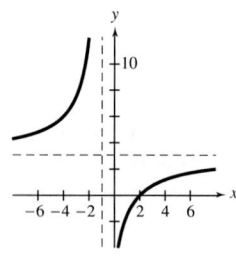

14. 2   **15.** $-\frac{2y^6}{3x^4}$ **16.** $t^{1/2}$ **17.** 1.6×10^7
18. $35\sqrt{5x}$ **19.** $2x - 6\sqrt{2x} + 9$ **20.** $\sqrt{10} + 2$
21. $-4 + 3\sqrt{2}i$ **22.** 6   **23.** 20 hr, 30 hr

Chapter 7

Chapter Opener   *(page 517)*

$t = 10$ sec

SECTION 7.1   *(page 526)*

Warm-Up   *(page 526)*

1. $(4x + 11)(4x - 11)$ **2.** $(3t - 4)^2$
3. $(x - 3)(5x + 2)$ **4.** $(x - 10)(x - 4)$
5. $y(2y - 1)(2y + 3)$ **6.** $4x(x^2 - 3x + 4)$
7. $-\frac{6}{5}$ **8.** $\frac{7}{2}$ **9.** $-9, \frac{15}{2}$ **10.** $\frac{8}{5}, 12$

---

1. 0, 3   **3.** 9, 12   **5.** $\pm\frac{5}{2}$ **7.** 6
9. -30 **11.** 1, 6   **13.** $\frac{1}{2}, -\frac{5}{6}$ **15.** $\frac{5}{3}, 6$
17. ± 3 **19.** ± 8 **21.** ± 8 **23.** $\pm\frac{4}{5}$
25. $\pm\frac{15}{2}$ **27.** 9, -17 **29.** 2.5, 3.5
31. $2 \pm \sqrt{7}$ **33.** $\frac{-1 \pm 5\sqrt{2}}{2}$ **35.** $\frac{3 \pm 7\sqrt{2}}{4}$
37. $\pm 6i$ **39.** $\pm 2i$ **41.** $3 \pm 5i$
43. $-\frac{4}{3} \pm 4i$ **45.** $-6 \pm \frac{11}{3}i$ **47.** $1 \pm 3\sqrt{3}i$
49. $-1 \pm 0.2i$ **51.** $\frac{2}{3} \pm \frac{1}{3}i$ **53.** $-\frac{7}{3} \pm \frac{\sqrt{38}}{3}i$
55. $\pm 2, \pm 4$ **57.** $\frac{1}{27}, -27$ **59.** $\pm 2\sqrt{2}$
61. ± 10 **63.** $-5, 15$ **65.** $5 \pm 10i$
67. $\pm\sqrt{15}, \pm\sqrt{5}i$ **69.** $\pm 2, \pm 2i$
71. $\pm\sqrt{3}, \pm 1$ **73.** $x^2 - 3x - 10 = 0$
75. $x^2 - 2x - 1 = 0$ **77.** 7.37 m   **79.** 10
81. 5.66   **83.** 4.33   **85.** 19.60 ft
87. 166 mi, 44 min   **89.** 3.31 mi   **91.** 9 sec
93. 1980

SECTION 7.2   *(page 537)*

Warm-Up   *(page 537)*

1. $x^2 + 6x + 8$ **2.** $x^2 + 20x + 150$
3. $u^2 - u + \frac{1}{4}$ **4.** $v^2 - 24v + 44$
5. $\pm\frac{2}{7}$ **6.** $\pm\frac{3}{4}i$ **7.** $4 \pm 6i$

8. $-2 \pm \sqrt[2]{5}$ **9.** $\frac{2}{3} \pm \frac{\sqrt{5}}{3}$ **10.** $\frac{3}{8}$, 12

---

1. 16   **3.** 100   **5.** $\frac{25}{4}$ **7.** $\frac{9}{25}$ **9.** $\frac{9}{100}$
11. 0.04   **13.** 0, 25   **15.** 1, 7   **17.** 4, -6
19. $-3, -4$ **21.** $-3, 6$ **23.** $\frac{3}{2}$, 4

25. $2 + \sqrt{7} \approx 4.65$
$2 - \sqrt{7} \approx -0.65$

27. $-2 + \sqrt{7} \approx 0.65$
$-2 - \sqrt{7} \approx -4.65$

29. $2 + \sqrt{3} \approx 3.73$
$2 - \sqrt{3} \approx 0.27$

31. $-1 + \sqrt{2}i \approx -1 + 1.41i$
$-1 - \sqrt{2}i \approx -1 - 1.41i$

33. $5 + 3\sqrt{3} \approx 10.20$
$5 - 3\sqrt{3} \approx -0.20$

35. $-10 + 3\sqrt{10} \approx -0.51$
$-10 - 3\sqrt{10} \approx -19.49$

37. $\frac{1 + 2\sqrt{7}}{3} \approx 2.10$
$\frac{1 - 2\sqrt{7}}{3} \approx -1.43$

39. $\frac{-5 + \sqrt{13}}{2} \approx -0.70$
$\frac{-5 - \sqrt{13}}{2} \approx -4.30$

41. $\frac{-3 + \sqrt{17}}{2} \approx 0.56$
$\frac{-3 - \sqrt{17}}{2} \approx -3.56$

43. $\frac{1}{2} + \frac{\sqrt{3}}{2}i \approx 0.5 + 0.87i$
$\frac{1}{2} - \frac{\sqrt{3}}{2}i \approx 0.5 - 0.87i$

45. $\frac{-4 + \sqrt{10}}{2} \approx -0.42$
$\frac{-4 - \sqrt{10}}{2} \approx -3.58$

47. $\frac{-9 + \sqrt{21}}{6} \approx -0.74$
$\frac{-9 - \sqrt{21}}{6} \approx -2.26$

49. $\frac{-1 + \sqrt{10}}{2} \approx 1.08$
$\frac{-1 - \sqrt{10}}{2} \approx -2.08$

51. $\frac{7 + \sqrt{57}}{2} \approx 7.27$
$\frac{7 - \sqrt{57}}{2} \approx -0.27$

53. $\frac{-5 + \sqrt{17}}{2} \approx -0.44$
$\frac{-5 - \sqrt{17}}{2} \approx -4.56$

55. 2.73, -0.73 **57.** 6.83
59. (a) $x^2 + 8x$ (b) $x^2 + 8x + 16$ (c) $(x + 4)^2$
61. 4 cm, 6 cm   **63.** 73.5 ft
65. 42 units, 58 units

SECTION 7.3   *(page 548)*

Warm-Up   *(page 548)*

1. 3, 7   **2.** 2, -16 **3.** 3, $-\frac{10}{3}$
4. $\frac{3}{5}$, 3   **5.** $\frac{-3 \pm \sqrt{17}}{2}$ **6.** $-1 \pm \frac{\sqrt{3}}{3}i$
7. 3   **8.** $\sqrt{11}i$ **9.** $4\sqrt{6}$ **10.** $3\sqrt{17}$

---

1. $2x^2 + 2x - 7 = 0$ **3.** $-x^2 + 10x - 5 = 0$
5. 4, 7   **7.** $-2, -4$ **9.** $-\frac{1}{2}$
11. $-\frac{3}{2}$ **13.** $-\frac{1}{2}, \frac{2}{3}$ **15.** $-15, 20$
17. 2 distinct imaginary solutions
19. 2 distinct irrational solutions
21. 2 distinct imaginary solutions
23. 1 (repeated) rational solution
25. $1 \pm \sqrt{5}$ **27.** $-2 \pm \sqrt{3}$ **29.** $-3 \pm 2\sqrt{3}$
31. $5 \pm \sqrt{2}$ **33.** $-\frac{3}{2} \pm \frac{\sqrt{3}}{2}i$ **35.** $\frac{1 \pm \sqrt{3}}{2}$
37. $\frac{-2 \pm \sqrt{10}}{2}$ **39.** $\frac{-1 \pm \sqrt{5}}{3}$ **41.** $\frac{1 \pm \sqrt{5}}{5}$
43. $\frac{-1 \pm \sqrt{10}}{5}$ **45.** ± 13 **47.** 0, -15
49. $\frac{9}{5}, \frac{21}{5}$ **51.** $-4 \pm 4i$ **53.** $\frac{-5 \pm 5\sqrt{17}}{12}$
55. 8, 16   **57.** $-4 \pm 3i$ **59.** 0.372, 3.228
61. 0.200, 99.800   **63.** 2.33, -1.08 **65.** 3.56
67. (a) $c < 9$ (b) $c = 9$ (c) $c > 9$
69. (a) $c < 16$ (b) $c = 16$ (c) $c > 16$
71. (a) 2.5 sec   (b) $\frac{5 + 5\sqrt{3}}{4} \approx 3.4$ sec
73. 1968

CHAPTER 7 MID-CHAPTER QUIZ   *(page 551)*

1. $0, 9$ **2.** $\pm\frac{5}{2}$ **3.** $4, -5$ **4.** $-\frac{2}{3}, \frac{3}{4}$

5. $\pm 2\sqrt{2}$ **6.** $1, 19$ **7.** $-3 \pm \sqrt{13}$

8. $\dfrac{3 \pm \sqrt{19}}{2}$ **9.** $\dfrac{1}{2}, 1$ **10.** $\dfrac{1 \pm \sqrt{11}i}{6}$

11. $-3 \pm 3i$ **12.** $\pm\sqrt{2}, \pm\sqrt{5}i$

13. $-3, 5$ **14.** $-1, 5$

15. (a) $c = \frac{49}{4}$ (b) $c < \frac{49}{4}$ (c) $c > \frac{49}{4}$

16. 200 units   **17.** 20 ft $\times$ 8 ft

USING A GRAPHING CALCULATOR   *(page 552)*

1. x-intercepts are -0.9 and 2.6

3. x-intercepts are -0.4 and -23.6

5. x-intercepts are -15.7 and 1.7

7. $10.9, 19.1$

SECTION 7.4   *(page 562)*

> ### Warm-Up   *(page 561)*
>
> **1.** -7 **2.** $0, \frac{2}{15}$ **3.** $\frac{5}{2}, -1$ **4.** $10, -12$
>
> **5.** $1, -17$ **6.** 1 **7.** $-\dfrac{3}{2} \pm \dfrac{\sqrt{11}}{2}i$
>
> **8.** $\dfrac{-3 \pm \sqrt{29}}{2}$ **9.** $6, -1$ **10.** $3, -1$

1. $15, 16$ **3.** $14, 16$ **5.** $12, 13$ **7.** 108 in.^{2}

9. 70 ft   **11.** 64 in.   **13.** 180 km^2 **15.** 440 m

17. 50 ft $\times$ 250 ft or 125 ft $\times$ 100 ft

19. Base: 24 in.; Height: 16 in.

21. No.

Area $= \frac{1}{2}(b_1 + b_2)h = \frac{1}{2}x\,[x + (550 - 2x)] = 43{,}560$

This equation has no real solution.

23. 8% **25.** 18 doz, $\$1.20$/doz   **27.** 48

29. 15.86 mi   **31.** 400 mph   **33.** 9.1 hr, 11.1 hr

35. 3 sec   **37.** 9.5 sec   **39.** 4.7 sec   **41.** 1988

SECTION 7.5   *(page 573)*

> ### Warm-Up   *(page 573)*
>
> **1.** 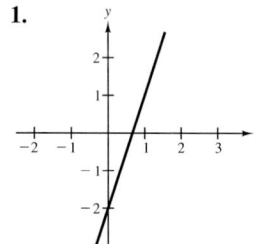   **2.**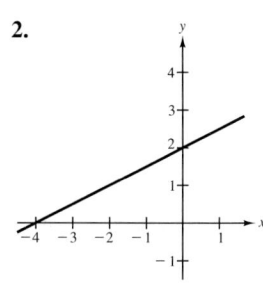
>
> **3.** 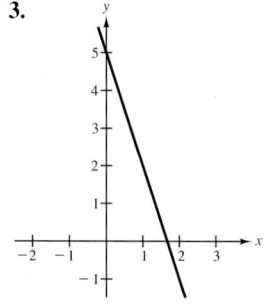   **4.**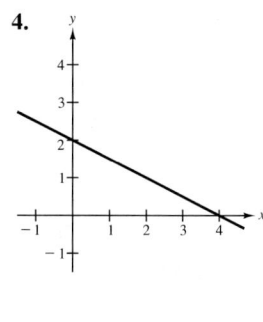
>
> **5.** $\frac{9}{2}$ **6.** 20 **7.** $0, -8$
>
> **8.** $\dfrac{-1 \pm \sqrt{41}}{4}$ **9.** $2x^2 + 20x + 20$
>
> **10.** $-3x^2 + 12x - 19$

1. e   **2.** f   **3.** b   **4.** c   **5.** d   **6.** a

7. Up, $(0, 2)$ **9.** Down, $(10, 4)$

11. Up, $(0, -6)$ **13.** Up, $(3, 0)$

15. $(-5, 0), (5, 0)$ **17.** $(0, 0), (9, 0)$

19. $\left(\frac{3}{2}, 0\right)$ **21.** No x-intercepts

23. $f(x) = (x - 2)^2 + 3, (2, 3)$

25. $f(x) = (x - 0)^2 + 2, (0, 2)$

27. $y = -(x - 1)^2 - 6, (1, -6)$

29. $y = 2\left[x - \left(-\frac{3}{2}\right)\right]^2 - \frac{5}{2}, \left(-\frac{3}{2}, -\frac{5}{2}\right)$

31.
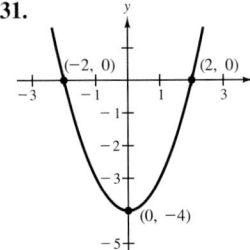

33.
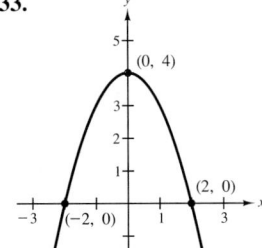

35.
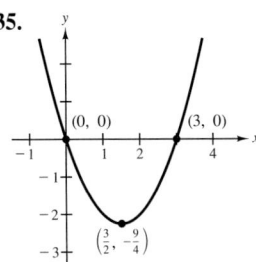

37.
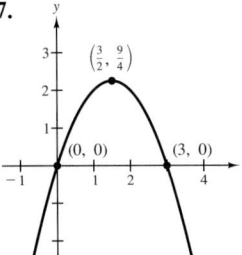

39.
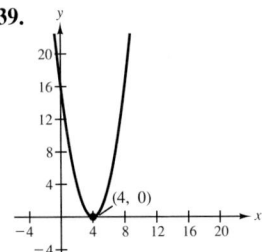

41.
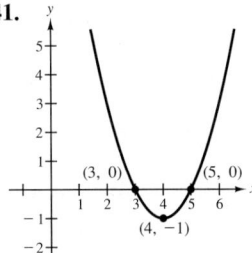

43.
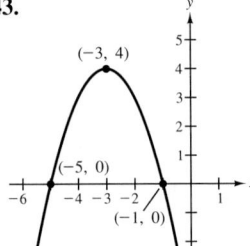

45.
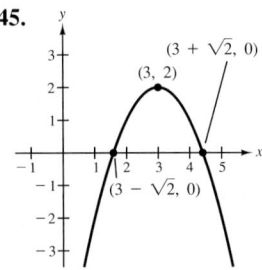

47.
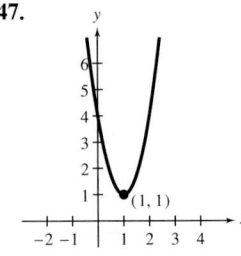

49.
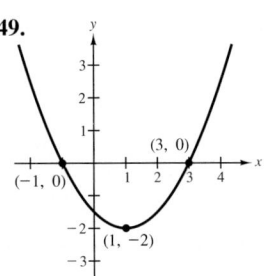

51.
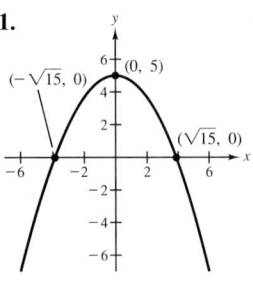

53.
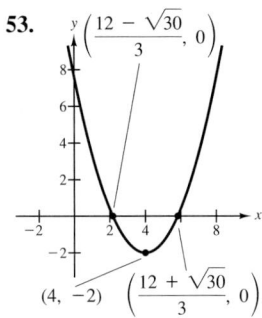

55. $x = -2, \ x = 2$

57. $x = 3 \pm \sqrt{2}$

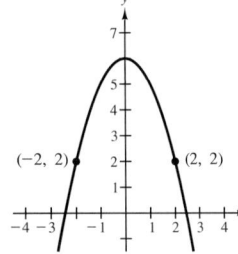

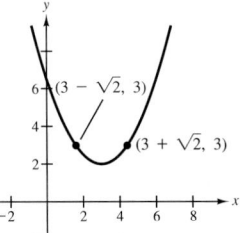

59. $y = -x^2 - 4x + 4$ **61.** $y = -x^2 - 4x$

63. $y = -2x^2 - 12x - 15$ **65.** $y = x^2 - 4x + 5$

67. $y = -x^2 - 6x - 5$ **69.** $y = x^2 - 4x$

71. $y = \frac{1}{2}x^2 - 3x + \frac{13}{2}$ **73.** $y = -4x^2 - 8x + 1$

75. $y = \frac{1}{25}x^2 - \frac{2}{5}x + 3$

77. (a)
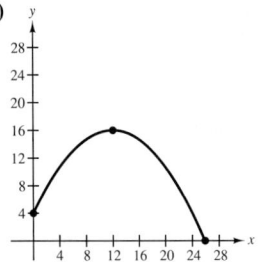

(b) 4 ft **(c)** 16 ft **(d)** $12 + 8\sqrt{3} \approx 25.9$ ft

79. (b) (50, 3375), 150 units

(c) Recommended only for orders between 100 and 150 units.

81. (a) 100 ft **(b)** 400 ft

(c)

x	± 100	± 200	± 300	± 400	± 500
y	16	64	144	256	400

83. Vertex: (2, 0.5)   **85.** Vertex: (−2, 4.9)

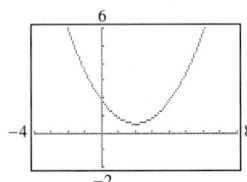   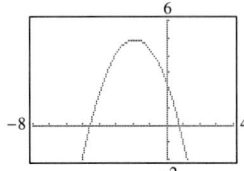

87. 50

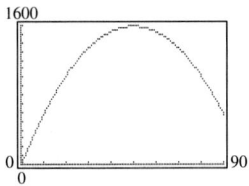

SECTION 7.6   *(page 588)*

Warm-Up   *(page 588)*

1. $x \le 5$

2. $x \ge -\frac{5}{3}$

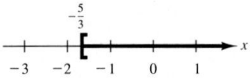

3. $x < \frac{3}{2}$

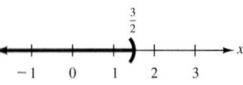

4. $x < 5$

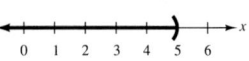

5. $1 < x < 5$

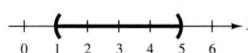

6. $x < 2,\ x > 8$

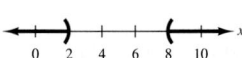

7. Negative   **8.** Zero
9. Positive   **10.** Negative

1. 3

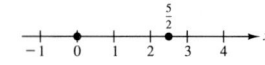

3. 0, $\frac{5}{2}$

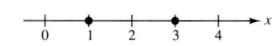

5. 1, 3

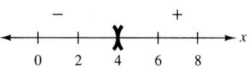

7. Negative: $(-\infty, 4)$
Positive: $(4, \infty)$

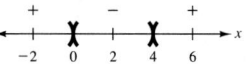

9. Negative: $(0, 4)$
Positive: $(-\infty, 0) \cup (4, \infty)$

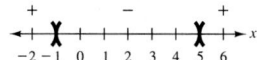

11. Negative: $(-1, 5)$
Positive: $(-\infty, -1) \cup (5, \infty)$

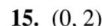

13. $[-3, \infty)$ **15.** $(0, 2)$

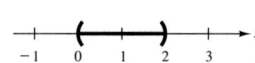

17. $(-\infty, 0) \cup (2, \infty)$ **19.** $(-\infty, -2] \cup [2, \infty)$

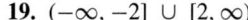

21. $[-5, 2]$ **23.** $\left(-\infty, -\frac{1}{2}\right) \cup (4, \infty)$

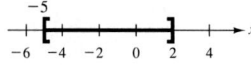   

25. No solution
27. $(-\infty, \infty)$

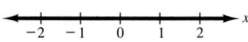

29. $(-\infty, 2 - \sqrt{2}) \cup (2 + \sqrt{2}, \infty)$

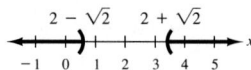

31. No solution

33. $(-\infty, 5 - \sqrt{6}) \cup (5 + \sqrt{6}, \infty)$

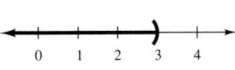

35. $(-2, 0) \cup (2, \infty)$

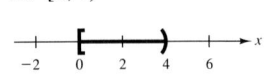

37. 3

39. 0, -5

41. $(3, \infty)$

43. $(-\infty, 3)$

45. $[0, 4)$

47. $(-\infty, 3] \cup \left[\frac{11}{2}, \infty\right)$

49. $(4, 7)$

51. $(-\infty, 3) \cup [6, \infty)$

53. $(3, 5)$ **55.** $(12, 20)$ **57.** $r > 7.24\%$

43.

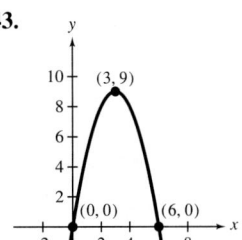

45.

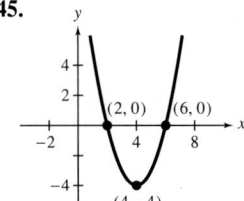

47. $y = -2x^2 + 12x - 13$ **49.** $y = \frac{1}{16}x^2 - \frac{5}{8}x + \frac{25}{16}$

51. $(-\infty, 3)$

53. $(0, 7)$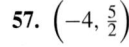

55. $(-\infty, -2] \cup [6, \infty)$ **57.** $\left(-4, \frac{5}{2}\right)$

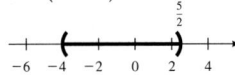

59. $(-\infty, 0] \cup \left(\frac{7}{2}, \infty\right)$

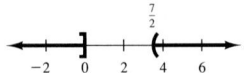

61. 11, 12   **63.** $\dfrac{5\sqrt{2}}{2}$ sec

65. (a) $P = 2l + 2w = 48$
$l + w = 24$
$w = 24 - l$

(b) $A = lw$
$w = 24 - l$
$A = l(24 - l)$

(c) 44, 80, 108, 128, 140, 144, 140, 128, 108

(d) The rectangle with the largest area has a length and width of 12. This rectangle is a square.

67. 15 people

69. $\dfrac{9 + 3\sqrt{17}}{2} \approx 10.68$ hr

$\dfrac{9 + 3\sqrt{17}}{2} + 3 \approx 13.68$ hr

71. 1993

CHAPTER 7 REVIEW EXERCISES   *(page 591)*

1. 0, -12 **3.** $-10, \frac{8}{3}$ **5.** $\pm\frac{1}{2}$

7. $-\frac{5}{2}$ **9.** $-4, 5$ **11.** $-9, 10$

13. ± 100 **15.** ± 1.5 **17.** $-4, 36$

19. $3 \pm 2\sqrt{3}$ **21.** $\dfrac{3}{2} \pm \dfrac{\sqrt{3}}{2}i$ **23.** $\dfrac{-5 \pm \sqrt{19}}{2}$

25. 5, -6 **27.** $3, -\dfrac{7}{2}$ **29.** $\dfrac{10}{3} \pm \dfrac{5\sqrt{2}}{3}i$

31. $3 \pm 5\sqrt{10}$ **33.** $-7, 12$ **35.** $-2, 20$

37. $3 \pm i$ **39.** $-\dfrac{1}{2}, -1$ **41.** $\dfrac{3 \pm \sqrt{17}}{2}$

CHAPTER 7 TEST   *(page 594)*

1. -5, 10 **2.** $-\frac{3}{8}$, 3 **3.** 1.7, 2.3

4. $-3 \pm 9i$ **5.** $\dfrac{9}{4}$ **6.** $\dfrac{3 \pm \sqrt{3}}{2}$

7. -56; two distinct imaginary solutions

8. $\dfrac{4 \pm \sqrt{7}}{3}$ **9.** 7.41 **10.** $x^2 - x - 20 = 0$

11.

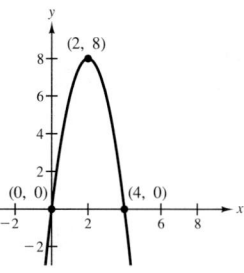

12. $y = \frac{2}{3}x^2 - 4x + 4$

13. $(0, 3)$

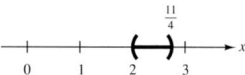

14. $(-\infty, -2] \cup [6, \infty)$

15. $\left(2, \frac{11}{4}\right)$

16. 14, 15 **17.** 12 ft $\times$ 20 ft

18. 50 mph   **19.** 1.58 sec   **20.** 120

Chapter 8

Chapter Opener   *(page 595)*

Width $= 150$ yd, Length $= 180$ yd

SECTION 8.1   *(page 602)*

Warm-Up   *(page 602)*

1. All real values x
2. All real values x such that $x \le 10$
3. All real values x such that $x \ne 5$
4. All real values x such that $x = 5$ and $x \ne -5$

5.

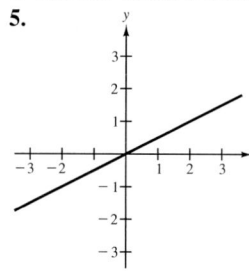

6.

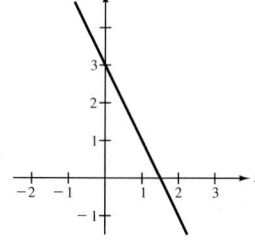

7.

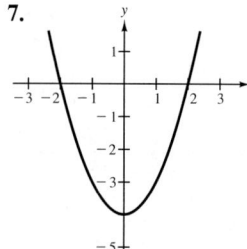

8.

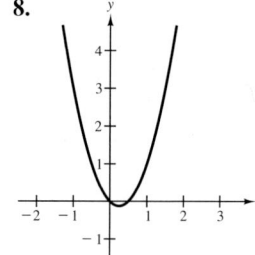

9.

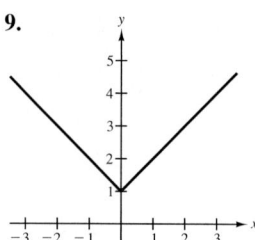

10.

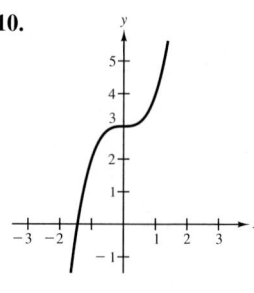

1. $(f + g)(x) = x^2 + 2x, \ (-\infty, \infty)$
$(f - g)(x) = 2x - x^2, \ (-\infty, \infty)$
$(fg)(x) = 2x^3, \ (-\infty, \infty)$
$\left(\dfrac{f}{g}\right)(x) = \dfrac{2}{x}, \ (-\infty, 0) \cup (0, \infty)$

3. $(f + g)(x) = x^2 + 4x - 12, \ (-\infty, \infty)$
$(f - g)(x) = -x^2 + 4x + 6, \ (-\infty, \infty)$
$(fg)(x) = 4x^3 - 3x^2 - 36x + 27, \ (-\infty, \infty)$
$\left(\dfrac{f}{g}\right)(x) = \dfrac{4x - 3}{x^2 - 9}, \ (-\infty, -3) \cup (-3, 3) \cup (3, \infty)$

5. $(f + g)(x) = \dfrac{1}{x} + 5x, \ (-\infty, 0) \cup (0, \infty)$
$(f - g)(x) = \dfrac{1}{x} - 5x, \ (-\infty, 0) \cup (0, \infty)$
$(fg)(x) = 5, \ (-\infty, 0) \cup (0, \infty)$
$\left(\dfrac{f}{g}\right)(x) = \dfrac{1}{5x^2}, \ (-\infty, 0) \cup (0, \infty)$

7. $(f + g)(x) = \sqrt{x + 4} + \sqrt{4 - x}, \ [-4, 4]$
$(f - g)(x) = \sqrt{x + 4} - \sqrt{4 - x}, \ [-4, 4]$
$(fg)(x) = \sqrt{16 - x^2}, \ [-4, 4]$
$\left(\dfrac{f}{g}\right)(x) = \dfrac{\sqrt{x + 4}}{\sqrt{4 - x}}, \ [-4, 4]$

9. (a) 11   **(b)** 0   **11. (a)** 19   **(b)** -79
13. (a) 10   **(b)** $\frac{2}{5}$
15. (a) $\frac{1}{3}$ **(b)** Undefined (division by zero)
17. (a) $x^2 - 3$ **(b)** $(x - 3)^2$ **(c)** 13   **(d)** 16
19. (a) $3|x - 1|$ **(b)** $3|x - 3|$ **(c)** 0   **(d)** 3
21. (a) $\sqrt{x + 5}$ **(b)** $\sqrt{x} + 5$ **(c)** 3   **(d)** 8
23. (a) $\dfrac{1}{\sqrt{x} - 3}$ **(b)** $\dfrac{1}{\sqrt{x - 3}}$ **(c)** $\dfrac{1}{4}$ **(d)** $\dfrac{1}{3}$
25. $f \circ g: (-\infty, \infty); \ g \circ f: (-\infty, \infty)$
27. $f \circ g: [2, \infty); \ g \circ f: [0, \infty)$
29. $f \circ g: (-\infty, \infty); \ g \circ f: (-\infty, -9) \cup (-9, \infty)$
31. $t^2 - 8t - 3$ **33.** $t^2 - 2t + 3$ **35.** 5
37. 10   **39.** $5y^2 - 15y + 3$
41. (a) -1 **(b)** -2 **(c)** -2
43. -1 **45.** -3
47.

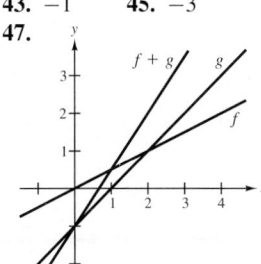

49.

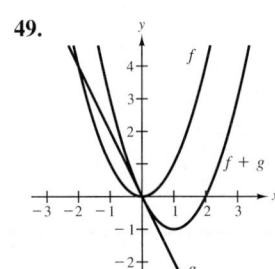

51. $T(x) = \frac{3}{4}x + \frac{1}{15}x^2$

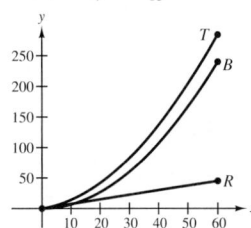

53. (a) $f(g(x)) = 0.02x - 200{,}000$
   (b) $g(f(x)) = 0.02(x - 200{,}000)$
   This part represents the bonus, because it gives 2% of sales over \$200,000.

55.

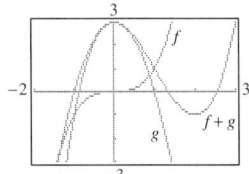

$f(x)$ is more significant in determining the magnitude of the sum.

SECTION 8.2   *(page 614)*

Warm-Up *(page 613)*

1. All real values x **2.** All real values x
3. Real values x such that $-4 \le x \le 4$
4. Real values x such that $x \ne -6$ and $x \ne 6$
5. $(f + g)(x) = x^2 + 3x - 9, \ (-\infty, \infty)$
   $(f - g)(x) = x^2 - 3x - 9, \ (-\infty, \infty)$
   $(fg)(x) = 3x^3 - 27x, \ (-\infty, \infty)$
   $\left(\dfrac{f}{g}\right)(x) = \dfrac{x^2 - 9}{3x}, \ (-\infty, 0) \cup (0, \infty)$
   $(f \circ g)(x) = 9x^2 - 9, \ (-\infty, \infty)$
   $(g \circ f)(x) = 3(x^2 - 9), \ (-\infty, \infty)$
6. $(f + g)(x) = \frac{1}{2}x^2 + 2x - 3, \ (-\infty, \infty)$
   $(f - g)(x) = -\frac{1}{2}x^2 + 2x - 3, \ (-\infty, \infty)$
   $(fg)(x) = x^3 - \frac{3}{2}x^2, \ (-\infty, \infty)$
   $\left(\dfrac{f}{g}\right)(x) = \dfrac{4x - 6}{x^2}, \ (-\infty, 0) \cup (0, \infty)$
   $(f \circ g)(x) = x^2 - 3, \ (-\infty, \infty)$
   $(g \circ f)(x) = \frac{1}{2}(2x - 3)^2, \ (-\infty, \infty)$

7. $(f+g)(x) = \sqrt[3]{x-2} + x^2,\ (-\infty, \infty)$

$(f-g)(x) = \sqrt[3]{x-2} - x^2,\ (-\infty, \infty)$

$(fg)(x) = x^2\sqrt[3]{x-2},\ (-\infty, \infty)$

$\left(\dfrac{f}{g}\right)(x) = \dfrac{\sqrt[3]{x-2}}{x^2},\ (-\infty, 0) \cup (0, \infty)$

$(f \circ g)(x) = \sqrt[3]{x^2 - 2},\ (-\infty, \infty)$

$(g \circ f)(x) = \sqrt[3]{(x-2)^2},\ (-\infty, \infty)$

8. $(f+g)(x) = \sqrt{x} + x^2 - 4,\ [0, \infty)$

$(f-g)(x) = \sqrt{x} - x^2 + 4,\ [0, \infty)$

$(fg)(x) = \sqrt{x}(x^2 - 4),\ [0, \infty)$

$\left(\dfrac{f}{g}\right)(x) = \dfrac{\sqrt{x}}{x^2 - 4},\ [0, 2) \cup (2, \infty)$

$(f \circ g)(x) = \sqrt{x^2 - 4},\ (-\infty, -2] \cup [2, \infty)$

$(g \circ f)(x) = x - 4,\ [0, \infty)$

9. $(f+g)(x) = \dfrac{2(x-1)}{x(x-2)},\ (-\infty, 0) \cup (0, 2) \cup (2, \infty)$

$(f-g)(x) = \dfrac{2}{x(2-x)},\ (-\infty, 0) \cup (0, 2) \cup (2, \infty)$

$(fg)(x) = \dfrac{1}{x(x-2)},\ (-\infty, 0) \cup (0, 2) \cup (2, \infty)$

$\left(\dfrac{f}{g}\right)(x) = \dfrac{x-2}{x},\ (-\infty, 0) \cup (0, 2) \cup (2, \infty)$

$(f \circ g)(x) = x - 2,\ (-\infty, 2) \cup (2, \infty)$

$(g \circ f)(x) = \dfrac{x}{1 - 2x},$

$\left(-\infty, \dfrac{1}{2}\right) \cup \left(\dfrac{1}{2}, 2\right) \cup (2, \infty)$

10. $(f+g)(x) = \dfrac{x^2 + 4x + 5}{x+2},\ (-\infty, -2) \cup (-2, \infty)$

$(f-g)(x) = \dfrac{x^2 + 4x + 3}{x+2},\ (-\infty, -2) \cup (-2, \infty)$

$(fg)(x) = 1,\ (-\infty, -2) \cup (-2, \infty)$

$\left(\dfrac{f}{g}\right)(x) = (x+2)^2,\ (-\infty, -2) \cup (-2, \infty)$

$(f \circ g)(x) = \dfrac{2x+5}{x+2},\ (-\infty, -2) \cup (-2, \infty)$

$(g \circ f)(x) = \dfrac{1}{x+4},\ (-\infty, -4) \cup (-4, \infty)$

1. $f^{-1}(x) = \dfrac{x}{5}$ **3.** $f^{-1}(x) = x - 10$

5. $f^{-1}(x) = \sqrt[7]{x}$ **7.** $f^{-1}(x) = x^3$

9. $f(g(x)) = f\left(\tfrac{1}{10}x\right) = 10\left(\tfrac{1}{10}x\right) = x$

$g(f(x)) = g(10x) = \tfrac{1}{10}(10x) = x$

11. $f(g(x)) = f(x - 15) = (x - 15) + 15 = x$

$g(f(x)) = g(x + 15) = (x + 15) - 15 = x$

13. $f(g(x)) = f\left(\dfrac{1-x}{2}\right)$

$\qquad = 1 - 2\left(\dfrac{1-x}{2}\right)$

$\qquad = 1 - (1 - x) = x$

$g(f(x)) = g(1 - 2x)$

$\qquad = \dfrac{1 - (1 - 2x)}{2}$

$\qquad = \dfrac{2x}{2} = x$

15. $f(g(x)) = f\left[\tfrac{1}{3}(2 - x)\right] = 2 - 3\left[\tfrac{1}{3}(2 - x)\right]$

$\qquad = 2 - (2 - x)$

$\qquad = x$

$g(f(x)) = g(2 - 3x) = \tfrac{1}{3}[2 - (2 - 3x)]$

$\qquad = \tfrac{1}{3}(3x)$

$\qquad = x$

17. $f(g(x)) = f(x^3 - 1) = \sqrt[3]{(x^3 - 1) + 1}$

$\qquad = \sqrt[3]{x^3}$

$\qquad = x$

$g(f(x)) = g\left(\sqrt[3]{x+1}\right) = \left(\sqrt[3]{x+1}\right)^3 - 1$

$\qquad = x + 1 - 1$

$\qquad = x$

19. $f(g(x)) = f\left(\dfrac{1}{x}\right) = \dfrac{1}{(1/x)} = x$

$g(f(x)) = g\left(\dfrac{1}{x}\right) = \dfrac{1}{(1/x)} = x$

21. b   **22.** c   **23.** d   **24.** a

25.

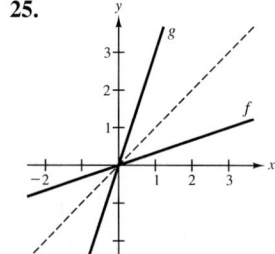

27.

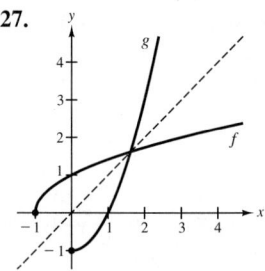

29.

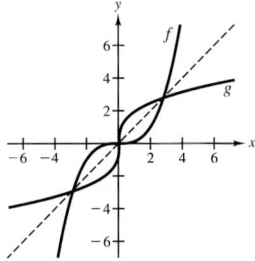

31. $f^{-1}(x) = \dfrac{x}{8}$ **33.** $g^{-1}(x) = x - 25$

35. $g^{-1}(x) = \dfrac{3 - x}{4}$ **37.** $h^{-1}(x) = x^2,\ x \geq 0$

39. $f^{-1}(t) = \sqrt[3]{t + 1}$ **41.** $g^{-1}(s) = \dfrac{5}{s}$

43. No   **45.** Yes   **47.** No
49. Yes   **51.** No   **53.** Yes
55. No   **57.** $x \geq 0,\ f^{-1}(x) = \sqrt[4]{x}$
59. $x \geq 2,\ f^{-1}(x) = \sqrt{x} + 2$

61.

x	0	1	3	4
f^{-1}	6	4	2	0

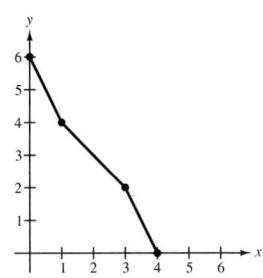

63.

x	-4	-2	2	3
f^{-1}	-2	-1	1	3

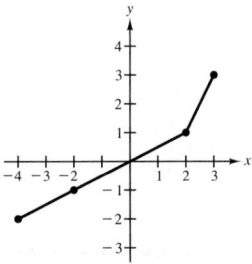

65. (a) $f^{-1}(x) = \dfrac{3 - x}{2}$ **(b)** $(f^{-1})^{-1}(x) = 3 - 2x$

67. $f^{-1}(x) = \sqrt[3]{x - 1}$

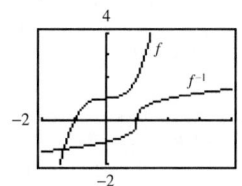

69. (a) $y = \frac{20}{13}(x - 9)$
(b) x: hourly wage; y: number of units produced
(c) 8

SECTION 8.3   *(page 625)*

Warm-Up   *(page 625)*

1. $\frac{1}{3}$ **2.** $\frac{10}{3}$ **3.** 0.92
4. 513   **5.** 40   **6.** $\frac{400}{7}$

7.

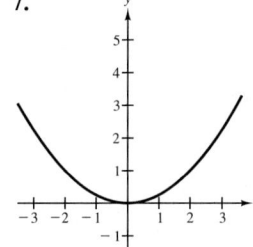

8.

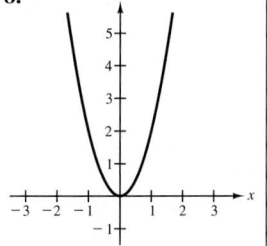

9.

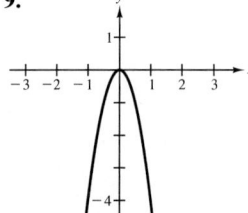

10.

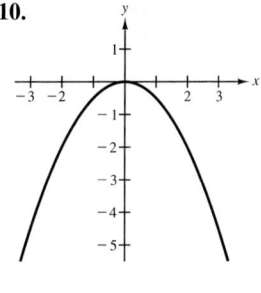

1. $I = kV$ **3.** $u = kv^2$ **5.** $p = \dfrac{k}{d}$

7. $P = \dfrac{k}{\sqrt{1+r}}$ **9.** $A = klw$

11. $P = \dfrac{k}{V}$ **13.** A varies jointly as b and h.

15. V varies jointly as the square of r and as h.

17. $s = 5t$ **19.** $F = \dfrac{5}{16}x^2$ **21.** $n = \dfrac{48}{m}$

23. $F = \dfrac{25}{6}xy$ **25.** $d = \dfrac{120x^2}{r}$ **27.** \$4921.25

29. (a) 2 in.   (b) 15 lb   **31.** 18 lb

33. $208\frac{1}{3}$ ft   **35.** 3072 watts   **37.** 324 lb

39. $p = \dfrac{114}{t}$, 17.5%   **41.** 667 units

43. $\frac{1}{4}$ **45.** 3125 lb

47.

x	2	4	6	8	10
$y = 1x^2$	4	16	36	64	100

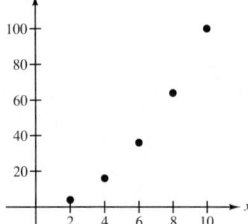

49.

x	2	4	6	8	10
$y = \frac{1}{2}x^2$	2	8	18	32	50

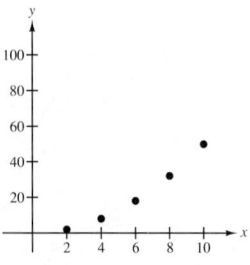

51.

x	2	4	6	8	10
$y = \dfrac{2}{x^2}$	$\dfrac{1}{2}$	$\dfrac{1}{8}$	$\dfrac{1}{18}$	$\dfrac{1}{32}$	$\dfrac{1}{50}$

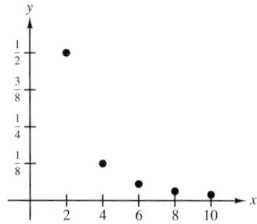

53.

x	2	4	6	8	10
$y = \dfrac{10}{x^2}$	$\dfrac{5}{2}$	$\dfrac{5}{8}$	$\dfrac{5}{18}$	$\dfrac{5}{32}$	$\dfrac{1}{10}$

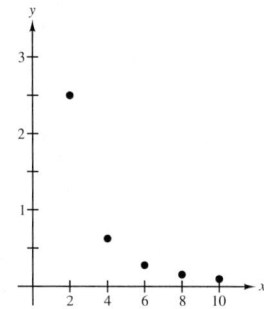

CHAPTER 8 MID-CHAPTER QUIZ   *(page 629)*

1. $-x^2 + 2x - 8$, $(-\infty, \infty)$ **2.** $x^2 + 2x - 8$, $(-\infty, \infty)$
3. $2x^3 - 8x^2$, $(-\infty, \infty)$ **4.** $x^2 - 2x + 8$, $(-\infty, \infty)$
5. $\dfrac{2x - 8}{x^2}$, $(-\infty, 0) \cup (0, \infty)$

6. $\dfrac{x^2}{2x - 8}$, $(-\infty, 4) \cup (4, \infty)$

7. $2x^2 - 8$, $(-\infty, \infty)$ **8.** $(2x - 8)^2$, $(-\infty, \infty)$
9. $C(x(t)) = 3000t + 750$, $\quad 0 \le t \le 40$
Cost of units produced in t hours

10. $f(g(x)) = 100 - 15\left[\frac{1}{15}(100 - x)\right]$

$= 100 - (100 - x) = x$

$g(f(x)) = \frac{1}{15}\left[100 - (100 - 15x)\right]$

$= \frac{1}{15}(15x) = x$

11. $f^{-1}(x) = \sqrt[3]{x + 8}$

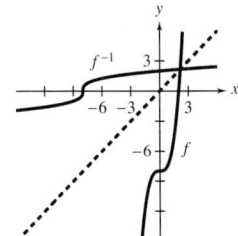

12. No. Let $x > 0$.　　**13.** $z = 3t$　　**14.** $m = \dfrac{14}{n}$

15. $S = 5hr^2$　　**16.** $N = \dfrac{15t^2}{s}$　　**17.** 30 min

SECTION 8.4　*(page 636)*

1. Horizontal line　　　**3.** Line with slope -3

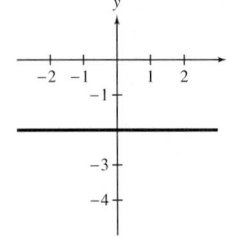　　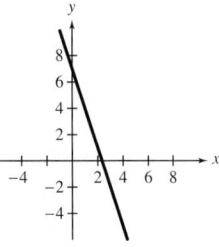

5. Parabola

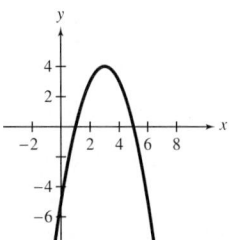

7. 4　　**9.** 1　　**11.** 3

13. Horizontal translation 5 units to the right

15. Vertical translation 3 units upward

17. Reflection in the x-axis

19.

(a) 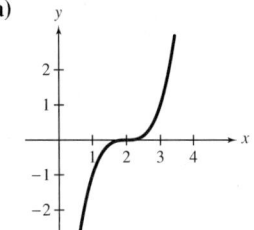　　**(b)**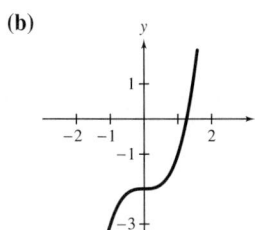

(c) 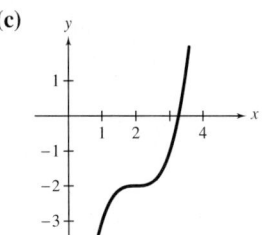　　**(d)**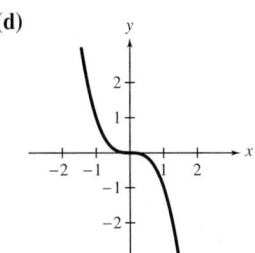

21. Rises to the left; Rises to the right

23. Falls to the left; Rises to the right

25. Rises to the left; Falls to the right

27. $(\pm 5, 0)$, $(0, -25)$　　**29.** $(3, 0)$, $(0, 9)$

31. $(-2, 0)$, $(1, 0)$, $(0, -2)$

33. $(2, 0)$, $(0, 0)$　　**35.** $(\pm 1, 0)$, $\left(0, -\frac{1}{2}\right)$

37. $(0, 10)$　　**39.** e　　**40.** c　　**41.** b

42. f　　**43.** a　　**44.** g　　**45.** d　　**46.** h

47.

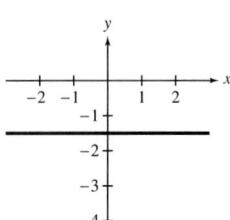

49.

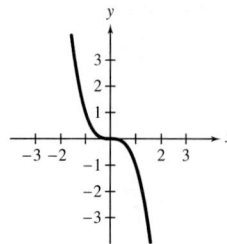

SECTION 8.5   *(page 650)*

Warm-Up   *(page 649)*

1. $y = \frac{1}{3}(9 - 2x)$ **2.** $y = \pm\sqrt{36 - x^2}$

3. $y = \pm\sqrt{\dfrac{6 - 2x^2}{3}}$ **4.** $y = \pm\sqrt{2x^2 - 3}$

5. $15\sqrt{2}$ **6.** $15\sqrt{6}$

7.

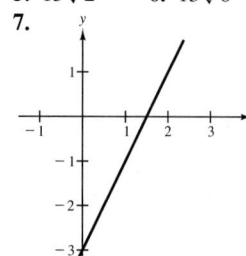

8.

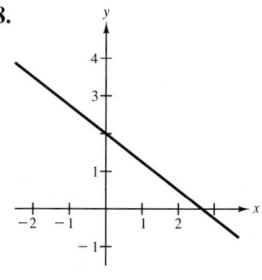

51.

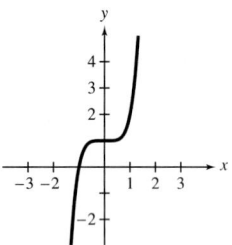

53.

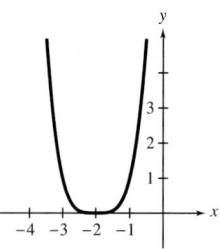

9.

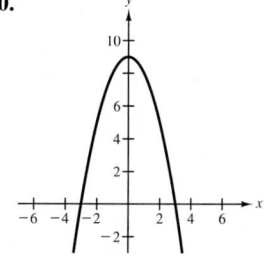

10.

55.

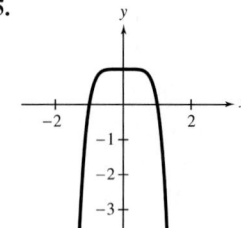

57.

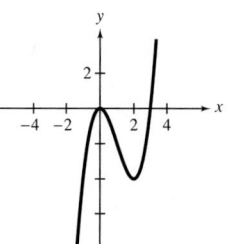

59.

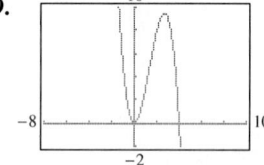

61.

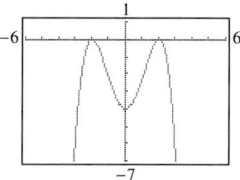

1. c   **2.** f   **3.** e   **4.** b   **5.** a   **6.** d
7. $x^2 + y^2 = 25$ **9.** $x^2 + y^2 = \frac{4}{9}$
11. $x^2 + y^2 = 64$ **13.** $x^2 + y^2 = 29$
15. Center: $(0, 0)$ **17.** Center: $(0, 0)$
     $r = 4$ $r = \frac{1}{2}$

63.

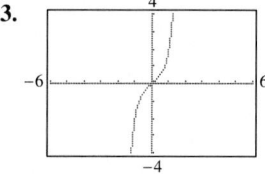

65. (b) $(0, 6)$
     (c)

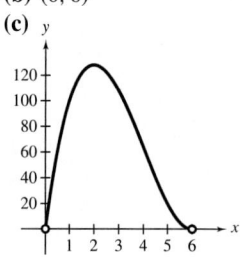

$x = 2$

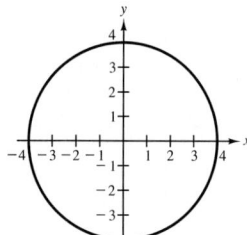

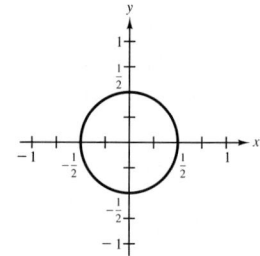

19. Center: (0, 0)

$r = \frac{12}{5}$

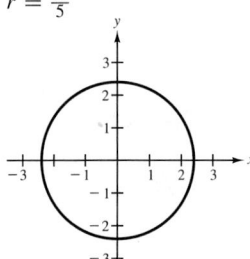

21. $\dfrac{x^2}{16} + \dfrac{y^2}{9} = 1$ **23.** $\dfrac{x^2}{9} + \dfrac{y^2}{16} = 1$

25. $\dfrac{x^2}{100} + \dfrac{y^2}{36} = 1$

27. Vertices: $(\pm 4, 0)$ **29.** Vertices: $(0, \pm 4)$
Co-vertices: $(0, \pm 2)$ Co-vertices: $(\pm 2, 0)$

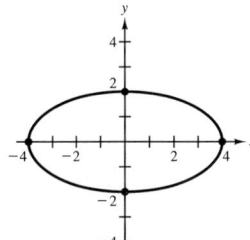

31. Vertices: $\left(\pm\frac{5}{3}, 0\right)$ **33.** Vertices: $(0, \pm 2)$
                                                   Co-vertices: $(\pm 1, 0)$
Co-vertices: $\left(0, \pm\frac{4}{3}\right)$

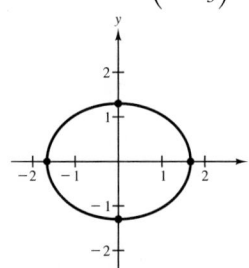 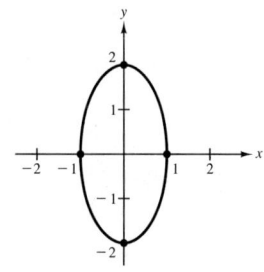

35. Vertices: $(\pm 3, 0)$ **37.** Vertices: $(0, \pm 3)$
Asymptotes: $y = \pm x$ Asymptotes: $y = \pm x$

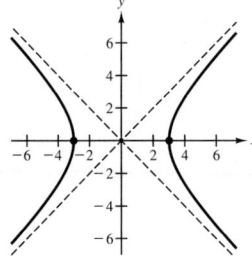 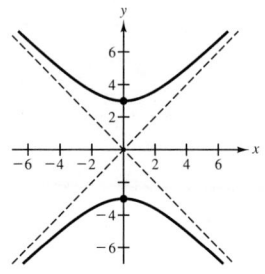

39. Vertices: $(\pm 3, 0)$ **41.** Vertices: $(0, \pm 3)$
Asymptotes: $y = \pm\frac{5}{3}x$ Asymptotes: $y = \pm\frac{3}{5}x$

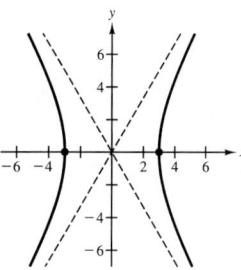

43. Vertices: $(\pm 1, 0)$ **45.** Vertices: $(\pm 4, 0)$
Asymptotes: $y = \pm\frac{3}{2}x$ Asymptotes: $y = \pm\frac{1}{2}x$

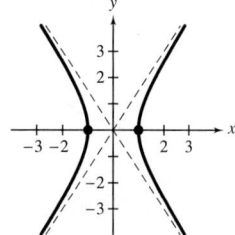 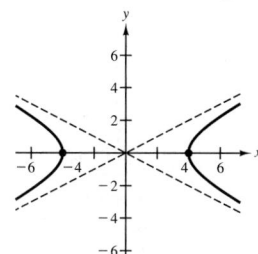

47. $\dfrac{x^2}{16} - \dfrac{y^2}{64} = 1$ **49.** $\dfrac{y^2}{16} - \dfrac{x^2}{64} = 1$ **51.** $\dfrac{x^2}{81} - \dfrac{y^2}{36} = 1$

53. Parabola **55.** Ellipse **57.** Hyperbola

59. Circle **61.** Line **63.** Parabola

65. $x^2 + y^2 = 4500^2 = 20{,}250{,}000$

67. Circle: $x^2 + y^2 = 25^2 = 625$

Height: $2y = 2\sqrt{625 - x^2}$

Area: $(2x)(2y) = 4x\sqrt{625 - x^2}$

69. $\sqrt{304} \approx 17.4$ ft

71.

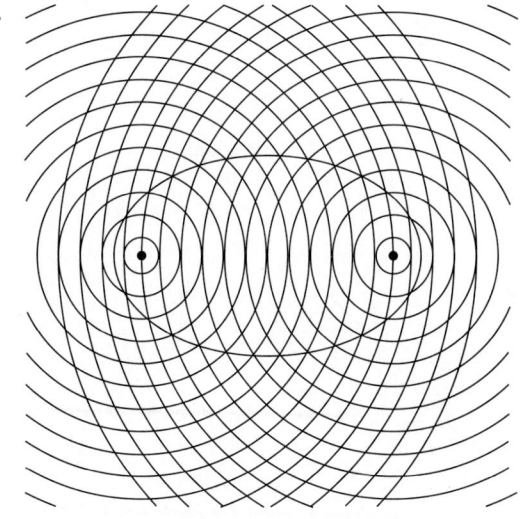

SECTION 8.6 *(page 660)*

Warm-Up *(page 660)*

1.
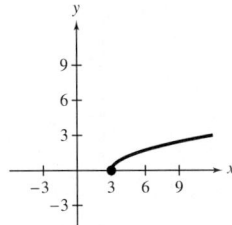

2.
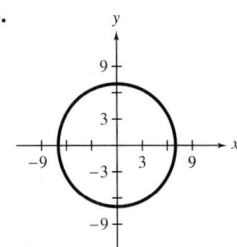

3.
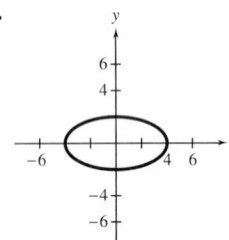

4.
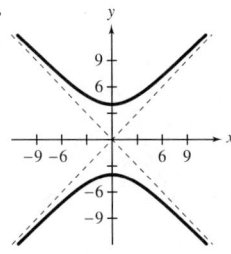

5.
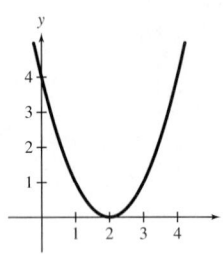

6.
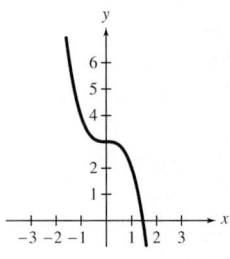

7. $\left(2, \frac{1}{2}\right)$ **8.** $(4, -2)$
9. $(1, -4, 3)$ **10.** $(4, 0, 6)$

1. $(-2, 4)$, $(1, 1)$ **3.** No real solution
5. $(0, 3)$, $(-2, 0)$
7. $(0, 0)$, $(1, 1)$, $(-1, -1)$ **9.** $(0, 0)$, $(2, 4)$

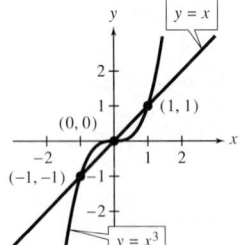

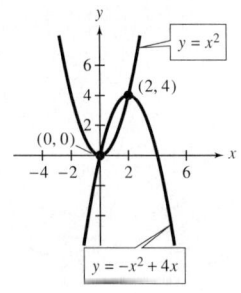

11. $(4, 2)$, $(9, 3)$

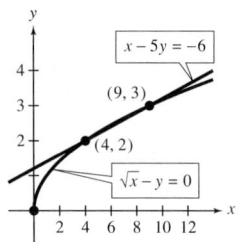

13. $(-3, 18)$, $(2, 8)$ **15.** $(2, 5)$, $(-3, 0)$
17. No real solution   **19.** $(0, 5)$, $(-4, -3)$
21. $(0, 2)$, $(3, 1)$ **23.** $(0, 4)$, $(3, 0)$
25. No real solution   **27.** $(10, 30)$, $(20, 15)$
29. $(\pm\sqrt{3}, -1)$ **31.** $(2, \pm2\sqrt{3})$, $(-1, \pm3)$

33. $(\pm2, \pm\sqrt{3})$ **35.** $\left(\pm\dfrac{2\sqrt{5}}{5}, \pm\dfrac{2\sqrt{5}}{5}\right)$

37. $(0, 0)$, $(1, 1)$

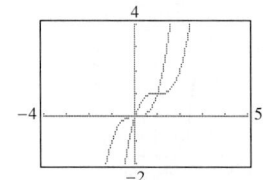

39. $\left(-\dfrac{3}{5}, -\dfrac{4}{5}\right)$ to $\left(\dfrac{4}{5}, -\dfrac{3}{5}\right)$ **41.** Late 1987

USING A GRAPHING CALCULATOR *(page 664)*

1.
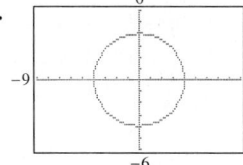

3.
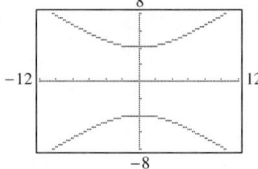

5.
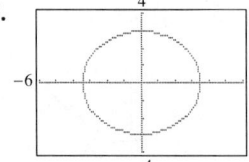

7. $(-3, -1)$, $(1, -3)$ **9.** $(0, 0)$, $(2, 1)$

CHAPTER 8 REVIEW EXERCISES *(page 665)*

1. (a) 0   (b) 5   (c) 12   (d) -1

3. (a) $-\dfrac{1}{3}$ (b) 83   (c) $-\dfrac{1}{48}$ (d) $-\dfrac{\sqrt{2}}{6}$

5. (a) $-\dfrac{8}{3}$ (b) -2 (c) 1   (d) $\dfrac{1}{10}$

7. (a) $x^2 + 2$ (b) $(x + 2)^2$ (c) 6   (d) 1

9. (a) $|x|$ (b) x (c) 5   (d) -1

11. $f \circ g:\ [2, \infty);\ g \circ f:\ [4, \infty)$

13. No   15. Yes   17. $f^{-1}(x) = 4x$

19. $h^{-1}(x) = x^2,\ x \geq 0$

21. Because it is possible to find a horizontal line that intersects the graph of f at more than one point, the function does not have an inverse.

23. $k = 6,\ y = 6\sqrt[3]{x}$ 25. $k = \dfrac{1}{18},\ T = \dfrac{1}{18}rs^2$

27. (a) $k = \dfrac{1}{8}$ (b) 1953.125 kilowatts

29. 150 lb   31. 945 units

33. Falls to the left; Falls to the right

35. Rises to the left; Rises to the right

37. $(2, 0),\ (0, 8)$ 39. $(0, 0),\ (-1, 0),\ (2, 0)$

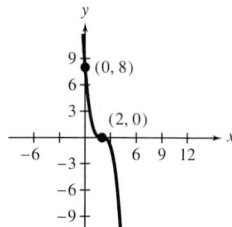   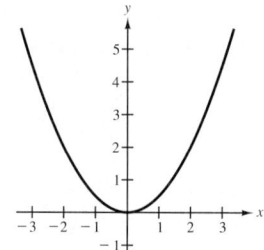

41. $(0, 0),\ (-3, 0)$ 43. Parabola

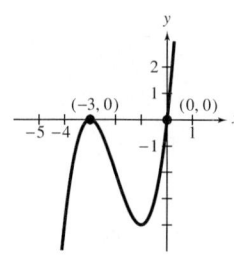   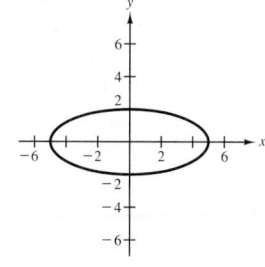

45. Hyperbola   47. Ellipse

49. Circle

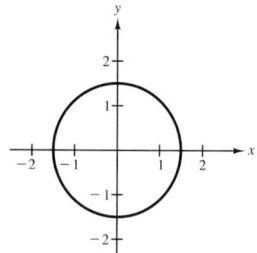

51. $\dfrac{x^2}{36} - \dfrac{y^2}{4} = 1$ 53. $\dfrac{x^2}{4} + \dfrac{y^2}{25} = 1$

55. $x^2 + y^2 = 400$ 57. $x^2 + y^2 = 25{,}000{,}000$

59. $(-1, 5),\ (-2, 20)$ 61. $(-1, 0),\ (0, -1)$

63. $(0, 2),\ \left(-\dfrac{16}{5}, -\dfrac{6}{5}\right)$ 65. $(\pm 5, 0)$

CHAPTER 8 TEST *(page 668)*

1. -13 2. -30 3. $\dfrac{1}{3}$ 4. 12

5. $[-5, 5]$ 6. $f^{-1}(x) = 2x + 2$

7. $S = \dfrac{kx^2}{y}$ 8. $v = \dfrac{1}{4}\sqrt{u}$

9. 240 m³

10. Rises to the left; Falls to the right

11. $(3, 0),\ (0, 9)$

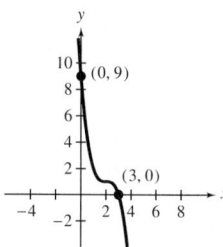

12. $x^2 + y^2 = 25$ 13. $\dfrac{x^2}{9} + \dfrac{y^2}{100} = 1$

14. $\dfrac{x^2}{9} - \dfrac{y^2}{\frac{9}{4}} = 1$

15.
(a)

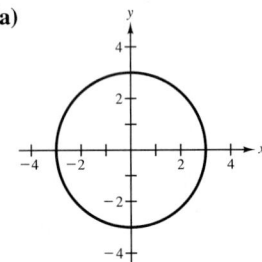

(b)

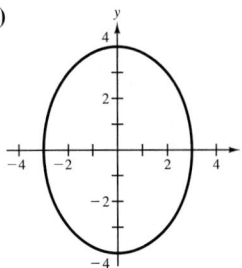

(c)

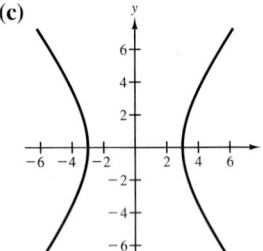

(d)

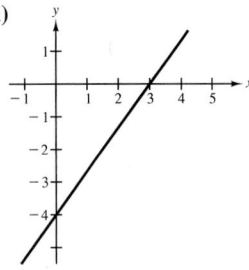

16. (a) $(2, -2)$, $(6, -18)$ **(b)** $(6, 8)$, $(8, 6)$

Chapter 9

Chapter Opener   *(page 669)*

$V_4 \approx 1.277$ $V_5 \approx 1.357$ Yes

SECTION 9.1   *(page 676)*

Warm-Up   *(page 676)*

1. $\frac{27}{64}$ **2.** 900   **3.** $\frac{1}{4}$ **4.** $\frac{1}{81}$ **5.** 5
6. $\frac{1}{9}$ **7.** 1   **8.** y^5 **9.** $13x$ **10.** $-\frac{8}{y^3}$

1. $\log_7 49 = 2$ **3.** $\log_3 \frac{1}{9} = -2$
5. $\log_8 4 = \frac{2}{3}$ **7.** $5^2 = 25$ **9.** $4^{-2} = \frac{1}{16}$
11. $36^{1/2} = 6$ **13.** 3   **15.** -2 **17.** 0
19. There is no power to which 2 can be raised to obtain -3.
21. 1   **23.** 3   **25.** $\frac{1}{2}$ **27.** $\frac{3}{4}$ **29.** 4
31. -2 **33.** 2   **35.** 0   **37.** 1   **39.** 1

41. 1.4914   **43.** -0.0706 **45.** 0.7335
47. 3.2189   **49.** -0.2877 **51.** 0.0757
53.

r	0.07	0.08	0.09
t	9.9	8.7	7.7

r	0.10	0.11	0.12
t	6.9	6.3	5.8

55. 53.4 in.

SECTION 9.2   *(page 684)*

Warm-Up   *(page 683)*

1. x^{20} **2.** $\frac{x^2}{16}$ **3.** $\frac{3}{2y^5}$ **4.** $\frac{4y^2}{9x^4}$
5. $\frac{1}{6}x^3 y$ **6.** 1   **7.** $4^3 = 64$ **8.** $3^{-4} = \frac{1}{81}$
9. $e^{2x} = e^{2x}$ **10.** $e^{1.6094 \cdots} = 5$

1. 2   **3.** 1   **5.** 0   **7.** -3 **9.** 0.862
11. 1.114   **13.** 0.252   **15.** 0.216   **17.** 2.407
19. 0   **21.** $\log_3 11 + \log_3 x$ **23.** $3 \ln y$
25. $\log_2 z - \log_2 17$ **27.** $-2 \log_5 x$
29. $\frac{1}{3}\log_3(x + 1)$ **31.** $\ln 3 + 2\ln x + \ln y$
33. $2\log_2 x - \log_2(x - 3)$ **35.** $\frac{1}{3}[\ln x + \ln(x + 5)]$
37. $\log_2 3x$ **39.** $\log_{10} \frac{4}{x}$ **41.** $\ln b^4$, $b > 0$
43. $\log_5(2x)^{-2}$, $x > 0$ **45.** $\ln \sqrt[3]{2x + 1}$
47. $\log_3 2\sqrt{y}$ **49.** $\ln \frac{x^2 y^3}{z}$, $x > 0$
51. $\ln(xy)^4$, $x > 0$, $y > 0$ **53.** True   **55.** False
57. False   **59.** $1 - \log_4 x$
61. $1 + \frac{1}{2}\log_5 2$ **63.** $2 + \ln 3$
65. $\ln 1 = 0$, $\ln 4 \approx 1.3862$, $\ln 6 \approx 1.7917$, $\ln 8 \approx 2.0793$,
$\ln 9 \approx 2.1972$, $\ln 10 \approx 2.3025$, $\ln 12 \approx 2.4848$,
$\ln 15 \approx 2.7080$, $\ln 16 \approx 2.7724$, $\ln 18 \approx 2.8903$,
$\ln 20 \approx 2.9956$
$\ln 7$, $\ln 11$, $\ln 13$, $\ln 14$, $\ln 17$, and $\ln 19$ cannot be approximated using $\ln 2$, $\ln 3$, and $\ln 5$.
67. $B = 10(\log_{10} I + 16)$, 60 decibels
69. $E = 1.4 \log_{10} \frac{C_2}{C_1}$

SECTION 9.3   *(page 694)*

Warm-Up   *(page 693)*

1. x^8 **2.** 1   **3.** z^3 **4.** $\frac{3}{4}(x-3)^4$

5. x^8y^4 **6.** $\dfrac{x^5}{32y^5}$ **7.** $\dfrac{x^4}{y^3}$ **8.** $\dfrac{1}{a^8}$

9. b^3 **10.** $2x$

---

1. (a) $\frac{1}{9}$ **(b)** 1   **(c)** 3

3. (a) 9   **(b)** 1   **(c)** $\frac{1}{3}$

5. (a) 500   **(b)** 250   **(c)** 56.657

7. (a) 1000   **(b)** 1628.895   **(c)** 2653.298

9. (a) 0.368   **(b)** 1   **(c)** 1.649

11. (a) 73.891   **(b)** 1.353   **(c)** 0.183

13. b   **14.** a   **15.** e   **16.** d   **17.** f   **18.** c

19. 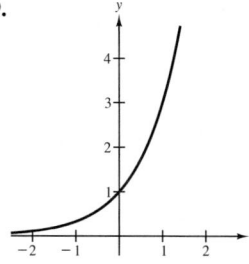   **21.**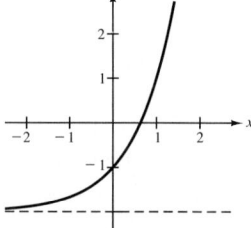

23. 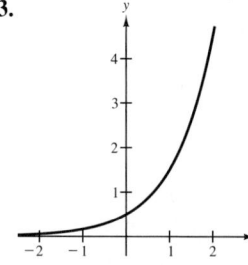   **25.**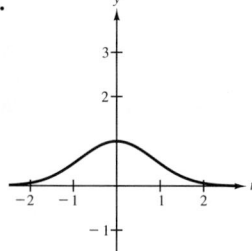

27. 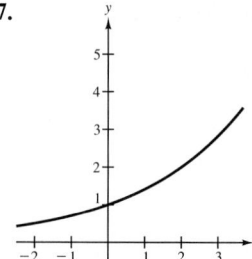   **29.**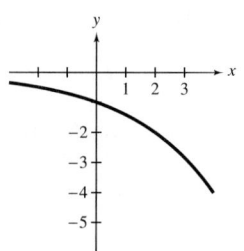

31. 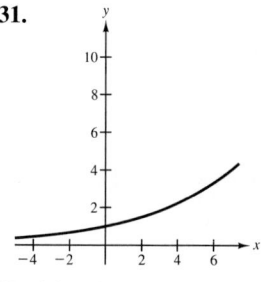   **33.**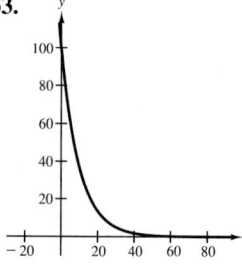

35. (a) 272.184 million   **(b)** 288.627 million

37. $9000

39. $32.50

41.

n	1	4	12
A	$466.10	$487.54	$492.68

n	365	Continuous compounding
A	$495.22	$495.30

43.

n	1	4	12
A	$226,296.28	$259,889.34	$268,503.32

n	365	Continuous compounding
A	$272,841.24	$272,990.75

45.

n	1	4	12
A	$152,203.13	$167,212.90	$170,948.62

n	365	Continuous compounding
A	$172,813.72	$172,877.82

47.

n	1	4	12
P	\$2541.75	\$2498.00	\$2487.98

n	365	Continuous compounding
P	\$2483.09	\$2482.93

49.

n	1	4	12
P	\$18,429.30	\$15,830.43	\$15,272.04

n	365	Continuous compounding
P	\$15,004.64	\$14,995.58

51.

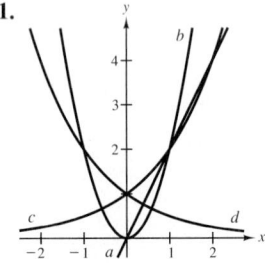

53.

t	0	25	50	75
h	2000 ft	1450 ft	950 ft	450 ft

Ground level: 97.5 sec

55. 2.155 million

57. 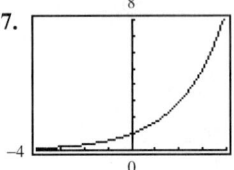   **59.**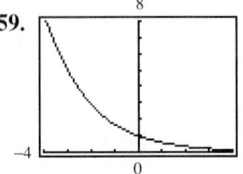

61. 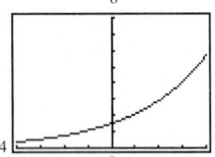   **63.**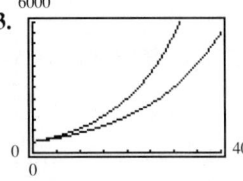

CHAPTER 9 MID-CHAPTER QUIZ *(page 698)*

1. (a) $\log_5 125 = 3$ (b) $\ln 1 = 0$
2. (a) $16^{1/2} = 4$ (b) $5^{-2} = \frac{1}{25}$
3. 2   **4.** -2 **5.** (a) 2.7126   (b) 7.6578
6. $3[\ln x - \ln(x + 3)]$ **7.** $\log_3 \left[x^2 (x + 5)^4 \right]$, $x > 0$
8. $3 - \log_5 x$ **9.** 864.665
10. (a)   (b)

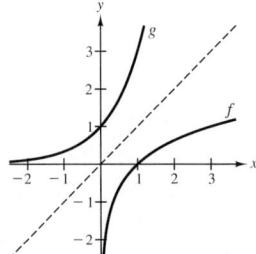

11. (a) \$2099.59   (b) \$2102.53   **12.** 4.6 lb/in.^{2}

SECTION 9.4 *(page 705)*

> **Warm-Up** *(page 704)*
>
> **1.** 4   **2.** 3   **3.** -3
> **4.** 3   **5.** 0.368   **6.** 1.649
> **7.** g is a vertical translation of f 4 units downward.
> **8.** g is a horizontal translation of f 4 units to the right.
> **9.** g is a reflection of f in the x-axis.
> **10.** g is a reflection of f in the y-axis.

1. f is the inverse of g.   **3.** f is the inverse of g.

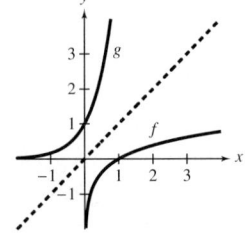

5. e   **6.** b   **7.** d   **8.** c   **9.** a   **10.** f

11.

13.

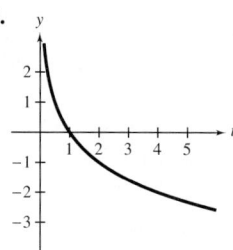

35.

37.

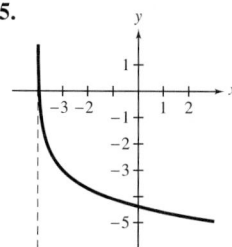
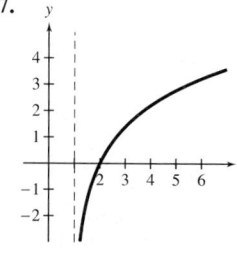

39. 2.3481   **41.** 1.7712   **43.** −0.4739
45. 2.6332   **47.** −2   **49.** 1.3481
51. $(0, \infty)$ **53.** $3 \le f(x) \le 4$ or [3, 4]
55. A factor of 10

15.

17.

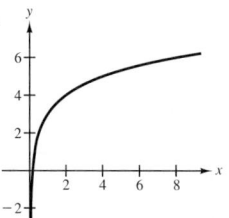
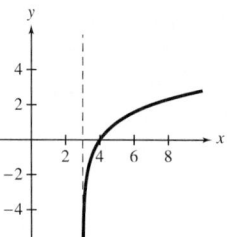

57.

59.

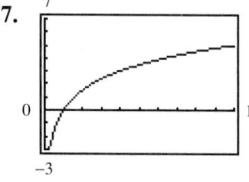
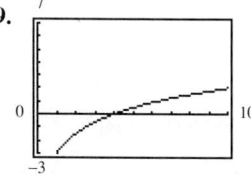

19.

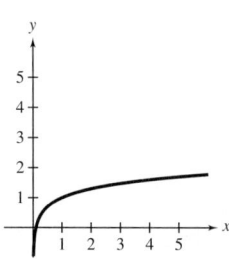

61.

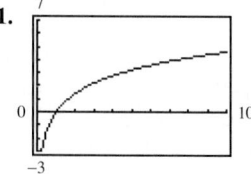

21. b   **22.** e   **23.** d   **24.** c   **25.** f   **26.** a

27.

29.

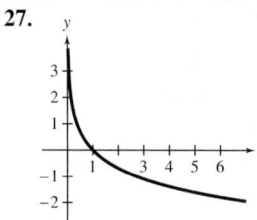
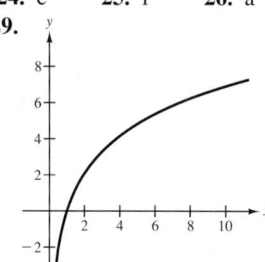

USING A GRAPHING CALCULATOR   *(page 710)*

1.

3.

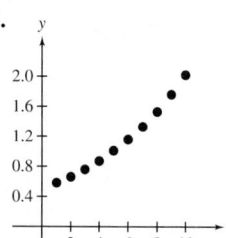
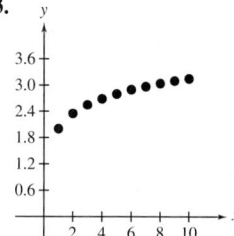

31.

33.

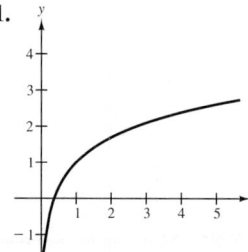
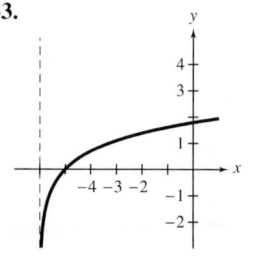

Model:
$$y = 0.5013(1.14960)^x$$

Model:
$$y = 2.008 + 0.4993 \ln x,$$
$$y = 2.008 + 0.49931 \ln x$$

5.

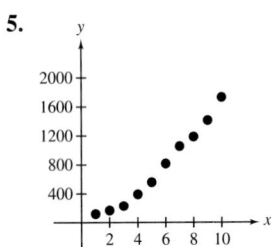

Model: $y = 100.487857(1.360675)^x$;
1991 sales: $2974.5 million

SECTION 9.5   *(page 717)*

Warm-Up   *(page 717)*

1. $\frac{5}{2}$ **2.** $-\frac{7}{2}$ **3.** $10, -2$ **4.** $2, 3$

5. $\dfrac{28}{\ln 5}$ **6.** 0 **7.** 3 **8.** 6

9. 0 **10.** 4

---

1. 5 **3.** 8 **5.** 3 **7.** -3 **9.** 4
11. $\frac{22}{5}$ **13.** 81 **15.** $500,000$ **17.** 5.49
19. 0.86 **21.** 3.28 **23.** 2.49 **25.** 0.39
27. 6.80 **29.** 0.80 **31.** 35.35 **33.** 21.82
35. 8.99 **37.** 1 **39.** 1000 **41.** 7.91
43. 8.17 **45.** -0.78 **47.** $31,622,771.6$
49. ± 20.09 **51.** $-2, 5$ **53.** 7.70 yr
55. $10^{-8.5}$ **57.** 1.0234 g/cm^3; 1.0241 g/cm^3

SECTION 9.6   *(page 726)*

Warm-Up   *(page 726)*

1. $\frac{33}{4}$ **2.** $\frac{3}{5}, -1$ **3.** 41 **4.** $\frac{11}{2}$
5. 18 **6.** 69 **7.** 4.16 **8.** 11.31
9. 139.32 **10.** -1.39

---

1. 7% **3.** 9% **5.** 8%
7. 8.75 yr   **9.** 6.60 yr   **11.** 9.24 yr
13. Continuous   **15.** Quarterly   **17.** 8.33%
19. 7.23% **21.** 5.39% **23.** \$1652.99
25. \$626.46   **27.** \$5496.57   **29.** \$320,250.81
31. Total deposits: \$7200.00; Total interest: \$10,529.42
33. $k = \frac{1}{2} \ln \frac{8}{3} \approx 0.4904$ **35.** $k = \frac{1}{3} \ln \frac{1}{2} \approx -0.2310$
37. $y = 9.6 e^{0.0072t}$; 11.1 million
39. $y = 3.3 e^{0.0452t}$; 8.1 million
41. 3.3 g   **43.** $15,700$ yr
45. (a) $S = 10(1 - e^{-0.0575x})$ (b) 3300 units
47. 501 times as great   **49.** 7.04
51. $10,000,000$ times   **53.** 8.995 billion
55. (a)

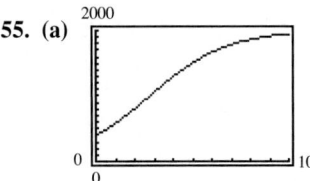

(b) 1298 units   (c) 3.2 yr   (d) 2000 units

CHAPTER 9 REVIEW EXERCISES   *(page 730)*

1. $\log_4 64 = 3$ **3.** 3 **5.** -2
7. 7 **9.** 0 **11.** $\log_4 6 + 4 \log_4 x$
13. $\frac{1}{5} \log_5 (x + 2)$ **15.** $\ln(x + 2) - \ln(x - 2)$
17. $\log_4 \dfrac{x}{10}$ **19.** $\log_8 32 x^3$
21. $\ln \dfrac{9}{4x^2}, \ x > 0$ **23.** True   **25.** False
27. True   **29.** 1.7959 **31.** -0.4307
33. 1.0293 **35.** (a) $\frac{1}{8}$ (b) 2 (c) 4
37. (a) 2.718 (b) 0.351 (c) 0.135
39. (a) 0 (b) 3 (c) -0.631
41. (a) 1 (b) -1.099 (c) 2.303
43. (a) -6 (b) 0 (c) 22.5
45. d   **46.** f   **47.** a   **48.** b   **49.** c   **50.** e
51.   **53.**

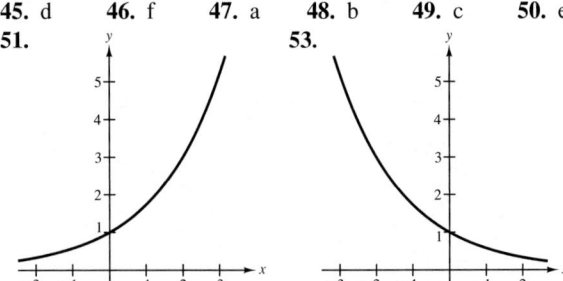

55.
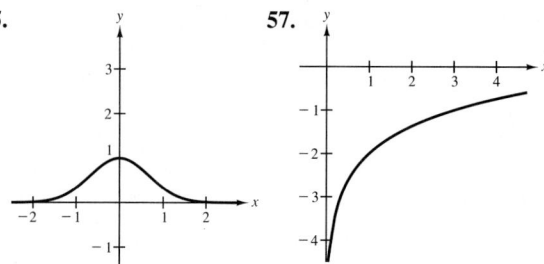

57.

59.
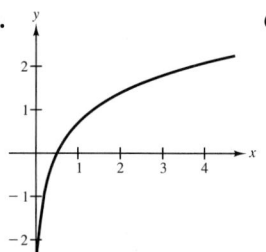

61.
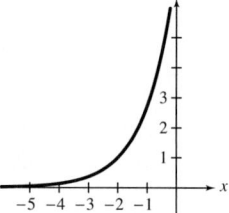

63. 1.585   **65.** 2.132   **67.** 6   **69.** 1
71. 243   **73.** 5.66   **75.** 6.23   **77.** 32.99
79. 15.81   **81.** 1408.10   **83.** 0.32
85.

n	1	4	12
A	\$3806.13	\$4009.59	\$4058.25

n	365	Continuous compounding
A	\$4082.26	\$4083.08

87.

n	1	4	12
A	\$67,275.00	\$72,095.68	\$73,280.74

n	365	Continuous compounding
A	\$73,870.32	\$73,890.56

89.

n	1	4	12
P	\$2301.55	\$2103.50	\$2059.87

n	365	Continuous compounding
P	\$2038.82	\$2038.11

91. 4.6 yrs   **93.** 150 units
95.

t	0	1	5	10
$p(t)$	200	227	346	472

t	20	50	75
$p(t)$	579	600	600

The limiting size of the herd is 600.

CHAPTER 9 TEST   *(page 734)*

1. $5^3 = 125$ **2.** $\log_4 \frac{1}{16} = -2$ **3.** $\frac{1}{3}$
4. $\log_4 5 + 2\log_4 x - \frac{1}{2}\log_4 y$ **5.** $\ln \frac{x}{y^4}$, $y > 0$
6. $3 + \log_5 6$
7.
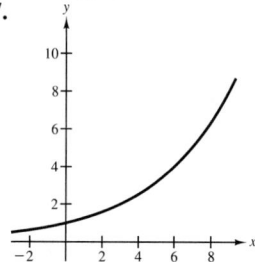

8. (a) \$8012.78   **(b)** \$8110.40
9. $g = f^{-1}$

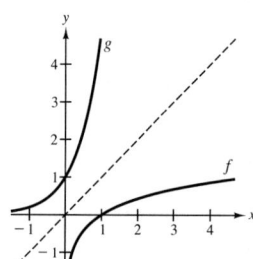

10. (a) 54   **(b)** 24   **(c)** 81   **(d)** 44.09
11. 64   **12.** 0.97   **13.** 13.73   **14.** 15.52
15. \$10,806.08   **16.** 7%   **17.** \$8469
18. (a) 600   **(b)** 1141   **(c)** 4.4 years

Chapter 10

Chapter Opener *(page 735)*

$320.71 after 20 years; $1842.02 after 50 years

SECTION 10.1 *(page 743)*

Warm-Up *(page 743)*

1. $\frac{1}{7}$ **2.** 24   **3.** $\frac{3}{4}$ **4.** $\frac{13}{12}$

5. n **6.** $n-1$ **7.** $\dfrac{2n-3}{n-5}$ **8.** $\dfrac{n+1}{n}$

9. $\dfrac{n+2}{n^2}$ **10.** $\dfrac{2n}{n^2-1}$

1. 2, 4, 6, 8, 10   **3.** $-2, 4, -6, 8, -10$

5. $\frac{1}{2}, \frac{1}{4}, \frac{1}{8}, \frac{1}{16}, \frac{1}{32}$ **7.** $\frac{1}{4}, -\frac{1}{8}, \frac{1}{16}, -\frac{1}{32}, \frac{1}{64}$

9. $\frac{1}{2}, \frac{1}{3}, \frac{1}{4}, \frac{1}{5}, \frac{1}{6}$ **11.** $\frac{2}{5}, \frac{1}{2}, \frac{6}{11}, \frac{4}{7}, \frac{10}{17}$

13. $-1, \frac{1}{4}, -\frac{1}{9}, \frac{1}{16}, -\frac{1}{25}$ **15.** $2, 2, \frac{4}{3}, \frac{2}{3}, \frac{4}{15}$

17. $\frac{9}{2}, \frac{19}{4}, \frac{39}{8}, \frac{79}{16}, \frac{159}{32}$ **19.** $0, 6, -6, 18, -30$

21. 5   **23.** $\dfrac{1}{132}$ **25.** $\dfrac{1}{n+1}$ **27.** $2n$

29. 63   **31.** 35   **33.** $\frac{47}{60}$ **35.** -48 **37.** $\frac{8}{9}$

39. $\frac{182}{243}$ **41.** 112   **43.** 852   **45.** 6.5793

47. $\displaystyle\sum_{k=1}^{5} k$ **49.** $\displaystyle\sum_{k=1}^{5} 2k$ **51.** $\displaystyle\sum_{k=1}^{10} \frac{1}{2k}$

53. $\displaystyle\sum_{k=1}^{20} \frac{1}{k^2}$ **55.** $\displaystyle\sum_{k=0}^{9} \frac{1}{(-3)^k}$ **57.** $\displaystyle\sum_{k=1}^{11} \frac{k}{k+1}$

59. $\displaystyle\sum_{k=0}^{6} k!$ **61.** 3.6   **63.** 0.8

65. (a) $502.92, $505.85, $508.80, $511.77,
$514.75, $517.76, $520.78, $523.82
(b) $2019.37

67. $a_5 = 108°$, $a_6 = 120°$
At the point where any two hexagons and a pentagon
meet, the sum of the three angles is
$$a_5 + 2a_6 = 348° < 360°.$$
Therefore, there is a gap of $12°$.

SECTION 10.2 *(page 752)*

Warm-Up *(page 752)*

1. $\frac{49}{20}$ **2.** $\frac{455}{72}$ **3.** 120   **4.** 40   **5.** 285

6. 1960   **7.** 77   **8.** $\frac{355}{2}$ **9.** 63   **10.** 144

1. 3   **3.** -6 **5.** $\frac{2}{3}$

7. Arithmetic, 2   **9.** Arithmetic, $\frac{3}{2}$

11. Arithmetic, -16 **13.** Not arithmetic

15. Arithmetic, 0.8   **17.** Not arithmetic

19. 7, 10, 13, 16, 19   **21.** 6, 4, 2, 0, -2

23. $\frac{3}{2}, 4, \frac{13}{2}, 9, \frac{23}{2}$ **25.** 4, $\frac{15}{4}, \frac{7}{2}, \frac{13}{4}, 3$

27. d   **28.** b   **29.** c   **30.** a

31. $a_n = \frac{1}{2}n + \frac{5}{2}$ **33.** $a_n = -25n + 1025$

35. $a_n = 4n + 16$ **37.** $a_n = -10n + 60$

39. 5, 8, 11, 14, 17   **41.** 9, 6, 3, 0, -3

43. $-10, -4, 2, 8, 14$ **45.** 210   **47.** 2700

49. 19,500   **51.** 62,625   **53.** 522   **55.** 900

57. 12,200   **59.** 243   **61.** 2850   **63.** 2550

65. $246,000   **67.** $25.43   **69.** 632 bales

71. 1024 ft

SECTION 10.3 *(page 762)*

Warm-Up *(page 762)*

1. $\frac{16}{81}$ **2.** $\frac{27}{125}$ **3.** $\frac{1}{72}$ **4.** 1   **5.** n

6. $n+1$ **7.** $60n^6$ **8.** $16n^5$

9. $\dfrac{n}{2}$ **10.** $\dfrac{9}{5}n^2$

1. 3   **3.** $-\frac{1}{2}$ **5.** $-\frac{3}{2}$ **7.** e

9. Not geometric   **11.** Geometric, $\frac{1}{2}$

13. Geometric, $-\frac{2}{3}$ **15.** Geometric, 1.02

17. 4, 8, 16, 32, 64   **19.** 6, 2, $\frac{2}{3}, \frac{2}{9}, \frac{2}{27}$

21. 1, $-\frac{1}{2}, \frac{1}{4}, -\frac{1}{8}, \frac{1}{16}$

23. 1000, 1010, 1020.1, 1030.301, 1040.60401

25. $a_n = 2(3)^{n-1}$ **27.** $a_n = 2^{n-1}$

29. $a_n = 4\left(-\frac{1}{2}\right)^{n-1}$ **31.** $a_n = 8\left(\frac{1}{4}\right)^{n-1}$

33. $\frac{3}{256}$ **35.** $48\sqrt{2}$ **37.** $\frac{81}{64}$ **39.** $\pm\frac{243}{32}$

41. 1023   **43.** 772.478   **45.** 2.250

47. 5727.500   **49.** −14,762   **51.** 16.000
53. 152.095   **55.** 1103.568
57. (a) $250,000(0.75)^n$ **(b)** \$59,326.17
59. \$3,623,993   **61.** \$19,496.56   **63.** \$105,428.44
65. (a) \$5,368,709.11   **(b)** \$10,737,418.23
67. 70.875 in.^{2} **69.** 1.476×10^{20}; No

33. $x^6 - 12x^5 + 60x^4 - 160x^3 + 240x^2 - 192x + 64$
35. 120   **37.** −1365   **39.** 1760   **41.** −4
43. $2035 + 828i$ **45.** $\frac{1}{32} + \frac{5}{32} + \frac{10}{32} + \frac{10}{32} + \frac{5}{32} + \frac{1}{32}$
47. $\frac{1}{256} + \frac{12}{256} + \frac{54}{256} + \frac{108}{256} + \frac{81}{256}$
49. 1.172   **51.** 510,568.785
53. True; $_nC_m = \dfrac{n!}{m!(n-m)!} = {_nC_{n-m}}$

CHAPTER 10 MID-CHAPTER QUIZ   *(page 765)*

1. 25, 20, 15, 10, 5   **2.** $1, -\frac{1}{4}, \frac{1}{16}, -\frac{1}{64}, \frac{1}{256}$

3. $\sum_{n=1}^{7} (6n - 5)$ **4.** $\sum_{n=0}^{4} \left(\frac{4}{3}\right)^n$ **5.** 132

6. $n(n-1)$ **7.** Geometric, $r = 4$
8. Arithmetic, $d = -\frac{1}{2}$ **9.** $a_n = 9n - 6$
10. $a_n = 5(-0.5)^{n-1}$ **11.** 36   **12.** 88
13. 5.864×10^{12} **14.** $\frac{153}{32}$
15. (a) \$10,066.67, \$10,133.78, \$10,201.34,
     \$10,269.35, \$10,337.81
   (b) \$22,196.40
16. \$2737.50   **17.** \$141,021.55 million

SECTION 10.4   *(page 772)*

Warm-Up   *(page 771)*

1. $-3x^2 + 6x$ **2.** $x^2 - 9$ **3.** $x^2 - 4x + 4$
4. $4x^2 + 12x + 9$ **5.** $\dfrac{5x^3}{y^2}$ **6.** $-8x^6 y^3$
7. 24   **8.** 8   **9.** 1320   **10.** 45

1. 15   **3.** 252   **5.** 1   **7.** 1225
9. 12,650   **11.** 455   **13.** 53,130
15. 1, 9, 36, 84, 126, 126, 84, 36, 9, 1
17. (a) 35   **(b)** 126
19. $x^4 + 4x^3 y + 6x^2 y^2 + 4xy^3 + y^4$
21. $x^5 + 5x^4 + 10x^3 + 10x^2 + 5x + 1$
23. $x^6 + 18x^5 + 135x^4 + 540x^3 + 1215x^2 + 1458x + 729$
25. $x^5 - 5x^4 y + 10x^3 y^2 - 10x^2 y^3 + 5xy^4 - y^5$
27. $u^3 - 6u^2 v + 12uv^2 - 8v^3$
29. $81a^4 + 216a^3 b + 216a^2 b^2 + 96ab^3 + 16b^4$
31. $x^8 + 8x^7 y + 28x^6 y^2 + 56x^5 y^3 + 70x^4 y^4 +$
$56x^3 y^5 + 28x^2 y^6 + 8xy^7 + y^8$

SECTION 10.5   *(page 781)*

Warm-Up   *(page 781)*

1. 10   **2.** 120   **3.** 84   **4.** 1   **5.** 576
6. 720   **7.** 870   **8.** 1260   **9.** 4845
10. 24

1. 10   **3.** 8   **5.** 6   **7.** 7   **9.** 6   **11.** 260
13. 6,760,000   **15.** 6   **17.** 48   **19.** 16
21. XYZ, XZY, YXZ, YZX, ZXY, ZYX
23. 120   **25.** 40,320
27. {A, B}, {A, C}, {A, D}, {A, E}, {A, F}, {B, C},
   {B, D}, {B, E}, {B, F}, {C, D}, {C, E}, {C, F},
   {D, E}, {D, F}, {E, F}
29. 1140   **31.** 126
33. (a) 3   **(b)** 6   **(c)** 15   **(d)** 28   **(e)** 45   **(f)** 66
35. 30   **37.** 84   **39.** 5   **41.** 20

SECTION 10.6   *(page 789)*

Warm-Up   *(page 789)*

1. $\frac{11}{12}$ **2.** $\frac{5}{8}$ **3.** $\frac{37}{30}$ **4.** $\frac{95}{144}$ **5.** $\frac{1}{24}$
6. $\frac{1}{120}$ **7.** $\frac{3}{14}$ **8.** $\frac{1}{6}$ **9.** $\frac{5}{8}$ **10.** $\frac{117}{125}$

1. {A, B, C, D, E, ..., X, Y, Z}
3. {AB, AC, AD, AE, BC, BD, BE, CD, CE, DE}
5. $\frac{3}{8}$ **7.** $\frac{7}{8}$ **9.** $\frac{1}{2}$ **11.** $\frac{3}{13}$ **13.** $\frac{1}{6}$ **15.** $\frac{5}{6}$
17. (a) $\frac{1}{5}$ **(b)** $\frac{1}{3}$ **(c)** 1
19. (a) $\frac{1}{20}$ **(b)** $\frac{2}{5}$ **(c)** $\frac{1}{2}$ **(d)** $\frac{1}{4}$

21. $\frac{14}{65}$ **23.** 0.9   **25.** 0.4375

27.

		Female	
		X	X
Male	X	XX	XX
	Y	XY	XY

Probability of a girl $= \frac{1}{2}$

Probability of a boy $= \frac{1}{2}$

29. $\frac{1}{24}$ **31.** $\frac{1}{45}$ **33.** $\frac{1}{54,145}$ **35.** $1 - p$

USING A GRAPHING CALCULATOR   *(page 795)*

1. 10; 40%   **3.** 1000; 5.5%

5. Experimental results will vary.

CHAPTER 10 REVIEW EXERCISES   *(page 796)*

1. $\displaystyle\sum_{k=1}^{4} (5k - 3)$ **3.** $\displaystyle\sum_{k=1}^{6} \frac{1}{3k}$ **5.** 380

7. $n(n-1)(n-2)$ **9.** 127, 122, 117, 112, 107

11. $\frac{5}{4}, 2, \frac{11}{4}, \frac{7}{2}, \frac{17}{4}$ **13.** 10, 30, 90, 270, 810

15. 100, -50, 25, -12.5, 6.25   **17.** $4n + 6$

19. $-50n + 1050$ **21.** $a_n = \left(-\frac{2}{3}\right)^{n-1}$

23. $a_n = 24(2)^{n-1}$ **25.** $a_n = 12\left(-\frac{1}{2}\right)^{n-1}$

27. 28   **29.** $\frac{4}{5}$ **31.** 486   **33.** $\frac{2525}{2}$

35. 8190   **37.** 2.571   **39.** 19.842

41. 115,019.345   **43.** 5100   **45.** 462

47. (a) $a_n = 85,000(1.012)^n$ **(b)** 154,328

49. $x^{10} + 10x^9 + 45x^8 + 120x^7 + 210x^6 + 252x^5 +$
$210x^4 + 120x^3 + 45x^2 + 10x + 1$

51. $y^6 - 12y^5 + 60y^4 - 160y^3 + 240y^2 - 192y + 64$

53. $x^8 - 4x^7 + 7x^6 - 7x^5 + \frac{35}{8}x^4 - \frac{7}{4}x^3 + \frac{7}{16}x^2 - \frac{1}{16}x + \frac{1}{256}$

55. 8   **57.** 3003   **59.** $\frac{1}{3}$ **61.** $\frac{1}{24}$ **63.** 0.346

CHAPTER 10 TEST   *(page 798)*

1. $1, -\frac{2}{3}, \frac{4}{9}, -\frac{8}{27}, \frac{16}{81}$ **2.** 35

3. $\displaystyle\sum_{k=1}^{12} \frac{2}{3k+1}$ **4.** 12, 16, 20, 24, 28

5. $a_n = -100n + 5100$ **6.** 3825

7. $-\frac{3}{2}$ **8.** $a_n = 4\left(\frac{1}{2}\right)^{n-1}$

9. 765   **10.** \$47,868.33   **11.** 1140

12. $x^5 - 10x^4 + 40x^3 - 80x^2 + 80x - 32$

13. 56   **14.** 26,000   **15.** 12,650

16. 0.25   **17.** $\frac{3}{26}$ **18.** $\frac{1}{6}$

CUMULATIVE TEST: CHAPTERS 7-10   *(page 799)*

1. $5 \pm 5\sqrt{2}i$ **2.** $\dfrac{-3 \pm \sqrt{3}}{3}$ **3.** 9

4. (a)

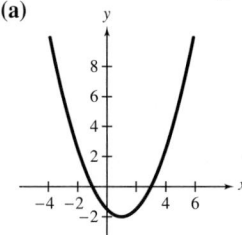

(b)

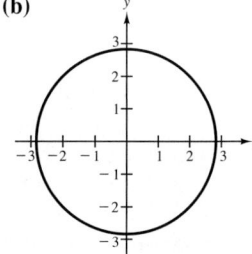

(c)

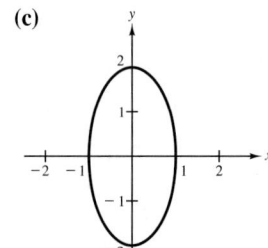

(d)

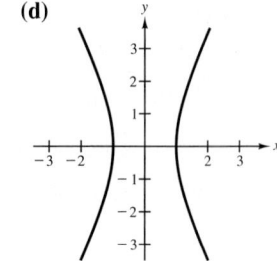

5. $y = \frac{2}{3}x^2 - 4x + 4$ **6.** 11 ft

7. (a) -5 **(b)** 51   **(c)** $\dfrac{\sqrt{3}}{5}$ **(d)** 3

8. $f^{-1}(x) = \frac{1}{3}(x + 2)$

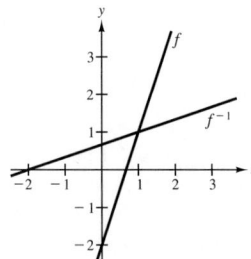

9. 128 ft

10. $(\pm\sqrt{3}, 1)$

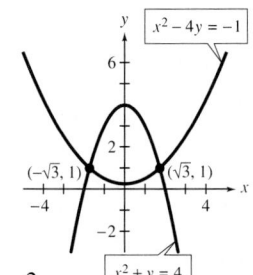

11. -2

12.

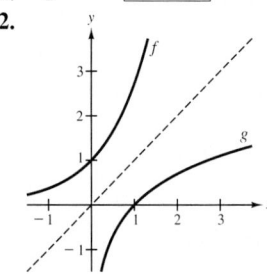

$g = f^{-1}$

13. $\log_2 \dfrac{(xy)^3}{z}$

14. (a) 3   (b) 12.182   (c) 18.013   (d) 0.867

15. 8.3%   **16.** 15.4 yr   **17.** 800

18. $\frac{4}{81}$ **19.** -35 **20.** 48

21. 120   **22.** 0.535, 0.465   **23.** $\frac{5}{18}$

Appendix   Introduction to Logic

SECTION A.1   *(page A5)*

1. Statement   **3.** Open statement

5. Open statement   **7.** Open statement

9. Nonstatement   **11.** Open statement

13. (a) True   (b) False   **15.** (a) True   (b) True

17. (a) False   (b) False   **19.** (a) True   (b) False

21. (a) The sun is not shining.
(b) It is not hot.
(c) The sun is shining and it is hot.
(d) The sun is shining or it is hot.

23. (a) Lions are not mammals.
(b) Lions are not carnivorous.
(c) Lions are mammals and lions are carnivorous.
(d) Lions are mammals or lions are carnivorous.

25. (a) The sun is not shining and it is hot.
(b) The sun is not shining or it is hot.
(c) The sun is shining and it is not hot.
(d) The sun is shining or it is not hot.

27. (a) Lions are not mammals and lions are carnivorous.
(b) Lions are not mammals or lions are carnivorous.
(c) Lions are mammals and lions are not carnivorous.
(d) Lions are mammals or lions are not carnivorous.

29. $p \wedge \sim q$ **31.** $\sim p \vee q$ **33.** $\sim p \vee \sim q$

35. $\sim p \wedge q$ **37.** The bus is blue.

39. x is not equal to 4.   **41.** The Earth is flat.

43.

p	q	$\sim p$	$\sim p \wedge q$
T	T	F	F
T	F	F	F
F	T	T	T
F	F	T	F

45.

p	q	$\sim p$	$\sim q$	$\sim p \vee \sim q$
T	T	F	F	F
T	F	F	T	T
F	T	T	F	T
F	F	T	T	T

47.

p	q	$\sim q$	$p \vee \sim q$
T	T	F	T
T	F	T	T
F	T	F	F
F	F	T	T

49. Not logically equivalent   **51.** Logically equivalent

53. Logically equivalent   **55.** Not logically equivalent

57. Logically equivalent   **59.** Not a tautology

61. A tautology

63.

p	q	$\sim p$	$\sim q$	$p \wedge q$	$\sim(p \wedge q)$	$\sim p \vee \sim q$
T	T	F	F	T	F	F
T	F	F	T	F	T	T
F	T	T	F	F	T	T
F	F	T	T	F	T	T

Identical (over columns $\sim(p \wedge q)$ and $\sim p \vee \sim q$)

SECTION A.2 *(page A13)*

1. **(a)** If the engine is running, then the engine is wasting gasoline.
   (b) If the engine is wasting gasoline, then the engine is running.
   (c) If the engine is not wasting gasoline, then the engine is not running.
   (d) If the engine is running, then the engine is not wasting gasoline.
3. **(a)** If the integer is even, then it is divisible by two.
   (b) If it is divisible by two, then the integer is even.
   (c) If it is not divisible by two, then the integer is not even.
   (d) If the integer is even, then it is not divisible by two.
5. $q \rightarrow p$ 7. $p \rightarrow q$ 9. $p \rightarrow q$ 11. True
13. True   15. False   17. True   19. True
21. Converse: If you can see the eclipse, then the sky is clear.
   Inverse: If the sky is not clear, then you cannot see the eclipse.
   Contrapositive: If you cannot see the eclipse, then the sky is not clear.
23. Converse: If the deficit increases, then taxes were raised.
   Inverse: If taxes are not raised, then the deficit will not increase.
   Contrapositive: If the deficit does not increase, then taxes were not raised.
25. Converse: It is necessary to apply for the visa to have a birth certificate.
   Inverse: It is not necessary to have a birth certificate to not apply for the visa.
   Contrapositive: It is not necessary to apply for the visa to not have a birth certificate.
27. Paul is not a junior and not a senior.
29. The temperature will increase and the metal rod will not expand.
31. We will go to the ocean and the weather forecast is not good.
33. No students are in extracurricular activities.
35. Some contact sports are not dangerous.
37. Some children are allowed at the concert.
39. None of the $20 bills is counterfeit.

41.

p	q	$\sim q$	$p \rightarrow q$	$p \rightarrow \sim q$	$\sim(p \rightarrow \sim q)$
T	T	F	T	F	T
T	F	T	F	T	F
F	T	F	T	T	F
F	F	T	T	T	F

43.

p	q	$p \rightarrow q$	$q \rightarrow p$	$\sim(q \rightarrow p)$	$\sim(q \rightarrow p) \wedge q$
T	T	T	T	F	F
T	F	F	T	F	F
F	T	T	F	T	T
F	F	T	T	F	F

45.

p	q	$\sim p$	$p \vee q$	$(p \vee q) \wedge (\sim p)$	$[(p \vee q) \wedge (\sim p)] \rightarrow q$
T	T	F	T	F	T
T	F	F	T	F	T
F	T	T	T	T	T
F	F	T	F	F	T

47.

p	q	$\sim p$	$\sim q$	$p \to \sim q$	$\sim q \to p$	$p \leftrightarrow \sim q$	$(p \leftrightarrow \sim q) \to \sim p$
T	T	F	F	F	T	F	T
T	F	F	T	T	T	T	F
F	T	T	F	T	T	T	T
F	F	T	T	T	F	F	T

49.

p	q	$\sim p$	$\sim q$	$q \to p$	$\sim p \to \sim q$
T	T	F	F	T	T
T	F	F	T	T	T
F	T	T	F	F	F
F	F	T	T	T	T

                     └─Identical─┘

51.

p	q	$\sim q$	$p \to q$	$\sim(p \to q)$	$p \wedge \sim q$
T	T	F	T	F	F
T	F	T	F	T	T
F	T	F	T	F	F
F	F	T	T	F	F

                         └─Identical─┘

53.

p	q	$\sim q$	$p \to q$	$(p \to q) \vee \sim q$	$p \vee \sim p$
T	T	F	T	T	T
T	F	T	F	T	T
F	T	F	T	T	T
F	F	T	T	T	T

                           └─Identical─┘

55.

p	q	$\sim p$	$\sim p \wedge q$	$p \to (\sim p \wedge q)$
T	T	F	F	F
T	F	F	F	F
F	T	T	T	T
F	F	T	F	T

                     └──── Identical ────┘

57. (c)   **59.** (a)

61.

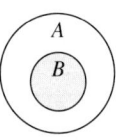

63.

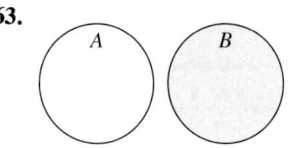

65.

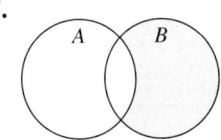

67.

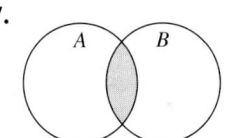

69.

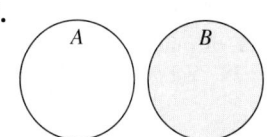

71. (a) Statement does not follow
    (b) Statement follows
73. (a) Statement does not follow
    (b) Statement does not follow

SECTION A.3   *(page A22)*

1.

p	q	$\sim p$	$\sim q$	$p \to \sim q$	$(p \to \sim q) \wedge q$	$[(p \to \sim q) \wedge q] \to \sim p$
T	T	F	F	F	F	T
T	F	F	T	T	F	T
F	T	T	F	T	T	T
F	F	T	T	T	F	T

3.

p	q	$\sim p$	$p \vee q$	$(p \vee q) \wedge \sim p$	$[(p \vee q) \wedge \sim p] \to q$
T	T	F	T	F	T
T	F	F	T	F	T
F	T	T	T	T	T
F	F	T	F	F	T

5.

p	q	$\sim p$	$\sim q$	$\sim p \to q$	$(\sim p \to q) \wedge p$	$[(\sim p \to q) \wedge p] \to \sim q$
T	T	F	F	T	T	F
T	F	F	T	T	T	T
F	T	T	F	T	F	T
F	F	T	T	F	F	T

7.

p	q	$p \vee q$	$(p \vee q) \wedge q$	$[(p \vee q) \wedge q] \to p$
T	T	T	T	T
T	F	T	F	T
F	T	T	T	F
F	F	F	F	T

9. Valid   **11.** Invalid   **13.** Valid   **15.** Valid
17. Invalid   **19.** Valid   **21.** Invalid   **23.** (b)
25. (c)   **27.** (b)   **29.** (c)
31. Valid

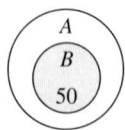

A = All numbers divisible by five
B = All numbers divisible by ten

33. Invalid

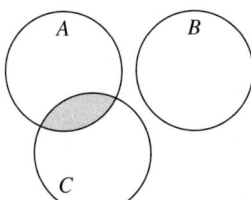

A = People eligible to vote
B = People under the age of 18
C = College students

35. Let p represent the statement "Sue drives to work," let q represent "She will stop at the grocery store," and let r represent "She'll buy milk."

First write:

Premise #1: $p \rightarrow q$
Premise #2: $q \rightarrow r$
Premise #3: p

Reorder the premises:

Premise #3: p
Premise #1: $p \rightarrow q$
Premise #2: $q \rightarrow r$
Conclusion: r

Then we can conclude r. That is, "Sue will get milk."

37. Let p represent "This is a good product," let q represent "We will buy it," and let r represent "The product was made by XYZ Corporation."

First write:

Premise #1: $p \rightarrow q$
Premise #2: $r \lor \sim q$
Premise #3: $\sim r$

Note that $p \rightarrow q \equiv \sim q \rightarrow \sim p$, and reorder the premises:

Premise #2: $r \lor \sim q$
Premise #3: $\sim r$
(Conclusion from Premise #2, Premise #3: $\sim q$)
Premise #1: $\sim q \rightarrow \sim p$
Conclusion: $\sim p$

Then we can conclude $\sim p$. That is, "It is not a good product."

Index of Applications

Biology and Life Science Applications

Air pollutant emissions, 228
Air pressure, 698
American elk, 677
Bats, 229
Deer herd, 733
Environmental hazard, 214
Forager honeybees, 113
Getting a workout, 351
Health risks, 200
Human memory model, 682, 685, 718
Number of ancestors, 764
Ocean temperatures, 627
Oceanography, 718
Oil spill, 627
Pollution level of a pond, 432
Pollution removal, 384, 443
Population growth, 443, 695, 722, 728, 763, 797
Population limit, 733
Population of animal species, 430, 447
Rattlesnake strikes, 238
Time of death, 682
Tree height, 202
U. S. blood types, 790, 791
World population, 728

Business Applications

Advertising effect, 622, 728
Annual revenue, 761
Average cost, 384, 427, 439
Bookstore sales, 275
Break-even analysis, 293, 299, 367
Cassette tape sales, 367
Children's book sales, 39
Company profits, 16, 590
Comparing revenues, 691
Computer inventory, 362
Cost, 126, 165, 274, 577
Cost and profit, 279
Declining balances depreciation, 472
Defective units, 200, 792
Demand for a product, 498, 628
Depreciation, 763, 797
Inventory costs, 369
Maintaining a profit level, 587
Maximum profit, 358, 365, 370, 576
Minimum cost, 366, 370
Monthly expenses, 230
Net earnings per share of stock, 126
Newspaper sales, 141
Operating costs, 228
Partnership costs, 443, 448
Price and demand, 621, 695, 732
Product design, 69
Profit, 45, 58, 228, 231
Projected expenses, 230
Revenue, 530, 539
Revenue and price, 69
Sales growth, 729, 750
Selling price, 563
Soft drink sales, 275
Straight-line depreciation, 141, 144, 274
Time management study, 241, 782
Units produced, 200
Weekly production cost, 629

Chemistry and Physics Applications

Acid mixture, 299, 316, 367
Acidity model, 728
Alcohol mixture, 207, 231, 299
Antifreeze coolant, 213
Boyle's Law, 625, 668
Carbon 14 dating, 728
Chemical reaction, 69
Compression ratio, 201
Crop spraying, 316
Earthquake intensity, 728
Falling distance, 495
Falling time, 592
Force to move a block, 489
Free-falling object, 52, 58, 102, 497, 529, 550, 755
Frictional force, 628
Fuel mixture, 213, 278
Fuel usage, 230
Gasoline mixture, 299
Gear ratio, 201
Gold alloys, 330
Height of a baseball, 564
Height of a projectile, 586, 590
Hooke's Law, 619, 626, 666
Intensity of sound, 677, 685, 718
Leaning Tower of Pisa, 259
Mass of Earth and Sun, 461
Maximum height of a dive, 576
Maximum height of an object, 102
Metal mixture, 212
Mining danger, 629
Model rocket launch, 560
Molecular transport, 685
Morse Code, 797
Newton's Law of Universal Gravitation, 625
Parachute drop, 696
Path of a ball, 576
Period of a pendulum, 482, 529
Piston pressure, 109
Power consumption for electric appliance, 495
Power generation, 627
Pressure vs. weight, 113
Radioactive decay, 689, 724, 728
Rope friction, 718
Satellite orbit, 652, 667
Strength of wooden beam, 489
Temperature conversion, 274
Velocity of a ball, 182
Velocity of a stream, 473, 627
Wind turbine power generation, 627

Construction Problems

Amount of roofing material, 44
Auditorium seating, 797
Best buy, 628
Bridge design, 577
Building a railroad, 193
Dimensions of a corral, 539
Dimensions of a house, 419
Dimensions of a picture, 555
Dimensions of an open box, 102
Dimensions of room, 200, 525
Gothic arch construction, 652
Height of a ladder, 528
Height of an arch, 652
Housing construction, 127

Length of a wire, 528
Length of fencing, 590
Lumber storage area, 562
Road grade, 259
Rocker arm design, 182
Shadow length, 202
Suspension bridge construction, 572
Swimming pool depth markers, 276
Vietnam Veterans Memorial, 300

Consumer Applications

Amount financed, 212
Auto repair bill, 204
Average price for car sales, 557
Cash register contents, 189
Commission rate, 212
Comparison shopping, 211, 277, 604
Cost of clean water, 430
Decreased price, 592
Depreciation, 695, 728, 729, 745
Discount price, 274, 601
Discount rate, 204
Fuel efficiency, 125
Gross pay, 192
Hourly wage, 126, 617
Inflation rate, 695, 732
Insurance premium, 212
Job offer, 736
Labor charges, 211
Lifetime salary, 760
Long distance rates, 211, 228
Markup rate, 203, 277
Monthly budget, 227
Monthly payment, 406
Overtime hours, 212, 231
Price inflation, 200
Price per pound, 211
Property tax, 201, 278
Property value, 695
Real estate commission, 193, 200
Recycling costs, 430
Reduced fare, 563
Reduced ticket price, 563
Reimbursed expenses, 274
Repair time, 200
Retail price, 277
Salary increase, 754, 763, 797
Sale price, 231, 277
Sales bonus, 604
Sales commission, 274
Sales goal, 30, 277

Sales tax, 19
Ski club trip, 557
State and local taxes, 164
State income tax, 230
Tip rate, 212, 277
Tire cost, 211
Total charge, 86
Travel expense, 227
Weekly salary, 190, 205, 212

Geometric Applications

Advertising space, 199
Area of a circular ring, 112
Area of a rectangle, 30, 57
Area of a region, 343
Area of a square, 57, 112, 164
Area of a trapezoid, 44, 107
Dimensions of a rectangle,
    102, 182, 199, 277, 562
Dimensions of a triangle, 112, 199
Dimensions of an open box, 102
Golden section, 487
Length of a rectangle, 106
Perimeter of a pond, 112
Perimeter of a rectangle, 43, 69
Perimeter of a square, 69, 164
Surface area of a cube, 164
Surface area of a soccer ball, 745
Surface area of Spaceship Earth, 88
Test chamber dimensions, 112
Volume of a cube, 164
Width of a rectangle, 69

Interest Rate Applications

Annual interest rate, 590
Annual percentage rate, 556, 719
Balance in an account, 15, 108,
    709, 727
Compound interest, 58, 112, 691,
    697, 716, 744
Continuously compounded
    interest, 692, 696
Doubling time, 677, 718, 720,
    726, 732
Effective yield, 724, 726
Increasing annuity, 761, 764
Investments, 292, 329, 362,
    365, 368, 563
Principal, 727

Savings plan, 15, 696
Simple interest, 69, 107, 112,
    213, 277, 299, 329, 624
Type of compounding, 721, 726

Miscellaneous Applications

Amount of illumination, 628
Annual evaluation of employees,
    778
Average life span of currency, 160
Baling hay, 755
Baseball batting average, 440, 448
Basketball lineup, 782
Bicycle chainwheel, 652
Charitable contributions, 635
Class election, 790
Clock chimes, 755
Coin problem, 212, 277
College enrollment, 274
Committee selection, 782, 788
Company layoff, 200
Computer software sales, 269
Conducting a survey, 787
Continuing education, 790
Course grade, 200
Cutting across the lawn, 538
Delivery van purchase, 291
Diet supplement, 363
Eight-track cartridge sales, 696
Exam scores, 125
Feed mixture, 212
Filling a swimming pool, 444
Finding ocean treasure, 348
Floral arrangement, 213
Fruit distribution, 369
Game show, 791
Hay mixture, 299
Height of a bridge, 513
Height of a picture frame, 112
Height of an attic, 259
Height of Eiffel Tower, 564
Height of Sears Tower, 564
High school survey, 784
License plate numbers, 776, 781
Making a test, 782
Menu selection, 782
Mountain bike owners, 564
Multiple choice test, 790
Number of stamps, 212
Nut mixture, 213, 299
Opinion poll, 212, 277

Photocopy rate, 443
Pizza and soda pop, 351
Pizza toppings, 782
Probability, 786
Public opinion poll, 201
Pumping rate, 443
Pumping time, 201
Random selection, 790
Recipe enlargement, 278
Recipe proportions, 201
Rental occupancy, 275
Renting videos, 430
Rope length, 367
School orchestra, 316
Seed mixture, 206
Slope of a beach, 675
Slope of sewer line, 270
Snowshoes, 627
Softball league, 782
Staffing choices, 781
Storage space, 351
Study questions, 790
Study time, 274
Swimming safety, 362
Telephone numbers, 776
Ticket sales, 212, 299, 363, 366, 751
Toboggan ride, 782
Tornadoes, 677
Trajectory of a ball, 179
Water flow rate, 439
Weather balloon, 112
Weight of an astronaut, 627, 628, 666
Work-rate problem, 102, 209, 214, 231, 277, 559, 563

Time and Distance Applications

Air speed, 299, 563
Average speed, 113, 208, 213, 277, 563, 592
Busing boundary, 662
Delivery route, 563
Distance problem, 113, 189, 228
Distance to the sun, 461
Distance traveled, 384
Driving distance, 330
Flying distance, 213
Flying speed, 368
Map of Wisconsin, 529
Map scale, 278
New York City Marathon, 113
Pulling a boat in to a dock, 539
Running time, 214
Speed of a ship, 469
Speed of light, 113
Spider and the fly, 128
Stopping time, 604, 626, 666
Travel distance, 529
Travel time, 213, 277
Wind speed, 443

U.S. Demographics Applications

American doctors, 242
Annual sales for Liz Claiborne, Inc., 711
Athletic shoe sales, 44
Average amount spent on swimwear in the U.S., 458
Average bus driver salary, 214
Average cafeteria worker salary, 215

Average temperatures in Duluth, 125
Car rental revenues, 444
Cellular phones, 564
Comparing average temperatures, 227
Comparing elevations, 228
Comparing populations, 662
Consumption of milk, 58
Cost of Medicare, 385
Dentists in the U.S., 144
Earnings per share for Xerox Corporation, 258
Energy use, 215
Engineering doctorates earned by women, 250
Federal debt, 462
Generation of electricity, 215
Law enforcement, 258
Life insurance, 708
Magazine circulation, 86
Marriage and divorce rates, 402
Number of African-American women holding elected office, 271
Number of practicing lawyers, 550
Per capita income in New Hampshire and Nevada, 303
Percentage of college graduates, 593
Pianos sold in the U.S., 444
Population of California, 116
Private school expenditures, 102
Recreational boats, 697
Running the IRS, 275
Seizure of illegal drugs, 447
Shrimp and lobster consumption in the U.S., 53

Index

A

Absolute value, 6
    equations, 232
    inequality, 236
    of real numbers, 7
Addition
    of fractions, 12
    of rational expressions with like
        denominators, 395
    of real numbers, 8
Addition and Subtraction Properties,
    219
Addition Property of Equality, 24
Additive Identity Property, 22, 32
Additive Inverse Property, 22, 32
Algebra, Basic Rules of, 32
Algebraic
    expression, 31
    inequalities, 216
Approximately equal to, 3
Area formula, 105
Area of a triangle, 338
Arithmetic sequence, 736, 746
    nth partial sum of, 748
    nth term of, 747
Associative Property of Addition,
    22, 32
Associative Property of Multiplication,
    22, 32
Asymptotes, 421, 422, 644
Augmented matrix, 317
Average rate of change, 250
Axiom, 24
Axis, 641, 644

B

Back-substitute, 286
Bar chart, 15
Base, 10
Basic graphing steps for the TI-81 and
    TI-82 graphing calculators,
    146

Basic Rules of Algebra, 32
Binomial, 47, 766
    coefficients, 766
    expansion, 770
    square of, 51
Binomial Theorem, 766, 767
Bounded intervals on the real number
    line, 216

C

Calculators, 467
Cancellation Property of Addition, 24
Cancellation Property of
    Multiplication, 25
Cartesian plane, 114
Center, 639, 641
Central rectangle, 644
Change of base formula, 702
Characteristics of a function, 154
Check a solution, 89
Circle, 639
Closure Property of Addition and
    Multiplication, 22
Coefficient, 31, 46
Coefficient matrix, 317
Collinear points, test for, 339
Combinations of n elements taken m
    at a time, 779, 780
Common formulas, 105
Common logarithmic function, 673
Common ratio, 755
Commutative Property of Addition,
    22, 32
Commutative Property of
    Multiplication, 22, 32
Complete the square, 531
Complex conjugate, 506
Complex number, 504
    imaginary part, 505
    real part, 505
Composition, 599
    of two functions, 599
Compound fraction, 390, 400
Compound inequality, 223

Compound interest, 691, 719
    formulas for, 106
Compression ratio, 201
Conditional equation, 89
Conic sections, 639
Consecutive integers, 190
Consistent, 285
Constant function, 170
Constant of proportionality, 618
Constant rate of change, 250
Constant term, 31, 46
Constraint linear inequalities, 355
Continuous compounding, 692
Coordinate system, 114
Co-vertices, 641
Cramer's Rule, 334, 335
Critical number, 578, 584
Cross-multiplication, 195
Cube root, 11

D

Definition
    of absolute value of real number, 7
    of an arithmetic sequence, 746
    of a complex number, 504
    of composition of two functions,
        599
    of distance between two real
        numbers, 6
    of exponential function, 686
    of factorial, 739
    of a function, 153
    of geometric sequence, 755
    of intercepts, 134
    of the inverse of a function, 606
    of linear equation, 91
    of logarithm, 670
    of nth root of a number, 462
    of order on the real number line, 5
    of permutation, 777
    of a polynomial in x, 46
    of a quadratic equation in x, 95
    of ratio, 194
    of rational exponents, 465

of rational expression, 374
of row-echelon form of a matrix, 321
of a sequence, 737
of sigma notation, 741
of the slope of a line, 245
of sum, difference, product, and quotient of functions, 596
of zero exponents and negative exponents, 451
Degree of a polynomial, 46
Denominator, 9
Dependent variable, 474
Depreciation, straight-line, 141
Determinant, 331
    of a 2×2 matrix, 331
Difference, 8
    of two cubes, 64
    of two squares, 62, 63
    of two terms, 51
Direct variation, 618
    as nth power, 620
Directly proportional, 618
Discriminant, 544
Distance between two real numbers, 6
Distance formula, 106, 122
Distributive Property, 22, 32
Dividend, 9
Division
    of a polynomial by a monomial, 408
    of radical expressions, 485
    of rational expressions, 389
    of real numbers, 10
    synthetic, 414
    by zero, 25
Division Property of Zero, 25
Divisor, 9
Does not equal, 3
Domain, 151, 158
Double solution, 518

E

Effective yield, 721
Elementary row operations, 319
Elimination, method of, 287, 288
Ellipse, 639, 641
Empty solution set, 89
Endpoints, 216
Equals approximately, 3

Equation(s), 89
    absolute value, 232
    of a circle, 640
    conditional, 89
    of an ellipse, 639
    equivalent, 91
    first-degree, 91
    graph of, 130
    of a line, point-slope form, 264
    linear, 91, 132
        row-echelon form, 304
        slope-intercept form, 249
    literal, 103
    of parabolas, 565
    position, 311
    quadratic, 95
    solving, 89
    summary of, 267
    system of, 283
    two-point form, 340
Equivalent equations, 90, 305
Evaluate a function, 156
Event, 783
Expand a binomial, 770
Expansion by minors, 332, 333
Exponent, 10
    properties of, 452, 465
    rational, 465
Exponential
    decay, 722
    form, 10
    function with base a, 686
    growth and decay, 722
    growth model, 722
Extract
    imaginary square roots, 522
    square roots, 520
Extraneous, 436
Extremes of the proportion, 195

F

Factor, 8
Factorial, definition of, 739
Factoring
    difference of two cubes, 64
    by grouping, 61
    polynomials, 59, 377
Feasible solutions, 355
Features of graphs of polynomial functions, 631

Finding test intervals for a polynomial, 578
Finding the inverse of a function, 608
First-degree equation, 91
Foci, 641, 644
Focus, 641
FOIL Method, 49, 507
Forming equivalent equations, 91
Formulas, 105
    for compound interest, 692
Fraction(s), 3
    addition of, 12
    compound, 390
    reduction law, 376
Function(s), 153
    characteristics of, 154
    composition of two, 599
    constant, 170
    evaluate, 156
    exponential, 686
    graph of, 166
    inverse of, 605, 606
    linear, 243
    logarithmic, 699, 702
    notation, 156
    rational, 420
    sum, difference, product, and quotient, 596
    vertical line test for, 168
Fundamental Counting Principle, 775

G

Gaussian
    elimination, 305
    elimination with back-substitution, 322
Gear ratio, 201
Geometric sequence, 736, 755
    nth partial sum of, 758
    nth term of, 756
Golden section, 487
Graph, 130, 345
    of an equation, 130
    of a function, 166
    of an inequality, 345
Graphic check of a solution, 138
Graphical interpretation of solutions, 285
Greater than, 5
Greater than or equal to, 5
Greatest common factor, 59

Guidelines
  for factoring polynomials, 78, 377
  for finding the least common
    multiple of polynomials,
    397
  for graphing rational functions, 424
  for limiting possible factorizations
    of a trinomial, 75
  for solving exponential and
    logarithmic equations, 714
  for solving a linear programming
    problem, 356

H

Half-planes, 345
Horizontal asymptote, 421, 422
Horizontal line test for inverse
    functions, 612
Horizontal shift, 172
Human memory model, 682
Hyperbola, 639, 644

I

Identity
  Additive Property, 22, 32
  Multiplicative Property, 22, 32
i-form, 502
Imaginary number, 504
Imaginary unit i, 502
Implied domain, 158
Inconsistent, 285
Independent variable, 155
Index, 11, 462
Index of summation, 741
Inequality
  absolute value, 236
  algebraic, 216
  compound, 223
  graph of, 216
  linear, 220, 344
    graph of, 345
  properties of, 219
  quadratic, 580
  solution of, 216
Infinite sequence, 737
Initial height, 52
Initial velocity, 52

Integers, 2
  consecutive, 190
Intercepts, definition of, 134
Inverse
  Additive Property, 22, 32
  Multiplicative Property, 22, 32
Inverse function, 605, 606
  horizontal line test for, 612
Inverse properties
  of nth powers and nth roots, 464
Inverse variation, 621
Irrational number, 3
Isolate, 92

J

Joint variation, 624

K

Kepler's Third Law, 462

L

Labels for integers, 191
Law of trichotomy, 7
Leading, 1, 321
Leading coefficient, 46
Least common denominator, 398
Least common multiple, 396
Less than, 5
Less than or equal to, 5
Like terms, 35
Linear equation, 91, 132
  row-echelon form, 304
  slope-intercept form, 249
  two-point form, 340
Linear extrapolation, 270
Linear function, 243
Linear inequality, 220, 344
  graph of, 345
Linear programming, 355
Literal equation, 103
Logarithm of x with base a, 670
Logarithmic function with base a, 699
Logarithms, properties of, 678
Long division, 409
Lower limit of summation, 741

M

Major axis, 641
Markup, 203
Markup rate, 203
Mathematical model, 191, 618
Mathematical system, 21
Matrix, 317
  augmented, 317
  coefficient, 317
  determinant of, 331
  elementary row operations, 319
  entry, 317
  order, 317
  row-echelon form, 321
  row-equivalent, 319
  row operations, 319
Means of the proportion, 195
Method
  of elimination, 287, 288
  of substitution, 286, 657
Minor axis, 641
Miscellaneous common formulas, 106
Monomial, 47
Multiplication
  of radical expressions, 484
  of rational expressions, 386
  of real numbers, 8
Multiplication and Division Properties
    (Negative quantities), 219
Multiplication and Division Properties
    (Positive quantities), 219
Multiplication Property of Equality,
    24
Multiplication Property of Zero, 25
Multiplicative Identity Property,
    22, 32
Multiplicative Inverse Property, 22, 32

N

Natural
  base, 673
  exponential function, 690
  logarithm, 673
  logarithmic function, 702
  numbers, 2
Negative, 4
  exponents, definition of, 451
  integers, 2
Negative infinity, 218

Nonlinear system of equations, 654
Nonnegative real number, 4
Notation, function, 156
nth partial sum
   of an arithmetic sequence, 748
   of a geometric sequence, 758
nth root, 462
nth row, 769
nth term
   of an arithmetic sequence, 747
   of a geometric sequence, 756
Number
   of combinations of n elements
      taken m at a time, 780
   of permutations of n elements, 778
   of solutions of a system of linear
      equations, 308
Numerator, 9

O

Objective function, 355
Operations with complex numbers,
   505
Opposites, 7
Optimization problems, 355
Order, 317
   of operations, 12
Ordered pair, 114
Ordered triple, 304
Origin, 4, 114

P

Parabola, 565, 639
   equation of, 565
   sketch of, 568
Parallel lines, 252
Pascal's Triangle, 768
Perfect square trinomials, 70
Perimeter formula, 105
Permutation, 777
Perpendicular lines, 253
Pie chart, 15
Pixels, 145
Plotting, 4
Plotting a point, 114

Point-plotting method of sketching
   a graph, 131
Points of intersection, 284
Point-slope form of the equation
   of a line, 264
Polynomial
   degree of, 46
   factoring, 59
   graph of, 631
   least common multiple of, 397
   in one variable, 46
   position, 52
   third-degree, 414
   in x, 46
Position
   equation, 311
   polynomial, 52
Positive
   infinity, 218
   integers, 2
Principal nth root, 462
Probability of an event, 783
Product, 8
   of the sum and difference of
      two terms, 51
Proofs, 24
Properties
   of equality, 24
   of exponential and logarithmic
      equations, 712
   of exponential functions, 686
   of exponents, 452, 465
   of inequalities, 219
   of logarithms, 672, 678
   of natural logarithms, 674
   of negation, 25
   of radicals, 474
   of real numbers, 21
   of zero, 25
Proportion, 195
Pythagorean Theorem, 121, 494

Q

Quadrant, 114
Quadratic equation, 95
   solving by factoring, 518
Quadratic Formula, 540
Quadratic inequalities, 580
Quadratic type, 522
Quotient, 9

R

Radical, 11, 462, 467
   properties of, 474
   simplifying, 474
Radical expressions
   dividing, 485
   multiplying, 484
   simplifying, 476
Radicand, 11, 462
Radius, 639
Raising a to the nth power, 10
Raising both sides of an equation
      to the nth power, 490
Range, 151
Rate of work, 209
Rate problems, 208
Ratio, 194
Rational exponents, 464, 465
Rational expression, 374
   addition of, 395
   division of, 389
   multiplication of, 386
   simplifying, 376
   subtraction of, 395
Rational function, 420
   guidelines for graphing, 424
Rational number, 3
Rationalize the denominator, 476
Real number line, 4
   bounded intervals on, 216
   unbounded intervals on, 218
Real numbers, 2
   addition of, 8
   division of, 10
   multiplication of, 8
   nonnegative, 4
   properties of, 21
   subtraction of, 8
Reciprocal, 9
Rectangular coordinate system, 114
Recursive formula, 748
Reduction law for fractions, 376
Reflection
   in the x-axis, 174
   in the y-axis, 174
Relation, 151
Repeated solution, 97, 518
Repeating number, 3
Resolution, 145
Roots
   cube, 10
   nth, 11, 462
   principal nth, 11, 462
   square, 10, 520

Rounding a decimal, 3
Row operations, 305, 319
Row-echelon form, 304, 321
Row-equivalent, 319

S

Sample space, 783
Scientific notation, 454
Sequence, 736, 737
Set, 2
Sigma notation, 741
Simple interest formula, 106
Simplify, 35
  radical expressions, 476
  rational expressions, 376
Sketch
  the graph of a linear inequality
    in two variables, 345
  a parabola, 568
Slope of a line, 244, 245
Slope-intercept form of the equation
  of a line, 249
Solution, 283, 344
  point, 117
  set, 89
  set of the inequality, 216
Solve
  an absolute value equation, 232
  an absolute value inequality, 236
  an inequality, 216
  a system of equations, 283
Square of a binomial, 51
Square root, 10, 520
  of a negative number, 502
Standard form, 96, 504, 565
  of the equation of a circle
    (center at origin), 640
  of the equation of a circle, 639
  of the equation of a hyperbola
    (center at origin), 644
  of the equation of an ellipse
    (center at origin), 642
Straight-line depreciation, 141
Strategy for solving word problems,
  196, 555
Subset, 2
Subtraction
  of fractions, 12
  of rational expressions with like
    denominators, 395

of real numbers, 8
Subtraction and Addition Properties,
  219
Sum
  or difference of two cubes, 64
  and difference of two terms, 51
  of two numbers, 8
Summary of equations of lines, 267
Summation
  index of, 741
  lower limit of, 741
  upper limit of, 741
Symmetric Property of Equality, 26
Synthetic division, 414
System of equations, 283
  consistent, 285
  equivalent, 90, 305
  inconsistent, 285
  nonlinear, 654
Systems of linear inequalities, 351

T

Temperature formula, 106
Term, 31, 736
  of the sum, 8
Terminating number, 3
Test for collinear points, 339
Test intervals, 578
Theorems, 24
Transitive Property, 219
  of equality, 25
Translating key words and phrases,
  186
Transverse axis, 644
Triangle, area of, 338
Trichotomy, law of, 7
Trinomial, perfect square, 70
Two-point form, 265
  of the equation of a line, 340

U

Unbounded intervals on the real
  number line, 217, 218
Union symbol, 237, 374
Upper limit of summation, 741
Using the discriminant, 544

V

Value of f, 156
Variable, 31
  dependent, 155
  independent, 155
Variation, 618
  constant of proportionality, 618
  direct, 618
  inverse, 621
  joint, 624
Vertex, 565
Vertical and horizontal shifts, 172
Vertical asymptote, 421, 422
Vertical line test for functions, 168
Volume formula, 106

W

Whole number, 2
Word problems, strategy for solving,
  196, 555

X

x-axis, 114
x-coordinate, 114
x-intercept, 134

Y

y-axis, 114
y-coordinate, 114
y-intercept, 134

Z

Zero, 2, 578
  Division Property of, 25
  Multiplication Property of, 25
Zero exponents, definition of, 451
Zero row, 769
Zero-Factor Property, 95

Graphing Linear Equations

A **linear equation in two variables** is an equation of first degree in both variables.

The **graph of an equation** is the set of all points in the rectangular coordinate system whose coordinates are solutions of the equation.

The **graph of a linear equation in two variables** is a straight line.

To find the **x-intercepts** (where the graph intersects the x-axis), let y be zero and solve the equation for x.

To find the **y-intercepts** (where the graph intersects the y-axis), let x be zero and solve the equation for y.

Slope, $m = \dfrac{y_2 - y_1}{x_2 - x_1} = \dfrac{\text{Change in } y}{\text{Change in } x}$.

Slope-intercept form of the equation of a line:
$y = mx + b$, m is the slope, and $(0, b)$ is the y-intercept.

Point-slope form of the equation of a line:
$y - y_1 = m(x - x_1)$, m is the slope, and (x_1, y_1) is a point on the line.

Graphs of Systems of Linear Equations

The **solution** of a system of linear equations is an ordered pair (a, b) that satisfies each of the equations. A system of linear equations may have no solution, exactly one solution, or infinitely many solutions.

Graphs of Quadratic Equations

Standard form of a quadratic equation:
$ax^2 + bx + c = 0$, a, b, and c are real numbers with $a \neq 0$. A quadratic equation can be solved by factoring, extracting square roots, completing the square, or using the Quadratic Formula.

Extracting square roots: If $u^2 = d$, where $d > 0$, then $u = \pm\sqrt{d}$.

Quadratic Formula: $x = \dfrac{-b \pm \sqrt{b^2 - 4ac}}{2a}$

Discriminant: $b^2 - 4ac$
If $b^2 - 4ac > 0$, then the equation has two real number solutions.
If $b^2 - 4ac = 0$, then the equation has one (repeated) real number solution.
If $b^2 - 4ac < 0$, then the equation has no real number solutions.

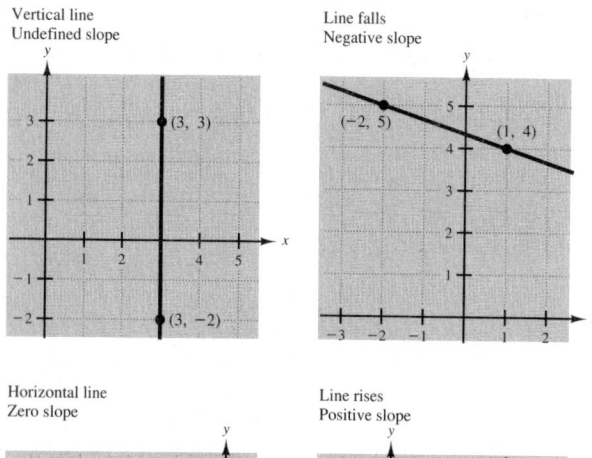

Vertical line
Undefined slope

Line falls
Negative slope

Horizontal line
Zero slope

Line rises
Positive slope

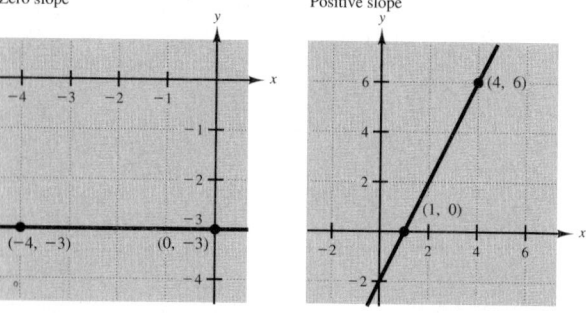

(a) Consistent (b) (c) Inconsistent

(a) Two lines that intersect at a single point
(b) Two lines that coincide with infinitely many points of intersection
(c) Two parallel lines with no point of intersection

The graph of $y = ax^2 + bx + c$, $a \neq 0$, is called a **parabola** which opens up if $a > 0$ and opens down if $a < 0$.

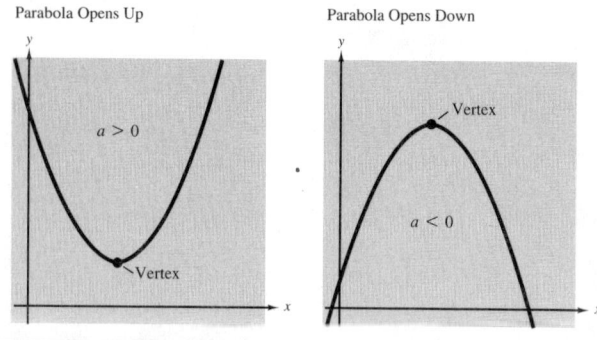

Parabola Opens Up

Parabola Opens Down